Advanced Modern Engineering Mathematics

Second Edition

Advanced Modern Engineering Mathematics

Second Edition

GLYN JAMES Coventry University

and

DAVID BURLEY University of Sheffield
DICK CLEMENTS University of Bristol
PHIL DYKE University of Plymouth
JOHN SEARL University of Edinburgh
NIGEL STEELE Coventry University
JERRY WRIGHT AT&T

Prentice Hall

An imprint of Pearson Education

Harlow, England · London · New York · Reading, Massachusetts · San Francisco · Toronto · Don Mills, Ontario · Sydney
Tokyo · Singapore · Hong Kong · Seoul · Taipei · Cape Town · Madrid · Mexico City · Amsterdam · Munich · Paris · Milan

Pearson Education Limited
Edinburgh Gate
Harlow
Essex CM20 2JE
England

and Associated Companies throughout the world

Visit us on the World Wide Web at:
http://www.pearsoneduc.com

Typeset in 10/12pt Times by 35
Produced by Pearson Education Asia (Pte) Ltd.
Printed in Singapore (MPM)

First edition printed 1993
This second edition printed 1999

ISBN 0–201–59621–0

British Library Cataloguing-in-Publication Data
A catalogue record for this book is available from the British Library

Library of Congress Cataloging-in-Publication Data
James, Glyn.
Advanced modern engineering mathematics/Glyn James . . . [et al.].
– 2nd ed.
p. cm.
Includes bibliographical references.
ISBN 0–201–59621–1 (alk. paper)
1. Engineering mathematics. I. Title.
TA330. J33 1999
510'.2462–dc21 99–18069
CIP

10 9 8 7 6 5 4
05 04 03 02 01

About the Authors

Glyn James is Professor and Dean of the School of Mathematical and Information Sciences at Coventry University. He graduated from the University College of Wales, Cardiff in the late fifties, obtaining first class honours degrees in both Mathematics and Chemistry. He obtained a PhD in Engineering Science in 1971 as an external student of the University of Warwick. He has been employed at Coventry since 1964 and held the position of the Head of Mathematics Department prior to his appointment as Dean in 1992. His current research interests are in control theory and its applications to industrial problems and he is co-director of a large research centre at the University. He also has a keen interest in mathematical education, particularly in relation to the teaching of engineering mathematics and mathematical modelling. He was co-chairman of the European Mathematics Working Group established by the European Society for Engineering Education (SEFI) in 1982, a past chairman of the Education Committee of the Institute of Mathematics and its Applications (IMA), and a member of the Royal Society Mathematics Education Subcommittee. In 1995 he was chairman of the Working Group that produced the report 'Mathematics Matters in Engineering' on behalf of the professional bodies in engineering and mathematics within the UK. He is also a member of the editorial/advisory board of three international journals. He has published numerous papers and is co-editor of five books on various aspects of mathematical modelling. He is a member of the Council and past Vice-President of the IMA and has also served a period as Honorary Secretary of the Institute. He is a Chartered Mathematician and a Fellow of the IMA.

David Burley has recently taken early retirement and now lectures part time in the University of Sheffield. He graduated in mathematics from King's College, University of London in 1955 and obtained his PhD in mathematical physics. After working in the University of Glasgow, he has spent most of his academic career in the University of Sheffield, being Head of Department for six years. He has long experience of teaching engineering students and has been particularly interested in encouraging students to construct mathematical models in physical and biological contexts to enhance their learning. His research work has ranged through statistical mechanics, optimization and fluid mechanics. Current interests involve the flow of molten glass in a variety of situations and the application of results in the glass industry.

Dick Clements is Reader in the Department of Engineering Mathematics at Bristol University. He read for the Mathematical Tripos at Christ's College, Cambridge in the late 1960s. He went on to take a PGCE at Leicester University School of

Education before returning to Cambridge to research a PhD in Aeronautical Engineering. In 1973 he was appointed Lecturer in Engineering Mathematics at Bristol University and has taught mathematics to engineering students ever since. He has undertaken research in a wide range of engineering topics but is particularly interested in mathematical modelling and the development of new ways of teaching mathematics to engineers. He has published numerous papers and one previous book, *Mathematical Modelling: A Case Study Approach*. He is a Chartered Engineer, a Member of the Royal Aeronautical Society, a Chartered Mathematician, a Fellow of the Institute of Mathematics and its Applications and a Member of the Royal Institute of Navigation.

Phil Dyke is Professor of Applied Mathematics and Head of School of Mathematics and Statistics at the University of Plymouth. After graduating with first class honours in Mathematics from the University of London, he gained a PhD in coastal sea modelling at Reading in 1972. Since then, Phil Dyke has been a full-time academic initially at Heriot-Watt University teaching engineers followed by a brief spell at Sunderland. He has been at Plymouth since 1984. He still engages in teaching and is actively involved in building mathematical models relevant to environmental issues.

John Searl is Director of the Edinburgh Centre for Mathematical Education at the University of Edinburgh. As well as lecturing on mathematical education, he teaches service courses for engineers and scientists. His current research concerns the development of learning environments that make for the effective learning of mathematics for 16–20 year olds. As an applied mathematician who has worked collaboratively with (among others) engineers, physicists, biologists and pharmacologists, he is keen to develop the problem solving skills of the students and to encourage them to think for themselves.

Jerry Wright is a Principal Member of Technical Staff at the AT&T Shannon Laboratory, New Jersey, USA. He graduated in Engineering (BSc and PhD at the University of Southampton) and in Mathematics (MSc at the University of London) and worked at the National Physical Laboratory before moving to the University of Bristol in 1978. There he acquired wide experience in the teaching of mathematics to students of engineering, and became Senior Lecturer in Engineering Mathematics. He held a Royal Society Industrial Fellowship for 1994, and is a Fellow of the Institute of Mathematics and its Applications. In 1996 he moved to AT&T Labs (originally part of Bell Labs) to continue his research in spoken language understanding and human/computer dialogue systems.

Nigel Steele is Head of Mathematics at Coventry University. He graduated in Mathematics from Southampton University in 1967, receiving his Master's degree in 1969. He has been a member of the Mathematics Department at Coventry University since 1968. He has a particular interest in teaching Engineering Mathematics, and is joint editor of the European Society for Engineering Education (SEFI) report 'A Common Core Curriculum in Mathematics for the European

Engineer'. He has published numerous papers and contributed to several books. His current research interests are in the application of neurocomputing techniques, fuzzy logic and in control theory. He is a Member of the Royal Aeronautical Society, a Chartered Mathematician and a Fellow and Council Member of the Institute of Mathematics and its Applications.

Contents

Preface to the First Edition

Throughout the course of history, engineering and mathematics have developed in parallel. All branches of engineering depend on mathematics for their description and there has been a steady flow of ideas and problems from engineering that has stimulated and sometimes initiated branches of mathematics. Thus it is vital that engineering students receive a thorough grounding in mathematics, with the treatment related to their interests and problems. This has been the motivation in the production of this book – a companion text to *Modern Engineering Mathematics*, which was designed to provide a first-level core studies course in mathematics for undergraduate programmes in all engineering disciplines.

Skills development

Although the pace of this book is at a somewhat more advanced level than *Modern Engineering Mathematics* the philosophy of learning by doing is retained with continuing emphasis on the development of students' ability to use mathematics with understanding to solve engineering problems.

Examples and Exercises

The book contains over 320 fully worked examples and more than 750 exercises, with answers to all the questions given. There are copious review exercises at the end of each chapter and these are designed both to reinforce understanding and to extend the range of topics included. In many areas hand-worked exercises can be checked using a suitable software package. It is strongly recommended that this is done whenever possible, not simply to increase confidence, but also to gain insight into the use and the performance of such packages. By this means the student will rapidly develop an appreciation of the need to have a good understanding of the underpinning mathematical theory and of computational problems such as the sensitivity of eigenvalue calculation to rounding errors.

Applications

Recognizing the growing importance of mathematical modelling in engineering practice, each chapter contains specific sections on engineering applications. As in *Modern Engineering Mathematics*, these sections form an ideal framework for individual, or group, case study assignments leading to a written report and/or oral presentation, thereby helping to reinforce the skills of mathematical modelling, which are seen as essential if engineers are to tackle the increasingly complex systems they are being called upon to analyse and design. Also, recognizing the increased use of numerical methods in problem solving, the treatment of numerical methods is integrated with the analytical work throughout the book. Algorithms are written in pseudocode and, therefore, are readily transferable to any specific language by the user.

Building on the foundations laid in *Modern Engineering Mathematics*, this book gives an extensive treatment of some of the more advanced areas of mathematics that have applications in various fields of engineering, particularly as tools for computer-based system modelling, analysis and design. In making the choice of material to be included, the authoring team have recognized the increasing importance of digital techniques and statistics in engineering practice. It is believed that the content is sufficiently broad to provide the necessary second-level core studies for most engineering studies, where in each case a selection of the material may be made.

Content

Chapter 1 is designed to develop an understanding of the standard techniques associated with functions of a complex variable which are still seen as an important component of a practising engineer's mathematical 'tool-kit'. The level of treatment recognizes the increasing emphasis in engineering practice on the use of computational methods that has led to less emphasis on the teaching of complex variable theory on engineering undergraduate courses. **Chapter 2 and 3** are devoted to Laplace and z transforms, which provide the necessary mathematical foundation for the frequency domain analysis and design of continuous- and discrete-time systems respectively. While Laplace transforms have traditionally been a component of most engineering undergraduate courses, the more recent inclusion of the z transform reflects the increased impact of digital signal processing in many areas of engineering. **Chapters 4 and 5** deal with Fourier analysis, which is central to many areas of applications in engineering, particularly in the analysis of signals and systems, with discrete Fourier transforms providing one of the key methods for discrete signal analysis and widely used in fields such as communications theory and speech and image processing.

Matrix techniques provide the framework for much of the developments in modern engineering. **Chapter 6** extends the introductory work on matrix algebra covered in *Modern Engineering Mathematics* to embrace the eigenvalue problem which arises naturally in many fields. In addition, the chapter develops the techniques of matrix analysis, which are used extensively in control and systems engineering among others. **Chapter 7** provides a basic introduction to vector calculus, which is seen as necessary to ensure that engineers have a good understanding of the mathematical tools they bring to their tasks when using software packages for the modelling of engineering problems in more than one dimension. Many physical processes fundamental to engineering, such as heat transfer and wave propagation, are governed by partial differential equations; **Chapter 8** is devoted to such equations with recognition given to the increasing emphasis on numerical solutions. Optimization, which is dealt with in **Chapter 9**, is a rapidly developing field that has taken full advantage of numerical methods and developments in computer technology. Indeed, many complex and practical optimization tasks have only become tractable in the last few years. The book concludes with **Chapter 10**, which is devoted to probability and statistics. In almost all branches of engineering, data from experiments have to be analysed and conclusions drawn, decisions have to be made, production and distribution need to be organized and monitored, and quality has to be controlled. In all these activities, probability and statistics have at least a supporting role to play, and sometimes a central one.

Acknowledgements

A number of individuals have contributed to the development of this book and it is a pleasure to acknowledge their contribution. In particular, the contribution of Dick Clements of Bristol University has been significant, the continuing support and enthusiasm for the project of Sarah Mallen (former development director at Addison-Wesley) has been an inspiration, and the comments and suggestions made by many reviewers during the preparation of the manuscript were extremely valuable. The authoring team also wish to acknowledge the many sources from which exercises have been drawn over many years. Again the authoring team have been fortunate in having an excellent production team at Addison-Wesley. The team wish to thank sincerely all those concerned, in particular the production editor Susan Keany; without her considerable effort the book would not have been completed. Finally, I would like to thank my wife, Dolan, for her full support throughout the preparation of this book.

Glyn James
Coventry
August 1993

Acknowledgments

Preface to the Second Edition

The revision in this edition reflect both the changes made in the companion text *Modern Engineering Mathematics* when producing its second edition and the feedback obtained from users of the text.

In general the feedback from users has been positive so changes have been kept to a minimum. Easier examples and exercises have been introduced where appropriate and introductory sections on matrix algebra and vector spaces incorporated in Chapter 6, *Matrix Analysis*. More significantly a new chapter, Chapter 8, on *Numerical Solution of Ordinary Differential Equations* has been introduced, reflecting reduced emphasis on this topic in the revised version of the companion text. For those adopting the book as a course text a Solutions Manual for all the exercises is available from the publishers.

Acknowledgements

The authoring team is extremely grateful to all the reviewers and users of the text who have provided valuable comments on the first edition of this book. Most of this was highly constructive and very much appreciated. The team has continued to enjoy the full support of the very enthusiastic production team at Addison Wesley Longman and wishes to thank all those concerned.

Glyn James
Coventry
December 1998

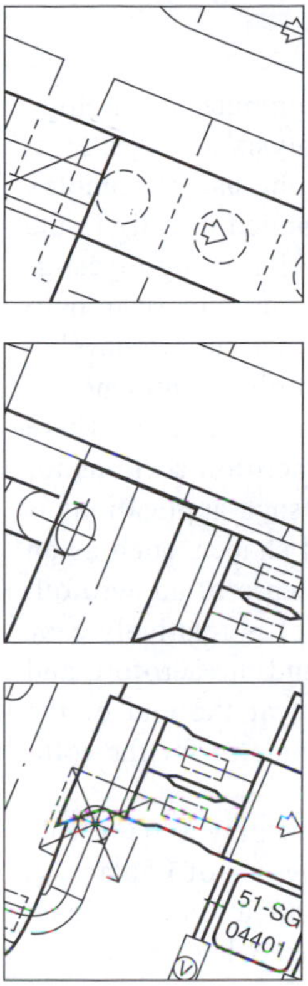

1

Functions of a Complex Variable

CONTENTS

1.1 Introduction

In the theory of alternating currents, the application of quantities such as complex impedance involves functions having complex numbers as independent variables. There are many other areas in engineering where this is the case; for example the motion of fluids, the transfer of heat or the processing of signals. Some of these applications are discussed later in this book.

Traditionally, complex variable techniques have been important, and extensively used, in a wide variety of engineering situations. This has been especially the case in areas such as electromagnetic and electrostatic field theory, fluid dynamics, aerodynamics and elasticity. With the rapid developments in computer technology and the consequential use of sophisticated algorithms for analysis and design in engineering there has, in recent years, been less emphasis on the use of complex variable techniques and a shift towards numerical techniques applied directly to the underlying full partial differential equations model of the situation being investigated. However, even when this is the case there is still considerable merit in having an analytical solution, possibly for an idealized model, in order both to develop a better understanding of the behaviour of the solution and to give confidence in numerical estimates for the solution of enhanced models. An example of where the theory has made a significant contribution is the design of aerofoil sections for aircraft and other lifting bodies. The strength of the theory in such applications is its ability to generate mappings which transform complicated shapes, such as an aerofoil section, into a simpler shape, typically a circle in the case of an aerofoil. The idealized airflow around the transformed shape (the circle) is relatively easy to calculate and by reversing the transformation the flow around the aerofoil, and hence its lifting capabilities, can be deduced. An application at the end of the chapter illustrates the technique applied to simplifying the geometry for the solution of a problem in the flow of heat.

Throughout engineering, transforms in one form or another play a major role in analysis and design. An area of continuing importance is the use of Laplace, z, Fourier and other transforms in areas such as control, communication and signal processing. Such transforms are considered later in this book, where it will be seen that functions of a complex variable play a key role. This chapter is devoted to developing an understanding of the standard techniques of complex variables so as to enable the reader to apply them with confidence in application areas.

1.2 Complex functions and mappings

The concept of a function involved two sets X and Y and a rule that assigns to each element x in the set X (written $x \in X$) precisely one element $y \in Y$. Whenever this situation arises, we say that there is a **function** f that **maps** the set X to the set Y, and represent this symbolically by

$$y = f(x) \quad (x \in X)$$

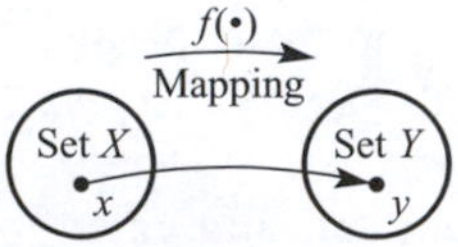

Figure 1.1 Real mapping $y = f(x)$.

Schematically we illustrate a function as in Figure 1.1. While x can take any value in the set X, the variable $y = f(x)$ depends on the particular element chosen for x. We therefore refer to x as the **independent** variable and y as the **dependent** variable. The set X is called the **domain** of the function, and the set of all images $y = f(x)$ $(x \in X)$ is called the **image set** or **range** of f. Previously we were concerned with real functions, so that x and y were real numbers. If the independent variable is a complex variable $z = x + jy$, where x and y are real and $j = \sqrt{(-1)}$,

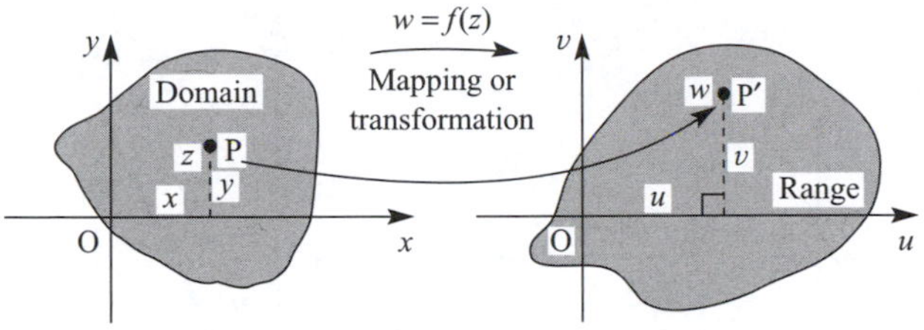

(a) z plane (b) w plane

Figure 1.2 Complex mapping $w = f(z)$.

then the function $f(z)$ of z will in general also be complex. For example, if $f(z) = z^2$ then, replacing z by $x + jy$ and expanding, we have

$$f(z) = (x + jy)^2 = (x^2 - y^2) + j2xy = u + jv \quad \text{(say)}$$

where u and v are real. Such a function $f(z)$ is called a **complex function**, and we write

$$w = f(z)$$

where, in general, the dependent variable $w = u + jv$ is also complex.

The reader will recall that a complex number $z = x + jy$ can be represented on a plane called the **Argand diagram**, as illustrated in Figure 1.2(a). However, we cannot plot the values of x, y and $f(z)$ on one set of axes, as we were able to do for real functions $y = f(x)$. We therefore represent the values of

$$w = f(z) = u + jv$$

on a second plane as illustrated in Figure 1.2(b). The plane containing the independent variable z is called the z plane and the plane containing the dependent variable w is called the w plane. Thus the complex function $w = f(z)$ may be regarded as a **mapping** or **transformation** of points P within a region in the z plane (called the **domain**) to corresponding image points P′ within a region in the w plane (called the **range**).

It is this facility for mapping that gives the theory of complex functions much of its application in engineering. In most useful mappings the entire z plane is mapped onto the entire w plane, except perhaps for isolated points. Throughout this chapter the domain will be taken to be the entire z plane (that is, the set of all complex numbers, denoted by $\mathbb{C}$). This is analogous, for real functions, to the domain being the entire real line (that is, the set of all real numbers $\mathbb{R}$). If this is not the case then the complex function is termed 'not well defined'. In contrast, as for real functions, the range of the complex function may well be a proper subset of $\mathbb{C}$.

EXAMPLE 1.1

Find the image in the w plane of the straight line $y = 2x + 4$ in the z plane, $z = x + jy$, under the mapping

$$w = 2z + 6$$

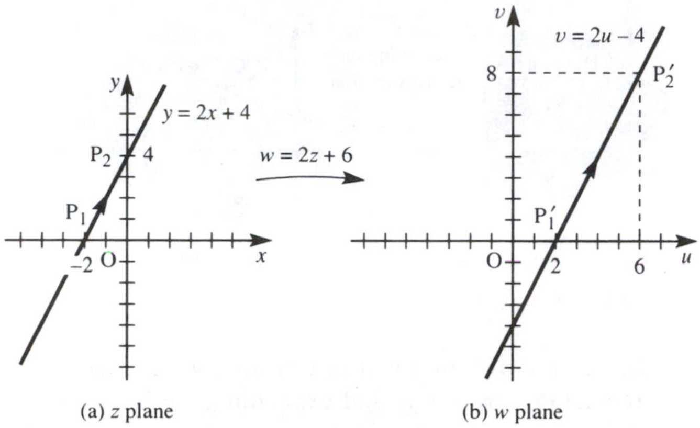

Figure 1.3 The mapping of Example 1.1.

Solution Writing $w = u + jv$, where u and v are real, the mapping becomes

$$w = u + jv = 2(x + jy) + 6$$

or

$$u + jv = (2x + 6) + j2y$$

Equating real and imaginary parts then gives

$$u = 2x + 6, \quad v = 2y \tag{1.1}$$

which, on solving for x and y, leads to

$$x = \tfrac{1}{2}(u - 6), \quad y = \tfrac{1}{2}v$$

Thus the image of the straight line

$$y = 2x + 4$$

in the z plane is represented by

$$\tfrac{1}{2}v = 2 \times \tfrac{1}{2}(u - 6) + 4$$

or

$$v = 2u - 4$$

which corresponds to a straight line in the w plane. The given line in the z plane and the mapped image line in the w plane are illustrated in Figures 1.3(a) and (b) respectively.

Note from (1.1) that, in particular, the point $P_1(-2 + j0)$ in the z plane is mapped to the point $P_1'(2 + j0)$ in the w plane, and that the point $P_2(0 + j4)$ in the z plane is mapped to the point $P_2'(6 + j8)$ in the w plane. Thus, as the point P moves from P_1 to P_2 along the line $y = 2x + 4$ in the z plane, the mapped point P′ moves from P_1' to P_2' along the line $v = 2u - 4$ in the w plane. It is usual to indicate this with the arrowheads as illustrated in Figure 1.3.

1.2.1 Linear mappings

The mapping $w = 2z + 6$ in Example 1.1 is a particular example of a mapping corresponding to the general complex linear function

$$w = \alpha z + \beta \qquad (1.2)$$

where w and z are complex-valued variables, and α and β are complex constants. In this section we shall investigate mappings of the z plane onto the w plane corresponding to (1.2) for different choices of the constants α and β. In so doing we shall also introduce some general properties of mappings.

Case (a) $\alpha = 0$

Letting $\alpha = 0$ (or $\alpha = 0 + \mathrm{j}0$) in (1.2) gives

$$w = \beta$$

which implies that $w = \beta$, no matter what the value of z. This is quite obviously a degenerate mapping, with the entire z plane being mapped onto the one point $w = \beta$ in the w plane. If nothing else, this illustrates the point made earlier in this section, that the image set may only be part of the entire w plane. In this particular case the image set is a single point. Since the whole of the z plane maps onto $w = \beta$, it follows that, in particular, $z = \beta$ maps to $w = \beta$. The point β is thus a **fixed point** in this mapping, which is a useful concept in helping us to understand a particular mapping. A further question of interest when considering mappings is that of whether, given a point in the w plane, we can tell from which point in the z plane it came under the mapping. If it is possible to get back to a unique point in the z plane then the mapping is said to have an **inverse mapping**. Clearly, for an inverse mapping $z = g(w)$ to exist, the point in the w plane has to be in the image set of the original mapping $w = f(z)$. Also, from the definition of a mapping, each point w in the w plane image set must lead to a single point z in the z plane under the inverse mapping $z = g(w)$. (Note the similarity to the requirements for the existence of an inverse function $f^{-1}(x)$ of a real function $f(x)$.) For the particular mapping $w = \beta$ considered here the image set is the single point $w = \beta$ in the w plane, and it is clear from Figure 1.4 that there is no way of getting back to just a single point in the z plane. Thus the mapping $w = \beta$ has no inverse.

Case (b) $\beta = 0$, $\alpha \neq 0$

With such a choice for the constants α and β, the mapping corresponding to (1.2) becomes

$$w = \alpha z$$

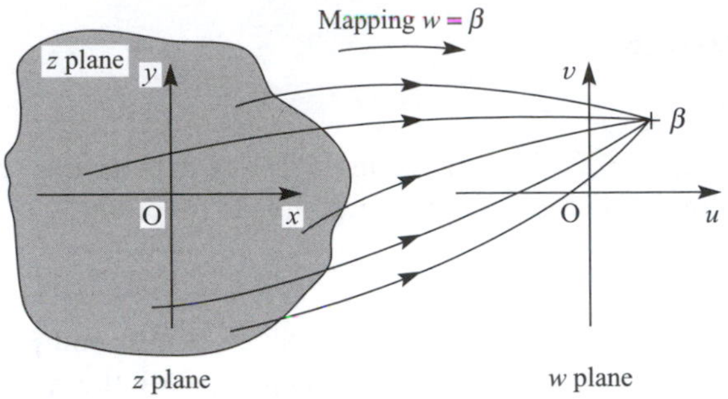

Figure 1.4 The degenerate mapping $w = \beta$.

Under this mapping, the origin is the only fixed point, there being no other fixed points that are finite. Also, in this case there exists an inverse mapping

$$z = \frac{1}{\alpha} w$$

that enables us to return from the w plane to the z plane to the very same point from which we started under $w = \alpha z$. To illustrate this mapping at work, let us choose $\alpha = 1 + \mathrm{j}$, so that

$$w = (1 + \mathrm{j})z \tag{1.3}$$

and consider what happens to a general point z_0 in the z plane under this mapping. In general, there are two ways of doing this. We can proceed as in Example 1.1 and split both z and w into real and imaginary parts, equate real and imaginary parts and hence find the image curves in the w plane to specific curves (usually the lines $\mathrm{Re}(z) =$ constant, $\mathrm{Im}(z) =$ constant) in the z plane. Alternatively, we can rearrange the expression for w and deduce the properties of the mapping directly. The former course of action, as we shall see in this chapter, is the one most frequently used. Here, however, we shall take the latter approach and write $\alpha = 1 + \mathrm{j}$ in polar form as

$$1 + \mathrm{j} = \sqrt{2}\mathrm{e}^{\mathrm{j}\pi/4}$$

Then, if

$$z = r\mathrm{e}^{\mathrm{j}\theta}$$

in polar form it follows from (1.3) that

$$w = r\sqrt{2}\mathrm{e}^{\mathrm{j}(\theta+\pi/4)} \tag{1.4}$$

We can then readily deduce from (1.4) what the mapping does. The general point in the z plane with modulus r and argument θ is mapped onto an image point w, with modulus $r\sqrt{2}$ and argument $\theta + \frac{1}{4}\pi$ in the w plane as illustrated in Figure 1.5.

It follows that in general the mapping

$$w = \alpha z$$

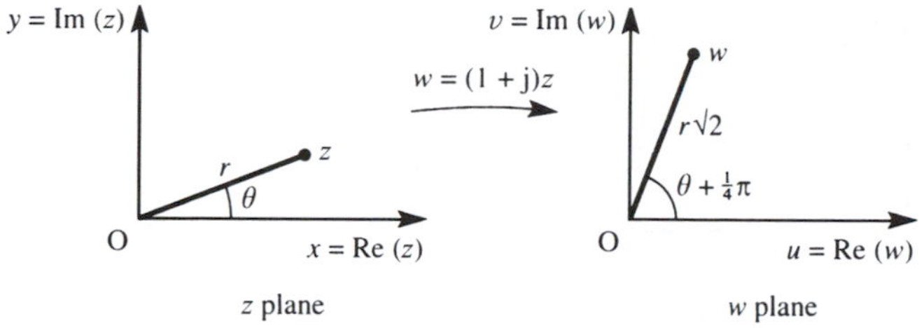

Figure 1.5 The mapping $w = (1 + j)z$.

maps the origin in the z plane to the origin in the w plane (fixed point), but effects an expansion by $|\alpha|$ and an anticlockwise rotation by arg α. Of course, arg α need not be positive, and indeed it could even be zero (corresponding to α being real). The mapping can be loosely summed up in the phrase 'magnification and rotation, but no translation'. Certain geometrical properties are also preserved, the most important being that straight lines in the z plane will be transformed to straight lines in the w plane. This is readily confirmed by noting that the equation of any straight line in the z plane can always be written in the form

$$|z - a| = |z - b|$$

where a and b are complex constants (this being the equation of the perpendicular bisector of the join of the two points representing a and b on the Argand diagram). Under the mapping $w = \alpha z$, the equation maps to

$$\left|\frac{w}{\alpha} - a\right| = \left|\frac{w}{\alpha} - b\right| \quad (\alpha \neq 0)$$

or

$$|w - a\alpha| = |w - b\alpha|$$

in the w plane, which is clearly another straight line.

We now return to the general linear mapping (1.2) and rewrite it in the form

$$w - \beta = \alpha z$$

This can be looked upon as two successive mappings: first,

$$\zeta = \alpha z$$

identical to $w = \alpha z$ considered earlier, but this time mapping points from the z plane to points in the ζ plane; secondly,

$$w = \zeta + \beta \tag{1.5}$$

mapping points in the ζ plane to points in the w plane. Elimination of ζ regains equation (1.2). The mapping (1.5) represents a translation in which the origin in the ζ plane is mapped to the point $w = \beta$ in the w plane, and the mapping of any other point in the ζ plane is obtained by adding β to the coordinates to obtain the equivalent point in the w plane. Geometrically, the mapping (1.5) is as if the ζ plane is picked up and, without rotation, the origin placed over the point β. The

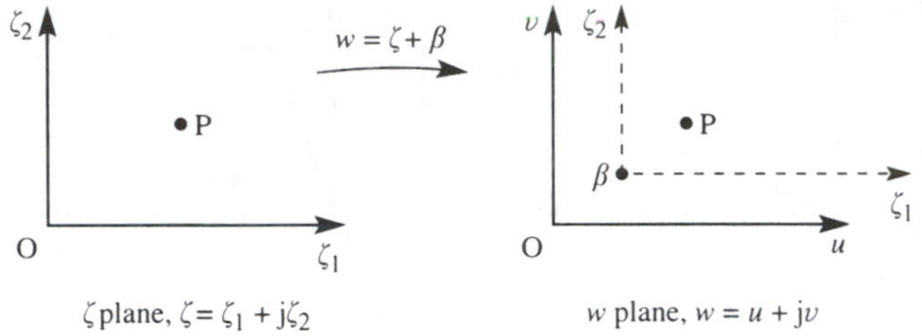

Figure 1.6 The mapping $w = \zeta + \beta$.

original axes then represent the w plane as illustrated in Figure 1.6. Obviously *all* curves, in particular straight lines, are preserved under this translation.

We are now in a position to interpret (1.2), the general linear mapping, geometrically as a combination of mappings that can be regarded as fundamental, namely

- translation
- rotation, and
- magnification

that is,

$$z \xrightarrow[\text{rotation}]{} e^{j\theta}z \xrightarrow[\text{magnification}]{} |\alpha|e^{j\theta}z \xrightarrow[\text{translation}]{} |\alpha|e^{j\theta}z + \beta = \alpha z + \beta = w$$

It clearly follows that a straight line in the z plane is mapped onto a corresponding straight line in the w plane under the linear mapping $w = \alpha z + \beta$. A second useful property of the linear mapping is that circles are mapped onto circles. To confirm this, consider the general circle

$$|z - z_0| = r$$

in the z plane, having the complex number z_0 as its centre and the real number r as its radius. Rearranging the mapping equation $w = \alpha z + \beta$ gives

$$z = \frac{w}{\alpha} - \frac{\beta}{\alpha} \quad (\alpha \neq 0)$$

so that

$$z - z_0 = \frac{w}{\alpha} - \frac{\beta}{\alpha} - z_0 = \frac{1}{\alpha}(w - w_0)$$

where $w_0 = \alpha z_0 + \beta$. Hence

$$|z - z_0| = r$$

implies

$$|w - w_0| = |\alpha|r$$

which is a circle, with centre w_0 given by the image of z_0 in the w plane and with radius $|\alpha|r$ given by the radius of the z plane circle magnified by $|\alpha|$.

We conclude this section by considering examples of linear mappings.

EXAMPLE 1.2 Examine the mapping

$$w = (1 + j)z + 1 - j$$

as a succession of fundamental mappings: translation, rotation and magnification.

Solution The linear mapping can be regarded as the following sequence of simple mappings:

$$z \xrightarrow[\substack{\text{rotation}\\ \text{anticlockwise}\\ \text{by } \frac{1}{4}\pi}]{} e^{j\pi/4}z \xrightarrow[\substack{\text{magnification}\\ \text{by } \sqrt{2}}]{} \sqrt{2}e^{j\pi/4}z \xrightarrow[\substack{\text{translation}\\ 0\to 1-j \text{ or}\\ (0,0)\to(1,-1)}]{} \sqrt{2}e^{j\pi/4}z + 1 - j = w$$

Figure 1.7 illustrates this process diagramatically. The shading in Figure 1.7 helps to identify how the z plane moves, turns and expands under this mapping. For example, the line joining the points $0 + j2$ and $1 + j0$ in the z plane has the cartesian equation

$$\tfrac{1}{2}y + x = 1$$

Taking $w = u + jv$ and $z = x + jy$, the mapping

$$w = (1 + j)z + 1 - j$$

becomes

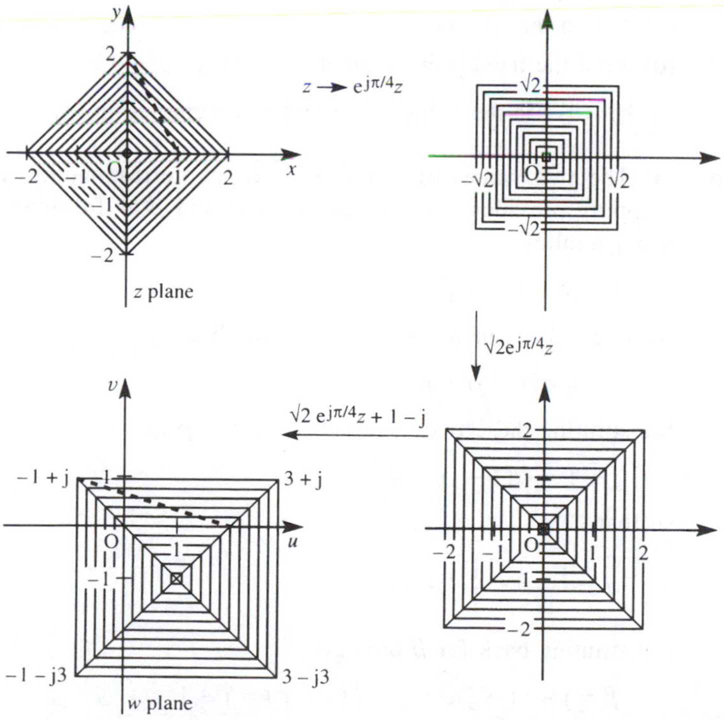

Figure 1.7 The mapping $w = (1 + j)z + 1 - j$.

$$u + jv = (1 + j)(x + jy) + 1 - j = (x - y + 1) + j(x + y - 1)$$

Equating real and imaginary parts then gives

$$u = x - y + 1, \qquad v = x + y - 1$$

which on solving for x and y gives

$$2x = u + v, \qquad 2y = v - u + 2$$

Substituting for x and y into the equation $\frac{1}{2}y + x = 1$ then gives the image of this line in the w plane as the line

$$3v + u = 2$$

which crosses the real axis in the w plane at 2 and the imaginary axis at $\frac{2}{3}$. Both lines are shown dashed, in the z and w planes respectively, in Figure 1.7.

EXAMPLE 1.3

The mapping $w = \alpha z + \beta$ (α, β constant complex numbers) maps the point $z = 1 + j$ to the point $w = j$, and the point $z = 1 - j$ to the point $w = -1$.

(a) Determine α and β.

(b) Find the region in the w plane corresponding to the right half-plane $\text{Re}(z) \geqslant 0$ in the z plane.

(c) Find the region in the w plane corresponding to the interior of the unit circle $|z| < 1$ in the z plane.

(d) Find the fixed point(s) of the mapping.

In (b)–(d) use the values of α and β determined in (a).

Solution

(a) The two values of z and w given as corresponding under the given linear mapping provides two equations for α and β as follows: $z = 1 + j$ mapping to $w = j$ implies

$$j = \alpha(1 + j) + \beta$$

while $z = 1 - j$ mapping to $w = -1$ implies

$$-1 = \alpha(1 - j) + \beta$$

Substituting these two equations in α and β gives

$$j + 1 = \alpha(1 + j) - \alpha(1 - j)$$

so that

$$\alpha = \frac{1 + j}{j2} = \frac{1}{2}(1 - j)$$

Substituting back for β then gives

$$\beta = j - (1 + j)\alpha = j - \tfrac{1}{2}(1 - j^2) = j - 1$$

so that

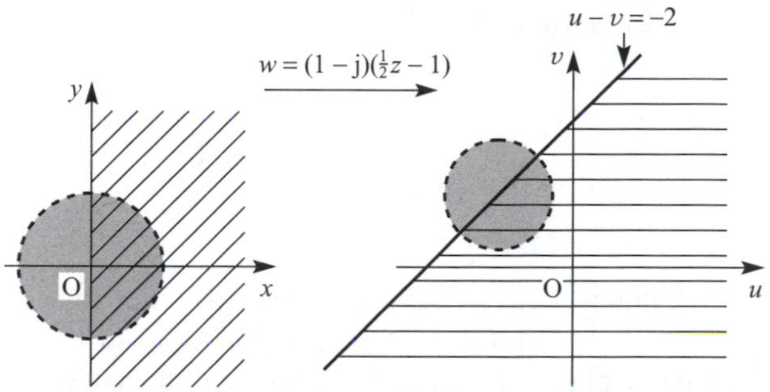

Figure 1.8 The mappings of Example 1.3.

$$w = \tfrac{1}{2}(1 - j)z + j - 1 = (1 - j)(\tfrac{1}{2}z - 1)$$

(b) The best way to find specific image curves in the w plane is first to express z (= $x + jy$) in terms of w (= $u + jv$) and then, by equating real and imaginary parts, to express x and y in terms of u and v. We have

$$w = (1 - j)(\tfrac{1}{2}z - 1)$$

which, on dividing by $1 - j$, gives

$$\frac{w}{1 - j} = \frac{1}{2}z - 1$$

Taking $w = u + jv$ and $z = x + jy$ and then rationalizing the left-hand side, we have

$$\tfrac{1}{2}(u + jv)(1 + j) = \tfrac{1}{2}(x + jy) - 1$$

Equating real and imaginary parts then gives

$$u - v = x - 2, \quad u + v = y \tag{1.6}$$

The first of these can be used to find the image of $x \geqslant 0$. It is $u - v \geqslant -2$, which is also a region bordered by the straight line $u - v = 2$ and shown in Figure 1.8. Pick one point in the right half of the z plane, say $z = 2$, and the mapping gives $w = 0$ as the image of this point. This allays any doubts about which side of $u - v = -2$ corresponds to the right half of the z plane, $x \geqslant 0$. The two corresponding regions are shown 'hatched' in Figure 1.8.

Note that the following is always true, although we shall not prove it here. If a curve cuts the z plane in two then the corresponding curve in the w plane also cuts the w plane in two, and, further, points in one of the two distinct sets of the z plane partitioned by the curve correspond to points in just one of the similarly partitioned sets in the w plane.

(c) In cartesian form, with $z = x + jy$, the equation of the unit circle $|z| = 1$ is

$$x^2 + y^2 = 1$$

Substituting for x and y from the mapping relationships (1.6) gives the image of this circle as

$$(u - v + 2)^2 + (u + v)^2 = 1$$

or

$$u^2 + v^2 + 2u - 2v + \tfrac{3}{2} = 0$$

which, on completing the squares, leads to

$$(u + 1)^2 + (v - 1)^2 = \tfrac{1}{2}$$

As expected, this is a circle, having in this particular case centre (−1, 1) and radius $\sqrt{\tfrac{1}{2}}$. If $x^2 + y^2 < 1$ then $(u + 1)^2 + (v - 1)^2 < \tfrac{1}{2}$, so the region inside the circle $|z| = 1$ in the z plane corresponds to the region inside its image circle in the w plane. Corresponding regions are shown shaded in Figure 1.8.

(d) The fixed point(s) of the mapping are obtained by putting $w = z$ in $w = \alpha z + \beta$, leading to

$$z = (\tfrac{1}{2}z - 1)(1 - \mathrm{j})$$

that is,

$$z = \tfrac{1}{2}z - \tfrac{1}{2}\mathrm{j}z - 1 + \mathrm{j}$$

so that

$$z = \frac{-1 + \mathrm{j}}{\tfrac{1}{2} + \tfrac{1}{2}\mathrm{j}} = \mathrm{j}2$$

is the only fixed point.

One final point is in order before we leave this example. In Figure 1.8 the images of $x = 0$ and $x^2 + y^2 = 1$ can also be seen in the context of translation, rotation (the line in Figure 1.8 rotates about $z = 2\mathrm{j}$) and magnification (in fact, shrinkage, as can be seen by the decrease in diameter of the circle compared with its image in the w plane).

1.2.2 Exercises

1 Find in the cartesian form $y = mx + c$ (m and c real constants) the equations of the following straight lines in the z plane, $z = x + \mathrm{j}y$:

(a) $|z - 2 + \mathrm{j}| = |z - \mathrm{j} + 3|$

(b) $|z + z^* + 4\mathrm{j}(z - z^*) = 6$

where * denotes the complex conjugate.

2 Find the point of intersection and the angle of intersection of the straight lines

$$|z - 1 - \mathrm{j}| = |z - 3 + \mathrm{j}|$$

$$|z - 1 + \mathrm{j}| = |z - 3 - \mathrm{j}|$$

3 The function $w = \mathrm{j}z + 4 - 3\mathrm{j}$ is a combination of translation and rotation. Show this diagramatically, following the procedure used in Example 1.2. Find the image of the line $6x + y = 22$ ($z = x + \mathrm{j}y$) in the w plane under this mapping.

4 Show that the mapping $w = (1 - \mathrm{j})z$, where $w = u + \mathrm{j}v$ and $z = x + \mathrm{j}y$, maps the region $y > 1$ in the z plane onto the region $u + v > 2$ in the w plane. Illustrate the regions in a diagram.

5 Under the mapping $w = \mathrm{j}z + \mathrm{j}$, where $w = u + \mathrm{j}v$ and $z = x + \mathrm{j}y$, show that the half-plane $x > 0$ in the

z plane maps onto the half-plane $v > 1$ in the w plane.

6 For $z = x + jy$ find the image region in the w plane corresponding to the semi-infinite strip $x > 0$, $0 < y < 2$ in the z plane under the mapping $w = jz + 1$. Illustrate the regions in both planes.

7 Find the images of the following curves under the mapping

$$w = (j + \sqrt{3})z + j\sqrt{3} - 1$$

(a) $y = 0$ (b) $x = 0$

(c) $x^2 + y^2 = 1$ (d) $x^2 + y^2 + 2y = 1$

where $z = x + jy$.

8 The mapping $w = \alpha z + \beta$ (a, β both constant complex numbers) maps the point $z = 1 + j$ to the point $w = j$ and the point $z = -1$ to the point $w = 1 + j$.

(a) Determine α and β.

(b) Find the region in the w plane corresponding to the upper half-plane $\text{Im}(z) > 0$ and illustrate diagrammatically.

(c) Find the region in the w plane corresponding to the disc $|z| < 2$ and illustrate diagrammatically.

(d) Find the fixed point(s) of the mapping.

In (b)–(d) use the values of α and β determined in (a).

1.2.3 Inversion

The inversion mapping is of the form

$$w = \frac{1}{z} \tag{1.7}$$

and in this subsection we shall consider the image of circles and straight lines in the z plane under such a mapping. Clearly, under this mapping the image in the w plane of the general circle

$$|z - z_0| = r$$

in the z plane, with centre at z_0 and radius r, is given by

$$\left|\frac{1}{w} - z_0\right| = r \tag{1.8}$$

but it is not immediately obvious what shaped curve this represents in the w plane. To investigate, we take $w = u + jv$ and $z_0 = x_0 + jy_0$ in (1.8), giving

$$\left|\frac{u - jv}{u^2 + v^2} - x_0 - jy_0\right| = r$$

Squaring we have

$$\left(\frac{u}{u^2 + v^2} - x_0\right)^2 + \left(\frac{v}{u^2 + v^2} + y_0\right)^2 = r^2$$

which on expanding leads to

$$\frac{u^2}{(u^2 + v^2)^2} - \frac{2ux_0}{u^2 + v^2} + x_0^2 + \frac{v^2}{(u^2 + v^2)^2} + \frac{2vy_0}{(u^2 + v^2)} + y_0^2 = r^2$$

or

$$\frac{u^2+v^2}{(u^2+v^2)^2} + \frac{2vy_0 - 2ux_0}{u^2+v^2} = r^2 - x_0^2 - y_0^2$$

so that

$$(u^2 + v^2)(r^2 - x_0^2 - y_0^2) + 2ux_0 - 2vy_0 = 1 \qquad \textbf{(1.9)}$$

The expression is a quadratic in u and v, with the coefficients of u^2 and v^2 equal and no term in uv. It therefore represents a circle, unless the coefficient of $u^2 + v^2$ is itself zero, which occurs when

$$x_0^2 + y_0^2 = r^2, \quad \text{or} \quad |z_0| = r$$

and we have

$$2ux_0 - 2vy_0 = 1$$

which represents a straight line in the w plane.

Summarizing, the inversion mapping $w = 1/z$ maps the circle $|z - z_0| = r$ in the z plane onto another circle in the w plane unless $|z_0| = r$, in which case the circle is mapped onto a straight line in the w plane that does not pass through the origin.

When $|z_0| \neq r$, we can divide the equation of the circle (1.9) in the w plane by the factor $r^2 - x_0^2 - y_0^2$ to give

$$u^2 + v^2 + \frac{2x_0u}{r^2 - x_0^2 - y_0^2} - \frac{2y_0v}{r^2 - x_0^2 - y_0^2} = \frac{1}{r^2 - x_0^2 - y_0^2}$$

which can be written in the form

$$(u - u_0)^2 + (v - v_0)^2 = R^2$$

where (u_0, v_0) are the coordinates of the centre and R the radius of the w plane circle. It is left as an exercise for the reader to show that

$$(u_0, v_0) = \left(-\frac{x_0}{r^2 - |z_0|^2}, \frac{y_0}{r^2 - |z_0|^2}\right), \qquad R = \frac{r}{r^2 - |z_0|^2}$$

Next we consider the general straight line

$$|z - a_1| = |z - a_2|$$

in the z plane, where a_1 and a_2 are constant complex numbers with $a_1 \neq a_2$. Under the mapping (1.7), this becomes the curve in the w plane represented by the equation

$$\left|\frac{1}{w} - a_1\right| = \left|\frac{1}{w} - a_2\right| \qquad \textbf{(1.10)}$$

Again, it is not easy to identify this curve, so we proceed as before and take

$$w = u + jv, \qquad a_1 = p + jq, \qquad a_2 = r + js$$

where p, q, r and s are real constants. Substituting in (1.10) and squaring both sides, we have

$$\left(\frac{u}{u^2+v^2} - p\right)^2 + \left(\frac{v}{u^2+v^2} + q\right)^2 = \left(\frac{u}{u^2+v^2} - r\right)^2 + \left(\frac{v}{u^2+v^2} + s\right)^2$$

Expanding out each term, the squares of $u/(u^2 + v^2)$ and $v/(u^2 + v^2)$ cancel, giving

$$-\frac{2up}{u^2 + v^2} + p^2 + \frac{2vq}{u^2 + v^2} + q^2 = -\frac{2ur}{u^2 + v^2} + r^2 + \frac{2vs}{u^2 + v^2} + s^2$$

which on rearrangement becomes

$$(u^2 + v^2)(p^2 + q^2 - r^2 - s^2) + 2u(r - p) + 2v(q - s) = 0 \qquad \textbf{(1.11)}$$

Again this represents a circle *through the origin* in the w plane, unless

$$p^2 + q^2 = r^2 + s^2$$

which implies $|a_1| = |a_2|$, when it represents a straight line, also through the origin, in the w plane. The algebraic form of the coordinates of the centre of the circle and its radius can be deduced from (1.11).

We can therefore make the important conclusion that the inversion mapping $w = 1/z$ takes circles or straight lines in the z plane onto circles or straight lines in the w plane. Further, since we have carried out the algebra, we can be more specific. If the circle in the z plane passes through the origin (that is $|z_0| = r$ in (1.9)) then it is mapped onto a straight line that does *not* pass through the origin in the w plane. If the straight line in the z plane passes through the origin ($|a_1| = |a_2|$ in (1.11)) then it is mapped onto a straight line through the origin in the w plane. Figure 1.9 summarizes these conclusions.

To see why this is the case, we first note that the fixed points of the mapping, determined by putting $w = z$, are

$$z = \frac{1}{z}, \quad \text{or} \quad z^2 = 1$$

so that $z = \pm 1$.

We also note that $z = 0$ is mapped to infinity in the w plane and $w = 0$ is mapped to infinity in the z plane and vice versa in both cases. Further, if we apply the mapping a second time, we get the identity mapping. That is, if

$$w = \frac{1}{z}, \quad \text{and} \quad \zeta = \frac{1}{w}$$

then

$$\zeta = \frac{1}{1/z} = z$$

which is the identity mapping.

The inside of the unit circle in the z plane, $|z| < 1$, is mapped onto $|1/w| < 1$ or $|w| > 1$, the outside of the unit circle in the w plane. By the same token, therefore, the outside of the unit circle in the z plane $|z| > 1$ is mapped onto $|1/w| > 1$ or $|w| < 1$, the inside of the unit circle in the w plane. Points actually on $|z| = 1$ in the z plane are mapped to points on $|w| = 1$ in the w plane, with ± 1 staying fixed, as already shown. Figure 1.10 summarizes this property.

It is left as an exercise for the reader to show that the top half-boundary of $|z| = 1$ is mapped onto the bottom half-boundary of $|w| = 1$.

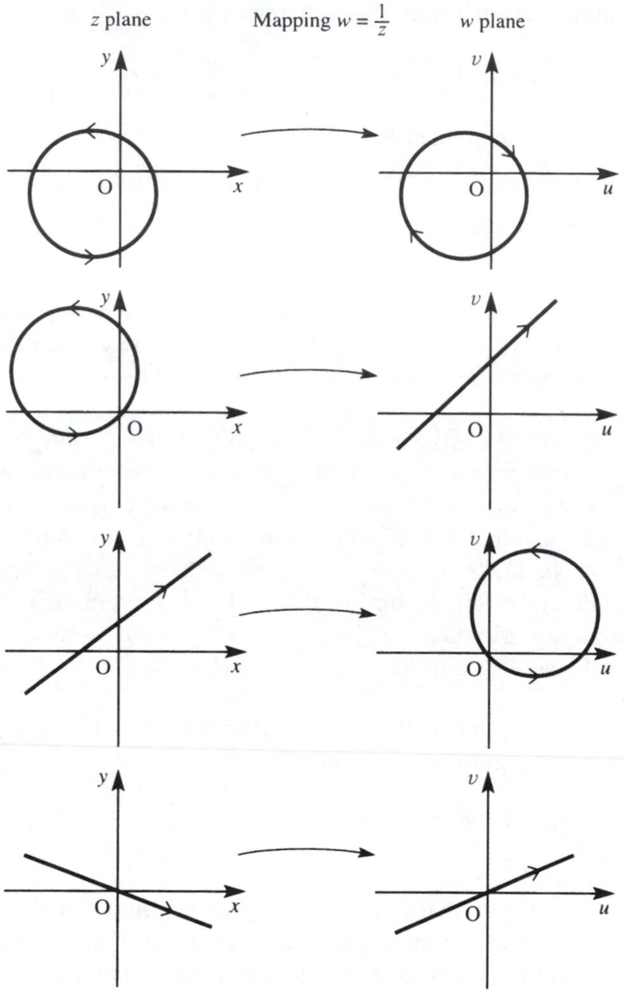

Figure 1.9 The inversion mapping $w = 1/z$.

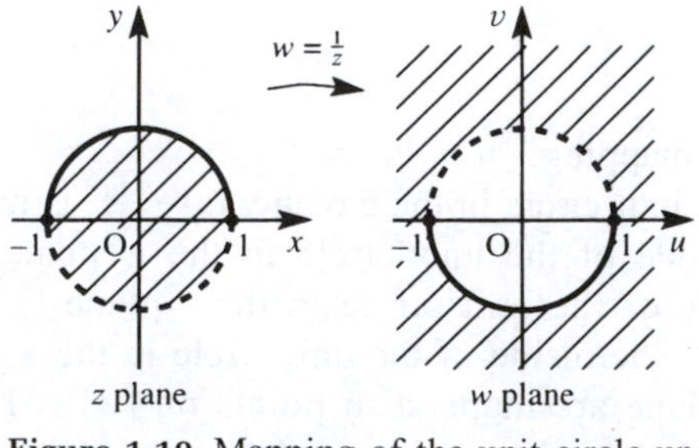

Figure 1.10 Mapping of the unit circle under $w = 1/z$.

For any point z_0 in the z plane the point $1/z_0$ is called the **inverse of z_0 with respect to the circle** $|z| = 1$; this is the reason for the name of the mapping. (Note the double meaning of inverse; here it means the reciprocal function and not the 'reverse' mapping.) The more general definition of inverse is that for any point z_0 in the z plane the point r^2/z_0 is the inverse of z_0 with respect to the circle $|z| = r$, where r is a real constant.

EXAMPLE 1.4

Determine the image path in the w plane corresponding to the circle $|z - 3| = 2$ in the z plane under the mapping $w = 1/z$. Sketch the paths in both the z and w planes and shade the region in the w plane corresponding to the region inside the circle in the z plane.

Solution The image in the w plane of the circle $|z - 3| = 2$ in the z plane under the mapping $w = 1/z$ is given by

$$\left|\frac{1}{w} - 3\right| = 2$$

which, on taking $w = u + \text{j}v$, gives

$$\left|\frac{u - \text{j}v}{u^2 + v^2} - 3\right| = 2$$

Squaring both sides, we then have

$$\left(\frac{u}{u^2 + v^2} - 3\right)^2 + \left(\frac{-v}{u^2 + v^2}\right)^2 = 4$$

or

$$\frac{u^2 + v^2}{(u^2 + v^2)^2} - \frac{6u}{u^2 + v^2} + 5 = 0$$

which reduces to

$$1 - 6u + 5(u^2 + v^2) = 0$$

or

$$(u - \tfrac{3}{5})^2 + v^2 = \tfrac{4}{25}$$

Thus the image in the w plane is a circle with centre $(\frac{3}{5}, 0)$ and radius $\frac{2}{5}$. The corresponding circles in the z and w planes are shown in Figure 1.11.

Taking $z = x + \text{j}y$, the mapping $w = 1/z$ becomes

$$u + \text{j}v = \frac{1}{x + \text{j}y} = \frac{x - \text{j}y}{x^2 + y^2}$$

which, on equating real and imaginary parts, gives

$$u = \frac{x}{x^2 + y^2}, \qquad v = \frac{-y}{x^2 + y^2}$$

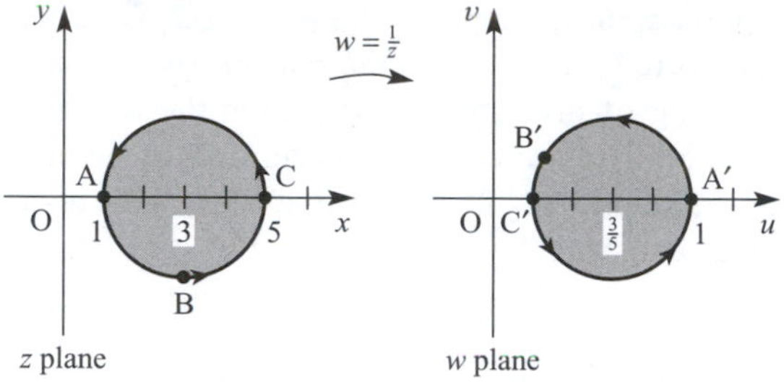

Figure 1.11 The mapping of Example 1.4.

We can now use these two relationships to determine the images of particular points under the mapping. In particular, the centre (3, 0) of the circle in the z plane is mapped onto the point $u = \frac{1}{3}$, $v = 0$ in the w plane, which is inside the mapped circle. Thus, under the mapping, the region inside the circle in the z plane is mapped onto the region inside the circle in the w plane.

Further, considering three sample points A(1 + j0), B(3 − j2) and C(5 + j0) on the circle in the z plane, we find that the corresponding image points on the circle in the w plane are A′(1, 0), B′($\frac{3}{13}$, $\frac{2}{13}$) and C′($\frac{1}{5}$, 0). Thus, as the point z traverses the circle in the z plane in an anticlockwise direction, the corresponding point w in the w plane will also traverse the mapped circle in an anticlockwise direction as indicated in Figure 1.11.

1.2.4 Bilinear mappings

A **bilinear mapping** is a mapping of the form

$$w = \frac{az + b}{cz + d} \tag{1.12}$$

where a, b, c and d are prescribed complex constants. It is called the bilinear mapping in z and w since it can be written in the form $Awz + Bw + Cz + D = 0$, which is linear in both z and w.

Clearly the bilinear mapping (1.12) is more complicated than the linear mapping given by (1.2). In fact, the general linear mapping is a special case of the bilinear mapping, since setting $c = 0$ and $d = 1$ in (1.12) gives (1.2). In order to investigate the bilinear mapping, we rewrite the right-hand side of (1.12) as follows:

$$w = \frac{az + b}{cz + d} = \frac{\frac{a}{c}(cz + d) - \frac{ad}{c} + b}{cz + d}$$

so that

$$w = \frac{a}{c} + \frac{bc - ad}{c(cz + d)} \tag{1.13}$$

This mapping clearly degenerates to $w = a/c$ unless we demand that $bc - ad \neq 0$. We therefore say that (1.12) represents a bilinear mapping provided the determinant

$$\begin{vmatrix} a & b \\ c & d \end{vmatrix} = ad - bc$$

is non-zero. This is sometimes referred to as the **determinant of the mapping**. When the condition holds, the inverse mapping

$$z = \frac{-dw + b}{cw - a}$$

obtained by rearranging (1.12), is also bilinear, since

$$\begin{vmatrix} -d & b \\ c & -a \end{vmatrix} = da - cb \neq 0$$

Renaming the constants so that $\lambda = a/c$, $\mu = bc - ad$, $\alpha = c^2$ and $\beta = cd$, (1.13) becomes

$$w = \lambda + \frac{\mu}{\alpha z + \beta}$$

and we can break the mapping down into three steps as follows:

$$z_1 = \alpha z + \beta$$

$$z_2 = \frac{1}{z_1}$$

$$w = \lambda + \mu z_2$$

The first and third of these steps are linear mappings as considered in Section 1.2.1, while the second is the inversion mapping considered in Section 1.2.3. The bilinear mapping (1.12) can thus be generated from the following elementary mappings:

$$z \xrightarrow[\text{rotation and magnification}]{} \alpha z \xrightarrow[\text{translation}]{} \alpha z + \beta \xrightarrow[\text{inversion}]{} \frac{1}{\alpha z + \beta}$$

$$\xrightarrow[\text{magnification and rotation}]{} \frac{\mu}{\alpha z + \beta} \xrightarrow[\text{translation}]{} \lambda + \frac{\mu}{\alpha z + \beta} = w$$

We saw in Section 1.2.1 that the general linear transformation $w = \alpha z + \beta$ does not change the shape of the curve being mapped from the z plane onto the w plane. Also, in Section 1.2.3 we saw that the inversion mapping $w = 1/z$ maps circles or straight lines in the z plane onto circles or straight lines in the w plane. It follows that the bilinear mapping also exhibits this important property, in that it also will map circles or straight lines in the z plane onto circles or straight lines in the w plane.

EXAMPLE 1.5 Investigate the mapping

$$w = \frac{z-1}{z+1}$$

by finding the images in the w plane of the lines $\text{Re}(z)$ = constant and $\text{Im}(z)$ = constant. Find the fixed points of the mapping.

Solution Since we are seeking specific image curves in the w plane, we first express z in terms of w and then express x and y in terms of u and v, where $z = x + jy$ and $w = u + jv$. Rearranging

$$w = \frac{z-1}{z+1}$$

gives

$$z = \frac{1+w}{1-w}$$

Taking $z = x + jy$ and $w = u + jv$, we have

$$\begin{aligned} x + jy &= \frac{1+u+jv}{1-u-jv} \\ &= \frac{1+u+jv}{1-u-jv}\,\frac{1-u+jv}{1-u+jv} \end{aligned}$$

which reduces to

$$x + jy = \frac{1-u^2-v^2}{(1-u)^2+v^2} + j\frac{2v}{(1-u)^2+v^2}$$

Equating real and imaginary parts then gives

$$x = \frac{1-u^2-v^2}{(1-u)^2+v^2} \tag{1.14a}$$

$$y = \frac{2v}{(1-u)^2+v^2} \tag{1.14b}$$

It follows from (1.14a) that the lines $\text{Re}(z) = x = c_1$, which are parallel to the imaginary axis in the z plane, correspond to the curves

$$c_1 = \frac{1-u^2-v^2}{(1-u)^2+v^2}$$

where c_1 is a constant, in the w plane. Rearranging this leads to

$$c_1(1 - 2u + u^2 + v^2) = 1 - u^2 - v^2$$

or, assuming that $1 + c_1 \neq 0$,

$$u^2 + v^2 - \frac{2c_1u}{1 + c_1} + \frac{c_1 - 1}{c_1 + 1} = 0$$

which, on completing squares, gives

$$\left(u - \frac{c_1}{1 + c_1}\right)^2 + v^2 = \left(\frac{1}{1 + c_1}\right)^2$$

It is now clear that the corresponding curve in the w plane is a circle, centre $(u = c_1/(1 + c_1), v = 0)$ and radius $(1 + c_1)^{-1}$.

In the algebraic manipulation we assumed that $c_1 \neq -1$, in order to divide by $1 + c_1$. In the exceptional case $c_1 = -1$, we have $u = 1$, and the mapped curve is a straight line in the w plane parallel to the imaginary axis.

Similarly, it follows from (1.14b) that the lines $\text{Im}(z) = y = c_2$, which are parallel to the imaginary axis in the z plane, correspond to the curves

$$c_2 = \frac{2v}{(1 - u)^2 + v^2}$$

where c_2 is a constant, in the w plane. Again, this usually represents a circle in the w plane, but exceptionally will represent a straight line. Rearranging the equation we have

$$(1 - u)^2 + v^2 = \frac{2v}{c_2}$$

provided that $c_2 \neq 0$. Completing the square then leads to

$$(u - 1)^2 + \left(v - \frac{1}{c_2}\right)^2 = \frac{1}{c_2^2}$$

which represents a circle in the w plane, centre $(u = 1, v = 1/c_2)$ and radius $1/c_2$.

In the exceptional case $c_2 = 0$, $v = 0$ and we see that the real axis $y = 0$ in the z plane maps onto the real axis $v = 0$ in the w plane.

Putting a sequence of values to c_1 and then to c_2, say -10 to $+10$ in steps of $+1$, enables us to sketch the mappings shown in Figure 1.12. The fixed points of the mapping are given by

$$z = \frac{z - 1}{z + 1}$$

that is,

$$z^2 = -1, \quad \text{or} \quad z = \pm \text{j}$$

In general, all bilinear mappings will have two fixed points. However, although there are mathematically interesting properties associated with particular mappings having coincident fixed points, they do not impinge on engineering applications, so they only deserve passing reference here.

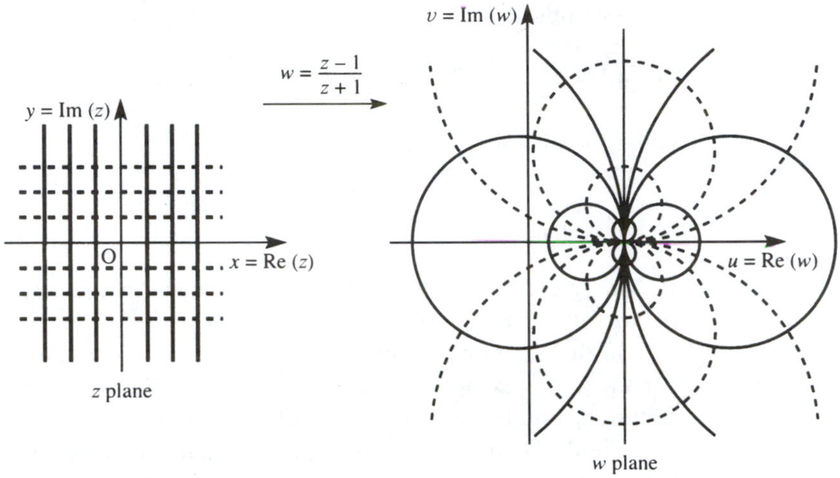

Figure 1.12 The mapping $w = (z - 1)/(z + 1)$.

EXAMPLE 1.6

Find the image in the w plane of the circle $|z| = 2$ in the z plane under the bilinear mapping

$$w = \frac{z - \mathrm{j}}{z + \mathrm{j}}$$

Sketch the curves in both the z and w planes and shade the region in the w plane corresponding to the region inside the circle in the z plane.

Solution Rearranging the transformation, we have

$$z = \frac{\mathrm{j}w + \mathrm{j}}{1 - w}$$

so that the image in the w plane of the circle $|z| = 2$ in the z plane is determined by

$$\left|\frac{\mathrm{j}w + \mathrm{j}}{1 - w}\right| = 2 \qquad \textbf{(1.15)}$$

One possible way of proceeding now is to put $w = u + \mathrm{j}v$ and proceed as in Example 1.4, but the algebra becomes a little messy. An alternative approach is to use the property of complex numbers that $|z_1/z_2| = |z_1|/|z_2|$, so that (1.15) becomes

$$|\mathrm{j}w + \mathrm{j}| = 2|1 - w|$$

Taking $w = u + \mathrm{j}v$ then gives

$$|-v + \mathrm{j}(u + 1)| = 2|(1 - u) - \mathrm{j}v|$$

which on squaring both sides leads to

$$v^2 + (1 + u)^2 = 4[(1 - u)^2 + v^2]$$

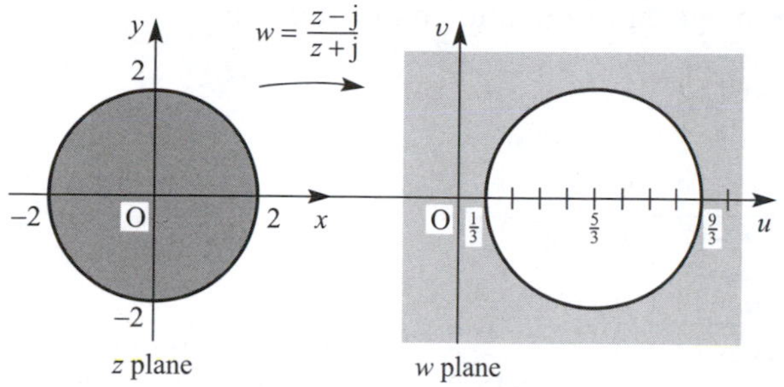

Figure 1.13 The mapping $w = (z - j)/(z + j)$.

or

$$u^2 + v^2 - \tfrac{10}{3}u + 1 = 0$$

Completing the square of the u term then gives

$$(u - \tfrac{5}{3})^2 + v^2 = \tfrac{16}{9}$$

indicating that the image curve in the w plane is a circle centre ($u = \tfrac{5}{3}$, $v = 0$) and radius $\tfrac{4}{3}$. The corresponding circles in the z and w planes are illustrated in Figure 1.13. To identify corresponding regions, we consider the mapping of the point $z = 0 + j0$ inside the circle in the z plane. Under the given mapping, this maps to the point

$$w = \frac{0 - j}{0 + j} = -1 + j0$$

in the w plane. It then follows that the region inside the circle $|z| = 2$ in the z plane maps onto the region outside the mapped circle in the w plane.

An interesting property of (1.12) is that there is just one bilinear transformation that maps three given distinct points z_1, z_2 and z_3 in the z plane onto three specified distinct points w_1, w_2 and w_3 respectively in the w plane. It is left as an exercise for the reader to show that the bilinear transformation is given by

$$\frac{(w - w_1)(w_2 - w_3)}{(w - w_3)(w_2 - w_1)} = \frac{(z - z_1)(z_2 - z_3)}{(z - z_3)(z_2 - z_1)} \tag{1.16}$$

The right-hand side of (1.16) is called the cross-ratio of z_1, z_2, z_3 and z. We shall illustrate with an example.

EXAMPLE 1.7

Find the bilinear transformation that maps the three points $z = 0$, $-j$ and -1 onto the three points $w = j$, 1, 0 respectively in the w plane.

Solution Taking the transformation to be

$$w = \frac{az + b}{cz + d}$$

on using the given information on the three pairs of corresponding points we have

$$\mathrm{j} = \frac{a(0) + b}{c(0) + d} = \frac{b}{d} \tag{1.17a}$$

$$1 = \frac{a(-\mathrm{j}) + b}{c(-\mathrm{j}) + d} \tag{1.17b}$$

$$0 = \frac{a(-1) + b}{c(-1) + d} \tag{1.17c}$$

From (1.17c) $a = b$; then from (1.17a)

$$d = \frac{b}{\mathrm{j}} = -\mathrm{j}b = -\mathrm{j}a$$

and from (1.17b) $c = \mathrm{j}a$. Thus

$$w = \frac{az + a}{\mathrm{j}az - \mathrm{j}a} = \frac{1}{\mathrm{j}}\frac{z + 1}{z - 1} = -\mathrm{j}\frac{z + 1}{z - 1}$$

Alternatively, using (1.16) we can obtain

$$\frac{(w - \mathrm{j})(1 - 0)}{(w - 0)(1 - \mathrm{j})} = \frac{(z - 0)(-\mathrm{j} + 1)}{(z + 1)(-\mathrm{j} - 0)}$$

or

$$w = -\mathrm{j}\frac{z + 1}{z - 1}$$

as before.

1.2.5 Exercises

9 Show that if $z = x + \mathrm{j}y$, the image of the half-plane $y > c$ (c constant) under the mapping $w = 1/z$ is the interior of a circle, provided that $c > 0$. What is the image when $c = 0$ and when $c < 0$? Illustrate with sketches in the w plane.

10 Determine the image in the w plane of the circle

$$\left|z + \tfrac{3}{4} + \mathrm{j}\right| = \tfrac{7}{4}$$

under the inversion mapping $w = 1/z$.

11 Show that the mapping $w = 1/z$ maps the circle $|z - a| = a$, with a being a positive real constant, onto a straight line in the w plane. Sketch the corresponding curves in the z and w planes, indicating the region onto which the interior of the circle in the z plane is mapped.

12 Find a bilinear mapping that maps $z = 0$ to $w = \mathrm{j}$, $z = -\mathrm{j}$ to $w = 1$ and $z = -1$ to $w = 0$. Hence sketch the mapping by finding the images in the w plane of the lines $\mathrm{Re}(z)$ = constant and $\mathrm{Im}(z)$ = constant in the z plane. Verify that $z = \frac{1}{2}(\mathrm{j} - 1)(-1 \pm \sqrt{3})$ are fixed points of the mapping.

13 The two complex variables w and z are related through the inverse mapping

$$w = \frac{1 + \mathrm{j}}{z}$$

(a) Find the images of the points $z = 1$, $1 - \mathrm{j}$ and 0 in the w plane.
(b) Find the region of the w plane corresponding to the interior of the unit circle $|z| < 1$ in the z plane.
(c) Find the curves in the w plane corresponding to the straight lines $x = y$ and $x + y = 1$ in the z plane.
(d) Find the fixed points of the mapping.

14 Given the complex mapping

$$w = \frac{z + 1}{z - 1}$$

where $w = u + \mathrm{j}v$ and $z = x + \mathrm{j}y$, determine the image curve in the w plane corresponding to the semicircular arc $x^2 + y^2 = 1$ ($x \leqslant 0$) described from the point (0, −1) to the point (0, 1).

15 (a) Map the region in the z plane ($z = x + \mathrm{j}y$) that lies between the lines $x = y$ and $y = 0$, with $x < 0$, onto the w plane under the bilinear mapping

$$w = \frac{z + \mathrm{j}}{z - 3}$$

(*Hint*: Consider the point $w = \frac{2}{3}$ to help identify corresponding regions.)
(b) Show that, under the same mapping as in (a), the straight line $3x + y = 4$ in the z plane corresponds to the unit circle $|w| = 1$ in the w plane and that the point $w = 1$ does not correspond to a finite value of z.

16 If $w = (z - \mathrm{j})/(z + \mathrm{j})$, find and sketch the image in the w plane corresponding to the circle $|z| = 2$ in the z plane.

17 Show that the bilinear mapping

$$w = \mathrm{e}^{\mathrm{j}\theta_0} \frac{z - z_0}{z - z_0^*}$$

where θ_0 is a real constant $0 \leqslant \theta_0 < 2\pi$, z_0 a fixed complex number and z_0^* its conjugate, maps the upper half of the z plane ($\mathrm{Im}(z) > 0$) onto the inside of the unit circle in the w plane ($|w| < 1$). Find the values of z_0 and θ_0 if $w = 0$ corresponds to $z = \mathrm{j}$ and $w = -1$ corresponds to $z = \infty$.

18 Show that, under the mapping

$$w = \frac{2\mathrm{j}z}{z + \mathrm{j}}$$

circular arcs or the straight line through $z = 0$ and $z = \mathrm{j}$ in the z plane are mapped onto circular arcs or the straight line through $w = 0$ and $w = \mathrm{j}$ in the w plane. Find the images of the regions $|z - \frac{1}{2}| < \frac{1}{2}$ and $|z| < |z - \mathrm{j}|$ in the w plane.

19 Find the most general bilinear mapping that maps the unit circle $|z| = 1$ in the z plane onto the unit circle $|w| = 1$ in the w plane and the point $z = z_0$ in the z plane to the origin $w = 0$ in the w plane.

1.2.6 The mapping $w = z^2$

There are a number of other mappings that are used by engineers. For example, in dealing with Laplace and z transforms, the subjects of Chapters 2 and 3 respectively, we are concerned with the polynomial mapping

$$w = a_0 + a_1 z + \ldots + a_n z^n$$

where $a_0, a_1, \ldots, a_n$ are complex constants, the rational function

$$w = \frac{P(z)}{Q(z)}$$

where P and Q are polynomials in z, and the exponential mapping

$$w = a\mathrm{e}^{bz}$$

where e = 2.71828 . . . , the base of natural logarithms. As is clear from the bilinear mapping in Section 1.2.4, even elementary mappings can be cumbersome to analyse. Fortunately, we have two factors on our side. First, very detailed tracing of specific curves and their images is not required, only images of points. Secondly, by using complex differentiation, the subject of Section 1.3, various facets of these more complicated mappings can be understood without lengthy algebra. As a prelude, in this subsection we analyse the mapping $w = z^2$, which is the simplest polynomial mapping.

EXAMPLE 1.8 Investigate the mapping $w = z^2$ by plotting the images on the w plane of the lines x = constant and y = constant in the z plane.

Solution There is some difficulty in inverting this mapping to get z as a function of w, since square roots lead to problems of uniqueness. However, there is no need to invert here, for taking $w = u + \mathrm{j}v$ and $z = x + \mathrm{j}y$, the mapping becomes

$$w = u + \mathrm{j}v = (x + \mathrm{j}y)^2 = (x^2 - y^2) + \mathrm{j}2xy$$

which, on taking real and imaginary parts, gives

$$\begin{aligned} u &= x^2 - y^2 \\ v &= 2xy \end{aligned} \tag{1.18}$$

If $x = \alpha$, a real constant, then (1.18) becomes

$$u = \alpha^2 - y^2, \qquad v = 2\alpha y$$

which, on eliminating y, gives

$$u = \alpha^2 - \frac{v^2}{4\alpha^2}$$

or

$$4\alpha^2 u = 4\alpha^4 - v^2$$

so that

$$v^2 = 4\alpha^4 - 4\alpha^2 u = 4\alpha^2(\alpha^2 - u)$$

This represents a parabola in the w plane, and, since the right-hand side must be positive, $\alpha^2 \geqslant u$ so the 'nose' of the parabola is at $u = \alpha^2$ on the positive real axis in the w plane.

If $y = \beta$, a real constant, then (1.18) becomes

$$u = x^2 - \beta^2, \qquad v = 2x\beta$$

which, on eliminating x, gives

$$u = \frac{v^2}{4\beta^2} - \beta^2$$

or

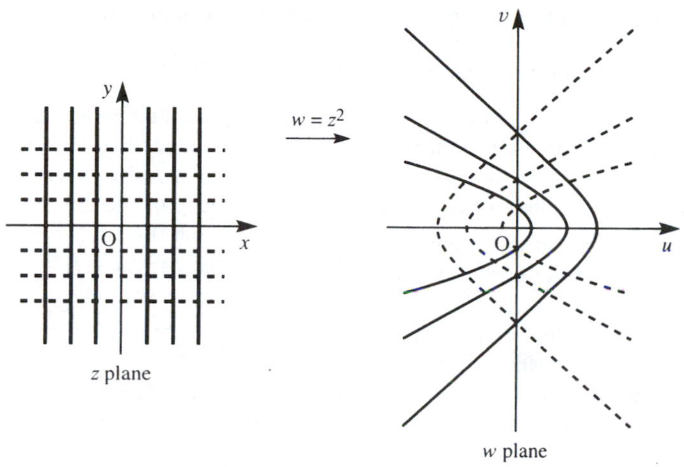

Figure 1.14 The mapping $w = z^2$.

$$4\beta^2 = v^2 - 4\beta^4$$

so that

$$v^2 = 4\beta^2 u + 4\beta^4 = 4\beta^2(u + \beta^2)$$

This is also a parabola, but pointing in the opposite direction. The right-hand side, as before, must be positive, so that $u > -\beta^2$ and the 'nose' of the parabola is on the negative real axis. These curves are drawn in Figure 1.14.

We shall not dwell further on the finer points of the mapping $w = z^2$. Instead, we note that in general it is extremely difficult to plot images of curves in the z plane, even the straight lines parallel to the axes, under polynomial mappings. We also note that we do not often need to do so, and that we have done it only as an aid to understanding.

The exercises that follow should also help in understanding this topic. We shall then return to examine polynomial, rational and exponential mappings in Section 1.3.4, after introducing complex differentiation.

1.2.7 Exercises

20 Find the image region in the w plane corresponding to the region inside the triangle in the z plane having vertices at $0 + j0$, $2 + j0$ and $0 + j2$ under the mapping $w = z^2$. Illustrate with sketches.

21 Find the images of the lines $y = x$ and $y = -x$ under the mapping $w = z^2$. Also find the image of the general line through the origin $y = mx$. By putting $m = \tan\theta_0$, deduce that straight lines intersecting at the origin in the z plane map onto lines intersecting at the origin in the w plane, but that the angle between these image lines is double that between the original lines.

22 Consider the mapping $w = z^n$, where n is an integer (a generalization of the mapping $w = z^2$). Use the polar representation of complex numbers to show that

(a) Circles centred at the origin in the z plane are mapped onto circles centred at the origin in the w plane.
(b) Straight lines passing through the origin intersecting with angle θ_0 in the z plane are mapped onto straight lines passing through the origin in the w plane but intersecting at an angle $n\theta_0$.

23 If the complex function

$$w = \frac{1 + z^2}{z}$$

is represented by a mapping from the z plane onto the w plane, find u in terms of x and y, and v in terms of x and y, where $z = x + jy$, $w = u + jv$. Find the image of the unit circle $|z| = 1$ in the w plane. Show that the circle centred at the origin, of radius r, in the z plane ($|z| = r$) is mapped onto the curve

$$\left(\frac{r^2 u}{r^2 + 1}\right)^2 + \left(\frac{r^2 v}{r^2 - 1}\right)^2 = r^2 \quad (r \neq 1)$$

in the w plane. What kind of curves are these? What happens for very large r?

1.3 Complex differentiation

The derivative of a real function $f(x)$ of a single real variable x at $x = x_0$ is given by the limit

$$f'(x_0) = \lim_{x \to x_0} \left[\frac{f(x) - f(x_0)}{x - x_0}\right]$$

Here, of course, x_0 is a real number and so can be represented by a single point on the real line. The point representing x can then approach the fixed x_0 either from the left or from the right along this line. Let us now turn to complex variables and functions depending on them. We know that a plane is required to represent complex numbers, so z_0 is now a fixed point in the Argand diagram, somewhere in the plane. The definition of the derivative of the function $f(z)$ of the complex variable z at the point z_0 will thus be

$$f'(z_0) = \lim_{z \to z_0} \left[\frac{f(z) - f(z_0)}{z - z_0}\right]$$

It may appear that if we merely exchange z for x, the rest of this section will follow similar lines to the differentiation of functions of real variables. For real variables taking the limit could only be done from the left or from the right, and the existence of a unique limit was not difficult to establish. For complex variables, however the point that represents the fixed complex number z_0 can be approached along an infinite number of curves in the z plane. The existence of a unique limit is thus a very stringent requirement. That most complex functions can be differentiated in the usual way is a remarkable property of the complex variable. Since $z = x + jy$, and x and y can vary independently, there are some connections with the calculus of functions of two real variables, but we shall not pursue this connection here.

Rather than use the word 'differentiable' to describe complex functions for which a derivative exists, if the function $f(z)$ has a derivative $f'(z)$ that exists at all points of a region R of the z plane then $f(z)$ is called **analytic** in R. Other terms such as **regular** or **holomorphic** are also used as alternatives to analytic. (Strictly, functions that have a power series expansion – see Section 1.4.1 – are called **analytic functions**. Since differentiable functions have a power series expansion they are referred to as analytic functions. However, there are examples of analytic functions that are not differentiable.)

1.3.1 Cauchy–Riemann equations

The following result is an important property of the analytic function.

If $z = x + \mathrm{j}y$ and $f(z) = u(x, y) + \mathrm{j}v(x, y)$, and $f(z)$ is analytic in some region R of the z plane, then the two equations

$$\frac{\partial u}{\partial x} = \frac{\partial v}{\partial y}, \quad \frac{\partial u}{\partial y} = -\frac{\partial v}{\partial x} \tag{1.19}$$

known as the **Cauchy–Riemann equations**, hold throughout R.

It is instructive to prove this result. Since $f'(z)$ exists at any point z_0 in R,

$$f'(z_0) = \lim_{z \to z_0} \left[\frac{f(z) - f(z_0)}{z - z_0} \right]$$

where z can tend to z_0 along any path within R. Examination of (1.19) suggests that we might choose paths parallel to the x direction and parallel to the y direction, since these will lead to partial derivatives with respect to x and y. Thus, choosing $z - z_0 = \Delta x$, a real path, we see that

$$f'(z_0) = \lim_{\Delta x \to 0} \left[\frac{f(z_0 + \Delta x) - f(z_0)}{\Delta x} \right]$$

Since $f(z) = u + \mathrm{j}v$, this means that

$$f'(z_0) = \lim_{\Delta x \to 0} \left[\frac{u(x_0 + \Delta x, y_0) + \mathrm{j}v(x_0 + \Delta x, y_0) - u(x_0, y_0) - \mathrm{j}v(x_0, y_0)}{\Delta x} \right]$$

or, on splitting into real and imaginary parts,

$$f'(z_0) = \lim_{\Delta x \to 0} \left[\frac{u(x_0 + \Delta x, y_0) - u(x_0, y_0)}{\Delta x} + \mathrm{j}\frac{v(x_0 + \Delta x, y_0) - v(x_0, y_0)}{\Delta x} \right]$$

giving

$$f'(z_0) = \left[\frac{\partial u}{\partial x} + j\frac{\partial v}{\partial x}\right]_{x=x_0,\ y=y_0} \tag{1.20}$$

Starting again from the definition of $f'(z_0)$, but this time choosing $z - z_0 = j\Delta y$ for the path parallel to the y axis, we obtain

$$f'(z_0) = \lim_{j\Delta y \to 0}\left[\frac{f(z_0 + j\Delta y) - f(z_0)}{j\Delta y}\right]$$

Once again, using $f(z) = u + jv$ and splitting into real and imaginary parts, we see that

$$f'(z_0) = \lim_{j\Delta y \to 0}\left[\frac{u(x_0, y_0 + \Delta y) + jv(x_0, y_0 + \Delta y) - u(x_0, y_0) - jv(x_0, y_0)}{j\Delta y}\right]$$

$$= \lim_{\Delta y \to 0}\left[\frac{1}{j}\frac{u(x_0, y_0 + \Delta y) - u(x_0, y_0)}{\Delta y} + \frac{v(x_0, y_0 + \Delta y) - v(x_0, y_0)}{\Delta y}\right]$$

giving

$$f'(z_0) = \left[\frac{1}{j}\frac{\partial u}{\partial y} + \frac{\partial v}{\partial y}\right]_{x=x_0,\ y=y_0} \tag{1.21}$$

Since $f'(z_0)$ must be the same no matter what path is followed, the two values obtained in (1.20) and (1.21) must be equal. Hence

$$\frac{\partial u}{\partial x} + j\frac{\partial v}{\partial x} = \frac{1}{j}\frac{\partial u}{\partial y} + \frac{\partial v}{\partial y} = -j\frac{\partial u}{\partial y} + \frac{\partial v}{\partial y}$$

Equating real and imaginary parts then gives the required Cauchy–Riemann equations

$$\frac{\partial u}{\partial x} = \frac{\partial v}{\partial y}, \quad \frac{\partial v}{\partial x} = -\frac{\partial u}{\partial y}$$

at the point $z = z_0$. However, z_0 is an arbitrarily chosen point in the region R; hence the Cauchy–Riemann equations hold throughout R, and we have thus proved the required result.

It is tempting to think that should we choose more paths along which to let $z - z_0$ tend to zero, we could derive more relationships along the same lines as the Cauchy–Riemann equations. It turns out, however, that we merely reproduce them or expressions derivable from them, and it is possible to prove that satisfaction of the Cauchy–Riemann equations (1.19) is a necessary condition for a function $f(z) = u(x, y) + jv(x, y)$, $z = x + jy$, to be analytic in a specified region. At points where $f'(z)$ exists it may be obtained from either (1.20) or (1.21) as

$$f'(z) = \frac{\partial u}{\partial x} + \mathrm{j}\frac{\partial v}{\partial x}$$

or

$$f'(z) = \frac{\partial v}{\partial y} - \mathrm{j}\frac{\partial u}{\partial y}$$

If z is given in the polar form $z = r\,\mathrm{e}^{\mathrm{j}\theta}$ then

$$f(z) = u(r, \theta) + \mathrm{j}v(r, \theta)$$

and the corresponding polar forms of the Cauchy–Riemann equations are

$$\frac{\partial u}{\partial r} = \frac{1}{r}\frac{\partial v}{\partial \theta}, \quad \frac{\partial v}{\partial r} = -\frac{1}{r}\frac{\partial u}{\partial \theta} \tag{1.22}$$

At points where $f'(z)$ exists it may be obtained from either of

$$f'(z) = \mathrm{e}^{-\mathrm{j}\theta}\left(\frac{\partial u}{\partial r} + \mathrm{j}\frac{\partial v}{\partial r}\right) \tag{1.23a}$$

or

$$f'(z) = \mathrm{e}^{-\mathrm{j}\theta}\left(\frac{1}{r}\frac{\partial v}{\partial \theta} - \frac{\mathrm{j}}{r}\frac{\partial u}{\partial \theta}\right) \tag{1.23b}$$

EXAMPLE 1.9 Verify that the function $f(z) = z^2$ satisfies the Cauchy–Riemann equations, and determine the derivative $f'(z)$.

Solution Since $z = x + \mathrm{j}y$, we have

$$f(z) = z^2 = (x + \mathrm{j}y)^2 = (x^2 - y^2) + \mathrm{j}2xy$$

so if $f(z) = u(x, y) + \mathrm{j}v(x, y)$ then

$$u = x^2 - y^2, \quad v = 2xy$$

giving the partial derivatives as

$$\frac{\partial u}{\partial x} = 2x, \quad \frac{\partial u}{\partial y} = -2y$$

$$\frac{\partial v}{\partial x} = 2y, \quad \frac{\partial v}{\partial y} = 2x$$

It is readily seen that the Cauchy–Riemann equations

$$\frac{\partial u}{\partial x} = \frac{\partial v}{\partial y}, \quad \frac{\partial u}{\partial y} = -\frac{\partial v}{\partial x}$$

are satisfied.

The derivative $f'(z)$ is then given by

$$f'(z) = \frac{\partial u}{\partial x} + j\frac{\partial v}{\partial x} = 2x + j2y = 2z$$

as expected.

EXAMPLE 1.10

Verify that the exponential function $f(z) = e^{\alpha z}$, where α is a constant, satisfies the Cauchy–Riemann equations, and show that $f'(z) = \alpha e^{\alpha z}$.

Solution

$$f(z) = u + jv = e^{\alpha z} = e^{\alpha(x+jy)} = e^{\alpha x} e^{j\alpha y}$$
$$= e^{\alpha x}(\cos \alpha y + j \sin \alpha y)$$

so, equating real and imaginary parts,

$$u = e^{\alpha x} \cos \alpha y, \quad v = e^{\alpha x} \sin \alpha y$$

The partial derivatives are

$$\frac{\partial u}{\partial x} = \alpha e^{\alpha x} \cos \alpha y, \quad \frac{\partial v}{\partial x} = \alpha e^{\alpha x} \sin \alpha y$$

$$\frac{\partial u}{\partial y} = -\alpha e^{\alpha x} \sin \alpha y, \quad \frac{\partial v}{\partial y} = \alpha e^{\alpha x} \cos \alpha y$$

confirming that the Cauchy–Riemann equations are satisfied. The derivative $f'(z)$ is then given by

$$f'(z) = \frac{\partial u}{\partial x} + j\frac{\partial v}{\partial x} = \alpha e^{\alpha x}(\cos \alpha y + j \sin \alpha y) = \alpha e^{\alpha z}$$

so that

$$\frac{d}{dz} e^{\alpha z} = \alpha e^{\alpha z} \tag{1.24}$$

As in the real variable case, we have

$$e^{jz} = \cos z + j \sin z \tag{1.25}$$

(Section 1.4.3), so that $\cos z$ and $\sin z$ may be expressed as

$$\left.\begin{aligned} \cos z &= \frac{e^{jz} + e^{-jz}}{2} \\ \sin z &= \frac{e^{jz} - e^{-jz}}{2j} \end{aligned}\right\} \tag{1.26a}$$

Using result (1.24) from Example 1.10, it is then readily shown that

$$\frac{d}{dz}(\sin z) = \cos z$$

$$\frac{d}{dz}(\cos z) = -\sin z$$

Similarly, we define the hyperbolic functions $\sinh z$ and $\cosh z$ by

$$\left.\begin{aligned}\sinh z &= \frac{e^z - e^{-z}}{2} = -j\sin jz \\ \cosh z &= \frac{e^z + e^{-z}}{2} = \cos jz\end{aligned}\right\} \qquad (1.26b)$$

from which, using (1.24), it is readily deduced that

$$\frac{d}{dz}(\sinh z) = \cosh z$$

$$\frac{d}{dz}(\cosh z) = \sinh z$$

We note from above that e^z has the following real and imaginary parts:

$$\text{Re}(e^z) = e^x \cos y$$

$$\text{Im}(e^z) = e^x \sin y$$

In real variables the exponential and circular functions are contrasted, one being monotonic, the other oscillatory. In complex variables, however, the real and imaginary parts of e^z are (two-variable) combinations of exponential and circular functions, which might seem surprising for an exponential function. Similarly, the circular functions of a complex variable have unfamiliar properties. For example, it is easy to see that $|\cos z|$ and $|\sin z|$ are unbounded for complex z by using the above relationships between circular and hyperbolic functions of complex variables. Contrast this with $|\cos x| \leqslant 1$ and $|\sin x| \leqslant 1$ for a real variable x.

In a similar way to the method adopted in Examples 1.9 and 1.10 it can be shown that the derivatives of the majority of functions $f(x)$ of a real variable x carry over to the complex variable case $f(z)$ at points where $f(z)$ is analytic. Thus, for example,

$$\frac{d}{dz} z^n = nz^{n-1}$$

for all z in the z plane, and

$$\frac{d}{dz} \ln z = \frac{1}{z}$$

for all z in the z plane except for points on the non-positive real axis, where $\ln z$ is non-analytic.

It can also be shown that the rules associated with derivatives of a function of a real variable, such as the sum, product, quotient and chain rules, carry over to the complex variable case. Thus,

$$\frac{\mathrm{d}}{\mathrm{d}z}[f(z) + g(z)] = \frac{\mathrm{d}f(z)}{\mathrm{d}z} + \frac{\mathrm{d}g(z)}{\mathrm{d}z}$$

$$\frac{\mathrm{d}}{\mathrm{d}z}[f(z)g(z)] = f(z)\frac{\mathrm{d}g(z)}{\mathrm{d}z} + \frac{\mathrm{d}f(z)}{\mathrm{d}z}g(z)$$

$$\frac{\mathrm{d}}{\mathrm{d}z}f(g(z)) = \frac{\mathrm{d}f}{\mathrm{d}g}\frac{\mathrm{d}g}{\mathrm{d}z}$$

$$\frac{\mathrm{d}}{\mathrm{d}z}\left[\frac{f(z)}{g(z)}\right] = \frac{g(z)f'(z) - f(z)g'(z)}{[g(z)]^2}$$

1.3.2 Conjugate and harmonic functions

A pair of functions $u(x, y)$ and $v(x, y)$ of the real variable x and y that satisfy the Cauchy–Riemann equations (1.19) are said to be **conjugate functions**. (Note here the different use of the word 'conjugate' to that used in complex number work, where $z^* = x - \mathrm{j}y$ is the complex conjugate of $z = x + \mathrm{j}y$.) Conjugate functions satisfy the orthogonality property in that the curves in the (x, y) plane defined by $u(x, y) = \text{constant}$ and $v(x, y) = \text{constant}$ are orthogonal curves. This follows since the gradient at any point on the curve $u(x, y) = \text{constant}$ is given by

$$\left[\frac{\mathrm{d}y}{\mathrm{d}x}\right]_u = -\frac{\partial u}{\partial y}\bigg/\frac{\partial u}{\partial x}$$

and the gradient at any point on the curve $v(x, y) = \text{constant}$ is given by

$$\left[\frac{\mathrm{d}y}{\mathrm{d}x}\right]_v = -\frac{\partial v}{\partial y}\bigg/\frac{\partial v}{\partial x}$$

It follows from the Cauchy–Riemann equations (1.19) that

$$\left[\frac{\mathrm{d}y}{\mathrm{d}x}\right]_u\left[\frac{\mathrm{d}y}{\mathrm{d}x}\right]_v = -1$$

so the curves are orthogonal.

A function that satisfies Laplace's equation in two dimensions is said to be **harmonic**; that is, $u(x, y)$ is a harmonic function if

$$\frac{\partial^2 u}{\partial x^2} + \frac{\partial^2 u}{\partial y^2} = 0$$

It is readily shown (see Example 1.12) that if $f(z) = u(x, y) + \mathrm{j}v(x, y)$ is analytic, so that the Cauchy–Riemann equations are satisfied, then both u and v are **harmonic** functions. Therefore u and v are **conjugate harmonic functions**. Harmonic functions have applications in such areas as stress analysis in plates, inviscid two-dimensional fluid flow and electrostatics.

EXAMPLE 1.11 Given $u(x, y) = x^2 - y^2 + 2x$, find the conjugate function $v(x, y)$ such that $f(z) = u(x, y) + jv(x, y)$ is an analytic function of z throughout the z plane.

Solution We are given $u(x, y) = x^2 - y^2 + 2x$, and, since $f(z) = u + jv$ is to be analytic, the Cauchy–Riemann equations must hold. Thus, from (1.19),

$$\frac{\partial v}{\partial y} = \frac{\partial u}{\partial x} = 2x + 2$$

Integrating this with respect to y gives

$$v = 2xy + 2y + F(x)$$

where $F(x)$ is an arbitrary function of x, since the integration was performed holding x constant. Differentiating v partially with respect to x gives

$$\frac{\partial v}{\partial x} = 2y + \frac{\mathrm{d}F}{\mathrm{d}x}$$

but this equals $-\partial u/\partial y$ by the second of the Cauchy–Riemann equations (1.19). Hence

$$\frac{\partial u}{\partial y} = -2y - \frac{\mathrm{d}F}{\mathrm{d}x}$$

But since $u = x^2 - y^2 + 2x$, $\partial u/\partial y = -2y$, and comparison yields $F(x) =$ constant. This constant is set equal to zero, since no conditions have been given by which it can be determined. Hence

$$u(x, y) + jv(x, y) = x^2 - y^2 + 2x + j(2xy + 2y)$$

To confirm that this is a function of z, note that $f(z)$ is $f(x + jy)$, and becomes just $f(x)$ if we set $y = 0$. Therefore we set $y = 0$ to obtain

$$f(x + j0) = f(x) = u(x, 0) + jv(x, 0) = x^2 + 2x$$

and it follows that

$$f(z) = z^2 + 2z$$

which can be easily checked by separation into real and imaginary parts.

EXAMPLE 1.12 Show that the real and imaginary parts $u(x, y)$ and $v(x, y)$ of a complex analytic function $f(z)$ are harmonic.

Solution Since

$$f(z) = u(x, y) + jv(x, y)$$

is analytic, the Cauchy–Riemann equations

$$\frac{\partial v}{\partial x} = -\frac{\partial u}{\partial y}, \quad \frac{\partial u}{\partial x} = \frac{\partial v}{\partial y}$$

are satisfied. Differentiating the first with respect to x gives

$$\frac{\partial^2 v}{\partial x^2} = -\frac{\partial^2 u}{\partial x \partial y} = -\frac{\partial^2 u}{\partial y \partial x} = -\frac{\partial}{\partial y}\left(\frac{\partial u}{\partial x}\right)$$

which is $-\partial^2 v/\partial y^2$, by the second Cauchy–Riemann equation. Hence

$$\frac{\partial^2 v}{\partial x^2} = -\frac{\partial^2 v}{\partial y^2}, \quad \text{or} \quad \frac{\partial^2 v}{\partial x^2} + \frac{\partial^2 v}{\partial y^2} = 0$$

and v is a harmonic function.

Similarly,

$$\frac{\partial^2 u}{\partial y^2} = -\frac{\partial^2 v}{\partial y \partial x} = -\frac{\partial}{\partial x}\left(\frac{\partial v}{\partial y}\right) = -\frac{\partial^2 u}{\partial x^2}$$

so that

$$\frac{\partial^2 v}{\partial x^2} + \frac{\partial^2 v}{\partial y^2} = 0$$

and u is also a harmonic function. We have assumed that both u and v have continuous second-order partial derivatives, so that

$$\frac{\partial^2 u}{\partial x \partial y} = \frac{\partial^2 u}{\partial y \partial x}, \quad \frac{\partial^2 v}{\partial x \partial y} = \frac{\partial^2 v}{\partial y \partial x}$$

1.3.3 Exercises

24 Determine whether the following functions are analytic, and find the derivative where appropriate:

(a) $z\,\mathrm{e}^z$ (b) $\sin 4z$

(c) zz^* (d) $\cos 2z$

25 Determine the constants a and b in order that

$$w = x^2 + ay^2 - 2xy + \mathrm{j}(bx^2 - y^2 + 2xy)$$

be analytic. For these values of a and b find the derivative of w, and express both w and $\mathrm{d}w/\mathrm{d}z$ as functions of $z = x + \mathrm{j}y$.

26 Find a function $v(x, y)$ such that, given $u = 2x(1 - y)$, $f(z) = u + \mathrm{j}v$ is analytic in z.

27 Show that $\phi(x, y) = \mathrm{e}^x(x\cos y - y\sin y)$ is a harmonic function, and find the conjugate harmonic function $\psi(x, y)$. Write $\phi(x, y) + \mathrm{j}\psi(x, y)$ as a function of $z = x + \mathrm{j}y$ only.

28 Show that $u(x, y) = \sin x \cosh y$ is harmonic. Find the harmonic conjugate $v(x, y)$ and express $w = u + \mathrm{j}v$ as a function of $z = x + \mathrm{j}y$.

29 Find the orthogonal trajectories of the following families of curves:

(a) $x^3y - xy^3 = \alpha$ (constant α)

(b) $\mathrm{e}^{-x}\cos y + xy = \alpha$ (constant α)

30 Find the real and imaginary parts of the functions

(a) $z^2 e^{2z}$ (b) $\sin 2z$

Verify that they are analytic and find their derivatives.

31 Give a definition of the inverse sine function $\sin^{-1} z$ for complex z. Find the real and imaginary parts of $\sin^{-1} z$. (*Hint*: put $z = \sin w$, split into real and imaginary parts, and with $w = u + jv$ and $z = x + jy$ solve for u and v in terms of x and y.) Is $\sin^{-1} z$ analytic? If so, what is its derivative?

32 Establish that if $z = x + jy$, $|\sinh y| \leqslant |\sin z| \leqslant \cosh y$.

1.3.4 Mappings revisited

In Section 1.2 we examined mappings from the z plane to the w plane, where in the main the relationship between w and z, $w = f(z)$ was linear or bilinear. There is an important property of mappings, hinted at in Example 1.8 when considering the mapping $w = z^2$. A mapping $w = f(z)$ that preserves angles is called **conformal**. Under such a mapping, the angle between two intersecting curves in the z plane is the same as the angle between the corresponding intersecting curves in the w plane. The sense of the angle is also preserved. That is, if θ is the angle between curves 1 and 2 taken in the anticlockwise sense in the z plane then θ is also the angle between the image of curve 1 and the image of curve 2 in the w plane, and it too is taken in the anticlockwise sense. Figure 1.15 should make the idea of a conformal mapping clearer. If $f(z)$ is analytic then $w = f(z)$ defines a conformal mapping except at points where the derivative $f'(z)$ is zero.

Clearly the linear mappings

$$w = \alpha z + \beta \quad (\alpha \neq 0)$$

are conformal everywhere, since $dw/dz = \alpha$ and is not zero for any point in the z plane. Bilinear mappings given by (1.12) are not so straightforward to check. However, as we saw in Section 1.2.4, (1.12) can be rearranged as

$$w = \lambda + \frac{\mu}{\alpha z + \beta} \quad (\alpha, \mu \neq 0)$$

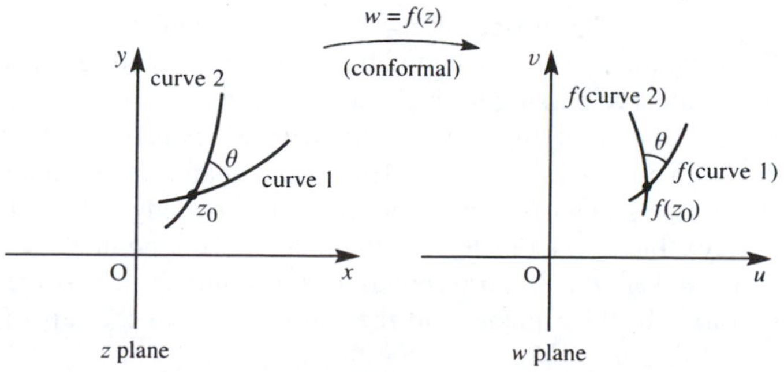

Figure 1.15 Conformal mappings.

Thus

$$\frac{\mathrm{d}w}{\mathrm{d}z} = -\frac{\mu\alpha}{(\alpha z + \beta)^2}$$

which again is never zero for any point in the z plane. In fact, the only mapping we have considered so far that has a point at which it is not conformal everywhere is $w = z^2$ (cf. Example 1.8), which is not conformal at $z = 0$.

EXAMPLE 1.13 Determine the points at which the mapping $w = z + 1/z$ is not conformal and demonstrate this by considering the image in the w plane of the real axis in the z plane.

Solution Taking $z = x + \mathrm{j}y$ and $w = u + \mathrm{j}v$, we have

$$w = u + \mathrm{j}v = x + \mathrm{j}y + \frac{x - \mathrm{j}y}{x^2 + y^2}$$

which, on equating real and imaginary parts, gives

$$u = x + \frac{x}{x^2 + y^2}$$

$$v = y - \frac{y}{x^2 + y^2}$$

The real axis, $y = 0$, in the z plane corresponds to $v = 0$, the real axis in the w plane. Note, however, that the fixed point of the mapping is given by

$$z = z + \frac{1}{z}$$

or $z = \infty$. From the Cauchy–Riemann equations it is readily shown that w is analytic everywhere except at $z = 0$. Also, $\mathrm{d}w/\mathrm{d}z = 0$ when

$$1 - \frac{1}{z^2} = 0, \quad \text{that is,} \quad z = \pm 1$$

which are both on the real axis. Thus the mapping fails to be conformal at $z = 0$ and $z = \pm 1$. The image of $z = 1$ is $w = 2$, and the image of $z = -1$ is $w = -2$. Consideration of the image of the real axis is therefore perfectly adequate, since this is a curve passing through each point where $w = z + 1/z$ fails to be conformal. It would be satisfying if we could analyse this mapping in the same manner as we did with $w = z^2$ in Example 1.8. Unfortunately, we cannot do this, because the algebra gets unwieldy (and, indeed, our knowledge of algebraic curves is also too scanty). Instead, let us look at the image of the point $z = 1 + \varepsilon$, where ε is a small real number. $\varepsilon > 0$ corresponds to the point Q just to the right of $z = 1$ on the real axis in the z plane, and the point P just to the left of $z = 1$ corresponds to $\varepsilon < 0$ (Figure 1.16).

If $z = 1 + \varepsilon$ then

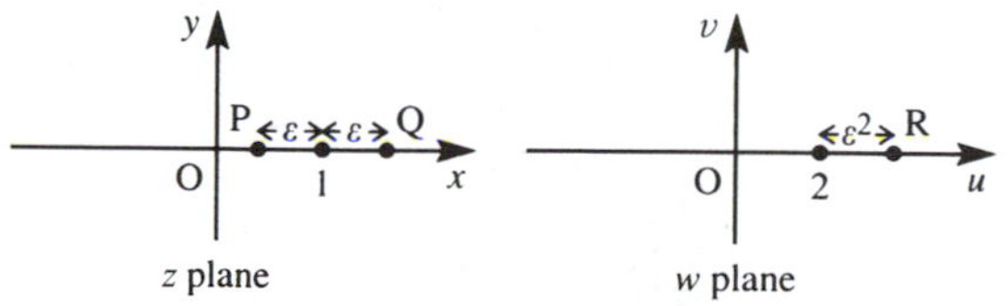

Figure 1.16 Image of $z = 1 + \varepsilon$ of Example 1.13.

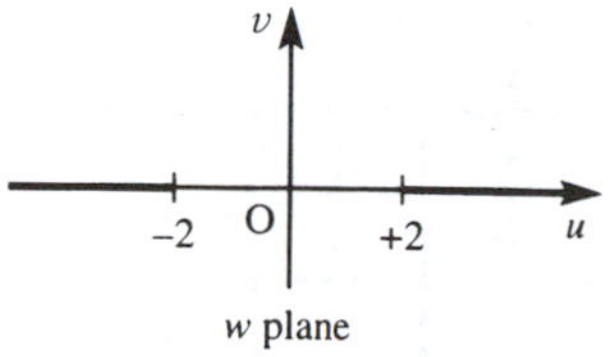

Figure 1.17 Image in w plane of the real axis in the z plane for Example 1.13.

$$\begin{aligned} w &= 1 + \varepsilon + \frac{1}{1 + \varepsilon} \\ &= 1 + \varepsilon + (1 + \varepsilon)^{-1} \\ &= 1 + \varepsilon + 1 - \varepsilon + \varepsilon^2 - \varepsilon^3 + \ldots \\ &\simeq 2 + \varepsilon^2 \end{aligned}$$

if $|\varepsilon|$ is much smaller than 1 (we shall discuss the validity of the power series expansion in Section 1.4). Whether ε is positive or negative, the point $w = 2 + \varepsilon^2$ is to the right of $w = 2$ in the w plane as indicated by the point R in Figure 1.16. Therefore, as $\varepsilon \to 0$, a curve (the real axis) that passes through $z = 1$ in the z plane making an angle $\theta = \pi$ corresponds to a curve (again the real axis) that approaches $w = 2$ in the w plane along the real axis from the right making an angle $\theta = 0$. Non-conformality has thus been confirmed. The treatment of $z = -1$ follows in an identical fashion, so the details are omitted. Note that when $y = 0$ ($v = 0$), $u = x + 1/x$ so, as the real axis in the z plane is traversed from $x = -\infty$ to $x = 0$, the real axis in the w plane is traversed from $u = -\infty$ to -2 *and back to* $u = -\infty$ *again* (when $x = -1$, u reaches 2). As the real axis in the z plane is traversed from $x = 0$ through $x = 1$ to $x = +\infty$, so the real axis in the w plane is traversed from $u = +\infty$ to $u = +2$ ($x = 1$) *back to* $u = \infty$ *again*. Hence the points on the real axis in the w plane in the range $-2 < u < 2$ do not correspond to real values of z. Solving $u = x + 1/x$ for x gives

$$x = \tfrac{1}{2}[u \pm \surd(u^2 - 4)]$$

which makes this point obvious. Figure 1.17 shows the image in the w plane of the real axis in the z plane. This mapping is very rich in interesting properties, but we shall not pursue it further here. Aeronautical engineers may well meet it again if they study the flow around an aerofoil in two dimensions, for this mapping takes circles centred at the origin in the z plane onto meniscus (lens-shaped) regions in the w plane, and only a slight alteration is required before these images become aerofoil-shaped.

EXAMPLE 1.14

Examine the mapping

$$w = e^z$$

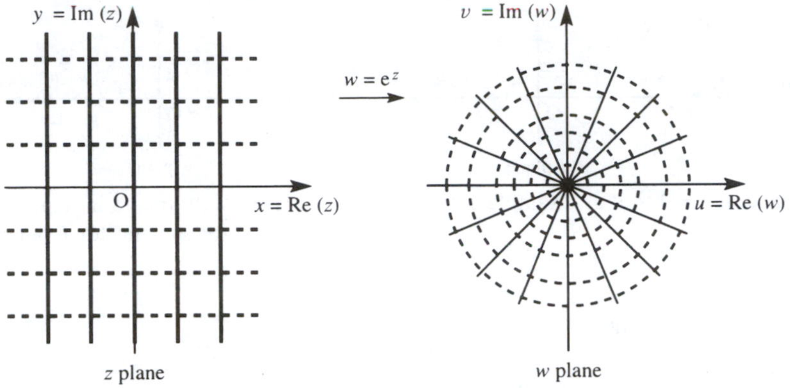

Figure 1.18 Mapping of lines under $w = e^z$.

by (a) finding the images in the w plane of the lines x = constant and y = constant into the z plane, and (b) finding the image in the w plane of the left half-plane ($x < 0$) in the z plane.

Solution Taking $z = x + jy$ and $w = u + jv$, for $w = e^z$ we have

$$u = e^x \cos y$$

$$v = e^x \sin y$$

Squaring and adding these two equations, we obtain

$$u^2 + v^2 = e^{2x}$$

On the other hand, dividing the two equations gives

$$\frac{v}{u} = \tan y$$

We can now tackle the questions.

(a) Since $u^2 + v^2 = e^{2x}$, putting x = constant shows that the lines parallel to the imaginary axis in the z plane correspond to circles centred at the origin in the w plane. The equation

$$\frac{v}{u} = \tan y$$

shows that the lines parallel to the real axis in the z plane correspond to straight lines through the origin in the w plane ($v = u \tan \alpha$ if $y = \alpha$, a constant). Figure 1.18 shows the general picture.

(b) Since $u^2 + v^2 = e^{2x}$, if $x = 0$ then $u^2 + v^2 = 1$, so the imaginary axis in the z plane corresponds to the unit circle in the w plane. If $x < 0$ then $e^{2x} < 1$, and as $x \to -\infty$, $e^{2x} \to 0$, so the left half of the z plane corresponds to the interior of the unit circle in the w plane, as illustrated in Figure 1.19.

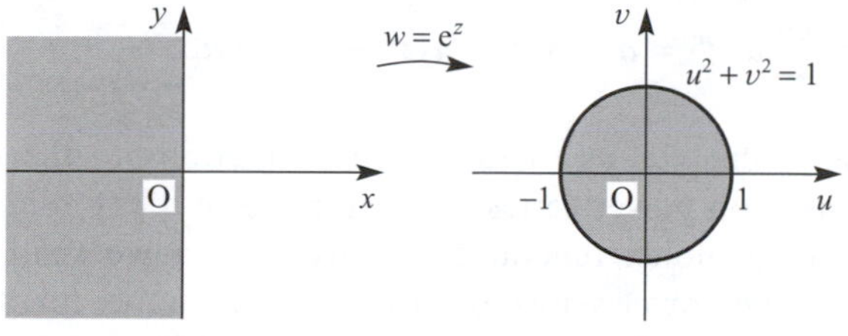

Figure 1.19 Mapping of half plane under $w = e^z$.

1.3.5 Exercises

33 Determine the points at which the following mappings are *not* conformal:

(a) $w = z^2 - 1$ (b) $w = 2z^3 - 21z^2 + 72z + 6$

(c) $w = 8z + \dfrac{1}{2z^2}$

34 Follow Example 1.13 for the mapping $w = z - 1/z$. Again determine the points at which the mapping is not conformal, but this time demonstrate this by looking at the image of the *imaginary* axis.

35 Find the region of the w plane corresponding to the following regions of the z plane under the exponential mapping $w = e^z$:

(a) $0 \leqslant x < \infty$ (b) $0 \leqslant x \leqslant 1,\ 0 \leqslant y \leqslant 1$

(c) $\frac{1}{2}\pi \leqslant y \leqslant \pi,\ 0 \leqslant x < \infty$

36 Consider the mapping $w = \sin z$. Determine the points at which the mapping is not conformal. By finding the images in the w plane of the lines x = constant and y = constant in the z plane ($z = x + jy$), draw the mapping along similar lines to Figures 1.14 and 1.18.

37 Show that the transformation

$$z = \zeta + \frac{a^2}{\zeta}$$

where $z = x + jy$ and $\zeta = R\,e^{j\theta}$ maps a circle, with centre at the origin and radius a, in the ζ plane, onto a straight line segment in the z plane. What is the length of the line? What happens if the circle in the ζ plane is centred at the origin but of radius b, where $b \neq a$?

1.4 Complex series

In *Modern Engineering Mathematics* we saw that there were distinct advantages in being able to express a function $f(x)$, such as the exponential, trigonometric and logarithmic functions, of a real variable x in terms of its power series expansion

$$f(x) = \sum_{n=0}^{\infty} a_n x^n = a_0 + a_1 x + a_2 x^2 + \ldots + a_r x^r + \ldots \qquad \textbf{(1.27)}$$

Power series are also very important in dealing with complex functions. In fact, any real function $f(x)$ which has a power series of the form in (1.27) has a corresponding complex function $f(z)$ having the same power series expansion, that is

$$f(z) = \sum_{n=0}^{\infty} a_n z^n = a_0 + a_1 z + a_2 z^2 + \ldots + a_r z^r + \ldots \tag{1.28}$$

This property enables us to extend real functions to the complex case, so that methods based on power series expansions have a key role to play in formulating the theory of complex functions. In this section we shall consider some of the properties of the power series expansion of a complex function by drawing, wherever possible, an analogy with the power series expansion of the corresponding real function.

1.4.1 Power series

A series having the form

$$\sum_{n=0}^{\infty} a_n (z - z_0)^n = a_0 + a_1(z - z_0) + a_2(z - z_0)^2 + \ldots + a_r(z - z_0)^r + \ldots \tag{1.29}$$

in which the coefficients a_r are real or complex and z_0 is a fixed point in the complex z plane is called a **power series** about z_0 or a power series centred on z_0. Where $z_0 = 0$, the series (1.29) reduces to the series (1.28), which is a power series centred at the origin. In fact, on making the change of variable $z' = z - z_0$, (1.29) takes the form (1.28), so there is no loss of generality in considering the latter below.

Tests for the convergence or divergence of complex power series are similar to those used for power series of a real variable. However, in complex series it is essential that the modulus $|a_n|$ be used. For example, the geometric series

$$\sum_{n=0}^{\infty} z^n$$

has a sum to N terms

$$S_N = \sum_{n=0}^{N-1} z^n = \frac{1 - z^N}{1 - z}$$

and converges, if $|z| < 1$, to the limit $1/(1 - z)$ as $N \to \infty$. If $|z| \geqslant 1$, the series diverges. These results appear to be identical with the requirement that $|x| < 1$ to ensure convergence of the real power series

$$\frac{1}{1 - x} = \sum_{n=0}^{\infty} x^n$$

However, in the complex case the geometrical interpretation is different in that the condition $|z| < 1$ implies that z lies inside the circle centred at the origin and radius 1 in the z plane. Thus the series $\sum_{n=0}^{\infty} z^n$ converges if z lies inside this circle and diverges if z lies on or outside it. The situation is illustrated in Figure 1.20.

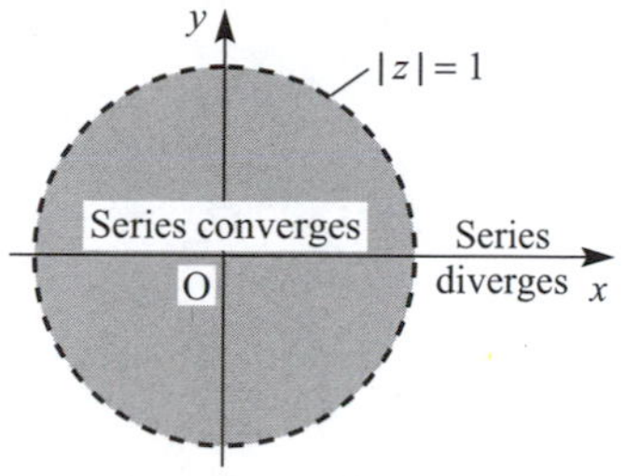

Figure 1.20 Region of convergence of $\sum_{n=0}^{\infty} z^n$.

The existence of such a circle leads to an important concept in that in general there exists a circle centred at the origin and of radius R such that the series

$$\sum_{n=0}^{\infty} a_n z^n \begin{cases} \text{converges if} & |z| < R \\ \text{diverges if} & |z| > R \end{cases}$$

The radius R is called the **radius of convergence** of the power series. What happens when $|z| = R$ is normally investigated as a special case?

We have introduced the radius of convergence based on a circle centred at the origin, while the concept obviously does not depend on the location of the centre of the circle. If the series is centred on z_0 as in (1.29) then the convergence circle would be centred on z_0. Indeed it could even be centred at infinity, when the power series becomes

$$\sum_{n=0}^{\infty} a_n z^{-n} = a_0 + \frac{a_1}{z} + \frac{a_2}{z^2} + \ldots + \frac{a_r}{z^r} + \ldots$$

which we shall consider further in Section 1.4.5.

In order to determine the radius of convergence R for a given series, various tests for convergence, such as those introduced in *Modern Engineering Mathematics* for real series, may be applied. In particular, using d'Alembert's ratio test, it can be shown that the radius of convergence R of the complex series $\sum_{n=0}^{\infty} a_n z^n$ is given by

$$R = \lim_{n\to\infty} \left| \frac{a_n}{a_{n+1}} \right| \qquad \textbf{(1.30)}$$

provided that the limit exists. Then the series is convergent within the disc $|z| < R$. In general, of course, the limit may not exist, and in such cases an alternative method must be used.

EXAMPLE 1.15 Find the power series, in the form indicated, representing the function $1/(z - 3)$ in the following three regions:

(a) $|z| < 3$; $\quad \sum_{n=0}^{\infty} a_n z^n$

(b) $|z - 2| < 1$; $\quad \sum_{n=0}^{\infty} a_n (z - 2)^n$

(c) $|z| > 3$; $\quad \sum_{n=0}^{\infty} \frac{a_n}{z^n}$

and sketch these regions on an Argand diagram.

Solution We know that the binomial series expansion

$$(1+z)^n = 1 + nz + \frac{n(n-1)}{2!}z^2 + \ldots + \frac{n(n-1)(n-2)\cdots(n-r+1)}{r!}z^r + \ldots$$

is valid for $|z| < 1$. To solve the problem, we exploit this result by expanding the function $1/(z - 3)$ in three different ways:

(a) $$\frac{1}{z-3} = \frac{-\frac{1}{3}}{1-\frac{1}{3}z} = -\frac{1}{3}\left(1-\frac{1}{3}z\right)^{-1} = -\frac{1}{3}\left[1+\frac{1}{3}z+\left(\frac{1}{3}z\right)^2+\ldots+\left(\frac{1}{3}z\right)^n+\ldots\right]$$

for $|\frac{1}{3}z| < 1$, that is $|z| < 3$, giving the power series

$$\frac{1}{z-3} = -\frac{1}{3} - \frac{1}{9}z - \frac{1}{27}z^2 - \ldots \quad (|z| < 3)$$

(b) $$\frac{1}{z-3} = \frac{1}{(z-2)-1} = [(z-2)-1]^{-1}$$

$$= -[1 + (z-2) + (z-2)^2 + \ldots] \quad (|z-2| < 1)$$

giving the power series

$$\frac{1}{z-3} = -1 - (z-2) - (z-2)^2 - \ldots \quad (|z-2| < 1)$$

(c) $$\frac{1}{z-3} = \frac{1/z}{1-3/z} = \frac{1}{z}\left[1+\frac{3}{z}+\left(\frac{3}{z}\right)^2+\ldots\right]$$

giving the power series

$$\frac{1}{z-3} = \frac{1}{z} + \frac{3}{z^2} + \frac{9}{z^3} + \ldots \quad (|z| > 3)$$

The three regions are sketched in Figure 1.21. Note that none of the regions include the point $z = 3$, which is termed a **singularity** of the function, a concept we shall discuss in Section 1.5.1.

In Example 1.15 the whole of the circle $|z| = 3$ was excluded from the three regions where the power series converge. In fact, it is possible to include any selected point in the z plane as a centre of the circle in which to define a power

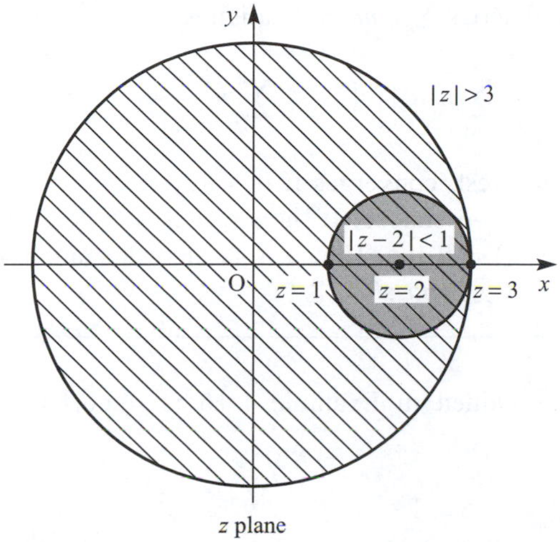

Figure 1.21 Regions of convergence for the series in Example 1.15.

series that converges to $1/(z-3)$ everywhere inside the circle, with the exception of the point $z = 3$. For example, the point $z = 4j$ would lead to the expansion of

$$\frac{1}{z-4j+4j-3} = \frac{1}{4j-3}\,\frac{1}{\dfrac{z-4j}{4j-3}+1}$$

in a binomial series in powers of $(z-4j)/(4j-3)$, which converges to $1/(z-3)$ inside the circle

$$|z-4j| = |4j-3| = \sqrt{(16+9)} = 5$$

We should not expect the point $z = 3$ to be included in any of the circles, since the function $1/(z-3)$ is infinite there and hence not defined.

EXAMPLE 1.16 Prove that both the power series $\sum_{n=0}^{\infty} a_n z^n$ and the corresponding series of derivatives $\sum_{n=0}^{\infty} na_n z^{n-1}$ have the same radius of convergence.

Solution Let R be the radius of convergence of the power series $\sum_{n=0}^{\infty} a_n z^n$. Since $\lim_{n\to\infty}(a_n z_0^n) = 0$ (otherwise the series has no chance of convergence), if $|z_0| < R$ for some complex number z_0 then it is always possible to choose

$$|a_n| < |z_0|^{-n}$$

for $n > N$, with N a fixed integer. We now use d'Alembert's ratio test, namely

$$\text{if} \quad \lim_{n\to\infty}\left|\frac{a_{n+1}}{a_n}\right| < 1 \quad \text{then} \quad \sum_{n=0}^{\infty} a_n z^n \quad \text{converges}$$

$$\text{if} \quad \lim_{n\to\infty}\left|\frac{a_{n+1}}{a_n}\right| > 1 \quad \text{then} \quad \sum_{n=0}^{\infty} a_n z^n \quad \text{diverges}$$

The differentiated series $\sum_{n=0}^{\infty} na_n z^{n-1}$ satisfies

$$\sum_{n=1}^{\infty} |na_n z^{n-1}| < \sum_{n=1}^{\infty} n|a_n||z|^{n-1} < \sum_{n=1}^{\infty} n\frac{|z|^{n-1}}{|z_0|^n}$$

which, by the ratio test, converges if $0 < |z_0| < R$, since $|z| < |z_0|$ and $|z_0|$ can be as close to R as we choose. If, however, $|z| > R$ then $\lim_{n\to\infty}(a_n z^n) \neq 0$ and thus $\lim_{n\to\infty}(na_n z^{n-1}) \neq 0$ too. Hence R is also the radius of convergence of the differentiated series $\sum_{n=1}^{\infty} na_n z^{n-1}$.

The result obtained in Example 1.16 is important, since if the complex function

$$f(z) = \sum_{n=0}^{\infty} a_n z^n$$

converges in $|z| < R$ then the derivative

$$f'(z) = \sum_{n=1}^{\infty} na_n z^{n-1}$$

also converges in $|z| < R$. We can go on differentiating $f(z)$ through its power series and be sure that the differentiated function and the differentiated power series are equal inside the circle of convergence.

1.4.2 Exercises

38 Find the power series representation for the function $1/(z - \mathrm{j})$ in the regions

(a) $|z| < 1$ (b) $|z| > 1$ (c) $|z - 1 - \mathrm{j}| < \sqrt{2}$

Deduce that the radius of convergence of the power series representation of this function is $|z_0 - \mathrm{j}|$, where $z = z_0$ is the centre of the circle of convergence ($z_0 \neq \mathrm{j}$).

39 Find the power series representation of the function

$$f(z) = \frac{1}{z^2 + 1}$$

in the disc $|z| < 1$. Use Example 1.16 to deduce the power series for

(a) $\dfrac{1}{(z^2 + 1)^2}$ (b) $\dfrac{1}{(z^2 + 1)^3}$

valid in this same disc.

1.4.3 Taylor series

In *Modern Engineering Mathematics* we introduced the Taylor series expansion

$$f(x+a) = f(a) + \frac{x}{1!}f^{(1)}(a) + \frac{x^2}{2!}f^{(2)}(a) + \ldots = \sum_{n=0}^{\infty} \frac{x^n}{n!}f^{(n)}(a) \tag{1.31}$$

of a function $f(x)$ of a real variable x about $x = a$ and valid within the interval of convergence of the power series. For the engineer the ability to express a function in such a power series expansion is seen to be particularly useful in the development of numerical methods and the assessment of errors. The ability to express a complex function as a Taylor series is also important to engineers in many fields of applications, such as control and communications theory. The form of the Taylor series in the complex case is identical with that of (1.31).

If $f(z)$ is a complex function analytic inside and on a simple closed curve C (usually a circle) in the z plane then it follows from Example 1.16 that the higher derivatives of $f(z)$ also exist inside C. If z_0 and $z_0 + h$ are two fixed points inside C then

$$f(z_0 + h) = f(z_0) + hf^{(1)}(z_0) + \frac{h^2}{2!}f^{(2)}(z_0) + \ldots + \frac{h^n}{n!}f^{(n)}(z_0) + \ldots$$

where $f^{(k)}(z_0)$ is the kth derivative of $f(z)$ evaluated at $z = z_0$. Normally, $z = z_0 + h$ is introduced so that $h = z - z_0$, and the series expansion then becomes

$$f(z) = f(z_0) + (z - z_0)f^{(1)}(z_0) + \frac{(z-z_0)^2}{2!}f^{(2)}(z_0) + \ldots$$
$$+ \frac{(z-z_0)^n}{n!}f^{(n)}(z_0) + \ldots = \sum_{n=0}^{\infty} \frac{(z-z_0)^n}{n!}f^{(n)}(z_0) \tag{1.32}$$

The power series expansion (1.32) is called the **Taylor series expansion** of the complex function $f(z)$ about z_0. The region of convergence of this series is $|z - z_0| < R$, a disc centred on $z = z_0$ and of radius R, the radius of convergence. Figure 1.22 illustrates the region of convergence. When $z_0 = 0$, as in real variables, the series expansion about the origin is often called a **Maclaurin series expansion**.

Since the proof of the Taylor series expansion does not add to our understanding of how to apply the result to the solution of engineering problems, we omit it at this stage.

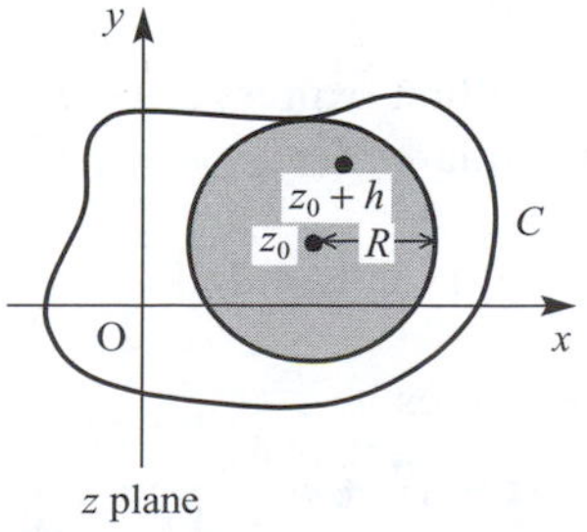

Figure 1.22 Region of convergence of the Taylor series.

EXAMPLE 1.17 Determine the Taylor series expansion of the function

$$f(z) = \frac{1}{z(z - 2\mathrm{j})}$$

about the point $z = \mathrm{j}$:

(a) directly up to the term $(z - \mathrm{j})^4$,

(b) using the binomial expansion.

Determine the radius of convergence.

Solution (a) The disadvantage with functions other than the most straightforward is that obtaining their derivatives is prohibitively complicated in terms of algebra. It is easier in this particular case to resolve the given function into partial fractions as

$$f(z) = \frac{1}{z(z - 2\mathrm{j})} = \frac{1}{2\mathrm{j}}\left(\frac{1}{z - 2\mathrm{j}} - \frac{1}{z}\right)$$

The right-hand side is now far easier to differentiate repeatedly. Proceeding to determine $f^{(k)}(\mathrm{j})$, we have

$$f(z) = \frac{1}{2\mathrm{j}}\left(\frac{1}{z - 2\mathrm{j}} - \frac{1}{z}\right), \quad \text{so that} \quad f(\mathrm{j}) = 1$$

$$f^{(1)}(z) = \frac{1}{2\mathrm{j}}\left[-\frac{1}{(z - 2\mathrm{j})^2} + \frac{1}{z^2}\right], \quad \text{so that} \quad f^{(1)}(\mathrm{j}) = 0$$

$$f^{(2)}(z) = \frac{1}{2\mathrm{j}}\left[\frac{2}{(z - 2\mathrm{j})^3} - \frac{2}{z^3}\right], \quad \text{so that} \quad f^{(2)}(\mathrm{j}) = -2$$

$$f^{(3)}(z) = \frac{1}{2\mathrm{j}}\left[-\frac{6}{(z - 2\mathrm{j})^4} + \frac{6}{z^4}\right], \quad \text{so that} \quad f^{(3)}(\mathrm{j}) = 0$$

$$f^{(4)}(z) = \frac{1}{2\mathrm{j}}\left[\frac{24}{(z - 2\mathrm{j})^5} - \frac{24}{z^5}\right], \quad \text{so that} \quad f^{(4)}(\mathrm{j}) = 24$$

leading from (1.32) to the Taylor series expansion

$$\frac{1}{z(z - 2\mathrm{j})} = 1 - \frac{2}{2!}(z - \mathrm{j})^2 + \frac{24}{4!}(z - \mathrm{j})^4 + \dots$$

$$= 1 - (z - \mathrm{j})^2 + (z - \mathrm{j})^4 + \dots$$

(b) To use the binomial expansion, we first express $z(z - 2\mathrm{j})$ as $(z - \mathrm{j} + \mathrm{j})(z - \mathrm{j} - \mathrm{j})$, which, being the difference of two squares $((z - \mathrm{j})^2 - \mathrm{j}^2)$, leads to

$$f(z) = \frac{1}{z(z - 2\mathrm{j})} = \frac{1}{(z - \mathrm{j})^2 + 1} = [1 + (z - \mathrm{j})^2]^{-1}$$

Use of the binomial expansion then gives

$$f(z) = 1 - (z - \mathrm{j})^2 + (z - \mathrm{j})^4 - (z - \mathrm{j})^6 + \dots$$

valid for $|z - \mathrm{j}| < 1$, so the radius of convergence is 1.

The points where $f(z)$ is infinite (its singularities) are precisely at distance 1 away from $z = j$, so this value for the radius of convergence comes as no surprise.

EXAMPLE 1.18 Suggest a function to represent the power series

$$1 + z + \frac{z^2}{2!} + \frac{z^3}{3!} + \ldots + \frac{z^n}{n!} + \ldots$$

and determine its radius of convergence.

Solution Set

$$f(z) = 1 + z + \frac{z^2}{2!} + \frac{z^3}{3!} + \ldots = \sum_{n=0}^{\infty} \frac{z^n}{n!}$$

Assuming we can differentiate the series for $f(z)$ term by term, we obtain

$$f'(z) = \sum_{n=1}^{\infty} \frac{nz^{n-1}}{n!} = \sum_{n=1}^{\infty} \frac{z^{n-1}}{(n-1)!} = f(z)$$

Hence $f(z)$ is its own derivative. Since e^x is its own derivative in real variables, and is the only such function, it seems sensible to propose that

$$f(z) = \sum_{n=0}^{\infty} \frac{z^n}{n!} = e^z \tag{1.33}$$

the complex exponential function. Indeed the complex exponential e^z is defined by the power series (1.33). According to d'Alembert's ratio test the series $\sum_{n=0}^{\infty} a_n$ is convergent if $|a_{n+1}/a_n| \to L < 1$ as $n \to \infty$, where L is a real constant. If $a_n = z^n/n!$ then $|a_{n+1}/a_n| = |z|/(n+1)$ which is less than unity for sufficiently large n, no matter how big $|z|$ is. Hence $\sum_{n=0}^{\infty} z^n/n!$ is convergent for *all* z and so has an infinite radius of convergence. Note that this is confirmed from (1.30). Such functions are called **entire**.

In the same way as we define the exponential function e^z by the power series expansion (1.31), we can define the circular functions $\sin z$ and $\cos z$ by the power series expansions

$$\sin z = z - \frac{z^3}{3!} + \frac{z^5}{5!} - \frac{z^7}{7!} + \ldots + (-1)^n \frac{z^{2n+1}}{(2n+1)!} + \ldots$$

$$\cos z = 1 - \frac{z^2}{2!} + \frac{z^4}{4!} - \frac{z^6}{6!} + \ldots + (-1)^n \frac{z^{2n}}{(2n)!} + \ldots$$

both of which are valid for all z. Using these power series definitions, we can readily prove the result (1.25), namely

$$e^{jz} = \cos z + j \sin z$$

1.4.4 Exercises

40 Find the first four non-zero terms of the Taylor series expansions of the following functions about the points indicated, and determine the radius of convergence in each case:

(a) $\dfrac{1}{1+z}$ $(z = 1)$ (b) $\dfrac{1}{z(z-4j)}$ $(z = 2j)$

(c) $\dfrac{1}{z^2}$ $(z = 1 + j)$

41 Find the Maclaurin series expansion of the function

$$f(z) = \frac{1}{1 + z + z^2}$$

up to and including the term in z^3.

42 Without explicitly finding each Taylor series expansion, find the radius of convergence of the function

$$f(z) = \frac{1}{z^4 - 1}$$

about the three points $z = 0$, $z = 1 + j$ and $z = 2 + 2j$. Why is there no Taylor series expansion of this function about $z = j$?

43 Determine a Taylor series expansion of $f(z) = \tan z$. What is its radius of convergence?

1.4.5 Laurent series

Let us now examine more closely the solution of Example 1.15(c), where the power series obtained was

$$\frac{1}{z-3} = \frac{1}{z} + \frac{3}{z^2} + \frac{9}{z^3} + \dots$$

valid for $|z| > 3$. In the context of the definition, this is a power series about '$z = \infty$', the 'point at infinity'. Some readers, quite justifiably, may not be convinced that there is a single unique point at infinity. Figure 1.23 shows what is termed the **Riemann sphere**. A sphere lies on the complex z plane, with the contact point at the origin O. Let O′ be the top of the sphere, at the diametrically opposite point to O. Now, for any arbitrarily chosen point P in the z plane, by joining O′ and P we determine a unique point P′ where the line O′P intersects the sphere. There is thus exactly one point P′ on the sphere corresponding to each P in the z plane. The point O′ itself is the only point on the sphere that does not have

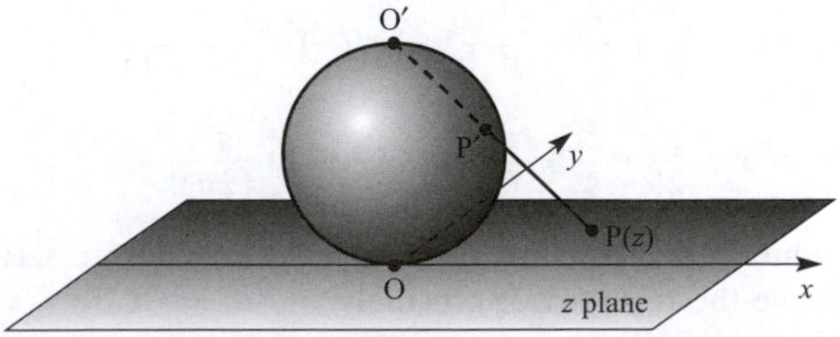

Figure 1.23 The Riemann sphere.

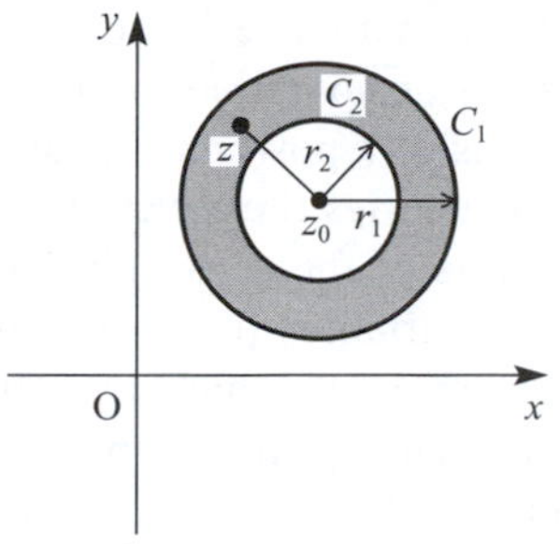

Figure 1.24 Region of validity of the Laurent series.

a corresponding point on the (finite) z plane; we therefore say it corresponds to the **point at infinity** on the z plane.

Returning to consider power series, we know that, inside the radius of convergence, a given function and its Taylor series expansion are identically equal. Points at which a function fails to be analytic are called **singularities**, which we shall discuss in Section 1.5.1. No Taylor series expansion is possible about a singularity. Indeed, a Taylor series expansion about a point z_0 at which a function is analytic is only valid within a circle, centre z_0, up to the nearest singularity. Thus all singularities must be excluded in any Taylor series consideration. The Laurent series representation includes (or at least takes note of) the behaviour of the function in the vicinity of a singularity.

If $f(z)$ is a complex function analytic on concentric circles C_1 and C_2 of radii r_1 and r_2 (with $r_2 < r_1$), centred at z_0, and also analytic throughout the region between the circles (that is, an annular region), then for each point z within the annulus (Figure 1.24) $f(z)$ may be represented by the **Laurent series**

$$\begin{aligned} f(z) &= \sum_{n=-\infty}^{\infty} c_n(z-z_0)^n \\ &= \ldots + \frac{c_{-r}}{(z-z_0)^r} + \frac{c_{-r+1}}{(z-z_0)^{r-1}} + \ldots + \frac{c_{-1}}{z-z_0} + c_0 \\ &\quad + c_1(z-z_0) + \ldots + c_r(z-z_0)^r + \ldots \end{aligned} \tag{1.34}$$

where in general the coefficients c_r are complex. The annular shape of the region is necessary in order to exclude the point $z = z_0$, which may be a singularity of $f(z)$, from consideration. If $f(z)$ is analytic at $z = z_0$ then $c_n = 0$ for $n = -1, -2, \ldots, -\infty$, and the Laurent series reduces to the Taylor series.

The Laurent series (1.34) for $f(z)$ may be written as

$$f(z) = \sum_{n=-\infty}^{-1} c_n(z-z_0)^n + \sum_{n=0}^{\infty} c_n(z-z_0)^n$$

and the first sum on the right-hand side, the 'non-Taylor' part, is called the **principal part** of the Laurent series.

Of course, we can seldom actually sum a series to infinity. There is therefore often more than theoretical interest in the so-called 'remainder terms', these being the difference between the first n terms of a power series and the exact value of the function. For both Taylor and Laurent series these remainder terms are expressed, as in the case of real variable, in terms of the $(n + 1)$th derivative of the function itself. For Laurent series in complex variables these derivatives can be expressed in terms of contour integrals (Section 1.6), which may be amenable to simple computation. Many of the details are outside the scope of this book, but there is some introductory material in Section 1.6.

EXAMPLE 1.19 For $f(z) = 1/z^2(z + 1)$ find the Laurent series expansion about (a) $z = 0$ and (b) $z = -1$. Determine the region of validity in each case.

Solution As with Example 1.15, problems such as this are tackled by making use of the binomial series expansion

$$(1+z)^n = 1 + nz + \frac{n(n-1)}{2!}z^2 + \ldots + \frac{n(n-1)(n-2)\ldots(n-r+1)}{r!}z^r + \ldots$$

provided that $|z| < 1$.

(a) In this case $z_0 = 0$, so we need a series in powers of z. Thus

$$\begin{aligned}\frac{1}{z^2(1+z)} &= \frac{1}{z^2}(1+z)^{-1}\\ &= \frac{1}{z^2}(1 - z + z^2 - z^3 + z^4 - \ldots) \quad (0 < |z| < 1)\end{aligned}$$

Thus the required Laurent series expansion is

$$\frac{1}{z^2(z+1)} = \frac{1}{z^2} - \frac{1}{z} + 1 - z + z^2 \ldots$$

valid for $0 < |z| < 1$. The value $z = 0$ must be excluded because of the first two terms of the series. The region $0 < |z| < 1$ is an example of a **punctured disc**, a common occurrence in this branch of mathematics.

(b) In this case $z_0 = -1$, so we need a series in powers of $(z + 1)$. Thus

$$\begin{aligned}\frac{1}{z^2(z+1)} &= \frac{1}{(z+1)}(z + 1 - 1)^{-2}\\ &= \frac{1}{(z+1)}[1 - (z + 1)]^{-2}\\ &= \frac{1}{(z+1)}[1 + 2(z + 1) + 3(z + 1)^2 + \ldots]\\ &= \frac{1}{z+1} + 2 + 3(z + 1) + 4(z + 1)^2 + \ldots\end{aligned}$$

valid for $0 < |z + 1| < 1$. Note that in a meniscus-shaped region (that is, the region of overlap between the two circular regions $|z| < 1$ and $|z + 1| < 1$) both Laurent series are simultaneously valid. This is quite typical, and not a cause for concern.

EXAMPLE 1.20 Determine the Laurent series expansions of

$$f(z) = \frac{1}{(z+1)(z+3)}$$

valid for

(a) $1 < |z| < 3$
(b) $|z| > 3$
(c) $0 < |z + 1| < 2$
(d) $|z| < 1$

Solution (a) Resolving into partial functions,

$$f(z) = \frac{1}{2}\left(\frac{1}{z+1}\right) - \frac{1}{2}\left(\frac{1}{z+3}\right)$$

Since $|z| > 1$ and $|z| < 3$, we express this as

$$\begin{aligned} f(z) &= \frac{1}{2z}\left(\frac{1}{1+1/z}\right) - \frac{1}{6}\left(\frac{1}{1+\frac{1}{3}z}\right) \\ &= \frac{1}{2z}\left(1+\frac{1}{z}\right)^{-1} - \frac{1}{6}\left(1+\frac{1}{3}z\right)^{-1} \\ &= \frac{1}{2z}\left(1 - \frac{1}{z} + \frac{1}{z^2} - \frac{1}{z^3} + \ldots\right) - \frac{1}{6}\left(1 - \frac{1}{3}z + \frac{1}{9}z^2 - \frac{1}{27}z^3 + \ldots\right) \\ &= \ldots - \frac{1}{2z^4} + \frac{1}{2z^3} - \frac{1}{2z^2} + \frac{1}{2z} - \frac{1}{6} + \frac{1}{18}z - \frac{1}{54}z^2 + \frac{1}{162}z^3 - \ldots \end{aligned}$$

(b) $$f(z) = \frac{1}{2}\left(\frac{1}{z+1}\right) - \frac{1}{2}\left(\frac{1}{z+3}\right)$$

Since $|z| > 3$, we express this as

$$\begin{aligned} f(z) &= \frac{1}{2z}\left(\frac{1}{1+1/z}\right) - \frac{1}{2z}\left(\frac{1}{1+3/z}\right) \\ &= \frac{1}{2z}\left(1+\frac{1}{z}\right)^{-1} - \frac{1}{2z}\left(1+\frac{3}{z}\right)^{-1} \\ &= \frac{1}{2z}\left(1 - \frac{1}{z} + \frac{1}{z^2} - \frac{1}{z^3} + \ldots\right) - \frac{1}{2z}\left(1 - \frac{3}{z} + \frac{9}{z^2} - \frac{27}{z^3} + \ldots\right) \\ &= \frac{1}{z^2} - \frac{4}{z^3} + \frac{13}{z^4} - \frac{40}{z^5} + \ldots \end{aligned}$$

(c) We can proceed as in Example 1.18. Alternatively, we can take $z + 1 = u$; then $0 < |u| < 2$ and

$$f(u) = \frac{1}{u(u+2)} = \frac{1}{2u(1+\frac{1}{2}u)}$$
$$= \frac{1}{2u}\left(1 - \frac{1}{2}u + \frac{1}{4}u^2 - \frac{1}{8}u^3 + \ldots\right)$$

giving

$$f(z) = \frac{1}{2(z+1)} - \frac{1}{4} + \frac{1}{8}(z+1) - \frac{1}{16}(z+1)^2 + \ldots$$

(d) $$f(z) = \frac{1}{2(z+1)} - \frac{1}{2(z+3)}$$

Since $|z| < 1$, we express this as

$$f(z) = \frac{1}{2(1+z)} - \frac{1}{6(1+\frac{1}{3}z)}$$
$$= \tfrac{1}{2}(1+z)^{-1} - \tfrac{1}{6}(1+\tfrac{1}{3}z)^{-1}$$
$$= \tfrac{1}{2}(1 - z + z^2 - z^3 + \ldots) - \tfrac{1}{6}(1 - \tfrac{1}{3}z + \tfrac{1}{9}z^2 - \tfrac{1}{27}z^3 + \ldots)$$
$$= \tfrac{1}{3} - \tfrac{4}{9}z + \tfrac{13}{27}z^2 - \tfrac{40}{81}z^3 + \ldots$$

EXAMPLE 1.21 Determine the Laurent series expansion of the function $f(z) = z^3 e^{1/z}$ about

(a) $z = 0$

(b) $z = a$, a finite, non-zero complex number

(c) $z = \infty$

Solution (a) From (1.33),

$$e^z = 1 + z + \frac{z^2}{2!} + \ldots \quad (0 \leqslant |z| < \infty)$$

Substituting $1/z$ for z, we obtain

$$e^{1/z} = 1 + \frac{1}{z} + \frac{1}{2!z^2} + \ldots \quad (0 < |z| \leqslant \infty)$$

so that

$$z^3 e^{1/z} = z^3 + z^2 + \frac{z}{2!} + \frac{1}{3!} + \frac{1}{4!z} + \frac{1}{5!z^2} + \ldots \quad (0 < |z| \leqslant \infty)$$

This series has infinitely many terms in its principal part, but stops at z^3 (it is written back to front). Series with never-ending principal parts are a problem, and fortunately are uncommon in engineering. Note also that the series is valid in an infinite punctured disc.

(b) The value of $f(a)$ must be $a^3 e^{1/a}$, which is not infinite since $a \neq 0$. Therefore $f(z)$ has a Taylor series expansion

$$f(z) = f(a) + (z - a)f^{(1)}(a) + \frac{(z-a)^2}{2!}f^{(2)}(a) + \dots$$

about $z = a$. We have

$$f^{(1)}(z) = \frac{d}{dz}(z^3 e^{1/z}) = 3z^2 e^{1/z} - z e^{1/z}$$

$$f^{(2)}(z) = \frac{d}{dz}(3z^2 e^{1/z} - z e^{1/z}) = 6z e^{1/z} - 4 e^{1/z} + \frac{1}{z^2} e^{1/z}$$

giving the series as

$$z^3 e^{1/z} = a^3 e^{1/a} + (z - a)(3a^2 e^{1/a} - a e^{1/a}) + \frac{1}{2!}(z-a)^2\left(6a e^{1/a} - 4e^{1/a} + \frac{1}{a^2} e^{1/a}\right) + \dots$$

which is valid in the region $|z - a| < R$, where R is the distance between the origin, where $f(z)$ is not defined, and the point a; hence $R = |a|$. Thus the region of validity for this Taylor series is the disc $|z - a| < |a|$.

(c) To expand about $z = \infty$, let $w = 1/z$, so that

$$f(z) = \frac{1}{w^3} e^w$$

Expanding about $w = 0$ then gives

$$f\left(\frac{1}{w}\right) = \frac{1}{w^3}\left(1 + w + \frac{w^2}{2!} + \frac{w^3}{3!} + \dots\right)$$

$$= \frac{1}{w^3} + \frac{1}{w^2} + \frac{1}{2!w} + \frac{1}{3!} + \frac{w}{4!} + \dots \quad (0 < |w| < \infty)$$

Note that this time there are only three terms in the principal part of $f(z)(=f(1/w))$.

1.4.6 Exercises

44 Determine the Laurent series expansion of

$$f(z) = \frac{1}{z(z-1)^2}$$

about (a) $z = 0$ and (b) $z = 1$, and specify the region of validity for each.

45 Determine the Laurent series expansion of the function

$$f(z) = z^2 \sin \frac{1}{z}$$

about the points

(a) $z = 0$ (b) $z = \infty$

(c) $z = a$, a finite non-zero complex number

(For (c), do *not* calculate the coefficients explicitly.)

46 Expand

$$f(z) = \frac{z}{(z-1)(2-z)}$$

in a Laurent series expansion valid for

(a) $|z| < 1$ (b) $1 < |z| < 2$ (c) $|z| > 2$

(d) $|z - 1| > 1$ (e) $0 < |z - 2| < 1$

1.5 Singularities, zeros and residues

1.5.1 Singularities and zeros

As indicated in Section 1.4.5 a **singularity** of a complex function $f(z)$ is a point of the z plane where $f(z)$ ceases to be analytic. Normally, this means $f(z)$ is infinite at such a point, but it can also mean that there is a choice of values, and it is not possible to pick a particular one. In this chapter we shall be mainly concerned with singularities at which $f(z)$ has an infinite value. A **zero** of $f(z)$ is a point in the z plane at which $f(z) = 0$.

Singularities can be classified in terms of the Laurent series expansion of $f(z)$ about the point in question. If $f(z)$ has a Taylor series expansion, that is a Laurent series expansion with zero principal part, about the point $z = z_0$, then z_0 is a **regular point** of $f(z)$. If $f(z)$ has a Laurent series expansion with only a finite number of terms in its principal part, for example

$$f(z) = \frac{a_{-m}}{(z - z_0)^m} + \ldots + \frac{a_{-1}}{(z - z_0)} + a_0 + a_1(z - z_0) + \ldots + a_m(z - z_0)^m + \ldots$$

then $f(z)$ has a singularity at $z = z_0$ called a **pole**. If there are m terms in the principal part, as in this example, then the pole is said to be of **order** m. Another way of defining this is to say that z_0 is a pole of order m if

$$\lim_{z \to z_0} (z - z_0)^m f(z) = a_{-m} \qquad \textbf{(1.35)}$$

where a_{-m} is finite and non-zero. If the principal part of the Laurent series for $f(z)$ at $z = z_0$ has infinitely many terms, which means that the above limit does not exist for any m, then $z = z_0$ is called an **essential singularity** of $f(z)$.

If $f(z)$ appears to be singular at $z = z_0$, but it turns out to be possible to define a Taylor series expansion there, then $z = z_0$ is called a **removable singularity**. The following examples illustrate these cases.

(a) $f(z) = z^{-1}$ has a pole of order one, called a **simple pole**, at $z = 0$.

(b) $f(z) = (z - 1)^{-3}$ has a pole of order three at $z = 1$.

(c) $f(z) = e^{1/(z-j)}$ has an essential singularity at $z = j$.

(d) The function

$$f(z) = \frac{z-1}{(z+2)(z-3)^2}$$

has a zero at $z = 1$, a simple pole at $z = -2$ and a pole of order two at $z = 3$.

(e) The function

$$f(z) = \frac{\sin z}{z}$$

is not defined at $z = 0$, and appears to be singular there. However, defining

$$\operatorname{sinc} z = \begin{cases} (\sin z)/z & (z \neq 0) \\ 1 & (z = 0) \end{cases}$$

gives a function having a Taylor series expansion

$$\operatorname{sinc} z = 1 - \frac{z^2}{3!} + \frac{z^4}{5!} - \dots$$

that is regular at $z = 0$. Therefore the (apparent) singularity at $z = 0$ has been removed, and thus $f(z) = (\sin z)/z$ has a removable singularity at $z = 0$.

Functions whose only singularities are poles are called **meromorphic** and, by and large, in engineering applications of complex variables most functions are meromorphic. To help familiarize the reader with these definitions, the following example should prove instructive.

EXAMPLE 1.22

Find the singularities and zeros of the following complex functions:

(a) $\dfrac{1}{z^4 - z^2(1+j) + j}$ (b) $\dfrac{z-1}{z^4 - z^2(1+j) + j}$

(c) $\dfrac{\sin(z-1)}{z^4 - z^2(1+j) + j}$ (d) $\dfrac{1}{[z^4 - z^2(1+j) + j]^3}$

Solution (a) For

$$f(z) = \frac{1}{z^4 - z^2(1+j) + j}$$

the numerator is never zero, and the denominator is only infinite when z is infinite. Thus $f(z)$ has no zeros in the finite z plane. The denominator is zero when

$$z^4 - z^2(1 + j) + j = 0$$

which factorizes to give

$$(z^2 - 1)(z^2 - j) = 0$$

leading to

$$z^2 = 1 \text{ or } j$$

so that the singularities are at

$$z = +1, -1, (1 + j)/\sqrt{2}, (1 - j)/\sqrt{2} \tag{1.36}$$

all of which are simple poles since none of the roots are repeated.

(b) The function

$$f(z) = \frac{z-1}{z^4 - z^2(1+j) + j}$$

is similar to $f(z)$ in (a), except that it has the additional term $z - 1$ in the numerator. Therefore, at first glance, it seems that the singularities are as in (1.36). However, a closer look indicates that $f(z)$ can be rewritten as

$$f(z) = \frac{z-1}{(z-1)(z+1)[z - \sqrt{\tfrac{1}{2}}(1+j)]\,[z - \sqrt{\tfrac{1}{2}}(1-j)]}$$

and the factor $z - 1$ cancels, rendering $z = 1$ a removable singularity, and reducing $f(z)$ to

$$f(z) = \frac{1}{(z+1)[z - \sqrt{\tfrac{1}{2}}(1+j)]\,[z - \sqrt{\tfrac{1}{2}}(1-j)]}$$

which has no (finite) zeros and $z = -1$, $\sqrt{\tfrac{1}{2}}(1 + j)$ and $\sqrt{\tfrac{1}{2}}(1 - j)$ as simple poles.

(c) In the case of

$$f(z) = \frac{\sin(z-1)}{z^4 - z^2(1+j) + j}$$

the function may be rewritten as

$$f(z) = \frac{\sin(z-1)}{z-1} \, \frac{1}{(z+1)[z - \sqrt{\tfrac{1}{2}}(1+j)]\,[z - \sqrt{\tfrac{1}{2}}(1-j)]}$$

Now

$$\frac{\sin(z-1)}{z-1} \to 1 \quad \text{as} \quad z \to 1$$

so once again $z = 1$ is a removable singularity. Also, as in (b), $z = -1$, $\sqrt{\tfrac{1}{2}}(1 + j)$ and $\sqrt{\tfrac{1}{2}}(1 - j)$ are simple poles and the only singularities. However,

$$\sin(z - 1) = 0$$

has the general solution $z = 1 + N\pi$ ($N = 0, \pm 1, \pm 2, \dots$). Thus, apart from $N = 0$, all of these are zeros of $f(z)$.

(d) For

$$f(z) = \frac{1}{[z^4 - z^2(1+j)+j]^3}$$

factorizing as in (b), we have

$$f(z) = \frac{1}{(z-1)^3(z+1)^3[z-\sqrt{\tfrac{1}{2}}(1+j)]^3\,[z-\sqrt{\tfrac{1}{2}}(1-j)]^3}$$

so -1, $+1$, $\sqrt{\tfrac{1}{2}}(1+j)$ and $\sqrt{\tfrac{1}{2}}(1-j)$ are still singularities, but this time they are triply repeated. Hence they are all poles of order three. There are no zeros.

1.5.2 Exercises

47 Determine the location of, and classify, the singularities and zeros of the following functions. Specify also any zeros that may exist.

(a) $\dfrac{\cos z}{z^2}$ (b) $\dfrac{1}{(z+j)^2(z-j)}$ (c) $\dfrac{z}{z^4-1}$

(d) $\coth z$ (e) $\dfrac{\sin z}{z^2+\pi^2}$ (f) $e^{z/(1-z)}$

(g) $\dfrac{z-1}{z^2+1}$ (h) $\dfrac{z+j}{(z+2)^3(z-3)}$

(i) $\dfrac{1}{z^2(z^2-4z+5)}$

48 Expand each of the following functions in a Laurent series about $z = 0$, and give the type of singularity (if any) in each case:

(a) $\dfrac{1-\cos z}{z}$

(b) $\dfrac{e^{z^2}}{z^3}$

(c) $z^{-1}\cosh z^{-1}$

(d) $\tan^{-1}(z^2 + 2z + 2)$

49 Show that if $f(z)$ is the ratio of two polynomials then it cannot have an essential singularity.

1.5.3 Residues

If a complex function $f(z)$ has a pole at the point $z = z_0$ then the coefficient a_{-1} of the term $1/(z - z_0)$ in the Laurent series expansion of $f(z)$ about $z = z_0$ is called the **residue** of $f(z)$ at the point $z = z_0$. The importance of residues will become apparent when we discuss integration in Section 1.6. Here we shall concentrate on efficient ways of calculating them, usually without finding the Laurent series expansion explicitly. However, experience and judgement are sometimes the only help in finding the easiest way of calculating residues. First let us consider the case when $f(z)$ has a simple pole at $z = z_0$. This implies, from the definition of a simple pole, that

$$f(z) = \frac{a_{-1}}{z-z_0} + a_0 + a_1(z-z_0) + \ldots$$

in an appropriate annulus $S < |z - z_0| < R$. Multiplying by $z - z_0$ gives

$$(z - z_0)f(z) = a_{-1} + a_0(z - z_0) + \ldots$$

which is a Taylor series expansion of $(z - z_0)f(z)$. If we let z approach z_0, we then obtain the result

$$\begin{matrix}\text{residue at a}\\ \text{simple pole } z_0\end{matrix} = \lim_{z\to z_0}[(z - z_0)f(z)] = a_{-1} \tag{1.37}$$

Hence this limit gives a way of calculating the residue at a simple pole.

EXAMPLE 1.23 Determine the residues of

$$f(z) = \frac{2z}{(z^2 + 1)(2z - 1)}$$

at each of its poles in the finite z plane.

Solution Factorizing the denominator, we have

$$f(z) = \frac{2z}{(z - \mathrm{j})(z + \mathrm{j})(2z - 1)}$$

so that $f(z)$ has simple poles at $z = \mathrm{j}, -\mathrm{j}$ and $\frac{1}{2}$. Using (1.37) then gives

$$\begin{matrix}\text{residue}\\ \text{at } z = \mathrm{j}\end{matrix} = \lim_{z\to \mathrm{j}}(z - \mathrm{j})\frac{2z}{(z - \mathrm{j})(z + \mathrm{j})(2z - 1)}$$

$$= \frac{2\mathrm{j}}{2\mathrm{j}(2\mathrm{j} - 1)} = -\frac{1 + 2\mathrm{j}}{5}$$

$$\begin{matrix}\text{residue}\\ \text{at } z = -\mathrm{j}\end{matrix} = \lim_{z\to -\mathrm{j}}(z + \mathrm{j})\frac{2z}{(z - \mathrm{j})(z + \mathrm{j})(2z - 1)}$$

$$= \frac{-2\mathrm{j}}{-2\mathrm{j}(-2\mathrm{j} - 1)} = -\frac{1 - 2\mathrm{j}}{5}$$

$$\begin{matrix}\text{residue}\\ \text{at } z = \frac{1}{2}\end{matrix} = \lim_{z\to \frac{1}{2}}(z - \tfrac{1}{2})\frac{z}{(z - \mathrm{j})(z + \mathrm{j})(z - \frac{1}{2})}$$

$$= \frac{\frac{1}{2}}{(\frac{1}{2} - \mathrm{j})(\frac{1}{2} + \mathrm{j})} = \frac{2}{5}$$

Note in this last case the importance of expressing $2z - 1$ as $2(z - \frac{1}{2})$.

EXAMPLE 1.24 Determine the residues of the function $1/(1 + z^4)$ at each of its poles in the finite z plane.

Solution The function $1/(1 + z^4)$ has poles where

$$1 + z^4 = 0$$

that is, at the points where

$$z^4 = -1 = e^{\pi j + 2\pi n j}$$

with n an integer. Recalling how to determine the roots of a complex number, these points are

$$z = e^{\pi j/4 + \pi j n/2} \quad (n = 0, 1, 2, 3)$$

that is

$$z = e^{\pi j/4},\ e^{3\pi j/4},\ e^{5\pi j/4},\ e^{7\pi j/4}$$

or

$$z = (1 + j)/\sqrt{2},\ (-1 + j)/\sqrt{2},\ (-1 - j)/\sqrt{2},\ (1 - j)/\sqrt{2}$$

To find the residue at the point z_0, we use (1.37), giving

$$\begin{matrix}\text{residue} \\ \text{at } z_0\end{matrix} = \lim_{z \to z_0} \left(\frac{z - z_0}{1 + z^4} \right)$$

where z_0 is one of the above roots of $z^4 = -1$. It pays to use L'Hôpital's rule before substituting for a particular z_0. This is justified since $(z - z_0)/(1 + z_4)$ is of the indeterminate form 0/0 at each of the four simple poles. Differentiating numerator and denominator gives

$$\lim_{z \to z_0} \left(\frac{z - z_0}{1 + z^4} \right) = \lim_{z \to z_0} \left(\frac{1}{4z^3} \right)$$

$$= \frac{1}{4z_0^3}$$

since $4z^3$ is not zero at any of the poles. $1/4z_0^3$ is thus the value of each residue at $z = z_0$. Substituting for the four values $(\pm 1 \pm j)/\sqrt{2}$ gives the following:

$$\begin{matrix}\text{residue} \\ \text{at } z = (1 + j)/\sqrt{2}\end{matrix} = \frac{1}{4(\sqrt{\frac{1}{2}})^3 (1 + j)^3} = -(1 + j)/4\sqrt{2}$$

$$\begin{matrix}\text{residue} \\ \text{at } z = (1 - j)/\sqrt{2}\end{matrix} = \frac{1}{4(\sqrt{\frac{1}{2}})^3 (1 - j)^3} = (-1 + j)/4\sqrt{2}$$

$$\begin{matrix}\text{residue} \\ \text{at } z = (-1 + j)/\sqrt{2}\end{matrix} = \frac{1}{4(\sqrt{\frac{1}{2}})^3 (-1 + j)^3} = (1 - j)/4\sqrt{2}$$

$$\begin{matrix}\text{residue} \\ \text{at } z = (-1 - j)/\sqrt{2}\end{matrix} = \frac{1}{4(\sqrt{\frac{1}{2}})^3 (-1 - j)^3} = (1 + j)/4\sqrt{2}$$

Finding each Laurent series for the four poles explicitly would involve far more difficult manipulation. However, the enthusiastic reader may like to check at least one of the above residues.

Next suppose that we have a pole of order two at $z = z_0$. The function $f(z)$ then has a Laurent series expansion of the form

$$f(z) = \frac{a_{-2}}{(z-z_0)^2} + \frac{a_{-1}}{z-z_0} + a_0 + a_1(z - z_0) + \dots$$

Again, we are only interested in isolating the residue a_{-1}. This time we cannot use (1.37). Instead, we multiply $f(z)$ by $(z - z_0)^2$ to obtain

$$(z - z_0)^2 f(z) = a_{-2} + a_{-1}(z - z_0) + a_0(z - z_0)^2 + \dots$$

and we differentiate to eliminate the unwanted a_{-2}:

$$\frac{d}{dz}[(z - z_0)^2 f(z)] = a_{-1} + 2a_0(z - z_0) + \dots$$

Letting z tend to z_0 then gives

$$\lim_{z\to z_0}\left[\frac{d}{dz}(z-z_0)^2 f(z)\right] = a_{-1}$$

the required residue.

We now have the essence of finding residues, so let us recapitulate and generalize. If $f(z)$ has a pole of order m at $z - z_0$, we first multiply $f(z)$ by $(z - z_0)^m$. If $m \geqslant 2$, we then need to differentiate as many times as it takes (that is, $m - 1$ times) to make a_{-1} the leading term, without the multiplying factor $z - z_0$. The general formula for the residue at a pole of order m is thus

$$\frac{1}{(m-1)!}\lim_{z\to z_0}\left\{\frac{d^{m-1}}{dz^{m-1}}[(z-z_0)^m f(z)]\right\} \qquad \textbf{(1.38)}$$

where the factor $(m - 1)!$ arises when the term $a_{-1}(z - z_0)^{m-1}$ is differentiated $m - 1$ times. This formula looks as difficult to apply as finding the Laurent series expansion directly. This indeed is often so; and hence experience and judgement are required. A few examples will help to decide on which way to calculate residues. A word of warning is in order here: a common source of error is confusion between the derivative in the formula for the residue, and the employment of L'Hôpital's rule to find the resulting limit.

EXAMPLE 1.25 Determine the residues of

$$f(z) = \frac{z^2 - 2z}{(z+1)^2(z^2+4)}$$

at each of its poles in the finite z plane.

Solution Factorizing the denominator gives

$$f(z) = \frac{z^2 - 2z}{(z+1)^2(z-2j)(z+2j)}$$

so that $f(z)$ has simple poles at $z = 2j$ and $z = -2j$ and a pole of order two at $z = -1$. Using (1.37),

$$\begin{aligned}
\text{residue at } z = 2j &= \lim_{z\to 2j} (z-2j)\frac{z^2-2z}{(z+1)^2(z-2j)(z+2j)} \\
&= \frac{-4-4j}{(2j+1)^2(4j)} = \frac{1}{25}(7+j) \\
\text{residue at } z = -2j &= \lim_{z\to -2j} (z+2j)\frac{z^2-2z}{(z+1)^2(z-2j)(z+2j)} \\
&= \frac{-4+4j}{(-2j+1)^2(-4j)} = \frac{1}{25}(7-j)
\end{aligned}$$

Using (1.38) with $m = 2$ we know that

$$\begin{aligned}
\text{residue at } z = -1 &= \frac{1}{1!}\lim_{z\to -1}\frac{d}{dz}\left[(z+1)^2\frac{z^2-2z}{(z+1)^2(z^2+4)}\right] \\
&= \lim_{z\to -1}\frac{(z^2+4)(2z-2)-(z^2-2z)(2z)}{(z^2+4)^2} = \frac{(5)(-4)-(3)(-2)}{25} = -\frac{14}{25}
\end{aligned}$$

EXAMPLE 1.26 Determine the residues of the following functions at the points indicated:

(a) $\dfrac{e^z}{(1+z^2)^2}$ $(z = j)$ (b) $\left(\dfrac{\sin z}{z^2}\right)^3$ $(z = 0)$ (c) $\dfrac{z^4}{(z+1)^3}$ $(z = -1)$

Solution (a) Since

$$\frac{e^z}{(z^2+1)^2} = \frac{e^z}{(z+j)^2(z-j)^2}$$

and e^z is regular at $z = j$, it follows that $z = j$ is a pole of order two. Thus, from (1.38),

$$\begin{aligned}
\text{residue} &= \lim_{z\to j}\frac{d}{dz}\left[(z-j)^2\frac{e^z}{(z+j)^2(z-j)^2}\right] \\
&= \lim_{z\to j}\left\{\frac{d}{dz}\left[\frac{e^z}{(z+j)^2}\right]\right\} = \lim_{z\to j}\frac{(z+j)^2e^z - 2(z+j)e^z}{(z+j)^4} \\
&= \frac{(2j)^2e^j - 2(2j)e^j}{(2j)^4} = -\frac{1}{4}(1+j)e^j
\end{aligned}$$

Since $e^j = \cos 1 + j\sin 1$, we calculate the residue at $z = j$ as $0.075 - j0.345$.

(b) The function $[(\sin z)/z^2]^3$ has a pole at $z = 0$, and, since $(\sin z/z) \to 1$ as $z \to 0$, $(\sin^2 z)/z^2$ may also be defined as 1 at $z = 0$. Therefore, since

$$\left(\frac{\sin z}{z^2}\right)^3 = \frac{\sin^3 z}{z^3}\frac{1}{z^3}$$

the singularity at $z = 0$ must be a pole of order three. We could use (1.38) to obtain the residue, which would involve determining the second derivative, but it is easier in this case to derive the coefficient of $1/z$ from the Laurent series expansion

$$\frac{\sin z}{z} = 1 - \frac{z^2}{3!} + \frac{z^4}{5!} - \dots$$

giving

$$\frac{\sin z}{z^2} = \frac{1}{z} - \frac{1}{6}z + \frac{1}{120}z^3 - \dots$$

Taking the cube of this series, we have

$$\left(\frac{\sin z}{z^2}\right)^3 = \left(\frac{1}{z} - \frac{1}{6}z + \frac{1}{120}z^3 - \dots\right)^3 = \frac{1}{z^3} - 3\frac{1}{z^2}\frac{z}{6} + \dots = \frac{1}{z^3} - \frac{1}{2z} + \dots$$

Hence the residue at $z = 0$ is $-\frac{1}{2}$.

(c) The function $z^4/(z + 1)^3$ has a triple pole at $z = -1$, so, using (1.38),

$$\text{residue} = \lim_{z\to-1}\left\{\frac{1}{2}\frac{\mathrm{d}^2}{\mathrm{d}z^2}\left[(z+1)^3\frac{z^4}{(z+1)^3}\right]\right\} = \lim_{z\to-1}\left[\frac{1}{2}\frac{\mathrm{d}^2}{\mathrm{d}z^2}(z^4)\right]$$

$$= \lim_{z\to-1} \tfrac{1}{2}\times 4\times 3z^2 = 6(-1)^2 = 6$$

Residues are sometimes difficult to calculate using (1.38), especially if circular functions are involved and the pole is of order three or more. In such cases direct calculation of the Laurent series expansion using the standard series for $\sin z$ and $\cos z$ together with the binomial series, as in Example 1.26(b), is the best procedure.

1.5.4 Exercises

50 Determine the residues of the following rational functions at each pole in the finite z plane:

(a) $\dfrac{2z+1}{z^2-z-2}$ (b) $\dfrac{1}{z^2(1-z)}$

(c) $\dfrac{3z^2+2}{(z-1)(z^2+9)}$ (d) $\dfrac{z^3-z^2+z-1}{z^3+4z}$

(e) $\dfrac{z^6+4z^4+z^3+1}{(z-1)^5}$ (f) $\left(\dfrac{z+1}{z-1}\right)^2$

(g) $\dfrac{z+1}{(z-1)^2(z+3)}$

(h) $\dfrac{3+4z}{z^3+3z^2+2z}$

51 Calculate the residues at the simple poles indicated of the following functions:

(a) $\dfrac{\cos z}{z}$ $(z = 0)$ (b) $\dfrac{\sin z}{z^4+z^2+1}$ $(z = e^{\pi j/3})$

(c) $\dfrac{z^4-1}{z^4+1}$ $(z = e^{\pi j/4})$

(d) $\dfrac{z}{\sin z}$ $(z = \pi)$ (e) $\dfrac{1}{(z^2+1)^2}$ $(z = j)$

52 The following functions have poles at the points indicated. Determine the order of the pole and the residue there.

(a) $\dfrac{\cos z}{z^3}$ $(z = 0)$

(b) $\dfrac{z^2-2z}{(z+1)^2(z^2+4)}$ $(z = -1)$

(c) $\dfrac{e^z}{\sin^2 z}$ $(z = n\pi$, n an integer$)$

(*Hint:* use $\lim_{u\to 0}(\sin u)/u = 1$ $(u = z - n\pi)$, after differentiating, to replace $\sin u$ by u under the limit.)

1.6 Contour integration

Consider the definite integral

$$\int_{z_1}^{z_2} f(z)\,dz$$

of the function $f(z)$ of a complex variable z, in which z_1 and z_2 are a pair of complex numbers. This implies that we evaluate the integral as z takes values, in the z plane, from the point z_1 to the point z_2. Since these are two points in a plane, it follows that to evaluate the definite integral we require that some path from z_1 to z_2 be defined. It is therefore clear that a definite integral of a complex function $f(z)$ is in fact a **line integral**.

Line integrals will be considered in Section 7.4.1. Briefly, for now, a line integral in the (x, y) plane, of the real variables x and y, is an integral of the form

$$\int_C [P(x, y)\,dx + Q(x, y)\,dy] \qquad \textbf{(1.39)}$$

where C denotes the path of integration between two points A and B in the plane. In the particular case when

$$\frac{\partial P}{\partial y} = \frac{\partial Q}{\partial x} \qquad \textbf{(1.40)}$$

the integrand $P(x, y)\,dx + Q(x, y)\,dy$ is a total differential, and the line integral is independent of the path C joining A and B.

In this section we introduce **contour integration**, which is the term used for evaluating line integrals in the complex plane.

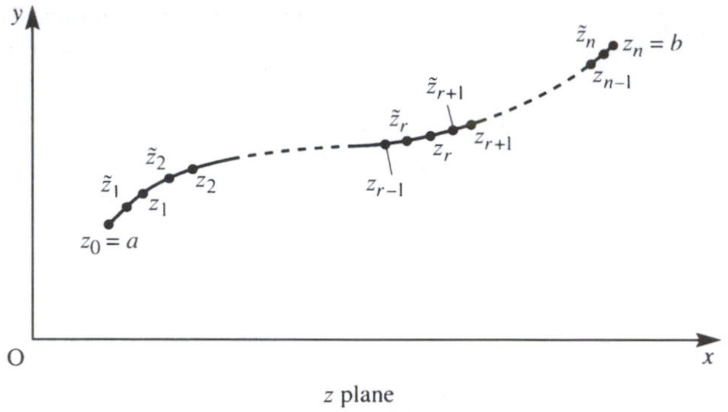

Figure 1.25 Partitioning of the curve C.

1.6.1 Contour integrals

Let $f(z)$ be a complex function that is continuous at all points of a simple curve C in the z plane that is of finite length and joins two points a and b. (We have not gone into great detail regarding the question of continuity for complex variables. Suffice it to say that the intuitive concepts described in Chapter 9 of *Modern Engineering Mathematics* for real variables carry over to the case of complex variables.) Subdivide the curve into n parts by the points $z_1, z_2, \ldots, z_{n-1}$, taking $z_0 = a$ and $z_n = b$ (Figure 1.25). On each arc joining z_{k-1} to z_k $(k = 1, \ldots, n)$ choose a point $\tilde{z}_k$. Form the sum

$$S_n = f(\tilde{z}_1)(z_1 - z_0) + f(\tilde{z}_2)(z_2 - z_1) + \ldots + f(\tilde{z}_n)(z_n - z_{n-1})$$

Then, writing $z_k - z_{k-1} = \Delta z_k$, S_n becomes

$$S_n = \sum_{k=1}^{n} f(\tilde{z}_k)\,\Delta z_k$$

If we let n increase in such a way that the largest of the chord lengths $|\Delta z_k|$ approaches zero then the sum S_n approaches a limit that does not depend on the mode of subdivision of the curve. We call this limit the **contour integral** of $f(z)$ along the curve C:

$$\int_C f(z)\,dz = \lim_{|\Delta z_k| \to 0} \sum_{k=1}^{n} f(\tilde{z}_k)\,\Delta z_k \tag{1.41}$$

If we take $z = x + jy$ and express $f(z)$ as

$$f(z) = u(x, y) + jv(x, y)$$

then it can be shown from (1.41) that

$$\int_C f(z)\,dz = \int_C [u(x,y) + jv(x,y)](dx + j\,dy)$$

or

$$\int_C f(z)\,dz = \int_C [u(x,y)\,dx - v(x,y)\,dy] + j\int_C [v(x,y)\,dx + u(x,y)\,dy] \qquad (1.42)$$

Both of the integrals on the right-hand side of (1.42) are real line integrals of the form (1.39), and can therefore be evaluated using the methods developed for such integrals.

EXAMPLE 1.27 Evaluate the contour integral $\int_C z^2 dz$ along the path C from $-1 + j$ to $5 + j3$ and composed of the two straight line segments, the first from $-1 + j$ to $5 + j$ and the second from $5 + j$ to $5 + j3$.

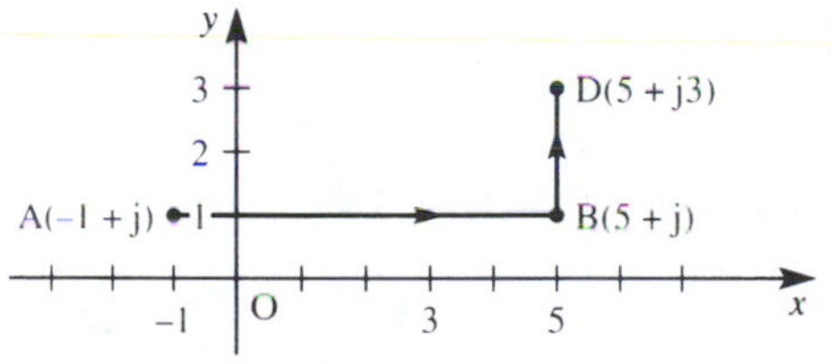

Figure 1.26 Path of integration for Example 1.27.

Solution The path of integration C is shown in Figure 1.26. Since

$$z^2 = (x + jy)^2 = (x^2 - y^2) + j2xy$$

it follows from (1.42) that

$$I = \int_C z^2\,dz = \int_C [(x^2 - y^2)dx - 2xy\,dy] + j\int_C [2xy\,dx + (x^2 - y^2)dy]$$

Along AB, $y = 1$ and $dy = 0$, so that

$$I_{AB} = \int_{-1}^{5} (x^2 - 1)\,dx + j\int_{-1}^{5} 2x\,dx$$

$$= [\tfrac{1}{3}x^3 - x]_{-1}^{5} + j[x^2]_{-1}^{5} = 36 + j24$$

Along BD, $x = 5$ and $dx = 0$, so that

$$I_{BD} = \int_1^3 -10y\,dy + j\int_1^3 (25 - y^2)\,dy$$
$$= [-5y^2]_1^3 + j[25y - \tfrac{1}{3}y^3]_1^3$$
$$= -40 + j\tfrac{124}{3}$$

Thus

$$\int_C z^2\,dz = I_{AB} + I_{BD} = (36 + j24) + (-40 + j\tfrac{124}{3}) = -4 + j\tfrac{196}{3}$$

EXAMPLE 1.28 Show that $\int_C (z + 1)\,dz = 0$, where C is the boundary of the square with vertices at $z = 0$, $z = 1 + j0$, $z = 1 + j1$ and $z = 0 + j1$.

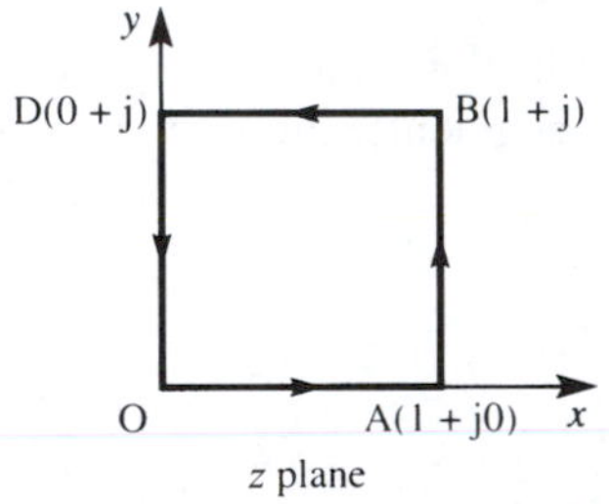

Figure 1.27 Path of integration for Example 1.28.

Solution The path of integration C is shown in Figure 1.27.

Since $z + 1 = (x + 1) + jy$, it follows from (1.42) that

$$I = \int_C (z + 1)\,dz = \int_C [(x + 1)\,dx - y\,dy] + j\int_C [y\,dx + (x + 1)\,dy]$$

Along OA, $y = 0$ and $dy = 0$, so that

$$I_{OA} = \int_0^1 (x + 1)\,dx = \tfrac{3}{2}$$

Along AB, $x = 1$ and $dx = 0$, so that

$$I_{AB} = \int_0^1 -y\,dy + j\int_0^1 2\,dy = -\tfrac{1}{2} + j2$$

Along BD, $y = 1$ and $dy = 0$, so that

$$I_{BD} = \int_1^0 (x + 1)\,dx + j\int_1^0 dx = -\tfrac{3}{2} - j$$

Along DO, $x = 0$ and $dx = 0$, so that

$$I_{DO} = \int_1^0 -y\,dy + j\int_1^0 dx = \tfrac{1}{2} - j$$

Thus

$$\int_C (z+1)\,dz = I_{OA} + I_{AB} + I_{BD} + I_{DO} = 0$$

1.6.2 Cauchy's theorem

The most important result in the whole of complex variable theory is called Cauchy's theorem and it provides the foundation on which the theory of integration with respect to a complex variable is based. The theorem may be stated as follows.

THEOREM 1.1 Cauchy's theorem

If $f(z)$ is an analytic function with derivative $f'(z)$ that is continuous at all points inside and on a simple closed curve C then

$$\oint_C f(z)dz = 0$$

(Note the use of the symbol $\oint_C$ to denote integration around a closed curve, with the convention being that the integral is evaluated travelling round C in the positive or anticlockwise direction.)

Proof To prove the theorem, we make use of **Green's theorem** in a plane, which will be introduced in Section 7.4.5. At this stage a statement of the theorem is sufficient.

If C is a simple closed curve enclosing a region A in a plane, and $P(x, y)$ and $Q(x, y)$ are continuous functions with continuous partial derivatives, then

$$\oint_C (P\mathrm{d}x + Q\mathrm{d}y) = \iint_A \left(\frac{\partial Q}{\partial x} - \frac{\partial P}{\partial y}\right) \mathrm{d}x\,\mathrm{d}y \tag{1.43}$$

Returning to the contour integral and taking

$$f(z) = u(x, y) + \mathrm{j}v(x, y), \quad z = x + \mathrm{j}y$$

we have from (1.42)

$$\oint_C f(z)\,\mathrm{d}z = \oint_C (u\,\mathrm{d}x - v\,\mathrm{d}y) + \mathrm{j}\oint_C (v\,\mathrm{d}x + u\,\mathrm{d}y) \tag{1.44}$$

Since $f(z)$ is analytic, the Cauchy–Riemann equations

$$\frac{\partial u}{\partial x} = \frac{\partial v}{\partial y}, \quad \frac{\partial v}{\partial x} = -\frac{\partial u}{\partial y}$$

are satisfied on C and within the region R enclosed by C.

Since $u(x, y)$ and $v(x, y)$ satisfy the conditions imposed on $P(x, y)$ and $Q(x, y)$ in Green's theorem, we can apply (1.43) to both integrals on the right-hand side of (1.44) to give

$$\oint_C f(z)\,\mathrm{d}z = \iint_R \left(-\frac{\partial v}{\partial x} - \frac{\partial u}{\partial y}\right)\mathrm{d}x\,\mathrm{d}y + \mathrm{j}\iint_R \left(\frac{\partial u}{\partial x} - \frac{\partial v}{\partial y}\right)\mathrm{d}x\,\mathrm{d}y$$

$$= 0 + \mathrm{j}0$$

by the Cauchy–Riemann equations. Thus

$$\oint_C f(z)\mathrm{d}z = 0$$

as required. □

In fact, the restriction in Cauchy's theorem that $f'(z)$ has to be continuous on C can be removed and so make the theorem applicable to a wider class of functions. A revised form of Theorem 1.1, with the restriction removed, is referred to as the **fundamental theorem of complex integration**. Since the proof that $f'(z)$ need not be continuous on C was first proposed by Goursat, the fundamental theorem is also sometimes referred to as the **Cauchy–Goursat theorem**. We shall not pursue the consequences of relaxation of this restriction any further in this book.

In practice, we frequently need to evaluate contour integrals involving functions such as

$$f_1(z) = \frac{1}{z-2}, \quad f_2(z) = \frac{z}{(z-3)^2(z+2)}$$

that have singularities associated with them. Since the function ceases to be analytic at such points, how do we accommodate for a singularity if it is inside the contour of integration? To resolve the problem the singularity is removed by **deforming the contour**.

First let us consider the case when the complex function $f(z)$ has a single isolated singularity at $z = z_0$ inside a closed curve C. To remove the singularity, we surround it by a circle γ, of radius ρ, and then cut the region between the circle and the outer contour C by a straight line AB. This leads to the deformed contour indicated by the arrows in Figure 1.28. In the figure the line linking the circle γ to the contour C is shown as a narrow channel in order to enable us to distinguish between the path A to B and the path B to A. The region inside this deformed contour is shown shaded in the figure (recall that the region inside a closed contour is the region on the left as we travel round it). Since this contains no singularities, we can apply Cauchy's theorem and write

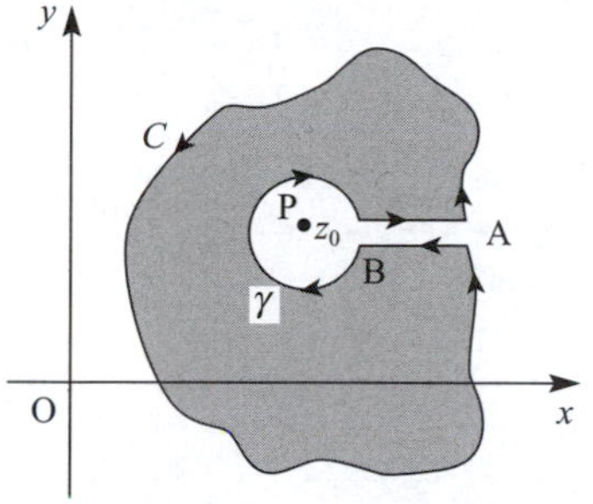

Figure 1.28 Deformed contour for an isolated singularity.

$$\oint_C f(z)\mathrm{d}z + \int_{AB} f(z)\,\mathrm{d}z + \oint_\gamma f(z)\,\mathrm{d}z + \oint_{BA} f(z)\,\mathrm{d}z = 0$$

Since

$$\oint_{BA} = -\oint_{AB}, \quad \text{and} \quad \oint_\gamma = -\oint_\gamma$$

this reduces to

$$\oint_C f(z)\,\mathrm{d}z = \oint_{\gamma^+} f(z)\,\mathrm{d}z \tag{1.45}$$

with the + indicating the change of sense from clockwise to anticlockwise around the circle γ.

EXAMPLE 1.29 Evaluate the integral $\oint_C \mathrm{d}z/z$ around

(a) any contour containing the origin;

(b) any contour not containing the origin.

Solution (a) $f(z) = 1/z$ has a singularity (a simple pole) at $z = 0$. Hence, using (1.45), the integral around any contour enclosing the origin is the same as the integral around a circle γ, centred at the origin and of radius ρ_0. We thus need to evaluate

$$\oint_\gamma \frac{1}{z}\,\mathrm{d}z$$

As can be seen from Figure 1.29, on the circle γ

$$z = \rho_0\,\mathrm{e}^{\mathrm{j}\theta}\ (0 \leqslant \theta < 2\pi)$$

so

$$\mathrm{d}z = \mathrm{j}\rho_0\,\mathrm{e}^{\mathrm{j}\theta}\,\mathrm{d}\theta$$

leading to

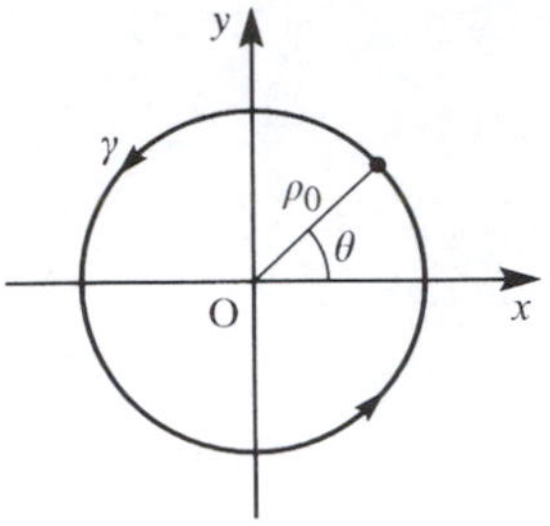

Figure 1.29 A circle of radius ρ_0 centred at the origin.

$$\oint_\gamma \frac{1}{z}\,dz = \int_0^{2\pi} \frac{j\rho_0\,e^{j\theta}}{\rho_0\,e^{j\theta}}\,d\theta = \int_0^{2\pi} j\,d\theta = 2\pi j$$

Hence if C encloses the origin then

$$\oint_C \frac{dz}{z} = 2\pi j$$

(b) If C does not enclose the origin then, by Cauchy's theorem

$$\oint_C \frac{dz}{z} = 0$$

since $1/z$ is analytic inside and on any curve that does not enclose the origin.

EXAMPLE 1.30 Generalize the result of Example 1.29 by evaluating

$$\oint_C \frac{dz}{z^n}$$

where n is an integer, around any contour containing the origin.

Solution If $n \leqslant 0$, we can apply Cauchy's theorem straight away (or evaluate the integral directly) to show the integral is zero. If $n > 1$, we proceed as in Example 1.29 and evaluate the integral around a circle, centred at the origin. Taking $z = \rho_0 e^{j\theta}$ as in Example 1.29, we have

$$\oint_C \frac{dz}{z^n} = \int_0^{2\pi} \frac{j\rho_0\,e^{j\theta}}{\rho_0^n\,e^{nj\theta}}\,d\theta$$

where ρ_0 is once more the radius of the circle. If $n \neq 1$,

$$\oint_C \frac{dz}{z^n} = j\int_0^{2\pi} \frac{d\theta}{\rho_0^{n-1}\,e^{(n-1)j\theta}} = j\rho_0^{1-n}\left[\frac{e^{(1-n)j\theta}}{(1-n)j}\right]_0^{2\pi} = \frac{\rho_0^{1-n}}{1-n}(e^{(1-n)2\pi j} - 1) = 0$$

since $e^{2\pi jN} = 1$ for any integer N. Hence

$$\oint_C \frac{dz}{z^n} = 0 \quad (n \neq 1)$$

In Examples 1.29 and 1.30 we have thus established the perhaps surprising result that if C is a contour containing the origin then

$$\oint_C \frac{dz}{z^n} = \begin{cases} 2\pi j & (n = 1) \\ 0 & (n \text{ any other integer}) \end{cases}$$

If C does not contain the origin, the integral is of course zero by Cauchy's theorem.

EXAMPLE 1.31

Evaluate the integral

$$\oint_C \frac{dz}{z-2-j}$$

around any contour C containing the point $z = 2 + j$.

Solution The function

$$f(z) = \frac{1}{z-2-j}$$

has a singularity (simple pole) at $z = 2 + j$. Hence, using (1.45), the integral around any contour C enclosing the point $z = 2 + j$ is the same as the integral around a circle γ centred at $z = 2 + j$ and of radius ρ. Thus we need to evaluate

$$\oint_\gamma \frac{dz}{z-2-j}$$

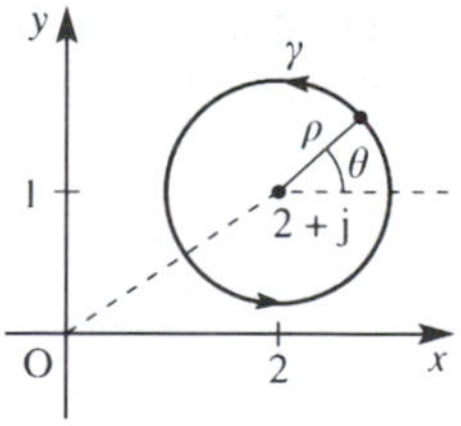

Figure 1.30 A circle of radius ρ centred at $2 + j$.

As can be seen from Figure 1.30, on the circle γ

$$z = (2 + j) + \rho e^{j\theta} \quad (0 \leqslant \theta < 2\pi)$$

$$dz = j\rho e^{j\theta} d\theta$$

leading to

$$\oint_\gamma \frac{dz}{z-2-j} = \int_0^{2\pi} \frac{j\rho e^{j\theta}}{\rho e^{j\theta}} d\theta$$

$$= \int_0^{2\pi} j d\theta = j2\pi$$

Hence if C encloses the point $z = 2 + j$ then

$$\oint_C \frac{dz}{z-2-j} = \pi j2$$

Compare this with the answer to Example 1.29.

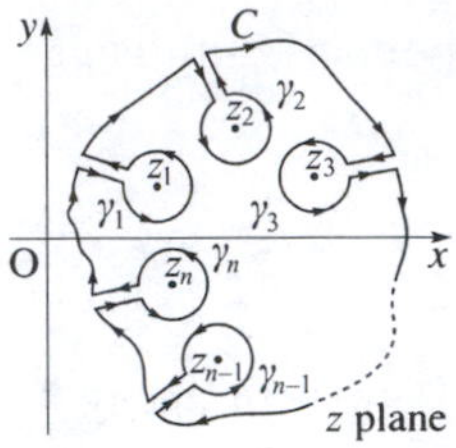

Figure 1.31
Deformed contour for n singularities.

So far we have only considered functions having a single singularity inside the closed contour C. The method can be extended to accommodate any finite number of singularities. If the function $f(z)$ has a finite number of singularities at $z = z_1, z_2, \dots, z_n$, inside a closed contour C then we can deform the latter by introducing n circles $\gamma_1, \gamma_2, \dots, \gamma_n$ to surround each of the singularities as shown in Figure 1.31. It is then readily shown that

$$\oint_C f(z)\,dz = \oint_{\gamma_1} f(z)\,dz + \oint_{\gamma_2} f(z)\,dz + \dots + \oint_{\gamma_n} f(z)\,dz \tag{1.46}$$

EXAMPLE 1.32 Evaluate the contour integral

$$\oint_C \frac{z\,dz}{(z-1)(z+2j)}$$

where C is

(a) any contour enclosing both the points $z = 1$ and $z = -2j$;
(b) any contour enclosing $z = -2j$ but excluding the point $z = 1$.

Solution The function

$$f(z) = \frac{z}{(z-1)(z+2j)}$$

has singularities at both $z = 1$ and $z = -2j$.

(a) Since the contour encloses both singularities, we need to evaluate the integrals around circles γ_1 and γ_2 of radii ρ_1 and ρ_2 surrounding the points $z = 1$ and $z = -2j$ respectively. Alternatively, we can resolve $f(z)$ into partial fractions as

$$f(z) = \frac{\frac{1}{3}(1-j2)}{z-1} + \frac{\frac{1}{5}(4+2j)}{z+2j}$$

and consider

$$I = \oint_C \frac{z\,dz}{(z-1)(2-2j)} = \frac{1}{3}(1-2j)\oint_C \frac{dz}{z-1} + \frac{1}{5}(4+2j)\oint_C \frac{dz}{z+2j} = I_1 + I_2$$

The integrand of I_1 has a single singularity at $z = 1$, and we simply need to evaluate it around the circle γ_1 of radius ρ_1 about $z = 1$ to give

$$I_1 = 2\pi j$$

Similarly, I_2 has a single singularity at $z = -2j$, and we evaluate it around the circle γ_2 to give

$$I_2 = 2\pi j$$

Then

$$I = \tfrac{1}{3}(1 - j2)2\pi j + \tfrac{1}{5}(4 + j2)2\pi j = 2\pi j(\tfrac{17}{15} - j\tfrac{4}{15})$$

Thus if the contour C contains both the singularities then

$$\oint_C \frac{z\,dz}{(z-1)(z+j2)} = 2\pi j\left(\frac{17}{15} - j\frac{4}{15}\right)$$

(b) If the contour C only contains the singularity $z = -2j$ then

$$\oint_C \frac{z\,dz}{(z-1)(z+2j)} = I_2 = 2\pi j\left(\frac{4}{5} + j\frac{2}{5}\right)$$

In Examples 1.29–1.32 we can note some similarity in the answers, with the common occurrence of the term $2\pi j$. It therefore appears that it may be possible to obtain some general results to assist in the evaluation of contour integrals. Indeed, this is the case, and such general results are contained in the **Cauchy integral theorem**.

THEOREM 1.2 Cauchy integral theorem

Let $f(z)$ be an analytic function within and on a simple closed contour C. If z_0 is any point in C then

$$\oint_C \frac{f(z)}{z - z_0}\,dz = 2\pi j\,f(z_0) \tag{1.47}$$

If we differentiate repeatedly n times with respect to z under the integral sign then it also follows that

$$\oint_C \frac{f(z)}{(z - z_0)^{n+1}}\,dz = \frac{2\pi j}{n!}f^{(n)}(z_0) \tag{1.48}$$

□

Note that (1.48) implies that if $f'(z)$ exists at $z = z_0$ so does $f^{(n)}(z)$ for all n, as predicted earlier in the observations following Example 1.16.

EXAMPLE 1.33

Evalute the contour integral

$$\oint_C \frac{2z}{(z-1)(z+2)(z+j)}\,dz$$

where C is a contour that includes the three points $z = 1$, $z = -2$ and $z = -j$.

Solution Since

$$f(z) = \frac{2z}{(z-1)(z+2)(z+\mathrm{j})}$$

has singularities at the points $z = 1$, $z = -2$ and $z = -\mathrm{j}$ inside the contour, it follows from (1.46) that

$$\oint_C f(z)\,\mathrm{d}z = \oint_{\gamma_1} f(z)\,\mathrm{d}z + \oint_{\gamma_2} f(z)\,\mathrm{d}z + \oint_{\gamma_3} f(z)\,\mathrm{d}z \tag{1.49}$$

where γ_1, γ_2 and γ_3 are circles centred at the singularities $z = 1$, $z = -2$ and $z = -\mathrm{j}$ respectively. In order to make use of the Cauchy integral theorem, (1.49) is written as

$$\oint_C f(z)\,\mathrm{d}z = \oint_{\gamma_1} \frac{\{2z/[(z+2)(z+\mathrm{j})]\}}{z-1}\,\mathrm{d}z + \oint_{\gamma_2} \frac{\{2z/[(z-1)(z+\mathrm{j})]\}}{z+2}\,\mathrm{d}z$$

$$+ \oint_{\gamma_3} \frac{\{2z/[(z-1)(z+2)]\}}{z+\mathrm{j}}\,\mathrm{d}z$$

$$= \oint_{\gamma_1} \frac{f_1(z)}{z-1}\,\mathrm{d}z + \oint_{\gamma_2} \frac{f_2(z)}{z+2}\,\mathrm{d}z + \oint_{\gamma_3} \frac{f_3(z)}{z+\mathrm{j}}\,\mathrm{d}z$$

Since $f_1(z)$, $f_2(z)$ and $f_3(z)$ are analytic within and on the circles γ_1, γ_2 and γ_3 respectively, it follows from (1.47) that

$$\oint_C f(z)\,\mathrm{d}z = 2\pi\mathrm{j}[f_1(1) + f_2(-2) + f_3(-\mathrm{j})]$$

$$= 2\pi\mathrm{j}\left[\frac{3}{2(1+\mathrm{j})} + \frac{-4}{(-3)(-2+\mathrm{j})} + \frac{-2\mathrm{j}}{(-\mathrm{j}-1)(-\mathrm{j}+2)}\right]$$

so that

$$\oint_C \frac{2z\,\mathrm{d}z}{(z-1)(z+2)(z+\mathrm{j})} = 0$$

EXAMPLE 1.34 Evaluate the contour integral

$$\oint_C \frac{z^4}{(z-1)^3}\,\mathrm{d}z$$

where the contour C encloses the point $z = 1$.

Solution Since $f(z) = z^4/(z-1)^3$ has a pole of order three at $z = 1$, it follows that

$$\oint_C f(z)\,\mathrm{d}z = \oint_{\gamma} \frac{z^4}{(z-1)^3}\,\mathrm{d}z$$

where γ is a circle centred at $z = 1$. Writing $f_1(z) = z^4$, then

$$\oint_C f(z)\,dz = \oint_\gamma \frac{f_1(z)}{(z-1)^3}\,dz$$

and, since $f_1(z)$ is analytic within and on the circle γ, it follows from (1.48) that

$$\oint_C f(z)\,dz = 2\pi j \frac{1}{2!}\left[\frac{d^2}{dz^2} f_1(z)\right]_{z=1} = \pi j(12z^2)_{z=1}$$

so that

$$\oint_C \frac{z^4}{(z-1)^3}\,dz = 12\pi j$$

1.6.3 Exercises

53 Evaluate $\int_C (z^2 + 3z)\,dz$ along the following contours C in the complex z plane:

(a) the straight line joining $2 + j0$ to $0 + j2$;
(b) the straight lines from $2 + j0$ to $2 + j2$ and then to $0 + j2$;
(c) the circle $|z| = 2$ from $2 + j0$ to $0 + j2$ in an anticlockwise direction.

54 Evaluate $\oint_C (5z^4 - z^3 + 2)\,dz$ around the following closed contours C in the z plane:

(a) the circle $|z| = 1$;
(b) the square with vertices at $0 + j0$, $1 + j0$, $1 + j1$ and $0 + j1$;
(c) the curve consisting of the parabolas $y = x^2$ from $0 + j0$ to $1 + j1$ and $y^2 = x$ from $1 + j$ to $0 + j0$.

55 Generalize the result of Example 1.30, and show that

$$\oint_C \frac{dz}{(z-z_0)^n} = \begin{cases} j2\pi & (n = 1) \\ 0 & (n \neq 1) \end{cases}$$

where C is a simple closed contour surrounding the point $z = z_0$.

56 Evaluate the contour integral

$$\oint_C \frac{dz}{z-4}$$

where C is any simple closed curve and $z = 4$ is

(a) outside C (b) inside C

57 Using the Cauchy integral theorem, evaluate the contour integral

$$\oint_C \frac{2z\,dz}{(2z-1)(z+2)}$$

where C is

(a) the circle $|z| = 1$
(b) the circle $|z| = 3$

58 Using the Cauchy integral theorem, evaluate the contour integral

$$\oint_C \frac{5z\,dz}{(z+1)(z-2)(z+4j)}$$

where C is

(a) the circle $|z| = 3$
(b) the circle $|z| = 5$

59 Using the Cauchy integral theorem, evaluate the following contour integrals:

(a) $$\oint_C \frac{z^3+z}{(2z+1)^3}\,dz$$

where C is the unit circle $|z| = 1$;

(b) $$\oint_C \frac{4z}{(z-1)(z+2)^2}\,dz$$

where C is the circle $|z| = 3$.

1.6.4 The residue theorem

This theorem draws together the theories of differentiation and integration of a complex function. It is concerned with the evaluation of the contour integral

$$I = \oint_C f(z)\,dz$$

where the complex function $f(z)$ has a finite number n of isolated singularities at $z_1, z_2, \dots, z_n$ inside the closed contour C. Defining the contour C as in Figure 1.31, we have as in (1.46) that

$$I = \oint_C f(z)\,dz = \oint_{\gamma_1} f(z)\,dz + \oint_{\gamma_2} f(z)\,dz + \dots + \oint_{\gamma_n} f(z)\,dz \qquad \textbf{(1.46)}$$

If we assume that $f(z)$ has a pole of order m at $z = z_i$ then it can be represented by the Laurent series expansion

$$f(z) = \frac{a_{-m}^{(i)}}{(z-z_i)^m} + \dots + \frac{a_{-1}^{(i)}}{z-z_i} + a_0^{(i)} + a_1^{(i)}(z-z_i) + \dots + a_m^{(i)}(z-z_i)^m + \dots$$

valid in the annulus $r_i < |z - z_i| < R_i$. If the curve C lies entirely within this annulus then, by Cauchy's theorem, (1.46) becomes

$$I = \oint_C f(z)\,dz = \oint_{\gamma_i} f(z)\,dz$$

Substituting the Laurent series expansion of $f(z)$, which we can certainly do since we are within the annulus of convergence, we obtain

$$\oint_{\gamma_i} f(z)\,dz = \oint_{\gamma_i} \left[\frac{a_{-m}^{(i)}}{(z-z_i)^m} + \dots + \frac{a_{-1}^{(i)}}{z-z_i} + a_0^{(i)} + a_1^{(i)}(z-z_i) + \dots + a_m^{(i)}(z-z_i)^m + \dots \right] dz$$

$$= a_{-m}^{(i)} \oint_{\gamma_i} \frac{dz}{(z-z_i)^m} + \dots + a_{-1}^{(i)} \oint_{\gamma_i} \frac{dz}{z-z_i} + a_0^{(i)} \oint_{\gamma_i} dz + a_1^{(i)} \oint (z-z_i)\,dz + \dots$$

Using the result from Exercise 55, all of these integrals are zero, except the one multiplying $a_{-1}^{(i)}$, the residue, which has the value $2\pi j$. We have therefore shown that

$$\oint_{\gamma_i} f(z)\,dz = 2\pi j a_{-1}^{(i)} = 2\pi j \times \text{residue at } z = z_i$$

This clearly generalizes, so that (1.46) becomes

$$I = \oint_C f(z)\,dz = 2\pi j \sum_{i=1}^{n} (\text{residue at } z = z_i)$$

$$= 2\pi j \times (\text{sum of residues inside } C)$$

Thus we have the following general result.

THEOREM 1.3 **The residue theorem**

> If $f(z)$ is an analytic function within and on a simple closed curve C, apart from a finite number of poles, then
>
> $$\oint_C f(z)\,dz = 2\pi j \times [\text{sum of residues of } f(z) \text{ at the poles inside } C]$$ □

This is quite a remarkable result in that it enables us to evaluate the contour integral $\oint_c f(z)\,dz$ by simply evaluating one coefficient of the Laurent series expansion of $f(z)$ at each of its singularities inside C.

EXAMPLE 1.35 Evaluate the contour integral $\oint_c dz/[z(1 + z)]$ if C is

(a) the circle $|z| = \frac{1}{2}$; (b) the circle $|z| = 2$.

Solution The singularities of $1/[z(1 + z)]$ are at $z = 0$ and -1. Evaluating the residues using (1.37), we have

$$\text{residue at } z = 0 = \lim_{z\to 0} z \frac{1}{z(1+z)} = 1$$

$$\text{residue at } z = -1 = \lim_{z\to -1} (z+1) \frac{1}{z(1+z)} = -1$$

(a) If C is $|z| = \frac{1}{2}$ then it contains the pole at $z = 0$, but *not* the pole at $z = -1$. Hence, by the residue theorem,

$$\oint_C \frac{dz}{z(z+1)} = 2\pi j \times (\text{residue at } z = 0) = 2\pi j$$

(b) If C is $|z| = 2$ then both poles are inside C. Hence, by the residue theorem,

$$\oint_C \frac{dz}{z(z+1)} = 2\pi j(1 - 1) = 0$$

EXAMPLE 1.36 Evaluate the contour integral $\displaystyle\oint_C \frac{z^3 - z^2 + z - 1}{z^3 + 4z}\,dz$ where C is

(a) $|z| = 1$ (b) $|z| = 3$

Solution The rational function

$$\frac{z^3 - z^2 + z - 1}{z^3 + 4z}$$

has poles at $z = 0$ and $\pm 2\mathrm{j}$. Evaluating the residues using (1.37) gives

$$\begin{array}{l}\text{residue} \\ \text{at } z = 0\end{array} = \lim_{z\to 0} \frac{z(z^3 - z^2 + z - 1)}{z(z^2 + 4)} = -\frac{1}{4}$$

$$\begin{array}{l}\text{residue} \\ \text{at } z = 2\mathrm{j}\end{array} = \lim_{z\to 2\mathrm{j}} \frac{(z - 2\mathrm{j})(z^3 - z^2 + z - 1)}{z(z - 2\mathrm{j})(z + 2\mathrm{j})} = -\frac{3}{8} + \frac{3}{4}\mathrm{j}$$

$$\begin{array}{l}\text{residue} \\ \text{at } z = -2\mathrm{j}\end{array} = \lim_{z\to -2\mathrm{j}} \frac{(z + 2\mathrm{j})(z^3 - z^2 + z - 1)}{z(z - 2\mathrm{j})(z + 2\mathrm{j})} = -\frac{3}{8} - \frac{3}{4}\mathrm{j}$$

(Note that these have been evaluated in Exercise 50(d).)

(a) If C is $|z| = 1$ then only the pole at $z = 0$ is inside the contour, so only the residue there is taken into account in the residue theorem, and

$$\oint_C \frac{z^3 - z^2 + z - 1}{z^3 + 4z}\,\mathrm{d}z = 2\pi\mathrm{j}\left(-\frac{1}{4}\right) = -\frac{1}{2}\pi\mathrm{j}$$

(b) If C is $|z| = 3$ then all the poles are inside the contour. Hence, by the residue theorem,

$$\oint_C \frac{z^3 - z^2 + z - 1}{z^3 + 4z}\,\mathrm{d}z = 2\pi\mathrm{j}\left(-\frac{1}{4} - \frac{3}{8} + \frac{3}{4}\mathrm{j} - \frac{3}{8} - \frac{3}{4}\mathrm{j}\right) = -2\pi\mathrm{j}$$

EXAMPLE 1.37 Evaluate the contour integral

$$\oint_C \frac{\mathrm{d}z}{z^3(z^2 + 2z + 2)}$$

where C is the circle $|z| = 3$.

Solution The poles of $1/z^3(z^2 + 2z + 2)$ are as follows: a pole of order three at $z = 0$, and two simple poles where $z^2 + 2z + 2 = 0$, that is at $z = -1 \pm \mathrm{j}$. All of these poles lie inside the contour C.

From (1.38), the residue at $z = 0$ is given by

$$\lim_{z\to 0} \frac{1}{2!}\frac{\mathrm{d}^2}{\mathrm{d}z^2}\left[\frac{1}{z^2 + 2z + 2}\right] = \lim_{z\to 0} \frac{1}{2}\frac{\mathrm{d}}{\mathrm{d}z}\left[\frac{-(2z+2)}{(z^2 + 2z + 2)^2}\right] = \lim_{z\to 0} \frac{\mathrm{d}}{\mathrm{d}z}\left[\frac{-(z+1)}{(z^2 + 2z + 2)^2}\right]$$

$$= \lim_{z\to 0} \frac{-(z^2 + 2z + 2)^2 + (z+1)2(z^2 + 2z + 2)(2z + 2)}{(z^2 + 2z + 2)^4} = \tfrac{1}{4}$$

From (1.37), the residue at $z = -1 - \mathrm{j}$ is

$$\lim_{z\to-1-j}(z+1+j)\frac{1}{z^3(z+1+j)(z+1-j)} = \lim_{z\to-1-j}\frac{1}{z^3(z+1-j)}$$

$$= \frac{1}{(-1-j)^3(-2j)} = \frac{1}{(1+j)^3 2j} = \frac{1}{(-2+2j)2j}$$

using $(1+j)^3 = 1 + 3j + 3j^2 + j^3 = -2 + 2j$. Hence

$$\begin{array}{l}\text{residue}\\ \text{at } z = -1-j\end{array} = \tfrac{1}{4}\frac{1}{-1-j} = -\tfrac{1}{4}\frac{1-j}{2} = \tfrac{1}{8}(-1+j)$$

Also, using (1.37),

$$\begin{array}{l}\text{residue}\\ \text{at } z = -1+j\end{array} = \lim_{z\to-1+j}(z+1-j)\frac{1}{z^3(z+1+j)(z+1-j)}$$

which is precisely the complex conjugate of the residue at $z = -1 - j$. Hence we can take a short cut with the algebra and state the residue as $\frac{1}{8}(-1-j)$.

The sum of the residues is

$$\tfrac{1}{4} + \tfrac{1}{8}(-1+j) + \tfrac{1}{8}(-1-j) = 0$$

so, by the residue theorem,

$$\oint_C \frac{dz}{z^3(z^2+2z+2)} = 2\pi j(0) = 0$$

1.6.5 Evaluation of definite real integrals

The evaluation of definite integrals is often achieved by using the residue theorem together with a suitable complex function $f(z)$ and a suitable closed contour C. In this section we shall briefly consider two of the most common types of real integrals that can be evaluated in this way.

Type 1: Infinite real integrals of the form

$$\int_{-\infty}^{\infty} f(x)\,dx$$

where $f(x)$ is a rational function of the real variable x

To evaluate such integrals we consider the contour integral

$$\oint_C f(z)\,\mathrm{d}z$$

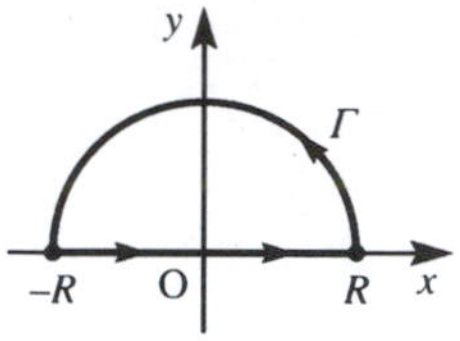

Figure 1.32 The closed contour for evaluating $\int_{-\infty}^{\infty} f(x)\,\mathrm{d}x$.

where C is the closed contour illustrated in Figure 1.32, consisting of the real axis from $-R$ to $+R$ and the semicircle Γ, of radius R, in the upper half z plane. Since $z = x$ on the real axis,

$$\oint_C f(z)\,\mathrm{d}z = \int_{-R}^{R} f(x)\,\mathrm{d}x + \int_\Gamma f(z)\,\mathrm{d}z$$

Then, provided that $\lim_{R\to\infty}\int_\Gamma f(z)\,\mathrm{d}z = 0$, taking $R \to \infty$ gives

$$\oint_C f(z)\,\mathrm{d}z = \int_{-\infty}^{\infty} f(x)\,\mathrm{d}x$$

On the semicircular path Γ, $z = R\mathrm{e}^{\mathrm{j}\theta}$ $(0 \leqslant \theta \leqslant \frac{1}{2}\pi)$, giving

$$\mathrm{d}z = \mathrm{j}R\,\mathrm{e}^{\mathrm{j}\theta}\,\mathrm{d}\theta$$

and

$$\int_\Gamma f(z)\,\mathrm{d}z = \int_0^{\pi/2} f(R\,\mathrm{e}^{\mathrm{j}\theta})\,\mathrm{j}R\,\mathrm{e}^{\mathrm{j}\theta}\,\mathrm{d}\theta$$

For this to tend to zero as $R \to \infty$, $|f(R\,\mathrm{e}^{\mathrm{j}\theta})|$ must decrease at least as rapidly as R^{-2}, implying that the degree of the denominator of the rational function $f(x)$ must be at least two more than the degree of the numerator. Thus, provided that this condition is satisfied, this approach may be used to calculate the infinite real integral $\int_{-\infty}^{\infty} f(x)\,\mathrm{d}x$. Note that if $f(x)$ is an even function of x then the same approach can also be used to evaluate $\int_0^{\infty} f(x)\,\mathrm{d}x$, since if $f(x)$ is even, it follows that

$$\int_{-\infty}^{\infty} f(x)\,\mathrm{d}x = 2\int_0^{\infty} f(x)\,\mathrm{d}x$$

EXAMPLE 1.38

Using contour integration, show that

$$\int_{-\infty}^{\infty} \frac{\mathrm{d}x}{(x^2+4)^2} = \frac{1}{16}\pi$$

Solution Consider the contour integral

$$I = \oint_C \frac{\mathrm{d}z}{(z^2+4)^2}$$

where C is the closed semicircular contour shown in Figure 1.32. The integrand $1/(z^2 + 4)^2$ has poles of order two at $z = \pm 2\mathrm{j}$. However, the only singularity inside the contour C is the double pole at $z = 2\mathrm{j}$. From (1.38),

$$\begin{aligned}\text{residue at } z = 2\text{j} &= \lim_{z\to 2\text{j}} \frac{1}{1!}\frac{\text{d}}{\text{d}z}(z-2\text{j})^2 \frac{1}{(z-2\text{j})^2(z+2\text{j})^2} \\ &= \lim_{z\to 2\text{j}} \frac{-2}{(z+2\text{j})^3} = \frac{-2}{(4\text{j})^3} = -\frac{1}{32}\text{j}\end{aligned}$$

so, by the residue theorem,

$$\oint_C \frac{\text{d}z}{(z^2+4)^2} = 2\pi\text{j}(-\tfrac{1}{32}\text{j}) = \frac{1}{16}\pi$$

Since

$$\oint_C \frac{\text{d}z}{(z^2+4)^2} = \int_{-R}^{R} \frac{\text{d}x}{(x^2+4)^2} + \int_{\Gamma} \frac{\text{d}z}{(z^2+4)^2}$$

letting $R \to \infty$, and noting that the second integral becomes zero, gives

$$\oint_C \frac{\text{d}z}{(z^2+4)^2} = \int_{-\infty}^{\infty} \frac{\text{d}x}{(x^2+4)^2} = \frac{1}{16}\pi$$

Note that in this particular case we could have evaluated the integral without using contour integration. Making the substitution $x = 2\tan\theta$, $\text{d}x = 2\sec^2\theta\,\text{d}\theta$ gives

$$\int_{-\infty}^{\infty} \frac{\text{d}x}{(x^2+4)^2} = \int_{-\pi/2}^{\pi/2} \frac{2\sec^2\theta\,\text{d}\theta}{(4\sec^2\theta)^2} = \frac{1}{8}\int_{-\pi/2}^{\pi/2} \cos^2\theta\,\text{d}\theta = \frac{1}{16}[\tfrac{1}{2}\sin 2\theta + \theta]_{-\pi/2}^{\pi/2} = \frac{1}{16}\pi$$

Type 2: Real integrals of the form

$$I = \int_0^{2\pi} G(\sin\theta, \cos\theta)\,\text{d}\theta$$

where G is a rational function of $\sin\theta$ *and* $\cos\theta$

We take $z = \text{e}^{\text{j}\theta}$, so that

$$\sin\theta = \frac{1}{2\text{j}}\left(z - \frac{1}{z}\right), \quad \cos\theta = \frac{1}{2}\left(z + \frac{1}{z}\right)$$

and

$$\text{d}z = \text{j}\,\text{e}^{\text{j}\theta}\,\text{d}\theta, \quad \text{or} \quad \text{d}\theta = \frac{\text{d}z}{\text{j}z}$$

On substituting back, the integral I becomes

$$I = \oint_C f(z)\,\text{d}z$$

where C is the unit circle $|z| = 1$ shown in Figure 1.33.

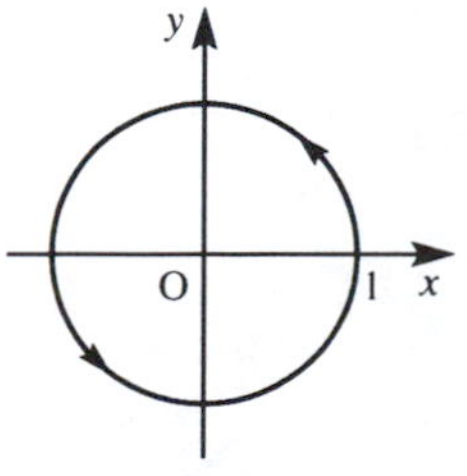

Figure 1.33 The unit-circle contour for evaluating $\int_0^{2\pi} G(\sin\theta, \cos\theta)\,\text{d}\theta$.

EXAMPLE 1.39 Using contour integration, evaluate

$$I = \int_0^{2\pi} \frac{d\theta}{2 + \cos\theta}$$

Solution Take $z = e^{j\theta}$, so that

$$\cos\theta = \frac{1}{2}\left(z + \frac{1}{z}\right), \quad d\theta = \frac{dz}{jz}$$

On substituting, the integral becomes

$$I = \oint_C \frac{dz}{jz[2 + \frac{1}{2}(z + 1/z)]} = \frac{2}{j}\oint_C \frac{dz}{z^2 + 4z + 1}$$

where C is the unit circle $|z| = 1$ shown in Figure 1.33. The integrand has singularities at

$$z^2 + 4z + 1 = 0$$

that is, at $z = -2 \pm \sqrt{3}$. The only singularity inside the contour C is the simple pole at $z = -2 + \sqrt{3}$. From (1.37),

$$\text{residue at } z = -2 + \sqrt{3}$$

$$= \lim_{z \to -2+\sqrt{3}} \left[\frac{2}{j}(z + 2 - \sqrt{3})\frac{1}{(z + 2 - \sqrt{3})(z + 2 + \sqrt{3})}\right] = \frac{2}{j}\frac{1}{2\sqrt{3}} = \frac{1}{j\sqrt{3}}$$

so, by the residue theorem,

$$I = 2\pi j\left(\frac{1}{j\sqrt{3}}\right) = \frac{2\pi}{\sqrt{3}}$$

Thus

$$\int_0^{2\pi} \frac{d\theta}{2 + \cos\theta} = \frac{2\pi}{\sqrt{3}}$$

1.6.6 Exercises

60 Evaluate the integral

$$\oint_C \frac{z\,dz}{z^2 + 1}$$

where C is

(a) the circle $|z| = \frac{1}{2}$ (b) the circle $|z| = 2$

61 Evaluate the integral

$$\oint_C \frac{z^2 + 3jz - 2}{z^3 + 9z}\,dz$$

where C is

(a) the circle $|z| = 1$ (b) the circle $|z| = 4$

62 Calculate the residues at all the poles of the function

$$f(z) = \frac{(z^2+2)(z^2+4)}{(z^2+1)(z^2+6)}$$

Hence calculate the integral

$$\oint_C f(z)\,dz$$

where C is

(a) the circle $|z| = 2$
(b) the circle $|z - j| = 1$
(c) the circle $|z| = 4$

63 Evaluate the integral

$$\oint_C \frac{dz}{z^2(1+z^2)^2}$$

where C is

(a) the circle $|z| = \frac{1}{2}$ (b) the circle $|z| = 2$

64 Using the residue theorem, evaluate the following contour integrals:

(a) $\oint_C \frac{(3z^2+2)\,dz}{(z-1)(z^2+4)}$,

where C is (i) the circle $|z-2| = 2$; (ii) the circle $|z| = 4$

(b) $\oint_C \frac{(z^2-2z)\,dz}{(z+1)^2(z^2+4)}$,

where C is (i) the circle $|z| = 3$; (ii) the circle $|z+j| = 2$

(c) $\oint_C \frac{dz}{(z+1)^3(z-1)(z-2)}$,

where C is (i) the circle $|z| = \frac{1}{2}$; (ii) the circle $|z+1| = 1$; (iii) the rectangle with vertices at $\pm j$, $3 \pm j$

(d) $\oint_C \frac{(z-1)\,dz}{(z^2-4)(z+1)^4}$,

where C is (i) the circle $|z| = \frac{1}{2}$; (ii) the circle $|z+\frac{3}{2}| = 2$; (iii) the triangle with vertices at $-\frac{3}{2}+j$, $-\frac{3}{2}-j$, $3+j0$

65 Using a suitable contour integral, evaluate the following real integrals:

(a) $\int_{-\infty}^{\infty} \frac{dx}{x^2+x+1}$ (b) $\int_{-\infty}^{\infty} \frac{dx}{(x^2+1)^2}$

(c) $\int_0^{\infty} \frac{dx}{(x^2+1)(x^2+4)^2}$

(d) $\int_0^{2\pi} \frac{\cos 3\theta}{5-4\cos\theta}\,d\theta$

(e) $\int_0^{2\pi} \frac{4\,d\theta}{5+4\sin\theta}$

(f) $\int_{-\infty}^{\infty} \frac{x^2\,dx}{(x^2+1)^2(x^2+2x+2)}$

(g) $\int_0^{2\pi} \frac{d\theta}{3-2\cos\theta+\sin\theta}$ (h) $\int_0^{\infty} \frac{dx}{x^4+1}$

(i) $\int_{-\infty}^{\infty} \frac{dx}{(x^2+4x+5)^2}$ (j) $\int_0^{2\pi} \frac{\cos\theta}{3+2\cos\theta}\,d\theta$

1.7 Engineering application: analysing AC circuits

In the circuit shown in Figure 1.34 we wish to find the variation in impedance Z and admittance Y as the capacitance C of the capacitor varies from 0 to ∞. Here

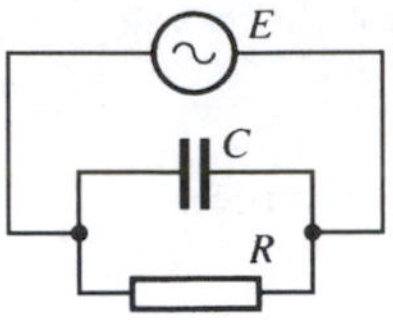

Figure 1.34 AC circuit of Section 1.7.

$$\frac{1}{Z} = \frac{1}{R} + j\omega C, \quad Y = \frac{1}{Z}$$

Writing

$$\frac{1}{Z} = \frac{1 + j\omega CR}{R}$$

we clearly have

$$Z = \frac{R}{1 + j\omega CR} \tag{1.50}$$

Equation (1.50) can be interpreted as a bilinear mapping with Z and C as the two variables. We examine what happens to the real axis in the C plane (C varies from 0 to ∞ and, of course, is real) under the inverse of the mapping given by (1.50). Rearranging (1.50), we have

$$C = \frac{R - Z}{j\omega RZ} \tag{1.51}$$

Taking $Z = x + jy$

$$C = \frac{R - x - jy}{j\omega R(x + jy)} = \frac{x + jy - R}{\omega R(y - jx)} = \frac{(x + jy - R)(y + jx)}{\omega R(x^2 + y^2)} \tag{1.52}$$

Equating imaginary parts, and remembering that C is real, gives

$$0 = x^2 + y^2 - Rx \tag{1.53}$$

which represents a circle, with centre at $(\frac{1}{2}R, 0)$ and of radius $\frac{1}{2}R$. Thus the real axis in the C plane is mapped onto the circle given by (1.53) in the Z plane. Of course, C is positive. If $C = 0$, (1.53) indicates that $Z = R$. The circuit of Figure 1.34 confirms that the impedance is R in this case. If $C \to \infty$ then $Z \to 0$, so the positive real axis in the plane is mapped onto either the upper or lower half of the circle. Equating real parts in (1.52) gives

$$C = \frac{-y}{\omega(x^2 + y^2)}$$

so $C > 0$ gives $y < 0$, implying that the lower half of the circle is the image in the Z plane of the positive real axis in the C plane, as indicated in Figure 1.35. A diagram such as Figure 1.35 gives an immediate visual impression of how the impedance Z varies as C varies.

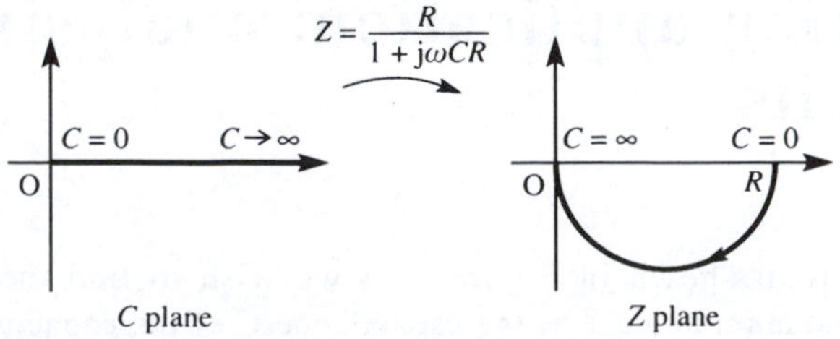

Figure 1.35 Mapping for the impedance Z.

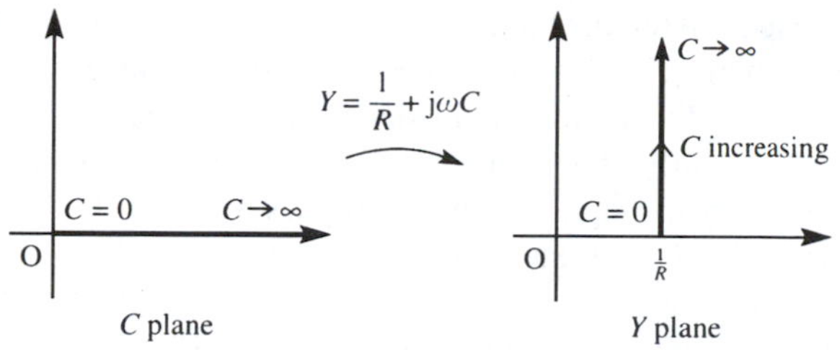

Figure 1.36 Mapping for the admittance Y.

The admittance $Y = 1/Z$ is given by

$$Y = \frac{1}{R} + j\omega C$$

which represents a linear mapping as shown in Figure 1.36.

1.8 Engineering application: use of harmonic functions

In this section we discuss two engineering applications where use is made of the properties of harmonic functions.

1.8.1 A heat transfer problem

We saw in Section 1.3.2 that every analytic function generates a pair of harmonic functions. The problem of finding a function that is harmonic in a specified region and satisfies prescribed boundary conditions is one of the oldest and most important problems in science-based engineering. Sometimes the solution can be found by means of a conformal mapping defined by an analytic function. This, essentially, is a consequence of the 'function of a function' rule of calculus, which implies that every harmonic function of x and y transforms into a harmonic function of u and v under the mapping

$$w = u + jv = f(x + jy) = f(z)$$

where $f(z)$ is analytic. Furthermore, the level curves of the harmonic function in the z plane are mapped onto corresponding level curves in the w plane, so that a harmonic function that has a constant value along part of the boundary of a region or has a zero normal derivative along part of the boundary is mapped onto a harmonic function with the same property in the w plane.

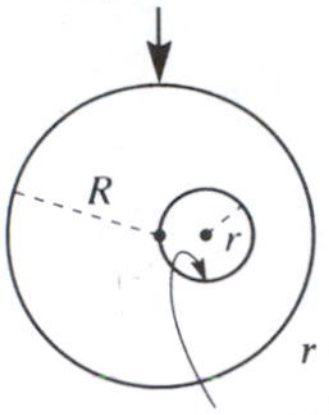

Figure 1.37 Schematic diagram of heat transfer problem.

For heat transfer problems the level curves of the harmonic function correspond to isotherms, and a zero normal derivative corresponds to thermal insulation. To illustrate these ideas, consider the simple steady-state heat transfer problem shown schematically in Figure 1.37. There is a cylindrical pipe with an offset cylindrical cavity through which steam passes at 100 °C. The outer temperature of the pipe is 0 °C. The radius of the inner circle is $\frac{3}{10}$ of that of the outer circle, so by choosing the outer radius as the unit of length the problem can be stated as that of finding a harmonic function $T(x, y)$ such that

$$\frac{\partial^2 T}{\partial x^2} + \frac{\partial^2 T}{\partial y^2} = 0$$

in the region between the circles $|z| = 1$ and $|z - 0.3| = 0.3$, and $T = 0$ on $|z| = 1$ and $T = 100$ on $|z - 0.3| = 0.3$.

The mapping

$$w = \frac{z - 3}{3z - 1}$$

transforms the circle $|z| = 1$ onto the circle $|w| = 1$ and the circle $|z - 0.3| = 0.3$ onto the circle $|w| = 3$ as shown in Figure 1.38. Thus the problem is transformed into the axially symmetric problem in the w plane of finding a harmonic function $T(u, v)$ such that $T(u, v) = 100$ on $|w| = 1$ and $T(u, v) = 0$ on $|w| = 3$. Harmonic functions with such axial symmetry have the general form

$$T(u, v) = A \ln (u^2 + v^2) + B$$

where A and B are constants.

Here we require, in addition to the axial symmetry, that $T(u, v) = 100$ on $u^2 + v^2 = 1$ and $T(u, v) = 0$ on $u^2 + v^2 = 9$. Thus $B = 100$ and $A = -100 \ln 9$, and the solution on the w plane is

$$T(u, v) = \frac{100\,[1 - \ln (u^2 + v^2)]}{\ln 9}$$

We need the solution on the z plane, which means in general we have to obtain u and v in terms of x and y. Here, however, it is a little easier, since $u^2 + v^2 = |w|^2$ and

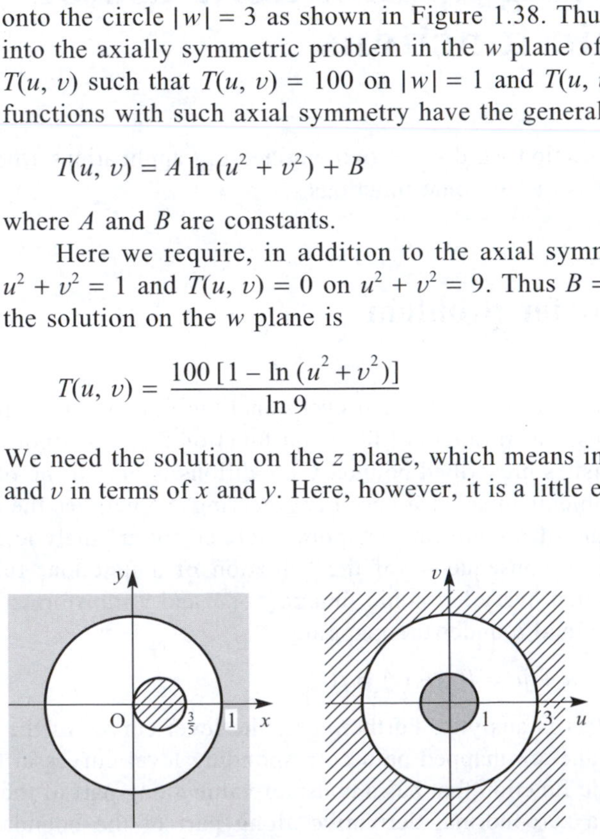

Figure 1.38 The mapping $w = (z - 3)/(3z - 1)$.

$$|w|^2 = \left|\frac{z-3}{3z-1}\right|^2 = \frac{|z-3|^2}{|3z-1|^2} = \frac{(x-3)^2+y^2}{(3x-1)^2+9y^2}$$

Thus

$$T(x, y) = \frac{100}{\ln 9}\{1 - \ln[(x-3)^2 + y^2] - \ln[(3x-1)^2 + 9y^2]\}$$

1.8.2 Current in a field-effect transistor

The fields (E_x, E_y) in an insulated-gate field-effect transistor are harmonic conjugates that satisfy a non-linear boundary condition. For the transistor shown schematically in Figure 1.39 we have

$$\frac{\partial E_y}{\partial x} = \frac{\partial E_x}{\partial y}, \quad \frac{\partial E_y}{\partial y} = \frac{-\partial E_x}{\partial x}$$

with conditions

$$E_x = 0 \quad \text{on the electrodes}$$

$$E_x\left(E_y + \frac{V_0}{h}\right) = -\frac{I}{2\mu\varepsilon_0\varepsilon_r} \quad \text{on the channel}$$

$$E_y \rightarrow -\frac{V_g}{h} \quad \text{as} \quad x \rightarrow -\infty \quad (0 < y < h)$$

$$E_y \rightarrow \frac{V_d - V_g}{h} \quad \text{as} \quad x \rightarrow \infty \quad (0 < y < h)$$

where V_0 is a constant with dimensions of potential, h is the insulator thickness, I is the current in the channel, which is to be found, μ, ε_0 and ε_r have their usual meanings, and the gate potential V_g and the drain potential V_d are taken with respect to the source potential.

The key to the solution of this problem is the observation that the non-linear boundary condition

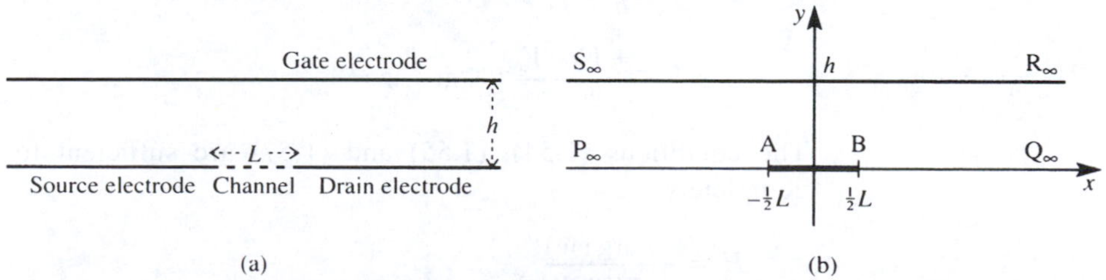

Figure 1.39 (a) Schematic diagram for an insulated-gate field-effect transistor; (b) an appropriate coordinate system for the application.

$$2E_x\left(E_y + \frac{V_0}{h}\right) = -\frac{I}{\mu\varepsilon_0\varepsilon_r}$$

contains the harmonic function (now of E_x and E_y)

$$H(E_x, E_y) = 2E_x\left(E_y + \frac{V_0}{h}\right)$$

A harmonic conjugate of H is the function

$$G(E_x, E_y) = \left(E_y + \frac{V_0}{h}\right)^2 - E_x^2$$

Since E_x and E_y are harmonic conjugates with respect to x and y, so are G and H. Thus the problem may be restated as that of finding harmonic conjugates G and H such that

$$H = 0 \quad \text{on the electrodes}$$

$$H = -\frac{I}{\mu\varepsilon_0\varepsilon_r} \quad \text{on the channel}$$

$$G \to \left(\frac{V_0 - V_g}{h}\right)^2 \quad \text{as} \quad x \to \infty \quad (0 < y < h)$$

$$G \to \left(\frac{V_0 + V_d - V_g}{h}\right)^2 \quad \text{as} \quad x \to -\infty \quad (0 < y < h)$$

Using the sequence of mappings shown in Figure 1.40, which may be composed into the single formula

$$w = \frac{a\,e^{bz} - a^2}{a\,e^{bz} - 1}$$

where $a = e^{bL/2}$ and $b = \pi/h$, the problem is transformed into finding harmonic-conjugate functions G and H (on the w plane) such that

$$H = 0 \quad \text{on} \quad v = 0 \quad (u > 0) \tag{1.54}$$

$$H = -\frac{I}{\mu\varepsilon_0\varepsilon_r} \quad \text{on} \quad v = 0 \quad (u < 0) \tag{1.55}$$

$$G = \left(\frac{V_0 - V_g}{h}\right)^2 \quad \text{at} \quad w = e^{bL} \tag{1.56}$$

$$G = \left(\frac{V_0 + V_d - V_g}{h}\right)^2 \quad \text{at} \quad w = 1 \tag{1.57}$$

The conditions (1.54), (1.55) and (1.57) are sufficient to determine H and G completely

$$H = -\frac{I\arg(w)}{\pi\mu\varepsilon_0\varepsilon_r}$$

$$G = \frac{I\ln|w|}{\pi\mu\varepsilon_0\varepsilon_r} + \left(\frac{V_0 + V_d - V_g}{h}\right)^2$$

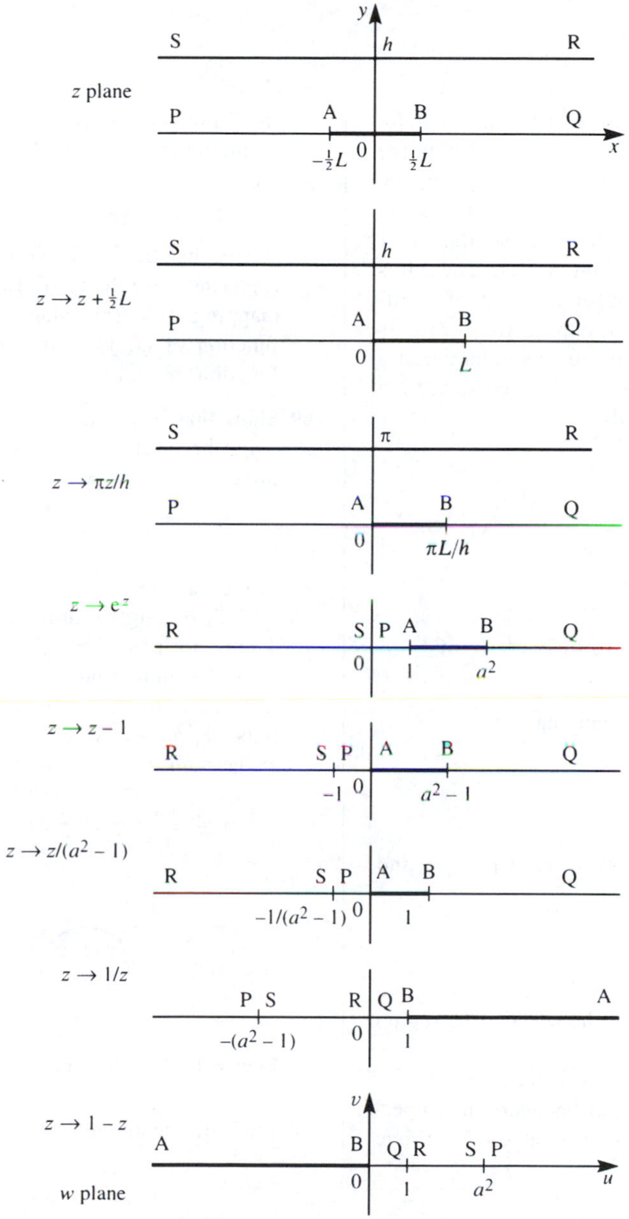

Figure 1.40 Sequence of mappings to simplify the problem.

while the condition (1.56) determines the values of I

$$I = \frac{\mu\varepsilon_0\varepsilon_r}{Lh}(2V_0 - 2V_g + V_d)V_d$$

This example shows the power of complex variable methods for solving difficult problems arising in engineering mathematics. The following exercises give some simpler examples for the reader to investigate.

1.8.3 Exercises

66 Show that the transformation $w = 1/z$, $w = u + jv$, $z = x + jy$, transforms the circle $x^2 + y^2 = 2ax$ in the z plane into the straight line $u = 1/2a$ in the w plane. Two long conducting wires of radius a are placed adjacent and parallel to each other, so that their cross-section appears as in Figure 1.41. The wires are separated at O by an insulating gap of negligible dimensions, and carry potentials $\pm V_0$ as indicated. Find an expression for the potential at a general point (x, y) in the plane of the cross-section and sketch the equipotentials.

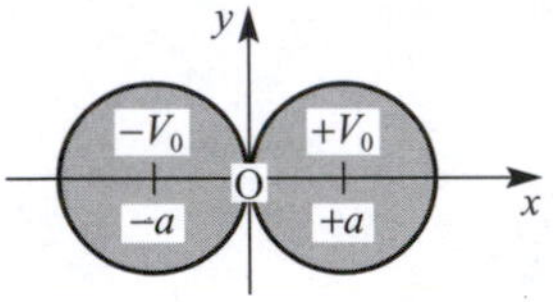

Figure 1.41 Conducting wires of Exercise 66.

67 Find the images under the mapping

$$w = \frac{z+1}{1-z}$$

$z = x + jy$, of

(a) the points A(−1, 0), B(0, 1), C($\frac{24}{25}$, $\frac{7}{25}$) and D($\frac{3}{4}$, 0) in the z plane,
(b) the straight line $y = 0$,
(c) the circle $x^2 + y^2 = 1$.

Illustrate your answer with a diagram showing the z and w planes and shade on the w plane the region corresponding to $x^2 + y^2 < 1$.

A semicircular disc of unit radius, $[(x, y): x^2 + y^2 \leqslant 1, y \geqslant 0]$, has its straight boundary at temperature 0 °C and its curved boundary at 100 °C. Prove that the temperature at the point (x, y) is

$$T = \frac{200}{\pi}\tan^{-1}\left(\frac{2y}{1-x^2-y^2}\right)$$

68 (a) Show that the function

$$G(x, y) = 2x(1 - y)$$

satisfies the Laplace equation and construct its harmonic conjugate $H(x, y)$ that satisfies $H(0, 0) = 0$. Hence obtain, in terms of z, where $z = x + jy$, the function F such that $W = F(z)$ where $W = G + jH$.

(b) Show that under the mapping $w = \ln z$, the harmonic function $G(x, y)$ defined in (a) is mapped into the function

$$G(u, v) = 2e^u \cos v - e^{2u} \sin 2v$$

Verify that $G(u, v)$ is harmonic.
(c) Generalize the result (b) to prove that under the mapping $w = f(z)$, where $f'(z)$ exists, a harmonic function of (x, y) is transformed into a harmonic function of (u, v).

69 Show that if $w = (z + 3)/(z - 3)$, $w = u + jv$, $z = x + jy$, the circle $u^2 + v^2 = k^2$ in the w plane is the image of the circle

$$x^2 + y^2 + 6\frac{1+k^2}{1-k^2}x + 9 = 0 \quad (k^2 \neq 1)$$

in the z plane.

Two long cylindrical wires, each of radius 4 mm, are placed parallel to each other with their axes 10 mm apart, so that their cross-section appears as in Figure 1.42. The wires carry potentials $\pm V_0$ as shown. Show that the potential $V(x, y)$ at the point (x, y) is given by

$$V = \frac{V_0}{\ln 4}\{\ln[(x + 3)^2 + y^2] - \ln[(x - 3)^2 + y^2]\}$$

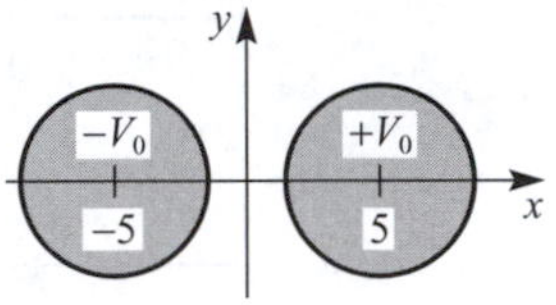

Figure 1.42 Cylindrical wires of Exercise 69.

70 Find the image under the mapping

$$w = \frac{j(1-z)}{1+z}$$

$z = x + jy$, $w = u + jv$, of

(a) the points A(1, 0), B(0, 1), C(0, −1) in the z plane,
(b) the straight line $y = 0$,
(c) the circle $x^2 + y^2 = 1$.

A circular plate of unit radius, $[((x, y): x^2 + y^2 \leqslant 1]$, has one half (with $y > 0$) of its rim, $x^2 + y^2 = 1$, at temperature 0 °C and the other half (with $y < 0$) at temperature 100 °C. Using the above mapping, prove that the steady-state temperature at the point (x, y) is

$$T = \frac{100}{\pi}\tan^{-1}\left(\frac{1 - x^2 - y^2}{2y}\right)$$

71 The problem shown schematically in Figure 1.43 arose during a steady-state heat transfer investigation. T is the temperature. By applying the successive mappings

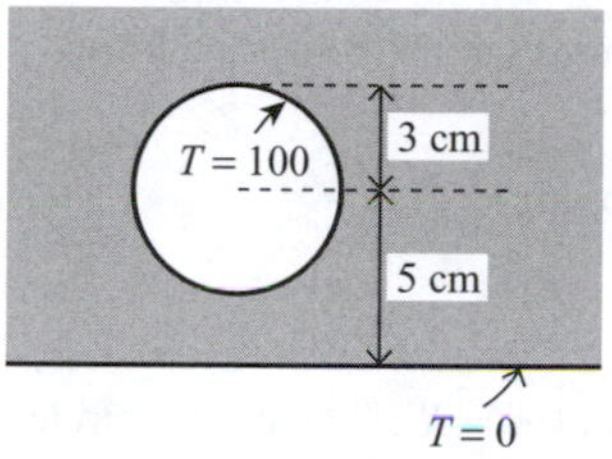

Figure 1.43 Schematic representation of Exercise 71.

$$z_1 = \frac{z + j4}{z - j4}, \quad w = \ln z_1$$

show that the temperature at the point (x, y) in the shaded region in the figure is given by

$$T(x, y) = \frac{50}{\ln 3}\ln\left[\frac{x^2 + (4 + y)^2}{x^2 + (4 - y)^2}\right]$$

72 The functions

$$w = z + \frac{1}{z}, \quad w = \frac{z + 1}{z - 1}$$

perform the mappings shown in Figure 1.44. A long bar of semicircular cross-section has the temperature of the part of its curved surface corresponding to the arc PQ in Figure 1.45 kept at 100 °C while the rest of the surface is kept at 0 °C. Show that the temperature T at the point (x, y) is given by

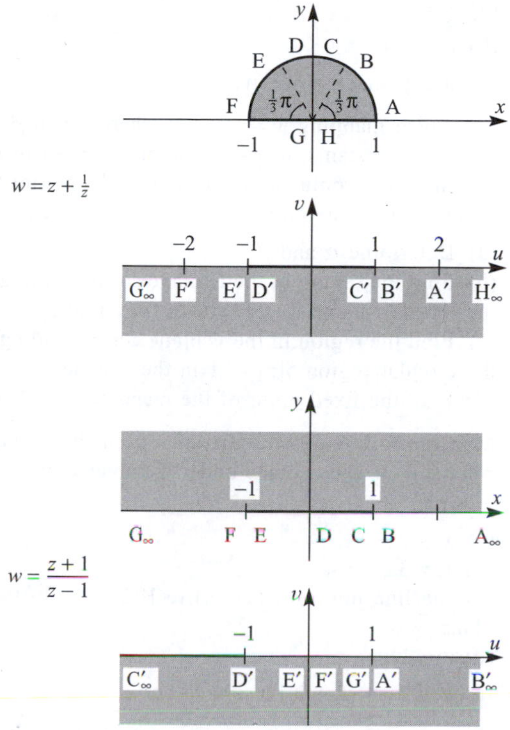

Figure 1.44 Mappings of Exercise 72.

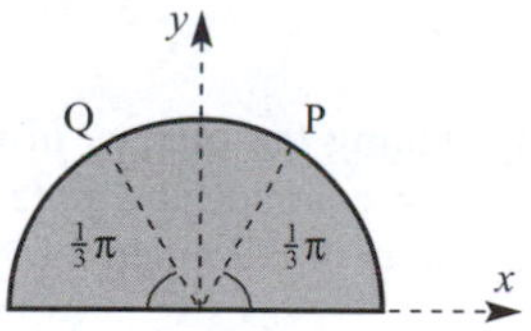

Figure 1.45 Cross-section of bar of Exercise 72.

$$T = \frac{100}{\pi}[\arg(z^2 + z + 1) - \arg(z^2 - z + 1)]$$

1.9 Review exercises (1–24)

1 Find the images of the following points under the mappings given:

(a) $z = 1 + j$ under $w = (1 + j)z + j$

(b) $z = 1 - j2$ under $w = j3z + j + 1$

(c) $z = 1$ under $w = \frac{1}{2}(1 - j)z + \frac{1}{2}(1 + j)$

(d) $z = j2$ under $w = \frac{1}{2}(1 - j)z + \frac{1}{2}(1 + j)$

2 Under each of the mappings given in Review exercise 1, find the images in the w plane of the two straight lines

(a) $y = 2x$
(b) $x + y = 1$
in the z plane, $z = x + jy$.

3 The linear mapping $w = \alpha z + \beta$, where α and β are complex constants, maps the point $z = 2 - j$ in the z plane to the point $w = 1$ in the w plane, and the point $z = 0$ to the point $w = 3 + j$.

(a) Determine α and β.
(b) Find the region in the w plane corresponding to the left half-plane $\text{Re}(z) \leqslant 0$ in the z plane.
(c) Find the region in the w plane corresponding to the circular region $5|z| \leqslant 1$ in the z plane.
(d) Find the fixed point of the mapping.

4 Map the following straight lines from the z plane, $z = x + jy$, to the w plane under the inverse mapping $w = j/z$:

(a) $x = y + 1$
(b) $y = 3x$
(c) the line joining A$(1 + j)$ to B$(2 + j3)$ in the z plane
(d) $y = 4$

In each case sketch the image curve.

5 Two complex variables w and z are related by the mapping

$$w = \frac{z+1}{z-1}$$

Sketch this mapping by finding the images in the w plane of the lines $\text{Re}(z) = $ constant and $\text{Im}(z) = $ constant. Find the fixed points of the mapping.

6 The mapping

$$w = \frac{1 - z^2}{z}$$

takes points from the z plane to the w plane. Find the fixed points of the mapping, and show that the circle of radius r with centre at the origin in the z plane is transformed to the ellipse

$$\left(\frac{ur^2}{r^2 - 1}\right)^2 + \left(\frac{vr^2}{r^2 + 1}\right)^2 = r^2$$

in the w plane, where $w = u + jv$. Investigate what happens when $r = 1$.

7 Find the real and imaginary parts of the complex function $w = z^3$, and verify the Cauchy–Riemann equations.

8 Find a function $v(x, y)$ such that, given

$$u(x, y) = x \sin x \cosh y - y \cos x \sinh y$$

$f(z) = u + jv$ is an analytic function of z ($f(0) = 0$).

9 Find the bilinear transformation that maps the three points $z = 0$, j and $\frac{1}{2}(1 + j)$ in the z plane to the three points $w = \infty$, $-j$ and $1 - j$ respectively in the w plane. Check that the transformation will map

(a) the lower half of the z plane onto the upper half of the w plane
(b) the interior of the circle with centre $z = j\frac{1}{2}$ and radius $\frac{1}{2}$ in the z plane onto the half-plane $\text{Im}(w) < -1$ in the w plane.

10 Show that the mapping

$$z = \zeta + \frac{a^2}{4\zeta}$$

where $z = x + jy$ and $\zeta = R\,e^{j\theta}$ maps the circle $R = $ constant in the ζ plane onto an ellipse in the z plane. Suggest a possible use for this mapping.

11 Find the power series representation of the function

$$\frac{1}{1 + z^3}$$

in the disc $|z| < 1$. Deduce the power series for

$$\frac{1}{(1 + z^3)^2}$$

valid in the same disc.

12 Find the first four non-zero terms of the Taylor series expansion of the following functions about the point indicated, and determine the radius of convergence of each:

(a) $\dfrac{1 - z}{1 + z}$ $(z = 0)$ (b) $\dfrac{1}{z^2 + 1}$ $(z = 1)$

(c) $\dfrac{z}{z + 1}$ $(z = j)$

13 Find the radius of convergence of each Taylor series expansion of the following function about the points indicated, *without* finding the series itself:

$$f(z) = \frac{1}{z(z^2 + 1)}$$

at the points $z = 1, -1, 1 + j, 1 + j\frac{1}{2}$ and $2 + j3$.

14 Determine the Laurent series expansion of the function

$$f(z) = \frac{1}{(z^2 + 1)z}$$

about the points (a) $z = 0$ and (b) $z = 1$, and determine the region of validity of each.

15 Find the Laurent series expansion of the function

$$f(z) = e^z \sin\left(\frac{1}{1-z}\right)$$

about (a) $z = 0$, (b) $z = 1$ and (c) $z = \infty$, indicating the range of validity in each case. (Do *not* find terms explicitly, indicate only the form of the principal part.)

16 Find the real and imaginary parts of the functions

(a) $e^z \sinh z$ (b) $\cos 2z$

(c) $\dfrac{\sin z}{z}$ (d) $\tan z$

17 Determine whether the following mappings are conformal, and, if not, find the non-conformal points:

(a) $w = \dfrac{1}{z^2}$

(b) $w = 2z^3 + 3z^2 + 6(1 - j)z + 1$

(c) $w = 64z + \dfrac{1}{z^3}$

18 Consider the mapping $w = \cos z$. Determine the points where the mapping is not conformal. By finding the images in the w plane of the lines $x =$ constant and $y =$ constant in the z plane ($z = x + jy$), draw the mapping similarly to Figures 1.14 and 1.18.

19 Determine the location of and classify the singularities of the following functions:

(a) $\dfrac{\sin z}{z^2}$ (b) $\dfrac{1}{(z^3-8)^2}$

(c) $\dfrac{z+1}{z^4-1}$ (d) $\operatorname{sech} z$

(e) $\sinh z$ (f) $\sin\left(\dfrac{1}{z}\right)$ (g) z^z

20 Find the residues of the following functions at the points indicated:

(a) $\dfrac{e^{2z}}{(1+z)^2}$ $(z = -1)$ (b) $\dfrac{\cos z}{2z-\pi}$ $(z = \frac{1}{2}\pi)$

(c) $\dfrac{\tan z}{2z-\pi}$ $(z = \frac{1}{2}\pi)$ (d) $\dfrac{z}{(z+8)^3}$ $(z = -8)$

21 Find the poles and zeros, and determine all the residues, of the rational function

$$f(z) = \frac{(z^2-1)(z^2+3z+5)}{z(z^4+1)}$$

22 Determine the residue of the rational function

$$\frac{z^7 + 6z^5 - 30z^4}{(z-1-j)^3}$$

23 Evaluate the following contour integrals along the circular paths indicated:

(a) $\displaystyle\oint_C \frac{z\,dz}{z^2+7z+6}$, where C is $|z| = 2$

(b) $\displaystyle\oint_C \frac{(z^2+1)(z^2+3)}{(z^2+9)(z^2+4)}\,dz$, where C is $|z| = 4$

(c) $\displaystyle\oint_C \frac{dz}{z^2(1-z^2)^2}$, where $\begin{cases}\text{(i) } C \text{ is } |z| = \frac{1}{2}\\ \text{(ii) } C \text{ is } |z| = 2\end{cases}$

(d) $\displaystyle\oint_C \frac{dz}{(2z-3j)(z+j)}$,

where $\begin{cases}\text{(i) } C \text{ is } |z| = 2\\ \text{(ii) } C \text{ is } |z-1| = 1\end{cases}$

(e) $\displaystyle\oint_C \frac{z^3\,dz}{(z^2+1)(z^2+z+1)}$, where C is $|z - j| = \frac{1}{2}$

(f) $\displaystyle\oint_C \frac{(z-1)\,dz}{z(z-2)^2(z-3)}$, where $\begin{cases}\text{(i) } C \text{ is } |z| = 1\\ \text{(ii) } C \text{ is } |z| = \frac{5}{2}\end{cases}$

24 Using a suitable contour integral, evaluate the following real integrals:

(a) $\displaystyle\int_{-\infty}^{\infty} \frac{x^2\,dx}{(x^2+1)^2(x^2+2x+2)}$

(b) $\displaystyle\int_0^{\infty} \frac{x^2\,dx}{x^4+16}$ (c) $\displaystyle\int_0^{2\pi} \frac{\sin^2\theta\,d\theta}{5+4\cos\theta}$

(d) $\displaystyle\int_0^{2\pi} \frac{\cos 2\theta\,d\theta}{5-4\cos\theta}$

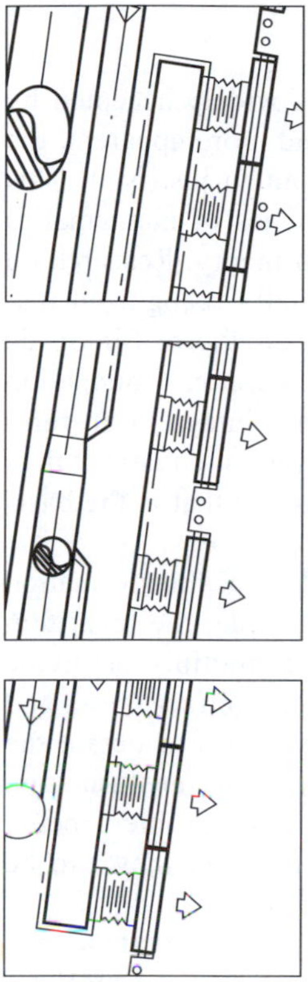

2

Laplace Transforms

CONTENTS

2.1 Introduction

Laplace transform methods have a key role to play in the modern approach to the analysis and design of engineering systems. The stimulus for developing these methods was the pioneering work of the English electrical engineer Oliver Heaviside (1850–1925) in developing a method for the systematic solution of ordinary differential equations with constant coefficients. Heaviside was concerned with solving practical problems, and his method was based mainly on intuition, lacking mathematical rigour: consequently it was frowned upon by theoreticians at the time. However, Heaviside himself was not concerned with rigorous proofs, and was satisfied that his method gave the correct results. Using his ideas, he was able to solve important practical problems that could not be dealt with using classical methods. This led to many new results in fields such as the propagation of currents and voltages along transmission lines.

Because it worked in practice, Heaviside's method was widely accepted by engineers. As its power for problem-solving became more and more apparent, the method attracted the attention of mathematicians, who set out to justify it. This provided the stimulus for rapid developments in many branches of mathematics including improper integrals, asymptotic series and transform theory. Research on the problem continued for many years before it was eventually recognized that an integral transform developed by the French mathematician Pierre Simon de Laplace (1749–1827) almost a century before provided a theoretical foundation for Heaviside's work. It was also recognized that the use of this integral transform provided a more systematic alternative for investigating differential equations than the method proposed by Heaviside. It is this alternative approach that is the basis of the **Laplace transform method**.

We have already come across instances where a mathematical transformation has been used to simplify the solution of a problem. For example, the logarithm is used to simplify multiplication and division problems. To multiply or divide two numbers, we transform them into their logarithms, add or subtract these, and then perform the inverse transformation (that is, the antilogarithm) to obtain the product or quotient of the original numbers. The purpose of using a transformation is to create a new domain in which it is easier to handle the problem being investigated. Once results have been obtained in the new domain, they can be inverse-transformed to give the desired results in the original domain.

The Laplace transform is an example of a class called **integral transforms**, and it takes a function $f(t)$ of one variable t (which we shall refer to as **time**) into a function $F(s)$ of another variable s (the **complex frequency**). Another integral transform widely used by engineers is the **Fourier transform**, which is dealt with in Chapter 5. The attraction of the Laplace transform is that it transforms *differential* equations in the t (time) domain into *algebraic* equations in the s (frequency) domain. Solving differential equations in the t domain therefore reduces to solving algebraic equations in the s domain. Having done the latter for the desired unknowns, their values as functions of time may be found by taking inverse transforms. Another advantage of using the Laplace transform for solving differential equations is that initial conditions play an essential role in the transformation process, so they are automatically incorporated into the solution. This constrasts with the classical approach considered in Chapter 10 of the companion text *Modern Engineering Mathematics*, where the initial conditions are only introduced when the unknown constants of integration are determined. The Laplace transform is therefore an ideal tool for solving initial-value problems such as those occurring in the investigation of electrical circuits and mechanical vibrations.

The Laplace transform finds particular application in the field of signals and linear systems analysis. A distinguishing feature of a system is that when it is subjected to an excitation (input), it produces a response (output). When the input $u(t)$ and output $x(t)$ are functions of a single variable t, representing time, it is normal to refer to them as **signals**. Schematically, a system may be represented as in Figure 2.1. The problem facing the engineer is that of determining the system output $x(t)$ when it is subjected to an input $u(t)$ applied at some instant of time, which we can take to be $t = 0$. The relationship between output and input is

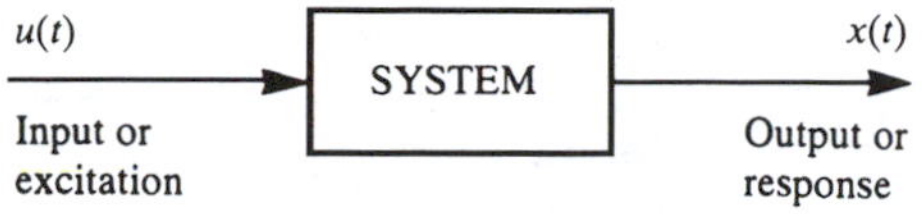

Figure 2.1 Schematic representation of a system.

determined by the laws governing the behaviour of the system. If the system is linear and time-invariant then the output is related to the input by a linear differential equation with constant coefficients, and we have a standard initial-value problem, which is amenable to solution using the Laplace transform.

While many of the problems considered in this chapter can be solved by the classical approach, the Laplace transform leads to a more unified approach and provides the engineer with greater insight into system behaviour. In practice, the input signal $u(t)$ may be a discontinuous or periodic function, or even a pulse, and in such cases the use of the Laplace transform has distinct advantages over the classical approach. Also, more often than not, an engineer is interested not only in system analysis but also in system synthesis or design. Consequently, an engineer's objective in studying a system's response to specific inputs is frequently to learn more about the system with a view to improving or controlling it so that it satisfies certain specifications. It is in this area that the use of the Laplace transform is attractive, since by considering the system response to particular inputs, such as a sinusoid, it provides the engineer with powerful graphical methods for system design that are relatively easy to apply and widely used in practice.

In modelling the system by a differential equation, it has been assumed that both the input and output signals can vary at any instant of time; that is, they are functions of a continuous time variable (note that this does not mean that the signals themselves have to be continuous functions of time). Such systems are called **continuous-time systems**, and it is for investigating these that the Laplace transform is best suited. With the introduction of computer control into system design, signals associated with a system may only change at discrete instants of time. In such cases the system is said to be a **discrete-time system**, and is modelled by a difference equation rather than a differential equation. Such systems are dealt with using the z transform considered in Chapter 3.

2.2 The Laplace transform

2.2.1 Definition and notation

We define the Laplace transform of a function $f(t)$ by the expression

$$\mathscr{L}\{f(t)\} = \int_0^\infty e^{-st} f(t)\, dt \tag{2.1}$$

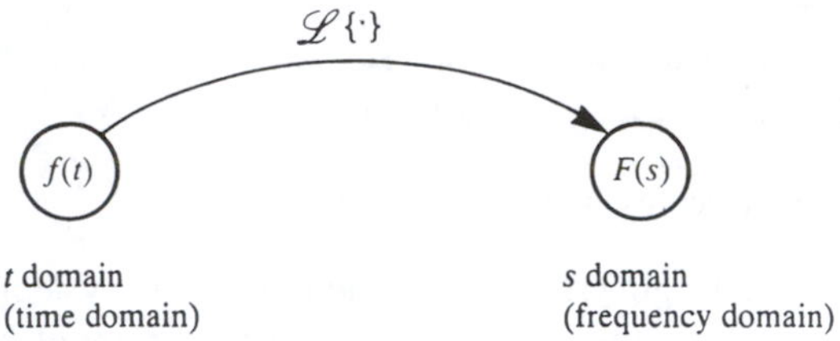

Figure 2.2 The Laplace transform operator.

where s is a complex variable and e^{-st} is called the **kernel** of the transformation.

It is usual to represent the Laplace transform of a function by the corresponding capital letter, so that we write

$$\mathscr{L}\{f(t)\} = F(s) = \int_0^\infty e^{-st} f(t)\,dt \tag{2.2}$$

An alternative notation in common use is to denote $\mathscr{L}\{f(t)\}$ by $\bar{f}(s)$ or simply $\bar{f}$.

Before proceeding, there are a few observations relating to the definition (2.2) worthy of comment.

(a) The symbol $\mathscr{L}$ denotes the **Laplace transform operator**; when it operates on a function $f(t)$, it transforms it into a function $F(s)$ of the complex variable s. We say the operator transforms the function $f(t)$ in the t domain (usually called the **time domain**) into the function $F(s)$ in the s domain (usually called the **complex frequency domain**, or simply the **frequency domain**). This relationship is depicted graphically in Figure 2.2, and it is usual to refer to $f(t)$ and $F(s)$ as a **Laplace transform pair**, written as $\{f(t), F(s)\}$.

(b) Because the upper limit in the integral is infinite, the domain of integration is infinite. Thus the integral is an example of an **improper integral**, as introduced in Section 9.3 of *Modern Engineering Mathematics*; that is,

$$\int_0^\infty e^{-st} f(t)\,dt = \lim_{T\to\infty} \int_0^T e^{-st} f(t)\,dt$$

This immediately raises the question of whether or not the integral converges, an issue we shall consider in Section 2.2.3.

(c) Because the lower limit in the integral is zero, it follows that when taking the Laplace transform, the behaviour of $f(t)$ for negative values of t is ignored or suppressed. This means that $F(s)$ contains information on the behaviour of $f(t)$ only for $t \geqslant 0$, so that the Laplace transform is not a suitable tool for investigating problems in which values of $f(t)$ for $t < 0$ are relevant. In most engineering applications this does not cause any problems, since we are then concerned with physical systems for which the functions we are dealing with vary with time t. An attribute of physical realizable systems is that they are **non-anticipatory** in the sense that there is no output (or response) until an input (or excitation) is applied. Because of this causal relationship between the input and output, we define a

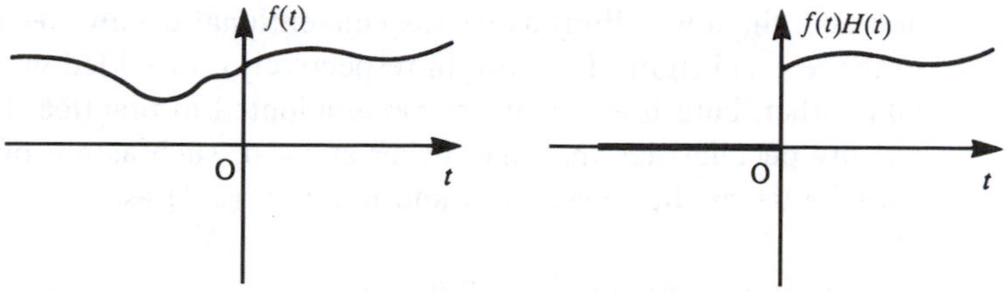

Figure 2.3 Graph of $f(t)$ and its causal equivalent function.

function $f(t)$ to be **causal** if $f(t) = 0(t < 0)$. In general, however, unless the domain is clearly specified, a function $f(t)$ is normally intepreted as being defined for all real values, both positive and negative, of t. Making use of the Heaviside unit step function $H(t)$ (see also Section 2.5.1), where

$$H(t) = \begin{cases} 0 & (t < 0) \\ 1 & (t \geqslant 0) \end{cases}$$

we have

$$f(t)H(t) = \begin{cases} 0 & (t < 0) \\ f(t) & (t \geqslant 0) \end{cases}$$

Thus the effect of multiplying $f(t)$ by $H(t)$ is to convert it into a causal function. Graphically, the relationship between $f(t)$ and $f(t)H(t)$ is as shown in Figure 2.3.

It follows that the corresponding Laplace transform $F(s)$ contains full information on the behaviour of $f(t)H(t)$. Consequently, strictly speaking one should refer to $\{f(t)H(t), F(s)\}$ rather than $\{f(t), F(s)\}$ as being a Laplace transform pair. However, it is common practice to drop the $H(t)$ and assume that we are dealing with causal functions. It should be noted that it is possible to take the Laplace transform of a non-causal function $f(t)$, recognizing that all information concerning its behaviour for $t < 0$ is lost under transformation.

(d) If the behaviour of $f(t)$ for $t < 0$ is of interest then we need to use the alternative **two-sided** or **bilateral Laplace transform** of the function $f(t)$, defined by

$$\mathscr{L}_B\{f(t)\} = \int_{-\infty}^{\infty} e^{-st} f(t)\,dt \tag{2.3}$$

The Laplace transform defined by (2.2), with lower limit zero, is sometimes referred to as the **one-sided** or **unilateral Laplace transform** of the function $f(t)$. In this chapter we shall concern ourselves only with the latter transform, and refer to it simply as the Laplace transform of the function $f(t)$. Note that when $f(t)$ is a causal function,

$$\mathscr{L}_B\{f(t)\} = \mathscr{L}\{f(t)\}$$

(e) Another issue concerning the lower limit of zero is the interpretation of $f(0)$ when $f(t)$ has a peculiarity at the origin. The question then arises as to whether or not we should include the peculiarity and take the lower limit as 0– or exclude it and take the lower limit as 0+ (as conventional 0− and 0+ denote values of t just to the left and right of the origin respectively). Provided we are consistent, we can take either, both interpretations being adopted in practice. In order to accommodate any peculiarities that may occur at $t = 0$, such as an impulse applied at $t = 0$, we take 0− as the lower limit and interpret (2.2) as

$$\mathscr{L}\{f(t)\} = F(s) = \int_{0-}^{\infty} e^{-st} f(t)\, dt \tag{2.4}$$

We shall return to this issue when considering the impulse response in Section 2.5.8.

2.2.2 Transforms of simple functions

In this section we obtain the Laplace transformations of some simple functions.

EXAMPLE 2.1 Determine the Laplace transform of the function

$$f(t) = c$$

where c is a constant.

Solution Using the definition (2.2),

$$\mathscr{L}(c) = \int_0^{\infty} e^{-st} c\, dt = \lim_{T\to\infty} \int_0^T e^{-st} c\, dt$$

$$= \lim_{T\to\infty} \left[-\frac{c}{s} e^{-st} \right]_0^T = \frac{c}{s}\left(1 - \lim_{T\to\infty} e^{-sT}\right)$$

Taking $s = \sigma + j\omega$, where σ and ω are real,

$$\lim_{T\to\infty} e^{-sT} = \lim_{T\to\infty} (e^{-(\sigma+j\omega)T}) = \lim_{T\to\infty} e^{-\sigma T}(\cos \omega T + j \sin \omega T)$$

A finite limit exists provided that $\sigma = \text{Re}(s) > 0$, when the limit is zero. Thus, provided that $\text{Re}(s) > 0$, the Laplace transform is

$$\mathscr{L}(c) = \frac{c}{s}, \quad \text{Re}(s) > 0$$

so that

$$\left.\begin{aligned} f(t) &= c \\ F(s) &= \frac{c}{s} \end{aligned}\right\} \quad \text{Re}(s) > 0 \tag{2.5}$$

constitute an example of a Laplace transform pair.

EXAMPLE 2.2 Determine the Laplace transform of the ramp function

$$f(t) = t$$

Solution From the definition (2.2),

$$\mathscr{L}\{t\} = \int_0^\infty e^{-st} t \, dt = \lim_{T\to\infty} \int_0^T e^{-st} t \, dt$$

$$= \lim_{T\to\infty} \left[-\frac{t}{s} e^{-st} - \frac{e^{-st}}{s^2} \right]_0^T$$

$$= \frac{1}{s^2} - \lim_{T\to\infty} \frac{Te^{-sT}}{s} - \lim_{T\to\infty} \frac{e^{-sT}}{s^2}$$

Following the same procedure as in Example 2.1, limits exist provided that $\mathrm{Re}(s) > 0$, when

$s = \sigma + j\omega$

$$\lim_{T\to\infty} \frac{Te^{-sT}}{s} = \lim_{T\to\infty} \frac{e^{-sT}}{s^2} = 0$$

Thus, provided that $\mathrm{Re}(s) > 0$,

$$\mathscr{L}\{t\} = \frac{1}{s^2}$$

giving us the Laplace transform pair

$$\left.\begin{aligned} f(t) &= t \\ F(s) &= \frac{1}{s^2} \end{aligned}\right\} \quad \mathrm{Re}(s) > 0 \tag{2.6}$$

EXAMPLE 2.3 Determine the Laplace transform of the one-sided exponential function

$$f(t) = e^{kt}$$

Solution The definition (2.2) gives

$$\mathscr{L}\{e^{kt}\} = \int_0^\infty e^{-st} e^{kt} \, dt = \lim_{T\to\infty} \int_0^T e^{-(s-k)t} \, dt$$

$$= \lim_{T\to\infty} \frac{-1}{s-k} [e^{-(s-k)t}]_0^T = \frac{1}{s-k} \left(1 - \lim_{T\to\infty} e^{-(s-k)T} \right)$$

Writing $s = \sigma + j\omega$, where σ and ω are real, we have

$$\lim_{T\to\infty} e^{-(s-k)T} = \lim_{T\to\infty} e^{-(\sigma-k)T} e^{j\omega T}$$ $(\cos \omega T + j \sin \omega T)$

If k is real, then, provided that $\sigma = \text{Re}(s) > k$, the limit exists, and is zero. If k is complex, say $k = a + \text{j}b$, then the limit will also exist, and be zero, provided that $\sigma > a$ (that is, $\text{Re}(s) > \text{Re}(k)$). Under these conditions, we then have

$$\mathscr{L}\{\text{e}^{kt}\} = \frac{1}{s-k}$$

giving us the Laplace transform pair

$$\left.\begin{aligned} f(t) &= \text{e}^{kt} \\ F(s) &= \frac{1}{s-k} \end{aligned}\right\} \quad \text{Re}(s) > \text{Re}(k) \tag{2.7}$$

EXAMPLE 2.4 Determine the Laplace transforms of the sine and cosine functions

$$f(t) = \sin at, \quad g(t) = \cos at$$

where a is a real constant.

Solution Since

$$\text{e}^{\text{j}at} = \cos at + \text{j} \sin at$$

we may write

$$f(t) = \sin at = \text{Im e}^{\text{j}at}$$

$$g(t) = \cos at = \text{Re e}^{\text{j}at}$$

Using this formulation, the required transforms may be obtained from the result

$$\mathscr{L}\{\text{e}^{kt}\} = \frac{1}{s-k}, \quad \text{Re}(s) > \text{Re}(k)$$

of Example 2.3.

Taking $k = \text{j}a$ in this result gives

$$\mathscr{L}\{\text{e}^{\text{j}at}\} = \frac{1}{s-\text{j}a}, \quad \text{Re}(s) > 0$$

or

$$\mathscr{L}\{\text{e}^{\text{j}at}\} = \frac{s+\text{j}a}{s^2+a^2}, \quad \text{Re}(s) > 0$$

Thus, equating real and imaginary parts and assuming s is real,

$$\mathscr{L}\{\sin at\} = \text{Im } \mathscr{L}\{\text{e}^{\text{j}at}\} = \frac{a}{s^2+a^2}$$

$$\mathscr{L}\{\cos at\} = \text{Re } \mathscr{L}\{\text{e}^{\text{j}at}\} = \frac{s}{s^2+a^2}$$

These results also hold when s is complex, giving us the Laplace transform pairs

$$\mathscr{L}\{\sin at\} = \frac{a}{s^2 + a^2}, \quad \text{Re}(s) > 0 \tag{2.8}$$

$$\mathscr{L}\{\cos at\} = \frac{s}{s^2 + a^2}, \quad \text{Re}(s) > 0 \tag{2.9}$$

2.2.3 Existence of the Laplace transform

Clearly, from the definition (2.2), the Laplace transform of a function $f(t)$ exists if and only if the improper integral in the definition converges for at least some values of s. The examples of Section 2.2.2 suggest that this relates to the boundedness of the function, with the factor e^{-st} in the transform integral acting like a convergence factor in that the allowed values of Re(s) are those for which the integral converges. In order to be able to state sufficient conditions on $f(t)$ for the existence of $\mathscr{L}\{f(t)\}$, we first introduce the definition of a function of exponential order.

DEFINITION 2.1

A function $f(t)$ is said to be of **exponential order** as $t \to \infty$ if there exists a real number σ and positive constants M and T such that

$$|f(t)| < M e^{\sigma t}$$

for all $t > T$.

What this definition tells us is that a function $f(t)$ is of exponential order if it does not grow faster than some exponential function of the form $Me^{\sigma t}$. Fortunately most functions of practical significance satisfy this requirement, and are therefore of exponential order. There are, however, functions that are not of exponential order, an example being e^{t^2}, since this grows more rapidly than $Me^{\sigma t}$ as $t \to \infty$ whatever the values of M and σ.

EXAMPLE 2.5 The function $f(t) = e^{3t}$ is of exponential order, with $\sigma \geqslant 3$.

EXAMPLE 2.6 Show that the function $f(t) = t^3$ $(t \geqslant 0)$ is of exponential order.

Solution Since

$$e^{\alpha t} = 1 + \alpha t + \tfrac{1}{2}\alpha^2 t^2 + \tfrac{1}{6}\alpha^3 t^3 + \dots$$

it follows that for any $\alpha > 0$

$$t^3 < \frac{6}{\alpha^3}\,\mathrm{e}^{\alpha t}$$

so that t^3 is of exponential order, with $\sigma > 0$.

It follows from Examples 2.5 and 2.6 that the choice of σ in Definition 2.1 is not unique for a particular function. For this reason, we define the greatest lower bound σ_c of the set of possible values of σ to be the **abscissa of convergence** of $f(t)$. Thus, in the case of the function $f(t) = \mathrm{e}^{3t}$, $\sigma_c = 3$, while in the case of the function $f(t) = t^3$, $\sigma_c = 0$.

Returning to the definition of the Laplace transform given by (2.2), it follows that if $f(t)$ is a continuous function and is also of exponential order with abscissa of convergence σ_c, so that

$$|f(t)| < M\mathrm{e}^{\sigma t}, \quad \sigma > \sigma_c$$

then, taking $T = 0$ in Definition 2.1,

$$|F(s)| = \left|\int_0^\infty \mathrm{e}^{-st} f(t)\,\mathrm{d}t\right| \leqslant \int_0^\infty |\mathrm{e}^{-st}|\,|f(t)|\,\mathrm{d}t$$

Writing $s = \sigma + \mathrm{j}\omega$, where σ and ω are real, since $|\mathrm{e}^{-\mathrm{j}\omega t}| = 1$, we have

$$|\mathrm{e}^{-st}| = |\mathrm{e}^{-\sigma t}|\,|\mathrm{e}^{-\mathrm{j}\omega t}| = |\mathrm{e}^{-\sigma t}| = \mathrm{e}^{-\sigma t}$$

so that

$$|F(s)| \leqslant \int_0^\infty \mathrm{e}^{-\sigma t}\,|f(t)|\,\mathrm{d}t \leqslant M\int_0^\infty \mathrm{e}^{-\sigma t}\,\mathrm{e}^{\sigma_d t}\,\mathrm{d}t, \quad \sigma_d > \sigma_c$$

$$= M\int_0^\infty \mathrm{e}^{-(\sigma-\sigma_d)t}\,\mathrm{d}t$$

This last integral is finite whenever $\sigma = \mathrm{Re}(s) > \sigma_d$. Since σ_d can be chosen arbitrarily such that $\sigma_d > \sigma_c$ we conclude that $F(s)$ exists for $\sigma > \sigma_c$. Thus a continuous function $f(t)$ of exponential order, with abscissa of convergence σ_c, has a Laplace transform

$$\mathscr{L}\{f(t)\} = F(s), \quad \mathrm{Re}(s) > \sigma_c$$

where the region of convergence is as shown in Figure 2.4.

In fact, the requirement that $f(t)$ be continuous is not essential, and may be relaxed to $f(t)$ being piecewise-continuous, as defined in Section 8.6.1 of *Modern Engineering Mathematics*; that is, $f(t)$ must have only a finite number of finite discontinuities, being elsewhere continuous and bounded.

We conclude this section by stating a theorem that ensures the existence of a Laplace transform.

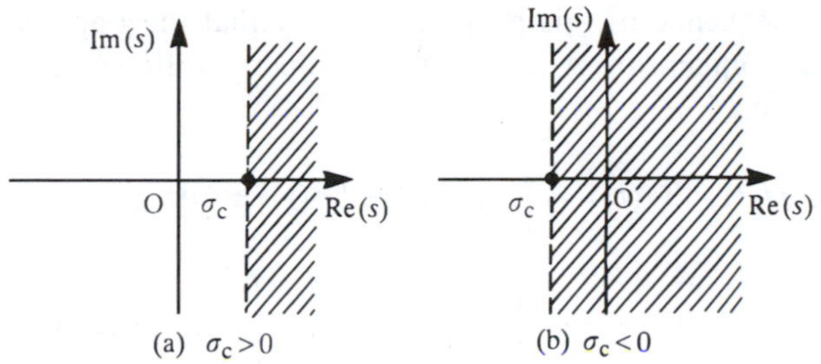

Figure 2.4 Region of convergence for $\mathscr{L}\{f(t)\}$; σ_c is the abscissa of convergence for $f(t)$.

THEOREM 2.1

Existence of Laplace transform

If the causal function $f(t)$ is piecewise-continuous on $[0, \infty]$ and is of exponential order, with abscissa of convergence σ_c, then its Laplace transform exists, with region of convergence $\mathrm{Re}(s) > \sigma_c$ in the s domain; that is,

$$\mathscr{L}\{f(t)\} = F(s) = \int_0^\infty e^{-st} f(t)\,dt, \quad Re(s) > \sigma_c \qquad \square$$

The conditions of this theorem are *sufficient* for ensuring the existence of the Laplace transform of a function. They do not, however, constitute *necessary* conditions for the existence of such a transform, and it does not follow that if the conditions are violated then a transform does not exist. In fact, the conditions are more restrictive than necessary, since there exist functions with infinite discontinuities that possess Laplace transforms.

2.2.4 Properties of the Laplace transform

In this section we consider some of the properties of the Laplace transform that will enable us to find further transform pairs $\{f(t), F(s)\}$ without having to compute them directly using the definition. Further properties will be developed in later sections when the need arises.

Property 2.1: The linearity property

A fundamental property of the Laplace transform is its linearity, which may be stated as follows:

If $f(t)$ and $g(t)$ are functions having Laplace transforms and if α and β are any constants then

$$\mathscr{L}\{\alpha f(t) + \beta g(t)\} = \alpha\mathscr{L}\{f(t)\} + \beta\mathscr{L}\{g(t)\}$$

As a consequence of this property, we say that the Laplace transform operator $\mathscr{L}$ is a **linear operator**. A proof of the property follows readily from the definition (2.2), since

$$\begin{aligned}
\mathscr{L}\{\alpha f(t)+\beta g(t)\} &= \int_0^\infty [\alpha f(t)+\beta g(t)]\,e^{-st}\,dt \\
&= \int_0^\infty \alpha f(t)\,e^{-st}\,dt + \int_0^\infty \beta g(t)\,e^{-st}\,dt \\
&= \alpha\int_0^\infty f(t)\,e^{-st}\,dt + \beta\int_0^\infty g(t)\,e^{-st}\,dt \\
&= \alpha\mathscr{L}\{f(t)\} + \beta\mathscr{L}\{g(t)\}
\end{aligned}$$

Regarding the region of convergence, if $f(t)$ and $g(t)$ have abscissae of convergence σ_f and σ_g respectively, and $\sigma_1 > \sigma_f$, $\sigma_2 > \sigma_g$, then

$$|f(t)| < M_1 e^{\sigma_1 t}, \quad |g(t)| < M_2 e^{\sigma_2 t}$$

It follows that

$$\begin{aligned}
|\alpha f(t) + \beta g(t)| &\leqslant |\alpha|\,|f(t)| + |\beta|\,|g(t)| \\
&\leqslant |\alpha| M_1 e^{\sigma_1 t} + |\beta| M_2 e^{\sigma_2 t} \\
&\leqslant (|\alpha| M_1 + |\beta| M_2)\,e^{\sigma t}
\end{aligned}$$

where $\sigma = \max(\sigma_1, \sigma_2)$, so that the abscissa of convergence of the linear sum $\alpha f(t) + \beta g(t)$ is less than or equal to the maximum of those for $f(t)$ and $g(t)$.

This linearity property may clearly be extended to a linear combination of any finite number of functions.

EXAMPLE 2.7 Determine $\mathscr{L}\{3t + 2\,e^{3t}\}$.

Solution Using the results given in (2.6) and (2.7),

$$\mathscr{L}\{t\} = \frac{1}{s^2}, \quad \mathrm{Re}(s) > 0$$

$$\mathscr{L}\{e^{3t}\} = \frac{1}{s-3}, \quad \mathrm{Re}(s) > 3$$

so, by the linearity property,

$$\begin{aligned}
\mathscr{L}\{3t + 2e^{3t}\} &= 3\mathscr{L}\{t\} + 2\mathscr{L}\{e^{3t}\} \\
&= \frac{3}{s^2} + \frac{2}{s-3}, \quad \mathrm{Re}(s) > \max\{0, 3\} \\
&= \frac{3}{s^2} + \frac{2}{s-3}, \quad \mathrm{Re}(s) > 3
\end{aligned}$$

EXAMPLE 2.8 Determine $\mathscr{L}\{5 - 3t + 4 \sin 2t - 6e^{4t}\}$.

Solution Using the results given in (2.5)–(2.8),

$$\mathscr{L}\{5\} = \frac{5}{s}, \quad \text{Re}(s) > 0 \qquad \mathscr{L}\{t\} = \frac{1}{s^2}, \quad \text{Re}(s) > 0$$

$$\mathscr{L}\{\sin 2t\} = \frac{2}{s^2 + 4}, \quad \text{Re}(s) > 0 \qquad \mathscr{L}\{e^{4t}\} = \frac{1}{s - 4}, \quad \text{Re}(s) > 4$$

so, by the linearity property,

$$\mathscr{L}\{5 - 3t + 4 \sin 2t - 6\,e^{4t}\} = \mathscr{L}\{5\} - 3\mathscr{L}\{t\} + 4\mathscr{L}\{\sin 2t\} - 6\mathscr{L}\{e^{4t}\}$$

$$= \frac{5}{s} - \frac{3}{s^2} + \frac{8}{s^2 + 4} - \frac{6}{s - 4}, \quad \text{Re}(s) > \max\{0, 4\}$$

$$= \frac{5}{s} - \frac{3}{s^2} + \frac{8}{s^2 + 4} - \frac{6}{s - 4}, \quad \text{Re}(s) > 4$$

The first shift property is another property that enables us to add more combinations to our repertoire of Laplace transform pairs. As with the linearity property, it will prove to be of considerable importance in our later discussions particularly when considering the inversion of Laplace transforms.

Property 2.2: The first shift property

The property is contained in the following theorem, commonly referred to as the **first shift theorem** or sometimes as the **exponential modulation theorem**.

THEOREM 2.2 **The first shift theorem**

If $f(t)$ is a function having Laplace transform $F(s)$, with $\text{Re}(s) > \sigma_c$, then the function $e^{at}f(t)$ also has a Laplace transform, given by

$$\mathscr{L}\{e^{at}f(t)\} = F(s - a), \quad \text{Re}(s) > \sigma_c + \text{Re}(a)$$

Proof A proof of the theorem follows directly from the definition of the Laplace transform, since

$$\mathscr{L}\{e^{at}f(t)\} = \int_0^\infty e^{at}f(t)\,e^{-st}\,dt = \int_0^\infty f(t)\,e^{-(s-a)t}\,dt$$

Then, since

$$\mathscr{L}\{f(t)\} = F(s) = \int_0^{\infty} f(t)\,\mathrm{e}^{-st}\,\mathrm{d}t, \quad \mathrm{Re}(s) > \sigma_c$$

we see that the last integral above is in structure exactly the Laplace transform of $f(t)$ itself, except that $s - a$ takes the place of s, so that

$$\mathscr{L}\{\mathrm{e}^{at} f(t)\} = F(s - a), \quad \mathrm{Re}(s - a) > \sigma_c$$

or

$$\mathscr{L}\{\mathrm{e}^{at} f(t)\} = F(s - a), \quad \mathrm{Re}(s) > \sigma_c + \mathrm{Re}(a) \qquad \square$$

An alternative way of expressing the result of Theorem 2.2, which may be found more convenient in application, is

$$\mathscr{L}\{\mathrm{e}^{at}f(t)\} = [\mathscr{L}\{f(t)\}]_{s \to s-a} = [F(s)]_{s \to s-a}$$

In other words, the theorem says that the Laplace transform of e^{at} times a function $f(t)$ is equal to the Laplace transform of $f(t)$ itself, with s replaced by $s - a$.

EXAMPLE 2.9 Determine $\mathscr{L}\{t\,\mathrm{e}^{-2t}\}$.

Solution From the result given in (2.6),

$$\mathscr{L}\{t\} = F(s) = \frac{1}{s^2}, \quad \mathrm{Re}(s) > 0$$

so, by the first shift theorem,

$$\mathscr{L}\{t\mathrm{e}^{-2t}\} = F(s + 2) = [F(s)]_{s \to s+2}, \quad \mathrm{Re}(s) > 0 - 2$$

that is,

$$\mathscr{L}\{t\,\mathrm{e}^{-2t}\} = \frac{1}{(s+2)^2}, \quad \mathrm{Re}(s) > -2$$

EXAMPLE 2.10 Determine $\mathscr{L}\{\mathrm{e}^{-3t}\sin 2t\}$.

Solution From the result (2.8),

$$\mathscr{L}\{\sin 2t\} = F(s) = \frac{2}{s^2 + 4}, \quad \mathrm{Re}(s) > 0$$

so, by the first shift theorem,

$$\mathscr{L}\{\mathrm{e}^{-3t}\sin 2t\} = F(s + 3) = [F(s)]_{s \to s+3}, \quad \mathrm{Re}(s) > 0 - 3$$

that is,

$$\mathscr{L}\{\mathrm{e}^{-3t}\sin 2t\} = \frac{2}{(s+3)^2 + 4} = \frac{2}{s^2 + 6s + 13}, \quad \mathrm{Re}(s) > -3$$

The function $e^{-3t}\sin 2t$ in Example 2.10 is a member of a general class of functions called **damped sinusoids**. These play an important role in the study of engineering systems, particularly in the analysis of vibrations. For this reason, we add the following two general members of the class to our standard library of Laplace transform pairs:

$$\mathscr{L}\{e^{-kt}\sin at\} = \frac{a}{(s+k)^2 + a^2}, \quad \text{Re}(s) > -k \tag{2.10}$$

$$\mathscr{L}\{e^{-kt}\cos at\} = \frac{s+k}{(s+k)^2 + a^2}, \quad \text{Re}(s) > -k \tag{2.11}$$

where in both cases k and a are real constants.

Property 2.3: Derivative-of-transform property

This property relates operations in the time domain to those in the transformed s domain, but initially we shall simply look upon it as a method of increasing our repertoire of Laplace transform pairs. The property is also sometimes referred to as the **multiplication-by-t** property. A statement of the property is contained in the following theorem.

THEOREM 2.3 Derivative of transform

If $f(t)$ is a function having Laplace transform

$$F(s) = \mathscr{L}\{f(t)\}, \quad \text{Re}(s) > \sigma_c$$

then the functions $t^n f(t)$ $(n = 1, 2, \ldots)$ also have Laplace transforms, given by

$$\mathscr{L}\{t^n f(t)\} = (-1)^n \frac{d^n F(s)}{ds^n}, \quad \text{Re}(s) > \sigma_c$$

Proof By definition,

$$\mathscr{L}\{f(t)\} = F(s) = \int_0^\infty e^{-st} f(t)\,dt$$

so that

$$\frac{d^n F(s)}{ds^n} = \frac{d^n}{ds^n}\int_0^\infty e^{-st} f(t)\,dt$$

Owing to the convergence properties of the improper integral involved, we can interchange the operations of differentiation and integration and differentiate with respect to s under the integral sign. Thus

$$\frac{d^n F(s)}{ds^n} = \int_0^\infty \frac{\partial^n}{\partial s^n}[e^{-st} f(t)]\,dt$$

which, on carrying out the repeated differentiation, gives

$$\frac{\mathrm{d}^n F(s)}{\mathrm{d}s^n} = (-1)^n \int_0^\infty \mathrm{e}^{-st} t^n f(t)\,\mathrm{d}t$$
$$= (-1)^n \mathscr{L}\{t^n f(t)\}, \quad \mathrm{Re}(s) > \sigma_c$$

the region of convergence remaining unchanged. □

In other words, Theorem 2.3 says that differentiating the transform of a function with respect to s is equivalent to multiplying the function itself by $-t$. As with the previous properties, we can now use this result to add to our list of Laplace transform pairs.

EXAMPLE 2.11 Determine $\mathscr{L}\{t \sin 3t\}$.

Solution Using the result (2.8),

$$\mathscr{L}\{\sin 3t\} = F(s) = \frac{3}{s^2 + 9}, \quad \mathrm{Re}(s) > 0$$

so, by the derivative theorem,

$$\mathscr{L}\{t \sin 3t\} = -\frac{\mathrm{d}F(s)}{\mathrm{d}s} = \frac{6s}{(s^2 + 9)^2}, \quad \mathrm{Re}(s) > 0$$

EXAMPLE 2.12 Determine $\mathscr{L}\{t^2 \mathrm{e}^t\}$.

Solution From the result (2.7),

$$\mathscr{L}\{\mathrm{e}^t\} = F(s) = \frac{1}{s - 1}, \quad \mathrm{Re}(s) > 1$$

so, by the derivative theorem,

$$\mathscr{L}\{t^2 \mathrm{e}^t\} = (-1)^2 \frac{\mathrm{d}^2 F(s)}{\mathrm{d}s^2} = (-1)^2 \frac{\mathrm{d}^2}{\mathrm{d}s^2}\left(\frac{1}{s-1}\right)$$
$$= (-1)\frac{\mathrm{d}}{\mathrm{d}s}\left(\frac{1}{(s-1)^2}\right) = \frac{2}{(s-1)^3}, \quad \mathrm{Re}(s) > 1$$

Note that the result is easier to deduce using the first shift theorem.

EXAMPLE 2.13 Determine $\mathscr{L}\{t^n\}$, where n is a positive integer.

Solution Using the result (2.5),

$$\mathscr{L}\{1\} = \frac{1}{s}, \quad \mathrm{Re}(s) > 0$$

so, by the derivative theorem,

$$\mathscr{L}\{t^n\} = (-1)^n \frac{\mathrm{d}^n}{\mathrm{d}s^n}\left(\frac{1}{s}\right) = \frac{n!}{s^{n+1}}, \quad \operatorname{Re}(s) > 0$$

2.2.5 Table of Laplace transforms

It is appropriate at this stage to draw together the results proved to date for easy access. This is done in the form of two short tables. Figure 2.5(a) lists some Laplace transform pairs and Figure 2.5(b) lists the properties already considered.

(a)

$f(t)$	$\mathscr{L}\{f(t)\} = F(s)$	Region of convergence
c, c a constant	$\dfrac{c}{s}$	$\operatorname{Re}(s) > 0$
t	$\dfrac{1}{s^2}$	$\operatorname{Re}(s) > 0$
t^n, n a positive integer	$\dfrac{n!}{s^{n+1}}$	$\operatorname{Re}(s) > 0$
e^{kt}, k a constant	$\dfrac{1}{s-k}$	$\operatorname{Re}(s) > \operatorname{Re}(k)$
$\sin at$, a a real constant	$\dfrac{a}{s^2+a^2}$	$\operatorname{Re}(s) > 0$
$\cos at$, a a real constant	$\dfrac{s}{s^2+a^2}$	$\operatorname{Re}(s) > 0$
$\mathrm{e}^{-kt}\sin at$, k and a real constants	$\dfrac{a}{(s+k)^2+a^2}$	$\operatorname{Re}(s) > -k$
$\mathrm{e}^{-kt}\cos at$, k and a real constants	$\dfrac{s+k}{(s+k)^2+a^2}$	$\operatorname{Re}(s) > -k$

(b)

$$\mathscr{L}\{f(t)\} = F(s), \quad \operatorname{Re}(s) > \sigma_1 \quad \text{and} \quad \mathscr{L}\{g(t)\} = G(s), \quad \operatorname{Re}(s) > \sigma_2$$

Linearity: $\mathscr{L}\{\alpha f(t) + \beta g(t)\} = \alpha F(s) + \beta G(s), \quad \operatorname{Re}(s) > \max(\sigma_1, \sigma_2)$

First shift theorem: $\mathscr{L}\{\mathrm{e}^{at} f(t)\} = F(s-a), \quad \operatorname{Re}(s) > \sigma_1 + \operatorname{Re}(a)$

Derivative of transform:

$$\mathscr{L}\{t^n f(t)\} = (-1)^n \frac{\mathrm{d}^n F(s)}{\mathrm{d}s^n}, \quad (n = 1, 2, \ldots), \operatorname{Re}(s) > \sigma_1$$

Figure 2.5 (a) Table of Laplace transform pairs; (b) some properties of the Laplace transform.

2.2.6 Exercises

1 Use the definition of the Laplace transform to obtain the transforms of $f(t)$ when $f(t)$ is given by

(a) $\cosh 2t$ (b) t^2 (c) $3 + t$ (d) $t e^{-t}$

stating the region of convergence in each case.

2 What are the abscissae of convergence for the following functions?

(a) e^{5t} (b) e^{-3t}
(c) $\sin 2t$ (d) $\sinh 3t$
(e) $\cosh 2t$ (f) t^4
(g) $e^{-5t} + t^2$ (h) $3\cos 2t - t^3$
(i) $3e^{2t} - 2e^{-2t} + \sin 2t$ (j) $\sinh 3t + \sin 3t$

3 Using the results shown in Figure 2.5, obtain the Laplace transforms of the following functions, stating the region of convergence:

(a) $5 - 3t$ (b) $7t^3 - 2\sin 3t$
(c) $3 - 2t + 4\cos 2t$ (d) $\cosh 3t$
(e) $\sinh 2t$ (f) $5e^{-2t} + 3 - 2\cos 2t$
(g) $4te^{-2t}$ (h) $2e^{-3t}\sin 2t$
(i) $t^2 e^{-4t}$ (j) $6t^3 - 3t^2 + 4t - 2$
(k) $2\cos 3t + 5\sin 3t$ (l) $t\cos 2t$
(m) $t^2 \sin 3t$ (n) $t^2 - 3\cos 4t$
(o) $t^2 e^{-2t} + e^{-t}\cos 2t + 3$

2.2.7 The inverse transform

The symbol $\mathscr{L}^{-1}\{F(s)\}$ denotes a causal function $f(t)$ whose Laplace transform is $F(s)$; that is,

$$\text{if} \quad \mathscr{L}\{f(t)\} = F(s) \quad \text{then} \quad f(t) = \mathscr{L}^{-1}\{F(s)\}$$

This correspondence between the functions $F(s)$ and $f(t)$ is called the **inverse Laplace transformation**, $f(t)$ being the **inverse transform** of $F(s)$, and $\mathscr{L}^{-1}$ being referred to as the **inverse Laplace transform operator**. These relationships are depicted in Figure 2.6.

As was pointed out in observation (c) of Section 2.2.1, the Laplace transform $F(s)$ only determines the behaviour of $f(t)$ for $t \geqslant 0$. Thus $\mathscr{L}^{-1}\{F(s)\} = f(t)$ only for $t \geqslant 0$. When writing $\mathscr{L}^{-1}\{F(s)\} = f(t)$, it is assumed that $t \geqslant 0$ so strictly speaking, we should write

$$\mathscr{L}^{-1}\{F(s)\} = f(t)H(t) \tag{2.12}$$

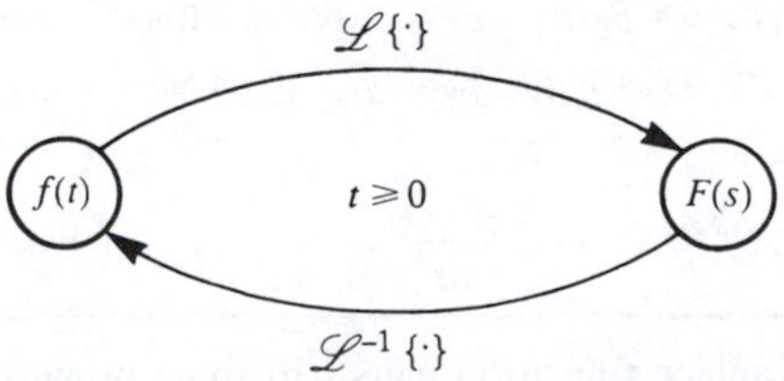

Figure 2.6 The Laplace transform and its inverse.

EXAMPLE 2.14 Since

$$\mathscr{L}\{e^{at}\} = \frac{1}{s-a}$$

if follows that

$$\mathscr{L}^{-1}\left\{\frac{1}{s-a}\right\} = e^{at}$$

EXAMPLE 2.15 Since

$$\mathscr{L}\{\sin \omega t\} = \frac{\omega}{s^2+\omega^2}$$

it follows that

$$\mathscr{L}^{-1}\left\{\frac{\omega}{s^2+\omega^2}\right\} = \sin \omega t$$

The linearity property for the Laplace transform (Property 2.1), states that if α and β are any constants then

$$\begin{aligned}\mathscr{L}\{\alpha f(t) + \beta g(t)\} &= \alpha\mathscr{L}\{f(t)\} + \beta\mathscr{L}\{g(t)\}\\ &= \alpha F(s) + \beta G(s)\end{aligned}$$

It then follows from the above definition that

$$\begin{aligned}\mathscr{L}^{-1}\{\alpha F(s) + \beta G(s)\} &= \alpha f(t) + \beta g(t)\\ &= \alpha\mathscr{L}^{-1}\{F(s)\} + \beta\mathscr{L}^{-1}\{G(s)\}\end{aligned}$$

so that the inverse Laplace transform operator $\mathscr{L}^{-1}$ is also a linear operator.

2.2.8 Evaluation of inverse transforms

The most obvious way of finding the inverse transform of the function $F(s)$ is to make use of a table of transforms such as that given in Figure 2.5. Sometimes it is possible to write down the inverse transform directly from the table, but more often than not it is first necessary to carry out some algebraic manipulation on $F(s)$. In particular, we frequently need to determine the inverse transform of a rational function of the form $p(s)/q(s)$, where $p(s)$ and $q(s)$ are polynomials in s. In such cases the procedure is first to resolve the function into partial fractions and then to use the table of transforms.

EXAMPLE 2.16 Find

$$\mathcal{L}^{-1}\left\{\frac{1}{(s+3)(s-2)}\right\}$$

Solution First $1/(s + 3)(s + 2)$ is resolved into partial fractions, giving

$$\frac{1}{(s+3)(s-2)} = \frac{-\frac{1}{5}}{s+3} + \frac{\frac{1}{5}}{s-2}$$

Then, using the result $\mathcal{L}^{-1}\{1/(s + a)\} = e^{-at}$ together with the linearity property, we have

$$\mathcal{L}^{-1}\left\{\frac{1}{(s+3)(s-2)}\right\} = -\frac{1}{5}\mathcal{L}^{-1}\left\{\frac{1}{s+3}\right\} + \frac{1}{5}\mathcal{L}^{-1}\left\{\frac{1}{s-2}\right\} = -\frac{1}{5}e^{-3t} + \frac{1}{5}e^{2t}$$

EXAMPLE 2.17 Find

$$\mathcal{L}^{-1}\left\{\frac{s+1}{s^2(s^2+9)}\right\}$$

Solution Resolving $(s + 1)/s^2(s^2 + 9)$ into partial fractions gives

$$\frac{s+1}{s^2(s^2+9)} = \frac{\frac{1}{9}}{s} + \frac{\frac{1}{9}}{s^2} - \frac{1}{9}\frac{s+1}{s^2+9}$$

$$= \frac{\frac{1}{9}}{s} + \frac{\frac{1}{9}}{s^2} - \frac{1}{9}\frac{s}{s^2+3^2} - \frac{1}{27}\frac{3}{s^2+3^2}$$

Using the results in Figure 2.5, together with the linearity property, we have

$$\mathcal{L}^{-1}\left\{\frac{s+1}{s^2(s^2+9)}\right\} = \frac{1}{9} + \frac{1}{9}t - \frac{1}{9}\cos 3t - \frac{1}{27}\sin 3t$$

2.2.9 Inversion using the first shift theorem

In Theorem 2.2 we saw that if $F(s)$ is the Laplace transform of $f(t)$ then, for a scalar a, $F(s - a)$ is the Laplace transform of $e^{at}f(t)$. This theorem normally causes little difficulty when used to obtain the Laplace transforms of functions, but it does frequently lead to problems when used to obtain inverse transforms. Expressed in the inverse form, the theorem becomes

$$\mathscr{L}^{-1}\{F(s-a)\} = \mathrm{e}^{at}f(t)$$

The notation

$$\mathscr{L}^{-1}\{[F(s)]_{s\to s-a}\} = \mathrm{e}^{at}[f(t)]$$

where $F(s) = \mathscr{L}\{f(t)\}$ and $[F(s)]_{s\to s-a}$ denotes that s in $F(s)$ is replaced by $s - a$, may make the relation clearer.

EXAMPLE 2.18 Find

$$\mathscr{L}^{-1}\left\{\frac{1}{(s+2)^2}\right\}$$

Solution

$$\frac{1}{(s+2)^2} = \left[\frac{1}{s^2}\right]_{s\to s+2}$$

and, since $1/s^2 = \mathscr{L}\{t\}$, the shift theorem gives

$$\mathscr{L}^{-1}\left\{\frac{1}{(s+2)^2}\right\} = t\,\mathrm{e}^{-2t}$$

EXAMPLE 2.19 Find

$$\mathscr{L}^{-1}\left\{\frac{2}{s^2+6s+13}\right\}$$

Solution

$$\frac{2}{s^2+6s+13} = \frac{2}{(s+3)^2+4} = \left[\frac{2}{s^2+2^2}\right]_{s\to s+3}$$

and, since $2/(s^2 + 2^2) = \mathscr{L}\{\sin 2t\}$, the shift theorem gives

$$\mathscr{L}^{-1}\left\{\frac{2}{s^2+6s+13}\right\} = \mathrm{e}^{-3t}\sin 2t$$

EXAMPLE 2.20 Find

$$\mathscr{L}^{-1}\left\{\frac{s+7}{s^2+2s+5}\right\}$$

Solution

$$\frac{s+7}{s^2+2s+5} = \frac{s+7}{(s+1)^2+4} = \frac{(s+1)}{(s+1)^2+4} + 3\frac{2}{(s+1)^2+4}$$

$$= \left[\frac{s}{s^2+2^2}\right]_{s\to s+1} + 3\left[\frac{2}{s^2+2^2}\right]_{s\to s+1}$$

Since $s/(s^2 + 2^2) = \mathscr{L}\{\cos 2t\}$ and $2/(s^2 + 2^2) = \mathscr{L}\{\sin 2t\}$, the shift theorem gives

$$\mathscr{L}^{-1}\left\{\frac{s+7}{s^2+2s+5}\right\} = e^{-t}\cos 2t + 3e^{-t}\sin 2t$$

EXAMPLE 2.21 Find

$$\mathscr{L}^{-1}\left\{\frac{1}{(s+1)^2(s^2+4)}\right\}$$

Solution Resolving $1/(s+1)^2(s^2+4)$ into partial fractions gives

$$\frac{1}{(s+1)^2(s^2+4)} = \frac{\frac{2}{25}}{s+1} + \frac{\frac{1}{5}}{(s+1)^2} - \frac{1}{25}\frac{2s+3}{s^2+4}$$

$$= \frac{\frac{2}{25}}{s+1} + \frac{1}{5}\left[\frac{1}{s^2}\right]_{s\to s+1} - \frac{2}{25}\frac{s}{s^2+2^2} - \frac{3}{50}\frac{2}{s^2+2^2}$$

Since $1/s^2 = \mathscr{L}\{t\}$, the shift theorem, together with the results in Figure 2.5, gives

$$\mathscr{L}^{-1}\left\{\frac{1}{(s+1)^2(s^2+4)}\right\} = \frac{2}{25}e^{-t} + \frac{1}{5}e^{-t}t - \frac{2}{25}\cos 2t - \frac{3}{50}\sin 2t$$

2.2.10 Exercise

4 Find $\mathscr{L}^{-1}\{F(s)\}$ when $F(s)$ is given by

(a) $\dfrac{1}{(s+3)(s+7)}$

(b) $\dfrac{s+5}{(s+1)(s-3)}$

(c) $\dfrac{s-1}{s^2(s+3)}$

(d) $\dfrac{2s+6}{s^2+4}$

(e) $\dfrac{1}{s^2(s^2+16)}$

(f) $\dfrac{s+8}{s^2+4s+5}$

(g) $\dfrac{s+1}{s^2(s^2+4s+8)}$

(h) $\dfrac{4s}{(s-1)(s+1)^2}$

(i) $\dfrac{s+7}{s^2+2s+5}$

(j) $\dfrac{3s^2-7s+5}{(s-1)(s-2)(s-3)}$

(k) $\dfrac{5s-7}{(s+3)(s^2+2)}$

(l) $\dfrac{s}{(s-1)(s^2+2s+2)}$

(m) $\dfrac{s-1}{s^2+2s+5}$

(n) $\dfrac{s-1}{(s-2)(s-3)(s-4)}$

(o) $\dfrac{3s}{(s-1)(s^2-4)}$

(p) $\dfrac{36}{s(s^2+1)(s^2+9)}$

(q) $\dfrac{2s^2+4s+9}{(s+2)(s^2+3s+3)}$

(r) $\dfrac{1}{(s+1)(s+2)(s^2+2s+10)}$

2.3 Solution of differential equations

We first consider the Laplace transforms of derivatives and integrals, and then apply these to the solution of differential equations.

2.3.1 Transforms of derivatives

If we are to use Laplace transform methods to solve differential equations, we need to find convenient expressions for the Laplace transforms of derivatives such as $\mathrm{d}f/\mathrm{d}t$, $\mathrm{d}^2f/\mathrm{d}t^2$ or, in general, $\mathrm{d}^nf/\mathrm{d}t^n$. By definition,

$$\mathscr{L}\left\{\frac{\mathrm{d}f}{\mathrm{d}t}\right\} = \int_0^\infty \mathrm{e}^{-st}\frac{\mathrm{d}f}{\mathrm{d}t}\,\mathrm{d}t$$

Integrating by parts, we have

$$\begin{aligned}\mathscr{L}\left\{\frac{\mathrm{d}f}{\mathrm{d}t}\right\} &= [\mathrm{e}^{-st}f(t)]_0^\infty + s\int_0^\infty \mathrm{e}^{-st}f(t)\,\mathrm{d}t\\ &= -f(0) + sF(s)\end{aligned}$$

that is,

$$\mathscr{L}\left\{\frac{\mathrm{d}f}{\mathrm{d}t}\right\} = sF(s) - f(0) \qquad \textbf{(2.13)}$$

In taking the Laplace transform of a derivative we have assumed that $f(t)$ is continuous at $t = 0$, so that $f(0^-) = f(0) = f(0^+)$. In Section 2.5.8, when considering the impulse function, $f(0^-) \neq f(0^+)$ and we have to revert to a more generalized calculus to resolve the problem.

The advantage of using the Laplace transform when dealing with differential equations can readily be seen, since it enables us to replace the operation of differentiation in the time domain by a simple algebraic operation in the s domain.

Note that to deduce the result (2.13), we have assumed that $f(t)$ is continuous, with a piecewise-continuous derivative $\mathrm{d}f/\mathrm{d}t$, for $t \geqslant 0$ and that it is also of exponential order as $t \to \infty$.

Likewise, if both $f(t)$ and $\mathrm{d}f/\mathrm{d}t$ are continuous on $t \geqslant 0$ and are of exponential order as $t \to \infty$, and $\mathrm{d}^2f/\mathrm{d}t^2$ is piecewise-continuous for $t \geqslant 0$, then

$$\mathscr{L}\left\{\frac{\mathrm{d}^2f}{\mathrm{d}t^2}\right\} = \int_0^\infty \mathrm{e}^{-st}\frac{\mathrm{d}^2f}{\mathrm{d}t^2}\,\mathrm{d}t = \left[\mathrm{e}^{-st}\frac{\mathrm{d}f}{\mathrm{d}t}\right]_0^\infty + s\int_0^\infty \mathrm{e}^{-st}\frac{\mathrm{d}f}{\mathrm{d}t}\,\mathrm{d}t = -\left[\frac{\mathrm{d}f}{\mathrm{d}t}\right]_{t=0} + s\mathscr{L}\left\{\frac{\mathrm{d}f}{\mathrm{d}t}\right\}$$

which, on using (2.12), gives

$$\mathscr{L}\left\{\frac{\mathrm{d}^2 f}{\mathrm{d}t^2}\right\} = -\left[\frac{\mathrm{d}f}{\mathrm{d}t}\right]_{t=0} + s[sF(s) - f(0)]$$

leading to the result

$$\mathscr{L}\left\{\frac{\mathrm{d}^2 f}{\mathrm{d}t^2}\right\} = s^2F(s) - sf(0) - \left[\frac{\mathrm{d}f}{\mathrm{d}t}\right]_{t=0} = s^2F(s) - sf(0) - f^{(1)}(0) \qquad \textbf{(2.14)}$$

Clearly, provided that $f(t)$ and its derivatives satisfy the required conditions, this procedure may be extended to obtain the Laplace transform of $f^{(n)}(t) = \mathrm{d}^n f/\mathrm{d}t^n$ in the form

$$\begin{aligned}\mathscr{L}\{f^{(n)}(t)\} &= s^nF(s) - s^{n-1}f(0) - s^{n-2}f^{(1)}(0) - \ldots - f^{(n-1)}(0)\\ &= s^nF(s) - \sum_{i=1}^{n} s^{n-i}f^{(i-1)}(0)\end{aligned} \qquad \textbf{(2.15)}$$

a result that may be readily proved by induction.

Again it is noted that in determining the Laplace transform of $f^{(n)}(t)$ we have assumed that $f^{(n-1)}(t)$ is continuous.

2.3.2 Transforms of integrals

In some applications the behaviour of a system may be represented by an **integro-differential equation**, which is an equation containing both derivatives and integrals of the unknown variable. For example, the current i in a series electrical circuit consisting of a resistance R, an inductance L and capacitance C, and subject to an applied voltage E, is given by

$$L\frac{\mathrm{d}i}{\mathrm{d}t} + iR + \frac{1}{C}\int_0^t i(\tau)\,\mathrm{d}\tau = E$$

To solve such equations directly, it is convenient to be able to obtain the Laplace transform of integrals such as $\int_0^t f(\tau)\,\mathrm{d}\tau$.

Writing

$$g(t) = \int_0^t f(\tau)\,\mathrm{d}\tau$$

we have

$$\frac{\mathrm{d}g}{\mathrm{d}t} = f(t), \quad g(0) = 0$$

Taking Laplace transforms,

$$\mathscr{L}\left\{\frac{dg}{dt}\right\} = \mathscr{L}\{f(t)\}$$

which, on using (2.13), gives

$$sG(s) = F(s)$$

or

$$\mathscr{L}\{g(t)\} = G(s) = \frac{1}{s}F(s) = \frac{1}{s}\mathscr{L}\{f(t)\}$$

leading to the result

$$\mathscr{L}\left\{\int_0^t f(\tau)\,d\tau\right\} = \frac{1}{s}\mathscr{L}\{f(t)\} = \frac{1}{s}F(s) \qquad (2.16)$$

EXAMPLE 2.22 Obtain

$$\mathscr{L}\left\{\int_0^t (\tau^3 + \sin 2\tau)\,d\tau\right\}$$

In this case $f(t) = t^3 + \sin 2t$, giving

$$F(s) = \mathscr{L}\{f(t)\} = \mathscr{L}\{t^3\} + \mathscr{L}\{\sin 2t\}$$

$$= \frac{6}{s^4} + \frac{2}{s^2 + 4}$$

so, by (2.16),

$$\mathscr{L}\left\{\int_0^t (\tau^3 + \sin 2\tau)\,d\tau\right\} = \frac{1}{s}F(s) = \frac{6}{s^5} + \frac{2}{s(s^2+4)}$$

2.3.3 Ordinary differential equations

Having obtained expressions for the Laplace transforms of derivatives, we are now in a position to use Laplace transform methods to solve ordinary linear differential equations with constant coefficients. To illustrate this, consider the general second-order linear differential equation

$$a\frac{d^2x}{dt^2} + b\frac{dx}{dt} + cx = u(t) \quad (t \geqslant 0) \qquad (2.17)$$

subject to the initial conditions $x(0) = x_0$, $\dot{x}(0) = v_0$ where as usual a dot denotes differentiaton with respect to time, t. Such a differential equation may model the dynamics of some system for which the variable $x(t)$ determines the **response** of the system to the **forcing** or **excitation** term $u(t)$. The terms **system input** and **system output** are also frequently used for $u(t)$ and $x(t)$ respectively. Since the differential equation is linear and has constant coefficients, a system characterized by such a model is said to be a **linear time-invariant system**.

Taking Laplace transforms of each term in (2.17) gives

$$a\mathscr{L}\left\{\frac{\mathrm{d}^2x}{\mathrm{d}t^2}\right\} + b\mathscr{L}\left\{\frac{\mathrm{d}x}{\mathrm{d}t}\right\} + c\mathscr{L}\{x\} = \mathscr{L}\{u(t)\}$$

which on using (2.13) and (2.14) leads to

$$a[s^2X(s) - sx(0) - \dot{x}(0)] + b[sX(s) - x(0)] + cX(s) = U(s)$$

Rearranging gives

$$(as^2 + bs + c)X(s) = U(s) + (as + b)x_0 + av_0$$

so that

$$X(s) = \frac{U(s) + (as + b)x_0 + av_0}{as^2 + bs + c} \tag{2.18}$$

Equation (2.18) determines the Laplace transform $X(s)$ of the response, from which, by taking the inverse transform, the desired time response $x(t)$ may be obtained.

Before considering specific examples, there are a few observations worth noting at this stage.

(a) As we have already noted in Section 2.3.1, a distinct advantage of using the Laplace transform is that it enables us to replace the operation of differentiation by an algebraic operation. Consequently, by taking the Laplace transform of each term in a differential equation, it is converted into an algebraic equation in the variable s. This may then be rearranged using algebraic rules to obtain an expression for the Laplace transform of the response; the desired time response is then obtained by taking the inverse transform.

(b) The Laplace transform method yields the complete solution to the linear differential equation, with the initial conditions automatically included. This contrasts with the classical approach, in which the general solution consists of two components, the **complementary function** and the **particular integral**, with the initial conditions determining the undetermined constants associated with the complementary function. When the solution is expressed in the general form (2.18), upon inversion the term involving $U(s)$ leads to a particular integral while that involving x_0 and v_0 gives a complementary function. A useful side issue is that an explicit solution for the transient is obtained that reflects the initial conditions.

(c) The Laplace transform method is ideally suited for solving initial-value problems; that is, linear differential equations in which all the initial conditions $x(0)$,

$\dot{x}(0)$, and so on, at time $t = 0$ are specified. The method is less attractive for boundary-value problems, when the conditions on $x(t)$ and its derivatives are not all specified at $t = 0$, but some are specified at other values of the independent variable. It is still possible, however, to use the Laplace transform method by assigning arbitrary constants to one or more of the initial conditions and then determining their values using the given boundary conditions.

(d) It should be noted that the denominator of the right-hand side of (2.18) is the left-hand side of (2.17) with the operator d/dt replaced by s. The denominator equated to zero also corresponds to the auxiliary equation or characteristic equation used in the classical approach. Given a specific initial-value problem, the process of obtaining a solution using Laplace transform methods is fairly straightforward, and is illustrated by Example 2.23.

EXAMPLE 2.23 Solve the differential equation

$$\frac{d^2x}{dt^2} + 5\frac{dx}{dt} + 6x = 2\,e^{-t} \quad (t \geqslant 0)$$

subject to the initial conditions $x = 1$ and $dx/dt = 0$ at $t = 0$.

Solution Taking Laplace transforms

$$\mathscr{L}\left\{\frac{d^2x}{dt^2}\right\} + 5\mathscr{L}\left\{\frac{dx}{dt}\right\} + 6\mathscr{L}\{x\} = 2\mathscr{L}\{e^{-t}\}$$

leads to the transformed equation

$$[s^2X(s) - sx(0) - \dot{x}(0)] + 5[sX(s) - x(0)] + 6X(s) = \frac{2}{s+1}$$

which on rearrangement gives

$$(s^2 + 5s + 6)X(s) = \frac{2}{s+1} + (s + 5)x(0) + \dot{x}(0)$$

Incorporating the given initial conditions $x(0) = 1$ and $\dot{x}(0) = 0$ leads to

$$(s^2 + 5s + 6)X(s) = \frac{2}{s+1} + s + 5$$

That is,

$$X(s) = \frac{2}{(s+1)(s+2)(s+3)} + \frac{s+5}{(s+3)(s+2)}$$

Resolving the rational terms into partial fractions gives

$$X(s) = \frac{1}{s+1} - \frac{2}{s+2} + \frac{1}{s+3} + \frac{3}{s+2} - \frac{2}{s+3}$$

$$= \frac{1}{s+1} + \frac{1}{s+2} - \frac{1}{s+3}$$

Taking inverse transforms gives the desired solution

$$x(t) = e^{-t} + e^{-2t} - e^{-3t} \quad (t \geqslant 0)$$

In principle the procedure adopted in Example 2.23 for solving a second-order linear differential equation with constant coefficients is readily carried over to higher-order differential equations. A general nth-order linear differential equation may be written as

$$a_n \frac{d^n x}{dt^n} + a_{n-1} \frac{d^{n-1} x}{dt^{n-1}} + \ldots + a_0 x = u(t) \quad (t \geqslant 0) \tag{2.19}$$

where $a_n, a_{n-1}, \ldots, a_0$ are constants, with $a_n \neq 0$. This may be written in the more concise form

$$q(D)x(t) = u(t) \tag{2.20}$$

where D denotes the operator d/dt and $q(D)$ is the polynomial

$$q(D) = \sum_{r=0}^{n} a_r D^r$$

The objective is then to determine the response $x(t)$ for a given forcing function $u(t)$ subject to the given set of initial conditions

$$D^r x(0) = \left[\frac{d^r x}{dt^r}\right]_{t=0} = c_r \quad (r = 0, 1, \ldots, n-1)$$

Taking Laplace transforms in (2.20) and proceeding as before leads to

$$X(s) = \frac{p(s)}{q(s)}$$

where

$$p(s) = U(s) + \sum_{r=0}^{n-1} c_r \sum_{i=r+1}^{n} a_i s^{i-r-1}$$

Then, in principle, by taking the inverse transform, the desired response $x(t)$ may be obtained as

$$x(t) = \mathscr{L}^{-1}\left\{\frac{p(s)}{q(s)}\right\}$$

For high-order differential equations the process of performing this inversion may prove to be rather tedious, and matrix methods may be used as indicated in Chapter 6.

To conclude this section, further worked examples are developed in order to help consolidate understanding of this method for solving linear differential equations.

EXAMPLE 2.24 Solve the differential equation

$$\frac{d^2x}{dt^2}+6\frac{dx}{dt}+9x = \sin t \quad (t \geqslant 0)$$

subject to the initial conditions $x = 0$ and $dx/dt = 0$ at $t = 0$.

Solution Taking the Laplace transforms

$$\mathscr{L}\left\{\frac{d^2x}{dt^2}\right\}+6\mathscr{L}\left\{\frac{dx}{dt}\right\}+9\mathscr{L}\{x\} = \mathscr{L}\{\sin t\}$$

leads to the equation

$$[s^2X(s) - sx(0) - \dot{x}(0)] + 6[sX(s) - x(0)] + 9X(s) = \frac{1}{s^2+1}$$

which on rearrangement gives

$$(s^2 + 6s + 9)X(s) = \frac{1}{s^2+1} + (s + 6)x(0) + \dot{x}(0)$$

Incorporating the given initial conditions $x(0) = \dot{x}(0) = 0$ leads to

$$X(s) = \frac{1}{(s^2+1)(s+3)^2}$$

Resolving into partial fractions gives

$$X(s) = \frac{3}{50}\frac{1}{s+3}+\frac{1}{10}\frac{1}{(s+3)^2}+\frac{2}{25}\frac{1}{s^2+1}-\frac{3}{50}\frac{s}{s^2+1}$$

that is,

$$X(s) = \frac{3}{50}\frac{1}{s+3}+\frac{1}{10}\left[\frac{1}{s^2}\right]_{s\to s+3}+\frac{2}{25}\frac{1}{s^2+1}-\frac{3}{50}\frac{s}{s^2+1}$$

Taking inverse transforms, using the shift theorem, leads to the desired solution

$$x(t) = \tfrac{3}{50}e^{-3t}+\tfrac{1}{10}te^{-3t}+\tfrac{2}{25}\sin t-\tfrac{3}{50}\cos t \quad (t \geqslant 0)$$

EXAMPLE 2.25 Solve the differential equation

$$\frac{d^3x}{dt^3}+5\frac{d^2x}{dt^2}+17\frac{dx}{dt}+13x = 1 \quad (t \geqslant 0)$$

subject to the initial conditions $x = dx/dt = 1$ and $d^2x/dt^2 = 0$ at $t = 0$.

Solution Taking Laplace transforms

$$\mathscr{L}\left\{\frac{d^3x}{dt^3}\right\}+5\mathscr{L}\left\{\frac{d^2x}{dt^2}\right\}+17\mathscr{L}\left\{\frac{dx}{dt}\right\}+13\mathscr{L}\{x\} = \mathscr{L}\{1\}$$

leads to the equation

$$s^3X(s) - s^2x(0) - s\dot{x}(0) - \ddot{x}(0) + 5[s^2X(s) - sx(0) - \dot{x}(0)]$$
$$+ 17[sX(s) - x(0)] + 13X(s) = \frac{1}{s}$$

which on rearrangement gives

$$(s^3 + 5s^2 + 17s + 13)X(s) = \frac{1}{s} + (s^2 + 5s + 17)x(0) + (s + 5)\dot{x}(0) + \ddot{x}(0)$$

Incorporating the given initial conditions $x(0) = \dot{x}(0) = 1$ and $\ddot{x}(0) = 0$ leads to

$$X(s) = \frac{s^3 + 6s^2 + 22s + 1}{s(s^3 + 5s^2 + 17s + 13)}$$

Clearly $s + 1$ is a factor of $s^3 + 5s^2 + 17s + 13$, and by algebraic division we have

$$X(s) = \frac{s^3 + 6s^2 + 22s + 1}{s(s + 1)(s^2 + 4s + 13)}$$

Resolving into partial fractions,

$$X(s) = \frac{\frac{1}{13}}{s} + \frac{\frac{8}{5}}{s+1} - \frac{1}{65}\frac{44s + 7}{s^2 + 4s + 13} = \frac{\frac{1}{13}}{s} + \frac{\frac{8}{5}}{s+1} - \frac{1}{65}\frac{44(s + 2) - 27(3)}{(s + 2)^2 + 3^2}$$

Taking inverse transforms, using the shift theorem, leads to the solution

$$x(t) = \tfrac{1}{13} + \tfrac{8}{5}e^{-t} - \tfrac{1}{65}e^{-2t}(44\cos 3t - 27\sin 3t) \quad (t \geq 0)$$

2.3.4 Simultaneous differential equations

In engineering we frequently encounter systems whose characteristics are modelled by a set of simultaneous linear differential equations with constant coefficients. The method of solution is essentially the same as that adopted in Section 2.3.3 for solving a single differential equation in one unknown. Taking Laplace transforms throughout, the system of simultaneous differential equations is transformed into a system of simultaneous algebraic equations, which are then solved for the transformed variables; inverse transforms then give the desired solutions.

EXAMPLE 2.26 Solve for $t \geq 0$ the simultaneous first-order differential equations

$$\frac{dx}{dt} + \frac{dy}{dt} + 5x + 3y = e^{-t} \quad \textbf{(2.21)}$$

$$2\frac{dx}{dt} + \frac{dy}{dt} + x + y = 3 \quad \textbf{(2.22)}$$

subject to the initial conditions $x = 2$ and $y = 1$ at $t = 0$.

Solution Taking Laplace transforms in (2.21) and (2.22) gives

$$sX(s) - x(0) + sY(s) - y(0) + 5X(s) + 3Y(s) = \frac{1}{s+1}$$

$$2[sX(s) - x(0)] + sY(s) - y(0) + X(s) + Y(s) = \frac{3}{s}$$

Rearranging and incorporating the given initial conditions $x(0) = 2$ and $y(0) = 1$ leads to

$$(s + 5)X(s) + (s + 3)Y(s) = 3 + \frac{1}{s+1} = \frac{3s+4}{s+1} \tag{2.23}$$

$$(2s + 1)X(s) + (s + 1)Y(s) = 5 + \frac{3}{s} = \frac{5s+3}{s} \tag{2.24}$$

Hence, by taking Laplace transforms, the pair of simultaneous differential equations (2.21) and (2.22) in $x(t)$ and $y(t)$ has been transformed into a pair of simultaneous algebraic equations (2.23) and (2.24) in the transformed variables $X(s)$ and $Y(s)$. These algebraic equations may now be solved simultaneously for $X(s)$ and $Y(s)$ using standard algebraic techniques.

Solving first for $X(s)$ gives

$$X(s) = \frac{2s^2 + 14s + 9}{s(s+2)(s-1)}$$

Resolving into partial fractions,

$$X(s) = -\frac{\frac{9}{2}}{s} - \frac{\frac{11}{6}}{s+2} + \frac{\frac{25}{3}}{s-1}$$

which on inversion gives

$$x(t) = -\tfrac{9}{2} - \tfrac{11}{6}e^{-2t} + \tfrac{25}{3}e^{t} \quad (t \geqslant 0) \tag{2.25}$$

Likewise, solving for $Y(s)$ gives

$$Y(s) = \frac{s^3 - 22s^2 - 39s - 15}{s(s+1)(s+2)(s-1)}$$

Resolving into partial fractions,

$$Y(s) = \frac{\frac{15}{2}}{s} + \frac{\frac{1}{2}}{s+1} + \frac{\frac{11}{2}}{s+2} - \frac{\frac{25}{2}}{s-1}$$

which on inversion gives

$$y(t) = \tfrac{15}{2} + \tfrac{1}{2}e^{-t} + \tfrac{11}{2}e^{-2t} - \tfrac{25}{2}e^{t} \quad (t \geqslant 0)$$

Thus the solution to the given pair of simultaneous differential equations is

$$\left.\begin{aligned} x(t) &= -\tfrac{9}{2} - \tfrac{11}{6}e^{-2t} + \tfrac{25}{3}e^{t} \\ y(t) &= \tfrac{15}{2} + \tfrac{1}{2}e^{-t} + \tfrac{11}{2}e^{-2t} - \tfrac{25}{2}e^{t} \end{aligned}\right\} \quad (t \geqslant 0)$$

Note: When solving a pair of first-order simultaneous differential equations such as (2.21) and (2.22), an alternative approach to obtaining the value of $y(t)$ having obtained $x(t)$ is to use (2.21) and (2.22) directly.

Eliminating dy/dt from (2.21) and (2.22) gives

$$2y = \frac{dx}{dt} - 4x - 3 + e^{-t}$$

Substituting the solution obtained in (2.25) for $x(t)$ gives

$$2y = (\tfrac{11}{3}e^{-2t} + \tfrac{25}{3}e^{t}) - 4(-\tfrac{9}{2} - \tfrac{11}{6}e^{-2t} + \tfrac{25}{3})e^{t} - 3 + e^{-t}$$

leading as before to the solution

$$y = \tfrac{15}{2} + \tfrac{1}{2}e^{-t} + \tfrac{11}{2}e^{-2t} - \tfrac{25}{2}e^{t}$$

A further alternative is to express (2.23) and (2.24) in matrix form and solve for $X(s)$ and $Y(s)$ using Gaussian elimination.

In principle, the same procedure as used in Example 2.26 can be employed to solve a pair of higher-order simultaneous differential equations or a larger system of differential equations involving more unknowns. However, the algebra involved can become quite complicated, and matrix methods are usually preferred, as indicated in Chapter 6.

2.3.5 Exercises

5 Using Laplace transform methods, solve for $t \geqslant 0$ the following differential equations, subject to the specified initial conditions:

(a) $\dfrac{dx}{dt} + 3x = e^{-2t}$

subject to $x = 2$ at $t = 0$

(b) $3\dfrac{dx}{dt} - 4x = \sin 2t$

subject to $x = \frac{1}{3}$ at $t = 0$

(c) $\dfrac{d^2x}{dt^2} + 2\dfrac{dx}{dt} + 5x = 1$

subject to $x = 0$ and $\dfrac{dx}{dt} = 0$ at $t = 0$

(d) $\dfrac{d^2y}{dt^2} + 2\dfrac{dy}{dt} + y = 4\cos 2t$

subject to $y = 0$ and $\dfrac{dy}{dt} = 2$ at $t = 0$

(e) $\dfrac{d^2x}{dt^2} - 3\dfrac{dx}{dt} + 2x = 2e^{-4t}$

subject to $x = 0$ and $\dfrac{dx}{dt} = 1$ at $t = 0$

(f) $\dfrac{d^2x}{dt^2} + 4\dfrac{dx}{dt} + 5x = 3e^{-2t}$

subject to $x = 4$ and $\dfrac{dx}{dt} = -7$ at $t = 0$

(g) $\dfrac{d^2x}{dt^2} + \dfrac{dx}{dt} - 2x = 5e^{-t}\sin t$

subject to $x = 1$ and $\dfrac{dx}{dt} = 0$ at $t = 0$

(h) $\dfrac{d^2y}{dt^2} + 2\dfrac{dy}{dt} + 3y = 3t$

subject to $y = 0$ and $\dfrac{dy}{dt} = 1$ at $t = 0$

(i) $\dfrac{d^2x}{dt^2} + 4\dfrac{dx}{dt} + 4x = t^2 + e^{-2t}$

subject to $x = \frac{1}{2}$ and $\dfrac{dx}{dt} = 0$ at $t = 0$

(j) $9\dfrac{d^2x}{dt^2} + 12\dfrac{dx}{dt} + 5x = 1$

subject to $x = 0$ and $\dfrac{dx}{dt} = 0$ at $t = 0$

(k) $\dfrac{d^2x}{dt^2} + 8\dfrac{dx}{dt} + 16x = 16 \sin 4t$

subject to $x = -\frac{1}{2}$ and $\dfrac{dx}{dt} = 1$ at $t = 0$

(l) $9\dfrac{d^2y}{dt^2} + 12\dfrac{dy}{dt} + 4y = e^{-t}$

subject to $y = 1$ and $\dfrac{dy}{dt} = 1$ at $t = 0$

(m) $\dfrac{d^3x}{dt^3} - 2\dfrac{d^2x}{dt^2} - \dfrac{dx}{dt} + 2x = 2 + t$

subject to $x = 0$, $\dfrac{dx}{dt} = 1$ and $\dfrac{d^2x}{dt^2} = 0$ at $t = 0$

(n) $\dfrac{d^3x}{dt^3} + \dfrac{d^2x}{dt^2} + \dfrac{dx}{dt} + x = \cos 3t$

subject to $x = 0$, $\dfrac{dx}{dt} = 1$ and $\dfrac{d^2x}{dt^2} = 1$ at $t = 0$

6 Using Laplace transform methods, solve for $t \geqslant 0$ the following simultaneous differential equations subject to the given initial conditions:

(a) $2\dfrac{dx}{dt} - 2\dfrac{dy}{dt} - 9y = e^{-2t}$

$2\dfrac{dx}{dt} + 4\dfrac{dy}{dt} + 4x - 37y = 0$

subject to $x = 0$ and $y = \frac{1}{4}$ at $t = 0$

(b) $\dfrac{dx}{dt} + 2\dfrac{dy}{dt} + x - y = 5 \sin t$

$2\dfrac{dx}{dt} + 3\dfrac{dy}{dt} + x - y = e^{t}$

subject to $x = 0$ and $y = 0$ at $t = 0$

(c) $\dfrac{dx}{dt} + \dfrac{dy}{dt} + 2x + y = e^{-3t}$

$\dfrac{dy}{dt} + 5x + 3y = 5e^{-2t}$

subject to $x = -1$ and $y = 4$ at $t = 0$

(d) $3\dfrac{dx}{dt} + 3\dfrac{dy}{dt} - 2x = e^{t}$

$\dfrac{dx}{dt} + 2\dfrac{dy}{dt} - y = 1$

subject to $x = 1$ and $y = 1$ at $t = 0$

(e) $3\dfrac{dx}{dt} + \dfrac{dy}{dt} - 2x = 3 \sin t + 5 \cos t$

$2\dfrac{dx}{dt} + \dfrac{dy}{dt} + y = \sin t + \cos t$

subject to $x = 0$ and $y = -1$ at $t = 0$

(f) $\dfrac{dx}{dt} + \dfrac{dy}{dt} + y = t$

$\dfrac{dx}{dt} + 4\dfrac{dy}{dt} + x = 1$

subject to $x = 1$ and $y = 0$ at $t = 0$

(g) $2\dfrac{dx}{dt} + 3\dfrac{dy}{dt} + 7x = 14t + 7$

$5\dfrac{dx}{dt} - 3\dfrac{dy}{dt} + 4x + 6y = 14t - 14$

subject to $x = y = 0$ at $t = 0$

(h) $\dfrac{d^2x}{dt^2} = y - 2x$

$\dfrac{d^2y}{dt^2} = x - 2y$

subject to $x = 4$, $y = 2$, $dx/dt = 0$ and $dy/dt = 0$ at $t = 0$

(i) $5\dfrac{d^2x}{dt^2} + 12\dfrac{d^2y}{dt^2} + 6x = 0$

$5\dfrac{d^2x}{dt^2} + 16\dfrac{d^2y}{dt^2} + 6y = 0$

subject to $x = \frac{7}{4}$, $y = 1$, $dx/dt = 0$ and $dy/dt = 0$ at $t = 0$

(j) $2\dfrac{d^2x}{dt^2} - \dfrac{d^2y}{dt^2} - \dfrac{dx}{dt} - \dfrac{dy}{dt} = 3y - 9x$

$2\dfrac{d^2x}{dt^2} - \dfrac{d^2y}{dt^2} + \dfrac{dx}{dt} + \dfrac{dy}{dt} = 5y - 7x$

subject to $x = dx/dt = 1$ and $y = dy/dt = 0$ at $t = 0$

2.4 Engineering applications: electrical circuits and mechanical vibrations

To illustrate the use of Laplace transforms, we consider here their application to the analysis of electrical circuits and vibrating mechanical systems. Since initial conditions are automatically taken into account in the transformation process, the Laplace transform is particularly attractive for examining the transient behaviour of such systems.

2.4.1 Electrical circuits

Passive electrical circuits are constructed of three basic elements: **resistors** (having resistance R, measured in ohms Ω), **capacitors** (having capacitance C, measured in farads F) and **inductors** (having inductance L, measured in henries H), with the associated variables being **current** $i(t)$ (measured in amperes A) and **voltage** $v(t)$ (measured in volts V). The current flow in the circuit is related to the charge $q(t)$ (measured in coulombs C) by the relationship

$$i = \frac{\mathrm{d}q}{\mathrm{d}t}$$

Conventionally, the basic elements are represented symbolically as in Figure 2.7.

The relationship between the flow of current $i(t)$ and the voltage drops $v(t)$ across these elements at time t are

$$\text{voltage drop across resistor} = Ri \text{ (Ohm's law)}$$

$$\text{voltage drop across capacitor} = \frac{1}{C}\int i\,\mathrm{d}t = \frac{q}{C}$$

The interaction between the individual elements making up an electrical circuit is determined by **Kirchhoff's laws**:

Law 1
The algebraic sum of all the currents entering any junction (or node) of a circuit is zero.

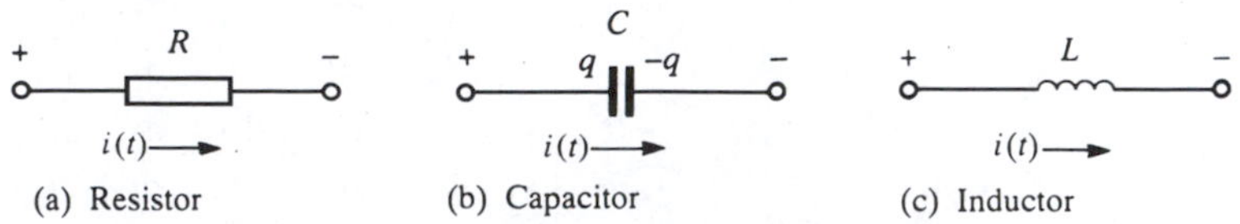

Figure 2.7 Constituent elements of an electrical circuit.

Law 2
The algebraic sum of the voltage drops around any closed loop (or path) in a circuit is zero.

Use of these laws leads to circuit equations, which may then be analysed using Laplace transform techniques.

EXAMPLE 2.27

The *LCR* circuit of Figure 2.8 consists of a resistor R, a capacitor C and an inductor L connected in series together with a voltage source $e(t)$. Prior to closing the switch at time $t = 0$, both the charge on the capacitor and the resulting current in the circuit are zero. Determine the charge $q(t)$ on the capacitor and the resulting current $i(t)$ in the circuit at time t given that $R = 160\Omega$, $L = 1\,\text{H}$, $C = 10^{-4}\,\text{F}$ and $e(t) = 20\text{V}$.

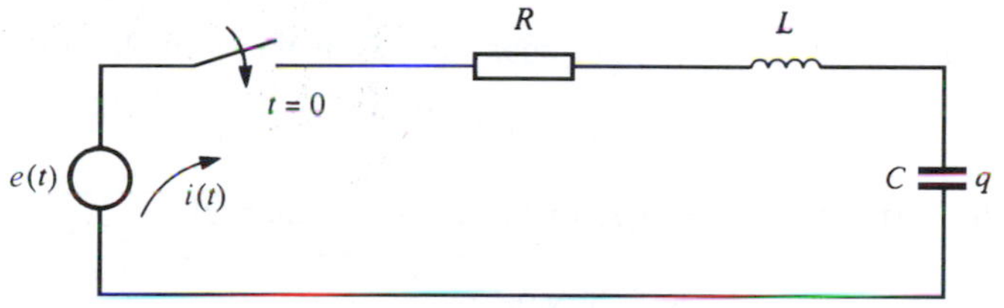

Figure 2.8 *LCR* circuit of Example 2.27.

Solution

Applying Kirchhoff's second law to the circuit of Figure 2.8 gives

$$Ri + L\frac{di}{dt} + \frac{1}{C}\int i\,dt = e(t) \qquad R\,I(s) + LsI(s) + i(0) \qquad (2.26)$$

$$\frac{10^4\,I(s)}{s}$$

or, using $i = dq/dt$,

$$L\frac{d^2q}{dt^2} + R\frac{dq}{dt} + \frac{1}{C}q = e(t)$$

Substituting the given values for L, R, C and $e(t)$ gives

$$\frac{d^2q}{dt^2} + 160\frac{dq}{dt} + 10^4 q = 20$$

Taking Laplace transforms throughout leads to the equation

$$(s^2 + 160s + 10^4)Q(s) = [sq(0) + \dot{q}(0)] + 160q(0) + \frac{20}{s}$$

where $Q(s)$ is the transform of $q(t)$. We are given that $q(0) = 0$ and $\dot{q}(0) = i(0) = 0$, so that this reduces to

$$(s^2 + 160s + 10^4)Q(s) = \frac{20}{s}$$

that is,

$$Q(s) = \frac{20}{s(s^2 + 160s + 10^4)}$$

Resolving into partial fractions gives

$$Q(s) = \frac{\frac{1}{500}}{s} - \frac{1}{500}\frac{s + 160}{s^2 + 160s + 10^4}$$

$$= \frac{1}{500}\left[\frac{1}{s} - \frac{(s + 80) + \frac{4}{3}(60)}{(s + 80)^2 + (60)^2}\right]$$

$$= \frac{1}{500}\left[\frac{1}{s} - \left[\frac{s + \frac{4}{3} \times 60}{s^2 + 60^2}\right]_{s \to s+80}\right]$$

Taking inverse transforms, making use of the shift theorem (Theorem 2.2), gives

$$q(t) = \tfrac{1}{500}(1 - e^{-80t}\cos 60t - \tfrac{4}{3}e^{-80t}\sin 60t)$$

The resulting current $i(t)$ in the circuit is then given by

$$i(t) = \frac{dq}{dt} = \frac{1}{3}e^{-80t}\sin 60t$$

Note that we could have determined the current by taking Laplace transforms in (2.26). Substituting the given values for L, R, C and $e(t)$ and using (2.26) leads to the transformed equation

$$160I(s) + sI(s) + \frac{10^4}{s}I(s) = \frac{20}{s}$$

that is,

$$I(s) = \frac{20}{(s^2 + 80)^2 + 60^2} \quad (= sQ(s) \quad \text{since} \quad q(0) = 0)$$

which, on taking inverse transforms, gives as before

$$i(t) = \tfrac{1}{3}e^{-80t}\sin 60t$$

EXAMPLE 2.28 In the parallel network of Figure 2.9 there is no current flowing in either loop prior to closing the switch at time $t = 0$. Deduce the currents $i_1(t)$ and $i_2(t)$ flowing in the loops at time t.

Solution Applying Kirchhoff's first law to node X gives

$$i = i_1 + i_2$$

Applying Kirchhoff's second law to each of the two loops in turn gives

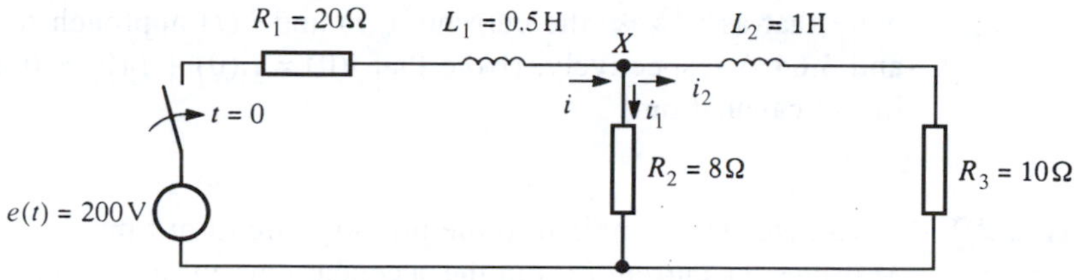

Figure 2.9 Parallel circuit of Example 2.28.

$$R_1(i_1 + i_2) + L_1\frac{\mathrm{d}}{\mathrm{d}t}(i_1 + i_2) + R_2 i_1 = 200$$

$$L_2\frac{\mathrm{d}i_2}{\mathrm{d}t} + R_3 i_2 - R_2 i_1 = 0$$

Substituting the given values for the resistances and inductances gives

$$\left.\begin{aligned}\frac{\mathrm{d}i_1}{\mathrm{d}t} + \frac{\mathrm{d}i_2}{\mathrm{d}t} + 56i_1 + 40i_2 &= 400\\ \frac{\mathrm{d}i_2}{\mathrm{d}t} - 8i_1 + 10i_2 &= 0\end{aligned}\right\} \tag{2.27}$$

Taking Laplace transforms and incorporating the initial conditions $i_1(0) = i_2(0) = 0$ leads to the transformed equations

$$(s + 56)I_1(s) + (s + 40)I_2(s) = \frac{400}{s} \tag{2.28}$$

$$-8I_1(s) + (s + 10)I_2(s) = 0 \tag{2.29}$$

Hence

$$I_2(s) = \frac{3200}{s(s^2 + 74s + 880)} = \frac{3200}{s(s + 59.1)(s + 14.9)}$$

Resolving into partial fractions gives

$$I_2(s) = \frac{3.64}{s} + \frac{1.22}{s + 59.1} - \frac{4.86}{s + 14.9}$$

which, on taking inverse transforms, leads to

$$i_2(t) = 3.64 + 1.22\,\mathrm{e}^{-59.1t} - 4.86\,\mathrm{e}^{-14.9t}$$

From (2.27),

$$i_1(t) = \tfrac{1}{8}\left(10i_2 + \frac{\mathrm{d}i_2}{\mathrm{d}t}\right)$$

that is,

$$i_1(t) = 4.55 - 7.49\,\mathrm{e}^{-59.1t} + 2.98\,\mathrm{e}^{-14.9t}$$

Note that as $t \to \infty$, the currents $i_1(t)$ and $i_2(t)$ approach the constant values 4.55 and 3.64 A respectively. (Note that $i(0) = i_1(0) + i_2(0) \neq 0$ due to rounding errors in the calculation.)

EXAMPLE 2.29 A voltage $e(t)$ is applied to the primary circuit at time $t = 0$, and mutual induction M drives the current $i_2(t)$ in the secondary circuit of Figure 2.10. If, prior to closing the switch, the currents in both circuits are zero, determine the induced current $i_2(t)$ in the secondary circuit at time t when $R_1 = 4\Omega$, $R_2 = 10\Omega$, $L_1 = 2\text{H}$, $L_2 = 8\text{H}$, $M = 2\text{H}$ and $e(t) = 28 \sin 2t$ V.

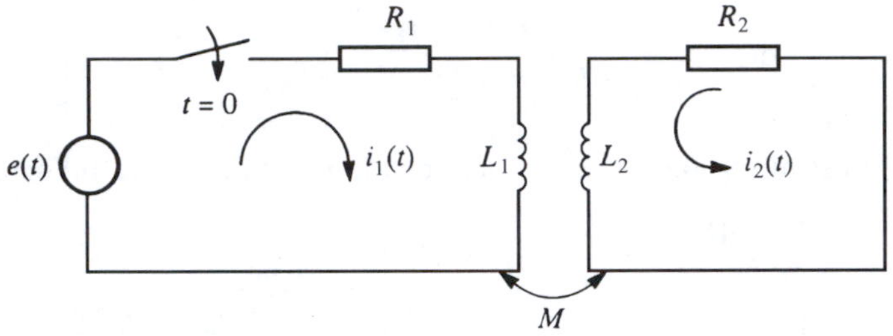

Figure 2.10 Circuit of Example 2.29.

Solution Applying Kirchhoff's second law to the primary and secondary circuits respectively gives

$$R_1 i_1 + L_1 \frac{di_1}{dt} + M\frac{di_2}{dt} = e(t)$$

$$R_2 i_2 + L_2 \frac{di_2}{dt} + M\frac{di_1}{dt} = 0$$

Substituting the given values for the resistances, inductances and applied voltage leads to

$$2\frac{di_1}{dt} + 4i_1 + 2\frac{di_2}{dt} = 28 \sin 2t$$

$$2\frac{di_1}{dt} + 8\frac{di_2}{dt} + 10i_2 = 0$$

Taking Laplace transforms and noting that $i_1(0) = i_2(0) = 0$ leads to the equations

$$(s + 2)I_1(s) + sI_2(s) = \frac{28}{s^2 + 4} \tag{2.30}$$

$$sI_1(s) + (4s + 5)I_2(s) = 0 \tag{2.31}$$

Solving for $I_2(s)$ yields

$$I_2(s) = -\frac{28s}{(3s + 10)(s + 1)(s^2 + 4)}$$

Resolving into partial fractions gives

$$I_2(s) = -\frac{\frac{45}{17}}{3s+10} + \frac{\frac{4}{5}}{s+1} + \frac{7}{85}\frac{s-26}{s^2+4}$$

Taking inverse Laplace transforms gives the current in the secondary circuit as

$$i_2(t) = \tfrac{4}{5}\mathrm{e}^{-t} - \tfrac{15}{17}\mathrm{e}^{-10t/3} + \tfrac{7}{85}\cos 2t - \tfrac{91}{85}\sin 2t$$

As $t \to \infty$, the current will approach the sinusoidal response

$$i_2(t) = \tfrac{7}{85}\cos 2t - \tfrac{91}{85}\sin 2t$$

2.4.2 Mechanical vibrations

Mechanical translational systems may be used to model many situations, and involve three basic elements: **masses** (having mass M, measured in kg), **springs** (having spring stiffness K, measured in $\mathrm{N\,m^{-1}}$) and **dampers** (having damping coefficient B, measured in $\mathrm{N\,s\,m^{-1}}$). The associated variables are **displacement** $x(t)$ (measured in m) and **force** $F(t)$ (measured in N). Conventionally, the basic elements are represented symbolically as in Figure 2.11.

Assuming we are dealing with ideal springs and dampers (that is, assuming that they behave linearly), the relationships between the forces and displacements at time t are

mass: $F = M\dfrac{\mathrm{d}^2x}{\mathrm{d}t^2} = M\ddot{x}$ (Newton's law)

spring: $F = K(x_2 - x_1)$ (Hooke's law)

damper: $F = B\left(\dfrac{\mathrm{d}x_2}{\mathrm{d}t} - \dfrac{\mathrm{d}x_1}{\mathrm{d}t}\right) = B(\dot{x}_2 - \dot{x}_1)$

Using these relationships leads to the system equations, which may then be analysed using Laplace transform techniques.

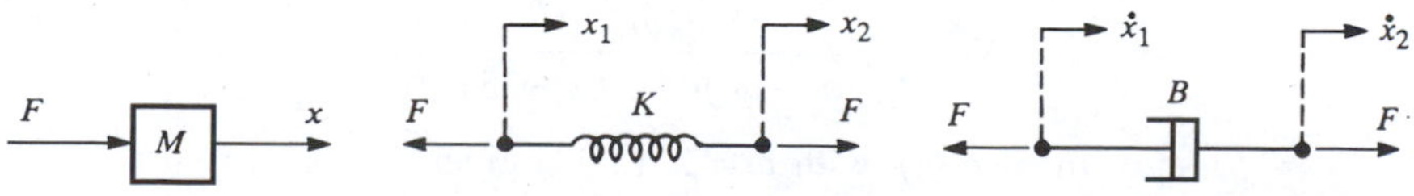

Figure 2.11 Constituent elements of a translational mechanical system.

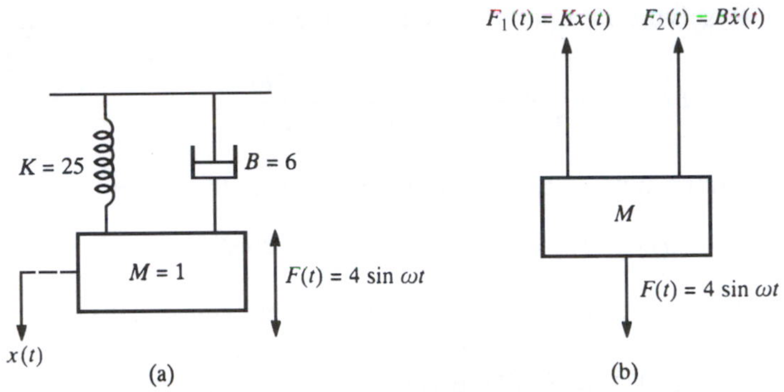

Figure 2.12 Mass–spring–damper system of Example 2.30.

EXAMPLE 2.30

The mass of the mass–spring–damper system of Figure 2.12(a) is subjected to an externally applied periodic force $F(t) = 4 \sin \omega t$ at time $t = 0$. Determine the resulting displacement $x(t)$ of the mass at time t, given that $x(0) = \dot{x}(0) = 0$, for the two cases

(a) $\omega = 2$ (b) $\omega = 5$

In the case $\omega = 5$, what would happen to the response if the damper were missing?

Solution As indicated in Figure 1.12(b), the forces acting on the mass M are the applied forces $F(t)$ and the restoring forces F_1 and F_2 due to the spring and damper respectively. Thus, by Newton's law,

$$M\ddot{x}(t) = F(t) - F_1(t) - F_2(t)$$

Since $M = 1$, $F(t) = 4 \sin \omega t$, $F_1(t) = Kx(t) = 25x(t)$ and $F_2(t) = B\dot{x}(t) = 6\dot{x}(t)$, this gives

$$\ddot{x}(t) + 6\dot{x}(t) + 25x(t) = 4 \sin \omega t \tag{2.32}$$

as the differential equation representing the motion of the system.

Taking Laplace transforms throughout in (2.32) gives

$$(s^2 + 6s + 25)X(s) = [sx(0) + \dot{x}(0)] + 6x(0) + \frac{4\omega}{s^2 + \omega^2}$$

where $X(s)$ is the transform of $x(t)$. Incorporating the given initial conditions $x(0) = \dot{x}(0) = 0$ leads to

$$X(s) = \frac{4\omega}{(s^2 + \omega^2)(s^2 + 6s + 25)} \tag{2.33}$$

In case (a), with $\omega = 2$, (2.33) gives

$$X(s) = \frac{8}{(s^2 + 4)(s^2 + 6s + 25)}$$

which, on resolving into partial fractions, leads to

$$X(s) = \frac{4}{195}\,\frac{-4s+14}{s^2+4} + \frac{2}{195}\,\frac{8s+20}{s^2+6s+25}$$

$$= \frac{4}{195}\,\frac{-4s+14}{s^2+4} + \frac{2}{195}\,\frac{8(s+3)-4}{(s+3)^2+16}$$

Taking inverse Laplace transforms gives the required response

$$x(t) = \tfrac{4}{195}(7\sin 2t - 4\cos 2t) + \tfrac{2}{195}\mathrm{e}^{-3t}(8\cos 4t - \sin 4t) \qquad \textbf{(2.34)}$$

In case (b), with $\omega = 5$, (2.33) gives

$$X(s) = \frac{20}{(s^2+25)(s^2+6s+25)} \qquad \textbf{(2.35)}$$

that is,

$$X(s) = \frac{-\frac{2}{15}s}{s^2+25} + \frac{1}{15}\,\frac{2(s+3)+6}{(s+3)^2+16}$$

which, on taking inverse Laplace transforms, gives the required response

$$x(t) = -\tfrac{2}{15}\cos 5t + \tfrac{1}{15}\mathrm{e}^{-3t}(2\cos 4t + \tfrac{3}{2}\sin 4t) \qquad \textbf{(2.36)}$$

If the damping term were missing then (2.35) would become

$$X(s) = \frac{20}{(s^2+25)^2} \qquad \textbf{(2.37)}$$

By Theorem 2.3,

$$\mathscr{L}\{t\cos 5t\} = -\frac{\mathrm{d}}{\mathrm{d}s}\mathscr{L}\{\cos 5t\} = -\frac{\mathrm{d}}{\mathrm{d}s}\left(\frac{s}{s^2+25}\right)$$

that is,

$$\mathscr{L}\{t\cos 5t\} = -\frac{1}{s^2+25} + \frac{2s^2}{(s^2+25)^2} = \frac{1}{s^2+25} - \frac{50}{(s^2+25)^2}$$

$$= \tfrac{1}{5}\mathscr{L}\{\sin 5t\} - \frac{50}{(s^2+25)^2}$$

Thus, by the linearity property (2.11),

$$\mathscr{L}\{\tfrac{1}{5}\sin 5t - t\cos 5t\} = \frac{50}{(s^2+25)^2}$$

so that taking inverse Laplace transforms in (2.37) gives the response as

$$x(t) = \tfrac{2}{25}(\sin 5t - 5t\cos 5t)$$

Because of the term $t \cos 5t$, the response $x(t)$ is unbounded as $t \to \infty$. This arises because in this case the applied force $F(t) = 4 \sin 5t$ is in **resonance** with the system (that is, the vibrating mass), whose natural oscillating frequency is $5/2\pi$ Hz, equal to that of the applied force. Even in the presence of damping, the amplitude of the system response is maximized when the applied force is approaching resonance with the system. (This is left as an exercise for the reader.) In the absence of damping we have the limiting case of **pure resonance**, leading to an unbounded response. As noted in Section 10.10.3 of *Modern Engineering Mathematics*, resonance is of practical importance, since, for example, it can lead to large and strong structures collapsing under what appears to be a relatively small force.

EXAMPLE 2.31

Consider the mechanical system of Figure 2.13(a), which consists of two masses $M_1 = 1$ and $M_2 = 2$, each attached to a fixed base by a spring, having constants $K_1 = 1$ and $K_3 = 2$ respectively, and attached to each other by a third spring having constant $K_2 = 2$. The system is released from rest at time $t = 0$ in a position in which M_1 is displaced 1 unit to the left of its equilibrium position and M_2 is displaced 2 units to the right of its equilibrium position. Neglecting all frictional effects, determine the positions of the masses at time t.

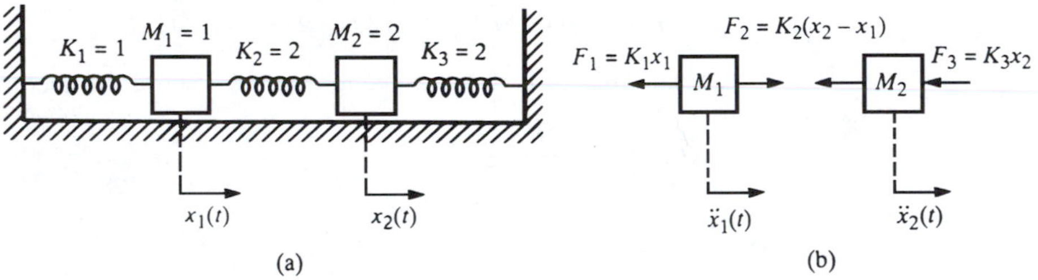

Figure 2.13 Two-mass system of Example 2.31.

Solution Let $x_1(t)$ and $x_2(t)$ denote the displacements of the masses M_1 and M_2 respectively from their equilibrium positions. Since frictional effects are neglected, the only forces acting on the masses are the restoring forces due to the springs, as shown in Figure 2.13(b). Applying Newton's law to the motions of M_1 and M_2 respectively gives

$$M_1\ddot{x}_1 = F_2 - F_1 = K_2(x_2 - x_1) - K_1x_1$$

$$M_2\ddot{x}_2 = -F_3 - F_2 = -K_3x_2 - K_2(x_2 - x_1)$$

which, on substituting the given values for M_1, M_2, K_1, K_2 and K_3, gives

$$\ddot{x}_1 + 3x_1 - 2x_2 = 0 \quad \textbf{(2.38)}$$

$$2\ddot{x}_2 + 4x_2 - 2x_1 = 0 \quad \textbf{(2.39)}$$

Taking Laplace transforms leads to the equations

$$(s^2 + 3)X_1(s) - 2X_2(s) = sx_1(0) + \dot{x}_1(0)$$

$$-X_1(s) + (s^2 + 2)X_2(s) = sx_2(0) + \dot{x}_2(0)$$

Since $x_1(t)$ and $x_2(t)$ denote displacements to the right of the equilibrium positions, we have $x_1(0) = -1$ and $x_2(0) = 2$. Also, the system is released from rest, so that $\dot{x}_1(0) = \dot{x}_2(0) = 0$. Incorporating these initial conditions, the transformed equations become

$$(s^2 + 3)X_1(s) - 2X_2(s) = -s \tag{2.40}$$

$$-X_1(s) + (s^2 + 2)X_2(s) = 2s \tag{2.41}$$

Hence

$$X_2(s) = \frac{2s^3 + 5s}{(s^2 + 4)(s^2 + 1)}$$

Resolving into partial fractions gives

$$X_2(s) = \frac{s}{s^2 + 1} + \frac{s}{s^2 + 4}$$

which, on taking inverse Laplace transforms, leads to the response

$$x_2(t) = \cos t + \cos 2t$$

Substituting for $x_2(t)$ in (2.39) gives

$$x_1(t) = 2x_2(t) + \ddot{x}_2(t)$$

$$= 2 \cos t + 2 \cos 2t - \cos t - 4 \cos 2t$$

that is,

$$x_1(t) = \cos t - 2 \cos 2t$$

Thus the positions of the masses at time t are

$$x_1(t) = \cos t - 2 \cos 2t$$

$$x_2(t) = \cos t + \cos 2t$$

2.4.3 Exercises

7 Use the Laplace transform technique to find the transforms $I_1(s)$ and $I_2(s)$ of the respective currents flowing in the circuit of Figure 2.14, where $i_1(t)$ is that through the capacitor and $i_2(t)$ that through the resistance. Hence, determine $i_2(t)$. (Initially, $i_1(0) = i_2(0) = q_1(0) = 0$.) Sketch $i_2(t)$ for large values of t.

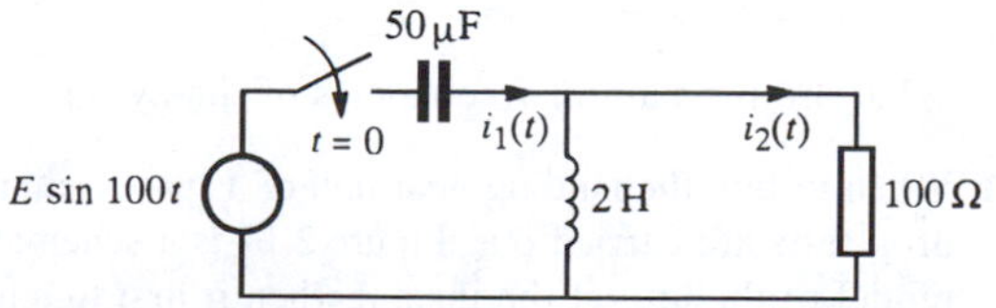

Figure 2.14 Circuit of Exercise 7.

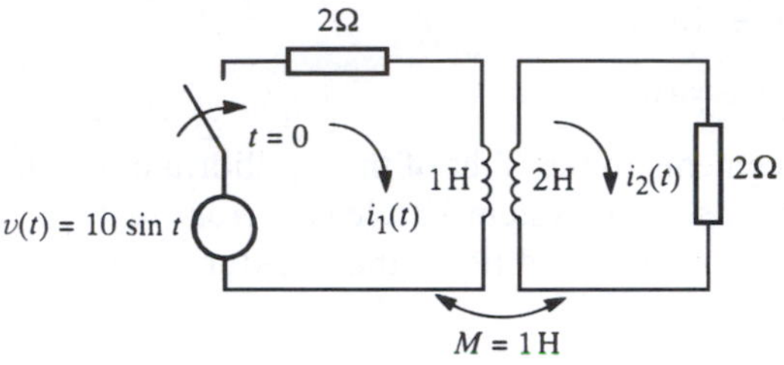

Figure 2.15 Circuit of Exercise 8.

8 At time $t = 0$, with no currents flowing, a voltage $v(t) = 10 \sin t$ is applied to the primary circuit of a transformer that has a mutual inductance of 1 H, as show in Figure 2.15. Denoting the current flowing at time t in the secondary circuit by $i_2(t)$, show that

$$\mathscr{L}\{i_2(t)\} = \frac{10s}{(s^2 + 7s + 6)(s^2 + 1)}$$

and deduce that

$$i_2(t) = -e^{-t} + \tfrac{12}{37}e^{-6t} + \tfrac{25}{37}\cos t + \tfrac{35}{37}\sin t$$

9 In the circuit of Figure 2.16 there is no energy stored (that is, there is no charge on the capacitors and no current flowing in the inductances) prior to the closure of the switch at time $t = 0$. Determine $i_1(t)$ for $t > 0$ for a constant applied voltage $E_0 = 10$ V.

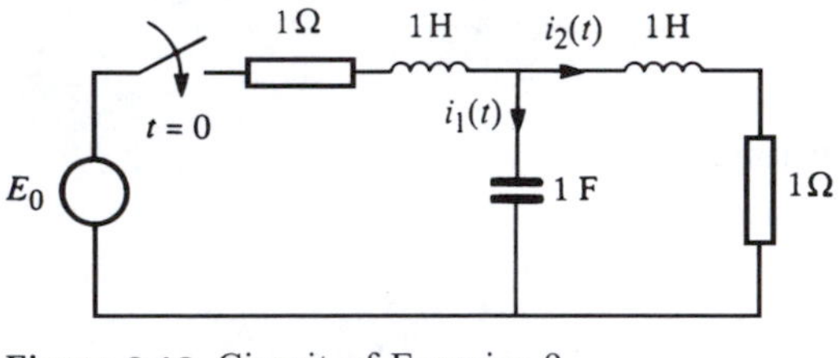

Figure 2.16 Circuit of Exercise 9.

10 Determine the displacements of the masses M_1 and M_2 in Figure 2.13 at time $t > 0$ when

$$M_1 = M_2 = 1$$
$$K_1 = 1,\ K_2 = 3 \quad \text{and} \quad K_3 = 9$$

What are the natural frequencies of the system?

11 When testing the landing-gear unit of a space vehicle, drop tests are carried out. Figure 2.17 is a schematic model of the unit at the instant when it first touches the ground. At this instant the spring is fully extended and the velocity of the mass is $\sqrt{(2gh)}$, where h is the height from which the unit has been dropped. Obtain the equation representing the displacement of the mass at time $t > 0$ when $M = 50$ kg, $B = 180$ N s m^{-1} and $K = 474.5$ N m^{-1}, and investigate the effects of different dropping heights h. (g is the acceleration due to gravity, and may be taken as 9.8 m s^{-2}.)

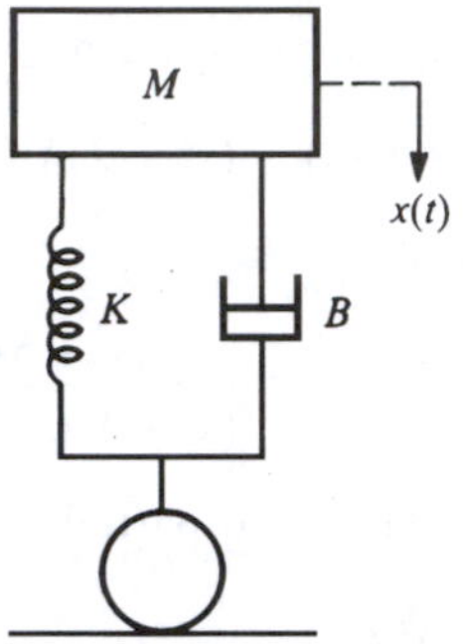

Figure 2.17 Landing-gear of Exercise 11.

12 Consider the mass–spring–damper system of Figure 2.18, which may be subject to two input forces $u_1(t)$ and $u_2(t)$. Show that the displacements $x_1(t)$ and $x_2(t)$ of the two masses are given by

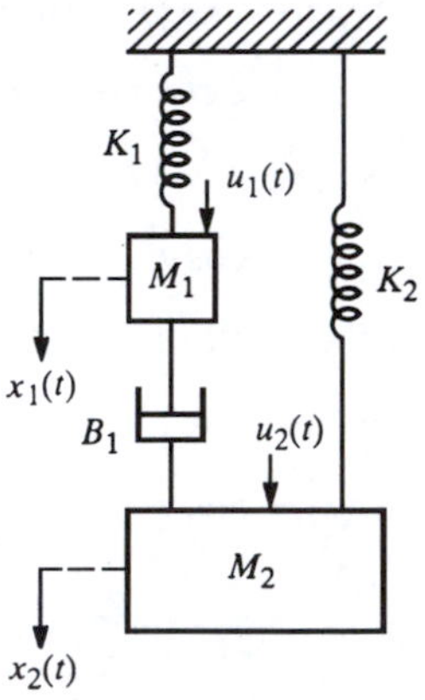

Figure 2.18 Mechanical system of Exercise 12.

$$x_1(t) = \mathscr{L}^{-1}\left\{\frac{M_2s^2 + B_1s + K_2}{\Delta}U_1(s) + \frac{B_1s}{\Delta}U_2(s)\right\}$$

$$x_2(t) = \mathscr{L}^{-1}\left\{\frac{B_1s}{\Delta}U_1(s) + \frac{M_1s^2 + B_1s + K_1}{\Delta}U_2(s)\right\}$$

where

$$\Delta = (M_1s^2 + B_1s + K_1)(M_2s^2 + B_1s + K_2) + B_1^2s^2$$

2.5 Step and impulse functions

2.5.1 The Heaviside step function

In Sections 2.3 and 2.4 we considered linear differential equations in which the forcing functions were continuous. In many engineering applications the forcing function may frequently be discontinuous, for example a square wave resulting from an on/off switch. In order to accommodate such discontinuous functions, we use the Heaviside unit step function $H(t)$, which, as we saw in Section 2.2.1, is defined by

$$H(t) = \begin{cases} 0 & (t < 0) \\ 1 & (t \geqslant 0) \end{cases}$$

and is illustrated graphically in Figure 2.19(a). The Heaviside function is also frequently referred to simply as the **unit step function**. A function representing a unit step at $t = a$ may be obtained by a horizontal translation of duration a. This is depicted graphically in Figure 2.19(b), and defined by

$$H(t-a) = \begin{cases} 0 & (t < a) \\ 1 & (t \geqslant a) \end{cases}$$

The product function $f(t)H(t - a)$ takes values

$$f(t)H(t-a) = \begin{cases} 0 & (t < a) \\ f(t) & (t \geqslant a) \end{cases}$$

so the function $H(t - a)$ may be interpreted as a device for 'switching on' the function $f(t)$ at $t = a$. In this way the unit step function may be used to write a concise formulation of piecewise-continuous functions. To illustrate this, consider the piecewise-continuous function $f(t)$ illustrated in Figure 2.20 and defined by

$$f(t) = \begin{cases} f_1(t) & (0 \leqslant t < t_1) \\ f_2(t) & (t_1 \leqslant t < t_2) \\ f_3(t) & (t \geqslant t_2) \end{cases}$$

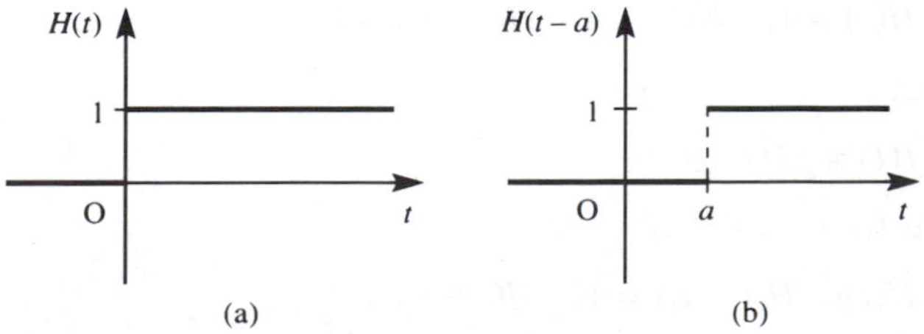

Figure 2.19 Heaviside unit step function.

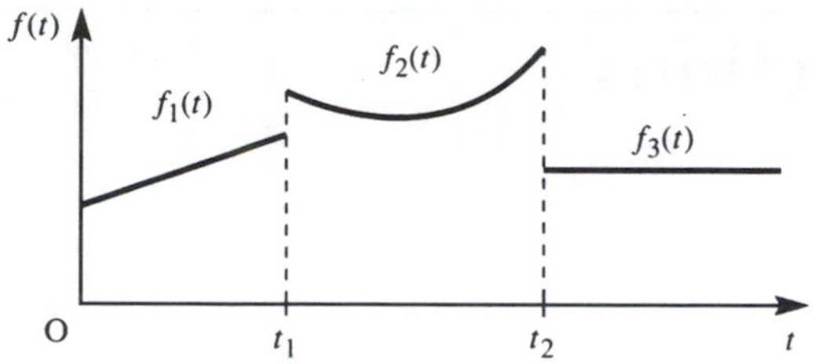

Figure 2.20 Piecewise-continuous function.

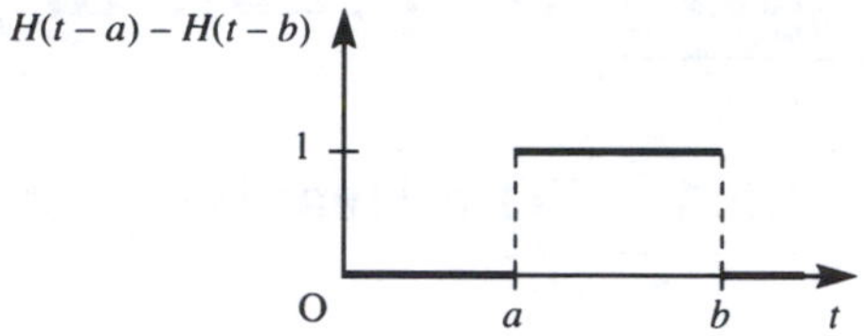

Figure 2.21 Top hat function.

To construct this function $f(t)$, we could use the following 'switching' operations:

(a) switch on the function $f_1(t)$ at $t = 0$;

(b) switch on the function $f_2(t)$ at $t = t_1$ and at the same time switch off the function $f_1(t)$;

(c) switch on the function $f_3(t)$ at $t = t_2$ and at the same time switch off the function $f_2(t)$.

In terms of the unit step function, the function $f(t)$ may thus be expressed as

$$f(t) = f_1(t)H(t) + [f_2(t) - f_1(t)]H(t - t_1) + [f_3(t) - f_2(t)]H(t - t_2)$$

Alternatively, $f(t)$ may be constructed using the **top hat function** $H(t - a) - H(t - b)$. Clearly,

$$H(t-a)-H(t-b) = \begin{cases} 1 & (a \leqslant t < b) \\ 0 & \text{otherwise} \end{cases} \tag{2.42}$$

Which, as illustrated in Figure 2.21, gives

$$f(t)[H(t-a)-H(t-b)] = \begin{cases} f(t) & (a \leqslant t < b) \\ 0 & \text{otherwise} \end{cases}$$

Using this approach, the function $f(t)$ of Figure 2.20 may be expressed as

$$f(t) = f_1(t)[H(t) - H(t - t_1)] + f_2(t)[H(t - t_1) - H(t - t_2)] + f_3(t)H(t - t_2)$$

giving, as before,

$$f(t) = f_1(t)H(t) + [f_2(t) - f_1(t)]H(t - t_1) + [f_3(t) - f_2(t)]H(t - t_2)$$

It is easily checked that this corresponds to the given formulation, since for $0 \leqslant t < t_1$

$$H(t) = 1, \quad H(t - t_1) = H(t - t_2) = 0$$

giving

$$f(t) = f_1(t) \quad (0 \leqslant t < t_1)$$

while for $t_1 \leqslant t < t_2$

$$H(t) = H(t - t_1) = 1, \quad H(t - t_2) = 0$$

giving

$$f(t) = f_1(t) + [f_2(t) - f_1(t)] = f_2(t) \quad (t_1 \leqslant t \leqslant t_2)$$

and finally for $t \geqslant t_2$

$$H(t) = H(t - t_1) = H(t - t_2) = 1$$

giving

$$f(t) = f_1(t) + [f_2(t) - f_1(t)] + [f_3(t) - f_2(t)] = f_3(t) \quad (t \geqslant t_2)$$

EXAMPLE 2.32 Express in terms of unit step functions the piecewise-continuous causal function

$$f(t) = \begin{cases} 2t^2 & (0 \leqslant t < 3) \\ t+4 & (3 \leqslant t < 5) \\ 9 & (t \geqslant 5) \end{cases}$$

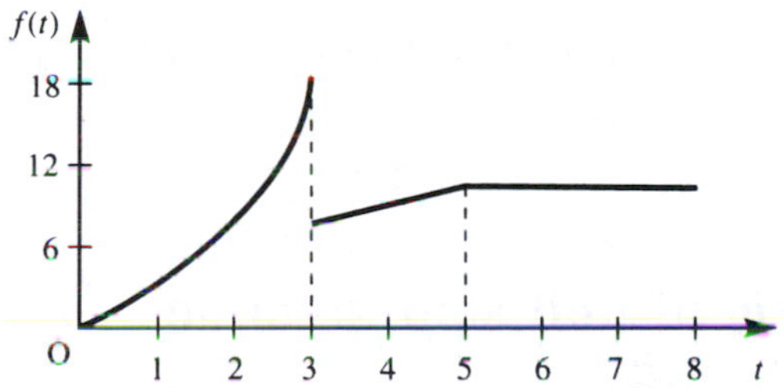

Figure 2.22 Piecewise-continuous function of Example 2.32.

Solution $f(t)$ is depicted graphically in Figure 2.22, and in terms of unit step functions it may be expressed as

$$f(t) = 2t^2H(t) + (t + 4 - 2t^2)H(t - 3) + (9 - t - 4)H(t - 5)$$

That is,

$$f(t) = 2t^2H(t) + (4 + t - 2t^2)H(t - 3) + (5 - t)H(t - 5)$$

EXAMPLE 2.33 Express in terms of unit step functions the piecewise-continuous causal function

$$f(t) = \begin{cases} 0 & (t < 1) \\ 1 & (1 \leqslant t < 3) \\ 3 & (3 \leqslant t < 5) \\ 2 & (5 \leqslant t < 6) \\ 0 & (t \geqslant 6) \end{cases}$$

H(t−1) + 2H(t−3) − H(t−5) +2H(t−6)

Solution $f(t)$ is depicted graphically in Figure 2.23, and in terms of unit step functions it may be expressed as

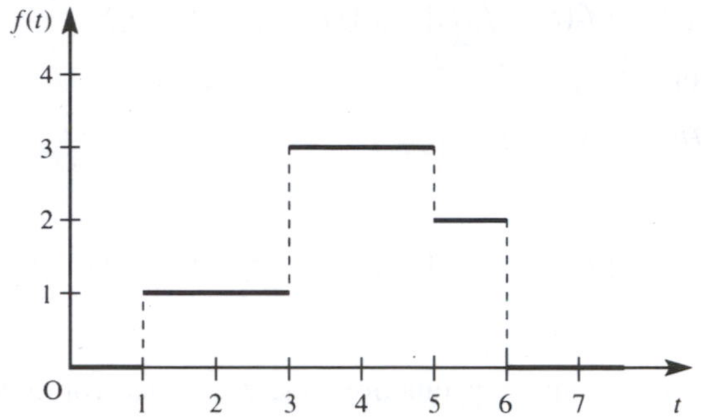

Figure 2.23 Piecewise-continuous function of Example 2.33.

$$f(t) = 1H(t-1) + (3-1)H(t-3) + (2-3)H(t-5) + (0-2)H(t-6)$$

That is,

$$f(t) = 1H(t-1) + 2H(t-3) - 1H(t-5) - 2H(t-6)$$

2.5.2 Laplace transform of unit step function

By definition of the Laplace transform, the transform of $H(t-a)$, $a \geqslant 0$, is given by

$$\begin{aligned}\mathscr{L}\{H(t-a)\} &= \int_0^{\infty} H(t-a)\,\mathrm{e}^{-st}\,\mathrm{d}t \\ &= \int_0^{a} 0\,\mathrm{e}^{-st}\,\mathrm{d}t + \int_a^{\infty} 1\,\mathrm{e}^{-st}\,\mathrm{d}t \\ &= \left[\frac{\mathrm{e}^{-st}}{-s}\right]_a^{\infty} = \frac{\mathrm{e}^{-as}}{s}\end{aligned}$$

That is,

$$\mathscr{L}\{H(t-a)\} = \frac{\mathrm{e}^{-as}}{s} \quad (a \geqslant 0) \tag{2.43}$$

and in the particular case of $a = 0$

$$\mathscr{L}\{H(t)\} = \frac{1}{s} \tag{2.44}$$

EXAMPLE 2.34 Determine the Laplace transform of the rectangular pulse

$$f(t) = \begin{cases} 0 & (t < a) \\ K & (a \leqslant t < b) \\ 0 & (t \geqslant b) \end{cases} \quad K \text{ constant}, \quad b > a > 0$$

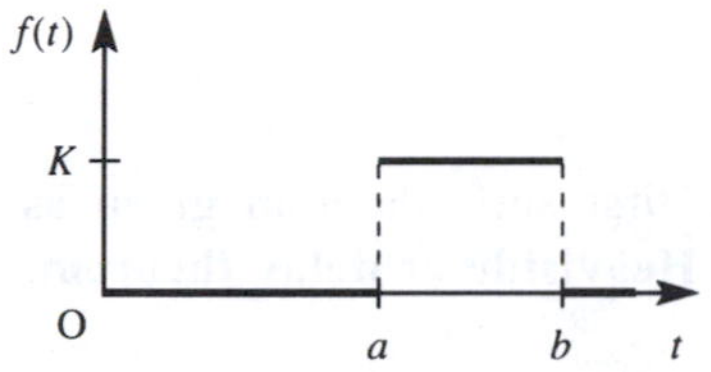

Figure 2.24 Rectangular pulse.

$$\mathcal{L}\{KH(t-a) - KH(t-b)\}$$
$$= \frac{Ke^{-as}}{s} - \frac{Ke^{-bs}}{s}$$
$$= \frac{K}{s}(e^{-as} - e^{-bs}).$$

Solution The pulse is depicted graphically in Figure 2.24. In terms of unit step functions, it may be expressed, using the top hat function, as

$$f(t) = K[H(t-a) - H(t-b)]$$

Then, taking Laplace transforms,

$$\mathcal{L}\{f(t)\} = K\mathcal{L}\{H(t-a)\} - K\mathcal{L}\{H(t-b)\}$$

which, on using the result (2.24), gives

$$\mathcal{L}\{f(t)\} = K\frac{e^{-as}}{s} - K\frac{e^{-bs}}{s}$$

That is,

$$\mathcal{L}\{f(t)\} = \frac{K}{s}(e^{-as} - e^{-bs})$$

EXAMPLE 2.35 Determine the Laplace transform of the piecewise-constant function $f(t)$ shown in Figure 2.23.

Solution From Example 2.33 $f(t)$ may be expressed as

$$f(t) = 1H(t-1) + 2H(t-3) - 1H(t-5) - 2H(t-6)$$

Taking Laplace transforms,

$$\mathcal{L}\{f(t)\} = 1\mathcal{L}\{H(t-1)\} + 2\mathcal{L}\{H(t-3)\} - 1\mathcal{L}\{H(t-5)\} - 2\mathcal{L}\{H(t-6)\}$$

which, on using the result (2.43), gives

$$\mathcal{L}\{f(t)\} = \frac{e^{-s}}{s} + 2\frac{e^{-3s}}{s} - \frac{e^{-5s}}{s} - 2\frac{e^{-6s}}{s}$$

That is,

$$\mathscr{L}\{f(t)\} = \frac{1}{s}(e^{-s} + 2e^{-3s} - e^{-5s} - 2e^{-6s})$$

2.5.3 The second shift theorem

This theorem is dual to the first shift theorem given as Theorem 2.2, and is sometimes referred to as the **Heaviside** or **delay theorem**.

THEOREM 2.4 If $\mathscr{L}\{f(t)\} = F(s)$ then for a positive constant a

$$\mathscr{L}\{f(t-a)H(t-a)\} = e^{-as}F(s)$$

Proof By definition,

$$\mathscr{L}\{f(t-a)H(t-a)\} = \int_0^\infty f(t-a)H(t-a)e^{-st}\,dt$$

$$= \int_a^\infty f(t-a)e^{-st}\,dt$$

Making the substitution $T = t - a$,

$$\mathscr{L}\{f(t-a)H(t-a)\} = \int_0^\infty f(T)e^{-s(T+a)}\,dT$$

$$= e^{-sa}\int_0^\infty f(T)e^{-sT}\,dT$$

Since $F(s) = \mathscr{L}\{f(t)\} = \int_0^\infty f(T)\,e^{-sT}$, it follows that

$$\mathscr{L}\{f(t-a)H(t-a)\} = e^{-as}F(s)$$ □

It is important to distinguish between the two functions $f(t)H(t-a)$ and $f(t-a)H(t-a)$. As we saw earlier, $f(t)H(t-a)$ simply indicates that the function $f(t)$ is 'switched on' at time $t = a$, so that

$$f(t)H(t-a) = \begin{cases} 0 & (t < a) \\ f(t) & (t \geqslant a) \end{cases}$$

On the other hand, $f(t-a)H(t-a)$ represents a translation of the function $f(t)$ by a units to the right (to the right, since $a > 0$), so that

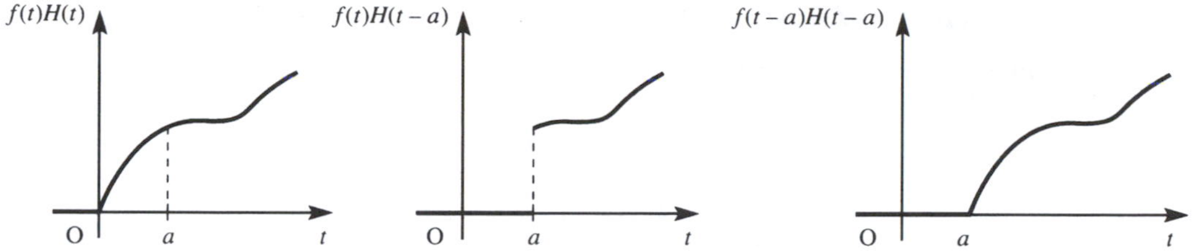

Figure 2.25 Illustration of $f(t-a)H(t-a)$.

$$f(t-a)H(t-a) = \begin{cases} 0 & (t < a) \\ f(t-a) & (t \geqslant a) \end{cases}$$

The difference between the two is illustrated graphically in Figure 2.25. $f(t-a)H(t-a)$ may be interpreted as representing the function $f(t)$ delayed in time by a units. Thus, when considering its Laplace transform $e^{-as}F(s)$, where $F(s)$ denotes the Laplace transform of $f(t)$, the component e^{-as} may be interpreted as a delay operator on the transform $F(s)$, indicating that the response of the system characterized by $F(s)$ will be delayed in time by a units. Since many practically important systems have some form of delay inherent in their behaviour, it is clear that the result of this theorem is very useful.

EXAMPLE 2.36 Determine the Laplace transform of the causal function $f(t)$ defined by

$$f(t) = \begin{cases} t & (0 \leqslant t < b) \\ 0 & (t \geqslant b) \end{cases}$$

$f(t) = tH(t) - tH(t-b)$

$= tH(t) - [(t-b)+b]H(t-b)$

$= \dfrac{1}{s^2} - \dfrac{e^{-bs}}{s^2} - \dfrac{b e^{-bs}}{s}$

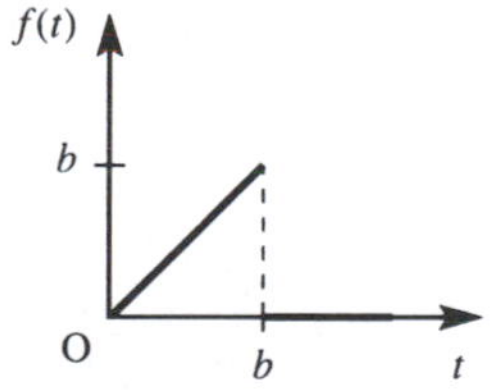

Figure 2.26 Sawtooth pulse.

Solution $f(t)$ is illustrated graphically in Figure 2.26, and is seen to characterize a sawtooth pulse of duration b. In terms of unit step functions,

$$f(t) = tH(t) - tH(t-b)$$

In order to apply the second shift theorem, each term must be rearranged to be of the form $f(t-a)H(t-a)$; that is, the time argument $t-a$ of the function must be the same as that of the associated step function. In this particular example this gives

$$f(t) = tH(t) - (t-b)H(t-b) - bH(t-b)$$

Taking Laplace transforms,

$$\mathscr{L}\{f(t)\} = \mathscr{L}\{tH(t)\} - \mathscr{L}\{(t-b)H(t-b)\} - b\mathscr{L}\{H(t-b)\}$$

which, on using Theorem 2.4, leads to

$$\mathscr{L}\{f(t)\} = \frac{1}{s^2} - e^{-bs}\mathscr{L}(t) - b\frac{e^{-bs}}{s}$$

$$= \frac{1}{s^2} - \frac{e^{-bs}}{s^2} - b\frac{e^{-bs}}{s}$$

giving

$$\mathscr{L}\{f(t)\} = \frac{1}{s^2}(1 - e^{-bs}) - \frac{b}{s}e^{-bs}$$

It should be noted that this result could have been obtained without the use of the second shift theorem, since, directly from the definition of the Laplace transform,

$$\mathscr{L}\{f(t)\} = \int_0^\infty f(t)\,e^{-st}\,dt = \int_0^b t\,e^{-st}\,dt + \int_b^\infty 0\,e^{-st}\,dt$$

$$= \left[-\frac{t\,e^{-st}}{s}\right]_0^b + \int_0^b \frac{e^{-st}}{s}dt$$

$$= \left[-\frac{t\,e^{-st}}{s} - \frac{e^{-st}}{s^2}\right]_0^b$$

$$= \left(-\frac{b\,e^{-sb}}{s} - \frac{e^{-sb}}{s^2}\right) - \left(-\frac{1}{s^2}\right)$$

$$= \frac{1}{s^2}(1 - e^{-bs}) - \frac{b}{s}e^{-bs}$$

as before.

EXAMPLE 2.37 Obtain the Laplace transform of the piecewise-continuous causal function

$$f(t) = \begin{cases} 2t^2 & (0 \leqslant t < 3) \\ t+4 & (3 \leqslant t < 5) \\ 9 & (t \geqslant 5) \end{cases}$$

considered in Example 2.32.

Solution In Example 2.32 we saw that $f(t)$ may be expressed in terms of unit step functions as

$$f(t) = 2t^2H(t) - (2t^2 - t - 4)H(t-3) - (t-5)H(t-5)$$

Before we can find $\mathscr{L}\{f(t)\}$, the function $2t^2 - t - 4$ must be expressed as a function of $t - 3$. This may be readily achieved as follows. Let $z = t - 3$. Then

$$\begin{aligned}2t^2 - t - 4 &= 2(z + 3)^2 - (z + 3) - 4\\ &= 2z^2 + 11z + 11\\ &= 2(t - 3)^2 + 11(t - 3) + 11\end{aligned}$$

Hence

$$f(t) = 2t^2H(t) - [2(t - 3)^2 + 11(t - 3) + 11]H(t - 3) - (t - 5)H(t - 5)$$

Taking Laplace transforms,

$$\begin{aligned}\mathscr{L}\{f(t)\} = 2\mathscr{L}\{t^2H(t)\} &- \mathscr{L}\{[2(t - 3)^2 + 11(t - 3) + 11]H(t - 3)\}\\ &- \mathscr{L}\{(t - 5)H(t - 5)\}\end{aligned}$$

which, on using Theorem 2.4, leads to

$$\begin{aligned}\mathscr{L}\{f(t)\} &= 2\frac{2}{s^3} - e^{-3s}\mathscr{L}\{2t^2 + 11t + 11\} - e^{-5s}\mathscr{L}\{t\}\\ &= \frac{4}{s^3} - e^{-3s}\left(\frac{4}{s^3} + \frac{11}{s^2} + \frac{11}{s}\right) - \frac{e^{-5s}}{s^2}\end{aligned}$$

Again this result could have been obtained directly from the definition of the Laplace transform, but in this case the required integration by parts is a little more tedious.

2.5.4 Inversion using the second shift theorem

We have seen in Examples 2.34 and 2.35 that, to obtain the Laplace transforms of piecewise-continuous functions, use of the second shift theorem could be avoided, since it is possible to obtain such transforms directly from the definition of the Laplace transform.

In practice, the importance of the theorem lies in determining *inverse* transforms, since, as indicated earlier, delays are inherent in most practical systems and engineers are interested in knowing how these influence the system response. Consequently, by far the most useful form of the second shift theorem is

$$\mathscr{L}^{-1}\{e^{-as}F(s)\} = f(t - a)H(t - a) \tag{2.45}$$

Comparing (2.45) with the result (2.12), namely

$$\mathscr{L}^{-1}\{F(s)\} = f(t)H(t)$$

we see that

$$\mathscr{L}^{-1}\{e^{-as}F(s)\} = [f(t)H(t)] \quad \text{with } t \text{ replaced by } t - a$$

indicating that the response $f(t)$ has been delayed in time by a units. This is why the theorem is sometimes called the delay theorem.

EXAMPLE 2.38 Determine $\mathscr{L}^{-1}\left\{\dfrac{4\,e^{-4s}}{s(s+2)}\right\}$.

Solution This may be written as $\mathscr{L}^{-1}\{e^{-4s}F(s)\}$, where

$$F(s) = \frac{4}{s(s+2)}$$

First we obtain the inverse transform $f(t)$ of $F(s)$. Resolving into partial fractions,

$$F(s) = \frac{2}{s} - \frac{2}{s+2}$$

which, on inversion, gives

$$f(t) = 2 - 2e^{-2t}$$

a graph of which is shown in Figure 2.27(a). Then, using (2.45), we have

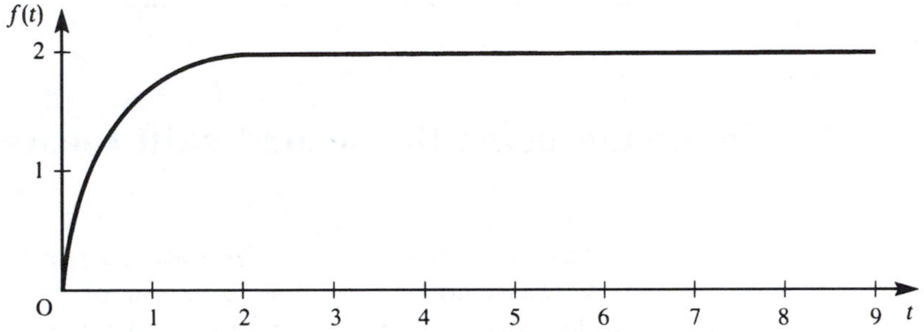

(a) Graph of $f(t)$

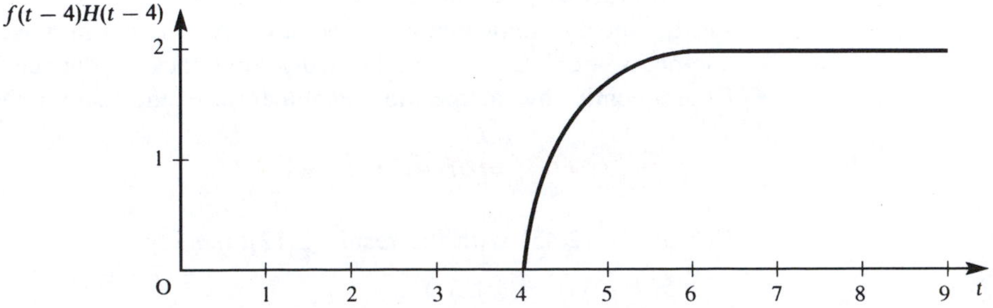

(b) Graph of $f(t-4)H(t-4)$

Figure 2.27 Inverse transforms of Example 2.38.

$$\mathscr{L}^{-1}\left\{e^{-4s}\frac{4}{s(s+2)}\right\} = \mathscr{L}^{-1}\{e^{-4s}F(s)\} = f(t-4)H(t-4)$$

$$= (2 - 2e^{-2(t-4)})H(t-4)$$

giving

$$\mathscr{L}^{-1}\left\{\frac{4e^{-4s}}{s(s+2)}\right\} = \begin{cases} 0 & (t < 4) \\ 2(1 - e^{-2(t-4)}) & (t \geqslant 4) \end{cases}$$

which is plotted in Figure 2.27(b).

EXAMPLE 2.39   Determine $\mathscr{L}^{-1}\left\{\dfrac{e^{-s\pi}(s+3)}{s(s^2+1)}\right\}$.

$$H(t-\pi)\,\mathscr{L}^{-1}\left\{\frac{s+3}{s(s^2+1)}\right\}$$

Solution   This may be written as $\mathscr{L}^{-1}\{e^{-s\pi}F(s)\}$, where

$$F(s) = \frac{s+3}{s(s^2+1)}$$

$$H(t-\pi)\,\mathscr{L}^{-1}\left\{\frac{3}{s} - \frac{3s}{s^2+1} + \frac{1}{s^2+1}\right\}$$

Resolving into partial fractions,

$$= (3 - 3\cos(t-\pi) + \sin(t-\pi))H(t-\pi)$$

$$F(s) = \frac{3}{s} - \frac{3s}{s^2+1} + \frac{1}{s^2+1}$$

$$= (3 + 3\cos t - \sin t)H(t-\pi)$$

which, on inversion, gives

$$f(t) = 3 - 3\cos t + \sin t$$

a graph of which is shown in Figure 2.28(a). Then, using (2.45), we have

$$\mathscr{L}^{-1}\left\{\frac{e^{-s\pi}(s+3)}{s(s^2+1)}\right\} = \mathscr{L}^{-1}\{e^{-s\pi}F(s)\} = f(t-\pi)H(t-\pi)$$

$$= [3 - 3\cos(t-\pi) + \sin(t-\pi)]H(t-\pi)$$

$$= (3 + 3\cos t - \sin t)H(t-\pi)$$

giving

$$\mathscr{L}^{-1}\left\{\frac{e^{-s\pi}(s+3)}{s(s^2+1)}\right\} = \begin{cases} 0 & (t < \pi) \\ 3 + 3\cos t - \sin t & (t \geqslant \pi) \end{cases}$$

which is plotted in Figure 2.28(b).

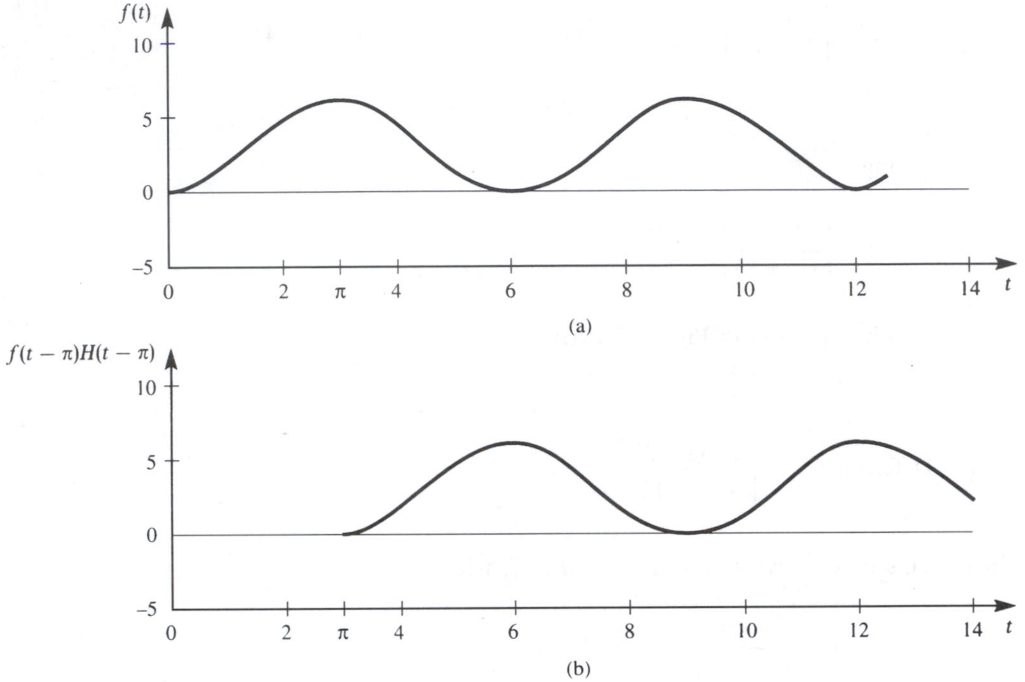

Figure 2.28 Inverse transforms of Example 2.39.

2.5.5 Differential equations

We now return to the solution of linear differential equations for which the forcing function $f(t)$ is piecewise-continuous, like that illustrated in Figure 2.20. One approach to solving a differential equation having such a forcing function is to solve it separately for each of the continuous components $f_1(t)$, $f_2(t)$, and so on, comprising $f(t)$, using the fact that in this equation all the derivatives, except the highest, must remain continuous so that values at the point of discontinuity provide the initial conditions for the next section. This approach is obviously rather tedious, and a much more direct one is to make use of Heaviside step functions to specify $f(t)$. Then the method of solution follows that used in Section 2.3, and we shall simply illustrate it by examples.

EXAMPLE 2.40 Obtain the solution $x(t)$, $t \geqslant 0$, of the differential equation

$$\frac{d^2x}{dt^2} + 5\frac{dx}{dt} + 6x = f(t) \tag{2.46}$$

where $f(t)$ is the pulse function

$$f(t) = \begin{cases} 3 & (0 \leqslant t < 6) \\ 0 & (t \geqslant 6) \end{cases}$$

and subject to the initial conditions $x(0) = 0$ and $\dot{x}(0) = 2$.

Solution To illustrate the advantage of using a step function formulation of the forcing function $f(t)$, we shall first solve separately for each of the time ranges.

Method 1 For $0 \leqslant t < 6$, (2.46) becomes

$$\frac{d^2x}{dt^2} + 5\frac{dx}{dt} + 6x = 3$$

with $x(0) = 0$ and $\dot{x}(0) = 2$.

Taking Laplace transforms gives

$$(s^2 + 5s + 6)X(s) = sx(0) + \dot{x}(0) + 5x(0) + \frac{3}{s} = 2 + \frac{3}{s}$$

That is,

$$X(s) = \frac{2s+3}{s(s+2)(s+3)} = \frac{\frac{1}{2}}{s} + \frac{\frac{1}{2}}{s+2} - \frac{1}{s+3}$$

which, on inversion, gives

$$x(t) = \tfrac{1}{2} + \tfrac{1}{2}e^{-2t} - e^{-3t} \quad (0 \leqslant t < 6)$$

We now determine the values of $x(6)$ and $\dot{x}(6)$ in order to provide the initial conditions for the next stage:

$$x(6) = \tfrac{1}{2} + \tfrac{1}{2}e^{-12} - e^{-18} = \alpha, \quad \dot{x}(6) = -e^{-12} + 3\,e^{-18} = \beta$$

For $t \geqslant 6$ we make the change of independent variable $T = t - 6$, whence (2.46) becomes

$$\frac{d^2x}{dT^2} + 5\frac{dx}{dT} + 6x = 0$$

subject to $x(T = 0) = \alpha$ and $\dot{x}(T = 0) = \beta$.

Taking Laplace transforms gives

$$\begin{aligned}(s^2 + 5s + 6)X(s) &= sx(T = 0) + \dot{x}(T = 0) + 5x(T = 0) \\ &= \alpha s + 5\alpha + \beta\end{aligned}$$

That is,

$$X(s) = \frac{\alpha s + 5\alpha + \beta}{(s+2)(s+3)} = \frac{\beta + 3\alpha}{s+2} - \frac{\beta + 2\alpha}{s+3}$$

which, on taking inverse transforms, gives

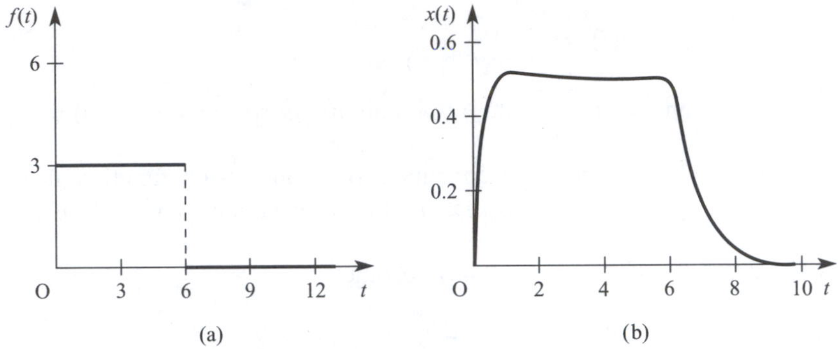

Figure 2.29 Forcing function and response of Example 2.40.

$$x(T) = (\beta + 3\alpha)e^{-2T} - (\beta + 2\alpha)e^{-3T}$$

Substituting the values of α and β and reverting to the independent variable t gives

$$x(t) = (\tfrac{3}{2} + \tfrac{1}{2}e^{-12})\,e^{-2(t-6)} - (1 + e^{-18})\,e^{-3(t-6)} \quad (t \geqslant 6)$$

That is,

$$x(t) = (\tfrac{1}{2}e^{-2t} - e^{-3t}) + (\tfrac{3}{2}e^{-2(t-6)} - e^{-3(t-6)}) \quad (t \geqslant 6)$$

Thus the solution of the differential equation is

$$x(t) = \begin{cases} \frac{1}{2} + \frac{1}{2}e^{-2t} - e^{-3t} & (0 \leqslant t < 6) \\ (\frac{1}{2}e^{-2t} - e^{-3t}) + (\frac{3}{2}e^{-2(t-6)} - e^{-3(t-6)}) & (t \geqslant 6) \end{cases}$$

The forcing function $f(t)$ and response $x(t)$ are shown in Figures 2.29(a) and (b) respectively.

Method 2 In terms of Heaviside step functions,

$$f(t) = 3H(t) - 3H(t - 6)$$

so that, using (2.43),

$$\mathscr{L}\{f(t)\} = \frac{3}{s} - \frac{3}{s}\,e^{-6s}$$

Taking Laplace transforms in (2.46) then gives

$$(s^2 + 5s + 6)X(s) = sx(0) + \dot{x}(0) + 5x(0) + \mathscr{L}\{f(t)\}$$

$$= 2 + \frac{3}{s} - \frac{3}{s}\,e^{-6s}$$

That is,

$$X(s) = \frac{2s+3}{s(s+2)(s+3)} - e^{-6s}\frac{3}{s(s+2)(s+3)}$$

$$= \left(\frac{\frac{1}{2}}{s} + \frac{\frac{1}{2}}{s+2} - \frac{1}{s+3}\right) - e^{-6s}\left(\frac{\frac{1}{2}}{s} - \frac{\frac{3}{2}}{s+2} + \frac{1}{s+3}\right)$$

Taking inverse Laplace transforms and using the result (2.45) gives

$$x(t) = (\tfrac{1}{2} + \tfrac{1}{2}e^{-2t} - e^{-3t}) - (\tfrac{1}{2} - \tfrac{3}{2}e^{-2(t-6)} + e^{-3(t-6)})H(t-6)$$

which is the required solution. This corresponds to that obtained in Method 1, since, using the definition of $H(t-6)$, it may be written as

$$x(t) = \begin{cases} \frac{1}{2} + \frac{1}{2}e^{-2t} - e^{-3t} & (0 \leqslant t < 6) \\ (\frac{1}{2}e^{-2t} - e^{-3t}) + (\frac{3}{2}e^{-2(t-6)} - e^{-3(t-6)}) & (t \geqslant 6) \end{cases}$$

This approach is clearly less tedious, since the initial conditions at the discontinuities are automatically taken account of in the solution.

EXAMPLE 2.41 Determine the solution $x(t)$, $t \geqslant 0$, of the differential equation

$$\frac{d^2x}{dt^2} + 2\frac{dx}{dt} + 5x = f(t) \qquad \textbf{(2.47)}$$

where

$$f(t) = \begin{cases} t & (0 \leqslant t \leqslant \pi) \\ 0 & (t \geqslant \pi) \end{cases}$$

and subject to the initial conditions $x(0) = 0$ and $\dot{x}(0) = 3$.

Solution Following the procedures of Example 2.36, we have

$$f(t) = tH(t) - tH(t-\pi)$$
$$= tH(t) - (t-\pi)H(t-\pi) - \pi H(t-\pi)$$

so that, using Theorem 2.4,

$$\mathscr{L}\{f(t)\} = \frac{1}{s^2} - \frac{e^{-\pi s}}{s^2} - \frac{\pi e^{-\pi s}}{s}$$

$$= \frac{1}{s^2} - e^{-\pi s}\left(\frac{1}{s^2} + \frac{\pi}{s}\right)$$

Taking Laplace transforms in (2.47) then gives

$$(s^2 + 2s + 5)X(s) = sx(0) + \dot{x}(0) + 2x(0) + \mathscr{L}\{f(t)\}$$

$$= 3 + \frac{1}{s^2} - e^{-\pi s}\left(\frac{1}{s^2} + \frac{\pi}{s}\right)$$

using the given initial conditions.

Thus

$$X(s) = \frac{3s^2+1}{s^2(s^2+2s+5)} - e^{-\pi s}\frac{1+s\pi}{s^2(s^2+2s+5)}$$

which, on resolving into partial fractions, leads to

$$X(s) = \tfrac{1}{25}\left[-\frac{2}{s}+\frac{5}{s^2}+\frac{2s+74}{(s+1)^2+4}\right] - \frac{e^{-\pi s}}{25}\left[\frac{5\pi-2}{s}+\frac{5}{s^2}-\frac{(5\pi-2)s+(10\pi+1)}{(s+1)^2+4}\right]$$

$$= \tfrac{1}{25}\left[-\frac{2}{s}+\frac{5}{s^2}+\frac{2(s+1)+72}{(s+1)^2+4}\right]$$

$$-\frac{e^{-\pi s}}{25}\left[\frac{5\pi-2}{s}+\frac{5}{s^2}-\frac{(5\pi-2)(s+1)+(5\pi+3)}{(s+1)^2+4}\right]$$

Taking inverse Laplace transforms and using (2.45) gives the desired solution:

$$x(t) = \tfrac{1}{25}(-2+5t+2\,e^{-t}\cos 2t+36\,e^{-t}\sin 2t)$$

$$-\tfrac{1}{25}[(5\pi-2)+5(t-\pi)-(5\pi-2)\,e^{-(t-\pi)}\cos 2(t-\pi)$$

$$-\tfrac{1}{2}(5\pi+3)\,e^{-(t-\pi)}\sin 2(t-\pi)]H(t-\pi)$$

That is,

$$x(t) = \tfrac{1}{25}[5t-2+2\,e^{-t}(\cos 2t+18\sin 2t)]$$

$$-\tfrac{1}{25}\{5t-2-e^{\pi}\,e^{-t}[(5\pi-2)\cos 2t+\tfrac{1}{2}(5\pi+3)\sin 2t]\}H(t-\pi)$$

or, in alternative form,

$$x(t) = \begin{cases} \tfrac{1}{25}[5t-2+2\,e^{-t}(\cos 2t+18\sin 2t)] & (0 \leqslant t < \pi) \\ \tfrac{1}{25}e^{-t}\{(2+(5\pi-2)\,e^{\pi})\cos 2t+[36+\tfrac{1}{2}(5\pi+3)\,e^{\pi}]\sin 2t\} & (t \geqslant \pi) \end{cases}$$

2.5.6 Periodic functions

We have already determined the Laplace transforms of periodic functions, such as $\sin \omega t$ and $\cos \omega t$, which are smooth (differentiable) continuous functions. In many engineering applications, however, one frequently encounters periodic functions that exhibit discontinuous behaviour. Examples of typical periodic functions of practical importance are shown in Figure 2.30.

Such periodic functions may be represented as infinite series of terms involving step functions; once expressed in such a form, the result (2.43) may then be used to obtain their Laplace transforms.

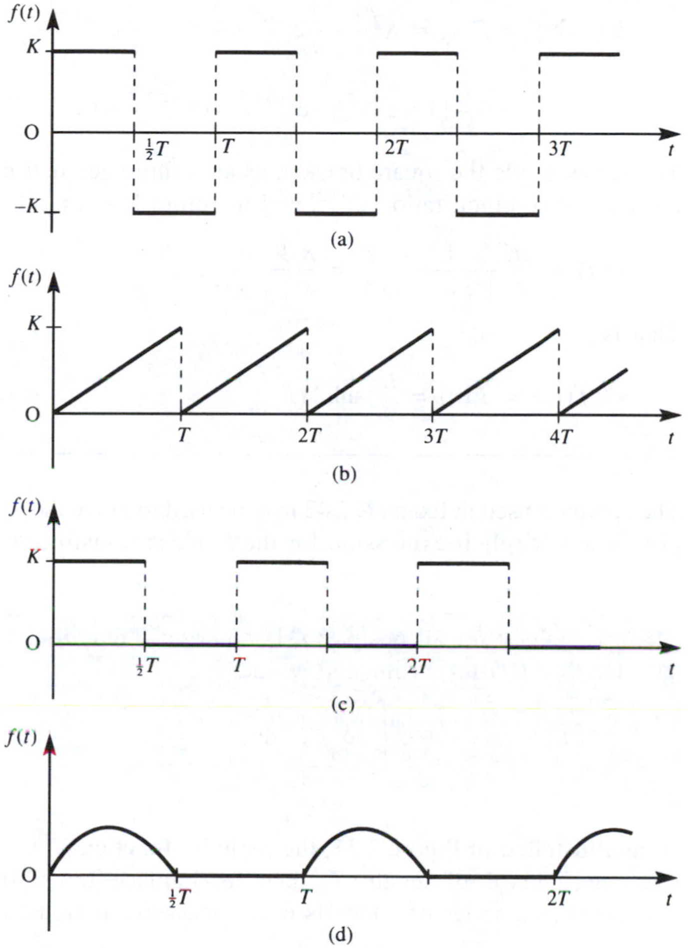

Figure 2.30 Typical practically important periodic functions: (a) square wave; (b) sawtooth wave; (c) repeated pulse wave; (d) half-wave rectifier.

EXAMPLE 2.42 Obtain the Laplace transform of the square wave illustrated in Figure 2.30(a).

Solution In terms of step functions, the square wave may be expressed in the form

$$\begin{aligned} f(t) &= KH(t) - 2KH(t - \tfrac{1}{2}T) + 2KH(t - T) - 2KH(t - \tfrac{3}{2}T) \\ &\quad + 2KH(t - 2T) + \ldots \\ &= K[H(t) - 2H(t - \tfrac{1}{2}T) + 2H(t - T) - 2H(t - \tfrac{3}{2}T) + 2H(t - 2T) + \ldots] \end{aligned}$$

Taking Laplace transforms and using the result (2.43) gives

$$\mathscr{L}\{f(t)\} = F(s) = K\left(\frac{1}{s} - \frac{2}{s}\mathrm{e}^{-sT/2} + \frac{2}{s}\mathrm{e}^{-sT} - \frac{2}{s}\mathrm{e}^{-3sT/2} + \frac{2}{s}\mathrm{e}^{-2sT} + \dots\right)$$

$$= \frac{2K}{s}[1 - \mathrm{e}^{-sT/2} + (\mathrm{e}^{-sT/2})^2 - (\mathrm{e}^{-sT/2})^3 + (\mathrm{e}^{-sT/2})^4 - \dots] - \frac{K}{s}$$

The series inside the square brackets is an infinite geometric progression with first term 1 and common ratio $-\mathrm{e}^{-sT/2}$, and therefore has sum $(1 + \mathrm{e}^{-sT/2})^{-1}$. Thus,

$$F(s) = \frac{2K}{s}\frac{1}{1+\mathrm{e}^{-sT/2}} - \frac{K}{s} = \frac{K}{s}\frac{1-\mathrm{e}^{-sT/2}}{1+\mathrm{e}^{-sT/2}}$$

That is,

$$\mathscr{L}\{f(t)\} = F(s) = \frac{K}{s}\tanh\tfrac{1}{4}sT$$

The approach used in Example 2.42 may be used to prove the following theorem, which provides an explicit expression for the Laplace transform of a periodic function.

THEOREM 2.5

If $f(t)$, defined for all positive t, is a periodic function with period T, that is, $f(t + nT) = f(t)$ for all integers n, then

$$\mathscr{L}\{f(t)\} = \frac{1}{1-\mathrm{e}^{-sT}}\int_0^T \mathrm{e}^{-st} f(t)\,\mathrm{d}t$$

Proof If, as illustrated in Figure 2.31, the periodic function $f(t)$ is piecewise-continuous over an interval of length T, then its Laplace transform exists and can be expressed as a series of integrals over successive periods; that is,

$$\mathscr{L}\{f(t)\} = \int_0^\infty f(t)\,\mathrm{e}^{-st}\,\mathrm{d}t$$

$$= \int_0^T f(t)\,\mathrm{e}^{-st}\,\mathrm{d}t + \int_T^{2T} f(t)\,\mathrm{e}^{-st}\,\mathrm{d}t + \int_{2T}^{3T} f(t)\,\mathrm{e}^{-st}\,\mathrm{d}t + \dots$$

$$+ \int_{(n-1)T}^{nT} f(t)\,\mathrm{e}^{-st}\,\mathrm{d}t + \dots$$

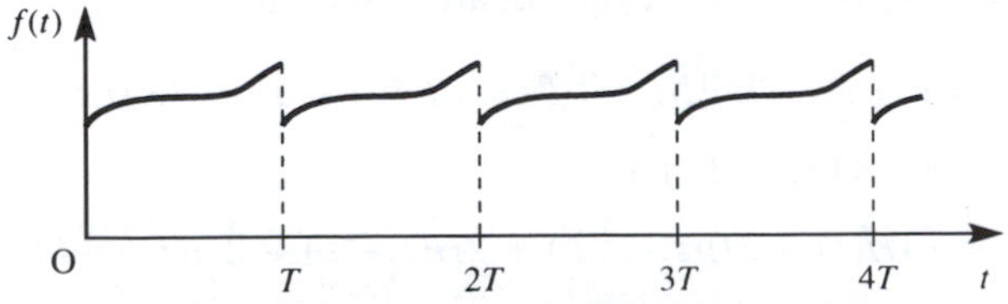

Figure 2.31 Periodic function having period T.

If in successive integrals we make the substitutions

$$t = \tau + nT \quad (n = 0, 1, 2, 3, \ldots)$$

then

$$\mathscr{L}\{f(t)\} = \sum_{n=0}^{\infty} \int_0^T f(\tau + nT)\,\mathrm{e}^{-s(\tau+nT)}\,\mathrm{d}\tau$$

Since $f(t)$ is periodic with period T,

$$f(\tau + nT) = f(t) \quad (n = 0, 1, 2, 3, \ldots)$$

so that

$$\mathscr{L}\{f(t)\} = \sum_{n=0}^{\infty} \int_0^T f(\tau)\,\mathrm{e}^{-s\tau}\,\mathrm{e}^{-snT}\,\mathrm{d}\tau = \left(\sum_{n=0}^{\infty} \mathrm{e}^{-snT}\right) \int_0^T f(\tau)\,\mathrm{e}^{-s\tau}\,\mathrm{d}\tau$$

The series $\sum_{n=0}^{\infty} \mathrm{e}^{-snT} = 1 + \mathrm{e}^{-sT} + \mathrm{e}^{-2sT} + \mathrm{e}^{-3sT} + \ldots$ is an infinite geometric progression with first term 1 and common ratio e^{-sT}. Its sum is given by $(1 - \mathrm{e}^{-sT})^{-1}$, so that

$$\mathscr{L}\{f(t)\} = \frac{1}{1 - \mathrm{e}^{-sT}} \int_0^T f(\tau)\,\mathrm{e}^{-s\tau}\,\mathrm{d}\tau$$

Since, within the integral, τ is a 'dummy' variable, it may be replaced by t to give the desired result. □

We note that, in terms of the Heaviside step function, Theorem 2.5 may be stated as follows:

If $f(t)$, defined for all positive t, is a periodic function with period T and

$$f_1(t) = f(t)(H(t) - H(t - T))$$

then

$$\mathscr{L}\{f(t)\} = (1 - \mathrm{e}^{-sT})^{-1}\mathscr{L}\{f_1(t)\}$$

This formulation follows since $f(t)$ is periodic and $f_1(t) = 0$ for $t > T$. For the periodic function $f(t)$ shown in Figure 2.31 the corresponding function $f_1(t)$ is shown in Figure 2.32. We shall see from the following examples that this formulation simplifies the process of obtaining Laplace transforms of periodic functions.

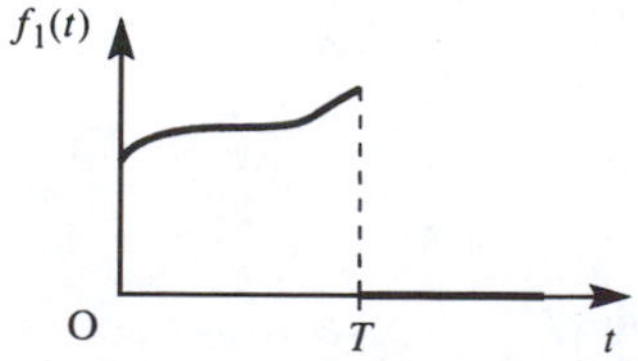

Figure 2.32 Plot of periodic function within one period.

EXAMPLE 2.43 Confirm the result obtained in Example 2.42 using Theorem 2.5.

Solution For the square wave $f(t)$ illustrated in Figure 2.30(a), $f(t)$ is defined over the period $0 < t < T$ by

$$f(t) = \begin{cases} K & (0 < t < \frac{1}{2}T) \\ -K & (\frac{1}{2}T < t < T) \end{cases}$$

Hence we can write $f_1(t) = K[H(t) - 2H(t - \frac{1}{2}T) + H(t - T)]$, and thus

$$\mathscr{L}\{f_1(t)\} = K\left(\frac{1}{s} - \frac{2}{s}e^{-sT/2} + \frac{1}{s}e^{-sT}\right) = \frac{K}{s}(1 - e^{-sT/2})^2$$

Using the result of Theorem 2.5,

$$\mathscr{L}\{f(t)\} = \frac{K(1 - e^{-sT/2})^2}{s(1 - e^{-sT})} = \frac{K(1 - e^{-sT/2})^2}{s(1 - e^{-sT/2})(1 + e^{-sT/2})}$$

$$= \frac{K}{s}\frac{1 - e^{-sT/2}}{1 + e^{-sT/2}} = \frac{K}{s}\tanh\frac{1}{4}sT$$

confirming the result obtained in Example 2.42.

EXAMPLE 2.44 Determine the Laplace transform of the rectified half-wave defined by

$$f(t) = \begin{cases} \sin \omega t & (0 < t < \pi/\omega) \\ 0 & (\pi/\omega < t < 2\pi/\omega) \end{cases}$$

$$f(t + 2n\pi/\omega) = f(t) \quad \text{for all integers } n$$

Solution $f(t)$ is illustrated in Figure 2.30(d), with $T = 2\pi/\omega$. We can express $f_1(t)$ as

$$f_1(t) = \sin \omega t[H(t) - H(t - \pi/\omega)]$$
$$= \sin \omega t H(t) + \sin \omega(t - \pi/\omega)H(t - \pi/\omega)$$

So

$$\mathscr{L}\{f_1(t)\} = \frac{\omega}{s^2 + \omega^2} + e^{-s\pi/\omega}\frac{\omega}{s^2 + \omega^2} = \frac{\omega}{s^2 + \omega^2}(1 + e^{-s\pi/\omega})$$

Then, by the result of Theorem 2.5,

$$\mathscr{L}\{f(t)\} = \frac{\omega}{s^2 + \omega^2}\frac{1 + e^{-s\pi/\omega}}{1 - e^{-2s\pi/\omega}}$$

$$= \frac{\omega}{(s^2 + \omega^2)(1 - e^{-s\pi/\omega})}$$

2.5.7 Exercises

13 A function $f(t)$ is defined by

$$f(t) = \begin{cases} t & (0 \leqslant t \leqslant 1) \\ 0 & (t > 1) \end{cases}$$

Express $f(t)$ in terms of Heaviside unit step functions and show that

$$\mathscr{L}\{f(t)\} = \frac{1}{s^2}(1 - e^{-s}) - \frac{1}{s}e^{-s}$$

14 Express in terms of Heaviside unit step functions the following piecewise-continuous causal functions. In each case obtain the Laplace transform of the function.

(a) $f(t) = \begin{cases} 3t^2 & (0 < t \leqslant 4) \\ 2t - 3 & (4 < t < 6) \\ 5 & (t > 6) \end{cases}$

(b) $g(t) = \begin{cases} t & (0 \leqslant t < 1) \\ 2 - t & (1 < t < 2) \\ 0 & (t > 2) \end{cases}$

15 Obtain the inverse Laplace transforms of the following:

(a) $\dfrac{e^{-5s}}{(s-2)^4}$ (b) $\dfrac{3\,e^{-2s}}{(s+3)(s+1)}$

(c) $\dfrac{s+1}{s^2(s^2+1)}e^{-s}$ (d) $\dfrac{s+1}{s^2+s+1}e^{-\pi s}$

(e) $\dfrac{s}{s^2+25}e^{-4\pi s/5}$ (f) $\dfrac{e^{-s}(1-e^{-s})}{s^2(s^2+1)}$

16 Given that $x = 0$ when $t = 0$, obtain the solution of the differential equation

$$\frac{dx}{dt} + x = f(t) \quad (t \geqslant 0)$$

where $f(t)$ is the function defined in Exercise 13. Sketch a graph of the solution.

17 Given that $x = 1$ and $dx/dt = 0$, obtain the solution of the differential equation

$$\frac{d^2x}{dt^2} + \frac{dx}{dt} + x = g(t) \quad (t \geqslant 0)$$

where $g(t)$ is the piecewise-continuous function defined in Exercise 14(b).

18 Show that the function

$$f(t) = \begin{cases} 0 & (0 \leqslant t < \frac{1}{2}\pi) \\ \sin t & (t \geqslant \frac{1}{2}\pi) \end{cases}$$

may be expressed in the form $f(t) = \cos(t - \frac{1}{2}\pi) H(t - \frac{1}{2}\pi)$, where $H(t)$ is the Heaviside unit step function. Hence solve the differential equation

$$\frac{d^2x}{dt^2} + 3\frac{dx}{dt} + 2x = f(t)$$

where $f(t)$ is given above, and $x = 1$ and $dx/dt = -1$ when $t = 0$.

19 Express the function

$$f(t) = \begin{cases} 3 & (0 \leqslant t < 4) \\ 2t - 5 & (t \geqslant 4) \end{cases}$$

in terms of Heaviside unit step functions and obtain its Laplace transform. Obtain the response of the harmonic oscillator

$$\ddot{x} + x = f(t)$$

to such a forcing function, given that $x = 1$ and $dx/dt = 0$ when $t = 0$.

20 The response $\theta_o(t)$ of a system to a forcing function $\theta_i(t)$ is determined by the second-order differential equation

$$\ddot{\theta}_o + 6\dot{\theta}_o + 10\theta_o = \theta_i \quad (t \geqslant 0)$$

Suppose that $\theta_i(t)$ is a constant stimulus applied for a limited period and characterized by

$$\theta_i(t) = \begin{cases} 3 & (0 \leqslant t < a) \\ 0 & (t \geqslant a) \end{cases}$$

determine the response of the system at time t given that the system was initially in a quiescent state. Show that the transient response at time T $(> a)$ is

$$-\tfrac{3}{10}e^{-3T}\{\cos T + 3\sin T - e^{3a}[\cos(T-a) + 3\sin(T-a)]\}$$

21 The input $\theta_i(t)$ and output $\theta_o(t)$ of a servomechanism are related by the differential equation

$$\ddot{\theta}_o + 8\dot{\theta}_o + 16\theta_o = \theta_i \quad (t \geqslant 0)$$

and initially $\theta_o(0) = \dot{\theta}_o(0) = 0$. For $\theta_i = f(t)$, where

$$f(t) = \begin{cases} 1 - t & (0 < t < 1) \\ 0 & (t > 1) \end{cases}$$

show that

$$\mathscr{L}\{\theta_i(t)\} = \frac{s-1}{s^2} + \frac{1}{s^2}e^{-s}$$

and hence obtain an expression for the response of the system at time t.

22 During the time interval t_1 to t_2, a constant electromotive force e_0 acts on the series RC circuit shown in Figure 2.33. Assuming that the circuit is initially in a quiescent state, show that the current in the circuit at time t is

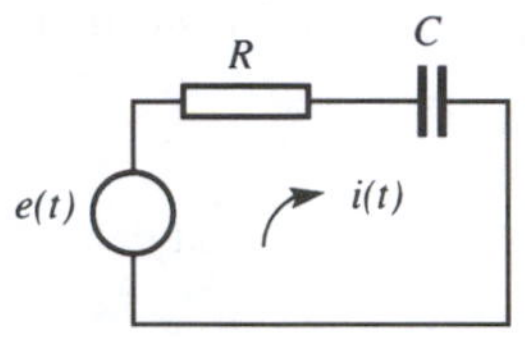

Figure 2.33 Circuit of Exercise 22.

$$i(t) = \frac{e_0}{R}[e^{-(t-t_1)/RC}H(t-t_1) - e^{-(t-t_2)/RC}H(t-t_2)]$$

Sketch this as a function of time.

23 A periodic function $f(t)$, with period 4 units, is defined within the interval $0 \leqslant t < 4$ by

$$f(t) = \begin{cases} 3t & (0 \leqslant t < 2) \\ 6 & (2 \leqslant t < 4) \end{cases}$$

Sketch a graph of the function for $0 \leqslant t < 12$ and obtain its Laplace transform.

24 Obtain the Laplace transform of the periodic sawtooth wave with period T, illustrated in Figure 2.30(b).

2.5.8 The impulse function

In many engineering applications we are interested in seeking the responses of systems to forcing functions that are applied suddenly but only for a very short time. These functions are known as **impulsive forces**. Mathematically, such forcing functions are idealized by the **impulse function**, which is a function whose total value is concentrated at one point. To develop a mathematical formulation of the impulse function and obtain some insight into its physical interpretation, consider the pulse function $\phi(t)$ defined by

$$\phi(t) = \begin{cases} 0 & (0 < t < a - \frac{1}{2}T) \\ A/T & (a - \frac{1}{2}T \leqslant t < a + \frac{1}{2}T) \\ 0 & (t \geqslant a + \frac{1}{2}T) \end{cases}$$

and illustrated in Figure 2.34(a). Since the height of the pulse is A/T and its duration (or width) is T, the area under the pulse is A; that is,

$$\int_{-\infty}^{\infty} \phi(t)\mathrm{d}t = \int_{a-T/2}^{a+T/2} \frac{A}{T}\,\mathrm{d}t = A$$

If we now consider the limiting process in which the duration of the pulse approaches zero, in such a way that the area under the pulse remains A, then we obtain a formulation of the impulse function of magnitude A occurring at time $t = a$. It is important to appreciate that the magnitude of the impulse function is measured by its area.

The impulse function whose magnitude is unity is called the **unit impulse function** or **Dirac delta function**. The unit impulse occurring at $t = a$ is the limiting case of the pulse $\phi(t)$ of Figure 2.34(a) with A having the value unity. It is denoted by $\delta(t - a)$ and has the properties

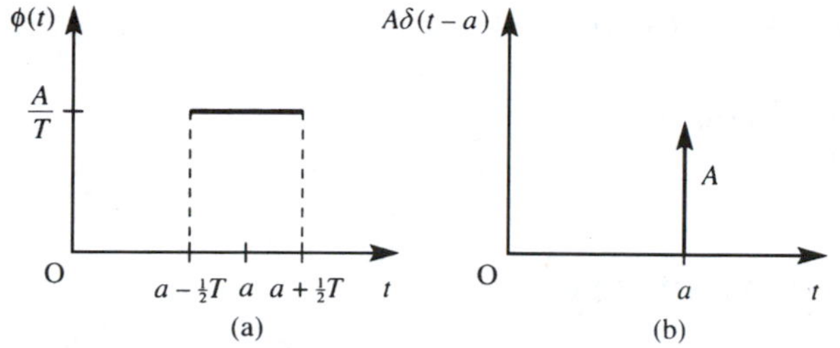

Figure 2.34 Impulse function.

$$\delta(t-a) = 0 \quad (t \neq a)$$

$$\int_{-\infty}^{\infty} \delta(t-a)\,\mathrm{d}t = 1$$

Likewise, an impulse function of magnitude A occurring at $t = a$ is denoted by $A\delta(t - a)$ and may be represented diagrammatically as in Figure 2.34(b).

An impulse function is not a function in the usual sense, but is an example of a class of what are called **generalized functions**, which may be analysed using the theory of **generalized calculus**. (It may also be regarded mathematically as a **distribution** and investigated using the **theory of distributions**.) However, its properties are such that, used with care, it can lead to results that have physical or practical significance and which in many cases cannot be obtained by any other method. In this context it provides engineers with an important mathematical tool. Although, clearly, an impulse function is not physically realizable, it follows from the above formulation that physical signals can be produced that closely approximate it.

We noted that the magnitude of the impulse function is determined by the area under the limiting pulse. The actual shape of the limiting pulse is not really important, provided that the area contained within it remains constant as its duration approaches zero. Physically, therefore, the unit impulse function at $t = a$ may equally well be regarded as the pulse $\phi_1(t)$ of Figure 2.35 in the limiting case as T approaches zero.

In some applications we need to consider a unit impulse function at time $t = 0$. This is denoted by $\delta(t)$ and is defined as the limiting case of the pulse $\phi_2(t)$ illustrated in Figure 2.36 as T approaches zero. It has the properties

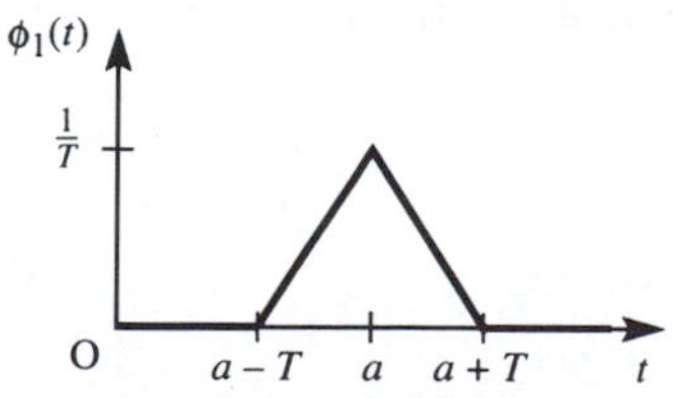

Figure 2.35 Approximation to a unit pulse.

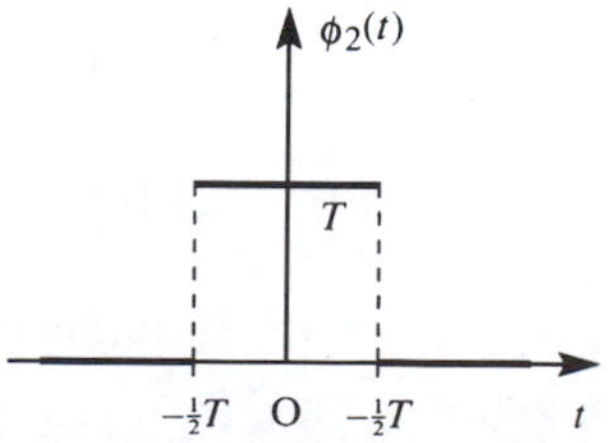

Figure 2.36 Pulse at the origin.

$$\delta(t) = 0 \quad (t \neq 0)$$

$$\int_{-\infty}^{\infty} \delta(t)\,dt = 1$$

2.5.9 The sifting property

An important property of the unit impulse function that is of practical significance is the so-called **sifting property**, which states that if $f(t)$ is continuous at $t = a$ then

$$\int_{-\infty}^{\infty} f(t)\delta(t-a)\,dt = f(a) \tag{2.48}$$

This is referred to as the sifting property because it provides a method of isolating, or sifting out, the value of a function at any particular point.

For theoretical reasons it is convenient to use infinite limits in (2.48), while in reality finite limits can be substituted. This follows since for $\alpha < a < \beta$, where α and β are constants,

$$\int_{\alpha}^{\beta} f(t)\delta(t-a)\,dt = f(a) \tag{2.49}$$

For example,

$$\int_{0}^{2\pi} \cos t\,\delta(t-\tfrac{1}{3}\pi)\,dt = \cos\tfrac{1}{3}\pi = \tfrac{1}{2}$$

2.5.10 Laplace transforms of impulse functions

By the definition of the Laplace transform, we have for any $a > 0$

$$\mathscr{L}\{\delta(t-a)\} = \int_{0}^{\infty} \delta(t-a)\,e^{-st}\,dt$$

which, using the sifting property, gives the important result

$$\mathscr{L}\{\delta(t-a)\} = e^{-as} \tag{2.50}$$

or, in terms of the inverse transform,

$$\mathscr{L}^{-1}\{e^{-as}\} = \delta(t-a) \tag{2.51}$$

As mentioned earlier, in many applications we may have an impulse function $\delta(t)$ at $t = 0$, and it is in order to handle such a function that we must carefully specify whether the lower limit in the Laplace integral defined in Section 2.2.1 is 0− or 0+. Adopting the notation

$$\mathscr{L}_{+}\{f(t)\} = \int_{0+}^{\infty} f(t)\,\mathrm{e}^{-st}\,\mathrm{d}t$$

$$\mathscr{L}_{-}\{f(t)\} = \int_{0-}^{\infty} f(t)\,\mathrm{e}^{-st}\,\mathrm{d}t$$

we have

$$\mathscr{L}_{-}\{f(t)\} = \int_{0-}^{0+} f(t)\,\mathrm{e}^{-st}\,\mathrm{d}t + \int_{0+}^{\infty} f(t)\,\mathrm{e}^{-st}\,\mathrm{d}t$$

If $f(t)$ does not involve an impulse function at $t = 0$ then clearly $\mathscr{L}_{+}\{f(t)\} = \mathscr{L}_{-}\{f(t)\}$. However, if $f(t)$ does involve an impulse function at $t = 0$ then

$$\int_{0-}^{0+} f(t)\,\mathrm{d}t \neq 0$$

and it follows that

$$\mathscr{L}_{+}\{f(t)\} \neq \mathscr{L}_{-}\{f(t)\}$$

In Section 2.2.1 we adopted the definition

$$\mathscr{L}\{f(t)\} = \mathscr{L}_{-}\{f(t)\}$$

so that (2.50) and (2.51) hold for $a = 0$, giving

$$\mathscr{L}\{\delta(t)\} = \int_{0-}^{\infty} \delta(t)\,\mathrm{e}^{-st}\,\mathrm{d}t = \mathrm{e}^{-s0} = 1$$

so that

$$\mathscr{L}\{\delta(t)\} = 1 \qquad \textbf{(2.52)}$$

or, in inverse form,

$$\mathscr{L}^{-1}\{1\} = \delta(t) \qquad \textbf{(2.53)}$$

EXAMPLE 2.45 Determine $\mathscr{L}^{-1}\left\{\dfrac{s^2}{s^2+4}\right\}$.

Solution Since

$$\frac{s^2}{s^2+4} = \frac{s^2+4-4}{s^2+4} = 1 - \frac{4}{s^2+4}$$

we have

$$\mathscr{L}^{-1}\left\{\frac{s^2}{s^2+4}\right\} = \mathscr{L}^{-1}\{1\} - \mathscr{L}^{-1}\left\{\frac{4}{s^2+4}\right\}$$

giving

$$\mathscr{L}^{-1}\left\{\frac{s^2}{s^2+4}\right\} = \delta(t) - 2\sin 2t$$

EXAMPLE 2.46 Determine the solution of the differential equation

$$\frac{d^2x}{dt^2} + 3\frac{dx}{dt} + 2x = 1 + \delta(t-4) \qquad \textbf{(2.54)}$$

subject to the initial conditions $x(0) = \dot{x}(0) = 0$.

Solution Taking Laplace transforms in (2.54) gives

$$[s^2X(s) - sx(0) - \dot{x}(0)] + 3[sX(s) - x(0)] + 2X(s) = \mathscr{L}\{1\} + \mathscr{L}\{\delta(t-4)\}$$

which, on incorporating the given initial conditions and using (2.50), leads to

$$(s^2 + 3s + 2)X(s) = \frac{1}{s} + e^{-4s}$$

giving

$$X(s) = \frac{1}{s(s+2)(s+1)} + e^{-4s}\frac{1}{(s+2)(s+1)}$$

Resolving into partial fractions, we have

$$X(s) = \tfrac{1}{2}\left(\frac{1}{s} + \frac{1}{s+2} - \frac{2}{s+1}\right) + e^{-4s}\left(\frac{1}{s+1} - \frac{1}{s+2}\right)$$

which, on taking inverse transforms and using the result (2.45), gives the required response:

$$x(t) = \tfrac{1}{2}(1 + e^{-2t} - 2\,e^{-t}) + (e^{-(t-4)} - e^{-2(t-4)})H(t-4)$$

or, in an alternative form,

$$x(t) = \begin{cases} \tfrac{1}{2}(1 + e^{-2t} - 2\,e^{-t}) & (0 \leqslant t < 4) \\ \tfrac{1}{2} + (e^4 - 1)\,e^{-t} - (e^8 - \tfrac{1}{2})\,e^{-2t} & (t \geqslant 4) \end{cases}$$

We note that, although the response $x(t)$ is continuous at $t = 4$, the consequence of the impulsive input at $t = 4$ is a step change in the derivative $\dot{x}(t)$.

2.5.11 Relationship between Heaviside step and impulse functions

From the definitions of $H(t)$ and $\delta(t)$, it can be argued that

$$H(t) = \int_{-\infty}^{t} \delta(\tau)\,d\tau \tag{2.55}$$

since the interval of integration contains zero if $t > 0$ but not if $t < 0$. Conversely, (2.55) may be written as

$$\delta(t) = \frac{d}{dt}H(t) = H'(t) \tag{2.56}$$

which expresses the fact that $H'(t)$ is zero everywhere except at $t = 0$, when the jump in $H(t)$ occurs.

While this argument may suffice in practice, since we are dealing with generalized functions a more formal proof requires the development of some properties of generalized functions. In particular, we need to define what is meant by saying that two generalized functions are equivalent.

One method of approach is to use the concept of a **test function** $\theta(t)$, which is a continuous function that has continuous derivatives of all orders and that is zero outside a finite interval. One class of testing function, adopted by R. R. Gabel and R. A. Roberts (*Signals and Linear Systems*, Wiley, New York, 1973), is

$$\theta(t) = \begin{cases} e^{-d^2/(d^2-t^2)} & (|t| < d), \quad \text{where } d = \text{constant} \\ 0 & \text{otherwise} \end{cases}$$

For a generalized function $g(t)$ the integral

$$G(\theta) = \int_{-\infty}^{\infty} \theta(t)g(t)\,dt$$

is evaluated. This integral assigns the number $G(\theta)$ to each function $\theta(t)$, so that $G(\theta)$ is a generalization of the concept of a function: it is a **linear functional** on the space of test functions $\theta(t)$. For example, if $g(t) = \delta(t)$ then

$$G(\theta) = \int_{-\infty}^{\infty} \theta(t)\delta(t)\,dt = \theta(0)$$

so that in this particular case, for each weighting function $\theta(t)$, the value $\theta(0)$ is assigned to $G(\theta)$.

We can now use the concept of a test function to define what is meant by saying that two generalized functions are equivalent or 'equal'.

DEFINITION 2.2: The equivalence property

If $g_1(t)$ and $g_2(t)$ are two generalized functions then $g_1(t) = g_2(t)$ if and only if

$$\int_{-\infty}^{\infty} \theta(t)g_1(t)\,\mathrm{d}t = \int_{-\infty}^{\infty} \theta(t)g_2(t)\,\mathrm{d}t$$

for all test functions $\theta(t)$ for which the integrals exist.

The test function may be regarded as a 'device' for examining the generalized function. Gabel and Roberts draw a rough parallel with the role of using the output of a measuring instrument to deduce properties about what is being measured. In such an analogy $g_1(t) = g_2(t)$ if the measuring instrument can detect no differences between them.

Using the concept of a test function $\theta(t)$, the Dirac delta function $\delta(t)$ may be defined in the generalized form

$$\int_{-\infty}^{\infty} \theta(t)\delta(t)\mathrm{d}t = \theta(0)$$

Interpreted as an ordinary integral, this has no meaning. The integral and the function $\delta(t)$ are merely defined by the number $\theta(0)$. In this sense we can handle $\delta(t)$ as if it were an ordinary function, except that we never talk about the value of $\delta(t)$; rather we talk about the value of integrals involving $\delta(t)$.

Using the equivalence property, we can now confirm the result (2.56), namely that

$$\delta(t) = \frac{\mathrm{d}}{\mathrm{d}t}H(t) = H'(t)$$

To prove this, we must show that

$$\int_{-\infty}^{\infty} \theta(t)\delta(t)\,\mathrm{d}t = \int_{-\infty}^{\infty} \theta(t)H'(t)\,\mathrm{d}t \tag{2.57}$$

Integrating the right-hand side of (2.57) by parts, we have

$$\int_{-\infty}^{\infty} \theta(t)H'(t)\,\mathrm{d}t = [H(t)\theta(t)]_{-\infty}^{\infty} - \int_{-\infty}^{\infty} H(t)\theta'(t)\,\mathrm{d}t$$

$$= 0 - \int_{-\infty}^{\infty} \theta'(t)\mathrm{d}t \quad \text{(by the definitions of } \theta(t) \text{ and } H(t))$$

$$= -[\theta(t)]_0^{\infty} = \theta(0)$$

Since the left-hand side of (2.57) is also $\theta(0)$, the equivalence of $\delta(t)$ and $H'(t)$ is proved.

Likewise, it can be shown that

$$\delta(t-a) = \frac{\mathrm{d}}{\mathrm{d}t}H(t-a) = H'(t-a) \tag{2.58}$$

The results (2.56) and (2.58) may be used to obtain the **generalized derivatives** of piecewise-continuous functions having jump discontinuities $d_1, d_2, \ldots, d_n$ at

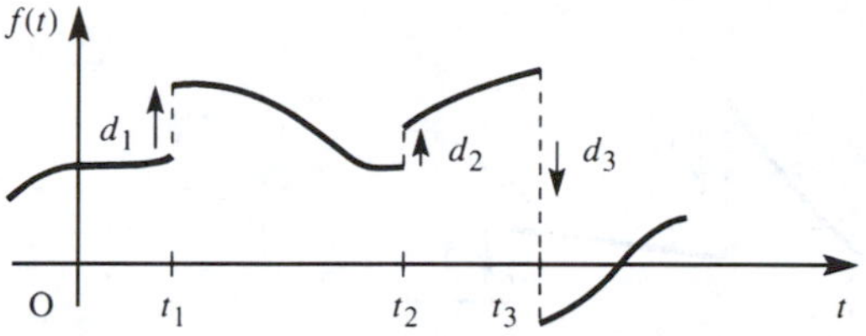

Figure 2.37 Piecewise-continuous function with jump discontinuities.

times $t_1, t_2, \ldots, t_n$ respectively, as illustrated in Figure 2.37. On expressing $f(t)$ in terms of Heaviside step functions as in Section 2.5.1, and differentiating using the product rule, use of (2.56) and (2.58) leads to the result

$$f'(t) = g'(t) + \sum_{i=1}^{n} d_i \delta(t - t_i) \tag{2.59}$$

where $g'(t)$ denotes the ordinary derivative of $f(t)$ where it exists. The result (2.59) tells us that the derivative of a piecewise-continuous function with jump discontinuities is the ordinary derivative where it exists plus the sum of delta functions at the discontinuities multiplied by the magnitudes of the respective jumps.

By the magnitude d_i of a jump in a function $f(t)$ at a point t_i, we mean the difference between the right-hand and left-hand limits of $f(t)$ at t_i; that is,

$$d_i = f(t_i + 0) - f(t_i - 0)$$

If follows that an upward jump, such as d_1 and d_2 in Figure 2.37, is positive, while a downward jump, such as d_3 in Figure 2.37, is negative.

The result (2.59) gives an indication as to why the use of differentiators in practical systems is not encouraged, since the introduction of impulses means that derivatives increase noise levels in signal reception. In contrast, integrators have a smoothing effect on signals, and are widely used.

EXAMPLE 2.47 Obtain the generalized derivative of the piecewise-continuous function

$$f(t) = \begin{cases} 2t^2 + 1 & (0 \leqslant t < 3) \\ t + 4 & (3 \leqslant t < 5) \\ 4 & (t \geqslant 5) \end{cases}$$

Solution $f(t)$ is depicted graphically in Figure 2.38, and it has jump discontinuities of magnitudes 1, −12 and −5 at times $t = 0$, 3 and 5 respectively. Using (2.59), the generalized derivative is

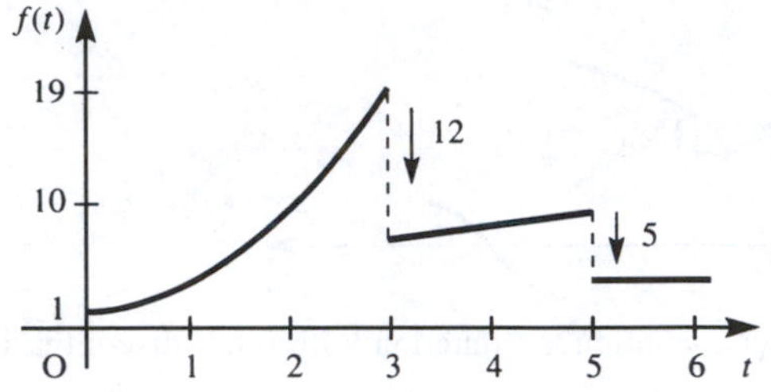

Figure 2.38 Piecewise-continuous function of Example 2.47.

$$f'(t) = g'(t) + 1\delta(t) - 12\delta(t-3) - 5\delta(t-5)$$

where

$$g'(t) = \begin{cases} 4t & (0 \leqslant t < 3) \\ 1 & (3 \leqslant t < 5) \\ 0 & (t \geqslant 5) \end{cases}$$

EXAMPLE 2.48 A system is characterized by the differential equation model

$$\frac{d^2x}{dt^2} + 5\frac{dx}{dt} + 6x = u + 3\frac{du}{dt} \tag{2.60}$$

Determine the response of the system to a forcing function $u(t) = e^{-t}$ applied at time $t = 0$, given that it was initially in a quiescent state.

Solution Since the system is initially in a quiescent state, the transformed equation corresponding to (2.60) is

$$(s^2 + 5s + 6)X(s) = (3s + 1)U(s)$$

giving

$$X(s) = \frac{3s+1}{s^2+5s+6}U(s)$$

In the particular case when $u(t) = e^{-t}$, $U(s) = 1/(s + 1)$, so that

$$X(s) = \frac{(3s+1)}{(s+1)(s+2)(s+3)} = \frac{-1}{s+1} + \frac{5}{s+2} - \frac{4}{s+3}$$

which, on taking inverse transforms, gives the desired response as

$$x(t) = -e^{-t} + 5e^{-2t} - 4e^{-3t} \quad (t \geqslant 0)$$

One might have been tempted to adopt a different approach and substitute for $u(t)$ directly in (2.60) before taking Laplace transforms. This leads to

$$\frac{d^2x}{dt^2} + 5\frac{dx}{dt} + 6x = e^{-t} - 3e^{-t} = -2e^{-t}$$

which, on taking Laplace transforms, leads to

$$(s^2 + 5s + 6)X(s) = -\frac{2}{s+1}$$

giving

$$X(s) = \frac{-2}{(s+1)(s+2)(s+3)} = \frac{-1}{s+1} + \frac{2}{s+2} - \frac{1}{s+3}$$

which, on inversion, gives

$$x(t) = -\mathrm{e}^{-t} + 2\,\mathrm{e}^{-2t} - \mathrm{e}^{-3t} \quad (t \geqslant 0)$$

Clearly this approach results in a different solution, and therefore appears to lead to a paradox. However, this apparent paradox can be resolved by noting that the second approach is erroneous in that it ignores the important fact that we are dealing with causal functions. Strictly speaking,

$$u(t) = \mathrm{e}^{-t}H(t)$$

and, when determining $\mathrm{d}u/\mathrm{d}t$, the product rule of differential calculus should be employed, giving

$$\frac{\mathrm{d}u}{\mathrm{d}t} = -\mathrm{e}^{-t}H(t) + \mathrm{e}^{-t}\frac{\mathrm{d}}{\mathrm{d}t}H(t)$$

$$= -\mathrm{e}^{-t}H(t) + \mathrm{e}^{-t}\delta(t)$$

Substituting this into (2.60) and taking Laplace transforms gives

$$(s^2 + 5s + 6)X(s) = \frac{1}{s+1} + 3\left(-\frac{1}{s+1} + 1\right)$$

$$= \frac{3s+1}{s+1}$$

That is,

$$X(s) = \frac{3s+1}{(s+1)(s^2+5s+6)}$$

leading to the same response

$$x(t) = -\mathrm{e}^{-t} + 5\,\mathrm{e}^{-2t} - 4\,\mathrm{e}^{-3t} \quad (t \geqslant 0)$$

as in the first approach above.

The differential equation used in Example 2.48 is of a form that occurs frequently in practice, so it is important that the causal nature of the forcing term be recognized.

The derivative $\delta'(t)$ of the delta function is also a generalized function, and, using the equivalence property, it is readily shown that

$$\int_{-\infty}^{\infty} f(t)\delta'(t)\,dt = -f'(0)$$

or, more generally,

$$\int_{-\infty}^{\infty} f(t)\delta'(t-a)\,dt = -f'(a)$$

provided that $f'(t)$ is continuous at $t = a$.

Likewise, the nth derivative satisfies

$$\int_{-\infty}^{\infty} f(t)\delta^{n}(t-a)\,dt = (-1)^{n}f^{(n)}(a)$$

provided that $f^{(n)}(t)$ is continuous at $t = a$.

Using the definition of the Laplace transform, it follows that

$$\mathscr{L}\{\delta^{(n)}(t-a)\} = s^{n}e^{-as}$$

and, in particular,

$$\mathscr{L}\{\delta^{(n)}(t)\} = s^{n} \tag{2.61}$$

2.5.12 Exercises

25 Obtain the inverse Laplace transforms of the following:

(a) $\dfrac{2s^2+1}{(s+2)(s+3)}$ (b) $\dfrac{s^2-1}{s^2+4}$ (c) $\dfrac{s^2+2}{s^2+2s+5}$

26 Solve for $t \geqslant 0$ the following differential equations, subject to the specified initial conditions:

(a) $\dfrac{d^2x}{dt^2} + 7\dfrac{dx}{dt} + 12x = 2 + \delta(t-2)$

subject to $x = 0$ and $\dfrac{dx}{dt} = 0$ at $t = 0$

(b) $\dfrac{d^2x}{dt^2} + 6\dfrac{dx}{dt} + 13x = \delta(t-2\pi)$

subject to $x = 0$ and $\dfrac{dx}{dt} = 0$ at $t = 0$

(c) $\dfrac{d^2x}{dt^2} + 7\dfrac{dx}{dt} + 12x = \delta(t-3)$

subject to $x = 1$ and $\dfrac{dx}{dt} = 1$ at $t = 0$

27 Obtain the generalized derivatives of the following piecewise-continuous functions:

(a) $f(t) = \begin{cases} 3t^2 & (0 \leqslant t < 4) \\ 2t-3 & (4 \leqslant t < 6) \\ 5 & (t \geqslant 6) \end{cases}$

(b) $g(t) = \begin{cases} t & (0 \leqslant t < 1) \\ 2-t & (1 \leqslant t < 2) \\ 0 & (t \geqslant 2) \end{cases}$

(c) $f(t) = \begin{cases} 2t+5 & (0 \leqslant t < 2) \\ 9-3t & (2 \leqslant t < 4) \\ t^2-t & (t \geqslant 4) \end{cases}$

28 Solve for $t \geqslant 0$ the differential equation

$$\frac{d^2x}{dt^2} + 7\frac{dx}{dt} + 10x = 2u + 3\frac{du}{dt}$$

subject to $x = 0$ and $dx/dt = 2$ at $t = 0$ and where $u(t) = e^{-2t}H(t)$.

29 A periodic function $f(t)$ is an infinite train of unit impulses at $t = 0$ and repeated at intervals of $t = T$. Show that

$$\mathscr{L}\{f(t)\} = \frac{1}{1-e^{-sT}}$$

The response of a harmonic oscillator to such a periodic stimulus is determined by the differential equation

$$\frac{d^2x}{dt^2} + \omega^2 x = f(t) \quad (t \geq 0)$$

Show that

$$x(t) = \frac{1}{\omega}\sum_{n=0}^{\infty} H(t-nT)\sin\omega(t-nT) \quad (t \geq 0)$$

and sketch the responses from $t = 0$ to $t = 6\pi/\omega$ for the two cases (a) $T = \pi/\omega$ and (b) $T = 2\pi/\omega$.

30 An impulse voltage $E\delta(t)$ is applied at time $t = 0$ to a circuit consisting of a resistor R, a capacitor C and an inductor L connected in series. Prior to application of this voltage, both the charge on the capacitor and the resulting current in the circuit are zero. Determine the charge $q(t)$ on the capacitor and the resulting current $i(t)$ in the circuit at time t.

2.5.13 Bending of beams

So far, we have considered examples in which Laplace transform methods have been used to solve initial-value-type problems. These methods may also be used to solve boundary-value problems, and, to illustrate, we consider in this section the application of Laplace transform methods to determine the transverse deflection of a uniform thin beam due to loading.

Consider a thin uniform beam of length l and let $y(x)$ be its transverse displacement, at distance x measured from one end, from the original position due to loading. The situation is illustrated in Figure 2.39, with the displacement measured upwards. Then, from the elementary theory of beams, we have

$$EI\frac{d^4y}{dx^4} = -W(x) \quad \textbf{(2.62)}$$

where $W(x)$ is the transverse force per unit length, with a downwards force taken to be positive, and EI is the flexural rigidity of the beam (E is Young's modulus of elasticity and I is the moment of inertia of the beam about its central axis). It is assumed that the beam has uniform elastic properties and a uniform cross-section over its length, so that both E and I are taken to be constants.

Equation (2.62) is sometimes written as

$$EI\frac{d^4y}{dx^4} = W(x)$$

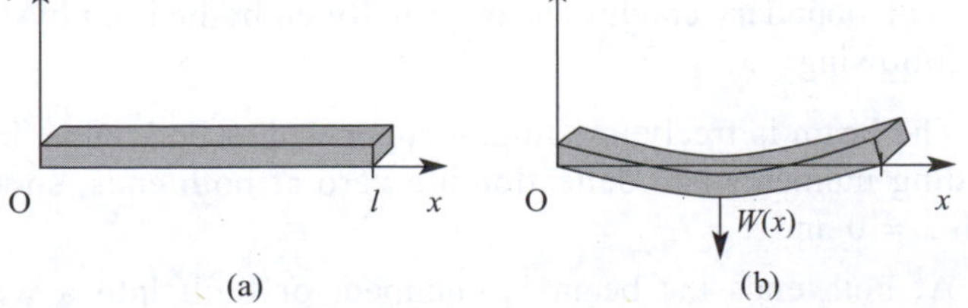

Figure 2.39 Transverse deflection of a beam: (a) initial position; (b) displaced position.

where $y(x)$ is the transverse displacement measured downwards and not upwards as in (2.62).

In cases when the loading is uniform along the full length of the beam, that is, $W(x)$ = constant, (2.62) may be readily solved by the normal techniques of integral calculus. However, when the loading is non-uniform, the use of Laplace transform methods has a distinct advantage, since by making use of Heaviside unit functions and impulse functions, the problem of solving (2.62) independently for various sections of the beam may be avoided.

Taking Laplace transforms throughout in (2.62) gives

$$EI[s^4Y(s) - s^3y(0) - s^2y_1(0) - sy_2(0) - y_3(0)] = -W(s) \tag{2.63}$$

where

$$y_1(0) = \left(\frac{dy}{dx}\right)_{x=0}, \quad y_2(0) = \left(\frac{d^2y}{dx^2}\right)_{x=0}, \quad y_3(0) = \left(\frac{d^3y}{dx^3}\right)_{x=0}$$

and may be interpreted physically as follows:

$EIy_3(0)$ is the shear at $x = 0$

$EIy_2(0)$ is the bending moment at $x = 0$

$y_1(0)$ is the slope at $x = 0$

$y(0)$ is the deflection at $x = 0$

Solving (2.63) for $y(s)$ leads to

$$Y(s) = -\frac{W(s)}{EIs^4} + \frac{y(0)}{s} + \frac{y_1(0)}{s^2} + \frac{y_2(0)}{s^3} + \frac{y_3(0)}{s^4} \tag{2.64}$$

Thus four boundary conditions need to be found, and ideally they should be the shear, bending moment, slope and deflection at $x = 0$. However, in practice these boundary conditions are not often available. While some of them are known, other boundary conditions are specified at points along the beam other than at $x = 0$, for example conditions at the far end, $x = l$, or conditions at possible points of support along the beam. That is, we are faced with a boundary-value problem rather than an initial-value problem.

To proceed, known conditions at $x = 0$ are inserted, while the other conditions among $y(0)$, $y_1(0)$, $y_2(0)$ and $y_3(0)$ that are not specified are carried forward as undetermined constants. Inverse transforms are taken throughout in (2.45) to obtain the deflection $y(x)$, and the outstanding undetermined constants are obtained using the boundary conditions specified at points along the beam other than at $x = 0$.

The boundary conditions are usually embodied in physical conditions such as the following:

(a) The beam is freely, or simply, supported at both ends, indicating that both the bending moments and deflection are zero at both ends, so that $y = d^2y/dx^2 = 0$ at both $x = 0$ and $x = l$.

(b) At both ends the beam is clamped, or built into a wall. Thus the beam is horizontal at both ends, so that $y = dy/dx = 0$ at both $x = 0$ and $x = l$.

(c) The beam is a cantilever with one end free (that is, fixed horizontally at one end, with the other end free). At the fixed end (say $x = 0$)

$$y = \frac{dy}{dx} = 0 \quad \text{at } x = 0$$

and at the free end ($x = l$), since both the shearing force and bending moment are zero,

$$\frac{d^2y}{dx^2} = \frac{d^3y}{dx^3} = 0 \quad \text{at } x = l$$

If the load is not uniform along the full length of the beam, use is made of Heaviside step functions and impulse functions in specifying $W(x)$ in (2.62). For example, a uniform load w per unit length over the portion of the beam $x = x_1$ to $x = x_2$ is specified as $wH(x - x_1) - wH(x - x_2)$, and a point load w at $x = x_1$ is specified as $w\delta(x - x_1)$.

EXAMPLE 2.49 Figure 2.40 illustrates a uniform beam of length l, freely supported at both ends, bending under uniformly distributed self-weight W and a concentrated point load P at $x = \frac{1}{3}l$. Determine the transverse deflection $y(x)$ of the beam.

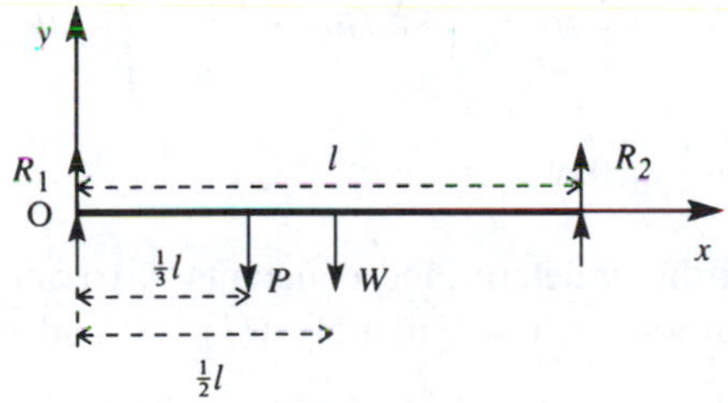

Figure 2.40 Loaded beam of Example 2.49.

Solution As in Figure 2.39, the origin is taken at the left-hand end of the beam, and the deflection $y(x)$ measured upwards from the horizontal at the level of the supports. The deflection $y(x)$ is then given by (2.62), with the force function $W(x)$ having contributions from the weight W, the concentrated load P and the support reactions R_1 and R_2. However, since we are interested in solving (2.62) for $0 \leqslant x \leqslant l$, point loads or reactions at the end $x = l$ may be omitted from the force function.

As a preliminary, we need to determine R_1. This is done by taking static moments about the end $x = l$, assuming the weight W to be concentrated at the centroid $x = \frac{1}{2}l$, giving

$$R_1 l = \tfrac{1}{2}Wl + P\tfrac{2}{3}l$$

or

$$R_1 = \tfrac{1}{2}W + \tfrac{2}{3}P$$

The force function $W(x)$ may then be expressed as

$$W(x) = \frac{W}{l}H(x) + P\delta(x - \tfrac{1}{3}l) - (\tfrac{1}{2}W + \tfrac{2}{3}P)\delta(x)$$

with a Laplace transform

$$W(s) = \frac{W}{ls} + P\,\mathrm{e}^{-ls/3} - (\tfrac{1}{2}W + \tfrac{2}{3}P)$$

Since the beam is freely supported at both ends, the deflection and bending moments are zero at both ends, so we take the boundary conditions as

$$y = 0 \quad \text{at } x = 0 \text{ and } x = l$$

$$\frac{\mathrm{d}^2y}{\mathrm{d}x^2} = 0 \quad \text{at } x = 0 \text{ and } x = l$$

The transformed equation (2.64) becomes

$$Y(s) = -\frac{1}{EI}\left[\frac{W}{ls^5} + \frac{P}{s^4}\mathrm{e}^{-ls/3} - \left(\frac{1}{2}W + \frac{2}{3}P\right)\frac{1}{s^4}\right] + \frac{y_1(0)}{s^2} + \frac{y_3(0)}{s^4}$$

Taking inverse transforms, making use of the second shift theorem (Theorem 2.4), gives the deflection $y(x)$ as

$$y(x) = -\frac{1}{EI}\left[\frac{1}{24}\frac{W}{l}x^4 + \frac{1}{6}P\left(x - \frac{1}{3}l\right)^3 H\left(x - \frac{1}{3}l\right) - \frac{1}{6}\left(\frac{1}{2}W + \frac{2}{3}P\right)x^3\right]$$

$$+ y_1(0)x + \frac{1}{6}y_3(0)x^3$$

To obtain the value of the undetermined constants $y_1(0)$ and $y_3(0)$, we employ the unused boundary conditions at $x = l$, namely $y(l) = 0$ and $y_2(l) = 0$. For $x > \frac{1}{3}l$

$$y(x) = -\frac{1}{EI}\left[\frac{1}{24}\frac{W}{l}x^4 + \frac{1}{6}P\left(x - \frac{1}{3}l\right)^3 - \frac{1}{6}\left(\frac{1}{2}W + \frac{2}{3}P\right)x^3\right] + y_1(0)x + \frac{1}{6}y_3(0)x^3$$

$$\frac{\mathrm{d}^2y}{\mathrm{d}x^2} = y_2(x) = -\frac{1}{EI}\left[\frac{Wx^2}{2l} + P\left(x - \frac{1}{3}l\right) - \left(\frac{1}{3}W + \frac{2P}{3}\right)x\right] + y_3(0)x$$

Thus taking $y_2(l) = 0$ gives $y_3(0) = 0$, and taking $y(l) = 0$ gives

$$-\frac{1}{EI}\left(\frac{1}{24}Wl^3 + \frac{4}{81}Pl^3 - \frac{1}{12}Wl^3 - \frac{1}{9}Pl^3\right) + y_1(0)l = 0$$

so that

$$y_1(0) = -\frac{l^2}{EI}\left(\frac{1}{24}W + \frac{5}{81}P\right)$$

Substituting back, we find that the deflection $y(x)$ is given by

$$y(x) = -\frac{W}{EI}\left(\frac{x^4}{24l} - \frac{1}{12}x^3 + \frac{1}{24}l^2x\right) - \frac{P}{EI}\left(\frac{5}{81}l^2x - \frac{1}{9}x^3\right) - \frac{P}{6EI}\left(x - \frac{1}{3}l\right)^3 H\left(x - \frac{1}{3}l\right)$$

or, for the two sections of the beam,

$$y(x) = \begin{cases} -\dfrac{W}{EI}\left(\dfrac{x}{24l} - \dfrac{1}{12}x^3 + \dfrac{1}{24}l^2x\right) - \dfrac{P}{EI}\left(\dfrac{5}{81}l^2x - \dfrac{1}{9}x^3\right) & (0 < x < \tfrac{1}{3}l) \\ -\dfrac{W}{EI}\left(\dfrac{x^4}{24l} - \dfrac{1}{12}x^3 + \dfrac{1}{24}l^2x\right) - \dfrac{P}{EI}\left(\dfrac{19}{162}l^2x + \dfrac{1}{18}x^3 - \dfrac{1}{6}x^2l - \dfrac{1}{162}l^3\right) & \left(\dfrac{1}{3}l < x < l\right) \end{cases}$$

2.5.14 Exercises

31 Find the deflection of a beam simply supported at its ends $x = 0$ and $x = l$, bending under a uniformly distributed self-weight M and a concentrated load W at $x = \frac{1}{2}l$.

32 A cantilever beam of negligible weight and of length l is clamped at the end $x = 0$. Determine the deflection of the beam when it is subjected to a load per unit length, w, over the section $x = x_1$ to $x = x_2$. What is the maximum deflection if $x_1 = 0$ and $x_2 = l$?

33 A uniform cantilever beam of length l is subjected to a concentrated load W at a point distance b from the fixed end. Determine the deflection of the beam, distinguishing between the sections $0 < x \leqslant b$ and $b < x \leqslant l$.

2.6 Transfer functions

2.6.1 Definitions

The **transfer function** of a linear time-invariant system is defined to be the ratio of the Laplace transform of the system output (or response function) to the Laplace transform of the system input (or forcing function), *under the assumption that all the initial conditions are zero* (that is, the system is initially in a **quiescent state**).

Transfer functions are frequently used in engineering to characterize the input–output relationships of linear time-invariant systems, and play an important role in the analysis and design of such systems.

Consider a linear time-invariant system characterized by the differential equation

$$a_n\frac{\mathrm{d}^nx}{\mathrm{d}t^n} + a_{n-1}\frac{\mathrm{d}^{n-1}x}{\mathrm{d}t^{n-1}} + \ldots + a_0x = b_m\frac{\mathrm{d}^mu}{\mathrm{d}t^m} + \ldots + b_0u \tag{2.65}$$

where $n \geqslant m$, the as and bs are constant coefficients, and $x(t)$ is the system response or output to the input or forcing term $u(t)$ applied at time $t = 0$. Taking Laplace transforms throughout in (2.65) will lead to the transformed equation.

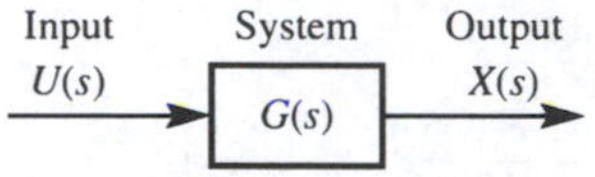

Figure 2.41 Transfer function block diagram.

Since all the initial conditions are assumed to be zero, we see from (2.15) that, in order to obtain the transformed equation, we simply replace d/dt by s, giving

$$(a_n s^n + a_{n-1}s^{n-1} + \ldots + a_0)X(s) = (b_m s^m + \ldots + b_0)U(s)$$

where $X(s)$ and $U(s)$ denote the Laplace transforms of $x(t)$ and $u(t)$ respectively.

The system transfer function $G(s)$ is then defined to be

$$G(s) = \frac{X(s)}{U(s)} = \frac{b_m s^m + \ldots + b_0}{a_n s^n + \ldots + a_0} \tag{2.66}$$

and the system may be represented diagrammatically by the operation box of Figure 2.41. This representation is referred to as the **input–output block diagram** representation of the system.

Writing

$$P(s) = b_m s^m + \ldots + b_0$$

$$Q(s) = a_n s^n + \ldots + a_0$$

the transfer function may be expressed as

$$G(s) = \frac{P(s)}{Q(s)}$$

where, in order to make the system physically realizable, the degrees m and n of the polynomials $P(s)$ and $Q(s)$ must be such that $n \geqslant m$. This is because it follows from (2.61) that if $m > n$ then the system response $x(t)$ to a realistic input $u(t)$ will involve impulses.

The equation $Q(s) = 0$ is called the **characteristic equation** of the system; its order determines the **order of the system**, and its roots are referred to as the **poles** of the transfer function. Likewise, the roots of $P(s) = 0$ are referred to as the **zeros** of the transfer function.

It is important to realize that, in general, a transfer function is only used to characterize a linear time-invariant system. It is a property of the system itself, and is independent of both system input and output.

Although the transfer function characterizes the dynamics of the system, it provides no information concerning the actual physical structure of the system, and in fact systems that are physically different may have identical transfer functions; for example, the mass–spring–damper system of Figure 2.12 and the *LCR* circuit of Figure 2.8 both have the transfer function

$$G(s) = \frac{X(s)}{U(s)} = \frac{1}{\alpha s^2 + \beta s + \gamma}$$

In the mass–spring–damper system $X(s)$ determines the displacement $x(t)$ of the mass and $U(s)$ represents the applied force $F(t)$, while α denotes the mass, β the damping coefficient and γ the spring constant. On the other hand, in the LCR circuit $X(s)$ determines the charge $q(t)$ on the condenser and $U(s)$ represents the applied emf $e(t)$, while α denotes the inductance, β the resistance and γ the reciprocal of the capacitance.

In practice, an overall system may be made up of a number of components each characterized by its own transfer function and related operation box. The overall system input–output transfer function is then obtained by the rules of **block diagram algebra**.

Since $G(s)$ may be written as

$$G(s) = \frac{b_m}{a_m} \frac{(s - z_1)(s - z_2) \ldots (s - z_m)}{(s - p_1)(s - p_2) \ldots (s - p_n)}$$

where the z_is and p_is are the transfer function zeros and poles respectively, we observe that $G(s)$ is known, apart from a constant factor, if the positions of all the poles and zeros are known. Consequently, a plot of the poles and zeros of $G(s)$ is often used as an aid in the graphical analysis of the transfer function (a common convention is to mark the position of a zero by a circle ○ and that of a pole by a cross ×). Since the coefficients of the polynomials $P(s)$ and $Q(s)$ are real, all complex roots always occur in complex conjugate pairs, so that the **pole–zero plot** is symmetrical about the real axis.

EXAMPLE 2.50 The response $x(t)$ of a system to a forcing function $u(t)$ is determined by the differential equation

$$9\frac{d^2x}{dt^2} + 12\frac{dx}{dt} + 13x = 2\frac{du}{dt} + 3u$$

(a) Determine the transfer function characterizing the system.

(b) Write down the characteristic equation of the system. What is the order of the system?

(c) Determine the transfer function poles and zeros, and illustrate them diagrammatically in the s plane.

Solution (a) Assuming all the initial conditions to be zero, taking Laplace transforms throughout in the differential equation

$$9\frac{d^2x}{dt^2} + 12\frac{dx}{dt} + 13x = 2\frac{du}{dt} + 3u$$

leads to

$$(9s^2 + 12s + 13)X(s) = (2s + 3)U(s)$$

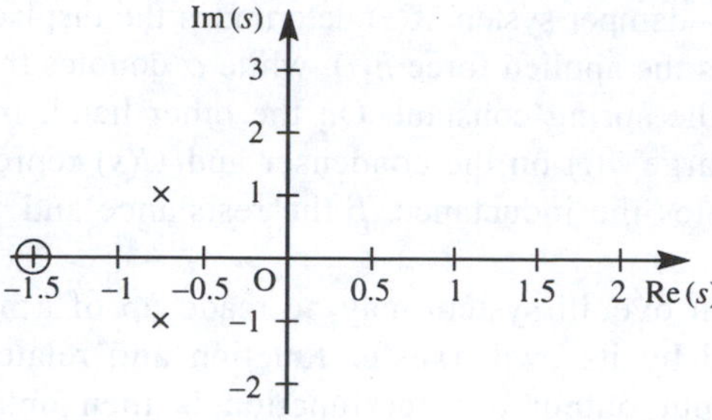

Figure 2.42 Pole (X)–zero (O) plot for Example 2.50.

so that the system transfer function is given by

$$G(s) = \frac{X(s)}{U(s)} = \frac{2s + 3}{9s^2 + 12s + 13}$$

(b) The characteristic equation of the system is

$$9s^2 + 12s + 13 = 0$$

and the system is of order 2.

(c) The transfer function poles are the roots of the characteristic equation

$$9s^2 + 12s + 13 = 0$$

which are

$$s = \frac{-12 \pm \sqrt{(144 - 468)}}{18} = \frac{-2 \pm j3}{3}$$

That is, the transfer function has simple poles at

$$s = -\tfrac{2}{3} + j \quad \text{and} \quad s = -\tfrac{2}{3} - j$$

The transfer function zeros are determined by equating the numerator polynomial $2s + 3$ to zero, giving a single zero at

$$s = -\tfrac{3}{2}$$

The corresponding pole–zero plot in the s plane is shown in Figure 2.42.

2.6.2 Stability

The stability of a system is a property of vital importance to engineers. Intuitively, we may regard a stable system as one that will remain at rest unless it is excited by an external source, and will return to rest if all such external influences are removed. Thus a stable system is one whose response, in the absence of an input, will approach zero as time approaches infinity. This then ensures that any bounded input produces a bounded output; this property is frequently taken to be the definition of a **stable linear system**.

Clearly, stability is a property of the system itself, and does not depend on the system input or forcing function. Since a system may be characterized in the s domain by its transfer function $G(s)$, it should be possible to use the transfer function to specify conditions for the system to be stable.

In considering the time response of

$$X(s) = G(s)U(s), \quad G(s) = \frac{P(s)}{Q(s)}$$

to any given input $u(t)$, it is necessary to factorize the denominator polynomial

$$Q(s) = a_n s^n + a_{n-1}s^{n-1} + \ldots + a_0$$

and various forms of factors can be involved.

Simple factor of the form $s + \alpha$, with α real

This corresponds to a simple pole at $s = -\alpha$, and will in the partial-fractions expansion of $G(s)$ lead to a term of the form $c/(s + \alpha)$ having corresponding time response $c\mathrm{e}^{-\alpha t}H(t)$, using the strict form of the inverse given in (2.12). If $\alpha > 0$, so that the pole is in the left half of the s plane, the time response will tend to zero as $t \to \infty$. If $\alpha < 0$, so that the pole is in the right half of the s plane, the time response will increase without bound as $t \to \infty$. It follows that a stable system must have real-valued simple poles of $G(s)$ in the left half of the s plane.

$\alpha = 0$ corresponds to a simple pole at the origin, having a corresponding time response that is a step $cH(t)$. A system having such a pole is said to be **marginally stable**; this does not ensure that a bounded input will lead to a bounded output, since, for example, if such a system has an input that is a step d applied at time $t = 0$ then the response will be a ramp $cdtH(t)$, which is unbounded as $t \to \infty$.

Repeated simple factors of the form $(s + \alpha)^n$, with α real

This corresponds to a multiple pole at $s = -\alpha$, and will lead in the partial-fractions expansion of $G(s)$ to a term of the form $c/(s + \alpha)^n$ having corresponding time response $[c/(n - 1)!]t^{n-1}\,\mathrm{e}^{-\alpha t}H(t)$. Again the response will decay to zero as $t \to \infty$ only if $\alpha > 0$, indicating that a stable system must have all real-valued repeated poles of $G(s)$ in the left half of the s plane.

Quadratic factors of the form $(s + \alpha)^2 + \beta^2$, with α and β real

This corresponds to a pair of complex conjugate poles at $s = -\alpha + j\beta$, $s = -\alpha - j\beta$, and will lead in the partial-fractions expansion of $G(s)$ to a term of the form

$$\frac{c(s+\alpha)+d\beta}{(s+\alpha)^2+\beta^2}$$

having corresponding time response

$$e^{-\alpha t}(c\cos\beta t + d\sin\beta t) \equiv A\,e^{-\alpha t}\sin(\beta t + \gamma)$$

where $A = \sqrt{(c^2 + d^2)}$ and $\gamma = \tan^{-1}(c/d)$.

Again we see that poles in the left half of the s plane (corresponding to $\alpha > 0$) have corresponding time responses that die away, in the form of an exponentially damped sinusoid, as $t \to \infty$. A stable system must therefore have complex conjugate poles located in the left half of the s plane; that is, all complex poles must have a negative real part.

If $\alpha = 0$, the corresponding time response will be a periodic sinusoid, which will not die away as $t \to \infty$. Again this corresponds to a marginally stable system, and will, for example, give rise to a response that increases without bound as $t \to \infty$ when the input is a sinusoid at the same frequency β.

A summary of the responses corresponding to the various types of poles is given in Figure 2.43.

The concept of stability may be expressed in the form of Definition 2.3.

DEFINITION 2.3

A physically realizable causal time-invariant linear system with transfer function $G(s)$ is stable provided that all the poles of $G(s)$ are in the left half of the s plane.

The requirement in the definition that the system be physically realizable, that is, $n \geqslant m$ in the transfer function $G(s)$ of (2.66), avoids terms of the form s^{m-n} in the partial-fractions expansion of $G(s)$. Such a term would correspond to differentiation of degree $m - n$, and were an input such as $\sin\omega t$ used to excite the system then the response would include a term such as $\omega^{m-n}\sin\omega t$ or $\omega^{m-n}\cos\omega t$, which could be made as large as desired by increasing the input frequency ω.

In terms of the poles of the transfer function $G(s)$, its abscissa of convergence σ_c corresponds to the real part of the pole located furthest to the right in the s plane. For example, if

$$G(s) = \frac{s+1}{(s+3)(s+2)}$$

then the abscissa of convergence $\sigma_c = -2$.

Poles of $G(s)$ in form $\sigma + j\omega$	Poles in complex s plane	Corresponding time response	Nature of response
$\sigma = \omega = 0$	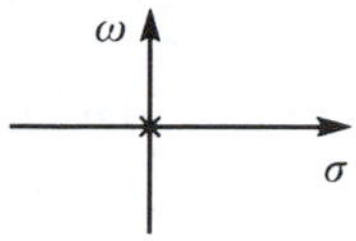	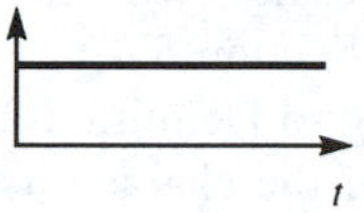	Constant
$\sigma = \omega = 0$ (multiplicity 2)	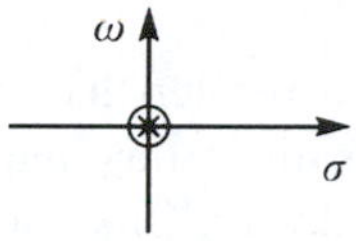	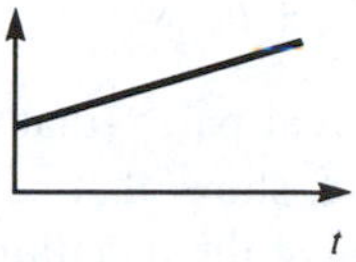	Ramp
$\sigma < 0, \omega = 0$	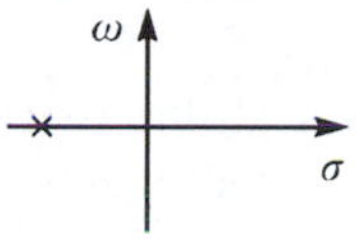	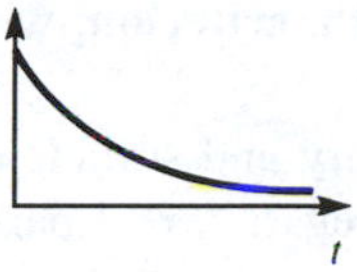	Exponential decay
$\sigma > 0, \omega = 0$	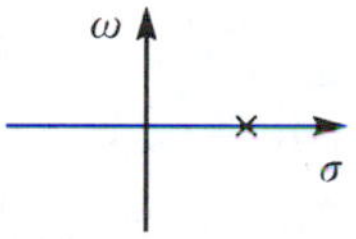	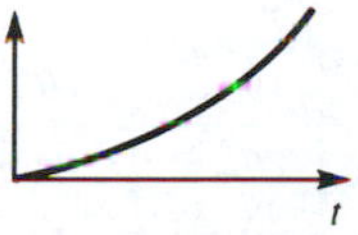	Exponential growth
$\sigma = 0, \omega > 0$	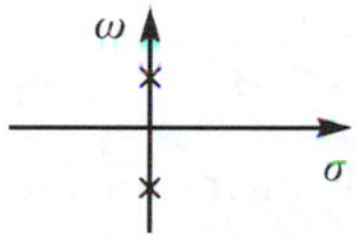	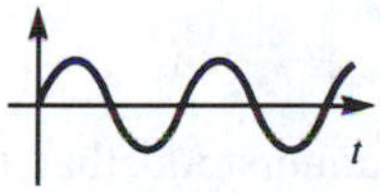	Sinusoidal
$\sigma = 0, \omega > 0$ (multiplicity 2)	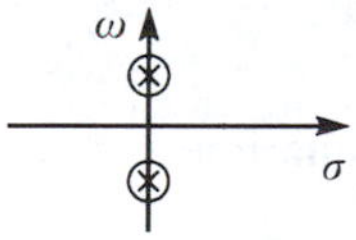	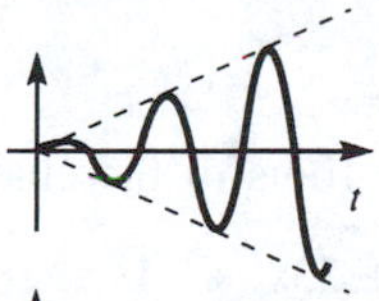	Linearly growing sinusoidal
$\sigma < 0, \omega > 0$	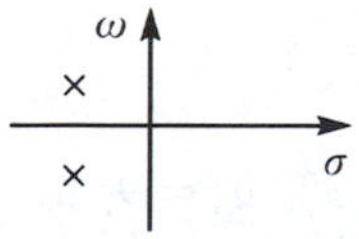	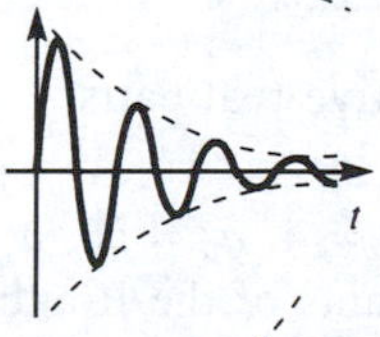	Exponentially decaying sinusoidal
$\sigma > 0, \omega > 0$	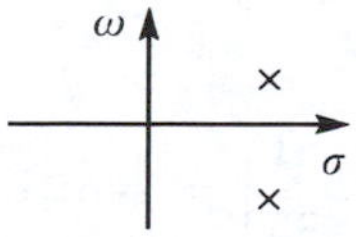	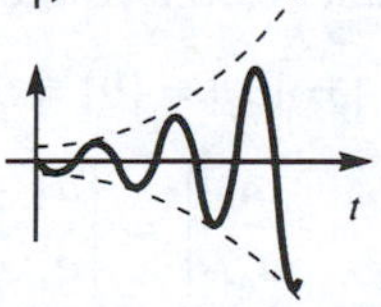	Exponentially growing sinusoidal

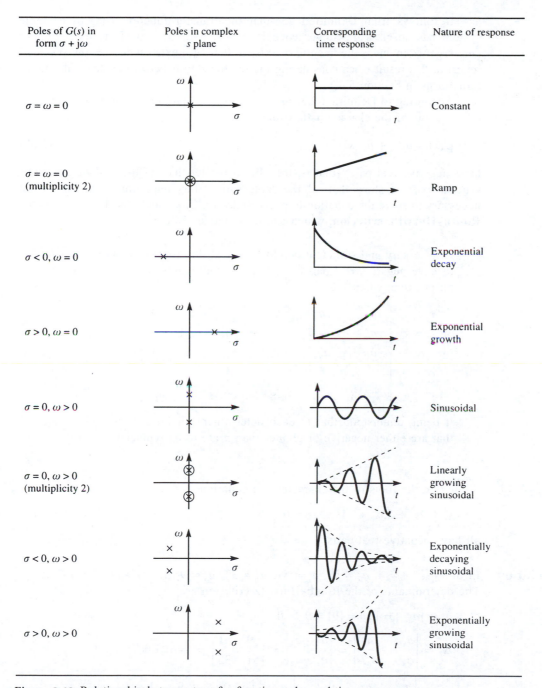

Figure 2.43 Relationship between transfer function poles and time response.

It follows from Definition 2.3 that the transfer function $G(s)$ of a stable system has abscissa of convergence $\sigma_c = -\alpha$, with $\alpha > 0$. Thus its region of convergence includes the imaginary axis, so that $G(s)$ exists when $s = j\omega$. We shall return to this result when considering the relationship between Laplace and Fourier transforms in Section 5.4.1.

According to Definition 2.3, in order to prove stability, we need to show that all the roots of the characteristic equation

$$Q(s) = a_n s^n + a_{n-1}s^{n-1} + \ldots + a_1 s + a_0 = 0 \tag{2.67}$$

have negative real parts (that is, they lie in the left half of the s plane). Various criteria exist to show that all the roots satisfy this requirement, and it is not necessary to solve the equation to prove stability. One widely used criterion is the **Routh–Hurwitz criterion**, which can be stated as follows:

A necessary and sufficient condition for all the roots of equation (2.67) to have negative real parts is that the determinants $\Delta_1, \Delta_2, \ldots, \Delta_n$ are all positive, where

$$\Delta_r = \begin{vmatrix} a_{n-1} & a_n & 0 & 0 & \cdots & 0 \\ a_{n-3} & a_{n-2} & a_{n-1} & a_n & \cdots & 0 \\ a_{n-5} & a_{n-4} & a_{n-3} & a_{n-2} & \cdots & 0 \\ \vdots & \vdots & \vdots & \vdots & & \vdots \\ a_{n-(2r-1)} & a_{n-2r} & a_{n-2r-1} & a_{n-2r-2} & \cdots & a_{n-r} \end{vmatrix} \tag{2.68}$$

it being understood that in each determinant all the as with subscripts that are either negative or greater than n are to be replaced by zero.

EXAMPLE 2.51 Show that the roots of the characteristic equation

$$s^4 + 9s^3 + 33s^2 + 51s + 26 = 0$$

all have negative real parts.

Solution In this case $n = 4$, $a_0 = 26$, $a_1 = 51$, $a_2 = 33$, $a_3 = 9$, $a_4 = 1$ and $a_r = 0$ $(r > 4)$. The determinants of the Routh–Hurwitz criterion are

$$\Delta_1 = |a_{n-1}| = |a_3| = |9| = 9 > 0$$

$$\Delta_2 = \begin{vmatrix} a_{n-1} & a_n \\ a_{n-3} & a_{n-2} \end{vmatrix} = \begin{vmatrix} a_3 & a_4 \\ a_1 & a_2 \end{vmatrix} = \begin{vmatrix} 9 & 1 \\ 51 & 33 \end{vmatrix} = 246 > 0$$

$$\Delta_3 = \begin{vmatrix} a_{n-1} & a_n & 0 \\ a_{n-3} & a_{n-2} & a_{n-1} \\ a_{n-5} & a_{n-4} & a_{n-3} \end{vmatrix} = \begin{vmatrix} a_3 & a_4 & 0 \\ a_1 & a_2 & a_3 \\ a_{-1} & a_0 & a_1 \end{vmatrix} = \begin{vmatrix} 9 & 1 & 0 \\ 51 & 33 & 9 \\ 0 & 26 & 51 \end{vmatrix} = 10\,440 > 0$$

$$\Delta_4 = \begin{vmatrix} a_{n-1} & a_n & 0 & 0 \\ a_{n-3} & a_{n-2} & a_{n-1} & a_n \\ a_{n-5} & a_{n-4} & a_{n-3} & a_{n-2} \\ a_{n-7} & a_{n-6} & a_{n-5} & a_{n-4} \end{vmatrix} = \begin{vmatrix} a_3 & a_4 & 0 & 0 \\ a_1 & a_2 & a_3 & a_4 \\ a_{-1} & a_0 & a_1 & a_2 \\ a_{-3} & a_{-2} & a_{-1} & a_0 \end{vmatrix}$$

$$= \begin{vmatrix} 9 & 1 & 0 & 0 \\ 51 & 33 & 9 & 1 \\ 0 & 26 & 51 & 37 \\ 0 & 0 & 0 & 26 \end{vmatrix} = 26\Delta_3 > 0$$

Thus $\Delta_1 > 0$, $\Delta_2 > 0$, $\Delta_3 > 0$ and $\Delta_4 > 0$, so that all the roots of the given characteristic equation have negative real parts. This is readily checked, since the roots are -2, -1, $-3 + j2$ and $-3 - j2$.

EXAMPLE 2.52 The steady motion of a steam-engine governor is modelled by the differential equations

$$m\ddot{\eta} + b\dot{\eta} + d\eta - e\omega = 0 \tag{2.69}$$

$$I_0\dot{\omega} = -f\eta \tag{2.70}$$

where η is a small fluctuation in the angle of inclination, ω a small fluctuation in the angular velocity of rotation, and m, b, d, e, f and I_0 are all positive constants. Show that the motion of the governor is stable provided that

$$\frac{bd}{m} > \frac{ef}{I_0}$$

Solution Differentiating (2.69) gives

$$m\dddot{\eta} + b\ddot{\eta} + d\dot{\eta} - e\dot{\omega} = 0$$

which, on using (2.70), leads to

$$m\dddot{\eta} + b\ddot{\eta} + d\dot{\eta} + \frac{ef}{I_0}\eta = 0$$

for which the corresponding characteristic equation is

$$ms^3 + bs^2 + ds + \frac{ef}{I_0} = 0$$

This is a cubic polynomial, so the parameters of (2.67) are

$$n = 3, \quad a_0 = \frac{ef}{I_0}, \quad a_1 = d, \quad a_2 = b, \quad a_3 = m \quad (a_r = 0,\ r > 3)$$

The determinants (2.68) of the Routh–Hurwitz criterion are

$$\Delta_1 = |a_2| = b > 0$$

$$\Delta_2 = \begin{vmatrix} a_2 & a_3 \\ a_0 & a_1 \end{vmatrix} = \begin{vmatrix} b & m \\ ef/I_0 & d \end{vmatrix} = bd - \frac{mef}{I_0}$$

(and so $\Delta_2 > 0$ provided that $bd - mef/I_0 > 0$ or $bd/m > ef/I_0$), and

$$\Delta_3 = \begin{vmatrix} a_2 & a_3 & 0 \\ a_0 & a_1 & a_2 \\ 0 & 0 & a_0 \end{vmatrix} = a_0\Delta_2 > 0 \quad \text{if } \Delta_2 > 0$$

Thus the action of the governor is stable provided that $\Delta_2 > 0$; that is,

$$\frac{bd}{m} > \frac{ef}{I_0}$$

2.6.3 Impulse response

From (2.66), we find that for a system having transfer function $G(s)$ the response $x(t)$ of the system, initially in a quiescent state, to an input $u(t)$ is determined by the transformed relationship

$$X(s) = G(s)U(s)$$

If the input $u(t)$ is taken to be the unit impulse function $\delta(t)$ then the system response will be determined by

$$X(s) = G(s)\mathscr{L}\{\delta(t)\} = G(s)$$

Taking inverse Laplace transforms leads to the corresponding time response $h(t)$, which is called the **impulse response** of the system (it is also sometimes referred to as the **weighting function** of the system); that is, the impulse response is given by

$$h(t) = \mathscr{L}^{-1}\{X(s)\} = \mathscr{L}^{-1}\{G(s)\} \tag{2.71}$$

We therefore have the following definition.

DEFINITION 2.4: IMPULSE RESPONSE

The impulse response $h(t)$ of a linear time-invariant system is the response of the system to a unit impulse applied at time $t = 0$ when all the initial conditions are zero. It is such that $\mathscr{L}\{h(t)\} = G(s)$, where $G(s)$ is the system transfer function.

Since the impulse response is the inverse Laplace transform of the transfer function, it follows that both the impulse response and the transfer function carry the same information about the dynamics of a linear time-invariant system. Theoretically, therefore, it is possible to determine the complete information about the system by exciting it with an impulse and measuring the response. For this reason, it is common practice in engineering to regard the transfer function as being the Laplace transform of the impulse response, since this places greater emphasis on the parameters of the system when considering system design.

We saw in Section 2.6.2 that, since the transfer function $G(s)$ completely characterizes a linear time-invariant system, it can be used to specify conditions for system stability, which are that all the poles of $G(s)$ lie in the left half of the s plane. Alternatively, characterizing the system by its impulse response, we can say that the system is stable provided that its impulse response decays to zero as $t \to \infty$.

EXAMPLE 2.53 Determine the impulse response of the linear system whose response $x(t)$ to an input $u(t)$ is determined by the differential equation

$$\frac{d^2x}{dt^2} + 5\frac{dx}{dt} + 6x = 5u(t) \tag{2.72}$$

Solution The impulse response $h(t)$ is the system response to $u(t) = \delta(t)$ when all the initial conditions are zero. It is therefore determined as the solution of the differential equation

$$\frac{d^2h}{dt^2} + 5\frac{dh}{dt} + 6h = 5\delta(t) \tag{2.73}$$

subject to the initial conditions $h(0) = \dot{h}(0) = 0$. Taking Laplace transforms in (2.73) gives

$$(s^2 + 5s + 6)H(s) = 5\mathscr{L}\{\delta(t)\} = 5$$

so that

$$H(s) = \frac{5}{(s+3)(s+2)} = \frac{5}{s+2} - \frac{5}{s+3}$$

which, on inversion, gives the desired impulse response

$$h(t) = 5(e^{-2t} - e^{-3t})$$

Alternatively, the transfer function $G(s)$ of the system determined by (2.72) is

$$G(s) = \frac{5}{s^2 + 5s + 6}$$

so that $h(t) = \mathscr{L}^{-1}\{G(s)\} = 5(e^{-2t} - e^{-3t})$ as before.

Note: This example serves to illustrate the necessity for incorporating 0– as the lower limit in the Laplace transform integral, in order to accommodate for an impulse applied at $t = 0$. The effect of the impulse is to cause a step change in $\dot{x}(t)$ at $t = 0$, with the initial condition accounting for what happens up to 0–.

2.6.4 Initial- and final-value theorems

The initial- and final-value theorems are two useful theorems that enable us to predict system behaviour as $t \to 0$ and $t \to \infty$ without actually inverting Laplace transforms.

THEOREM 2.6 The initial-value theorem

If $f(t)$ and $f'(t)$ are both Laplace-transformable and if $\lim_{s\to\infty} sF(s)$ exists then

$$\lim_{t\to 0+} f(t) = f(0+) = \lim_{s\to\infty} sF(s)$$

Proof From (2.13),

$$\mathcal{L}\{f'(t)\} = \int_{0-}^{\infty} f'(t)\,\mathrm{e}^{-st}\,\mathrm{d}t = sF(s) - f(0-)$$

where we have highlighted the fact that the lower limit is 0–. Hence

$$\lim_{s\to\infty} [sF(s) - f(0-)] = \lim_{s\to\infty} \int_{0-}^{\infty} f'(t)\,\mathrm{e}^{-st}\,\mathrm{d}t$$

$$= \lim_{s\to\infty} \int_{0-}^{0+} f'(t)\,\mathrm{e}^{-st}\,\mathrm{d}t + \lim_{s\to\infty} \int_{0+}^{\infty} f'(t)\,\mathrm{e}^{-st}\,\mathrm{d}t \tag{2.74}$$

If $f(t)$ is discontinuous at the origin, so that $f(0+) \neq f(0-)$, then, from (2.59), $f'(t)$ contains an impulse term $[f(0+) - f(0-)]\delta(t)$, so that

$$\lim_{s\to\infty} \int_{0-}^{0+} f'(t)\,\mathrm{e}^{-st}\,\mathrm{d}t = f(0+) - f(0-)$$

Also, since the Laplace transform of $f'(t)$ exists, it is of exponential order and we have

$$\lim_{s\to\infty} \int_{0+}^{\infty} f'(t)\,\mathrm{e}^{-st}\,\mathrm{d}t = 0$$

so that (2.74) becomes

$$\lim_{s\to\infty} sF(s) - f(0-) = f(0+) - f(0-)$$

giving the required result:

$$\lim_{s\to\infty} sF(s) = f(0+)$$

If $f(t)$ is continuous at the origin then $f'(t)$ does not contain an impulse term, and the right-hand side of (2.74) is zero, giving

$$\lim_{s\to\infty} sF(s) = f(0-) = f(0+)$$ □

It is important to recognize that the initial-value theorem does not give the initial value $f(0-)$ used when determining the Laplace transform, but rather gives the value of $f(t)$ as $t \to 0+$. This distinction is highlighted in the following example.

EXAMPLE 2.54 The circuit of Figure 2.44 consists of a resistance R and a capacitance C connected in series together with constant voltage source E. Prior to closing the switch at time $t = 0$, both the charge on the capacitor and the resulting current in the circuit are zero. Determine the current $i(t)$ in the circuit at time t after the switch is closed, and investigate the use of the initial-value theorem.

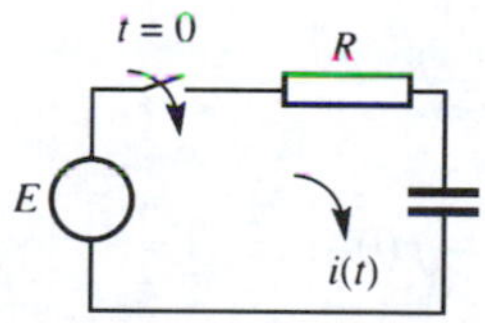

Figure 2.44 *RC* circuit of Example 2.54.

Solution Applying Kirchhoff's law to the circuit of Figure 2.44, we have

$$Ri + \frac{1}{C}\int i\,\mathrm{d}t = E_0$$

which, on taking Laplace transforms, gives the transformed equation

$$RI(s) + \frac{1}{c}\frac{I(s)}{s} = \frac{E_0}{s}$$

Therefore

$$I(s) = \frac{E_0/R}{s + 1/RC}$$

Taking inverse transforms gives the current $i(t)$ at $t \geqslant 0$ as

$$i(t) = \frac{E_0}{R}\mathrm{e}^{-t/RC} \tag{2.75}$$

Applying the initial-value theorem,

$$\lim_{t\to 0+} i(t) = \lim_{s\to\infty} sI(s) = \lim_{s\to\infty} \frac{sE_0/R}{s + 1/RC}$$

$$= \lim_{s\to\infty} \frac{E_0/R}{1 + 1/RCs} = \frac{E_0}{R}$$

That is,

$$i(0+) = \frac{E_0}{R}$$

a result that is readily confirmed by allowing $t \to 0+$ in (2.75). We note that this is not the same as the initial state $i(0) = 0$ owing to the fact that there is a step change in $i(t)$ at $t = 0$.

THEOREM 2.7 The final-value theorem

If $f(t)$ and $f'(t)$ are both Laplace-transformable and $\lim_{t\to\infty} f(t)$ exists then

$$\lim_{t\to\infty} f(t) = \lim_{s\to 0} sF(s)$$

Proof From (2.13),

$$\mathcal{L}\{f'(t)\} = \int_{0-}^{\infty} f'(t)\,\mathrm{e}^{-st}\,\mathrm{d}t = sF(s) - f(0-)$$

Taking limits, we have

$$\lim_{s\to 0} [sF(s) - f(0-)] = \lim_{s\to 0} \int_{0-}^{\infty} f'(t)\,\mathrm{e}^{-st}\,\mathrm{d}t$$

$$= \int_{0-}^{\infty} f'(t)\,\mathrm{d}t = [f(t)]_{0-}^{\infty}$$

$$= \lim_{t\to\infty} f(t) - f(0-)$$

giving the required result:

$$\lim_{t\to\infty} f(t) = \lim_{s\to 0} sF(s)$$ □

The restriction that $\lim_{t\to\infty} f(t)$ must exist means that the theorem does not hold for functions such as e^t, which tends to infinity as $t \to \infty$, or $\sin \omega t$, whose limit is undefined. Since in practice the final-value theorem is used to obtain the behaviour of $f(t)$ as $t \to \infty$ from knowledge of the transform $F(s)$, it is more common to express the restriction in terms of restrictions on $F(s)$, which are that $sF(s)$ must have all its poles in the left half of the s plane; that is, $sF(s)$ must represent a stable transfer function. It is important that the theorem be used with caution and that this restriction be fully recognized, since the existence of $\lim_{s\to 0} sF(s)$ does *not* imply that $f(t)$ has a limiting value as $t \to \infty$.

EXAMPLE 2.55 Investigate the application of the final-value theorem to the transfer function

$$F(s) = \frac{1}{(s+2)(s-3)} \tag{2.76}$$

Solution

$$\lim_{s\to 0} sF(s) = \lim_{s\to 0} \frac{s}{(s+2)(s-3)} = 0$$

so the use of the final-value theorem implies that for the time function $f(t)$ corresponding to $F(s)$ we have

$$\lim_{t\to\infty} f(t) = 0$$

However, taking inverse transforms in (2.76) gives

$$f(t) = \frac{1}{s}(e^{3t} - e^{-2t})$$

implying that $f(t)$ tends to infinity as $t \to \infty$. This implied contradiction arises since the theorem is not valid in this case. Although $\lim_{s\to 0} sF(s)$ exists, $sF(s)$ has a pole at $s = 3$, which is not in the left half of the s plane.

The final-value theorem provides a useful vehicle for determining a system's **steady-state gain (SSG)** and the **steady-state errors**, or **offsets**, in feedback control systems, both of which are important features in control system design.

The SSG of a stable system is the system's steady-state response, that is, the response as $t \to \infty$, to a unit step input. For a system with transfer function $G(s)$ we have, from (2.66), that its response $x(t)$ is related to the input $u(t)$ by the transformed equation

$$X(s) = G(s)U(s)$$

For a unit step input

$$u(t) = 1H(t) \quad \text{giving} \quad U(s) = \frac{1}{s}$$

so that

$$X(s) = \frac{G(s)}{s}$$

From the final-value theorem, the steady-state gain is

$$\text{SSG} = \lim_{t\to\infty} x(t) = \lim_{s\to 0} sX(s) = \lim_{s\to 0} G(s)$$

EXAMPLE 2.56 Determine the steady-state gain of a system having transfer function

$$G(s) = \frac{20(1+3s)}{s^2+7s+10}$$

Solution The response $x(t)$ to a unit step input $u(t) = 1H(t)$ is given by the transformed equation

$$X(s) = G(s)U(s)$$

$$= \frac{20(1+3s)}{s^2+7s+10}\,\frac{1}{s}$$

Then, by the final-value theorem, the steady-state gain is given by

$$\text{SSG} = \lim_{t\to\infty} x(t) = \lim_{s\to 0} sX(s)$$

$$= \lim_{s\to 0} \frac{20(1+3s)}{s^2+7s+10} = 2$$

Note that for a step input of magnitude K, that is, $u(t) = KH(t)$, the steady-state response will be $\lim_{s\to 0} kG(s) = 2K$; that is,

steady-state response to step input = SSG × magnitude of step input

---

A unity feedback control system having forward-path transfer function $G(s)$, reference input or desired output $r(t)$ and actual output $x(t)$ is illustrated by the block diagram of Figure 2.45. Defining the error to be $e(t) = r(t) - x(t)$, it follows that

$$G(s)E(s) = X(s) = R(s) - E(s)$$

giving

$$E(s) = \frac{R(s)}{1+G(s)}$$

Thus, from the final-value theorem, the steady-state error (SSE) is

$$\text{SSE} = \lim_{t\to\infty} e(t) = \lim_{s\to 0} sE(s) = \lim_{s\to 0} \frac{sR(s)}{1+G(s)} \tag{2.77}$$

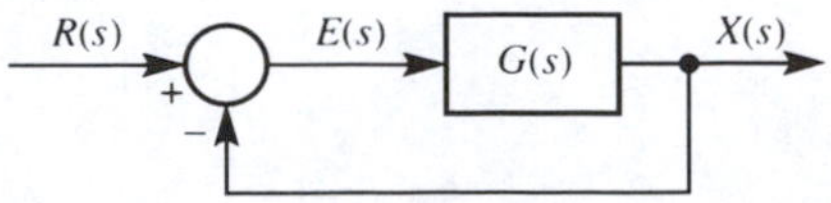

Figure 2.45 Unity feedback control system.

EXAMPLE 2.57 Determine the SSE for the system of Figure 2.45 when $G(s)$ is the same as in Example 2.50 and $r(t)$ is a step of magnitude K.

Solution Since $r(t) = KH(t)$, we have $R(s) = K/s$, so, using (2.77),

$$\text{SSE} = \lim_{s\to 0} \frac{sK/s}{1+G(s)} = \frac{K}{1+\text{SSG}}$$

where SSG = 2 as determined in Example 2.56. Thus

$$\text{SSE} = \tfrac{1}{3}K$$

It is clear from Example 2.57 that if we are to reduce the SSE, which is clearly desirable in practice, then the SSG needs to be increased. However, such an increase could lead to an undesirable transient response, and in system design a balance must be achieved. Detailed design techniques for alleviating such problems are not considered here; for such a discussion the reader is referred to specialist texts (see for example J. Schwarzenbach and K. F. Gill, *System Modelling and Control*, Edward Arnold, London, 1984).

2.6.5 Exercises

34 The response $x(t)$ of a system to a forcing function $u(t)$ is determined by the differential equation model

$$\frac{d^2x}{dt^2} + 2\frac{dx}{dt} + 5x = 3\frac{du}{dt} + 2u$$

(a) Determine the transfer function characterizing the system.
(b) Write down the characteristic equation of the system. What is the order of the system?
(c) Determine the transfer function poles and zeros, and illustrate them diagrammatically in the s plane.

35 Repeat Exercise 34 for a system whose response $x(t)$ to an input $u(t)$ is determined by the differential equation

$$\frac{d^3x}{dt^3} + 5\frac{d^2x}{dt^2} + 17\frac{dx}{dt} + 13x = \frac{d^2u}{dt^2} + 5\frac{du}{dt} + 6$$

36 Which of the following transfer functions represent stable systems and which represent unstable systems?

(a) $\dfrac{s-1}{(s+2)(s^2+4)}$ (b) $\dfrac{(s+2)(s-2)}{(s+1)(s-1)(s+4)}$

(c) $\dfrac{s-1}{(s+2)(s+4)}$ (d) $\dfrac{6}{(s^2+s+1)(s+1)^2}$

(e) $\dfrac{5(s+10)}{(s+5)(s^2-s+10)}$

37 Which of the following characteristic equations are representative of stable systems?

(a) $s^2 - 4s + 13 = 0$
(b) $5s^3 + 13s^2 + 31s + 15 = 0$
(c) $s^3 + s^2 + s + 1 = 0$
(d) $24s^4 + 11s^3 + 26s^2 + 45s + 36 = 0$
(e) $s^3 + 2s^2 + 2s + 1 = 0$

38 The differential equation governing the motion of a mass–spring–damper system with controller is

$$m\frac{d^3x}{dt^3} + c\frac{d^2x}{dt^2} + K\frac{dx}{dt} + Krx = 0$$

where m, c, K and r are positive constants. Show that the motion of the system is stable provided that $r < c/m$.

39 The behaviour of a system having a gain controller is characterized by the characteristic equation

$$s^4 + 2s^3 + (K + 2)s^2 + 7s + K = 0$$

where K is the controller gain. Show that the system is stable provided that $K > 2.1$.

40 A feedback control system has characteristic equation

$$s^3 + 15Ks^2 + (2K - 1)s + 5K = 0$$

where K is a constant gain factor. Determine the range of positive values of K for which the system will be stable.

41 Determine the impulse responses of the linear systems whose response $x(t)$ to an input $u(t)$ is determined by the following differential equations:

(a) $\frac{d^2x}{dt^2} + 15\frac{dx}{dt} + 56x = 3u(t)$

(b) $\frac{d^2x}{dt^2} + 8\frac{dx}{dt} + 25x = u(t)$

(c) $\frac{d^2x}{dt^2} - 2\frac{dx}{dt} - 8x = 4u(t)$

(d) $\frac{d^2x}{dt^2} - 4\frac{dx}{dt} + 13x = u(t)$

What can be said about the stability of each of the systems?

42 The response of a given system to a unit step $u(t) = 1H(t)$ is given by

$$x(t) = 1 - \tfrac{7}{3}e^{-t} + \tfrac{3}{2}e^{-2t} - \tfrac{1}{6}e^{-4t}$$

What is the transfer function of the system?

43 Verify the initial-value theorem for the functions

(a) $2 - 3\cos t$ (b) $(3t - 1)^2$

(c) $t + 3\sin 2t$

44 Verify the final-value theorem for the functions

(a) $1 + 3e^{-t}\sin 2t$ (b) $t^2 e^{-2t}$

(c) $3 - 2e^{-3t} + e^{-t}\cos 2t$

45 Using the final-value theorem, check the value obtained for $i_2(t)$ as $t \to \infty$ for the circuit of Example 2.28.

46 Discuss the applicability of the final-value theorem for obtaining the value of $i_2(t)$ as $t \to \infty$ for the circuit of Example 2.29.

47 Use the initial- and final-value theorems to find the jump at $t = 0$ and the limiting value as $t \to \infty$ for the solution of the initial-value problem

$$7\frac{dy}{dt} + 5y = 4 + e^{-3t} + 2\delta(t)$$

with $y(0-) = -1$.

2.6.6 Convolution

Convolution is a useful concept that has many applications in various fields of engineering. In Section 2.6.7 we shall use it to obtain the response of a linear system to any input in terms of the impulse response.

DEFINITION 2.5: Convolution

Given two piecewise-continuous functions $f(t)$ and $g(t)$, the **convolution** of $f(t)$ and $g(t)$, denoted by $f * g(t)$, is defined as

$$f * g(t) = \int_{-\infty}^{\infty} f(\tau)g(t - \tau)\,d\tau$$

In the particular case when $f(t)$ and $g(t)$ are causal functions

$$f(\tau) = g(\tau) = 0 \quad (\tau < 0), \qquad g(t - \tau) = 0 \quad (\tau > t)$$

and we have

$$f * g(t) = \int_0^t f(\tau)g(t - \tau)\,d\tau \tag{2.78}$$

The notation $f * g(t)$ indicates that the convolution $f * g$ is a function of t; that is, it could also be written as $(f * g)(t)$. The integral $\int_{-\infty}^{\infty} f(\tau)g(t - \tau)\,d\tau$ is called the **convolution integral**. Alternative names are the **superposition integral**, **Duhamel integral**, **folding integral** and **faltung integral**.

Convolution can be considered as a generalized function, and as such it has many of the properties of multiplication. In particular, the commutative law is satisfied, so that

$$f * g(t) = g * f(t)$$

or, for causal functions,

$$\int_0^t f(\tau)g(t - \tau)\,d\tau = \int_0^t f(t - \tau)g(\tau)\,d\tau \tag{2.79}$$

This means that the convolution can be evaluated by time-shifting either of the two functions. The result (2.79) is readily proved, since by making the substitution $\tau_1 = t - \tau$ in (2.78) we obtain

$$f * g(t) = \int_t^0 f(t - \tau_1)g(\tau_1)(-d\tau_1) = \int_0^t f(t - \tau_1)g(\tau_1)\,d\tau_1 = g * f(t)$$

EXAMPLE 2.58

For the two causal functions

$$f(t) = tH(t), \quad g(t) = \sin 2t\,H(t)$$

show that $f * g(t) = g * f(t)$.

Solution

$$f * g(t) = \int_0^t f(\tau)g(t - \tau)\,d\tau = \int_0^t \tau \sin 2(t - \tau)\,d\tau$$

Integrating by parts gives

$$f * g(t) = [\tfrac{1}{2}\tau \cos 2(t - \tau) + \tfrac{1}{4}\sin 2(t - \tau)]_0^t = \tfrac{1}{2}t - \tfrac{1}{4}\sin 2t$$

$$g * f(t) = \int_0^t f(t - \tau)g(\tau)\,d\tau = \int_0^t (t - \tau)\sin 2\tau\,d\tau$$

$$= [-\tfrac{1}{2}(t - \tau)\cos 2\tau - \tfrac{1}{4}\sin 2\tau]_0^t = \tfrac{1}{2}t - \tfrac{1}{4}\sin 2t$$

so that $f * g(t) = g * f(t)$.

The importance of convolution in Laplace transform work is that it enables us to obtain the inverse transform of the product of two transforms. The necessary result for doing this is contained in the following theorem.

THEOREM 2.8 Convolution theorem for Laplace transforms

If $f(t)$ and $g(t)$ are of exponential order σ, piecewise-continuous on $t \geqslant 0$ and have Laplace transforms $F(s)$ and $G(s)$ respectively, then, for $s > \sigma$

$$\mathscr{L}\left\{\int_0^t f(\tau)g(t-\tau)\,\mathrm{d}\tau\right\} = \mathscr{L}\{f * g(t)\} = F(s)G(s)$$

or, in the more useful inverse form,

$$\mathscr{L}^{-1}\{F(s)G(s)\} = f * g(t) \tag{2.80}$$

Proof By definition,

$$F(s)G(s) = \mathscr{L}\{f(t)\}\mathscr{L}\{g(t)\} = \left[\int_0^\infty \mathrm{e}^{-sx} f(x)\,\mathrm{d}x\right]\left[\int_0^\infty \mathrm{e}^{-sy} g(y)\,\mathrm{d}y\right]$$

where we have used the 'dummy' variables x and y, rather than t, in the integrals to avoid confusion. This may now be expressed in the form of the double integral

$$F(s)G(s) = \int_0^\infty\int_0^\infty \mathrm{e}^{-s(x+y)} f(x)g(y)\,\mathrm{d}x\,\mathrm{d}y = \iint_R \mathrm{e}^{-s(x+y)} f(x)g(y)\,\mathrm{d}x\,\mathrm{d}y$$

where R is the the first quadrant in the (x, y) plane, as shown in Figure 2.46(a). On making the substitution

$$x + y = t, \qquad y = \tau$$

the double integral is transformed into

$$F(s)G(s) = \iint_{R_1} \mathrm{e}^{-st} f(t-\tau)g(\tau)\,\mathrm{d}t\,\mathrm{d}\tau$$

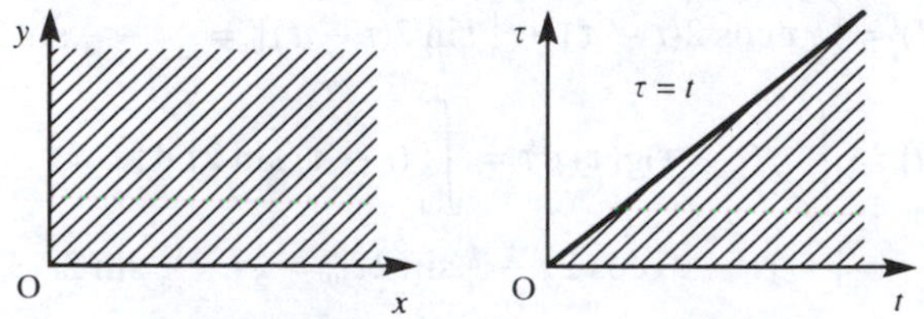

(a) Region R (b) Region R_1

Figure 2.46 Regions of integration.

where R_1 is the semi-infinite region in the (τ, t) plane bounded by the lines $\tau = 0$ and $\tau = t$, as shown in Figure 2.46(b). This may be written as

$$F(s)G(s) = \int_0^\infty e^{-st}\left(\int_0^t f(t-\tau)g(\tau)\,d\tau\right)dt$$

$$= \int_0^\infty e^{-st}[g * f(t)]\,dt$$

$$= \mathcal{L}\{g * f(t)\}$$

and, since convolution is commutative, we may write this as

$$F(s)G(s) = \mathcal{L}\{f * g(t)\}$$

which concludes the proof.   □

EXAMPLE 2.59   Using the convolution theorem, determine $\mathcal{L}^{-1}\left\{\dfrac{1}{s^2(s+2)^2}\right\}$.

Solution   We express $1/s^2(s+2)^2$ as $(1/s^2)[1/(s+2)^2]$; then, since

$$\mathcal{L}\{t\} = \frac{1}{s^2}, \quad \mathcal{L}\{t\,e^{-2t}\} = \frac{1}{(s+2)^2}$$

$$f(s) = \frac{1}{s^2}$$
$$g(s) = \frac{1}{(s+2)^2}$$

taking $f(t) = t$ and $g(t) = t\,e^{-2t}$ in the convolution theorem gives

$$\mathcal{L}^{-1}\left\{\frac{1}{s^2}\frac{1}{(s+2)^2}\right\} = \int_0^t f(t-\tau)g(\tau)\,d\tau = \int_0^t (t-\tau)\tau\,e^{-2\tau}\,d\tau$$

which on integration by parts gives

$$\mathcal{L}^{-1}\left\{\frac{1}{s^2}\frac{1}{(s+2)^2}\right\} = [-\tfrac{1}{2}e^{-2\tau}[(t-\tau)\tau + \tfrac{1}{2}(t-2\tau) - \tfrac{1}{2}]]_0^t$$

$$= \tfrac{1}{4}[t - 1 + (t+1)e^{-2t}]$$

$$f(t) = t$$
$$g(t) = e^{-2t}\,t$$
$$f * g(t) = \int_0^t \tau\,e^{-2t+2\tau}(t-\tau)\,d\tau$$

We can check this result by first expressing the given transform in partial-fractions form and then inverting to give

$$\frac{1}{s^2(s+2)^2} = \frac{-\tfrac{1}{4}}{s} + \frac{\tfrac{1}{4}}{s^2} + \frac{\tfrac{1}{4}}{s+2} + \frac{\tfrac{1}{4}}{(s+2)^2}$$

$$\int_0^t f(\tau)\,g(t-\tau) = \mathcal{L}^{-1}\{f(s)g(s)\}$$

so that

$$\mathcal{L}^{-1}\left\{\frac{1}{s^2(s+2)^2}\right\} = -\frac{1}{4} + \frac{1}{4}t + \frac{1}{4}e^{-2t} + \frac{1}{4}t\,e^{-2t} = \frac{1}{4}[t - 1 + (t+1)e^{-2t}]$$

$$\int_0^t f(\tau)\,g(t-\tau) = \mathcal{L}^{-1}\{f(s)g(s)\}$$

as before.

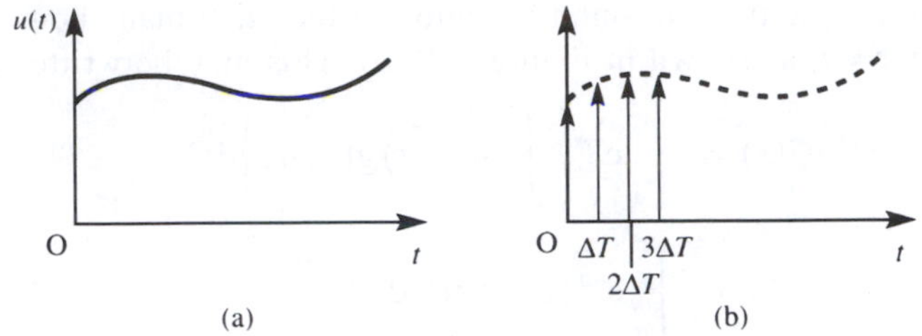

Figure 2.47 Approximation to a continuous input.

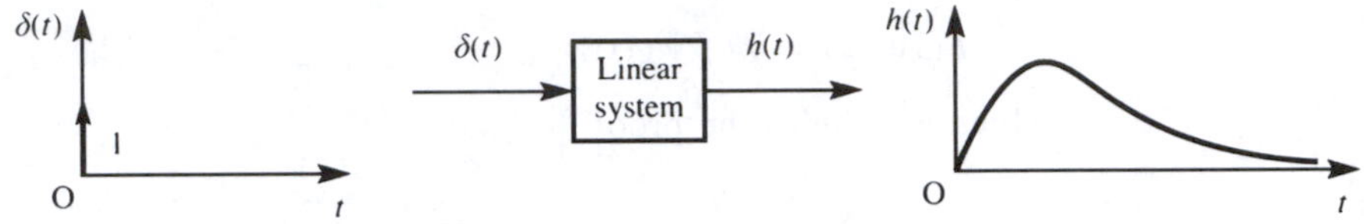

Figure 2.48 Impulse response of a linear system.

2.6.7 System response to an arbitrary input

The impulse response of a linear time-invariant system is particularly useful in practice in that it enables us to obtain the response of the system to an arbitrary input using the convolution integral. This provides engineers with a powerful approach to the analysis of dynamical systems.

Let us consider a linear system characterized by its impulse response $h(t)$. Then we wish to determine the response $x(t)$ of the system to an arbitrary input $u(t)$ such as that illustrated in Figure 2.47(a). We first approximate the continuous function $u(t)$ by an infinite sequence of impulses of magnitude $u(n\,\Delta T)$, $n = 0, 1, 2, \dots,$ as shown in Figure 2.47(b). This approximation for $u(t)$ may be written as

$$u(t) \simeq \sum_{n=0}^{\infty} u(n\Delta T)\delta(t - n\Delta T)\,\Delta T \tag{2.81}$$

Since the system is linear, the **principle of superposition** holds, so that the response of the system to the sum of the impulses is equal to the sum of the responses of the system to each of the impulses acting separately. Depicting the impulse response $h(t)$ of the linear system by Figure 2.48, the responses due to the individual impulses forming the sum in (2.81) are illustrated in the sequence of plots in Figure 2.49.

Summing the individual responses, we find that the response due to the sum of the impulses is

$$\sum_{n=0}^{\infty} u(n\Delta T)h(t - n\Delta T)\,\Delta T \tag{2.82}$$

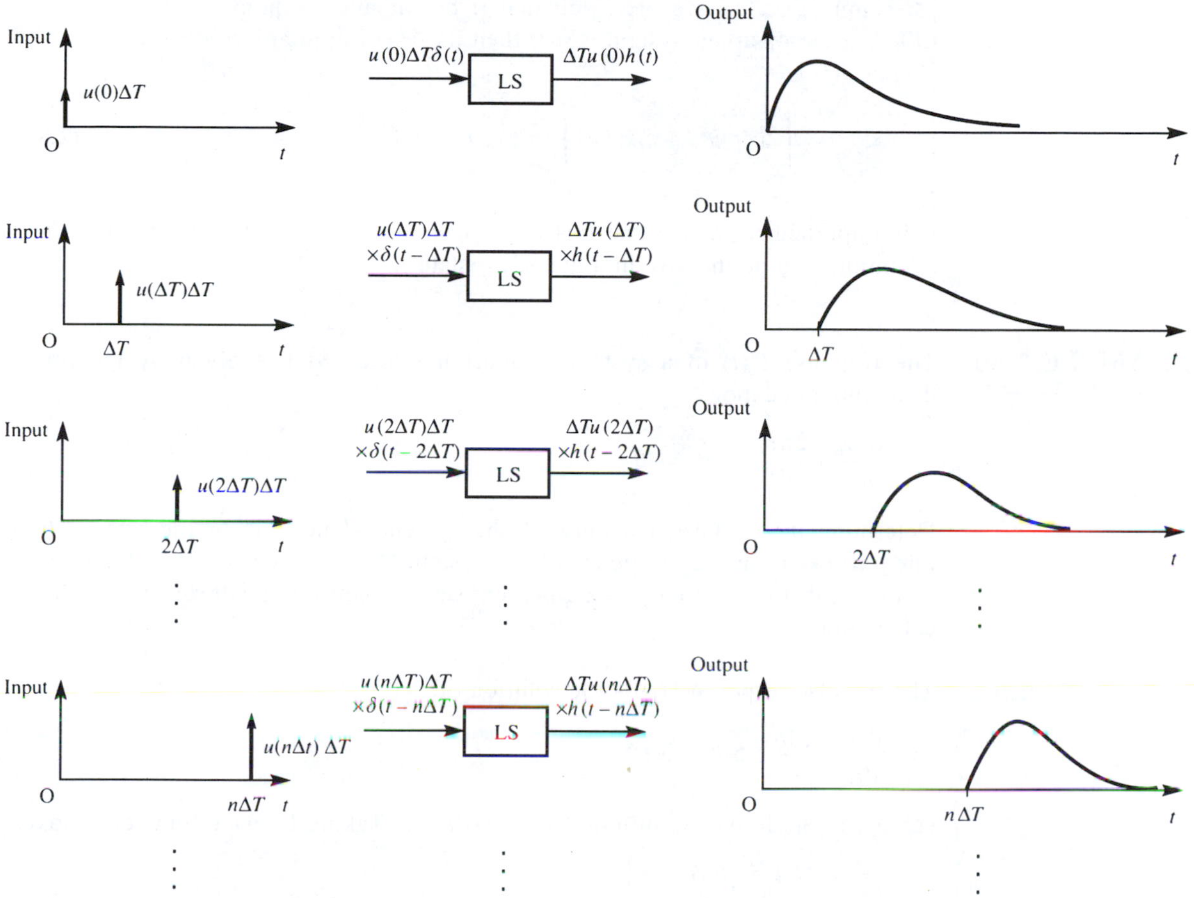

Figure 2.49 Responses due to individual impulses.

Allowing $\Delta T \to 0$, so that $n\,\Delta T$ approaches a continuous variable τ, the above sum will approach an integral that will be representative of the system response $x(t)$ to the continuous input $u(t)$. Thus

$$x(t) = \int_0^\infty u(\tau)h(t-\tau)\,\mathrm{d}\tau = \int_0^t u(\tau)h(t-\tau)\,\mathrm{d}\tau \quad \text{(since } h(t) \text{ is a causal function)}$$

That is,

$$x(t) = u * h(t)$$

Since convolution is commutative, we may also write

$$x(t) = h * u(t) = \int_0^t h(\tau)u(t-\tau)\,\mathrm{d}\tau$$

In summary, we have the result that if the impulse response of a linear time-invariant system is $h(t)$ then its response to an arbitrary input $u(t)$ is

$$x(t) = \int_0^t u(\tau)h(t-\tau)\,d\tau = \int_0^t h(\tau)u(t-\tau)\,d\tau \tag{2.83}$$

It is important to realize that this is the response of the system to the input $u(t)$ assuming it to be initially in a quiescent state.

EXAMPLE 2.60 The response $\theta_o(t)$ of a system to a driving force $\theta_i(t)$ is given by the linear differential equation

$$\frac{d^2\theta_o}{dt^2} + \frac{2d\theta_o}{dt} + 5\theta_o = \theta_i$$

Determine the impulse response of the system. Hence, using the convolution integral, determine the response of the system to a unit step input at time $t = 0$, assuming that it is initially in a quiescent state. Confirm this latter result by direct calculation.

Solution The impulse response $h(t)$ is the solution of

$$\frac{d^2h}{dt^2} + 2\frac{dh}{dt} + 5h = \delta(t)$$

subject to the initial conditions $h(0) = \dot{h}(0) = 0$. Taking Laplace transforms gives

$$(s^2 + 2s + 5)H(s) = \mathscr{L}\{\delta(t)\} = 1$$

so that

$$H(s) = \frac{1}{s^2 + 2s + 5} = \frac{1}{2}\frac{2}{(s+1)^2 + 2^2}$$

which, on inversion, gives the impulse response as

$$h(t) = \tfrac{1}{2}e^{-t}\sin 2t$$

Using the convolution integral

$$\theta_o(t) = \int_0^t h(\tau)\theta_i(t-\tau)\,d\tau$$

with $\theta_i(t) = 1H(t)$ gives the response to the unit step as

$$\theta_o(t) = \tfrac{1}{2}\int_0^t e^{-\tau}\sin 2\tau\,d\tau$$

Integrating by parts twice gives

$$\theta_o(t) = -\tfrac{1}{2}e^{-t}\sin 2t - e^{-t}\cos 2t + 1 - 2\int_0^t e^{-\tau}\sin 2\tau\,d\tau$$
$$= -\tfrac{1}{2}e^{-t}\sin 2t - e^{-t}\cos 2t + 1 - 4\theta_o(t)$$

Hence

$$\theta_o(t) = \tfrac{1}{5}(1 - e^{-t}\cos 2t - \tfrac{1}{2}e^{-t}\sin 2t)$$

(Note that in this case, because of the simple form of $\theta_i(t)$, the convolution integral $\int_0^t h(\tau)\theta_i(t-\tau)\,d\tau$ is taken in preference to $\int_0^t \theta_i(\tau)h(t-\tau)\,d\tau$.)

To obtain the step response directly, we need to solve for $t \geq 0$ the differential equation

$$\frac{d^2\theta_o}{dt^2} + 2\frac{d\theta_o}{dt} + 5\theta_o = 1$$

subject to the initial conditions $\theta_o(0) = \dot{\theta}_o(0) = 0$. Taking Laplace transforms gives

$$(s^2 + 2s + 5)\Theta(s) = \frac{1}{s}$$

so that

$$\Theta = \frac{1}{s(s^2 + 2s + 5)}$$
$$= \frac{\frac{1}{5}}{s} - \frac{1}{5}\frac{s+2}{(s+1)^2 + 4}$$

which, on inversion, gives

$$\theta_o(t) = \tfrac{1}{5} - \tfrac{1}{5}e^{-t}(\cos 2t + \tfrac{1}{2}\sin 2t)$$
$$= \tfrac{1}{5}(1 - e^{-t}\cos 2t - \tfrac{1}{2}e^{-t}\sin 2t)$$

confirming the previous result.

We therefore see that a linear time-invariant system may be characterized in the frequency domain (or s domain) by its transfer function $G(s)$ or in the time domain by its impulse response $h(t)$, as depicted in Figures 2.50(a) and (b) respectively. The response in the frequency domain is obtained by algebraic multiplication, while the time-domain response involves a convolution. This equivalence of the operation of convolution in the time domain with algebraic multiplication in

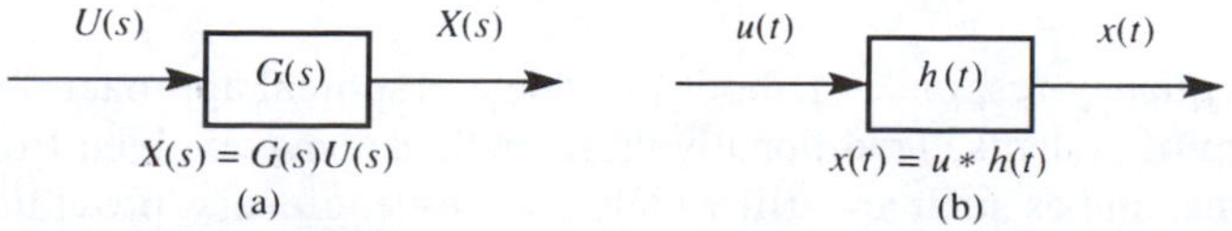

Figure 2.50 (a) Frequency-domain and (b) time-domain representations of a linear time-invariant system.

the frequency domain is clearly a powerful argument for the use of frequency-domain techniques in engineering design.

2.6.8 Exercises

48 For the following pairs of causal functions $f(t)$ and $g(t)$ show that $f * g(t) = g * f(t)$:

(a) $f(t) = t, \quad g(t) = \cos 3t$

(b) $f(t) = t + 1, \quad g(t) = e^{-2t}$

(c) $f(t) = t^2, \quad g(t) = \sin 2t$

(d) $f(t) = e^{-t}, \quad g(t) = \sin t$

49 Using the convolution theorem, determine the following inverse Laplace transforms. Check your results by first expressing the given transform in partial-fractions form and then inverting using the standard results

(a) $\mathscr{L}^{-1}\left\{\dfrac{1}{s(s+3)^3}\right\}$ (b) $\mathscr{L}^{-1}\left\{\dfrac{1}{(s-2)^2(s+3)^2}\right\}$

(c) $\mathscr{L}^{-1}\left\{\dfrac{1}{s^2(s+4)}\right\}$

50 Taking $f(\lambda) = \lambda$ and $g(\lambda) = e^{-\lambda}$, use the inverse form (2.80) of the convolution theorem to show that the solution of the integral equation

$$y(t) = \int_0^t \lambda e^{-(t-\lambda)} d\lambda$$

is

$$y(t) = (t - 1) + e^{-t}.$$

51 Find the impulse response of the system characterized by the differential equation

$$\frac{d^2x}{dt^2} + 7\frac{dx}{dt} + 12x = u(t)$$

and hence find the response of the system to the pulse input $u(t) = A[H(t) - H(t - T)]$, assuming that it is initially in a quiescent state.

52 The response $\theta_o(t)$ of a servomechanism to a driving force $\theta_i(t)$ is given by the second-order differential equation

$$\frac{d^2\theta_o}{dt^2} + 4\frac{d\theta_o}{dt} + 5\theta_o = \theta_i \quad (t \geqslant 0)$$

Determine the impulse response of the system, and hence, using the convolution integral, obtain the response of the servomechanism to a unit step driving force, applied at time $t = 0$, given that the system is initially in a quiescent state.

Check your answer by directly solving the differential equation

$$\frac{d^2\theta_o}{dt^2} + 4\frac{d\theta_o}{dt} + 5\theta_o = 1$$

subject to the initial conditions $\theta_o = \dot{\theta}_o = 0$ when $t = 0$.

2.7 Engineering application: frequency response

Frequency-response methods provide a graphical approach for the analysis and design of systems. Traditionally these methods have evolved from practical considerations, and as such are still widely used by engineers, providing tremendous insight into overall system behaviour. In this section we shall illustrate how the frequency response can be readily obtained from the system transfer function $G(s)$ by simply replacing s by $j\omega$. Methods of representing it graphically will also be considered.

Consider the system depicted in Figure 2.41, with transfer function

$$G(s) = \frac{K(s-z_1)(s-z_2)\cdots(s-z_m)}{(s-p_1)(s-p_2)\cdots(s-p_n)} \quad (m \leqslant n) \tag{2.84}$$

When the input is the sinusoidally varying signal

$$u(t) = A \sin \omega t$$

applied at time $t = 0$, the system response $x(t)$ for $t \geqslant 0$ is determined by

$$X(s) = G(s)\mathscr{L}\{A \sin \omega t\}$$

That is,

$$\begin{aligned} X(s) &= G(s)\frac{A\omega}{s^2+\omega^2} \\ &= \frac{KA\omega(s-z_1)(s-z_2)\cdots(s-z_m)}{(s-p_1)(s-p_2)\cdots(s-p_n)(s-\mathrm{j}\omega)(s+\mathrm{j}\omega)} \end{aligned}$$

which, on expanding in partial fractions, gives

$$X(s) = \frac{\alpha_1}{s-\mathrm{j}\omega} + \frac{\alpha_2}{s+\mathrm{j}\omega} + \sum_{i=1}^{n} \frac{\beta_i}{s-p_i}$$

where α_1, α_2, β_1, β_2, . . . , β_n are constants. Here the first two terms in the summation are generated by the input and determine the steady-state response, while the remaining terms are generated by the transfer function and determine the system transient response.

Taking inverse Laplace transforms, the system response $x(t)$, $t \geqslant 0$, is given by

$$x(t) = \alpha_1 \mathrm{e}^{\mathrm{j}\omega t} + \alpha_2 \mathrm{e}^{-\mathrm{j}\omega t} + \sum_{i=1}^{n} \beta_i \mathrm{e}^{p_i t} \quad (t \geqslant 0)$$

In practice we are generally concerned with systems that are stable, for which the poles p_i, $i = 1, 2, \ldots, n$, of the transfer function $G(s)$ lie in the left half of the s plane. Consequently, for practical systems the time-domain terms $\beta_i \mathrm{e}^{p_i t}$, $i = 1, 2, \ldots, n$, decay to zero as t increases, and will not contribute to the steady-state response $x_{ss}(t)$ of the system. Thus for stable linear systems the latter is determined by the first two terms as

$$x_{ss}(t) = \alpha_1 \mathrm{e}^{\mathrm{j}\omega t} + \alpha_2 \mathrm{e}^{-\mathrm{j}\omega t}$$

Using the 'cover-up' rule for determining the coefficients α_1 and α_2 in the partial-fraction expansions gives

$$\alpha_1 = \left[\frac{(s-\mathrm{j}\omega)G(s)A\omega}{(s-\mathrm{j}\omega)(s+\mathrm{j}\omega)}\right]_{s=\mathrm{j}\omega} = \frac{A}{2\mathrm{j}}G(\mathrm{j}\omega)$$

$$\alpha_2 = \left[\frac{(s+\mathrm{j}\omega)G(s)A\omega}{(s-\mathrm{j}\omega)(s+\mathrm{j}\omega)}\right]_{s=-\mathrm{j}\omega} = -\frac{A}{2\mathrm{j}}G(-\mathrm{j}\omega)$$

so that the steady-state response becomes

$$x_{ss}(t) = \frac{A}{2\mathrm{j}} G(\mathrm{j}\omega)\, \mathrm{e}^{\mathrm{j}\omega t} - \frac{A}{2\mathrm{j}} G(-\mathrm{j}\omega)\, \mathrm{e}^{-\mathrm{j}\omega t} \tag{2.85}$$

$G(\mathrm{j}\omega)$ can be expressed in the polar form

$$G(\mathrm{j}\omega) = |G(\mathrm{j}\omega)|\, \mathrm{e}^{\mathrm{j} \arg G(\mathrm{j}\omega)}$$

where $|G(\mathrm{j}\omega)|$ denotes the magnitude (or modulus) of $G(\mathrm{j}\omega)$. (Note that both the magnitude and argument vary with frequency ω.) Then, assuming that the system has real parameters,

$$G(-\mathrm{j}\omega) = |G(\mathrm{j}\omega)|\, \mathrm{e}^{-\mathrm{j} \arg G(\mathrm{j}\omega)}$$

and the steady-state response (2.85) becomes

$$\begin{aligned} x_{ss}(t) &= \frac{A}{2\mathrm{j}} [|G(\mathrm{j}\omega)|\mathrm{e}^{\mathrm{j} \arg G(\mathrm{j}\omega)}]\mathrm{e}^{\mathrm{j}\omega t} - \frac{A}{2\mathrm{j}} [|G(\mathrm{j}\omega)|\mathrm{e}^{-\mathrm{j} \arg G(\mathrm{j}\omega)}]\mathrm{e}^{-\mathrm{j}\omega t} \\ &= \frac{A}{2\mathrm{j}} |G(\mathrm{j}\omega)| [\mathrm{e}^{\mathrm{j}[\omega t + \arg G(\mathrm{j}\omega)]} - \mathrm{e}^{-\mathrm{j}[\omega t + \arg G(\mathrm{j}\omega)]}] \end{aligned}$$

That is,

$$x_{ss}(t) = A\, |G(\mathrm{j}\omega)| \sin[\omega t + \arg G(\mathrm{j}\omega)] \tag{2.86}$$

This indicates that if a stable linear system with transfer function $G(s)$ is subjected to a sinusoidal input then

(a) the steady-state system response is also a sinusoid having the same frequency ω as the input;

(b) the amplitude of this response is $|G(\mathrm{j}\omega)|$ times the amplitude A of the input sinusoid; the input is said to be **amplified** if $|G(\mathrm{j}\omega)| > 1$ and **attenuated** if $|G(\mathrm{j}\omega)| < 1$;

(c) the phase shift between input and output is $\arg G(\mathrm{j}\omega)$. The system is said to **lead** if $\arg G(\mathrm{j}\omega) > 0$ and **lag** if $\arg G(\mathrm{j}\omega) < 0$.

The variations in both the magnitude $|G(\mathrm{j}\omega)|$ and argument $\arg G(\mathrm{j}\omega)$ as the frequency ω of the input sinusoid is varied constitute the **frequency response of the system**, the magnitude $|G(\mathrm{j}\omega)|$ representing the **amplitude gain** or **amplitude ratio** of the system for sinusoidal input with frequency ω, and the argument $\arg G(\mathrm{j}\omega)$ representing the **phase shift**.

The result (2.86) implies that the function $G(\mathrm{j}\omega)$ may be found experimentally by subjecting a system to sinusoidal excitations and measuring the amplitude gain and phase shift between output and input as the input frequency is varied over the range $0 < \omega < \infty$. In principle, therefore, frequency-response measurements may be used to determine the system transfer function $G(s)$.

In Chapters 4 and 5, dealing with Fourier series and Fourier transforms, we shall see that most functions can be written as sums of sinusoids, and consequently the response of a linear system to almost any input can be deduced in the form of the corresponding sinusoidal responses. It is important, however, to appreciate that the term 'response' in the expression 'frequency response' only relates to the steady-state response behaviour of the system.

The information contained in the system frequency response may be conveniently displayed in graphical form. In practice it is usual to represent it by two graphs: one showing how the amplitude $|G(j\omega)|$ varies with frequency and one showing how the phase shift arg $G(j\omega)$ varies with frequency.

EXAMPLE 2.61 Determine the frequency response of the *RC* filter shown in Figure 2.51. Sketch the amplitude and phase-shift plots.

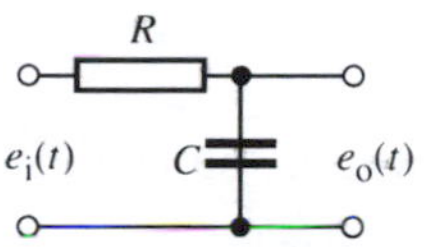

Figure 2.51 *RC* filter.

Solution The input–output relationship is given by

$$E_o(s) = \frac{1}{RCs+1}E_i(s)$$

so that the filter is characterized by the transfer function

$$G(s) = \frac{1}{RCs+1}$$

Therefore

$$G(j\omega) = \frac{1}{RCj\omega+1} = \frac{1-jRC\omega}{1+R^2C^2\omega^2}$$

$$= \frac{1}{1+R^2C^2\omega^2} - j\frac{RC\omega}{1+R^2C^2\omega^2}$$

giving the frequency-response characteristics

$$\text{amplitude ratio} = |G(j\omega)| = \sqrt{\left[\frac{1}{(1+R^2C^2\omega^2)^2} + \frac{R^2C^2\omega^2}{(1+R^2C^2\omega^2)^2}\right]}$$

$$= \frac{1}{\sqrt{(1+R^2C^2\omega^2)}}$$

$$\text{phase shift} = \arg G(j\omega) = -\tan^{-1}(RC\omega)$$

Note that for $\omega = 0$

$$|G(j\omega)| = 1, \quad \arg G(j\omega) = 0$$

and as $\omega \to \infty$

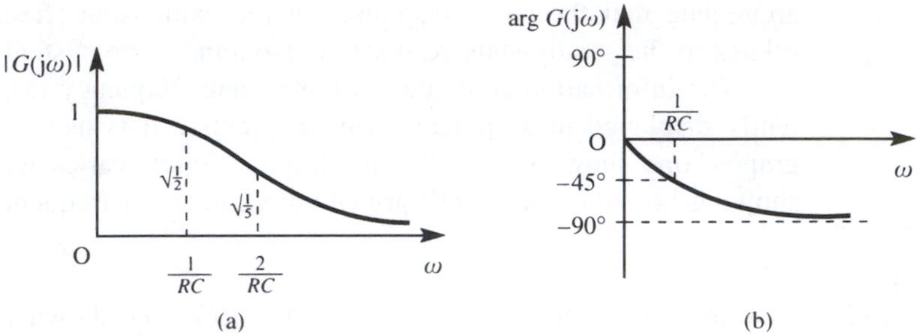

Figure 2.52 Frequency-response plots for Example 2.61: (a) amplitude plot; (b) phase-shift plot.

$$|G(j\omega)| \to 0, \quad \arg G(j\omega) \to -\tfrac{1}{2}\pi$$

Plots of the amplitude and phase-shift curves are shown in Figures 2.52(a) and (b) respectively.

For the simple transfer function of Example 2.61, plotting the amplitude and phase-shift characteristics was relatively easy. For higher-order transfer functions it can be a rather tedious task, and it would be far more efficient to use a suitable computer package. However, to facilitate the use of frequency-response techniques in system design, engineers adopt a different approach, making use of **Bode plots** to display the relevant information. This approach is named after H. W. Bode, who developed the techniques at the Bell Laboratories in the late 1930s. Again it involves drawing separate plots of amplitude and phase shift, but in this case on semi-logarithmic graph paper, with frequency plotted on the horizontal logarithmic axis and amplitude, or phase, on the vertical linear axis. It is also normal to express the amplitude gain in decibels (dB); that is,

$$\text{amplitude gain in dB} = 20 \log |G(j\omega)|$$

and the phase shift $\arg G(j\omega)$ in degrees. Thus the Bode plots consist of

(a) a plot of amplitude in decibels versus $\log \omega$, and

(b) a plot of phase shift in degrees versus $\log \omega$.

Note that with the amplitude gain measured in decibels, the input signal will be amplified if the gain is greater than zero and attenuated if it is less than zero.

The advantage of using Bode plots is that the amplitude and phase information can be obtained from the constituent parts of the transfer function by graphical addition. It is also possible to make simplifying approximations in which curves can be replaced by straight-line asymptotes. These can be drawn relatively quickly, and provide sufficient information to give an engineer a 'feel' for the system behaviour. Desirable system characteristics are frequently specified in terms of frequency-response behaviour, and since the approximate Bode plots permit quick determination of the effect of changes, they provide a good test for the system designer.

EXAMPLE 2.62 Draw the approximate Bode plots corresponding to the transfer function

$$G(s) = \frac{4 \times 10^3(5+s)}{s(100+s)(20+s)} \tag{2.87}$$

Solution First we express the transfer function in what is known as the **standard form**, namely

$$G(s) = \frac{10(1+0.2s)}{s(1+0.01s)(1+0.05s)}$$

giving

$$G(j\omega) = \frac{10(1+j0.2\omega)}{j\omega(1+j0.01\omega)(1+j0.05\omega)}$$

Taking logarithms to base 10,

$$20 \log|G(j\omega)| = 20 \log 10 + 20 \log|1 + j0.2\omega| - 20 \log|j\omega| - 20 \log|1 + j0.01\omega| - 20 \log|1 + j0.05\omega|$$

$$\arg G(j\omega) = \arg 10 + \arg(1 + j0.2\omega) - \arg j\omega - \arg(1 + j0.01\omega) - \arg(1 + j0.05\omega) \tag{2.88}$$

The transfer function involves constituents that are again a simple zero and simple poles (including one at the origin). We shall now illustrate how the Bode plots can be built up from those of the constituent parts.

Consider first the amplitude gain plot, which is a plot of $20 \log|G(j\omega)|$ versus $\log \omega$:

(a) for a simple gain k a plot of $20 \log k$ is a horizontal straight line, being above the 0 dB axis if $k > 1$ and below it if $k < 1$;

(b) for a simple pole at the origin a plot of $-20 \log \omega$ is a straight line with slope -20 dB/decade and intersecting the 0 dB axis at $\omega = 1$;

(c) for a simple zero or pole not at the origin we see that

$$20 \log|1 + j\tau\omega| \to \begin{cases} 0 & \text{as } \omega \to 0 \\ 20 \log \tau\omega = 20 \log \omega - 20 \log(1/\tau) & \text{as } \omega \to \infty \end{cases}$$

Note that the graph of $20 \log \tau\omega$ is a straight line with slope 20 dB/decade and intersecting the 0 dB axis at $\omega = 1/\tau$. Thus the plot of $20 \log|1 + j\tau\omega|$ may be approximated by two straight lines: one for $\omega < 1/\tau$ and one for $\omega > 1/\tau$. The frequency at intersection $\omega = 1/\tau$ is called the **breakpoint** or **corner frequency**; here $|1 + j\tau\omega| = \sqrt{2}$, enabling the true curve to be indicated at this frequency. Using this approach, straight-line approximations to the amplitude plots of a simple zero and a simple pole, neither at zero, are shown in Figures 2.53(a) and (b) respectively (actual plots are also shown).

Using the approximation plots for the constituent parts as indicated in (a)–(c) earlier, we can build up the approximate amplitude gain plot corresponding to

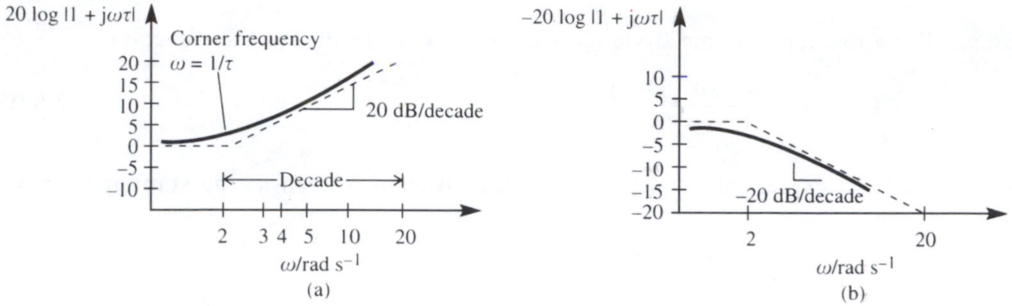

Figure 2.53 Straight-line approximations to Bode amplitude plots: (a) simple zero; (b) simple pole.

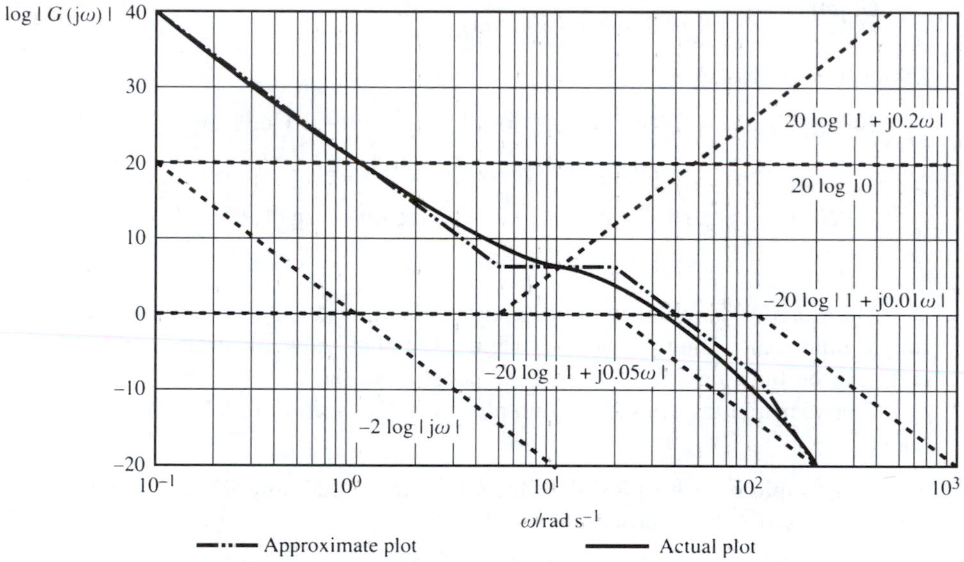

Figure 2.54 Amplitude Bode plots for the $G(s)$ of Example 2.62.

(2.87) by graphical addition as indicated in Figure 2.54. The actual amplitude gain plot, produced using a software package, is also shown.

The idea of using asymptotes can also be used to draw the phase-shift Bode plots; again taking account of the accumulated effects of the individual components making up the transfer function, namely that

(i) the phase shift associated with a constant gain k is zero;

(ii) the phase shift associated with a simple pole or zero at the origin is +90° or −90° respectively;

(iii) for a simple zero or pole not at the origin

$$\tan^{-1}(\omega\tau) \to \begin{cases} 0 & \text{as } \omega \to 0 \\ 90^\circ & \text{as } \omega \to \infty \end{cases}$$

$$\tan^{-1}(\omega\tau) = 45^\circ \quad \text{when } \omega\tau = 1$$

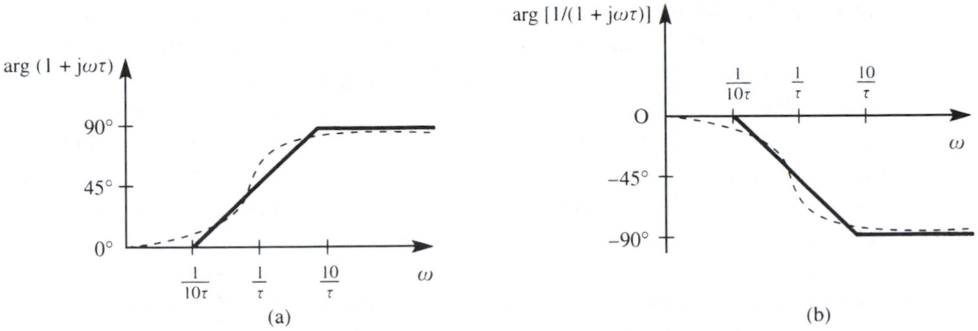

Figure 2.55 Approximate Bode phase-shift plots: (a) simple zero; (b) simple pole.

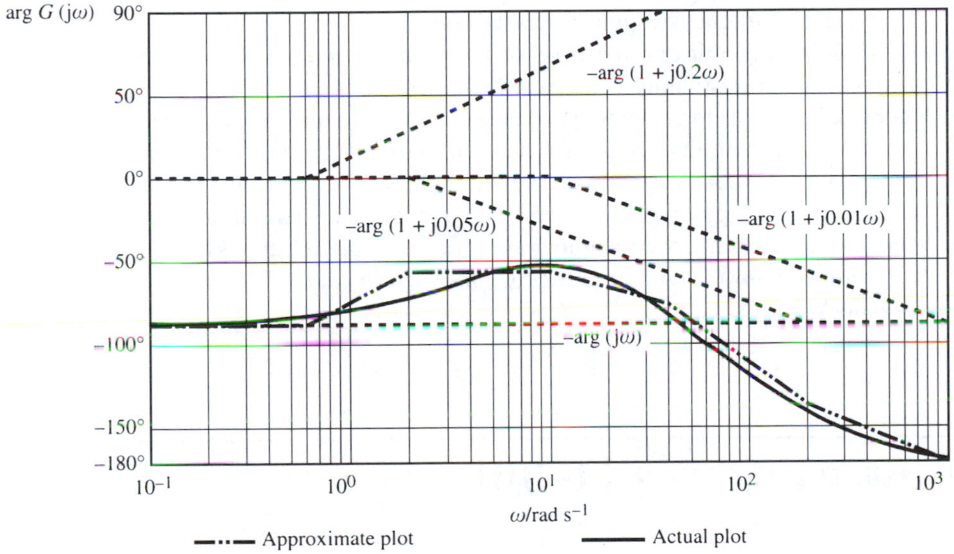

Figure 2.56 Phase-shift Bode plot for the $G(s)$ of Example 2.62.

With these observations in mind, the following approximations are made. For frequencies ω less than one-tenth of the corner frequency $\omega = 1/\tau$ (that is, for $\omega < 1/10\tau$) the phase shift is assumed to be 0°, and for frequencies greater than ten times the corner frequency (that is, for $\omega > 10/\tau$) the phase shift is assumed to be $\pm 90^\circ$. For frequencies between these limits (that is, $1/10\tau < \pi < 10/\tau$) the phase-shift plot is taken to be a straight line that passes through 0° at $\omega = 1/10\tau$, $\pm 45^\circ$ at $\omega = 1/\tau$, and $\pm 90^\circ$ at $\omega = 10/\tau$. In each case the plus sign is associated with a zero and the minus sign with a pole. With these assumptions, straight-line approximations to the phase-shift plots for a simple zero and pole, neither located at the origin, are shown in Figures 2.55(a) and (b) respectively (the actual plots are represented by the broken curves).

Using these approximations, a straight-line approximate phase-gain plot corresponding to (2.88) is shown in Figure 2.56. Again, the actual phase-gain plot, produced using a software package, is shown.

In the graphical approach adopted in this section, separate plots of amplitude gain and phase shift versus frequency have been drawn. It is also possible to represent the frequency response graphically using only one plot. When this is done using the pair of polar coordinates $(|G(j\omega)|, \arg G(j\omega))$ and allowing the frequency ω to vary, the resulting Argand diagram is referred to as the **polar plot** or **frequency-response plot**. Such a graphical representation of the transfer function forms the basis of the **Nyquist approach** to the analysis of feedback systems. In fact, the main use of frequency-response methods in practice is in the analysis and design of closed-loop control systems. For the unity feedback system of Figure 2.45 the frequency-response plot of the forward-path transfer function $G(s)$ is used to infer overall closed-loop system behaviour. The Bode plots are perhaps the quickest plots to construct, especially when straight-line approximations are made, and are useful when attempting to estimate a transfer function from a set of physical frequency-response measurements. Other plots used in practice are the **Nichols diagram** and the **inverse Nyquist** (or **polar**) **plot**, the first of these being useful for designing feedforward compensators and the second for designing feedback compensators. Although there is no simple mathematical relationship, it is also worth noting that transient behaviour may also be inferred from the various frequency-response plots. For example, the reciprocal of the inverse M circle centred on the −1 point in the inverse Nyquist plot gives an indication of the peak overshoot in the transient behaviour (see, for example, G. Franklin, D. Powell and A. Naeini-Emami (1986), *Feedback Control of Dynamic Systems*, Reading, MA: Addison-Wesley).

2.8 Review exercises (1–30)

1 Solve, using Laplace transforms, the following differential equations:

(a) $$\frac{d^2x}{dt^2} + 4\frac{dx}{dt} + 5x = 8\cos t$$

subject to $x = \frac{dx}{dt} = 0$ at $t = 0$

(b) $$5\frac{d^2x}{dt^2} - 3\frac{dx}{dt} - 2x = 6$$

subject to $x = 1$ and $\frac{dx}{dt} = 1$ at $t = 0$

2 (a) Find the inverse Laplace transform of

$$\frac{1}{(s+1)(s+2)(s^2+2s+2)}$$

(b) A voltage source $V e^{-t}\sin t$ is applied across a series LCR circuit with $L = 1$, $R = 3$ and $C = \frac{1}{2}$. Show that the current $i(t)$ in the circuit satisfies the differential equation

$$\frac{d^2i}{dt^2} + 3\frac{di}{dt} + 2i = V e^{-t}\sin t$$

Find the current $i(t)$ in the circuit at time $t \geqslant 0$ if $i(t)$ satisfies the initial conditions $i(0) = 1$ and $(di/dt)(0) = 2$.

3 Use Laplace transform methods to solve the simultaneous differential equations

$$\frac{d^2x}{dt^2} - x + 5\frac{dy}{dt} = t$$

$$\frac{d^2y}{dt^2} - 4y - 2\frac{dx}{dt} = -2$$

subject to $x = y = \frac{dx}{dt} = \frac{dy}{dt} = 0$ at $t = 0$.

4 Solve the differential equation

$$\frac{d^2x}{dt^2} + 2\frac{dx}{dt} + 2x = \cos t$$

subject to the initial conditions $x = x_0$ and $dx/dt = x_1$ at $t = 0$. Identify the steady-state and transient solutions. Find the amplitude and phase shift of the steady-state solution.

5 Resistors of 5 and 20 Ω are connected to the primary and secondary coils of a transformer with inductances as shown in Figure 2.57. At time $t = 0$, with no currents flowing, a voltage $E = 100$ V is applied to the primary circuit. Show that subsequently the current in the secondary circuit is

$$\frac{20}{\sqrt{41}}(e^{-(11+\sqrt{41})t/2} - e^{-(11-\sqrt{41})t/2})$$

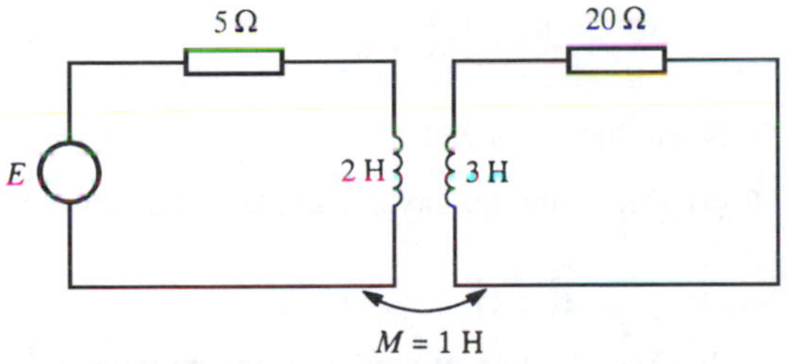

Figure 2.57 Circuit of Review exercise 5.

6 (a) Find the Laplace transforms of

(i) $\cos(\omega t + \phi)$ (ii) $e^{-\omega t}\sin(\omega t + \phi)$

(b) Using Laplace transform methods, solve the differential equation

$$\frac{d^2x}{dt^2} + 4\frac{dx}{dt} + 8x = \cos 2t$$

given that $x = 2$ and $dx/dt = 1$ when $t = 0$.

7 (a) Find the inverse Laplace transform of

$$\frac{s-4}{s^2+4s+13}$$

(b) Solve using Laplace transforms the differential equation

$$\frac{dy}{dt} + 2y = 2(2 + \cos t + 2\sin t)$$

given that $y = -3$ when $t = 0$.

8 Using Laplace transforms, solve the simultaneous differential equations

$$\frac{dx}{dt} + 5x + 3y = 5\sin t - 2\cos t$$

$$\frac{dy}{dt} + 3y + 5x = 6\sin t - 3\cos t$$

where $x = 1$ and $y = 0$ when $t = 0$.

9 The charge q on a capacitor in an inductive circuit is given by the differential equation

$$\frac{d^2q}{dt^2} + 300\frac{dq}{dt} + 2\times10^4 q = 200\sin 100t$$

and it is also known that both q and dq/dt are zero when $t = 0$. Use the Laplace transform method to find q. What is the phase difference between the steady-state component of the current dq/dt and the applied emf $200\sin 100t$ to the nearest half-degree?

10 Use Laplace transforms to find the value of x given that

$$4\frac{dx}{dt} + 6x + y = 2\sin 2t$$

$$\frac{d^2x}{dt^2} + x - \frac{dy}{dt} = 3e^{-2t}$$

and that $x = 2$ and $dx/dt = -2$ when t = 0.

11 (a) Use Laplace transforms to solve the differential equation

$$\frac{d^2\theta}{dt^2} + 8\frac{d\theta}{dt} + 16\theta = \sin 2t$$

given that $\theta = 0$ and $d\theta/dt = 0$ when $t = 0$.

(b) Using Laplace transforms, solve the simultaneous differential equations

$$\frac{di_1}{dt} + 2i_1 + 6i_2 = 0$$

$$i_1 + \frac{di_2}{dt} - 3i_2 = 0$$

given that $i_1 = 1$, $i_2 = 0$ when $t = 0$.

12 The terminals of a generator producing a voltage V are connected through a wire of resistance R and a coil of inductance L (and negligible resistance). A capacitor of capacitance C is connected in parallel with the resistance R as shown in Figure 2.58.

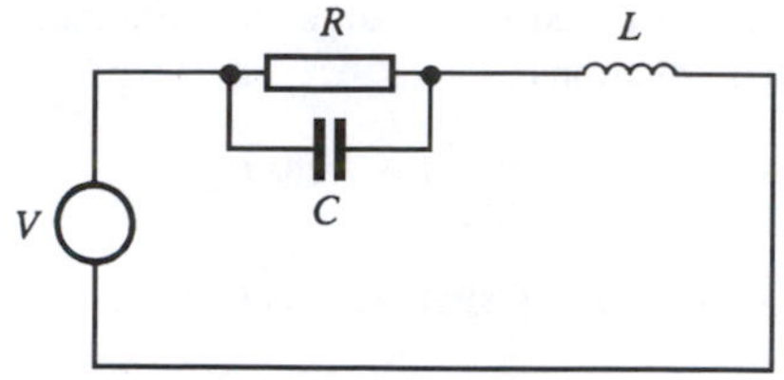

Figure 2.58 Circuit of Review exercise 12.

Show that the current i flowing through the resistance R is given by

$$LCR\frac{d^2i}{dt^2} + L\frac{di}{dt} + Ri = V$$

Suppose that

(i) $V = 0$ for $t < 0$ and $V = E$ (constant) for $t \geq 0$
(ii) $L = 2R^2C$
(iii) $CR = 1/2n$

and show that the equation reduces to

$$\frac{d^2i}{dt^2} + 2n\frac{di}{dt} + 2n^2i = 2n^2\frac{E}{R}$$

Hence, assuming that $i = 0$ and $di/dt = 0$ when $t = 0$, use Laplace transforms to obtain an expression for i in terms of t.

13 Show that the currents in the coupled circuits of Figure 2.59 are determined by the simultaneous differential equations

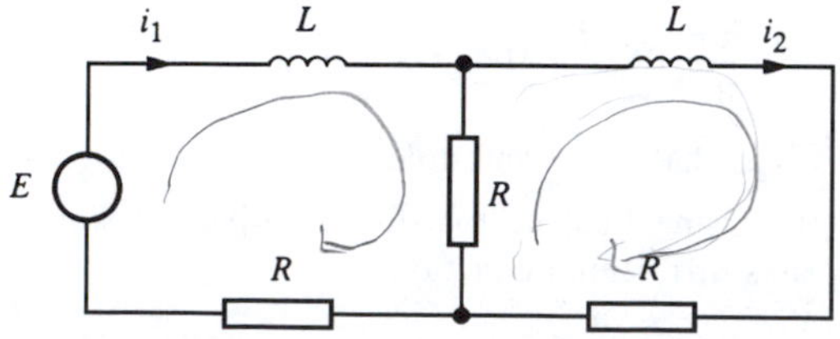

Figure 2.59 Circuit of Review exercise 13.

$$E = L\frac{di}{dt} + R(i_1 - i_2) + Ri_1$$
$$(i_2 - i_1)R + L\frac{di_2}{dt} + Ri_2 = 0$$

$$L\frac{di_1}{dt} + R(i_1 - i_2) + Ri_1 = E$$

$$L\frac{di_2}{dt} + Ri_2 - R(i_1 - i_2) = 0$$

Find i_1 in terms of t, L, E and R, given that $i_1 = 0$ and $di_1/dt = E/L$ at $t = 0$, and show that $i_1 \simeq \frac{2}{3}E/R$ for large t. What does i_2 tend to for large t?

14 A system consists of two unit masses lying in a straight line on a smooth surface and connected together to two fixed points by three springs. When a sinusoidal force is applied to the system, the displacements $x_1(t)$ and $x_2(t)$ of the respective masses from their equilibrium positions satisfy the equations

$$\frac{d^2x_1}{dt^2} = x_2 - 2x_1 + \sin 2t$$

$$\frac{d^2x_2}{dt^2} = -2x_2 + x_1$$

Given that the system is initially at rest in the equilibrium position ($x_1 = x_2 = 0$), use the Laplace transform method to solve the equations for $x_1(t)$ and $x_2(t)$.

15 (a) Obtain the inverse Laplace transforms of

(i) $\dfrac{s+4}{s^2+2s+10}$ (ii) $\dfrac{s-3}{(s-1)^2(s-2)}$

(b) Use Laplace transforms to solve the differential equation

$$\frac{d^2y}{dt^2} + 2\frac{dy}{dt} + y = 3t\,e^{-t}$$

given that $y = 4$ and $dy/dt = 2$ when $t = 0$.

16 (a) Determine the inverse Laplace transform of

$$\frac{5}{s^2 - 14s + 53}$$

(b) The equation of motion of the moving coil of a galvanometer when a current i is passed thought it is of the form

$$\frac{d^2\theta}{dt^2} + 2K\frac{d\theta}{dt} + n^2\theta = \frac{n^2i}{K}$$

where θ is the angle of deflection from the 'no-current' position and n and K are positive constants. Given that i is a constant and $\theta = 0 = d\theta/dt$ when $t = 0$, obtain an expression for the Laplace transform of θ.

In constructing the galvanometer, it is desirable to have it critically damped (that is, $n = K$). Use the Laplace transform method to solve the differential equation in this case, and sketch the graph of θ against t for positive values of t.

17 (a) Given that α is a positive constant, use the second shift theorem to

(i) show that the Laplace transform of $\sin t\, H(t - \alpha)$ is

$$e^{-\alpha s}\frac{\cos\alpha + s\sin\alpha}{s^2+1}$$

(ii) find the inverse transform of

$$\frac{s\,e^{-\alpha s}}{s^2+2s+5}$$

(b) Solve the differential equation

$$\frac{d^2y}{dt^2}+2\frac{dy}{dt}+5y = \sin t - \sin t\,H(t-\pi)$$

given that $y = dy/dt = 0$ when $t = 0$.

18 Show that the Laplace transform of the voltage $v(t)$, with period T, defined by

$$v(t) = \begin{cases} 1 & (0 \leqslant t < \frac{1}{2}T) \\ -1 & (\frac{1}{2}T \leqslant t < T) \end{cases} \qquad v(t+T) = v(t)$$

is

$$V(s) = \frac{1}{s}\frac{1-e^{-sT/2}}{1+e^{-sT/2}}$$

This voltage is applied to a capacitor of 100 μF and a resistor of 250 Ω in series, with no charge initially on the capacitor. Show that the Laplace transform $I(s)$ of the current $i(t)$ flowing, for $t \geqslant 0$, is

$$I(s) = \frac{1}{250(s+40)}\frac{1-e^{-sT/2}}{1+e^{-sT/2}}$$

and give an expression, involving Heaviside step functions, for $i(t)$ where $0 \leqslant t \leqslant 2T$. For $T = 10^{-3}$ s, is this a good representation of the steady-state response of the circuit? Briefly give a reason for your answer.

19 The response $x(t)$ of a control system to a forcing term $u(t)$ is given by the differential equation

$$\frac{d^2x}{dt^2}+2\frac{dx}{dt}+2x = u(t) \quad (t \geqslant 0)$$

Determine the impulse response of the system, and hence, using the convolution integral, obtain the response of the system to a unit step $u(t) = 1H(t)$ applied at $t = 0$, given that initially the system is in a quiescent state. Check your solution by directly solving the differential equation

$$\frac{d^2x}{dt^2}+2\frac{dx}{dt}+2x = 1 \quad (t \geqslant 0)$$

with $x = dx/dt = 0$ at $t = 0$.

20 A light horizontal beam, of length 5 m and constant flexural rigidity EI, built in at the left-hand end $x = 0$, is simply supported at the point $x = 4$ m and carries a distributed load with density function

$$W(x) = \begin{cases} 12\text{ kN m}^{-1} & (0 < x < 4) \\ 24\text{ kN m}^{-1} & (4 < x < 5) \end{cases}$$

Write down the fourth-order boundary-value problem satisfied by the deflection $y(x)$. Solve this problem to determine $y(x)$, and write down the resulting expressions for $y(x)$ for the cases $0 \leqslant x \leqslant 4$ and $4 \leqslant x \leqslant 5$. Calculate the end reaction and moment by evaluating appropriate derivatives of $y(x)$ at $x = 0$. Check that your results satisfy the equation of equilibrium for the beam as a whole.

21 (a) Sketch the function defined by

$$f(t) = \begin{cases} 0 & (0 \leqslant t < 1) \\ 1 & (1 \leqslant t < 2) \\ 0 & (t > 2) \end{cases}$$

Express $f(t)$ in terms of Heaviside step functions, and use the Laplace transform to solve the differential equation

$$\frac{dx}{dt}+x = f(t)$$

given that $x = 0$ at $t = 0$.

(b) The Laplace transform $I(s)$ of the current $i(t)$ in a certain circuit is given by

$$I(s) = \frac{E}{s[Ls + R/(1+Cs)]}$$

where E, L, R and C are positive constants. Determine (i) $\lim_{t\to 0} i(t)$ and (ii) $\lim_{t\to\infty} i(t)$.

22 Show that the Laplace transform of the half-rectified sine-wave function

$$v(t) = \begin{cases} \sin t & (0 \leqslant t \leqslant \pi) \\ 0 & (\pi \leqslant t \leqslant 2\pi) \end{cases}$$

of period 2π, is

$$\frac{1}{(1+s^2)(1-e^{-\pi s})}$$

Such a voltage $v(t)$ is applied to a 1 Ω resistor and a 1 H inductor connected in series. Show that the resulting current, initially zero, is $\sum_{n=0}^{\infty} f(t - n\pi)$, where $f(t) = (\sin t - \cos t + e^{-t})H(t)$. Sketch a graph of the function $f(t)$.

23 (a) Find the inverse Laplace transform of $1/s^2(s+1)^2$ by writing the expression in the form $(1/s^2)[1/(s+1)^2]$ and using the convolution theorem.

(b) Use the convolution theorem to solve the integral equation

$$y(t) = t + 2\int_0^t y(u)\cos(t-u)\,du$$

and the integro-differential equation

$$\int_0^t y''(u)y'(t-u)\,du = y(t)$$

where $y(0) = 0$ and $y'(0) = y_1$. Comment on the solution of the second equation.

24 A beam of negligible weight and length $3l$ carries a point load W at a distance l from the left-hand end. Both ends are clamped horizontally at the same level. Determine the equation governing the deflection of the beam. If, in addition, the beam is now subjected to a load per unit length, w, over the shorter part of the beam, what will then be the differential equation determining the deflection?

25 (a) Using Laplace transforms, solve the differential equation

$$\frac{d^2x}{dt^2} - 3\frac{dx}{dt} + 3x = H(t-a) \quad (a > 0)$$

where $H(t)$ is the Heaviside unit step function, given that $x = 0$ and $dx/dt = 0$ at $t = 0$.

(b) The output $x(t)$ from a stable linear control system with input $\sin \omega t$ and transfer function $G(s)$ is determined by the relationship

$$X(s) = G(s)\mathscr{L}\{\sin \omega t\}$$

where $X(s) = \mathscr{L}\{x(t)\}$. Show that, after a long time t, the output approaches $x_s(t)$, where

$$x_s(t) = \text{Re}\left(\frac{e^{j\omega t}G(j\omega)}{j}\right)$$

26 Consider the feedback system of Figure 2.60, where K is a constant feedback gain.

(a) In the absence of feedback (that is, $K = 0$) is the system stable?

(b) Write down the transfer function $G_1(s)$ for the overall feedback system.

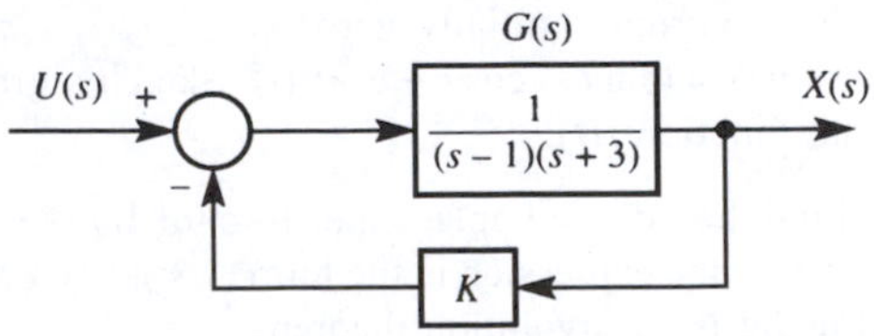

Figure 2.60 Feedback system of Review exercise 26.

(c) Plot the locus of poles of $G_1(s)$ in the s plane for both positive and negative values of K.

(d) From the plots in (c), specify for what range of values of K the feedback system is stable.

(e) Confirm your answer to (d) using the Routh–Hurwitz criterion.

27 (a) For the feedback control system of Figure 2.61(a) it is known that the impulse response is $h(t) = 2\,e^{-2t}\sin t$. Use this to determine the value of the parameter α.

(b) Consider the control system of Figure 2.61(b), having both proportional and rate feedback. Determine the critical value of the gain K for stability of the closed-loop system.

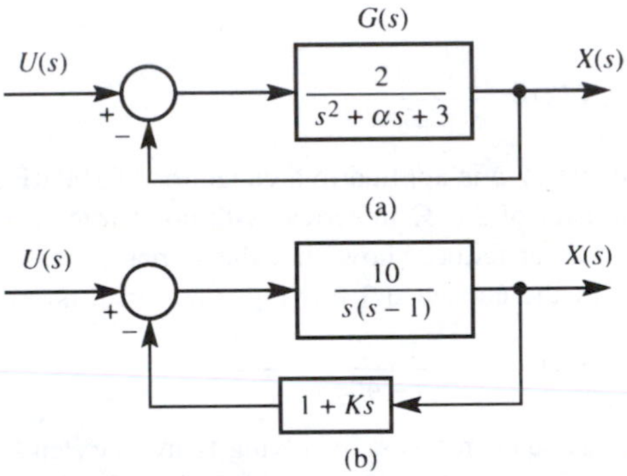

Figure 2.61 Feedback control systems of Review exercise 27.

28 (An extended problem) The transient response of a practical control system to a unit step input often exhibits damped oscillations before reaching steady-state. The following properties are some of those used to specify the transient response characteristics of an underdamped system:

rise time, the time required for the response to rise from 0 to 100% of its final value;

peak time, the time required for the response to reach the first peak of the overshoot;

settling time, the time required for the response curve to reach and stay within a range about the final value of size specified by an absolute percentage of the final value (usually 2% or 5%);

maximum overshoot, the maximum peak value of the response measured from unity.

Consider the feedback control system of Figure 2.62 having both proportional and derivative feedback.

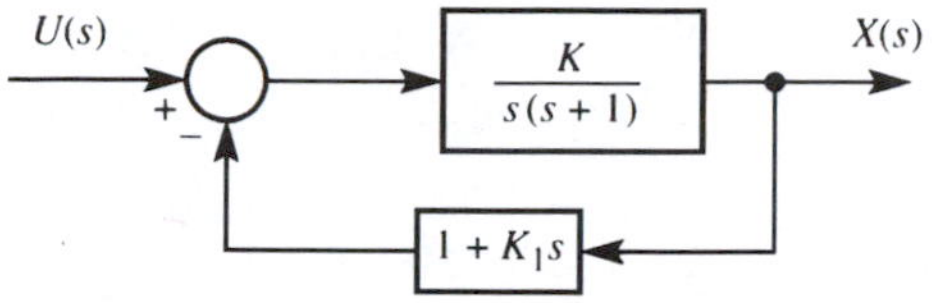

Figure 2.62 Feedback control system of Review exercise 28.

It is desirable to choose the values of the gains K and K_1 so that the system unit step response has a maximum overshoot of 0.2 and a peak time of 1 s.

(a) Obtain the overall transfer function of the closed-loop system.
(b) Show that the unit step response of the system, assuming zero initial conditions, may be written in the form

$$x(t) = 1 - e^{-\omega_n \xi t}\left[\cos \omega_d t + \frac{\xi}{\sqrt{(1-\xi^2)}} \sin \omega_d t\right] \quad (t \geqslant 0)$$

where $\omega_d = \omega_n\sqrt{(1-\xi^2)}$, $\omega_n^2 = K$ and $2\omega_n\xi = 1 + KK_1$.
(c) Determine the values of the gains K and K_1 so that the desired characteristics are achieved.
(d) With these values of K and K_1, determine the rise time and settling time, comparing both the 2% and 5% criteria for the latter.

29 (An extended problem) The mass M_1 of the mechanical system of Figure 2.63(a) is subjected to a harmonic forcing term $\sin \omega t$. Determine the steady-state response of the system.

It is desirable to design a vibration absorber to absorb the steady-state oscillations so that in the

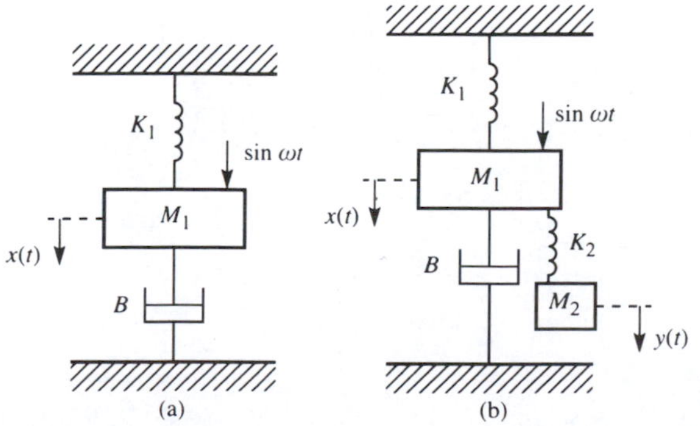

Figure 2.63 Vibration absorber of Review exercise 29.

steady state $x(t) \equiv 0$. To achieve this, a secondary system is attached as illustrated in Figure 2.63(b).

(a) Show that, with an appropriate choice of M_2 and K_2, the desired objective may be achieved.
(b) What is the corresponding steady-state motion of the mass M_2?
(c) Comment on the practicality of your design.

30 (An extended problem) The electronic amplifier of Figure 2.64 has open-loop transfer function $G(s)$ with the following characteristics: a low-frequency gain of 120 dB and simple poles at 1 MHz, 10 MHz and 25 MHz. It may be assumed that the amplifier is ideal, so that $K/(1 + K\beta) \simeq 1/\beta$, where β is the feedback gain and K the steady-state gain associated with $G(s)$.

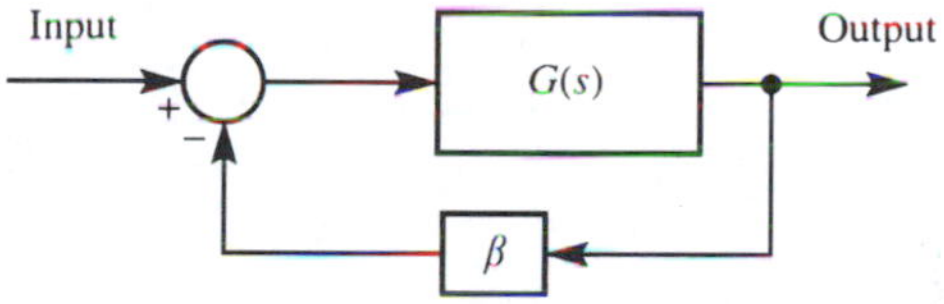

Figure 2.64 Electronic amplifier of Review exercise 30.

(a) Construct the magnitude versus log frequency and phase versus log frequency plots (Bode plots) for the open-loop system.
(b) Determine from the Bode plots whether or not the system is stable in the case of unity feedback (that is, $\beta = 1$).
(c) Determine the value of β for marginal stability, and hence the corresponding value of the closed-loop low-frequency gain.
(d) Feedback is now applied to the amplifier to reduce the overall closed-loop gain at low frequencies to 100 dB. Determine the gain and phase margin corresponding to this closed-loop configuration.
(e) Using the given characteristics, express $G(s)$ in the form

$$G(s) = \frac{K}{(1 + s\tau_1)(1 + s\tau_2)(1 + s\tau_3)}$$

and hence obtain the input–output transfer function for the amplifier.
(f) Write down the characteristic equation for the closed-loop system and, using the Routh–Hurwitz criterion, reconsider parts (b) and (c).

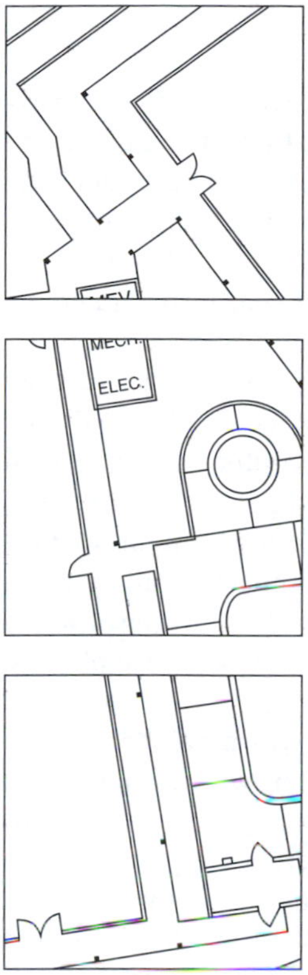

3

The z Transform

CONTENTS

3.1 Introduction

In this chapter we focus attention on discrete-(time) processes. With the advent of fast and cheap digital computers, there has been renewed emphasis on the analysis and design of digital systems, which represent a major class of engineering systems. The main thrust of this chapter will be in this direction. However, it is a mistake to believe that the mathematical basis of this area of work is of such recent vintage. The first comprehensive text in English dealing with difference equations was *The Treatise of the Calculus of Finite Differences* due to George Boole and published in 1860. Much of the early impetus for the **finite calculus** was due to the need to carry out interpolation and to approximate derivatives and integrals. Later, numerical methods for the solution of differential equations were devised, many of which were based on **finite difference methods**, involving the approximation of the derivative terms to produce a **difference equation**. The underlying idea in each case so far discussed is some form of approximation of an underlying continuous function or continuous-time process. There are situations, however, where it is more appropriate to propose a discrete-time model from the start.

Digital systems operate on digital signals, which are usually generated by **sampling** a continuous-time signal, that is, a signal defined for every instant of a possibly infinite time interval. The sampling process generates a **discrete-time signal**, defined only at the instants when sampling takes place so that a digital sequence is generated. After processing by a computer, the output digital signal may be used to construct a new continuous-time signal, perhaps by the use of a **zero-order hold** device, and this in turn might be used to control a plant or process. Digital signal processing devices have made a major impact in many areas of engineering, as well as in the home. For example, many readers will have heard the significantly improved reproduction quality available from compact disc players, which operate using digital teachnology.

We have seen in Chapter 2 that the Laplace transform was a valuable aid in the analysis of continuous-time systems, and in this chapter we develop the z transform, which will perform the same task for discrete-time systems. We introduce the transform in connection with the solution of difference equations, and later we show how difference equations arise as discrete-time system models.

The chapter includes two engineering applications. The first is on the design of digital filters, and highlights one of the major applications of transform methods as a design tool. It may be expected that whenever sampling is involved, performance will improve as sampling rate is increased. Engineers have found that this is not the full story, and the second application deals with some of the problems encountered. This leads on to an introduction to the unifying concept of the $\mathscr{D}$ transform, which brings together the theories of the Laplace and z transforms.

3.2 The *z* transform

Since z transforms relate to sequences, we first review the notation associated with sequences, which were considered in more detail in Chapter 7 of *Modern Engineering Mathematics*. A finite sequence $\{x_k\}_0^n$ is an ordered set of $n + 1$ real or complex numbers:

$$\{x_k\}_0^n = \{x_0, x_1, x_2, \dots, x_n\}$$

Note that the set of numbers is ordered so that position in the sequence is important. The position is identified by the position index k, where k is an integer. If the number of elements in the set is infinite then this leads to the **infinite sequence**

$$\{x_k\}_0^\infty = \{x_0, x_1, x_2, \dots\}$$

When dealing with sampled functions of time t, it is necessary to have a means of allowing for $t < 0$. To do this, we allow the sequence of numbers to extend to infinity on both sides of the initial position x_0, and write

$$\{x_k\}_{-\infty}^{\infty} = \{\ldots, x_{-2}, x_{-1}, x_0, x_1, x_2, \ldots\}$$

Sequences $\{x_k\}_{-\infty}^{\infty}$ for which $x_k = 0$ $(k < 0)$ are called **causal sequences**, by analogy with continuous-time causal functions $f(t)H(t)$ defined in Section 2.2.1 as

$$f(t)H(t) = \begin{cases} 0 & (t < 0) \\ f(t) & (t \geqslant 0) \end{cases}$$

While for some finite sequences it is possible to specify the sequence by listing all the elements of the set, it is normally the case that a sequence is specified by giving a formula for its general element x_k.

3.2.1 Definition and notation

The z **transform** of a sequence $\{x_k\}_{-\infty}^{\infty}$ is defined in general as

$$\mathscr{Z}\{x_k\}_{-\infty}^{\infty} = X(z) = \sum_{k=-\infty}^{\infty} \frac{x_k}{z^k} \tag{3.1}$$

whenever the sum exists and where z is a complex variable, as yet undefined.

The process of taking the z transform of a sequence thus produces a function of a complex variable z, whose form depends upon the sequence itself. The symbol $\mathscr{Z}$ denotes the **z-transform operator**; when it operates on a sequence $\{x_k\}$ it transforms the latter into the function $X(z)$ of the complex variable z. It is usual to refer to $\{x_k\}$, $X(z)$ as a **z-transform pair**, which is sometimes written as $\{x_k\} \leftrightarrow X(z)$. Note the similarity to obtaining the Laplace transform of a function in Section 2.2.1. We shall return to consider the relationship between Laplace and z transforms in Section 3.7.

For sequences $\{x_k\}_{-\infty}^{\infty}$ that are *causal*, that is

$$x_k = 0 \quad (k < 0)$$

the z transform given in (3.1) reduces to

$$\mathscr{Z}\{x_k\}_{0}^{\infty} = X(z) = \sum_{k=0}^{\infty} \frac{x_k}{z^k} \tag{3.2}$$

In this chapter we shall be concerned with causal sequences, and so the definition given in (3.2) will be the one that we shall use henceforth. We shall therefore from now on take $\{x_k\}$ to denote $\{x_k\}_0^\infty$. Non-causal sequences, however, are of importance, and arise particularly in the field of digital image processing, among others.

EXAMPLE 3.1 Determine the z transform of the sequence

$$\{x_k\} = \{2^k\} \quad (k \geqslant 0)$$

Solution From the definition (3.2),

$$\mathscr{Z}\{2^k\} = \sum_{k=0}^{\infty} \frac{2^k}{z^k} = \sum_{k=0}^{\infty} \left(\frac{2}{z}\right)^k$$

which we recognize as a geometric series, with common ratio $r = 2/z$ between successive terms. The series thus converges for $|z| > 2$, when

$$\sum_{k=0}^{\infty} \left(\frac{2}{z}\right)^k = \lim_{k\to\infty} \frac{1-(2/z)^k}{1-2/z} = \frac{1}{1-2/z}$$

leading to

$$\mathscr{Z}\{2^k\} = \frac{z}{z-2} \quad (|z| > 2) \tag{3.3}$$

so that

$$\left.\begin{aligned} \{x_k\} &= \{2^k\} \\ X(z) &= \frac{z}{z-2} \end{aligned}\right\}$$

is an example of a z-transform pair.

From Example 3.1, we see that the z transform of the sequence $\{2^k\}$ exists provided that we restrict the complex variable z so that it lies outside the circle $|z| = 2$ in the z plane. From another point of view, the function

$$X(z) = \frac{z}{z-2} \quad (|z| > 2)$$

may be thought of as a **generating function** for the sequence $\{2^k\}$, in the sense that the coefficient of z^{-k} in the expansion of $X(z)$ in powers of $1/z$ *generates* the kth term of the sequence $\{2^k\}$. This can easily be verified, since

$$\frac{z}{z-2} = \frac{1}{1-2/z} = \left(1 - \frac{2}{z}\right)^{-1}$$

and, since $|z| > 2$, we can expand this as

$$\left(1 - \frac{2}{z}\right)^{-1} = 1 + \frac{2}{z} + \left(\frac{2}{z}\right)^2 + \ldots + \left(\frac{2}{z}\right)^k + \ldots$$

and we see that the coefficient of z^{-k} is indeed 2^k, as expected.

We can generalize the result (3.3) in an obvious way to determine $\mathscr{Z}\{a^k\}$, the z transform of the sequence $\{a^k\}$, where a is a real or complex constant. At once

$$\mathscr{Z}\{a^k\} = \sum_{k=0}^{\infty} \frac{a^k}{z^k} = \frac{1}{1 - a/z} \quad (|z| > |a|)$$

so that

$$\mathscr{Z}\{a^k\} = \frac{z}{z - a} \quad (|z| > |a|) \tag{3.4}$$

EXAMPLE 3.2 Show that

$$\mathscr{Z}\{(-\tfrac{1}{2})^k\} = \frac{2z}{2z + 1} \quad (|z| > \tfrac{1}{2})$$

Solution Taking $a = -\frac{1}{2}$ in (3.4), we have

$$\mathscr{Z}\{(-\tfrac{1}{2})^k\} = \sum_{k=0}^{\infty} \frac{(-\frac{1}{2})^k}{z^k} = \frac{z}{z - (-\frac{1}{2})} \quad (|z| > \tfrac{1}{2})$$

so that

$$\mathscr{Z}\{(-\tfrac{1}{2})^k\} = \frac{2z}{2z + 1} \quad (|z| > \tfrac{1}{2})$$

Further z-transform pairs can be obtained from (3.4) by formally differentiating with respect to a, which for the moment we regard as a parameter. This gives

$$\frac{\mathrm{d}}{\mathrm{d}a}\mathscr{Z}\{a^k\} = \mathscr{Z}\left\{\frac{\mathrm{d}a^k}{\mathrm{d}a}\right\} = \frac{\mathrm{d}}{\mathrm{d}a}\left(\frac{z}{z - a}\right)$$

leading to

$$\mathscr{Z}\{ka^{k-1}\} = \frac{z}{(z - a)^2} \quad (|z| > |a|) \tag{3.5}$$

In the particular case $a = 1$ this gives

$$\mathscr{Z}\{k\} = \frac{z}{(z - 1)^2} \quad (|z| > 1) \tag{3.6}$$

EXAMPLE 3.3 Find the z transform of the sequence

$$\{2k\} = \{0, 2, 4, 6, 8, \dots\}$$

From (3.6),

$$\mathscr{Z}\{k\} = \mathscr{Z}\{0, 1, 2, 3, \dots\} = \sum_{k=0}^{\infty} \frac{k}{z^k} = \frac{z}{(z-1)^2}$$

Using the definition (3.1),

$$\mathscr{Z}\{0, 2, 4, 6, 8, \dots\} = 0 + \frac{2}{z} + \frac{4}{z^2} + \frac{6}{z^3} + \frac{8}{z^4} + \dots = 2\sum_{k=0}^{\infty} \frac{k}{z^k}$$

so that

$$\mathscr{Z}\{2k\} = 2\mathscr{Z}\{k\} = \frac{2z}{(z-1)^2} \tag{3.7}$$

Example 3.3 demonstrates the 'linearity' property of the z transform, which we shall consider further in Section 3.3.1.

A sequence of particular importance is the **unit pulse** or **impulse** sequence

$$\{\delta_k\} = \{1\} = \{1, 0, 0, \dots\}$$

It follows directly from the definition (3.4) that

$$\mathscr{Z}\{\delta_k\} = 1 \tag{3.8}$$

3.2.2 Sampling: a first introduction

Sequences are often generated in engineering applications through the sampling of continuous-time signals, described by functions $f(t)$ of a continuous-time variable t. Here we shall not discuss the means by which a signal is sampled, but merely suppose this to be possible in idealized form.

Figure 3.1 illustrates the idealized sampling process in which a continuous-time signal $f(t)$ is sampled instantaneously and perfectly at uniform intervals T, the **sampling interval**. The idealized sampling process generates the sequence

$$\{f(kT)\} = \{f(0), f(T), f(2T), \dots, f(nT), \dots\} \tag{3.9}$$

Using the definition (3.1), we can take the z transform of the sequence (3.9) to give

$$\mathscr{Z}\{f(kT)\} = \sum_{k=0}^{\infty} \frac{f(kT)}{z^k} \tag{3.10}$$

whenever the series converges. This idea is simply demonstrated by an example.

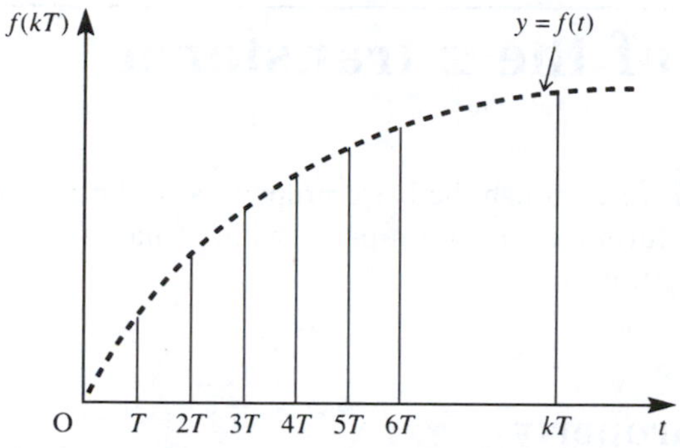

Figure 3.1 Sampling of a continuous-time signal.

EXAMPLE 3.4 The signal $f(t) = e^{-t}H(t)$ is sampled at intervals T. What is the z transform of the resulting sequence of samples?

Solution Sampling the causal function $f(t)$ generates the sequence

$$\{f(kT)\} = \{f(0), f(T), f(2T), \dots, f(nT), \dots\}$$
$$= \{1, e^{-T}, e^{-2T}, e^{-3T}, \dots, e^{-nT}, \dots\}$$

Then, using (3.1),

$$\mathscr{Z}\{f(kT)\} = \sum_{k=0}^{\infty} \frac{e^{-kT}}{z^k} = \sum_{k=0}^{\infty} \left(\frac{e^{-T}}{z}\right)^k$$

so that

$$\mathscr{Z}\{e^{-kT}\} = \frac{z}{z - e^{-T}} \quad (|z| > e^{-T}) \tag{3.11}$$

It is important to note in Example 3.4 that the region of convergence depends on the sampling interval T.

3.2.3 Exercises

1 Calculate the z transform of the following sequences, stating the region of convergence in each case:

(a) $\{(\frac{1}{4})^k\}$ (b) $\{3^k\}$ (c) $\{(-2)^k\}$

(d) $\{-(2^k)\}$ (e) $\{3k\}$

2 The continuous-time signal $f(t) = e^{-2\omega t}$, where ω is a real constant, is sampled when $t \geqslant 0$ at intervals T. Write down the general term of the sequence of samples, and calculate the z transform of the sequence.

3.3 Properties of the z transform

In this section we establish the basic properties of the z transform that will enable us to develop further z-transform pairs, without having to compute them directly using the definition.

3.3.1 The linearity property

As for Laplace transforms, a fundamental property of the z transform is its linearity, which may be stated as follows.

If $\{x_k\}$ and $\{y_k\}$ are sequences having z transforms $X(z)$ and $Y(z)$ respectively and if α and β are any constants, real or complex then

$$\mathscr{Z}\{\alpha x_k + \beta y_k\} = \alpha\mathscr{Z}\{x_k\} + \beta\mathscr{Z}\{y_k\} = \alpha X(z) + \beta Y(z) \tag{3.12}$$

As a consequence of this property, we say that the z-transform operator $\mathscr{Z}$ is a **linear operator**. A proof of the property follows readily from the definition (3.4), since

$$\begin{aligned}\mathscr{Z}\{\alpha x_k + \beta y_k\} &= \sum_{k=0}^{\infty}\frac{\alpha x_k + \beta y_k}{z^k} = \alpha\sum_{k=0}^{\infty}\frac{x_k}{z^k} = \beta\sum_{k=0}^{\infty}\frac{y_k}{z^k}\\ &= \alpha X(z) + \beta Y(z)\end{aligned}$$

The region of existence of the z transform, in the z plane, of the linear sum will be the intersection of the regions of existence (that is, the region common to both) of the individual z transforms $X(z)$ and $Y(z)$.

EXAMPLE 3.5 The continuous-time function $f(t) = \cos \omega t\, H(t)$, ω a constant, is sampled in the idealized sense at intervals T to generate the sequence $\{\cos k\omega T\}$. Determine the z transform of the sequence.

Solution Using the result $\cos k\omega T = \frac{1}{2}(e^{jk\omega T} + e^{-jk\omega T})$ and the linearity property, we have

$$\mathscr{Z}\{\cos k\omega T\} = \mathscr{Z}\{\tfrac{1}{2}e^{jk\omega T} + \tfrac{1}{2}e^{-jk\omega T}\} = \tfrac{1}{2}\mathscr{Z}\{e^{jk\omega T}\} + \tfrac{1}{2}\mathscr{Z}\{e^{-jk\omega T}\}$$

Using (3.7) and noting that $|e^{jk\omega T}| = |e^{-jk\omega T}| = 1$ gives

$$\begin{aligned}\mathscr{Z}\{\cos k\omega T\} &= \frac{1}{2}\frac{z}{z - e^{j\omega T}} + \frac{1}{2}\frac{z}{z - e^{-j\omega T}} \qquad (|z| > 1)\\ &= \frac{1}{2}\frac{z(z - e^{-j\omega T}) + z(z - e^{j\omega T})}{z^2 - (e^{j\omega T} + e^{-j\omega T})z + 1}\end{aligned}$$

leading to the z-transform pair

$$\mathscr{Z}\{\cos k\omega T\} = \frac{z(z - \cos \omega T)}{z^2 - 2z \cos \omega T + 1} \quad (|z| > 1) \tag{3.13}$$

In a similar manner to Example 3.5, we can verify the z-transform pair

$$\mathscr{Z}\{\sin k\omega T\} = \frac{z \sin \omega T}{z^2 - 2z \cos \omega T + 1} \quad (|z| > 1) \tag{3.14}$$

and this is left as an exercise for the reader (see Exercise 3).

3.3.2 The first shift property (delaying)

In this and the next section we introduce two properties relating the z transform of a sequence to the z transform of a shifted version of the same sequence. In this section we consider a delayed version of the sequence $\{x_k\}$, denoted by $\{y_k\}$, with

$$y_k = x_{k-k_0}$$

Here k_0 is the number of steps in the delay; for example, if $k_0 = 2$ then $y_k = x_{k-2}$, so that

$$y_0 = x_{-2}, \quad y_1 = x_{-1}, \quad y_2 = x_0, \quad y_3 = x_1$$

and so on. Thus the sequence $\{y_k\}$ is simply the sequence $\{x_k\}$ moved backward, or delayed, by two steps. From the definition (3.1),

$$\mathscr{Z}\{y_k\} = \sum_{k=0}^{\infty} \frac{y_k}{z^k} = \sum_{k=0}^{\infty} \frac{x_{k-k_0}}{z^k} = \sum_{p=-k_0}^{\infty} \frac{x_p}{z^{p+k_0}}$$

where we have written $p = k - k_0$. If $\{x_k\}$ is a causal sequence, so that $x_p = 0$ $(p < 0)$, then

$$\mathscr{Z}\{y_k\} = \sum_{p=0}^{\infty} \frac{x_p}{z^{p+k_0}} = \frac{1}{z^{k_0}} \sum_{p=0}^{\infty} \frac{x_p}{z^p} = \frac{1}{z^{k_0}} X(z)$$

where $X(z)$ is the z transform of $\{x_k\}$.

We therefore have the result

$$\mathscr{Z}\{x_{k-k_0}\} = \frac{1}{z^{k_0}} \mathscr{Z}\{x_k\} \tag{3.15}$$

which is referred to as the **first shift property** of z transforms.

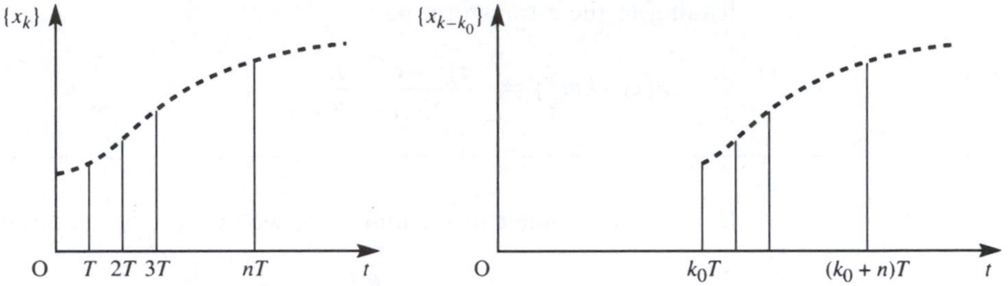

Figure 3.2 Sequence and its shifted form.

If $\{x_k\}$ represents the sampled form, with uniform sampling interval T, of the continuous signal $x(t)$ then $\{x_{k-k_0}\}$ represents the sampled form of the continuous signal $x(t-k_0T)$ which, as illustrated in Figure 3.2, is the signal $x(t)$ delayed by a multiple k_0 of the sampling interval T. The reader will find it of interest to compare this result with the results for the Laplace transforms of integrals (2.16).

EXAMPLE 3.6 The causal sequence $\{x_k\}$ is generated by

$$x_k = (\tfrac{1}{2})^k \quad (k \geqslant 0)$$

Determine the z transform of the shifted sequence $\{x_{k-2}\}$.

Solution By the first shift property,

$$\mathscr{Z}\{x_{k-2}\} = \frac{1}{z^2}\mathscr{Z}\left\{\left(\frac{1}{2}\right)^k\right\}$$

which, on using (3.4), gives

$$\mathscr{Z}\{x_{k-2}\} = \frac{1}{z^2}\frac{z}{z-\frac{1}{2}} \quad \left(|z| > \frac{1}{2}\right) = \frac{1}{z^2}\frac{2z}{2z-1} = \frac{2}{z(2z-1)} \quad \left(|z| > \frac{1}{2}\right)$$

We can confirm this result by direct use of the definition (3.1). From this, and the fact that $\{x_k\}$ is a causal sequence,

$$\{x_{k-2}\} = \{x_{-2}, x_{-1}, x_0, x_1, \ldots\} = \{0, 0, 1, \tfrac{1}{2}, \tfrac{1}{4}, \ldots\}$$

Thus,

$$\mathscr{Z}\{x_{k-2}\} = 0 + 0 + \frac{1}{z^2} + \frac{1}{2z^3} + \frac{1}{4z^4} + \ldots = \frac{1}{z^2}\left(1 + \frac{1}{2z} + \frac{1}{4z^2} + \ldots\right)$$

$$= \frac{1}{z^2}\frac{z}{z-\frac{1}{2}} \quad \left(|z| > \frac{1}{2}\right) = \frac{z}{z(2z-1)} \quad \left(|z| > \frac{1}{2}\right)$$

3.3.3 The second shift property (advancing)

In this section we seek a relationship between the z transform of an advanced version of a sequence and that of the original sequence. First we consider a single-step advance. If $\{y_k\}$ is the single-step advanced version of the sequence $\{x_k\}$ then $\{y_k\}$ is generated by

$$y_k = x_{k+1} \quad (k \geqslant 0)$$

Then

$$\mathscr{Z}\{y_k\} = \sum_{k=0}^{\infty} \frac{y_k}{z^k} = \sum_{k=0}^{\infty} \frac{x_{k+1}}{z^k} = z\sum_{k=0}^{\infty} \frac{x_{k+1}}{z^{k+1}}$$

and putting $p = k + 1$ gives

$$\mathscr{Z}\{y_k\} = z\sum_{p=1}^{\infty} \frac{x_p}{z^p} = z\left(\sum_{p=0}^{\infty} \frac{x_p}{z^p} - x_0\right) = zX(z) - zx_0$$

where $X(z)$ is the z transform of $\{x_k\}$.

We therefore have the result

$$\mathscr{Z}\{x_{k+1}\} = zX(z) - zx_0 \tag{3.16}$$

In a similar manner it is readily shown that for a two-step advanced sequence $\{x_{k+2}\}$

$$\mathscr{Z}\{x_{k+2}\} = z^2X(z) - z^2x_0 - zx_1 \tag{3.17}$$

Note the similarity in structure between (3.16) and (3.17) on the one hand and those for the Laplace transforms of first and second derivatives (Section 2.3.1). In general, it is readily proved by induction that for a k_0-step advanced sequence $\{x_{k+k_0}\}$

$$\mathscr{Z}\{x_{k+k_0}\} = z^{k_0}X(z) - \sum_{n=0}^{k_0-1} x_n z^{k_0-n} \tag{3.18}$$

In Section 3.5.2 we shall use these results to solve difference equations.

3.3.4 Some further properties

In this section we shall state some further useful properties of the z transform, leaving their verification to the reader as Exercises 9 and 10.

(i) *Multiplication by a^k*

If $Z\{x_k\} = X(z)$ then for a constant a

$$\mathscr{Z}\{a^k x_k\} = X(a^{-1}z) \tag{3.19}$$

(ii) *Multiplication by k^n*

If $\mathscr{Z}\{x_k\} = X(z)$ then for a positive integer n

$$\mathscr{Z}\{k^n x_k\} = \left(-z\frac{\mathrm{d}}{\mathrm{d}z}\right)^n X(z) \tag{3.20}$$

Note that in (3.20) the operator $-z\,\mathrm{d}/\mathrm{d}z$ means 'first differentiate with respect to z and then multiply by $-z$'. Raising to the power of n means 'repeat the operation n times'.

(iii) *Initial-value theorem*

If $\{x_k\}$ is a sequence with z transform $X(z)$ then the initial-value theorem states that

$$\lim_{z\to\infty} X(z) = x_0 \tag{3.21}$$

(iv) *Final-value theorem*

If $\{x_k\}$ is a sequence with z transform $X(z)$ then the final-value theorem states that

$$\lim_{k\to\infty} x_k = \lim_{z\to 1}(1 - z^{-1})X(z) \tag{3.22}$$

provided that the poles of $(1 - z^{-1})X(z)$ are inside the unit circle.

3.3.5 Table of z transforms

It is appropriate at this stage to draw together the results proved so far for easy access. This is done in the form of a table in Figure 3.3.

$\{x_k\}$ $(k \geqslant 0)$	$\mathscr{Z}\{x_k\}$	Region of existence
$x_k = \begin{cases} 1 & (k = 0) \\ 0 & (k > 0) \end{cases}$ (unit pulse sequence)	1	All z
$x_k = 1$ (unit step sequence)	$\frac{z}{z-1}$	$\lvert z\rvert > 1$
$x_k = a^k$ (a constant)	$\frac{z}{z-a}$	$\lvert z\rvert > \lvert a\rvert$
$x_k = k$	$\frac{z}{(z-1)^2}$	$\lvert z\rvert > 1$
$x_k = ka^{k-1}$ (a constant)	$\frac{z}{(z-a)^2}$	$\lvert z\rvert > a$
$x_k = e^{-kT}$ (T constant)	$\frac{z}{z-e^{-T}}$	$\lvert z\rvert > e^{-T}$
$x_k = \cos k\omega T$ (ω, T constants)	$\frac{z(z-\cos\omega T)}{z^2 - 2z\cos\omega T + 1}$	$\lvert z\rvert > 1$
$x_k = \sin k\omega T$ (ω, T constants)	$\frac{z\sin\omega T}{z^2 - 2z\cos\omega T + 1}$	$\lvert z\rvert > 1$

Figure 3.3 A short table of z transforms.

3.3.6 Exercises

3 Use the method of Example 3.5 to confirm (3.14), namely

$$\mathscr{Z}\{\sin k\omega T\} = \frac{z\sin\omega T}{z^2 - 2z\cos\omega T + 1}$$

where ω and T are constants.

4 Use the first shift property to calculate the z transform of the sequence $\{y_k\}$, with

$$y_k = \begin{cases} 0 & (k < 3) \\ x_k & (k \geqslant 3) \end{cases}$$

where $\{x_k\}$ is causal and $x_k = (\frac{1}{2})^k$. Confirm your result by direct evaluation of $\mathscr{Z}\{y_k\}$ using the definition of the z transform.

5 Determine the z transforms of the sequences

(a) $\{(-\frac{1}{5})^k\}$ (b) $\{\cos k\pi\}$

6 Determine $\mathscr{Z}\{(\frac{1}{2})^k\}$. Using (3.6), obtain the z transform of the sequence $\{k(\frac{1}{2})^k\}$.

7 Show that for a constant α

(a) $\mathscr{Z}\{\sinh k\alpha\} = \dfrac{z\sinh\alpha}{z^2 - 2z\cosh\alpha + 1}$

(b) $\mathscr{Z}\{\cosh k\alpha\} = \dfrac{z^2 - z\cosh\alpha}{z^2 - 2z\cosh\alpha + 1}$

8 Sequences are generated by sampling a causal continuous-time signal $u(t)$ $(t \geqslant 0)$ at uniform intervals T. Write down an expression for u_k, the general term of the sequence, and calculate the corresponding z transform when $u(t)$ is

(a) e^{-4t} (b) $\sin t$ (c) $\cos 2t$

9 Prove the initial- and final-value theorems given in (3.21) and (3.22).

10 Prove the multiplication properties given in (3.19) and (3.20).

3.4 The inverse z transform

In this section we consider the problem of recovering a causal sequence $\{x_k\}$ from knowledge of its z transform $X(z)$. As we shall see, the work on the inversion of Laplace transforms in Section 2.2.7 will prove a valuable asset for this task.

> Formally the symbol $\mathscr{Z}^{-1}[X(z)]$ denotes a causal sequence $\{x_k\}$ whose z transform is $X(z)$; that is,
>
> $$\text{if } \mathscr{Z}\{x_k\} = X(z) \quad \text{then} \quad \{x_k\} = \mathscr{Z}^{-1}[X(z)]$$

This correspondence between $X(z)$ and $\{x_k\}$ is called the **inverse *z* transformation**, $\{x_k\}$ being the **inverse transform** of $X(z)$, and $\mathscr{Z}^{-1}$ being referred to as the **inverse *z*-transform operator**.

As for the Laplace transforms in Section 2.2.8, the most obvious way of finding the inverse transform of $X(z)$ is to make use of a table of transforms such as that given in Figure 3.3. Sometimes it is possible to write down the inverse transform directly from the table, but more often than not it is first necessary to carry out some algebraic manipulation on $X(z)$. In particular, we frequently need to determine the inverse transform of a rational expression of the form $P(z)/Q(z)$, where $P(z)$ and $Q(z)$ are polynomials in z. In such cases the procedure, as for Laplace transforms, is first to resolve the expression, or a revised form of the expression, into partial fractions and then to use the table of transforms. We shall now illustrate the approach through some examples.

3.4.1 Inverse techniques

EXAMPLE 3.7 Find

$$\mathscr{Z}^{-1}\left[\frac{z}{z-2}\right]$$

Solution From Figure 3.3, we see that $z/(z-2)$ is a special case of the transform $z/(z-a)$, with $a = 2$. Thus

$$\mathscr{Z}^{-1}\left[\frac{z}{z-2}\right] = \{2^k\}$$

EXAMPLE 3.8 Find

$$\mathscr{Z}^{-1}\left[\frac{z}{(z-1)(z-2)}\right]$$

Solution Guided by our work on Laplace transforms, we might attempt to resolve

$$Y(z) = \frac{z}{(z-1)(z-2)}$$

into partial fractions. This approach does produce the correct result, as we shall show later. However, we notice that most of the entries in Figure 3.3 contain a factor z in the numerator of the transform. We therefore resolve

$$\frac{Y(z)}{z} = \frac{1}{(z-1)(z-2)}$$

into partial fractions, as

$$\frac{Y(z)}{z} = \frac{1}{z-2} - \frac{1}{z-1}$$

so that

$$Y(z) = \frac{z}{z-2} - \frac{z}{z-1}$$

Then using the result $\mathscr{Z}^{-1}[z/(z-a)] = \{a^k\}$ together with the linearity property, we have

$$\mathscr{Z}^{-1}[Y(z)] = \mathscr{Z}^{-1}\left(\frac{z}{z-2} - \frac{z}{z-1}\right) = \mathscr{Z}^{-1}\left(\frac{z}{z-2}\right) - \mathscr{Z}^{-1}\left(\frac{z}{z-1}\right)$$

$$= \{2^k\} - \{1^k\} \quad (k \geqslant 0)$$

so that

$$\mathscr{Z}^{-1}\left[\frac{z}{(z-1)(z-2)}\right] = \{2^k - 1\} \quad (k \geqslant 0) \qquad \textbf{(3.23)}$$

Suppose that in Example 3.8 we had not thought so far ahead and we had simply resolved $Y(z)$, rather than $Y(z)/z$, into partial fractions. Would the result be the same? The answer of course is 'yes', as we shall now show. Resolving

$$Y(z) = \frac{z}{(z-1)(z-2)}$$

into partial fractions gives

$$Y(z) = \frac{2}{z-2} - \frac{1}{z-1}$$

which may be written as

$$Y(z) = \frac{1}{z}\frac{2z}{z-2} - \frac{1}{z}\frac{z}{z-1}$$

Since

$$\mathscr{Z}^{-1}\left[\frac{2z}{z-2}\right] = 2\mathscr{Z}^{-1}\left(\frac{z}{z-2}\right) = 2\{2^k\}$$

it follows from the first shift property (3.15) that

$$\mathscr{Z}^{-1}\left[\frac{1}{z}\frac{2z}{z-2}\right] = \begin{cases} \{2 \cdot 2^{k-1}\} & (k > 0) \\ 0 & (k = 0) \end{cases}$$

Similarly,

$$\mathscr{Z}^{-1}\left[\frac{1}{z}\frac{z}{z-1}\right] = \begin{cases} \{1^{k-1}\} = \{1\} & (k > 0) \\ 0 & (k = 0) \end{cases}$$

Combining these last two results, we have

$$\begin{aligned} \mathscr{Z}^{-1}[Y(z)] &= \mathscr{Z}^{-1}\left[\frac{1}{z}\frac{2z}{z-2}\right] - \mathscr{Z}^{-1}\left[\frac{1}{z}\frac{z}{z-1}\right] \\ &= \begin{cases} \{2^k - 1\} & (k > 0) \\ 0 & (k = 0) \end{cases} \end{aligned}$$

which, as expected, is in agreement with the answer obtained in Example 3.8.

We can see that adopting this latter approach, while producing the correct result, involved extra effort in the use of a shift theorem. When possible, we avoid this by 'extracting' the factor z as in Example 3.8, but of course this is not always possible, and recourse may be made to the shift property, as the following example illustrates.

EXAMPLE 3.9 Find

$$\mathscr{Z}^{-1}\left[\frac{2z+1}{(z+1)(z-3)}\right]$$

Solution In this case there is no factor z available in the numerator, and so we must resolve

$$Y(z) = \frac{2z+1}{(z+1)(z-3)}$$

into partial fractions, giving

$$Y(z) = \frac{1}{4}\frac{1}{z+1} + \frac{7}{4}\frac{1}{z-3} = \frac{1}{4}\frac{1}{z}\frac{z}{z+1} + \frac{7}{4}\frac{1}{z}\frac{z}{z-3}$$

Since

$$\mathscr{Z}^{-1}\left[\frac{z}{z+1}\right] = \{(-1)^k\} \quad (k \geqslant 0)$$

$$\mathscr{Z}^{-1}\left[\frac{z}{z-3}\right] = \{3^k\} \quad (k \geqslant 0)$$

it follows from the first shift property (3.15) that

$$\mathscr{Z}^{-1}\left[\frac{1}{z}\frac{z}{z+1}\right] = \begin{cases} \{(-1)^{k-1}\} & (k > 0) \\ 0 & (k = 0) \end{cases}$$

$$\mathscr{Z}^{-1}\left[\frac{1}{z}\frac{z}{z-3}\right] = \begin{cases} 3^{k-1} & (k > 0) \\ 0 & (k = 0) \end{cases}$$

Then, from the linearity property

$$\mathscr{Z}^{-1}[Y(z)] = \frac{1}{4}\mathscr{Z}^{-1}\left[\frac{1}{z}\frac{z}{z+1}\right] + \frac{7}{4}\mathscr{Z}^{-1}\left[\frac{1}{z}\frac{z}{z-3}\right]$$

giving

$$\mathscr{Z}^{-1}\left[\frac{2z+1}{(z+1)(z-3)}\right] = \begin{cases} \{\frac{1}{4}(-1)^{k-1} + \frac{7}{4}3^{k-1}\} & (k > 0) \\ 0 & (k = 0) \end{cases}$$

It is often the case that the rational function $P(z)/Q(z)$ to be inverted has a quadratic term in the denominator. Unfortunately, in this case there is nothing resembling the first shift theorem of the Laplace transform which, as we saw in Section 2.2.9, proved so useful in similar circumstances. Looking at Figure 3.3, the only two transforms with quadratic terms in the denominator are those associated with the sequences $\{\cos k\omega T\}$ and $\{\sin k\omega T\}$. In practice these prove difficult to apply in the inverse form, and a 'first principles' approach is more appropriate. We illustrate this with two examples, demonstrating that all that is really required is the ability to handle complex numbers at the stage of resolution into partial fractions.

EXAMPLE 3.10 Invert the z transform

$$Y(z) = \frac{z}{z^2 + a^2}$$

where a is a real constant.

Solution In view of the factor z in the numerator, we resolve $Y(z)/z$ into partial fractions, giving

$$\frac{Y(z)}{z} = \frac{1}{z^2 + a^2} = \frac{1}{(z+\mathrm{j}a)(z-\mathrm{j}a)} = \frac{1}{\mathrm{j}2a}\frac{1}{(z-\mathrm{j}a)} - \frac{1}{\mathrm{j}2a}\frac{1}{(z+\mathrm{j}a)}$$

That is

$$Y(z) = \frac{1}{\mathrm{j}2a}\left(\frac{z}{z - \mathrm{j}a} - \frac{z}{z + \mathrm{j}a}\right)$$

Using the result $\mathscr{Z}^{-1}[z/(z - a)] = \{a^k\}$, we have

$$\mathscr{Z}^{-1}\left[\frac{z}{z - \mathrm{j}a}\right] = \{(\mathrm{j}a)^k\} = \{\mathrm{j}^k a^k\}$$

$$\mathscr{Z}^{-1}\left[\frac{z}{z + \mathrm{j}a}\right] = \{(-\mathrm{j}a)^k\} = \{(-\mathrm{j})^k a^k\}$$

From the relation $\mathrm{e}^{\mathrm{j}\theta} = \cos\theta + \mathrm{j}\sin\theta$, we have

$$\mathrm{j} = \mathrm{e}^{\mathrm{j}\pi/2}, \qquad -\mathrm{j} = \mathrm{e}^{-\mathrm{j}\pi/2}$$

so that

$$\mathscr{Z}^{-1}\left[\frac{z}{z - \mathrm{j}a}\right] = \{a^k(\mathrm{e}^{\mathrm{j}\pi/2})^k\} = \{a^k\mathrm{e}^{\mathrm{j}k\pi/2}\} = \{a^k(\cos\tfrac{1}{2}k\pi + \mathrm{j}\sin\tfrac{1}{2}k\pi)\}$$

$$\mathscr{Z}^{-1}\left[\frac{z}{z + \mathrm{j}a}\right] = \{a^k(\cos\tfrac{1}{2}k\pi - \mathrm{j}\sin\tfrac{1}{2}k\pi)\}$$

The linearity property then gives

$$\begin{aligned}\mathscr{Z}^{-1}[Y(z)] &= \left\{\frac{a^k}{\mathrm{j}2a}(\cos\tfrac{1}{2}k\pi + \mathrm{j}\sin\tfrac{1}{2}k\pi - \cos\tfrac{1}{2}k\pi + \mathrm{j}\sin\tfrac{1}{2}k\pi)\right\}\\ &= \{a^{k-1}\sin\tfrac{1}{2}k\pi\}\end{aligned}$$

EXAMPLE 3.11 Invert

$$Y(z) = \frac{z}{z^2 - z + 1}$$

Solution The denominator of the transform may be factorized as

$$z^2 - z + 1 = \left(z - \frac{1}{2} - \mathrm{j}\frac{\sqrt{3}}{2}\right)\left(z - \frac{1}{2} + \mathrm{j}\frac{\sqrt{3}}{2}\right)$$

In exponential form we have $\frac{1}{2} \pm \mathrm{j}\frac{1}{2}\sqrt{3} = \mathrm{e}^{\pm\mathrm{j}\pi/3}$, so the denominator may be written as

$$z^2 - z + 1 = (z - \mathrm{e}^{\mathrm{j}\pi/3})(z - \mathrm{e}^{-\mathrm{j}\pi/3})$$

We then have

$$\frac{Y(z)}{z} = \frac{1}{(z - e^{j\pi/3})(z - e^{-j\pi/3})}$$

which can be resolved into partial fractions as

$$\frac{Y(z)}{z} = \frac{1}{e^{j\pi/3} - e^{-j\pi/3}} \frac{1}{z - e^{j\pi/3}} + \frac{1}{e^{-j\pi/3} - e^{j\pi/3}} \frac{1}{z - e^{-j\pi/3}}$$

Noting that $\sin\theta = (e^{j\theta} - e^{-j\theta})/j2$, this reduces to

$$\frac{Y(z)}{z} = \frac{1}{j2\sin\frac{1}{3}\pi} \frac{z}{z - e^{j\pi/3}} - \frac{1}{j2\sin\frac{1}{3}\pi} \frac{z}{z - e^{-j\pi/3}}$$

$$= \frac{1}{j\sqrt{3}} \frac{z}{z - e^{j\pi/3}} - \frac{1}{j\sqrt{3}} \frac{z}{z - e^{-j\pi/3}}$$

Using the result $\mathscr{Z}^{-1}[z/(z-a)] = \{a^k\}$, this gives

$$\mathscr{Z}^{-1}[Y(z)] = \frac{1}{j\sqrt{3}}(e^{jk\pi/3} - e^{-jk\pi/3}) = \left\{2\sqrt{\frac{1}{3}}\sin\frac{1}{3}k\pi\right\}$$

We conclude this section with two further examples, illustrating the inversion technique applied to frequently occurring transform types.

EXAMPLE 3.12 Find the sequence whose z transform is

$$F(z) = \frac{z^3 + 2z^2 + 1}{z^3}$$

Solution $F(z)$ is unlike any z transform treated so far in the examples. However, it is readily expanded in a power series in z^{-1} as

$$F(z) = 1 + \frac{2}{z} + \frac{1}{z^3}$$

Using (3.4), it is then apparent that

$$\mathscr{Z}^{-1}[F(z)] = \{f_k\} = \{1, 2, 0, 1, 0, 0, \dots\}$$

EXAMPLE 3.13 Find $\mathscr{Z}^{-1}[G(z)]$ where

$$G(z) = \frac{z(1 - e^{-aT})}{(z-1)(z - e^{-aT})}$$

where a and T are positive constants.

Solution Resolving into partial fractions,

$$\frac{G(z)}{z} = \frac{1}{z-1} - \frac{1}{z-\mathrm{e}^{-aT}}$$

giving

$$G(z) = \frac{1}{z-1} - \frac{1}{z-\mathrm{e}^{-aT}}$$

Using the result $\mathscr{Z}^{-1}[z/(z-a)] = \{a^k\}$, we have

$$\mathscr{Z}^{-1}[G(z)] = \{(1 - \mathrm{e}^{-akT})\} \quad (k \geq 0)$$

In this particular example $G(z)$ is the z transform of a sequence derived by sampling the continuous-time signal

$$f(t) = 1 - \mathrm{e}^{-at}$$

at intervals T.

3.4.2 Exercises

11 Invert the following z transforms. Give the general term of the sequence in each case.

(a) $\dfrac{z}{z-1}$ (b) $\dfrac{z}{z+1}$ (c) $\dfrac{z}{z-\frac{1}{2}}$

(d) $\dfrac{z}{3z+1}$ (e) $\dfrac{z}{z-\mathrm{j}}$ (f) $\dfrac{z}{z+\mathrm{j}\sqrt{2}}$

(g) $\dfrac{1}{z-1}$ (h) $\dfrac{z+2}{z+1}$

12 By first resolving $Y(z)/z$ into partial fractions, find $\mathscr{Z}^{-1}[Y(z)]$ when $Y(z)$ is given by

(a) $\dfrac{z}{(z-1)(z+2)}$ (b) $\dfrac{z}{(2z+1)(z-3)}$

(c) $\dfrac{z^2}{(2z+1)(z-1)}$ (d) $\dfrac{2z}{2z^2+z-1}$

(e) $\dfrac{z}{z^2+1}$ [*Hint:* $z^2+1=(z+\mathrm{j})(z-\mathrm{j})$]

(f) $\dfrac{z}{z^2-2\sqrt{3}z+4}$ (g) $\dfrac{2z^2-7z}{(z-1)^2(z-3)}$

(h) $\dfrac{z^2}{(z-1)^2(z^2-z+1)}$

13 Find $\mathscr{Z}^{-1}[Y(z)]$ when $Y(z)$ is given by

(a) $\dfrac{1}{z}+\dfrac{2}{z^7}$ (b) $1+\dfrac{3}{z^2}-\dfrac{2}{z^9}$

(c) $\dfrac{3z+z^2+5z^5}{z^5}$ (d) $\dfrac{1+z}{z^3}+\dfrac{3z}{3z+1}$

(e) $\dfrac{2z^3+6z^2+5z+1}{z^2(2z+1)}$ (f) $\dfrac{2z^2-7z+7}{(z-1)^2(z-2)}$

(g) $\dfrac{z-3}{z^2-3z+2}$

3.5 Discrete-time systems and difference equations

In Chapter 2 the Laplace transform technique was examined, first as a method for solving differential equations, then as a way of characterizing a continuous-time system. In fact, much could be deduced concerning the behaviour of the system and its properties by examining its transform-domain representation, without looking for specific time-domain responses at all. In this section we shall discuss the idea of a linear discrete-time system and its model, a **difference equation**. Later we shall see that the z transform plays an analogous role to the Laplace transform for such systems, by providing a transform-domain representation of the system.

3.5.1 Difference equations

First we illustrate the motivation for studying difference equations by means of an example.

Suppose that a sequence of observations $\{x_k\}$ is being recorded and we receive observation x_k at (time) step or index k. We might attempt to process (for example smooth or filter) this sequence of observations $\{x_k\}$ using the discrete-time feedback system illustrated in Figure 3.4. At time step k the observation x_k enters the system as an input, and, after combination with the 'feedback' signal at the summing junction S, proceeds to the block labelled D. This block is a unit delay block, and its function is to hold its input signal until the 'clock' advances one step, to step $k + 1$. At this time the input signal is passed without alteration to become the signal y_{k+1}, the $(k + 1)$th member of the output sequence $\{y_k\}$. At the same time this signal is fed back through a scaling block of amplitude α to the summing junction S. This process is instantaneous, and at S the feedback signal is subtracted from the next input observation x_{k+1} to provide the next input to the delay block D. The process then repeats at each 'clock' step.

To analyse the system, let $\{r_k\}$ denote the sequence of input signals to D; then, owing to the delay action of D, we have

$$y_{k+1} = r_k$$

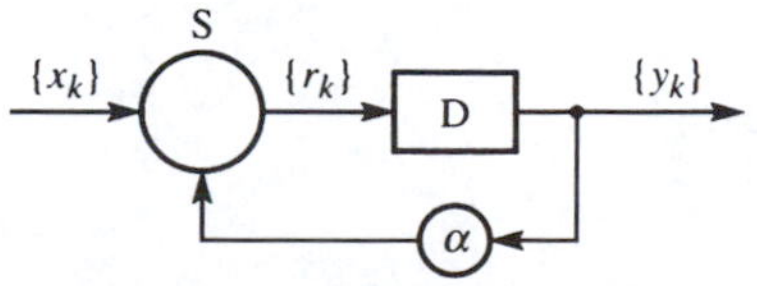

Figure 3.4 Discrete-time signal processing system.

Also, owing to the feedback action,

$$r_k = x_k - \alpha y_k$$

where α is the feedback gain. Combining the two expressions gives

$$y_{k+1} = x_k - \alpha y_k$$

or

$$y_{k+1} + \alpha y_k = x_k \tag{3.24}$$

Equation (3.24) is an example of a first-order difference equation, and it relates adjacent members of the sequence $\{y_k\}$ to each other and to the input sequence $\{x_k\}$.

A solution of the difference equation (3.24) is a formula for y_k, the general term of the output sequence $\{y_k\}$, and this will depend on both k and the input sequence $\{x_k\}$ as well as, in this case, the feedback gain α.

EXAMPLE 3.14 Find a difference equation to represent the system shown in Figure 3.5, having input and output sequences $\{x_k\}$ and $\{y_k\}$ respectively, where D is the unit delay block and a and b are constant feedback gains.

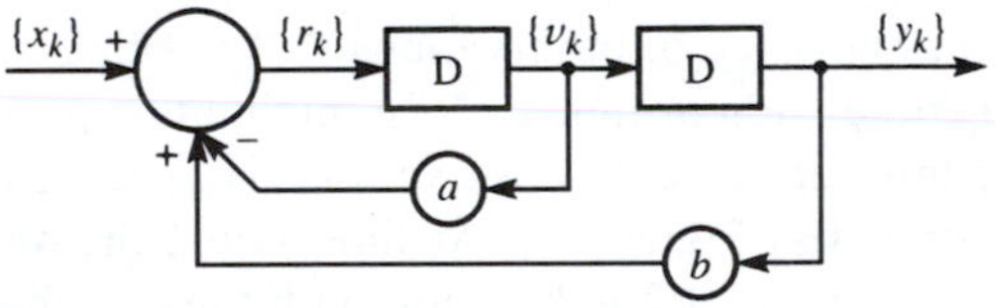

Figure 3.5 The system for Example 3.14.

Solution Introducing intermediate signal sequences $\{r_k\}$ and $\{v_k\}$ as shown in Figure 3.5, at each step the outputs of the delay blocks are

$$y_{k+1} = v_k \tag{3.25}$$

$$v_{k+1} = r_k \tag{3.26}$$

and at the summing junction

$$r_k = x_k - av_k + by_k \tag{3.27}$$

From (3.25),

$$y_{k+2} = v_{k+1}$$

which on using (3.26) gives

$$y_{k+2} = r_k$$

Substituting for r_k from (3.27) then gives

$$y_{k+2} = x_k - av_k + by_k$$

which on using (3.25) becomes

$$y_{k+2} = x_k - ay_{k+1} + by_k$$

Rearranging this gives

$$y_{k+2} + ay_{k+1} - by_k = x_k \tag{3.28}$$

as the difference equation representing the system.

The difference equation (3.28) is an example of a second-order linear constant-coefficient difference equation, and there are strong similarities between this and a second-order linear constant-coefficient differential equation. It is of second order because the term involving the greatest shift of the $\{y_k\}$ sequence is the term in y_{k+2}, implying a shift of two steps. As demonstrated by Example 3.14, the degree of shift, or the order of the equation, is closely related to the number of delay blocks in the block diagram.

3.5.2 The solution of difference equations

Difference equations arise in a variety of ways, sometimes from the direct modelling of systems in discrete time or as an approximation to a differential equation describing the behaviour of a system modelled as a continuous-time system. We do not discuss this further here; rather we restrict ourselves to the technique of solution but examples of applications will be apparent from the exercises. The z-transform method is based upon the second shift property (Section 3.3.3), and it will quickly emerge as a technique almost identical to the Laplace transform method for ordinary differential equations introduced in Section 2.3.3. We shall introduce the method by means of an example.

EXAMPLE 3.15 If in Example 3.14, $a = 1$, $b = 2$ and the input sequence $\{x_k\}$ is the unit step sequence $\{1\}$, solve the resulting difference equation (3.28).

Solution Substituting for a, b and $\{x_k\}$ in (3.28) leads to the difference equation

$$y_{k+2} + y_{k+1} - 2y_k = 1 \quad (k \geqslant 0) \tag{3.29}$$

Taking z transforms throughout in (3.29) gives

$$\mathscr{Z}\{y_{k+2} + y_{k+1} - 2y_k\} = \mathscr{Z}\{1, 1, 1, \dots\}$$

which, on using the linearity property and the result $\mathscr{Z}\{1\} = z/(z-1)$, may be written as

$$\mathscr{Z}\{y_{k+2}\} + \mathscr{Z}\{y_{k+1}\} - 2\mathscr{Z}\{y_k\} = \frac{z}{z-1}$$

Using (3.16) and (3.17) then gives

$$[z^2Y(z) - z^2y_0 - zy_1] + [zY(z) - zy_0] - 2Y(z) = \frac{z}{z-1}$$

which on rearranging leads to

$$(z^2 + z - 2)Y(z) = \frac{z}{z-1} + z^2y_0 + z(y_1 + y_0) \tag{3.30}$$

To proceed, we need some further information, namely the first and second terms y_0 and y_1 of the solution sequence $\{y_k\}$. Without this additional information, we cannot find a unique solution. As we saw in Section 2.3.3, this compares with the use of the Laplace transform method to solve second-order differential equations, where the values of the solution and its first derivative at time $t = 0$ are required.

Suppose that we know (or are given) that

$$y_0 = 0, \qquad y_1 = 1$$

Then (3.30) becomes

$$(z^2 + z - 2)Y(z) = z + \frac{z}{z-1}$$

or

$$(z + 2)(z - 1)Y(z) = z + \frac{z}{z-1}$$

and solving for $Y(z)$ gives

$$Y(z) = \frac{z}{(z+2)(z-1)} + \frac{z}{(z+2)(z-1)^2} = \frac{z^2}{(z+2)(z-1)^2} \tag{3.31}$$

To obtain the solution sequence $\{y_k\}$, we must take the inverse transform in (3.31). Proceeding as in Section 3.4, we resolve $Y(z)/z$ into partial fractions as

$$\frac{Y(z)}{z} = \frac{z}{(z+2)(z-1)^2} = \frac{1}{3}\frac{z}{(z-1)^2} + \frac{2}{9}\frac{1}{z-1} - \frac{2}{9}\frac{z}{z+2}$$

and so

$$Y(z) = \frac{1}{3}\frac{z}{(z-1)^2} + \frac{2}{9}\frac{z}{z-1} - \frac{2}{9}\frac{1}{z+2}$$

Using the results $\mathscr{Z}^{-1}[z/(z-a)] = \{a^k\}$ and $\mathscr{Z}^{-1}[z/(z-1)^2] = \{k\}$ from Figure 3.3, we obtain

$$\{y_k\} = \{\tfrac{1}{3}k + \tfrac{2}{9} - \tfrac{2}{9}(-2)^k\} \quad (k \geqslant 0)$$

as the solution sequence for the difference equation satisfying the conditions $y_0 = 0$ and $y_1 = 1$.

The method adopted in Example 3.15 is called the **z-transform method for solving linear constant-coefficient difference equations**, and is analogous to the Laplace transform method for solving linear constant-coefficient differential equations.

To conclude this section, two further examples are given to help consolidate understanding of the method.

EXAMPLE 3.16 Solve the difference equation

$$8y_{k+2} - 6y_{k+1} + y_k = 9 \quad (k \geqslant 0)$$

given that $y_0 = 1$ and $y_1 = \frac{3}{2}$.

Solution Taking z transforms

$$8\mathscr{Z}\{y_{k+2}\} - 6\mathscr{Z}\{y_{k+1}\} + \mathscr{Z}\{y_k\} = 9\mathscr{Z}\{1\}$$

Using (3.16) and (3.17) and the result $\mathscr{Z}\{1\} = z/(z-1)$ gives

$$8[z^2Y(z) - z^2y_0 - zy_1] - 6[zY(z) - zy_0] + Y(z) = \frac{9z}{z-1}$$

which on rearranging leads to

$$(8z^2 - 6z + 1)Y(z) = 8z^2y_0 + 8zy_1 - 6zy_0 + \frac{9z}{z-1}$$

We are given that $y_0 = 1$ and $y_1 = \frac{3}{2}$, so

$$(8z^2 - 6z + 1)Y(z) = 8z^2 + 6z + \frac{9z}{z-1}$$

or

$$\frac{Y(z)}{z} = \frac{8z+6}{(4z-1)(2z-1)} + \frac{9}{(4z-1)(2z-1)(z-1)}$$

$$= \frac{z+\frac{3}{4}}{(z-\frac{1}{4})(z-\frac{1}{2})} + \frac{\frac{9}{8}}{(z-\frac{1}{4})(z-\frac{1}{2})(z-1)}$$

Resolving into partial fractions gives

$$\frac{Y(z)}{z} = \frac{5}{z-\frac{1}{2}} - \frac{4}{z-\frac{1}{4}} + \frac{6}{z-\frac{1}{4}} - \frac{9}{z-\frac{1}{2}} + \frac{3}{z-1}$$

$$= \frac{2}{z-\frac{1}{4}} - \frac{4}{z-\frac{1}{2}} + \frac{3}{z-1}$$

and so

$$Y(z) = \frac{2z}{z-\frac{1}{4}} - \frac{4z}{z-\frac{1}{2}} + \frac{3z}{z-1}$$

Using the result $\mathscr{Z}^{-1}\{z/(z-a)\} = \{a^k\}$ from Figure 3.3, we take inverse transforms, to obtain

$$\{y_k\} = \{2(\tfrac{1}{4})^k - 4(\tfrac{1}{2})^k + 3\} \quad (k \geqslant 0)$$

as the required solution.

EXAMPLE 3.17 Solve the difference equation

$$y_{k+2} + 2y_k = 0 \quad (k \geqslant 0)$$

given that $y_0 = 1$ and $y_1 = \sqrt{2}$.

Solution Taking z transforms, we have

$$[z^2Y(z) - z^2y_0 - zy_1] + 2Y(z) = 0$$

and substituting the given values of y_0 and y_1 gives

$$z^2Y(z) - z^2 - \sqrt{2}z + 2Y(z) = 0$$

or

$$(z^2 + 2)Y(z) = z^2 + \sqrt{2}z$$

Resolving $Y(z)/z$ into partial fractions gives

$$\frac{Y(z)}{z} = \frac{z+\sqrt{2}}{z^2+2} = \frac{z+\sqrt{2}}{(z+\mathrm{j}\sqrt{2})(z-\mathrm{j}\sqrt{2})}$$

Following the approach adopted in Example 3.13, we write

$$\mathrm{j}\sqrt{2} = \sqrt{2}\,\mathrm{e}^{\mathrm{j}\pi/2}, \qquad -\mathrm{j}\sqrt{2} = \sqrt{2}\,\mathrm{e}^{-\mathrm{j}\pi/2}$$

$$\frac{Y(z)}{z} = \frac{z+\sqrt{2}}{(z-\sqrt{2}\,\mathrm{e}^{\mathrm{j}\pi/2})(z-\sqrt{2}\,\mathrm{e}^{-\mathrm{j}\pi/2})} = \frac{(1+\mathrm{j})/\mathrm{j}2}{z-\sqrt{2}\,\mathrm{e}^{\mathrm{j}\pi/2}} - \frac{(1-\mathrm{j})/\mathrm{j}2}{z-\sqrt{2}\,\mathrm{e}^{-\mathrm{j}\pi/2}}$$

Thus

$$Y(z) = \frac{1}{\mathrm{j}2}\left[(1+\mathrm{j})\frac{z}{z-\sqrt{2}\,\mathrm{e}^{\mathrm{j}\pi/2}} - (1-\mathrm{j})\frac{z}{z-\sqrt{2}\,\mathrm{e}^{-\mathrm{j}\pi/2}}\right]$$

which on taking inverse transforms gives

$$\begin{aligned}\{y_k\} &= \left\{\frac{2^{k/2}}{\mathrm{j}2}[(1+\mathrm{j})\,\mathrm{e}^{\mathrm{j}k\pi/2} - (1-\mathrm{j})\,\mathrm{e}^{-\mathrm{j}k\pi/2}]\right\} \\ &= \{2^{k/2}(\cos\tfrac{1}{2}k\pi + \sin\tfrac{1}{2}k\pi)\} \quad (k \geqslant 0)\end{aligned}$$

as the required solution.

The solution in Example 3.17 was found to be a real-valued sequence, and this comes as no surprise because the given difference equation and the 'starting' values y_0 and y_1 involved only real numbers. This observation provides a useful check on the algebra when complex partial fractions are involved.

3.5.3 Exercises

14 Find difference equations representing the discrete-time systems shown in Figure 3.6.

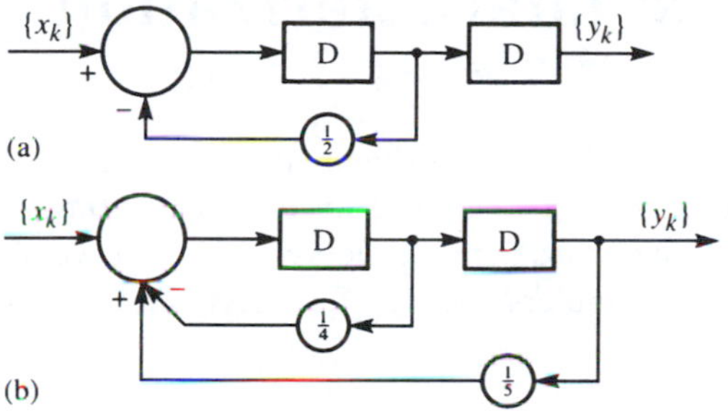

Figure 3.6 The systems for Exercise 14.

15 Using z-transform methods, solve the following difference equations:

(a) $y_{k+2} - 2y_{k+1} + y_k = 0$ subject to $y_0 = 0$, $y_1 = 1$

(b) $y_{n+2} - 8y_{n+1} - 9y_n = 0$ subject to $y_0 = 2$, $y_1 = 1$

(c) $y_{k+2} + 4y_k = 0$ subject to $y_0 = 0$, $y_1 = 1$

(d) $2y_{k+2} - 5y_{k+1} - 3y_k = 0$ subject to $y_0 = 3$, $y_1 = 2$

16 Using z-transform methods, solve the following difference equations:

(a) $6y_{k+2} + 5y_{k+1} - y_k = 5$ subject to $y_0 = y_1 = 0$

(b) $y_{k+2} - 5y_{k+1} + 6y_k = 5$ subject to $y_0 = 0$, $y_1 = 1$

(c) $y_{n+2} - 5y_{n+1} + 6y_n = (\frac{1}{2})^n$ subject to $y_0 = y_1 = 0$

(d) $y_{n+2} - 3y_{n+1} + 3y_n = 1$ subject to $y_0 = 1$, $y_1 = 0$

(e) $2y_{n+2} - 3y_{n+1} - 2y_n = 6n + 1$ subject to $y_0 = 1$, $y_1 = 2$

(f) $y_{n+2} - 4y_n = 3n - 5$ subject to $y_0 = y_1 = 0$

17 A person's capital at the beginning, and expenditure during, a given year k are denoted by C_k and E_k respectively, and satisfy the difference equations

$$C_{k+1} = 1.5C_k - E_k$$
$$E_{k+1} = 0.21C_k + 0.5E_k$$

(a) Show that eventually the person's capital grows at 20% per annum.

(b) If the capital at the beginning of year 1 is £6000 and the expenditure during year 1 is £3720 then find the year in which the expenditure is a minimum and the capital at the beginning of that year.

18 The dynamics of a discrete-time system are determined by the difference equation

$$y_{k+2} - 5y_{k+1} + 6y_k = u_k$$

Determine the response of the system to the unit step input

$$u_k = \begin{cases} 0 & (k < 0) \\ 1 & (k \geq 0) \end{cases}$$

given that $y_0 = y_1 = 1$.

19 As a first attempt to model the national economy, it is assumed that the national income I_k at year k is given by

$$I_k = C_k + P_k + G_k$$

where C_k is the consumer expenditure, P_k is private investment and G_k is government expenditure. It is also assumed that the consumer spending is proportional to the national income in the previous year, so that

$$C_k = aI_{k-1} \quad (0 < a < 1)$$

It is further assumed that private investment is proportional to the change in consumer spending over the previous year, so that

$$P_k = b(C_k - C_{k-1}) \quad (0 < b \leq 1)$$

Show that under these assumptions the national income I_k is determined by the difference equation

$$I_{k+2} - a(1 + b)I_{k+1} + abI_k = G_{k+2}$$

If $a = \frac{1}{2}$, $b = 1$, government spending is at a constant level (that is, $G_k = G$ for all k) and $I_0 = 2G$, $I_1 = 3G$, show that

$$I_k = 2[1 + (\tfrac{1}{2})^{k/2} \sin \tfrac{1}{4} k\pi]G$$

Discuss what happens as $k \to \infty$.

20 The difference equation for current in a particular ladder network of N loops is

$$R_1 i_{n+1} + R_2(i_{n+1} - i_n) + R_2(i_{n+1} - i_{n+2}) = 0 \quad (0 \leqslant n \leqslant N-2)$$

where i_n is the current in the $(n + 1)$th loop, and R_1 and R_2 are constant resistors.

(a) Show that this may be written as

$$i_{n+2} - 2 \cosh \alpha i_{n+1} + i_n = 0 \quad (0 \leqslant n \leqslant N-2)$$

where

$$\alpha = \cosh^{-1}\left(1 + \frac{R_1}{2R_2}\right)$$

(b) By solving the equation in (a), show that

$$i_n = \frac{i_1 \sinh n\alpha - i_0 \sinh (n-1)\alpha}{\sinh \alpha} \quad (2 \leqslant n \leqslant N)$$

3.6 Discrete linear systems: characterization

In this section we examine the concept of a discrete-time linear system and its difference equation model. Ideas developed in Chapter 2 for continuous-time system modelling will be seen to carry over to discrete-time systems, and we shall see that the z transform is the key to the understanding of such systems.

3.6.1 *z* transfer functions

In Section 2.6, when considering continuous-time linear systems modelled by differential equations, we introduced the concept of the system (Laplace) transfer function. This is a powerful tool in the description of such systems, since it contains all the information on system stability and also provides a method of calculating the response to an arbitrary input signal using a convolution integral. In the same way, we can identify a z transfer function for a discrete-time linear time-invariant system modelled by a difference equation, and we can arrive at results analogous to those of Chapter 2.

Let us consider the general linear constant-coefficient difference equation model for a linear time-invariant system, with input sequence $\{u_k\}$ and output sequence $\{y_k\}$. Both $\{u_k\}$ and $\{y_k\}$ are causal sequences throughout. Such a difference equation model takes the form

$$\begin{aligned} &a_n y_{k+n} + a_{n-1} y_{k+n-1} + a_{n-2} y_{k+n-2} + \ldots + a_0 y_k \\ &= b_m u_{k+m} + b_{m-1} u_{k+m-1} + b_{m-2} u_{k+m-2} + \ldots + b_0 u_k \end{aligned} \tag{3.32}$$

where $k \geqslant 0$ and n, m (with $n \geqslant m$) are positive integers and the a_i and b_j are constants. The difference equation (3.32) differs in one respect from the examples considered in Section 3.5 in that the possibility of delayed terms in the input sequence $\{u_k\}$ is also allowed for. The order of the difference equation is n if $a_n \neq 0$, and for the system to be physically realizable, $n \geqslant m$.

Assuming the system to be initially in a quiescent state, we take z transforms throughout in (3.32) to give

$$(a_n z^n + a_{n-1} z^{n-1} + \ldots + a_0)Y(z) = (b_m z^m + b_{m-1} z^{m-1} + \ldots + b_0)U(z)$$

where $Y(z) = \mathscr{Z}\{y_k\}$ and $U(z) = \mathscr{Z}\{u_k\}$. The **system discrete** or **z transfer function** $G(z)$ is defined as

$$G(z) = \frac{Y(z)}{U(z)} = \frac{b_m z^m + b_{m-1} z^{m-1} + \ldots + b_0}{a_n z^n + a_{n-1} z^{n-1} + \ldots + a_0} \tag{3.33}$$

and is normally rearranged (by dividing numerator and denominator by a_n) so that the coefficient of z^n in the denominator is 1. In deriving $G(z)$ in this form, we have assumed that the system was initially in a quiescent state. This assumption is certainly valid for the system (3.32) if

$$y_0 = y_1 = \ldots = y_{n-1} = 0$$

$$u_0 = u_1 = \ldots = u_{m-1} = 0$$

This is not the end of the story, however, and we shall use the term 'quiescent' to mean that no non-zero values are stored on the delay elements before the initial time.

On writing

$$P(z) = b_m z^m + b_{m-1} z^{m-1} + \ldots + b_0$$

$$Q(z) = a_n z^n + a_{n-1} z^{n-1} + \ldots + a_0$$

the discrete transfer function may be expressed as

$$G(z) = \frac{P(z)}{Q(z)}$$

As for the continuous model in Section 2.6.1, the equation $Q(z) = 0$ is called the **characteristic equation** of the discrete system, its order, n, determines the **order of the system**, and its roots are referred to as the **poles** of the discrete transfer function. Likewise, the roots of $P(z) = 0$ are referred to as the **zeros** of the discrete transfer function.

EXAMPLE 3.18 Draw a block diagram to represent the system modelled by the difference equation

$$y_{k+2} + 3y_{k+1} - y_k = u_k \tag{3.34}$$

and find the corresponding z transfer function.

Solution The difference equation may be thought of as a relationship between adjacent members of the solution sequence $\{y_k\}$. Thus at each time step k we have from (3.34)

$$y_{k+2} = -3y_{k+1} + y_k + u_k \tag{3.35}$$

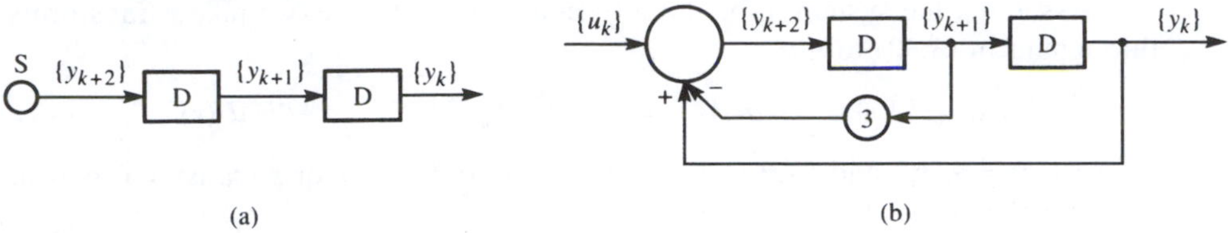

Figure 3.7 (a) The basic second-order block diagram substructure; (b) block diagram representation of (3.34).

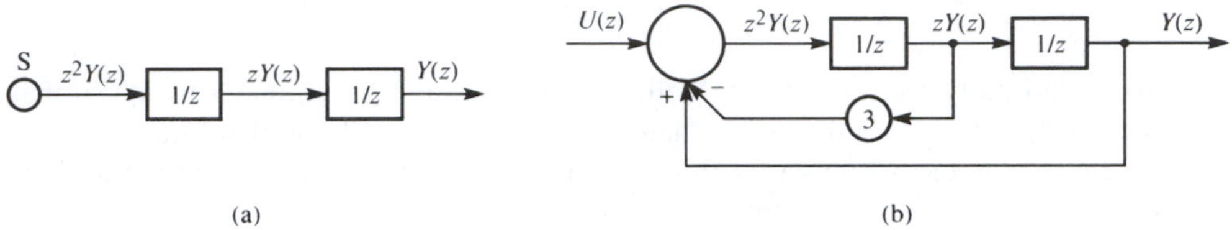

Figure 3.8 (a) The z-transform domain basic second-order block diagram substructure; (b) the z-transform domain block diagram representation of (3.34).

which provides a formula for y_{k+2} involving y_k, y_{k+1} and the input u_k. The structure shown in Figure 3.7(a) illustrates the generation of the sequence $\{y_k\}$ from $\{y_{k+2}\}$ using two delay blocks.

We now use (3.35) as a prescription for generating the sequence $\{y_{k+2}\}$ and arrange for the correct combination of signals to be formed at each step k at the input summing junction S of Figure 3.7(a). This leads to the structure shown in Figure 3.7(b), which is the required block diagram.

We can of course produce a block diagram in the z-transform domain, using a similar process. Taking the z transform throughout in (3.34), under the assumption of a quiescent initial state, we obtain

$$z^2Y(z) + 3zY(z) - Y(z) = U(z) \tag{3.36}$$

or

$$z^2Y(z) = -3zY(z) + Y(z) + U(z) \tag{3.37}$$

The representation (3.37) is the transform domain version of (3.35), and the z-transform domain basic structure corresponding to the time-domain structure of Figure 3.7(a) is shown in Figure 3.8(a).

The unit delay blocks, labelled D in Figure 3.7(a), become '$1/z$' elements in the z-transform domain diagram, in line with the first shift property (3.15), where a number k_0 of delay steps involves multiplication by z^{-k_0}.

It is now a simple matter to construct the 'signal' transform $z^2Y(z)$ from (3.37) and arrange for it to be available at the input to the summing junction S in Figure 3.8(a). The resulting block diagram is shown in Figure 3.8(b).

The z transfer function follows at once from (3.36) as

$$G(z) = \frac{Y(z)}{U(z)} = \frac{1}{z^2 + 3z - 1} \tag{3.38}$$

EXAMPLE 3.19 A system is specified by its z transfer function

$$G(z) = \frac{z-1}{z^2 + 3z + 2}$$

What is the order n of the system? Can it be implemented using only n delay elements? Illustrate this.

Solution If $\{u_k\}$ and $\{y_k\}$ denote respectively the input and output sequences to the system then

$$G(z) = \frac{Y(z)}{U(z)} = \frac{z-1}{z^2 + 3z + 2}$$

so that

$$(z^2 + 3z + 2)Y(z) = (z - 1)U(z)$$

Taking inverse transforms, we obtain the corresponding difference equation model assuming the system is initially in a quiescent state

$$y_{k+2} + 3y_{k+1} + 2y_k = u_{k+1} - u_k \tag{3.39}$$

The difference equation (3.39) has a more complex right-hand side than the difference equation (3.34) considered in Example 3.18. This results from the existence of z terms in the numerator of the transfer function. By definition, the order of the difference equation (3.39) is still 2. However, realization of the system with two delay blocks is not immediately apparent, although this can be achieved, as we shall now illustrate.

Introduce a new signal sequence $\{r_k\}$ such that

$$(z^2 + 3z + 2)R(z) = U(z) \tag{3.40}$$

where $R(z) = \mathscr{Z}\{r_k\}$. In other words, $\{r_k\}$ is the output of the system having transfer function $1/(z^2 + 3z + 2)$.

Multiplying both sides of (3.40) by z, we obtain

$$z(z^2 + 3z + 2)R(z) = zU(z)$$

or

$$(z^2 + 3z + 2)zR(z) = zU(z) \tag{3.41}$$

Subtracting (3.40) from (3.41) we have

$$(z^2 + 3z + 2)zR(z) - (z^2 + 3z + 2)R(z) = zU(z) - U(z)$$

giving

$$(z^2 + 3z + 2)[zR(z) - R(z)] = (z - 1)U(z)$$

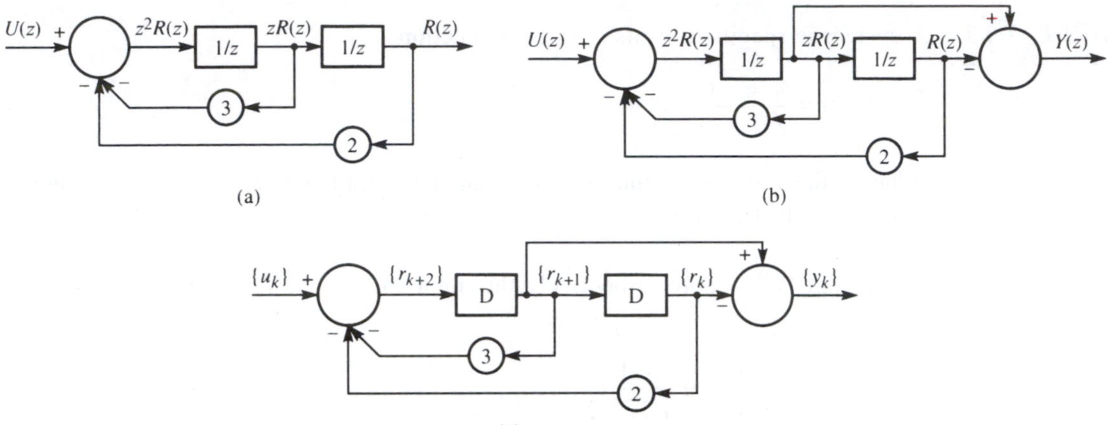

Figure 3.9 The z-transform block diagrams for (a) the system (3.40), (b) the system (3.39), and (c) the time-domain realization of the system in Example 3.19.

Finally, choosing

$$Y(z) = zR(z) - R(z) \tag{3.42}$$

$$(z^2 + 3z + 2)Y(z) = (z - 1)U(z)$$

which is a realization of the given transfer function.

To construct a block diagram realization of the system, we first construct a block diagram representation of (3.40) as in Figure 3.9(a). We now 'tap-off' appropriate signals to generate $Y(z)$ according to (3.42) to construct a block diagram representation of the specified system. The resulting block diagram is shown in Figure 3.9(b).

In order to implement the system, we must exhibit a physically realizable time-domain structure, that is, one containing only D elements. Clearly, since Figure 3.9(b) contains only '$1/z$' blocks, we can immediately produce a realizable time-domain structure as shown in Figure 3.9(c), where, as before, D is the unit delay block.

EXAMPLE 3.20

A system is specified by its z transfer function

$$G(z) = \frac{z}{z^2 + 0.3z + 0.02}$$

Draw a block diagram to illustrate a time-domain realization of the system. Find a second structure that also implements the system.

Solution We know that if $\mathscr{Z}\{u_k\} = U\{z\}$ and $\mathscr{Z}\{y_k\} = Y(z)$ are the z transforms of the input and output sequences respectively then, by definition,

$$G(z) = \frac{Y(z)}{U(z)} = \frac{z}{z^2 + 0.3z + 0.02} \tag{3.43}$$

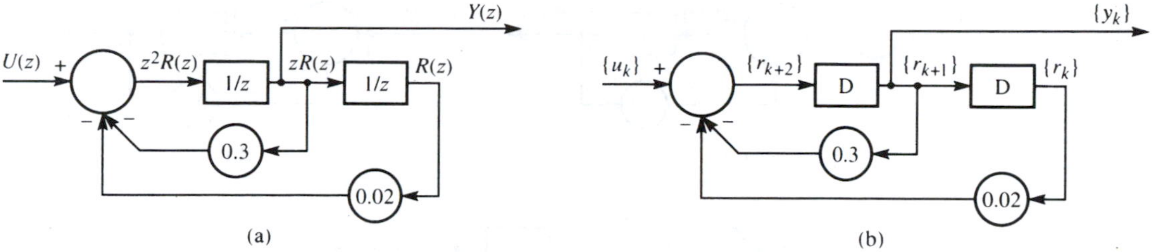

Figure 3.10 (a) The z-transform block diagram for the system of Example 3.20; and (b) the time-domain implementation of (a).

which may be rewritten as

$$(z^2 + 0.3z + 0.02)Y(z) = zU(z)$$

Noting the presence of the factor z on the right-hand side, we follow the procedure of Example 3.19 and consider the system

$$(z^2 + 0.3z + 0.02)R(z) = U(z) \tag{3.44}$$

Multiplying both sides by z, we have

$$(z^2 + 0.3z + 0.02)zR(z) = zU(z)$$

and so, if the output $Y(z) = zR(z)$ is extracted from the block diagram corresponding to (3.44), we have the block diagram representation of the given system (3.43). This is illustrated in Figure 3.10(a), with the corresponding time-domain implementation shown in Figure 3.10(b).

To discover a second form of time-domain implementation, note that

$$G(z) = \frac{z}{z^2 + 0.3z + 0.02} = \frac{2}{z + 0.2} - \frac{1}{z + 0.1}$$

We may therefore write

$$Y(z) = G(z)U(z) = \left(\frac{2}{z + 0.2} - \frac{1}{z + 0.1}\right)U(z)$$

so that

$$Y(z) = R_1(z) - R_2(z)$$

where

$$R_1(z) = \frac{2}{z + 0.2}U(z) \tag{3.45a}$$

$$R_2(z) = \frac{1}{z + 0.1}U(z) \tag{3.45b}$$

From (3.45a), we have

$$(z + 0.2)R_1(z) = 2U(z)$$

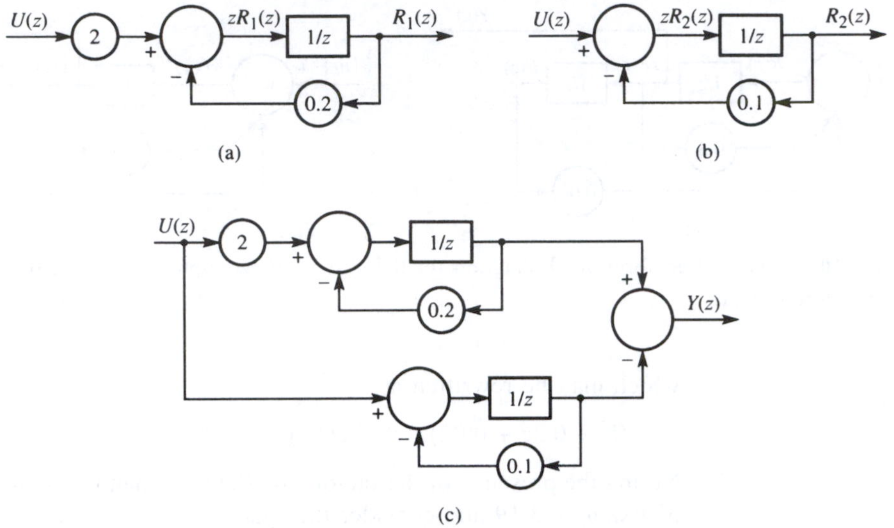

Figure 3.11 The block diagrams for (a) the subsystem (3.45a), (b) the subsystem (3.45b), and (c) an alternative z-transform block diagram for the system of Example 3.20.

which can be represented by the block diagram shown in Figure 3.11(a). Likewise, (3.45b) may be represented by the block diagram shown in Figure 3.11(b).

Recalling that $Y(z) = R_1(z) - R_2(z)$, it is clear that the given system can be represented and then implemented by an obvious coupling of the two subsystems represented by (3.45a, b). The resulting z-transform block diagram is shown in Figure 3.11(c). The time-domain version is readily obtained by replacing the '$1/z$' blocks by D and the transforms $U(z)$ and $Y(z)$ by their corresponding sequences $\{u_k\}$ and $\{y_k\}$ respectively.

3.6.2 The impulse response

In Example 3.20 we saw that two quite different realizations were possible for the same transfer function $G(z)$, and others are possible. Whichever realization of the transfer function is chosen, however, when presented with the same input sequence $\{u_k\}$, the same output sequence $\{y_k\}$ will be produced. Thus we identify the system as characterized by its transfer function as the key concept, rather than any particular implementation. This idea is reinforced when we consider the impulse response sequence for a discrete-time linear time-invariant system, and its role in convolution sums.

Consider the sequence

$$\{\delta_k\} = \{1, 0, 0, \dots\}$$

that is, the sequence consisting of a single 'pulse' at $k = 0$, followed by a train of zeros. As we saw in Section 3.2.1, the z transform of this sequence is easily found from the definition (3.1) as

$$\mathscr{Z}\{\delta_k\} = 1 \tag{3.46}$$

The sequence $\{\delta_k\}$ is called the **impulse sequence**, by analogy with the continuous-time counterpart $\delta(t)$, the impulse function. The analogy is perhaps clearer on considering the transformed version (3.46). In continuous-time analysis, using Laplace transform methods, we observed that $\mathscr{L}\{\delta(t)\} = 1$, and (3.46) shows that the 'entity' with z transform equal to unity is the sequence $\{\delta_k\}$. It is in fact the property that $\mathscr{Z}\{\delta_k\} = 1$ that makes the impulse sequence of such great importance.

Consider a system with transfer function $G(z)$, so that the z transform $Y(z)$ of the output sequence $\{y_k\}$ corresponding to an input sequence $\{u_k\}$ with z transform $U(z)$ is

$$Y(z) = G(z)U(z) \tag{3.47}$$

If the input sequence $\{y_k\}$ is the impulse sequence $\{\delta_k\}$ and the system is initially quiescent, then the output sequence $\{y_{\delta_k}\}$ is called the impulse response of the system. Hence

$$\mathscr{Z}\{y_{\delta_k}\} = Y_\delta(z) = G(z) \tag{3.48}$$

That is, the z transfer function of the system is the z transform of the impulse response. Alternatively, we can say that the impulse response of a system is the inverse z transform of the system transfer function. This compares with the definition of the impulse response for continuous systems given in Section 2.6.3.

Substituting (3.48) into (3.47), we have

$$Y(z) = Y_\delta(z)U(z) \tag{3.49}$$

Thus the z transform of the system output in response to any input sequence $\{u_k\}$ is the product of the transform of the input sequence with the transform of the system impulse response. The result (3.49) shows the underlying relationship between the concepts of impulse response and transfer function, and explains why the impulse response (or the transfer function) is thought of as characterizing a system. In simple terms, if either of these is known then we have all the information about the system for any analysis we may wish to do.

EXAMPLE 3.21 Find the impulse response of the system with z transfer function

$$G(z) = \frac{z}{z^2 + 3z + 2}$$

Solution Using (3.48),

$$Y_\delta(z) = \frac{z}{z^2 + 3z + 2} = \frac{z}{(z + 2)(z + 1)}$$

Resolving $Y_\delta(z)/z$ into partial fractions gives

$$\frac{Y_\delta(z)}{z} = \frac{1}{(z+2)(z+1)} = \frac{1}{z+1} - \frac{1}{z+2}$$

which on inversion gives the impulse response sequence

$$\{Y_{\delta_k}\} = \mathscr{Z}^{-1}\left[\frac{z}{z+1} - \frac{z}{z+2}\right] = \{(-1)^k - (2^k)\} \quad (k \geqslant 0)$$

EXAMPLE 3.22 A system has the impulse response sequence

$$\{y_{\delta_k}\} = \{a^k - 0.5^k\}$$

where $a > 0$ is a real constant. What is the nature of this response when (a) $a = 0.4$, (b) $a = 1.2$? Find the step response of the system in both cases.

Solution When $a = 0.4$

$$\{y_{\delta_k}\} = \{0.4^k - 0.5^k\}$$

and, since both $0.4^k \to 0$ as $k \to \infty$ and $0.5^k \to 0$ as $k \to \infty$, we see that the terms of the impulse response sequence go to zero as $k \to \infty$.

On the other hand, when $a = 1.2$, since $(1.2)^k \to \infty$ as $k \to \infty$, we see that in this case the impulse response sequence terms become unbounded, implying that the system 'blows up'.

In order to calculate the step response, we first determine the system transfer function $G(z)$, using (3.48), as

$$G(z) = Y_\delta(z) = \mathscr{Z}\{a^k - 0.5^k\}$$

giving

$$G(z) = \frac{z}{z-a} - \frac{z}{z-0.5}$$

The system step response is the system response to the unit step sequence $\{h_k\} = \{1, 1, 1, \ldots\}$ which, from Figure 3.3, has z transform

$$\mathscr{Z}\{h_k\} = \frac{z}{z-1}$$

Hence, from (3.46), the step response is determined by

$$Y(z) = G(z)\mathscr{Z}\{h_k\} = \left(\frac{z}{z-a} - \frac{z}{z-0.5}\right)\frac{z}{z-1}$$

so that

$$\frac{Y(z)}{z} = \frac{z}{(z-a)(z-1)} - \frac{z}{(z-0.5)(z-1)}$$

$$= \frac{a}{a-1}\frac{1}{z-a} - \frac{1}{z-0.5} + \left(-2 + \frac{1}{1-a}\right)\frac{1}{z-1}$$

giving

$$Y(z) = \frac{a}{a-1}\frac{z}{z-a} - \frac{z}{z-0.5} + \left(-2 + \frac{1}{1-a}\right)\frac{z}{z-1}$$

which on taking inverse transforms gives the step response as

$$\{y_k\} = \left\{\frac{a}{a-1}a^k - (0.5)^k + \left(-2 + \frac{1}{1-a}\right)\right\} \tag{3.50}$$

Considering the output sequence (3.50), we see that when $a = 0.4$, since $(0.4)^k \to 0$ as $k \to \infty$ (and $(0.5)^k \to 0$ as $k \to \infty$), the output sequence terms tend to the constant value

$$-2 + \frac{1}{1-0.4} = 0.3333$$

In the case of $a = 1.2$, since $(1.2)^k \to \infty$ as $k \to \infty$, the output sequence is unbounded, and again the system 'blows up'.

3.6.3 Stability

Example 3.22 illustrated the concept of system stability for discrete systems. When $a = 0.4$, the impulse response decayed to zero with increasing k, and we observed that the step response remained bounded (in fact, the terms of the sequence approached a constant limiting value). However, when $a = 1.2$, the impulse response became unbounded, and we observed that the step response also increased without limit. In fact, as we saw for continuous systems in Section 2.6.3, a linear constant-coefficient discrete-time system is stable provided that its impulse response goes to zero as $t \to \infty$. As for the continuous case, we can relate this definition to the poles of the system transfer function

$$G(z) = \frac{P(z)}{Q(z)}$$

As we saw in Section 3.6.1, the system poles are determined as the n roots of its characteristic equation

$$Q(z) = a_n z^n + a_{n-1}z^{n-1} + \ldots + a_0 = 0 \tag{3.51}$$

For instance, in Example 3.19 we considered a system with transfer function

$$G(z) = \frac{z-1}{z^2 + 3z + 2}$$

having poles determined by $z^2 + 3z + 2 = 0$, that is poles at $z = -1$ and $z = -2$. Since the impulse response is the inverse transform of $G(z)$, we expect this system to 'blow up' or rather, be unstable, because its impulse response sequence would be expected to contain terms of the form $(-1)^k$ and $(-2)^k$, neither of which goes to zero as $k \to \infty$. (Note that the term in $(-1)^k$ neither blows up nor goes to zero, simply alternating between +1 and −1; however, $(-2)^k$ certainly becomes unbounded as $k \to \infty$.) On the other hand, in Example 3.20 we encountered a system with transfer function

$$G(z) = \frac{z}{z^2 + 0.3z + 0.02}$$

having poles determined by

$$Q(z) = z^2 + 0.3z + 0.02 = (z + 0.2)(z + 0.1) = 0$$

that is poles at $z = -0.2$ and $z = -0.1$. Clearly, this system is stable, since its impulse response contains terms in $(-0.2)^k$ and $(-0.1)^k$, both of which go to zero as $k \to \infty$.

Both of these illustrative examples gave rise to characteristic polynomials $Q(z)$ that were quadratic in form and that had real coefficients. More generally, $Q(z) = 0$ gives rise to a polynomial equation of order n, with real coefficients. From the theory of polynomial equations, we know that $Q(z) = 0$ has n roots α_i $(i = 1, 2, \dots, n)$, which may be real or complex (with complex roots occurring in conjugate pairs).

Hence the characteristic equation may be written in the form

$$Q(z) = a_n(z - \alpha_1)(z - \alpha_2) \dots (z - \alpha_n) = 0 \tag{3.52}$$

The system poles α_i $(i = 1, 2, \dots, n)$ determined by (3.52) may be expressed in the polar form

$$\alpha_i = r_i \mathrm{e}^{\mathrm{j}\theta_i} \quad (i = 1, 2, \dots, n)$$

where $\theta_i = 0$ or π if α_i is real. From the interpretation of the impulse response as the inverse transform of the transfer function $G(z) = P(z)/Q(z)$, it follows that the impulse response sequence of the system will contain terms in

$$r_1^k \mathrm{e}^{\mathrm{j}k\theta_1}, \; r_2^k \mathrm{e}^{\mathrm{j}k\theta_2}, \; \dots, \; r_n^k \mathrm{e}^{\mathrm{j}k\theta_n}$$

Since, for stability, terms in the impulse response sequence must tend to zero as $k \to \infty$, it follows that a system having characteristic equation $Q(z) = 0$ will be stable provided that

$$r_i < 1 \quad \text{for} \quad i = 1, 2, \dots, n$$

Therefore a linear constant-coefficient discrete-time system with transfer function $G(z)$ is stable if and only if all the poles of $G(z)$ lie within the unit circle $|z| < 1$ in the complex z plane, as illustrated in Figure 3.12. If one or more poles lie outside this unit circle then the system will be unstable. If one or more distinct poles lie on the unit circle $|z| = 1$, with all the other poles inside, then the system is said to be **marginally stable**.

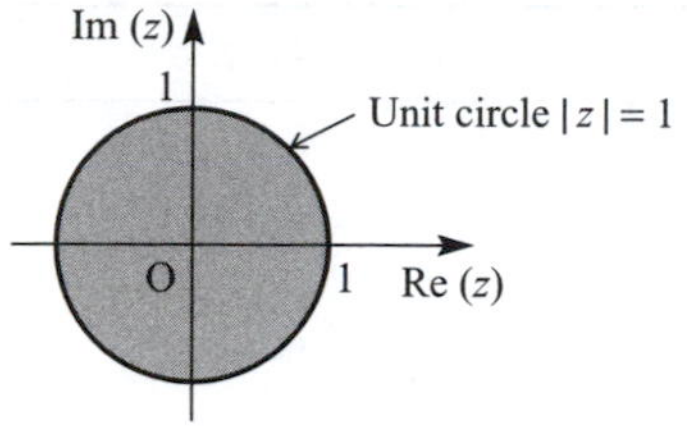

Figure 3.12 Region of stability in the z plane.

EXAMPLE 3.23 Which of the following systems, specified by their transfer function $G(z)$, are stable?

(a) $G(z) = \dfrac{1}{z + 0.25}$ (b) $G(z) = \dfrac{z}{z^2 - z + 0.5}$ (c) $G(z) = \dfrac{z^2}{z^3 - 3z^2 + 2.5z - 1}$

Solution (a) The single pole is at $z = -0.25$, so $r_1 = 0.25 < 1$, and the system is stable.

(b) The system poles are determined by

$$z^2 - z + 0.5 = [z - 0.5(1 + j)][z - 0.5(1 - j)] = 0$$

giving the poles as the conjugate pair $z_1 = 0.5(1 + j)$, $z_2 = 0.5(1 - j)$. The amplitudes $r_1 = r_2 = 0.707 < 1$, and again the system is stable.

(c) The system poles are determined by

$$z^3 - 3z^2 + 2.5z - 1 = (z - 2)[z - 0.5(1 + j)][z - 0.5(1 - j)]$$

giving the poles as $z_1 = 2$, $z_2 = 0.5(1 + j)$, $z_3 = 0.5(1 - j)$, and so their amplitudes are $r_1 = 2$, $r_2 = r_3 = 0.707$. Since $r_1 > 1$, it follows that the system is unstable.

According to our definition, it follows that to prove stability we must show that all the roots of the characteristic equation

$$Q(z) = z^n + a_{n-1}z^{n-1} + \dots + a_0 = 0 \quad \textbf{(3.53)}$$

lie within the unit circle $|z| = 1$ (note that for convenience we have arranged for the coefficient of z^n to be unity in (3.53)). Many mathematical criteria have been developed to test for this property. One such method, widely used in practice, is the **Jury stability criterion** introduced by E. I. Jury in 1963. This procedure gives necessary and sufficient conditions for the polynomial equation (3.53) to have all its roots inside the unit circle $|z| = 1$.

The first step in the procedure is to set up a table as in Figure 3.13 using information from the given polynomial equation (3.53) and where

Row	z^n	z^{n-1}	z^{n-2}	...	z^{n-k}	...	z^2	z^1	z^0
1	1	a_{n-1}	a_{n-2}	...	a_{n-k}	...	a_2	a_1	a_0
2	a_0	a_1	a_2	...	a_k	...	a_{n-2}	a_{n-1}	1
3	$\Delta_1 = b_0$	b_1	b_2	...	b_k	...	b_{n-2}	b_{n-1}	
4	b_{n-1}	b_{n-2}	b_{n-3}	...	b_{n-1-k}	...	b_1	b_0	
5	$\Delta_2 = c_0$	c_1	c_2	...	c_k	...	c_{n-2}		
6	c_{n-2}	c_{n-3}	c_{n-4}	...	c_{n-2-k}	...	c_0		
7	$\Delta_3 = d_0$	d_1	d_2	...	d_k	...			
8	d_{n-3}	d_{n-4}	d_{n-5}	...	d_{n-3-k}	...			
⋮									
⋮									
$2n-5$	$\Delta_{n-3} = s_0$	s_1	s_2	s_3					
$2n-4$	s_3	s_2	s_1	s_0					
$2n-3$	$\Delta_{n-2} = r_0$	r_1	r_2						
$2n-2$	r_2	r_1	r_0						
$2n-1$	$\Delta_{n-1} = t_0$								

Figure 3.13 Jury stability table for the polynomial equation (3.53).

$$b_k = \begin{vmatrix} 1 & a_k \\ a_0 & a_{n-k} \end{vmatrix}, \quad c_k = \begin{vmatrix} b_0 & b_{n-1-k} \\ b_{n-1} & b_k \end{vmatrix}, \quad d_k = \begin{vmatrix} c_0 & c_{n-2-k} \\ c_{n-2} & c_k \end{vmatrix}, \quad \ldots,$$

$$t_0 = \begin{vmatrix} r_0 & r_2 \\ r_2 & r_0 \end{vmatrix}$$

Note that the elements of row $2j + 2$ consist of the elements of row $2j + 1$ written in the reverse order for $j = 0, 1, 2, \ldots, n$; that is, the elements of the even rows consist of the elements of the odd rows written in reverse order. Necessary and sufficient conditions for the polynomial equation (3.53) to have all its roots inside the unit circle $|z| = 1$ are then given by

$$\text{(i)}\quad Q(1) > 0, \quad (-1)^n Q(-1) > 0 \qquad \textbf{(3.54)}$$

$$\text{(ii)}\quad \Delta_1 > 0, \quad \Delta_2 > 0, \quad \Delta_3 > 0, \quad \ldots, \quad \Delta_{n-2} > 0, \quad \Delta_{n-1} > 0$$

EXAMPLE 3.24 Show that all the roots of the polynomial equation

$$F(z) = z^3 + \tfrac{1}{3}z^2 - \tfrac{1}{4}z - \tfrac{1}{12} = 0$$

lie within the unit circle $|z| = 1$.

Solution The corresponding Jury stability table is shown in Figure 3.14. In this case

$$\text{(i)}\quad F(1) = 1 + \tfrac{1}{3} - \tfrac{1}{4} - \tfrac{1}{12} > 0$$

$$(-1)^n F(-1) = (-1)^3(-1 + \tfrac{1}{3} + \tfrac{1}{4} - \tfrac{1}{12}) > 0$$

$$\text{(ii)}\quad \Delta_1 = \tfrac{143}{144} > 0, \quad \Delta_2 = (\tfrac{143}{144})^2 - \tfrac{4}{81} > 0$$

Row	z^3	z^2	z^1	z^0
1	1	$\frac{1}{3}$	$-\frac{1}{4}$	$-\frac{1}{12}$
2	$-\frac{1}{12}$	$-\frac{1}{4}$	$\frac{1}{3}$	1
3	$\Delta_1 = \begin{vmatrix} 1 & -\frac{1}{12} \\ -\frac{1}{12} & 1 \end{vmatrix} = \frac{143}{144}$	$\begin{vmatrix} 1 & -\frac{1}{4} \\ -\frac{1}{12} & \frac{1}{3} \end{vmatrix} = \frac{5}{16}$	$\begin{vmatrix} 1 & \frac{1}{3} \\ -\frac{1}{12} & -\frac{1}{4} \end{vmatrix} = -\frac{2}{9}$	
4	$-\frac{2}{9}$	$\frac{5}{16}$	$\frac{143}{144}$	
5	$\Delta_2 = \begin{vmatrix} \frac{143}{144} & -\frac{2}{9} \\ -\frac{2}{9} & \frac{143}{144} \end{vmatrix} = 0.936\,78$			

Figure 3.14 Jury stability table for Example 3.24.

Thus, by the criteria (3.54), all the roots lie within the unit circle. In this case this is readily confirmed, since the polynomial $F(z)$ may be factorized as

$$F(z) = (z - \tfrac{1}{2})(z + \tfrac{1}{2})(z + \tfrac{1}{3}) = 0$$

So the roots are $z_1 = \frac{1}{2}$, $z_2 = -\frac{1}{2}$ and $z_3 = -\frac{1}{3}$.

The Jury stability table may also be used to determine how many roots of the polynomial equation (3.53) lie outside the unit circle. The number of such roots is determined by the number of changes in sign in the sequence

$$1,\quad \Delta_1,\quad \Delta_2,\quad \ldots,\quad \Delta_{n-1}$$

EXAMPLE 3.25

Show that the polynomial equation

$$F(z) = z^3 - 3z^2 - \tfrac{1}{4}z + \tfrac{3}{4} = 0$$

has roots that lie outside the unit circle $|z| = 1$. Determine how many such roots there are.

Solution The corresponding Jury stability table is shown in Figure 3.15. Hence, in this case

$$F(z) = 1 - 3 - \tfrac{1}{4} + \tfrac{3}{4} = -\tfrac{3}{2}$$

$$(-1)^n F(-1) = (-1)^3(-1 - 3 + \tfrac{1}{4} + \tfrac{3}{4}) = 3$$

As $F(1) < 0$, it follows from (3.54) that the polynomial equation has roots outside the unit circle $|z| = 1$. From Figure 3.15, the sequence $1, \Delta_1, \Delta_2$ is $1, \frac{7}{16}, -\frac{15}{16}$, and since there is only one sign change in the sequence, it follows that one root lies outside the unit circle. Again this is readily confirmed, since $F(z)$ may be factorized as

Row	z^3	z^2	z^1	z^0
1	1	-3	$-\frac{1}{4}$	$\frac{3}{4}$
2	$\frac{3}{4}$	$-\frac{1}{4}$	-3	1
3	$\Delta_1 = \frac{7}{16}$	$-\frac{45}{16}$	2	
4	2	$-\frac{45}{16}$	$\frac{7}{16}$	
5	$\Delta_2 = -\frac{5}{16}$			

Figure 3.15 Jury stability table for Example 3.25.

$$F(z) = (z - \tfrac{1}{2})(z + \tfrac{1}{2})(z - 3) = 0$$

showing that there is indeed one root outside the unit circle at $z = 3$.

3.6.4 Convolution

Here we shall briefly extend the concept of convolution introduced in Section 2.6.6 to discrete-time systems. From (3.45), for an initially quiescent system with an impulse response sequence $\{y_{\delta_k}\}$ with z transform $Y_\delta(z)$, the z transform $Y(z)$ of the output sequence $\{y_k\}$ in response to an input sequence $\{u_k\}$ with z transform $U(z)$ is given by

$$Y(z) = Y_\delta(z)U(z) \tag{3.49}$$

For the purposes of solving a particular problem, the best approach to determining $\{y_k\}$ for a given $\{u_k\}$ is to invert the right-hand side of (3.49) as an ordinary z transform with no particular thought as to its structure. However, to understand more of the theory of linear systems in discrete time, it is worth exploring the general situation a little further. To do this, we revert to the time domain.

Suppose that a linear discrete-time time-invariant system has impulse response sequence $\{y_{\delta_k}\}$, and suppose that we wish to find the system response $\{y_k\}$ to an input sequence $\{u_k\}$, with the system initially in a quiescent state. First we express the input sequence

$$\{u_k\} = \{u_0, u_1, u_2, \dots u_n, \dots\} \tag{3.55}$$

as

$$\{u_k\} = u_0\{\delta_k\} + u_1\{\delta_{k-1}\} + u_2\{\delta_{k-2}\} + \dots + u_n\{\delta_{k-n}\} + \dots \tag{3.56}$$

where

$$\delta_{k-j} = \begin{cases} 0 & (k \neq j) \\ 1 & (k = j) \end{cases}$$

In other words, $\{\delta_{k-j}\}$ is simply an impulse sequence with the pulse shifted to $k = j$. Thus, in going from (3.55) to (3.56), we have decomposed the input sequence $\{u_k\}$

into a weighted sum of shifted impulse sequences. Under the assumption of an initially quiescent system, linearity allows us to express the response $\{y_k\}$ to the input sequence $\{u_k\}$ as the appropriately weighted sum of shifted impulse responses. Thus, since the impulse response is $\{y_{\delta_k}\}$, the response to the shifted impulse sequence $\{\delta_{k-j}\}$ will be $\{y_{\delta_{k-j}}\}$, and the response to the weighted impulse sequence $u_j\{\delta_{k-j}\}$ will be simply $u_j\{y_{\delta_{k-j}}\}$. Summing the contributions from all the sequences in (3.56), we obtain

$$\{y_k\} = \sum_{j=0}^{\infty} u_j\{y_{\delta_{k-j}}\} \tag{3.57}$$

as the response of the system to the input sequence $\{u_k\}$. Expanding (3.57), we have

$$\begin{aligned}
\{y_k\} &= u_0\{y_{\delta_k}\} + u_1\{y_{\delta_{k-1}}\} + \ldots + u_j\{y_{\delta_{k-j}}\} + \ldots \\
&= u_0\{y_{\delta_0}, y_{\delta_1}, y_{\delta_2}, \ldots, \quad y_{\delta_h}, \ldots\} \\
&\quad + u_1\{0, y_{\delta_0}, y_{\delta_1}, \ldots, \quad y_{\delta_{h-1}}, \ldots\} \\
&\quad + u_2\{0, 0, y_{\delta_0}, \ldots, \quad y_{\delta_{h-2}}, \ldots\} \\
&\quad \vdots \\
&\quad + u_h\{0, 0, 0, \quad \ldots, 0, \underset{\substack{\uparrow \\ h\text{th position}}}{y_{\delta_0}}, y_{\delta_1}, \ldots\} \\
&\quad + \ldots
\end{aligned}$$

From this expansion, we find that the hth term of the output sequence is determined by

$$y_h = \sum_{j=0}^{h} u_j y_{\delta_{h-j}} \tag{3.58}$$

That is,

$$\{y_k\} = \left\{\sum_{j=0}^{k} u_j y_{\delta_{k-j}}\right\} \tag{3.59}$$

The expression (3.58) is called the **convolution sum**, and the result (3.59) is analogous to (2.83) for continuous systems.

EXAMPLE 3.26 A system has z transfer function

$$G(z) = \frac{z}{z + \frac{1}{2}}$$

What is the system step response? Verify the result using (3.59).

Solution From (3.46), the system step response is

$$Y(z) = G(z)\mathscr{Z}\{h_k\}$$

where $\{h_k\} = \{1, 1, 1, \dots\}$. From Figure 3.3, $\mathscr{Z}\{h_k\} = z/(z-1)$, so

$$Y(z) = \frac{z}{z+\frac{1}{2}}\frac{z}{z-1}$$

Resolving $Y(z)/z$ into partial fractions gives

$$\frac{Y(z)}{z} = \frac{z}{(z+\frac{1}{2})(z-1)} = \frac{2}{3}\frac{1}{z-1} + \frac{1}{3}\frac{1}{z+\frac{1}{2}}$$

so

$$Y(z) = \frac{2}{3}\frac{z}{z-1} + \frac{1}{3}\frac{z}{z+\frac{1}{2}}$$

Taking inverse transforms then gives the step response as

$$\{y_k\} = \{\tfrac{2}{3} + \tfrac{1}{3}(-\tfrac{1}{2})^k\}$$

Using (3.59), we first have to find the impulse response, which, from (3.48), is given by

$$\{y_{\delta_k}\} = \mathscr{Z}^{-1}[G(z)] = \mathscr{Z}^{-1}\left[\frac{z}{z+\frac{1}{2}}\right]$$

so that

$$\{y_{\delta_k}\} = \{(-\tfrac{1}{2})^k\}$$

Taking $\{u_k\}$ to be the unit step sequence $\{h_k\}$, where $h_k = 1$ $(k \geqslant 0)$, the step response may then be determined from (3.59) as

$$\{y_k\} = \left\{\sum_{j=0}^{k} u_j y_{\delta_{k-j}}\right\} = \left\{\sum_{j=0}^{k} 1 \cdot (-\tfrac{1}{2})^{k-j}\right\}$$

$$= \left\{(-\tfrac{1}{2})^k \sum_{j=0}^{k} (-\tfrac{1}{2})^{-j}\right\} = \left\{(-\tfrac{1}{2})^k \sum_{j=0}^{k} (-2)^j\right\}$$

Recognizing the sum as the sum to $k+1$ terms of a geometric series with common ratio -2, we have

$$\{y_k\} = \left\{\left(-\frac{1}{2}\right)^k \frac{1-(-2)^{k+1}}{1-(-2)}\right\} = \left\{\frac{1}{3}\left(\left(-\frac{1}{2}\right)^k + 2\right)\right\} = \left\{\frac{2}{3} + \frac{1}{3}\left(-\frac{1}{2}\right)^k\right\}$$

which concurs with the sequence obtained by direct evaluation.

Example 3.26 reinforces the remark made earlier that the easiest approach to obtaining the response is by direct inversion of (3.32). However, (3.59), together with the argument leading to it, provides a great deal of insight into the way in which the response sequence $\{y_k\}$ is generated. It also serves as a useful 'closed form' for the output of the system, and readers should consult specialist texts on signals and systems for a full discussion (P. Kraniauskas, *Transforms in Signals and Systems*, Addison-Wesley, Wokingham, 1992).

The astute reader will recall that we commenced this section by suggesting that we were about to study the implications of the input–output relationship (3.49), namely

$$Y(z) = Y_\delta(z)U(z)$$

We have in fact explored the time-domain input–output relationship for a linear system, and we now proceed to link this approach with our work in the transform domain. By definition,

$$U(z) = \sum_{k=0}^{\infty} u_k z^{-k} = u_0 + \frac{u_1}{z} + \frac{u_2}{z^2} + \ldots + \frac{u_k}{z^k} + \ldots$$

$$Y_\delta(z) = \sum_{k=0}^{\infty} y_{\delta_k} z^{-k} = y_{\delta_0} + \frac{y_{\delta_1}}{z} + \frac{y_{\delta_2}}{z^2} + \ldots + \frac{y_{\delta_k}}{z^k} + \ldots$$

so

$$Y_\delta(z)U(z) = u_0 y_{\delta_0} + (u_0 y_{\delta_1} + u_1 y_{\delta_0})\frac{1}{z} + (u_0 y_{\delta_2} + u_1 y_{\delta_1} + u_2 y_{\delta_0})\frac{1}{z^2} + \ldots \quad \textbf{(3.60)}$$

Considering the kth term of (3.60), we see that the coefficient of z^{-k} is simply

$$\sum_{j=0}^{k} u_j y_{\delta_{k-j}}$$

However, by definition, since $Y(z) = Y_\delta(z)U(z)$, this is also $y(k)$, the kth term of the output sequence, so that the latter is

$$\{y_k\} = \left\{\sum_{j=0}^{k} u_j y_{\delta_{k-j}}\right\}$$

as found in (3.59). We have thus shown that the time-domain and transform-domain approaches are equivalent, and, in passing, we have established the z transform of the convolution sum as

$$\mathscr{Z}\left\{\sum_{j=0}^{k} u_j v_{k-j}\right\} = U(z)V(z) \quad \textbf{(3.61)}$$

where

$$\mathscr{Z}\{u_k\} = U(z), \quad \mathscr{Z}\{v_k\} = V(z)$$

Putting $p = k - j$ in (3.61) shows that

$$\sum_{j=0}^{k} u_j v_{k-j} = \sum_{p=0}^{k} u_{k-p} v_p \quad \textbf{(3.62)}$$

confirming that the convolution process is commutative.

3.6.5 Exercises

21 Find the transfer functions of each of the following discrete-time systems, given that the system is initially in a quiescent state:

(a) $y_{k+2} - 3y_{k+1} + 2y_k = u_k$

(b) $y_{k+2} - 3y_{k+1} + 2y_k = u_{k+1} - u_k$

(c) $y_{k+3} - y_{k+2} + 2y_{k+1} + y_k = u_k + u_{k-1}$

22 Draw a block diagram representing the discrete-time system

$$y_{k+2} + 0.5y_{k+1} + 0.25y_k = u_k$$

Hence find a block diagram representation of the system

$$y_{k+2} + 0.5y_{k+1} + 0.25y_k = u_k - 0.6u_{k+1}$$

23 Find the impulse response for the systems with z transfer function

(a) $\dfrac{z}{8z^2 + 6z + 1}$ (b) $\dfrac{z^2}{z^2 - 3z + 3}$

(c) $\dfrac{z^2}{z^2 - 0.2z - 0.08}$ (d) $\dfrac{5z^2 - 12z}{z^2 - 6z + 8}$

24 Obtain the impulse response for the systems of Exercises 21(a, b).

25 Which of the following systems are stable?

(a) $9y_{k+2} + 9y_{k+1} + 2y_k = u_k$

(b) $9y_{k+2} - 3y_{k+1} - 2y_k = u_k$

(c) $2y_{k+2} - 2y_{k+1} + y_k = u_{k+1} - u_k$

(d) $2y_{k+2} + 3y_{k+1} - y_k = u_k$

(e) $4y_{k+2} - 3y_{k+1} - y_k = u_{k+1} - 2u_k$.

26 Use the method of Example 3.26 to calculate the step response of the system with transfer function

$$\frac{z}{z - \frac{1}{2}}$$

Verify the result by direct calculation.

27 A sampled data system described by the difference equation

$$y_{n+1} - y_n = u_n$$

is controlled by making the input u_n proportional to the previous error according to

$$u_n = K\left(\frac{1}{2^n} - y_{n-1}\right)$$

where K is a positive gain. Determine the range of values of K for which the system is stable. Taking $K = \frac{2}{9}$, determine the response of the system given $y_0 = y_1 = 0$.

28 Show that the system

$$y_{n+2} + 2y_{n+1} + 2y_n = u_{n+1} \quad (n \geqslant 0)$$

has transfer function

$$D(z) = \frac{z}{z^2 + 2z + 2}$$

Show that the poles of the system are at $z = -1 + \mathrm{j}$ and $z = -1 - \mathrm{j}$. Hence show that the impulse response of the system is given by

$$h_n = \mathscr{Z}^{-1}D(z) = 2^{n/2} \sin \tfrac{3}{4} n\pi$$

3.7 The relationship between Laplace and z transforms

Throughout this chapter we have attempted to highlight similarities, where they occur, between results in Laplace transform theory and those for z transforms. In this section we take a closer look at the relationship between the two transforms. In Section 3.2.2 we introduced the idea of sampling a continuous-time signal $f(t)$ instantaneously at uniform intervals T to produce the sequence

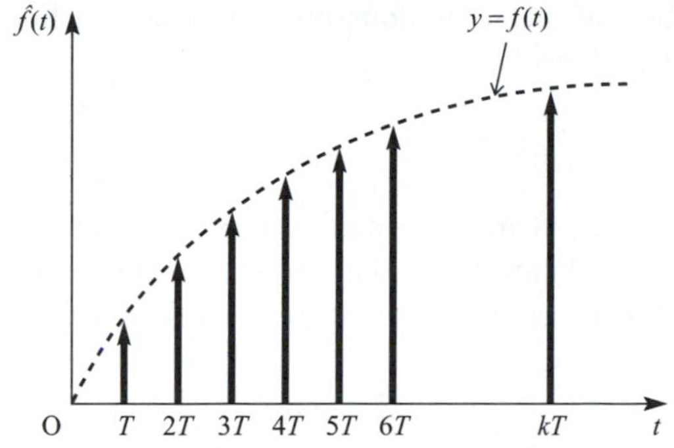

Figure 3.16 Sampled function $f(t)$.

$$\{f(nT)\} = \{f(0), f(T), f(2T), \ldots, f(nT), \ldots\} \tag{3.63}$$

An alternative way of representing the sampled function is to define the continuous-time sampled version of $f(t)$ as $\hat{f}(t)$ where

$$\begin{aligned}\hat{f}(t) &= \sum_{n=0}^{\infty} f(t)\delta(t - nT)\\ &= \sum_{n=0}^{\infty} f(nT)\delta(t - nT)\end{aligned} \tag{3.64}$$

The representation (3.64) may be interpreted as defining a row of impulses located at the sampling points and weighted by the appropriate sampled values (as illustrated in Figure 3.16). Taking the Laplace transform of $\hat{f}(t)$, following the results of Section 2.5.10, we have

$$\begin{aligned}\mathscr{L}\{\hat{f}(t)\} &= \int_{0-}^{\infty}\left[\sum_{k=0}^{\infty} f(kT)\delta(t - kT)\right]\mathrm{e}^{-st}\,\mathrm{d}t\\ &= \sum_{k=0}^{\infty} f(kT)\int_{0-}^{\infty}\delta(t - kT)\,\mathrm{e}^{-st}\,\mathrm{d}t\end{aligned}$$

giving

$$\mathscr{L}\{\hat{f}(t)\} = \sum_{k=0}^{\infty} f(kT)\,\mathrm{e}^{-ksT}$$

Making the change of variable $z = \mathrm{e}^{sT}$ in (3.65) leads to the result

$$\mathscr{L}\{\hat{f}(t)\} = \sum_{k=0}^{\infty} f(kT)z^{-k} = F(z) \tag{3.66}$$

where, as in (3.10), $F(z)$ denotes the z transform of the sequence $\{f(kT)\}$. We can therefore view the z transform of a sequence of samples in discrete time as the

Laplace transform of the continuous-time sampled function $\hat{f}(t)$ with an appropriate change of variable

$$z = e^{sT} \quad \text{or} \quad s = \frac{1}{T}\ln z$$

In Chapter 1 we saw that under this transformation the left half of the s plane, $\text{Re}(s) < 0$, is mapped onto the region inside the unit circle in the z plane, $|z| < 1$. This is consistent with our stability criteria in the s and z domains.

3.8 Engineering application: design of discrete-time systems

An important development in many areas of modern engineering is the replacement of analogue devices by digital ones. Perhaps the most widely known example is the compact disc player, in which mechanical transcription followed by analogue signal processing has been superseded by optical technology and digital signal processing. There are other examples in many fields of engineering, particularly where automatic control is employed.

3.8.1 Analogue filters

At the centre of most signal processing applications are **filters**. These have the effect of changing the spectrum of input signals, that is, attenuating components of signals by an amount depending on the frequency of the component. For example, an analogue **ideal low-pass filter** passes without attenuation all signal components at frequencies less than a critical frequency $\omega = \omega_c$ say. The amplitude of the frequency response $|G(j\omega)|$ (see Section 2.7) of such an ideal filter is shown in Figure 3.17.

One class of analogue filters whose frequency response approximates that of the ideal low-pass filter comprises those known as **Butterworth filters**. As well

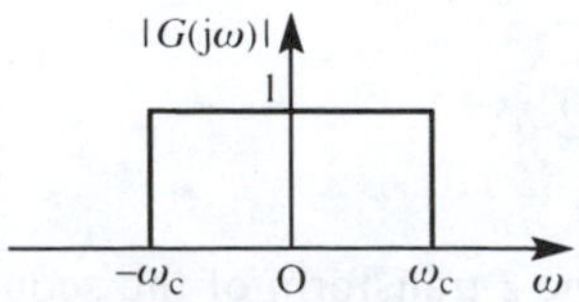

Figure 3.17 Amplitude response for an ideal low-pass filter.

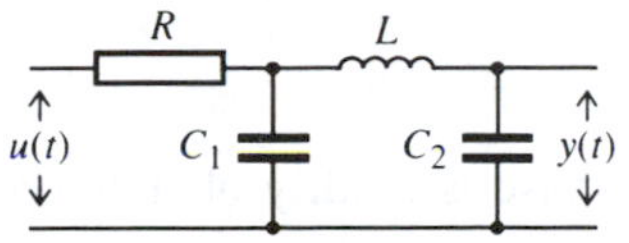

Figure 3.18 *LCR* network for implementing a second-order Butterworth filter.

as having 'good' characteristics, these can be implemented using a network as illustrated in Figure 3.18 for the second-order filter.

It can be shown (see M. J. Chapman, D. P. Goodall and N. C. Steele, *Signal Processing in Electronic Communication*, Horwood Publishing, Chichester, 1997) that the transfer function $G_n(s)$ of the nth-order filter is

$$G_n(s) = \frac{1}{B_n(x)} \quad \text{where} \quad B_n(x) = \sum_{k=0}^{n} a_k x^k$$

with

$$x = \frac{s}{\omega_c}, \quad a_k = \prod_{r=1}^{k} \frac{\cos(r-1)\alpha}{\sin r\alpha}, \quad \alpha = \frac{\pi}{2n}$$

Using these relations, it is readily shown that

$$G_2(s) = \frac{\omega_c^2}{s^2 + \sqrt{2}\,\omega_c s + \omega_c^2} \tag{3.67}$$

$$G_3(s) = \frac{\omega_c^3}{s^3 + 2\omega_c s^2 + 2\omega_c^2 s + \omega_c^3} \tag{3.68}$$

and so on. On sketching the amplitudes of the frequency responses $G_n(j\omega)$, it becomes apparent that increasing n improves the approximation to the response of the ideal low-pass filter of Figure 3.17.

3.8.2 Designing a digital replacement filter

Suppose that we now wish to design a discrete-time system, to operate on samples taken from an input signal, that will operate in a similar manner to a Butterworth filter. We shall assume that the input signal $u(t)$ and the output signal $y(t)$ of the analogue filter are both sampled at the same intervals T to generate the input sequence $\{u(kT)\}$ and the output sequence $\{y(kT)\}$ respectively. Clearly, we need to specify what is meant by 'operate in a similar manner'. In this case, we shall select as our design strategy a method that matches the impulse response sequence of the digital design with a sequence of samples, drawn at the appropriate instants T from the impulse response of an analogue 'prototype'. We shall select the prototype from one of the Butterworth filters discussed in Section 3.8.1, although there are many other possibilities.

Let us select the first-order filter, with cut-off frequency ω_c, as our prototype. Then the first step is to calculate the impulse response of this filter. The Laplace transfer function of the filter is

$$G(s) = \frac{\omega_c}{s + \omega_c}$$

So, from (2.71), the impulse response is readily obtained as

$$h(t) = \omega_c \mathrm{e}^{-\omega_c T} \quad (t \geqslant 0) \tag{3.69}$$

Next, we sample this response at intervals T to generate the sequence

$$\{h(kT)\} = \{\omega_c \mathrm{e}^{-\omega_c kT}\}$$

which on taking the z transform, gives

$$\mathscr{Z}\{h(kT)\} = H(z) = \omega_c \frac{z}{z - \mathrm{e}^{-\omega_c T}}$$

Finally, we choose $H(z)$ to be the transfer function of our digital system. This means simply that the input–output relationship for the design of the digital system will be

$$Y(z) = H(z)U(z)$$

where $Y(z)$ and $U(z)$ are the z transforms of the output and input sequences $\{y(kT)\}$ and $\{u(kT)\}$ respectively. Thus we have

$$Y(z) = \omega_c \frac{z}{z - \mathrm{e}^{-\omega_c T}} U(z) \tag{3.70}$$

Our digital system is now defined, and we can easily construct the corresponding difference equation model of the system as

$$(z - \mathrm{e}^{-\omega_c T})Y(z) = \omega_c z U(z)$$

that is

$$zY(z) - \mathrm{e}^{-\omega_c T} Y(z) = \omega_c z U(z)$$

Under the assumption of zero initial conditions, we can take inverse transforms to obtain the first-order difference equation model

$$y(k + 1) - \mathrm{e}^{-\omega_c T} y(k) = \omega_c u(k + 1) \tag{3.71}$$

A block diagram implementation of (3.71) is shown in Figure 3.19.

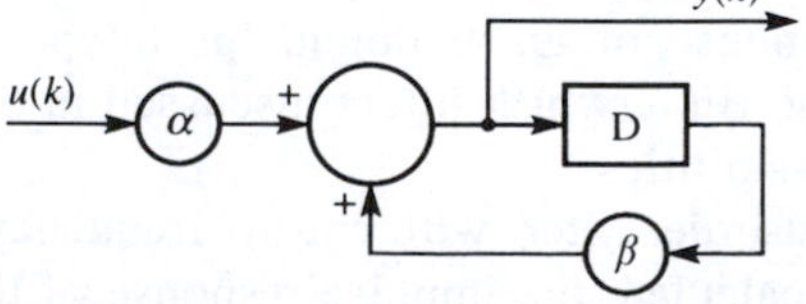

Figure 3.19 Block diagram for the digital replacement filter, $\alpha = k\omega_c$, $\beta = \mathrm{e}^{-\omega_c t}$.

3.8.3 Possible developments

The design method we have considered is called the **impulse invariant technique**, and is only one of many available. The interested reader may develop this study in various ways:

(1) Write a computer program to evaluate the sequence generated by (3.71) with $\omega_c = 1$, and compare with values obtained at the sampling instants for the impulse response (3.69) of the prototype analogue filter.

(2) Repeat the design process for the second-order Butterworth filter.

(3) By setting $s = j\omega$ in the Laplace transfer function of the prototype, and $z = e^{j\omega T}$ in the z transfer function of the digital design, compare the amplitude of the frequency responses in both cases. For an explanation of the results obtained, see Chapter 5.

(4) An alternative design strategy is to replace s in the Laplace transfer function with

$$\frac{2}{T}\frac{z-1}{z+1}$$

(this is a process that makes use of the trapezoidal method of approximate integration). Design alternative digital filters using this technique.

(5) Show that filters designed using either of these techniques will be stable provided that the prototype design is itself stable.

3.9 Engineering application: the delta operator and the $\mathscr{D}$ transform

3.9.1 Introduction

In recent years, sampling rates for digital systems have increased many-fold, and traditional model formulations based on the z transform have produced unsatisfactory results in some applications. It is beyond the scope of this text to describe this situation in detail, but it is possible to give a brief introduction to the problem and to suggest an approach to the solution. For further details see R. M. Middleton and G. C. Goodwin, *Digital Control and Estimation, A Unified Approach* (Prentice-Hall, Englewood Cliffs, NJ, 1990) or W. Forsythe and R. M. Goodall, *Digital Control*

(Macmillan, London, 1991). The contribution of Colin Paterson to the development of this application is gratefully acknowledged.

3.9.2 The q or shift operator and the δ operator

In the time domain we define the shift operator q in terms of its effect on a sequence $\{x_k\}$ as

$$q\{x_k\} = \{x_{k+1}\}$$

That is, the effect of the shift operator is to shift the sequence by one position, so that the kth term of the new sequence is the $(k + 1)$th term of the original sequence. It is then possible to write the difference equation

$$y_{k+2} + 2y_{k+1} + 5y_k = u_{k+1} - u_k$$

as

$$q^2y_k + 2qy_k + 5y_k = qu_k - u_k$$

or

$$(q^2 + 2q + 5)y_k = (q - 1)u_k \tag{3.72}$$

Note that if we had taken the z transform of the difference equation, with an initially quiescent system, we would have obtained

$$(z^2 + 2z + 5)Y(z) = (z - 1)U(z)$$

We see at once the correspondence between the time-domain q operator and the z-transform operator $\mathscr{Z}$.

The next step is to introduce the δ operator, defined as

$$\delta = \frac{q - 1}{\Delta}$$

where Δ has the dimensions of time and is often chosen as the sampling period T. Note that

$$\delta y_k = \frac{(q - 1)y_k}{\Delta} = \frac{y_{k+1} - y_k}{\Delta}$$

so that if $\Delta = T$ then, in the limit of rapid sampling,

$$\delta y_k \simeq \frac{dy}{dt}$$

Solving for q we see that

$$q = 1 + \Delta\delta$$

The difference equation (3.72) can thus be written as

$$((1 + \Delta\delta)^2 + 2(1 + \Delta\delta) + 5)y_k = [(1 + \Delta\delta) - 1]u_k$$

or

$$[(\Delta\delta)^2 + 4\Delta\delta + 8]y_k = \Delta\delta u_k$$

or, finally, as

$$\left(\delta^2 + \frac{4\delta}{\Delta} + \frac{8}{\Delta^2}\right)y_k = \frac{\delta}{\Delta}u_k$$

3.9.3 Constructing a discrete-time system model

So far, we have simply demonstrated a method of rewriting a difference equation in an alternative form. We now examine the possible advantages of constructing discrete-time system models using the δ operator. To do this, we consider a particular example, in which we obtain two different discrete-time forms of the second-order Butterworth filter, both based on the bilinear transform method, sometimes known as **Tustin's method**. This method has its origins in the trapezoidal approximation to the integration process; full details are given in M. J. Chapman, D. P. Goodall and N. C. Steele, Signal Processing in Electronic Communication (Horwood Publishing, Chichester, 1997).

The continuous-time second-order Butterworth filter with cut-off frequency $\omega_c = 1$ is modelled, as indicated by (3.67), by the differential equation

$$\frac{d^2y}{dt^2} + 1.414\,21\frac{dy}{dt} + y = u(t) \tag{3.73}$$

where $u(t)$ is the input and $y(t)$ the filter response. Taking Laplace transforms throughout on the assumption of quiescent initial conditions, that is, $y(0) = (dy/dt)(0) = 0$, we obtain the transformed equation

$$(s^2 + 1.414\,21s + 1)Y(s) = U(s) \tag{3.74}$$

This represents a stable system, since the system poles, given by

$$s^2 + 1.414\,21s + 1 = 0$$

are located at $s = -0.707\,10 \pm j0.707\,10$ and thus lie in the left half-plane of the complex s plane.

We now seek a discrete-time version of the differential equation (3.73). To do this, we first transform (3.74) into the z domain using the **bilinear transform method**, which involves replacing s by

$$\frac{2}{T}\frac{z - 1}{z + 1}$$

Equation (3.74) then becomes

$$\left[\frac{4}{T^2}\left(\frac{z-1}{z+1}\right)^2 + 1.414\,21\,\frac{2}{T}\left(\frac{z-1}{z+1}\right) + 1\right] Y(z) = U(z)$$

or

$$[(\tfrac{1}{4}T^2 + 1.414\,21 \times \tfrac{1}{2}T + 4)z^2 + (\tfrac{1}{2}T^2 - 8)z + \tfrac{1}{4}T^2 - 1.414\,21 \times \tfrac{1}{2}T + 4]Y(z)$$
$$= \tfrac{1}{4}T^2(z^2 + 2z + 1)U(z) \qquad \textbf{(3.75)}$$

We can now invert this transformed equation to obtain the time-domain model

$$(\tfrac{1}{4}T^2 + 1.414\,21 \times \tfrac{1}{2}T + 4)y_{k+2} + (\tfrac{1}{2}T^2 - 8)y_{k+1} + (\tfrac{1}{4}T^2 - 1.414\,21 \times \tfrac{1}{2}T + 4)y_k$$
$$= \tfrac{1}{4}T^2(u_{k+2} + 2u_{k+1} + u_k) \qquad \textbf{(3.76)}$$

For illustrative purposes we set $T = 0.1\text{s}$ in (3.76) to obtain

$$4.073\,21y_{k+2} - 7.995\,00y_{k+1} + 3.931\,79y_k = 0.025\,00(u_{k+2} + 2u_{k+1} + u_k)$$

Note that the roots of the characteristic equation have modulus of about 0.9825, and are thus quite close to the stability boundary.

When $T = 0.01\text{s}$, (3.76) becomes

$$4.007\,10y_{k+2} - 7.999\,95y_{k+1} + 3.992\,95y_k = 0.000\,03(u_{k+2} + 2u_{k+1} + u_k)$$

In this case the roots have modulus of about 0.9982, and we see that increasing the sampling rate has moved them even closer to the stability boundary, and that *high accuracy in the coefficients is essential*, thus adding to the expense of implementation.

An alternative method of proceeding is to avoid the intermediate stage of obtaining the *z*-domain model (3.75) and to proceed directly to a discrete-time representation from (3.73), using the transformation

$$s \rightarrow \frac{2}{T}\frac{\mathrm{q}-1}{\mathrm{q}+1}$$

leading to the same result as in (3.76). Using the δ operator instead of the shift operator q, noting that $\mathrm{q} = 1 + \Delta\delta$, we make the transformation

$$s \rightarrow \frac{2}{T}\frac{\Delta\delta}{2+\Delta\delta}$$

or, if $T = \Delta$, the transformation

$$s \rightarrow \frac{2\delta}{2+\Delta\delta}$$

in (3.74), which becomes

$$[\delta^2 + 1.414\,21 \times \tfrac{1}{2}\delta(2 + \Delta\delta) + \tfrac{1}{4}(2 + \Delta\delta)^2]y_k = \tfrac{1}{4}(2 + \Delta\delta)^2 u_k$$

Note that in this form it is easy to see that in the limit as $\Delta \rightarrow 0$ (that is, as sampling becomes very fast) we regain the original differential equation model. Rearranging this equation, we have

$$\left[\delta^2 + \frac{(1.414\,21 + \Delta)}{(1 + 1.414\,21 \times \frac{1}{2}\Delta + \frac{1}{4}\Delta^2)}\delta + \frac{1}{(1 + 1.414\,21 \times \frac{1}{2}\Delta + \frac{1}{4}\Delta^2)}\right]y_k$$

$$= \frac{(2 + \Delta\delta)^2}{4(1 + 1.414\,21 \times \frac{1}{2}\Delta + \frac{1}{4}\Delta^2)}u_k \tag{3.77}$$

In order to assess stability, it is helpful to introduce a transform variable γ associated with the δ operator. This is achieved by defining γ in terms of z as

$$\gamma = \frac{z - 1}{\Delta}$$

The region of stability in the z plane, $|z| < 1$, thus becomes

$$|1 + \Delta\gamma| < 1$$

or

$$\left|\frac{1}{\Delta} + \gamma\right| < \frac{1}{\Delta} \tag{3.78}$$

This corresponds to a circle in the γ domain, centre $(-1/\Delta, 0)$ and radius $1/\Delta$. As $\Delta \to 0$, we see that this circle expands in such a way that the stability region is the entire open left half-plane, and coincides with the stability region for continuous-time systems.

Let us examine the pole locations for the two cases previously considered, namely $T = 0.1$ and $T = 0.01$. With $\Delta = T = 0.1$, the characteristic equation has the form

$$\gamma^2 + 1.410\,92\gamma + 0.931\,78 = 0$$

with roots, corresponding to poles of the system, at $-0.705\,46 \pm \text{j}0.658\,87$. The centre of the circular stability region is now at $-1/0.1 = -10$, with radius 10, and these roots lie at a radial distance of about 9.3178 from this centre. Note that the distance of the poles from the stability boundary is just less than 0.7. The poles of the original continuous-time model were also at about this distance from the appropriate boundary, and we observe the sharp contrast from our first discretized model, when the discretization process itself moved the pole locations very close to the stability boundary. In that approach the situation became exacerbated when the sampling rate was increased, to $T = 0.01$, and the poles moved nearer to the boundary. Setting $T = 0.01$ in the new formulation, we find that the characteristic equation becomes

$$\gamma^2 + 1.414\,13\gamma + 0.992\,95 = 0$$

with roots at $-0.707\,06 \pm \text{j}0.702\,14$. The stability circle is now centred at -100, with radius 100, and the radial distance of the poles is about 99.2954. Thus the distance from the boundary remains at about 0.7. Clearly, in the limit as $\Delta \to 0$, the pole locations become those of the continuous-time model, with the stability circle enlarging to become the entire left half of the complex γ plane.

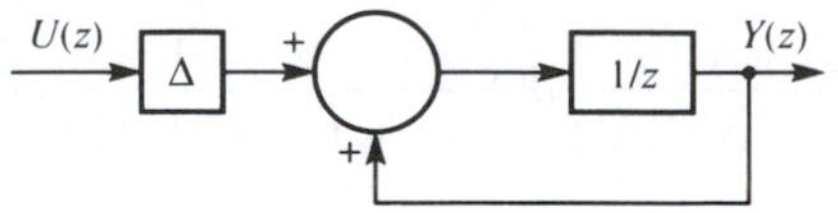

Figure 3.20 The δ^{-1} block.

3.9.4 Implementing the design

The discussion so far serves to demonstrate the utility of the δ operator formulation, but the problem of implementation of the design remains. It is possible to construct a δ^{-1} block based on delay or $1/z$ blocks, as shown in Figure 3.20. Systems can be realized using these structures in cascade or otherwise, and simulation studies have produced successful results. An alternative approach is to make use of the **state-space form** of the system model (see Section 6.10). We demonstrate this approach again for the case $T = 0.01$, when, with $T = \Delta = 0.01$, (3.77) becomes

$$(\delta^2 + 1.414\,13\delta + 0.992\,95)y_k = (0.000\,02\delta^2 + 0.009\,30\delta + 0.992\,95)u_k \quad \textbf{(3.79a)}$$

Based on (3.79a) we are led to consider the equation

$$(\delta^2 + 1.414\,13\delta + 0.992\,95)p_k = u_k \quad \textbf{(3.79b)}$$

Defining the state variables

$$x_{1,k} = p_k, \quad x_{2,k} = \delta p_k$$

equation (3.79b) can be represented by the pair of equations

$$\delta x_{1,k} = x_{2,k}$$

$$\delta x_{2,k} = -0.992\,95x_{1,k} - 1.414\,13x_{2,k} + u_k$$

Choosing

$$y_k = 0.992\,95p_k + 0.009\,30\delta p_k + 0.000\,002\delta^2 p_k \quad \textbf{(3.79c)}$$

equations (3.79b) and (3.79c) are equivalent to (3.79a). In term of the state variables we see that

$$y_k = 0.992\,93x_{1,k} + 0.009\,72x_{2,k} + 0.000\,02u_k$$

Defining the vectors $\boldsymbol{x}_k = [x_{1,k} \quad x_{2,k}]^{\mathrm{T}}$ and $\delta\boldsymbol{x}_k = [\delta x_{1,k} \quad \delta x_{2,k}]^{\mathrm{T}}$, equation (3.79a) can be represented in matrix form as

$$\delta\boldsymbol{x}_k = \begin{bmatrix} 0 & 1 \\ -0.992\,95 & -1.414\,13 \end{bmatrix}\boldsymbol{x}_k + \begin{bmatrix} 0 \\ 1 \end{bmatrix}u_k \quad \textbf{(3.80a)}$$

with

$$y_k = [0.992\,93 \quad 0.009\,72]\boldsymbol{x}_k + 0.000\,02u_k \tag{3.80b}$$

We now return to the q form to implement the system. Recalling that $\delta = (\mathrm{q} - 1)/\Delta$, (3.80a) becomes

$$\mathrm{q}\boldsymbol{x}_k = \boldsymbol{x}_{k+1} = \boldsymbol{x}_k + \Delta\left(\begin{bmatrix} 0 & 1 \\ -0.992\,95 & -1.414\,13 \end{bmatrix}\boldsymbol{x}_k + \begin{bmatrix} 0 \\ 1 \end{bmatrix}u_k\right) \tag{3.81}$$

with (3.80b) remaining the same and where $\Delta = 0.01$, in this case. Equations (3.81) and (3.80b) may be expressed in the vector–matrix form

$$\boldsymbol{x}_{k+1} = \boldsymbol{x}_k + \Delta[\boldsymbol{A}(\Delta)\boldsymbol{x}_k + \boldsymbol{b}u_k]$$

$$y = \boldsymbol{c}^{\mathrm{T}}(\Delta)\boldsymbol{x}_k + d(\Delta)u_k$$

This matrix difference equation can now be implemented without difficulty using standard delay blocks, and has a form similar to the result of applying a simple Euler discretization of the original continuous-time model expressed in state-space form.

3.9.5 The $\mathscr{D}$ transform

In Section 3.9.3 we introduced a transform variable

$$\gamma = \frac{z-1}{\Delta}$$

The purpose of this was to enable us to analyse the stability of systems described in the δ form. We now define a transform in terms of the z transform using the notation given by R. M. Middleton and G. C. Goodwin, *Digital Control and Estimation, A Unified Approach* (Prentice-Hall, Englewood Cliffs, NJ, 1990). Let the sequence $\{f_k\}$ have z transform $F(z)$; then the new transform is given by

$$F'_\Delta(\gamma) = F(z)|_{z=\Delta\gamma+1}$$

$$= \sum_{k=0}^{\infty} \frac{f_k}{(1+\Delta\gamma)^k}$$

The $\mathscr{D}$ transform is formally defined as a slight modification to this form, as

$$\mathscr{D}(f_k) = F_\Delta(\gamma) = \Delta F'_\Delta(\gamma) = \Delta\sum_{k=0}^{\infty} \frac{f_k}{(1+\Delta\gamma)^k}$$

The purpose of this modification is to permit the construction of a *unified theory of transforms* encompassing both continuous- and discrete-time models in the same structure. These developments are beyond the scope of the text, but may be pursued by the interested reader in the reference given above. We conclude the discussion with an example to illustrate the ideas. The ramp sequence $\{u_k\} = \{k\Delta\}$ can be obtained by sampling the continuous-time function $f(t) = t$ at intervals Δ. This sequence has z transform

$$U(z) = \frac{\Delta z}{(z-1)^2}$$

and the corresponding $\mathscr{D}$ transform is then

$$\Delta\, U'_\Delta(\gamma) = \frac{1 + \Delta\gamma}{\gamma^2}$$

Note that on setting $\Delta = 0$ and $\gamma = s$ one recovers the Laplace transform of $f(t)$.

3.9.6 Exercises

29 A continuous-time system having input $y(t)$ and output $y(t)$ is defined by its transfer function

$$H(s) = \frac{1}{(s+1)(s+2)}$$

Use the methods described above to find the q and δ form of the discrete-time system model obtained using the transformation

$$s \to \frac{2}{\Delta}\frac{z-1}{z+1}$$

where Δ is the sampling interval. Examine the stability of the original system and that of the discrete-time systems when $\Delta = 0.1$ and when $\Delta = 0.01$.

30 Use the formula in equation (3.68) to obtain the transfer function of the third-order Butterworth filter with $\omega_c = 1$, and obtain the corresponding δ form discrete-time system when $T = \Delta$.

31 Make the substitution

$$x_1(t) = y(t)$$

$$x_2(t) = \frac{dy(t)}{dt}$$

in Exercise 29 to obtain the state-space form of the system model,

$$\dot{\boldsymbol{x}}(t) = \boldsymbol{A}\boldsymbol{x}(t) + \boldsymbol{b}u(t)$$

$$y(t) = \boldsymbol{c}^{\mathrm{T}}\boldsymbol{x}(t) + du(t)$$

The **Euler discretization technique** replaces $\dot{\boldsymbol{x}}(t)$ by

$$\frac{\boldsymbol{x}((k+1)\Delta) - \boldsymbol{x}(k\Delta)}{\Delta}$$

Show that this corresponds to the model obtained above with $\boldsymbol{A} = \boldsymbol{A}(0)$, $\boldsymbol{c} = \boldsymbol{c}(0)$ and $d = d(0)$.

32 The discretization procedure used in Section 3.9.3 has been based on the bilinear transform method, derived from the trapezoidal approximation to the integration process. An alternative approximation is the Adams–Bashforth procedure, and it can be shown that this means that we should make the transformation

$$s \to \frac{12}{\Delta}\frac{z^2 - z}{5z^2 + 8z - 1}$$

where Δ is the sampling interval (see W. Forsythe and R. M. Goodall, *Digital Control*, Macmillan, London, 1991). Use this transformation to discretize the system given by

$$H(s) = \frac{s}{s+1}$$

when $\Delta = 0.1$ in

(a) the z form, and

(b) the γ form.

3.10 Review exercises (1–16)

1 The signal $f(t) = t$ is sampled at intervals T to generate the sequence $\{f(kT)\}$. Show that

$$\mathscr{Z}\{f(kT)\} = \frac{Tz}{(z-1)^2}$$

2 Show that

$$\mathscr{Z}\{a^k \sin k\omega\} = \frac{az \sin \omega}{z^2 - 2az \cos \omega + a^2} \quad (a > 0)$$

3 Show that

$$\mathscr{Z}\{k^2\} = \frac{z(z+1)}{(z-1)^3}$$

4 Find the impulse response for the system with transfer function

$$H(z) = \frac{(3z^2 - z)}{z^2 - 2z + 1}$$

5 Calculate the step response for the system with transfer function

$$H(z) = \frac{1}{z^2 + 3z + 2}$$

6 A process with Laplace transfer function $H(s) = 1/(s+1)$ is in cascade with a zero-order hold device with Laplace transfer function $G(s) = (1 - e^{-sT})/s$. The overall transfer function is then

$$\frac{1 - e^{-sT}}{s(s+1)}$$

Write $F(s) = 1/s(s+1)$, and find $f(t) = \mathscr{L}^{-1}\{F(s)\}$. Sample $f(t)$ at intervals T to produce the sequence $\{f(kT)\}$ and find $\tilde{F}(z) = \mathscr{Z}\{f(kT)\}$. Deduce that

$$e^{-sT}F(s) \rightarrow \frac{1}{z}\tilde{F}(z)$$

and hence show that the overall z transfer function for the process and zero-order hold is

$$\frac{1 - e^{-T}}{z - e^{-T}}$$

7 A system has Laplace transfer function

$$H(s) = \frac{s+1}{(s+2)(s+3)}$$

Calculate the impulse response, and obtain the z transform of this response when sampled at intervals T.

8 It can be established that if $X(z)$ is the z transform of the sequence $\{x_n\}$ then the general term of that sequence is given by

$$x_n = \frac{1}{j2\pi} \oint_C X(z)z^{n-1}\,dz$$

where C is any closed contour containing all the sigularities of $X(z)$. If we assume that all the singularities of $X(z)$ are poles located within a circle of finite radius then it is an easy application of the residue theorem to show that

$$x_n = \sum[\text{residues of } X(z)z^{n-1} \text{ at poles of } X(z)]$$

(a) Let $X(z) = z/(z-a)(z-b)$, with a and b real. Where are the poles of $X(z)$? Calculate the residues of $z^{n-1}X(z)$, and hence invert the transform to obtain $\{x_n\}$.
(b) Use the residue method to find

(i) $\mathscr{Z}^{-1}\left\{\dfrac{z}{(z-3)^2}\right\}$ (ii) $\mathscr{Z}^{-1}\left\{\dfrac{z}{z^2 - z + 1}\right\}$

9 The impulse response of a certain discrete-time system is $\{(-1)^k - 2^k\}$. What is the step response?

10 A discrete-time system has transfer function

$$H(z) = \frac{z^2}{(z+1)(z-1)}$$

Find the response to the sequence $\{1, -1, 0, 0, \ldots\}$.

11 Show that the response of the second-order system with transfer function

$$\frac{z^2}{(z-\alpha)(z-\beta)}$$

to the input $(1, -(\alpha+\beta), \alpha\beta, 0, 0, 0, \ldots\}$ is

$$\{\delta_k\} = \{1, 0, 0, \ldots\}$$

Deduce that the response of the system

$$\frac{z}{(z-\alpha)(z-\beta)}$$

to the same input will be

$$\{\delta_{k-1}\} = \{0, 1, 0, 0, \ldots\}$$

12 A system is specified by its Laplace transfer function

$$H(s) = \frac{s}{(s+1)(s+2)}$$

Calculate the impulse response $y_\delta(t) = \mathscr{L}^{-1}\{H(s)\}$, and show that if this response is sampled at intervals T to generate the sequence $\{y_\delta(nT)\}$ $(n = 0, 1, 2, \dots)$ then

$$D(z) = \mathscr{Z}\{y_\delta(nT)\} = \frac{2z}{z - e^{-2T}} - \frac{z}{z - e^{-T}}$$

A discrete-time system is now constructed so that

$$Y(z) = TD(z)X(z)$$

where $X(z)$ is the z transform of the input sequence $\{x_n\}$ and $Y(z)$ that of the output sequence $\{y_n\}$, with $x_n = x(nT)$ and $y_n = y(nT)$. Show that if $T = 0.5$s then the difference equation governing the system is

$$y_{n+2} - 0.9744y_{n+1} + 0.2231y_n = 0.5x_{n+2} - 0.4226x_{n+1}$$

Sketch a block diagram for the discrete-time system modelled by the difference equation

$$p_{n+2} - 0.9744p_{n+1} + 0.2231p_n = x_n$$

and verify that the signal y_n, as defined above, is generated by taking $y_n = 0.5p_{n+2} - 0.4226p_{n+1}$ as output.

13 In a discrete-time position-control system the position y_n satisfies the difference equation

$$y_{n+1} = y_n + av_n \quad (a \text{ constant})$$

where v_n and u_n satisfy the difference equations

$$v_{n+1} = v_n + bu_n \quad (b \text{ constant})$$

$$u_n = k_1(x_n - y_n) - k_2v_n \quad (k_1, k_2 \text{ constants})$$

(a) Show that if $k_1 = 1/4ab$ and $k_2 = 1/b$ then the z transfer function of the system is

$$\frac{Y(z)}{X(z)} = \frac{1}{(1 - 2z)^2}$$

where $Y(z) = \mathscr{Z}\{y_n\}$ and $X(z) = \mathscr{Z}\{x_n\}$.

(b) If also $x_n = A$ (where A is a constant), determine the response sequence $\{y_n\}$ given that $y_0 = y_1 = 0$.

14 The step response of a continuous-time system is modelled by the differential equation

$$\frac{d^2y}{dt^2} + 3\frac{dy}{dt} + 2y = 1 \quad (t \geqslant 0)$$

with $y(0) = \dot{y}(0) = 0$. Use the backward-difference approximation

$$\frac{dy}{dt} \simeq \frac{y_k - y_{k-1}}{T}$$

$$\frac{d^2y}{dt^2} \simeq \frac{y_k - 2y_{k-1} + y_{k-2}}{T^2}$$

to show that this differential equation may be approximated by

$$\frac{y_k - 2y_{k-1} + y_{k-2}}{T^2} + 3\frac{y_k - y_{k-1}}{T} + 2y_k = 1$$

Take the z transform of this difference equation, and show that the system poles are at

$$z = \frac{1}{1+T}, \quad z = \frac{1}{1+2T}$$

Deduce that the general solution is thus

$$y_k = \alpha\left(\frac{1}{1+T}\right)^k + \beta\left(\frac{1}{1+2T}\right)^k + \gamma$$

Show that $\gamma = \frac{1}{2}$ and, noting that the initial conditions $y(0) = 0$ and $\dot{y}(0) = 0$ imply $y_0 = y_{-1} = 0$, deduce that

$$y_k = \frac{1}{2}\left[\left(\frac{1}{1+2T}\right)^k - 2\left(\frac{1}{1+T}\right)^k + 1\right]$$

Note that the z-transform method could be used to obtain this result if we redefine $\mathscr{Z}\{y_k\} = \sum_{j=-1}^{\infty}(y_j/z^j)$, with appropriate modifications to the formulae for $\mathscr{Z}\{y_{k+1}\}$ and $\mathscr{Z}\{y_{k+2}\}$.

Explain why the calculation procedure is always stable in theory, but note the pole locations for very small T.

Finally, verify that the solution of the differential equation is

$$y(t) = \tfrac{1}{2}(e^{-2t} - 2e^{-t} + 1)$$

and plot graphs of the exact and approximate solutions with $T = 0.1$s and $T = 0.05$s.

15 Again consider the step response of the system modelled by the differential equation

$$\frac{d^2y}{dt^2} + 3\frac{dy}{dt} + 2y = 1 \quad (t \geqslant 0)$$

with $y(0) = \dot{y}(0) = 0$. Now discretize using the bilinear transform method, that is, take the Laplace transform and make the transformation

$$s \to \frac{2}{T}\frac{z-1}{z+1}$$

where T is the sampling interval. Show that the poles of the resulting z transfer function are at

$$z = \frac{1-T}{1+T}, \quad z = \frac{2-T}{2+T}$$

Deduce that the general solution is then

$$y_k = \alpha\left(\frac{1-T}{1+T}\right)^k + \beta\left(\frac{2-T}{2+T}\right)^k + \gamma$$

Deduce that $\gamma = \frac{1}{2}$ and, using the conditions $y_0 = y_{-1} = 0$, show that

$$y_k = \frac{1}{2}\left[(1-T)\left(\frac{1-T}{1+T}\right)^k - (2-T)\left(\frac{2-T}{2+T}\right)^k + 1\right]$$

Plot graphs to illustrate the exact solution and the approximate solution when $T = 0.1$s and $T = 0.05$s.

16 Show that the z transform of the sampled version of the signal $f(t) = t^2$ is

$$F(z) = \frac{z(z+1)\Delta^2}{(z-1)^3}$$

where Δ is the sampling interval. Verify that the $\mathscr{D}$ transform is then

$$\frac{(1+\Delta v)(2+\Delta v)}{v^3}$$

4

Fourier Series

CONTENTS

4.1 Introduction

The representation of a function in the form of a series is fairly common practice in mathematics. Probably the most familiar expansions are power series of the form

$$f(x) = \sum_{n=0}^{\infty} a_n x^n$$

in which the resolved components or **base set** comprise the power functions

$$1, x, x^2, x^3, \ldots, x^n, \ldots$$

For example, we recall that the exponential function may be represented by the infinite series

$$e^x = 1 + x + \frac{x^2}{2!} + \frac{x^3}{3!} + \ldots + \frac{x^n}{n!} + \ldots = \sum_{n=0}^{\infty} \frac{x^n}{n!}$$

There are frequently advantages in expanding a function in such a series, since the first few terms of a good approximation are easy to deal with. For example, term-by-term integration or differentiation may be applied or suitable function approximations can be made.

Power functions comprise only one example of a base set for the expansion of functions: a number of other base sets may be used. In particular, a **Fourier series** is an expansion of a periodic function $f(t)$ of period $T = 2\pi/\omega$ in which the base set is the set of sine functions, giving an expanded representation of the form

$$f(t) = A_0 + \sum_{n=1}^{\infty} A_n \sin(n\omega t + \phi_n)$$

Although the idea of expanding a function in the form of such a series had been used by Bernoulli, D'Alembert and Euler (*c*. 1750) to solve problems associated with the vibration of strings, it was Joseph Fourier (1768–1830) who developed the approach to a stage where it was generally useful. Fourier, a French physicist, was interested in heat-flow problems: given an initial temperature at all points of a region, he was concerned with determining the change in the temperature distribution over time. When Fourier postulated in 1807 that an arbitrary function $f(x)$ could be represented by a trigonometric series of the form

$$\sum_{n=0}^{\infty} (A_n \cos nkx + B_n \sin nkx)$$

the result was considered so startling that it met considerable opposition from the leading mathematicians of the time, notably Laplace, Poisson and, more significantly, Lagrange, who is regarded as one of the greatest mathematicians of all time. They questioned his work because of its lack of rigour, and it was probably this opposition that delayed the publication of Fourier's work, his classic text *Théorie Analytique de la Chaleur* (The Analytical Theory of Heat) not appearing until 1822. This text has since become the source for the modern methods of solving practical problems associated with partial differential equations subject to prescribed boundary conditions. In addition to heat flow, this class of problems includes structural vibrations, wave propagation and diffusion, which are discussed in Chapter 9. The task of giving Fourier's work a more rigorous mathematical underpinning was undertaken later by Dirichlet (*c*. 1830) and subsequently Riemann, his successor at the University of Göttingen.

In addition to its use in solving boundary-value problems associated with partial differential equations, Fourier series analysis is central to many other applications in engineering. In Chapter 2 we saw how the frequency response of a dynamical system, modelled by a linear differential equation with constant coefficients, is readily determined and the role that it plays in both system analysis and design. In such cases the frequency response, being the steady-state response to a sinusoidal input signal $A \sin \omega t$, is also a sinusoid having the same frequency as the input signal. As mentioned in Section 2.5.6, periodic functions, which are not purely sinusoidal, frequently occur as input signals in engineering applications, particularly in electrical engineering, since many electrical sources of practical value, such as electronic rectifiers, generate non-sinusoidal periodic waveforms. Fourier series provide the ideal framework for analysing the steady-state response to such periodic input signals, since they enable us to represent the signals as infinite sums of sinusoids. The steady-state response due to each sinusoid can then be determined as in Section 2.7, and, because of the linear character of the system, the desired steady-state response can be determined as the sum of the individual responses. As the Fourier series expansion will consist of sinusoids having frequencies $n\omega$ that are multiples of the input signal frequency ω, the steady-state response will also have components having such frequencies. If one of the multiple frequencies $n\omega$ happens to be close in value to the natural oscillating frequency of the system, then it will resonate with the system, and the component at this frequency will dominate the steady-state response. Thus a distinction of significant practical interest between a non-sinusoidal periodic input signal and a sinusoidal input signal is that although the signal may have a frequency considerably lower than the natural frequency of the system, serious problems can still arise owing to resonance. A Fourier series analysis helps to identify such a possibility.

In Chapter 5 we shall illustrate how Fourier series analysis may be extended to aperiodic functions by the use of Fourier transforms. The discrete versions of such transforms provide one of the most advanced methods for discrete signal analysis, and are widely used in such fields as communications theory and speech and image processing. Applications to boundary-value problems are considered in Chapter 9.

4.2 Fourier series expansion

In this section we develop the Fourier series expansion of periodic functions and discuss how closely they approximate the functions. We also indicate how symmetrical properties of the function may be taken advantage of in order to reduce the amount of mathematical manipulation involved in determining the Fourier series.

4.2.1 Periodic functions

A function $f(t)$ is said to be **periodic** if its image values are repeated at regular intervals in its domain. Thus the graph of a periodic function can be divided into

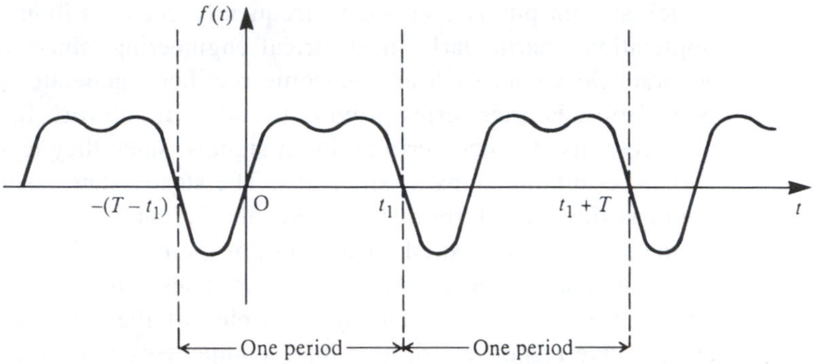

Figure 4.1 A periodic function with period T.

'vertical strips' that are replicas of each other, as illustrated in Figure 4.1. The interval between two successive replicas is called the **period** of the function. We therefore say that a function $f(t)$ is periodic with period T if, for all its domain values t,

$$f(t + mT) = f(t)$$

for any integer m.

To provide a measure of the number of repetitions per unit of t, we define the **frequency** of a periodic function to be the reciprocal of its period, so that

$$\text{frequency} = \frac{1}{\text{period}} = \frac{1}{T}$$

The term **circular frequency** is also used in engineering, and is defined by

$$\text{circular frequency} = 2\pi \times \text{frequency} = \frac{2\pi}{T}$$

and is measured in radians per second. It is common to drop the term 'circular' and refer to this simply as the frequency when the context is clear.

4.2.2 Fourier's theorem

This theorem states that a periodic function that satisfies certain conditions can be expressed as the sum of a number of sine functions of different amplitudes, phases and periods. That is, if $f(t)$ is a periodic function with period T then

$$f(t) = A_0 + A_1 \sin(\omega t + \phi_1) + A_2 \sin(2\omega t + \phi_2) + \ldots$$
$$+ A_n \sin(n\omega t + \phi_n) + \ldots \quad \textbf{(4.1)}$$

where the As and ϕs are constants and $\omega = 2\pi/T$ is the frequency of $f(t)$. The term $A_1 \sin(\omega t + \phi_1)$ is called the **first harmonic** or the **fundamental mode**, and it has the same frequency ω as the parent function $f(t)$. The term $A_n \sin(n\omega t + \phi_n)$ is called the ***n*th harmonic**, and it has frequency $n\omega$, which is n times that of the fundamental. A_n denotes the **amplitude** of the nth harmonic and ϕ_n is its **phase angle**, measuring the lag or lead of the nth harmonic with reference to a pure sine wave of the same frequency.

Since

$$A_n \sin(n\omega t + \phi_n) \equiv (A_n \cos\phi_n)\sin n\omega t + (A_n \sin\phi_n)\cos n\omega t$$
$$\equiv b_n \sin \omega t + a_n \cos n\omega t$$

where

$$b_n = A_n \cos\phi_n, \quad a_n = A_n \sin\phi_n \tag{4.2}$$

the expansion (4.1) may be written as

$$f(t) = \tfrac{1}{2}a_0 + \sum_{n=1}^{\infty} a_n \cos n\omega t + \sum_{n=1}^{\infty} b_n \sin n\omega t \tag{4.3}$$

where $a_0 = 2A_0$ (we shall see later that taking the first term as $\frac{1}{2}a_0$ rather than a_0 is a convenience that enables us to make a_0 fit a general result). The expansion (4.3) is called the **Fourier series expansion** of the function $f(t)$, and the as and bs are called the **Fourier coefficients**. In electrical engineering it is common practice to refer to a_n and b_n respectively as the **in-phase** and **phase quadrature components** of the nth harmonic, this terminology arising from the use of the phasor notation $e^{jn\omega t} = \cos n\omega t + j\sin n\omega t$. Clearly, (4.1) is an alternative representation of the Fourier series with the amplitude and phase of the nth harmonic being determined from (4.2) as

$$A_n = \sqrt{(a_n^2 + b_n^2)}, \quad \phi_n = \tan^{-1}\left(\frac{a_n}{b_n}\right)$$

with care being taken over choice of quadrant.

4.2.3 The Fourier coefficients

Before proceeding to evaluate the Fourier coefficients, we first state the following integrals, in which $T = 2\pi/\omega$:

$$\int_d^{d+T} \cos n\omega t \, dt = \begin{cases} 0 & (n \neq 0) \\ T & (n = 0) \end{cases} \tag{4.4}$$

$$\int_{d}^{d+T} \sin n\omega t \,\mathrm{d}t = 0 \quad (\text{all } n) \tag{4.5}$$

$$\int_{d}^{d+T} \sin m\omega t \sin n\omega t \,\mathrm{d}t = \begin{cases} 0 & (m \neq n) \\ \frac{1}{2}T & (m = n \neq 0) \end{cases} \tag{4.6}$$

$$\int_{d}^{d+T} \cos m\omega t \cos n\omega t \,\mathrm{d}t = \begin{cases} 0 & (m \neq n) \\ \frac{1}{2}T & (m = n \neq 0) \end{cases} \tag{4.7}$$

$$\int_{d}^{d+T} \cos m\omega t \sin n\omega t \,\mathrm{d}t = 0 \ (\text{all } m \text{ and } n) \tag{4.8}$$

The results (4.4)–(4.8) constitute the **orthogonality relations** for sine and cosine functions, and show that the set of functions

$$\{1, \cos \omega t, \cos 2\omega t, \ldots, \cos n\omega t, \sin \omega t, \sin 2\omega t, \ldots, \sin n\omega t\}$$

is an orthogonal set of functions on the interval $d \leqslant t \leqslant d + T$. The choice of d is arbitrary in these results, it only being necessary to integrate over a period of duration T.

Integrating the series (4.3) with respect to t over the period $t = d$ to $t = d + T$, and using (4.4) and (4.5), we find that each term on the right-hand side is zero except for the term involving a_0; that is, we have

$$\begin{aligned} \int_{d}^{d+T} f(t)\,\mathrm{d}t &= \tfrac{1}{2}a_0 \int_{d}^{d+T} \mathrm{d}t + \sum_{n=1}^{\infty} \left(a_n \int_{d}^{d+T} \cos n\omega t \,\mathrm{d}t + b_n \int_{d}^{d+T} \sin n\omega t \,\mathrm{d}t \right) \\ &= \tfrac{1}{2}a_0(T) + \sum_{n=1}^{\infty} [a_n(0) + b_n(0)] \\ &= \tfrac{1}{2}Ta_0 \end{aligned}$$

Thus

$$\frac{1}{2}a_0 = \frac{1}{T}\int_{d}^{d+T} f(t)\,\mathrm{d}t$$

and we can see that the constant term $\frac{1}{2}a_0$ in the Fourier series expansion represents the mean value of the function $f(t)$ over one period. For an electrical signal it represents the bias level or DC (direct current) component. Hence

$$a_0 = \frac{2}{T}\int_{d}^{d+T} f(t)\,\mathrm{d}t \tag{4.9}$$

To obtain this result, we have assumed that term-by-term integration of the series (4.3) is permissible. This is indeed so because of the convergence properties of the series – its validity is discussed in detail in more advanced texts.

To obtain the Fourier coefficient a_n $(n \neq 0)$, we multiply (4.3) throughout by $\cos m\omega t$ and integrate with respect to t over the period $t = d$ to $t = d + T$, giving

$$\int_d^{d+T} f(t)\cos m\omega t\,\mathrm{d}t = \tfrac{1}{2}a_0\int_d^{d+T}\cos m\omega t\,\mathrm{d}t + \sum_{n=1}^{\infty} a_n\int_d^{d+T}\cos n\omega t\cos m\omega t\,\mathrm{d}t$$

$$+ \sum_{n=1}^{\infty} b_n\int_d^{d+T}\cos m\omega t\sin n\omega t\,\mathrm{d}t$$

Assuming term-by-term integration to be possible, and using (4.4), (4.7) and (4.8), we find that, when $m \neq 0$, the only non-zero integral on the right-hand side is the one that occurs in the first summation when $n = m$. That is, we have

$$\int_d^{d+T} f(t)\cos m\omega t\,\mathrm{d}t = a_m\int_d^{d+T}\cos m\omega t\cos m\omega t\,\mathrm{d}t = \tfrac{1}{2}a_m T$$

giving

$$a_m = \frac{2}{T}\int_d^{d+T} f(t)\cos m\omega t\,\mathrm{d}t$$

which, on replacing m by n, gives

$$a_n = \frac{2}{T}\int_d^{d+T} f(t)\cos n\omega t\,\mathrm{d}t \tag{4.10}$$

The value of a_0 given in (4.9) may be obtained by taking $n = 0$ in (4.10), so that we may write

$$a_n = \frac{2}{T}\int_d^{d+T} f(t)\cos m\omega t\,\mathrm{d}t \quad (n = 0, 1, 2, \ldots) \tag{4.11}$$

This explains why the constant term in the Fourier series expansion was taken as $\frac{1}{2}a_0$ and not a_0, since this ensures compatibility of the results (4.9) and (4.10). Although a_0 and a_n satisfy the same formula, it is usually safer to work them out separately.

Finally, to obtain the Fourier coefficients b_n, we multiply (4.3) throughout by $\sin m\omega t$ and integrate with respect to t over the period $t = d$ to $t = d + T$, giving

$$\int_d^{d+T} f(t)\sin m\omega t\,\mathrm{d}t = \tfrac{1}{2}a_0\int_d^{d+T}\sin m\omega t\,\mathrm{d}t$$

$$+ \sum_{n=1}^{\infty}\left(a_n\int_d^{d+T}\sin m\omega t\cos n\omega t\,\mathrm{d}t + b_n\int_t^{d+T}\sin m\omega t\sin n\omega t\,\mathrm{d}t\right)$$

Assuming term-by-term integration to be possible, and using (4.5), (4.6) and (4.8), we find that the only non-zero integral on the right-hand side is the one that occurs in the second summation when $m = n$. That is, we have

$$\int_d^{d+T} f(t)\sin m\omega t\,\mathrm{d}t = b_m\int_d^{d+T}\sin m\omega t\sin m\omega t\,\mathrm{d}t = \tfrac{1}{2}b_m T$$

giving, on replacing m by n,

$$b_n = \frac{2}{T}\int_d^{d+T} f(t)\sin m\omega t\,\mathrm{d}t \quad (n = 1, 2, 3, \ldots) \tag{4.12}$$

The equations (4.11) and (4.12) giving the Fourier coefficients are known as **Euler's formulae**.

Summary

In summary, we have shown that if a periodic function $f(t)$ of period $T = 2\pi/\omega$ can be expanded as a Fourier series then that series is given by

$$f(t) = \tfrac{1}{2}a_0 + \sum_{n=1}^{\infty} a_n \cos n\omega t + \sum_{n=1}^{\infty} b_n \sin n\omega t \tag{4.3}$$

where the coefficients are given by the Euler formulae

$$a_n = \frac{2}{T}\int_d^{d+T} f(t)\cos n\omega t\,\mathrm{d}t \quad (n = 0, 1, 2, \ldots) \tag{4.11}$$

$$b_n = \frac{2}{T}\int_d^{d+T} f(t)\sin n\omega t\,\mathrm{d}t \quad (n = 1, 2, 3, \ldots) \tag{4.12}$$

The limits of integration in Euler's formulae may be specified over any period, so that the choice of d is arbitrary, and may be made in such a way as to help in the calculation of a_n and b_n. In practice, it is common to specify $f(t)$ over either the period $-\frac{1}{2}T < t < \frac{1}{2}T$ or the period $0 < t < T$, leading respectively to the limits of integration being $-\frac{1}{2}T$ and $\frac{1}{2}T$ (that is, $d = -\frac{1}{2}T$) or 0 and T (that is, $d = 0$).

It is also worth noting that an alternative approach may simplify the calculation of a_n and b_n. Using the formula

$$\mathrm{e}^{\mathrm{j}n\omega t} = \cos n\omega t + \mathrm{j}\sin n\omega t$$

we have

$$a_n + \mathrm{j}b_n = \frac{2}{T}\int_d^{d+T} f(t)\,\mathrm{e}^{\mathrm{j}n\omega t}\,\mathrm{d}t \tag{4.13}$$

Evaluating this integral and equating real and imaginary parts on each side gives the values of a_n and b_n. This approach is particularly useful when only the amplitude $|a_n + \mathrm{j}b_n|$ of the nth harmonic is required.

4.2.4 Functions of period 2π

If the period T of the periodic function $f(t)$ is taken to be 2π then $\omega = 1$, and the series (4.3) becomes

$$f(t) = \tfrac{1}{2}a_0 + \sum_{n=1}^{\infty} a_n \cos nt + \sum_{n=1}^{\infty} b_n \sin nt \tag{4.14}$$

with the coefficients given by

$$a_n = \frac{1}{\pi}\int_d^{d+2\pi} f(t)\cos nt\,\mathrm{d}t \quad (n = 0, 1, 2, \ldots) \tag{4.15}$$

$$b_n = \frac{1}{\pi}\int_d^{d+2\pi} f(t)\sin nt\,\mathrm{d}t \quad (n = 1, 2, \ldots) \tag{4.16}$$

While a unit frequency may rarely be encountered in practice, consideration of this particular case reduces the amount of mathematical manipulation involved in determining the coefficients a_n and b_n. Also, there is no loss of generality in considering this case, since if we have a function $f(t)$ of period T, we may write $t_1 = 2\pi t/T$, so that

$$f(t) \equiv f\left(\frac{Tt_1}{2\pi}\right) \equiv F(t_1)$$

where $F(t_1)$ is a function of period 2π. That is, by a simple change of variable, a periodic function $f(t)$ of period T may be transformed into a periodic function $F(t_1)$ of period 2π. Thus, in order to develop an initial understanding and to discuss some of the properties of Fourier series, we shall first consider functions of period 2π, returning to functions of period other than 2π in Section 4.2.10.

EXAMPLE 4.1 Obtain the Fourier series expansion of the periodic function $f(t)$ of period 2π defined by

$$f(t) = t \quad (0 < t < 2\pi), \qquad f(t) = f(t + 2\pi)$$

Solution A sketch of the function $f(t)$ over the interval $-4\pi < t < 4\pi$ is shown in Figure 4.2. Since the function is periodic we only need to sketch it over one period, the pattern being repeated for other periods. Using (4.15) to evaluate the Fourier coefficients a_0 and a_n gives

$$a_0 = \frac{1}{\pi}\int_0^{2\pi} f(t)\,\mathrm{d}t = \frac{1}{\pi}\int_0^{2\pi} t\,\mathrm{d}t = \frac{1}{\pi}\left[\frac{t^2}{2}\right]_0^{2\pi} = 2\pi$$

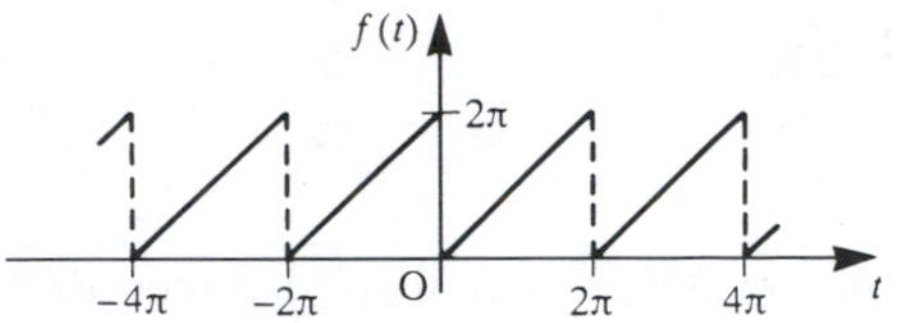

Figure 4.2 Sawtooth wave of Example 4.1.

and

$$a_n = \frac{1}{\pi}\int_0^{2\pi} f(t)\cos nt\,dt \quad (n = 1, 2, \dots)$$

$$= \frac{1}{\pi}\int_0^{2\pi} t\cos nt\,dt$$

which, on integration by parts, gives

$$a_n = \frac{1}{\pi}\left[t\frac{\sin nt}{n} + \frac{\cos nt}{n^2}\right]_0^{2\pi} = \frac{1}{\pi}\left(\frac{2\pi}{n}\sin 2n\pi + \frac{1}{n^2}\cos 2n\pi - \frac{\cos 0}{n^2}\right) = 0$$

since $\sin 2n\pi = 0$ and $\cos 2n\pi = \cos 0 = 1$. Note the need to work out a_0 separately from a_n in this case. The formula (4.16) for b_n gives

$$b_n = \frac{1}{\pi}\int_0^{2\pi} f(t)\sin nt\,dt \quad (n = 1, 2, \dots)$$

$$= \frac{1}{\pi}\int_0^{2\pi} t\sin nt\,dt$$

which, on integration by parts, gives

$$b_n = \frac{1}{\pi}\left[-\frac{t}{n}\cos nt + \frac{\sin nt}{n^2}\right]_0^{2\pi}$$

$$= \frac{1}{\pi}\left(-\frac{2\pi}{n}\cos 2n\pi\right) \quad (\text{since } \sin 2n\pi = \sin 0 = 0)$$

$$= -\frac{2}{n} \quad (\text{since } \cos 2n\pi = 1)$$

Hence from (4.14) the Fourier series expansion of $f(t)$ is

$$f(t) = \pi - \sum_{n=1}^{\infty} \frac{2}{n}\sin nt$$

or, in expanded form,

$$f(t) = \pi - 2\left(\sin t + \frac{\sin 2t}{2} + \frac{\sin 3t}{3} + \dots + \frac{\sin nt}{n} + \dots\right)$$

EXAMPLE 4.2 As period function $f(t)$ with period 2π is defined by

$$f(t) = t^2 + t \quad (-\pi < t < \pi), \qquad f(t) = f(t + 2\pi)$$

Sketch a graph of the function $f(t)$ for values of t from $t = -3\pi$ to $t = 3\pi$ and obtain a Fourier series expansion of the function.

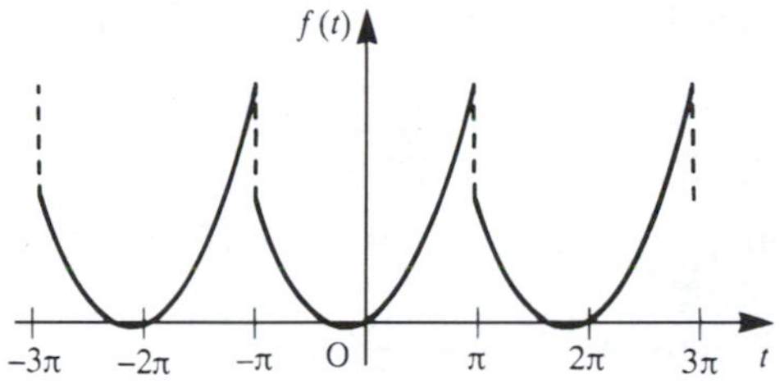

Figure 4.3 Graph of the function $f(t)$ of Example 4.2.

Solution A graph of the function $f(t)$ for $-3\pi < t < 3\pi$ is shown in Figure 4.3. From (4.15) we have

$$a_0 = \frac{1}{\pi}\int_{-\pi}^{\pi} f(t)\,dt = \frac{1}{\pi}\int_{-\pi}^{\pi} (t^2 + t)\,dt = \tfrac{2}{3}\pi^2$$

and

$$a_n = \frac{1}{\pi}\int_{-\pi}^{\pi} f(t)\cos nt\,dt \quad (n = 1, 2, 3, \ldots)$$

n starts from 1 not 0 !!

$$= \frac{1}{\pi}\int_{-\pi}^{\pi} (t^2 + t)\cos nt\,dt$$

which, on integration by parts, gives

$$a_n = \frac{1}{\pi}\left[\frac{t^2}{n}\sin nt + \frac{2t}{n^2}\cos nt - \frac{2}{n^3}\sin nt + \frac{t}{n}\sin nt + \frac{1}{n^2}\cos nt\right]_{-\pi}^{\pi}$$

$$= \frac{1}{\pi}\frac{4\pi}{n^2}\cos n\pi \qquad \left(\text{since } \sin n\pi = 0 \text{ and } \left[\frac{1}{n^2}\cos nt\right]_{-\pi}^{\pi} = 0\right)$$

$$= \frac{4}{n^2}(-1)^n \qquad (\text{since } \cos n\pi = (-1)^n)$$

From (4.16)

$$b_n = \frac{1}{\pi}\int_{-\pi}^{\pi} f(t)\sin nt\,dt \quad (n = 1, 2, 3, \ldots)$$

$$= \frac{1}{\pi}\int_{-\pi}^{\pi} (t^2 + t)\sin nt\,dt$$

which, on integration by parts, gives

$$b_n = \frac{1}{\pi}\left[-\frac{t^2}{n}\cos nt + \frac{2t}{n^2}\sin nt + \frac{2}{n^3}\cos nt - \frac{t}{n}\cos nt + \frac{1}{n^2}\sin nt\right]_{-\pi}^{\pi}$$

$$= -\frac{2}{n}\cos n\pi = -\frac{2}{n}(-1)^n \quad (\text{since } \cos n\pi = (-1)^n)$$

Hence from (4.14) the Fourier series expansion of $f(t)$ is

$$f(t) = \frac{1}{3}\pi^2 + \sum_{n=1}^{\infty}\frac{4}{n^2}(-1)^n\cos nt - \sum_{n=1}^{\infty}\frac{2}{n}(-1)^n\sin nt$$

or, in expanded form,

$$f(t) = \frac{1}{3}\pi^2 + 4\left(-\cos t + \frac{\cos 2t}{2^2} - \frac{\cos 3t}{3^2} + \ldots\right)$$

$$+ 2\left(\sin t - \frac{\sin 2t}{2} + \frac{\sin 3t}{3}\ldots\right)$$

To illustrate the alternative approach, using (4.13) gives

$$a_n + \mathrm{j}b_n = \frac{1}{\pi}\int_{-\pi}^{\pi} f(t)\mathrm{e}^{\mathrm{j}nt}\,\mathrm{d}t = \frac{1}{\pi}\int_{-\pi}^{\pi}(t^2+t)\mathrm{e}^{\mathrm{j}nt}\,\mathrm{d}t$$

$$= \frac{1}{\pi}\left(\left[\frac{t^2+t}{\mathrm{j}n}\mathrm{e}^{\mathrm{j}nt}\right]_{-\pi}^{\pi} - \int_{-\pi}^{\pi}\frac{2t+1}{\mathrm{j}n}\mathrm{e}^{\mathrm{j}nt}\,\mathrm{d}t\right)$$

$$= \frac{1}{\pi}\left[\frac{t^2+t}{\mathrm{j}n}\mathrm{e}^{\mathrm{j}nt} - \frac{2t+1}{(\mathrm{j}n)^2}\mathrm{e}^{\mathrm{j}nt} + \frac{2\mathrm{e}^{\mathrm{j}nt}}{(\mathrm{j}n)^3}\right]_{-\pi}^{\pi}$$

Since

$$\mathrm{e}^{\mathrm{j}n\pi} = \cos n\pi + \mathrm{j}\sin n\pi = (-1)^n$$

$$\mathrm{e}^{-\mathrm{j}n\pi} = \cos n\pi - \mathrm{j}\sin n\pi = (-1)^n$$

and

$$1/\mathrm{j} = -\mathrm{j}$$

$$a_n + \mathrm{j}b_n = \frac{(-1)^n}{\pi}\left(-\mathrm{j}\frac{\pi^2+\pi}{n} + \frac{2\pi+1}{n^2} + \frac{\mathrm{j}2}{n^3} + \mathrm{j}\frac{\pi^2-\pi}{n} - \frac{1-2\pi}{n^2} - \frac{\mathrm{j}2}{n^3}\right)$$

$$= (-1)^n\left(\frac{4}{n^2} - \mathrm{j}\frac{2}{n}\right)$$

Equating real and imaginary parts gives, as before,

$$a_n = \frac{4}{n^2}(-1)^n, \quad b_n = -\frac{2}{n}(-1)^n$$

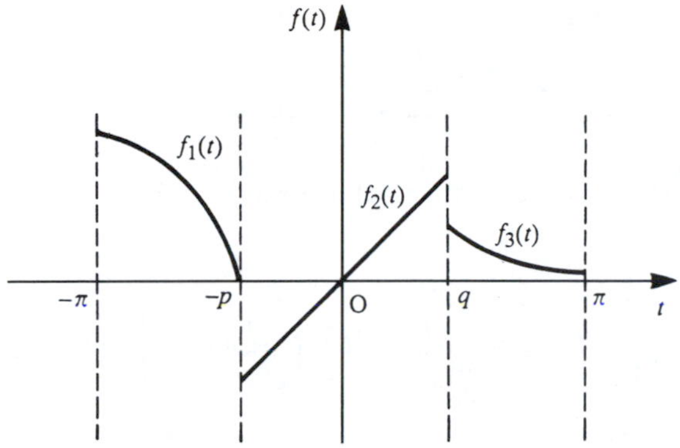

Figure 4.4 Piecewise-continuous function over a period.

A periodic function $f(t)$ may be specified in a piecewise fashion over a period, or, indeed, it may only be piecewise-continuous over a period, as illustrated in Figure 4.4. In order to calculate the Fourier coefficients in such cases, it is necessary to break up the range of integration in the Euler formulae to correspond to the various components of the function. For example, for the function shown in Figure 4.4, $f(t)$ is defined in the interval $-\pi < t < \pi$ by

$$f(t) = \begin{cases} f_1(t) & (-\pi < t < -p) \\ f_2(t) & (-p < t < q) \\ f_3(t) & (q < t < \pi) \end{cases}$$

and is periodic with period 2π. The Euler formulae (4.15) and (4.16) for the Fourier coefficients become

$$a_n = \frac{1}{\pi}\left[\int_{-\pi}^{-p} f_1(t)\cos nt\,dt + \int_{-p}^{q} f_2(t)\cos nt\,dt + \int_{q}^{\pi} f_3(t)\cos nt\,dt\right]$$

$$b_n = \frac{1}{\pi}\left[\int_{-\pi}^{-p} f_1(t)\sin nt\,dt + \int_{-p}^{q} f_2(t)\sin nt\,dt + \int_{q}^{\pi} f_3(t)\sin nt\,dt\right]$$

EXAMPLE 4.3 A periodic function $f(t)$ of period 2π is defined within the period $0 \leqslant t \leqslant 2\pi$ by

$$f(t) = \begin{cases} t & (0 \leqslant t \leqslant \frac{1}{2}\pi) \\ \frac{1}{2}\pi & (\frac{1}{2}\pi \leqslant t \leqslant \pi) \\ \pi - \frac{1}{2}t & (\pi \leqslant t \leqslant 2\pi) \end{cases}$$

Sketch a graph of $f(t)$ for $-2\pi \leqslant t \leqslant 3\pi$ and find a Fourier series expansion of it.

Solution A graph of the function $f(t)$ for $-2\pi \leqslant t \leqslant 3\pi$ is shown in Figure 4.5. From (4.15),

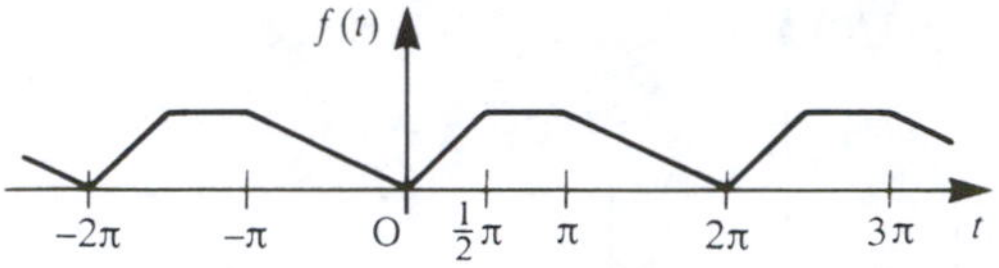

Figure 4.5 Graph of the function $f(t)$ of Example 4.3.

$$a_0 = \frac{1}{\pi}\int_0^{2\pi} f(t)\mathrm{d}t = \frac{1}{\pi}\left[\int_0^{\pi/2} t\,\mathrm{d}t + \int_{\pi/2}^{\pi} \frac{1}{2}\pi\,\mathrm{d}t + \int_{\pi}^{2\pi}\left(\pi - \frac{1}{2}t\right)\mathrm{d}t\right] = \frac{5}{8}\pi$$

and

$$a_n = \frac{1}{\pi}\int_0^{2\pi} f(t)\cos nt\,\mathrm{d}t \quad (n = 1,2,3,\ldots)$$

$$= \frac{1}{\pi}\left[\int_0^{\pi/2} t\cos nt\,\mathrm{d}t + \int_{\pi/2}^{\pi}\frac{1}{2}\pi\cos nt\,\mathrm{d}t + \int_{\pi}^{2\pi}\left(\pi - \frac{1}{2}t\right)\cos nt\,\mathrm{d}t\right]$$

$$= \frac{1}{\pi}\left(\left[\frac{t}{n}\sin nt + \frac{\cos nt}{n^2}\right]_0^{\pi/2} + \left[\frac{\pi}{2n}\sin nt\right]_{\pi/2}^{\pi} + \left[\frac{2\pi - t}{2}\frac{\sin nt}{n} - \frac{\cos nt}{2n^2}\right]_{\pi}^{2\pi}\right)$$

$$= \frac{1}{\pi}\left(\frac{\pi}{2n}\sin\frac{1}{2}n\pi + \frac{1}{n^2}\cos\frac{1}{2}n\pi - \frac{1}{n^2} - \frac{\pi}{2n}\sin\frac{1}{2}n\pi - \frac{1}{2n^2} + \frac{1}{2n^2}\cos n\pi\right)$$

$$= \frac{1}{2\pi n^2}(2\cos\tfrac{1}{2}n\pi - 3 + \cos n\pi)$$

that is,

$$a_n = \begin{cases} \dfrac{1}{\pi n^2}[(-1)^{n/2} - 1] & (\text{even } n) \\ -\dfrac{2}{\pi n^2} & (\text{odd } n) \end{cases}$$

From (4.16),

$$b_n = \frac{1}{\pi}\int_0^{2\pi} f(t)\sin nt\,\mathrm{d}t \quad (\mathrm{n} = 1,2,3,\ldots)$$

$$= \frac{1}{\pi}\left[\int_0^{\pi/2} t\sin nt\,\mathrm{d}t + \int_{\pi/2}^{\pi}\frac{1}{2}\pi\sin nt\,\mathrm{d}t + \int_{\pi}^{2\pi}\left(\pi - \frac{1}{2}t\right)\sin nt\,\mathrm{d}t\right]$$

$$= \frac{1}{\pi}\left(\left[-\frac{t}{n}\cos nt + \frac{1}{n^2}\sin nt\right]_0^{\pi/2} + \left[-\frac{\pi}{2n}\cos nt\right]_{\pi/2}^{\pi}\right.$$

$$\left. + \left[\frac{t - 2\pi}{2n}\cos nt - \frac{1}{2n^2}\sin nt\right]_{\pi}^{2\pi}\right)$$

$$= \frac{1}{\pi}\left(-\frac{\pi}{2n}\cos\tfrac{1}{2}n\pi + \frac{1}{n^2}\sin\frac{1}{2}n\pi - \frac{\pi}{2n}\cos n\pi + \frac{\pi}{2n}\cos\frac{1}{2}n\pi + \frac{\pi}{2n}\cos n\pi\right)$$

$$= \frac{1}{\pi n^2}\sin\frac{1}{2}n\pi$$

$$= \begin{cases} 0 & (\text{even } n) \\ \dfrac{(-1)^{(n-1)/2}}{\pi n^2} & (\text{odd } n) \end{cases}$$

Hence from (4.14) the Fourier series expansion of $f(t)$ is

$$f(t) = \frac{5}{16}\pi - \frac{2}{\pi}\left(\cos t + \frac{\cos 3t}{3^2} + \frac{\cos 5t}{5^2} + \ldots\right)$$

$$- \frac{2}{\pi}\left(\frac{\cos 2t}{2^2} + \frac{\cos 6t}{6^2} + \frac{\cos 10t}{10^2} + \ldots\right)$$

$$+ \frac{1}{\pi}\left(\sin t - \frac{\sin 3t}{3^2} + \frac{\sin 5t}{5^2} - \frac{\sin 7t}{7^2} + \ldots\right)$$

4.2.5 Even and odd functions

Noting that a particular function possesses certain symmetrical properties enables us both to tell which terms are absent from a Fourier series expansion of the function and to simplify the expressions determining the remaining coefficients. In this section we consider even and odd function symmetries, while in Section 4.2.6 we shall consider symmetry due to even and odd harmonics.

First we consider the properties of even and odd functions that are useful for determining the Fourier coefficients. If $f(t)$ is an even function then $f(t) = f(-t)$ for all t, and the graph of the function is symmetrical about the vertical axis as illustrated in Figure 4.6(a). From the definition of integration, it follows that if $f(t)$ is an even function then

$$\int_{-a}^{a} f(t)\,dt = 2\int_{0}^{a} f(t)\,dt$$

If $f(t)$ is an odd function then $f(t) = -f(-t)$ for all t, and the graph of the function is symmetrical about the origin; that is, there is opposite-quadrant symmetry, as illustrated in Figure 4.6(b). It follows that if $f(t)$ is an odd function then

$$\int_{-a}^{a} f(t)\,dt = 0$$

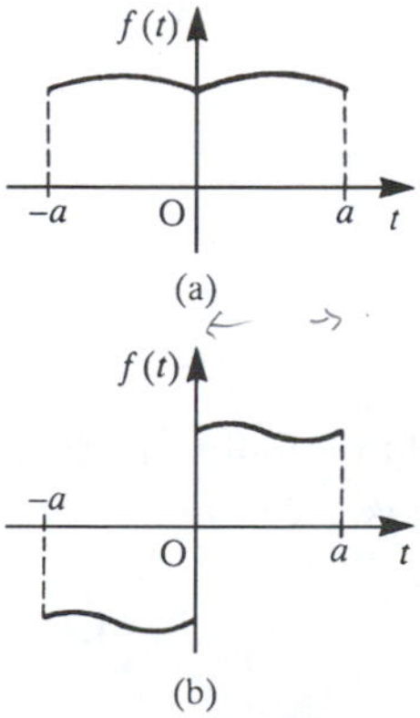

Figure 4.6 Graphs of (a) an even function and (b) an odd function.

The following properties of even and odd functions are also useful for our purposes:

(a) the *sum* of two (or more) *odd* functions is an *odd* function;

(b) the *product* of two *even* functions is an *even* function;

(c) the *product* of two *odd* functions is an *even* function;

(d) the *product* of an *odd* and an *even* function is an *odd* function;

(e) the *derivative* of an *even* function is an *odd* function;

(f) the *derivative* of an *odd* function is an *even* function.

(Noting that t^{even} is even and t^{odd} is odd helps one to remember (a)–(f).)

Using these properties, and taking $d = -\frac{1}{2}T$ in (4.11) and (4.12), we have the following:

(i) If $f(t)$ is an *even* periodic function of period T then

$$a_n = \frac{2}{T}\int_{-T/2}^{T/2} f(t)\cos n\omega t\,\mathrm{d}t = \frac{4}{T}\int_0^{T/2} f(t)\cos n\omega t\,\mathrm{d}t$$

using property (b), and

$$b_n = \frac{2}{T}\int_{-T/2}^{T/2} f(t)\sin n\omega t\,\mathrm{d}t = 0$$

using property (d). Thus the Fourier series expansion of an even periodic function $f(t)$ with period T consists of cosine terms only and, from (4.3), is given by

$$f(t) = \tfrac{1}{2}a_0 + \sum_{n=1}^{\infty} a_n \cos n\omega t \qquad \textbf{(4.17)}$$

with

$$a_n = \frac{4}{T}\int_0^{T/2} f(t)\cos n\omega t \quad (n = 0, 1, 2, \ldots) \qquad \textbf{(4.18)}$$

(ii) If $f(t)$ is an *odd* periodic function of period T then

$$a_n = \frac{2}{T}\int_{-T/2}^{T/2} f(t)\cos n\omega t\,\mathrm{d}t = 0$$

using property (d), and

$$b_n = \frac{2}{T}\int_{-T/2}^{T/2} f(t)\sin n\omega t\,\mathrm{d}t = \frac{4}{T}\int_0^{T/2} f(t)\sin n\omega t\,\mathrm{d}t$$

using property (c). Thus the Fourier series expansion of an odd periodic function $f(t)$ with period T consists of sine terms only and, from (4.3), is given by

$$f(t) = \sum_{n=1}^{\infty} b_n \sin n\omega t \qquad \textbf{(4.19)}$$

with

$$b_n = \frac{4}{T}\int_0^{T/2} f(t)\sin n\omega t\,\mathrm{d}t \quad (n = 1, 2, 3, \ldots) \qquad \textbf{(4.20)}$$

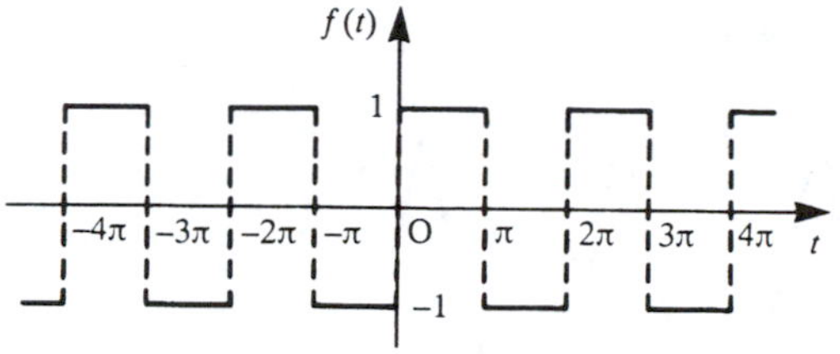

Figure 4.7 Square wave of Example 4.4.

EXAMPLE 4.4 A periodic function $f(t)$ with period 2π is defined within the period $-\pi < t < \pi$ by

$$f(t) = \begin{cases} -1 & (-\pi < t < 0) \\ 1 & (0 < t < \pi) \end{cases}$$

Find its Fourier series expansion.

Solution A sketch of the function $f(t)$ over the interval $-4\pi < t < 4\pi$ is shown in Figure 4.7. Clearly $f(t)$ is an odd function of t, so that its Fourier series expansion consists of sine terms only. Taking $T = 2\pi$, that is $\omega = 1$, in (4.19) and (4.20), the Fourier series expansion is given by

$$f(t) = \sum_{n=1}^{\infty} b_n \sin nt$$

with

$$\begin{aligned} b_n &= \frac{2}{\pi}\int_0^{\pi} f(t)\sin nt \,\mathrm{d}t \quad (n = 1, 2, 3, \ldots) \\ &= \frac{2}{\pi}\int_0^{\pi} 1 \sin nt \,\mathrm{d}t = \frac{2}{\pi}\left[-\frac{1}{n}\cos nt\right]_0^{\pi} \\ &= \frac{2}{n\pi}(1 - \cos n\pi) = \frac{2}{n\pi}[1 - (-1)^n] \\ &= \begin{cases} 4/n\pi & (\text{odd } n) \\ 0 & (\text{even } n) \end{cases} \end{aligned}$$

Thus the Fourier series expansion of $f(t)$ is

$$f(t) = \frac{4}{\pi}\left(\sin t + \frac{1}{3}\sin 3t + \frac{1}{5}\sin 5t + \ldots\right) = \frac{4}{\pi}\sum_{n=1}^{\infty}\frac{\sin(2n-1)t}{2n-1} \tag{4.21}$$

EXAMPLE 4.5 A periodic function $f(t)$ with period 2π is defined as

$$f(t) = t^2 \quad (-\pi < t < \pi), \qquad f(t) = f(t + 2\pi)$$

Obtain a Fourier series expansion for it.

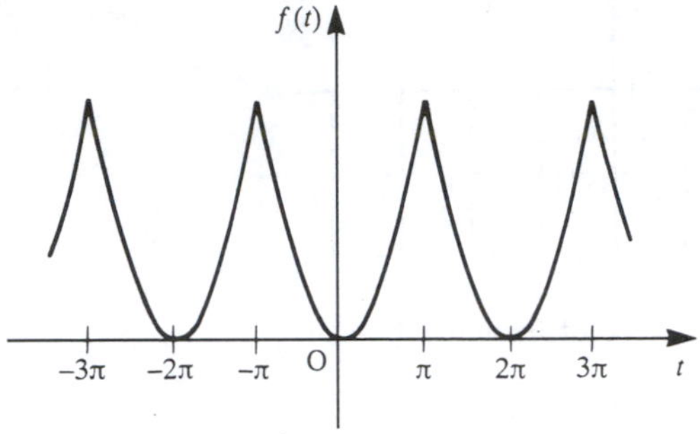

Figure 4.8 The function $f(t)$ of Example 4.5.

Solution A sketch of the function $f(t)$ over the interval $-3\pi < t < 3\pi$ is shown is Figure 4.8. Clearly, $f(t)$ is an even function of t, so that its Fourier series expansion consists of cosine terms only. Taking $T = 2\pi$, that is $\omega = 1$, in (4.17) and (4.18), the Fourier series expansion is given by

$$f(t) = \tfrac{1}{2}a_0 + \sum_{n=1}^{\infty} a_n \cos nt$$

with

$$a_0 = \frac{2}{\pi}\int_0^{\pi} f(t)\,\mathrm{d}t = \frac{2}{\pi}\int_0^{\pi} t^2\,\mathrm{d}t = \frac{2}{3}\pi^2$$

and

$$\begin{aligned} a_n &= \frac{2}{\pi}\int_0^{\pi} f(t)\cos nt\,\mathrm{d}t \quad (n = 1, 2, 3, \dots) \\ &= \frac{2}{\pi}\int_0^{\pi} t^2 \cos nt\,\mathrm{d}t \\ &= \frac{2}{\pi}\left[\frac{t^2}{n}\sin nt + \frac{2t}{n^2}\cos nt - \frac{2}{n^3}\sin nt\right]_0^{\pi} \\ &= \frac{2}{\pi}\left(\frac{2\pi}{n^2}\cos n\pi\right) = \frac{4}{n^2}(-1)^n \end{aligned}$$

since $\sin n\pi = 0$ and $\cos n\pi = (-1)^n$. Thus the Fourier series expansion of $f(t) = t^2$ is

$$f(t) = \frac{1}{3}\pi^2 + 4\sum_{n=1}^{\infty}\frac{(-1)^n}{n^2}\cos nt \tag{4.22}$$

or, writing out the first few terms,

$$f(t) = \tfrac{1}{3}\pi^2 - 4\cos t + \cos 2t - \tfrac{4}{9}\cos 3t + \dots$$

4.2.6 Even and odd harmonics

In this section we consider types of symmetry that can be identified in order to eliminate terms from the Fourier series expansion having even values of n (including $n = 0$) or odd values of n.

(a) If a periodic function $f(t)$ is such that

$$f(t + \tfrac{1}{2}T) = f(t)$$

then it has period $T/2$ and frequency $\omega = 2(2\pi/T)$ so only even harmonics are present in its Fourier series expansion. For even n we have

$$a_n = \frac{4}{T}\int_0^{T/2} f(t)\cos n\omega t \,\mathrm{d}t \tag{4.23}$$

$$b_n = \frac{4}{T}\int_0^{T/2} f(t)\sin n\omega t \,\mathrm{d}t \tag{4.24}$$

An example of such a function is given in Figure 4.9(a).

(b) If a periodic function $f(t)$ with period T is such that

$$f(t + \tfrac{1}{2}T) = -f(t)$$

then only odd harmonics are present in its Fourier series expansion. For odd n

$$a_n = \frac{4}{T}\int_0^{T/2} f(t)\cos n\omega t \,\mathrm{d}t \tag{4.25}$$

$$b_n = \frac{4}{T}\int_0^{T/2} f(t)\sin n\omega t \,\mathrm{d}t \tag{4.26}$$

An example of such a function is shown in Figure 4.9(b).

The square wave of Example 4.4 is such that $f(t + \pi) = -f(t)$, so that, from (b), its Fourier series expansion consists of only odd harmonics. Since it is also an odd function, it follows that its Fourier series expansion consists only of odd-harmonic sine terms, which is confirmed by the result (4.21).

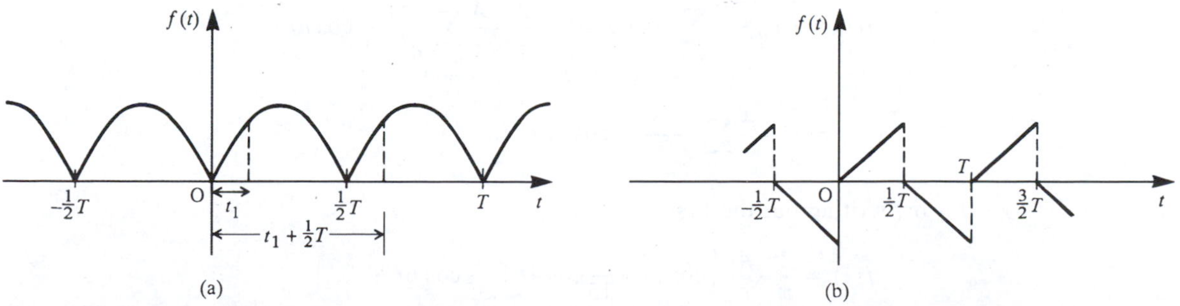

Figure 4.9 Functions having Fourier series with (a) only even harmonics and (b) only odd harmonics.

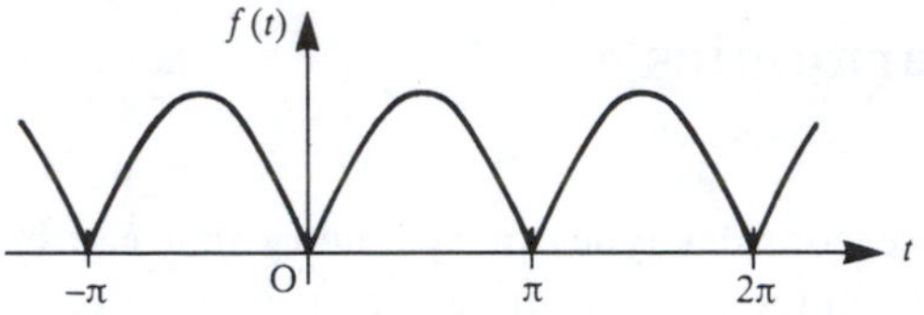

Figure 4.10 Rectified wave $f(t) = |\sin t|$.

EXAMPLE 4.6 Obtain the Fourier series expansion of the rectified sine wave

$$f(t) = |\sin t|$$

Solution A sketch of the wave over the interval $-\pi < t < 2\pi$ is shown in Figure 4.10. Clearly, $f(t + \pi) = f(t)$ so that only even harmonics are present in the Fourier series expansion. Since the function is also an even function of t, it follows that the Fourier series expansion will consist only of even-harmonic cosine terms. Taking $T = 2\pi$, that is $\omega = 1$, in (4.23), the coefficients of the even harmonics are given by

$$a_n = \frac{2}{\pi}\int_0^{\pi} f(t)\cos nt \quad (\text{even } n) \quad = \frac{2}{\pi}\int_0^{\pi} \sin t \cos nt \, \mathrm{d}t$$

$$= \frac{1}{\pi}\int_0^{\pi} [\sin (n+1)t - \sin (n-1)t] \, \mathrm{d}t$$

$$= \frac{1}{\pi}\left[-\frac{\cos (n+1)t}{n+1} + \frac{\cos (n-1)t}{n-1}\right]_0^{\pi}$$

Since both $n + 1$ and $n - 1$ are odd when n is even,

$$\cos (n+1)\pi = \cos (n-1)\pi = -1$$

so that

$$a_n = \frac{1}{\pi}\left[\left(\frac{1}{n+1} - \frac{1}{n-1}\right) - \left(-\frac{1}{n+1} + \frac{1}{n-1}\right)\right] = -\frac{4}{\pi}\frac{1}{n^2-1}$$

Thus the Fourier series expansion of $f(t)$ is

$$f(t) = \frac{1}{2}a_0 + \sum_{\substack{n=2 \\ (n \text{ even})}}^{\infty} a_n \cos nt = \frac{2}{\pi} - \frac{4}{\pi}\sum_{\substack{n=2 \\ (n \text{ even})}}^{\infty} \frac{1}{n^2-1}\cos nt$$

$$= \frac{2}{\pi} - \frac{4}{\pi}\sum_{n=1}^{\infty} \frac{1}{4n^2-1}\cos 2nt$$

or, writing out the first few terms,

$$f(t) = \frac{2}{\pi} - \frac{4}{\pi}\left(\frac{1}{3}\cos 2t + \frac{1}{15}\cos 4t + \frac{1}{35}\cos 6t + \ldots\right)$$

4.2.7 Linearity property

The linearity property as applied to Fourier series may be stated in the form of the following theorem.

THEOREM 4.1 If $f(t) = lg(t) + mh(t)$, where $g(t)$ and $h(t)$ are periodic functions of period T and l and m are arbitrary constants, then $f(t)$ has a Fourier series expansion in which the coefficients are the sums of the coefficients in the Fourier series expansions of $g(t)$ and $h(t)$ multiplied by l and m respectively.

Proof Clearly $f(t)$ is periodic with period T. If the Fourier series expansions of $g(t)$ and $h(t)$ are

$$g(t) = \tfrac{1}{2}a_0 + \sum_{n=1}^{\infty} a_n \cos n\omega t + \sum_{n=1}^{\infty} b_n \sin n\omega t$$

$$h(t) = \tfrac{1}{2}\alpha_0 + \sum_{n=1}^{\infty} \alpha_n \cos n\omega t + \sum_{n=1}^{\infty} \beta_n \sin n\omega t$$

then, using (4.11) and (4.12), the Fourier coefficients in the expansion of $f(t)$ are

$$\begin{aligned} A_n &= \frac{2}{T}\int_d^{d+T} f(t)\cos n\omega t\,\mathrm{d}t = \frac{2}{T}\int_d^{d+T} [lg(t) + mh(t)]\cos n\omega t\,\mathrm{d}t \\ &= \frac{2l}{T}\int_d^{d+T} g(t)\cos n\omega t\,\mathrm{d}t + \frac{2m}{T}\int_d^{d+T} h(t)\cos n\omega t\,\mathrm{d}t \\ &= la_n + m\alpha_n \end{aligned}$$

and

$$\begin{aligned} B_n &= \frac{2}{T}\int_d^{d+T} f(t)\sin n\omega t\,\mathrm{d}t = \frac{2l}{T}\int_d^{d+T} g(t)\sin n\omega t\,\mathrm{d}t + \frac{2m}{T}\int_d^{d+T} h(t)\sin n\omega t\,\mathrm{d}t \\ &= lb_n + m\beta_n \end{aligned}$$

confirming that the Fourier series expansion of $f(t)$ is

$$f(t) = \tfrac{1}{2}(la_0 + m\alpha_0) + \sum_{n=1}^{\infty}(la_n + m\alpha_n)\cos n\omega t + \sum_{n=1}^{\infty}(lb_n + m\beta_n)\sin n\omega t \qquad \square$$

EXAMPLE 4.7 Suppose that $g(t)$ and $h(t)$ are periodic functions of period 2π and are defined within the period $-\pi < t < \pi$ by

$$g(t) = t^2, \quad h(t) = t$$

Determine the Fourier series expansions of both $g(t)$ and $h(t)$ and use the linearity property to confirm the expansion obtained in Example 4.2 for the periodic function $f(t)$ defined within the period $-\pi < t < \pi$ by $f(t) = t^2 + t$.

Solution The Fourier series of $g(t)$ is given by (4.22) as

$$g(t) = \frac{1}{3}\pi^2 + 4\sum_{n=1}^{\infty} \frac{(-1)^n}{n^2} \cos nt$$

Recognizing that $h(t) = t$ is an odd function of t, we find, taking $T = 2\pi$ and $\omega = 1$ in (4.19) and (4.20), that its Fourier series expansion is

$$h(t) = \sum_{n=1}^{\infty} b_n \sin nt$$

where

$$b_n = \frac{2}{\pi}\int_0^{\pi} h(t) \sin nt \, \mathrm{d}t \quad (n = 1, 2, 3, \ldots)$$

$$= \frac{2}{\pi}\int_0^{\pi} t \sin nt \, \mathrm{d}t = \frac{2}{\pi}\left[-\frac{t}{n}\cos nt + \frac{\sin nt}{n^2}\right]_0^{\pi}$$

$$= -\frac{2}{n}(-1)^n$$

recognizing again that $\cos n\pi = (-1)^n$ and $\sin n\pi = 0$. Thus the Fourier series expansion of $h(t) = t$ is

$$h(t) = -2\sum_{n=1}^{\infty} \frac{(-1)^n}{n} \sin nt \tag{4.27}$$

Using the linearity property, we find, by combining (4.12) and (4.27), that the Fourier series expansion of $f(t) = g(t) + h(t) = t^2 + t$ is

$$f(t) = \frac{1}{3}\pi^2 + 4\sum_{n=1}^{\infty} \frac{(-1)^n}{n^2} \cos nt - 2\sum_{n=1}^{\infty} \frac{(-1)^n}{n} \sin nt$$

which conforms to the series obtained in Example 4.2.

4.2.8 Convergence of the Fourier series

So far we have concentrated our attention on determining the Fourier series expansion corresponding to a given periodic function $f(t)$. In reality, this is an exercise in integration, since we merely have to compute the coefficients a_n and b_n using Euler's formulae (4.11) and (4.12) and then substitute these values into (4.3). We have not yet considered the question of whether or not the Fourier series thus obtained is a valid representation of the periodic function $f(t)$. It should not be assumed that the existence of the coefficients a_n and b_n in itself implies that the associated series converges to the function $f(t)$.

A full discussion of the convergence of a Fourier series is beyond the scope of this book and we shall confine ourselves to simply stating a set of conditions which ensures that $f(t)$ has convergent Fourier series expansion. These conditions, known as **Dirichlet's conditions**, may be stated in the form of Theorem 4.2.

THEOREM 4.2 **Dirichlet's conditions**

If $f(t)$ is a bounded periodic function that in any period has

(a) a finite number of isolated maxima and minima, and

(b) a finite number of points of finite discontinuity

then the Fourier series expansion of $f(t)$ converges to $f(t)$ at all points where $f(t)$ is continuous and to the average of the right- and left-hand limits of $f(t)$ at points where $f(t)$ is discontinuous (that is, to the mean of the discontinuity). □

EXAMPLE 4.8 Give reasons why the functions

(a) $\dfrac{1}{3-t}$ (b) $\sin\left(\dfrac{1}{t-2}\right)$

do not satisfy Dirichlet's conditions in the interval $0 < t < 2\pi$.

Solution (a) The function $f(t) = 1/(3 - t)$ has an infinite discontinuity at $t = 3$, which is within the interval, and therefore does not satisfy the condition that $f(t)$ must only have *finite* discontinuities within a period (i.e. it is bounded).

(b) The function $f(t) = \sin[1/(t - 2)]$ has an infinite number of maxima and minima in the neighbourbood of $t = 2$, which is within the interval, and therefore does not satisfy the requirement that $f(t)$ must have only a finite number of isolated maxima and minima within one period.

The conditions of Theorem 4.2 are sufficient to ensure that a representative Fourier series expansion of $f(t)$ exists. However, they are not necessary conditions for convergence, and it does not follow that a representative Fourier series does not exist if they are not satisfied. Indeed, necessary conditions on $f(t)$ for the existence of a convergent Fourier series are not yet known. In practice, this does not cause any problems, since for almost all conceivable practical applications the functions that are encoutered satisfy the conditions of Theorem 4.2 and therefore have representative Fourier series.

Another issue of importance in practical applications is the rate of convergence of a Fourier series, since this is an indication of how many terms must be taken in the expansion in order to obtain a realistic approximation to the function $f(t)$ it represents. Obviously, this is determined by the coefficients a_n and b_n of the Fourier series and the manner in which these decrease as n increases.

In an example, such as Example 4.1, in which the function $f(t)$ is only piecewise-continuous, exhibiting jump discontinuities, the Fourier coefficients decrease as $1/n$, and it may be necessary to include a large number of terms to obtain an adequate approximation to $f(t)$. In an example, such as Example 4.3, in which the function is a continuous function but has discontinuous first derivatives (owing to the sharp corners), the Fourier coefficients decrease as $1/n^2$, and so one would expect the series to converge more rapidly. Indeed, this argument applies in general, and we may summarize as follows:

(a) if $f(t)$ is only piecewise-continuous then the coefficients in its Fourier series representation decrease as $1/n$;

(b) if $f(t)$ is continuous everywhere but has discontinuous first derivatives then the coefficients in its Fourier series representation decrease as $1/n^2$;

(c) if $f(t)$ and all its derivatives up to that of the rth order are continuous but the $(r + 1)$th derivative is discontinuous then the coefficients in its Fourier series representation decrease as $1/n^{r+2}$.

These observations are not surprising, since they simply tell us that the smoother the function $f(t)$, the more rapidly will its Fourier series representation converge.

To illustrate some of these issues related to convergence we return to Example 4.4, in which the Fourier series (4.21) was obtained as a representation of the square wave of Figure 4.7.

Since (4.21) is an infinite series, it is clearly not possible to plot a graph of the result. However, by considering finite partial sums, it is possible to plot graphs of approximations to the series. Denoting the sum of the first N terms in the infinite series by $f_N(t)$, that is

$$f_N(t) = \frac{4}{\pi}\sum_{n=1}^{N}\frac{\sin(2n-1)t}{2n-1} \tag{4.28}$$

the graphs of $f_N(t)$ for $N = 1, 2, 3$ and 20 are as shown in Figure 4.11. It can be seen that at points where $f(t)$ is continuous the approximation of $f(t)$ by $f_N(t)$ improves as N increases, confirming that the series converges to $f(t)$ at all such points. It can also be seen that at points of discontinuity of $f(t)$, which occur at $t = \pm n\pi$ $(n = 0, 1, 2, \dots)$ the series converges to the mean value of the discontinuity, which in this particular example is $\frac{1}{2}(-1 + 1) = 0$. As a consequence, the equality sign in (4.21) needs to be interpreted carefully. Although such use may be acceptable, in the sense that the series converges to $f(t)$ for values of t where $f(t)$ is continuous, this is not so at points of discontinuity. To overcome this problem, the symbol ~ (read as 'behaves as' or 'represented by') rather than = is frequently used in the Fourier series representation of a function $f(t)$, so that (4.21) is often written as

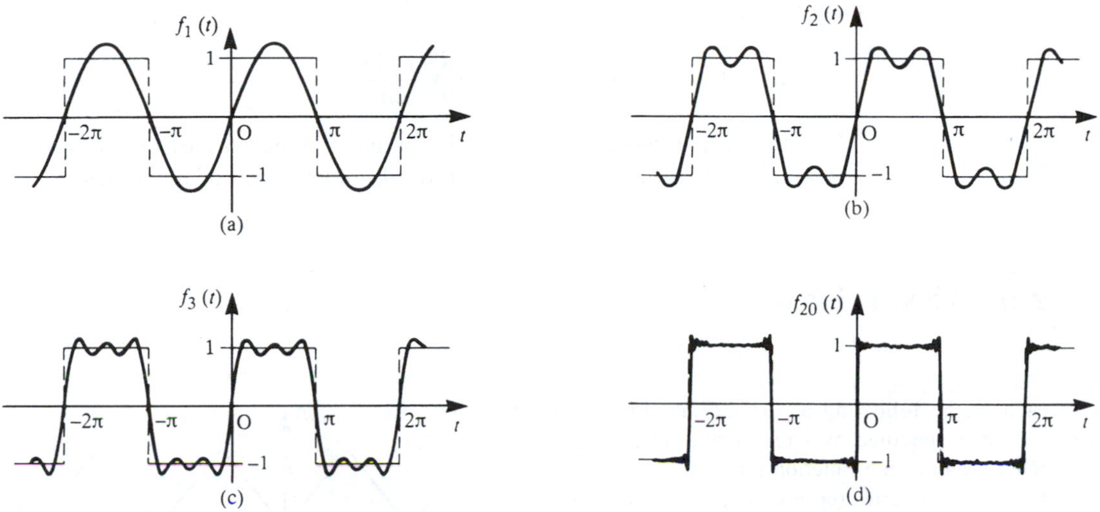

Figure 4.11 Plots of $f_N(t)$ for a square wave; (a) $N = 1$; (b) 2; (c) 3; (d) 20.

$$f(t) \sim \sum_{n=1}^{\infty} \frac{\sin(2n-1)t}{2n-1}$$

In Section 4.7.3 it is shown that the Fourier series converges to $f(t)$ in the sense that the integral of the square of the difference between $f(t)$ and $f_N(t)$ is minimized and tends to zero as $N \to \infty$.

We note that convergence of the Fourier series is slowest near a point of discontinuity, such as the one that occurs at $t = 0$. Although the series does converge to the mean value of the discontinuity (namely zero) at $t = 0$, there is, as indicated in Figure 4.11(d), an undershoot at $t = 0-$ (that is, just to the left of $t = 0$) and an overshoot at $t = 0+$ (that is, just to the right of $t = 0$). This non-smooth convergence of the Fourier series leading to the occurrence of an undershoot and an overshoot at points of discontinuity of $f(t)$ is a characteristic of all Fourier series representing discontinuous functions, not only that of the square wave of Example 4.4, and is known as **Gibbs' phenomenon** after the American physicist J. W. Gibbs (1839–1903). The magnitude of the undershoot/overshoot does not diminish as $N \to \infty$ in (4.28), but simply gets 'sharper' and 'sharper', tending to a spike. In general, the magnitude of the undershoot and overshoot together amount to about 18% of the magnitude of the discontinuity (that is, the difference in the values of the function $f(t)$ to the left and right of the discontinuity). It is important that the existence of this phenomenon be recognized, since in certain practical applications these spikes at discontinuities have to be suppressed by using appropriate smoothing factors.

Theoretically, we can use the series (4.21) to obtain an approximation to π. This is achieved by taking $t = \frac{1}{2}\pi$, when $f(t) = 1$; (4.21) then gives

$$1 = \frac{4}{\pi} \sum_{n=1}^{\infty} \frac{\sin\frac{1}{2}(2n-1)\pi}{2n-1}$$

leading to

$$\pi = 4(1 - \tfrac{1}{3} + \tfrac{1}{5} - \tfrac{1}{7} + \ldots) = 4\sum_{n=1}^{\infty} \frac{(-1)^{n+1}}{2n-1}$$

For practical purposes, however, this is not a good way of obtaining an approximation to π, because of the slow rate of convergence of the series.

4.2.9 Exercises

1 In each of the following a periodic function $f(t)$ of period 2π is specified over one period. In each case sketch a graph of the function for $-4\pi \leqslant t \leqslant 4\pi$ and obtain a Fourier series representation of the function.

(a) $f(t) = \begin{cases} -\pi & (-\pi < t < 0) \\ t & (0 < t < \pi) \end{cases}$

(b) $f(t) = \begin{cases} t+\pi & (-\pi < t < 0) \\ 0 & (0 < t < \pi) \end{cases}$

(c) $f(t) = 1 - \dfrac{t}{\pi} \quad (0 \leqslant t \leqslant 2\pi)$

(d) $f(t) = \begin{cases} 0 & (-\pi \leqslant t \leqslant -\frac{1}{2}\pi) \\ 2\cos t & (-\frac{1}{2}\pi \leqslant t \leqslant \frac{1}{2}\pi) \\ 0 & (\frac{1}{2}\pi \leqslant t \leqslant \pi) \end{cases}$

(e) $f(t) = \cos\frac{1}{2}t \quad (-\pi < t < \pi)$

(f) $f(t) = |t| \quad (-\pi < t < \pi)$

(g) $f(t) = \begin{cases} 0 & (-\pi \leqslant t \leqslant 0) \\ 2t-\pi & (0 < t \leqslant \pi) \end{cases}$

(h) $f(t) = \begin{cases} -t + e^t & (-\pi \leqslant t < 0) \\ t + e^t & (0 \leqslant t < \pi) \end{cases}$

2 Obtain the Fourier series expansion of the periodic function $f(t)$ of period 2π defined over the period $0 \leqslant t \leqslant 2\pi$ by

$$f(t) = (\pi - t)^2 \quad (0 \leqslant t \leqslant 2\pi)$$

Use the Fourier series to show that

$$\frac{1}{12}\pi^2 = \sum_{n=1}^{\infty} \frac{(-1)^{n+1}}{n^2}$$

3 The charge $q(t)$ on the plates of a capacitor at time t is as shown in Figure 4.12. Express $q(t)$ as a Fourier series expansion.

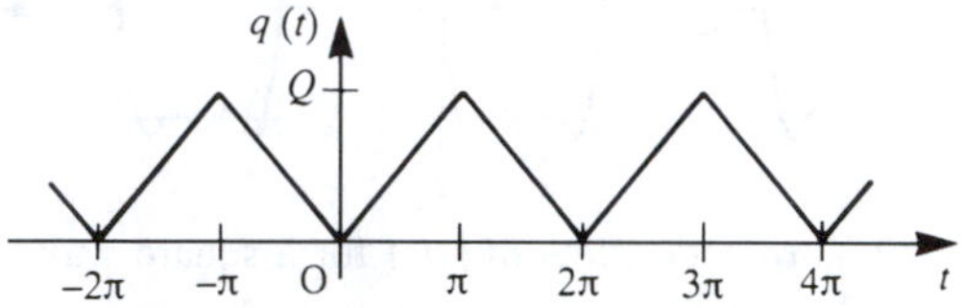

Figure 4.12 Plot of the charge $q(t)$ in Exercise 3.

4 The clipped response of a half-wave rectifier is the periodic function $f(t)$ of period 2π defined over the period $0 \leqslant t \leqslant 2\pi$ by

$$f(t) = \begin{cases} 5\sin t & (0 \leqslant t \leqslant \pi) \\ 0 & (\pi \leqslant t \leqslant 2\pi) \end{cases}$$

Express $f(t)$ as a Fourier series expansion.

5 Show that the Fourier series representing the periodic function $f(t)$, where

$$f(t) = \begin{cases} \pi^2 & (-\pi < t < 0) \\ (t-\pi)^2 & (0 < t < \pi) \end{cases}$$

$$f(t + 2\pi) = f(t)$$

is

$$f(t) = \frac{2}{3}\pi^2 + \sum_{n=1}^{\infty}\left[\frac{2}{n^2}\cos nt + \frac{(-1)^n}{n}\pi \sin nt\right] - \frac{4}{\pi}\sum_{n=1}^{\infty} \frac{\sin(2n-1)t}{(2n-1)^3}$$

Use this result to show that

(a) $\displaystyle\sum_{n=1}^{\infty} \frac{1}{n^2} = \frac{1}{6}\pi^2$ (b) $\displaystyle\sum_{n=1}^{\infty} \frac{(-1)^{n+1}}{n^2} = \frac{1}{12}\pi^2$

6 A periodic function $f(t)$ of period 2π is defined within the domain $0 \leqslant t \leqslant \pi$ by

$$f(t) = \begin{cases} t & (0 \leqslant t \leqslant \frac{1}{2}\pi) \\ \pi - t & (\frac{1}{2}\pi \leqslant t \leqslant \pi) \end{cases}$$

Sketch a graph of $f(t)$ for $-2\pi < t < 4\pi$ for the two cases where

(a) $f(t)$ is an even function
(b) $f(t)$ is an odd function

Find the Fourier series expansion that represents the even function for all values of t, and use it to show that

$$\frac{1}{8}\pi^2 = \sum_{n=1}^{\infty} \frac{1}{(2n-1)^2}$$

7 A periodic function $f(t)$ of period 2π is defined within the period $0 \leqslant t \leqslant 2\pi$ by

$$f(t) = \begin{cases} 2 - t/\pi & (0 \leqslant t \leqslant \pi) \\ t/\pi & (\pi \leqslant t \leqslant 2\pi) \end{cases}$$

Draw a graph of the function for $-4\pi \leqslant t \leqslant 4\pi$ and obtain its Fourier series expansion.

By replacing t by $t - \frac{1}{2}\pi$ in your answer, show that the periodic function $f(t - \frac{1}{2}\pi) - \frac{3}{2}$ is represented by a sine series of odd harmonics.

4.2.10 Functions of period T

Although all the results have been related to periodic functions having period T, all the examples we have considered so far have involved periodic functions of period 2π. This was done primarily for ease of manipulation in determining the Fourier coefficients while becoming acquainted with Fourier series. As mentioned in Section 4.2.4, functions having unit frequency (that is, of period 2π) are rarely encountered in practice, and in this section we consider examples of periodic functions having periods other than 2π.

EXAMPLE 4.9

A periodic function $f(t)$ of period 4 (that is, $f(t + 4) = f(t)$) is defined in the range $-2 < t < 2$ by

$$f(t) = \begin{cases} 0 & (-2 < t < 0) \\ 1 & (0 < t < 2) \end{cases}$$

Sketch a graph of $f(t)$ for $-6 \leqslant t \leqslant 6$ and obtain a Fourier series expansion for the function.

Solution A graph of $f(t)$ for $-6 \leqslant t \leqslant 6$ is shown is Figure 4.13. Taking $T = 4$ in (4.11) and (4.12), we have

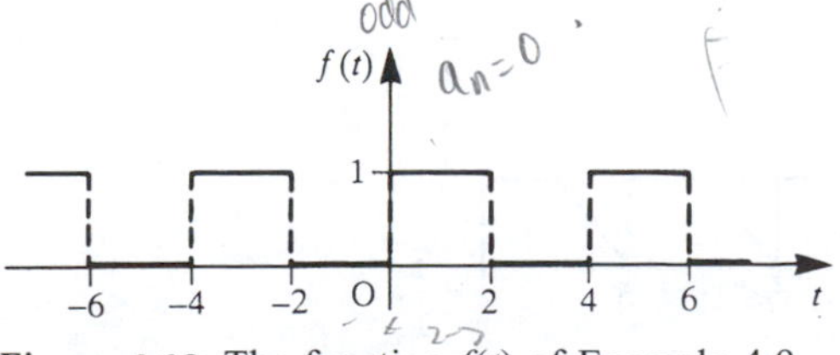

Figure 4.13 The function $f(t)$ of Example 4.9.

$$a_0 = \tfrac{1}{2}\int_{-2}^{2} f(t)\mathrm{d}t = \frac{1}{2}\left(\int_{-2}^{0} 0\,\mathrm{d}t + \int_{0}^{2} 1\,\mathrm{d}t\right) = 1$$

$$a_n = \tfrac{1}{2}\int_{-2}^{2} f(t)\cos\tfrac{1}{2}n\pi t\,\mathrm{d}t \quad (n = 1, 2, 3, \dots)$$

$$= \tfrac{1}{2}\left(\int_{-2}^{0} 0\,\mathrm{d}t + \int_{0}^{2} \cos\tfrac{1}{2}n\pi t\,\mathrm{d}t\right) = 0$$

and

$$b_n = \tfrac{1}{2}\int_{-2}^{2} f(t)\sin\tfrac{1}{2}n\pi t\,\mathrm{d}t \quad (n = 1, 2, 3, \dots)$$

$$= \frac{1}{2}\left(\int_{-2}^{0} 0\,\mathrm{d}t + \int_{0}^{2} \sin\frac{1}{2}n\pi t\,\mathrm{d}t\right) = \frac{1}{n\pi}(1 - \cos n\pi) = \frac{1}{n\pi}[1 - (-1)^n]$$

$$= \begin{cases} 0 & (\text{even } n) \\ 2/n\pi & (\text{odd } n) \end{cases}$$

Thus, from (4.10), the Fourier series expansion of $f(t)$ is

$$f(t) = \frac{1}{2} + \frac{2}{\pi}\left(\sin\frac{1}{2}\pi t + \frac{1}{3}\sin\frac{3}{2}\pi t + \frac{1}{5}\sin\frac{5}{2}\pi t + \dots\right)$$

$$= \frac{1}{2} + \frac{2}{\pi}\sum_{n=1}^{\infty}\frac{1}{2n-1}\sin\frac{1}{2}(2n-1)\pi t$$

EXAMPLE 4.10 A periodic function $f(t)$ of period 2 is defined by

$$f(t) = \begin{cases} 3t & (0 < t < 1) \\ 3 & (1 < t < 2) \end{cases}$$

$$f(t + 2) = f(t)$$

Sketch a graph of $f(t)$ for $-4 \leqslant t \leqslant 4$ and determine a Fourier series expansion for the function.

Solution A graph of $f(t)$ for $-4 \leqslant t \leqslant 4$ is shown in Figure 4.14. Taking $T = 2$ in (4.11) and (4.12), we have

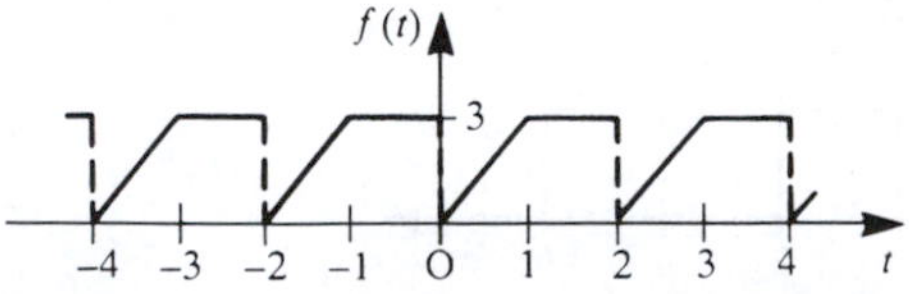

Figure 4.14 The function $f(t)$ of Example 4.10.

$$a_0 = \frac{2}{2}\int_0^2 f(t)\mathrm{d}t = \int_0^1 3t\,\mathrm{d}t + \int_1^2 3\,\mathrm{d}t = \tfrac{9}{2}$$

$$a_n = \frac{2}{2}\int_0^2 f(t)\cos\frac{n\pi t}{1}\,\mathrm{d}t \quad (n = 1, 2, 3, \ldots)$$

$$= \int_0^1 3t\cos n\pi t\,\mathrm{d}t + \int_1^2 3\cos n\pi t\,\mathrm{d}t$$

$$= \left[\frac{3t\sin n\pi t}{n\pi} + \frac{3\cos n\pi t}{(n\pi)^2}\right]_0^1 + \left[\frac{3\sin n\pi t}{n\pi}\right]_1^2$$

$$= \frac{3}{(n\pi)^2}(\cos n\pi - 1)$$

$$= \begin{cases} 0 & (\text{even } n) \\ -6/(n\pi)^2 & (\text{odd } n) \end{cases}$$

and

$$b_n = \frac{2}{2}\int_0^2 f(t)\sin\frac{n\pi t}{1}\,\mathrm{d}t \quad (n = 1, 2, 3, \ldots)$$

$$= \int_0^1 3t\sin n\pi t\,\mathrm{d}t + \int_1^2 3\sin n\pi t\,\mathrm{d}t$$

$$= \left[-\frac{3\cos n\pi t}{n\pi} + \frac{3\sin n\pi t}{(n\pi)^2}\right]_0^1 + \left[-\frac{3\cos n\pi t}{n\pi}\right]_1^2$$

$$= -\frac{3}{n\pi}\cos 2n\pi = -\frac{3}{n\pi}$$

Thus, from (4.10), the Fourier series expansion of $f(t)$ is

$$f(t) = \frac{9}{4} - \frac{6}{\pi^2}\left(\cos\pi t + \frac{1}{9}\cos 3\pi t + \frac{1}{25}\cos 5\pi t + \ldots\right)$$

$$- \frac{3}{\pi}\left(\sin\pi t + \frac{1}{2}\sin 2\pi t + \frac{1}{3}\sin 3\pi t + \ldots\right)$$

$$= \frac{9}{4} - \frac{6}{\pi^2}\sum_{n=1}^{\infty}\frac{\cos(2n-1)\pi t}{(2n-1)^2} - \frac{3}{\pi}\sum_{n=1}^{\infty}\frac{\sin n\pi t}{n}$$

4.2.11 Exercises

8 Find a Fourier series expansion of the periodic function

$$f(t) = t \ (-l < t < l)$$
$$f(t + 2l) = f(t)$$

9 A periodic function $f(t)$ of period $2l$ is defined over one period by

$$f(t) = \begin{cases} -\dfrac{K}{l}(l+t) & (-l < t < 0) \\ \dfrac{K}{l}(l-t) & (0 < t < l) \end{cases}$$

Determine its Fourier series expansion and illustrate graphically for $-3l < t < 3l$.

10 A periodic function of period 10 is defined within the period $-5 < t < 5$ by

$$f(t) = \begin{cases} 0 & (-5 < t < 0) \\ 3 & (0 < t < 5) \end{cases}$$

Determine its Fourier series expansion and illustrate graphically for $-12 < t < 12$.

11 Passing a sinusoidal voltage $A \sin \omega t$ through a half-wave rectifier produces the clipped sine wave shown in Figure 4.15. Determine a Fourier series expansion of the rectified wave.

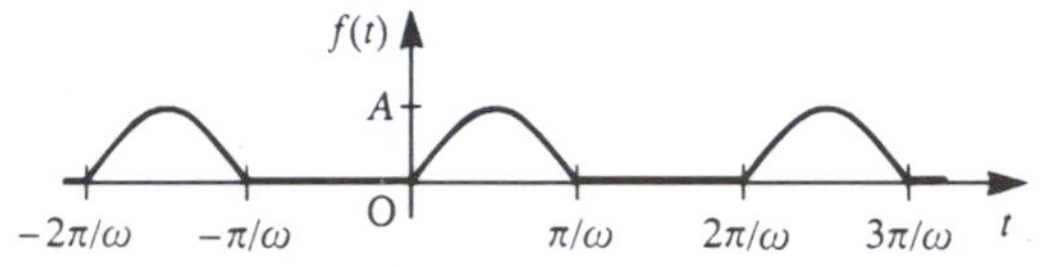

Figure 4.15 Rectified sine wave of Exercise 11.

12 Obtain a Fourier series expansion of the periodic function

$$f(t) = t^2 \quad (-T < t < T)$$
$$f(t + 2T) = f(t)$$

and illustrate graphically for $-3T < t < 3T$.

13 Determine a Fourier series representation of the periodic voltage $e(t)$ shown in Figure 4.16.

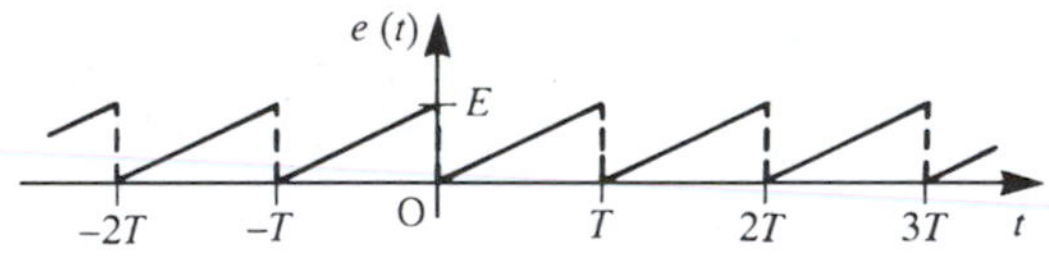

Figure 4.16 Voltage $e(t)$ of Exercise 13.

4.3 Functions defined over a finite interval

One of the requirements of Fourier's theorem is that the function to be expanded be periodic. Therefore a function $f(t)$ that is not periodic cannot have a Fourier series representation that converges to it *for all values* of t. However, we can obtain a Fourier series expansion that represents a *non-periodic* function $f(t)$ that is defined only over a finite time interval $0 \leqslant t \leqslant \tau$. This is a facility that is frequently used to solve problems in practice, particularly boundary-value problems involving partial differential equations, such as the consideration of heat flow along a bar or the vibrations of a string. Various forms of Fourier series representations of $f(t)$, valid only in the interval $0 \leqslant t \leqslant \tau$, are possible, including series consisting of cosine terms only or series consisting of sine terms only. To obtain these, various periodic extensions of $f(t)$ are formulated.

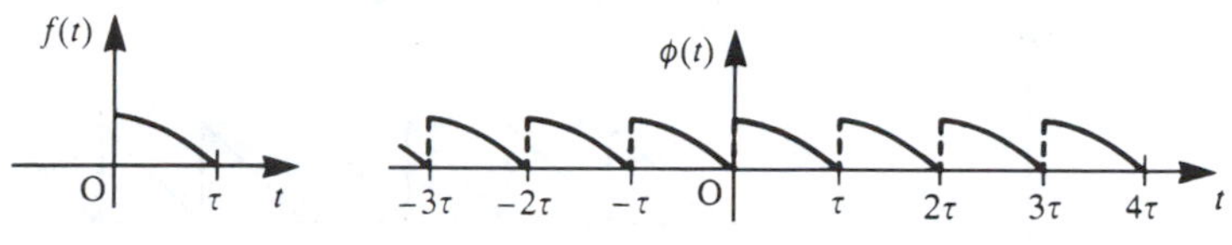

Figure 4.17 Graphs of a function defined only over (a) a finite interval $0 \leqslant t \leqslant \tau$ and (b) its periodic extension.

4.3.1 Full-range series

Suppose the given function $f(t)$ is defined only over the finite time interval $0 \leqslant t \leqslant \tau$. Then, to obtain a full-range Fourier series representation of $f(t)$ (that is, a series consisting of both cosine and sine terms), we define the **periodic extension** $\phi(t)$ of $f(t)$ by

$$\phi(t) = f(t) \quad (0 < t < \tau)$$

$$\phi(t + \tau) = \phi(t)$$

The graphs of a possible $f(t)$ and its periodic extension $\phi(t)$ are shown in Figures 4.17(a) and (b) respectively.

Provided that $f(t)$ satisfies Dirichlet's conditions in the interval $0 \leqslant t \leqslant \tau$, the new function $\phi(t)$, of period τ, will have a convergent Fourier series expansion. Since, within the particular period $0 < t < \tau$, $\phi(t)$ is identical with $f(t)$, it follows that this Fourier series expansion of $\phi(t)$ will be representative of $f(t)$ within this interval.

EXAMPLE 4.11 Find a full-range Fourier series expansion of $f(t) = t$ valid in the finite interval $0 < t < 4$. Draw graphs of both $f(t)$ and the periodic function represented by the Fourier series obtained.

Solution Define the periodic function $\phi(t)$ by

$$\phi(t) = f(t) = t \quad (0 < t < 4)$$

$$\phi(t + 4) = \phi(t)$$

Then the graphs of $f(t)$ and its periodic extension $\phi(t)$ are as shown in Figures 4.18(a) and (b) respectively. Since $\phi(t)$ is a periodic function with period 4, it has a convergent Fourier series expansion. Taking $T = 4$ in (4.11) and (4.12), the Fourier coefficients are determined as

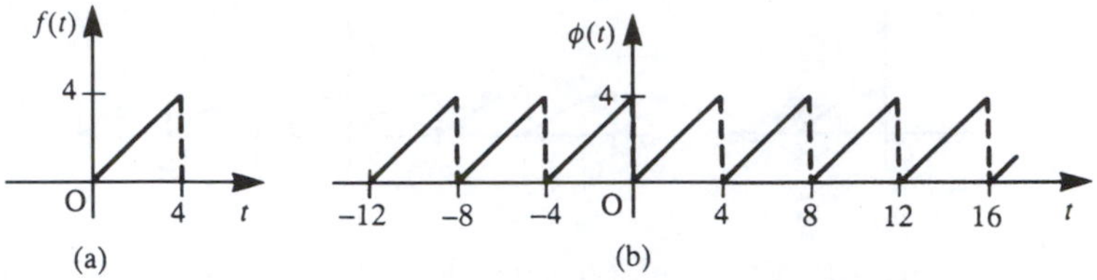

Figure 4.18 The functions $f(t)$ and $\phi(t)$ of Example 4.11.

$$a_0 = \tfrac{1}{2}\int_0^4 f(t)\,\mathrm{d}t = \tfrac{1}{2}\int_0^4 t\,\mathrm{d}t = 4$$

$$a_n = \tfrac{1}{2}\int_0^4 f(t)\cos\tfrac{1}{2}n\pi t\,\mathrm{d}t \quad (n = 1, 2, 3, \dots)$$

$$= \frac{1}{2}\int_0^4 t\cos\frac{1}{2}n\pi t\,\mathrm{d}t = \frac{1}{2}\left[\frac{2t}{n\pi}\sin\tfrac{1}{2}n\pi t + \frac{4}{(n\pi)^2}\cos\frac{1}{2}n\pi t\right]_0^4 = 0$$

and

$$b_n = \tfrac{1}{2}\int_0^4 f(t)\sin\tfrac{1}{2}n\pi t\,\mathrm{d}t \quad (n = 1, 2, 3, \dots)$$

$$= \frac{1}{2}\int_0^4 t\sin\frac{1}{2}n\pi t\,\mathrm{d}t = \frac{1}{2}\left[-\frac{2t}{n\pi}\cos\frac{1}{2}n\pi t + \frac{4}{(n\pi)^2}\sin\frac{1}{2}n\pi t\right]_0^4 = -\frac{4}{n\pi}$$

Thus, by (4.10), the Fourier series expansion of $\phi(t)$ is

$$\phi(t) = 2 - \frac{4}{\pi}\left(\sin\frac{1}{2}\pi t + \frac{1}{2}\sin\pi t + \frac{1}{3}\sin\frac{3}{2}\pi t + \frac{1}{4}\sin 2t + \frac{1}{5}\sin\frac{5}{2}\pi t + \dots\right)$$

$$= 2 - \frac{4}{\pi}\sum_{n=1}^{\infty}\frac{1}{n}\sin\frac{1}{2}n\pi t$$

Since $\phi(t) = f(t)$ for $0 < t < 4$, it follows that this Fourier series is representative of $f(t)$ within this interval, so that

$$f(t) = t = 2 - \frac{4}{\pi}\sum_{n=1}^{\infty}\frac{1}{n}\sin\frac{1}{2}n\pi t \quad (0 < t < 4) \tag{4.29}$$

It is important to appreciate that this series converges to t only within the interval $0 < t < 4$. For values of t outside this interval it converges to the periodic extended function $\phi(t)$. Again convergence is to be interpreted in the sense of Theorem 4.2, so that at the end points $t = 0$ and $t = 4$ the series does not converge to t but to the mean of the discontinuity in $\phi(t)$, namely the value 2.

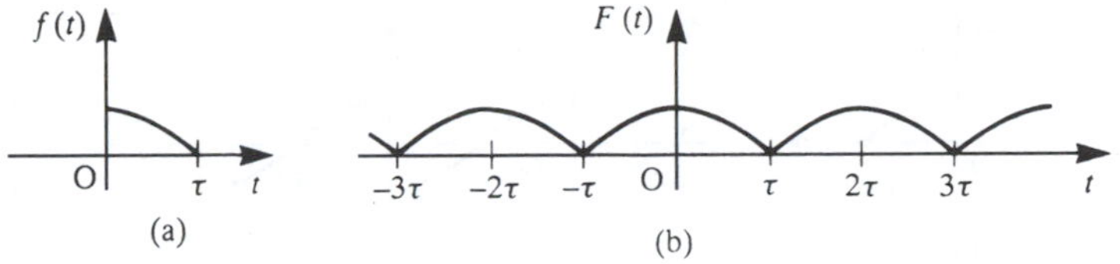

Figure 4.19 (a) A function $f(t)$; (b) its even periodic extension $F(t)$.

4.3.2 Half-range cosine and sine series

Rather than develop the periodic extension $\phi(t)$ of $f(t)$ as in Section 4.3.1, it is possible to formulate periodic extensions that are either even or odd functions, so that the resulting Fourier series of the extended periodic functions consist either of cosine terms only or sine terms only.

For a function $f(t)$ defined only over the finite interval $0 \leqslant t \leqslant \tau$ its **even periodic extension** $F(t)$ is the even periodic function defined by

$$F(t) = \begin{cases} f(t) & (0 < t < \tau) \\ f(-t) & (-\tau < t < 0) \end{cases}$$

$$F(t + 2\tau) = f(t)$$

As an illustration, the even periodic extension $F(t)$ of the function $f(t)$ shown in Figure 4.17(a) (redrawn in Figure 4.19a) is shown in Figure 4.19(b).

Provided that $f(t)$ satisfies Dirichlet's conditions in the interval $0 < t < \tau$, since it is an even function of period 2τ, it follows from Section 4.2.5 that the even periodic extension $F(t)$ will have a convergent Fourier series representation consisting of cosine terms only and given by

$$F(t) = \tfrac{1}{2}a_0 + \sum_{n=1}^{\infty} a_n \cos \frac{n\pi t}{\tau} \tag{4.30}$$

where

$$a_n = \frac{2}{\tau}\int_0^{\tau} f(t) \cos \frac{n\pi t}{\tau}\,dt \quad (n = 0, 1, 2, \ldots) \tag{4.31}$$

Since, within the particular interval $0 < t < \tau$, $F(t)$ is identical with $f(t)$, it follows that the series (4.30) also converges to $f(t)$ within this interval.

For a function $f(t)$ defined only over the finite interval $0 \leqslant t \leqslant \tau$, its **odd periodic extension** $G(t)$ is the odd periodic function defined by

$$G(t) = \begin{cases} f(t) & (0 < t < \tau) \\ -f(-t) & (-\tau < t < 0) \end{cases}$$

$$G(t + 2\tau) = G(t)$$

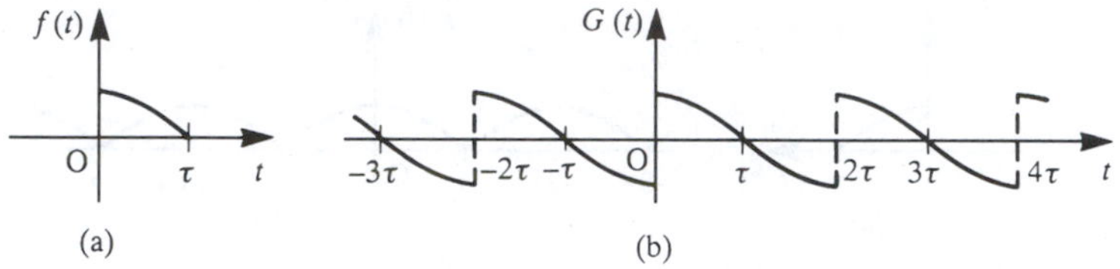

Figure 4.20 (a) A function $f(t)$; (b) its odd periodic extension $G(t)$.

Again, as an illustration, the odd periodic extension $G(t)$ of the function $f(t)$ shown in Figure 4.17(a) (redrawn in Figure 4.20a) is shown in Figure 4.20(b).

Provided that $f(t)$ satisfies Dirichlet's conditions in the interval $0 < t < \tau$, since it is an odd function of period 2τ, it follows from Section 4.2.5 that the odd periodic extension $G(t)$ will have a convergent Fourier series representation consisting of sine terms only and given by

$$G(t) = \sum_{n=1}^{\infty} b_n \sin \frac{n\pi t}{\tau} \tag{4.32}$$

where

$$b_n = \frac{2}{\tau}\int_0^{\tau} f(t) \sin \frac{n\pi t}{\tau}\, dt \quad (n = 1, 2, 3, \dots) \tag{4.33}$$

Again, since, within the particular interval $0 < t < \tau$, $G(t)$ is identical with $f(t)$, it follows that the series (4.32) also converges to $f(t)$ within this interval.

We note that both the even and odd periodic extensions $F(t)$ and $G(t)$ are of period 2τ, which is twice the length of the interval over which $f(t)$ is defined. However, the resulting Fourier series (4.30) and (4.32) are based only on the function $f(t)$, and for this reason are called the **half-range Fourier series expansions** of $f(t)$. In particular, the even half-range expansion $F(t)$, (4.30), is called the **half-range cosine series expansion** of $f(t)$, while the odd half-range expansion $G(t)$, (4.32), is called the **half-range sine series expansion** of $f(t)$.

EXAMPLE 4.12

For the function $f(t) = t$ defined only in the interval $0 < t < 4$, and considered in Example 4.11, obtain

(a) a half-range cosine series expansion

(b) a half-range sine series expansion.

Draw graphs of $f(t)$ and of the periodic functions represented by the two series obtained for $-20 < t < 20$.

Solution (a) *Half-range cosine series*. Define the periodic function $F(t)$ by

$$F(t) = \begin{cases} f(t) = t & (0 < t < 4) \\ f(-t) = -t & (-4 < t < 0) \end{cases}$$

$$F(t + 8) = F(t)$$

Then, since $F(t)$ is an even periodic function with period 8, it has a convergent Fourier series expansion given by (4.30). Taking $\tau = 4$ in (4.31), we have

$$a_0 = \tfrac{2}{4}\int_0^4 f(t)\,\mathrm{d}t = \tfrac{1}{2}\int_0^4 t\,\mathrm{d}t = 4$$

$$a_n = \tfrac{2}{4}\int_0^4 f(t)\cos\tfrac{1}{4}n\pi t\,\mathrm{d}t \quad (n = 1, 2, 3, \dots)$$

$$= \frac{1}{2}\int_0^4 t\cos\frac{1}{4}n\pi t\,\mathrm{d}t = \frac{1}{2}\left[\frac{4t}{n\pi}\sin\frac{1}{4}n\pi t + \frac{16}{(n\pi)^2}\cos\frac{1}{4}n\pi t\right]_0^4$$

$$= \frac{8}{(n\pi)^2}(\cos n\pi - 1) = \begin{cases} 0 & (\text{even } n) \\ -16/(n\pi)^2 & (\text{odd } n) \end{cases}$$

Then, by (4.30), the Fourier series expansion of $F(t)$ is

$$F(t) = 2 - \frac{16}{\pi^2}\left(\cos\frac{1}{4}\pi t + \frac{1}{3^2}\cos\frac{3}{4}\pi t + \frac{1}{5^2}\cos\frac{5}{4}\pi t + \dots\right)$$

or

$$F(t) = 2 - \frac{16}{\pi^2}\sum_{n=1}^{\infty}\frac{1}{(2n-1)^2}\cos\frac{1}{4}(2n-1)\pi t$$

Since $F(t) = f(t)$ for $0 < t < 4$, it follows that this Fourier series is representative of $f(t)$ within this interval. Thus the half-range cosine series expansion of $f(t)$ is

$$f(t) = t = 2 - \frac{16}{\pi^2}\sum_{n=1}^{\infty}\frac{1}{(2n-1)^2}\cos\frac{1}{4}(2n-1)\pi t \quad (0 < t < 4) \qquad \textbf{(4.34)}$$

(b) *Half-range sine series*. Define the periodic function $G(t)$ by

$$G(t) = \begin{cases} f(t) = t & (0 < t < 4) \\ -f(-t) = t & (-4 < t < 0) \end{cases}$$

$$G(t + 8) = G(t)$$

Then, since $G(t)$ is an odd periodic function with period 8, it has a convergent Fourier series expansion given by (4.32). Taking $\tau = 4$ in (4.33), we have

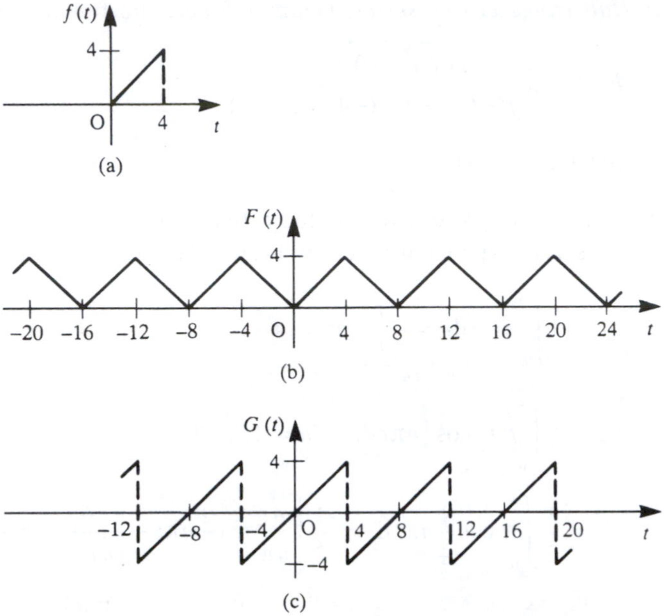

Figure 4.21 The functions $f(t)$, $F(t)$ and $G(t)$ of Example 4.12.

$$\begin{aligned} b_n &= \tfrac{2}{4}\int_0^4 f(t)\sin\tfrac{1}{4}n\pi t\,\mathrm{d}t \quad (n = 1, 2, 3, \ldots) \\ &= \frac{1}{2}\int_0^4 t\sin\frac{1}{4}n\pi t\,\mathrm{d}t = \frac{1}{2}\left[-\frac{4t}{n\pi}\cos\frac{1}{4}n\pi t + \frac{16}{(n\pi)^2}\sin\frac{1}{4}n\pi t\right]_0^4 \\ &= -\frac{8}{n\pi}\cos n\pi = -\frac{8}{n\pi}(-1)^n \end{aligned}$$

Thus, by (4.32), the Fourier series expansion of $G(t)$ is

$$G(t) = \frac{8}{\pi}\left(\sin\frac{1}{4}\pi t - \frac{1}{2}\sin\frac{1}{2}\pi t + \frac{1}{3}\sin\frac{3}{4}\pi t - \ldots\right)$$

or

$$G(t) = \frac{8}{\pi}\sum_{n=1}^{\infty}\frac{(-1)^{n+1}}{n}\sin\frac{1}{4}n\pi t$$

Since $G(t) = f(t)$ for $0 < t < 4$, it follows that this Fourier series is representative of $f(t)$ within this interval. Thus the half-range sine series expansion of $f(t)$ is

$$f(t) = t = \frac{8}{\pi}\sum_{n=1}^{\infty}\frac{(-1)^{n+1}}{n}\sin\frac{1}{4}n\pi t \quad (0 < t < 4) \tag{4.35}$$

Graphs of the given function $f(t)$ and of the even and odd periodic expansions $F(t)$ and $G(t)$ are given in Figures 4.21(a), (b) and (c) respectively.

It is important to realize that the three different Fourier series representations (4.29), (4.34) and (4.35) are representative of the function $f(t) = t$ only within the defined interval $0 < t < 4$. Outside this interval the three Fourier series converge to the three different functions $\phi(t)$, $F(t)$ and $G(t)$, illustrated in Figures 4.18(b), 4.21(b) and 4.21(c) respectively.

4.3.3 Exercises

14 Show that the half-range Fourier sine series expansion of the function $f(t) = 1$, valid for $0 < t < \pi$, is

$$f(t) = \frac{4}{\pi}\sum_{n=1}^{\infty}\frac{\sin(2n-1)t}{2n-1} \quad (0 < t < \pi)$$

Sketch the graphs of both $f(t)$ and the periodic function represented by the series expansion for $-3\pi < t < 3\pi$.

15 Determine the half-range cosine series expansion of the function $f(t) = 2t - 1$, valid for $0 < t < 1$. Sketch the graphs of both $f(t)$ and the periodic function represented by the series expansion for $-2 < t < 2$.

16 The function $f(t) = 1 - t^2$ is to be represented by a Fourier series expansion over the finite interval $0 < t < 1$. Obtain a suitable

(a) full-range series expansion,
(b) half-range sine series expansion,
(c) half-range cosine series expansion.

Draw graphs of $f(t)$ and of the periodic functions represented by each of the three series for $-4 < t < 4$.

17 A function $f(t)$ is defined by

$$f(t) = \pi t - t^2 \quad (0 \leqslant t \leqslant \pi)$$

and is to be represented by either a half-range Fourier sine series or a half-range Fourier cosine series. Find both of these series and sketch the graphs of the functions represented by them for $-2\pi < t < 2\pi$.

18 A tightly stretched flexible uniform string has its ends fixed at the points $x = 0$ and $x = l$. The mid-point of the string is displaced a distance a, as shown in Figure 4.22. If $f(x)$ denotes the displaced profile of the string, express $f(x)$ as a Fourier series expansion consisting only of sine terms.

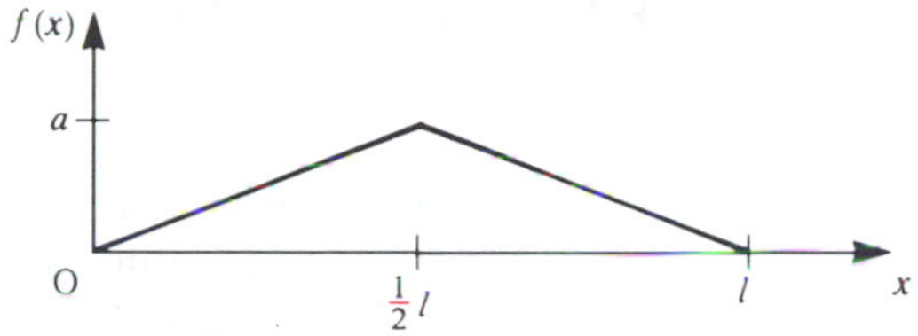

Figure 4.22 Displaced string of Exercise 18.

19 Repeat Exercise 18 for the case where the displaced profile of the string is as shown in Figure 4.23.

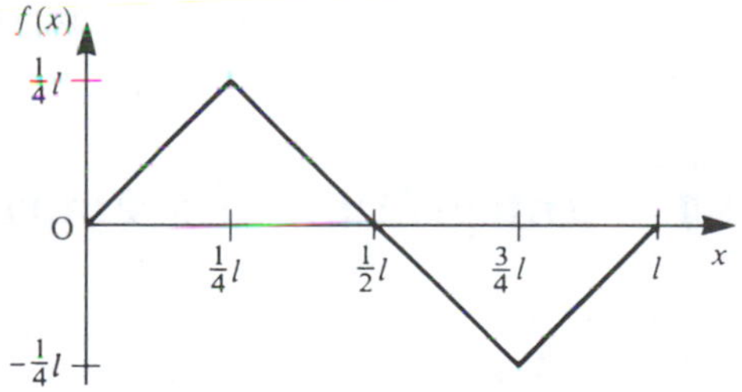

Figure 4.23 Displaced string of Exercise 19.

20 A function $f(t)$ is defined on $0 \leqslant t \leqslant \pi$ by

$$f(t) = \begin{cases} \sin t & (0 \leqslant t < \frac{1}{2}\pi) \\ 0 & (\frac{1}{2}\pi \leqslant t \leqslant \pi) \end{cases}$$

Find a half-range Fourier series expansion of $f(t)$ on this interval. Sketch a graph of the function represented by the series for $-2\pi \leqslant t \leqslant 2\pi$.

21 A function $f(t)$ is defined on the interval $-l \leqslant x \leqslant l$ by

$$f(x) = \frac{A}{l}(|x| - l)$$

Obtain a Fourier series expansion of $f(x)$ and sketch a graph of the function represented by the series for $-3l \leqslant x \leqslant 3l$.

22 The temperature distribution $T(x)$ at a distance x, measured from one end, along a bar of length L is given by

$$T(x) = Kx(L - x) \quad (0 \leqslant x \leqslant L), \quad K = \text{constant}$$

Express $T(x)$ as a Fourier series expansion consisting of sine terms only.

23 Find the Fourier series expansion of the function $f(t)$ valid for $-1 < t < 1$, where

$$f(t) = \begin{cases} 1 & (-1 < t < 0) \\ \cos \pi t & (0 < t < 1) \end{cases}$$

To what value does this series converge when $t = 1$?

4.4 Differentiation and integration of Fourier series

It is inevitable that the desire to obtain the derivative or the integral of a Fourier series will arise in some applications. Since the smoothing effects of the integration process tend to eliminate discontinuities, whereas the process of differentiation has the opposite effect, it is not surprising that the integration of a Fourier series is more likely to be possible than its differentiation. We shall not pursue the theory in depth here; rather we shall state, without proof, two theorems concerned with the term-by-term integration and differentiation of Fourier series, and make some observations on their use.

4.4.1 Integration of a Fourier series

THEOREM 4.3 A Fourier series expansion of a periodic function $f(t)$ that satisfies Dirichlet's conditions may be integrated term by term, and the integrated series converges to the integral of the function $f(t)$. □

According to this theorem, if $f(t)$ satisfies Dirichlet's conditions in the interval $-\pi \leqslant t \leqslant \pi$ and has a Fourier series expansion

$$f(t) = \tfrac{1}{2}a_0 + \sum_{n=1}^{\infty} (a_n \cos nt + b_n \sin nt)$$

then for $-\pi \leqslant t_1 < t \leqslant \pi$

$$\int_{t_1}^{t} f(t)\,dt = \int_{t_1}^{t} \tfrac{1}{2}a_0\,dt + \sum_{n=1}^{\infty} \int_{t_1}^{t} (a_n \cos nt + b_n \sin nt)\,dt$$

$$= \tfrac{1}{2}a_0(t - t_1) + \sum_{n=1}^{\infty} \left[\frac{b_n}{n}(\cos nt_1 - \cos nt) + \frac{a_n}{n}(\sin nt - \sin nt_1)\right]$$

Because of the presence of the term $\frac{1}{2}a_0 t$ on the right-hand side, this is clearly not a Fourier series expansion of the integral on the left-hand side. However, the result can be rearranged to be a Fourier series expansion of the function

$$g(t) = \int_{t_1}^{t} f(t)\,\mathrm{d}t - \tfrac{1}{2}a_0 t$$

Example 4.13 serves to illustrate this process. Note also that the Fourier coefficients in the new Fourier series are $-b_n/n$ and a_n/n, so, from the observations made in Section 4.2.8, the integrated series converges faster than the original series for $f(t)$. If the given function $f(t)$ is piecewise-continuous, rather than continuous, over the interval $-\pi \leqslant t \leqslant \pi$ then care must be taken to ensure that the integration process is carried out properly over the various subintervals. Again, Example 4.14 serves to illustrate this point.

EXAMPLE 4.13

From Example 4.5, the Fourier series expansion of the function

$$f(t) = t^2 \quad (-\pi \leqslant t \leqslant \pi), \qquad f(t + 2\pi) = f(\pi)$$

is

$$t^2 = \frac{1}{3}\pi^2 + 4\sum_{n=1}^{\infty} \frac{(-1)^n \cos nt}{n^2} \quad (-\pi \leqslant t \leqslant \pi)$$

Integrating this result between the limits $-\pi$ and t gives

$$\int_{-\pi}^{t} t^2\,\mathrm{d}t = \int_{-\pi}^{t} \frac{1}{3}\pi^2\,\mathrm{d}t + 4\sum_{n=1}^{\infty} \int_{-\pi}^{t} \frac{(-1)^n \cos nt}{n^2}\,\mathrm{d}t$$

that is,

$$\frac{1}{3}t^3 = \frac{1}{3}\pi^2 t + 4\sum_{n=1}^{\infty} \frac{(-1)^n \sin nt}{n^3} \quad (-\pi \leqslant t \leqslant \pi)$$

Because of the term $\frac{1}{3}\pi^2 t$ on the right-hand side, this is clearly not a Fourier series expansion. However, rearranging, we have

$$t^3 - \pi^2 t = 12\sum_{n=1}^{\infty} \frac{(-1)^n \sin nt}{n^2}$$

and now the right-hand side may be taken to be the Fourier series expansion of the function

$$g(t) = t^3 - \pi^2 t \quad (-\pi \leqslant t \leqslant \pi)$$

$$g(t + 2\pi) = g(t)$$

EXAMPLE 4.14 Integrate term by term the Fourier series expansion obtained in Example 4.4 for the square wave

$$f(t) = \begin{cases} -1 & (-\pi < t < 0) \\ 1 & (0 < t < \pi) \end{cases}$$

$$f(t + 2\pi) = f(t)$$

illustrated in Figure 4.7.

Solution From (4.21), the Fourier series expansion for $f(t)$ is

$$f(t) = \frac{4}{\pi}\frac{\sin(2n-1)t}{2n-1}$$

We now need to integrate between the limits $-\pi$ and t and, owing to the discontinuity in $f(t)$ at $t = 0$, we must consider separately values of t in the intervals $-\pi < t < 0$ and $0 \leqslant t \leqslant \pi$.

Case (i), interval $-\pi < t < 0$. Integrating (4.21) term by term, we have

$$\int_{-\pi}^{t} (-1)\,\mathrm{d}t = \frac{4}{\pi}\sum_{n=1}^{\infty}\int_{-\pi}^{t}\frac{\sin(2n-1)t}{(2n-1)}\,\mathrm{d}t$$

that is,

$$-(t+\pi) = -\frac{4}{\pi}\sum_{n=1}^{\infty}\left[\frac{\cos(2n-1)t}{(2n-1)^2}\right]_{-\pi}^{t}$$

$$= -\frac{4}{\pi}\left[\sum_{n=1}^{\infty}\frac{\cos(2n-1)t}{(2n-1)^2} + \sum_{n=1}^{\infty}\frac{1}{(2n-1)^2}\right]$$

It can be shown that

$$\sum_{n=1}^{\infty}\frac{2}{(2n-1)^2} = \frac{1}{8}\pi^2$$

(see Exercise 6), so that the above simplifies to

$$-t = \frac{1}{2}\pi - \frac{4}{\pi}\sum_{n=1}^{\infty}\frac{\cos(2n-1)t}{(2n-1)^2} \quad (-\pi < t < 0) \tag{4.36}$$

Case (ii), interval $0 < t < \pi$. Integrating (4.21) term by term, we have

$$\int_{-\pi}^{0} (-1)\,\mathrm{d}t + \int_{0}^{t} 1\,\mathrm{d}t = \frac{4}{\pi}\sum_{n=1}^{\infty}\int_{-\pi}^{t}\frac{\sin(2n-1)t}{(2n-1)}\,\mathrm{d}t$$

giving

$$t = \frac{1}{2}\pi - \frac{4}{\pi}\sum_{n=1}^{\infty}\frac{\cos(2n-1)t}{(2n-1)^2} \quad (0 < t < \pi) \tag{4.37}$$

Taking (4.36) and (4.37) together, we find that the function

$$g(t) = |t| = \begin{cases} -t & (-\pi < t < 0) \\ t & (0 < t < \pi) \end{cases}$$

$$g(t + 2\pi) = g(t)$$

has a Fourier series expansion

$$g(t) = |t| = \frac{1}{2}\pi - \frac{4}{\pi}\sum_{n=1}^{\infty}\frac{\cos(2n-1)t}{(2n-1)^2}$$

4.4.2 Differentiation of a Fourier series

THEOREM 4.4

If $f(t)$ is a periodic function that satisfies Dirichlet's conditions then its derivative $f'(t)$, wherever it exists, may be found by term-by-term differentiation of the Fourier series of $f(t)$ if and only if the function $f(t)$ is continuous everywhere and the function $f'(t)$ has a Fourier series expansion (that is, $f'(t)$ satisfies Dirichlet's conditions). □

It follows from Theorem 4.4 that if the Fourier series expansion of $f(t)$ is differentiable term by term then $f(t)$ must be periodic at the end points of a period (owing to the condition that $f(t)$ must be continuous everywhere). Thus, for example, if we are dealing with a function $f(t)$ of period 2π and defined in the range $-\pi < t < \pi$ then we must have $f(-\pi) = f(\pi)$. To illustrate this point, consider the Fourier series expansion of the function

$$f(t) = t \quad (-\pi < t < \pi)$$

$$f(t + 2\pi) = f(t)$$

which, from Example 4.7, is given by

$$f(t) = 2(\sin t - \tfrac{1}{2}\sin 2t + \tfrac{1}{3}\sin 3t - \tfrac{1}{4}\sin 4t + \dots)$$

Differentiating term by term, we have

$$f'(t) = 2(\cos t - \cos 2t + \cos 3t - \cos 4t + \dots)$$

If this differentiation process is valid then $f'(t)$ must be equal to unity for $-\pi < t < \pi$. Clearly this is not the case, since the series on the right-hand side does not converge for any value of t. This follows since the nth term of the series is $2(-1)^{n+1}\cos nt$ and does not tend to zero as $n \to \infty$.

If $f(t)$ is continuous everywhere and has a Fourier series expansion

$$f(t) = \tfrac{1}{2}a_0 + \sum_{n=1}^{\infty} (a_n \cos nt + b_n \sin nt)$$

then, from Theorem 4.4, provided that $f'(t)$ satisfies the required conditions, its Fourier series expansion is

$$f'(t) = \sum_{n=1}^{\infty} (nb_n \cos nt - na_n \sin nt)$$

In this case the Fourier coefficients of the derived expansion are nb_n and na_n, so, in contrast to the integrated series, the derived series will converge more slowly than the original series expansion for $f(t)$.

EXAMPLE 4.15 Consider the process of differentiating term by term the Fourier series expansion of the function

$$f(t) = t^2 \quad (-\pi \leqslant t \leqslant \pi), \qquad f(t + 2\pi) = f(t)$$

Solution From Example 4.5, the Fourier series expansion of $f(t)$ is

$$t^2 = \frac{1}{3}\pi^2 + 4\sum_{n=1}^{\infty} \frac{(-1)^n \cos nt}{n^2} \quad (-\pi \leqslant t \leqslant \pi)$$

Since $f(t)$ is continuous within and at the end points of the interval $-\pi \leqslant t \leqslant \pi$, we may apply Theorem 4.4 to obtain

$$t = 2\sum_{n=1}^{\infty} \frac{(-1)^{n+1} \sin nt}{n} \quad (-\pi \leqslant t \leqslant \pi)$$

which conforms with the Fourier series expansion obtained for the function

$$f(t) = t \quad (-\pi < t < \pi), \qquad f(t + 2\pi) = f(t)$$

in Example 4.7.

4.4.3 Coefficients in terms of jumps at discontinuities

For periodic functions that, within a period, are piecewise polynomials and exhibit jump discontinuities, the Fourier coefficients may be determined in terms of the magnitude of the jumps and those of derived functions. This method is useful for determining describing functions (see Section 4.8) for nonlinear characteristics in control engineering, where only the fundamental component of the Fourier series is important; this applies particularly to the case of multivalued nonlinearities.

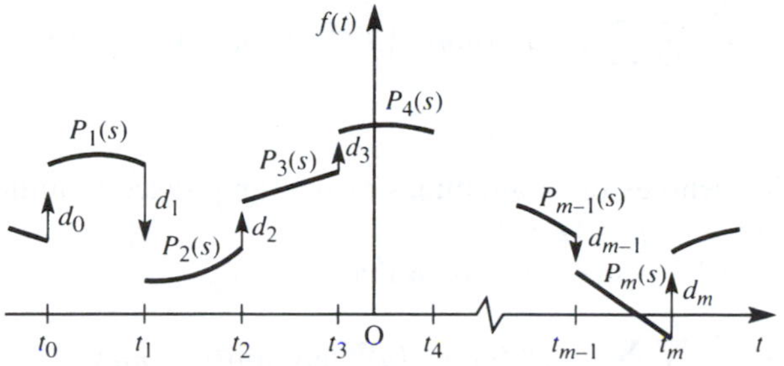

Figure 4.24 Piecewise polynomial periodic function exhibiting jump discontinuities.

Consider a periodic function $f(t)$, of period T, having within the time interval $-\frac{1}{2}T \leqslant t \leqslant \frac{1}{2}T$ a finite number $(m + 1)$ of jump discontinuities $d_0, d_1, \ldots, d_m$ at times $t_0, t_1, \ldots t_m$, with $t_0 = \frac{1}{2}T$ and $t_m = \frac{1}{2}T$. Furthermore, within the interval $t_{s-1} < t < t_s$ $(s = 1, 2, \ldots, m)$ let $f(t)$ be represented by polynomial functions $P_s(t)$ $(s = 1, 2, \ldots, m)$, as illustrated in Figure 4.24. If $f(t)$ is to be represented in terms of the Fourier series

$$f(t) = \tfrac{1}{2}a_0 + \sum_{n=1}^{\infty} a_n \cos n\omega t + \sum_{n=1}^{\infty} b_n \sin n\omega t$$

then, from (4.11),

$$a_n = \frac{2}{T}\sum_{s=1}^{m}\int_{t_{s-1}}^{t_s} P_s(t) \cos n\omega t \, dt$$

Defining the magnitude of the jump discontinuities as in Section 2.5.11, namely

$$d_i = f(t_i + 0) - f(t_i - 0)$$

and noting that $t_0 = -\frac{1}{2}T$ and $t_m = \frac{1}{2}T$, integration by parts and summation gives

$$a_n = -\frac{1}{n\pi}\sum_{s=1}^{m}\left[d_s \sin n\omega t_s + \int_{t_{s-1}}^{t_s} P_s^{(1)}(t) \sin n\omega t \, dt\right] \tag{4.38}$$

where $P_s^{(1)}(t)$ denotes the piecewise components of the derivative $f^{(1)}(t) \equiv f'(t)$ in the generalized sense of (2.59).

In a similar manner the integral terms of (4.38) may be expressed as

$$\sum_{s=1}^{m}\int_{t_{s-1}}^{t_s} P_s^{(1)} \sin n\omega t \, dt = \frac{1}{n\omega}\sum_{s=1}^{m}\left[d_s^{(1)} \cos n\omega t + \int_{t_{s-1}}^{t_s} P_s^{(2)}(t) \cos n\omega t \, dt\right]$$

where $d_s^{(1)}$ $(s = 1, 2, \ldots, m)$ denotes the magnitude of the jump discontinuities in the derivative $f^{(1)}(t)$.

Continuing in this fashion, integrals involving higher derivatives may be obtained. However, since all $P_s(t)$ $(s = 1, 2, \ldots, m)$ are polynomials, a stage is reached when all the integrals vanish. If the degree of $P_s(t)$ is less than or equal to N for $s = 1, 2, \ldots, m$ then

$$a_n = \frac{1}{n\pi}\sum_{s=1}^{m}\sum_{r=0}^{N}(-1)^{r+1}(n\omega)^{-2r}[d_s^{(2r)}\sin n\omega t_s + (n\omega)^{-1}d_s^{(2r+1)}\cos n\omega t_s]$$

$$(n \neq 0) \qquad \textbf{(4.39)}$$

where $d_s^{(r)}$ denotes the magnitudes of the jump discontinuities in the rth derivative of $f(t)$ according to (2.59).

Similarly, it may be shown that

$$b_n = \frac{1}{n\pi}\sum_{s=1}^{m}\sum_{r=0}^{N}(-1)^{r}(n\omega)^{-2r}[d_s^{(2r)}\cos n\omega t_s - (n\omega)^{-1}d_s^{(2r+1)}\sin n\omega t_s] \qquad \textbf{(4.40)}$$

and the coefficient a_0 is found by direct integration of the corresponding Euler formula

$$a_0 = \frac{2}{T}\int_{-T/2}^{T/2} f(t)\,\mathrm{d}t \qquad \textbf{(4.41)}$$

EXAMPLE 4.16 Using (4.39)–(4.41), obtain the Fourier series expansion of the periodic function $f(t)$ defined by

$$f(t) = \begin{cases} t^2 & (-\pi < t < 0) \\ -2 & (0 < t < \pi) \end{cases}$$

$$f(t + 2\pi) = f(t)$$

Solution In this case $N = 2$, and the graphs of $f(t)$ together with those of its first two derivatives are shown in Figure 4.25.

Jump discontinuities occur at $t = -\pi$, 0 and π, so that $m = 2$. The piecewise polynomials involved and the corresponding jump discontinuities are

(a) $P_1(t) = t^2$, $P_2(t) = -2$
$d_1 = -2$, $d_2 = \pi^2 + 2$

(b) $P_1^{(1)}(t) = 2t$ $P_2^{(1)}(t) = 0$
$d_1^{(1)} = 0$ $d_2^{(1)} = -2\pi$

(c) $P_1^{(2)}(t) = 2$, $P_2^{(2)}(t) = 0$
$d_1^{(2)} = -2$ $d_2^{(2)} = 2$

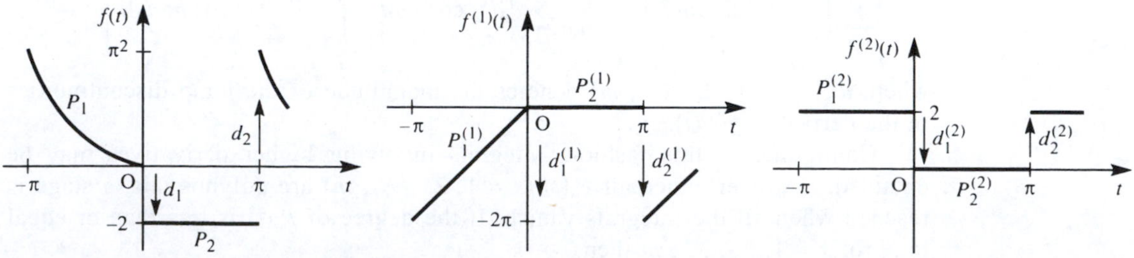

Figure 4.25 $f(t)$, $f^{(1)}(t)$, $f^{(2)}(t)$ of Example 4.16.

with $d_1^{(r)} = d_2^{(r)} = 0$ for $r > 2$.

Taking $\omega = 1$ (since $T = 2\pi$) in (4.39) gives

$$a_n = \frac{1}{n\pi}\left(-\sum_{s=1}^{2} d_s \sin nt_s - \frac{1}{n}\sum_{s=1}^{2} d_s^{(1)} \cos nt_s + \frac{1}{n^2}\sum_{s=1}^{2} d_s^{(2)} \sin nt_s\right)$$

Since $t_1 = 0$, $t_2 = \pi$, $\sin 0 = \sin n\pi = 0$, $\cos 0 = 1$ and $\cos n\pi = (-1)^n$, we have

$$a_n = \frac{2}{n^2}(-1)^n \quad (n = 1, 2, 3, \dots)$$

Likewise, from (4.40),

$$b_n = \frac{1}{n\pi}\left(\sum_{s=1}^{2} d_s \cos nt_s - \frac{1}{n}\sum_{s=1}^{2} d_s^{(1)} \sin nt_s - \frac{1}{n^2}\sum_{s=1}^{2} d_s^{(2)} \cos nt_s\right)$$

$$= \frac{1}{n\pi}\left\{-2 + (\pi^2 + 2)(-1)^n - \frac{1}{n^2}[-2 + 2(-1)^n]\right\}$$

$$= \frac{1}{n\pi}\left\{\left(\frac{2}{n^2} - 2\right)[1 - (-1)^n] + \pi^2(-1)^n\right\} \quad (n = 1, 2, 3, \dots)$$

and, from (4.41),

$$a_0 = \frac{1}{\pi}\left[\int_{-\pi}^{0} t^2 \,\mathrm{d}t + \int_{0}^{\pi} (-2)\,\mathrm{d}t\right] = \frac{1}{3}\pi^2 - 2$$

Thus the Fourier expansion for $f(t)$ is

$$f(t) = \left(\frac{1}{6}\pi^2 - 1\right) + \sum_{n=1}^{\infty} \frac{2}{n^2}(-1)^n \cos nt$$

$$+ \sum_{n=1}^{\infty} \frac{1}{n\pi}\left\{\left(\frac{2}{n^2} - 2\right)[1 - (-1)^n] + \pi^2(-1)^n\right\} \sin nt$$

4.4.4 Exercises

24 Show that the periodic function

$$f(t) = t \quad (-T < t < T)$$

$$f(t + 2T) = f(t)$$

has a Fourier series expansion

$$f(t) = \frac{2T}{\pi}\left(\sin\frac{\pi t}{T} - \frac{1}{2}\sin\frac{2\pi t}{T} + \frac{1}{3}\sin\frac{3\pi t}{T} - \frac{1}{4}\sin\frac{4\pi t}{T} + \dots\right)$$

By term-by-term integration of this series, show that the periodic function

$$g(t) = t^2 \quad (-T < t < T)$$

$$g(t + 2T) = g(t)$$

has a Fourier series expansion

$$g(t) = \frac{1}{3}T^2 - \frac{4T^2}{\pi^2}\left(\cos\frac{\pi t}{T} - \frac{1}{2^2}\cos\frac{2\pi t}{T} + \frac{1}{3^2}\cos\frac{3\pi t}{T} - \frac{1}{4^2}\cos\frac{4\pi t}{T} + \dots\right)$$

(*Hint*: A constant of integration must be introduced; it may be evaluated as the mean value over a period.)

25 The periodic function

$$h(t) = \pi^2 - t^2 \quad (-\pi < t < \pi)$$
$$h(t + 2\pi) = h(t)$$

has a Fourier series expansion

$$h(t) = \frac{2}{3}\pi^2 + 4\left(\cos t - \frac{1}{2^2}\cos 2t + \frac{1}{3^2}\cos 3t \ldots\right)$$

By term-by-term differentiation of this series, confirm the series obtained for $f(t)$ in Exercise 24 for the case when $T = \pi$.

26 (a) Suppose that the derivative $f'(t)$ of a periodic function $f(t)$ of period 2π has a Fourier series expansion

$$f'(t) = \tfrac{1}{2}A_0 + \sum_{n=1}^{\infty} A_n \cos nt + \sum_{n=1}^{\infty} B_n \sin nt$$

Show that

$$A_0 = \frac{1}{n}[f(\pi_-) - f(-\pi_+)]$$
$$A_n = (-1)^n A_0 + nb_n$$
$$B_n = -na_n$$

where a_0, a_n and b_n are the Fourier coefficients of the function $f(t)$.

(b) In Example 4.7 we saw that the periodic function

$$f(t) = t^2 + t \quad (-\pi < t < \pi)$$
$$f(t + 2\pi) = f(t)$$

has a Fourier series expansion

$$f(t) = \tfrac{1}{3}\pi^2 + \sum_{n=1}^{\infty} \frac{4}{n^2}(-1)^n \cos nt - \sum_{n=1}^{\infty} \frac{2}{n}(-1)^n \sin nt$$

Differentiate this series term by term, and explain why it is not a Fourier expansion of the periodic function

$$g(t) = 2t + 1 \quad (-\pi < t < \pi)$$
$$g(t + 2\pi) = g(t)$$

(c) Use the results of (a) to obtain the Fourier series expansion of $g(t)$ and confirm your solution by direct evaluation of the coefficients using Euler's formulae.

27 Using (4.39)–(4.41), confirm the following Fourier series expansions:

(a) (4.21) for the square wave of Example 4.4;

(b) the expansion obtained in Example 4.1 for the sawtooth wave;

(c) the expansion obtained for the piecewise-continuous function $f(t)$ of Example 4.3.

28 Consider the periodic function

$$f(t) = \begin{cases} 0 & (-\pi < t < -\tfrac{1}{2}\pi) \\ \pi + 2t & (-\tfrac{1}{2}\pi < t < 0) \\ \pi - 2t & (0 < t < \tfrac{1}{2}\pi) \\ 0 & (\tfrac{1}{2}\pi < t < \pi) \end{cases}$$

$$f(t + 2\pi) = f(t)$$

(a) Sketch a graph of the function for $-4\pi < t < 4\pi$.

(b) Use (4.39)–(4.41) to obtain the Fourier series expansion

$$f(t) = \frac{1}{4}\pi - \frac{4}{\pi}\sum_{n=1}^{\infty} \frac{1}{n^2}\left(\cos\frac{1}{2}n\pi - 1\right)\cos nt$$

and write out the first 10 terms of this series. (*Note*: Although the function $f(t)$ itself has no jump discontinuities, the method may be used since the derivative does have jump discontinuities.)

29 Use the method of Section 4.4.3 to obtain the Fourier series expansions for the following periodic functions:

(a) $f(t) = \begin{cases} 0 & (-\pi < t < 0) \\ t^2 & (0 < t < \pi) \end{cases}$

$$f(t + 2\pi) = f(t)$$

(b) $f(t) = \begin{cases} 2 & (-\pi < t < -\tfrac{1}{2}\pi) \\ t^3 & (-\tfrac{1}{2}\pi < t < \tfrac{1}{2}\pi) \\ -2 & (\tfrac{1}{2}\pi < t < \pi) \end{cases}$

$$f(t + 2\pi) = f(t)$$

(c) $f(t) = \begin{cases} t & (0 < t < 1) \\ 1 - t & (1 < t < 2) \end{cases}$

$$f(t + 2) = f(t)$$

(d) $f(t) = \begin{cases} \tfrac{1}{2} + t & (-\tfrac{1}{2} < t < 0) \\ \tfrac{1}{2} - t & (0 < t < \tfrac{1}{2}) \end{cases}$

$$f(t + 1) = f(t)$$

4.5 Engineering application: frequency response and oscillating systems

4.5.1 Response to periodic input

In Section 2.7 we showed that the frequency response, defined as the steady-state response to a sinusoidal input $A \sin \omega t$, of a stable linear system having a transfer function $G(s)$ is given by (2.86) as

$$x_{ss}(t) = A|G(j\omega)| \sin [\omega t + \arg G(j\omega)] \tag{4.42}$$

By employing a Fourier series expansion, we can use this result to determine the steady-state response of a stable linear system to a non-sinusoidal periodic input. For a stable linear system having a transfer function $G(s)$, let the input be a periodic function $P(t)$ of period $2T$ (that is, one having frequency $\omega = \pi/T$ in rad s^{-1}). From (4.21), $P(t)$ may be expressed in the form of the Fourier series expansion

$$P(t) = \tfrac{1}{2}a_0 + \sum_{n=1}^{\infty} A_n \sin (n\omega t + \phi_n) \tag{4.43}$$

where A_n and ϕ_n are defined as in Section 4.2.1. The steady-state response to each term in the series expansion (4.43) may be obtained using (4.42). Since the system is linear, the principle of superposition holds, so that the steady-state response to the periodic input $P(t)$ may be obtained as the sum of the steady-state responses to the individual sinusoids comprising the sum in (4.43). Thus the steady-state response to the input $P(t)$ is

$$x_{ss}(t) = \tfrac{1}{2}a_0 G(0) + \sum_{n=1}^{\infty} A_n|G(j\omega n)| \sin [n\omega t + \phi_n + \arg G(j\omega n)] \tag{4.44}$$

There are two issues related to this steady-state response that are worthy of note.

(a) For practical systems $|G(j\omega)| \to 0$ as $\omega \to \infty$, so that $|G(j\omega n)| \to 0$ as $n \to \infty$ in (4.44). As a consequence, the Fourier series representation of the steady-state response $x_{ss}(t)$ converges more rapidly than the Fourier series representation of the periodic input $P(t)$. From a practical point of view, this is not surprising, since it is a consequence of the smoothing action of the system (that is, as indicated in Section 4.4, integration is a 'smoothing' operation).

(b) There is a significant difference between the steady-state response (4.44) to a non-sinusoidal periodic input of frequency ω and the steady-state response (4.41) to a pure sinusoid at the same frequency. As indicated in (4.42), in the case of a sinusoidal input at frequency ω the steady-state response is also a sinusoid at the same frequency ω. However, for a non-sinusoidal periodic input $P(t)$ at frequency ω the steady-state response (4.44) is no longer at the same frequency; rather it comprises an infinite sum of sinusoids having frequencies $n\omega$ that are integer multiples of the input frequency ω. This clearly has important practical implications, particularly when considering the responses of oscillating or vibrating systems. If the frequency $n\omega$ of one of the harmonics in (4.44) is close to the natural oscillating frequency of an underdamped system then the phenomenon of **resonance** will arise.

To someone unfamiliar with the theory, it may seem surprising that a practical system may resonate at a frequency much higher than that of the input. As indicated in Example 2.30, the phenomenon of resonance is important in practice, and it is therefore important that engineers have some knowledge of the theory associated with Fourier series, so that the possible dominance of a system re-sponse by one of the higher harmonics, rather than the fundamental, may be properly interpreted.

EXAMPLE 4.17

The mass–spring–damper system of Figure 4.26(a) is initially at rest in a position of equilibrium. Determine the steady-state response of the system when the mass is subjected to an externally applied periodic force $P(t)$ having the form of the square wave shown in Figure 4.26(b).

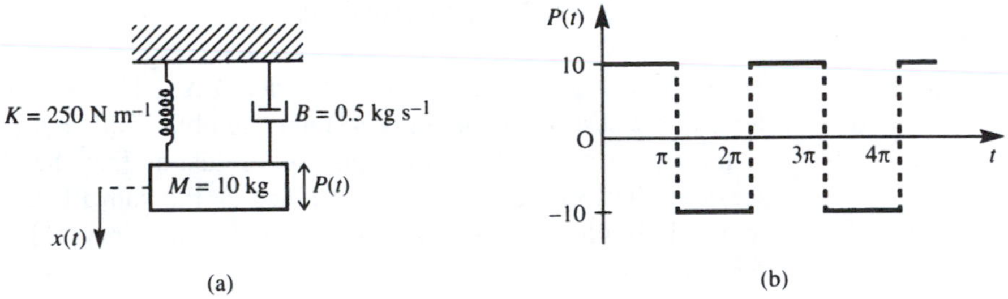

Figure 4.26 (a) System and (b) input for Example 4.17.

Solution From Newton's law, the displacement $x(t)$ of the mass at time t is given by

$$M\frac{d^2x}{dt^2} + B\frac{dx}{dt} + Kx = P(t) \tag{4.45}$$

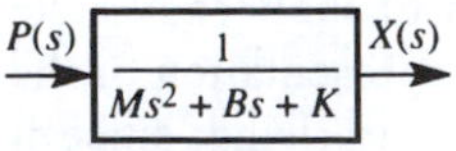

Figure 4.27 Block diagram for the system of Figure 4.26.

so that the system may be represented by the block diagram of Figure 4.27. Thus the system transfer function is

$$G(s) = \frac{1}{Ms^2 + Bs + K} \tag{4.46}$$

From Example 4.4, the Fourier series expansion for the square wave $P(t)$ is

$$P(t) = \frac{40}{\pi}\left[\sin t + \frac{\sin 3t}{3} + \frac{\sin 5t}{5} + \ldots + \frac{\sin (2n-1)t}{2n-1} + \ldots\right]$$

that is,

$$P(t) = u_1(t) + u_2(t) + u_3(t) + \ldots + u_n(t) + \ldots \tag{4.47}$$

where

$$u_n(t) = \frac{40}{\pi}\frac{\sin(2n-1)t}{2n-1} \tag{4.48}$$

Substituting the given values for M, B and K, the transfer function (4.46) becomes

$$G(s) = \frac{1}{10s^2 + 0.5s + 250}$$

Thus

$$G(j\omega) = \frac{1}{-10\omega^2 + 0.5j\omega + 250} = \frac{250 - 10\omega^2}{D} - j\frac{0.5\omega}{D}$$

where $D = (250 - 10\omega^2)^2 + 0.25\omega^2$, so that

$$|G(j\omega)| = \sqrt{\left[\frac{(250 - 10\omega^2)^2 + 0.25\omega^2}{D^2}\right]}$$

$$= \frac{1}{\sqrt{D}} = \frac{1}{\sqrt{[(250 - 10\omega^2)^2 + 0.25\omega^2]}} \tag{4.49}$$

$$\arg G(j\omega) = -\tan^{-1}\left(\frac{0.5\omega}{250 - 10\omega^2}\right) \tag{4.50}$$

Using (4.42), the steady-state response of the system to the nth harmonic $u_n(t)$ given by (4.48) is

$$x_{ssn}(t) = \frac{40}{\pi(2n-1)}|G(j(2n-1))| \sin[(2n-1)t + \arg G(j(2n-1))] \tag{4.51}$$

where $|G(j\omega)|$ and $\arg G(j\omega)$ are given by (4.49) and (4.50) respectively. The steady-state response $x_{ss}(t)$ of the system to the square-wave input $P(t)$ is then determined as the sum of the steady-state responses due to the individual harmonics in (4.47); that is,

$$x_{ss}(t) = \sum_{n=1}^{\infty} x_{ssn}(t) \tag{4.52}$$

where $x_{ssn}(t)$ is given by (4.51).

Evaluating the first few terms of the response (4.52), we have

$$x_{ss1}(t) = \frac{40}{\pi}\frac{1}{\sqrt{[(250-10)^2 + 0.25]}} \sin\left[t - \tan^{-1}\left(\frac{0.5}{240}\right)\right]$$

$$= 0.053 \sin(t - 0.003)$$

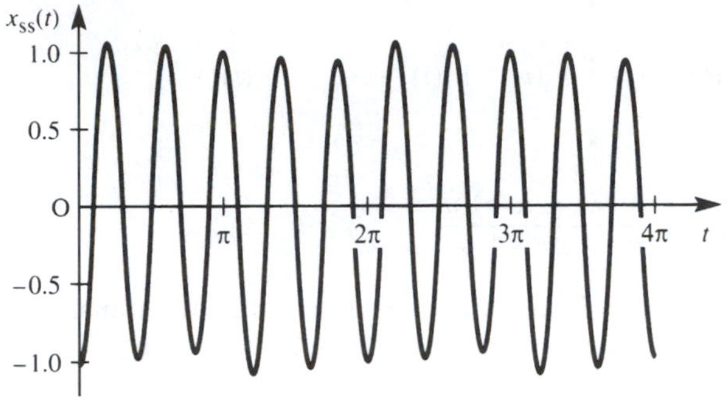

Figure 4.28 Steady-state response of system of Figure 4.26.

$$\begin{aligned}
x_{ss2}(t) &= \frac{40}{3\pi}\frac{1}{\sqrt{[(250-90)^2+2.25]}}\sin\left[3t-\tan^{-1}\left(\frac{1.5}{160}\right)\right]\\
&= 0.027\sin(3t-0.009)\\
x_{ss3}(t) &= \frac{40}{5\pi}\frac{1}{\sqrt{(6.25)}}\sin\left[5t-\tan^{-1}\left(\frac{2.5}{0}\right)\right]\\
&= 1.02\sin(5t-\tfrac{1}{2}\pi)\\
x_{ss4}(t) &= \frac{40}{7\pi}\frac{1}{\sqrt{[(250-490)^2+12.25]}}\sin\left[7t-\tan^{-1}\left(\frac{3.5}{-240}\right)\right]\\
&= 0.0076\sin(7t-3.127)
\end{aligned}$$

Thus a good approximation to the steady-state response (4.52) is

$$\begin{aligned}x_{ss}(t) \simeq\ & 0.053\sin(t-0.003)+0.027\sin(3t-0.54)+1.02\sin(5t-\tfrac{1}{2}\pi)\\ &+0.0076\sin(7t-3.127)\end{aligned} \tag{4.53}$$

The graph of this displacement is shown in Figure 4.28, and it appears from this that the response has a frequency about five times that of the input. This is because the term $1.02\sin(5t-\frac{1}{2}\pi)$ dominates in the response (4.53); this is a consequence of the fact that the natural frequency of oscillation of the system is $\sqrt{(K/M)} = 5\text{ rad s}^{-1}$, so that it is in resonance with this particular harmonic.

In conclusion, it should be noted that it was not essential to introduce transfer functions to solve this problem. Alternatively, by determining the particular integral of the differential equation (4.45), the steady-state response to an input $A\sin\omega t$ is determined as

$$x_{ss}(t) = \frac{A\sin(\omega t-\alpha)}{\sqrt{[(K-M\omega^2)^2+B^2\omega^2]}},\qquad \tan\alpha = \frac{\omega B}{K-M\omega^2}$$

giving $x_{ssn}(t)$ as in (4.52). The solution then proceeds as before.

4.5.2 Exercises

30 Determine the steady-state current in the circuit of Figure 4.29(a) as a result of the applied periodic voltage shown in Figure 4.29(b).

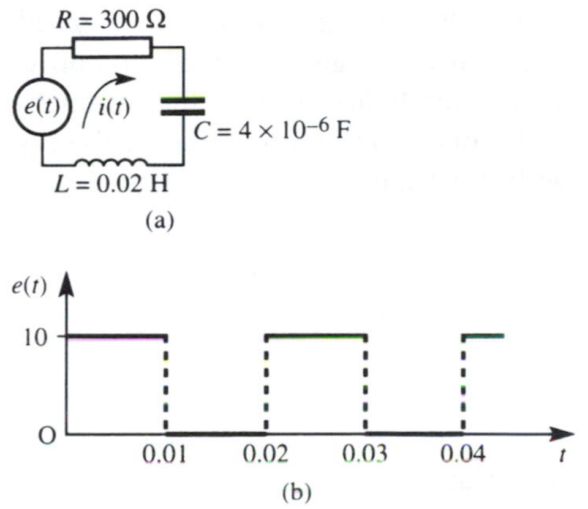

Figure 4.29 (a) Circuit of Exercise 30; (b) applied voltage.

31 Determine the steady-state response of the mass–spring–damper system of Figure 4.30(a) when the mass is subjected to the externally applied periodic force $f(t)$ shown in Figure 4.30(b).

What frequency dominates the response, and why?

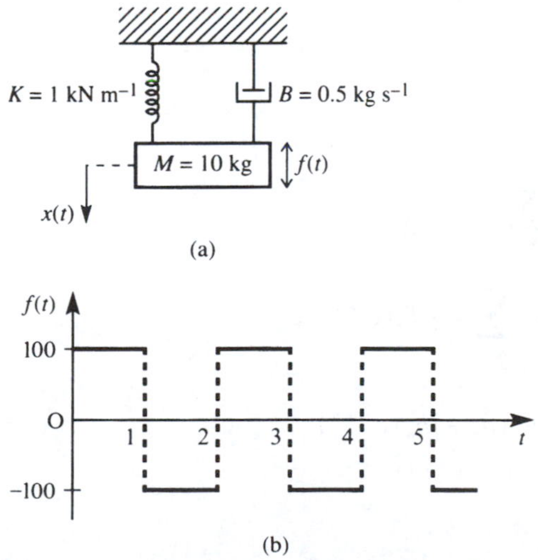

Figure 4.30 (a) Mass–spring–damper system of Exercise 31; (b) applied force.

32 Determine the steady-state motion of the mass of Figure 4.31(a) when it is subjected to the externally applied force of Figure 4.31(b).

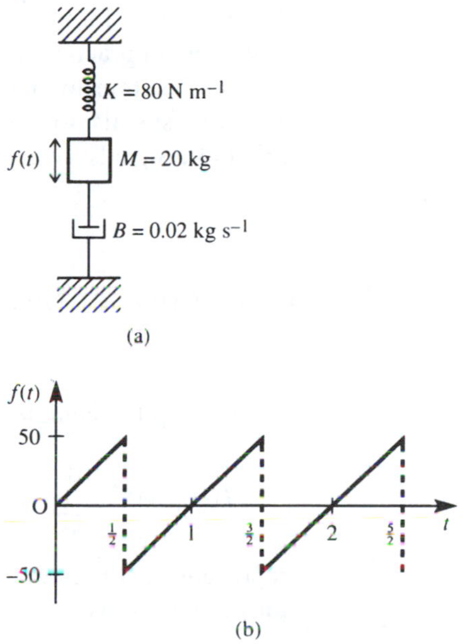

Figure 4.31 (a) Mass–spring–damper system of Exercise 32; (b) applied force.

33 Determine the steady-state current in the circuit shown in Figure 4.32(a) when the applied voltage is of the form shown in Figure 4.32(b).

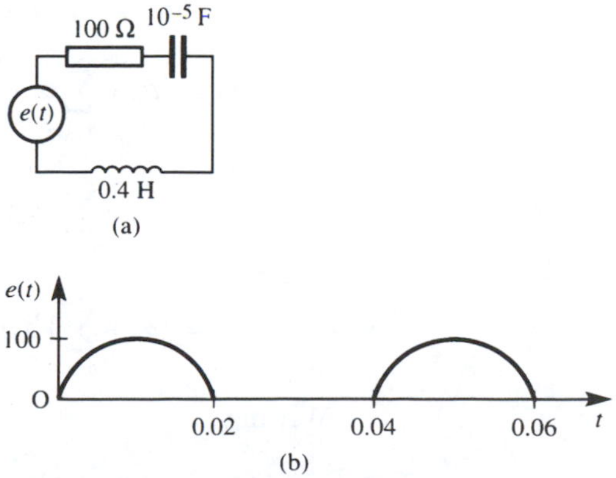

Figure 4.32 (a) Circuit of Exercise 33; (b) applied voltage.

4.6 Complex form of Fourier series

An alternative to the trigonometric form of the Fourier series considered so far is the complex or exponential form. As a result of the properties of the exponential function, this form is easily manipulated mathematically. It is widely used by engineers in practice, particularly in work involving signal analysis, and provides a smoother transition from the consideration of Fourier series for dealing with periodic signals to the consideration of Fourier transforms for dealing with aperiodic signals, which will be dealt with in Chapter 5.

4.6.1 Complex representation

To develop the complex form of the Fourier series

$$f(t) = \tfrac{1}{2}a_0 + \sum_{n=1}^{\infty} a_n \cos n\omega t + \sum_{n=1}^{\infty} b_n \sin n\omega t \tag{4.54}$$

representing a periodic function $f(t)$ of period T, we proceed as follows. Substituting the results

$$\sin n\omega t = \frac{1}{2\mathrm{j}}(\mathrm{e}^{\mathrm{j}n\omega t} - \mathrm{e}^{-\mathrm{j}n\omega t})$$

$$\cos n\omega t = \tfrac{1}{2}(\mathrm{e}^{\mathrm{j}n\omega t} + \mathrm{e}^{-\mathrm{j}n\omega t})$$

into (4.54) gives

$$\begin{aligned} f(t) &= \tfrac{1}{2}a_0 + \sum_{n=1}^{\infty} a_n \frac{\mathrm{e}^{\mathrm{j}n\omega t} + \mathrm{e}^{-\mathrm{j}n\omega t}}{2} + \sum_{n=1}^{\infty} b_n \frac{\mathrm{e}^{\mathrm{j}n\omega t} - \mathrm{e}^{-\mathrm{j}n\omega t}}{2\mathrm{j}} \\ &= \tfrac{1}{2}a_0 + \sum_{n=1}^{\infty} \tfrac{1}{2}a_n(\mathrm{e}^{\mathrm{j}n\omega t} + \mathrm{e}^{-\mathrm{j}n\omega t}) + \sum_{n=1}^{\infty} -\mathrm{j}\tfrac{1}{2}b_n(\mathrm{e}^{\mathrm{j}n\omega t} - \mathrm{e}^{-\mathrm{j}n\omega t}) \\ &= \tfrac{1}{2}a_0 + \sum_{n=1}^{\infty} [\tfrac{1}{2}(a_n - \mathrm{j}b_n)\,\mathrm{e}^{\mathrm{j}n\omega t} + \tfrac{1}{2}(a_n + \mathrm{j}b_n)\,\mathrm{e}^{-\mathrm{j}n\omega t}] \end{aligned} \tag{4.55}$$

Writing

$$c_0 = \tfrac{1}{2}a_0, \quad c_n = \tfrac{1}{2}(a_n - \mathrm{j}b_n), \quad c_{-n} = c_n^* = \tfrac{1}{2}(a_n + \mathrm{j}b_n) \tag{4.56}$$

(4.55) becomes

$$f(t) = c_0 + \sum_{n=1}^{\infty} c_n \, e^{jn\omega t} + \sum_{n=1}^{\infty} c_{-n} \, e^{-jn\omega t} = c_0 + \sum_{n=1}^{\infty} c_n \, e^{jn\omega t} + \sum_{n=-1}^{-\infty} c_n \, e^{jn\omega t}$$

$$= \sum_{n=-\infty}^{\infty} c_n \, e^{jn\omega t}, \quad \text{since} \quad c_0 \, e^0 = c_0$$

Thus the Fourier series (4.54) becomes simply

$$f(t) = \sum_{n=-\infty}^{\infty} c_n \, e^{jn\omega t} \tag{4.57}$$

which is referred to as the **complex** or **exponential form** of the Fourier series expansion of the function $f(t)$.

In order that we can apply this result directly, it is necessary to obtain a formula for calculating the complex coefficients c_n. To do this, we incorporate the Euler formulae (4.11) and (4.12) into the definitions given in (4.56), leading to

$$c_0 = \tfrac{1}{2}a_0 = \frac{1}{T}\int_d^{d+T} f(t)\, dt \tag{4.58}$$

$$c_n = \tfrac{1}{2}(a_n - jb_n) = \frac{1}{T}\left[\int_d^{d+T} f(t)\cos n\omega t \, dt - j\int_d^{d+T} f(t)\sin n\omega t\, dt\right]$$

$$= \frac{1}{T}\int_d^{d+T} f(t)(\cos n\omega t - j\sin n\omega t)\, dt$$

$$= \frac{1}{T}\int_d^{d+T} f(t)\, e^{-jn\omega t}\, dt \tag{4.59}$$

$$c_{-n} = \tfrac{1}{2}(a_n + jb_n) = \frac{1}{T}\int_d^{d+T} f(t)(\cos n\omega t + j\sin n\omega t)\, dt$$

$$= \frac{1}{T}\int_d^{d+T} f(t)\, e^{jn\omega t}\, dt \tag{4.60}$$

From (4.58)–(4.60), it is readily seen that for all values of n

$$c_n = \frac{1}{T}\int_d^{d+T} f(t)\, e^{-jn\omega t}\, dt \tag{4.61}$$

In summary, the complex form of the Fourier series expansion of a periodic function $f(t)$, of period T, is

$$f(t) = \sum_{n=-\infty}^{\infty} c_n \, e^{jn\omega t} \tag{4.57}$$

where

$$c_n = \frac{1}{T}\int_d^{d+T} f(t)\, e^{-jn\omega t}\, dt \quad (n = 0, \pm 1, \pm 2, \dots) \tag{4.61}$$

from zero.

In general the coefficients c_n $(n = 0, \pm 1, \pm 2, \ldots)$ are complex, and may be expressed in the form

$$c_n = |c_n| e^{j\phi_n}$$

where $|c_n|$, the magnitude of c_n, is given from the definitions (4.56) by

$$|c_n| = \sqrt{[(\tfrac{1}{2}a_n)^2 + (\tfrac{1}{2}b_n)^2]} = \tfrac{1}{2}\sqrt{(a_n^2 + b_n^2)}$$

so that $2|c_n|$ is the amplitude of the nth harmonic. The argument ϕ_n of c_n is related to the phase of the nth harmonic.

EXAMPLE 4.18 Find the complex form of the Fourier series expansion of the periodic function $f(t)$ defined by

$$f(t) = \cos \tfrac{1}{2}t \quad (-\pi < t < \pi), \qquad f(t + 2\pi) = f(t)$$

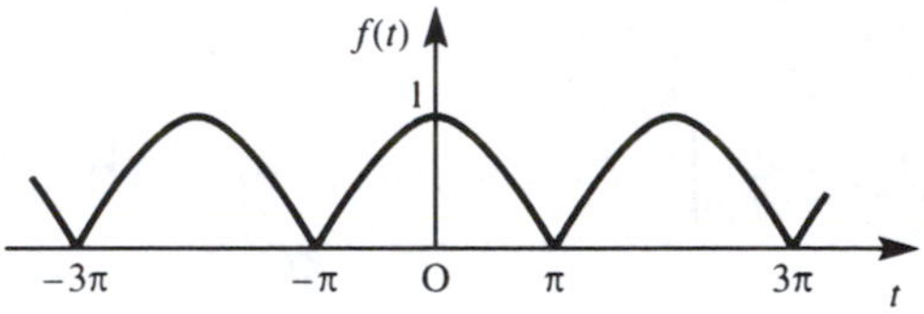

Figure 4.33 Function $f(t)$ of Example 4.18.

Solution A graph of the function $f(t)$ over the interval $-3\pi \leqslant t \leqslant 3\pi$ is shown in Figure 4.33. Here the period T is 2π, so from (4.61) the complex coefficients c_n are given by

$$\begin{aligned}
c_n &= \frac{1}{2\pi}\int_{-\pi}^{\pi} \cos \tfrac{1}{2}t\, e^{-jnt}\, dt = \frac{1}{4\pi}\int_{-\pi}^{\pi} (e^{jt/2} + e^{-jt/2})\, e^{-jnt}\, dt \\
&= \frac{1}{4\pi}\int_{-\pi}^{\pi} (e^{-j(n-1/2)t} + e^{-j(n+1/2)t})\, dt \\
&= \frac{1}{4\pi}\left[\frac{-2\,e^{-j(2n-1)t/2}}{j(2n-1)} - \frac{2\,e^{-j(2n+1)t/2}}{j(2n+1)}\right]_{-\pi}^{\pi} \\
&= \frac{j}{2\pi}\left[\left(\frac{e^{-jn\pi}e^{j\pi/2}}{2n-1} + \frac{e^{-jn\pi}e^{-j\pi/2}}{2n+1}\right) - \left(\frac{e^{jn\pi}e^{-j\pi/2}}{2n-1} + \frac{e^{jn\pi}e^{j\pi/2}}{2n+1}\right)\right]
\end{aligned}$$

Now $e^{j\pi/2} = \cos \tfrac{1}{2}\pi + j \sin \tfrac{1}{2}\pi = j$, $e^{-j\pi/2} = -j$ and $e^{jn\pi} = e^{-jn\pi} = \cos n\pi = (-1)^n$, so that

$$\begin{aligned}
c_n &= \frac{j}{2\pi}\left(\frac{j}{2n-1} - \frac{j}{2n+1} + \frac{j}{2n-1} - \frac{j}{2n+1}\right)(-1)^n \\
&= \frac{(-1)^n}{\pi}\left(\frac{1}{2n+1} - \frac{1}{2n-1}\right) = \frac{-2(-1)^n}{(4n^2-1)\pi}
\end{aligned}$$

Note that in this case c_n is real, which is as expected, since the function $f(t)$ is an even function of t.

From (4.57), the complex Fourier series expansion for $f(t)$ is

$$f(t) = \sum_{n=-\infty}^{\infty} \frac{2(-1)^{n+1}}{(4n^2 - 1)\pi} \mathrm{e}^{\mathrm{j}nt}$$

This may readily be converted back to the trigonometric form, since, from the definitions (4.56),

$$a_0 = 2c_0, \qquad a_n = c_n + c_n^*, \qquad b_n = \mathrm{j}(c_n - c_n^*)$$

so that in this particular case

$$a_0 = \frac{4}{\pi}, \qquad a_n = 2\left[\frac{2}{\pi}\frac{(-2)^{n+1}}{4n^2 + 1}\right] = \frac{4}{\pi}\frac{(-1)^{n+1}}{4n^2 - 1}, \qquad b_n = 0$$

Thus the trigonometric form of the Fourier series is

$$f(t) = \frac{2}{\pi} + \frac{4}{\pi}\sum_{n=1}^{\infty} \frac{(-1)^{n+1}}{4n^2 - 1} \cos nt$$

which corresponds to the solution to Exercise 1(e).

EXAMPLE 4.19 Obtain the complex form of the Fourier series of the sawtooth function $f(t)$ defined by

$$f(t) = \frac{2t}{T} \quad (0 < t < 2T), \qquad f(t + 2T) = f(t)$$

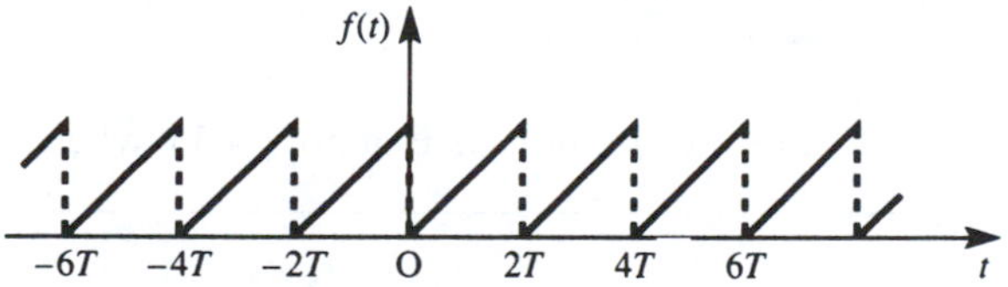

Figure 4.34 Function $f(t)$ of Example 4.19.

Solution A graph of the function $f(t)$ over the interval $-6T < t < 6T$ is shown in Figure 4.34. Here the period is $2T$, that is, $\omega = \pi/T$, so from (4.61) the complex coefficients c_n are given by

$$c_n = \frac{1}{2T}\int_0^{2T} f(t)\,\mathrm{e}^{-\mathrm{j}n\pi t/T}\,\mathrm{d}t = \frac{1}{2T}\int_0^{2T} \frac{2}{T} t\,\mathrm{e}^{-\mathrm{j}n\pi t/T}\,\mathrm{d}t$$

$$= \frac{1}{T^2}\left[\frac{Tt}{-\mathrm{j}n\pi}\mathrm{e}^{-\mathrm{j}n\pi t/T} - \frac{T^2}{(\mathrm{j}n\pi)^2}\mathrm{e}^{-\mathrm{j}n\pi t/T}\right]_0^{2T} \qquad (n \neq 0)$$

Now $e^{-jn2\pi} = e^{-j0} = 1$, so

$$c_n = \frac{1}{T^2}\left[\frac{2T^2}{-jn\pi} + \frac{T^2}{(n\pi)^2} - \frac{T^2}{(n\pi)^2}\right] = \frac{j2}{n\pi} \quad (n \neq 0)$$

In the particular case $n = 0$

$$c_0 = \frac{1}{2T}\int_0^{2T} f(t)\,dt = \frac{1}{2T}\int_0^{2T} \frac{2t}{T}\,dt = \frac{1}{T^2}\left[\frac{1}{2}t^2\right]_0^{2T} = 2$$

Thus from (4.57) the complex form of the Fourier series expansion of $f(t)$ is

$$f(t) = 2 + \sum_{n=-\infty}^{-1} \frac{j2}{n\pi} e^{jn\pi t/T} + \sum_{n=1}^{\infty} \frac{j2}{n\pi} e^{jn\pi t/T} = 2 + \sum_{\substack{n=-\infty \\ n\neq 0}}^{\infty} \frac{j2}{n\pi} e^{jn\pi t/T}$$

Noting that $j = e^{j\pi/2}$, this result may also be written in the form

$$f(t) = 2 + \frac{2}{\pi}\sum_{\substack{n=-\infty \\ n\neq 0}}^{\infty} \frac{1}{n} e^{j(n\pi t/T + \pi/2)}$$

As in Example 4.18, the Euler coefficients in the corresponding trigonometric series are

$$a_0 = 2c_0 = 4, \qquad a_n = c_n + c_n^* = 0, \qquad b_n = j(c_n + c_n^*) = j\left(\frac{2j}{n\pi} + \frac{2j}{n\pi}\right) = -\frac{4}{n\pi}$$

so that the corresponding trigonometric Fourier series expansion of $f(t)$ is

$$f(t) = 2 - \frac{4}{\pi}\sum_{n=1}^{\infty} \frac{1}{n} \sin\frac{n\pi t}{T}$$

which corresponds to the solution of Example 4.11 when $T = 2$.

4.6.2 The multiplication theorem and Parseval's theorem

Two useful results, particularly in the application of Fourier series to signal analysis, are the **multiplication theorem** and **Parseval's theorem**. The multiplication theorem enables us to write down the mean value of the product of two periodic functions over a period in terms of the coefficients of their Fourier series expansions, while Parseval's theorem enables us to write down the mean square value of a periodic function, which, as we will see in Section 4.6.4, determines the power spectrum of the function.

THEOREM 4.5 The multiplication theorem

If $f(t)$ and $g(t)$ are two periodic functions having the same period T then

$$\frac{1}{T}\int_{c}^{c+T} f(t)g(t)\,\mathrm{d}t = \sum_{n=-\infty}^{\infty} c_n d_n^* \tag{4.62}$$

where the c_n and d_n are the coefficients in the complex Fourier series expansions of $f(t)$ and $g(t)$ respectively.

Proof Let $f(t)$ and $g(t)$ have complex Fourier series given by

$$f(t) = \sum_{n=-\infty}^{\infty} c_n\,\mathrm{e}^{\mathrm{j}n2\pi t/T} \tag{4.63a}$$

with

$$c_n = \frac{1}{T}\int_{c}^{c+T} f(t)\,\mathrm{e}^{-\mathrm{j}n2\pi t/T}\,\mathrm{d}T \tag{4.63b}$$

and

$$g(t) = \sum_{n=-\infty}^{\infty} d_n\,\mathrm{e}^{\mathrm{j}n2\pi t/T} \tag{4.64a}$$

with

$$d_n = \frac{1}{T}\int_{c}^{c+T} g(t)\,\mathrm{e}^{-\mathrm{j}n2\pi t/T}\,\mathrm{d}t \tag{4.64b}$$

Then

$$\frac{1}{T}\int_{c}^{c+T} f(t)g(t)\,\mathrm{d}t = \frac{1}{T}\int_{c}^{c+T}\left(\sum_{n=-\infty}^{\infty} c_n\,\mathrm{e}^{\mathrm{j}n2\pi t/T}\right)g(t)\,\mathrm{d}t \quad \text{using (4.63a)}$$

$$= \sum_{n=-\infty}^{\infty} c_n\left[\frac{1}{T}\int_{c}^{c+T} g(t)\,\mathrm{e}^{\mathrm{j}n2\pi t/T}\,\mathrm{d}t\right] \quad \text{assuming term-by-term integration is possible}$$

$$= \sum_{n=-\infty}^{\infty} c_n d_{-n} \quad \text{using (4.64b)}$$

Since $d_{-n} = d_n^*$, the complex conjugate of d_n, this reduces to the required result:

$$\frac{1}{T}\int_{c}^{c+T} f(t)g(t)\,\mathrm{d}t = \sum_{n=-\infty}^{\infty} c_n d_n^*$$

□

In terms of the real coefficients a_n, b_n and α_n, β_n of the corresponding trigonometric Fourier series expansions of $f(t)$ and $g(t)$,

$$f(t) = \tfrac{1}{2}a_0 + \sum_{n=1}^{\infty} a_n \cos\left(\frac{n2\pi t}{T}\right) + \sum_{n=1}^{\infty} b_n \sin\left(\frac{n2\pi t}{T}\right)$$

$$g(t) = \tfrac{1}{2}\alpha_0 + \sum_{n=1}^{\infty} \alpha_n \cos\left(\frac{n2\pi t}{T}\right) + \sum_{n=1}^{\infty} \beta_n \sin\left(\frac{n2\pi t}{T}\right)$$

and using the definitions (4.56), the multiplication theorem result (4.62) reduces to

$$\begin{aligned}\frac{1}{T}\int_c^{c+T} f(t)g(t)\,\mathrm{d}t &= \sum_{n=1}^{\infty} c_{-n}d_n + c_0d_0 + \sum_{n=1}^{\infty} c_nd_{-n} \\ &= \tfrac{1}{4}\alpha_0a_0 + \tfrac{1}{4}\sum_{n=1}^{\infty}[(a_n - \mathrm{j}b_n)(\alpha_n + \mathrm{j}\beta_n) \\ &\qquad + (a_n + \mathrm{j}b_n)(\alpha_n - \mathrm{j}\beta_n)]\end{aligned}$$

giving

$$\frac{1}{T}\int_c^{c+T} f(t)g(t)\,\mathrm{d}t = \tfrac{1}{4}\alpha_0a_0 + \tfrac{1}{2}\sum_{n=1}^{\infty}(a_n\alpha_n + b_n\beta_n)$$

THEOREM 4.6 Parseval's theorem

If $f(t)$ is a periodic function with period T then

$$\frac{1}{T}\int_c^{c+T} [f(t)]^2\,\mathrm{d}t = \sum_{n=-\infty}^{\infty} c_nc_n^* = \sum_{n=-\infty}^{\infty} |c_n|^2 \tag{4.65}$$

where the c_n are the coefficients in the complex Fourier series expansion of $f(t)$.

Proof This result follows from the multiplication theorem, since, taking $g(t) = f(t)$ in (4.62), we obtain

$$\frac{1}{T}\int_c^{c+T} [f(t)]^2\,\mathrm{d}t = \sum_{n=-\infty}^{\infty} c_nc_n^* = \sum_{n=-\infty}^{\infty} |c_n|^2$$ □

Using (4.60), Parseval's theorem may be written in terms of the real coefficients a_n and b_n of the trigonometric Fourier series expansion of the function $f(t)$ as

$$\frac{1}{T}\int_c^{c+T} [f(t)]^2\,\mathrm{d}t = \tfrac{1}{4}a_0^2 + \tfrac{1}{2}\sum_{n=1}^{\infty}(a_n^2 + b_n^2) \tag{4.66}$$

The **root mean square (RMS)** value f_{RMS} of a periodic function $f(t)$ of period T, defined by

$$f_{RMS}^2 = \frac{1}{T}\int_c^{c+T} [f(t)]^2\,dt$$

may therefore be expressed in terms of the Fourier coefficients using (4.65) or (4.66).

EXAMPLE 4.20 By applying Parseval's theorem to the function

$$f(t) = \frac{2t}{T} \quad (0 < t < T), \qquad f(t + 2T) = f(t)$$

considered in Example 4.19, show that

$$\tfrac{1}{6}\pi^2 = \sum_{n=1}^{\infty} \frac{1}{n^2}$$

Solution From Example 4.19, the coefficients of the complex Fourier series expansion of $f(t)$ are

$$c_0 = 2,\ c_n = \frac{j2}{n\pi} \quad (n \neq 0)$$

Thus, applying the Parseval's theorem result (4.65), noting that the period in this case is $2T$, we obtain

$$\frac{1}{2T}\int_0^{2T} [f(t)]^2\,dt = c_0^2 + \sum_{n=-\infty}^{-1} |c_n|^2 + \sum_{n=1}^{\infty} |c_n|^2$$

giving

$$\frac{1}{2T}\int_0^{2T} \frac{4t^2}{T^2}\,dt = 4 + 2\sum_{n=1}^{\infty}\left(\frac{2}{n\pi}\right)^2$$

which reduces to

$$\tfrac{16}{3} = 4 + \sum_{n=1}^{\infty} \frac{8}{n^2\pi^2}$$

leading to the required result

$$\tfrac{1}{6}\pi^2 = \sum_{n=1}^{\infty} \frac{1}{n^2}$$

4.6.3 Discrete frequency spectra

In expressing a periodic function $f(t)$ by its Fourier series expansion, we are decomposing the function into its **harmonic** or **frequency components**. We have seen that if $f(t)$ is of period T then it has frequency components at frequencies

$$\omega_n = \frac{n2\pi}{T} = n\omega_0 \quad (n = 1, 2, 3, \ldots) \tag{4.67}$$

where ω_0 is the frequency of the parent function $f(t)$. (All frequencies here are measured in rad s^{-1}.)

A Fourier series may therefore be interpreted as constituting a **frequency spectrum** of the periodic function $f(t)$, and provides an alternative representation of the function to its time-domain waveform. This frequency spectrum is often displayed by plotting graphs of both the amplitudes and phases of the various harmonic components against angular frequency ω_n. A plot of amplitude against angular frequency is called the **amplitude spectrum**, while that of phase against angular frequency is called the **phase spectrum**. For a periodic function $f(t)$, of period T, harmonic components only occur at discrete frequencies ω_n, given by (4.63), so that these spectra are referred to as **discrete frequency spectra** or **line spectra**. In Chapter 5 Fourier transforms will be used to define continuous spectra for aperiodic functions. With the growing ability to process signals digitally, the representation of signals by their corresponding spectra is an approach widely used in almost all branches of engineering, especially electrical engineering, when considering topics such as filtering and modulation. An example of the use of a discrete spectral representation of a periodic function is in distortion measurements on amplifiers, where the harmonic content of the output, measured digitally, to a sinusoidal input provides a measure of the distortion.

If the Fourier series expansion of a periodic function $f(t)$, with period T, has been obtained in the trigonometric form

$$f(t) = \tfrac{1}{2}a_0 + \sum_{n=1}^{\infty} a_n \cos\left(\frac{n2\pi t}{T}\right) + \sum_{n=1}^{\infty} b_n \sin\left(\frac{n2\pi t}{T}\right)$$

then, as indicated in Section 4.2.2, this may be expressed in terms of the various harmonic components as

$$f(t) = A_0 + \sum_{n=1}^{\infty} A_n \sin\left(\frac{n2\pi t}{T} + \phi_n\right) \tag{4.68}$$

where

$$A_0 = \tfrac{1}{2}a_0, \quad A_n = \sqrt{(a_n^2 + b_n^2)}$$

and the ϕ_n are determined by

$$\sin\phi_n = \frac{b_n}{A_n}, \quad \cos\phi_n = \frac{a_n}{A_n}$$

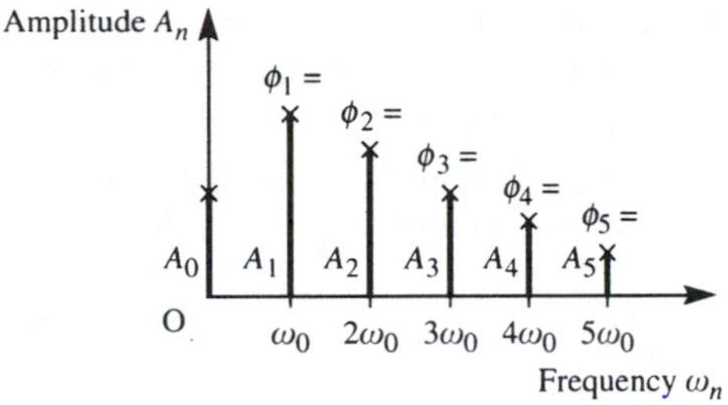

Figure 4.35 Real discrete frequency spectrum.

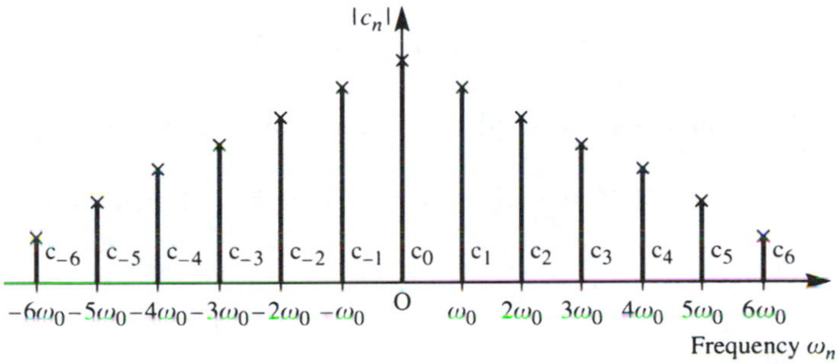

Figure 4.36 Complex form of the amplitude spectrum.

In this case a plot of A_n against angular frequency ω_n will constitute the amplitude spectrum and that of ϕ_n against ω_n the phase spectrum. These may be incorporated in the same graph by indicating the various phases on the amplitude spectrum as illustrated in Figure 4.35. It can be seen that the amplitude spectrum consists of a series of equally spaced vertical lines whose lengths are proportional to the amplitudes of the various harmonic components making up the function $f(t)$. Clearly the trigonometric form of the Fourier series does not in general lend itself to the plotting of the discrete frequency spectrum, and the amplitudes A_n and phases ϕ_n must first be determined from the values of a_n and b_n previously determined.

In work on signal analysis it is much more common to use the complex form of the Fourier series. For a periodic function $f(t)$, of period T, this is given by (4.57), with the complex coefficients being given by

$$c_n = |c_n|\,e^{j\phi_n} \quad (n = 0, \pm 1, \pm 2, \dots)$$

in which $|c_n|$ and ϕ_n denote the magnitude and argument of c_n respectively. Since in general c_n is a complex quantity, we need two line spectra to determine the discrete frequency spectrum; the amplitude spectrum being a plot of $|c_n|$ against ω_n and the phase spectrum that of ϕ_n against ω_n. In cases where c_n is real a single spectrum may be used to represent the function $f(t)$. Since $|c_{-n}| = |c_n^*| = |c_n|$, the amplitude spectrum will be symmetrical about the vertical axis, as illustrated in Figure 4.36.

Note that in the complex form of the discrete frequency spectrum we have components at the discrete frequencies $0, \pm\omega_0, \pm 2\omega_0, \pm 3\omega_0, \dots$; that is, both positive and negative discrete frequencies are involved. Clearly signals having negative frequencies are not physically realizable, and have been introduced for mathematical convenience. At frequency $n\omega_0$ we have the component $e^{jn\omega_0 t}$, which in itself is not a physical signal; to obtain a physical signal, we must consider this alongside the corresponding component $e^{-jn\omega_0 t}$ at the frequency $-n\omega_0$, since then we have

$$e^{jn\omega_0 t} + e^{-jn\omega_0 t} = 2\cos n\omega_0 t \tag{4.69}$$

EXAMPLE 4.21 Plot the discrete amplitude and phase spectra for the periodic function

$$f(t) = \frac{2t}{T} \quad (0 < t < 2T), \qquad f(t + 2T) = f(t)$$

of Example 4.19. Consider both complex and real forms.

Solution In Example 4.19 the complex coefficients were determined as

$$c_0 = 2, \qquad c_n = \frac{j2}{n\pi} \quad (n = \pm 1, \pm 2, \pm 3, \dots)$$

Thus

$$|c_n| = \begin{cases} 2/n\pi & (n = 1, 2, 3, \dots) \\ -2/n\pi & (n = -1, -2, -3, \dots) \end{cases}$$

$$\phi_n = \arg c_n = \begin{cases} \tfrac{1}{2}\pi & (n = 1, 2, 3, \dots) \\ -\tfrac{1}{2}\pi & (n = -1, -2, -3, \dots) \end{cases}$$

The corresponding amplitude and phase spectra are shown in Figures 4.37(a) and (b) respectively.

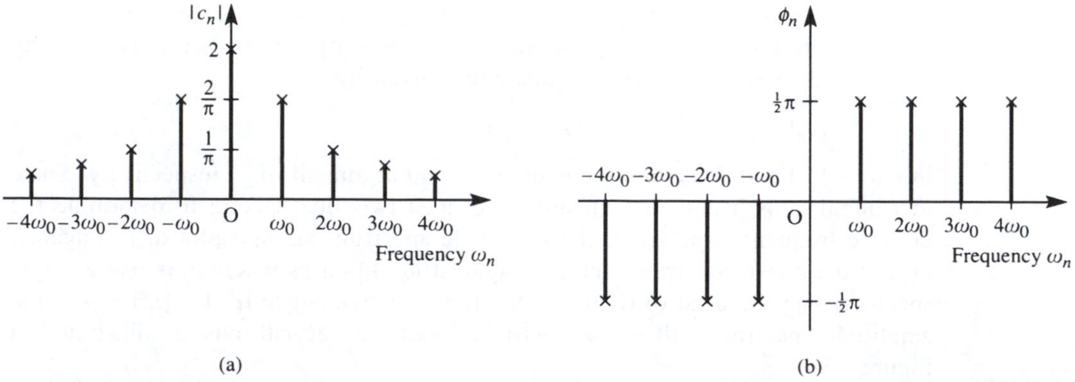

Figure 4.37 Complex discrete frequency spectra for Example 4.21, with $\omega_0 = \pi/T$: (a) amplitude spectrum; (b) phase spectrum.

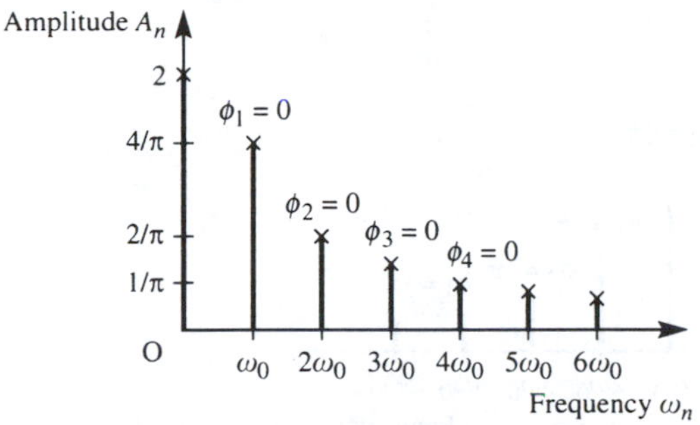

Figure 4.38 Real discrete frequency spectrum for Example 4.21 (corresponding to sinusoidal expansion).

In Example 4.19 we saw that the coefficients in the trigonometric form of the Fourier series expansion of $f(t)$ are

$$a_0 = 4, \quad a_n = 0, \quad b_n = -\frac{4}{n\pi}$$

so that the amplitude coefficients in (4.67) are

$$A_0 = 2, \quad A_n = \frac{4}{n\pi} \quad (n = 1, 2, 3, \ldots)$$

leading to the real discrete frequency spectrum of Figure 4.38.

Since $|c_n| = \frac{1}{2}\sqrt{(a_n^2 + b_n^2)} = \frac{1}{2}A_n$, the amplitude spectrum lines in the complex form (Figure 4.37) are, as expected, halved in amplitude relative to those in the real representation (Figure 4.38), the other half-value being allocated to the corresponding negative frequency. In the complex representation the phases at negative frequencies (Figure 4.37b) are the negatives of those at the corresponding positive frequencies. In our particular representation (4.68) of the real form the phases at positive frequencies differ by $\frac{1}{2}\pi$ between the real and complex form. Again this is not surprising, since from (4.69) we see that combining positive and negative frequencies in the complex form leads to a cosinusoid at that frequency rather than a sinusoid. In order to maintain equality of the phases at positive frequencies between the complex and real representations, a cosinusoidal expansion

$$f(t) = A_0 + \sum_{n=1}^{\infty} A_n \cos\left(\frac{n\pi t}{T} + \phi_n\right) \tag{4.70}$$

of the real Fourier series is frequently adopted as an alternative to the sinusoidal series expansion (4.68). Taking (4.70), the amplitude spectrum will remain the same as for (4.68), but the phase spectrum will be determined by

$$\sin \phi_n = -\frac{b_n}{A_n}, \quad \cos \phi_n = \frac{a_n}{A_n}$$

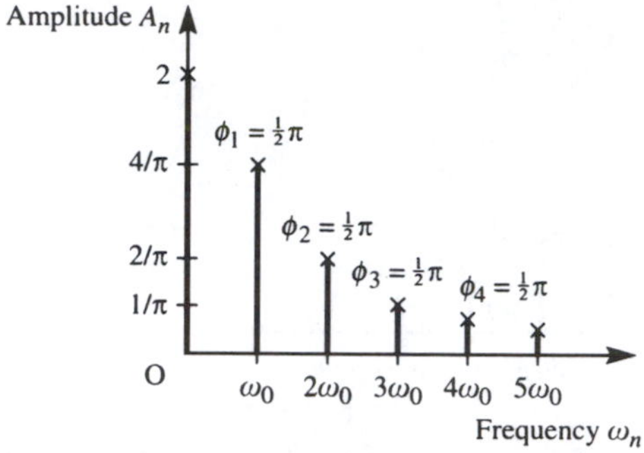

Figure 4.39 Real discrete frequency spectrum for Example 4.21 (corresponding to cosinusoidal expansion).

showing a phase shift of $\frac{1}{2}\pi$ from that of (4.68). Adopting the real representation (4.70), the corresponding real discrete frequency spectrum for the function $f(t)$ of Example 4.21 is as illustrated in Figure 4.39.

EXAMPLE 4.22

Determine the complex form of the Fourier series expansion of the periodic (period $2T$) infinite train of identical rectangular pulses of magnitude A and duration $2d$ illustrated in Figure 4.40. Draw the discrete frequency spectrum in the particular case when $d = \frac{1}{10}$ and $T = \frac{1}{2}$.

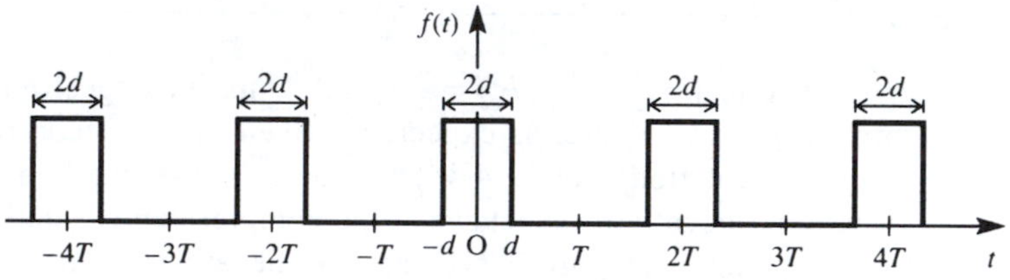

Figure 4.40 Infinite train of rectangular pulses of Example 4.22.

Solution Over one period $-T < t < T$ the function $f(t)$ representing the train is expressed as

$$f(t) = \begin{cases} 0 & (-T < t < -d) \\ A & (-d < t < d) \\ 0 & (d < t < T) \end{cases}$$

From (4.61), the complex coefficients c_n are given by

$$c_n = \frac{1}{2T}\int_{-T}^{T} f(t)\,\mathrm{e}^{-\mathrm{j}n\pi t/T}\,\mathrm{d}t = \frac{1}{2T}\int_{-d}^{d} A\,\mathrm{e}^{-\mathrm{j}n\pi t/T}\,\mathrm{d}t = \frac{A}{2T}\left[\frac{-T}{\mathrm{j}n\pi}\mathrm{e}^{-\mathrm{j}n\pi t/T}\right]_{-d}^{d} \quad (n \neq 0)$$

$$= \frac{A}{n\pi}\,\frac{\mathrm{e}^{\mathrm{j}n\pi d/T} - \mathrm{e}^{-\mathrm{j}n\pi d/T}}{\mathrm{j}2} = \frac{A}{n\pi}\sin\left(\frac{n\pi d}{T}\right) = \frac{Ad}{T}\,\frac{\sin(n\pi d/T)}{n\pi d/T} \quad (n = \pm 1, \pm 2, \ldots)$$

In the particular case when $n = 0$

$$c_0 = \frac{1}{2T}\int_{-T}^{T} f(t)\,\mathrm{d}t = \frac{1}{2T}\int_{-d}^{d} A\,\mathrm{d}t = \frac{Ad}{T}$$

so that

$$c_n = \frac{Ad}{T}\operatorname{sinc}\left(\frac{n\pi d}{T}\right) \quad (n = 0, \pm 1, \pm 2, \dots)$$

where the **sinc function** is defined by

$$\operatorname{sinc} t = \begin{cases} \dfrac{\sin t}{t} & (t \neq 0) \\ 1 & (t = 0) \end{cases}$$

Thus from (4.57) the complex Fourier series expansion for the infinite train of pulses $f(t)$ is

$$f(t) = \sum_{n=-\infty}^{\infty} \frac{Ad}{T}\operatorname{sinc}\left(\frac{n\pi d}{T}\right)\mathrm{e}^{\mathrm{j}n\pi t/T}$$

As expected, since $f(t)$ is an even function, c_n is real, so we need only plot the discrete amplitude spectrum to represent $f(t)$. Since the amplitude spectrum is a plot of $|c_n|$ against frequency $n\omega_0$, with $\omega_0 = \pi/T$, it will only take values at the discrete frequency values

$$0, \pm\frac{\pi}{T}, \pm\frac{2\pi}{7}, \pm\frac{3\pi}{7}, \dots$$

In the particular case $d = \frac{1}{10}$, $T = \frac{1}{2}$, $\omega_0 = 2\pi$ the amplitude spectrum will only exist at frequency values

$$0, \pm 2\pi, \pm 4\pi, \dots$$

Since in this case

$$c_n = \tfrac{1}{5}A \operatorname{sinc} \tfrac{1}{5}n\pi \quad (n = 0, \pm 1, \pm 2, \dots)$$

noting that $\operatorname{sinc} \frac{1}{5}n\pi = 0$ when $\frac{1}{5}n\pi = m\pi$ or $n = 5m$ $(m = \pm 1, \pm 2, \dots)$, the spectrum is as shown in Figure 4.41.

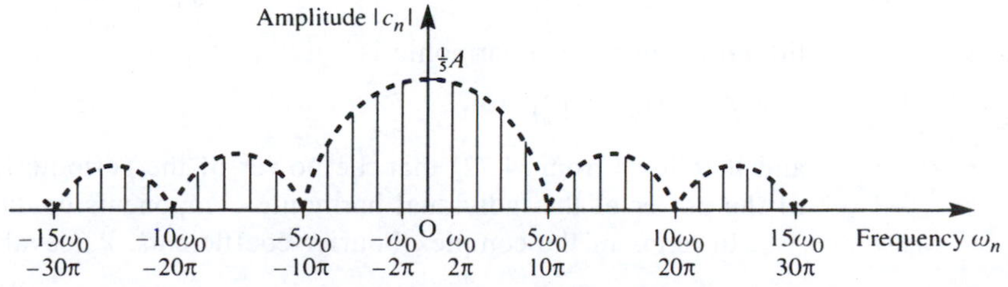

Figure 4.41 Discrete amplitude spectrum for an infinite train of pulses when $d = \frac{1}{10}$ and $T = \frac{1}{2}$.

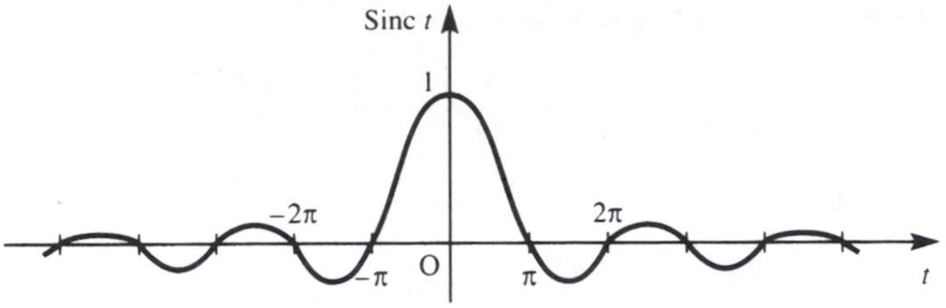

Figure 4.42 Graph of sinc t.

As we will see in Chapter 5, the sinc function $\text{sinc}\, t = (\sin t)/t$ plays an important role in signal analysis, and it is sometimes referred to as the **sampling function**. A graph of sinc t is shown in Figure 4.42, and it is clear that the function oscillates over intervals of length 2π and decreases in amplitude with increasing t. Note also that the function has zeros at $t = \pm n\pi$ $(n = 1, 2, 3, \dots)$.

4.6.4 Power spectrum

The **average power** P associated with a periodic signal $f(t)$, of period T, is defined as the mean square value; that is,

$$P = \frac{1}{T}\int_d^{d+T} [f(t)]^2 \,\mathrm{d}t \tag{4.71}$$

For example, if $f(t)$ represents a voltage waveform applied to a resistor then P represents the average power, measured in watts, dissipated by a $1\,\Omega$ resistor.

By Parseval's theorem (Theorem 4.6),

$$P = \tfrac{1}{4}a_0^2 + \tfrac{1}{2}\sum_{n=1}^{\infty}(a_n^2 + b_n^2) \tag{4.72}$$

Since

$$\frac{1}{T}\int_d^{d+T}\left[a_n \cos\left(\frac{2n\pi t}{T}\right)\right]^2 \mathrm{d}t = \frac{1}{2}a_n^2, \qquad \frac{1}{T}\int_d^{d+T}\left[b_n \cos\left(\frac{2n\pi t}{T}\right)\right]^2 \mathrm{d}t = \frac{1}{2}b_n^2$$

the power in the nth harmonic is

$$P_n = \tfrac{1}{2}(a_n^2 + b_n^2) \tag{4.73}$$

and it follows from (4.72) that the power of the periodic function $f(t)$ is the sum of the power of the individual harmonic components contained in $f(t)$.

In terms of the complex Fourier coefficients, Parseval's theorem gives

$$P = \sum_{n=-\infty}^{\infty} |c_n|^2 \tag{4.74}$$

As discussed in Section 4.6.3, the component $e^{jn\omega_0 t}$ at frequency $\omega_n = n\omega_0$, $\omega_0 = 2\pi/T$, must be considered alongside the component $e^{-jn\omega_0 t}$ at the corresponding negative frequency $-\omega_n$ in order to form the actual nth harmonic component of the function $f(t)$. Since $|c_{-n}|^2 = |c_n^*|^2 = |c_n|^2$, it follows that the power associated with the nth harmonic is the sum of the power associated with $e^{jn\omega_0 t}$ and $e^{-jn\omega_0 t}$; that is,

$$P_n = 2|c_n|^2 \tag{4.75}$$

which, since $|c_n| = \frac{1}{2}\sqrt{(a_n^2 + b_n^2)}$, corresponds to (4.73). Thus in the complex form half the power of the nth harmonic is associated with the positive frequency and half with the negative frequency.

Since the total power of a periodic signal is the sum of the power associated with each of the harmonics of which the signal is composed, it is again useful to consider a spectral representation, and a plot of $|c_n|^2$ against angular frequency ω_n is called the **power spectrum** of the function $f(t)$. Clearly such a spectrum is readily deduced from the discrete amplitude spectrum of $|c_n|$ against angular frequency ω_n.

EXAMPLE 4.23

For the spectrum of the infinite train of rectangular pulses shown in Figure 4.40, determine the percentage of the total power contained within the frequency band up to the first zero value (called the **zero crossing** of the spectrum) at 10π rad s^{-1}.

Solution From (4.71), the total power associated with the infinite train of rectangular pulses $f(t)$ is

$$P = \frac{1}{2T}\int_{-T}^{T} [f(t)]^2\,dt = \frac{1}{2T}\int_{-d}^{d} A^2\,dt$$

which in the particular case when $d = \frac{1}{10}$ and $T = \frac{1}{2}$ becomes

$$P = \int_{-1/10}^{1/10} A^2\,dt = \tfrac{1}{5}A^2$$

The power contained in the frequency band up to the first zero crossing at 10π rad s^{-1} is

$$P_1 = c_0^2 + 2(c_1^2 + c_2^2 + c_3^2 + c_4^2)$$

where

$$c_n = \tfrac{1}{5}A \operatorname{sinc} \tfrac{1}{5}n\pi$$

That is,

$$\begin{aligned} P_1 &= \tfrac{1}{25}A^2 + \tfrac{2}{25}A^2(\operatorname{sinc}^2 \tfrac{1}{5}\pi + \operatorname{sinc}^2 \tfrac{2}{5}\pi + \operatorname{sinc}^2 \tfrac{3}{5}\pi + \operatorname{sinc}^2 \tfrac{4}{5}\pi) \\ &= \tfrac{1}{25}A^2[1 + 2(0.875 + 0.756 + 0.255 + 0.055)] \\ &= \tfrac{1}{5}A^2(0.976) \end{aligned}$$

Thus $P_1 = 0.976P$, so that approximately 97.6% of the total power associated with $f(t)$ is contained in the frequency band up to the first zero crossing at 10π rad s^{-1}.

Suppose that a periodic voltage $v(t)$, of period T, applied to a linear circuit, results in a corresponding current $i(t)$, having the same period T. Then, given the Fourier series representation of both the voltage and current at a pair of terminals, we can use the multiplication theorem (Theorem 4.5) to obtain an expression for the average power P at the terminals. Thus, given

$$v(t) = \sum_{n=-\infty}^{\infty} c_n e^{j2n\pi t/T}, \qquad i(t) = \sum_{n=-\infty}^{\infty} d_n e^{j2n\pi t/T}$$

the instantaneous power at the terminals is vi and the average power is

$$P = \frac{1}{T}\int_d^{d+T} vi\,dt = \sum_{n=-\infty}^{\infty} c_n d_n^*$$

or, in terms of the corresponding trigonometric Fourier series coefficients a_n, b_n and α_n, β_n

$$P = \tfrac{1}{4}\alpha_0\beta_0 + \tfrac{1}{2}\sum_{n=1}^{\infty}(a_n\alpha_n + b_n\beta_n)$$

4.6.5 Exercises

34 Show that the complex form of the Fourier series expansion of the periodic function

$$f(t) = t^2 \quad (-\pi < t < \pi)$$
$$f(t + 2\pi) = f(t)$$

is

$$f(t) = \frac{\pi^2}{6} + \sum_{n=0}^{\infty} \frac{2}{n^2}(-1)^n e^{jnt}$$

Using (4.56), obtain the corresponding trigonometric series and check with the series obtained in Example 4.5.

35 Obtain the complex form of the Fourier series expansion of the square wave

$$f(t) = \begin{cases} 0 & (-2 < t < 0) \\ 1 & (0 < t < 2) \end{cases}$$

$$f(t + 4) = f(t)$$

Using (4.56), obtain the corresponding trigonometric series and check with the series obtained in Example 4.9.

36 Obtain the complex form of the Fourier series expansion of the following periodic functions.

(a) $f(t) = \begin{cases} \pi & (-\pi < t < 0) \\ t & (0 < t < \pi) \end{cases}$

$f(t + 2\pi) = f(t)$

(b) $f(t) = \begin{cases} a \sin \omega t & (0 < t < \tfrac{1}{2}T) \\ 0 & (\tfrac{1}{2}T < t < T) \end{cases}$

$f(t + T) = f(t), \quad T = 2\pi/\omega$

(c) $f(t) = \begin{cases} 2 & (-\pi < t < 0) \\ 1 & (0 < t < \pi) \end{cases}$

$f(t + 2\pi) = f(t)$

(d) $f(t) = |\sin t| \quad (-\pi < t < \pi)$

$f(t + 2\pi) = f(t)$

37 A periodic function $f(t)$, of period 2π, is defined within the period $-\pi < t < \pi$ by

$$f(t) = \begin{cases} 0 & (-\pi < t < 0) \\ 1 & (0 < t < \pi) \end{cases}$$

Using the Fourier coefficients of $f(t)$, together with Parseval's theorem, show that

$$\sum_{n=1}^{\infty} \frac{1}{(2n-1)^2} = \frac{1}{8}\pi^2$$

(*Note*: The Fourier coefficients may be deduced from Example 4.9 or Exercise 35.)

38 (a) Show that the Fourier series expansion of the periodic function

$$f(t) = 500\pi t \quad (0 < t < \tfrac{1}{50})$$

$$f(t + \tfrac{1}{50}) = f(t)$$

may be expressed as

$$f(t) = 5\pi - 10 \sum_{n=1}^{\infty} \frac{1}{n} \sin 100n\pi t$$

(b) Using (4.66), estimate the RMS value of $f(t)$ by (i) using the first four terms of the Fourier series; (ii) using the first eight terms of the Fourier series.

(c) Obtain the true RMS value of $f(t)$, and hence determine the percentage errors in the estimated values obtained in (b).

39 A periodic voltage $v(t)$ (in V) of period 5 ms and specified by

$$v(t) = \begin{cases} 60 & (0 < t < 1.25 \text{ ms}) \\ 0 & (1.25 \text{ ms} < t < 5 \text{ ms}) \end{cases}$$

$$v(t + 5 \text{ ms}) = v(t)$$

is applied across the terminals of a 15 Ω resistor.

(a) Obtain expressions for the coefficients c_n of the complex Fourier series representation of $v(t)$, and write down the values of the first five non-zero terms.

(b) Calculate the power associated with each of the first five non-zero terms of the Fourier expansion.

(c) Calculate the total power delivered to the 15 Ω resistor.

(d) What is the percentage of the total power delivered to the resistor by the first five non-zero terms of the Fourier series?

4.7 Orthogonal functions

As was noted in Section 4.2.3, the fact that the set of functions $\{1, \cos \omega t, \sin \omega t, \ldots, \cos n\omega t, \sin n\omega t, \ldots\}$ is an orthogonal set of functions on the interval $d \leqslant t \leqslant d + T$ was crucial in the evaluation of the coefficient in the Fourier series expansion of a function $f(t)$. It is natural to ask whether it is possible to express the function $f(t)$ as a series expansion in other sets of functions. In the case of periodic functions $f(t)$ there is no natural alternative, but if we are concerned with representing a function $f(t)$ only in a finite interval $t_1 \leqslant t \leqslant t_2$ then a variety of other possibilities exist. These possibilities are drawn from a class of functions called **orthogonal functions**, of which the trigonometric set $\{1, \cos \omega t, \sin \omega t, \ldots, \cos n\omega t, \sin n\omega t\}$ is a particular example.

4.7.1 Definitions

Two real functions $f(t)$ and $g(t)$ that are piecewise-continuous in the interval $t_1 \leqslant t \leqslant t_2$ are said to be **orthogonal** in this interval if

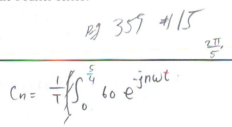

$$\int_{t_1}^{t_2} f(t)g(t)\,dt = 0$$

A set of real functions $\phi_1(t), \phi_2(t), \ldots \equiv \{\phi_n(t)\}$, each of which is piecewise-continuous on $t_1 \leqslant t \leqslant t_2$, is said to be an **orthogonal set** on this interval if $\phi_n(t)$ and $\phi_m(t)$ are orthogonal for each pair of distinct indices n, m; that is, if

$$\int_{t_1}^{t_2} \phi_n(t)\phi_m(t)\,dt = 0 \quad (n \neq m) \tag{4.76}$$

We shall also assume that no member of the set $\{\phi_n(t)\}$ is identically zero except at a finite number of points, so that

$$\int_{t_1}^{t_2} \phi_m^2(t)\,dt = \gamma_m \quad (m = 1, 2, 3, \ldots) \tag{4.77}$$

where γ_m $(m = 1, 2, \ldots)$ are all non-zero constants.

An orthogonal set $\{\phi_n(t)\}$ is said to be **orthonormal** if each of its components is also normalized; that is, $\gamma_m = 1$ $(m = 1, 2, 3, \ldots)$. We note that any orthogonal set $\{\phi_n(t)\}$ can be converted into an orthonormal set by dividing each member $\phi_m(t)$ of the set by $\sqrt{\gamma_m}$.

EXAMPLE 4.24 Since (4.4)–(4.8) hold,

$$\{1, \cos t, \sin t, \cos 2t, \sin 2t, \ldots, \cos nt, \sin nt\}$$

is an orthogonal set on the interval $d \leqslant t \leqslant d + 2\pi$, while the set

$$\left\{\frac{1}{\sqrt{(2\pi)}}, \frac{\cos t}{\sqrt{\pi}}, \frac{\sin t}{\sqrt{\pi}}, \ldots, \frac{\cos nt}{\sqrt{\pi}}, \frac{\sin nt}{\sqrt{\pi}}\right\}$$

forms an orthonormal set on the same interval.

The latter follows since

$$\int_d^{d+2\pi} \left[\frac{1}{\sqrt{(2\pi)}}\right]^2 dt = 1$$

$$\int_d^{d+2\pi} \left(\frac{\cos nt}{\sqrt{\pi}}\right)^2 dt = \int_d^{d+2\pi} \left(\frac{\sin nt}{\sqrt{\pi}}\right)^2 dt = 1 \quad (n = 1, 2, 3, \ldots)$$

The definition of orthogonality considered so far applies to real functions, and has to be amended somewhat if members of the set $\{\phi_n(t)\}$ are complex functions of the real variable t. In such a case the set $\{\phi_n(t)\}$ is said to be an orthogonal set on the interval $t_1 \leqslant t \leqslant t_2$ if

$$\int_{t_1}^{t_2} \phi_n(t)\phi_m^*(t)\,\mathrm{d}t = \begin{cases} 0 & (n \neq m) \\ \gamma & (n = m) \end{cases} \tag{4.78}$$

where $\phi_m^*(t)$ denotes the complex conjugate of $\phi(t)$.

EXAMPLE 4.25 Verify that the set of complex exponential functions

$$\{\mathrm{e}^{\mathrm{j}n\pi t/T}\} \quad (n = 0, \pm 1, \pm 2, \pm 3, \ldots)$$

used in the complex representation of the Fourier series is an orthogonal set on the interval $0 \leqslant t \leqslant 2T$.

Solution First,

$$\int_0^{2T} \mathrm{e}^{\mathrm{j}n\pi t/T}\, 1\, \mathrm{d}t = \left[\frac{T}{\mathrm{j}n\pi}\mathrm{e}^{\mathrm{j}n\pi t/T}\right]_0^{2T} = 0 \quad (n \neq 0)$$

since $\mathrm{e}^{\mathrm{j}2n\pi} = \mathrm{e}^0 = 1$. Secondly,

$$\int_0^{2T} \mathrm{e}^{\mathrm{j}n\pi t/T}(\mathrm{e}^{\mathrm{j}m\pi t/T})^*\, \mathrm{d}t = \int_0^{2T} \mathrm{e}^{\mathrm{j}(n-m)\pi t/T}\, \mathrm{d}t$$

$$= \left[\frac{T}{\mathrm{j}(n-m)\pi}\mathrm{e}^{\mathrm{j}(n-m)\pi t/T}\right]_0^{2T} = 0 \quad (n \neq m)$$

and, when $n = m$,

$$\int_0^{2T} \mathrm{e}^{\mathrm{j}n\pi t/T}(\mathrm{e}^{\mathrm{j}n\pi t/T})^*\, \mathrm{d}t = \int_0^{2T} 1\, \mathrm{d}t = 2T$$

Thus

$$\int_0^{2T} \mathrm{e}^{\mathrm{j}n\pi t/T}\, 1\, \mathrm{d}t = 0 \quad (n \neq 0)$$

$$\int_0^{2T} \mathrm{e}^{\mathrm{j}n\pi t/T}(\mathrm{e}^{\mathrm{j}m\pi t/T})^*\, \mathrm{d}t = \begin{cases} 0 & (n \neq m) \\ 2T & (n = m) \end{cases}$$

and, from (4.78), the set is an orthogonal set on the interval $0 \leqslant t \leqslant 2T$.

The trigonometric and exponential sets are examples of orthogonal sets that we have already used in developing the work on Fourier series. Examples of other sets of orthogonal functions that are widely used in practice are Legendre polynomials, Bessel functions, Hermite polynomials, Lagurre polynomials, Jacobi polynomials, Tchebyshev (sometimes written as Chebyshev) polynomials and Walsh functions.

4.7.2 Generalized Fourier series

Let $\{\phi_n(t)\}$ be an orthogonal set on the interval $t_1 \leqslant t \leqslant t_2$ and suppose that we wish to represent the piecewise-continuous function $f(t)$ in terms of this set within this interval. Following the Fourier series development, suppose that it is possible to express $f(t)$ as a series expansion of the form

$$f(t) = \sum_{n=1}^{\infty} c_n \phi_n(t) \tag{4.79}$$

We now wish to determine the coefficients c_n, and to do so we again follow the Fourier series development. Multiplying (4.79) throughout by $\phi_m(t)$ and integrating term by term, we obtain

$$\int_{t_1}^{t_2} f(t)\phi_m(t)\,\mathrm{d}t = \sum_{n=1}^{\infty} c_n \int_{t_1}^{t_2} \phi_m(t)\phi_n(t)\,\mathrm{d}t$$

which, on using (4.76) and (4.77), reduces to

$$\int_{t_1}^{t_2} f(t)\phi_n(t)\,\mathrm{d}t = c_n \gamma_n$$

giving

$$c_n = \frac{1}{\gamma_n}\int_{t_1}^{t_2} f(t)\phi_n(t)\,\mathrm{d}t \quad (n = 1, 2, 3, \dots) \tag{4.80}$$

Summarizing, if $f(t)$ is a piecewise-continuous function on the interval $t_1 \leqslant t \leqslant t_2$ and $\{\phi_n(t)\}$ is an orthogonal set on this interval then the series

$$f(t) = \sum_{n=1}^{\infty} c_n \phi_n(t)$$

is called the **generalized Fourier series** of $f(t)$ with respect to the basis set $\{\phi_n(t)\}$, and the coefficients c_n, given by (4.80), are called the **generalized Fourier coefficients** with respect to the same basis set.

A parallel can be drawn between a generalized Fourier series expansion of a function $f(t)$ with respect to an orthogonal basis set of functions $\{\phi_n(t)\}$ and the representation of a vector $\boldsymbol{f}$ in terms of an orthogonal basis set of vectors $\boldsymbol{v}_1, \boldsymbol{v}_2, \dots, \boldsymbol{v}_n$ as

$$\boldsymbol{f} = \alpha_1 \boldsymbol{v}_1 + \dots + \alpha_n \boldsymbol{v}_n$$

where

$$\alpha_i = \frac{\boldsymbol{f} \cdot \boldsymbol{v}_i}{\boldsymbol{v}_i \cdot \boldsymbol{v}_i} = \frac{\boldsymbol{f} \cdot \boldsymbol{v}_i}{|\boldsymbol{v}_i|^2}$$

There is clearly a similarity between this pair of results and the pair (4.79)–(4.80).

4.7.3 Convergence of generalized Fourier series

As in the case of a Fourier series expansion, partial sums of the form

$$F_N(t) = \sum_{n=1}^{N} c_n \phi_n(t) \tag{4.81}$$

can be considered, and we wish this representation to be, in some sense, a 'close approximation' to the parent function $f(t)$. The question arises when considering such a partial sum as to whether choosing the coefficients c_n as the generalized Fourier coefficients (4.80) leads to the 'best' approximation. Defining the **mean square error** E_N between the actual value of $f(t)$ and the approximation $F_N(t)$ as

$$E_N = \frac{1}{t_2 - t_1} \int_{t_1}^{t_2} [f(t) - F_N(t)]^2 \, \mathrm{d}t$$

it can be shown that E_N is minimized, for all N, when the coefficients c_n are chosen according to (4.80). Thus in this sense the finite generalized Fourier series gives the best approximation.

To verify this result, assume, for convenience, that the set $\{\phi_n(t)\}$ is orthonormal, and consider the Nth partial sum

$$F_N(t) = \sum_{n=1}^{N} \tilde{c}_n \phi_n(t)$$

where the $\tilde{c}_n$ are to be chosen in order to minimize the mean square error E_N. Now

$$\begin{aligned}(t_2 - t_1)E_N &= \int_{t_1}^{t_2} \left[f(t) - \sum_{n=1}^{N} \tilde{c}_n \phi_n(t) \right]^2 \mathrm{d}t \\ &= \int_{t_1}^{t_2} f^2(t)\,\mathrm{d}t - 2\sum_{n=1}^{N} \tilde{c}_n \int_{t_1}^{t_2} f(t)\phi_n(t)\,\mathrm{d}t + \sum_{n=1}^{N} \tilde{c}_n^2 \int_{t_1}^{t_2} \phi_n^2(t)\,\mathrm{d}t \\ &= \int_{t_1}^{t_2} f^2(t)\,\mathrm{d}t - 2\sum_{n=1}^{N} \tilde{c}_n c_n + \sum_{n=1}^{N} \tilde{c}_n^2\end{aligned}$$

since $\{\phi_n(t)\}$ is an orthonormal set. That is,

$$(t_2 - t_1)E_n = \int_{t_1}^{t_2} f^2(t)\,\mathrm{d}t - \sum_{n=1}^{N} c_n^2 + \sum_{n=1}^{N} (\tilde{c}_n - c_n)^2 \tag{4.82}$$

which is clearly minimized when $\tilde{c}_n = c_n$.

Taking $\tilde{c}_n = c_n$ in (4.82), the mean square error E_N in approximating $f(t)$ by $F_N(t)$ of (4.77) is given by

$$E_N = \frac{1}{t_2 - t_1} \left[\int_{t_1}^{t_2} f^2(t)\,\mathrm{d}t - \sum_{n=1}^{N} c_n^2 \right]$$

if the set $\{\phi_n(t)\}$ is orthonormal, and is given by

$$E_N = \frac{1}{t_2 - t_1}\left[\int_{t_1}^{t_2} f^2(t)\,\mathrm{d}t - \sum_{n=1}^{N} \gamma_n c_n^2\right] \tag{4.83}$$

if the set $\{\phi_n(t)\}$ is orthogonal.

Since, by definition, E_N is non-negative, it follows from (4.83) that

$$\int_{t_1}^{t_2} f^2(t)\,\mathrm{d}t \geqslant \sum_{n=1}^{N} \gamma_n c_n^2 \tag{4.84}$$

a result known as **Bessel's inequality**. The question that arises in practice is whether or not $E_N \to 0$ as $N \to \infty$, indicating that the sum

$$\sum_{n=1}^{N} c_n \phi_n(t)$$

converges to the function $f(t)$. If this were the case then, from (4.83),

$$\int_{t_1}^{t_2} f^2(t)\,\mathrm{d}t = \sum_{n=1}^{\infty} \gamma_n c_n^2 \tag{4.85}$$

which is the **generalized form of Parseval's theorem**, and the set $\{\phi_n(t)\}$ is said to be complete. Strictly speaking, the fact that Parseval's theorem holds ensures that the partial sum $F_N(t)$ converges in the mean to the parent function $f(t)$ as $N \to \infty$, and this does not necessarily guarantee convergence at any particular point. In engineering applications, however, this distinction may be overlooked, since for the functions met in practice convergence in the mean also ensures pointwise convergence at points where $f(t)$ is convergent, and convergence to the mean of the discontinuity at points where $f(t)$ is discontinuous.

EXAMPLE 4.26

The set $\{1, \cos t, \sin t, \ldots, \cos nt, \sin nt\}$ is a complete orthogonal set in the interval $d \leqslant t \leqslant d + 2\pi$. Following the same argument as above, it is readily shown that for a function $f(t)$ that is piecewise-continuous on $d \leqslant t \leqslant d + 2\pi$ the mean square error between $f(t)$ and the finite Fourier series

$$F_N(t) = \tfrac{1}{2}\tilde{a}_0 + \sum_{n=1}^{N} \tilde{a}_n \cos nt + \sum_{n=1}^{N} \tilde{b}_n \sin nt$$

is minimized when $\tilde{a}_0$, $\tilde{a}_n$ and $\tilde{b}_n$ $(n = 1, 2, 3, \ldots)$ are equal to the corresponding Fourier coefficients a_0, a_n and b_n $(n = 1, 2, 3, \ldots)$ determined using (4.11) and (4.12). In this case the mean square error E_N is given by

$$E_N = \frac{1}{2\pi}\left[\int_{d}^{d+2\pi} f^2(t)\,\mathrm{d}t - \pi\left[\frac{1}{2}a_0^2 + \sum_{n=1}^{N}(a_n^2 + b_n^2)\right]\right]$$

Bessel's inequality (4.84) becomes

$$\int_d^{d+2\pi} f^2(t)\,\mathrm{d}t \geqslant \pi\left[\frac{1}{2}a_0^2 + \sum_{n=1}^{N}(a_n^2 + b_n^2)\right]$$

and Parseval's theorem (4.85) reduces to

$$\frac{1}{2\pi}\int_d^{d+2\pi} f^2(t) = \frac{1}{4}a_0^2 + \frac{1}{2}\sum_{n=1}^{\infty}(a_n^2 + b_n^2)$$

which conforms with (4.66). Since, in this case, the basis set is complete, Parseval's theorem holds, and the Fourier series converges to $f(t)$ in the sense discussed above.

4.7.4 Exercises

40 The Fourier series expansion for the periodic square wave

$$f(t) = \begin{cases} -1 & (-\pi < t < 0) \\ 1 & (0 < t < \pi) \end{cases}$$

$$f(t + 2\pi) = f(t)$$

is

$$f(t) = \sum_{n=1}^{\infty} \frac{4}{\pi(2n-1)} \sin(2n-1)t$$

Determine the mean square error corresponding to approximations to $f(t)$ based on the use of one term, two terms and three terms respectively in the series expansion.

41 The Legendre polynomials $P_n(t)$ are generated by the formula

$$P_n(t) = \frac{1}{2^n n!}\frac{\mathrm{d}^n}{\mathrm{d}t^n}(t^2 - 1)^n \quad (n = 0, 1, 2, \dots)$$

and satisfy the recurrence relationship

$$nP_n(t) = (2n-1)tP_{n-1}(t) - (n-1)P_{n-2}(t)$$

(a) Deduce that

$$P_0(t) = 1, \quad P_1(t) = t$$
$$P_2(t) = \tfrac{1}{2}(3t^2 - 1), \quad P_3(t) = \tfrac{1}{2}(5t^3 - 3t)$$

(b) Show that the polynomials form an orthogonal set on the interval (−1, 1) and, in particular, that

$$\int_{-1}^{1} P_m(t)P_n(t)\,\mathrm{d}t = \begin{cases} 0 & (n \neq m) \\ 2/(2n+1) & (n = m;\ m = 0, 1, 2, \dots) \end{cases}$$

(c) Given that the function

$$f(t) = \begin{cases} -1 & (-1 < t < 0) \\ 0 & (t = 0) \\ 1 & (0 < t < 1) \end{cases}$$

is expressed as a Fourier–Legendre series expansion

$$f(t) = \sum_{r=0}^{\infty} c_r P_r(t)$$

determine the values of c_0, c_1, c_2 and c_3.

(d) Plot graphs to illustrate convergence of the series obtained in (c), and compare the mean square error with that of the corresponding Fourier series expansion.

42 Repeat parts (c) and (d) of Exercise 41 for the function

$$f(x) = \begin{cases} 0 & (-1 < x < 0) \\ x & (0 < x < 1) \end{cases}$$

43 Lagurre polynomials $L_n(t)$ are generated by the formula

$$L_n(t) = \mathrm{e}^t \frac{\mathrm{d}^n}{\mathrm{d}t^n}(t^n \mathrm{e}^{-t}) \quad (n = 0, 1, 2, \dots)$$

and satisfy the recurrence relation

$$L_n(t) = (2n - 1 - t)L_{n-1}(t) - (n - 1)^2 L_{n-2}(t) \quad (n = 2, 3, \dots)$$

These polynomials are orthogonal on the interval $0 \leq t < \infty$ with respect to the weighting function e^{-t}, so that

$$\int_0^{\infty} e^{-t} L_n(t) L_m(t)\, dt = \begin{cases} 0 & (n \neq m) \\ (n!)^2 & (n = m) \end{cases}$$

(a) Deduce that

$$L_0(t) = 1, \quad L_1(t) = 1 - t$$
$$L_2(t) = 2 - 4t + t^2$$
$$L_3(t) = 6 - 18t + 9t^2 - t^3$$

(b) Confirm the above orthogonality result in the case of L_0, L_1, L_2 and L_3.

(c) Given that the function $f(t)$ is to be approximated over the interval $0 \leq t < \infty$ by

$$f(t) = \sum_{r=0}^{\infty} c_r L_r(t)$$

show that

$$c_r = \frac{1}{(r!)^2} \int_0^{\infty} f(t)\, e^{-t} L_r(t)\, dt \quad (r = 0, 1, 2, \dots)$$

(*Note*: Lagurre polynomials are of particular importance to engineers, since they can be generated as the impulse responses of relatively simple networks.)

44 Hermite polynomials $H_n(t)$ are generated by the formula

$$H_n(t) = (-1)^n e^{t^2/2} \frac{d^n}{dt^n} e^{-t^2/2} \quad (n = 0, 1, 2, \dots)$$

and satisfy the recurrence relationship

$$H_n(t) = tH_{n-1}(t) - (n - 1)H_{n-2}(t) \quad (n = 2, 3, \dots)$$

The polynomials are orthogonal on the interval $-\infty < t < \infty$ with respect to the weighting function $e^{-t^2/2}$, so that

$$\int_{-\infty}^{\infty} e^{-t^2/2} H_n(t) H_m(t)\, dt = \begin{cases} 0 & (n \neq m) \\ \sqrt{(2\pi)}\, n! & (n = m) \end{cases}$$

(a) Deduce that

$$H_0(t) = 1, \quad H_1(t) = t$$
$$H_2(t) = t^2 - 1 \quad H_3(t) = t^3 - 3t$$
$$H_4(t) = t^4 - 6t^2 + 3$$

(b) Confirm the above orthogonality result for H_0, H_1, H_2 and H_3.

(c) Given that the function $f(t)$ is to be approximated over the interval $-\infty < t < \infty$ by

$$f(t) = \sum_{r=0}^{\infty} c_r H_r(t)$$

show that

$$c_r = \frac{1}{r!\sqrt{\pi}} \int_{-\infty}^{\infty} e^{-t^2/2} f(t) H_r(t)\, dt \quad (r = 0, 1, \dots)$$

45 Tchebyshev polynomials $T_n(t)$ are generated by the formula

$$T_n(t) = \cos(n \cos^{-1} t) \quad (n = 0, 1, 2, \dots)$$

or

$$T_n(t) = \sum_{r=0}^{[n/2]} (-1)^r \frac{n!}{(2r)!(n-2r)!} (1 - t^2)^r t^{n-2r} \quad (n = 0, 1, 2, \dots)$$

where

$$[n/2] = \begin{cases} n/2 & (\text{even } n) \\ (n-1)/2 & (\text{odd } n) \end{cases}$$

They also satisfy the recurrence relationship

$$T_n(t) = 2tT_{n-1}(t) - T_{n-2}(t) \quad (n = 2, 3, \dots)$$

and are orthogonal on the interval $-1 \leq t \leq 1$ with respect to the weighting function $1/\sqrt{(1 - t^2)}$, so that

$$\int_{-1}^{1} \frac{T_n(t) T_m(t)}{\sqrt{(1 - t^2)}}\, dt = \begin{cases} 0 & (m \neq n) \\ \frac{1}{2}\pi & (m = n \neq 0) \\ \pi & (m = n = 0) \end{cases}$$

(a) Deduce that

$$T_0(t) = 1, \quad T_1(t) = t$$
$$T_2(t) = 2t^2 - 1, \quad T_3(t) = 4t^3 - 3t$$
$$T_4(t) = 8t^4 - 8t^2 + 1$$
$$T_5(t) = 16t^5 - 20t^3 + 5t$$

(b) Confirm the above orthogonality result for T_0, T_1, T_2 and T_3.

(c) Given that the function $f(t)$ is to be approximated over the interval $-1 \leq t \leq 1$ by

$$f(t) = \sum_{r=0}^{\infty} c_r T_r(t)$$

show that

$$c_0 = \frac{1}{\pi} \int_{-1}^{1} \frac{f(t) T_0(t)}{\sqrt{(1 - t^2)}}\, dt$$

$$c_r = \frac{2}{\pi}\int_{-1}^{1} \frac{f(t)T_r(t)}{\sqrt{(1-t^2)}}\,dt \quad (r = 1, 2, \dots)$$

46 With developments in digital techniques, Walsh functions $W_n(t)$ have become of considerable importance in practice, since they are so easily generated by digital logic circuitry. The first four Walsh functions may be defined on the interval $0 \leqslant t \leqslant T$ by

$$W_0(t) = \frac{1}{\sqrt{T}} \quad (0 \leqslant t \leqslant T)$$

$$W_1(t) = \begin{cases} 1/\sqrt{T} & (0 \leqslant t < \frac{1}{2}T) \\ -1/\sqrt{T} & (\frac{1}{2}T < t \leqslant T) \end{cases}$$

$$W_2(t) = \begin{cases} 1/\sqrt{T} & (0 \leqslant t < \frac{1}{4}T, \frac{3}{4}T < t \leqslant T) \\ -1/\sqrt{T} & (\frac{1}{4}T < t < \frac{3}{4}T) \end{cases}$$

$$W_3(t) = \begin{cases} 1/\sqrt{T} & (0 \leqslant t < \frac{1}{8}T, \frac{3}{8}T < t < \frac{5}{8}T, \frac{7}{8}T < t \leqslant T) \\ -1/\sqrt{T} & (\frac{1}{8}T < t < \frac{3}{8}T, \frac{5}{8}T < t < \frac{7}{8}T) \end{cases}$$

(a) Plot graphs of the functions $W_0(t)$, $W_1(t)$, $W_2(t)$ and $W_3(t)$, and show that they are orthonormal on the interval $0 \leqslant t \leqslant T$. Write down an expression for $W_n(t)$.

(b) The Walsh functions may be used to obtain a Fourier–Walsh series expansion for a function $f(t)$, over the interval $0 \leqslant t \leqslant T$, in the form

$$f(t) = \sum_{r=0}^{\infty} c_r W_r(t)$$

Illustrate this for the square wave of Exercise 40. What is the corresponding mean square error? Comment on your answer.

4.8 Engineering application: describing functions

Many control systems containing a nonlinear element may be represented by the block diagram of Figure 4.43. In practice, describing function techniques are used to analyse and design such control systems. Essentially the method involves replacing the nonlinearity by an equivalent gain N and then using the techniques developed for linear systems, such as the frequency response methods of Section 2.7. If the nonlinear element is subjected to a sinusoidal input $e(t) = X \sin \omega t$ then its output $z(t)$ may be represented by the Fourier series expansion

$$\begin{aligned} z(t) &= \tfrac{1}{2}a_0 + \sum_{n=1}^{\infty} a_n \cos n\omega t + \sum_{n=1}^{\infty} b_n \sin n\omega t \\ &= \tfrac{1}{2}a_0 + \sum_{n=1}^{\infty} A_n \sin(n\omega t + \phi_n) \end{aligned}$$

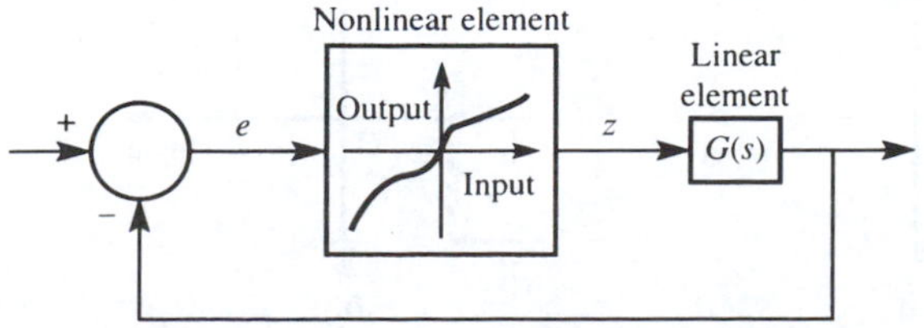

Figure 4.43 Nonlinear control system.

with $A_n = \sqrt{(a_n^2 + b_n^2)}$ and $\phi_n = \tan^{-1}(a_n/b_n)$.

The **describing function** $N(X)$ of the nonlinear element is then defined to be the complex ratio of the fundamental component of the output to the input; that is,

$$N(X) = \frac{A_1}{X} e^{j\phi_1}$$

with $N(X)$ being independent of the input frequency ω if the nonlinear element is memory-free.

Having determined the describing function, the behaviour of the closed-loop system is then determined by the characteristic equation

$$1 + N(X)G(j\omega) = 0$$

If a combination of X and ω can be found to satisfy this equation then the system is capable of sustained oscillations at that frequency and magnitude; that is, the system exhibits **limit-cycle behaviour**. In general, more than one combination can be found, and the resulting oscillations can be a stable or unstable limit cycle.

Normally the characteristic equation is investigated graphically by plotting $G(j\omega)$ and $-1/N(X)$, for all values of X, on the same polar diagram. Limit cycles then occur at frequencies and amplitudes corresponding to points of intersection of the curves. Sometimes plotting can be avoided by calculating the maximum value of $N(X)$ and hence the value of the gain associated with $G(s)$ that will just cause limit cycling to occur.

Using this background information, the following investigation is left as an exercise for the reader to develop.

(a) Show that the describing functions $N_1(X)$ and $N_2(X)$ corresponding respectively to the relay (on–off nonlinearity) of Figure 4.44(a) and the relay with dead zone of Figure 4.44(b) are

$$N_1(X) = \frac{4L}{\pi X}, \qquad N_2(X) = \frac{4L}{\pi X}\sqrt{\left[1 - \left(\frac{h}{X}\right)^2\right]}$$

(b) For the system of Figure 4.45 show that a limit cycle exists when the nonlinearity is the relay of Figure 4.44(a) with $L = 1$. Determine the amplitude and frequency of this limit cycle.

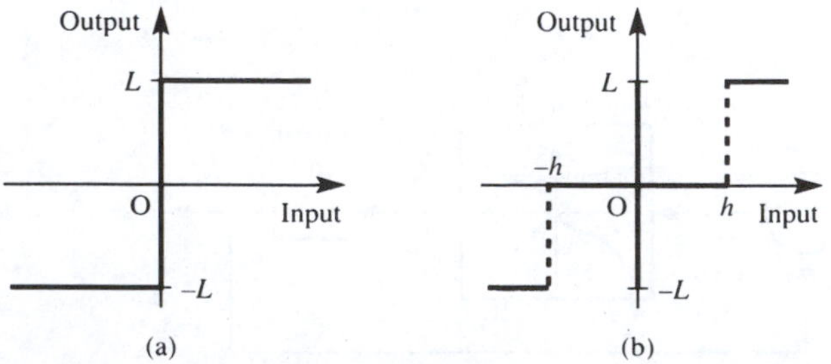

Figure 4.44 (a) Relay; (b) relay with dead zone.

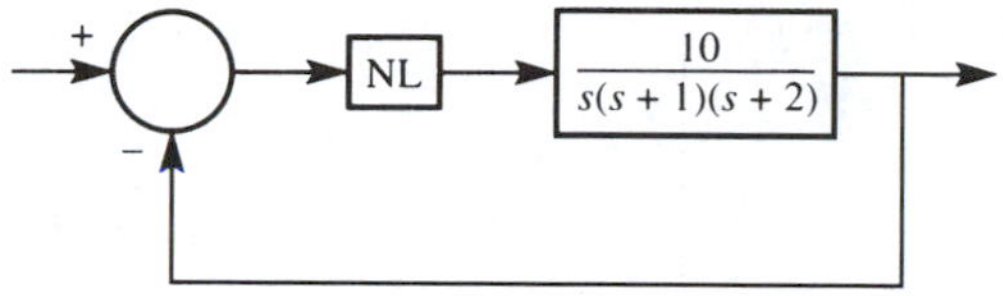

Figure 4.45 Nonlinear system of exercise.

In an attempt to eliminate the limit-cycle oscillation, the relay is replaced by the relay with dead zone illustrated in Figure 4.44(b), again with $L = 1$. Show that this allows our objective to be achieved provided that $h > 10/3\pi$.

4.9 Review exercises (1–20)

1 A periodic function $f(t)$ is defined by

$$f(t) = \begin{cases} t^2 & (0 \leqslant t < \pi) \\ 0 & (\pi < t \leqslant 2\pi) \end{cases}$$

$$f(t + 2\pi) = f(t)$$

Obtain a Fourier series expansion of $f(t)$ and deduce that

$$\tfrac{1}{6}\pi^2 = \sum_{r=1}^{\infty} \frac{1}{r^2}$$

2 Determine the full-range Fourier series expansion of the even function $f(t)$ of period 2π defined by

$$f(t) = \begin{cases} \tfrac{2}{3}t & (0 \leqslant t \leqslant \tfrac{1}{3}\pi) \\ \tfrac{1}{3}(\pi - t) & (\tfrac{1}{3}\pi \leqslant t \leqslant \pi) \end{cases}$$

To what value does the series converge at $t = \tfrac{1}{3}\pi$?

3 A function $f(t)$ is defined for $0 \leqslant t \leqslant \tfrac{1}{2}T$ by

$$f(t) = \begin{cases} t & (0 \leqslant t \leqslant \tfrac{1}{4}T) \\ \tfrac{1}{2}T - t & (\tfrac{1}{4}T \leqslant t \leqslant \tfrac{1}{2}T) \end{cases}$$

Sketch odd and even functions that have a period T and are equal to $f(t)$ for $0 \leqslant t \leqslant \tfrac{1}{2}T$.

(a) Find the half-range Fourier sine series of $f(t)$.

(b) To what value will the series converge for $t = -\tfrac{1}{4}T$?

(c) What is the sum of the following series?

$$S = \sum_{r=1}^{\infty} \frac{1}{(2r - 1)^2}$$

4 Prove that if $g(x)$ is an odd function and $f(x)$ an even function of x, the product $g(x)[c + f(x)]$ is an odd function if c is a constant.

A periodic function with period 2π is defined by

$$F(\theta) = \tfrac{1}{12}\theta(\pi^2 - \theta^2)$$

in the interval $-\pi \leqslant \theta \leqslant \pi$. Show that the Fourier series representation of the function is

$$F(\theta) = \sum_{n=1}^{\infty} \frac{(-1)^{n+1}}{n^3} \sin n\theta$$

5 A repeating waveform of period 2π is described by

$$f(t) = \begin{cases} \pi + t & (-\pi \leqslant t \leqslant -\tfrac{1}{2}\pi) \\ -t & (-\tfrac{1}{2}\pi \leqslant t \leqslant \tfrac{1}{2}\pi) \\ t - \pi & (\tfrac{1}{2}\pi \leqslant t \leqslant \pi) \end{cases}$$

Sketch the waveform over the range $t = -2\pi$ to $t = 2\pi$ and find the Fourier series representation of $f(t)$, making use of any properties of the waveform that you can identify before any integration is performed.

6 A function $f(x)$ is defined in the interval $-1 \leqslant x \leqslant 1$ by

$$f(x) = \begin{cases} 1/2\varepsilon & (-\varepsilon < x < \varepsilon) \\ 0 & (1 \leqslant x < -\varepsilon;\ \varepsilon < x \leqslant 1) \end{cases}$$

Sketch a graph of $f(x)$ and show that a Fourier series expansion of $f(x)$ valid in the interval $-1 \leq x \leq 1$ is given by

$$f(x) = \tfrac{1}{2} + \sum_{n=1}^{\infty} \frac{\sin n\pi\varepsilon}{n\pi\varepsilon} \cos n\pi x$$

7 Show that the half-range Fourier sine series for the function

$$f(t) = \left(1 - \frac{t}{\pi}\right)^2 \quad (0 \leq t \leq \pi)$$

is

$$f(t) = \sum_{n=1}^{\infty} \frac{2}{n\pi}\left\{1 - \frac{2}{n^2\pi^2}[1-(-1)^n]\right\} \sin nt$$

8 Find a half-range Fourier sine and Fourier cosine series for $f(x)$ valid in the interval $0 < x < \pi$ when $f(x)$ is defined by

$$f(x) = \begin{cases} x & (0 \leq x \leq \tfrac{1}{2}\pi) \\ \pi - x & (\tfrac{1}{2}\pi \leq x \leq \pi) \end{cases}$$

Sketch the graph of the Fourier series obtained for $-2\pi < x \leq 2\pi$.

9 A function $f(x)$ is periodic of period 2π and is defined by $f(x) = e^x$ $(-\pi < x < \pi)$. Sketch the graph of $f(x)$ from $x = -2\pi$ to $x = 2\pi$ and prove that

$$f(x) = \frac{2\sinh\pi}{\pi}\left[\frac{1}{2} + \sum_{n=1}^{\infty} \frac{(-1)^n}{1+n^2}(\cos nx - n\sin nx)\right]$$

10 A function $f(t)$ is defined on $0 < t < \pi$ by

$$f(t) = \pi - t$$

Find

(a) a half-range Fourier sine series, and
(b) a half-range Fourier cosine series

for $f(t)$ valid for $0 < t < \pi$.

Sketch the graphs of the functions represented by each series for $-2\pi < t < 2\pi$.

11 Show that the Fourier series

$$\frac{1}{2}\pi - \frac{4}{\pi}\sum_{n=1}^{\infty} \frac{\cos(2n-1)t}{(2n-1)^2}$$

represents the function $f(t)$, of period 2π, given by

$$f(t) = \begin{cases} t & (0 \leq t \leq \pi) \\ -t & (-\pi \leq t \leq 0) \end{cases}$$

Deduce that, apart from a transient component (that is, a complementary function that dies away as $t \to \infty$), the differential equation

$$\frac{dx}{dt} + x = f(t)$$

has the solution

$$x = \frac{1}{2}\pi - \frac{4}{\pi}\sum_{n=1}^{\infty} \frac{\cos(2n-1)t + (2n-1)\sin(2n-1)t}{(2n-1)^2[1+(2n-1)^2]}$$

12 Show that if $f(t)$ is a periodic function of period 2π and

$$f(t) = \begin{cases} t/\pi & (0 < t < \pi) \\ (2\pi - t)/\pi & (\pi < t < 2\pi) \end{cases}$$

then

$$f(t) = \frac{1}{2} - \frac{4}{\pi^2}\sum_{n=0}^{\infty} \frac{\cos(2n+1)t}{(2n+1)^2}$$

Show also that, when ω is not an integer,

$$y = \frac{1}{2\omega^2}(1 - \cos\omega t) - \frac{4}{\pi^2}\sum_{n=1}^{\infty} \frac{\cos(2n+1)t - \cos\omega t}{(2n+1)^2[\omega^2 - (2n+1)^2]}$$

satisfies the differential equation

$$\frac{d^2y}{dt^2} + \omega^2 y = f(t)$$

subject to the initial conditions $y = dy/dt = 0$ at $t = 0$.

13 (a) A periodic function $f(t)$, of period 2π, is defined in $-\pi \leq t \leq \pi$ by

$$f(t) = \begin{cases} -t & (-\pi \leq t \leq 0) \\ t & (0 \leq t \leq \pi) \end{cases}$$

Obtain a Fourier series expansion for $f(t)$, and from it, using Parseval's theorem, deduce that

$$\frac{1}{96}\pi^4 = \sum_{n=1}^{\infty} \frac{1}{(2n-1)^4}$$

(b) By formally differentiating the series obtained in (a), obtain the Fourier series expansion of the periodic square wave

$$g(t) = \begin{cases} -1 & (-\pi < t < 0) \\ 0 & (t = 0) \\ 1 & (0 < t < \pi) \end{cases}$$

$$g(t + 2\pi) = g(t)$$

Check the validity of your result by determining directly the Fourier series expansion of $g(t)$.

14 A periodic function $f(t)$, of period 2π, is defined in the range $-\pi < t < \pi$ by

$$f(t) = \sin \tfrac{1}{2}t$$

Show that the complex form of the Fourier series expansion for $f(t)$ is

$$f(t) = \sum_{n=-\infty}^{\infty} \frac{j4n(-1)^n}{\pi(4n^2 - 1)} e^{jnt}$$

15 (a) Find the Fourier series expansion of the voltage $v(t)$ represented by the half-wave rectified sine wave

$$v(t) = \begin{cases} 10 \sin (2\pi t/T) & (0 < t < \tfrac{1}{2}T) \\ 0 & (\tfrac{1}{2}T < t < T) \end{cases}$$

$$v(t + T) = v(t)$$

(b) If the voltage $v(t)$ in (a) is applied to a 10 Ω resistor, what is the total average power delivered to the resistor? What percentage of the total power is carried by the second-harmonic component of the voltage?

16 The periodic waveform $f(t)$ shown in Figure 4.46 may be written as

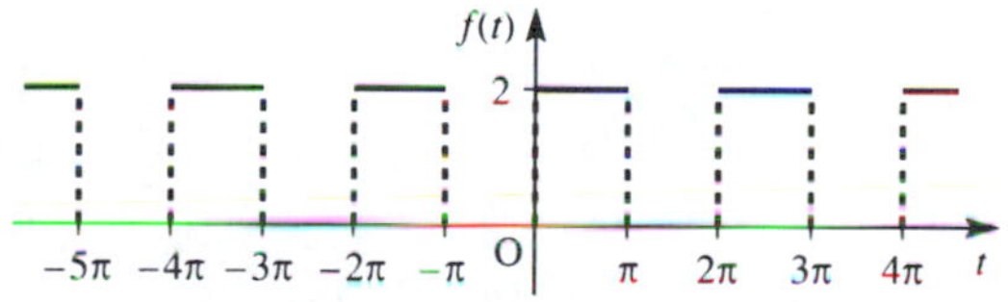

Figure 4.46 Waveform $f(t)$ of Review exercise 16.

$$f(t) = 1 + g(t)$$

where $g(t)$ represents an odd function.

(a) Sketch the graph of $g(t)$.

(b) Obtain the Fourier series expansion for $g(t)$, and hence write down the Fourier series expansion for $f(t)$.

17 Show that the complex Fourier series expansion for the periodic function

$$f(t) = t \quad (0 < t < 2\pi)$$

$$f(t + 2\pi) = f(t)$$

is

$$f(t) = \pi + \sum_{\substack{n=-\infty \\ n \neq 0}}^{\infty} \frac{j e^{jnt}}{n}$$

18 (a) A square-wave voltage $v(t)$ of period T is defined by

$$v(t) = \begin{cases} -1 & (-\tfrac{1}{2}T < t < 0) \\ 1 & (0 < t < \tfrac{1}{2}T) \end{cases}$$

$$v(t + T) = v(t)$$

Show that its Fourier series expansion is given by

$$v(t) = \frac{4}{\pi} \sum_{n=1}^{\infty} \frac{\sin [(4n - 2)\pi t/T]}{2n - 1}$$

(b) Find the steady-state response of the circuit shown in Figure 4.47 to the sinusoidal input voltage

$$v_\omega(t) = \sin \omega t$$

and hence write down the Fourier series expansion of the circuit's steady-state response to the square-wave voltage $v(t)$ in (a).

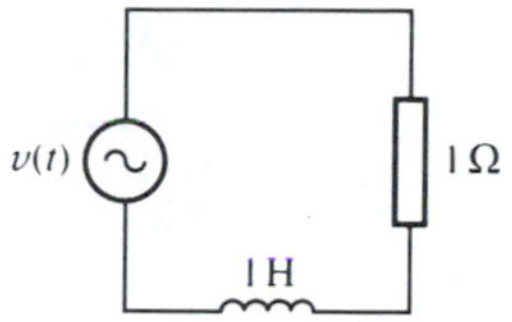

Figure 4.47 Circuit of Review exercise 18.

19 (a) Defining the nth Tchebyshev polynomial by

$$T_n(t) = \cos (n \cos^{-1} t)$$

use Euler's formula $\cos \theta = \tfrac{1}{2}(e^{j\theta} + e^{-j\theta})$ to obtain the expansions of t^{2k} and t^{2k+1} in Tchebyshev polynomials, where k is a positive integer.

(b) Establish the recurrence relation

$$T_n(t) = 2tT_{n-1}(t) - T_{n-2}(t)$$

(c) Write down the values of $T_0(t)$ and $T_1(t)$ from the definition, and then use (b) to find $T_2(t)$ and $T_3(t)$.

(d) Express $t^5 - 5t^4 + 7t^3 + 6t - 8$ in Tchebyshev polynomials.

(e) Find the cubic polynomial that approximates to

$$t^5 - 5t^4 + 7t^3 + 6t - 8$$

over the interval (−1, 1) with the smallest maximum error. Give an upper bound for this error. Is there a value of t for which this upper bound is attained?

20 The relationship between the input and output of a relay with a dead zone Δ and no hysteresis is shown in Figure 4.48. Show that the describing function is

$$N(x_i) = \frac{4M}{\pi x_i}\left[1 - \left(\frac{\Delta}{2x_i}\right)^2\right]^{1/2}$$

for an input amplitude x_i.

If this relay is used in the forward path of the on–off positional control system shown in Figure 4.49, where the transfer function

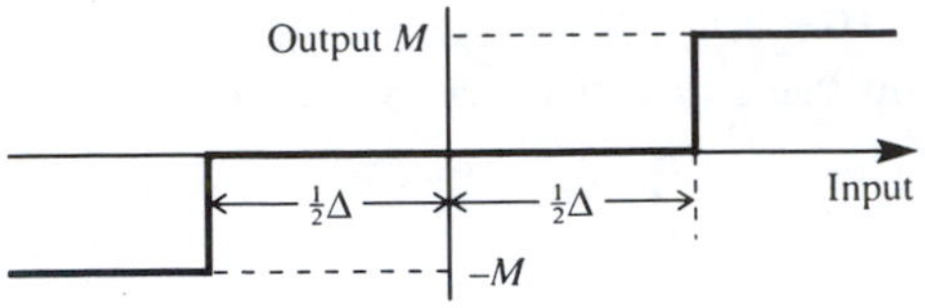

Figure 4.48 Relay with dead zone of Review exercise 20.

$$\frac{K}{s(T_1 s+1)(T_2 s+1)}$$

characterizes the time constant of the servo-motor, and the inertia and viscous damping of the load,

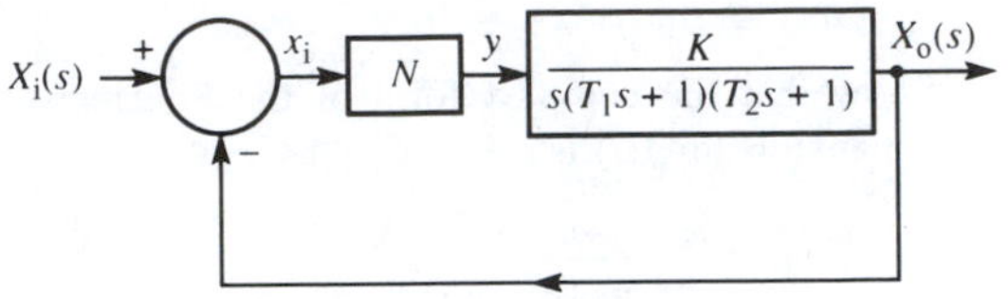

Figure 4.49 Positional control system of Review exercise 20.

show that a limit-cycle oscillation will not occur provided that the dead zone in the relay is such that

$$\Delta > \frac{4MK}{\pi}\frac{T_1 T_2}{T_1 + T_2}$$

5

The Fourier Transform

CONTENTS

5.1 Introduction

In Chapter 4 we saw how Fourier series provided an ideal framework for analysing the steady-state response of systems to a periodic input signal. In this chapter we extend the ideas of Fourier analysis to deal with non-periodic functions. We do this through the introduction of the Fourier transform. As the theory develops, we shall see how the complex exponential form of the Fourier series representation of a periodic function emerges as a special case of the Fourier transform. Similarities between the transform and the Laplace transform, discussed in Chapter 2, will also be highlighted.

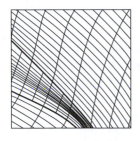

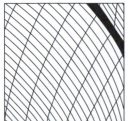

While Fourier transforms first found most application in the solution of partial differential equations, it is probably true to say that today Fourier-transform methods are most heavily used in the analysis of signals and systems. This chapter is therefore developed with such applications in mind, and its main aim is to develop an understanding of the underlying mathematics as a preparation for a specialist study of application areas in various branches of engineering.

Throughout this book we draw attention to the impact of digital computers on engineering and thus on the mathematics required to understand engineering concepts. While much of the early work on signal analysis was implemented using analogue devices, the bulk of modern equipment exploits digital technology. In Chapter 2 we developed the Laplace transform as an aid to the analysis and design of continuous-time systems while in Chapter 3 we introduced the z and $\mathcal{D}$ transforms to assist with the analysis and design of discrete-time systems. In this chapter the frequency-domain analysis introduced in Chapter 2 for continuous-time systems is consolidated and then extended to provide a framework for the frequency-domain description of discrete-time systems through the introduction of discrete Fourier transforms. These discrete transforms provide one of the most advanced methods for discrete signal analysis, and are widely used in such fields as communications theory and speech and image processing. In practice, the computational aspects of the work assume great importance, and the use of appropriate computational algorithms for the calculation of the discrete Fourier transform is essential. For this reason we have included an introduction to the fast Fourier transform algorithm, based on the pioneering work of J. W. Cooley and J. W. Tukey published in 1965, which it is hoped will serve the reader with the necessary understanding for progression to the understanding of specialist engineering applications.

5.2 The Fourier transform

5.2.1 The Fourier integral

In Chapter 4 we saw how Fourier series methods provided a technique for the frequency-domain representation of periodic functions. As indicated in Section 4.6.3, in expressing a function as its Fourier series expansion we are decomposing the function into its harmonic or frequency components. Thus a periodic function $f(t)$, of period T', has frequency components at discrete frequencies

$$\omega_n = \frac{2\pi n}{T'} = n\omega_0 \quad (n = 0, 1, 2, 3, \ldots)$$

where ω_0 is the fundamental frequency, that is, the frequency of the parent function $f(t)$. Consequently we were able to interpret a Fourier series as constituting a **discrete frequency spectrum** of the periodic function $f(t)$, thus providing an

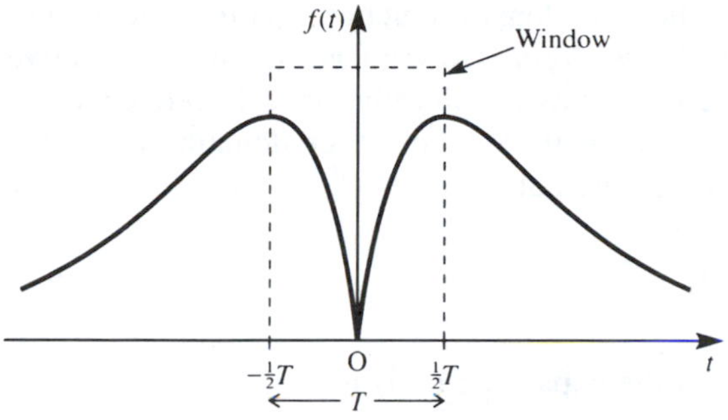

Figure 5.1 The view of $f(t)$ through a window of length T.

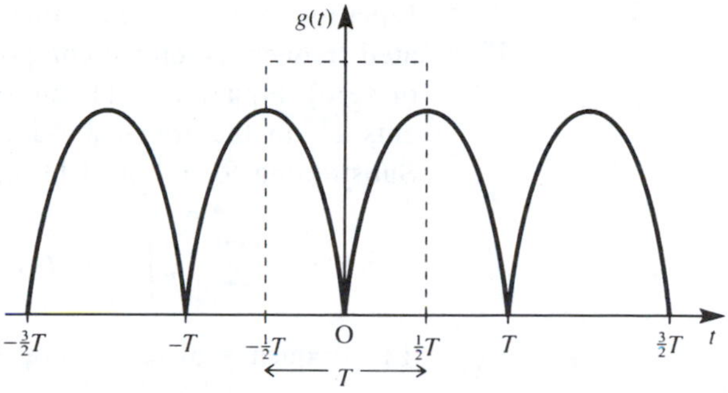

Figure 5.2 The periodic function $g(t)$ based on the 'windowed' view of $f(t)$.

alternative frequency-domain representation of the function to its time-domain waveform. However, not all functions are periodic and so we need to develop an approach that will give a similar representation for non-periodic functions, defined on $-\infty < t < \infty$. One way of achieving this is to look at a portion of a non-periodic function $f(t)$ over an interval T, by imagining that we are looking at a graph of $f(t)$ through a 'window' of length T, and then to consider what happens as T gets larger.

Figure 5.1 depicts this situation, with the window placed symmetrically about the origin. We could now concentrate only on the 'view through the window' and carry out a Fourier series development based on that portion of $f(t)$ alone. Whatever the behaviour of $f(t)$ outside the window, the Fourier series thus generated would represent the periodic function defined by

$$g(t) = \begin{cases} f(t) & (|t| < \frac{1}{2}T) \\ f(t - nT) & (\frac{1}{2}(2n-1)T < |t| < \frac{1}{2}(2n+1)T) \end{cases}$$

Figure 5.2 illustrates $g(t)$, and we can see that the graphs of $f(t)$ and $g(t)$ agree on the interval $(-\frac{1}{2}T, \frac{1}{2}T)$. Note that this approach corresponds to the one adopted in Section 4.3 to obtain the Fourier series expansion of functions defined over a finite interval.

Using the complex or exponential form of the Fourier series expansion, we have from (4.57) and (4.61) that

$$g(t) = \sum_{n=-\infty}^{\infty} G_n \, e^{jn\omega_0 t} \tag{5.1}$$

with

$$G_n = \frac{1}{T}\int_{-T/2}^{T/2} g(t)\, e^{-jn\omega_0 t}\, dt \tag{5.2}$$

and where

$$\omega_0 = 2\pi/T \tag{5.3}$$

Equation (5.2) in effect *transforms* the time-domain function $g(t)$ into the associated frequency-domain components G_n, where n is any integer (positive, negative or zero). Equation (5.1) can also be viewed as transforming the discrete components G_n in the frequency-domain representation to the time-domain form $g(t)$. Substituting for G_n in (5.1), using (5.2), we obtain

$$g(t) = \sum_{n=-\infty}^{\infty} \left[\frac{1}{T} \int_{-T/2}^{T/2} g(\tau)\, \mathrm{e}^{-\mathrm{j}n\omega_0\tau}\, \mathrm{d}\tau \right] \mathrm{e}^{\mathrm{j}n\omega_0 t} \tag{5.4}$$

The frequency of the general term in the expansion (5.4) is

$$\frac{2\pi n}{T} = n\omega_0 = \omega_n$$

and so the difference in frequency between successive terms is

$$\frac{2\pi}{T}[(n+1) - n] = \frac{2\pi}{T} = \Delta\omega$$

Since $\Delta\omega = \omega_0$, we can express (5.4) as

$$g(t) = \sum_{n=\infty}^{\infty} \left[\frac{1}{2\pi} \int_{-T/2}^{T/2} g(\tau)\, \mathrm{e}^{-j\omega_n\tau}\, \mathrm{d}\tau \right] \mathrm{e}^{\mathrm{j}\omega_n t} \Delta\omega \tag{5.5}$$

Defining $G(\mathrm{j}\omega)$ as

$$G(\mathrm{j}\omega) = \int_{-T/2}^{T/2} g(\tau)\, \mathrm{e}^{-j\omega t}\, \mathrm{d}\tau \tag{5.6}$$

we have

$$g(t) = \frac{1}{2\pi} \sum_{n=-\infty}^{\infty} \mathrm{e}^{\mathrm{j}\omega_n t} G(\mathrm{j}\omega_n) \Delta\omega \tag{5.7}$$

As $T \to \infty$, our window widens, so that $g(t) = f(t)$ everywhere and $\Delta\omega \to 0$. Since we also have

$$\lim_{\Delta\omega\to 0} \frac{1}{2\pi} \sum_{n=-\infty}^{\infty} \mathrm{e}^{\mathrm{j}\omega_n t} G(\mathrm{j}\omega_n) \Delta\omega = \frac{1}{2\pi} \int_{-\infty}^{\infty} \mathrm{e}^{\mathrm{j}\omega t} G(\mathrm{j}\omega)\, \mathrm{d}\omega$$

it follows from (5.7) and (5.6) that

$$f(t) = \int_{-\infty}^{\infty} \left[\frac{1}{2\pi} \mathrm{e}^{\mathrm{j}\omega t} \int_{-\infty}^{\infty} f(\tau)\, \mathrm{e}^{-\mathrm{j}\omega\tau}\, \mathrm{d}\tau \right] \mathrm{d}\omega \tag{5.8}$$

The result (5.8) is known as the **Fourier integral representation** of $f(t)$. A set of conditions that are sufficient for the existence of the Fourier integral is a revised form of Dirichlet's conditions for Fourier series, contained in Theorem 4.2. These conditions may be stated in the form of Theorem 5.1.

THEOREM 5.1 **Dirichlet's conditions for the Fourier integral**

If the function $f(t)$ is such that

(a) it is absolutely integrable, so that

$$\int_{-\infty}^{\infty} |f(t)|\,\mathrm{d}t < \infty$$

(that is, the integral is finite), and

(b) it has at most a finite number of maxima and minima and a finite number of discontinuities in any finite interval

then the Fourier integral representation of $f(t)$, given in (5.8), converges to $f(t)$ at all points where $f(t)$ is continuous and to the average of the right- and left-hand limits of $f(t)$ where $f(t)$ is discontinuous (that is, to the mean of the discontinuity). □

As was indicated in Section 4.2.8 for Fourier series, the use of the equality sign in (5.8) must be interpreted carefully because of the non-convergence to $f(t)$ at points of discontinuity. Again the symbol ~ (read as 'behaves as' or 'represented by') rather than = is frequently used.

The absolute integrable condition (a) of Theorem 5.1 implies that the absolute area under the graph of $y = f(t)$ is finite. Clearly this is so if $f(t)$ decays sufficiently fast with time. However, in general the condition seems to imply a very tight constraint on the nature of $f(t)$, since clearly functions of the form $f(t) = \text{constant}$, $f(t) = \mathrm{e}^{at}$, $f(t) = \mathrm{e}^{-at}$, $f(t) = \sin \omega t$, and so on, defined for $-\infty < t < \infty$, do not meet the requirement. In practice, however, signals are usually causal and do not last for ever (that is, they only exist for a finite time). Also, in practice no signal amplitude goes to infinity, so consequently no **practical signal** $f(t)$ can have an infinite area under its graph $y = f(t)$. Thus for practical signals the integral in (5.8) exists.

To obtain the trigonometric (or real) form of the Fourier integral, we substitute

$$\mathrm{e}^{-\mathrm{j}\omega(\tau-t)} = \cos \omega(\tau - t) - \mathrm{j} \sin \omega(\tau - t)$$

in (5.8) to give

$$f(t) = \frac{1}{2\pi}\int_{-\infty}^{\infty}\int_{-\infty}^{\infty} f(\tau)[\cos \omega(\tau - t) - \mathrm{j}\sin \omega(\tau - t)]\,\mathrm{d}\tau\,\mathrm{d}\omega$$

Since $\sin \omega(\tau - t)$ is an odd function of ω, this reduces to

$$f(t) = \frac{1}{2\pi}\int_{-\infty}^{\infty}\int_{-\infty}^{\infty} f(\tau) \cos \omega(\tau - t)\,\mathrm{d}\tau\,\mathrm{d}\omega$$

which, on noting that the integrand is an even function of ω, reduces further to

$$f(t) = \frac{1}{\pi}\int_{0}^{\infty} \mathrm{d}\omega \int_{-\infty}^{\infty} f(\tau) \cos \omega(\tau - t)\,\mathrm{d}\tau \qquad \textbf{(5.9)}$$

The representation (5.9) is then the required trigonometric form of the Fourier integral.

If $f(t)$ is either an odd function or an even function then further simplifications of (5.9) are possible. Detailed calculations are left as an exercise for the reader, and we shall simply quote the results.

(a) If $f(t)$ is an even function then (5.9) reduces to

$$f(t) = \frac{2}{\pi}\int_0^{\infty}\int_0^{\infty} f(\tau)\cos\omega\tau\cos\omega t\,d\tau\,d\omega \tag{5.10}$$

which is referred to as the **Fourier cosine integral**.

(b) If $f(t)$ is an odd function then (5.9) reduces to

$$f(t) = \frac{2}{\pi}\int_0^{\infty}\int_0^{\infty} f(\tau)\sin\omega\tau\sin\omega t\,d\tau\,d\omega \tag{5.11}$$

which is referred to as the **Fourier sine integral**.

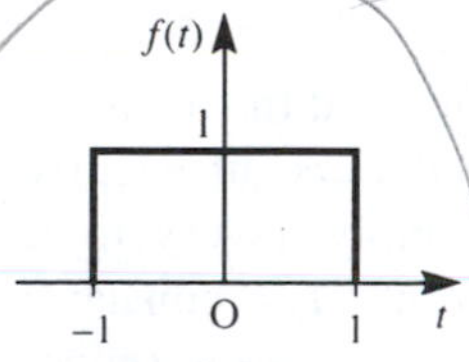

Figure 5.3 Rectangular pulse $f(t) = \begin{cases} 1 & (|t| \leq 1) \\ 0 & (|t| > 1) \end{cases}$

In the case of the Fourier series representation of a periodic function it was a matter of some interest to determine how well the first few terms of the expansion represented the function. The corresponding problem in the non-periodic case is to investigate how well the Fourier integral represents a function when only the components in the lower part of the (continuous) frequency range are taken into account. To illustrate, consider the rectangular pulse of Figure 5.3 given by

$$f(t) = \begin{cases} 1 & (|t| \leq 1) \\ 0 & (|t| > 1) \end{cases}$$

This is clearly an even function, so from (5.10) its Fourier integral is

$$f(t) = \frac{2}{\pi}\int_0^{\infty}\int_0^{1} 1\cos\omega\tau\cos\omega t\,d\tau\,d\omega = \frac{2}{\pi}\int_0^{\infty}\frac{\cos\omega t\sin\omega}{\omega}\,d\omega$$

An elementary evaluation of this integral is not possible, so we consider frequencies $\omega < \omega_0$, when

$$\begin{aligned} f(t) &\simeq \frac{2}{\pi}\int_0^{\omega_0}\frac{\cos\omega t\sin\omega}{\omega}\,d\omega \\ &= \frac{1}{\pi}\int_0^{\omega_0}\frac{\sin\omega(t+1)}{\omega}\,d\omega - \frac{1}{\pi}\int_0^{\omega_0}\frac{\sin\omega(t-1)}{\omega}\,d\omega \\ &= \frac{1}{\pi}\int_0^{\omega_0(t+1)}\frac{\sin u}{u}\,du - \frac{1}{\pi}\int_0^{\omega_0(t-1)}\frac{\sin u}{u}\,du \end{aligned}$$

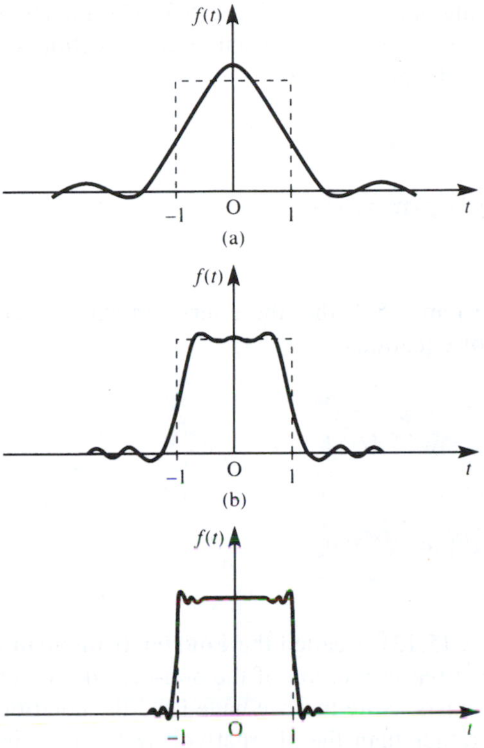

Figure 5.4 Plot of (5.12): (a) $\omega_0 = 4$; (b) $\omega_0 = 8$; (c) $\omega_0 = 16$.

> The integral
>
> $$\text{Si}(x) = \int_0^x \frac{\sin u}{u}\,du \quad (x \geqslant 0)$$
>
> occurs frequently, and it can be shown that
>
> $$\text{Si}(x) = \sum_{n=0}^{\infty} \frac{(-1)^n x^{2n+1}}{(2n+1)(2n+1)!}$$

Its values have been tabulated (see for example L. Rade and B. Westergren, *Beta Mathematics Handbook*, Chartwell-Bratt Ltd, 1990). Thus

$$f(t) \simeq \text{Si}(\omega_0(t+1)) - \text{Si}(\omega_0(t-1)) \tag{5.12}$$

This has been plotted for $\omega_0 = 4$, 8 and 16, and the responses are shown in Figures 5.4(a), (b) and (c) respectively. Physically, these responses describe the output of an ideal low-pass filter, cutting out all frequencies $\omega > \omega_0$, when the input signal is the rectangular pulse of Figure 5.3. The reader will no doubt note the similarities with the Fourier series discussion of Section 4.2.8 and the continuing existence of the Gibbs phenomenon.

5.2.2 The Fourier transform pair

We note from (5.6) and (5.7) that the Fourier integral (5.8) may be written in the form of the pair of equations

$$F(\mathrm{j}\omega) = \int_{-\infty}^{\infty} f(t)\mathrm{e}^{-\mathrm{j}\omega t}\,\mathrm{d}t \tag{5.13}$$

$$f(t) = \frac{1}{2\pi}\int_{-\infty}^{\infty} F(\mathrm{j}\omega)\mathrm{e}^{\mathrm{j}\omega t}\,\mathrm{d}\omega \tag{5.14}$$

$F(\mathrm{j}\omega)$ as defined by (5.13) is called the **Fourier transform** of $f(t)$, and it provides a frequency-domain representation of the non-periodic function $f(t)$, whenever the integral in (5.13) exists. Note that we have used the notation $F(\mathrm{j}\omega)$ for the Fourier transform of $f(t)$ rather than the alternative $F(\omega)$, which is also in common use. The reason for this choice is a consequence of the relationship between the Fourier and Laplace transforms, which will emerge later in Section 5.4.1. We stress that this is a *choice* that we have made, but the reader should have no difficulty in using either form, provided that once the choice has been made it is then adhered to. Equation (5.14) then provides us with a way of reconstructing $f(t)$ if we know its Fourier transform $F(\mathrm{j}\omega)$.

A word of caution is in order here regarding the scaling factor $1/2\pi$ in (5.14). Although the convention that we have adopted here is fairly standard, some authors associate the factor $1/2\pi$ with (5.13) rather than (5.14), while others associate a factor of $(2\pi)^{-1/2}$ with each of (5.13) and (5.14). In all cases the pair combine to give the Fourier integral (5.8). We could overcome this possible confusion by measuring the frequency in cycles per second or hertz rather than in radians per second, this being achieved using the substitution $f = \omega/2\pi$, where f is in hertz and ω is in radians per second. We have not adopted this approach, since ω is so widely used by engineers.

In line with our notation for Laplace transforms in Chapter 2, we introduce the symbol $\mathscr{F}$ to denote the Fourier transform operator. Then from (5.13) the Fourier transform $\mathscr{F}\{f(t)\}$ of a function $f(t)$ is defined by

$$\mathscr{F}\{f(t)\} = F(\mathrm{j}\omega) = \int_{-\infty}^{\infty} f(t)\mathrm{e}^{-\mathrm{j}\omega t}\,\mathrm{d}t \tag{5.15}$$

whenever the integral exists. Similarly, using (5.14), we define the inverse Fourier transform of $G(j\omega)$ as

$$\mathscr{F}^{-1}\{G(j\omega)\} = g(t) = \frac{1}{2\pi}\int_{-\infty}^{\infty} G(j\omega)e^{j\omega t}\,d\omega \tag{5.16}$$

whenever the integral exists. The relations (5.15) and (5.16) together constitute the **Fourier transform pair**, and they provide a pathway between the time- and frequency-domain representations of a function. Equation (5.15) expresses $f(t)$ in the frequency domain, and is analogous to resolving it into harmonic components with a continuously varying frequency ω. This contrasts with a Fourier series representation of a periodic function, where the resolved frequencies take discrete values.

The conditions for the existence of the Fourier transform $F(j\omega)$ of the function $f(t)$ are Dirichlet's conditions (Theorem 5.1). Corresponding trigonometric forms of the Fourier transform pair may be readily written down from (5.9), (5.10) and (5.11).

EXAMPLE 5.1

Does the function

$$f(t) = 1 \quad (-\infty < t < \infty)$$

have a Fourier transform representation?

Solution Since the area under the curve of $y = f(t)$ $(-\infty < t < \infty)$ is infinite, it follows that $\int_{-\infty}^{\infty}|f(t)|\,dt$ is unbounded, so the conditions of Theorem 5.1 are not satisfied. We can confirm that the Fourier transform does not exist from the definition (5.15). We have

$$\begin{aligned}\int_{-\infty}^{\infty} 1\,e^{-j\omega t}\,dt &= \lim_{\alpha\to\infty}\int_{-\alpha}^{\alpha} e^{-j\omega t}\,dt \\ &= \lim_{\alpha\to\infty}\left[-\frac{1}{j\omega}(e^{-j\omega\alpha} - e^{j\omega\alpha})\right] \\ &= \lim_{\alpha\to\infty}\frac{2\sin\omega\alpha}{\omega}\end{aligned}$$

Since this last limit does not exist, we conclude that $f(t) = 1$ $(-\infty < t < \infty)$ does not have a Fourier transform representation.

It is clear, using integration by parts, that $f(t) = t$ $(-\infty < t < \infty)$ does not have a Fourier transform, nor indeed does $f(t) = t^n$ ($n > 1$, an integer; $-\infty < t < \infty$).

While neither e^{at} nor e^{-at} $(a > 0)$ has a Fourier transform, when we consider the causal signal $f(t) = H(t)\,e^{-at}$ $(a > 0)$, we do obtain a transform.

EXAMPLE 5.2 Find the Fourier transform of the one-sided exponential function

$$f(t) = H(t)\,\mathrm{e}^{-at} \quad (a > 0)$$

where $f(t)$ is the Heaviside unit step function.

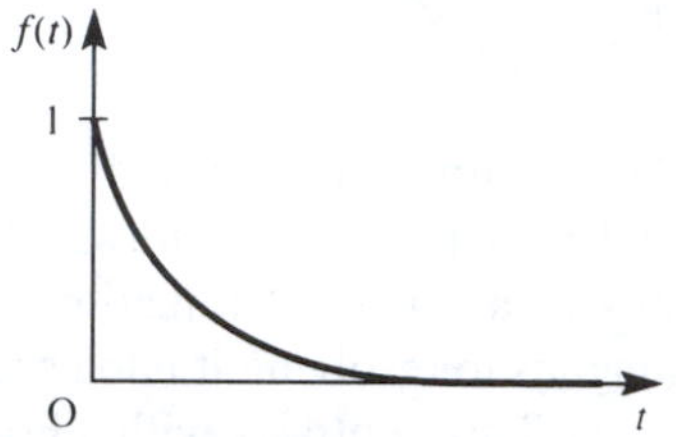

Figure 5.5 The 'one-sided' exponential function $f(t) = H(t)\,\mathrm{e}^{-at}$ $(a > 0)$.

Solution The graph of $f(t)$ is shown in Figure 5.5, and we can show that the area under the graph is bounded. Hence, by Theorem 5.1, a Fourier transform exists. Using the definition (5.15), we have

$$\mathscr{F}\{f(t)\} = \int_{-\infty}^{\infty} H(t)\,\mathrm{e}^{-at}\,\mathrm{e}^{-\mathrm{j}\omega t}\,\mathrm{d}t \quad (a > 0)$$

$$= \int_{0}^{\infty} \mathrm{e}^{-(a+\mathrm{j}\omega)t}\,\mathrm{d}t = \left[-\frac{\mathrm{e}^{-(a+\mathrm{j}\omega)t}}{a + \mathrm{j}\omega}\right]_0^{\infty}$$

so that

$$\mathscr{F}\{H(t)\,\mathrm{e}^{-at}\} = \frac{1}{a + \mathrm{j}\omega} \tag{5.17}$$

EXAMPLE 5.3 Calculate the Fourier transform of the rectangular pulse

$$f(t) = \begin{cases} A & (|t| \leqslant T) \\ 0 & (|t| > T) \end{cases}$$

Solution The graph of $f(t)$ is shown in Figure 5.6, and since the area under it is finite, a Fourier transform exists. From the definition (5.15), we have

$$\mathscr{F}\{f(t)\} = \int_{-T}^{T} A\,\mathrm{e}^{-\mathrm{j}\omega t}\,\mathrm{d}t = \begin{cases} \left[-\dfrac{A}{\mathrm{j}\omega}\mathrm{e}^{-\mathrm{j}\omega t}\right]_{-T}^{T} & \omega \neq 0 \\ 2A & \omega = 0 \end{cases}$$

$$= \frac{2A}{\omega}\sin \omega T = 2AT \operatorname{sinc} \omega T$$

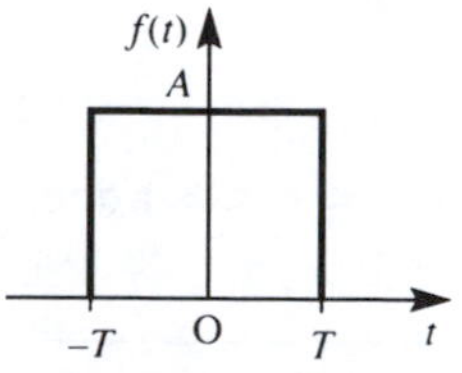

Figure 5.6 The rectangular pulse $f(t) = \begin{cases} A & (|t| \leqslant T) \\ 0 & (|t| > T) \end{cases}$.

where $\operatorname{sinc} x$ is defined, as in Example 4.22, by

$$\operatorname{sinc} x = \begin{cases} \dfrac{\sin x}{x} & (x \neq 0) \\ 1 & (x = 0) \end{cases}$$

$f(t)$	$\mathscr{F}\{f(t)\} = \int_{-\infty}^{\infty} f(t)\mathrm{e}^{-\mathrm{j}\omega t}\,\mathrm{d}t$
$\mathrm{e}^{-at}H(t) \quad (a > 0)$	$\dfrac{1}{a + \mathrm{j}\omega}$
$t\mathrm{e}^{-at}H(t) \quad (a > 0)$	$\dfrac{1}{(a + \mathrm{j}\omega)^2}$
$\begin{cases} A & (\lvert t \rvert \le T) \\ 0 & (\lvert t \rvert > T) \end{cases}$	$2AT \operatorname{sinc} \omega T$
$\mathrm{e}^{-a\lvert t \rvert} \quad (a > 0)$	$\dfrac{2a}{a^2 + \omega^2}$

Figure 5.7 A brief table of Fourier transforms.

By direct use of the definition (5.15), we can, as in Examples 5.2 and 5.3, determine the Fourier transforms of some standard functions. A brief table of transforms in given in Figure 5.7.

5.2.3 The continuous Fourier spectra

From Figure 5.7, it is clear that Fourier transforms are generally complex-valued functions of the real frequency variable ω. If $\mathscr{F}\{f(t)\} = F(\mathrm{j}\omega)$ is the Fourier transform of the signal $f(t)$ then $F(\mathrm{j}\omega)$ is also known as the **(complex) frequency spectrum** of $f(t)$. Writing $F(\mathrm{j}\omega)$ in the exponential form

$$F(\mathrm{j}\omega) = |F(\mathrm{j}\omega)|\mathrm{e}^{\mathrm{j}\arg F(\mathrm{j}\omega)}$$

plots of $|F(\mathrm{j}\omega)|$ and $\arg F(\mathrm{j}\omega)$, which are both real-valued functions of ω, are called the amplitude and phase spectra respectively of the signal $f(t)$. These two spectra represent the **frequency-domain portrait** of the signal $f(t)$. In contrast to the situation when $f(t)$ was periodic, where (as shown in Section 4.6.3) the amplitude and phase spectra were defined only at discrete values of ω, we now see that both spectra are defined for all values of the continuous variable ω.

EXAMPLE 5.4 Determine the amplitude and phase spectra of the causal signal

$$f(t) = \mathrm{e}^{-at}H(t) \quad (a > 0)$$

and plot their graphs.

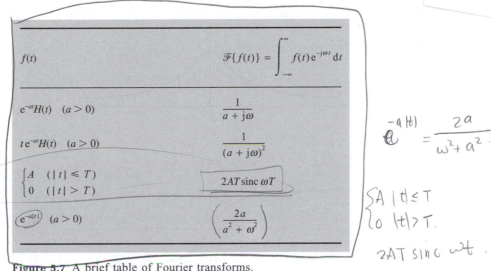

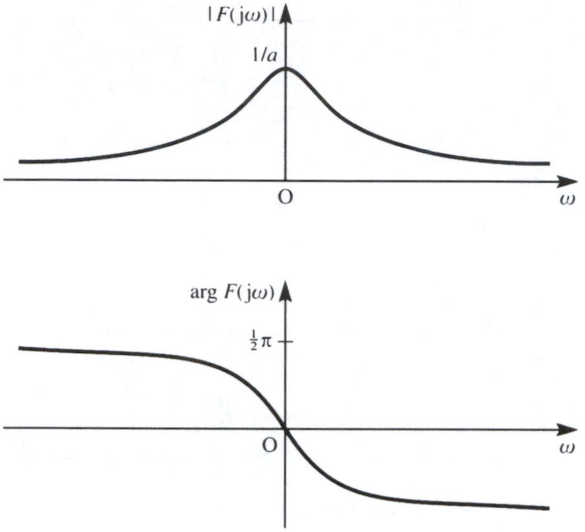

Figure 5.8 Amplitude and phase spectra of the one-sided exponential function $f(t) = e^{-at}H(t)$ $(a > 0)$.

Solution From (5.17),

$$\mathscr{F}\{f(t)\} = F(j\omega) = \frac{1}{a + j\omega}$$

Thus the amplitude and argument of $F(j\omega)$ are

$$|F(j\omega)| = \frac{1}{\sqrt{(a^2 + \omega^2)}} \quad \textbf{(5.18)}$$

$$\arg F(j\omega) = \tan^{-1}(1) - \tan^{-1}\left(\frac{\omega}{a}\right) = -\tan^{-1}\left(\frac{\omega}{a}\right) \quad \textbf{(5.19)}$$

These are the amplitude and phase spectra of $f(t)$, and are plotted in Figure 5.8.

Generally, as we have observed, the Fourier transform and thus the frequency spectrum are complex-valued quantities. In some cases, as for instance in Example 5.3, the spectrum is purely real. In Example 5.3 we found that the transform of the pulse illustrated in Figure 5.6 was

$$F(j\omega) = 2AT \operatorname{sinc} \omega T$$

where

$$\operatorname{sinc} \omega T = \begin{cases} \dfrac{\sin \omega T}{\omega T} & (\omega \neq 0) \\ 1 & (\omega = 0) \end{cases}$$

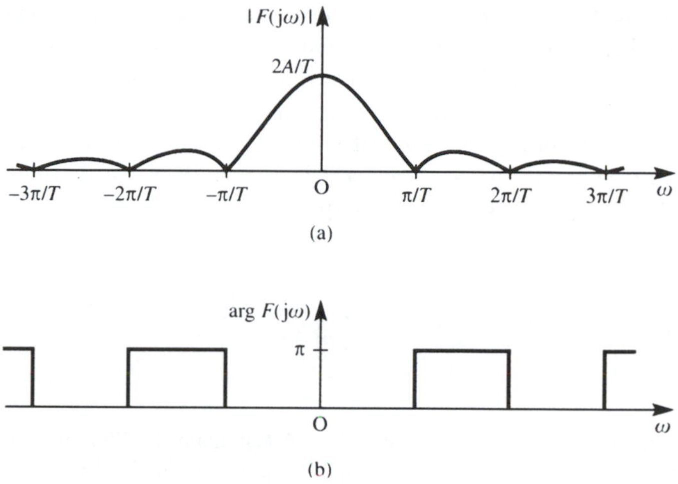

Figure 5.9 (a) Amplitude and (b) spectra of the pulse $f(t) = \begin{cases} A & (|t| \leqslant T) \\ 0 & (|t| > T) \end{cases}$.

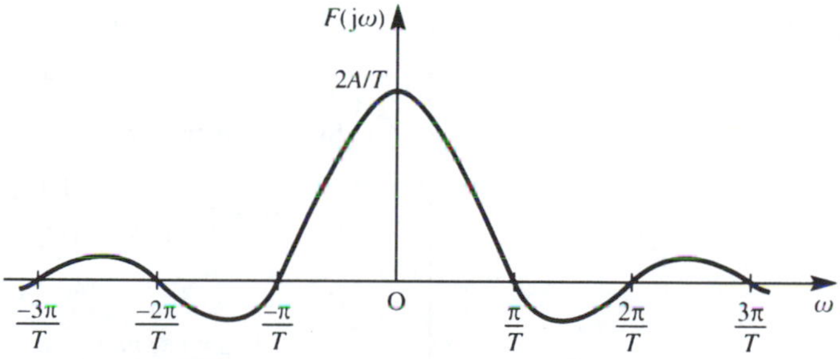

Figure 5.10 Frequency spectrum (real-valued) of the pulse $f(t) = \begin{cases} A & (|t| \leqslant T) \\ 0 & (|t| > T) \end{cases}$.

is an even function of ω, taking both positive and negative values. In this case the amplitude and phase spectra are given by

$$|F(\mathrm{j}\omega)| = 2AT\,|\mathrm{sinc}\;\omega T| \tag{5.20}$$

$$\arg F(\mathrm{j}\omega) = \begin{cases} 0 & (\mathrm{sinc}\;\omega T \geqslant 0) \\ \pi & (\mathrm{sinc}\;\omega T < 0) \end{cases} \tag{5.21}$$

with corresponding graphs shown in Figure 5.9.

In fact, when the Fourier transform is a purely real-valued function, we can plot all the information on a single frequency spectrum of $F(\mathrm{j}\omega)$ versus ω. For the rectangular pulse of Figure 5.6 the resulting graph is shown in Figure 5.10.

From Figure 5.7, we can see that the Fourier transforms discussed so far have two properties in common. First, the amplitude spectra are even functions of the frequency variable ω. This is always the case when the time signal $f(t)$ is real, that is, loosely speaking, a consequence of the fact that we have decomposed, or analysed $f(t)$, relative to complex exponentials rather than real-valued sines and cosines. The second point to note is that all the amplitude spectra decrease rapidly as ω increases. This means that most of the information concerning the 'shape' of the signal $f(t)$ is contained in a fairly small interval of the frequency axis around $\omega = 0$. From another point of view, we see that a device capable of passing signals of frequencies up to about $\omega = 3\pi/T$ would pass a reasonably accurate version of the rectangular pulse of Example 5.3.

5.2.4 Exercises

1 Calculate the Fourier transform of the two-sided exponential pulse given by

$$f(t) = \begin{cases} e^{at} & (t \leqslant 0) \\ e^{-at} & (t > 0) \end{cases} \quad (a > 0)$$

2 Determine the Fourier transform of the 'on–off' pulse shown in Figure 5.11.

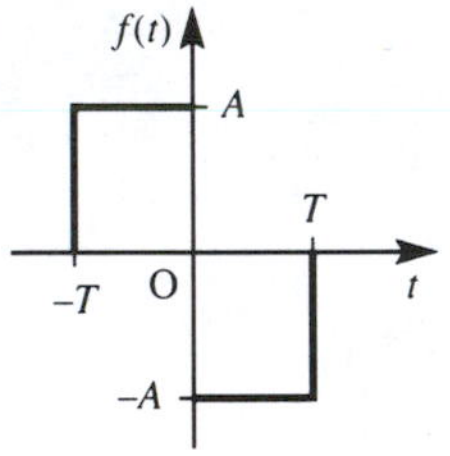

Figure 5.11 The 'on–off' pulse.

3 A triangular pulse is defined by

$$f(t) = \begin{cases} (A/T)t + A & (-T \leqslant t \leqslant 0) \\ (-A/T)t + A & (0 < t \leqslant T) \end{cases}$$

Sketch $f(t)$ and determine its Fourier transform. What is the relationship between this pulse and that of Exercise 2?

4 Determine the Fourier transforms of

$$f(t) = \begin{cases} 2K & (|t| \leqslant 2) \\ 0 & (|t| > 2) \end{cases}$$

$$g(t) = \begin{cases} K & (|t| \leqslant 1) \\ 0 & (|t| > 1) \end{cases}$$

Sketch the function $h(t) = f(t) - g(t)$ and determine its Fourier transform.

5 Calculate the Fourier transform of the 'off–on–off' pulse $f(t)$ defined by

$$f(t) = \begin{cases} 0 & (t < -2) \\ -1 & (-2 \leqslant t < -1) \\ 1 & (-1 \leqslant t \leqslant 1) \\ -1 & (1 < t \leqslant 2) \\ 0 & (t > 2) \end{cases}$$

6 Show that the Fourier transform of

$$f(t) = \begin{cases} \sin at & (|t| \leqslant \pi/a) \\ 0 & (|t| > \pi/a) \end{cases}$$

is

$$\frac{j2a \sin(\pi\omega/a)}{\omega^2 - a^2}$$

7 Calculate the Fourier transform of

$$f(t) = e^{-at} \sin \omega_0 t \, H(t)$$

8 Based on (5.10) and (5.11), define the **Fourier sine transform** as

$$F_s(x) = \int_0^\infty f(t) \sin xt \, dt$$

and the **Fourier cosine transform** as

$$F_c(x) = \int_0^\infty f(t) \cos xt \, dt$$

Show that

$$f(t) = \begin{cases} 0 & (t < 0) \\ \cos at & (0 \leqslant t \leqslant a) \\ 0 & (t > a) \end{cases}$$

has Fourier cosine transform

$$\frac{1}{2}\left[\frac{\sin(1+x)a}{1+x}+\frac{\sin(1-x)a}{1-x}\right]$$

9 Show that the Fourier sine and cosine transforms of

$$f(t)=\begin{cases}0 & (t<0)\\ 1 & (0\leqslant t\leqslant a)\\ 0 & (t>a)\end{cases}$$

are

$$\frac{1-\cos xa}{x},\quad \frac{\sin xa}{x}$$

respectively.

10 Find the sine and cosine transforms of $f(t) = \mathrm{e}^{-at}H(t)$ $(a > 0)$.

5.3 Properties of the Fourier transform

In this section we establish some of the properties of the Fourier transform that allow its use as a practical tool in system analysis and design.

5.3.1 The linearity property

Linearity is a fundamental property of the Fourier transform, and may be stated as follows.

If $f(t)$ and $g(t)$ are functions having Fourier transforms $F(\mathrm{j}\omega)$ and $G(\mathrm{j}\omega)$ respectively, and if α and β are constants then

$$\mathcal{F}\{\alpha f(t)+\beta g(t)\}=\alpha\mathcal{F}\{f(t)\}+\beta\mathcal{F}\{g(t)\}=\alpha F(\mathrm{j}\omega)+\beta G(\mathrm{j}\omega) \tag{5.22}$$

As a consequence of this, we say that the Fourier transform operator $\mathcal{F}$ is a **linear operator**. The proof of this property follows readily from the definition (5.15), since

$$\begin{aligned}\mathcal{F}\{\alpha f(t)+\beta g(t)\}&=\int_{-\infty}^{\infty}[\alpha f(t)+\beta g(t)]\mathrm{e}^{-\mathrm{j}\omega t}\,\mathrm{d}t\\ &=\alpha\int_{-\infty}^{\infty}f(t)\mathrm{e}^{-\mathrm{j}\omega t}\,\mathrm{d}t+\beta\int_{-\infty}^{\infty}g(t)\mathrm{e}^{-\mathrm{j}\omega t}\,\mathrm{d}t\\ &=\alpha F(\mathrm{j}\omega)+\beta G(\mathrm{j}\omega)\end{aligned}$$

Clearly the linearity property also applies to the inverse transform operator $\mathcal{F}^{-1}$.

5.3.2 Time-differentiation property

If the function $f(t)$ has a Fourier transform $F(j\omega)$ then, by (5.16),

$$f(t) = \frac{1}{2\pi}\int_{-\infty}^{\infty} F(j\omega)e^{j\omega t}\,d\omega$$

Differentiating with respect to t gives

$$\frac{df}{dt} = \frac{1}{2\pi}\int_{-\infty}^{\infty} \frac{\partial}{\partial t}[F(j\omega)e^{j\omega t}]\,d\omega = \frac{1}{2\pi}\int_{-\infty}^{\infty} (j\omega)F(j\omega)e^{j\omega t}\,d\omega$$

implying that the time signal df/dt is the inverse Fourier transform of $(j\omega)F(j\omega)$. In other words

$$\mathscr{F}\left\{\frac{df}{dt}\right\} = (j\omega)F(j\omega)$$

Repeating the argument n times, it follows that

$$\mathscr{F}\left\{\frac{d^n f}{dt^n}\right\} = (j\omega)^n F(j\omega) \tag{5.23}$$

The result (5.23) is referred to as the **time-differentiation property**, and may be used to obtain frequency-domain representations of differential equations.

EXAMPLE 5.5 Show that if the time signals $y(t)$ and $u(t)$ have Fourier transforms $Y(j\omega)$ and $U(j\omega)$ respectively, and if

$$\frac{d^2y(t)}{dt^2} + 3\frac{dy(t)}{dt} + 7y(t) = 3\frac{du(t)}{dt} + 2u(t) \tag{5.24}$$

then $Y(j\omega) = G(j\omega)U(j\omega)$ for some function $G(j\omega)$.

Solution Taking Fourier transforms throughout in (5.24), we have

$$\mathscr{F}\left\{\frac{d^2y(t)}{dt^2} + 3\frac{dy(t)}{dt} + 7y(t)\right\} = \mathscr{F}\left\{3\frac{du(t)}{dt} + 2u(t)\right\}$$

which, on using the linearity property (5.22), reduces to

$$\mathscr{F}\left\{\frac{d^2y(t)}{dt^2}\right\} + 3\mathscr{F}\left\{\frac{dy(t)}{dt}\right\} + 7\mathscr{F}\{y(t)\} = 3\mathscr{F}\left\{\frac{du(t)}{dt}\right\} + 2\mathscr{F}\{u(t)\}$$

Then, from (5.23),

$$(j\omega)^2Y(j\omega) + 3(j\omega)Y(j\omega) + 7Y(j\omega) = 3(j\omega)U(j\omega) + 2U(j\omega)$$

that is,

$$(-\omega^2 + j3\omega + 7)Y(j\omega) = (j3\omega + 2)U(j\omega)$$

giving

$$Y(j\omega) = G(j\omega)U(j\omega)$$

where

$$G(j\omega) = \frac{2 + j3\omega}{7 - \omega^2 + j3\omega}$$

The reader may at this stage be fearing that we are about to propose yet *another* method for solving differential equations. This is not the idea! Rather, we shall show that the Fourier transform provides an essential tool for the analysis (and synthesis) of linear systems from the viewpoint of the frequency domain.

5.3.3 Time-shift property

If a function $f(t)$ has Fourier transform $F(j\omega)$ then what is the Fourier transform of the shifted version of $f(\tau)$, defined by $g(t) = f(t - \tau)$? From the definition (5.15),

$$\mathscr{F}\{g(t)\} = \int_{-\infty}^{\infty} g(t)e^{-j\omega t}\,dt = \int_{-\infty}^{\infty} f(t - \tau)e^{-j\omega t}\,dt$$

Making the substitution $x = t - \tau$, we have

$$\mathscr{F}\{g(t)\} = \int_{-\infty}^{\infty} f(x)e^{-j\omega(x+\tau)}\,dx = e^{-j\omega\tau}\int_{-\infty}^{\infty} f(x)e^{-j\omega x}\,dx = e^{-j\omega\tau}F(j\omega)$$

that is,

$$\mathscr{F}\{f(t - \tau)\} = e^{-j\omega\tau}F(j\omega) \tag{5.25}$$

The result (5.25) is known as the **time-shift property**, and implies that delaying a signal by a time τ causes its Fourier transform to be multiplied by $e^{-j\omega\tau}$.

Since

$$|e^{-j\omega\tau}| = |\cos\omega\tau - j\sin\omega\tau| = |\sqrt{(\cos^2\omega\tau + \sin^2\omega\tau)}| = 1$$

we have

$$|e^{-j\omega\tau}F(j\omega)| = |F(j\omega)|$$

indicating that the amplitude spectrum of $f(t - \tau)$ is identical with that of $f(t)$. However,

$$\arg[e^{-j\omega t}F(j\omega)] = \arg F(j\omega) - \arg e^{j\omega\tau} = \arg F(j\omega) - \omega\tau$$

indicating that each frequency component is shifted by an amount proportional to its frequency ω.

EXAMPLE 5.6 Determine the Fourier transform of the rectangular pulse $f(t)$ shown in Figure 5.12.

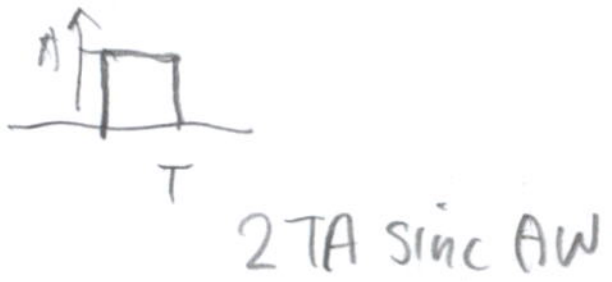

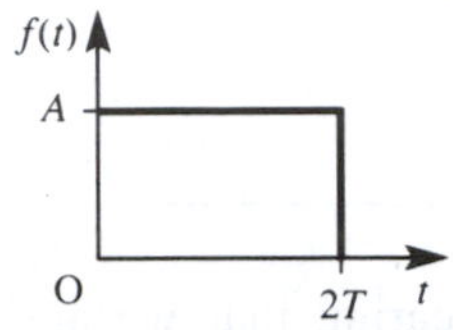

Figure 5.12 Rectangular pulse of Example 5.6.

Solution This is just the pulse of Example 5.3 (shown in Figure 5.6), delayed by T. The pulse of Example 5.3 had a Fourier transform $2AT \operatorname{sinc} \omega T$, and so, using the shift property (5.25) with $\tau = T$, we have

$$\mathcal{F}\{f(t)\} = F(\mathrm{j}\omega) = \mathrm{e}^{-\mathrm{j}\omega T} 2AT \operatorname{sinc} \omega T = 2AT\, \mathrm{e}^{-\mathrm{j}\omega T} \operatorname{sinc} \omega T$$

5.3.4 Frequency-shift property

Suppose that a function $f(t)$ has Fourier transform $F(\mathrm{j}\omega)$. Then, from the definition (5.15), the Fourier transform of the related function $g(t) = \mathrm{e}^{\mathrm{j}\omega_0 t} f(t)$ is

$$\mathcal{F}\{g(t)\} = \int_{-\infty}^{\infty} \mathrm{e}^{\mathrm{j}\omega_0 t} f(t) \mathrm{e}^{-\mathrm{j}\omega t}\, \mathrm{d}t = \int_{-\infty}^{\infty} f(t) \mathrm{e}^{-\mathrm{j}(\omega - \omega_0)t}\, \mathrm{d}t$$

$$= \int_{-\infty}^{\infty} f(t) \mathrm{e}^{-\mathrm{j}\tilde{\omega} t}\, \mathrm{d}t, \quad \text{where } \tilde{\omega} = \omega - \omega_0$$

$$= F(\mathrm{j}\tilde{\omega}), \quad \text{by definition}$$

Thus

$$\mathcal{F}\{\mathrm{e}^{\mathrm{j}\omega_0 t} f(t)\} = F(\mathrm{j}(\omega - \omega_0)) \tag{5.26}$$

The result (5.26) is known as the **frequency-shift property**, and indicates that multiplication by $\mathrm{e}^{\mathrm{j}\omega_0 t}$ simply shifts the spectrum of $f(t)$ so that it is centred on the point $\omega = \omega_0$ in the frequency domain. This phenomena is the mathematical foundation for the process of **modulation** in communication theory, illustrated in Example 5.7.

EXAMPLE 5.7 Determine the frequency spectrum of the signal $g(t) = f(t)\cos\omega_c t$.

Solution Since $\cos\omega_c t = \frac{1}{2}(e^{j\omega_c t} + e^{-j\omega_c t})$, it follows, using the linearity property (5.22), that

$$\begin{aligned}\mathcal{F}\{g(t)\} &= \mathcal{F}\{\tfrac{1}{2}f(t)(e^{j\omega_c t} + e^{-j\omega_c t})\}\\ &= \tfrac{1}{2}\mathcal{F}\{f(t)e^{j\omega_c t}\} + \tfrac{1}{2}\mathcal{F}\{f(t)e^{-j\omega_c t}\}\end{aligned}$$

If $\mathcal{F}\{f(t)\} = F(j\omega)$ then, using (5.26),

$$\mathcal{F}\{f(t)\cos\omega_c t\} = \mathcal{F}\{g(t)\} = \tfrac{1}{2}F(j(\omega - \omega_c)) + \tfrac{1}{2}F(j(\omega + \omega_c))$$

The effect of multiplying the signal $f(t)$ by the **carrier signal** $\cos\omega_c t$ is thus to produce a signal whose spectrum consists of two (scaled) versions of $F(j\omega)$, the spectrum of $f(t)$; one centred on $\omega = \omega_c$ and the other on $\omega = -\omega_c$. The carrier signal $\cos\omega_c t$ is said to be modulated by the signal $f(t)$.

Demodulation is considered in Exercise 5, Section 5.9, and the ideas of modulation and demodulation are developed in Section 5.8.

5.3.5 The symmetry property

From the definition of the transform pair (5.15) and (5.16) it is apparent that there is some symmetry of structure in relation to the variables t and ω. We can establish the exact form of this symmetry as follows. From (5.16),

$$f(t) = \frac{1}{2\pi}\int_{-\infty}^{\infty} F(j\omega)e^{j\omega t}\,d\omega$$

or, equivalently, by changing the 'dummy' variable in the integration,

$$2\pi f(t) = \int_{-\infty}^{\infty} F(jy)e^{jyt}\,dy$$

so that

$$2\pi f(-t) = \int_{-\infty}^{\infty} F(jy)e^{-jyt}\,dy$$

or, on replacing t by ω,

$$2\pi f(-\omega) = \int_{-\infty}^{\infty} F(jy)e^{-jy\omega}\,dy \qquad \textbf{(5.27)}$$

The right-hand side of (5.27) is simply the definition (5.15) of the Fourier transform of $F(jt)$, with the integration variable t replaced by y. We therefore conclude that

$$\mathcal{F}\{F(jt)\} = 2\pi f(-\omega) \qquad \textbf{(5.28a)}$$

given that

$$\mathcal{F}\{f(t)\} = F(j\omega) \qquad \textbf{(5.28b)}$$

What (5.28) tells us is that if $f(t)$ and $F(j\omega)$ form a Fourier transform pair then $F(jt)$ and $2\pi f(-\omega)$ also form a Fourier transform pair. This property is referred to as the **symmetry property of Fourier transforms**. It is also sometimes referred to as the **duality property**.

EXAMPLE 5.8 Determine the Fourier transform of the signal

$$g(t) = C \operatorname{sinc} at = \begin{cases} \dfrac{C \sin at}{at} & (t \neq 0) \\ C & (t = 0) \end{cases} \tag{5.29}$$

Solution From Example 5.3, we know that if

$$f(t) = \begin{cases} A & (|t| \leqslant T) \\ 0 & (|t| > T) \end{cases} \tag{5.30}$$

then

$$\mathscr{F}\{f(t)\} = F(j\omega) = 2AT \operatorname{sinc} \omega T$$

Thus, by the symmetry property (5.28), $F(jt)$ and $2\pi f(-\omega)$ are also a Fourier transform pair. In this case

$$F(jt) = 2AT \operatorname{sinc} tT$$

and so, choosing $T = a$ and $A = C/2a$ to correspond to (5.29), we see that

$$F(jt) = C \operatorname{sinc} at = g(t)$$

has Fourier transform $2\pi f(-\omega)$. Rewriting (5.30), we find that, since $|\omega| = |-\omega|$,

$$\mathscr{F}\{C \operatorname{sinc} at\} = \begin{cases} 2\pi C/2a & (|\omega| \leqslant a) \\ 0 & (|\omega| > a) \end{cases} = \begin{cases} \pi C/a & (|\omega| \leqslant a) \\ 0 & (|\omega| > a) \end{cases}$$

A graph of $g(t)$ and its Fourier transform $G(j\omega) = 2\pi f(-\omega)$ is shown in Figure 5.13.

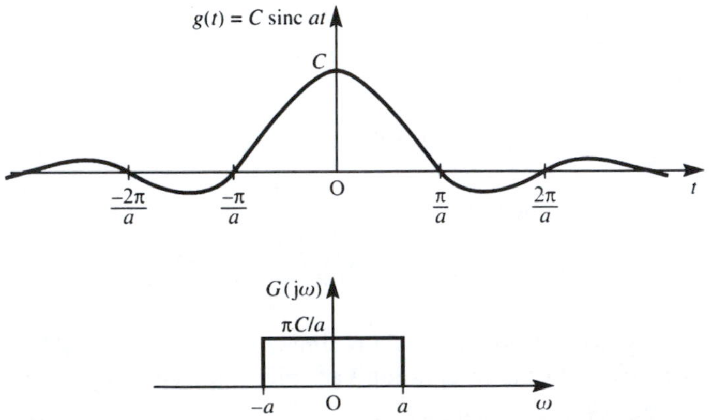

Figure 5.13 The Fourier transform pair $g(t)$ and $G(j\omega)$ of Example 5.8.

5.3.6 Exercises

11 Use the linearity property to verify the result in Exercise 4.

12 If $y(t)$ and $u(t)$ are signals with Fourier transforms $Y(j\omega)$ and $U(j\omega)$ respectively, and

$$\frac{d^2y(t)}{dt^2} + 3\frac{dy(t)}{dt} + y(t) = u(t)$$

show that $Y(j\omega) = H(j\omega)U(j\omega)$ for some function $H(j\omega)$. What is $H(j\omega)$?

13 Use the time-shift property to calculate the Fourier transform of the double pulse defined by

$$f(t) = \begin{cases} 1 & (1 \leq |t| \leq 2) \\ 0 & (\text{otherwise}) \end{cases}$$

14 Calculate the Fourier transform of the windowed cosine function

$$f(t) = \cos\omega_0 t[H(t + \tfrac{1}{2}T) - H(t - \tfrac{1}{2}T)]$$

15 Find the Fourier transform of the shifted form of the windowed cosine function

$$g(t) = \cos\omega_0 t[H(t) - H(t - T)]$$

16 Calculate the Fourier transform of the windowed sine function

$$f(t) = \sin 2t[H(t + 1) - H(t - 1)]$$

5.4 The frequency response

In this section we first consider the relationship between the Fourier and Laplace transforms, and then proceed to consider the frequency response in terms of the Fourier transform.

5.4.1 Relationship between Fourier and Laplace transforms

The differences between the Fourier and Laplace transforms are quite subtle. At first glance it appears that to obtain the Fourier transform from the Laplace transform we merely write $j\omega$ for s, and that the difference ends there. This is true in some cases, but not in all. Strictly, the Fourier and Laplace transforms are distinct, and neither is a generalization of the other.

Writing down the defining integrals, we have

The Fourier transform

$$\mathscr{F}\{f(t)\} = \int_{-\infty}^{\infty} f(t)e^{-j\omega t}\,dt \qquad (5.31)$$

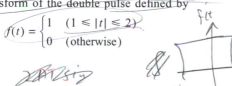

The bilateral Laplace transform

$$\mathscr{L}_B\{f(t)\} = \int_{-\infty}^{\infty} f(t)\mathrm{e}^{-st}\,\mathrm{d}t \tag{5.32}$$

The unilateral Laplace transform

$$\mathscr{L}\{f(t)\} = \int_{0-}^{\infty} f(t)\mathrm{e}^{-st}\,\mathrm{d}t \tag{5.33}$$

There is an obvious structural similarity between (5.31) and (5.32), while the connection with (5.33) is not so clear in view of the lower limit of integration. In the Laplace transform definitions recall that s is a complex variable, and may be written as

$$s = \sigma + \mathrm{j}\omega \tag{5.34}$$

where σ and ω are real variables. We can then interpret (5.31), the Fourier transform of $f(t)$, as a special case of (5.32), when $\sigma = 0$, provided that the Laplace transform exists when $\sigma = 0$, or equivalently when $s = \mathrm{j}\omega$ (that is, s describes the imaginary axis in the s plane). If we restrict our attention to causal functions, that is, functions (or signals) that are zero whenever $t < 0$, the bilateral Laplace transform (5.32) is identical with the unilateral Laplace transform (5.33). The Fourier transform can thus be regarded as a special case of the unilateral Laplace transform for causal functions, provided again that the unilateral Laplace transform exists on the imaginary axis $s = \mathrm{j}\omega$.

The next part of the story is concerned with a class of time signals $f(t)$ whose Laplace transforms do exist on the imaginary axis $s = \mathrm{j}\omega$. Recall from (2.71) that a causal linear time-invariant system with Laplace transfer function $G(s)$ has an impulse response $h(t)$ given by

$$h(t) = \mathscr{L}^{-1}\{G(s)\} = g(t)H(t), \quad \text{say} \tag{5.35}$$

Furthermore, if the system is stable then all the poles of $G(s)$ are in the left half-plane, implying that $g(t)H(t) \to 0$ as $t \to \infty$. Let the pole locations of $G(s)$ be

$$p_1, p_2, \dots, p_n$$

where

$$p_k = -a_k^2 + \mathrm{j}b_k$$

in which a_k, b_k are real and $a_k \neq 0$ for $k = 1, 2, \dots, n$. Examples of such poles are illustrated in Figure 5.14, where we have assumed that $G(s)$ is the transfer function of a real system so that poles that do not lie on the real axis occur in conjugate pairs. As indicated in Section 2.2.3, the Laplace transfer function $G(s)$ will exist in the shaded region of Figure 5.14 defined by

$$\mathrm{Re}(s) > -c^2$$

where $-c^2$ is the abscissa of convergence and is such that

$$0 < c^2 < \min a_k^2$$

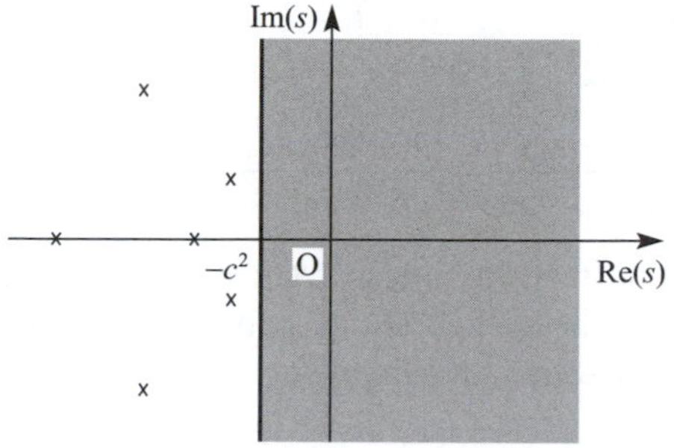

Figure 5.14 Pole locations for $G(s)$ and the region of existence of $G(s)$.

The important conclusion is that for such systems $G(s)$ always exists on the imaginary axis $s = j\omega$, and so $h(t) = g(t)H(t)$ always has a Fourier transform. In other words, we have demonstrated that the impulse response function $h(t)$ of a *stable causal*, linear time-invariant system always has a Fourier transform. Moreover, we have shown that this can be found by evaluating the Laplace transform on the imaginary axis, that is, by putting $s = j\omega$ in the Laplace transform. We have thus established that Fourier transforms exist for a significant class of useful signals; this knowledge will be used in Section 5.4.2.

EXAMPLE 5.9

Which of the following causal time-invariant systems have impulse responses that possess Fourier transforms? Find the latter when they exist.

(a) $\dfrac{d^2y(t)}{dt^2} + 3\dfrac{dy(t)}{dt} + 2y(t) = u(t)$

(b) $\dfrac{d^2y(t)}{dt^2} + \omega^2 y(t) = u(t)$

(c) $\dfrac{d^2y(t)}{dt^2} + \dfrac{dy(t)}{dt} + y(t) = 2u(t) + \dfrac{du(t)}{dt}$

Solution Assuming that the systems are initially in a quiescent state when $t < 0$, taking Laplace transforms gives

(a) $Y(s) = \dfrac{1}{s^2 + 3s + 2}U(s) = G_1(s)U(s)$

(b) $Y(s) = \dfrac{1}{s^2 + \omega^2}U(s) = G_2(s)U(s)$

(c) $Y(s) = \dfrac{s + 2}{s^2 + s + 1}U(s) = G_3(s)U(s)$

In case (a) the poles of $G_1(s)$ are at $s = -1$ and $s = -2$, so the system is stable and the impulse response has a Fourier transform given by

$$G_1(j\omega) = \left.\frac{1}{s^2 + 3s + 2}\right|_{s=j\omega} = \frac{1}{2 - \omega^2 + j3\omega}$$

$$= \frac{2 - \omega^2 - j3\omega}{(2 - \omega^2)^2 + 9\omega^2} = \frac{(2 - \omega^2) - j3\omega}{\omega^4 + 5\omega^2 + 4}$$

In case (b) we find that the poles of $G_2(s)$ are at $s = j\omega$ and $s = -j\omega$, that is, on the imaginary axis. The system is not stable (notice that the impulse response does not decay to zero), and the impulse response does not possess a Fourier transform.

In case (c) the poles of $G_3(s)$ are at $s = -\frac{1}{2} + j\frac{1}{2}\sqrt{3}$ and $s = -\frac{1}{2} - j\frac{1}{2}\sqrt{3}$. Since these are in the left half-plane, $\mathrm{Re}(s) < 0$, we conclude that the system is stable. The Fourier transform of the impulse response is then

$$G_3(j\omega) = \frac{2 + j\omega}{1 - \omega^2 + j\omega}$$

5.4.2 The frequency response

For a linear time-invariant system, initially in a quiescent state, having a Laplace transfer function $G(s)$, the response $y(t)$ to an input $u(t)$ is given in (2.66) as

$$Y(s) = G(s)U(s) \tag{5.36}$$

where $Y(s)$ and $U(s)$ are the Laplace transforms of $y(t)$ and $u(t)$ respectively. In Section 2.7 we saw that, subject to the system being stable, the steady-state response $y_{ss}(t)$ to a sinusoidal input $u(t) = A\sin\omega t$ is given by (2.86) as

$$y_{ss}(t) = A|G(j\omega)|\sin[\omega t + \arg G(j\omega)] \tag{5.37}$$

That is, the steady-state response is also sinusoidal, with the same frequency as the input signal but having an amplitude gain $|G(j\omega)|$ and a phase shift $\arg G(j\omega)$.

More generally, we could have taken the input to be the complex sinusoidal signal

$$u(t) = A\,e^{j\omega t}$$

and, subject to the stability requirement, showed that the steady-state response is

$$y_{ss}(t) = AG(j\omega)\,e^{j\omega t} \tag{5.38}$$

or

$$y_{ss}(t) = A|G(j\omega)|\,e^{j[\omega t + \arg G(j\omega)]} \tag{5.39}$$

As before, $|G(j\omega)|$ and $\arg G(j\omega)$ are called the amplitude gain and phase shift respectively. Both are functions of the real frequency variable ω, and their plots versus ω constitute the **system frequency response**, which, as we saw in

Section 2.7, characterizes the behaviour of the system. Note that taking imaginary parts throughout in (5.39) leads to the sinusoidal response (5.37).

We note that the steady-state response (5.38) is simply the input signal $Ae^{j\omega t}$ multiplied by the Fourier transform $G(j\omega)$ of the system's impulse response. Consequently $G(j\omega)$ is called the **frequency transfer function** of the system. Therefore if the system represented in (5.36) is stable, so that $G(j\omega)$ exists as the Fourier transform of its impulse response, and the input $u(t) = \mathscr{L}^{-1}\{U(s)\}$ has a Fourier transform $U(j\omega)$ then we may represent the system in terms of the frequency transfer function as

$$Y(j\omega) = G(j\omega)U(j\omega) \tag{5.40}$$

Equation (5.40) thus determines the Fourier transform of the system output, and can be used to determine the frequency spectrum of the output from that of the input. This means that both the amplitude and phase spectra of the output are available, since

$$|Y(j\omega)| = |G(j\omega)|\,|U(j\omega)| \tag{5.41a}$$

$$\arg Y(j\omega) = \arg G(j\omega) + \arg U(j\omega) \tag{5.41b}$$

We shall now consider an example that will draw together both these and some earlier ideas which serve to illustrate the relevance of this material in the communications industry.

EXAMPLE 5.10 A signal $f(t)$ consists of two components:

(a) a symmetric rectangular pulse of duration 2π (see Example 5.3) and

(b) a second pulse, also of duration 2π (that is a copy of (a)), modulating a signal with carrier frequency $\omega_0 = 3$ (the process of modulation was introduced in Section 5.3.4).

Write down an expression for $f(t)$ and illustrate its amplitude spectrum. Describe the amplitude spectrum of the output signal if $f(t)$ is applied to a stable causal system with a Laplace transfer function

$$G(s) = \frac{1}{s^2 + \sqrt{2}s + 1}$$

Solution Denoting the pulse of Example 5.3, with $T = \pi$, by $P_\pi(t)$, and noting the use of the term 'carrier signal' in Example 5.7, we have

$$f(t) = P_\pi(t) + (\cos 3t)P_\pi(t)$$

From Example 5.3,

$$\mathscr{F}\{P_\pi(t)\} = 2\pi \operatorname{sinc} \omega\pi$$

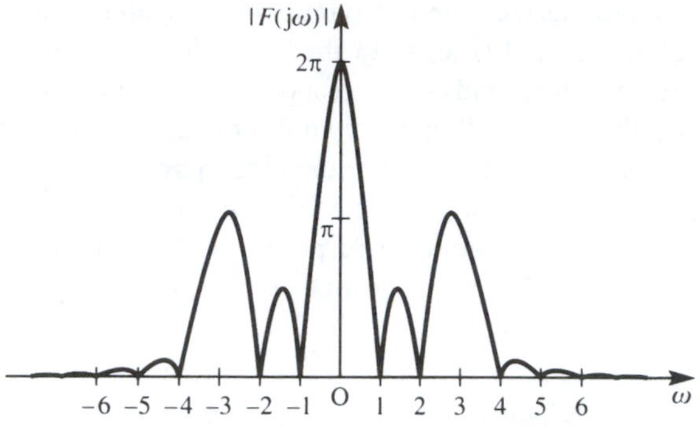

Figure 5.15 Amplitude spectrum of the signal $P_\pi(t) + (\cos 3t)P_\pi(t)$.

so, using the result of Example 5.7, we have

$$\mathscr{F}\{f(t)\} = F(j\omega) = 2\pi \operatorname{sinc} \omega\pi + \tfrac{1}{2}[2\pi \operatorname{sinc}(\omega - 3)\pi + 2\pi \operatorname{sinc}(\omega + 3)\pi]$$

The corresponding amplitude spectrum obtained by plotting $|F(j\omega)|$ versus ω is illustrated in Figure 5.15.

Since the system with transfer function

$$G(s) = \frac{1}{s^2 + \sqrt{2}s + 1}$$

is stable and causal, it has a frequency transfer function

$$G(j\omega) = \frac{1}{1 - \omega^2 + j\sqrt{2}\omega}$$

so that its amplitude gain is

$$|G(j\omega)| = \frac{1}{\sqrt{(\omega^4 + 1)}}$$

The amplitude spectrum of the output signal $|Y(j\omega)|$ when the input is $f(t)$ is then obtained from (5.41a) as the product of $|F(j\omega)|$ and $|G(j\omega)|$. Plots of both the amplitude gain spectrum $|G(j\omega)|$ and the output amplitude spectrum $|Y(j\omega)|$ are shown in Figures 5.16(a) and (b) respectively. Note from Figure 5.16(b) that we have a reasonably good copy of the amplitude spectrum of $P_\pi(t)$ (see Figure 5.8a). However, the second element of $f(t)$ has effectively vanished. Our system has 'filtered out' this latter component while 'passing' an almost intact version of the first. Examination of the time-domain response would show that the first component does in fact experience some 'smoothing', which, roughly speaking, consists of rounding of the sharp edges. The system considered here is a second-order 'low-pass' Butterworth filter (introduced in Section 3.8.1).

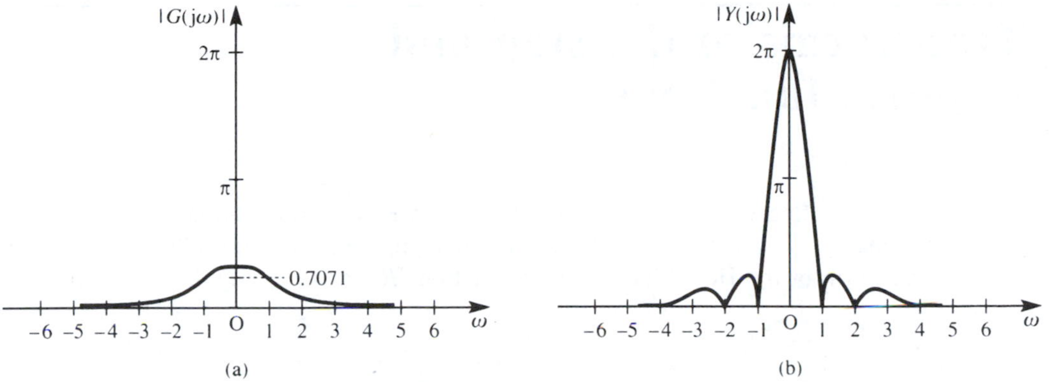

Figure 5.16 (a) Amplitude gain spectrum of the system with $G(s) = 1/(s^2 + \sqrt{2}\,s + 1)$; (b) amplitude spectrum of the output signal $|Y(j\omega)|$ of Example 5.10.

5.4.3 Exercises

17 Find the impulse response of systems (a) and (c) of Example 5.9. Calculate the Fourier transform of each using the definition (5.15), and verify the results given in Example 5.9.

18 Use the time-shift property to calculate the Fourier transform of the double rectangular pulse $f(t)$ illustrated in Figure 5.17.

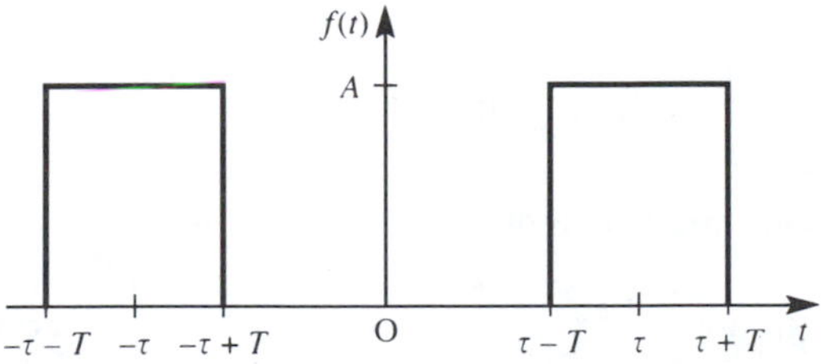

Figure 5.17 The double rectangular pulse of Exercise 18.

19 The system with transfer function

$$G(s) = \frac{1}{s^2 + \sqrt{2}s + 1}$$

was discussed in Example 5.10. Make a transformation

$$s \to \frac{1}{s'}$$

and write down $G(s')$. Examine the frequency response of a system with transfer function $G(s')$ and in particular find the amplitude response when $\omega = 0$ and as $\omega \to \infty$. How would you describe such a system?

20 Use the symmetry property, and the result of Exercise 1 to calculate the Fourier transform of

$$f(t) = \frac{1}{a^2 + t^2}$$

Sketch $f(t)$ and its transform (which is real).

21 Using the results of Examples 5.3 and 5.7, calculate the Fourier transform of the pulse-modulated signal

$$f(t) = P_T(t)\cos\omega_0 t$$

where

$$P_T(t) = \begin{cases} 1 & (|t| \leq T) \\ 0 & (|t| > T) \end{cases}$$

is the pulse of duration $2T$.

5.5 Transforms of the step and impulse functions

In this section we consider the application of Fourier transforms to the concepts of energy, power and convolution. In so doing, we shall introduce the Fourier transform of the Heaviside unit step function $H(t)$ and the impulse function $\delta(t)$.

5.5.1 Energy and power

In Section 4.6.4 we introduced the concept of the power spectrum of a periodic signal and found that it enabled us to deduce useful information relating to the latter. In this section we define two quantities associated with time signals $f(t)$, defined for $-\infty < t < \infty$, namely signal energy and signal power. Not only are these important quantities in themselves, but, as we shall see, they play an important role in characterizing signal types.

The total **energy** associated with the signal $f(t)$ is defined as

$$E = \int_{-\infty}^{\infty} [f(t)]^2 \, dt \tag{5.42}$$

If $f(t)$ has a Fourier transform $F(j\omega)$, so that, from (5.16),

$$f(t) = \frac{1}{2\pi} \int_{-\infty}^{\infty} F(j\omega) e^{j\omega t} \, d\omega$$

then (5.42) may be expressed as

$$E = \int_{-\infty}^{\infty} f(t)f(t) \, dt = \int_{-\infty}^{\infty} f(t) \left[\frac{1}{2\pi} \int_{-\infty}^{\infty} F(j\omega) e^{j\omega t} \, d\omega \right] dt$$

On changing the order of integration, this becomes

$$E = \frac{1}{2\pi} \int_{-\infty}^{\infty} F(j\omega) \left[\int_{-\infty}^{\infty} f(t) e^{j\omega t} \, dt \right] d\omega \tag{5.43}$$

From the defining integral (5.15) for $F(j\omega)$, we recognize the part of the integrand within the square brackets as $F(-j\omega)$, which, if $f(t)$ is real, is such that $F(-j\omega) = F^*(j\omega)$, where $F^*(j\omega)$ is the complex conjugate of $F(j\omega)$. Thus (5.43) becomes

$$E = \frac{1}{2\pi} \int_{-\infty}^{\infty} F(j\omega) F^*(j\omega) \, d\omega$$

so that

$$E = \int_{-\infty}^{\infty} [f(t)]^2 \, dt = \frac{1}{2\pi}\int_{-\infty}^{\infty} |F(j\omega)|^2 \, d\omega \qquad (5.44)$$

Equation (5.44) relates the total energy of the signal $f(t)$ to the integral over all frequencies of $|F(j\omega)|^2$. For this reason, $|F(j\omega)|^2$ is called the **energy spectral density**, and a plot of $|F(j\omega)|^2$ versus ω is called the **energy spectrum** of the signal $f(t)$. The result (5.44) is called **Parseval's theorem**, and is an extension of the result contained in Theorem 4.6 for periodic signals.

EXAMPLE 5.11

Determine the energy spectral densities of

(a) the one-sided exponential function $f(t) = e^{-at}H(t)$ $(a > 0)$,

(b) the rectangular pulse of Figure 5.6.

Solution (a) From (5.17), the Fourier transform of $f(t)$ is

$$F(j\omega) = \frac{a - j\omega}{a^2 + \omega^2}$$

The energy spectral density of the function is therefore

$$|F(j\omega)|^2 = F(j\omega)F^*(j\omega) = \frac{a - j\omega}{a^2 + \omega^2}\frac{a + j\omega}{a^2 + \omega^2}$$

that is,

$$|F(j\omega)|^2 = \frac{1}{a^2 + \omega^2}$$

(b) From Example 5.3, the Fourier transform $F(j\omega)$ of the rectangular pulse is

$$F(j\omega) = 2AT \operatorname{sinc} \omega T$$

Thus the energy spectral density of the pulse is

$$|F(j\omega)|^2 = 4A^2T^2 \operatorname{sinc}^2 \omega T$$

There are important signals $f(t)$, defined in general for $-\infty < t < \infty$, for which the integral $\int_{-\infty}^{\infty} [f(t)]^2 dt$ in (5.42) is either unbounded (that is, it becomes infinite) or does not converge to a finite limit; for example, $\sin t$. For such signals, instead of considering energy, we consider the average power P, frequently referred to as the **power** of the signal. This is defined by

$$P = \lim_{T\to\infty} \frac{1}{T}\int_{-T/2}^{T/2} [f(t)]^2 \, dt \qquad (5.45)$$

Note that for signals that satisfy the Dirichlet conditions (Theorem 5.1) the integral in (5.42) exists and, since in (5.45) we divide by the signal duration, it follows that such signals have zero power associated with them.

We now pose the question: 'Are there other signals which possess Fourier transforms?' As you may expect, the answer is 'Yes', although the manner of obtaining the transforms will be different from our procedure so far. We shall see that the transforms so obtained, on using the inversion integral (5.16) yield some very 'ordinary' signals so far excluded from our discussion.

We begin by considering the Fourier transform of the generalized function $\delta(t)$, the Dirac delta function introduced in Section 2.5.8. Recall from (2.49) that $\delta(t)$ satisfies the sifting property; that is, for a continuous function $g(t)$,

$$\int_a^b g(t)\delta(t-c)\,\mathrm{d}t = \begin{cases} g(c) & (a < c < b) \\ 0 & \text{otherwise} \end{cases}$$

Using the defining integral (5.15), we readily obtain the following two Fourier transforms:

$$\mathscr{F}\{\delta(t)\} = \int_{-\infty}^{\infty} \delta(t)\mathrm{e}^{-\mathrm{j}\omega t}\,\mathrm{d}t = 1 \qquad \textbf{(5.46)}$$

$$\mathscr{F}\{\delta(t-t_0)\} = \int_{-\infty}^{\infty} \delta(t-t_0)\mathrm{e}^{-\mathrm{j}\omega t}\,\mathrm{d}t = \mathrm{e}^{-\mathrm{j}\omega t_0} \qquad \textbf{(5.47)}$$

These two transforms are, by now, unremarkable, and, noting that $|\mathrm{e}^{-\mathrm{j}\omega t_0}| = 1$, we illustrate the signals and their spectra in Figure 5.18.

We now depart from the definition of the Fourier transform given in (5.15) and seek new transform pairs based on (5.46) and (5.47). Using the symmetry (duality) property of Section 5.3.5, we deduce from (5.46) that

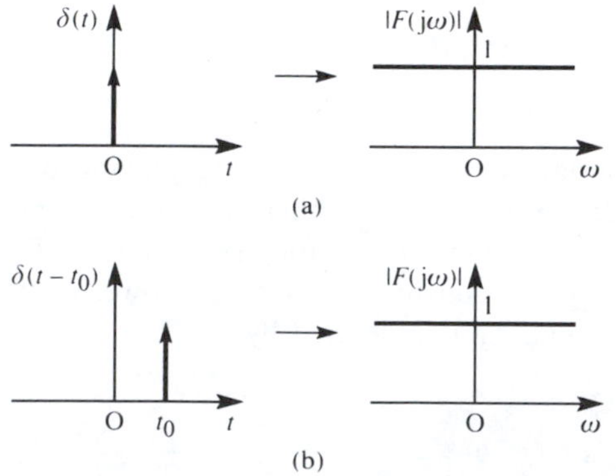

Figure 5.18 (a) $\delta(t)$ and its amplitude spectrum; (b) $\delta(t-t_0)$ and its amplitude spectrum.

$$1 \quad \text{and} \quad 2\pi\delta(-\omega) = 2\pi\delta(\omega) \tag{5.48}$$

is another Fourier transform pair. Likewise, from (5.47), we deduce that

$$e^{-jt_0t} \quad \text{and} \quad 2\pi\delta(-\omega - t_0)$$

is also a Fourier transform pair. Substituting $t_0 = -\omega_0$ into the latter, we have

$$e^{j\omega_0t} \quad \text{and} \quad 2\pi\delta(\omega_0 - \omega) = 2\pi\delta(\omega - \omega_0) \tag{5.49}$$

as another Fourier transform pair.

We are thus claiming that in (5.48) and (5.49) that $f_1(t) = 1$ and $f_2(t) = e^{j\omega_0t}$, which do not have 'ordinary' Fourier transforms as defined by (5.15), actually do have **'generalized' Fourier transforms** given by

$$F_1(j\omega) = 2\pi\delta(\omega) \tag{5.50}$$

$$F_2(j\omega) = 2\pi\delta(\omega - \omega_0) \tag{5.51}$$

respectively.

The term 'generalized' has been used because the two transforms contain the generalized functions $\delta(\omega)$ and $\delta(\omega - \omega_0)$. Let us now test our conjecture that (5.50) and (5.51) are Fourier transforms of $f_1(t)$ and $f_2(t)$ respectively. If (5.50) and (5.51) really are Fourier transforms then their time-domain images $f_1(t)$ and $f_2(t)$ respectively should reappear via the inverse transform (5.16). Substituting $F_1(j\omega)$ from (5.50) into (5.16), we have

$$\mathscr{F}^{-1}\{F_1(j\omega)\} = \frac{1}{2\pi}\int_{-\infty}^{\infty} F_1(j\omega)e^{j\omega t}\,d\omega = \frac{1}{2\pi}\int_{-\infty}^{\infty} 2\pi\delta(\omega)e^{j\omega t}\,d\omega = 1$$

so $f_1(t) = 1$ is recovered.

Similarly, using (5.51), we have

$$\mathscr{F}^{-1}\{F_2(j\omega)\} = \frac{1}{2\pi}\int_{-\infty}^{\infty} 2\pi\delta(\omega - \omega_0)e^{j\omega t}\,d\omega = e^{j\omega_0t}$$

so that $f_2(t) = e^{j\omega_0t}$ is also recovered.

Our approach has therefore been successful, and we do indeed have a way of generating new pairs of transforms. We shall therefore use the approach to find generalized Fourier transforms for the signals

$$f_3(t) = \cos\omega_0t, \qquad f_4(t) = \sin\omega_0t$$

Since

$$f_3(t) = \cos\omega_0t = \tfrac{1}{2}(e^{j\omega_0t} + e^{-j\omega_0t})$$

the linearity property (5.22) gives

$$\mathscr{F}\{f_3(t)\} = \tfrac{1}{2}\mathscr{F}\{e^{j\omega_0t}\} + \tfrac{1}{2}\mathscr{F}\{e^{-j\omega_0t}\}$$

which, on using (5.49), leads to the generalized Fourier transform pair

$$\mathcal{F}\{\cos \omega_0 t\} = \pi[\delta(\omega - \omega_0) + \delta(\omega + \omega_0)] \tag{5.52}$$

Likewise, we deduce the generalized Fourier transform pair

$$\mathcal{F}\{\sin \omega_0 t\} = \mathrm{j}\pi[\delta(\omega + \omega_0) - \delta(\omega - \omega_0)] \tag{5.53}$$

The development of (5.53) and the verification that both (5.52) and (5.53) invert correctly using the inverse transform (5.16) is left as an exercise for the reader.

It is worth noting at this stage that defining the Fourier transform $\mathcal{F}\{f(t)\}$ of $f(t)$ in (5.15) as

$$\mathcal{F}\{f(t)\} = \int_{-\infty}^{\infty} f(t)\mathrm{e}^{-\mathrm{j}\omega t}\,\mathrm{d}t$$

whenever the integral exists does not preclude the existence of other Fourier transforms, such as the generalized one just introduced, defined by other means.

It is clear that the total energy

$$E = \int_{-\infty}^{\infty} \cos^2 \omega_0 t\,\mathrm{d}t$$

associated with the signal $f_3(t) = \cos \omega_0 t$ is unbounded. However, from (5.45), we can calculate the power associated with the signal as

$$P = \lim_{T\to\infty} \frac{1}{T}\int_{-T/2}^{T/2} \cos^2 \omega_0 t\,\mathrm{d}t = \lim_{T\to\infty} \frac{1}{T}\left[t + \frac{1}{2\omega_0}\sin 2\omega_0 t\right]_{-T/2}^{T/2} = \tfrac{1}{2}$$

Thus, while the signal $f_3(t) = \cos \omega_0 t$ has unbounded energy associated with it, its power content is $\frac{1}{2}$. Signals whose associated energy is finite, for example $f(t) = \mathrm{e}^{-at}H(t)$ $(a > 0)$ are sometimes called **energy signals**, while those whose associated energy is unbounded but whose total power is finite are known as **power signals**. The concepts of power signals and power spectral density are important in the analysis of random signals, and the interested reader should consult specialized texts.

EXAMPLE 5.12

Suppose that a periodic function $f(t)$, defined on $-\infty < t < \infty$, may be expanded in a Fourier series having exponential form

$$f(t) = \sum_{n=-\infty}^{\infty} F_n \mathrm{e}^{\mathrm{j}n\omega_0 t}$$

What is the (generalized) Fourier transform of $f(t)$?

Solution From the definition,

$$\mathcal{F}\{f(t)\} = \mathcal{F}\left\{\sum_{n=-\infty}^{\infty} F_n \mathrm{e}^{\mathrm{j}n\omega_0 t}\right\} = \sum_{n=-\infty}^{\infty} F_n \mathcal{F}\{\mathrm{e}^{\mathrm{j}n\omega_0 t}\}$$

which, on using (5.49), gives

$$\mathscr{F}\{f(t)\} = \sum_{n=-\infty}^{\infty} F_n 2\pi\delta(\omega - n\omega_0)$$

That is,

$$\mathscr{F}\{f(t)\} = 2\pi \sum_{n=-\infty}^{\infty} F_n \delta(\omega - n\omega_0)$$

where F_n $(-\infty < n < \infty)$ are the coefficients of the exponential form of the Fourier series representation of $f(t)$.

EXAMPLE 5.13 Use the result of Example 5.12 to verify the Fourier transform of $f(t) = \cos \omega_0 t$ given in (5.52).

Solution Since

$$f(t) = \cos \omega_0 t = \tfrac{1}{2}\mathrm{e}^{\mathrm{j}\omega_0 t} + \tfrac{1}{2}\mathrm{e}^{-\mathrm{j}\omega_0 t}$$

the F_n of Example 5.12 are

$$F_{-1} = F_1 = \tfrac{1}{2}$$

$$F_n = 0 \quad (n \neq \pm 1)$$

Thus, using the result

$$\mathscr{F}\{f(t)\} = 2\pi \sum_{n=-\infty}^{\infty} F_n \delta(\omega - \omega_0)$$

we have

$$\begin{aligned}\mathscr{F}\{\cos \omega_0 t\} &= 2\pi F_{-1}\delta(\omega + \omega_0) + 2\pi F_1 \delta(\omega - \omega_0) \\ &= \pi[\delta(\omega + \omega_0) + \delta(\omega - \omega_0)]\end{aligned}$$

in agreement with (5.52).

EXAMPLE 5.14 Determine the (generalized) Fourier transform of the periodic 'sawtooth' function, defined by

$$f(t) = \frac{2t}{T} \quad (0 < t < 2T)$$

$$f(t + 2T) = f(t)$$

Solution In Example 4.19 we saw that the exponential form of the Fourier series representation of $f(t)$ is

$$f(t) = \sum_{n=-\infty}^{\infty} F_n \mathrm{e}^{\mathrm{j}n\omega_0 t}$$

with

$$\omega_0 = \frac{2\pi}{2T} = \frac{\pi}{T}$$

$$F_0 = 2$$

$$F_n = \frac{j2}{n\pi} \quad (n \neq 0)$$

It follows from Example 5.12 that the Fourier transform $\mathscr{F}\{f(t)\}$ is

$$\mathscr{F}\{f(t)\} = F(j\omega) = 4\pi\delta(\omega) + \sum_{\substack{n=-\infty \\ n\neq 0}}^{\infty} j\frac{4}{n}\delta(\omega - n\omega_0)$$

$$= 4\pi\delta(\omega) + j4\sum_{\substack{n=-\infty \\ n\neq 0}}^{\infty} \frac{1}{n}\delta\left(\omega - \frac{n\pi}{T}\right)$$

Thus we see that the amplitude spectrum simply consists of pulses located at integer multiples of the fundamental frequency $\omega_0 = \pi/T$. The discrete line spectra obtained via the exponential form of the Fourier series for this periodic function is thus reproduced, now with a scaling factor of 2π.

EXAMPLE 5.15 Determine the (generalized) Fourier transform of the unit impulse train $f(t) = \sum_{n=-\infty}^{\infty} \delta(t - nT)$ shown symbolically in Figure 5.19.

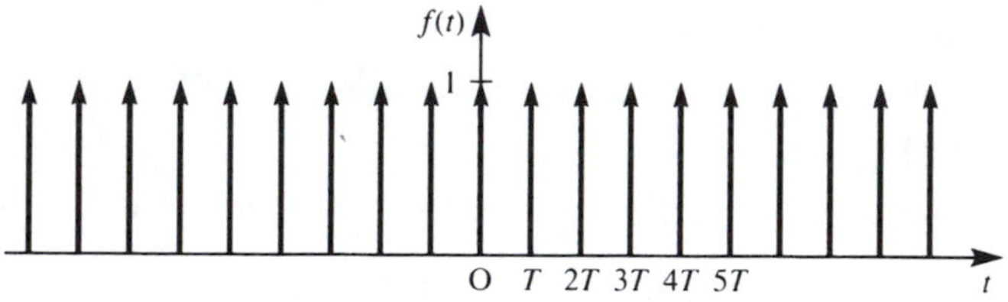

Figure 5.19 Unit impulse train $f(t) = \sum_{n=-\infty}^{\infty} \delta(t - nT)$.

Solution Although $f(t)$ is a generalized function, and not a function in the ordinary sense, it follows that since

$$f(t + kT) = \sum_{n=-\infty}^{\infty} \delta(t + (k - n)T) \quad (k \text{ an integer})$$

$$= \sum_{m=-\infty}^{\infty} \delta(t - mT) \quad (m = n - k)$$

$$= f(t)$$

it is periodic, with period T. Moreover, we can formally expand $f(t)$ as a Fourier series

$$f(t) = \sum_{n=-\infty}^{\infty} F_n e^{jn\omega_0 t} \quad \left(\omega_0 = \frac{2\pi}{T}\right)$$

with

$$F_n = \frac{1}{T}\int_{-T/2}^{T/2} f(t)\mathrm{e}^{-\mathrm{j}n\omega_0 t}\,\mathrm{d}t = \frac{1}{T}\int_{-T/2}^{T/2} \delta(t)\mathrm{e}^{-\mathrm{j}n\omega_0 t}\,\mathrm{d}t = \frac{1}{T} \quad \text{for all } n$$

It follows from Example 5.12 that

$$\mathscr{F}\{f(t)\} = 2\pi\sum_{n=-\infty}^{\infty}\frac{1}{T}\delta(\omega - n\omega_0) = \omega_0\sum_{n=-\infty}^{\infty}\delta(\omega - n\omega_0)$$

Thus we have shown that

$$\mathscr{F}\left\{\sum_{n=-\infty}^{\infty}\delta(t - nT)\right\} = \omega_0\sum_{n=-\infty}^{\infty}\delta(\omega - n\omega_0) \tag{5.54}$$

where $\omega_0 = 2\pi/T$. That is, the time-domain impulse train has another impulse train as its transform. We shall see in Section 5.6.4 that this result is of particular importance in dealing with sampled time signals.

Following our successful hunt for generalized Fourier transforms, we are led to consider the possibility that the Heaviside unit step function $H(t)$ defined in Section 2.5.1 may have a transform in this sense. Recall from (2.56) that if

$$f(t) = H(t)$$

then

$$\frac{\mathrm{d}f(t)}{\mathrm{d}t} = \delta(t)$$

From the time-differentiation property (5.23), we might expect that if

$$\mathscr{F}\{H(t)\} = \bar{H}(\mathrm{j}\omega)$$

then

$$(\mathrm{j}\omega)\bar{H}(\mathrm{j}\omega) = \mathscr{F}\{\delta(t)\} = 1 \tag{5.55}$$

Equation (5.55) suggests that a candidate for $\bar{H}(\mathrm{j}\omega)$ might be $1/\mathrm{j}\omega$, but this is not the case, since inversion using (5.16) does not give $H(t)$ back. Using (5.16) and complex variable techniques, it can be shown that

$$\mathscr{F}^{-1}\left\{\frac{1}{\mathrm{j}\omega}\right\} = \frac{1}{2\pi}\int_{-\infty}^{\infty}\frac{\mathrm{e}^{\mathrm{j}\omega t}}{\mathrm{j}\omega}\,\mathrm{d}\omega = \left\{\begin{matrix} \frac{1}{2} & (t > 0) \\ 0 & (t = 0) \\ -\frac{1}{2} & (t < 0)\end{matrix}\right\} = \tfrac{1}{2}\operatorname{sgn}(t)$$

where $\operatorname{sgn}(t)$ is the **signum function**, defined by

$$\operatorname{sgn}(t) = \begin{cases} 1 & (t > 0) \\ 0 & (t = 0) \\ -1 & (t < 0)\end{cases}$$

However, we note that (5.55) is also satisfied by

$$\bar{H}(j\omega) = \frac{1}{j\omega} + c\delta(\omega) \tag{5.56}$$

where c is a constant. This follows from the equivalence property (see Definition 2.2) $f(\omega)\delta(\omega) = f(0)\delta(\omega)$ with $f(\omega) = j\omega$, which gives

$$(j\omega)\bar{H}(j\omega) = 1 + (j\omega)c\delta(\omega) = 1$$

Inverting (5.56) using (5.16), we have

$$g(t) = \mathscr{F}^{-1}\left\{\frac{1}{j\omega} + c\delta(\omega)\right\} = \frac{1}{2\pi}\int_{-\infty}^{\infty}\left[\frac{1}{j\omega} + c\delta(\omega)\right] e^{j\omega t}\,d\omega$$

$$= \begin{cases} c/2\pi + \frac{1}{2} & (t > 0) \\ c/2\pi & (t = 0) \\ c/2\pi - \frac{1}{2} & (t < 0) \end{cases}$$

and, choosing $c = \pi$, we have

$$g(t) = \begin{cases} 1 & (t > 0) \\ \frac{1}{2} & (t = 0) \\ 0 & (t < 0) \end{cases}$$

Thus we have (almost) recovered the step function $H(t)$. Here $g(t)$ takes the value $\frac{1}{2}$ at $t = 0$, but this is not surprising in view of the convergence of the Fourier integral at points of discontinuity as given in Theorem 5.1. With this proviso, we have shown that

$$\bar{H}(j\omega) = \mathscr{F}\{H(t)\} = \frac{1}{j\omega} + \pi\delta(\omega) \tag{5.57}$$

We must confess to having made an informed guess as to what additional term to add in (5.56) to produce the Fourier transform (5.57). We could instead have chosen $c\delta(k\omega)$ with k a constant as an additional term. While it is possible to show that this would not lead to a different result, proving uniqueness is not trivial and is beyond the scope of this book.

5.5.2 Convolution

In Section 2.6.6 we saw that the convolution integral, in conjunction with the Laplace transform, provided a useful tool for *discussing* the nature of the solution of a differential equation, although it was not perhaps the most efficient way of evaluating the solution to a particular problem. As the reader may now have come to expect, in view of the duality between time and frequency domains, there are two convolution results involving the Fourier transform.

Convolution in time

Suppose that

$$\mathscr{F}\{u(t)\} = U(\mathrm{j}\omega) = \int_{-\infty}^{\infty} u(t)\,\mathrm{e}^{-\mathrm{j}\omega t}\,\mathrm{d}t$$

$$\mathscr{F}\{v(t)\} = V(\mathrm{j}\omega) = \int_{-\infty}^{\infty} v(t)\,\mathrm{e}^{-\mathrm{j}\omega t}\,\mathrm{d}t$$

then the Fourier transform of the convolution

$$y(t) = \int_{-\infty}^{\infty} u(\tau)v(t-\tau)\,\mathrm{d}\tau = u(t) * v(t) \tag{5.58}$$

is

$$\mathscr{F}\{y(t)\} = Y(\mathrm{j}\omega) = \int_{-\infty}^{\infty} \mathrm{e}^{-\mathrm{j}\omega t}\left[\int_{-\infty}^{\infty} u(\tau)v(t-\tau)\,\mathrm{d}\tau\right]\mathrm{d}t$$

$$= \int_{-\infty}^{\infty} u(\tau)\left[\int_{-\infty}^{\infty} \mathrm{e}^{-\mathrm{j}\omega t}v(t-\tau)\,\mathrm{d}t\right]\mathrm{d}\tau$$

Introducing the change of variables $z \to t - \tau$, $\tau \to \tau$ and following the procedure for change of variable from Section 2.6.6, the transform can be expressed as

$$Y(\mathrm{j}\omega) = \int_{-\infty}^{\infty} u(\tau)\left[\int_{-\infty}^{\infty} v(z)\,\mathrm{e}^{-\mathrm{j}\omega(z+\tau)}\,\mathrm{d}z\right]\mathrm{d}\tau$$

$$= \int_{-\infty}^{\infty} u(\tau)\,\mathrm{e}^{-\mathrm{j}\omega\tau}\,\mathrm{d}\tau \int_{-\infty}^{\infty} v(z)\,\mathrm{e}^{-\mathrm{j}\omega z}\,\mathrm{d}z$$

so that

$$Y(\mathrm{j}\omega) = U(\mathrm{j}\omega)V(\mathrm{j}\omega) \tag{5.59}$$

That is,

$$\mathscr{F}\{u(t) * v(t)\} = \mathscr{F}\{v(t) * u(t)\} = U(\mathrm{j}\omega)V(\mathrm{j}\omega) \tag{5.60}$$

indicating that a convolution in the time domain is transformed into a product in the frequency domain.

Convolution in frequency

If

$$\mathscr{F}\{u(t)\} = U(\mathrm{j}\omega), \qquad \text{with } u(t) = \frac{1}{2\pi}\int_{-\infty}^{\infty} U(\mathrm{j}\omega)\mathrm{e}^{\mathrm{j}\omega t}\,\mathrm{d}\omega$$

$$\mathscr{F}\{v(t)\} = V(\mathrm{j}\omega), \qquad \text{with } v(t) = \frac{1}{2\pi}\int_{-\infty}^{\infty} V(\mathrm{j}\omega)\mathrm{e}^{\mathrm{j}\omega t}\,\mathrm{d}\omega$$

then the inverse transform of the convolution

$$U(\mathrm{j}\omega) * V(\mathrm{j}\omega) = \int_{-\infty}^{\infty} U(\mathrm{j}y)V(\mathrm{j}(\omega - y))\,\mathrm{d}y$$

is given by

$$\begin{aligned}\mathscr{F}^{-1}\{U(\mathrm{j}\omega) * V(\mathrm{j}\omega)\} &= \frac{1}{2\pi}\int_{-\infty}^{\infty} \mathrm{e}^{\mathrm{j}\omega t}\left[\int_{-\infty}^{\infty} U(\mathrm{j}y)V(\mathrm{j}(\omega - y))\,\mathrm{d}y\right]\mathrm{d}\omega \\ &= \frac{1}{2\pi}\int_{-\infty}^{\infty} U(\mathrm{j}y)\left[\int_{-\infty}^{\infty} V(\mathrm{j}(\omega - y))\,\mathrm{e}^{\mathrm{j}\omega t}\,\mathrm{d}\omega\right]\mathrm{d}y\end{aligned}$$

A change of variable $z \to \omega - y$, $\omega \to \omega$ leads to

$$\begin{aligned}\mathscr{F}^{-1}\{U(\mathrm{j}\omega) * V(\mathrm{j}\omega)\} &= \frac{1}{2\pi}\int_{-\infty}^{\infty} U(\mathrm{j}y)\left[\int_{-\infty}^{\infty} V(\mathrm{j}z)\,\mathrm{e}^{\mathrm{j}(z+y)t}\,\mathrm{d}z\right]\mathrm{d}y \\ &= \frac{1}{2\pi}\int_{-\infty}^{\infty} U(\mathrm{j}y)\,\mathrm{e}^{\mathrm{j}yt}\,\mathrm{d}y\int_{-\infty}^{\infty} V(\mathrm{j}z)\,\mathrm{e}^{\mathrm{j}zt}\,\mathrm{d}z \\ &= 2\pi\, u(t)v(t)\end{aligned}$$

That is,

$$\mathscr{F}\{u(t)v(t)\} = \frac{1}{2\pi}U(\mathrm{j}\omega) * V(\mathrm{j}\omega) \tag{5.61}$$

and thus multiplication in the time domain corresponds to convolution in the frequency domain (subject to the scaling factor $1/(2\pi)$).

EXAMPLE 5.16

Suppose that $f(t)$ has a Fourier transform $F(\mathrm{j}\omega)$. Find an expression for the Fourier transform of $g(t)$, where

$$g(t) = \int_{-\infty}^{t} f(\tau)\,\mathrm{d}\tau$$

Solution Since

$$H(t-\tau) = \begin{cases} 1 & (\tau \leqslant t) \\ 0 & (\tau > t) \end{cases}$$

we can write

$$g(t) = \int_{-\infty}^{\infty} f(\tau)H(t-\tau)\,\mathrm{d}\tau = f(t) * H(t)$$

the convolution of $g(t)$ and $H(t)$. Then, using (5.60),

$$\mathscr{F}\{g(t)\} = G(\mathrm{j}\omega) = F(\mathrm{j}\omega)\bar{H}(\mathrm{j}\omega)$$

which, on using the expression for $\bar{H}(\mathrm{j}\omega)$ from (5.57), gives

$$G(\mathrm{j}\omega) = \frac{F(\mathrm{j}\omega)}{\mathrm{j}\omega} + \pi F(\mathrm{j}\omega)\delta(\omega)$$

so that

$$G(\mathrm{j}\omega) = \frac{F(\mathrm{j}\omega)}{\mathrm{j}\omega} + \pi F(0)\delta(\omega) \tag{5.62}$$

5.5.3 Exercises

22 Verify that $\mathscr{F}^{-1}\{\pi[\delta(\omega-\omega_0)+\delta(\omega+\omega_0)]\} = \cos\omega_0 t$.

23 Show that $\mathscr{F}\{\sin\omega_0 t\} = \mathrm{j}\pi[\delta(\omega+\omega_0)-\delta(\omega-\omega_0)]$. Use (5.16) to verify that

$$\mathscr{F}^{-1}\{\mathrm{j}\pi[\delta(\omega+\omega_0)-\delta(\omega-\omega_0)]\} = \sin\omega_0 t$$

24 Suppose that $f(t)$ and $g(t)$ have Fourier transforms $F(\mathrm{j}\omega)$ and $G(\mathrm{j}\omega)$ respectively, defined in the 'ordinary' sense (that is, using (5.15)), and show that

$$\int_{-\infty}^{\infty} f(t)G(\mathrm{j}t)\,\mathrm{d}t = \int_{-\infty}^{\infty} F(\mathrm{j}t)g(t)\,\mathrm{d}t$$

This result is known as **Parseval's formula**.

25 Use the results of Exercise 24 and the symmetry property to show that

$$\int_{-\infty}^{\infty} f(t)g(t)\,\mathrm{d}t = \frac{1}{2\pi}\int_{-\infty}^{\infty} F(\mathrm{j}\omega)G(-\mathrm{j}\omega)\,\mathrm{d}\omega$$

26 Use the convolution result in the frequency domain to obtain $\mathscr{F}\{H(t)\sin\omega_0 t\}$.

27 Calculate the exponential form of the Fourier series for the periodic pulse train shown in Figure 5.20. Hence show that

$$\mathscr{F}\{f(t)\} = \frac{2\pi A d}{T}\sum_{n=-\infty}^{\infty} \operatorname{sinc}\left(\frac{n\pi d}{T}\right)\delta(\omega - n\omega_0)$$

($\omega_0 = 2\pi/T$), and A is the height of the pulse.

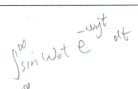

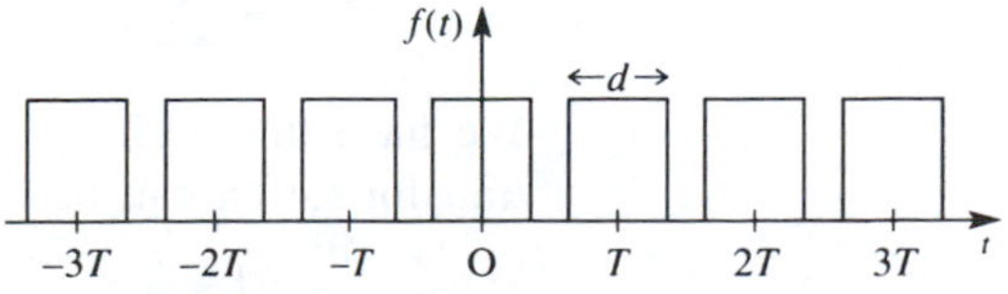

Figure 5.20 Periodic pulse train of Exercise 27.

5.6 The Fourier transform in discrete time

5.6.1 Introduction

The earlier sections of this chapter have discussed the Fourier transform of signals defined as functions of the continuous-time variable t. We have seen that a major area of application is in the analysis of signals in the frequency domain, leading to the concept of the frequency response of a linear system. In Chapter 4 we considered signals defined at discrete-time instants, together with linear systems modelled by difference equations. There we found that in system analysis the z transform plays a role similar to that of the Laplace transform for continuous-time systems. We now attempt to develop a theory of Fourier analysis to complement that for continuous-time systems, and then consider the problem of estimating the continuous-time Fourier transform in a form suitable for computer execution.

5.6.2 A Fourier transform for sequences

First we return to our work on Fourier series and write down the exponential form of the Fourier series representation for the periodic function $F(\mathrm{e}^{\mathrm{j}\theta})$ of period 2π. Writing $\theta = \omega t$, we infer from (4.57) and (4.61) that

$$F(\mathrm{e}^{\mathrm{j}\theta}) = \sum_{n=-\infty}^{\infty} f_n \mathrm{e}^{\mathrm{j}n\theta} \tag{5.63}$$

where

$$f_n = \frac{1}{2\pi}\int_{-\pi}^{\pi} F(\mathrm{e}^{\mathrm{j}\theta})\mathrm{e}^{-\mathrm{j}n\theta}\,\mathrm{d}\theta \tag{5.64}$$

Thus the operation has generated a sequence of numbers $\{f_n\}$ from the periodic function $F(\mathrm{e}^{\mathrm{j}\theta})$ of the continuous variable θ. Let us reverse the process and imagine that we *start* with a sequence $\{g_k\}$ and use (5.63) to *define* a periodic function $\tilde{G}'(\mathrm{e}^{\mathrm{j}\theta})$ such that

$$\tilde{G}'(\mathrm{e}^{\mathrm{j}\theta}) = \sum_{n=-\infty}^{\infty} g_n \mathrm{e}^{\mathrm{j}n\theta} \tag{5.65}$$

We have thus defined a transformation from the sequence $\{g_k\}$ to $\tilde{G}'(\mathrm{e}^{\mathrm{j}\theta})$. This transformation can be inverted, since, from (5.64),

$$g_k = \frac{1}{2\pi}\int_{-\pi}^{\pi} \tilde{G}'(\mathrm{e}^{\mathrm{j}\theta})\mathrm{e}^{-\mathrm{j}k\theta}\,\mathrm{d}\theta \tag{5.66}$$

and we recover the terms of the sequence $\{g_k\}$ from $\tilde{G}'(\mathrm{e}^{\mathrm{j}\theta})$.

It is convenient for our later work if we modify the definition slightly, defining the Fourier transform of the sequence $\{g_k\}$ as

$$\mathscr{F}\{g_k\} = G(e^{j\theta}) = \sum_{n=-\infty}^{\infty} g_n e^{-jn\theta} \tag{5.67}$$

whenever the series converges. The inverse transform is then given from (5.66), by

$$g_k = \frac{1}{2\pi}\int_{-\pi}^{\pi} G(e^{j\theta})e^{jk\theta}\,d\theta \tag{5.68}$$

The results (5.67) and (5.68) thus constitute the Fourier transform pair for the sequence $\{g_k\}$. Note that $G(e^{j\theta})$ is a function of the continuous variable θ, and since it is a function of $e^{j\theta}$ it is periodic (with a period of at most 2π), irrespective of whether or not the sequence $\{g_k\}$ is periodic.

Note that we have adopted the notation $G(e^{j\theta})$ rather than $G(\theta)$ for the Fourier transform, similar to our use of $F(j\omega)$ rather than $F(\omega)$ in the case of continuous-time signals. In the present case we shall be concerned with the relationship with the z transform of Chapter 3, where $z = r\,e^{j\theta}$, and the significance of our choice will soon emerge.

EXAMPLE 5.17 Find the transform of the sequence $\{g_k\}_{-\infty}^{\infty}$, where $g_0 = 2$, $g_2 = g_{-2} = 1$ and $g_k = g_{-k} = 0$ for $k \neq 0$ or 2.

Solution From the definition (5.67),

$$\begin{aligned}\mathscr{F}\{g_k\} &= G(e^{j\theta}) = \sum_{n=-\infty}^{\infty} g_n e^{-jn\theta}\\ &= g_{-2}\,e^{j2\theta} + g_0 1 + g_2\,e^{-j2\theta} = e^{j2\theta} + 2 + e^{-j2\theta}\\ &= 2(1 + \cos 2\theta) = 4\cos^2\theta\end{aligned}$$

In this particular case the transform is periodic of period π, rather than 2π. This is because $g_1 = g_{-1} = 0$, so that $\cos\theta$ does not appear in the transform. Since $G(e^{j\theta})$ is purely real, we may plot the transform as in Figure 5.21.

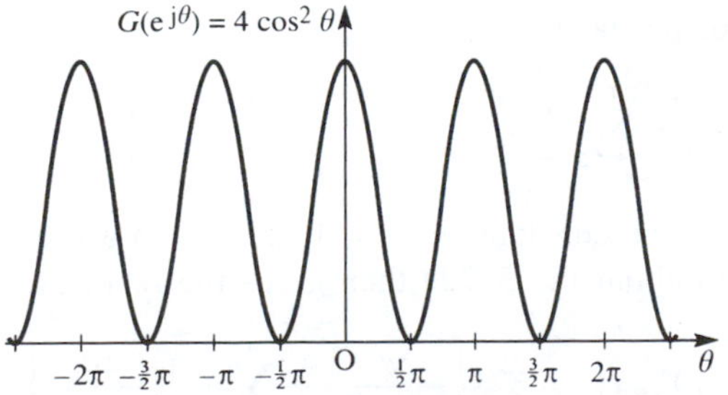

Figure 5.21 Transform of the sequence of Example 5.17.

Having defined a Fourier transform for sequences, we now wish to link it to the frequency response of discrete-time systems. In Section 5.4.2 the link between frequency responses and the Fourier transforms of continuous-time systems was established using the Laplace transform. We suspect therefore that the z transform should yield the necessary link for discrete-time systems. Indeed, the argument follows closely that of Section 5.4.2.

For a causal linear time-invariant discrete-time system with z transfer function $G(z)$ the relationship between the input sequence $\{u_k\}$ and output sequence $\{y_k\}$ in the transform domain is given from Section 3.6.1 by

$$Y(z) = G(z)U(z) \tag{5.69}$$

where $U(z) = \mathscr{Z}\{u_k\}$ and $Y(z) = \mathscr{Z}\{y_k\}$.

To investigate the system frequency response, we seek the output sequence corresponding to an input sequence

$$\{u_k\} = \{A\,\mathrm{e}^{\mathrm{j}\omega kT}\} = \{A\,\mathrm{e}^{\mathrm{j}k\theta}\}, \quad \theta = \omega T \tag{5.70}$$

which represents samples drawn, at equal intervals T, from the continuous-time complex sinusoidal signal $\mathrm{e}^{\mathrm{j}\omega t}$.

The frequency response of the discrete-time system is then its steady-state response to the sequence $\{u_k\}$ given in (5.70). As for the continuous-time case (Section 5.4.2), the complex form $\mathrm{e}^{\mathrm{j}\omega t}$ is used in order to simplify the algebra, and the steady-state sinusoidal response is easily recovered by taking imaginary parts, if necessary.

From Figure 3.3, we see that

$$\mathscr{Z}\{A\,\mathrm{e}^{\mathrm{j}k\theta}\} = \mathscr{Z}\{A(\mathrm{e}^{\mathrm{j}\theta})^k\} = \frac{Az}{z - \mathrm{e}^{\mathrm{j}\theta}}$$

so, from (5.69), the response of the system to the input sequence (5.70) is determined by

$$Y(z) = G(z)\frac{Az}{z - \mathrm{e}^{\mathrm{j}\theta}} \tag{5.71}$$

Taking the system to be of order n, and under the assumption that the n poles p_r $(r = 1, 2, \ldots, n)$ of $G(z)$ are distinct and none is equal to $\mathrm{e}^{\mathrm{j}\theta}$, we can expand $Y(z)/z$ in terms of partial fractions to give

$$\frac{Y(z)}{z} = \frac{c}{z - \mathrm{e}^{\mathrm{j}\theta}} + \sum_{r=1}^{n} \frac{c_r}{z - p_r} \tag{5.72}$$

where, in general, the constants c_r $(r = 1, 2, \ldots, n)$ are complex. Taking inverse z transforms throughout in (5.72) then gives the response sequence as

$$\{y_k\} = \mathscr{Z}^{-1}\{Y(z)\} = \mathscr{Z}^{-1}\left\{\frac{zc}{z - \mathrm{e}^{\mathrm{j}\theta}}\right\} + \sum_{r=1}^{n} \mathscr{Z}^{-1}\left\{\frac{zc_r}{z - p_r}\right\}$$

that is,

$$\{y_k\} = c\{e^{jk\theta}\} + \sum_{r=1}^{n} c_r\{p_r^k\} \tag{5.73}$$

If the transfer function $G(z)$ corresponds to a stable discrete-time system then all its poles p_r $(r = 1, 2, \ldots, n)$ lie within the unit circle $|z| < 1$, so that all the terms under the summation sign in (5.73) tend to zero as $k \to \infty$. This is clearly seen by expressing p_r in the form $p_r = |p_r|e^{j\phi_r}$ and noting that if $|p_r| < 1$ then $|p_r|^k \to 0$ as $k \to \infty$. Consequently, for stable systems the steady-state response corresponding to (5.73) is

$$\{y_{k_{ss}}\} = c\{e^{jk\theta}\}$$

Using the 'cover-up' rule for partial fractions, the constant c is readily determined from (5.71) as

$$c = AG(e^{j\theta})$$

so that the steady-state response becomes

$$\{y_{k_{ss}}\} = AG(e^{j\theta})\{e^{jk\theta}\} \tag{5.74}$$

We have assumed that the poles of $G(z)$ are distinct in order to simplify the algebra. Extending the development to accommodate multiple poles is readily accomplished, leading to the same steady-state response as given in (5.74).

The result (5.74) corresponds to (5.38) for continuous-time systems, and indicates that the steady-state response sequence is simply the input sequence with each term multiplied by $G(e^{j\theta})$. Consequently $G(e^{j\theta})$ is called the **frequency transfer function** of the discrete-time system and, as for the continuous case, it characterizes the system's frequency response. Clearly $G(e^{j\theta})$ is simply $G(z)$, the z transfer function, with $z = e^{j\theta}$, and so we are simply evaluating the z transfer function around the unit circle $|z| = 1$. The z transfer function $G(z)$ will exist on $|z| = 1$ if and only if the system is stable, and thus the result is the exact analogue of the result for continuous-time systems in Section 5.4.2, where the Laplace transfer function was evaluated along the imaginary axis to yield the frequency response of a stable linear continuous-time system.

To complete the analogy with continuous-time systems, we need one further result. From Section 3.6.2, the impulse response of the linear causal discrete-time system with z transfer function $G(z)$ is

$$\{y_{k_\delta}\} = \mathscr{Z}^{-1}\{G(z)\} = \{g_k\}_{k=0}^{\infty}, \quad \text{say}$$

Taking inverse transforms then gives

$$G(z) = \sum_{k=0}^{\infty} g_k z^{-k} = \sum_{k=-\infty}^{\infty} g_k z^{-k}$$

since $g_k = 0$ $(k < 0)$ for a causal system. Thus

$$G(e^{j\theta}) = \sum_{k=-\infty}^{\infty} g_k e^{-jk\theta}$$

and we conclude from (5.67) that $G(e^{j\theta})$ is simply the Fourier transform of the sequence $\{g_k\}$. Therefore the discrete-time frequency transfer function $G(e^{j\theta})$ is the Fourier transform of the impulse response sequence.

EXAMPLE 5.18 Determine the frequency transfer function of the causal discrete-time system shown in Figure 5.22 and plot its amplitude spectrum.

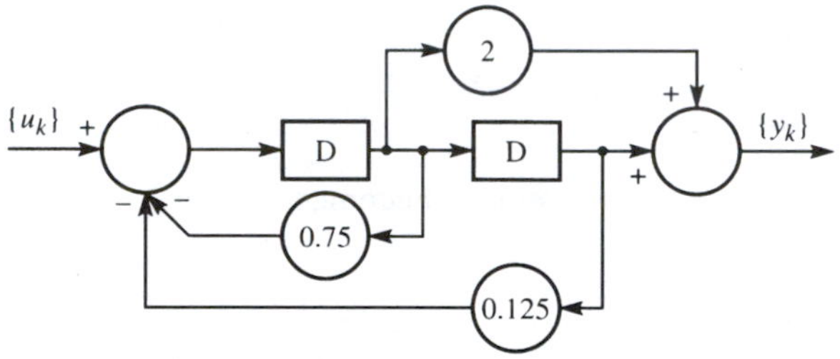

Figure 5.22 Discrete-time system of Example 5.18.

Solution Using the methods of Section 3.6.1, we readily obtain the z transfer function as

$$G(z) = \frac{2z + 1}{z^2 + 0.75z + 0.125}$$

Next we check for system stability. Since $z^2 + 0.75z + 0.125 = (z + 0.5)(z + 0.25)$, the poles of $G(z)$ are at $p_1 = -0.5$ and $p_2 = -0.25$, and since both are inside the unit circle $|z| = 1$, the system is stable. The frequency transfer function may then be obtained as $G(e^{j\theta})$, where

$$G(e^{j\theta}) = \frac{2\,e^{j\theta} + 1}{e^{j2\theta} + 0.75\,e^{j\theta} + 0.125}$$

To determine the amplitude spectrum, we evaluate $|G(e^{j\theta})|$ as

$$|G(e^{j\theta})| = \frac{|2\,e^{j\theta} + 1|}{|e^{j2\theta} + 0.75\,e^{j\theta} + 0.125|}$$

$$= \frac{\sqrt{(5 + 4\cos\theta)}}{\sqrt{(1.578 + 1.688\cos\theta + 0.25\cos 2\theta)}}$$

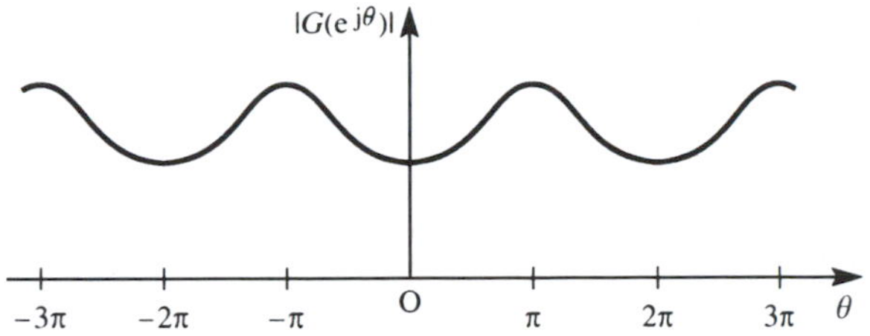

Figure 5.23 Amplitude spectrum of the system of Example 5.18.

A plot of $|G(e^{j\theta})|$ versus θ then leads to the amplitude spectrum of Figure 5.23.

In Example 5.18 we note the periodic behaviour of the amplitude spectrum, which is inescapable when discrete-time signals and systems are concerned. Note, however, that the periodicity is in the variable $\theta = \omega T$ and that we may have control over the choice of T, the time between samples of our input signal.

5.6.3 The discrete Fourier transform

The Fourier transform of sequences discussed in Section 5.6.2 transforms a sequence $\{g_k\}$ into a continuous function $G(e^{j\theta})$ of a frequency variable θ, where $\theta = \omega T$ and T is the time between signal samples. In this section, with an eye to computer requirements, we look at the implications of sampling $G(e^{j\theta})$. The overall operation will have commenced with samples of a time signal $\{g_k\}$ and proceeded via a Fourier transformation process, finally producing a sequence $\{G_k\}$ of samples drawn from the frequency-domain image $G(e^{j\theta})$ of $\{g_k\}$.

Suppose that we have a sequence $\{g_k\}$ of N samples drawn from a continuous-time signal $g(t)$, at equal intervals T; that is,

$$\{g_k\} = \{g(kT)\}_{k=0}^{N-1}$$

Using (5.67), the Fourier transform of this sequence is

$$\mathscr{F}\{g_k\} = G(e^{j\theta}) = \sum_{n=-\infty}^{\infty} g_n e^{-jn\theta} \tag{5.75}$$

where $g_k = 0$ $(k \notin [0, N-1])$. Then, with $\theta = \omega T$, we may write (5.75) as

$$G(e^{j\omega T}) = \sum_{n=0}^{N-1} g_n e^{-jn\omega T} \tag{5.76}$$

We now sample this transform $G(e^{j\omega T})$ at intervals $\Delta\omega$ in such a way as to create N samples spread equally over the interval $0 \leqslant \theta \leqslant 2\pi$, that is, over one period of the essentially periodic function $G(e^{j\theta})$. We then have

$$N\Delta\theta = 2\pi$$

where $\Delta\theta$ is the normalized frequency spacing. Since $\theta = \omega T$ and T is a constant such that $\Delta\theta = T\Delta\omega$, we deduce that

$$\Delta\omega = \frac{2\pi}{NT} \tag{5.77}$$

Sampling (5.76) at intervals $\Delta\omega$ produces the sequence

$$\{G_k\}_{k=0}^{N-1}, \quad \text{where} \quad G_k = \sum_{n=0}^{N-1} g_n \, \mathrm{e}^{-\mathrm{j}nk\Delta\omega T} \tag{5.78}$$

Since

$$\begin{aligned} G_{k+N} &= \sum_{n=0}^{N-1} g_n \, \mathrm{e}^{-\mathrm{j}n(k+N)\Delta\omega T} \\ &= \sum_{n=0}^{N-1} g_n \, \mathrm{e}^{-\mathrm{j}nk\Delta\omega T} \mathrm{e}^{-\mathrm{j}n2\pi}, \quad \text{using (5.77)} \\ &= \sum_{n=0}^{N-1} g_n \, \mathrm{e}^{-\mathrm{j}nk\Delta\omega T} = G_k \end{aligned}$$

it follows that the sequence $\{G_k\}_{-\infty}^{\infty}$ is periodic, with period N. We have therefore generated a sequence of samples in the frequency domain that in some sense represents the spectrum of the underlying continuous-time signal. We shall postpone the question of the exact nature of this representation for the moment, but as the reader will have guessed, it is crucial to the purpose of this section. First, we consider the question of whether, from knowledge of the sequence $\{G_k\}_{k=0}^{N-1}$ of (5.78), we can recover the original sequence $\{g_n\}_{n=0}^{N-1}$. To see how this can be achieved, consider a sum of the form

$$S_r = \sum_{k=0}^{N-1} G_k \, \mathrm{e}^{-\mathrm{j}kr\Delta\omega T}, \; (N-1) \leqslant r \leqslant 0 \tag{5.79}$$

Substituting for G_k from (5.78), we have

$$S_r = \sum_{k=0}^{N-1} \left(\sum_{m=0}^{N-1} g_m \, \mathrm{e}^{-\mathrm{j}mk\Delta\omega T} \right) \mathrm{e}^{-\mathrm{j}kr\Delta\omega T} = \sum_{k=0}^{N-1} \sum_{m=0}^{N-1} g_m \, \mathrm{e}^{-\mathrm{j}k\Delta\omega(m+r)T}$$

That is, on interchanging the order of integration,

$$S_r = \sum_{m=0}^{N-1} g_m \sum_{k=0}^{N-1} \mathrm{e}^{-\mathrm{j}k\Delta\omega(m+r)T} \tag{5.80}$$

Now

$$\sum_{k=0}^{N-1} \mathrm{e}^{-\mathrm{j}k\Delta\omega(m+r)T}$$

is a geometric progression with first term $\mathrm{e}^0 = 1$ and common ratio $\mathrm{e}^{-\mathrm{j}\Delta\omega(m+r)T}$, and so the sum to N terms is thus

$$\sum_{k=0}^{N-1} \mathrm{e}^{-\mathrm{j}k\Delta\omega(m+r)T} = \frac{1 - \mathrm{e}^{-\mathrm{j}\Delta\omega(m+r)NT}}{1 - \mathrm{e}^{-\mathrm{j}\Delta\omega(m+r)T}} = \frac{1 - \mathrm{e}^{-\mathrm{j}(m+r)2\pi}}{1 - \mathrm{e}^{-\mathrm{j}\Delta\omega(m+r)T}} = 0 \quad (m \neq -r + nN)$$

When $m = -r$

$$\sum_{k=0}^{N-1} \mathrm{e}^{-\mathrm{j}k\Delta\omega(m+r)T} = \sum_{k=0}^{N-1} 1 = N$$

Thus

$$\sum_{k=0}^{N-1} \mathrm{e}^{-\mathrm{j}k\Delta\omega(m+r)T} = N\delta_{m,-r} \qquad \textbf{(5.81)}$$

where δ_{ij} is the Kronecker delta defined by

$$\delta_{ij} = \begin{cases} 1 & (i = j) \\ 0 & (i \neq j) \end{cases}$$

Substituting (5.81) into (5.80), we have

$$S_r = N\sum_{m=0}^{N-1} g_m \delta_{m,-r} = Ng_{-r}$$

Returning to (5.79) and substituting for S_r we see that

$$g_{-r} = \frac{1}{N}\sum_{k=0}^{N-1} G_k \mathrm{e}^{-\mathrm{j}kr\Delta\omega T}$$

which on taking $n = -r$ gives

$$g_n = \frac{1}{N}\sum_{k=0}^{N-1} G_k \mathrm{e}^{\mathrm{j}kn\Delta\omega T} \qquad \textbf{(5.82)}$$

Thus (5.82) allows us to determine the members of the sequence

$$\{g_n\}_{n=0}^{N-1}$$

that is, it enables us to recover the time-domain samples from the frequency-domain samples *exactly*.

The relations

$$G_k = \sum_{n=0}^{N-1} g_n \mathrm{e}^{-\mathrm{j}nk\Delta\omega T} \qquad \textbf{(5.78)}$$

$$g_n = \frac{1}{N}\sum_{k=0}^{N-1} G_k \mathrm{e}^{\mathrm{j}nk\Delta\omega T} \qquad \textbf{(5.82)}$$

with $\Delta\omega = 2\pi/NT$, between the time- and frequency-domain sequences $\{g_n\}_{n=0}^{N-1}$ and $\{G_k\}_{k=0}^{N-1}$ define the **discrete Fourier transform (DFT)** pair. The pair provide pathways between time and frequency domains for discrete-time signals in exactly the same sense that (5.15) and (5.16) defined similar pathways for continuous-time signals. It should be stressed again that, whatever the properties of the sequences $\{g_n\}$ and $\{G_k\}$ on the right-hand sides of (5.78) and (5.82), the sequences generated on the left-hand sides will be periodic, with period N.

EXAMPLE 5.19 The sequence $\{g_k\}_{k=0}^{2} = \{1, 2, 1\}$ is generated by sampling a time signal $g(t)$ at intervals with $T = 1$. Determine the discrete Fourier transform of the sequence, and verify that the sequence can be recovered exactly from its transform.

Solution From (5.78), the discrete Fourier transform sequence $\{G_k\}_{k=0}^{2}$ is generated by

$$G_k = \sum_{n=0}^{2} g_n \,\mathrm{e}^{-\mathrm{j}kn\Delta\omega T} \quad (k = 0, 1, 2)$$

In this case $T = 1$ and, with $N = 3$, (5.77) gives

$$\Delta\omega = \frac{2\pi}{3 \times 1} = \frac{2}{3}\pi$$

Thus

$$G_0 = \sum_{n=0}^{2} g_n \,\mathrm{e}^{-\mathrm{j}n\times 0\times 2\pi/3} = \sum_{n=0}^{2} g_n = g_0 + g_1 + g_2 = 1 + 2 + 1 = 4$$

$$G_1 = \sum_{n=0}^{2} g_n \,\mathrm{e}^{-\mathrm{j}n\times 1\times 2\pi/3} = g_0\mathrm{e}^0 + g_1\mathrm{e}^{-\mathrm{j}2\pi/3} + g_2\mathrm{e}^{-\mathrm{j}4\pi/3} = 1 + 2\mathrm{e}^{-\mathrm{j}2\pi/3} + 1\,\mathrm{e}^{-\mathrm{j}4\pi/3}$$

$$= \mathrm{e}^{-\mathrm{j}2\pi/3}(\mathrm{e}^{\mathrm{j}2\pi/3} + 2 + \mathrm{e}^{-\mathrm{j}2\pi/3}) = 2\,\mathrm{e}^{-\mathrm{j}2\pi/3}(1 + \cos\tfrac{2}{3}\pi) = \mathrm{e}^{-\mathrm{j}2\pi/3}$$

$$G_2 = \sum_{n=0}^{2} g_n \,\mathrm{e}^{-\mathrm{j}n\times 2\times 2\pi/3} = \sum_{n=0}^{2} g_n \,\mathrm{e}^{-\mathrm{j}n4\pi/3} = g_0\mathrm{e}^0 + g_1\mathrm{e}^{-\mathrm{j}4\pi/3} + g_2\mathrm{e}^{-\mathrm{j}8\pi/3}$$

$$= \mathrm{e}^{-\mathrm{j}4\pi/3}[\mathrm{e}^{\mathrm{j}4\pi/3} + 2 + \mathrm{e}^{-\mathrm{j}4\pi/3}] = 2\,\mathrm{e}^{-\mathrm{j}4\pi/3}(1 + \cos\tfrac{4}{3}\pi) = \mathrm{e}^{-\mathrm{j}4\pi/3}$$

Thus

$$\{G_k\}_{k=0}^{2} = \{4, \mathrm{e}^{-\mathrm{j}2\pi/3}, \mathrm{e}^{-\mathrm{j}4\pi/3}\}$$

We must now show that use of (5.82) will recover the original sequence $\{g_k\}_{k=0}^{2}$. From (5.82), the inverse transform of $\{G_k\}_{k=0}^{2}$ is given by

$$\tilde{g}_n = \frac{1}{N}\sum_{k=0}^{N-1} G_k \,\mathrm{e}^{\mathrm{j}kn\,\Delta\omega\, T}$$

again with $T = 1$, $\Delta\omega = \frac{2}{3}\pi$ and $N = 3$. Thus

$$\tilde{g}_0 = \tfrac{1}{3}\sum_{k=0}^{2} G_k \,\mathrm{e}^{\mathrm{j}k\times 0\times 2\pi/3} = \tfrac{1}{3}\sum_{k=0}^{2} G_k = \tfrac{1}{3}(4 + \mathrm{e}^{-\mathrm{j}2\pi/3} + \mathrm{e}^{-\mathrm{j}4\pi/3})$$

$$= \tfrac{1}{3}[4 + \mathrm{e}^{-\mathrm{j}\pi}(\mathrm{e}^{\mathrm{j}\pi/3} + \mathrm{e}^{-\mathrm{j}\pi/3})] = \tfrac{1}{3}(4 - 2\cos\tfrac{1}{3}\pi) = 1$$

$$\tilde{g}_1 = \tfrac{1}{3}\sum_{k=0}^{2} G_k \,\mathrm{e}^{\mathrm{j}k\times 1\times 2\pi/3} = \tfrac{1}{3}(G_0 + G_1\,\mathrm{e}^{\mathrm{j}2\pi/3} + G_2\,\mathrm{e}^{\mathrm{j}4\pi/3})$$

$$= \tfrac{1}{3}(4 + 1 + 1) = 2$$

$$\tilde{g}_2 = \tfrac{1}{3}\sum_{k=0}^{2} G_k \,\mathrm{e}^{\mathrm{j}k\times 2\times 2\pi/3} = \tfrac{1}{3}(G_0 + G_1\,\mathrm{e}^{\mathrm{j}4\pi/3} + G_2\,\mathrm{e}^{\mathrm{j}8\pi/3})$$

$$= \tfrac{1}{3}[4 + \mathrm{e}^{\mathrm{j}\pi}(\mathrm{e}^{\mathrm{j}\pi/3} + \mathrm{e}^{-\mathrm{j}\pi/3})] = \tfrac{1}{3}(4 - 2\cos\tfrac{1}{3}\pi) = 1$$

That is

$$\{\tilde{g}_n\}_{n=0}^{2} = \{1, 2, 1\} = \{g_k\}_{k=0}^{2}$$

and thus the original sequence has been recovered exactly from its transform.

We see from Example 5.19 that the operation of calculating N terms of the transformed sequence involved $N \times N = N^2$ multiplications and $N(N-1)$ summations, all of which are operations involving complex numbers in general. The computation of the discrete Fourier transform in this direct manner is thus said to be a computation of complexity N^2. Such computations rapidly become impossible as N increases, owing to the time required for this execution.

5.6.4 Estimation of the continuous Fourier transform

We saw in Section 5.4.2 that the continuous Fourier transform provides a means of examining the frequency response of a stable linear time-invariant *continuous*-time system. Similarly, we saw in Section 5.6.2 how a discrete-time Fourier transform could be developed that allows examination of the frequency response of a stable linear time-invariant *discrete*-time system. By sampling this latter transform, we developed the discrete Fourier transform itself. Why did we do this? First we have found a way (at least in theory) of involving the computer in our efforts. Secondly, as we shall now show, we can use the discrete Fourier transform to estimate the continuous Fourier transform of a continuous-time signal. To see how this is done, let us first examine what happens when we sample a continuous-time signal.

Suppose that $f(t)$ is a non-periodic continuous-time signal, a portion of which is shown in Figure 5.24(a). Let us sample the signal at equal intervals T, to generate the sequence

$$\{f(0), f(T), \ldots, f(nT), \ldots\}$$

as shown in Figure 5.24(b). Imagine now that each of these samples is presented in turn, at the appropriate instant, as the input to a continuous linear time-invariant system with impulse response $h(t)$. The output would then be, from Section 2.6.6,

$$\begin{aligned} y(t) &= \int_{-\infty}^{\infty} h(t-\tau)f(0)\delta(\tau)\,d\tau + \int_{-\infty}^{\infty} h(t-\tau)f(\tau)\delta(\tau - T)\,d\tau \\ &\quad + \ldots + \int_{-\infty}^{\infty} h(t-\tau)f(nT)\delta(\tau - nT)\,d\tau + \ldots \\ &= \int_{-\infty}^{\infty} h(t-\tau)\sum_{k=0}^{\infty} f(kT)\delta(\tau - kT)\,d\tau \end{aligned}$$

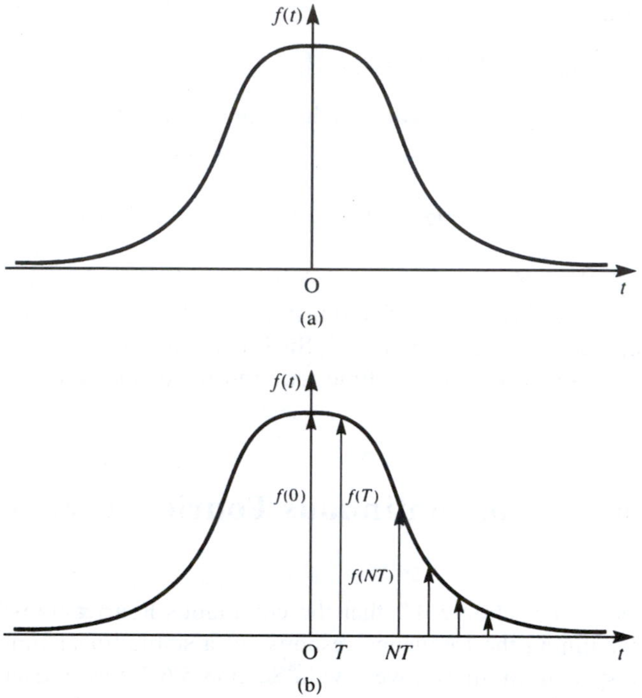

Figure 5.24 (a) Continuous-time signal $f(t)$; (b) samples drawn from $f(t)$.

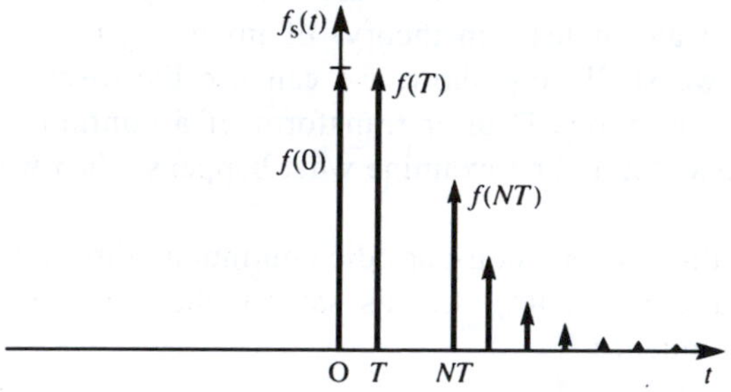

Figure 5.25 Visualization of $f_s(t)$ defined in (5.84).

Thus

$$y(t) = \int_{-\infty}^{\infty} h(t - \tau) f_s(\tau)\, d\tau \tag{5.83}$$

where

$$f_s(t) = \sum_{k=0}^{\infty} f(kT)\delta(t - kT) = f(t) \sum_{k=0}^{\infty} \delta(t - kT) \tag{5.84}$$

which we identify as a 'continuous-time' representation of the sampled version of $f(t)$. We are thus lead to picture $f_s(t)$ as in Figure 5.25.

In order to admit the possibility of signals that are non-zero for $t < 0$, we can generalize (5.84) slightly by allowing in general that

$$f_s(t) = f(t) \sum_{k=-\infty}^{\infty} \delta(t - kT) \tag{5.85}$$

We can now use convolution to find the Fourier transform $F_s(j\omega)$ of $f_s(t)$. Using the representation (5.85) for $f_s(t)$, we have

$$F_s(j\omega) = \mathscr{F}\{f_s(t)\} = \mathscr{F}\left\{f(t) \sum_{k=-\infty}^{\infty} \delta(t - kT)\right\}$$

which, on using (5.61), leads to

$$F_s(j\omega) = \frac{1}{2\pi} F(j\omega) * \mathscr{F}\left\{\sum_{k=-\infty}^{\infty} \delta(t - kT)\right\} \tag{5.86}$$

where

$$\mathscr{F}\{f(\mathrm{t})\} = F(j\omega)$$

From (5.54),

$$\mathscr{F}\left\{\sum_{k=-\infty}^{\infty} \delta(t - kT)\right\} = \frac{2\pi}{T} \sum_{k=-\infty}^{\infty} \delta\left(\omega - \frac{2\pi k}{T}\right)$$

so that, assuming the interchange of the order of integration and summation to be possible, (5.86) becomes

$$\begin{aligned}
F_s(j\omega) &= \frac{1}{2\pi} F(j\omega) * \frac{2\pi}{T} \sum_{k=-\infty}^{\infty} \delta\left(\omega - \frac{2\pi k}{T}\right) \\
&= \frac{1}{T} \int_{-\infty}^{\infty} F(j[\omega - \omega']) \sum_{k=-\infty}^{\infty} \delta\left(\omega' - \frac{2\pi k}{T}\right) d\omega' \\
&= \frac{1}{T} \sum_{k=-\infty}^{\infty} \int_{-\infty}^{\infty} F(j[\omega - \omega'])\delta\left(\omega' - \frac{2\pi k}{T}\right) d\omega' \\
&= \frac{1}{T} \sum_{k=-\infty}^{\infty} F\left(j\left(\omega - \frac{2\pi k}{T}\right)\right)
\end{aligned}$$

Thus

$$F_s(j\omega) = \frac{1}{T} \sum_{k=-\infty}^{\infty} F(j[\omega - k\omega_0]), \quad \omega_0 = \frac{2\pi}{T} \tag{5.87}$$

Examining (5.87), we see that the spectrum $F_s(j\omega)$ of the sampled version $f_s(t)$ of $f(t)$ consists of repeats of the spectrum $F(j\omega)$ of $f(t)$ scaled by a factor $1/T$, these repeats being spaced at intervals $\omega_0 = 2\pi/T$ apart. Figure 5.26(a) shows the

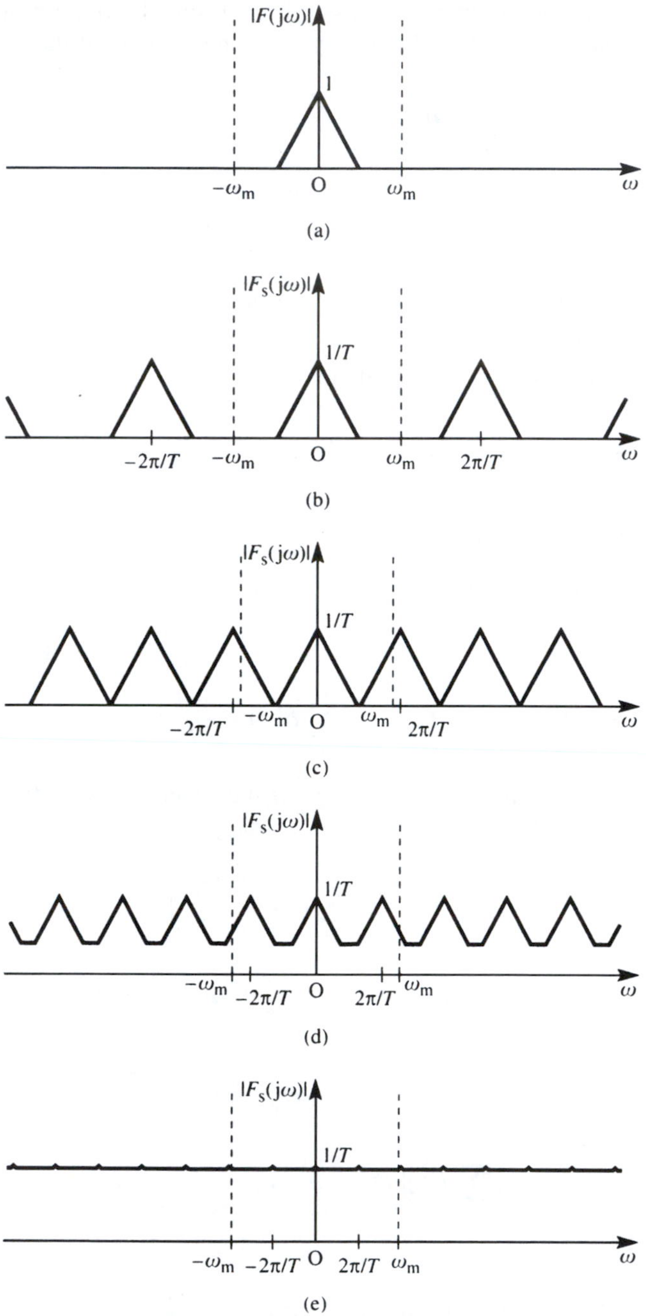

Figure 5.26 (a) Amplitude spectrum of a band-limited signal $f(t)$. (b)–(e) amplitude spectrum $|F_s(j\omega)|$ of $f_s(t)$, showing periodic repetition of $|F_s(j\omega)|$ and interaction effects as T increases.

amplitude spectrum $|F(j\omega)|$ of a band-limited signal $f(t)$, that is, a signal whose spectrum is zero for $|\omega| > \omega_m$. Figures 5.26(b–e) show the amplitude spectrum $|F_s(j\omega)|$ of the sampled version for increasing values of the sampling interval T. Clearly, as T increases, the spectrum of $F(j\omega)$, as observed using $|F_s(j\omega)|$ in $-\omega_m < \omega < \omega_m$, becomes more and more misleading because of 'interaction' from neighbouring copies.

As we saw in Section 5.6.2, the periodicity in the amplitude spectrum $|F_s(j\omega)|$ of $f_s(t)$ is inevitable as a consequence of the sampling process, and ways have to be found to minimize the problems it causes. The interaction observed in Figure 5.26 between the periodic repeats is known as **aliasing error**, and it is clearly essential to minimize this effect. This can be achieved in an obvious way if the original unsampled signal $f(t)$ is band-limited as in Figure 5.26(a). It is apparent that we must arrange that the periodic repeats of $|F(j\omega)|$ be far enough apart to prevent interaction between the copies. This implies that we have

$$\omega_0 \geqslant 2\omega_m$$

at an absolute (and impractical!) minimum. Since $\omega_0 = 2\pi/T$, the constraint implies that

$$T \leqslant \pi/\omega_m$$

where T is the interval between samples. The minimum time interval allowed is

$$T_{min} = \pi/\omega_m$$

which is known as the **Nyquist interval** and we have in fact deduced a form of the **Nyquist–Shannon sampling theorem**. If $T < T_{min}$ then the 'copies' of $F(j\omega)$ are isolated from each other, and we can focus on just one copy, either for the purpose of signal reconstruction, or for the purposes of the estimation of $F(j\omega)$ itself. Here we are concerned only with the latter problem. Basically, we have established a condition under which the spectrum of the samples of the band-limited signal $f(t)$, that is, the spectrum of $f_s(t)$, can be used to estimate $F(j\omega)$.

Suppose we have drawn N samples from a continuous signal $f(t)$ at intervals T, in accordance with the Nyquist criterion, as in Figure 5.27. We then consider

$$f_s(t) = \sum_{k=0}^{N-1} f(kT)\delta(t - kT)$$

or equivalently, the sequence

$$\{f_k\}_{k=0}^{N-1}, \quad \text{where} \quad f_k = f(kT)$$

Note that

$$f_s(t) = 0 \quad (t > (N-1)T)$$

so that

$$f_k = 0 \quad (k > N-1)$$

The Fourier transform of $f_s(t)$ is

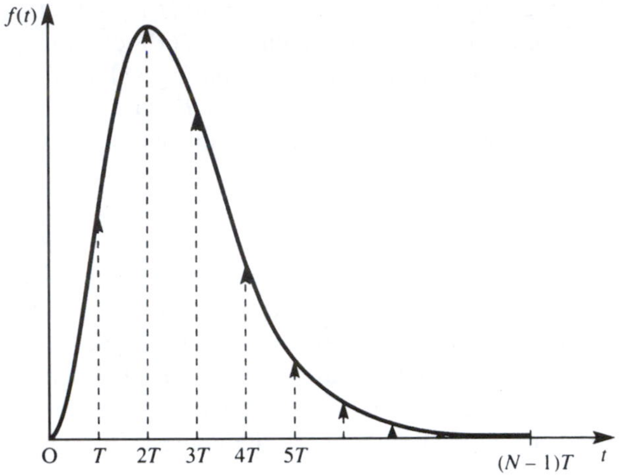

Figure 5.27 Sampling of a continuous-time signal.

$$
\begin{aligned}
F_s(j\omega) &= \int_{-\infty}^{\infty} f_s(t)\,e^{-j\omega t}\,dt \\
&= \int_{-\infty}^{\infty} \sum_{k=0}^{N-1} f(kT)\delta(t-kT)\,e^{-j\omega t}\,dt \\
&= \sum_{k=0}^{N-1} \int_{-\infty}^{\infty} f(kT)\delta(t-kT)\,e^{-j\omega t}\,dt \\
&= \sum_{k=0}^{N-1} f(kT)\,e^{-j\omega kT} = \sum_{k=0}^{N-1} f_k\,e^{-j\omega kT}
\end{aligned}
\tag{5.88}
$$

The transform in (5.88) is a function of the continuous variable ω, so, as in (5.78), we must now sample the continuous spectrum $F_s(j\omega)$ to permit computer evaluation.

We chose N samples to represent $f(t)$ in the time domain, and for this reason we also choose N samples in the frequency domain to represent $F(j\omega)$. Thus we sample (5.88) at intervals $\Delta\omega$, to generate the sequence

$$\{F_s(jn\,\Delta\omega)\}_{n=0}^{N-1} \tag{5.89a}$$

where

$$F_s(jn\,\Delta\omega) = \sum_{k=0}^{N-1} f_k\,e^{-jkn\,\Delta\omega\,T} \tag{5.89b}$$

We must now choose the frequency-domain sampling interval $\Delta\omega$. To see how to do this, recall that the sampled spectrum $F_s(j\omega)$ consisted of repeats of $F(j\omega)$, spaced at intervals $2\pi/T$ apart. Thus to sample just one copy in its entirety, we should choose

$$N\Delta\omega = 2\pi/T$$

or

$$\Delta\omega = 2\pi/NT \tag{5.90}$$

Note that the resulting sequence, defined outside $0 \leq n \leq N-1$, is periodic, as we should expect. However, note also that, following our discussion in Section 5.6, the process of recovering a time signal from samples of its spectrum will result in a periodic waveform, whatever the nature of the original time signal. We should not be surprised by this, since it is exactly in accordance with our introductory discussion in Section 5.1.

In view of the scaling factor $1/T$ in (5.87), our estimate of the Fourier transform $F(\mathrm{j}\omega)$ of $f(t)$ over the interval

$$0 \leq t \leq (N-1)T$$

will from (5.89), be the sequence of samples

$$\{TF_s(\mathrm{j}n\Delta\omega)\}_{n=0}^{N-1}$$

where

$$TF_s(\mathrm{j}n\Delta\omega) = T\sum_{k=0}^{N-1} f_k \mathrm{e}^{-\mathrm{j}kn\Delta\omega T}$$

which, from the definition of the discrete Fourier transform in (5.78), gives

$$TF_s(\mathrm{j}n\Delta\omega) = T \times \mathrm{DFT}\,\{f_k\}$$

where DFT $\{f_k\}$ is the discrete Fourier transform of the sequence $\{f_k\}$. We illustrate the use of this estimate in Example 5.20.

EXAMPLE 5.20

The delayed triangular pulse $f(t)$ is as illustrated in Figure 5.28. Estimate its Fourier transform using 10 samples and compare with the exact values.

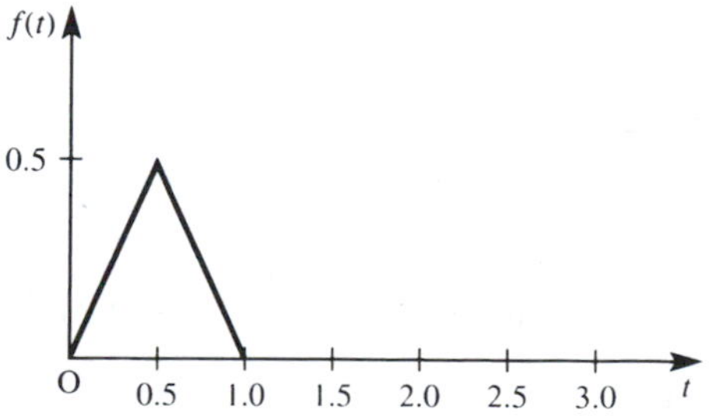

Figure 5.28 The delayed triangular pulse.

Solution Using $N = 10$ samples at intervals $T = 0.2$s, we generate the sequence

$$\{f_k\}_{k=0}^{9} = \{f(0), f(0.2), f(0.4), f(0.6), f(0.8), f(1.0), f(1.2), f(1.4), f(1.6), f(1.8)\}$$

Clearly, from Figure 5.28, we can express the continuous function $f(t)$ as

$$f(t) = \begin{cases} t & (0 \leqslant t \leqslant 0.5) \\ 1 - t & (0.5 < t < 1) \\ 0 & (t \geqslant 1) \end{cases}$$

and so

$$\{f_k\}_{k=0}^{9} = \{0, 0.2, 0.4, 0.4, 0.2, 0, 0, 0, 0, 0\}$$

Using (5.78), the discrete Fourier transform $\{F_n\}_{n=0}^{9}$ of the sequence $\{f_k\}_{k=0}^{9}$ is generated by

$$F_n = \sum_{k=0}^{9} f_k \, e^{-jkn\,\Delta\omega\,T}, \quad \text{where} \quad \Delta\omega = \frac{2\pi}{NT} = \frac{2\pi}{10 \times 0.2} = \pi$$

That is,

$$F_n = \sum_{k=0}^{9} f_k \, e^{-jkn(0.2\pi)}$$

or, since $f_0 = f_5 = f_6 = f_7 = f_8 = f_9 = 0$,

$$F_n = \sum_{k=1}^{4} f_k \, e^{-jnk(0.2\pi)}$$

The estimate of the Fourier transform, also based on $N = 10$ samples, is then the sequence

$$\{TF_n\}_{n=0}^{9} = \{0.2F_n\}_{n=0}^{9}$$

We thus have 10 values representing the Fourier transform at

$$\omega = n\Delta\omega \quad (n = 0, 1, 2, \ldots, 9)$$

or since $\Delta\omega = 2\pi/NT$

$$\omega = 0, \pi, 2\pi, \ldots, 9\pi$$

At $\omega = \pi$, corresponding to $n = 1$, our estimate is

$$\begin{aligned} 0.2F_1 &= 0.2\sum_{k=1}^{4} f_k \, e^{-jk(0.2\pi)} \\ &= 0.2[0.2\,e^{-j(0.2\pi)} + 0.4(e^{-j(0.4\pi)} + e^{-j(0.6\pi)}) + 0.2\,e^{-j(0.8\pi)}] \\ &= -0.1992j \end{aligned}$$

At $\omega = 2\pi$, corresponding to $n = 2$, our estimate is

$$\begin{aligned} 0.2F_2 &= 0.2\sum_{k=1}^{4} f_k \, e^{-jk(0.4\pi)} \\ &= 0.2[0.2\,e^{-j(0.4\pi)} + 0.4(e^{-j(0.8\pi)} + e^{-j(1.2\pi)}) + 0.2\,e^{-j(1.6\pi)}] \\ &= -0.1047 \end{aligned}$$

ω	Exact $F(j\omega)$	DFT estimate	$\lvert F(j\omega)\rvert$	$\lvert$DFT estimate$\rvert$	% error
0	0.2500	0.2400	0.2500	0.2400	4%
π	−0.2026j	−0.1992j	0.2026	0.1992	1.7%
2π	−0.1013	−0.1047	0.1013	0.1047	3.2%
3π	0.0225j	0.0180j	0.0225	0.0180	20%
4π	0	−0.0153	0	0.0153	–
5π	−0.0081j	0	0.0081	0	–
6π	−0.0113	−0.0153	0.0113	0.0153	–
7π	0.0041	−0.0180j	0.0041	0.0180	–
8π	0	−0.1047	0	0.1047	–
9π	−0.0025j	0.1992j	0.0025	0.1992	–

Figure 5.29 Comparison of exact results and DFT estimate for the amplitude spectrum of the signal of Example 5.20.

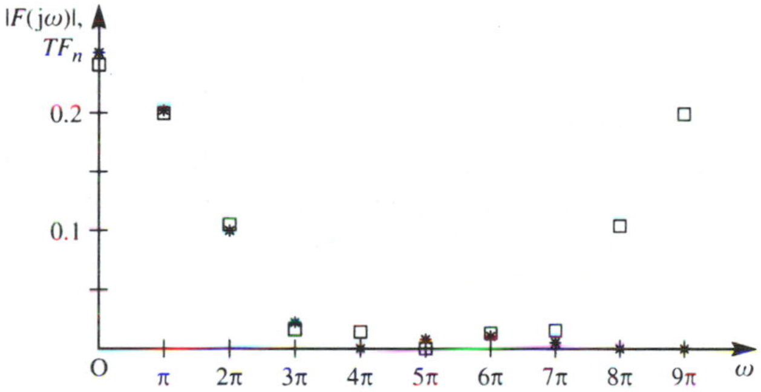

Figure 5.30 Exact result $|F(j\omega)|$ (∗) and DFT estimate TF_n (□) of the Fourier transform in Example 5.20.

Continuing in this manner, we compute the sequence

$$\{0.2F_0, 0.2F_1, \ldots, 0.2F_n\}$$

as

$$\{0.2400, -0.1992j, -0.1047, 0.0180j, -0.0153, 0, -0.0153, -0.0180j, -0.1047, 0.1992j\}$$

This then represents the estimate of the Fourier transform of the continuous function $f(t)$. The exact value of the Fourier transform of $f(t)$ is easily computed by direct use of the definition (5.15) as

$$F(j\omega) = \mathscr{F}\{f(t)\} = \tfrac{1}{4}\,e^{-j\omega/2}\,\text{sinc}^2\,\tfrac{1}{4}\,\omega$$

which we can use to examine the validity of our result. The comparison is shown in Figure 5.29 and illustrated graphically in Figure 5.30.

From the Nyquist–Shannon sampling theorem, with $T = 0.2$ s, we deduce that our results will be completely accurate if the original signal $f(t)$ is band-limited with a zero spectrum for $|\omega| > |\omega_m| = 5\pi$. Our signal is not strictly band-limited in this way, and we thus expect to observe some error in our results, particularly near $\omega = 5\pi$, because of the effects of aliasing. The estimate obtained is satisfactory at $\omega = 0, \pi, 2\pi$, but begins to lose accuracy at $\omega = 3\pi$. Results obtained above $\omega = 5\pi$ are seen to be images of those obtained for values below $\omega = 5\pi$, and this is to be expected owing to the periodicity of the DFT. In our calculation the DFT sequence will be periodic, with period $N = 10$; thus, for example,

$$|TF_7| = |TF_{7-10}| = |TF_{-3}| = T|F_{-3}|$$

As we have seen many times, for a real signal the amplitude spectrum is symmetric about $\omega = 0$. Thus $|F_{-3}| = |F_3|$, $|F_{-5}| = |F_5|$, and so on, and the effects of the symmetry are apparent in Figure 5.29. It is perhaps worth observing that if we had calculated (say) $\{TF_{-4}, TF_{-3}, \dots, TF_0, TF_1, \dots, TF_5\}$, we should have obtained a 'conventional' plot, with the right-hand portion, beyond $\omega = 5\pi$, translated to the left of the origin. However, using the plot of the amplitude spectrum in the chosen form does highlight the source of error due to aliasing.

In this section we have discussed a method by which Fourier transforms can be estimated numerically, at least in theory. It is apparent, though, that the amount of labour involved is significant, and as we observed in Section 5.6.3 an algorithm based on this approach is in general prohibitive in view of the amount of computing time required. The next section gives a brief introduction to a method of overcoming this problem.

5.6.5 The fast Fourier transform

The calculation of a discrete Fourier transform based on N sample values requires, as we have seen, N^2 complex multiplications and $N(N - 1)$ summations. For real signals, symmetry can be exploited, but for large N, $\frac{1}{2}N^2$ does not represent a significant improvement over N^2 for the purposes of computation. In fact, a totally new approach to the problem was required before the discrete Fourier transform could become a practical engineering tool. In 1965 Cooley and Tukey introduced the **fast Fourier transform (FFT)** in order to reduce the computational complexity (J. W. Cooley and J. W. Tukey, An algorithm for the machine computation of complex Fourier series, *Mathematics of Computation* **19** (1965) 297–301). We shall briefly introduce their approach in this section: for a full discussion see E. E. Brigham, *The Fast Fourier Transform* (Prentice-Hall, Englewood Cliffs, NJ, 1974), whose treatment is similar to that adopted here.

We shall restrict ourselves to the situation where $N = 2^{\gamma}$ for some integer γ, and, rather than examine the general case, we shall focus on a particular value of γ. In proceeding in this way, the idea should be clear and the extension to other values of γ appear credible. We can summarize the approach as being in three stages:

(a) matrix formulation;

(b) matrix factorization; and, finally,

(c) rearranging.

We first consider a matrix formulation of the DFT. From (5.78), the Fourier transform sequence $\{G_k\}_{k=0}^{N-1}$ of the sequence $\{g_n\}_{n=0}^{N-1}$ is generated by

$$G_k = \sum_{n=0}^{N-1} g_n \, e^{-j2\pi nk/N} \quad (k = 0, 1, \ldots, N-1) \tag{5.91}$$

We shall consider the particular case when $\gamma = 2$ (that is, $N = 2^2 = 4$), and define

$$W = e^{-j2\pi/N} = e^{-j\pi/2}$$

so that (5.91) becomes

$$G_k = \sum_{n=0}^{N-1} g_n W^{nk} = \sum_{n=0}^{3} g_n W^{nk} \quad (k = 0, 1, 2, 3)$$

Writing out the terms of the transformed sequence, we have

$$G_0 = g_0 W^0 + g_1 W^0 + g_2 W^0 + g_3 W^0$$
$$G_1 = g_0 W^0 + g_1 W^1 + g_2 W^2 + g_3 W^3$$
$$G_2 = g_0 W^0 + g_1 W^2 + g_2 W^4 + g_3 W^6$$
$$G_3 = g_0 W^0 + g_1 W^3 + g_2 W^6 + g_3 W^9$$

which may be expressed in the vector–matrix form

$$\begin{bmatrix} G_0 \\ G_1 \\ G_2 \\ G_3 \end{bmatrix} = \begin{bmatrix} W^0 & W^0 & W^0 & W^0 \\ W^0 & W^1 & W^2 & W^3 \\ W^0 & W^2 & W^4 & W^6 \\ W^0 & W^3 & W^6 & W^9 \end{bmatrix} \begin{bmatrix} g_0 \\ g_1 \\ g_2 \\ g_3 \end{bmatrix} \tag{5.92}$$

or, more generally, as

$$\boldsymbol{G}_k = \boldsymbol{W}^{nk} \boldsymbol{g}_n$$

where the vectors $\boldsymbol{G}_k$ and $\boldsymbol{g}_n$ and the square matrix $\boldsymbol{W}^{nk}$ are defined as in (5.92). The next step relates to the special properties of the entries in the matrix $\boldsymbol{W}^{nk}$. Note that $W^{nk} = W^{nk+pN}$, where p is an integer, and so,

$$W^4 = W^0 = 1$$
$$W^6 = W^2$$
$$W^9 = W^1$$

Thus (5.92) becomes

$$\begin{bmatrix} G_0 \\ G_1 \\ G_2 \\ G_3 \end{bmatrix} = \begin{bmatrix} 1 & 1 & 1 & 1 \\ 1 & W^1 & W^2 & W^3 \\ 1 & W^2 & W^0 & W^2 \\ 1 & W^3 & W^2 & W^1 \end{bmatrix} \begin{bmatrix} g_0 \\ g_1 \\ g_2 \\ g_3 \end{bmatrix} \tag{5.93}$$

Equation (5.93) is the end of the first stage of the development. In fact, we have so far only made use of the properties of the Nth roots of unity. Stage two involves the factorization of a matrix, the details of which will be explained later.

Note that

$$\begin{bmatrix} 1 & W^0 & 0 & 0 \\ 1 & W^2 & 0 & 0 \\ 0 & 0 & 1 & W^1 \\ 0 & 0 & 1 & W^3 \end{bmatrix} \begin{bmatrix} 1 & 0 & W^0 & 0 \\ 0 & 1 & 0 & W^0 \\ 1 & 0 & W^2 & 0 \\ 0 & 1 & 0 & W^2 \end{bmatrix} = \begin{bmatrix} 1 & 1 & 1 & 1 \\ 1 & W^2 & W^0 & W^2 \\ 1 & W^1 & W^2 & W^3 \\ 1 & W^3 & W^2 & W^1 \end{bmatrix} \tag{5.94}$$

where we have used $W^5 = W^1$ and $W^0 = 1$ (in the top row). The matrix on the right-hand side of (5.94) is the coefficient matrix of (5.93), *but with rows 2 and 3 interchanged*. Thus we can write (5.93) as

$$\begin{bmatrix} G_0 \\ G_2 \\ G_1 \\ G_3 \end{bmatrix} = \begin{bmatrix} 1 & W^0 & 0 & 0 \\ 1 & W^2 & 0 & 0 \\ 0 & 0 & 1 & W^1 \\ 0 & 0 & 1 & W^3 \end{bmatrix} \begin{bmatrix} 1 & 0 & W^0 & 0 \\ 0 & 1 & 0 & W^0 \\ 1 & 0 & W^2 & 0 \\ 0 & 1 & 0 & W^2 \end{bmatrix} \begin{bmatrix} g_0 \\ g_1 \\ g_2 \\ g_3 \end{bmatrix} \tag{5.95}$$

We now define a vector $\boldsymbol{g}'$ as

$$\boldsymbol{g}' = \begin{bmatrix} g'_0 \\ g'_1 \\ g'_2 \\ g'_3 \end{bmatrix} = \begin{bmatrix} 1 & 0 & W^0 & 0 \\ 0 & 1 & 0 & W^0 \\ 1 & 0 & W^2 & 0 \\ 0 & 1 & 0 & W^2 \end{bmatrix} \begin{bmatrix} g_0 \\ g_1 \\ g_2 \\ g_3 \end{bmatrix} \tag{5.96}$$

It then follows from (5.96) that

$$g'_0 = g_0 + W^0 g_2$$

$$g'_1 = g_1 + W^0 g_3$$

so that g'_0 and g'_1 are each calculated by one complex multiplication and one addition. Of course, in this special case, since $W^0 = 1$, the multiplication is unnecessary, but we are attempting to infer the general situation. For this reason, W^0 has not been replaced by 1.

Also, it follows from (5.96) that

$$g'_2 = g_0 + W^2 g_2$$

$$g'_3 = g_1 + W^2 g_3$$

and, since $W^2 = -W^0$, the computation of the pair g'_2 and g'_3 can make use of the computations of $W^0 g_2$ and $W^0 g_3$, with one further addition in each case. Thus the vector $\boldsymbol{g}'$ is determined by a total of four complex additions and two complex multiplications.

To complete the calculation of the transform, we return to (5.95), and rewrite it in the form

$$\begin{bmatrix} G_0 \\ G_2 \\ G_1 \\ G_3 \end{bmatrix} = \begin{bmatrix} 1 & W^0 & 0 & 0 \\ 1 & W^2 & 0 & 0 \\ 0 & 0 & 1 & W^1 \\ 0 & 0 & 1 & W^3 \end{bmatrix} \begin{bmatrix} g_0' \\ g_1' \\ g_2' \\ g_3' \end{bmatrix} \tag{5.97}$$

It then follows from (5.97) that

$$G_0 = g_0' + W^0 g_1'$$

$$G_2 = g_0' + W^2 g_1'$$

and we see that G_0 is determined by one complex multiplication and one complex addition. Furthermore, because $W^2 = -W^0$, G_2 follows after one further complex addition.

Similarly, it follows from (5.97) that

$$G_1 = g_2' + W^1 g_3'$$

$$G_3 = g_2' + W^3 g_3'$$

and, since $W^3 = -W^1$, a total of one further complex multiplication and two further additions are required to produce the re-ordered transform vector

$$[G_0 \quad G_2 \quad G_1 \quad G_3]^{\mathrm{T}}$$

Thus the total number of operations required to generate the (re-ordered) transform is four complex multiplications and eight complex additions. Direct calculation would have required $N^2 = 16$ complex multiplications and $N(N-1) = 12$ complex additions. Even with a small value of N, these savings are significant, and, interpreting computing time requirements as being proportional to the number of complex multiplications involved, it is easy to see why the FFT algorithm has become an essential tool for computational Fourier analysis. When $N = 2^\gamma$, the FFT algorithm is effectively a procedure for producing γ $N \times N$ matrices of the form (5.94). Extending our ideas, it is possible to see that generally the FFT algorithm, when $N = 2^\gamma$, will require $\frac{1}{2}N\gamma$ (four, when $N = 2^2 = 4$) complex multiplications and $N\gamma$ (eight, when $N = 4$) complex additions. Since

$$\gamma = \log_2 N$$

the demands of the FFT algorithm in terms of computing time, estimated on the basis of the number of complex multiplications, is often given as about $N \log_2 N$, as opposed to N^2 for the direct evaluation of the transform. This completes the second stage of our task, and we are only left with the problem of rearrangement of our transform vector into 'natural' order.

The means by which this is achieved is most elegant. Instead of indexing G_0, G_1, G_2, G_3 in decimal form, an alternative binary notation is used, and $[G_0 \quad G_1 \quad G_2 \quad G_3]^{\mathrm{T}}$ becomes

$$[G_{00} \quad G_{01} \quad G_{10} \quad G_{11}]^{\mathrm{T}}$$

The process of 'bit reversal' means rewriting a binary number with its bits or digits in reverse order. Applying this process to $[G_{00} \quad G_{01} \quad G_{10} \quad G_{11}]^{\mathrm{T}}$ yields

$$[G_{00} \quad G_{10} \quad G_{01} \quad G_{11}]^{\mathrm{T}} = [G_0 \quad G_2 \quad G_1 \quad G_3]^{\mathrm{T}}$$

with decimal labelling. This latter form is exactly the one obtained at the end of the FFT calculation, and we see that the natural order can easily be recovered by rearranging the output on the basis of bit reversal of the binary indexed version.

We have now completed our introduction to the fast Fourier transform. We shall now consider an example to illustrate the ideas discussed here. We shall then conclude by considering in greater detail the matrix factorization process used in the second stage.

EXAMPLE 5.21 Use the method of the FFT algorithm to compute the Fourier transform of the sequence

$$\{g_n\}_{n=0}^{3} = \{1, 2, 1, 0\}$$

Solution In this case $N = 4 = 2^2$, and we begin by computing the vector $\boldsymbol{g}'_n = [g'_0 \quad g'_1 \quad g'_2 \quad g'_3]^{\mathrm{T}}$, which, from (5.96), is given by

$$\boldsymbol{g}'_n = \begin{bmatrix} 1 & 0 & W^0 & 0 \\ 0 & 1 & 0 & W^0 \\ 1 & 0 & W^2 & 0 \\ 0 & 1 & 0 & W^2 \end{bmatrix} \begin{bmatrix} g_0 \\ g_1 \\ g_2 \\ g_3 \end{bmatrix}$$

For $N = 4$

$$W^n = (\mathrm{e}^{-\mathrm{j}2\pi/4})^n = \mathrm{e}^{-\mathrm{j}n\pi/2}$$

and so

$$\boldsymbol{g}'_n = \begin{bmatrix} 1 & 0 & 1 & 0 \\ 0 & 1 & 0 & 1 \\ 1 & 0 & -1 & 0 \\ 0 & 1 & 0 & -1 \end{bmatrix} \begin{bmatrix} 1 \\ 2 \\ 1 \\ 0 \end{bmatrix} = \begin{bmatrix} 2 \\ 2 \\ 0 \\ 2 \end{bmatrix}$$

Next, we compute the 'bit-reversed' order transform vector $\boldsymbol{G}'$, say, which from (5.97) is given by

$$\boldsymbol{G}' = \begin{bmatrix} 1 & W^0 & 0 & 0 \\ 1 & W^2 & 0 & 0 \\ 0 & 0 & 1 & W^1 \\ 0 & 0 & 1 & W^3 \end{bmatrix} \begin{bmatrix} g'_0 \\ g'_1 \\ g'_2 \\ g'_3 \end{bmatrix}$$

or, in this particular case,

$$G' = \begin{bmatrix} G_{00} \\ G_{10} \\ G_{01} \\ G_{11} \end{bmatrix} \begin{bmatrix} 1 & 1 & 0 & 0 \\ 1 & -1 & 0 & 0 \\ 0 & 0 & 1 & -\mathrm{j} \\ 0 & 0 & 1 & \mathrm{j} \end{bmatrix} \begin{bmatrix} 2 \\ 2 \\ 0 \\ 2 \end{bmatrix} = \begin{bmatrix} 4 \\ 0 \\ -2\mathrm{j} \\ 2\mathrm{j} \end{bmatrix} \tag{5.98}$$

Finally, we recover the transform vector $\boldsymbol{G} = [G_0 \quad G_1 \quad G_2 \quad G_3]^{\mathrm{T}}$ as

$$\boldsymbol{G} = \begin{bmatrix} 4 \\ -2\mathrm{j} \\ 0 \\ 2\mathrm{j} \end{bmatrix}$$

and we have thus established the Fourier transform of the sequence {1, 2, 1, 0} as the sequence

{4, −2j, 0, 2j}

It is interesting to compare the labour involved in this calculation with that in Example 5.19.

To conclude this section, we reconsider the matrix factorization operation, which is at the core of the process of calculating the fast Fourier transform. In a book of this nature it is not appropriate to reproduce a proof of the validity of the algorithm for any N of the form $N = 2^\gamma$. Rather, we shall illustrate how the factorization we introduced in (5.94) was obtained. The factored form of the matrix will not be generated in any calculation: what actually happens is that the various summations are performed using their structural properties.

From (5.91), with $W = \mathrm{e}^{-\mathrm{j}2\pi/N}$, we wish to calculate the sums

$$G_k = \sum_{n=0}^{N-1} g_n W^{nk} \quad k = 0, 1, \ldots, N-1 \tag{5.99}$$

In the case $N = 4$, $\gamma = 2$ we see that k and n take only the values 0, 1, 2 and 3, so we can represent both k and n using two-digit binary numbers; in general γ-digit binary numbers will be required.

We write $k = k_1k_0$ and $n = n_1n_0$, where k_0, k_1, n_0 and n_1 may take the values 0 or 1 only. For example, $k = 3$ becomes $k = 11$ and $n = 2$ becomes $n = 10$. The decimal form can always be recovered easily as $k = 2k_1 + k_0$ and $n = 2n_1 + n_0$.

Using binary notation, we can write (5.99) as

$$G_{k_1k_0} = \sum_{n_0=0}^{1} \sum_{n_1=0}^{1} g_{n_1n_0} W^{(2n_1+n_0)(2k_1+k_0)} \tag{5.100}$$

The single summation of (5.99) is now replaced, when $\gamma = 2$, by two summations. Again we see that for the more general case with $N = 2^\gamma$ a total of γ summations replaces the single sum of (5.99).

The matrix factorization operation with which we are concerned is now achieved by considering the term

$$W^{(2n_1+n_0)(2k_1+k_0)}$$

in (5.100). Expanding gives

$$\begin{aligned} W^{(2n_1+n_0)(2k_1+k_0)} &= W^{(2k_1+k_0)2n_1} W^{(2k_1+k_0)n_0} \\ &= W^{4n_1k_1} W^{2n_1k_0} W^{(2k_1+k_0)n_0} \end{aligned} \quad \textbf{(5.101)}$$

Since $W = e^{-j2\pi/N}$, and $N = 4$ in this case, the leading term in (5.101) becomes

$$\begin{aligned} W^{4n_1k_1} &= (e^{-j2\pi/4})^{4n_1k_1} = (e^{-j2\pi})^{n_1k_1} \\ &= 1^{n_1k_1} = 1 \end{aligned}$$

Again we observe that in the more general case such a factor will always emerge.

Thus (5.101) can be written as

$$W^{(2n_1+n_0)(2k_1+k_0)} = W^{2n_1k_0} W^{(2k_1+k_0)n_0}$$

so that (5.100) becomes

$$G_{k_1k_0} = \sum_{n_0=0}^{1} \left[\sum_{n_1=0}^{1} g_{n_1n_0} W^{2n_1k_0} \right] W^{(2k_1+k_0)n_0} \quad \textbf{(5.102)}$$

which is the required matrix factorization. This can be seen by writing

$$g'_{k_0n_0} = \sum_{n_1=0}^{1} g_{n_1n_0} W^{2n_1k_0} \quad \textbf{(5.103)}$$

so that the sum in the square brackets in (5.102) defines the four relations

$$\left. \begin{aligned} g'_{00} &= g_{00}W^{2.0.0} + g_{10}W^{2.1.0} = g_{00} + g_{10}W^0 \\ g'_{01} &= g_{01}W^{2.0.0} + g_{11}W^{2.1.0} = g_{01} + g_{11}W^0 \\ g'_{10} &= g_{00}W^{2.0.1} + g_{10}W^{2.1.1} = g_{00} + g_{10}W^2 \\ g'_{11} &= g_{01}W^{2.0.1} + g_{11}W^{2.1.1} = g_{01} + g_{11}W^2 \end{aligned} \right\} \quad \textbf{(5.104)}$$

which, in matrix form, becomes

$$\begin{bmatrix} g'_{00} \\ g'_{01} \\ g'_{10} \\ g'_{11} \end{bmatrix} = \begin{bmatrix} 1 & 0 & W^0 & 0 \\ 0 & 1 & 0 & W^0 \\ 1 & 0 & W^2 & 0 \\ 0 & 1 & 0 & W^2 \end{bmatrix} \begin{bmatrix} g_{00} \\ g_{01} \\ g_{10} \\ g_{11} \end{bmatrix} \quad \textbf{(5.105)}$$

and we see that we have re-established the system of equations (5.96), this time with binary indexing. Note that in (5.104) and (5.105) we distinguished between terms in W^0 depending on how the zero is generated. When the zero is generated through the value of the summation index (that is, when $n_1 = 0$ and thus a zero will always be generated whatever the value of γ) we replace W^0 by 1. When the index is zero because of the value of k_0, we maintain W^0 as an aid to generalization.

The final stage of the factorization appears when we write the outer summation of (5.102) as

$$G'_{k_0k_1} = \sum_{n_0=0}^{1} g'_{k_0n_0} W^{(2k_1+k_0)n_0} \tag{5.106}$$

which, on writing out in full, gives

$$G'_{00} = g'_{00}W^{0.0} + g'_{01}W^{0.1} = g'_{00} + g'_{01}W^0$$

$$G'_{01} = g'_{00}W^{2.0} + g'_{01}W^{2.1} = g'_{00} + g'_{01}W^2$$

$$G'_{10} = g'_{10}W^{1.0} + g'_{11}W^{1.1} = g'_{10} + g'_{11}W^1$$

$$G'_{11} = g'_{10}W^{3.0} + g'_{11}W^{3.1} = g'_{10} + g'_{11}W^3$$

or, in matrix form,

$$\begin{bmatrix} G'_{00} \\ G'_{01} \\ G'_{10} \\ G'_{11} \end{bmatrix} = \begin{bmatrix} 1 & W^0 & 0 & 0 \\ 1 & W^2 & 0 & 0 \\ 0 & 0 & 1 & W^1 \\ 0 & 0 & 1 & W^3 \end{bmatrix} \begin{bmatrix} g'_{00} \\ g'_{01} \\ g'_{10} \\ g'_{11} \end{bmatrix} \tag{5.107}$$

The matrix in (5.107) is exactly that of (5.97), and we have completed the factorization process as we intended. Finally, to obtain the transform in a natural order, we must carry out the bit-reversal operation. From (5.102) and (5.105), we achieve this by simply writing

$$G_{k_1k_0} = G'_{k_0k_1} \tag{5.108}$$

We can therefore summarize the Cooley–Tukey algorithm for the fast Fourier transform for the case $N = 4$ by the three relations (5.103), (5.106) and (5.108), that is,

$$g'_{k_0n_0} = \sum_{n_1=0}^{1} g_{n_1n_0} W^{2n_1k_0}$$

$$G'_{k_0k_1} = \sum_{n_0=0}^{1} g'_{k_0n_0} W^{(2k_1+k_0)n_0}$$

$$G_{k_1k_0} = G'_{k_0k_1}$$

The evaluation of these three relationships is equivalent to the matrix factorization process together with the bit-reversal procedure discussed above.

The fast Fourier transform is essentially a computer-orientated algorithm and highly efficient codes are available in software libraries, usually requiring a simple subroutine call for their implementation. The interested reader who would prefer to produce 'home-made' code may find listings in the textbook by Brigham quoted at the beginning of this section, as well as elsewhere.

5.6.6 Exercises

28 Calculate directly the discrete Fourier transform of the sequence

$$\{1, 0, 1, 0\}$$

using the methods of Section 5.6.3 (see Example 5.19).

29 Use the fast Fourier transform method to calculate the transform of the sequence of Exercise 28 (follow Example 5.21).

30 Use the FFT algorithm implemented on a computer to improve the experiment with the estimation of the spectrum of the signal of Example 5.20.

31 Derive an FFT algorithm for $N = 2^3 = 8$ points. Work from (5.99), writing

$$k = 4k_2 + 2k_1 + k_0, \quad k_i = 0 \text{ or } 1 \quad \text{for all } i$$

$$n = 4n_2 + 2n_1 + n_0, \quad n_i = 0 \text{ or } 1 \quad \text{for all } i$$

to show that

$$g'_{k_0 n_1 n_0} = \sum_{n_2=0}^{1} g_{n_2 n_1 n_0} W^{4k_0 n_2}$$

$$g''_{k_0 k_1 n_0} = \sum_{n_1=0}^{1} g'_{k_0 n_1 n_0} W^{(2k_1+k_0)2n_1}$$

$$G'_{k_0 k_1 k_2} = \sum_{n_0=0}^{1} g''_{k_0 k_1 n_0} W^{(4k_2+2k_1+k_0)n_0}$$

$$G_{k_2 k_1 k_0} = G'_{k_0 k_1 k_2}$$

5.7 Engineering application: the design of analogue filters

In this section we explore the ideas of mathematical design or synthesis. We shall express in mathematical form the desired performance of a system, and, utilizing the ideas we have developed, produce a system design.

This chapter has been concerned with the frequency-domain representation of signals and systems, and the system we shall design will operate on input signals to produce output signals with specific frequency-domain properties. In Figure 5.31 we illustrate the amplitude response of an ideal low-pass filter. This filter passes perfectly signals, or components of signals, at frequencies less than the cut-off frequency ω_c. Above ω_c, attenuation is perfect, meaning that signals above this frequency are not passed by this filter.

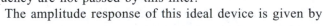

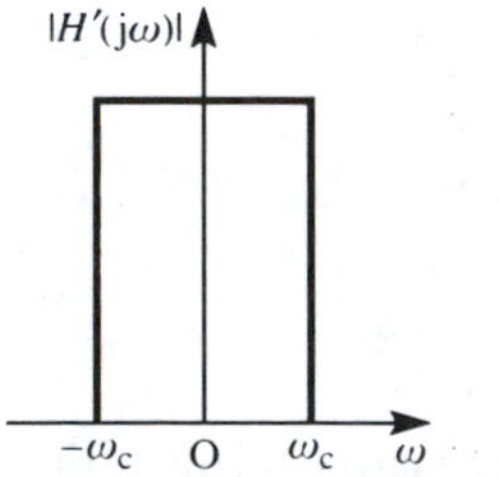

Figure 5.31 Amplitude response of an ideal low-pass filter.

The amplitude response of this ideal device is given by

$$|H'(j\omega)| = \begin{cases} 1 & (|\omega| \leq \omega_c) \\ 0 & (|\omega| > \omega_c) \end{cases}$$

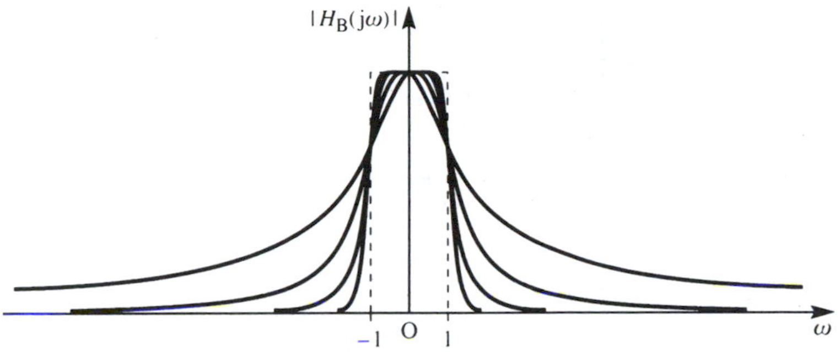

Figure 5.32 Amplitude responses of the Butterworth filters.

Such an ideal response cannot be attained by a real analogue device, and our design problem is to approximate this response to an acceptable degree using a system that can be constructed. A class of functions whose graphs resemble that of Figure 5.31 is the set

$$|H_B(j\omega)| = \frac{1}{\sqrt{[1 + (\omega/\omega_c)^{2n}]}}$$

and we see from Figure 5.32, which corresponds to $\omega_c = 1$, that, as n increases, the graph approaches the ideal response. This particular approximation is known as the **Butterworth approximation**, and is only one of a number of possibilities.

To explore this approach further, we must ask the question whether such a response could be obtained as the frequency response of a realizable, stable linear system. We assume that it can, although if our investigation leads to the opposite conclusion then we shall have to abandon this approach and seek another. If $H_B(j\omega)$ is the frequency response of such a system then it will have been obtained by replacing s with $j\omega$ in the system Laplace transfer function. This is at least possible since, by assumption, we are dealing with a stable system. Now

$$|H_B(j\omega)|^2 = \frac{1}{1 + (j\omega/j\omega_c)^{2n}}$$

where $|H_B(j\omega)|^2 = H_B(j\omega)H_B^*(j\omega)$. If $H_B(s)$ is to have real coefficients, and thus be realizable, then we must have $H_B^*(j\omega) = H(-j\omega)$. Thus

$$H_B(j\omega)H_B(-j\omega) = \frac{1}{1 + (\omega/\omega_c)^{2n}} = \frac{1}{1 + (j\omega/j\omega_c)^{2n}}$$

and we see that the response could be obtained by setting $s = j\omega$ in

$$H_B(s)H_B(-s) = \frac{1}{1 + (s/j\omega_c)^{2n}}$$

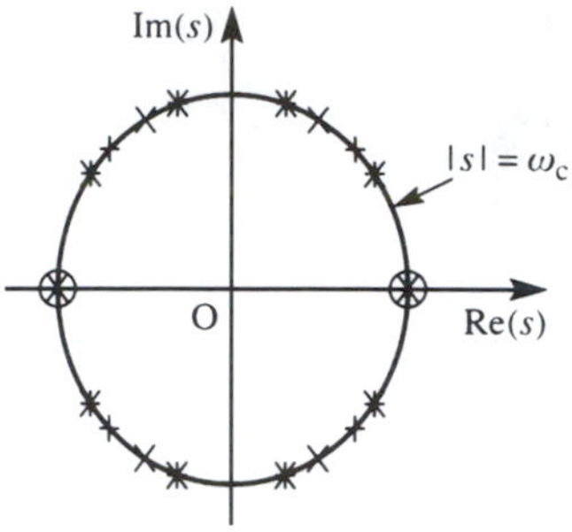

Figure 5.33 Pole locations for the Butterworth filters: (○) $n = 1$; (+) $n = 2$; (×) $n = 3$; (∗) $n = 5$.

Our task is now to attempt to separate $H_B(s)$ from $H_B(-s)$ in such a way that $H_B(s)$ represents the transfer function of a stable system. To do this, we solve the equation

$$1 + (s/j\omega_c)^{2n} = 0$$

to give the poles of $H_B(s)H_B(-s)$ as

$$s = \omega_c e^{j[(2k+1)\pi/2n+\pi/2]} \quad (k = 0, 1, 2, 3, \dots) \tag{5.109}$$

Figure 5.33 shows the pole locations for the cases $n = 1$, 2, 3 and 5. The important observations that we can make from this figure are that in each case there are $2n$ poles equally spaced around the circle of radius ω_c in the Argand diagram, and that there are no poles on the imaginary axis. If $s = s_1$ is a pole of $H_B(s)H_B(-s)$ then so is $s = -s_1$, and we can thus select as poles for the transfer function $H_B(s)$ those lying in the left half-plane. The remaining poles are then those of $H_B(-s)$. By this procedure, we have generated a stable transfer function $H_B(s)$ for our filter design.

The transfer function that we have generated from the frequency-domain specification of system behaviour must now be related to a real system, and this is the next step in the design process. The form of the transfer function for the filter of order n can be shown to be

$$H_B(s) = \frac{\omega_c^n}{(s - s_1)(s - s_2)\dots(s - s_n)}$$

where $s_1, s_2, \dots, s_n$ are the stable poles generated by (5.109). The reader is invited to show that the second-order Butterworth filter has transfer function

$$H_B(s) = \frac{\omega_c^2}{s^2 + \sqrt{2}\omega_c s + \omega_c^2}$$

Writing $Y(s) = H_B(s)U(s)$, with $H_B(s)$ as above, we obtain

$$Y(s) = \frac{\omega_c^2}{s^2 + \sqrt{2}\omega_c s + \omega_c^2} U(s)$$

or

$$(s^2 + \sqrt{2}\omega_c s + \omega_c^2)Y(s) = \omega_c^2\, U(s) \tag{5.110}$$

If we assume that all initial conditions are zero then (5.110) represents the Laplace transform of the differential equation

$$\frac{d^2y(t)}{dt^2} + \sqrt{2}\omega_c\frac{dy(t)}{dt} + \omega_c^2 y(t) = \omega_c^2 u(t) \tag{5.111}$$

This step completes the mathematical aspect of the design exercise. It is possible to show that a system whose behaviour is modelled by this differential equation can be constructed using elementary circuit components, and the specification of such a circuit would complete the design. For a fuller treatment of the subject the interested reader could consult M. J. Chapman, D. P. Goodall and N. C. Steele, *Signal Processing in Electronic Communications*, Horwood Publishing, Chichester, 1997.

To appreciate the operation of this filter, the use of a dynamic simulation package is recommended. After setting the cut-off frequency ω_c, at 4 for example, the output of the system $y(t)$ corresponding to an input signal $u(t) = \sin t + \sin 10t$ will demonstrate the almost-perfect transmission of the low-frequency ($\omega = 1$) term, with nearly total attenuation of the high-frequency ($\omega = 10$) signal. As an extension to this exercise, the differential equation to represent the third- and fourth-order filters should be obtained, and the responses compared. Using a simulation package and an FFT coding, it is possible to investigate the operation of such devices from the viewpoint of the frequency domain by examining the spectrum of samples drawn from both input and output signals.

5.8 Engineering application: modulation, demodulation and frequency-domain filtering

5.8.1 Introduction

In this section we demonstrate the practical implementation of modulation, demodulation and frequency-domain filtering. These are the processes by which an information-carrying signal can be combined with others for transmission along a channel, with the signal subsequently being recovered so that the transmitted information can be extracted. When a number of signals have to be transmitted along a single channel at the same time, one solution is to use the method of **amplitude modulation** as described in Section 5.3.4. We assume that the channel is 'noisy', so that the received signal contains noise, and this signal is then cleaned and demodulated using **frequency-domain filtering** techniques. This idea is easy to describe and to implement, but cannot usually be performed on-line in view of the heavy computational requirements. Our filtering operations are carried out on the frequency-domain version of the signal, and this is generated using the fast Fourier transform algorithm. Use is made of the **MATLAB package**, which has become almost a standard requirement for engineers, and is available in a student version (*The Student Edition of MATLAB*, Prentice-Hall, Englewood Cliffs, NJ, 1992). A MATLAB M-file demonstrating frequency-domain filtering is shown in Figure 5.34. It can be used with current versions of the package. (Note that in this figure, i is used instead of j to represent $\sqrt{-1}$.) Readers without access to MATLAB should find it fairly easy to interpret this file for the production of their own code.

```
% Demonstration of frequency domain filtering using the FFT.
%
%
% Some MATLAB housekeeping to prevent memory problems!
clear
clg
%
% Select a value of N for the number of samples to be taken.
% Make a selection by adding or removing % symbols.
% N must be a power of 2.
%N = 512;
N = 1024;
%N = 2048;
%N = 4096;
%N = 8192;
%
% T is the sampling interval and the choice of N determines the
% interval over which the signal is processed. Also, if
% N frequency domain values are to be produced the resolution
% is determined.
T = 0.001;
t = 0:T:(N - 1)*T;
delw = 2*pi/(T*N);
%
% Generate the 'information'
f = t .*exp(-t/2);
%
% Set the frequency of the carriers, wc is the carrier which
% will be modulated.
wc = 2*pi*50;
wca = 2*pi*120;
%
% Perform the modulation ...
```

Figure 5.34 'MATLAB' M-file demonstrating frequency-domain filtering using the fast Fourier transform.

```
x = f. *cos(wc*t) + cos(wca*t);
%
% ... and add channel noise here
nfac = 0.2;
rand('normal');
x = x + nfac*rand(t);
%
% Plot the 'received' time signal
plot(t,x)
title('The time signal, modulated carrier and noise if added')
xlabel('time, t')
ylabel('x(t)')
pause
%
% Calculate the DFT using the FFT algorithm ...
y = fft(x);
z = T*abs(y);
w = 0:delw:(N - 1)*delw;
%
% ... and plot the amplitude spectrum.
plot(w,z)
title('The amplitude spectrum. Spikes at frequencies of carriers')
xlabel('frequency, w')
ylabel('amplitude')
pause
%
% Construct a filter to isolate the information-bearing carrier.
%
% 2*hwind + 1 is the length of the filter 'window'.
% Set ffac to a value less than 1.0 ffac = 0.5 gives a filter
% of half length wc/2 where wc is frequency of carrier. Don't
% exceed a value of 0.95!
ffac = 0.5;
hwind = round(ffac*wc/delw);
l = 2*hwind + 1;
%
% Set the centre of the window at peak corresponding to wc.
% Check this is ok by setting l = 1!
l1 = round(wc/delw) - hwind;
%
% Remember that we must have both ends of the filter!
mask = [zeros(1,l1),ones(1,l),zeros(1,N - (2*l + 2*l1 - 1)),ones(1,l),zeros(1,l1 - 1)];
%
% Do the frequency domain filtering ...
zz = mask.*y;
%
% ... and calculate the inverse DFT
yya = ifft(zz);
%
% Remove rounding errors ... it is real!
yy = 0.5*(yya + conj(yya));
%
```

Figure 5.34 *continued*

```
% Plot the 'cleaned' spectrum with only lower carrier present.
plot(w,T*abs(zz))
title('Upper carrier eliminated and noise reduced')
xlabel('frequency, w')
ylabel('amplitude')
pause
%
% Now the signal is cleaned but needs demodulating so
% form the product with 2*carrier signal ...
dem = yy.*cos(wc*t);
dem = 2*dem;
%
% ... and take the DFT.
demft = fft(dem);
%
% Use a low-pass filter on the result, the length is llp.
% The same factor is used as before!
llp = round(ffac*wc/delw);
masklp = [ones(1,llp),zeros(1,N - (2*llp - 1)),ones(1,llp - 1)];
%
% Carry out the filtering...
op = masklp.*demft;
%
% ... and plot the DFT of filtered signal.
plot(w,T*abs(op))
title('Result of demodulation and low-pass filtering')
xlabel('frequency, w')
ylabel('amplitude')
pause
%
% Return to the time domain...
opta = ifft(op);
opt = 0.5*(opta + conj(opta));
act = f;
vp = N;
% ... and finally plot the extracted signal vs the original.
plot(t(1:vp),opt(1:vp),'-',t(1:vp),act(1:vp),':');
title('The extracted signal, with original')
xlabel('time, t')
ylabel('f(t)')
pause
%
% Clean-up ...
clg
clear
%
% ... but responsibly!
i = sqrt(-1);
home
```

Figure 5.34 *continued.*

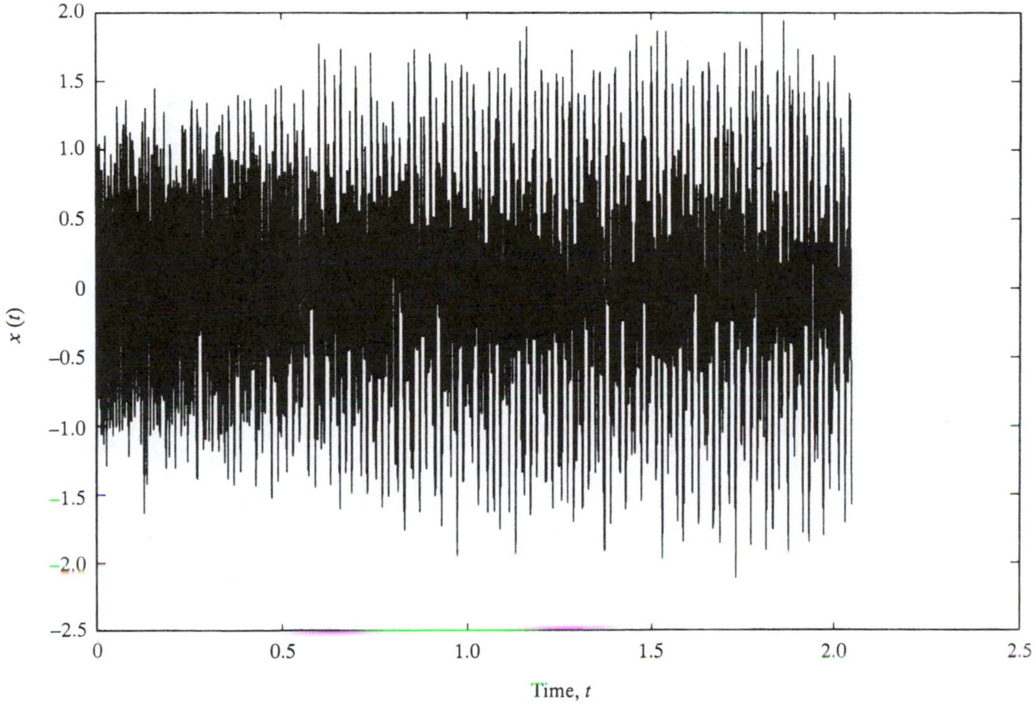

Figure 5.35 Time-domain version of noisy signal.

5.8.2 Modulation and transmission

We suppose that our 'information' consists of samples from the system $f(t) = t\,e^{-t/2}$, taken at intervals $T = 0.001$ s. This signal, or more correctly, data sequence, will be used to modulate the carrier signal cos (50*2*π*t). A second carrier signal is given by cos (120*2*π*t), and this can be thought of as carrying the signal $f(t) = 1$. We combine these two signals and add 'white noise' to represent the action of the channel. This part of the exercise corresponds to the signal generation and transmission part of the overall process, and Figure 5.35 shows the time-domain version of the resulting signal.

5.8.3 Identification and isolation of the information-carrying signal

Here we begin the signal-processing operations. The key to this is Fourier analysis, and we make use of the fast Fourier transform algorithm to perform the necessary

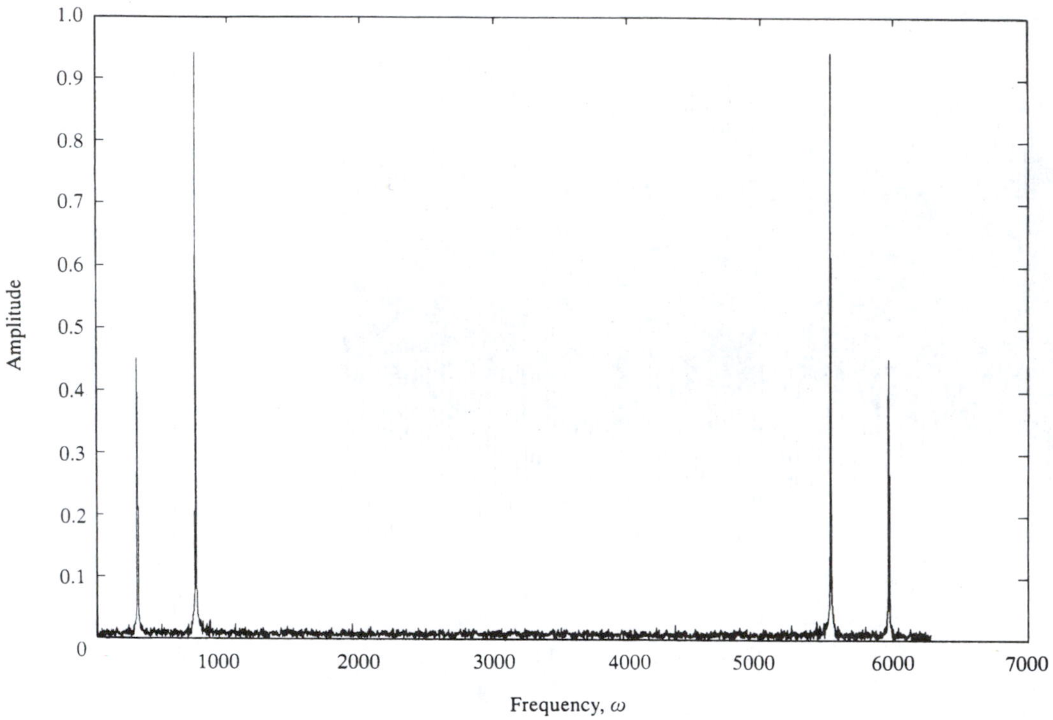

Figure 5.36 Spectrum of received signal. Spikes and frequencies of carriers.

transforms and their inverses. First we examine the spectrum of the received signal, shown in Figure 5.36. We immediately see two spikes corresponding to the carrier signals, and we know that the lower one is carrying the signal we wish to extract. We must design a suitable filter to operate in the frequency domain for the isolation of the selected carrier wave before using the demodulation operation to extract the information. To do this, we simply mask the transformed signal, multiplying by 1 those components we wish to pass, and by 0 those we wish to reject. Obviously we want to pass the carrier-wave frequency component itself, but we must remember that the spectrum of the information signal is centred on this frequency, and so we must pass a band of frequencies around this centre frequency. Again a frequency-domain filter is constructed. We thus have to construct a bandpass filter of suitable bandwidth to achieve this, and moreover, we must remember to include the right-hand half of the filter! There are no problems here with the Nyquist frequency – at first glance we simply need to avoid picking up the second carrier wave. However, the larger the bandwidth we select, the more noise we shall pass, and so a compromise has to be found between the necessary width for good signal recovery and noise elimination. Obviously, since we *know* the bandwidth of our information signal in this case, we could make our choice based on this knowledge. This, however, would be cheating, because usually the exact nature of the transmitted information is not known in advance: if it were, there would be little point in sending it! In the M file we have set the half-length of the filter to be a

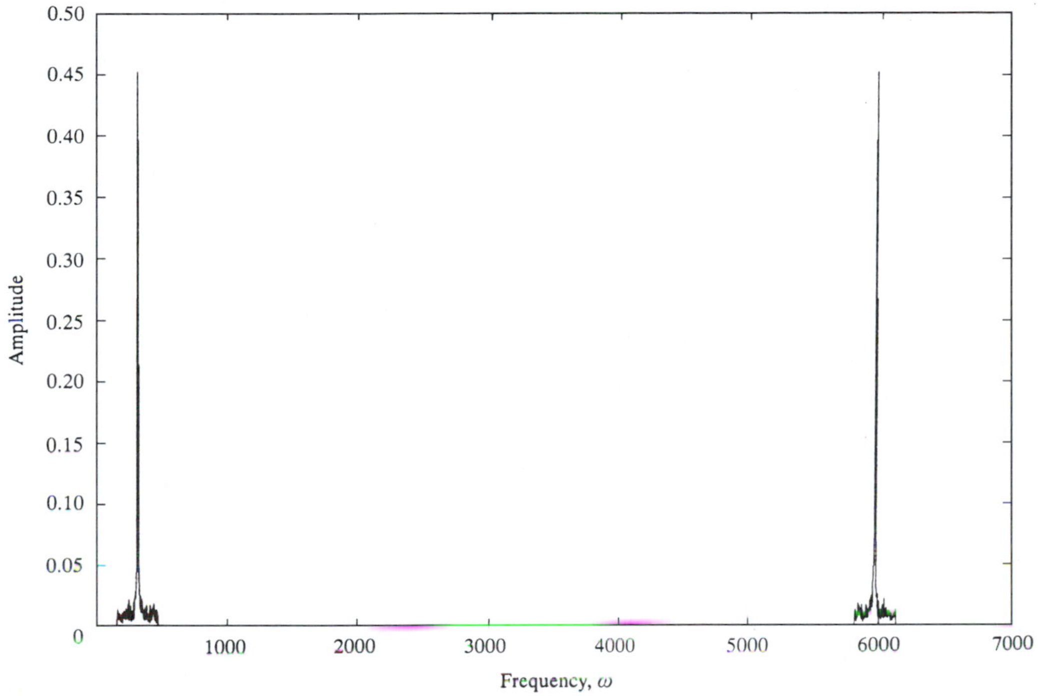

Figure 5.37 Spectrum after application of bandpass filter.

fraction of the carrier frequency. The carrier frequency ω_c represents the maximum possible channel bandwidth, and in practice a channel would have a specified maximum bandwidth associated with it. Figure 5.37 shows the resulting spectrum after application of the bandpass filter, with a bandwidth less than ω_c.

5.8.4 Demodulation stage

The purpose of this operation is to extract the information from the carrier wave, and it can be shown that multiplying the time signal by cos $\omega_c T$, where ω_c is the frequency of the carrier wave, has the effect of shifting the spectrum of the modulating signal so that it is again centred on the origin. To perform the multiplication operation, we have to return to the time domain, and this is achieved by using the inverse FFT algorithm. In the frequency-domain representation of the demodulated signal there are also copies of the spectrum of the modulating signal present, centred at higher frequencies ($2\omega_c$, $4\omega_c$), and so we must perform a final low-pass filtering operation on the demodulated signal. To do this, we return to the frequency domain using the FFT algorithm again. The result of the demodulation and low-pass filtering operations is shown in Figure 5.38.

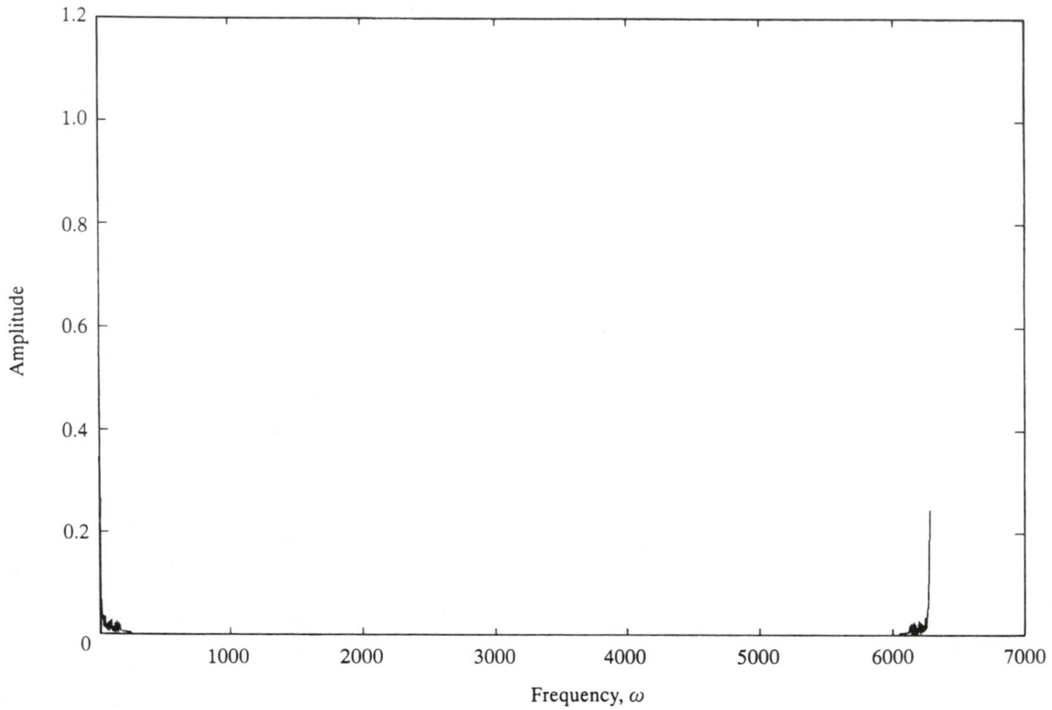

Figure 5.38 Result of demodulation and low-pass filtering operations.

5.8.5 Final signal recovery

The last operation to be performed is to return to the time domain to examine what we have achieved. After calling the inverse FFT routine, the extracted signal is plotted together with the original for comparison. The results with a fairly low value for the added noise are shown in Figure 5.39. If the process is carried out in the absence of noise altogether, excellent signal recovery is achieved, except for the characteristic 'ringing' due ot the sharp edges of the filters.

5.8.6 Further developments

Readers are invited to develop this case study to increase their understanding. Try adding a second information signal modulating the second carrier wave, and extract both signals after 'transmission'. Also add more carrier waves and modulating signals, and investigate signal recovery. If information signal bandwidths

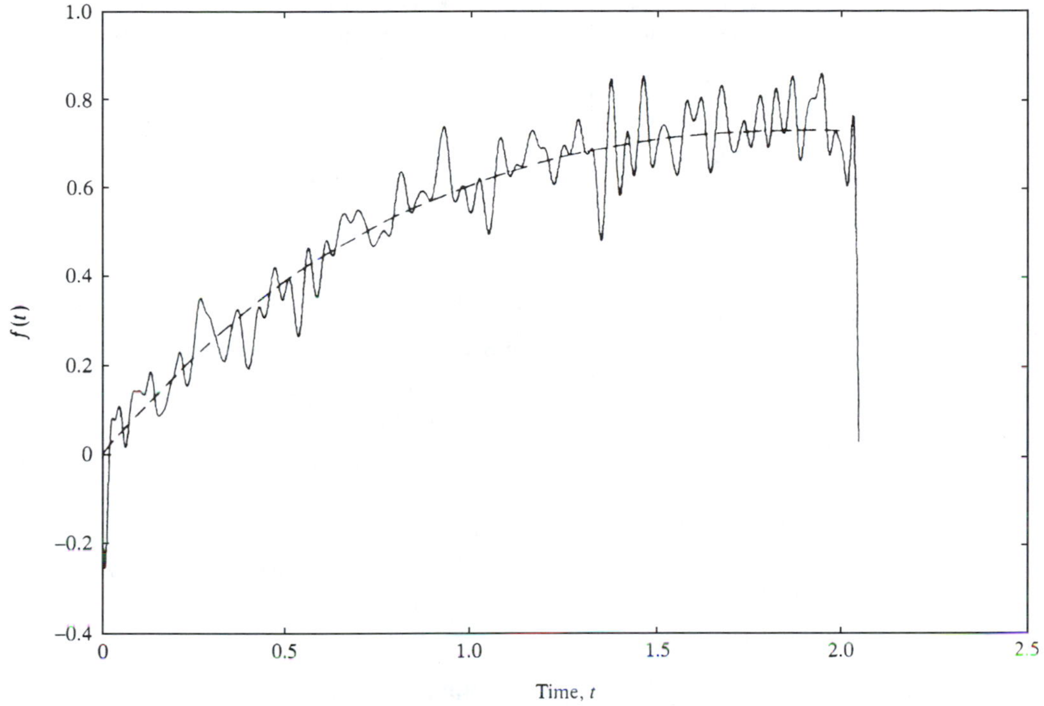

Figure 5.39 Extracted signal, shown together with the original signal.

are limited to a fixed value, how many signals can be transmitted and recovered satisfactorily? What happens if T is altered? Can the 'ringing' effect be reduced by smoothing the transition from the string of ones to the string of zeros in the filter masks? Seek references to various **window functions** in signal-processing texts to assist in resolving this question.

5.9 Review exercises (1–25)

1 Calculate the Fourier sine transform of the causal function $f(t)$ defined by

$$f(t) = \begin{cases} t & (0 \leqslant t \leqslant 1) \\ 1 & (1 < t \leqslant 2) \\ 0 & (t > 2) \end{cases}$$

2 Show that if $\mathscr{F}\{f(t)\} = F(\mathrm{j}\omega)$ then $\mathscr{F}\{f(-t)\} = F(-\mathrm{j}\omega)$. Show also that

$$\mathscr{F}\{f(-t-a)\} = \mathrm{e}^{\mathrm{j}a\omega}F(-\mathrm{j}\omega)$$

where a is real and positive.

Find $\mathscr{F}\{f(t)\}$ when

$$f(t) = \begin{cases} -\frac{1}{2}\pi & (t < -2) \\ \frac{1}{4}\pi t & (-2 \leqslant t \leqslant 2) \\ \frac{1}{2}\pi & (t > 2) \end{cases}$$

3 Use the result

$$\mathscr{F}[H(t + \tfrac{1}{2}T) - H(t - \tfrac{1}{2}T)] = T \operatorname{sinc} \tfrac{1}{2}\omega T$$

and the frequency convolution result to verify that the Fourier transform of the windowed cosine function

$$f(t) = \cos \omega_0 t\,[H(t + \tfrac{1}{2}T) - H(t - \tfrac{1}{2}T)]$$

is

$$\tfrac{1}{2}T\,[\operatorname{sinc}\tfrac{1}{2}(\omega - \omega_0)T + \operatorname{sinc}\tfrac{1}{2}(\omega + \omega_0)T]$$

4 Show that

$$\delta(t - t_1) * \delta(t - t_2) = \delta(t - (t_1 + t_2))$$

and hence show that

$$\mathscr{F}\{\cos \omega_0 t\, H(t)\} = \tfrac{1}{2}\pi[\delta(\omega + \omega_0) + \delta(\omega - \omega_0)] + \frac{\mathrm{j}\omega}{\omega_0^2 - \omega^2}$$

5 Establish the demodulation property,

$$\mathscr{F}\{f(t)\cos \omega_0 t \cos \omega_0 t\} = \tfrac{1}{2}F(\mathrm{j}\omega) + \tfrac{1}{4}[F(\mathrm{j}\omega + 2\mathrm{j}\omega_0) + F(\mathrm{j}\omega + 2\mathrm{j}\omega_0)]$$

6 Use the result $\mathscr{F}\{H(t + T) - H(t - T)\} = 2T \operatorname{sinc} \omega T$ and the symmetry property to show that

$$\mathscr{F}\{\operatorname{sinc} t\} = \pi[H(\omega + 1) - \mathrm{H}(\omega - 1)]$$

Check your result by use of the inversion integral.

7 For a wide class of frequently occurring Laplace transforms it is possible to deduce an inversion integral based on the Fourier inversion integral. If $X(s) = \mathscr{L}\{x(t)\}$ is such a transform, we have

$$x(t) = \frac{1}{\mathrm{j}2\pi}\int_{\gamma-\mathrm{j}\infty}^{\gamma+\mathrm{j}\infty} X(s)\,\mathrm{e}^{st}\,\mathrm{d}s$$

where Re $(s) = \gamma$, with γ real, defines a line in the s plane to the right of all the poles of $X(s)$. Usually the integral can be evaluated using the residue theorem, and we then have

$$x(t) = \sum \text{residues of } X(s)\,\mathrm{e}^{st} \text{ at all poles of } X(s)$$

(a) Write down the poles for the transform

$$X(s) = \frac{1}{(s-a)(s-b)}$$

where a and b are real. Calculate the residues of $X(s)\,\mathrm{e}^{st}$ at these poles and invert the transform.

(b) Calculate

(i) $\mathscr{L}^{-1}\left\{\dfrac{1}{(s-2)^2}\right\}$ (ii) $\mathscr{L}^{-1}\left\{\dfrac{1}{s^2(s+1)}\right\}$

(c) Show that

$$\mathscr{L}^{-1}\left\{\frac{2s}{(s^2+1)^2}\right\} = t \sin t$$

8 A linear system has impulse response $h(t)$, so that the output corresponding to an input $u(t)$ is

$$y(t) = \int_{-\infty}^{\infty} h(t-\tau)\,u(\tau)\,\mathrm{d}\tau$$

When $u(t) = \cos \omega_0 t$, $y(t) = -\sin \omega_0 t (\omega_0 \geqslant 0)$. Find the output when $u(t)$ is given by

(a) $\cos \omega_0(t + \frac{1}{4}\pi)$ (b) $\sin \omega_0 t$

(c) $\mathrm{e}^{\mathrm{j}\omega_0 t}$ (d) $\mathrm{e}^{-\mathrm{j}\omega_0 t}$

This system is known as a **Hilbert transformer**.

9 In Section 5.5.1 we established that

$$\mathscr{F}^{-1}\left\{\frac{1}{\mathrm{j}\omega}\right\} = \tfrac{1}{2}\operatorname{sgn}(t)$$

where sgn(t) is the signum function. Deduce that

$$\mathscr{F}\{\operatorname{sgn}(t)\} = \frac{2}{\mathrm{j}\omega}$$

and use the symmetry result to demonstrate that

$$\mathscr{F}\left\{-\frac{1}{\pi t}\right\} = \mathrm{j}\operatorname{sgn}(\omega)$$

10 The **Hilbert transform** of a signal $f(t)$ is defined by

$$F_{\mathrm{Hi}}(x) = \mathscr{H}\{f(t)\} = \frac{1}{\pi}\int_{-\infty}^{\infty}\frac{f(\tau)}{\tau - x}\,\mathrm{d}\tau$$

Show that the operation of taking the Hilbert transform is equivalent to the convolution

$$-\frac{1}{\pi t} * f(t)$$

and hence deduce that the Hilbert-transformed signal has an amplitude spectrum $F_{\mathrm{Hi}}(\mathrm{j}\omega)$ identical with $f(t)$. Show also that the phase of the transformed signal is changed by $\pm\frac{1}{2}\pi$, depending on the sign of ω.

11 Show that

$$\frac{t}{(t^2+a^2)(t-x)} = \frac{1}{x^2+a^2}\left(\frac{a^2}{t^2+a^2} + \frac{x}{t-x} - \frac{xt}{t^2+a^2}\right)$$

Hence show that the Hilbert transform of

$$f(t) = \frac{t}{t^2+a^2} \quad (a > 0)$$

is

$$\frac{a}{x^2+a^2}$$

12 If $F_{Hi}(x) = \mathscr{H}\{f(t)\}$ is the Hilbert transform of $f(t)$, establish the following properties:

(a) $\mathscr{H}\{f(a+t)\} = F_{Hi}(x+a)$

(b) $\mathscr{H}\{f(at)\} = F_{Hi}(ax) \quad (a > 0)$

(c) $\mathscr{H}\{f(-at)\} = -F_{Hi}(-ax) \quad (a > 0)$

(d) $\mathscr{H}\left\{\dfrac{df}{dt}\right\} = \dfrac{d}{dx}F_{Hi}(x)$

(e) $\mathscr{H}\{tf(t)\} = xF_{Hi}(x) + \dfrac{1}{\pi}\displaystyle\int_{-\infty}^{\infty} f(t)\,dt$

13 Show that

$$f(t) = -\frac{1}{\pi}\int_{-\infty}^{\infty} \frac{F_{Hi}(x)}{x-t}\,dx$$

14 Define the **analytic signal** associated with the real signal $f(t)$ as

$$f_a(t) = f(t) - jF_{Hi}(t)$$

where $F_{Hi}(t)$ is the Hilbert transform of $f(t)$. Use the method of Review exercise 3 to show that

$$\mathscr{F}\{f_a(t)\} = F_a(j\omega) = \begin{cases} 2F(j\omega) & (\omega > 0) \\ 0 & (\omega < 0) \end{cases}$$

15 Use the result $\mathscr{F}\{H(t)\} = 1/j\omega + \pi\delta(\omega)$ and the symmetry property to show that

$$\mathscr{F}\{H(\omega)\} = \tfrac{1}{2}\delta(t) + \frac{j}{2\pi t}$$

(*Hint*: $H(-\omega) = 1 - H(\omega)$.)

Hence show that if $\hat{f}(t)$ is defined by $\mathscr{F}\{\hat{f}(t)\} = 2H(\omega)F(j\omega)$ then $\hat{f}(t) = f(t) - F_{Hi}(t)$, the analytic signal associated with $f(t)$, where $F(j\omega) = \mathscr{F}\{f(t)\}$ and $F_{Hi}(t) = \mathscr{H}\{f(t)\}$.

If $f(t) = \cos\omega_0 t$ $(\omega_0 > 0)$, find $\mathscr{F}\{f(t)\}$ and hence $\hat{f}(t)$. Deduce that

$$\mathscr{H}\{\cos\omega_0 t\} = -\sin\omega_0 t$$

By considering the signal $g(t) = \sin\omega_0 t$ $(\omega_0 > 0)$, show that

$$\mathscr{H}\{\sin\omega_0 t\} = \cos\omega_0 t$$

16 A causal system has impulse response $\bar{h}(t)$, where $\bar{h}(t) = 0$ $(t < 0)$. Define the even part $\bar{h}_e(t)$ of $\bar{h}(t)$ as

$$\bar{h}_e(t) = \tfrac{1}{2}[\bar{h}(t) + \bar{h}(-t)]$$

and the odd part $h_o(t)$ as

$$\bar{h}_o(t) = \tfrac{1}{2}[\bar{h}(t) - \bar{h}(-t)]$$

Deduce that if $\bar{h}(t) = 0$ $(t < 0)$ then

$$\bar{h}_o(t) = \operatorname{sgn}(t)\bar{h}_e(t)$$

and that

$$\bar{h}(t) = \bar{h}_e(t) + \operatorname{sgn}(t)\bar{h}_e(t) \quad \text{for all } t$$

Verify this result for $\bar{h}(t) = \sin t\,H(t)$. Take the Fourier transform of this result to establish that

$$\bar{H}(j\omega) = \bar{H}_e(j\omega) + j\mathscr{H}\{\bar{H}_e(j\omega)\}$$

Let $\bar{h}(t) = e^{-at}H(t)$ be such a causal impulse response. By taking the Fourier transform, deduce the Hilbert transform pair

$$\mathscr{H}\left\{\frac{a}{a^2+t^2}\right\} = -\frac{x}{a^2+x^2}$$

Use the result

$$\mathscr{H}\{tf(t)\} = x\mathscr{H}\{f(t)\} + \frac{1}{\pi}\int_{-\infty}^{\infty} f(t)\,dt$$

to show that

$$\mathscr{H}\left\{\frac{t}{a^2+t^2}\right\} = \frac{a}{x^2+a^2}$$

17 The **Hartley transform** is defined as

$$F_H(s) = \mathrm{H}\{f(t)\} = \int_{-\infty}^{\infty} f(t)\operatorname{cas} 2\pi st\,dt$$

where $\operatorname{cas} t = \cos t + \sin t$. Find the Hartley transform of the functions

(a) $f(t) = e^{-at}H(t) \quad (a > 0)$

(b) $f(t) = \begin{cases} 0 & (|t| > T) \\ 1 & (|t| \leqslant T) \end{cases}$

18 An alternative form of the Fourier transform pair is given by

$$F(jp) = \int_{-\infty}^{\infty} f(t)e^{-j2\pi pt}\,dt$$

$$g(t) = \int_{-\infty}^{\infty} G(jp)e^{j2\pi pt}\,dt$$

where the frequency p is now measured in hertz. Define the even part of the Hartley transform as

$$E(s) = \tfrac{1}{2}[F_H(s) + F_H(-s)]$$

and the odd part as

$$O(s) = \tfrac{1}{2}[F_H(s) - F_H(-s)]$$

Show that the Fourier transform of $f(t)$ is given by

$$F(jp) = E(p) - jO(p)$$

and confirm your result for $f(t) = e^{-2t}H(t)$.

19 Prove the **time-shift result** for the Hartley transform in the form

$$F_H\{f(t-T)\} = \sin 2\pi T\,F_H(-s) + \cos 2\pi T\,F_H(s)$$

20 Using the alternative form of the Fourier transform given in Review exercise 18, it can be shown that the Fourier transform of the Heaviside step function is

$$\mathscr{F}\{H(t)\} = \frac{1}{\mathrm{j}p\pi} + \tfrac{1}{2}\delta(p)$$

Show that the Hartley transform of $H(t)$ is then

$$\tfrac{1}{2}\delta(s) + \frac{1}{\mathrm{j}s\pi}$$

and deduce that the Hartley transform of $H(t - \frac{1}{2})$ is

$$\tfrac{1}{2}\delta(s) + \frac{\cos \pi s - \sin \pi s}{s\pi}$$

21 Show that $F_{\mathrm{H}}\{\delta(t)\} = 1$ and deduce that $F_{\mathrm{H}}\{1\} = \delta(s)$. Show also that $F_{\mathrm{H}}\{\delta(t - t_0)\} = \operatorname{cas} 2\pi s t_0$ and that

$$\begin{aligned} F_{\mathrm{H}}\{\operatorname{cas} 2\pi s_0 t\} &= F_{\mathrm{H}}\{\cos 2\pi s_0 t\} + F_{\mathrm{H}}\{\sin 2\pi s_0 t\} \\ &= \delta(s - s_0) \end{aligned}$$

22 Prove the Hartley transform **modulation theorem** in the form

$$F_{\mathrm{H}}\{f(t) \cos 2\pi s_0 t\} = \tfrac{1}{2} F_{\mathrm{H}}(s - s_0) + \tfrac{1}{2} F_{\mathrm{H}}(s + s_0)$$

Hence show that

$$F_{\mathrm{H}}\{\cos 2\pi s_0 t\} = \tfrac{1}{2}[\delta(s - s_0) + \delta(s + s_0)]$$
$$F_{\mathrm{H}}\{\sin 2\pi s_0 t\} = \tfrac{1}{2}[\delta(s - s_0) - \delta(s + s_0)]$$

23 Show that

$$\mathscr{F}\{\tan^{-1} t\} = \frac{\pi \mathrm{e}^{-|\omega|}}{\mathrm{j}\omega}$$

$\left(\textit{Hint:}\text{ Consider } \int_{-\infty}^{t} (1 + t^2)^{-1}\, \mathrm{d}t.\right)$

24 Show that

$$x(t) = \tfrac{1}{2}(1 + \cos \omega_0 t)[H(t + \tfrac{1}{2}T) - H(t - \tfrac{1}{2}T)]$$

has Fourier transform

$$T[\operatorname{sinc} \omega + \tfrac{1}{2} \operatorname{sinc}(\omega - \omega_0) + \tfrac{1}{2} \operatorname{sinc}(\omega + \omega_0)]$$

25 The **discrete Hartley transform** of the sequence $\{f(r)\}_{r=0}^{N-1}$ is defined by

$$\mathsf{H}(v) = \frac{1}{N} \sum_{r=0}^{N-1} f(r) \operatorname{cas}\left(\frac{2\pi v r}{N}\right) \qquad (v = 0, 1, \dots, N - 1)$$

The inverse transform is

$$f(r) = \sum_{v=0}^{N-1} \mathsf{H}(v) \operatorname{cas}\left(\frac{2\pi v r}{N}\right) \qquad (r = 0, \dots, N - 1)$$

Show that in the case $N = 4$,

$$\mathbf{H} = \boldsymbol{T f}$$
$$\mathbf{H} = [\mathsf{H}(0) \quad \mathsf{H}(1) \quad \mathsf{H}(2) \quad \mathsf{H}(3)]^{\mathrm{T}}$$
$$\boldsymbol{f} = [f(0) \quad f(1) \quad f(2) \quad f(3)]^{\mathrm{T}}$$
$$\boldsymbol{T} = \frac{1}{4}\begin{bmatrix} 1 & 1 & 1 & 1 \\ 1 & 1 & -1 & -1 \\ 1 & -1 & 1 & -1 \\ 1 & -1 & -1 & 1 \end{bmatrix}$$

Hence compute the discrete Hartley transform of the sequence $\{1, 2, 3, 4\}$. Show that $\boldsymbol{T}^2 = \frac{1}{4}\boldsymbol{I}$ and hence that $\boldsymbol{T}^{-1} = 4\boldsymbol{T}$, and verify that applying the $\boldsymbol{T}^{-1}$ operator regains the original sequence.

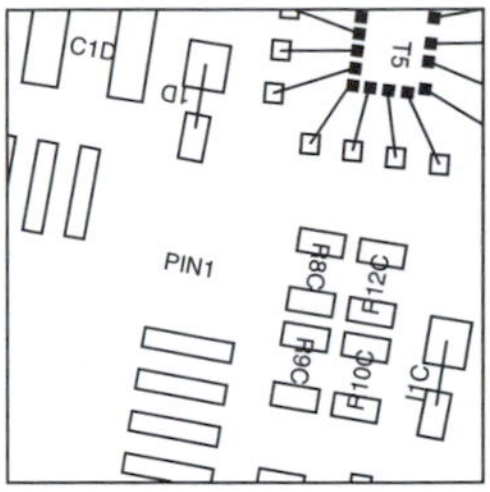

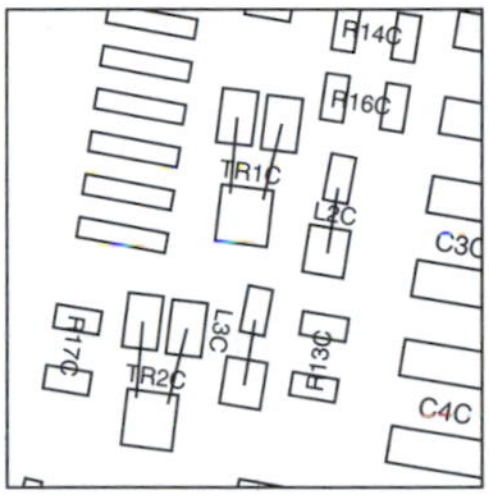

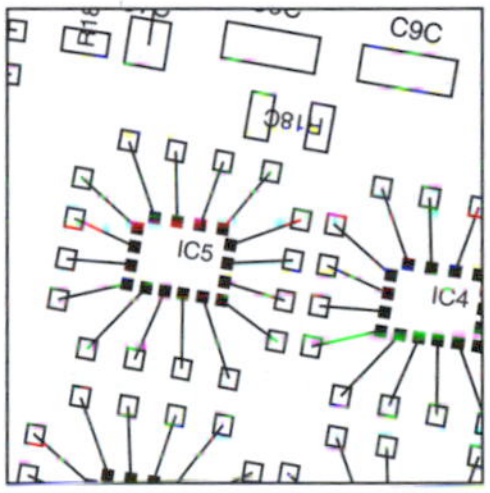

6

Matrix Analysis

CONTENTS

6.1 Introduction

In this chapter we turn our attention again to matrices, first considered in Chapter 5 of *Modern Engineering Mathematics*, and their applications in engineering. At the outset of the chapter we review the basic results of matrix algebra and briefly introduce vector spaces.

As the reader will be aware, matrices are arrays of real or complex numbers, and have a special, but not exclusive, relationship with systems of linear equations. An (incorrect) initial impression often formed by users of mathematics is that mathematicians have something of an obsession with these systems and their solution. However, such systems occur quite naturally in the process of numerical solution of ordinary differential equations used to model everyday engineering processes. In Chapter 9 we shall see that they also occur in numerical methods for the solution of partial differential equations, for example those modelling the flow of a fluid or the transfer of heat. Systems of linear first-order differential equations with constant coefficients are at the core of the **state-space** representation of linear system models. Identification, analysis and indeed design of such systems can conveniently be performed in the state-space representation, with this form assuming a particular importance in the case of multivariable systems.

In all these areas it is convenient to use a matrix representation for the systems under consideration, since this allows the system model to be manipulated following the rules of matrix algebra. A particularly valuable type of manipulation is **simplification** in some sense. Such a simplification process is an example of a system transformation, carried out by the process of matrix multiplication. At the heart of many transformations are the **eigenvalues** and **eigenvectors** of a square matrix. In addition to providing the means by which simplifying transformations can be deduced, system eigenvalues provide vital information on system stability, fundamental frequencies, speed of decay and long-term system behaviour. For this reason, we devote a substantial amount of space to the process of their calculation, both by hand and by numerical means when necessary. Our treatment of numerical methods is intended to be purely indicative rather than complete, because a comprehensive matrix algebra computational tool kit is now part of the essential armoury of all serious users of mathematics.

In addition to developing the use of matrix algebra techniques, we also demonstrate the techniques and applications of matrix analysis, focusing on the state-space system model widely used in control and systems engineering. Here we encounter the idea of a function of a matrix, in particular the matrix exponential, and we see again the role of the eigenvalues in its calculation. We also highlight the use of transform methods in the calculation of functions of matrices, linking with our work on the Laplace and z transforms in Chapters 2 and 3.

6.2 Review of matrix algebra

This section contains a summary of the definitions and properties associated with matrices and determinants. A full account can be found in chapters of ***Modern Engineering Mathematics*** or elsewhere. It is assumed that readers, prior to embarking on this chapter have a fairly thorough understanding of the material summarized in this section.

6.2.1 Definitions

(a) An array of real numbers

$$A = \begin{bmatrix} a_{11} & a_{12} & a_{13} & \dots & a_{1n} \\ a_{21} & a_{22} & a_{23} & \dots & a_{2n} \\ \vdots & \vdots & \vdots & \vdots & \vdots \\ a_{m1} & a_{m2} & a_{m3} & \dots & a_{mn} \end{bmatrix}$$

is called an $m \times n$ **matrix** with m rows and n columns. The a_{ij} is referred to as the i, jth **element** and denotes the element in the ith row and jth column. If $m = n$ then A is called a **square matrix** of order n. If the matrix has one column or one row then it is called a **column vector** or a **row vector** respectively.

(b) In a square matrix A of order n the diagonal containing the elements a_{11}, a_{22}, . . . , a_{nn} is called the **principal** or **leading** diagonal. The sum of the elements in this diagonal is called the **trace** of A, that is

$$\text{trace}\, A = \sum_{i=1}^{n} a_{ii}$$

(c) A **diagonal matrix** is a square matrix that has its only non-zero elements along the leading diagonal. A special case of a diagonal matrix is the **unit** or **identity** matrix I for which $a_{11} = a_{22} = \dots = a_{nn} = 1$.

(d) A **zero** or **null** matrix **0** is a matrix with every element zero.

(e) The **transposed matrix** A^T is the matrix A with rows and columns interchanged, its i, jth element being a_{ji}.

(f) A square matrix A is called a **symmetric matrix** if $A^T = A$. It is called **skew symmetric** if $A^T = -A$.

6.2.2 Basic operations on matrices

In what follows the matrices A, B and C are assumed to have the i, jth elements a_{ij}, b_{ij} and c_{ij} respectively.

Equality

The matrices A and B are **equal**, that is $A = B$, if they are of the same order $m \times n$ and

$$a_{ij} = b_{ij}, \quad 1 \leq i \leq m, \quad 1 \leq j \leq n$$

Multiplication by a scalar

If λ is a scalar then the matrix $\lambda\boldsymbol{A}$ has elements λa_{ij}.

Addition

We can only add an $m \times n$ matrix $\boldsymbol{A}$ to another $m \times n$ matrix $\boldsymbol{B}$ and the elements of the sum $\boldsymbol{A} + \boldsymbol{B}$ are

$$a_{ij} + b_{ij}, \quad 1 \leqslant i \leqslant m, \quad 1 \leqslant j \leqslant n$$

Properties of addition

(i) commutative law: $\boldsymbol{A} + \boldsymbol{B} = \boldsymbol{B} + \boldsymbol{A}$

(ii) associative law: $(\boldsymbol{A} + \boldsymbol{B}) + \boldsymbol{C} = \boldsymbol{A} + (\boldsymbol{B} + \boldsymbol{C})$

(iii) distributive law: $\lambda(\boldsymbol{A} + \boldsymbol{B}) = \lambda\boldsymbol{A} + \lambda\boldsymbol{B}$, λ scalar

Matrix multiplication

If $\boldsymbol{A}$ is a $m \times p$ matrix and $\boldsymbol{B}$ a $p \times n$ matrix then we define the product $\boldsymbol{C} = \boldsymbol{AB}$ as the $m \times n$ matrix with elements

$$c_{ij} = \sum_{k=1}^{p} a_{ik}b_{kj}, \quad i = 1, 2, \ldots, m; \quad j = 1, 2, \ldots, n$$

Properties of multiplication

(i) The commutative law is **not satisfied** in general; that is, in general $\boldsymbol{AB} \neq \boldsymbol{BA}$. Order matters and we distinguish between $\boldsymbol{AB}$ and $\boldsymbol{BA}$ by the terminology: **pre**-multiplication of $\boldsymbol{B}$ by $\boldsymbol{A}$ to form $\boldsymbol{AB}$ and **post**-multiplication of $\boldsymbol{B}$ by $\boldsymbol{A}$ to form $\boldsymbol{BA}$

(ii) Associative law: $\boldsymbol{A}(\boldsymbol{BC}) = (\boldsymbol{AB})\boldsymbol{C}$

(iii) If λ is a scalar then

$$(\lambda\boldsymbol{A})\boldsymbol{B} = \boldsymbol{A}(\lambda\boldsymbol{B}) = \lambda\boldsymbol{AB}$$

(iv) Distributive law over addition:

$$(\boldsymbol{A} + \boldsymbol{B})\boldsymbol{C} = \boldsymbol{AC} + \boldsymbol{BC}$$

$$\boldsymbol{A}(\boldsymbol{B} + \boldsymbol{C}) = \boldsymbol{AB} + \boldsymbol{AC}$$

Note the importance of maintaining order of multiplication.

(v) If $\boldsymbol{A}$ is an $m \times n$ matrix and if $\boldsymbol{I}_m$ and $\boldsymbol{I}_n$ are the unit matrices of order m and n respectively then

$$\boldsymbol{I}_m\boldsymbol{A} = \boldsymbol{AI}_n = \boldsymbol{A}$$

Properties of the transpose

If $\boldsymbol{A}^{\mathrm{T}}$ is the transposed matrix of $\boldsymbol{A}$ then

(i) $(\mathbf{A} + \mathbf{B})^{\mathrm{T}} = \mathbf{A}^{\mathrm{T}} + \mathbf{B}^{\mathrm{T}}$
(ii) $(\mathbf{A}^{\mathrm{T}})^{\mathrm{T}} = \mathbf{A}$
(iii) $(\mathbf{AB})^{\mathrm{T}} = \mathbf{B}^{\mathrm{T}}\mathbf{A}^{\mathrm{T}}$

6.2.3 Determinants

The determinant of a square $n \times n$ matrix $\mathbf{A}$ is denoted by det $\mathbf{A}$ or $|\mathbf{A}|$.

If we take a determinant and delete row i and column j then the determinant remaining is called the **minor** M_{ij} of the i, jth element. In general we can take any row i (or column) and evaluate an $n \times n$ determinant $|\mathbf{A}|$ as

$$|\mathbf{A}| = \sum_{j=1}^{n} (-1)^{i+j} a_{ij} M_{ij}$$

A minor multiplied by the appropriate sign is called the **cofactor** A_{ij} of the i, jth element so $A_{ij} = (-1)^{i+j} M_{ij}$ and thus

$$|\mathbf{A}| = \sum_{j=1}^{n} a_{ij} A_{ij}$$

Some useful properties

(i) $|\mathbf{A}^{\mathrm{T}}| = |\mathbf{A}|$
(ii) $|\mathbf{AB}| = |\mathbf{A}||\mathbf{B}|$
(iii) A square matrix $\mathbf{A}$ is said to be **non-singular** if $|\mathbf{A}| \neq 0$ and **singular** if $|\mathbf{A}| = 0$.

6.2.4 Adjoint and inverse matrices

Adjoint matrix

The **adjoint** of a square matrix $\mathbf{A}$ is the transpose of the matrix of cofactors, so for a 3×3 matrix $\mathbf{A}$

$$\text{adj}\,\mathbf{A} = \begin{bmatrix} A_{11} & A_{12} & A_{13} \\ A_{21} & A_{22} & A_{23} \\ A_{31} & A_{32} & A_{33} \end{bmatrix}^{\mathrm{T}}$$

Properties

(i) $\mathbf{A}(\text{adj}\,\mathbf{A}) = |\mathbf{A}|\mathbf{I}$
(ii) $|\text{adj}\,\mathbf{A}| = |\mathbf{A}|^{n-1}$, n being the order of $\mathbf{A}$
(iii) $\text{adj}\,(\mathbf{AB}) = (\text{adj}\,\mathbf{B})(\text{adj}\,\mathbf{A})$

Inverse matrix

Given a square matrix $\boldsymbol{A}$ if we can construct a square matrix $\boldsymbol{B}$ such that

$$\boldsymbol{BA} = \boldsymbol{AB} = \boldsymbol{I}$$

then we call $\boldsymbol{B}$ the inverse of $\boldsymbol{A}$ and write it as $\boldsymbol{A}^{-1}$.

Properties

(i) If $\boldsymbol{A}$ is non-singular then $|\boldsymbol{A}| \neq 0$ and $\boldsymbol{A}^{-1} = (\text{adj } \boldsymbol{A})/|\boldsymbol{A}|$

(ii) If $\boldsymbol{A}$ is singular then $|\boldsymbol{A}| = 0$ and $\boldsymbol{A}^{-1}$ does not exist

(iii) $(\boldsymbol{AB})^{-1} = \boldsymbol{B}^{-1}\boldsymbol{A}^{-1}$.

6.2.5 Linear equations

In this section we reiterate some definitive statements about the solution of the system of simultaneous linear equations

$$\begin{aligned} a_{11}x_1 + a_{12}x_2 + \ldots + a_{1n}x_n &= b_1 \\ a_{21}x_1 + a_{22}x_2 + \ldots + a_{2n}x_n &= b_2 \\ \vdots \qquad\qquad\qquad & \vdots \\ a_{n1}x_1 + a_{n2}x_2 + \ldots + a_{nn}x_n &= b_n \end{aligned}$$

or, in matrix notation

$$\begin{bmatrix} a_{11} & a_{12} & \ldots & a_{1n} \\ a_{21} & a_{22} & \ldots & a_{2n} \\ \vdots & \vdots & & \vdots \\ a_{n1} & a_{n2} & \ldots & a_{nn} \end{bmatrix} \begin{bmatrix} x_1 \\ x_2 \\ \vdots \\ x_n \end{bmatrix} = \begin{bmatrix} b_1 \\ b_2 \\ \vdots \\ b_n \end{bmatrix}$$

that is,

$$\boldsymbol{A}\boldsymbol{x} = \boldsymbol{b} \tag{6.1}$$

where $\boldsymbol{A}$ is the matrix of coefficients and $\boldsymbol{x}$ is the vector of unknowns. If $\boldsymbol{b} = 0$ the equations are called **homogeneous**, while if $\boldsymbol{b} \neq 0$ they are called **non-homogeneous** (or **inhomogeneous**). Considering individual cases:

Case (i)

If $\boldsymbol{b} \neq 0$ and $|\boldsymbol{A}| \neq 0$ then we have a unique solution $\boldsymbol{x} = \boldsymbol{A}^{-1}\boldsymbol{b}$.

Case (ii)

If $\boldsymbol{b} = 0$ and $|\boldsymbol{A}| \neq 0$ we have the trivial solution $\boldsymbol{x} = 0$.

Case (iii)

If $\boldsymbol{b} \neq 0$ and $|\boldsymbol{A}| = 0$ then we have two possibilities: **either** the equations are inconsistent and we have no solution **or** we have infinitely many solutions.

Case (iv)

If $\boldsymbol{b} = 0$ and $|\boldsymbol{A}| = 0$ then we have infinitely many solutions.

Case (iv) is one of the most important, since from it we can deduce the important result that **the homogeneous equation $\boldsymbol{A}x = 0$ has a non-trivial solution if and only if $|\boldsymbol{A}| = 0$.**

6.2.6 Rank of a matrix

The most commonly used definition of the **rank**, rank $\boldsymbol{A}$, of a matrix $\boldsymbol{A}$ is that it is the order of the largest square submatrix of $\boldsymbol{A}$ with a non-zero determinant. A square submatrix being formed by deleting rows and columns to form a square matrix. Unfortunately it is not always easy to compute the rank using this definition and an alternative definition, which provides a constructive approach to calculating the rank, is often adopted. First, using elementary row operations, the matrix $\boldsymbol{A}$ is reduced to **echelon form**

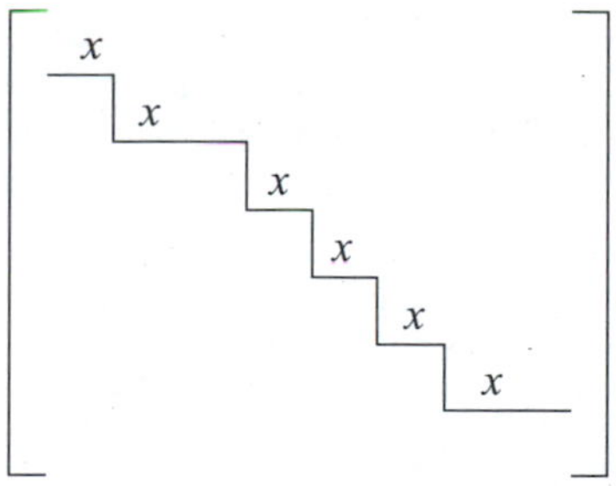

in which all the entries below the line are zero, and the leading element, marked x, in each row above the line is non-zero. The number of non-zero rows in the echelon form is equal to rank $\boldsymbol{A}$.

When considering the solution of equations (6.1) we saw that provided the determinant of the matrix $\boldsymbol{A}$ was not zero we could obtain explicit solutions in terms of the inverse matrix. However, when we looked at cases with zero determinant the results were much less clear. The idea of the rank of a matrix helps to make these results more precise. Defining the **augmented matrix** $(\boldsymbol{A} : \boldsymbol{b})$ for (6.1) as the matrix $\boldsymbol{A}$ with the column $\boldsymbol{b}$ added to it then we can state the results of cases (iii) and (iv) of Section 6.2.5 more clearly as follows:

If $\boldsymbol{A}$ and $(\boldsymbol{A} : \boldsymbol{b})$ have different rank then we have no solution to (6.1). If the two matrices have the same rank then a solution exists, and furthermore the solution will contain a number of free parameters equal to n-rank $\boldsymbol{A}$.

6.3 Vector spaces

Vectors and matrices form part of a more extensive formal structure called a vector space. The theory of vector spaces underpins many modern approaches to numerical methods and the approximate solution of many of the equations that arise in engineering analysis. In this section we shall, very briefly, introduce some of the basic ideas of vector spaces necessary for later work in this chapter.

DEFINITION

A **real vector space** V is a set of objects called **vectors** together with rules for addition and multiplication by real numbers. For any three vectors $\boldsymbol{a}$, $\boldsymbol{b}$ and $\boldsymbol{c}$ in V and any real numbers α and β the sum $\boldsymbol{a} + \boldsymbol{b}$ and the product $\alpha\boldsymbol{a}$ also belong to V and satisfy the following axioms:

(a) $\boldsymbol{a} + \boldsymbol{b} = \boldsymbol{b} + \boldsymbol{a}$

(b) $\boldsymbol{a} + (\boldsymbol{b} + \boldsymbol{c}) = (\boldsymbol{a} + \boldsymbol{b}) + \boldsymbol{c}$

(c) there exists a zero vector $\boldsymbol{0}$ such that

$$\boldsymbol{a} + \boldsymbol{0} = \boldsymbol{a}$$

(d) for each $\boldsymbol{a}$ in V there is an element $-\boldsymbol{a}$ in V such that

$$\boldsymbol{a} + (-\boldsymbol{a}) = \boldsymbol{0}$$

(e) $\alpha(\boldsymbol{a} + \boldsymbol{b}) = \alpha\boldsymbol{a} + \alpha\boldsymbol{b}$

(f) $(\alpha + \beta)\boldsymbol{a} = \alpha\boldsymbol{a} + \beta\boldsymbol{a}$

(g) $(\alpha\beta)\boldsymbol{a} = \alpha(\beta\boldsymbol{a})$

(h) $1\boldsymbol{a} = \boldsymbol{a}$

It is clear that the real numbers form a vector space. The properties given are also satisfied by vectors and by $m \times n$ matrices so vectors and matrices also form vector spaces. The space of all quadratics $a + bx + cx^2$ forms a vector space, as can be established by checking the axioms, (a)–(h). Many other common sets of objects also form vector spaces. If we can obtain useful information from the general structure then this will be of considerable use in specific cases.

6.3.1 Linear independence

The idea of linear dependence is a general one for any vector space. The vector $\boldsymbol{x}$ is said to be **linearly dependent** on $\boldsymbol{x}_1, \boldsymbol{x}_2, \ldots, \boldsymbol{x}_m$ if it can be written as

$$\boldsymbol{x} = \alpha_1\boldsymbol{x}_1 + \alpha_2\boldsymbol{x}_2 + \ldots + \alpha_m\boldsymbol{x}_m$$

for some scalars $\alpha_1, \ldots, \alpha_m$. The **set of vectors** $\boldsymbol{y}_1, \boldsymbol{y}_2, \ldots, \boldsymbol{y}_m$ are said to be **linearly independent** if and only if

$$\beta_1\boldsymbol{y}_1 + \beta_2\boldsymbol{y}_2 + \ldots + \beta_m\boldsymbol{y}_m = 0$$

implies that $\beta_1 = \beta_2 = \ldots = \beta_m = 0$.

Let us now take a linearly independent set of vectors $\boldsymbol{x}_1, \boldsymbol{x}_2, \ldots, \boldsymbol{x}_m$ in V and construct a set consisting of all vectors of the form

$$\boldsymbol{x} = \alpha_1\boldsymbol{x}_1 + \alpha_2\boldsymbol{x}_2 + \ldots + \alpha_m\boldsymbol{x}_m$$

We shall call this set $S(\boldsymbol{x}_1, \boldsymbol{x}_2, \ldots, \boldsymbol{x}_m)$. It is clearly a vector space, since all the axioms are satisfied.

EXAMPLE 6.1 Show that

$$\boldsymbol{e}_1 = \begin{bmatrix} 1 \\ 0 \\ 0 \end{bmatrix} \quad \text{and} \quad \boldsymbol{e}_2 = \begin{bmatrix} 0 \\ 1 \\ 0 \end{bmatrix}$$

form a linearly independent set and describe $S(\boldsymbol{e}_1, \boldsymbol{e}_2)$ geometrically.

Solution We have that

$$0 = \alpha\boldsymbol{e}_1 + \beta\boldsymbol{e}_2 = \begin{bmatrix} \alpha \\ \beta \\ 0 \end{bmatrix}$$

is only satisfied if $\alpha = \beta = 0$, and hence $\boldsymbol{e}_1$ and $\boldsymbol{e}_2$ are linearly independent.

$S(\boldsymbol{e}_1, \boldsymbol{e}_2)$ is the set of all vectors of the form $\begin{bmatrix} \alpha \\ \beta \\ 0 \end{bmatrix}$, which is just the (x_1, x_2) plane and is a subset of the three-dimensional Euclidean space.

If we can find a set B of linearly independent vectors $\boldsymbol{x}_1, \boldsymbol{x}_2, \ldots, \boldsymbol{x}_n$ in V such that

$$S(\boldsymbol{x}_1, \boldsymbol{x}_2, \ldots, \boldsymbol{x}_n) = V$$

then B is called a **basis** of the vector space V. Such a basis forms a crucial part of the theory, since every vector $\boldsymbol{x}$ in V can be written uniquely as

$$\boldsymbol{x} = \alpha_1\boldsymbol{x}_1 + \alpha_2\boldsymbol{x}_2 + \ldots + \alpha_n\boldsymbol{x}_n$$

The definition of B implies that $\boldsymbol{x}$ must take this form. To establish uniqueness, let us assume that we can also write $\boldsymbol{x}$ as

$$\boldsymbol{x} = \beta_1\boldsymbol{x}_1 + \beta_2\boldsymbol{x}_2 + \ldots + \beta_n\boldsymbol{x}_n$$

Then, on subtracting,

$$0 = (\alpha_1 - \beta_1)\boldsymbol{x}_1 + \ldots + (\alpha_n - \beta_n)\boldsymbol{x}_n$$

and since $\boldsymbol{x}_1, \ldots, \boldsymbol{x}_n$ are linearly independent, the only solution is $\alpha_1 = \beta_1$, $\alpha_2 = \beta_2$, . . . ; hence the two expressions for $\boldsymbol{x}$ are the same.

It can also be shown that any other basis for V must also contain n vectors and that any $n + 1$ vectors must be linearly dependent. Such a vector space is said to have **dimension** n (or **infinite dimension** if no finite n can be found). In a three-dimensional Euclidean space

$$\boldsymbol{e}_1 = \begin{bmatrix} 1 \\ 0 \\ 0 \end{bmatrix}, \quad \boldsymbol{e}_2 = \begin{bmatrix} 0 \\ 1 \\ 0 \end{bmatrix}, \quad \boldsymbol{e}_3 = \begin{bmatrix} 0 \\ 0 \\ 1 \end{bmatrix}$$

form an obvious basis, and

$$\boldsymbol{d}_1 = \begin{bmatrix} 1 \\ 0 \\ 0 \end{bmatrix}, \quad \boldsymbol{d}_2 = \begin{bmatrix} 1 \\ 1 \\ 0 \end{bmatrix}, \quad \boldsymbol{d}_3 = \begin{bmatrix} 1 \\ 1 \\ 1 \end{bmatrix}$$

is also a perfectly good basis. While the basis can change, the number of vectors in the basis, three in this case, is an intrinsic property of the vector space. If we consider the vector space of quadratics then the sets of functions $\{1, x, x^2\}$ and $\{1, x - 1, x(x - 1)\}$ are both bases for the space, since every quadratic can be written as $a + bx + cx^2$ or as $A + B(x - 1) + Cx(x - 1)$. We note that this space is three-dimensional.

6.3.2 Transformations between bases

Since any basis of a particular space contains the same number of vectors, we can look at transformations from one basis to another. We shall consider a three-dimensional space, but the results are equally valid in any number of dimensions. Let $\boldsymbol{e}_1$, $\boldsymbol{e}_2$, $\boldsymbol{e}_3$ and $\boldsymbol{e}'_1$, $\boldsymbol{e}'_2$, $\boldsymbol{e}'_3$ be two bases of a space. From the definition of a basis, the vectors $\boldsymbol{e}'_1$, $\boldsymbol{e}'_2$ and $\boldsymbol{e}'_3$ can be written in terms of $\boldsymbol{e}_1$, $\boldsymbol{e}_2$ and $\boldsymbol{e}_3$ as

$$\left.\begin{aligned} \boldsymbol{e}_1' &= a_{11}\boldsymbol{e}_1 + a_{21}\boldsymbol{e}_2 + a_{31}\boldsymbol{e}_3 \\ \boldsymbol{e}_1' &= a_{12}\boldsymbol{e}_2 + a_{22}\boldsymbol{e}_2 + a_{32}\boldsymbol{e}_3 \\ \boldsymbol{e}_3' &= a_{13}\boldsymbol{e}_3 + a_{23}\boldsymbol{e}_2 + a_{33}\boldsymbol{e}_3 \end{aligned}\right\} \tag{6.2}$$

Taking a typical vector $\boldsymbol{x}$ in V, which can be written both as

$$\boldsymbol{x} = x_1\boldsymbol{e}_1 + x_2\boldsymbol{e}_2 + x_3\boldsymbol{e}_3 \tag{6.3}$$

and as

$$\boldsymbol{x} = x_1'\boldsymbol{e}_1' + x_2'\boldsymbol{e}_2' + x_3'\boldsymbol{e}_3'$$

we can use the transformation (6.2) to give

$$\begin{aligned} \boldsymbol{x} &= x_1'(a_{11}\boldsymbol{e}_1 + a_{21}\boldsymbol{e}_2 + a_{31}\boldsymbol{e}_3) + x_2'(a_{12}\boldsymbol{e}_1 + a_{22}\boldsymbol{e}_2 + a_{32}\boldsymbol{e}_3) + x_3'(a_{13}\boldsymbol{e}_1 + a_{23}\boldsymbol{e}_2 + a_{33}\boldsymbol{e}_3) \\ &= (x_1'a_{11} + x_2'a_{12} + x_3'a_{13})\boldsymbol{e}_1 + (x_1'a_{21} + x_2'a_{22} + x_3'a_{23})\boldsymbol{e}_2 + (x_1'a_{31} + x_2'a_{32} + x_3'a_{33})\boldsymbol{e}_3 \end{aligned}$$

On comparing with (6.3) we see that

$$\begin{aligned} x_1 &= a_{11}x_1' + a_{12}x_2' + a_{13}x_3' \\ x_2 &= a_{21}x_1' + a_{22}x_2' + a_{23}x_3' \\ x_3 &= a_{31}x_1' + a_{32}x_2' + a_{33}x_3' \end{aligned}$$

or

$$X = \boldsymbol{A}X'$$

Thus changing from one basis to another is equivalent to transforming the coordinates by multiplication by a matrix, and we thus have another interpretation of matrices. Successive transformations to a third basis will just give $X' = \boldsymbol{B}X'$, and hence the composite transformation is $X = (\boldsymbol{AB})X''$ and is obtained through the standard matrix rules.

For convenience of working it is usual to take mutually orthogonal vectors as a basis, so that $\boldsymbol{e}_i^{\mathrm{T}}\boldsymbol{e}_j = \delta_{ij}$ and $\boldsymbol{e}_i'^{\mathrm{T}}\boldsymbol{e}_j' = \delta_{ij}$, where δ_{ij} is the Kronecker delta

$$\delta_{ij} = \begin{cases} 1 & \text{if} \quad i = j \\ 0 & \text{if} \quad i \neq j \end{cases}$$

Using (6.2) and multiplying out these orthogonality relations, we have

$$\boldsymbol{e}_i'^{\mathrm{T}}\boldsymbol{e}_j' = \sum_k a_{ki}\boldsymbol{e}_k^{\mathrm{T}} \sum_p a_{pj}\boldsymbol{e}_p = \sum_k\sum_p a_{ki}a_{pj}\boldsymbol{e}_k^{\mathrm{T}}\boldsymbol{e}_p = \sum_k\sum_p a_{ki}a_{pj}\delta_{kp} = \sum_k a_{ki}a_{kj}$$

Hence

$$\sum_k a_{ki}a_{kj} = \delta_{ij}$$

or in matrix form

$$\boldsymbol{A}^{\mathrm{T}}\boldsymbol{A} = \boldsymbol{I}$$

It should be noted that such a matrix $\boldsymbol{A}$ with $\boldsymbol{A}^{-1} = \boldsymbol{A}^{\mathrm{T}}$ is called an **orthogonal matrix**.

6.3.3 Exercises

1 Which of the following sets form a basis for a three-dimensional Euclidean space:

(a) $\begin{bmatrix}1\\0\\0\end{bmatrix}, \begin{bmatrix}1\\2\\0\end{bmatrix}, \begin{bmatrix}1\\2\\3\end{bmatrix}$ (b) $\begin{bmatrix}1\\0\\1\end{bmatrix}, \begin{bmatrix}1\\2\\3\end{bmatrix}, \begin{bmatrix}3\\2\\5\end{bmatrix}$

(c) $\begin{bmatrix}1\\0\\0\end{bmatrix}, \begin{bmatrix}1\\1\\0\end{bmatrix}, \begin{bmatrix}2\\1\\0\end{bmatrix}$

2 Given the unit vectors

$$e_1 = \begin{bmatrix}1\\0\\0\end{bmatrix}, \quad e_2 = \begin{bmatrix}0\\1\\0\end{bmatrix}, \quad e_3 = \begin{bmatrix}0\\0\\1\end{bmatrix}$$

find the transformation that takes these to the vectors

$$e_1' = \frac{1}{\sqrt{2}}\begin{bmatrix}1\\1\\0\end{bmatrix}, \quad e_2' = \frac{1}{\sqrt{2}}\begin{bmatrix}1\\-1\\0\end{bmatrix}, \quad e_3' = \begin{bmatrix}0\\0\\1\end{bmatrix}$$

Under this, how does the vector $\boldsymbol{x} = x_1\boldsymbol{e}_1 + x_2\boldsymbol{e}_2 + x_3\boldsymbol{e}_3$ transform and what is the geometrical interpretation? What lines transform into scalar multiples of themselves?

3 Show that the set of all cubic polynomials forms a vector space. Which of the following sets of functions are bases of that space?

(a) $\{1, x, x^2, x^3\}$

(b) $\{1 - x, 1 + x, 1 - x^3, 1 + x^3\}$

(c) $\{1 - x, 1 + x, x^2(1 - x), x^2(1 + x)\}$

(d) $\{x(1 - x), x(1 + x), 1 - x^3, 1 + x^3\}$

(e) $\{1 + 2x, 2x + 3x^2, 3x^2 + 4x^3, 4x^3 + 1\}$

4 Describe the vector space

$$S(x + 2x^3, 2x - 3x^5, x + x^3)$$

What is its dimension?

6.4 The eigenvalue problem

A problem that leads to a concept of crucial importance in many branches of mathematics and its applications is that of seeking non-trivial solutions $\boldsymbol{x} \neq 0$ to the matrix equation

$$\boldsymbol{A}\boldsymbol{x} = \lambda\boldsymbol{x}$$

This is referred to as the eigenvalue problem; values of the scalar λ for which non-trivial solutions exist are called **eigenvalues** and the corresponding solutions $\boldsymbol{x} \neq 0$ are called the **eigenvectors**. Such problems arise naturally in many branches of engineering. For example, in vibrations the eigenvalues and eigenvectors describe the frequency and mode of vibration respectively, while in mechanics they represent principal stresses and the principal axes of stress in bodies subjected to external forces. In Section 6.12 we shall see that eigenvalues also play an important role in the stability analysis of dynamical systems.

6.4.1 The characteristic equation

The set of simultaneous equations

$$\boldsymbol{A}\boldsymbol{x} = \lambda \boldsymbol{x} \tag{6.4}$$

where $\boldsymbol{A}$ is an $n \times n$ matrix and $\boldsymbol{x} = [x_1 \quad x_2 \quad \dots \quad x_n]^{\mathrm{T}}$ is an $n \times 1$ column vector can be written in the form

$$(\lambda \boldsymbol{I} - \boldsymbol{A})\boldsymbol{x} = 0 \tag{6.5}$$

where $\boldsymbol{I}$ is the identity matrix. The matrix equation (6.5) represents simply a set of homogeneous equations, and we know that a non-trivial solution exists if

$$c(\lambda) = |\lambda \boldsymbol{I} - \boldsymbol{A}| = 0 \tag{6.6}$$

Here $c(\lambda)$ is the expansion of the determinant and is a polynomial of degree n in λ, called the **characteristic polynomial** of $\boldsymbol{A}$. Thus

$$c(\lambda) = \lambda^n + c_{n-1}\lambda^{n-1} + c_{n-2}\lambda^{n-2} + \dots + c_1\lambda + c_0$$

and the equation $c(\lambda) = 0$ is called the **characteristic equation** of $\boldsymbol{A}$. We note that this equation can be obtained just as well by evaluating $|\boldsymbol{A} - \lambda \boldsymbol{I}| = 0$; however, the form (6.6) is preferred for the definition of the characteristic equation, since the coefficient of λ^n is then always +1.

In many areas of engineering, particularly in those involving vibration or the control of processes, the determination of those values of λ for which (6.5) has a non-trivial solution (that is, a solution for which $\boldsymbol{x} \neq 0$) is of vital importance. These values of λ are precisely the values that satisfy the characteristic equation, and are called the eigenvalues of $\boldsymbol{A}$.

EXAMPLE 6.2 Find the characteristic equation for the matrix

$$\boldsymbol{A} = \begin{bmatrix} 1 & 1 & -2 \\ -1 & 2 & 1 \\ 0 & 1 & -1 \end{bmatrix}$$

Solution By (6.6), the characteristic equation for $\boldsymbol{A}$ is the cubic equation

$$c(\lambda) = \begin{vmatrix} \lambda - 1 & -1 & 2 \\ 1 & \lambda - 2 & -1 \\ 0 & -1 & \lambda + 1 \end{vmatrix} = 0$$

Expanding the determinant along the first column gives

$$c(\lambda) = (\lambda - 1)\begin{vmatrix} \lambda - 2 & -1 \\ -1 & \lambda + 1 \end{vmatrix} - \begin{vmatrix} -1 & 2 \\ -1 & \lambda + 1 \end{vmatrix}$$

$$= (\lambda - 1)[(\lambda - 2)(\lambda + 1) - 1] - [2 - (\lambda + 1)]$$

Thus

$$c(\lambda) = \lambda^3 - 2\lambda^2 - \lambda + 2 = 0$$

is the required characteristic equation.

For matrices of large order, determining the characteristic polynomial by direct expansion of $|\lambda \boldsymbol{I} - \boldsymbol{A}|$ is unsatisfactory in view of the large number of terms involved in the determinant expansion. Alternative procedures are available to reduce the amount of calculation, and that due to Faddeev may be stated as follows.

The method of Faddeev

If the characteristic polynomial of an $n \times n$ matrix $\boldsymbol{A}$ is written as

$$\lambda^n - p_1\lambda^{n-1} - \ldots - p_{n-1}\lambda - p_n$$

then the coefficients $p_1, p_2, \ldots, p_n$ can be computed using

$$p_r = \frac{1}{r}\operatorname{trace} \boldsymbol{A}_r \quad (r = 1, 2, \ldots, n)$$

where

$$\boldsymbol{A}_r = \begin{cases} \boldsymbol{A} & (r = 1) \\ \boldsymbol{A}\boldsymbol{B}_{r-1} & (r = 2, 3, \ldots, n) \end{cases}$$

and

$\boldsymbol{B}_r = \boldsymbol{A}_r - p_r\boldsymbol{I}$, where $\boldsymbol{I}$ is the $n \times n$ identity matrix

The calculations may be checked using the result that

$\boldsymbol{B}_n = \boldsymbol{A}_n - p_n\boldsymbol{I}$ must be the zero matrix

EXAMPLE 6.3 Using the method of Faddeev, obtain the characteristic equation of the matrix $\boldsymbol{A}$ of Example 6.2.

Solution

$$\boldsymbol{A} = \begin{bmatrix} 1 & 1 & -2 \\ -1 & 2 & 1 \\ 0 & 1 & -1 \end{bmatrix}$$

Let the characteristic equation be

$$c(\lambda) = \lambda^3 - p_1\lambda^2 - p_2\lambda - p_3$$

Then, following the procedure described above,

$$p_1 = \text{trace}\,\boldsymbol{A} = (1 + 2 - 1) = 2$$

$$\boldsymbol{B}_1 = \boldsymbol{A} - 2\boldsymbol{I} = \begin{bmatrix} -1 & 1 & -2 \\ -1 & 0 & 1 \\ 0 & 1 & -3 \end{bmatrix}$$

$$\boldsymbol{A}_2 = \boldsymbol{A}\boldsymbol{B}_1 = \begin{bmatrix} -2 & -1 & 5 \\ -1 & 0 & 1 \\ -1 & -1 & 4 \end{bmatrix}$$

$$p_2 = \tfrac{1}{2}\text{trace}\,\boldsymbol{A}_2 = \tfrac{1}{2}(-2 + 0 + 4) = 1$$

$$\boldsymbol{B}_2 = \boldsymbol{A}_2 - \boldsymbol{I} = \begin{bmatrix} -3 & -1 & 5 \\ -1 & -1 & 1 \\ -1 & -1 & 3 \end{bmatrix}$$

$$\boldsymbol{A}_3 = \boldsymbol{A}\boldsymbol{B}_2 = \begin{bmatrix} -2 & 0 & 0 \\ 0 & -2 & 0 \\ 0 & 0 & -2 \end{bmatrix}$$

$$p_3 = \tfrac{1}{3}\text{trace}\,\boldsymbol{A}_3 = \tfrac{1}{3}(-2 - 2 - 2) = -2$$

Then, the characteristic polynomial of $\boldsymbol{A}$ is

$$c(\lambda) = \lambda^3 - 2\lambda^2 - \lambda + 2$$

in agreement with the result of Example 6.2. In this case, however, a check may be carried out on the computation, since

$$\boldsymbol{B}_3 = \boldsymbol{A}_3 + 2\boldsymbol{I} = 0$$

as required.

6.4.2 Eigenvalues and eigenvectors

The roots of the characteristic equation (6.6) are called the eigenvalues of the matrix $\boldsymbol{A}$ (the terms latent roots, proper roots and characteristic roots are also sometimes used). By the Fundamental Theorem of Algebra, a polynomial equation of degree n has exactly n roots, so that the matrix $\boldsymbol{A}$ has exactly n eigenvalues λ_i, $i = 1, 2, \ldots, n$. These eigenvalues may be real or complex, and not necessarily distinct. Corresponding to each eigenvalue λ_i, there is a non-zero solution $\boldsymbol{x} = \boldsymbol{e}_i$ of (6.5); $\boldsymbol{e}_i$ is called the eigenvector of $\boldsymbol{A}$ corresponding to the eigenvalue λ_i. (Again the terms latent vector, proper vector and characteristic vector are sometimes seen, but are generally obsolete.) We note that if $\boldsymbol{x} = \boldsymbol{e}_i$ satisfies (6.5) then any scalar multiple $\beta_i \boldsymbol{e}_i$ of $\boldsymbol{e}_i$ also satisfies (6.5), so that the eigenvector $\boldsymbol{e}_i$ may only be determined to within a scalar multiple.

EXAMPLE 6.4 Determine the eigenvalues and eigenvectors for the matrix $\boldsymbol{A}$ of Example 6.2.

Solution

$$\boldsymbol{A} = \begin{bmatrix} 1 & 1 & -2 \\ -1 & 2 & 1 \\ 0 & 1 & -1 \end{bmatrix}$$

The eigenvalues λ_i of $\boldsymbol{A}$ satisfy the characteristic equation $c(\lambda) = 0$, and this has been obtained in Examples 6.2 and 6.3 as the cubic

$$\lambda^3 - 2\lambda^2 - \lambda + 2 = 0$$

which can be solved to obtain the eigenvalues λ_1, λ_2 and λ_3.

Alternatively, it may be possible, using the determinant form $|\lambda \boldsymbol{I} - \boldsymbol{A}|$, or indeed (as we often do when seeking the eigenvalues) the form $|\boldsymbol{A} - \lambda \boldsymbol{I}|$, by carrying out suitable row and/or column operations to factorize the determinant.

In this case

$$|\boldsymbol{A} - \lambda \boldsymbol{I}| = \begin{vmatrix} 1-\lambda & 1 & -2 \\ -1 & 2-\lambda & 1 \\ 0 & 1 & -1-\lambda \end{vmatrix}$$

and adding column 1 to column 3 gives

$$\begin{vmatrix} 1-\lambda & 1 & -1-\lambda \\ -1 & 2-\lambda & 0 \\ 0 & 1 & -1-\lambda \end{vmatrix} = -(1+\lambda)\begin{vmatrix} 1-\lambda & 1 & 1 \\ -1 & 2-\lambda & 0 \\ 0 & 1 & 1 \end{vmatrix}$$

Subtracting row 3 from row 1 gives

$$-(1+\lambda)\begin{vmatrix} 1-\lambda & 0 & 0 \\ -1 & 2-\lambda & 0 \\ 0 & 1 & 1 \end{vmatrix} = -(1+\lambda)(1-\lambda)(2-\lambda)$$

Setting $|\boldsymbol{A} - \lambda \boldsymbol{I}| = 0$ gives the eigenvalues as $\lambda_1 = 2$, $\lambda_2 = 1$ and $\lambda_3 = -1$. The order in which they are written is arbitrary, but for consistency we shall adopt the convention of taking $\lambda_1, \lambda_2, \ldots, \lambda_n$ in decreasing order.

Having obtained the eigenvalues λ_i $(i = 1, 2, 3)$, the corresponding eigenvectors $\boldsymbol{e}_i$ are obtained by solving the appropriate homogeneous equations

$$(\boldsymbol{A} - \lambda_i \boldsymbol{I})\boldsymbol{e}_i = 0 \tag{6.7}$$

When $i = 1$, $\lambda_i = \lambda_1 = 2$ and (6.7) is

$$\begin{bmatrix} -1 & 1 & -2 \\ -1 & 0 & 1 \\ 0 & 1 & -3 \end{bmatrix}\begin{bmatrix} e_{11} \\ e_{12} \\ e_{13} \end{bmatrix} = 0$$

that is,

$$-e_{11} + e_{12} - 2e_{13} = 0$$

$$-e_{11} + 0e_{12} + e_{13} = 0$$

$$0e_{11} + e_{12} - 3e_{13} = 0$$

leading to the solution

$$\frac{e_{11}}{-1} = \frac{-e_{12}}{3} = \frac{e_{13}}{-1} = \beta_1$$

where β_1 is an arbitrary non-zero scalar. Thus the eigenvector $\boldsymbol{e}_1$ corresponding to the eigenvalue $\lambda_1 = 2$ is

$$\boldsymbol{e}_1 = \beta_1[1 \quad 3 \quad 1]^{\mathrm{T}}$$

As a check, we can compute

$$\boldsymbol{Ae}_1 = \beta_1\begin{bmatrix} 1 & 1 & -2 \\ -1 & 2 & 1 \\ 0 & 1 & -1 \end{bmatrix}\begin{bmatrix} 1 \\ 3 \\ 1 \end{bmatrix} = \beta_1\begin{bmatrix} 2 \\ 6 \\ 2 \end{bmatrix} = 2\beta_1\begin{bmatrix} 1 \\ 3 \\ 1 \end{bmatrix} = \lambda_1\boldsymbol{e}_1$$

and thus conclude that our calculation was correct.

When $i = 2$, $\lambda_i = \lambda_2 = 1$ and we have to solve

$$\begin{bmatrix} 0 & 1 & -2 \\ -1 & 1 & 1 \\ 0 & 1 & -2 \end{bmatrix}\begin{bmatrix} e_{21} \\ e_{22} \\ e_{23} \end{bmatrix} = 0$$

that is,

$$0e_{21} + e_{22} - 2e_{23} = 0$$

$$-e_{21} + e_{22} + e_{23} = 0$$

$$0e_{21} + e_{22} - 2e_{23} = 0$$

leading to the solution

$$\frac{e_{21}}{-3} = \frac{-e_{22}}{2} = \frac{e_{23}}{-1} = \beta_2$$

where β_2 is an arbitrary scalar. Thus the eigenvector $\boldsymbol{e}_2$ corresponding to the eigenvalue $\lambda_2 = 1$ is

$$\boldsymbol{e}_2 = \beta_2\,[3 \quad 2 \quad 1]^{\mathrm{T}}$$

Again a check could be made by computing $\boldsymbol{Ae}_2$.

Finally, when $i = 3$, $\lambda_i = \lambda_3 = -1$ and we obtain from (6.7)

$$\begin{bmatrix} 2 & 1 & -2 \\ -1 & 3 & 1 \\ 0 & 1 & 0 \end{bmatrix}\begin{bmatrix} e_{31} \\ e_{32} \\ e_{33} \end{bmatrix} = 0$$

that is,

$$2e_{31} + e_{32} - 2e_{33} = 0$$

$$-e_{31} + 3e_{32} + e_{33} = 0$$

$$0e_{31} + e_{32} + 0e_{33} = 0$$

and hence

$$\frac{e_{31}}{-1} = \frac{e_{32}}{0} = \frac{e_{33}}{-1} = \beta_3$$

Here again β_3 is an arbitrary scalar, and the eigenvector $\boldsymbol{e}_3$ corresponding to the eigenvalue λ_3 is

$$\boldsymbol{e}_3 = \beta_3[1 \quad 0 \quad 1]^{\mathrm{T}}$$

The calculation can be checked as before. Thus we have found that the eigenvalues of the matrix $\boldsymbol{A}$ are 2, 1 and −1, with corresponding eigenvectors

$$\beta_1[1 \quad 3 \quad 1]^{\mathrm{T}}, \quad \beta_2[3 \quad 2 \quad 1]^{\mathrm{T}} \quad \text{and} \quad \beta_3[1 \quad 0 \quad 1]^{\mathrm{T}}$$

respectively.

Since in Example 6.4 the β_i, $i = 1, 2, 3$, are arbitrary, it follows that there are an infinite number of eigenvectors, scalar multiples of each other, corresponding to each eigenvalue. Sometimes it is convenient to scale the eigenvectors according to some convention. A convention frequently adopted is to **normalize** the eigenvectors so that they are uniquely determined up to a scale factor of ±1. The normalized form of an eigenvector $\boldsymbol{e} = [e_1 \quad e_2 \quad \dots \quad e_n]^{\mathrm{T}}$ is denoted by $\hat{\boldsymbol{e}}$ and is given by

$$\hat{\boldsymbol{e}} = \frac{\boldsymbol{e}}{|\boldsymbol{e}|}$$

where

$$|\boldsymbol{e}| = \sqrt{(e_1^2 + e_2^2 + \dots + e_n^2)}$$

For example, for the matrix $\boldsymbol{A}$ of Example 6.4, the normalized forms of the eigenvectors are

$$\hat{\boldsymbol{e}}_1 = [1/\sqrt{11} \quad 3/\sqrt{11} \quad 1/\sqrt{11}]^{\mathrm{T}}, \quad \hat{\boldsymbol{e}}_2 = [3/\sqrt{14} \quad 2/\sqrt{14} \quad 1/\sqrt{14}]^{\mathrm{T}}$$

and

$$\hat{\boldsymbol{e}}_3 = [1/\sqrt{2} \quad 0 \quad 1/\sqrt{2}]^{\mathrm{T}}$$

However, throughout the text, unless otherwise stated, the eigenvectors will always be presented in their 'simplest' form, so that for the matrix of Example 6.4 we take $\beta_1 = \beta_2 = \beta_3 = 1$ and write

$$\boldsymbol{e}_1 = [1 \quad 3 \quad 1]^{\mathrm{T}}, \quad \boldsymbol{e}_2 = [3 \quad 2 \quad 1]^{\mathrm{T}} \quad \text{and} \quad \boldsymbol{e}_3 = [1 \quad 0 \quad 1]^{\mathrm{T}}$$

EXAMPLE 6.5 Find the eigenvalues and eigenvectors of

$$\boldsymbol{A} = \begin{bmatrix} \cos\theta & -\sin\theta \\ \sin\theta & \cos\theta \end{bmatrix}$$

Solution Now

$$|\lambda\boldsymbol{I} - \boldsymbol{A}| = \begin{vmatrix} \lambda - \cos\theta & \sin\theta \\ -\sin\theta & \lambda - \cos\theta \end{vmatrix}$$

$$= \lambda^2 - 2\lambda\cos\theta + \cos^2\theta + \sin^2\theta$$

$$= \lambda^2 - 2\lambda\cos\theta + 1$$

So the eigenvalues are the roots of

$$\lambda^2 - 2\lambda\cos\theta + 1 = 0$$

that is,

$$\lambda = \cos\theta \pm \mathrm{j}\sin\theta$$

Solving for the eigenvectors as in Example 6.4, we obtain

$$\boldsymbol{e}_1 = [1 \quad -\mathrm{j}]^{\mathrm{T}} \quad \text{and} \quad \boldsymbol{e}_2 = [1 \quad \mathrm{j}]^{\mathrm{T}}$$

In Example 6.5 we see that eigenvalues can be complex numbers, and that the eigenvectors may have complex components. This situation arises when the characteristic equation has complex (conjugate) roots.

6.4.3 Exercises

5 Using the method of Faddeev, obtain the characteristic polynomials of the matrices

(a) $\begin{bmatrix} 3 & 2 & 1 \\ 4 & 5 & -1 \\ 2 & 3 & 4 \end{bmatrix}$

(b) $\begin{bmatrix} 2 & -1 & 1 & 2 \\ 0 & 1 & 1 & 0 \\ -1 & 1 & 1 & 1 \\ 1 & 1 & 1 & 0 \end{bmatrix}$

6 Find the eigenvalues and corresponding eigenvectors of the matrices

(a) $\begin{bmatrix} 1 & 1 \\ 1 & 1 \end{bmatrix}$ (b) $\begin{bmatrix} 1 & 2 \\ 3 & 2 \end{bmatrix}$

(c) $\begin{bmatrix} 1 & 0 & -4 \\ 0 & 5 & 4 \\ -4 & 4 & 3 \end{bmatrix}$ (d) $\begin{bmatrix} 1 & 1 & 2 \\ 0 & 2 & 2 \\ -1 & 1 & 3 \end{bmatrix}$

(e) $\begin{bmatrix} 5 & 0 & 6 \\ 0 & 11 & 6 \\ 6 & 6 & -2 \end{bmatrix}$ (f) $\begin{bmatrix} 1 & -1 & 0 \\ 1 & 2 & 1 \\ -2 & 1 & -1 \end{bmatrix}$

(g) $\begin{bmatrix} 4 & 1 & 1 \\ 2 & 5 & 4 \\ -1 & -1 & 0 \end{bmatrix}$ (h) $\begin{bmatrix} 1 & -4 & -2 \\ 0 & 3 & 1 \\ 1 & 2 & 4 \end{bmatrix}$

6.4.4 Repeated eigenvalues

In the examples considered so far the eigenvalues λ_i ($i = 1, 2, \dots$) of the matrix $\boldsymbol{A}$ have been distinct, and in such cases the corresponding eigenvectors have been linearly independent. The matrix $\boldsymbol{A}$ is then said to have a full set of linearly independent eigenvectors. It is clear that the roots of the characteristic equation $c(\lambda)$ may not all be distinct; and when $c(\lambda)$ has $p \leqslant n$ distinct roots, $c(\lambda)$ may be factorized as

$$c(\lambda) = (\lambda - \lambda_1)^{m_1}(\lambda - \lambda_2)^{m_2} \dots (\lambda - \lambda_p)^{m_p}$$

indicating that the root $\lambda = \lambda_i$, $i = 1, 2, \dots, p$, is a root of order m_i, where the integer m_i is called the **algebraic multiplicity** of the eigenvalue λ_i. Clearly $m_1 + m_2 + \dots + m_p = n$. When a matrix $\boldsymbol{A}$ has repeated eigenvalues, the question arises as to whether it is possible to obtain a full set of linearly independent eigenvectors for $\boldsymbol{A}$. We first consider two examples to illustrate the situation.

EXAMPLE 6.6 Determine the eigenvalues and corresponding eigenvectors of the matrix

$$\boldsymbol{A} = \begin{bmatrix} 3 & -3 & 2 \\ -1 & 5 & -2 \\ -1 & 3 & 0 \end{bmatrix}$$

Solution We find the eigenvalues from

$$\begin{vmatrix} 3-\lambda & -3 & 2 \\ -1 & 5-\lambda & -2 \\ -1 & 3 & -\lambda \end{vmatrix} = 0$$

as $\lambda_1 = 4$, $\lambda_2 = \lambda_3 = 2$.

The eigenvectors are obtained from

$$(\boldsymbol{A} - \lambda \boldsymbol{I})\boldsymbol{e}_i = 0 \qquad \textbf{(6.8)}$$

and when $\lambda = \lambda_1 = 4$, we obtain from (6.8)

$$\boldsymbol{e}_1 = [1 \quad -1 \quad -1]^{\mathrm{T}}$$

When $\lambda = \lambda_2 = \lambda_3 = 2$, (6.8) becomes

$$\begin{bmatrix} 1 & -3 & 2 \\ -1 & 3 & -2 \\ -1 & 3 & -2 \end{bmatrix} \begin{bmatrix} e_{21} \\ e_{22} \\ e_{23} \end{bmatrix} = 0$$

so that the corresponding eigenvector is obtained from the single equation

$$e_{21} - 3e_{22} + 2e_{23} = 0 \qquad \textbf{(6.9)}$$

Clearly we are free to choose any two of the components e_{21}, e_{22} or e_{23} at will, with the remaining one determined by (6.9). Suppose we set $e_{22} = \alpha$ and $e_{23} = \beta$; then (6.9) means that $e_{21} = 3\alpha - 2\beta$, and thus

$$\begin{aligned} \boldsymbol{e}_2 &= [3\alpha - 2\beta \quad \alpha \quad \beta]^{\mathrm{T}} \\ &= \alpha\begin{bmatrix} 3 \\ 1 \\ 0 \end{bmatrix} + \beta\begin{bmatrix} -2 \\ 0 \\ 1 \end{bmatrix} \end{aligned} \tag{6.10}$$

Now $\lambda = 2$ is an eigenvalue of multiplicity 2, and we seek, if possible, two linearly independent eigenvectors defined by (6.10). Setting $\alpha = 1$ and $\beta = 0$ yields

$$\boldsymbol{e}_2 = [3 \quad 1 \quad 0]^{\mathrm{T}}$$

and setting $\alpha = 0$ and $\beta = 1$ gives a second vector

$$\boldsymbol{e}_3 = [-2 \quad 0 \quad 1]^{\mathrm{T}}$$

These two vectors are linearly independent and of the form defined by (6.10), and it is clear that many other choices are possible. However, any other choices of the form (6.10) will be linear combinations of $\boldsymbol{e}_2$ and $\boldsymbol{e}_3$ as chosen above. For example, $\boldsymbol{e} = [1 \quad 1 \quad 1]$ satisfies (6.10), but $\boldsymbol{e} = \boldsymbol{e}_2 + \boldsymbol{e}_3$.

In this example, although there was a repeated eigenvalue of algebraic multiplicity 2, it was possible to construct two linearly independent eigenvectors corresponding to this eigenvalue. Thus the matrix $\boldsymbol{A}$ has three and only three linearly independent eigenvectors.

EXAMPLE 6.7 Determine the eigenvalues and corresponding eigenvectors for the matrix

$$\boldsymbol{A} = \begin{bmatrix} 1 & 2 & 2 \\ 0 & 2 & 1 \\ -1 & 2 & 2 \end{bmatrix}$$

Solution Solving $|\boldsymbol{A} - \lambda \boldsymbol{I}| = 0$ gives the eigenvalues as $\lambda_1 = \lambda_2 = 2$, $\lambda_3 = 1$. The eigenvector corresponding to the non-repeated or simple eigenvalue $\lambda_3 = 1$ is easily found as

$$\boldsymbol{e}_3 = [1 \quad 1 \quad -1]^{\mathrm{T}}$$

When $\lambda = \lambda_1 = \lambda_2 = 2$, the corresponding eigenvector is given by

$$(\boldsymbol{A} - 2\boldsymbol{I})\boldsymbol{e}_1 = 0$$

that is, as the solution of

$$-e_{11} + 2e_{12} + 2e_{13} = 0 \tag{i}$$

$$e_{13} = 0 \tag{ii}$$

$$-e_{11} + 2e_{12} = 0 \tag{iii}$$

From (ii) we have $e_{13} = 0$, and from (i) and (ii) it follows that $e_{11} = 2e_{12}$. We deduce that there is only one linearly independent eigenvector corresponding to the repeated eigenvalue $\lambda = 2$, namely

$$e_1 = [2 \quad 1 \quad 0]^T$$

and in this case the matrix $\boldsymbol{A}$ does not possess a full set of linearly independent eigenvectors.

We see from Examples 6.6 and 6.7 that if an $n \times n$ matrix $\boldsymbol{A}$ has repeated eigenvalues then a full set of n linearly independent eigenvectors may or may not exist. The number of linearly independent eigenvectors associated with a repeated eigenvalue λ_i of algebraic multiplicity m_i is given by the **nullity** q_i of the matrix $\boldsymbol{A} - \lambda_i \boldsymbol{I}$, where

$$q_i = n - \text{rank}\,(\boldsymbol{A} - \lambda_i \boldsymbol{I}), \quad \text{with} \quad 1 \leqslant q_i \leqslant m_i \tag{6.11}$$

q_i is sometimes referred to as the **degeneracy** of the matrix $\boldsymbol{A} - \lambda_i \boldsymbol{I}$ or the **geometric multiplicity** of the eigenvalue λ_i, since it determines the dimension of the space spanned by the corresponding eigenvector(s) $\boldsymbol{e}_i$.

EXAMPLE 6.8

Confirm the findings of Examples 6.6 and 6.7 concerning the number of linearly independent eigenvectors found.

Solution In Example 6.6, we had an eigenvalue $\lambda_2 = 2$ of algebraic multiplicity 2. Correspondingly,

$$\boldsymbol{A} - \lambda_2 \boldsymbol{I} = \begin{bmatrix} 3-2 & -3 & 2 \\ -1 & 5-2 & -2 \\ -1 & 3 & -2 \end{bmatrix} = \begin{bmatrix} 1 & -3 & 2 \\ -1 & 3 & -2 \\ -1 & 3 & -2 \end{bmatrix}$$

and performing the row operation of adding row 1 to rows 2 and 3 yields

$$\begin{bmatrix} 1 & -3 & 2 \\ 0 & 0 & 0 \\ 0 & 0 & 0 \end{bmatrix}$$

Adding 3 times column 1 to column 2 followed by subtracting 2 times column 1 from column 3 gives finally

$$\begin{bmatrix} 1 & 0 & 0 \\ 0 & 0 & 0 \\ 0 & 0 & 0 \end{bmatrix}$$

indicating a rank of 1. Then from (6.11) the nullity $q_2 = 3 - 1 = 2$, confirming that corresponding to the eigenvalue $\lambda = 2$ there are two linearly independent eigenvectors, as found in Example 6.6.

In Example 6.7 we again had a repeated eigenvalue $\lambda_1 = 2$ of algebraic multiplicity 2. Then

$$\boldsymbol{A} - 2\boldsymbol{I} = \begin{bmatrix} 1-2 & 2 & 2 \\ 0 & 2-2 & 1 \\ -1 & 2 & 2-2 \end{bmatrix} = \begin{bmatrix} -1 & 2 & 2 \\ 0 & 0 & 1 \\ -1 & 2 & 0 \end{bmatrix}$$

Performing row and column operations as before produces the matrix

$$\begin{bmatrix} -1 & 0 & 0 \\ 0 & 0 & 1 \\ 0 & 0 & 0 \end{bmatrix}$$

this time indicating a rank of 2. From (6.11) the nullity $q_1 = 3 - 2 = 1$, confirming that there is one and only one linearly independent eigenvector associated with this eigenvalue, as found in Example 6.7.

6.4.5 Exercises

7 Obtain the eigenvalues and corresponding eigenvectors of the matrices

(a) $\begin{bmatrix} 2 & 2 & 1 \\ 1 & 3 & 1 \\ 1 & 2 & 2 \end{bmatrix}$ (b) $\begin{bmatrix} 0 & -2 & -2 \\ -1 & 1 & 2 \\ -1 & -1 & 2 \end{bmatrix}$

(c) $\begin{bmatrix} 4 & 6 & 6 \\ 1 & 3 & 2 \\ -1 & -5 & -2 \end{bmatrix}$

(d) $\begin{bmatrix} 7 & -2 & -4 \\ 3 & 0 & -2 \\ 6 & -2 & -3 \end{bmatrix}$

8 Given that $\lambda = 1$ is a three-times repeated eigenvalue of the matrix

$$\boldsymbol{A} = \begin{bmatrix} -3 & -7 & -5 \\ 2 & 4 & 3 \\ 1 & 2 & 2 \end{bmatrix}$$

using the concept of rank, determine how many linearly independent eigenvectors correspond to this value of λ. Determine a corresponding set of linearly independent eigenvectors.

9 Given that $\lambda = 1$ is a twice-repeated eigenvalue of the matrix

$$\boldsymbol{A} = \begin{bmatrix} 2 & 1 & -1 \\ -1 & 0 & 1 \\ -1 & -1 & 2 \end{bmatrix}$$

how many linearly independent eigenvectors correspond to this value of λ? Determine a corresponding set of linearly independent eigenvectors.

6.4.6 Some useful properties of eigenvalues

The following basic properties of the eigenvalues $\lambda_1, \lambda_2, \ldots, \lambda_n$ of an $n \times n$ matrix $\boldsymbol{A}$ are sometimes useful. The results are readily proved from either the definition of eigenvalues as the values of λ satisfying (6.4), or by comparison of corresponding characteristic polynomials (6.6). Consequently, the proofs are left to Exercise 10.

Property 6.1

The sum of the eigenvalues of $\boldsymbol{A}$ is

$$\sum_{i=1}^{n} \lambda_i = \text{trace}\,\boldsymbol{A} = \sum_{i=1}^{n} a_{ii}$$

Property 6.2

The product of the eigenvalues of $\boldsymbol{A}$ is

$$\prod_{i=1}^{n} \lambda_i = \det \boldsymbol{A}$$

where det $\boldsymbol{A}$ denotes the determinant of the matrix $\boldsymbol{A}$.

Property 6.3

The eigenvalues of the inverse matrix $\boldsymbol{A}^{-1}$, provided it exists, are

$$\frac{1}{\lambda_1}, \ \frac{1}{\lambda_2}, \ \ldots, \ \frac{1}{\lambda_n}$$

Property 6.4

The eigenvalues of the transposed matrix $\boldsymbol{A}^{\mathrm{T}}$ are

$$\lambda_1, \lambda_2, \ldots, \lambda_n$$

as for the matrix $\boldsymbol{A}$.

Property 6.5

If k is a scalar then the eigenvalues of $k\boldsymbol{A}$ are

$$k\lambda_1, \ k\lambda_2, \ \ldots, \ k\lambda_n$$

Property 6.6

If k is a scalar and $\boldsymbol{I}$ the $n \times n$ identity (unit) matrix then the eigenvalues of $\boldsymbol{A} \pm k\boldsymbol{I}$ are respectively

$$\lambda_1 \pm k, \quad \lambda_2 \pm k, \quad \ldots, \quad \lambda_n \pm k$$

Property 6.7

If k is a positive integer then the eigenvalues of $\boldsymbol{A}^k$ are

$$\lambda_1^k, \quad \lambda_2^k, \quad \ldots, \quad \lambda_n^k$$

6.4.7 Symmetric matrices

A square matrix $\boldsymbol{A}$ is said to be **symmetric** if $\boldsymbol{A}^{\mathrm{T}} = \boldsymbol{A}$. Such matrices form an important class and arise in a variety of practical situations. Two important results concerning the eigenvalues and eigenvectors of such matrices are

(a) the eigenvalues of a real symmetric matrix are real;

(b) for an $n \times n$ real symmetric matrix it is always possible to find n linearly independent eigenvectors $\boldsymbol{e}_1, \boldsymbol{e}_2, \ldots, \boldsymbol{e}_n$ that are mutually orthogonal so that $\boldsymbol{e}_i^{\mathrm{T}}\boldsymbol{e}_j = 0$ for $i \neq j$.

If the orthogonal eigenvectors of a symmetric matrix are normalized as

$$\hat{\boldsymbol{e}}_1, \hat{\boldsymbol{e}}_2, \ldots, \hat{\boldsymbol{e}}_n$$

then the **inner (scalar) product** is

$$\hat{\boldsymbol{e}}_i^{\mathrm{T}}\hat{\boldsymbol{e}}_j = \delta_{ij} \quad (i, j = 1, 2, \ldots, n)$$

where δ_{ij} is the Kronecker delta defined in Section 6.3.1.

The set of normalized eigenvectors of a symmetric matrix therefore form an orthonormal set (that is, they form a mutually orthogonal normalized set of vectors).

EXAMPLE 6.9

Obtain the eigenvalues and corresponding orthogonal eigenvectors of the symmetric matrix

$$\boldsymbol{A} = \begin{bmatrix} 2 & 2 & 0 \\ 2 & 5 & 0 \\ 0 & 0 & 3 \end{bmatrix}$$

and show that the normalized eigenvectors form an orthonormal set.

Solution The eigenvalues of $\boldsymbol{A}$ are $\lambda_1 = 6$, $\lambda_2 = 3$ and $\lambda_3 = 1$, with corresponding eigenvectors

$$e_1 = [1 \quad 2 \quad 0]^T, \quad e_2 = [0 \quad 0 \quad 1]^T, \quad e_3 = [-2 \quad 1 \quad 0]^T$$

which in normalized form are

$$\hat{e}_1 = [1 \quad 2 \quad 0]^T/\sqrt{5}, \quad \hat{e}_2 = [0 \quad 0 \quad 1]^T, \quad \hat{e}_3 = [-2 \quad 1 \quad 0]^T/\sqrt{5}$$

Evaluating the inner products, we see that, for example,

$$\hat{e}_1^T\hat{e}_1 = \tfrac{1}{5} + \tfrac{4}{5} + 0 = 1, \quad \hat{e}_1^T\hat{e}_3 = -\tfrac{2}{5} + \tfrac{2}{5} + 0 = 0$$

and that

$$\hat{e}_i^T\hat{e}_j = \delta_{ij} \quad (i, j = 1, 2, 3)$$

confirming that the eigenvectors form an orthonormal set.

6.4.8 Exercises

10 Verify Properties 6.1–6.7 of Section 6.4.6.

11 Given that the eigenvalues of the matrix

$$\boldsymbol{A} = \begin{bmatrix} 4 & 1 & 1 \\ 2 & 5 & 4 \\ -1 & -1 & 0 \end{bmatrix}$$

are 5, 3 and 1:

(a) confirm Properties 6.1–6.4 of Section 6.4.6.

(b) taking $k = 2$, confirm Properties 6.5–6.7 of Section 6.4.6.

12 Determine the eigenvalues and corresponding eigenvectors of the symmetric matrix

$$\boldsymbol{A} = \begin{bmatrix} -3 & -3 & -3 \\ -3 & 1 & -1 \\ -3 & -1 & 1 \end{bmatrix}$$

and verify that the eigenvectors are mutually orthogonal.

13 The 3×3 symmetric matrix $\boldsymbol{A}$ has eigenvalues 6, 3 and 2. The eigenvectors corresponding to the eigenvalues 6 and 3 are $[1 \quad 1 \quad 2]^T$ and $[1 \quad 1 \quad -1]^T$ respectively. Find an eigenvector corresponding to the eigenvalue 2.

6.5 Numerical methods

In practice we may well be dealing with matrices whose elements are decimal numbers or with matrices of high orders. In order to determine the eigenvalues and eigenvectors of such matrices, it is necessary that we have numerical algorithms at our disposal.

6.5.1 The power method

Consider a matrix $\boldsymbol{A}$ having n distinct eigenvalues $\lambda_1, \lambda_2, \ldots, \lambda_n$ and corresponding n linearly independent eigenvectors $\boldsymbol{e}_1, \boldsymbol{e}_2, \ldots, \boldsymbol{e}_n$. Taking this set of vectors as the basis, we can write any vector $\boldsymbol{x} = [x_1 \quad x_2 \quad \ldots \quad x_n]^{\mathrm{T}}$ as a linear combination in the form

$$\boldsymbol{x} = \alpha_1\boldsymbol{e}_1 + \alpha_2\boldsymbol{e}_2 + \ldots + \alpha_n\boldsymbol{e}_n = \sum_{i=1}^{n} \alpha_i\boldsymbol{e}_i$$

Then, since $\boldsymbol{A}\boldsymbol{e}_i = \lambda_i\boldsymbol{e}_i$ for $i = 1, 2, \ldots, n$,

$$\boldsymbol{A}\boldsymbol{x} = \boldsymbol{A}\sum_{i=1}^{n} \alpha_i\boldsymbol{e}_i = \sum_{i=1}^{n} \alpha_i\lambda_i\boldsymbol{e}_i$$

and, for any positive integer k,

$$\boldsymbol{A}^k\boldsymbol{x} = \sum_{i=1}^{n} \alpha_i\lambda_i^k\boldsymbol{e}_i$$

or

$$\boldsymbol{A}^k\boldsymbol{x} = \lambda_1^k\left[\alpha_1\boldsymbol{e}_1 + \sum_{i=2}^{n} \alpha_i\left(\frac{\lambda_i}{\lambda_1}\right)^k\boldsymbol{e}_i\right] \tag{6.12}$$

Assuming that the eigenvalues are ordered such that

$$|\lambda_1| > |\lambda_2| > \ldots > |\lambda_n|$$

and that $\alpha_1 \neq 0$, we have from (6.12)

$$\lim_{k\to\infty} \boldsymbol{A}^k\boldsymbol{x} = \lambda_1^k\alpha_1\boldsymbol{e}_1 \tag{6.13}$$

since all the other terms inside the square brackets tend to zero. The eigenvalue λ_1 and its corresponding eigenvector $\boldsymbol{e}_1$ are referred to as the **dominant** eigenvalue and eigenvector respectively. The other eigenvalues and eigenvectors are called **subdominant**.

Thus if we introduce the iterative process

$$\boldsymbol{x}^{(k+1)} = \boldsymbol{A}\boldsymbol{x}^{(k)} \quad (k = 0, 1, 2, \ldots)$$

starting with some arbitrary vector $\boldsymbol{x}^{(0)}$ not orthogonal to $\boldsymbol{e}_1$, it follows from (6.13) that

$$\boldsymbol{x}^{(k)} = \boldsymbol{A}^k\boldsymbol{x}^{(0)}$$

will converge to the dominant eigenvector of $\boldsymbol{A}$.

A clear disadvantage with this scheme is that if $|\lambda_1|$ is large then $\boldsymbol{A}^k\boldsymbol{x}^{(0)}$ will become very large, and computer overflow can occur. This can be avoided by

```
{read in x^T = [x_1 x_2 ... x_n]}
  m ← 0
  repeat
    mold ← m
    {evaluate y = Ax}
    {find m = max (y_i)}
    {x^T = [y_1/m y_2/m ... y_n/m]}
  until abs(m - mold) < tolerance
{write (results)}
```

Figure 6.1 Outline pseudocode program for power method to calculate the maximum eigenvalue.

scaling the vector $x^{(k)}$ after each iteration. The standard approach is to make the largest element of $x^{(k)}$ unity using the scaling factor max ($x^{(k)}$), which represents the element of $x^{(k)}$ having the largest modulus.

Thus in practice we adopt the iterative process

$$\begin{aligned} y^{(k+1)} &= Ax^{(k)} \\ x^{(k+1)} &= \frac{y^{(k+1)}}{\max(y^{(k+1)})} \quad (k = 0, 1, 2, \dots) \end{aligned} \tag{6.14}$$

and it is common to take $x^{(0)} = [1 \quad 1 \quad \dots \quad 1]^T$.

Corresponding to (6.12), we have

$$x^{(k)} = R\lambda_1^k\left[\alpha_1 e_1 + \sum_{i=2}^{n} \alpha_i\left(\frac{\lambda_i}{\lambda_1}\right)^k e_i\right]$$

where

$$R = [\max(y^{(1)})\max(y^{(2)}) \dots \max(y^{(k)})]^{-1}$$

Again we see that $x^{(k)}$ converges to a multiple of the dominant eigenvector e_1. Also, since $Ax^{(k)} \to \lambda_1 x^{(k)}$, we have $y^{(k+1)} \to \lambda_1 x^{(k)}$, and since the largest element of $x^{(k)}$ is unity, it follows that the scaling factors max($y^{(k+1)}$) converge to the dominant eigenvalue λ_1. The **rate of convergence** depends primarily on the ratios

$$\left|\frac{\lambda_2}{\lambda_1}\right|, \left|\frac{\lambda_3}{\lambda_1}\right|, \dots, \left|\frac{\lambda_n}{\lambda_1}\right|$$

The smaller these ratios, the faster the rate of convergence. The iterative process represents the simplest form of the **power method**, and a pseudocode for the basic algorithm is given in Figure 6.1.

EXAMPLE 6.10

Use the power method to find the dominant eigenvalue and the corresponding eigenvector of the matrix

$$\boldsymbol{A} = \begin{bmatrix} 1 & 1 & -2 \\ -1 & 2 & 1 \\ 0 & 1 & -1 \end{bmatrix}$$

Solution Taking $\boldsymbol{x}^{(0)} = [1 \quad 1 \quad 1]^{\mathrm{T}}$ in (6.14), we have

$$\boldsymbol{y}^{(1)} = \boldsymbol{A}\boldsymbol{x}^{(0)} = \begin{bmatrix} 1 & 1 & -2 \\ -1 & 2 & 1 \\ 0 & 1 & -1 \end{bmatrix}\begin{bmatrix} 1 \\ 1 \\ 1 \end{bmatrix} = \begin{bmatrix} 0 \\ 2 \\ 0 \end{bmatrix} = 2\begin{bmatrix} 0 \\ 1 \\ 0 \end{bmatrix}; \quad \lambda_1^{(1)} = 2$$

$$\boldsymbol{x}^{(1)} = \tfrac{1}{2}\boldsymbol{y}^{(1)} = \begin{bmatrix} 0 \\ 1 \\ 0 \end{bmatrix}$$

$$\boldsymbol{y}^{(2)} = \boldsymbol{A}\boldsymbol{x}^{(1)} = \begin{bmatrix} 1 & 1 & -2 \\ -1 & 2 & 1 \\ 0 & 1 & -1 \end{bmatrix}\begin{bmatrix} 0 \\ 1 \\ 0 \end{bmatrix} = \begin{bmatrix} 1 \\ 2 \\ 1 \end{bmatrix} = 2\begin{bmatrix} 0.5 \\ 1 \\ 0.5 \end{bmatrix}; \quad \lambda_2^{(2)} = 2$$

$$\boldsymbol{x}^{(2)} = \tfrac{1}{2}\boldsymbol{y}^{(2)} = \begin{bmatrix} \frac{1}{2} \\ 1 \\ \frac{1}{2} \end{bmatrix}$$

$$\boldsymbol{y}^{(3)} = \boldsymbol{A}\boldsymbol{x}^{(2)} = \begin{bmatrix} 1 & 1 & -2 \\ -1 & 2 & 1 \\ 0 & 1 & -1 \end{bmatrix}\begin{bmatrix} \frac{1}{2} \\ 1 \\ \frac{1}{2} \end{bmatrix} = \begin{bmatrix} \frac{1}{2} \\ 2 \\ \frac{1}{2} \end{bmatrix} = 2\begin{bmatrix} 0.25 \\ 1 \\ 0.25 \end{bmatrix}; \quad \lambda_3^{(2)} = 2$$

$$\boldsymbol{x}^{(3)} = \begin{bmatrix} 0.25 \\ 1 \\ 0.25 \end{bmatrix}$$

Continuing with the process, we have

$$\boldsymbol{y}^{(4)} = 2[0.375 \quad 1 \quad 0.375]^{\mathrm{T}}$$

$$\boldsymbol{y}^{(5)} = 2[0.312 \quad 1 \quad 0.312]^{\mathrm{T}}$$

$$\boldsymbol{y}^{(6)} = 2[0.344 \quad 1 \quad 0.344]^{\mathrm{T}}$$

$$\boldsymbol{y}^{(7)} = 2[0.328 \quad 1 \quad 0.328]^{\mathrm{T}}$$

$$\boldsymbol{y}^{(8)} = 2[0.336 \quad 1 \quad 0.336]^{\mathrm{T}}$$

Clearly $\boldsymbol{y}^{(k)}$ is approaching the vector $2[\frac{1}{3} \quad 1 \quad \frac{1}{3}]^{\mathrm{T}}$, so that the dominant eigenvalue is 2 and the corresponding eigenvector is $[\frac{1}{3} \quad 1 \quad \frac{1}{3}]^{\mathrm{T}}$, which conforms to the answer obtained in Example 6.4.

EXAMPLE 6.11 Find the dominant eigenvalue of

$$\boldsymbol{A} = \begin{bmatrix} 1 & 0 & -1 & 0 \\ 0 & 1 & 1 & 0 \\ -1 & 1 & 2 & 1 \\ 0 & 0 & 1 & -1 \end{bmatrix}$$

Solution Starting with $\boldsymbol{x}^{(0)} = [1 \quad 1 \quad 1 \quad 1]^{\mathrm{T}}$, the iterations give the following:

Iteration k		1	2	3	4	5	6	7
Eigenvalue	–	3	2.6667	3.3750	3.0741	3.2048	3.1636	3.1642
$x_1^{(k)}$	1	0	−0.3750	−0.4074	−0.4578	−0.4549	−0.4621	−0.4621
$x_2^{(k)}$	1	0.6667	0.6250	0.4815	0.4819	0.4624	0.4621	0.4621
$x_3^{(k)}$	1	1	1	1	1	1	1	1
$x_4^{(k)}$	1	0	0.3750	0.1852	0.2651	0.2293	0.2403	0.2401

This indicates that the dominant eigenvalue is aproximately 3.16, with corresponding eigenvector $[-0.46 \quad 0.46 \quad 1 \quad 0.24]^{\mathrm{T}}$.

The power method is suitable for obtaining the dominant eigenvalue and corresponding eigenvector of a matrix $\boldsymbol{A}$ having real distinct eigenvalues. The smallest eigenvalue, provided it is non-zero, can be obtained by using the same method on the inverse matrix $\boldsymbol{A}^{-1}$ when it exists. This follows since if $\boldsymbol{Ax} = \lambda\boldsymbol{x}$ then $\boldsymbol{A}^{-1}\boldsymbol{x} = \lambda^{-1}\boldsymbol{x}$. To find the subdominant eigenvalue using this method the dominant eigenvalue must first be removed from the matrix using **deflation methods**. We shall illustrate such a method for symmetric matrices only.

Let $\boldsymbol{A}$ be a symmetric matrix having real eigenvalues $\lambda_1, \lambda_2, \ldots, \lambda_n$. Then, by result (b) of Section 6.4.7, it has n corresponding mutually orthogonal normalized eigenvectors $\hat{\boldsymbol{e}}_1, \hat{\boldsymbol{e}}_2, \ldots, \hat{\boldsymbol{e}}_n$ such that

$$\hat{\boldsymbol{e}}_i^{\mathrm{T}}\hat{\boldsymbol{e}}_j = \delta_{ij} \quad (i, j = 1, 2, \ldots, n)$$

Let λ_1 be the dominant eigenvalue and consider the matrix

$$\boldsymbol{A}_1 = \boldsymbol{A} - \lambda_1\hat{\boldsymbol{e}}_1\hat{\boldsymbol{e}}_1^{\mathrm{T}}$$

which is such that

$$\boldsymbol{A}_1\hat{\boldsymbol{e}}_1 = (\boldsymbol{A} - \lambda_1\hat{\boldsymbol{e}}_1\hat{\boldsymbol{e}}_1^{\mathrm{T}})\hat{\boldsymbol{e}}_1 = \boldsymbol{A}\hat{\boldsymbol{e}}_1 - \lambda_1\hat{\boldsymbol{e}}_1(\hat{\boldsymbol{e}}_1^{\mathrm{T}}\hat{\boldsymbol{e}}_1) = \lambda_1\hat{\boldsymbol{e}}_1 - \lambda_1\hat{\boldsymbol{e}}_1 = 0$$

$$\boldsymbol{A}_1\hat{\boldsymbol{e}}_2 = \boldsymbol{A}\hat{\boldsymbol{e}}_2 - \lambda_1\hat{\boldsymbol{e}}_1(\hat{\boldsymbol{e}}_1^{\mathrm{T}}\hat{\boldsymbol{e}}_2) = \lambda_2\hat{\boldsymbol{e}}_2$$

$$\boldsymbol{A}_1\hat{\boldsymbol{e}}_3 = \boldsymbol{A}\hat{\boldsymbol{e}}_3 - \lambda_1\hat{\boldsymbol{e}}_1(\hat{\boldsymbol{e}}_1^{\mathrm{T}}\hat{\boldsymbol{e}}_3) = \lambda_3\hat{\boldsymbol{e}}_3$$

$$\vdots$$

$$\boldsymbol{A}_1\hat{\boldsymbol{e}}_n = \boldsymbol{A}\hat{\boldsymbol{e}}_n - \lambda_1\hat{\boldsymbol{e}}_1(\hat{\boldsymbol{e}}_1^{\mathrm{T}}\hat{\boldsymbol{e}}_n) = \lambda_n\hat{\boldsymbol{e}}_n$$

Thus the matrix $\mathbf{A}_1$ has the same eigenvalues and eigenvectors as the matrix $\mathbf{A}$, except that the eigenvalue corresponding to λ_1 is now zero. The power method can then be applied to the matrix $\mathbf{A}_1$ to obtain the subdominant eigenvalue λ_2 and its corresponding eigenvector $\boldsymbol{e}_2$. By repeated use of this technique, we can determine all the eigenvalues and corresponding eigenvectors of $\mathbf{A}$.

EXAMPLE 6.12 Given that the symmetric matrix

$$\mathbf{A} = \begin{bmatrix} 2 & 2 & 0 \\ 2 & 5 & 0 \\ 0 & 0 & 3 \end{bmatrix}$$

has a dominant eigenvalue $\lambda_1 = 6$ with corresponding normalized eigenvector $\hat{\boldsymbol{e}}_1 = [1 \quad 2 \quad 0]^{\mathrm{T}}/\sqrt{5}$, find the subdominant eigenvalue λ_2 and corresponding eigenvector $\hat{\boldsymbol{e}}_2$.

Solution Following the above procedure,

$$\mathbf{A}_1 = \mathbf{A} - \lambda_1 \hat{\boldsymbol{e}}_1 \hat{\boldsymbol{e}}_1^{\mathrm{T}}$$

$$= \begin{bmatrix} 2 & 2 & 0 \\ 2 & 5 & 0 \\ 0 & 0 & 3 \end{bmatrix} - \tfrac{6}{5} \begin{bmatrix} 1 \\ 2 \\ 0 \end{bmatrix} [1 \quad 2 \quad 0] = \begin{bmatrix} \frac{4}{5} & -\frac{2}{5} & 0 \\ -\frac{2}{5} & \frac{1}{5} & 0 \\ 0 & 0 & 3 \end{bmatrix}$$

Applying the power method procedure (6.14), with $\boldsymbol{x}^{(0)} = [1 \quad 1 \quad 1]^{\mathrm{T}}$, gives

$$\boldsymbol{y}^{(1)} = \mathbf{A}_1 \boldsymbol{x}^{(0)} = \begin{bmatrix} \frac{2}{5} \\ -\frac{1}{5} \\ 3 \end{bmatrix} = 3 \begin{bmatrix} \frac{2}{15} \\ -\frac{1}{15} \\ 1 \end{bmatrix}; \quad \lambda_2^{(1)} = 3$$

$$\boldsymbol{x}^{(1)} = \begin{bmatrix} \frac{2}{15} \\ -\frac{1}{15} \\ 1 \end{bmatrix} = \begin{bmatrix} 0.133 \\ -0.133 \\ 1 \end{bmatrix}$$

$$\boldsymbol{y}^{(2)} = \mathbf{A}_1 \boldsymbol{x}^{(1)} = \begin{bmatrix} \frac{2}{15} \\ -\frac{1}{15} \\ 3 \end{bmatrix} = 3 \begin{bmatrix} \frac{2}{45} \\ -\frac{2}{45} \\ 1 \end{bmatrix}; \quad \lambda_2^{(2)} = 3$$

$$\boldsymbol{x}^{(2)} = \begin{bmatrix} \frac{2}{45} \\ -\frac{2}{45} \\ 1 \end{bmatrix} = \begin{bmatrix} 0.044 \\ -0.044 \\ 1 \end{bmatrix}$$

$$y^{(3)} = \mathbf{A}_1 x^{(2)} = \begin{bmatrix} \frac{2}{45} \\ -\frac{2}{45} \\ 3 \end{bmatrix} = 3 \begin{bmatrix} \frac{2}{135} \\ -\frac{2}{135} \\ 1 \end{bmatrix}; \quad \lambda_2^{(2)} = 3$$

$$x^{(3)} = \begin{bmatrix} 0.015 \\ -0.015 \\ 1 \end{bmatrix}$$

Clearly the subdominant eigenvalue of $\mathbf{A}$ is $\lambda_2 = 3$, and a few more iterations confirms the corresponding normalized eigenvector is $\hat{e}_2 = [0 \quad 0 \quad 1]^T$. This is confirmed by the solution of Example 6.9. Note that the third eigenvalue may then be obtained using Property 6.1 of Section 6.4.6, since

$$\text{trace } \mathbf{A} = 10 = \lambda_1 + \lambda_2 + \lambda_3 = 6 + 3 + \lambda_3$$

giving $\lambda_3 = 1$. Alternatively, λ_3 and $\hat{e}_3$ can be obtained by applying the power method to the matrix $\mathbf{A}_2 = \mathbf{A}_1 - \lambda_2 \hat{e}_2 \hat{e}_2^T$.

Although it is good as an illustration of the principles underlying iterative methods for evaluating eigenvalues and eigenvectors, the power method is of little practical importance, except possibly when dealing with large sparse matrices. In order to evaluate all the eigenvalues and eigenvectors of a matrix, including those with repeated eigenvalues, more sophisticated methods are required. Many of the numerical methods available, such as the **Jacobi** and **Householder methods**, are only applicable to symmetric matrices, and involve reducing the matrix to a special form so that its eigenvalues can be readily calculated. Analogous methods for non-symmetric matrices are the **LR** and **QR methods**. It is methods such as these, together with others based on the inverse iterative method, that form the basis of the algorithms that exist in modern software libraries. Such methods will not be pursued further here, and the interested reader is referred to specialist texts on numerical analysis.

6.5.2 Gerschgorin circles

In many engineering applications it is not necessary to obtain accurate approximations to the eigenvalues of a matrix. All that is often required are bounds on the eigenvalues. For example, when considering the stability of continuous- or discrete-time systems (see Sections 6.8–6.10), we are concerned as to whether the eigenvalues lie in the negative half-plane or within the unit circle in the complex plane. (Note that the eigenvalues of a non-symmetric matrix can be complex.) The Gerschgorin theorems often provide a quick method to answer such questions without the need for detailed calculations. These theorems may be stated as follows.

THEOREM 6.1 **First Gerschgorin theorem**

Every eigenvalue of the matrix $\mathbf{A} = [a_{ij}]$, of order n, lies inside at least one of the circles (called **Gerschgorin circles**) in the complex plane with centre a_{ii} and radii $r_i = \sum_{j=1, j\neq i}^{n} |a_{ij}|$ $(i = 1, 2, \ldots, n)$. Expressed in another form, all the eigenvalues of the matrix $\mathbf{A} = [a_{ij}]$ lie in the union of the discs

$$|z - a_{ii}| \leqslant r_i = \sum_{\substack{j=1 \\ j\neq i}}^{n} |a_{ij}| \quad (i = 1, 2, \ldots, n)$$

in the complex z plane. □

THEOREM 6.2 **Second Gerschgorin theorem**

If the union of s of the Gerschgorin circles forms a connected region isolated from the remaining circles then exactly s of the eigenvalues lie within this region. □

Since the disc $|z - a_{ii}| \leqslant r_i$ is contained within the disc

$$|z| \leqslant |a_{ii}| + r_i = \sum_{j=1}^{n} |a_{ij}|$$

centred at the origin, we have a less precise but more easily applied criterion that all the eigenvalues of the matrix $\mathbf{A}$ lie within the disc

$$|z| \leqslant \max_i \left\{ \sum_{j=1}^{n} |a_{ij}| \right\} \quad (i = 1, 2, \ldots, n) \tag{6.15}$$

centred at the origin.

The **spectral radius** $\rho(\mathbf{A})$ of a matrix $\mathbf{A}$ is the modulus of its dominant eigenvalue; that is,

$$\rho(\mathbf{A}) = \max\{|\lambda_i|\} \quad (i = 1, 2, \ldots, n) \tag{6.16}$$

where $\lambda_1, \lambda_2, \ldots, \lambda_n$ are the eigenvalues of $\mathbf{A}$. Geometrically, $\rho(\mathbf{A})$ is the radius of the smallest circle centred at the origin in the complex plane such that all the eigenvalues of $\mathbf{A}$ lie inside the circle. It follows from (6.15) that

$$\rho(\mathbf{A}) \leqslant \max_i \left\{ \sum_{j=1}^{n} |a_{ij}| \right\} \quad (i = 1, 2, \ldots, n) \tag{6.17}$$

EXAMPLE 6.13 Draw the Gerschgorin circles corresponding to the matrix

$$\mathbf{A} = \begin{bmatrix} 10 & -1 & 0 \\ -1 & 2 & 2 \\ 0 & 2 & 3 \end{bmatrix}$$

What can be concluded about the eigenvalues of $\mathbf{A}$?

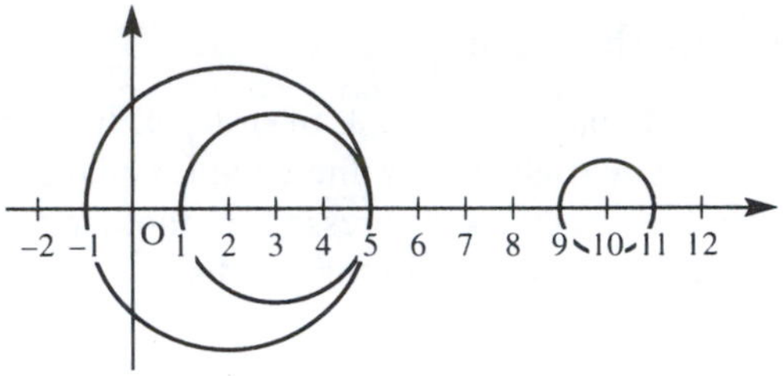

z plane

Figure 6.2 Gerschgorin circles for the matrix $\boldsymbol{A}$ of Example 6.13.

Solution The three Gerschgorin circles are

(i) $|z - 10| = |-1| + 0 = 1$

(ii) $|z - 2| = |-1| + |2| = 3$

(iii) $|z - 3| = |2| = 2$

and are illustrated in Figure 6.2.

It follows from Theorem 6.2 that one eigenvalue lies within the circle centred (10, 0) of radius 1, and two eigenvalues lie within the union of the other two circles; that is, within the circle centred at (2, 0) of radius 3. Since the matrix $\boldsymbol{A}$ is symmetric, it follows from result (a) of Section 6.4.7 that the eigenvalues are real. Hence

$$9 < \lambda_1 < 11$$

$$-1 < \{\lambda_2, \lambda_3\} < 5$$

6.5.3 Exercises

14 Use the power method to estimate the dominant eigenvalue and its corresponding eigenvector for the matrix

$$\boldsymbol{A} = \begin{bmatrix} 4 & 3 & 2 \\ 3 & 5 & 2 \\ 2 & 2 & 1 \end{bmatrix}$$

Stop when you consider the eigenvalue estimate is correct to two decimal places.

15 Repeat Exercise 14 for the matrices

(a) $\boldsymbol{A} = \begin{bmatrix} 2 & 1 & 0 \\ 1 & 2 & 1 \\ 1 & 1 & 2 \end{bmatrix}$ (b) $\boldsymbol{A} = \begin{bmatrix} 3 & 0 & 1 \\ 2 & 2 & 2 \\ 4 & 2 & 5 \end{bmatrix}$

(c) $\boldsymbol{A} = \begin{bmatrix} 2 & -1 & 0 & 0 \\ -1 & 2 & -1 & 0 \\ 0 & -1 & 2 & -1 \\ 0 & 0 & -1 & 2 \end{bmatrix}$

16 The symmetric matrix

$$\boldsymbol{A} = \begin{bmatrix} 3 & 1 & 1 \\ 1 & 3 & 1 \\ 1 & 1 & 5 \end{bmatrix}$$

has dominant eigenvector $\boldsymbol{e}_1 = [1 \quad 1 \quad 2]^{\mathrm{T}}$. Obtain the matrix

$$\boldsymbol{A}_1 = \boldsymbol{A} - \lambda_1 \hat{\boldsymbol{e}}_1 \hat{\boldsymbol{e}}_1^{\mathrm{T}}$$

where λ_1 is the eigenvalue corresponding to the eigenvector $\boldsymbol{e}_1$. Using the deflation method, obtain the subdominant eigenvalue λ_2 and corresponding eigenvector $\boldsymbol{e}_2$ correct to two decimal places, taking $[1 \quad 1 \quad 1]^T$ as a first approximation to $\boldsymbol{e}_2$. Continue the process to obtain the third eigenvalue λ_3 and its corresponding eigenvector $\boldsymbol{e}_3$.

17 Draw the Gerschgorin circles corresponding to the matrix

$$\boldsymbol{A} = \begin{bmatrix} 5 & 1 & -1 \\ 1 & 0 & 1 \\ -1 & 1 & -5 \end{bmatrix}$$

and hence show that the three eigenvalues are such that

$$3 < \lambda_1 < 7, \quad -2 < \lambda_2 < 2, \quad -7 < \lambda_3 < -3$$

18 Show that the characteristic equation of the matrix

$$\boldsymbol{A} = \begin{bmatrix} 10 & -1 & 0 \\ -1 & 2 & 2 \\ 0 & 2 & 3 \end{bmatrix}$$

of Example 6.13 is

$$f(\lambda) = \lambda^3 - 15\lambda^2 + 51\lambda - 17 = 0$$

Using the Newton–Raphson iterative procedure

$$\lambda_{n+1} = \lambda_n - \frac{f(\lambda_n)}{f'(\lambda_n)}$$

determine the eigenvalue identified in Example 6.13 to lie in the interval $9 < \lambda < 11$, correct to three decimal places.

Using Properties 6.1 and 6.2 of Section 6.4.6, determine the other two eigenvalues of $\boldsymbol{A}$ to the same approximation.

19 (a) If the eigenvalues of the $n \times n$ matrix $\boldsymbol{A}$ are

$$\lambda_1 > \lambda_2 > \lambda_3 \ldots \lambda_n \geq 0$$

show that the eigenvalue λ_n can be found by applying the power method to the matrix $k\boldsymbol{I} - \boldsymbol{A}$, where $\boldsymbol{I}$ is the identity matrix and $k \geq \lambda_1$.

(b) By considering the Gerschgorin circles, show that the eigenvalues of the matrix

$$\boldsymbol{A} = \begin{bmatrix} 2 & -1 & 0 \\ -1 & 2 & -1 \\ 0 & -1 & 2 \end{bmatrix}$$

satisfy the inequality

$$0 \leq \lambda \leq 4$$

Hence, using the result proved in (a), determine the smallest modulus eigenvalue of $\boldsymbol{A}$ correct to two decimal places.

6.6 Reduction to canonical form

In this section we examine the process of reduction of a matrix to **canonical form**. Specifically, we examine methods by which certain square matrices can be reduced or transformed into diagonal form. The process of transformation can be thought of as a change of system coordinates, with the new coordinate axes chosen in such a way that the system can be expressed in a simple form. The simplification may, for example, be a transformation to principal axes or a decoupling of system equations.

We will see that not all matrices can be reduced to diagonal form. In some cases we can only achieve the so-called Jordan canonical form, but many of the advantages of the diagonal form can be extended to this case as well.

The transformation to diagonal form is just one example of a **similarity** transform. Other such transforms exist, but, in common with the transformation to diagonal form, their purpose is usually that of simplifying the system model in some way.

6.6.1 Reduction to diagonal form

For an $n \times n$ matrix $\boldsymbol{A}$ possessing a full set of n linearly independent eigenvectors $\boldsymbol{e}_1, \boldsymbol{e}_2, \ldots, \boldsymbol{e}_n$ we can write down a **modal matrix** $\boldsymbol{M}$ having the n eigenvectors as its columns:

$$\boldsymbol{M} = [\boldsymbol{e}_1 \quad \boldsymbol{e}_2 \quad \boldsymbol{e}_3 \quad \ldots \quad \boldsymbol{e}_n]$$

The diagonal matrix having the eigenvalues of $\boldsymbol{A}$ as its diagonal elements is called the **spectral matrix** corresponding to the modal matrix $\boldsymbol{M}$ of $\boldsymbol{A}$, often denoted by $\boldsymbol{\Lambda}$. That is

$$\boldsymbol{\Lambda} = \begin{bmatrix} \lambda_1 & & & 0 \\ & \lambda_2 & & \\ & & \ddots & \\ 0 & & & \lambda_n \end{bmatrix}$$

with the ijth element being given by $\lambda_i \delta_{ij}$, where δ_{ij} is the Kronecker delta and $i, j = 1, 2, \ldots, n$. It is important in the work that follows that the pair of matrices $\boldsymbol{M}$ and $\boldsymbol{\Lambda}$ are written down correctly. If the ith column of $\boldsymbol{M}$ is the eigenvector $\boldsymbol{e}_i$ then the element in the (i, i) position in $\boldsymbol{\Lambda}$ must be λ_i, the eigenvalue corresponding to the eigenvector $\boldsymbol{e}_i$.

EXAMPLE 6.14 Obtain a modal matrix and the corresponding spectral matrix for the matrix $\boldsymbol{A}$ of Example 6.4.

Solution

$$\boldsymbol{A} = \begin{bmatrix} 1 & 1 & -2 \\ -1 & 2 & 1 \\ 0 & 1 & -1 \end{bmatrix}$$

having eigenvalues $\lambda_1 = 2$, $\lambda_2 = 1$ and $\lambda_3 = -1$, with corresponding eigenvectors

$$\boldsymbol{e}_1 = [1 \quad 3 \quad 1]^{\mathrm{T}}, \boldsymbol{e}_2 = [3 \quad 2 \quad 1]^{\mathrm{T}}, \boldsymbol{e}_3 = [1 \quad 0 \quad 1]^{\mathrm{T}}$$

Choosing as modal matrix $\boldsymbol{M} = [\boldsymbol{e}_1 \quad \boldsymbol{e}_2 \quad \boldsymbol{e}_3]^{\mathrm{T}}$ gives

$$\boldsymbol{M} = \begin{bmatrix} 1 & 3 & 1 \\ 3 & 2 & 0 \\ 1 & 1 & 1 \end{bmatrix}$$

The corresponding spectral matrix is

$$\boldsymbol{\Lambda} = \begin{bmatrix} 2 & 0 & 0 \\ 0 & 1 & 0 \\ 0 & 0 & -1 \end{bmatrix}$$

Returning to the general case, if we premultiply the matrix $\boldsymbol{M}$ by $\boldsymbol{A}$, we obtain

$$\begin{aligned}\boldsymbol{AM} &= \boldsymbol{A}[\boldsymbol{e}_1 \quad \boldsymbol{e}_2 \quad \dots \quad \boldsymbol{e}_n] = [\boldsymbol{Ae}_1 \quad \boldsymbol{Ae}_2 \quad \dots \quad \boldsymbol{Ae}_n] \\ &= [\lambda_1\boldsymbol{e}_1 \quad \lambda_2\boldsymbol{e}_2 \quad \dots \quad \lambda_n\boldsymbol{e}_n]\end{aligned}$$

so that

$$\boldsymbol{AM} = \boldsymbol{M\Lambda} \tag{6.18}$$

Since the n eigenvectors $\boldsymbol{e}_1, \boldsymbol{e}_2, \dots, \boldsymbol{e}_n$ are linearly independent, the matrix $\boldsymbol{M}$ is non-singular, so that $\boldsymbol{M}^{-1}$ exists. Thus premultiplying by $\boldsymbol{M}^{-1}$ gives

$$\boldsymbol{M}^{-1}\boldsymbol{AM} = \boldsymbol{M}^{-1}\boldsymbol{M\Lambda} = \boldsymbol{\Lambda} \tag{6.19}$$

indicating that the similarity transformation $\boldsymbol{M}^{-1}\boldsymbol{AM}$ reduces the matrix $\boldsymbol{A}$ to the **diagonal** or **canonical form** $\boldsymbol{\Lambda}$. Thus a matrix $\boldsymbol{A}$ possessing a full set of linearly independent eigenvectors is reducible to diagonal form, and the reduction process is often referred to as the **diagonalization** of the matrix $\boldsymbol{A}$. Since

$$\boldsymbol{A} = \boldsymbol{M\Lambda M}^{-1} \tag{6.20}$$

it follows that $\boldsymbol{A}$ is uniquely determined once the eigenvalues and corresponding eigenvectors are known. Note that knowledge of the eigenvalues and eigenvectors alone is not sufficient: in order to structure $\boldsymbol{M}$ and $\boldsymbol{\Lambda}$ correctly, the association of eigenvalues and the *corresponding* eigenvectors must also be known.

EXAMPLE 6.15 Verify results (6.19) and (6.20) for the matrix $\boldsymbol{A}$ of Example 6.14.

Solution Since

$$\boldsymbol{M} = \begin{bmatrix} 1 & 3 & 1 \\ 3 & 2 & 0 \\ 1 & 1 & 1 \end{bmatrix} \quad \text{we have} \quad \boldsymbol{M}^{-1} = \frac{1}{6}\begin{bmatrix} -2 & 2 & 2 \\ 3 & 0 & -3 \\ -1 & -2 & 7 \end{bmatrix}$$

Taking

$$\boldsymbol{\Lambda} = \begin{bmatrix} 2 & 0 & 0 \\ 0 & 1 & 0 \\ 0 & 0 & -1 \end{bmatrix}$$

matrix multiplication confirms the results

$$\boldsymbol{M}^{-1}\boldsymbol{AM} = \boldsymbol{\Lambda}, \ \boldsymbol{A} = \boldsymbol{M\Lambda M}^{-1}$$

For an $n \times n$ symmetric matrix $\boldsymbol{A}$ it follows, from result (b) of Section 6.4.7, that to the n real eigenvalues $\lambda_1, \lambda_2, \dots, \lambda_n$ there correspond n linearly independent normalized eigenvectors $\hat{\boldsymbol{e}}_1, \hat{\boldsymbol{e}}_2, \dots, \hat{\boldsymbol{e}}_n$ that are mutually orthogonal so that

$$\hat{e}_i^{\mathrm{T}}\hat{e}_j = \delta_{ij} \quad (i, j = 1, 2, \dots, n)$$

The corresponding modal matrix

$$\hat{\boldsymbol{M}} = [\hat{e}_1 \quad \hat{e}_2 \quad \dots \quad \hat{e}_n]$$

is then such that

$$\hat{\boldsymbol{M}}^{\mathrm{T}}\hat{\boldsymbol{M}} = \begin{bmatrix} \hat{e}_1^{\mathrm{T}} \\ \hat{e}_2^{\mathrm{T}} \\ \vdots \\ \hat{e}_n^{\mathrm{T}} \end{bmatrix} [\hat{e}_1 \quad \hat{e}_2 \quad \dots \quad \hat{e}_n] = \begin{bmatrix} \hat{e}_1^{\mathrm{T}}\hat{e}_1 & \hat{e}_1^{\mathrm{T}}\hat{e}_2 & \dots & \hat{e}_1^{\mathrm{T}}\hat{e}_n \\ \hat{e}_2^{\mathrm{T}}\hat{e}_1 & \hat{e}_2^{\mathrm{T}}\hat{e}_2 & \dots & \hat{e}_2^{\mathrm{T}}\hat{e}_n \\ \vdots & \vdots & & \vdots \\ \hat{e}_n^{\mathrm{T}}\hat{e}_1 & \hat{e}_n^{\mathrm{T}}\hat{e}_2 & \dots & \hat{e}_n^{\mathrm{T}}\hat{e}_n \end{bmatrix}$$

$$= \begin{bmatrix} 1 & 0 & \dots & 0 \\ 0 & 1 & \dots & 0 \\ \vdots & \vdots & & \vdots \\ 0 & 0 & \dots & 1 \end{bmatrix} = \boldsymbol{I}$$

That is, $\hat{\boldsymbol{M}}^{\mathrm{T}}\hat{\boldsymbol{M}} = \boldsymbol{I}$ and so $\hat{\boldsymbol{M}}^{\mathrm{T}} = \hat{\boldsymbol{M}}^{-1}$. Thus $\hat{\boldsymbol{M}}$ is an **orthogonal matrix** (the term **orthonormal matrix** would be more appropriate, but the nomenclature is long established).

It follows from (6.19) that a symmetric matrix $\boldsymbol{A}$ can be reduced to diagonal form $\boldsymbol{\Lambda}$ using the orthogonal transformation

$$\hat{\boldsymbol{M}}^{\mathrm{T}}\boldsymbol{A}\hat{\boldsymbol{M}} = \boldsymbol{\Lambda} \tag{6.21}$$

EXAMPLE 6.16 For the symmetric matrix $\boldsymbol{A}$ considered in Example 6.9 write down the corresponding orthogonal modal matrix $\hat{\boldsymbol{M}}$ and show that $\hat{\boldsymbol{M}}^{\mathrm{T}}\boldsymbol{A}\boldsymbol{M} = \boldsymbol{\Lambda}$, where $\boldsymbol{\Lambda}$ is the spectral matrix.

Solution From Example 6.9 the eigenvalues are $\lambda_1 = 6$, $\lambda_2 = 3$ and $\lambda_3 = 1$, with corresponding normalized eigenvectors

$$\hat{e}_1 = [1 \quad 2 \quad 0]^{\mathrm{T}}/\sqrt{5}, \quad \hat{e}_2 = [0 \quad 0 \quad 1]^{\mathrm{T}}, \quad \hat{e}_3 = [-2 \quad 1 \quad 0]^{\mathrm{T}}/\sqrt{5}$$

The corresponding modal matrix is

$$\hat{\boldsymbol{M}} = \begin{bmatrix} \sqrt{\tfrac{1}{5}} & 0 & -2\sqrt{\tfrac{1}{5}} \\ 2\sqrt{\tfrac{1}{5}} & 0 & \sqrt{\tfrac{1}{5}} \\ 0 & 1 & 0 \end{bmatrix}$$

and, by matrix multiplication,

$$\hat{\boldsymbol{M}}^{\mathrm{T}}\boldsymbol{A}\hat{\boldsymbol{M}} = \begin{bmatrix} 6 & 0 & 0 \\ 0 & 3 & 0 \\ 0 & 0 & 1 \end{bmatrix} = \boldsymbol{\Lambda}$$

6.6.2 The Jordan canonical form

If an $n \times n$ matrix $\boldsymbol{A}$ does not possess a full set of linearly independent eigenvectors then it cannot be reduced to diagonal form using the similarity transformation $\boldsymbol{M}^{-1}\boldsymbol{AM}$. In such a case, however, it is possible to reduce $\boldsymbol{A}$ to a **Jordan canonical form**, making use of 'generalized' eigenvectors.

As indicated in (6.11), if a matrix $\boldsymbol{A}$ has an eigenvalue λ_i of algebraic multiplicity m_i and geometric multiplicity q_i, with $1 \leqslant q_i \leqslant m_i$, then there are q_i linearly independent eigenvectors corresponding to λ_i. Consequently, we need to generate $m_i - q_i$ generalized eigenvectors in order to produce a full set. To obtain these, we first obtain the q_i linearly independent eigenvectors by solving

$$(\boldsymbol{A} - \lambda_i \boldsymbol{I})\boldsymbol{e}_i = \boldsymbol{0}$$

Then for each of these vectors we try to construct a generalized eigenvector $\boldsymbol{e}_i^*$ such that

$$(\boldsymbol{A} - \lambda_i \boldsymbol{I})\boldsymbol{e}_i^* = \boldsymbol{e}_i$$

If the resulting vector $\boldsymbol{e}_i^*$ is linearly independent of all the eigenvectors (and generalized eigenvectors) already found then it is a valid additional generalized eigenvector. If further generalized eigenvectors corresponding to λ_i are needed, we then repeat the process using

$$(\boldsymbol{A} - \lambda_i \boldsymbol{I})\boldsymbol{e}_i^{**} = \boldsymbol{e}_i^*$$

and so on until sufficient vectors are found.

EXAMPLE 6.17 Obtain a generalized eigenvector corresponding to the eigenvalue $\lambda = 2$ of Example 6.7.

Solution For

$$\boldsymbol{A} = \begin{bmatrix} 1 & 2 & 2 \\ 0 & 2 & 1 \\ -1 & 2 & 2 \end{bmatrix}$$

we found in Example 6.7 that corresponding to the eigenvalue $\lambda_i = 2$ there was only one linearly independent eigenvector

$$\boldsymbol{e}_1 = [2 \quad 1 \quad 0]^{\mathrm{T}}$$

and we need to find a generalized eigenvector to produce a full set. To obtain the generalized eigenvector $\boldsymbol{e}_1^*$, we solve

$$(\boldsymbol{A} - 2\boldsymbol{I})\boldsymbol{e}_1^* = \boldsymbol{e}_1$$

that is, we solve

$$\begin{bmatrix} -1 & 2 & 2 \\ 0 & 0 & 1 \\ -1 & 2 & 0 \end{bmatrix} \begin{bmatrix} e_{11}^* \\ e_{12}^* \\ e_{13}^* \end{bmatrix} = \begin{bmatrix} 2 \\ 1 \\ 0 \end{bmatrix}$$

At once, we have $e_{13}^* = 1$ and $e_{11}^* = 2e_{12}^*$, and so

$$e_1^* = [2 \quad 1 \quad 1]^{\mathrm{T}}$$

Thus, by including generalized eigenvectors, we have a full set of eigenvectors for the matrix $\boldsymbol{A}$ given by

$$\boldsymbol{e}_1 = [2 \quad 1 \quad 0]^{\mathrm{T}}, \quad \boldsymbol{e}_2 = [2 \quad 1 \quad 1]^{\mathrm{T}}, \quad \boldsymbol{e}_3 = [1 \quad -1 \quad 1]^{\mathrm{T}}$$

If we include the generalized eigenvectors, it is always possible to obtain for an $n \times n$ matrix $\boldsymbol{A}$ a modal matrix $\boldsymbol{M}$ with n linearly independent columns $\boldsymbol{e}_1, \boldsymbol{e}_2, \ldots, \boldsymbol{e}_n$. Corresponding to (6.18), we have

$$\boldsymbol{AM} = \boldsymbol{MJ}$$

where $\boldsymbol{J}$ is called the **Jordan form** of $\boldsymbol{A}$. Premultiplying by $\boldsymbol{M}^{-1}$ then gives

$$\boldsymbol{M}^{-1}\boldsymbol{AM} = \boldsymbol{J} \tag{6.22}$$

The process of reducing $\boldsymbol{A}$ to $\boldsymbol{J}$ is known as the **reduction** of $\boldsymbol{A}$ to its Jordan normal, or canonical, form.

If $\boldsymbol{A}$ has p distinct eigenvalues then the matrix $\boldsymbol{J}$ is of the block-diagonal form

$$\boldsymbol{J} = [\boldsymbol{J}_1 \quad \boldsymbol{J}_2 \quad \ldots \quad \boldsymbol{J}_p]$$

where each submatrix $\boldsymbol{J}_i$ $(i = 1, 2, \ldots, p)$ is associated with the corresponding eigenvalue λ_i. The submatrix $\boldsymbol{J}_i$ will have λ_i as its leading diagonal elements, with zeros elsewhere except on the diagonal above the leading diagonal. On this diagonal the entries will have the value 1 or 0, depending on the number of generalized eigenvectors used and how they were generated. To illustrate this, suppose that $\boldsymbol{A}$ is a 7×7 matrix with eigenvalues $\lambda_1 = 1$, $\lambda_2 = 2$ (occurring twice), $\lambda_3 = 3$ (occurring four times), and suppose that the number of linearly independent eigenvectors generated in each case is

$\lambda_1 = 1$, 1 eigenvector

$\lambda_2 = 2$, 1 eigenvector

$\lambda_3 = 3$, 2 eigenvectors

with one further generalized eigenvector having been determined for $\lambda_2 = 2$ and two more for $\lambda_3 = 3$.

Corresponding to $\lambda_1 = 1$, the Jordan block $\boldsymbol{J}_1$ will be just [1], while that corresponding to $\lambda_2 = 2$ will be

$$\boldsymbol{J}_2 = \begin{bmatrix} 2 & 1 \\ 0 & 2 \end{bmatrix}$$

corresponding to $\lambda_3 = 3$, the Jordan block $\boldsymbol{J}_3$ can take one of the two forms

$$\boldsymbol{J}_{3,1} = \left[\begin{array}{ccc:c} \lambda_3 & 1 & 0 & 0 \\ 0 & \lambda_3 & 1 & 0 \\ 0 & 0 & \lambda_3 & 0 \\ \hdashline 0 & 0 & 0 & \lambda_3 \end{array}\right] \quad \text{or} \quad \boldsymbol{J}_{3,2} = \left[\begin{array}{cc:cc} \lambda_3 & 1 & 0 & 0 \\ 0 & \lambda_3 & 0 & 0 \\ \hdashline 0 & 0 & \lambda_3 & 1 \\ 0 & 0 & 0 & \lambda_3 \end{array}\right]$$

depending on how the generalized eigenvectors are generated. Corresponding to $\lambda_3 = 3$, we had two linearly independent eigenvectors $\boldsymbol{e}_{3,1}$ and $\boldsymbol{e}_{3,2}$. If both generalized eigenvectors are generated from *one* of these vectors then $\boldsymbol{J}_3$ will take the form $\boldsymbol{J}_{3,1}$, whereas if one generalized eigenvector has been generated from each eigenvector then $\boldsymbol{J}_3$ will take the form $\boldsymbol{J}_{3,2}$.

EXAMPLE 6.18

Obtain the Jordan canonical form of the matrix $\boldsymbol{A}$ of Example 6.17, and show that $\boldsymbol{M}^{-1}\boldsymbol{A}\boldsymbol{M} = \boldsymbol{J}$ where $\boldsymbol{M}$ is a modal matrix that includes generalized eigenvectors.

Solution For

$$\boldsymbol{A} = \begin{bmatrix} 1 & 2 & 2 \\ 0 & 2 & 1 \\ -1 & 2 & 2 \end{bmatrix}$$

from Example 6.17 we know that the eigenvalues of $\boldsymbol{A}$ are $\lambda_1 = 2$ (twice) and $\lambda_3 = 1$. The eigenvector corresponding to $\lambda_3 = 1$ has been determined as $\boldsymbol{e}_3 = [1 \quad 1 \quad -1]^{\mathrm{T}}$ in Example 6.7 and corresponding to $\lambda_1 = 2$ we found one linearly independent eigenvector $\boldsymbol{e}_1 = [2 \quad 1 \quad 0]^{\mathrm{T}}$ and a generalized eigenvector $\boldsymbol{e}_1^* = [2 \quad 1 \quad 1]^{\mathrm{T}}$. Thus the modal matrix including this generalized eigenvector is

$$\boldsymbol{M} = \begin{bmatrix} 2 & 2 & 1 \\ 1 & 1 & 1 \\ 0 & 1 & -1 \end{bmatrix}$$

and the corresponding Jordan canonical form is

$$\boldsymbol{J} = \left[\begin{array}{cc:c} 2 & 1 & 0 \\ 0 & 2 & 0 \\ \hdashline 0 & 0 & 1 \end{array}\right]$$

To check this result, we compute $\boldsymbol{M}^{-1}$ as

$$\boldsymbol{M}^{-1} = \begin{bmatrix} 2 & -3 & -1 \\ -1 & 2 & 1 \\ -1 & 2 & 0 \end{bmatrix}$$

and, forming $\boldsymbol{M}^{-1}\boldsymbol{A}\boldsymbol{M}$, we obtain $\boldsymbol{J}$ as expected.

6.6.3 Exercises

20 Show that the eigenvalues of the matrix

$$\boldsymbol{A} = \begin{bmatrix} -1 & 6 & -12 \\ 0 & -13 & 30 \\ 0 & -9 & 20 \end{bmatrix}$$

are 5, 2 and −1. Obtain the corresponding eigenvectors. Write down the modal matrix $\boldsymbol{M}$ and spectral matrix Λ. Evaluate $\boldsymbol{M}^{-1}$ and show that $\boldsymbol{M}^{-1}\boldsymbol{AM} = \Lambda$.

21 Using the eigenvalues and corresponding eigenvectors of the symmetric matrix

$$\boldsymbol{A} = \begin{bmatrix} 2 & 2 & 0 \\ 2 & 5 & 0 \\ 0 & 0 & 3 \end{bmatrix}$$

obtained in Example 6.9, verify that $\hat{\boldsymbol{M}}^{\mathrm{T}}\boldsymbol{A}\hat{\boldsymbol{M}} = \Lambda$ where $\hat{\boldsymbol{M}}$ and Λ are respectively a normalized modal matrix and a spectral matrix of $\boldsymbol{A}$.

22 Given

$$\boldsymbol{A} = \begin{bmatrix} 5 & 10 & 8 \\ 10 & 2 & -2 \\ 8 & -2 & 11 \end{bmatrix}$$

find its eigenvalues and corresponding eigenvectors. Normalize the eigenvectors and write down the corresponding normalized modal matrix $\hat{\boldsymbol{M}}$. Write down $\hat{\boldsymbol{M}}^{\mathrm{T}}$ and show that $\hat{\boldsymbol{M}}^{\mathrm{T}}\boldsymbol{A}\hat{\boldsymbol{M}} = \Lambda$, where Λ is the spectral matrix of $\boldsymbol{A}$.

23 Determine the eigenvalues and corresponding eigenvectors of the matrix

$$\boldsymbol{A} = \begin{bmatrix} 1 & 1 & -2 \\ -1 & 2 & 1 \\ 0 & 1 & -1 \end{bmatrix}$$

Write down the modal matrix $\boldsymbol{M}$ and spectral matrix Λ. Confirm that $\boldsymbol{M}^{-1}\boldsymbol{AM} = \Lambda$ and that $\boldsymbol{A} = \boldsymbol{M}\Lambda\boldsymbol{M}^{-1}$.

24 Determine the eigenvalues and corresponding eigenvectors of the symmetric matrix

$$\boldsymbol{A} = \begin{bmatrix} 3 & -2 & 4 \\ -2 & -2 & 6 \\ 4 & 6 & -1 \end{bmatrix}$$

Verify that the eigenvectors are orthogonal, and write down an orthogonal matrix $\boldsymbol{L}$ such that $\boldsymbol{L}^{\mathrm{T}}\boldsymbol{AL} = \Lambda$, where Λ is the spectral matrix of $\boldsymbol{A}$.

25 A 3 × 3 symmetric matrix $\boldsymbol{A}$ has eigenvalues 6, 3 and 1. The eigenvectors corresponding to the eigenvalues 6 and 1 are $[1 \quad 2 \quad 0]^{\mathrm{T}}$ and $[-2 \quad 1 \quad 0]^{\mathrm{T}}$ respectively. Find the eigenvector corresponding to the eigenvalue 3, and hence determine the matrix $\boldsymbol{A}$.

26 Given that $\lambda = 1$ is a thrice-repeated eigenvalue of the matrix

$$\boldsymbol{A} = \begin{bmatrix} -3 & -7 & -5 \\ 2 & 4 & 3 \\ 1 & 2 & 2 \end{bmatrix}$$

use the nullity, given by (6.11), of a suitable matrix to show that there is only one corresponding linearly independent eigenvector. Obtain two further generalized eigenvectors, and write down the corresponding modal matrix $\boldsymbol{M}$. Confirm that $\boldsymbol{M}^{-1}\boldsymbol{AM} = \boldsymbol{J}$, where $\boldsymbol{J}$ is the appropriate Jordan matrix.

27 Show that the eigenvalues of the matrix

$$\boldsymbol{A} = \begin{bmatrix} 1 & 0 & 0 & -3 \\ 0 & 1 & -3 & 0 \\ -0.5 & -3 & 1 & 0.5 \\ -3 & 0 & 0 & 1 \end{bmatrix}$$

are −2, −2, 4 and 4. Using the nullity, given by (6.11), of appropriate matrices, show that there are two linearly independent eigenvectors corresponding to the repeated eigenvalue −2 and only one corresponding to the repeated eigenvalue 4. Obtain a further generalized eigenvector corresponding to the eigenvalue 4. Write down the Jordan canonical form of $\boldsymbol{A}$.

6.6.4 Quadratic forms

A **quadratic form** in n independent variables $x_1, x_2, \ldots, x_n$ is a homogeneous second-degree polynomial of the form

$$\begin{aligned} V(x_1, x_2, \ldots, x_n) &= \sum_{i=1}^{n}\sum_{j=1}^{n} a_{ij}x_ix_j \\ &= a_{11}x_1^2 + a_{12}x_1x_2 + \ldots + a_{1n}x_1x_n \\ &\quad + a_{21}x_2x_1 + a_{22}x_2^2 + \ldots + a_{2n}x_2x_n \\ &\quad \vdots \\ &\quad + a_{n1}x_nx_1 + a_{n2}x_nx_2 + \ldots + a_{nn}x_n^2 \end{aligned} \tag{6.23}$$

Defining the vector $\boldsymbol{x} = [x_1 \quad x_2 \quad \ldots \quad x_n]^{\mathrm{T}}$ and the matrix

$$\boldsymbol{A} = \begin{bmatrix} a_{11} & a_{12} & \ldots & a_{1n} \\ a_{21} & a_{22} & \ldots & a_{2n} \\ \vdots & \vdots & & \vdots \\ a_{n1} & a_{n2} & \ldots & a_{nn} \end{bmatrix}$$

the quadratic form (6.23) may be written in the form

$$V(\boldsymbol{x}) = \boldsymbol{x}^{\mathrm{T}}\boldsymbol{A}\boldsymbol{x} \tag{6.24}$$

The matrix $\boldsymbol{A}$ is referred to as the matrix of the quadratic form and the determinant of $\boldsymbol{A}$ is called the **discriminant** of the quadratic form.

Now a_{ij} and a_{ji} in (6.23) are both coefficients of the term x_ix_j $(i \neq j)$, so that for $i \neq j$ the coefficient of the term x_ix_j is $a_{ij} + a_{ji}$. By defining new coefficients a'_{ij} and a'_{ji} for x_ix_j and x_jx_i respectively, such that $a'_{ij} = a'_{ji} = \frac{1}{2}(a_{ij} + a_{ji})$, the matrix $\boldsymbol{A}$ associated with the quadratic form $V(\boldsymbol{x})$ may be taken to be symmetric. Thus for real quadratic forms we can, without loss of generality, consider the matrix $\boldsymbol{A}$ to be a symmetric matrix.

EXAMPLE 6.19 Find the real symmetric matrix corresponding to the quadratic form

$$V(x_1, x_2, x_3) = x_1^2 + 3x_2^2 - 4x_3^2 - 3x_1x_2 + 2x_1x_3 - 5x_2x_3$$

Solution If $\boldsymbol{x} = [x_1 \quad x_2 \quad x_3]^{\mathrm{T}}$, we have

$$V(x_1, x_2, x_3) = [x_1 \quad x_2 \quad x_3]\begin{bmatrix} 1 & -\frac{3}{2} & \frac{2}{2} \\ -\frac{3}{2} & 3 & -\frac{5}{2} \\ \frac{2}{2} & -\frac{5}{2} & -4 \end{bmatrix}\begin{bmatrix} x_1 \\ x_2 \\ x_3 \end{bmatrix} = \boldsymbol{x}^{\mathrm{T}}\boldsymbol{A}\boldsymbol{x}$$

where the matrix of the quadratic form is

$$\boldsymbol{A} = \begin{bmatrix} 1 & -\frac{3}{2} & 1 \\ -\frac{3}{2} & 3 & -\frac{5}{2} \\ 1 & -\frac{5}{2} & -4 \end{bmatrix}$$

In Section 6.6.1 we saw that a real symmetric matrix $\boldsymbol{A}$ can always be reduced to the diagonal form

$$\hat{\boldsymbol{M}}^{\mathrm{T}}\boldsymbol{A}\hat{\boldsymbol{M}} = \boldsymbol{\Lambda}$$

where $\hat{\boldsymbol{M}}$ is the normalized orthogonal modal matrix of $\boldsymbol{A}$ and $\boldsymbol{\Lambda}$ is its spectral matrix. Thus for a real quadratic form we can specify a change of variables

$$\boldsymbol{x} = \hat{\boldsymbol{M}}\boldsymbol{y}$$

where $\boldsymbol{y} = [y_1 \quad y_2 \quad \dots \quad y_n]^{\mathrm{T}}$, such that

$$V = \boldsymbol{x}^{\mathrm{T}}\boldsymbol{A}\boldsymbol{x} = \boldsymbol{y}^{\mathrm{T}}\hat{\boldsymbol{M}}^{\mathrm{T}}\boldsymbol{A}\hat{\boldsymbol{M}}\boldsymbol{y} = \boldsymbol{y}^{\mathrm{T}}\boldsymbol{\Lambda}\boldsymbol{y}$$

giving

$$V = \lambda_1 y_1^2 + \lambda_2 y_2^2 + \dots + \lambda_n y_n^2 \qquad \textbf{(6.25)}$$

Hence the quadratic form $\boldsymbol{x}^{\mathrm{T}}\boldsymbol{A}\boldsymbol{x}$ may be reduced to the sum of squares by the transformation $\boldsymbol{x} = \hat{\boldsymbol{M}}\boldsymbol{y}$, where $\hat{\boldsymbol{M}}$ is the normalized modal matrix of $\boldsymbol{A}$. The resulting form given in (6.25) is called the **canonical form** of the quadratic form V given in (6.24). The reduction of a quadratic form to its canonical form has many applications in engineering, particularly in stress analysis.

EXAMPLE 6.20

Find the canonical form of the quadratic form

$$V = 2x_1^2 + 5x_2^2 + 3x_3^2 + 4x_1x_2$$

Can V take negative values for any values of x_1, x_2 and x_3?

Solution At once, we have

$$V = \boldsymbol{x}^{\mathrm{T}}\begin{bmatrix} 2 & 2 & 0 \\ 2 & 5 & 0 \\ 0 & 0 & 3 \end{bmatrix}\boldsymbol{x} = \boldsymbol{x}^{\mathrm{T}}\boldsymbol{A}\boldsymbol{x}$$

where

$$\boldsymbol{x} = [x_1 \quad x_2 \quad x_3]^{\mathrm{T}}, \quad \boldsymbol{A} = \begin{bmatrix} 2 & 2 & 0 \\ 2 & 5 & 0 \\ 0 & 0 & 3 \end{bmatrix}$$

The real symmetric matrix $\boldsymbol{A}$ is the matrix of Example 6.16, where we found the normalized orthogonal modal matrix $\hat{\boldsymbol{M}}$ and spectral matrix $\boldsymbol{\Lambda}$ to be

$$\hat{\boldsymbol{M}} = \begin{bmatrix} \sqrt{\frac{1}{5}} & 0 & -2\sqrt{\frac{1}{5}} \\ 2\sqrt{\frac{1}{5}} & 0 & \sqrt{\frac{1}{5}} \\ 0 & 1 & 0 \end{bmatrix}, \quad \boldsymbol{\Lambda} = \begin{bmatrix} 6 & 0 & 0 \\ 0 & 3 & 0 \\ 0 & 0 & 1 \end{bmatrix}$$

such that $\hat{\boldsymbol{M}}^{\mathrm{T}}\boldsymbol{A}\hat{\boldsymbol{M}} = \boldsymbol{\Lambda}$. Thus, setting $\boldsymbol{x} = \hat{\boldsymbol{M}}\boldsymbol{y}$, we obtain

$$V = \boldsymbol{y}^{\mathrm{T}}\hat{\boldsymbol{M}}^{\mathrm{T}}\boldsymbol{A}\hat{\boldsymbol{M}}\boldsymbol{y} = \boldsymbol{y}^{\mathrm{T}}\begin{bmatrix} 6 & 0 & 0 \\ 0 & 3 & 0 \\ 0 & 0 & 1 \end{bmatrix}\boldsymbol{y} = 6y_1^2 + 3y_2^2 + y_3^2$$

as the required canonical form.

Clearly V is non-negative for all y_1, y_2 and y_3. Since $\boldsymbol{x} = \hat{\boldsymbol{M}}\boldsymbol{y}$ and $\hat{\boldsymbol{M}}$ is an orthogonal matrix it follows that $\boldsymbol{y} = \hat{\boldsymbol{M}}^{\mathrm{T}}\boldsymbol{x}$, so for all $\boldsymbol{x}$ there is a corresponding $\boldsymbol{y}$. It follows that V cannot take negative values for any values of x_1, x_2 and x_3.

The quadratic form of Example (6.20) was seen to be non-negative for any vector $\boldsymbol{x}$, and is positive provided that $\boldsymbol{x} \neq 0$. Such a quadratic form $\boldsymbol{x}^{\mathrm{T}}\boldsymbol{A}\boldsymbol{x}$ is called a **positive-definite** quadratic form, and, by reducing to canonical form, we have seen that this property depends only on the eigenvalues of the real symmetric matrix $\boldsymbol{A}$. This leads us to classify quadratic forms $V = \boldsymbol{x}^{\mathrm{T}}\boldsymbol{A}\boldsymbol{x}$, where $\boldsymbol{x} = [x_1 \quad x_2 \quad \dots \quad x_n]^{\mathrm{T}}$ in the following manner.

(a) V is **positive-definite**, that is, $V > 0$ for all vectors $\boldsymbol{x}$ except $\boldsymbol{x} = 0$, if and only if all the eigenvalues of $\boldsymbol{A}$ are positive.

(b) V is **positive-semidefinite**, that is, $V \geqslant 0$ for all vectors $\boldsymbol{x}$ and $V = 0$ for at least one vector $\boldsymbol{x} \neq 0$, if and only if all the eigenvalues of $\boldsymbol{A}$ are non-negative and at least one of the eigenvalues is zero.

(c) V is **negative-definite** if $-V$ is positive-definite, with a corresponding condition on the eigenvalues of $-\boldsymbol{A}$.

(d) V is **negative-semidefinite** if $-V$ is positive-semidefinite, with a corresponding condition on the eigenvalues of $-\boldsymbol{A}$.

(e) V is **indefinite**, that is, V takes at least one positive value and at least one negative value, if and only if the matrix $\boldsymbol{A}$ has both positive and negative eigenvalues.

Since the classification of a real quadratic form $\boldsymbol{x}^{\mathrm{T}}\boldsymbol{A}\boldsymbol{x}$ depends entirely on the location of the eigenvalues of the symmetric matrix $\boldsymbol{A}$, it may be viewed as a property of $\boldsymbol{A}$ itself. For this reason, it is common to talk of positive-definite, positive-semidefinite, and so on, symmetric matrices without reference to the underlying quadratic form.

EXAMPLE 6.21 Classify the following quadratic forms:

(a) $3x_1^2 + 2x_2^2 + 3x_3^2 - 2x_1x_2 - 2x_2x_3$

(b) $7x_1^2 + x_2^2 + x_3^2 - 4x_1x_2 - 4x_1x_3 + 8x_2x_3$

(c) $-3x_1^2 - 5x_2^2 - 3x_3^2 + 2x_1x_2 + 2x_2x_3 - 2x_1x_3$

(d) $4x_1^2 + x_2^2 + 15x_3^2 - 4x_1x_2$

Solution (a) The matrix corresponding to the quadratic form is

$$\boldsymbol{A} = \begin{bmatrix} 3 & -1 & 0 \\ -1 & 2 & -1 \\ 0 & -1 & 3 \end{bmatrix}$$

The eigenvalues of $\boldsymbol{A}$ are 4, 3 and 1, so the quadratic form is positive-definite.

(b) The matrix corresponding to the quadratic form is

$$\boldsymbol{A} = \begin{bmatrix} 7 & -2 & -2 \\ -2 & 1 & 4 \\ -2 & 4 & 1 \end{bmatrix}$$

The eigenvalues of $\boldsymbol{A}$ are 9, 3 and −3, so the quadratic form is indefinite.

(c) The matrix corresponding to the quadratic form is

$$\boldsymbol{A} = \begin{bmatrix} -3 & 1 & -1 \\ 1 & -5 & 1 \\ -1 & 1 & -3 \end{bmatrix}$$

The eigenvalues of $\boldsymbol{A}$ are −6, −3 and −2, so the quadratic form is negative-definite.

(d) The matrix corresponding to the quadratic form is

$$\boldsymbol{A} = \begin{bmatrix} 4 & -2 & 0 \\ -2 & 1 & 0 \\ 0 & 0 & 15 \end{bmatrix}$$

The eigenvalues of $\boldsymbol{A}$ are 15, 5 and 0, so the quadratic form is positive-semidefinite.

In Example 6.21 classifying the quadratic forms involved determining the eigenvalues of $\boldsymbol{A}$. If $\boldsymbol{A}$ contains one or more parameters then the task becomes difficult, if not impossible, even with the use of a symbolic algebra computer package. Frequently in engineering, particularly in stability analysis, it is necessary to determine the range of values of a parameter k, say, for which a quadratic form remains definite or at least semidefinite in sign. J. J. Sylvester determined criteria for the classification of quadratic forms (or the associated real symmetric matrix) that do not require the computation of the eigenvalues. These criteria are known as **Sylvester's conditions**, which we shall briefly discuss without proof.

In order to classify the quadratic form $\boldsymbol{x}^{\mathrm{T}}\boldsymbol{A}\boldsymbol{x}$ Sylvester's conditions involve consideration of the principal minors of $\boldsymbol{A}$. A **principal minor** P_i of order i ($i = 1, 2, \ldots, n$) of an $n \times n$ square matrix $\boldsymbol{A}$ is the determinant of the submatrix, of order i, whose principal diagonal is part of the principal diagonal of $\boldsymbol{A}$. Note that when $i = n$ the principal minor is det $\boldsymbol{A}$. In particular, the **leading principal minors** of $\boldsymbol{A}$ are

$$D_1 = |a_{11}|, \quad D_2 = \begin{vmatrix} a_{11} & a_{12} \\ a_{21} & a_{22} \end{vmatrix}, \quad D_3 = \begin{vmatrix} a_{11} & a_{12} & a_{13} \\ a_{21} & a_{22} & a_{23} \\ a_{31} & a_{32} & a_{33} \end{vmatrix}, \ldots, D_n = \det \boldsymbol{A}$$

EXAMPLE 6.22 Determine all the principal minors of the matrix

$$\boldsymbol{A} = \begin{bmatrix} 1 & k & 0 \\ k & 2 & 0 \\ 0 & 0 & 5 \end{bmatrix}$$

and indicate which are the leading principal minors.

Solution (a) The principal minor of order three is

$$P_3 = \det \boldsymbol{A} = 5(2 - k^2) \quad \text{(leading principal minor } D_3\text{)}$$

(b) The principal minors of order two are

(i) deleting row 1 and column 1,

$$P_{21} = \begin{vmatrix} 2 & 0 \\ 0 & 5 \end{vmatrix} = 10$$

(ii) deleting row 2 and column 2,

$$P_{22} = \begin{vmatrix} 1 & 0 \\ 0 & 5 \end{vmatrix} = 5$$

(iii) deleting row 3 and column 3,

$$P_{23} = \begin{vmatrix} 1 & k \\ k & 2 \end{vmatrix} = 2 - k^2 \quad \text{(leading principal minor } D_2\text{)}$$

(c) The principal minors of order one are

(i) deleting rows 1 and 2 and columns 1 and 2,

$$P_{11} = |5| = 5$$

(ii) deleting rows 1 and 3 and columns 1 and 3,

$$P_{12} = |2| = 2$$

(iii) deleting rows 2 and 3 and columns 2 and 3,

$$P_{13} = |1| = 1 \quad \text{(leading principal minor } D_1\text{)}$$

Sylvester's conditions

These state that the quadratic form $\mathbf{x}^T\mathbf{A}\mathbf{x}$, where $\mathbf{A}$ is an $n \times n$ real symmetric matrix, is

(a) **positive-definite** if and only if all the leading principal minors of $\mathbf{A}$ are positive; that is, $D_i > 0$ $(i = 1, 2, \ldots, n)$;

(b) **negative-definite** if and only if the leading principal minors of $\mathbf{A}$ alternate in sign with $a_{11} < 0$; that is, $(-1)^i D_i > 0$ $(i = 1, 2, \ldots, n)$;

(c) **positive-semidefinite** if and only if det $\mathbf{A} = 0$ and *all* the principal minors of $\mathbf{A}$ are non-negative; that is, det $\mathbf{A} = 0$ and $P_i \geqslant 0$ for *all* principal minors;

(d) **negative-semidefinite** if and only if det $\mathbf{A} = 0$ and $(-1)^i P_i \geqslant 0$ for *all* principal minors.

EXAMPLE 6.23 For what values of k is the matrix $\mathbf{A}$ of Example 6.22 positive-definite?

Solution The leading principal minors of $\mathbf{A}$ are

$$D_1 = 1, \quad D_2 = 2 - k^2, \quad D_3 = 5(2 - k^2)$$

These will be positive provided that $2 - k^2 > 0$, so the matrix will be positive-definite provided that $k^2 < 2$, that is, $-\sqrt{2} < k < \sqrt{2}$.

EXAMPLE 6.24 Using Sylvester's conditions, confirm the conclusions of Example 6.21.

Solution (a) The matrix of the quadratic form is

$$\mathbf{A} = \begin{bmatrix} 3 & -1 & 0 \\ -1 & 2 & -1 \\ 0 & -1 & 3 \end{bmatrix}$$

and its leading principal minors are

$$3, \quad \begin{vmatrix} 3 & -1 \\ -1 & 2 \end{vmatrix} = 5, \quad \det \mathbf{A} = 12$$

Thus, by Sylvester's condition (a), the quadratic form is positive-definite.

(b) The matrix of the quadratic form is

$$\mathbf{A} = \begin{bmatrix} 7 & -2 & -2 \\ -2 & 1 & 4 \\ -2 & 4 & 1 \end{bmatrix}$$

and its leading principal minors are

$$7, \quad \begin{vmatrix} 7 & -2 \\ -2 & 1 \end{vmatrix} = 3, \quad \det \boldsymbol{A} = -81$$

Thus none of Sylvester's conditions can be satisfied, and the quadratic form is indefinite.

(c) The matrix of the quadratic form is

$$\boldsymbol{A} = \begin{bmatrix} -3 & 1 & -1 \\ 1 & -5 & 1 \\ -1 & 1 & -3 \end{bmatrix}$$

and its leading principal minors are

$$-3, \quad \begin{vmatrix} -3 & 1 \\ 1 & -5 \end{vmatrix} = 14, \quad \det \boldsymbol{A} = -36$$

Thus, by Sylvester's condition (b), the quadratic form is negative-definite.

(d) The matrix of the quadratic form is

$$\boldsymbol{A} = \begin{bmatrix} 4 & -2 & 0 \\ -2 & 1 & 0 \\ 0 & 0 & 15 \end{bmatrix}$$

and its leading principal minors are

$$4, \quad \begin{vmatrix} 4 & -2 \\ -2 & 1 \end{vmatrix} = 0, \quad \det \boldsymbol{A} = 0$$

We therefore need to evaluate all the principal minors to see if the quadratic form is positive-semidefinite. The principal minors are

$$4, \quad 1, \quad 15, \quad \begin{vmatrix} 4 & -2 \\ -2 & 1 \end{vmatrix} = 0, \quad \begin{vmatrix} 1 & 0 \\ 0 & 15 \end{vmatrix} = 15, \quad \begin{vmatrix} 4 & 0 \\ 0 & 15 \end{vmatrix} = 60, \quad \det \boldsymbol{A} = 0$$

Thus, by Sylvester's condition (c), the quadratic form is positive-semidefinite.

6.6.5 Exercises

28 Classify the quadratic forms

(a) $x_1^2 + 2x_2^2 + 7x_3^2 - 2x_1x_2 + 4x_1x_3 - 2x_2x_3$

(b) $x_1^2 + 2x_2^2 + 5x_3^2 - 2x_1x_2 + 4x_1x_3 - 2x_2x_3$

(c) $x_1^2 + 2x_2^2 + 4x_3^2 - 2x_1x_2 + 4x_1x_3 - 2x_2x_3$

29 (a) Show that $ax_1^2 - 2bx_1x_2 + cx_2^2$ is positive-definite if and only if $a > 0$ and $ac > b^2$.

(b) Find inequalities that must be satisfied by a and b to ensure that $2x_1^2 + ax_2^2 + 3x_3^2 - 2x_1x_2 + 2bx_2x_3$ is positive-definite.

30 Evaluate the definiteness of the matrix

$$\boldsymbol{A} = \begin{bmatrix} 2 & 1 & -1 \\ 1 & 2 & 1 \\ -1 & 1 & 2 \end{bmatrix}$$

(a) by obtaining the eigenvalues;
(b) by evaluating the principal minors.

31 Determine the exact range of k for which the quadratic form

$$Q(x, y, z) = k(x^2 + y^2) + 2xy + z^2 + 2xz - 2yz$$

is positive-definite in x, y and z. What can be said about the definiteness of Q when $k = 2$?

32 Determine the minimum value of the constant a such that the quadratic form

$$\boldsymbol{x}^{\mathrm{T}}\begin{bmatrix} 3+a & 1 & 1 \\ 1 & a & 2 \\ 1 & 2 & a \end{bmatrix}\boldsymbol{x}$$

where $\boldsymbol{x} = [x_1 \quad x_2 \quad x_3]^{\mathrm{T}}$, is positive-definite.

33 Express the quadratic form

$$Q = x_1^2 + 4x_1x_2 - 4x_1x_3 - 6x_2x_3 + \lambda(x_2^2 + x_3^2)$$

in the form $\boldsymbol{x}^{\mathrm{T}}\boldsymbol{A}\boldsymbol{x}$, where $\boldsymbol{x} = [x_1 \quad x_2 \quad x_3]^{\mathrm{T}}$ and $\boldsymbol{A}$ is a symmetric matrix. Hence determine the range of values of λ for which Q is positive-definite.

6.7 Functions of a matrix

Let $\boldsymbol{A}$ be an $n \times n$ constant square matrix, so that

$$\boldsymbol{A}^2 = \boldsymbol{AA}, \ \boldsymbol{A}^3 = \boldsymbol{AA}^2 = \boldsymbol{A}^2\boldsymbol{A}, \qquad \text{and so on}$$

are all defined. We can then define a function $f(\boldsymbol{A})$ of the matrix $\boldsymbol{A}$ using a power series representation. For example,

$$f(\boldsymbol{A}) = \sum_{r=0}^{p} \beta_r \boldsymbol{A}^r = \beta_0 \boldsymbol{I} + \beta_1 \boldsymbol{A} + \ldots + \beta_p \boldsymbol{A}^p \tag{6.26}$$

where we have interpreted $\boldsymbol{A}^0$ as the $n \times n$ identity matrix $\boldsymbol{I}$.

EXAMPLE 6.25 Given the 2×2 square matrix

$$\boldsymbol{A} = \begin{bmatrix} 1 & -1 \\ 2 & 3 \end{bmatrix}$$

determine $f(\boldsymbol{A}) = \sum_{r=0}^{2} \beta_r \boldsymbol{A}^r$ when $\beta_0 = 1$, $\beta_1 = -1$ and $\beta_2 = 3$.

Solution Now

$$f(\boldsymbol{A}) = \beta_0\boldsymbol{I} + \beta_1\boldsymbol{A} + \beta_2\boldsymbol{A}^2 = 1\begin{bmatrix} 1 & 0 \\ 0 & 1 \end{bmatrix} - 1\begin{bmatrix} 1 & -1 \\ 2 & 3 \end{bmatrix} + 3\begin{bmatrix} -1 & -4 \\ 8 & 7 \end{bmatrix}$$

$$= \begin{bmatrix} -3 & -11 \\ 22 & 19 \end{bmatrix}$$

Note that $\boldsymbol{A}$ is a 2×2 matrix and $f(\boldsymbol{A})$ is another 2×2 matrix.

Suppose that in (6.26) we let $p \to \infty$, so that

$$f(\mathbf{A}) = \sum_{r=0}^{\infty} \beta_r \mathbf{A}^r$$

We can attach meaning to $f(\mathbf{A})$ in this case if the matrices

$$f_p(\mathbf{A}) = \sum_{r=0}^{p} \beta_r \mathbf{A}^r$$

tend to a constant $n \times n$ matrix in the limit as $p \to \infty$.

EXAMPLE 6.26 For the matrix

$$\mathbf{A} = \begin{bmatrix} 1 & 0 \\ 0 & 1 \end{bmatrix}$$

using a computer and larger and larger values of p, we infer that

$$f(\mathbf{A}) = \lim_{p \to \infty} \sum_{r=0}^{p} \frac{A^r}{r!} \simeq \begin{bmatrix} 2.718\,28 & 0 \\ 0 & 2.718\,28 \end{bmatrix}$$

indicating that

$$f(\mathbf{A}) = \begin{bmatrix} \mathrm{e} & 0 \\ 0 & \mathrm{e} \end{bmatrix}$$

What would be the corresponding results if

$$\text{(a)}\ \mathbf{A} = \begin{bmatrix} -1 & 0 \\ 0 & 1 \end{bmatrix}, \qquad \text{(b)}\ \mathbf{A} = \begin{bmatrix} -t & 0 \\ 0 & t \end{bmatrix}?$$

Solution (a) The computer will lead to the prediction

$$f(\mathbf{A}) \simeq \begin{bmatrix} (2.718\,28)^{-1} & 0 \\ 0 & 2.718\,28 \end{bmatrix}$$

indicating that

$$f(\mathbf{A}) = \begin{bmatrix} \mathrm{e}^{-1} & 0 \\ 0 & \mathrm{e} \end{bmatrix}$$

(b) The computer is of little help in this case. However, hand calculation shows that we are generating the matrix

$$f(\mathbf{A}) = \begin{bmatrix} 1 - t + \frac{1}{2}t^2 - \frac{1}{6}t^3 + \ldots & 0 \\ 0 & 1 + t + \frac{1}{2}t^2 + \frac{1}{6}t^3 + \ldots \end{bmatrix}$$

indicating that

$$f(\boldsymbol{A}) = \begin{bmatrix} e^{-t} & 0 \\ 0 & e^{t} \end{bmatrix}$$

By analogy with the definition of the scalar exponential function

$$e^{at} = 1 + at + \frac{a^2t^2}{2!} + \ldots + \frac{a^rt^r}{r!} + \ldots = \sum_{r=0}^{\infty} \frac{(at)^r}{r!}$$

It is natural to define the matrix function $e^{\boldsymbol{A}t}$, where t is a scalar parameter, by the power series

$$f(\boldsymbol{A}) = \sum_{r=0}^{\infty} \frac{\boldsymbol{A}^r}{r!} t^r \tag{6.27}$$

In fact the matrix in part (b) of Example 6.26 illustrates that this definition is reasonable.

In Example 6.26 we were able to spot the construction of the matrix $f(\boldsymbol{A})$, but this will not be the case when $\boldsymbol{A}$ is a general $n \times n$ square matrix. In order to overcome this limitation and generate a method that will not rely on our ability to 'spot' a closed form of the limiting matrix, we make use of the Cayley–Hamilton theorem, which may be stated as follows.

THEOREM 6.3 Cayley–Hamilton theorem

A square matrix $\boldsymbol{A}$ satisfies its own characteristic equation; that is, if

$$\lambda^n + c_{n-1}\lambda^{n-1} + c_{n-2}\lambda^{n-2} + \ldots + c_1\lambda + c_0 = 0$$

is the characteristic equation of an $n \times n$ matrix $\boldsymbol{A}$ then

$$\boldsymbol{A}^n + c_{n-1}\boldsymbol{A}^{n-1} + c_{n-2}\boldsymbol{A}^{n-2} + \ldots + c_1\boldsymbol{A} + c_0\boldsymbol{I} = \boldsymbol{0} \tag{6.28}$$

where $\boldsymbol{I}$ is the $n \times n$ identity matrix. □

The proof of this theorem is not trivial, and is not included here. We shall illustrate the theorem using a simple example.

EXAMPLE 6.27

Verify the Cayley–Hamilton theorem for the matrix

$$\boldsymbol{A} = \begin{bmatrix} 3 & 4 \\ 1 & 2 \end{bmatrix}$$

Solution The characteristic equation of $\boldsymbol{A}$ is

$$\begin{vmatrix} 3-\lambda & 4 \\ 1 & 2-\lambda \end{vmatrix} = 0 \quad \text{or} \quad \lambda^2 - 5\lambda + 2 = 0$$

Since

$$A^2 = \begin{bmatrix} 3 & 4 \\ 1 & 2 \end{bmatrix}\begin{bmatrix} 3 & 4 \\ 1 & 2 \end{bmatrix} = \begin{bmatrix} 13 & 20 \\ 5 & 8 \end{bmatrix}$$

we have

$$A^2 - 5A + 2I = \begin{bmatrix} 13 & 20 \\ 5 & 8 \end{bmatrix} - 5\begin{bmatrix} 3 & 4 \\ 1 & 2 \end{bmatrix} + 2\begin{bmatrix} 1 & 0 \\ 0 & 1 \end{bmatrix} = \mathbf{0}$$

thus verifying the validity of the Cayley–Hamilton theorem for this matrix.

In the particular case when A is a 2×2 matrix with characteristic equation

$$c(\lambda) = \lambda^2 + a_1\lambda + a_2 = 0 \quad \textbf{(6.29)}$$

it follows from the Cayley–Hamilton theorem that

$$c(A) = A^2 + a_1A + a_2I = \mathbf{0}$$

The significance of this result for our present purposes begins to appear when we rearrange to give

$$A^2 = -a_1A - a_2I$$

This means that A^2 can be written in terms of A and $A^0 = I$. Moreover, multiplying by A gives

$$A^3 = -a_1A^2 - a_2A = -a_1(-a_1A - a_2I) - a_2A$$

Thus A^3 can also be expressed in terms of A and $A^0 = I$; that is, in terms of powers of A less than $n = 2$, the order of the matrix A in this case. It is clear that we could continue the process of multiplying by A and substituting A^2 for as long as we could manage the algebra. However, we can quickly convince ourselves that for any integer $r \geqslant n$

$$A^r = \alpha_0 I + \alpha_1 A \quad \textbf{(6.30)}$$

where α_0 and α_1 are constants whose values will depend on r.

This is a key result deduced from the Cayley–Hamilton theorem, and the determination of the α_i $(i = 0, 1)$ is not as difficult as it might appear. To see how to perform the calculations, we use the characteristic equation of A itself. If we assume that the eigenvalues λ_1 and λ_2 of A are distinct then it follows from (6.29) that

$$c(\lambda_i) = \lambda_i^2 + a_1\lambda_i + a_2 = 0 \ (i = 1, 2)$$

Thus we can write

$$\lambda_i^2 = -a_1\lambda_i - a_2$$

in which a_1 and a_2 are the same constants as in (6.29). Then, for $i = 1, 2$,

$$\lambda_i^3 = -a_1\lambda_i^2 - a_2\lambda_i = -a_1(-a_1\lambda_i - a_2) - a_2\lambda_i$$

Proceeding in this way, we deduce that for each of the eigenvalues λ_1 and λ_2 we can write

$$\lambda_i^r = \alpha_0 + \alpha_1\lambda_i$$

with the same α_0 and α_1 as in (6.30). This therefore provides us with a procedure for the calculation of $\boldsymbol{A}^r$ when $r \geqslant n$ (the order of the matrix) is an integer.

EXAMPLE 6.28 Given that the matrix

$$\boldsymbol{A} = \begin{bmatrix} 0 & 1 \\ -2 & -3 \end{bmatrix}$$

has eigenvalues $\lambda_1 = -1$ and $\lambda_2 = -2$ calculate $\boldsymbol{A}^5$ and $\boldsymbol{A}^r$, where r is an integer greater than 2.

Solution Since $\boldsymbol{A}$ is a 2×2 square matrix, it follows from (6.30) that

$$\boldsymbol{A}^5 = \alpha_0 I + \alpha_1 \boldsymbol{A}$$

and for each eigenvalue λ_i $(i = 1, 2)$ α_0 and α_1 satisfy

$$\lambda_i^5 = \alpha_0 + \alpha_1\lambda_i$$

Substituting $\lambda_1 = -1$ and $\lambda_2 = -2$ leads to the following pair of simultaneous equations:

$$(-1)^5 = \alpha_0 + \alpha_1(-1), \qquad (-2)^5 = \alpha_0 + \alpha_1(-2)$$

which can be solved for α_0 and α_1 to give

$$\alpha_0 = 2(-1)^5 - (-2)^5, \quad \alpha_1 = (-1)^5 - (-2)^5$$

Then

$$\boldsymbol{A}^5 = [2(-1)^5 - (-2)^5]\begin{bmatrix} 1 & 0 \\ 0 & 1 \end{bmatrix} + [(-1)^5 - (-2)^5]\begin{bmatrix} 0 & 1 \\ -2 & -3 \end{bmatrix}$$

$$= \begin{bmatrix} 2(-1)^5 - (-2)^5 & (-1)^5 - (-2)^5 \\ (-2)((-1)^5 - (-2)^5) & 2(-2)^5 - (-1)^5 \end{bmatrix} = \begin{bmatrix} 30 & 31 \\ -62 & -63 \end{bmatrix}$$

Replacing the exponent 5 by the general value r, the algebra is identical, and it is easy to see that

$$\boldsymbol{A}^r = \begin{bmatrix} 2(-1)^r - (-2)^r & (-1)^r - (-2)^r \\ -2((-1)^r - (-2)^r) & 2(-2)^r - (-1)^r \end{bmatrix}$$

To evaluate α_0 and α_1 in (6.27), we assumed that the matrix $\boldsymbol{A}$ had distinct eigenvalues λ_1 and λ_2, leading to a pair of simultaneous equations for α_0 and α_1. What happens if the 2×2 matrix $\boldsymbol{A}$ has a repeated eigenvalue so that $\lambda_1 = \lambda_2 = \lambda$, say? We shall apparently have just a single equation to determine the two constants α_0 and α_1. However, we can obtain a second equation by differentiating with respect to λ, as illustrated in Example 6.29.

EXAMPLE 6.29 Given that the matrix

$$\boldsymbol{A} = \begin{bmatrix} 0 & 1 \\ -1 & -2 \end{bmatrix}$$

has eigenvalues $\lambda_1 = \lambda_2 = -1$, determine $\boldsymbol{A}^r$, where r is an integer greater than 2.

Solution Since $\boldsymbol{A}$ is a 2×2 matrix, it follows from (6.30) that

$$\boldsymbol{A} = \alpha_0 \boldsymbol{I} + \alpha_1 \boldsymbol{A}$$

with α_0 and α_1 satisfying

$$\lambda^r = \alpha_0 + \alpha_1 \lambda \tag{6.31}$$

Since in this case we have only one value of λ, namely $\lambda = -1$, we differentiate (6.31) with respect to λ, to obtain

$$r\lambda^{r-1} = \alpha_1 \tag{6.32}$$

Substituting $\lambda = -1$ in (6.31) and (6.32) leads to

$$\alpha_1 = (-1)^{r-1} r, \qquad \alpha_0 = (-1)^r + \alpha_1 = (1 - r)(-1)^r$$

giving

$$\boldsymbol{A}^r = (1 - r)(-1)^r \begin{bmatrix} 1 & 0 \\ 0 & 1 \end{bmatrix} - r(-1)^r \begin{bmatrix} 0 & 1 \\ -1 & -2 \end{bmatrix}$$

$$= \begin{bmatrix} (1 - r)(-1)^r & -r(-1)^r \\ r(-1)^r & (1 + r)(-1)^r \end{bmatrix}$$

Having found a straightforward way of expressing any positive integer power of the 2×2 square matrix $\boldsymbol{A}$ we see that the same process could be used for each of the terms in (6.26) for $r \geqslant 2$. Thus, for a 2×2 matrix $\boldsymbol{A}$ and some α_0 and α_1,

$$f(\boldsymbol{A}) = \sum_{r=0}^{p} \beta_r \boldsymbol{A}^r = \alpha_0 \boldsymbol{I} + \alpha_1 \boldsymbol{A}$$

If, as $p \to \infty$,

$$f(\boldsymbol{A}) = \lim_{p \to \infty} \sum_{r=0}^{p} \beta_r \boldsymbol{A}^r$$

exists, that is, it is a 2×2 matrix with finite entries independent of p, then we may write

$$f(\mathbf{A}) = \sum_{r=0}^{\infty} \beta_r \mathbf{A}^r = \alpha_0 \mathbf{I} + \alpha_1 \mathbf{A} \tag{6.33}$$

We are now in a position to check the results of our computer experiment with the matrix $\mathbf{A} = \begin{bmatrix} 1 & 0 \\ 0 & 1 \end{bmatrix}$ of Example 6.26. We have defined

$$f(\mathbf{A}) = e^{\mathbf{A}t} = \sum_{r=0}^{\infty} \frac{\mathbf{A}^r}{r!} t^r$$

so we can write

$$e^{\mathbf{A}t} = \alpha_0 \mathbf{I} + \alpha_1 \mathbf{A}$$

Since $\mathbf{A}$ has repeated eigenvalue $\lambda = 1$, we adopt the method of Example 6.29 to give

$$e^t = \alpha_0 + \alpha_1, \qquad t\,e^t = \alpha_1$$

leading to

$$\alpha_1 = t\,e^t, \qquad \alpha_0 = (1 - t)\,e^t$$

Thus

$$e^{\mathbf{A}t} = (1 - t)e^t\mathbf{I} + t\,e^t\mathbf{A} = e^t\mathbf{I} = \begin{bmatrix} e^t & 0 \\ 0 & e^t \end{bmatrix}$$

Setting $t = 1$ confirms our inference in Example 6.26.

EXAMPLE 6.30 Calculate $e^{\mathbf{A}t}$ and $\sin \mathbf{A}t$ when

$$\mathbf{A} = \begin{bmatrix} 1 & -1 \\ 0 & 1 \end{bmatrix}$$

Solution Again $\mathbf{A}$ has repeated eigenvalues, with $\lambda_1 = \lambda_2 = 1$. Thus for $e^{\mathbf{A}t}$ we have

$$e^{\mathbf{A}t} = \alpha_0 \mathbf{I} + \alpha_1 \mathbf{A}$$

with

$$e^t = \alpha_0 + \alpha_1, \quad . \quad t\,e^t = \alpha_1$$

leading to

$$e^{\mathbf{A}t} = \begin{bmatrix} e^t & -t\,e^t \\ 0 & e^t \end{bmatrix}$$

Similarly,

$$\sin \mathbf{A}t = \alpha_0 \mathbf{I} + \alpha_1 \mathbf{A}$$

with

$$\sin t = \alpha_0 + \alpha_1, \qquad t \cos t = \alpha_1$$

leading to

$$\sin \boldsymbol{A}t = \begin{bmatrix} \sin t & -t \cos t \\ 0 & \sin t \end{bmatrix}$$

Although we have worked so far with 2 × 2 matrices, nothing in our development restricts us to this case. The Cayley–Hamilton theorem allows us to express positive integer powers of any $n \times n$ square matrix $\boldsymbol{A}$ in terms of powers of $\boldsymbol{A}$ up to $n - 1$. That is, if $\boldsymbol{A}$ is an $n \times n$ matrix and $p \geqslant n$ then

$$\boldsymbol{A}^p = \sum_{r=0}^{n-1} \beta_r \boldsymbol{A}^r = \beta_0 \boldsymbol{I} + \beta_1 \boldsymbol{A} + \ldots + \beta_{n-1} \boldsymbol{A}^{n-1}$$

From this we can deduce that for an $n \times n$ matrix $\boldsymbol{A}$ we may write

$$f(\boldsymbol{A}) = \sum_{r=0}^{\infty} \beta_r \boldsymbol{A}^r$$

as

$$f(\boldsymbol{A}) = \sum_{r=0}^{n-1} \alpha_r \boldsymbol{A}^r \qquad \textbf{(6.34a)}$$

which generalizes the result (6.33). Again the coefficients $\alpha_0, \alpha_1, \ldots, \alpha_{n-1}$ are obtained by solving the n equations

$$f(\lambda_i) = \sum_{r=0}^{n-1} \alpha_r \lambda_i^r \quad (i = 1, 2, \ldots, n) \qquad \textbf{(6.34b)}$$

where $\lambda_1, \lambda_2, \ldots, \lambda_n$ are the eigenvalues of $\boldsymbol{A}$. If $\boldsymbol{A}$ has repeated eigenvalues, we differentiate as before, noting that if λ_i is an eigenvalue of multiplicity m then the first $m - 1$ derivatives

$$\frac{\mathrm{d}^k}{\mathrm{d}\lambda_i^k} f(\lambda_i) = \frac{\mathrm{d}^k}{\mathrm{d}\lambda_i^k} \sum_{r=0}^{n-1} \alpha_r \lambda_i^r \quad (k = 1, 2, \ldots, m - 1)$$

are also satisfied by λ_i.

Sometimes it is advantageous to use an alternative approach to evaluate

$$f(\mathbf{A}) = \sum_{r=0}^{p} \beta_r \mathbf{A}^r$$

If $\mathbf{A}$ possesses n linearly independent eigenvectors then there exists a modal matrix $\mathbf{M}$ and spectral matrix $\mathbf{\Lambda}$ such that

$$\mathbf{M}^{-1}\mathbf{AM} = \mathbf{\Lambda} = \text{diag}(\lambda_1, \lambda_2, \ldots, \lambda_n)$$

Now

$$\begin{aligned}
\mathbf{M}^{-1}f(\mathbf{A})\mathbf{M} &= \sum_{r=0}^{p} \beta_r(\mathbf{M}^{-1}\mathbf{A}^r\mathbf{M}) = \sum_{r=0}^{p} \beta_r(\mathbf{M}^{-1}\mathbf{AM})^r \\
&= \sum_{r=0}^{p} \beta_r \mathbf{\Lambda}^r \\
&= \sum_{r=0}^{p} \beta_r \,\text{diag}(\lambda_1^r, \lambda_2^r, \ldots, \lambda_n^r) \\
&= \text{diag}\left(\sum_{r=0}^{p} \beta_r\lambda_1^r, \sum_{r=0}^{p} \beta_r\lambda_2^r, \ldots, \sum_{r=0}^{p} \beta_r\lambda_n^r\right) \\
&= \text{diag}(f(\lambda_1), f(\lambda_2), \ldots, f(\lambda_n))
\end{aligned}$$

This gives us a second method of computing functions of a square matrix, since we see that

$$f(\mathbf{A}) = \mathbf{M}\,\text{diag}(f(\lambda_1), f(\lambda_2), \ldots, f(\lambda_n))\mathbf{M}^{-1} \tag{6.35}$$

EXAMPLE 6.31 Using the result (6.35), calculate $\mathbf{A}^k$ for the matrix

$$\mathbf{A} = \begin{bmatrix} 0 & 1 \\ -2 & -3 \end{bmatrix}$$

of Example 6.28.

Solution $\mathbf{A}$ has eigenvalues $\lambda_1 = -1$ and $\lambda_2 = -2$ with corresponding eigenvectors

$$\mathbf{e}_1 = [1 \quad -1]^T, \quad \mathbf{e}_2 = [1 \quad -2]^T$$

Thus a modal matrix $\mathbf{M}$ and corresponding spectral matrix $\mathbf{\Lambda}$ are

$$\mathbf{M} = \begin{bmatrix} 1 & 1 \\ -1 & -2 \end{bmatrix}, \quad \mathbf{\Lambda} = \begin{bmatrix} -1 & 0 \\ 0 & -2 \end{bmatrix}$$

Clearly

$$\mathbf{M}^{-1} = \begin{bmatrix} 2 & 1 \\ -1 & -1 \end{bmatrix}$$

Taking $f(\mathbf{A}) = \mathbf{A}^k$, we have

$$\text{diag}\,(f(-1), f(-2)) = \text{diag}\,((-1)^k, (-2)^k)$$

Thus, from (6.35),

$$f(\mathbf{A}) = \mathbf{M}\begin{bmatrix} (-1)^k & 0 \\ 0 & (-2)^k \end{bmatrix}\mathbf{M}^{-1} = \begin{bmatrix} 2(-1)^k - (-2)^k & (-1)^k - (-2)^k \\ 2((-2)^k - (-1)^k) & 2(-2)^k - (-1)^k \end{bmatrix}$$

as determined in Example 6.28.

Example 6.31 demonstrates a second approach to the calculation of a function of a matrix. There is little difference in the labour associated with each method, perhaps the only comment we should make is that each approach gives a different perspective on the construction of the matrix function either from powers of the matrix itself or from its spectral and modal matrices.

Later in this chapter we need to make use of some properties of the exponential matrix $e^{\mathbf{A}t}$, where $\mathbf{A}$ is a constant $n \times n$ square matrix. These are now briefly discussed. First, we have

$$\frac{d}{dt}(e^{\mathbf{A}t}) = \mathbf{A}e^{\mathbf{A}t} = e^{\mathbf{A}t}\mathbf{A} \tag{6.36}$$

which follows from the power series definition given in (6.27), and the proof is left as an exercise. Secondly,

$$e^{\mathbf{A}(t_1+t_2)} = e^{\mathbf{A}t_1}e^{\mathbf{A}t_2} \tag{6.37}$$

Although this property is true in general we shall illustrate its validity for the particular case when $\mathbf{A}$ has n linearly independent eigenvectors. Then, from (6.35),

$$e^{\mathbf{A}t_1} = \mathbf{M}\,\text{diag}\,(e^{\lambda_1 t_1}, e^{\lambda_2 t_1}, \dots, e^{\lambda_n t_1})\mathbf{M}^{-1}$$

$$e^{\mathbf{A}t_2} = \mathbf{M}\,\text{diag}\,(e^{\lambda_1 t_2}, e^{\lambda_2 t_2}, \dots, e^{\lambda_n t_2})\mathbf{M}^{-1}$$

so that

$$e^{\mathbf{A}t_1}e^{\mathbf{A}t_2} = \mathbf{M}\,\text{diag}\,(e^{\lambda_1(t_1+t_2)}, e^{\lambda_2(t_1+t_2)}, \dots, e^{\lambda_n(t_1+t_2)})\mathbf{M}^{-1} = e^{\mathbf{A}(t_1+t_2)}$$

It is important to note that in general

$$e^{\mathbf{A}t}\,e^{\mathbf{B}t} \neq e^{(\mathbf{A}+\mathbf{B})t}$$

It follows from the series definition that

$$e^{\mathbf{A}t}\,e^{\mathbf{B}t} = e^{(\mathbf{A}+\mathbf{B})t} \tag{6.38}$$

if and only if the matrices $\mathbf{A}$ and $\mathbf{B}$ commute, that is, if $\mathbf{AB} = \mathbf{BA}$.

To conclude this section we consider the derivative and integral of a matrix $\boldsymbol{A}(t) = [a_{ij}(t)]$, whose elements $a_{ij}(t)$ are functions of t. The derivative and integral of $\boldsymbol{A}(t)$ are defined respectively by

$$\frac{d}{dt}\boldsymbol{A}(t) = \left[\frac{d}{dt}a_{ij}(t)\right] \tag{6.39}$$

and

$$\int \boldsymbol{A}(t)\,dt = \left[\int a_{ij}(t)\,dt\right] \tag{6.40}$$

that is, each element of the matrix is differentiated or integrated as appropriate.

EXAMPLE 6.32 Evaluate $d\boldsymbol{A}/dt$ and $\int \boldsymbol{A}\,dt$ for the matrix

$$\begin{bmatrix} t^2+1 & t-3 \\ 2 & t^2+2t-1 \end{bmatrix}$$

Solution Using (6.39),

$$\frac{d\boldsymbol{A}}{dt} = \begin{bmatrix} \frac{d}{dt}(t^2+1) & \frac{d}{dt}(t-3) \\ \frac{d}{dt}(2) & \frac{d}{dt}(t^2+2t-1) \end{bmatrix} = \begin{bmatrix} 2t & 1 \\ 0 & 2t+2 \end{bmatrix}$$

Using (6.40),

$$\int \boldsymbol{A}\,dt = \begin{bmatrix} \int (t^2+1)\,dt & \int (t-3)\,dt \\ \int 2\,dt & \int (t^2+2t-1)\,dt \end{bmatrix}$$

$$= \begin{bmatrix} \frac{1}{3}t^3+t+c_{11} & \frac{1}{2}t^2-3t+c_{12} \\ 2t+c_{21} & \frac{1}{3}t^3+t^2-t+c_{22} \end{bmatrix}$$

$$= \begin{bmatrix} \frac{1}{3}t^3+t & \frac{1}{2}t^2-3t \\ 2t & \frac{1}{3}t^3+t^2-t \end{bmatrix} + \begin{bmatrix} c_{11} & c_{12} \\ c_{21} & c_{22} \end{bmatrix}$$

$$= \begin{bmatrix} \frac{1}{3}t^3+t & \frac{1}{2}t^2-3t \\ 2t & \frac{1}{3}t^3+t^2-t \end{bmatrix} + \boldsymbol{C}$$

where $\boldsymbol{C}$ is a constant matrix.

From the basic definitions, it follows that for constants α and β

$$\frac{d}{dt}(\alpha \boldsymbol{A} + \beta \boldsymbol{B}) = \alpha \frac{d\boldsymbol{A}}{dt} + \beta \frac{d\boldsymbol{B}}{dt} \tag{6.41}$$

$$\int (\alpha \boldsymbol{A} + \beta \boldsymbol{B})\, dt = \alpha \int \boldsymbol{A}\, dt + \beta \int \boldsymbol{B}\, dt \tag{6.42}$$

$$\frac{d}{dt}(\boldsymbol{AB}) = \boldsymbol{A}\frac{d\boldsymbol{B}}{dt} + \frac{d\boldsymbol{A}}{dt}\boldsymbol{B} \tag{6.43}$$

Note in (6.43) that order is important, since in general $\boldsymbol{AB} \neq \boldsymbol{BA}$.

Note that in general

$$\frac{d}{dt}[\boldsymbol{A}(t)]^n \neq n\boldsymbol{A}^{n-1}\frac{d\boldsymbol{A}}{dt}$$

6.7.1 Exercises

34 Show that the matrix

$$\boldsymbol{A} = \begin{bmatrix} 5 & 6 \\ 2 & 3 \end{bmatrix}$$

satisfies its own characteristic equation.

35 Given

$$\boldsymbol{A} = \begin{bmatrix} 1 & 2 \\ 1 & 1 \end{bmatrix}$$

use the Cayley–Hamilton theorem to evaluate

(a) $\boldsymbol{A}^2$ (b) $\boldsymbol{A}^3$ (c) $\boldsymbol{A}^4$

36 The characteristic equation of an $n \times n$ matrix $\boldsymbol{A}$ is

$$\lambda^n + c_{n-1}\lambda^{n-1} + c_{n-2}\lambda^{n-2} + \ldots + c_1\lambda + c_0 = 0$$

so, by the Cayley–Hamilton theorem,

$$\boldsymbol{A}^n + c_{n-1}\boldsymbol{A}^{n-1} + c_{n-2}\boldsymbol{A}^{n-2} + \ldots + c_1\boldsymbol{A} + c_0\boldsymbol{I} = \boldsymbol{0}$$

If $\boldsymbol{A}$ is non-singular then every eigenvalue is non-zero, so $c_0 \neq 0$ and

$$\boldsymbol{I} = -\frac{1}{c_0}(\boldsymbol{A}^n + c_{n-1}\boldsymbol{A}^{n-1} + \ldots + c_1\boldsymbol{A})$$

which on multiplying throughout by $\boldsymbol{A}^{-1}$ gives

$$\boldsymbol{A}^{-1} = -\frac{1}{c_0}(\boldsymbol{A}^{n-1} + c_{n-1}\boldsymbol{A}^{n-2} + \ldots + c_1\boldsymbol{I}) \tag{6.44}$$

(a) Using (6.44) find the inverse of the matrix

$$\boldsymbol{A} = \begin{bmatrix} 2 & 1 \\ 1 & 2 \end{bmatrix}$$

(b) Show that the characteristic equation of the matrix

$$\boldsymbol{A} = \begin{bmatrix} 1 & 1 & 2 \\ 3 & 1 & 1 \\ 2 & 3 & 1 \end{bmatrix}$$

is

$$\lambda^3 - 3\lambda^2 - 7\lambda - 11 = 0$$

Evaluate $\boldsymbol{A}^2$ and, using (6.44), determine $\boldsymbol{A}^{-1}$.

37 Given

$$\boldsymbol{A} = \begin{bmatrix} 2 & 3 & 1 \\ 3 & 1 & 2 \\ 1 & 2 & 3 \end{bmatrix}$$

compute $\boldsymbol{A}^2$ and, using the Cayley–Hamilton theorem, compute

$$\boldsymbol{A}^7 - 3\boldsymbol{A}^6 + \boldsymbol{A}^4 + 3\boldsymbol{A}^3 - 2\boldsymbol{A}^2 + 3\boldsymbol{I}$$

38 Evaluate $e^{\boldsymbol{A}t}$ for

(a) $\boldsymbol{A} = \begin{bmatrix} 1 & 0 \\ 1 & 1 \end{bmatrix}$ (b) $\boldsymbol{A} = \begin{bmatrix} 1 & 0 \\ 1 & 2 \end{bmatrix}$

39 Given

$$A = \frac{\pi}{2}\begin{bmatrix} 2 & 0 & 0 \\ 0 & 1 & 1 \\ 0 & 0 & 1 \end{bmatrix}$$

show that

$$\sin A = \frac{4}{\pi}A - \frac{4}{\pi^2}A^2 = \begin{bmatrix} 0 & 0 & 0 \\ 0 & 1 & 0 \\ 0 & 0 & 1 \end{bmatrix}$$

40 Given

$$A = \begin{bmatrix} t^2+1 & 2t-3 \\ 5-t & t^2-t+3 \end{bmatrix}$$

evaluate

(a) $\dfrac{dA}{dt}$ (b) $\displaystyle\int_1^2 A\,dt$

41 Given

$$A = \begin{bmatrix} t^2+1 & t-1 \\ 5 & 0 \end{bmatrix}$$

evaluate A^2 and show that

$$\frac{d}{dt}(A^2) \neq 2A\frac{dA}{dt}$$

6.8 State-space representation

In Section 2.3 we applied Laplace transform methods to obtain the response of a linear time-invariant system to a given input. It was noted at the time that the process of taking the inverse transform, to obtain the system response, could prove to be rather tedious for high-order systems. In this section we shall apply matrix techniques to obtain the response of such systems.

6.8.1 Single-input–single-output (SISO) systems

First let us consider the **single-input–single-output (SISO) system** characterized by the nth-order linear differential equation

$$a_n\frac{d^ny}{dt^n} + a_{n-1}\frac{d^{n-1}y}{dt^{n-1}} + \ldots + a_1\frac{dy}{dt} + a_0y = u(t) \qquad \textbf{(6.45)}$$

where, as in (2.19), the coefficients a_i ($i = 0, 1, \ldots, n$) are constants with $a_n \neq 0$ and it is assumed that the initial conditions $y(0), y^{(1)}(0), \ldots, y^{(n-1)}(0)$ are known.

We introduce the n variables $x_1(t), x_2(t), \ldots, x_n(t)$ defined by

$$x_1(t) = y(t)$$

$$x_2(t) = \frac{dy}{dt} = \dot{x}_1(t)$$

$$x_3(t) = \frac{d^2y}{dt^2} = \dot{x}_2(t)$$

$$\vdots$$

$$x_{n-1}(t) = \frac{d^{n-2}y}{dt^{n-2}} = \dot{x}_{n-2}(t)$$

$$x_n(t) = \frac{d^{n-1}y}{dt^{n-1}} = \dot{x}_{n-1}(t)$$

then, by substituting in (6.45), we have

$$a_n\dot{x}_n + a_{n-1}x_n + a_{n-2}x_{n-1} + \ldots + a_1x_2 + a_0x_1 = u(t)$$

giving

$$\dot{x}_n = -\frac{a_{n-1}}{a_n}x_n - \frac{a_{n-2}}{a_n}x_{n-1} - \ldots - \frac{a_1}{a_n}x_2 - \frac{a_0}{a_n}x_1 + \frac{1}{a_n}u$$

Thus, we can represent (6.45) as a system of n simultaneous first-order differential equations

$$\dot{x}_1 = x_2$$

$$\dot{x}_2 = x_3$$

$$\vdots$$

$$\dot{x}_{n-1} = x_n$$

$$\dot{x}_n = -\frac{a_0}{a_n}x_1 - \frac{a_1}{a_n}x_2 - \ldots - \frac{a_{n-1}}{a_n}x_n + \frac{1}{a_n}u$$

which may be written as the **vector–matrix differential equation**

$$\begin{bmatrix} \dot{x}_1 \\ \dot{x}_2 \\ \vdots \\ \dot{x}_{n-1} \\ \dot{x}_n \end{bmatrix} = \begin{bmatrix} 0 & 1 & 0 & \ldots & 0 & 0 \\ 0 & 0 & 1 & \ldots & 0 & 0 \\ \vdots & \vdots & \vdots & & \vdots & \vdots \\ 0 & 0 & 0 & \ldots & 0 & 1 \\ \dfrac{-a_0}{a_n} & \dfrac{-a_1}{a_n} & \dfrac{-a_2}{a_n} & \ldots & \dfrac{-a_{n-2}}{a_n} & \dfrac{-a_{n-1}}{a_n} \end{bmatrix} \begin{bmatrix} x_1 \\ x_2 \\ \vdots \\ x_{n-1} \\ x_n \end{bmatrix} + \begin{bmatrix} 0 \\ 0 \\ \vdots \\ 0 \\ \dfrac{1}{a_n} \end{bmatrix} u(t) \qquad \textbf{(6.46)}$$

This may be written in the more concise form

$$\dot{\boldsymbol{x}} = \boldsymbol{A}\boldsymbol{x} + \boldsymbol{b}u \qquad \textbf{(6.47)}$$

The vector $\boldsymbol{x}(t)$ is called the system **state vector**, and it contains all the information that one needs to know about the behaviour of the system. Its components are the n **state variables** $x_1, x_2, \ldots, x_n$, which may be considered as representing a set of coordinate axes in the n-dimensional coordinate space over which $\boldsymbol{x}(t)$ ranges. This is referred to as the **state space**, and as time increases the state vector $\boldsymbol{x}(t)$ will describe a locus in this space called a **trajectory**. In two dimensions the state space reduces to the **phase plane**. The matrix $\boldsymbol{A}$ is called the system **matrix** and the particular form adopted in (6.46) is known as the **companion form**, which is widely adopted in practice. Equation (6.47) is referred to as the system **state equation**.

It is important to realize that the choice of state variables $x_1, x_2, \ldots, x_n$ is not unique. For example, for the system represented by (6.45) we could also take

$$x_1 = \frac{\mathrm{d}^{n-1}y}{\mathrm{d}t^{n-1}}, \quad x_2 = \frac{\mathrm{d}^{n-2}y}{\mathrm{d}t^{n-2}}, \quad \ldots, \quad x_n = y$$

giving a state equation (6.47) with

$$\boldsymbol{A} = \begin{bmatrix} -\dfrac{a_{n-1}}{a_n} & -\dfrac{a_{n-2}}{a_n} & \ldots & -\dfrac{a_1}{a_n} & -\dfrac{a_0}{a_n} \\ 1 & 0 & \ldots & 0 & 0 \\ 0 & 1 & \ldots & 0 & 0 \\ \vdots & \vdots & & \vdots & \vdots \\ 0 & 0 & \ldots & 1 & 0 \end{bmatrix}, \quad \boldsymbol{b} = \begin{bmatrix} \dfrac{1}{a_n} \\ 0 \\ \vdots \\ 0 \end{bmatrix}$$

The output, or response, of the system determined by (6.45) is given by y, which in terms of the state variables is determined by x_1. Thus

$$y = [1 \quad 0 \quad \ldots \quad 0] \begin{bmatrix} x_1 \\ x_2 \\ \vdots \\ x_n \end{bmatrix}$$

or, more concisely,

$$y = \boldsymbol{c}^{\mathrm{T}}\boldsymbol{x} \tag{6.48}$$

where $\boldsymbol{c} = [1 \quad 0 \quad \ldots \quad 0]^{\mathrm{T}}$.

We shall see later that, in contrast with the transform methods adopted in Chapters 2 and 3, which are generally applied to SISO systems, a distinct advantage of the vector–matrix approach is that it is applicable to multivariable (that is, multi-input–multi-output MIMO) systems. In such cases it is particularly important to distinguish between the system state variables and the system outputs, which, in general, are linear combinations of the state variables.

Together the pair of equations (6.47) and (6.48) in the form

$$\dot{\boldsymbol{x}} = \boldsymbol{A}\boldsymbol{x} + \boldsymbol{b}u \tag{6.49a}$$

$$y = \boldsymbol{c}^{\mathrm{T}}\boldsymbol{x} \tag{6.49b}$$

constitute the **dynamic equations** of the system and are commonly referred to as the state-space model representation of the system. Such a representation forms the basis of the so-called 'modern approach' to the analysis and design of control systems in engineering. An obvious advantage of adopting the vector–matrix representation (6.49) is the compactness of the notation.

More generally the output y could be a linear combination of both the state and input, so that the more general form of the system dynamic equations (6.49) is

$$\left.\begin{aligned}\dot{x} &= \boldsymbol{A}x + bu \\ y &= c^{\mathrm{T}}x + du\end{aligned}\right\}\qquad \begin{aligned}&(6.50a)\\&(6.50b)\end{aligned}$$

EXAMPLE 6.33 Obtain a state-space representation of the system characterized by the third-order differential equation

$$\frac{d^3y}{dt^3} + 3\frac{d^2y}{dt^2} + 2\frac{dy}{dt} - 4y = e^{-t} \tag{6.51}$$

Solution Writing

$$x_1 = y, \qquad x_2 = \frac{dy}{dt} = \dot{x}_1, \qquad x_3 = \frac{d^2y}{dt^2} = \dot{x}_2$$

we have, from (6.51),

$$\dot{x}_3 = \frac{d^3y}{dt^3} = 4y - 2\frac{dy}{dt} - 3\frac{d^2y}{dt^2} + e^{-t} = 4x_1 - 2x_2 - 3x_3 + e^{-t}$$

Thus the corresponding state equation is

$$\begin{bmatrix}\dot{x}_1\\ \dot{x}_2\\ \dot{x}_3\end{bmatrix} = \begin{bmatrix}0 & 1 & 0\\ 0 & 0 & 1\\ 4 & -2 & -3\end{bmatrix}\begin{bmatrix}x_1\\ x_2\\ x_3\end{bmatrix} + \begin{bmatrix}0\\ 0\\ 1\end{bmatrix}e^{-t}$$

with the output y being given by

$$y = x_1 = [1 \quad 0 \quad 0]\begin{bmatrix}x_1\\ x_2\\ x_3\end{bmatrix}$$

These two equations then constitute the state-space representation of the system.

We now proceed to consider the more general SISO system characterized by the differential equation

$$\frac{d^ny}{dt^n} + a_n\frac{d^{n-1}y}{dt^{n-1}} + \ldots + a_0y = b_m\frac{d^mu}{dt^m} + \ldots + b_0u \quad (m \leqslant n) \tag{6.52}$$

in which the input involves derivative terms. As we saw in (2.66), this system may be characterized by the transfer function

$$G(s) = \frac{b_ms^m + \ldots + b_0}{s^n + a_{n-1}s^{n-1} + \ldots + a_0} \tag{6.53}$$

depicted by the input–output block diagram of Figure 2.41.

Again there are various ways of representing (6.53) in the state-space form, depending on the choice of the state variables. As an illustration, we shall consider one possible approach, introducing others in the exercises.

We define $\boldsymbol{A}$ and $\boldsymbol{b}$ as in (6.46); that is, we take $\boldsymbol{A}$ to be the companion matrix of the left-hand side of (6.51) (or the companion matrix of the denominator in the transfer-function model (6.53)), giving

$$\boldsymbol{A} = \begin{bmatrix} 0 & 1 & 0 & \dots & 0 & 0 \\ 0 & 0 & 1 & & & \\ \vdots & \vdots & \vdots & & \vdots & \vdots \\ 0 & 0 & 0 & \dots & 0 & 1 \\ -a_0 & -a_1 & -a_2 & \dots & -a_{n-2} & -a_{n-1} \end{bmatrix}$$

and we take $\boldsymbol{b} = [0 \quad 0 \quad \dots \quad 0 \quad 1]^{\mathrm{T}}$. In order to achieve the desired response, the vector $\boldsymbol{c}$ is then chosen to be

$$\boldsymbol{c} = [b_0 \quad b_1 \quad \dots \quad b_m \quad 0 \quad \dots \quad 0]^{\mathrm{T}} \tag{6.54}$$

To confirm that this structure is appropriate, it is convenient to adopt Laplace transform notation. Taking

$$X_1(s) = \frac{1}{s^n + a_{n-1}s^{n-1} + \dots + a_0} U(s)$$

we have

$$X_2(s) = sX_1(s),\ X_3(s) = sX_2(s) = s^2X_1(s),\ \dots,\ X_n(s) = sX_{n-1}(s) = s^{n-1}X_1(s)$$

so that

$$\begin{aligned} Y(s) = \boldsymbol{c}^{\mathrm{T}}\boldsymbol{X}(s) &= b_0X_1(s) + b_1X_2(s) + \dots + b_mX_{m+1}(s) \\ &= \frac{b_0 + b_1s + b_2s^2 + \dots + b_ms^m}{s^n + a_{n-1}s^{n-1} + \dots + a_0} \end{aligned}$$

which confirms (6.54).

EXAMPLE 6.34 For the system characterized by the differential equation model

$$\frac{\mathrm{d}^3y}{\mathrm{d}t^3} + 6\frac{\mathrm{d}^2y}{\mathrm{d}t^2} + 11\frac{\mathrm{d}y}{\mathrm{d}t} + 3y = 5\frac{\mathrm{d}^2u}{\mathrm{d}t^2} + \frac{\mathrm{d}u}{\mathrm{d}t} + u \tag{6.55}$$

obtain

(a) a transfer-function model; (b) a state-space model.

Solution (a) Assuming all initial conditions to be zero, taking Laplace transforms throughout in (6.55) leads to

$$(s^3 + 6s^2 + 11s + 3)Y(s) = (5s^2 + s + 1)U(s)$$

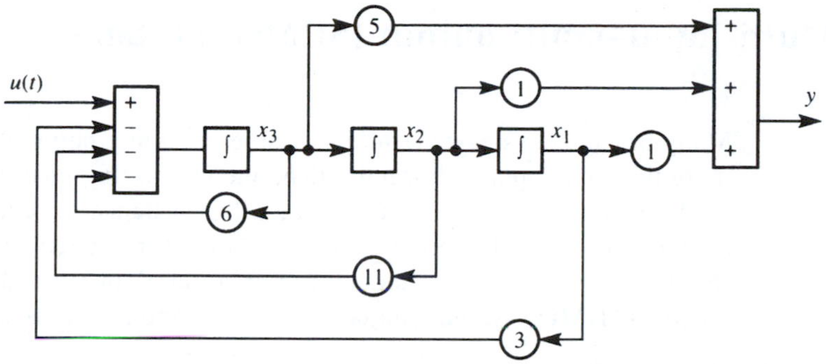

Figure 6.3 Block diagram for the state-space model of Example 6.34.

so that the transfer-function model is given by

$$G(s) = \frac{Y(s)}{U(s)} = \frac{5s^2 + s + 1}{s^3 + 6s^2 + 11s + 3} \quad \begin{matrix} \leftarrow \boldsymbol{c} \\ \leftarrow \boldsymbol{A} \end{matrix}$$

(b) Taking $\boldsymbol{A}$ to be the companion matrix of the left-hand side in (6.55)

$$\boldsymbol{A} = \begin{bmatrix} 0 & 1 & 0 \\ 0 & 0 & 1 \\ -3 & -11 & -6 \end{bmatrix} \quad \text{and} \quad \boldsymbol{b} = [0 \quad 0 \quad 1]^{\mathrm{T}}$$

we have, from (6.54),

$$\boldsymbol{c} = [1 \quad 1 \quad 5]^{\mathrm{T}}$$

(Note that the last row of $\boldsymbol{A}$ and $\boldsymbol{c}$ may be obtained from the transfer function in (a) by reading the coefficients of the denominator and numerator backwards, as indicated by the arrows, and negating those in the denominator.) Then from (6.49) the state-space model becomes

$$\dot{\boldsymbol{x}} = \boldsymbol{A}\boldsymbol{x} + \boldsymbol{b}u, \qquad y = \boldsymbol{c}^{\mathrm{T}}\boldsymbol{x}$$

This model structure may be depicted by the block diagram of Figure 6.3. It provides an ideal model for simulation studies, with the state variables being the outputs of the various integrators involved.

A distinct advantage of this approach to obtaining the state-space model is that $\boldsymbol{A}$, $\boldsymbol{b}$ and $\boldsymbol{c}$ are readily written down. A possible disadvantage in some applications is that the output y itself is not a state variable. An approach in which y is a state variable is developed in Exercise 47. In practice, it is also fairly common to choose the state variables from a physical consideration, as is illustrated in Example 6.35.

6.8.2 Multi-input–multi-output (MIMO) systems

Many practical systems are multivariable in nature, being characterized by having more than one input and/or more than one output. In general terms, the state-space model is similar to that in (6.50) for SISO systems, except that the input is now a vector $\boldsymbol{u}(t)$ as is the output $\boldsymbol{y}(t)$. Thus the more general form, corresponding to (6.50), of the state-space model representation of an nth-order **multi-input–multi-output (MIMO) system** subject to r inputs and l outputs is

$$\dot{\boldsymbol{x}} = \boldsymbol{A}\boldsymbol{x} + \boldsymbol{B}\boldsymbol{u} \quad (6.56a)$$

$$\boldsymbol{y} = \boldsymbol{C}\boldsymbol{x} + \boldsymbol{D}\boldsymbol{u} \quad (6.56b)$$

where $\boldsymbol{x}$ is the n-state vector, $\boldsymbol{u}$ is the r-input vector, $\boldsymbol{y}$ is the l-output vector, $\boldsymbol{A}$ is the $n \times n$ system matrix, $\boldsymbol{B}$ is the $n \times r$ control (or input) matrix, and $\boldsymbol{C}$ and $\boldsymbol{D}$ are respectively $l \times n$ and $l \times r$ output matrices.

EXAMPLE 6.35 Obtain the state-space model representation characterizing the two-input–one-output parallel network shown in Figure 6.4 in the form

$$\dot{\boldsymbol{x}} = \boldsymbol{A}\boldsymbol{x} + \boldsymbol{B}\boldsymbol{u}, \qquad y = \boldsymbol{c}^{\mathrm{T}}\boldsymbol{x} + \boldsymbol{d}^{\mathrm{T}}\boldsymbol{u}$$

where the elements x_1, x_2, x_3 of $\boldsymbol{x}$ and u_1, u_2 of $\boldsymbol{u}$ are as indicated in the figure, and the output y is the voltage drop across the inductor L_1 (v_C denotes the voltage drop across the capacitor C).

Solution Applying Kirchhoff's second law to each of the two loops in turn gives

$$R_1 i_1 + L_1 \frac{\mathrm{d}i_1}{\mathrm{d}t} + v_C = e_1 \quad (6.57)$$

$$L_2 \frac{\mathrm{d}i_2}{\mathrm{d}t} + v_C = e_2 \quad (6.58)$$

The voltage drop v_C across the capacitor C is given by

$$\dot{v}_C = \frac{1}{C}(i_1 + i_2) \quad (6.59)$$

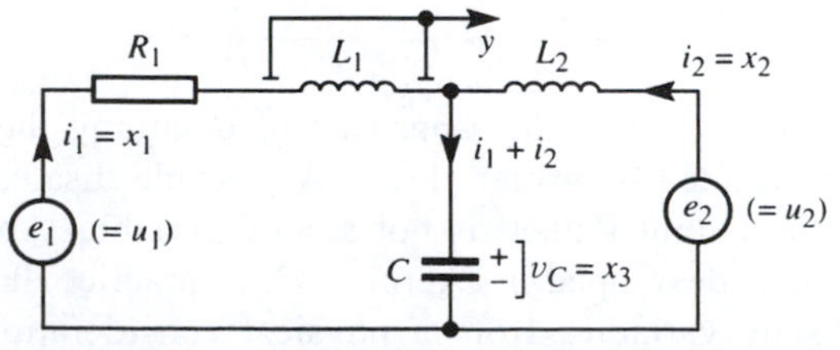

Figure 6.4 Parallel circuit of Example 6.35.

The output y, being the voltage drop across the inductor L_1, is given by

$$y = L_1 \frac{di_1}{dt}$$

which, using (6.57), gives

$$y = -R_1 i_1 - v_C + e_1 \tag{6.60}$$

Writing $x_1 = i_1$, $x_2 = i_2$, $x_3 = v_C$, $u_1 = e_1$ and $u_2 = e_2$, (6.57)–(6.60) give the state-space representation as

$$\begin{bmatrix} \dot{x}_1 \\ \dot{x}_2 \\ \dot{x}_3 \end{bmatrix} = \begin{bmatrix} -\frac{R_1}{L_1} & 0 & -\frac{1}{L_1} \\ 0 & 0 & -\frac{1}{L_2} \\ \frac{1}{C} & \frac{1}{C} & 0 \end{bmatrix} \begin{bmatrix} x_1 \\ x_2 \\ x_3 \end{bmatrix} + \begin{bmatrix} \frac{1}{L_1} & 0 \\ 0 & \frac{1}{L_2} \\ 0 & 0 \end{bmatrix} \begin{bmatrix} u_1 \\ u_2 \end{bmatrix}$$

$$y = [-R_1 \quad 0 \quad -1] \begin{bmatrix} x_1 \\ x_2 \\ x_3 \end{bmatrix} + [1 \quad 0] \begin{bmatrix} u_1 \\ u_2 \end{bmatrix}$$

which is of the required form

$$\dot{\boldsymbol{x}} = \boldsymbol{A}\boldsymbol{x} + \boldsymbol{B}\boldsymbol{u}$$
$$y = \boldsymbol{c}^{\mathrm{T}}\boldsymbol{x} + \boldsymbol{d}^{\mathrm{T}}\boldsymbol{u}$$

6.8.3 Exercises

42 Obtain the state-space forms of the differential equations

(a) $\dfrac{d^3y}{dt^3} + 4\dfrac{d^2y}{dt^2} + 5\dfrac{dy}{dt} + 4y = u(t)$

(b) $\dfrac{d^4y}{dt} + 2\dfrac{d^2y}{dt^2} + 4\dfrac{dy}{dt} = 5u(t)$

using the companion form of the system matrix in each case.

43 Obtain the dynamic equations in state-space form for the systems having transfer-function models

(a) $\dfrac{s^2 + 3s + 5}{s^3 + 6s^2 + 5s + 7}$ (b) $\dfrac{s^2 + 3s + 2}{s^3 + 4s^2 + 3s}$

using the companion form of the system matrix in each case.

44 Obtain the state-space model of the single-input–single-output network system of Figure 6.5 in the form $\dot{\boldsymbol{x}} = \boldsymbol{A}\boldsymbol{x} + \boldsymbol{b}u$, $y = \boldsymbol{c}^{\mathrm{T}}\boldsymbol{x}$, where u, y and the elements x_1, x_2, x_3 of $\boldsymbol{x}$ are as indicated.

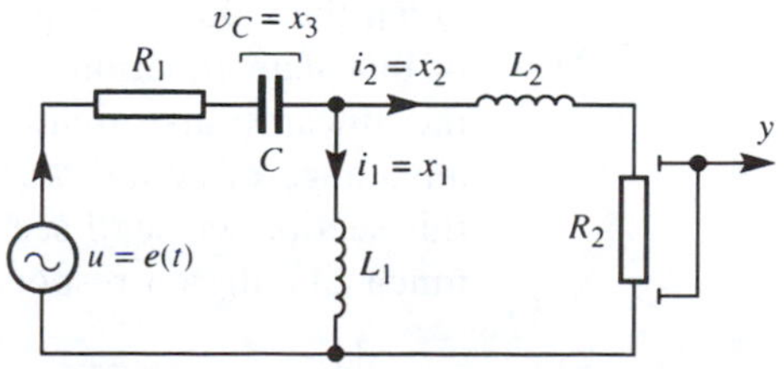

Figure 6.5 Network of Exercise 44.

45 The mass–spring–damper system of Figure 6.6 models the suspension system of a quarter-car. Obtain a state-space model in which the output represents the body mass vertical movement y and the input represents the tyre vertical movement $u(t)$ due to the road surface. All displacements are measured from equilibrium positions.

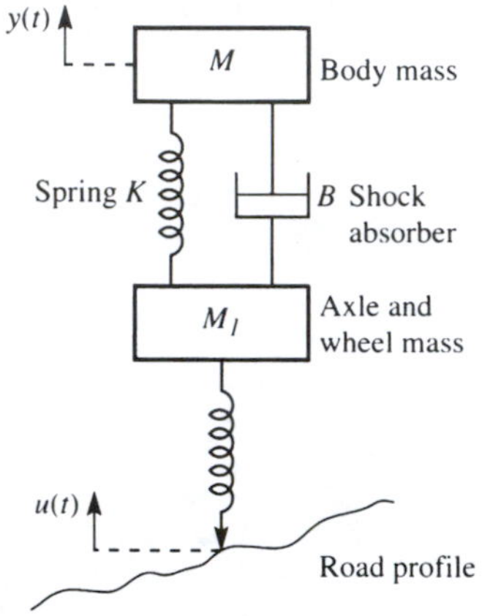

Figure 6.6 Quarter-car suspension model of Exercise 45.

46 Obtain the state-space model, in the form $\dot{x} = \boldsymbol{A}x + bu$, $y = \boldsymbol{C}x + d^{\mathrm{T}}u$ of the one-input–two-output network illustrated in Figure 6.7. The elements x_1, x_2 of the state vector x and y_1, y_2 of the output vector y are as indicated. If $R_1 = 1\ \mathrm{k\Omega}$, $R_2 = 5\ \mathrm{k\Omega}$,

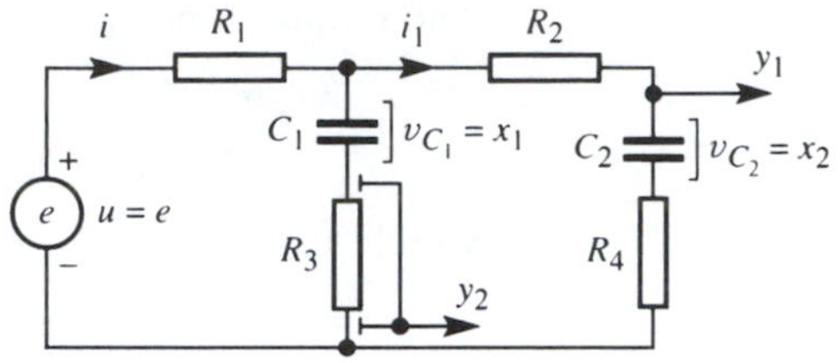

Figure 6.7 Network of Exercise 46.

$R_3 = R_4 = 3\ \mathrm{k\Omega}$, $C_1 = C_2 = 1\ \mathrm{\mu F}$ calculate the eigenvalues of the system matrix $\boldsymbol{A}$.

47 It was noted in Section 6.8.1 that a possible disadvantage of the state-space model representation (6.54) for the transfer function model (6.53) is that the output y is not a state variable. Suppose we wish the output y to be the state variable x_1; that is, in the state-space representation (6.49) we take $c^{\mathrm{T}} = [1 \quad 0 \quad \ldots \quad 0]^{\mathrm{T}}$. If $\boldsymbol{A}$ is again taken to be the companion matrix of the denominator then it can be shown that the coefficients $b_1, b_2, \ldots, b_n$ of the vector b are determined as the first n coefficients in the series in s^{-1} obtained by dividing the denominator of the transfer function (6.53) into the numerator. Illustrate this approach for the transfer-function model of Figure 6.8.

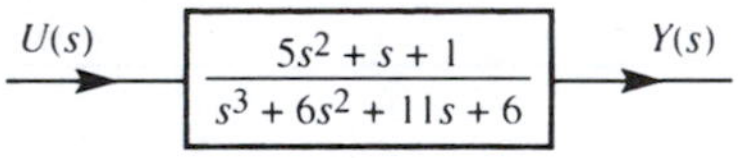

Figure 6.8 Transfer-function model of Exercise 47.

6.9 Solution of the state equation

In this section we are concerned with seeking the solution of the state equation

$$\dot{x} = \boldsymbol{A}x + \boldsymbol{B}u \tag{6.61}$$

given the value of x at some initial time t_0 to be x_0. Having obtained the solution of this state equation, a system response y may then be readily written down from the linear transformation (6.56b). As mentioned in Section 6.8.1, an obvious advantage of adopting the vector–matrix notation of (6.61) is its compactness. In this section we shall see that another distinct advantage is that (6.61) behaves very much like the corresponding first-order scalar differential equation

$$\frac{\mathrm{d}x}{\mathrm{d}t} = ax + bu, \quad x(t_0) = x_0 \tag{6.62}$$

6.9.1 Direct form of the solution

Before considering the nth-order system represented by (6.61), let us first briefly review the solution of (6.62). When the input u is zero, (6.62) reduces to the homogeneous equation

$$\frac{dx}{dt} = ax \tag{6.63}$$

which, by separation of variables,

$$\int_{x_0}^{x} \frac{dx}{x} = \int_{t_0}^{t} a\, dt$$

gives

$$\ln x - \ln x_0 = a(t - t_0)$$

leading to the solution

$$x = x_0 e^{a(t-t_0)} \tag{6.64}$$

for the unforced system.

If we consider the nonhomogeneous equation (6.62) directly, a solution can be obtained by first multiplying throughout by the integrating factor e^{-at} to obtain

$$e^{-at}\left(\frac{dx}{dt} - ax\right) = e^{-at} bu(t)$$

or

$$\frac{d}{dt}(e^{-at} x) = e^{-at} bu(t)$$

which on integration gives

$$e^{-at} x - e^{-at_0} x_0 = \int_{t_0}^{t} e^{-a\tau} bu(\tau)\, d\tau$$

leading to the solution

$$x(t) = e^{a(t-t_0)} x_0 + \int_{t_0}^{t} e^{a(t-\tau)} bu(\tau)\, d\tau \tag{6.65}$$

The first term of the solution, which corresponds to the solution of the unforced system, is a **complementary function**, while the convolution integral constituting the second term, which is dependent on the forcing function $u(t)$, is a **particular integral**.

Returning to (6.61), we first consider the unforced homogeneous system

$$\dot{\boldsymbol{x}} = \boldsymbol{A}\boldsymbol{x}, \quad \boldsymbol{x}(t_0) = \boldsymbol{x}_0 \tag{6.66}$$

which represents the situation when the system is 'relaxing' from an initial state.

The solution is completely analogous to the solution (6.64) of the scalar equation (6.63), and is of the form

$$x = e^{A(t-t_0)}x_0 \tag{6.67}$$

It is readily shown that this is a solution of (6.66). Using (6.36), differentiation of (6.67) gives

$$\dot{x} = A e^{A(t-t_0)}x_0 = Ax$$

so that (6.66) is satisfied. Also, from (6.67),

$$x(t_0) = e^{A(t_0-t_0)}x_0 = Ix_0 = x_0$$

using $e^0 = I$. Thus, since (6.67) satisfies the differential equation and the initial conditions, it represents the unique solution of (6.66).

Likewise, the nonhomogeneous equation (6.61) may be solved in an analogous manner to that used for solving (6.62). Premultiplying (6.61) throughout by e^{-At}, we obtain

$$e^{-At}(\dot{x} - Ax) = e^{At}Bu(t)$$

or using (6.36),

$$\frac{d}{dt}(e^{-At}x) = e^{-At}Bu(t)$$

Integration then gives

$$e^{-At}x(t) - e^{-At_0}x_0 = \int_{t_0}^{t} e^{-A\tau}Bu(\tau)\,d\tau$$

leading to the solution

$$x(t) = e^{A(t-t_0)}x_0 + \int_{t_0}^{t} e^{A(t-\tau)}Bu(\tau)\,d\tau \tag{6.68}$$

This is analogous to the solution given in (6.65) for the scalar equation (6.62). Again it contains two terms: one dependent on the initial state and corresponding to the solution of the unforced system, and one a convolution integral arising from the input. Having obtained the solution of the state equation, the system output $y(t)$ is then readily obtained from equation (6.56b).

6.9.2 The transition matrix

The matrix exponential $e^{A(t-t_0)}$ is referred to as the **fundamental** or **transition matrix** and is frequently denoted by $\Phi(t, t_0)$; so that (6.66) is written as

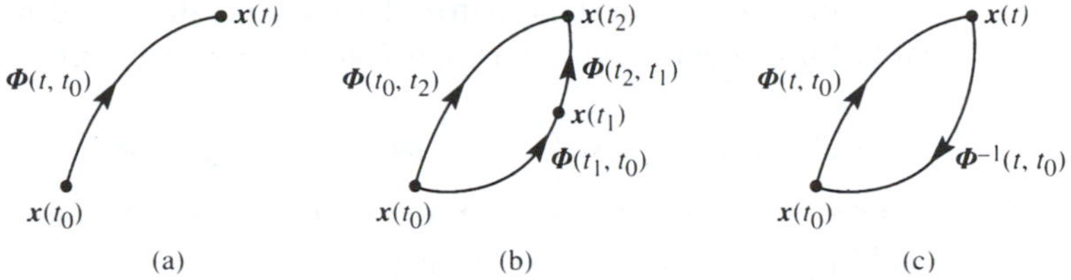

Figure 6.9 (a) Transition matrix $\Phi(t, t_0)$. (b) The transition property. (c) The inverse $\Phi^{-1}(t, t_0)$.

$$\boldsymbol{x}(t) = \boldsymbol{\Phi}(t, t_0)\boldsymbol{x}_0 \tag{6.69}$$

This is an important matrix, which can be used to characterize a linear system, and in the absence of any input it maps a given state $\boldsymbol{x}_0$ at any time t_0 to the state $\boldsymbol{x}(t)$ at any time t, as illustrated in Figure 6.9(a).

Using the properties of the exponential matrix given in Section 6.7, certain properties of the transition matrix may be deduced. From

$$e^{\boldsymbol{A}(t_1+t_2)} = e^{\boldsymbol{A}t_1} e^{\boldsymbol{A}t_2}$$

it follows that $\boldsymbol{\Phi}(t, t_0)$ satisfies the **transition property**

$$\boldsymbol{\Phi}(t_2, t_0) = \boldsymbol{\Phi}(t_2, t_1)\boldsymbol{\Phi}(t_1, t_0) \tag{6.70}$$

for any t_0, t_1 and t_2, as illustrated in Figure 6.9(b). From

$$e^{\boldsymbol{A}t} e^{-\boldsymbol{A}t} = \boldsymbol{I}$$

it follows that the inverse $\boldsymbol{\Phi}^{-1}(t, t_0)$ of the transition matrix is obtained by negating time, so that

$$\boldsymbol{\Phi}^{-1}(t, t_0) = \boldsymbol{\Phi}(-t, -t_0) = \boldsymbol{\Phi}(t_0, t) \tag{6.71}$$

for any t_0 and t, as illustrated in Figure 6.9(c).

6.9.3 Evaluating the transition matrix

Since, when dealing with time invariant systems, there is no loss of generality in taking $t_0 = 0$, we shall, for convenience, consider the evaluation of the transition matrix

$$\boldsymbol{\Phi}(t) = \boldsymbol{\Phi}(t, 0) = e^{\boldsymbol{A}t}$$

Clearly, methods of evaluating this are readily applicable to the evaluation of

$$\boldsymbol{\Phi}(t, \tau) = e^{\boldsymbol{A}(t-\tau)}$$

Indeed, since $\boldsymbol{A}$ is a constant matrix,

$$\boldsymbol{\Phi}(t, \tau) = \boldsymbol{\Phi}(t - \tau, 0)$$

so, having obtained $\boldsymbol{\Phi}(t)$, we can write down $\boldsymbol{\Phi}(t, \tau)$ by simply replacing t by $t - \tau$.

Since $\boldsymbol{A}$ is a constant matrix the methods discussed in Section 6.7 are applicable for evaluating the transition matrix. From (6.34a),

$$e^{\boldsymbol{A}t} = \alpha_0(t)\boldsymbol{I} + \alpha_1(t)\boldsymbol{A} + \alpha_2(t)\boldsymbol{A}^2 + \ldots + \alpha_{n-1}(t)\boldsymbol{A}^{n-1} \tag{6.72a}$$

where, using (6.34b), the $\alpha_i(t)$ $(i = 0, 1, \ldots, n-1)$ are obtained by solving simultaneously the n equations

$$e^{\lambda_j t} = \alpha_0(t) + \alpha_1(t)\lambda_j + \alpha_2(t)\lambda_j^2 + \ldots + \alpha_{n-1}(t)\lambda_j^{n-1} \tag{6.72b}$$

where λ_j $(j = 1, 2, \ldots, n)$ are the eigenvalues of $\boldsymbol{A}$. As in Section 6.7, if $\boldsymbol{A}$ has repeated eigenvalues then derivatives of $e^{\lambda t}$, with respect to λ, will have to be used.

EXAMPLE 6.36

A system is characterized by the state equation

$$\begin{bmatrix} \dot{x}_1(t) \\ \dot{x}_2(t) \end{bmatrix} = \begin{bmatrix} -1 & 0 \\ 1 & -3 \end{bmatrix} \begin{bmatrix} x_1(t) \\ x_2(t) \end{bmatrix} + \begin{bmatrix} 1 \\ 1 \end{bmatrix} u(t)$$

Given that the input is the unit step function

$$u(t) = H(t) = \begin{cases} 0 & (t < 0) \\ 1 & (t \geqslant 0) \end{cases}$$

and initially

$$x_1(0) = x_2(0) = 1$$

deduce the state $\boldsymbol{x}(t) = [x_1(t) \quad x_2(t)]^{\mathrm{T}}$ of the system at subsequent time t.

Solution From (6.68), the solution is given by

$$\boldsymbol{x}(t) = e^{\boldsymbol{A}t}\boldsymbol{x}(0) + \int_0^t e^{\boldsymbol{A}(t-\tau)}\boldsymbol{b}u(\tau)\,d\tau \tag{6.73}$$

where

$$\boldsymbol{A} = \begin{bmatrix} -1 & 0 \\ 1 & -3 \end{bmatrix}, \quad \boldsymbol{b} = [1 \quad 1]^{\mathrm{T}}$$

Since $\boldsymbol{A}$ is a 2×2 matrix, it follows from (6.72a) that

$$e^{\boldsymbol{A}t} = \alpha_0(t)\boldsymbol{I} + \alpha_1(t)\boldsymbol{A}$$

The eigenvalues of $\boldsymbol{A}$ are $\lambda_1 = -1$ and $\lambda_2 = -3$, so, using (6.72), we have

$$\alpha_0(t) = \tfrac{1}{2}(3e^{-t} - e^{-3t}), \quad \alpha_1(t) = \tfrac{1}{2}(e^{-t} - e^{-3t})$$

giving

$$e^{\boldsymbol{A}t} = \begin{bmatrix} e^{-t} & 0 \\ \tfrac{1}{2}(e^{-t} - e^{-3t}) & e^{-3t} \end{bmatrix}$$

Thus the first term in (6.73) becomes

$$e^{\boldsymbol{A}t}\boldsymbol{x}(0) = \begin{bmatrix} e^{-t} & 0 \\ \frac{1}{2}(e^{-t} - e^{-3t}) & e^{-3t} \end{bmatrix}\begin{bmatrix} 1 \\ 1 \end{bmatrix} = \begin{bmatrix} e^{-t} \\ \frac{1}{2}(e^{-t} + e^{-3t}) \end{bmatrix}$$

and the second term is

$$\int_0^t e^{\boldsymbol{A}(t-\tau)}\boldsymbol{b}u(\tau)\,d\tau = \int_0^t \begin{bmatrix} e^{-(t-\tau)} & 0 \\ \frac{1}{2}(e^{-(t-\tau)} - e^{-3(t-\tau)}) & e^{-3(t-\tau)} \end{bmatrix}\begin{bmatrix} 1 \\ 1 \end{bmatrix} 1\,d\tau$$

$$= \int_0^t \begin{bmatrix} e^{-(t-\tau)} \\ \frac{1}{2}(e^{-(t-\tau)} + e^{-3(t-\tau)}) \end{bmatrix} d\tau = \begin{bmatrix} e^{-(t-\tau)} \\ \frac{1}{2}(e^{-(t-\tau)} + \frac{1}{3}e^{-3(t-\tau)}) \end{bmatrix}_0^t$$

$$= \begin{bmatrix} e^{-0} \\ \frac{1}{2}(e^{-0} + \frac{1}{3}e^{-0}) \end{bmatrix} - \begin{bmatrix} e^{-t} \\ \frac{1}{2}(e^{-t} + \frac{1}{3}e^{-3t}) \end{bmatrix} = \begin{bmatrix} 1 - e^{-t} \\ \frac{2}{3} - \frac{1}{2}e^{-t} - \frac{1}{6}e^{-3t} \end{bmatrix}$$

Substituting back in (6.73) gives the required solution

$$\boldsymbol{x}(t) = \begin{bmatrix} e^{-t} \\ \frac{1}{2}(e^{-t} + e^{-3t}) \end{bmatrix} + \begin{bmatrix} 1 - e^{-t} \\ \frac{2}{3} - \frac{1}{2}e^{-t} - \frac{1}{6}e^{-3t} \end{bmatrix} = \begin{bmatrix} 1 \\ \frac{2}{3} + \frac{1}{3}e^{-3t} \end{bmatrix}$$

That is,

$$x_1(t) = 1, \quad x_2(t) = \tfrac{2}{3} + \tfrac{1}{3}e^{-3t}$$

6.9.4 Exercises

48 Obtain the transition matrix $\boldsymbol{\Phi}(t)$ of the system

$$\dot{\boldsymbol{x}} = \boldsymbol{A}\boldsymbol{x}$$

where

$$\boldsymbol{A} = \begin{bmatrix} 1 & 0 \\ 1 & 1 \end{bmatrix}$$

Verify that $\boldsymbol{\Phi}(t)$ has the following properties:

(a) $\boldsymbol{\Phi}(0) = \boldsymbol{I}$;

(b) $\boldsymbol{\Phi}(t_2) = \boldsymbol{\Phi}(t_2 - t_1)\boldsymbol{\Phi}(t_1)$;

(c) $\boldsymbol{\Phi}^{-1}(t) = \boldsymbol{\Phi}(-t)$.

49 Writing $x_1 = y$ and $x_2 = dy/dt$ express the differential equation

$$\frac{d^2y}{dt^2} + 2\frac{dy}{dt} + y = 0$$

in the vector–matrix form $\dot{\boldsymbol{x}} = \boldsymbol{A}\boldsymbol{x}$, $\boldsymbol{x} = [x_1 \quad x_2]^T$. Obtain the transition matrix and hence solve the differential equation given that $y = dy/dt = 1$ when $t = 0$. Confirm your answer by direct solution of the second-order differential equation.

50 Solve

$$\dot{\boldsymbol{x}} = \begin{bmatrix} \dot{x}_1 \\ \dot{x}_2 \end{bmatrix} = \begin{bmatrix} 1 & 0 \\ 1 & 1 \end{bmatrix}\begin{bmatrix} x_1 \\ x_2 \end{bmatrix}$$

subject to $\boldsymbol{x}(0) = [1 \quad 1]^T$.

51 Find the solution of

$$\dot{\boldsymbol{x}} = \begin{bmatrix} \dot{x}_1 \\ \dot{x}_2 \end{bmatrix} = \begin{bmatrix} 0 & 1 \\ -6 & -5 \end{bmatrix}\begin{bmatrix} x_1 \\ x_2 \end{bmatrix} + \begin{bmatrix} 0 \\ 6 \end{bmatrix}u(t) \quad (t \geqslant 0)$$

where $u(t) = 2$ and $\boldsymbol{x}(0) = [1 \quad -1]^T$.

52 Using (6.68), find the response for $t \geqslant 0$ of the system

$$\dot{x}_1 = x_2 + 2u$$

$$\dot{x}_2 = -2x_1 - 3x_2$$

to an input $u(t) = e^{-t}$ and subject to the initial conditions $x_1(0) = 0$, $x_2(0) = 1$.

53 A system is governed by the vector–matrix differential equation

$$\dot{\boldsymbol{x}}(t) = \begin{bmatrix} 3 & 4 \\ 2 & 1 \end{bmatrix} \boldsymbol{x}(t) + \begin{bmatrix} 0 & 1 \\ 1 & 1 \end{bmatrix} \boldsymbol{u}(t) \quad (t \geqslant 0)$$

where $\boldsymbol{x}(t)$ and $\boldsymbol{u}(t)$ are respectively the state and input vectors of the system. Determine the transition matrix of this system, and hence obtain an explicit expression for $\boldsymbol{x}(t)$ for the input $\boldsymbol{u}(t) = [4 \quad 3]^{\mathrm{T}}$ and subject to the initial condition $\boldsymbol{x}(0) = [1 \quad 2]^{\mathrm{T}}$.

6.9.5 Laplace transform solution

Defining

$$\mathscr{L}\{\boldsymbol{x}(t)\} = \begin{bmatrix} \mathscr{L}\{x_1(t)\} \\ \mathscr{L}\{x_2(t)\} \\ \vdots \\ \mathscr{L}\{x_n(t)\} \end{bmatrix} = \begin{bmatrix} X_1(s) \\ X_2(s) \\ \vdots \\ X_n(s) \end{bmatrix} = \boldsymbol{X}(s)$$

$$\mathscr{L}\{\boldsymbol{u}(t)\} = \begin{bmatrix} \mathscr{L}\{u_1(t)\} \\ \mathscr{L}\{u_2(t)\} \\ \vdots \\ \mathscr{L}\{u_r(t)\} \end{bmatrix} = \begin{bmatrix} U_1(s) \\ U_2(s) \\ \vdots \\ U_r(s) \end{bmatrix} = \boldsymbol{U}(s)$$

and then taking Laplace transforms throughout in the state equation

$$\dot{\boldsymbol{x}}(t) = \boldsymbol{A}\boldsymbol{x}(t) + \boldsymbol{B}\boldsymbol{u}(t)$$

gives

$$s\boldsymbol{X}(s) - \boldsymbol{x}(0) = \boldsymbol{A}\boldsymbol{X}(s) + \boldsymbol{B}\boldsymbol{U}(s)$$

which on rearranging gives

$$(s\boldsymbol{I} - \boldsymbol{A})\boldsymbol{X}(s) = x(0) + \boldsymbol{B}\boldsymbol{U}(s)$$

where $\boldsymbol{I}$ is the identity matrix. Premultiplying throughout by $(s\boldsymbol{I} - \boldsymbol{A})^{-1}$ gives

$$\boldsymbol{X}(s) = (s\boldsymbol{I} - \boldsymbol{A})^{-1}\boldsymbol{x}(0) + (s\boldsymbol{I} - \boldsymbol{A})^{-1}\boldsymbol{B}\boldsymbol{U}(s) \tag{6.74}$$

which on taking inverse Laplace transforms gives the response as

$$\boldsymbol{x}(t) = \mathscr{L}^{-1}\{(s\boldsymbol{I} - \boldsymbol{A})^{-1}\}\boldsymbol{x}(0) + \mathscr{L}^{-1}\{(s\boldsymbol{I} - \boldsymbol{A})^{-1}\boldsymbol{B}\boldsymbol{U}(s)\} \tag{6.75}$$

On comparing the solution (6.75) with that given in (6.68), we find that the transition, matrix $\boldsymbol{\Phi}(t) = e^{\boldsymbol{A}t}$ may also be written in the form

$$\boldsymbol{\Phi}(t) = \mathscr{L}^{-1}\{(s\boldsymbol{I} - \boldsymbol{A})^{-1}\}$$

As mentioned in Section 6.7.3, having obtained $\boldsymbol{\Phi}(t)$,

$$\boldsymbol{\Phi}(t, t_0) = \mathrm{e}^{\boldsymbol{A}(t-t_0)}$$

may be obtained on simply replacing t by $t - t_0$.

EXAMPLE 6.37

Using the Laplace transform approach, obtain an expression for the state $\boldsymbol{x}(t)$ of the system characterized by the state equation

$$\dot{\boldsymbol{x}}(t) = \begin{bmatrix} \dot{x}_1(t) \\ \dot{x}_2(t) \end{bmatrix} = \begin{bmatrix} -1 & 0 \\ 1 & -3 \end{bmatrix} \begin{bmatrix} x_1(t) \\ x_2(t) \end{bmatrix} + \begin{bmatrix} 1 \\ 1 \end{bmatrix} u(t)$$

when the input $u(t)$ is the unit step function

$$u(t) = H(t) = \begin{cases} 0 & (t < 0) \\ 1 & (t \geqslant 0) \end{cases}$$

and subject to the initial condition $\boldsymbol{x}(0) = [1 \quad 1]^{\mathrm{T}}$

Solution In this case

$$\boldsymbol{A} = \begin{bmatrix} -1 & 0 \\ 1 & -3 \end{bmatrix}, \qquad \boldsymbol{b} = \begin{bmatrix} 1 \\ 1 \end{bmatrix}, \qquad u(t) = H(t), \qquad \boldsymbol{x}_0 = [1 \quad 1]^{\mathrm{T}}$$

Thus

$$s\boldsymbol{I} - \boldsymbol{A} = \begin{bmatrix} s+1 & 0 \\ -1 & s+3 \end{bmatrix}, \qquad \det(s\boldsymbol{I} - \boldsymbol{A}) = (s + 1)(s + 3)$$

giving

$$(s\boldsymbol{I} - \boldsymbol{A})^{-1} = \frac{1}{(s+1)(s+3)} \begin{bmatrix} s+3 & 0 \\ 1 & s+1 \end{bmatrix} = \begin{bmatrix} \dfrac{1}{s+1} & 0 \\ \dfrac{1}{2(s+1)} - \dfrac{1}{2(s-3)} & \dfrac{1}{s+3} \end{bmatrix}$$

which, on taking inverse transforms, gives the transition matrix as

$$\mathrm{e}^{\boldsymbol{A}t} = \mathscr{L}^{-1}\{(s\boldsymbol{I} - \boldsymbol{A})^{-1}\} = \begin{bmatrix} \mathrm{e}^{-t} & 0 \\ \frac{1}{2}\mathrm{e}^{-t} - \frac{1}{2}\mathrm{e}^{-3t} & \mathrm{e}^{-3t} \end{bmatrix}$$

so that the first term in the solution (6.75) becomes

$$\mathscr{L}^{-1}\{(s\boldsymbol{I} - \boldsymbol{A})^{-1}\}\boldsymbol{x}_0 = \begin{bmatrix} \mathrm{e}^{-t} & 0 \\ \frac{1}{2}\mathrm{e}^{-t} - \frac{1}{2}\mathrm{e}^{-3t} & \mathrm{e}^{-3t} \end{bmatrix} \begin{bmatrix} 1 \\ 1 \end{bmatrix} = \begin{bmatrix} \mathrm{e}^{-t} \\ \frac{1}{2}\mathrm{e}^{-t} + \frac{1}{2}\mathrm{e}^{-3t} \end{bmatrix} \tag{6.76}$$

Since $U(s) = \mathscr{L}\{H(t)\} = 1/s$,

$$(s\boldsymbol{I}-\boldsymbol{A})^{-1}\boldsymbol{b}U(s) = \frac{1}{(s+1)(s+3)}\begin{bmatrix} s+3 & 0 \\ 1 & s+1 \end{bmatrix}\begin{bmatrix} 1 \\ 1 \end{bmatrix}\frac{1}{s}$$

$$= \frac{1}{s(s+1)(s+3)}\begin{bmatrix} s+3 \\ s+2 \end{bmatrix}$$

$$= \begin{bmatrix} \dfrac{1}{s} - \dfrac{1}{s+1} \\ \dfrac{2}{3s} - \dfrac{1}{2(s+1)} - \dfrac{1}{6(s+3)} \end{bmatrix}$$

so that the second term in (6.75) becomes

$$\mathscr{L}^{-1}\{(s\boldsymbol{I}-\boldsymbol{A})^{-1}\boldsymbol{b}U(s)\} = \begin{bmatrix} 1-\mathrm{e}^{-t} \\ \frac{2}{3}-\frac{1}{2}\mathrm{e}^{-t}-\frac{1}{6}\mathrm{e}^{-3t} \end{bmatrix} \tag{6.77}$$

Combining (6.76) and (6.77), the response $\boldsymbol{x}(t)$ is given by

$$\boldsymbol{x}(t) = \begin{bmatrix} \mathrm{e}^{-t} \\ \frac{1}{2}\mathrm{e}^{-t}+\frac{1}{2}\mathrm{e}^{-3t} \end{bmatrix} + \begin{bmatrix} 1-\mathrm{e}^{-t} \\ \frac{2}{3}-\frac{1}{2}\mathrm{e}^{-t}-\frac{1}{6}\mathrm{e}^{-3t} \end{bmatrix} = \begin{bmatrix} 1 \\ \frac{2}{3}+\frac{1}{3}\mathrm{e}^{-3t} \end{bmatrix}$$

Note that throughout the solution corresponds to that obtained in Example 6.36.

Again, having obtained an expression for the system state $\boldsymbol{x}(t)$, its output, or response, $\boldsymbol{y}(t)$ may be obtained from the linear transformation (6.56b).

We can also use the Laplace transform formulation to obtain the input–output transfer-function matrix for a multivariable system. Taking Laplace transforms throughout in (6.56b) gives

$$\boldsymbol{Y}(s) = \boldsymbol{C}\boldsymbol{X}(s) + \boldsymbol{D}\boldsymbol{U}(s) \tag{6.78}$$

where $\boldsymbol{Y}(s) = \mathscr{L}\{\boldsymbol{y}(t)\}$. Assuming zero initial conditions in (6.74), we have

$$\boldsymbol{X}(s) = (s\boldsymbol{I}-\boldsymbol{A})^{-1}\boldsymbol{B}\boldsymbol{U}(s)$$

which on substituting in (6.78) gives the system input–output relationship

$$\boldsymbol{Y}(s) = [\boldsymbol{C}(s\boldsymbol{I}-\boldsymbol{A})^{-1}\boldsymbol{B}+\boldsymbol{D}]\boldsymbol{U}(s)$$

so that the system transfer function matrix for a multi-input–multi-output system is

$$\boldsymbol{G}(s) = \boldsymbol{C}(s\boldsymbol{I}-\boldsymbol{A})^{-1}\boldsymbol{B}+\boldsymbol{D} \tag{6.79}$$

EXAMPLE 6.38

(a) Obtain the state-space model characterizing the network of Figure 6.10. Take the inductor current and the voltage drop across the capacitor as the state variables, take the input variable to be the output of the voltage source, and take the output variables to be the currents through L and R_2 respectively.

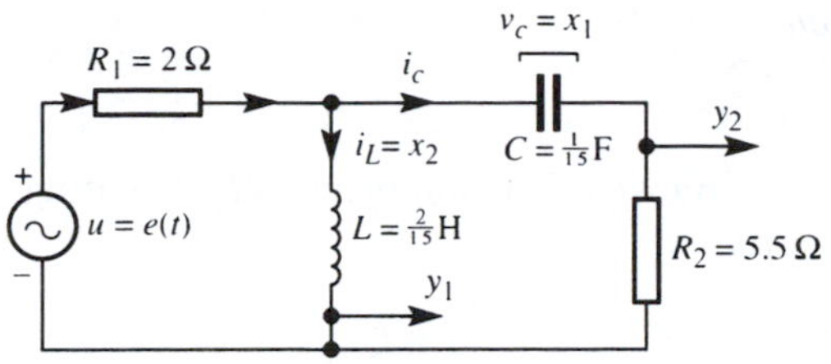

Figure 6.10 Network of Example 6.38.

(b) Find the transfer-function matrix relating the output variables y_1 and y_2 to the input variable u. Thus find the system response to the unit step $u(t) = H(t)$, assuming that the circuit is initially in a quiescent state.

Solution (a) The current i_C in the capacitor is given by

$$i_C = C\dot{v}_C = C\dot{x}_1$$

Applying Kirchhoff's second law to the outer loop gives

$$e = R_1(i_L + i_C) + v_C + R_2 i_C = R_1(x_2 + C\dot{x}_1) + x_1 + R_2 C\dot{x}_1$$

leading to

$$\dot{x}_1 = -\frac{1}{C(R_1 + R_2)}x_1 - \frac{R_1}{C(R_1 + R_2)}x_2 + \frac{e}{C(R_1 + R_2)}$$

Applying Kirchhoff's second law to the left-hand loop gives

$$e = R_1(i_L + i_C) + L\dot{i}_L = R_1(x_2 + C\dot{x}_1) + L\dot{x}_2$$

leading to

$$\dot{x}_2 = \frac{R_1}{L(R_1 + R_2)}x_1 - \frac{R_1 R_2}{L(R_1 + R_2)}x_2 + \frac{e}{L}\frac{R_2}{R_1 + R_2}$$

Also,

$$y_1 = x_2$$

$$y_2 = C\dot{x}_1 = -\frac{1}{R_1 + R_2}x_1 - \frac{R_1}{R_1 + R_2}x_2 + \frac{e}{R_1 + R_2}$$

Substituting the given parameter values leads to the state-space representation

$$\begin{bmatrix}\dot{x}_1\\ \dot{x}_2\end{bmatrix} = \begin{bmatrix}-2 & -4\\ 2 & -11\end{bmatrix}\begin{bmatrix}x_1\\ x_2\end{bmatrix} + \begin{bmatrix}2\\ \frac{11}{2}\end{bmatrix}u$$

$$\begin{bmatrix}y_1\\ y_2\end{bmatrix} = \begin{bmatrix}0 & 1\\ -\frac{2}{15} & -\frac{4}{15}\end{bmatrix}\begin{bmatrix}x_1\\ x_2\end{bmatrix} + \begin{bmatrix}0\\ \frac{2}{15}\end{bmatrix}u$$

which is of the standard form

$$\dot{x} = \mathbf{A}x + bu$$

$$y = \mathbf{C}x + du$$

(b) From (6.79), the transfer-function matrix $\mathbf{G}(s)$ relating the output variables y_1 and y_2 to the input u is

$$\mathbf{G}(s) = \mathbf{C}(s\mathbf{I} - \mathbf{A})^{-1}b + d$$

Now

$$s\mathbf{I} - \mathbf{A} = \begin{bmatrix} s+2 & 4 \\ -2 & s+11 \end{bmatrix}$$

giving

$$(s\mathbf{I} - \mathbf{A})^{-1} = \frac{1}{(s+3)(s+10)}\begin{bmatrix} s+11 & -4 \\ 2 & s+2 \end{bmatrix}$$

$$\mathbf{C}(s\mathbf{I} - \mathbf{A})^{-1}b = \frac{1}{(s+3)(s+10)}\begin{bmatrix} 0 & 1 \\ -\frac{2}{15} & -\frac{4}{15} \end{bmatrix}\begin{bmatrix} s+11 & -4 \\ 2 & s+2 \end{bmatrix}\begin{bmatrix} 2 \\ \frac{11}{2} \end{bmatrix}$$

$$= \frac{1}{(s+3)(s+10)}\begin{bmatrix} \frac{11}{2}s + 15 \\ -\frac{26}{15}s - 4 \end{bmatrix}$$

so that

$$\mathbf{G}(s) = \frac{1}{(s+3)(s+10)}\begin{bmatrix} \frac{11}{2}s + 15 \\ -\frac{26}{15}s - 4 \end{bmatrix} + \begin{bmatrix} 0 \\ \frac{2}{15} \end{bmatrix} = \begin{bmatrix} \dfrac{\frac{11}{2}s + 15}{(s+3)(s+10)} \\ \dfrac{-\frac{26}{15}s - 4}{(s+3)(s+10)} + \dfrac{2}{15} \end{bmatrix}$$

The output variables y_1 and y_2 are then given by the inverse Laplace transform of

$$Y(s) = \mathbf{G}(s)U(s)$$

where $U(s) = \mathscr{L}[u(t)] = \mathscr{L}[H(t)] = 1/s$; that is,

$$Y(s) = \begin{bmatrix} \dfrac{\frac{11}{2}s + 15}{s(s+3)(s+10)} \\ \dfrac{-\frac{26}{15}s - 4}{s(s+3)(s+10)} + \dfrac{2}{15s} \end{bmatrix} = \begin{bmatrix} \dfrac{\frac{1}{2}}{s} + \dfrac{\frac{1}{14}}{s+3} - \dfrac{\frac{4}{7}}{s+10} \\ \dfrac{-\frac{2}{15}}{s} - \dfrac{\frac{2}{35}}{s+3} + \dfrac{\frac{4}{21}}{s+10} + \dfrac{\frac{2}{15}}{s} \end{bmatrix}$$

which on taking inverse Laplace transforms gives the output variables as

$$\begin{bmatrix} y_1 \\ y_2 \end{bmatrix} = \begin{bmatrix} \frac{1}{2} + \frac{1}{14}\mathrm{e}^{-3t} - \frac{4}{7}\mathrm{e}^{-10t} \\ -\frac{2}{35}\mathrm{e}^{-3t} + \frac{4}{21}\mathrm{e}^{-10t} \end{bmatrix} \quad (t \geqslant 0)$$

6.9.6 Exercises

54 A system is governed by the vector–matrix differential equation

$$\dot{\boldsymbol{x}}(t) = \begin{bmatrix} 3 & 4 \\ 2 & 1 \end{bmatrix} \boldsymbol{x}(t) + \begin{bmatrix} 0 & 1 \\ 1 & 1 \end{bmatrix} u(t) \quad (t \geqslant 0)$$

where $\boldsymbol{x}(t)$ and $\boldsymbol{u}(t)$ are respectively the state and input vectors of the system. Use Laplace transforms to obtain the state vector $\boldsymbol{x}(t)$ for the input $u(t) = [4 \quad 3]^T$ and subject to the initial condition $\boldsymbol{x}(0) = [1 \quad 2]^T$. Compare with the answer to Exercise 53.

55 Given that the differential equations modelling a certain control system are

$$\dot{x}_1 = x_1 - 3x_2 + u$$
$$\dot{x}_2 = 2x_1 - 4x_2 + u$$

use (6.75) to determine the state vector $\boldsymbol{x} = [x_1 \quad x_2]^T$ for the control input $u = e^{-3t}$, applied at time $t = 0$, given that $x_1 = x_2 = 1$ at time $t = 0$.

56 Determine the response $y = x_1$ of the system governed by the differential equations

$$\left.\begin{aligned} \dot{x}_1 &= -2x_2 + u_1 - u_2 \\ \dot{x}_2 &= x_1 - 3x_2 + u_1 + u_2 \end{aligned}\right\} \quad (t \geqslant 0)$$

to an input $\boldsymbol{u} = [u_1 \quad u_2]^T = [1 \quad t]^T$ and subject to the initial conditions $\boldsymbol{x}_1(0) = 0$, $\boldsymbol{x}_2(0) = 1$.

57 Using the Laplace transform approach, obtain an expression for the state $\boldsymbol{x}(t)$ of the system characterized by the state equation

$$\dot{\boldsymbol{x}} = \begin{bmatrix} \dot{x}_1 \\ \dot{x}_2 \end{bmatrix} = \begin{bmatrix} 0 & 1 \\ -2 & -3 \end{bmatrix} \begin{bmatrix} \dot{x}_1 \\ \dot{x}_2 \end{bmatrix} + \begin{bmatrix} 2 \\ 0 \end{bmatrix} u \quad (t \geqslant 0)$$

where the input is

$$\boldsymbol{u}(t) = \begin{cases} 0 & (t < 0) \\ e^{-t} & (t \geqslant 0) \end{cases}$$

and subject to the initial condition $\boldsymbol{x}(0) = [1 \quad 0]^T$.

6.9.7 Spectral representation of response

We first consider the unforced system

$$\dot{\boldsymbol{x}}(t) = \boldsymbol{A}\boldsymbol{x}(t) \tag{6.80}$$

with the initial state $\boldsymbol{x}(t_0)$ at time t_0 given, and assume that the matrix $\boldsymbol{A}$ has as distinct eigenvalues λ_i $(i = 1, 2, \ldots, n)$ corresponding to n linearly independent eigenvectors $\boldsymbol{e}_i$ $(i = 1, 2, \ldots, n)$. Since the n eigenvectors are linearly independent, they may be used as a basis for the n-dimensional state space, so that the system state $\boldsymbol{x}(t)$ may be written as a linear combination in the form

$$\boldsymbol{x}(t) = c_1(t)\boldsymbol{e}_1 + \ldots + c_n(t)\boldsymbol{e}_n \tag{6.81}$$

where, since the eigenvectors are constant, the time-varying nature of $\boldsymbol{x}(t)$ is reflected in the coefficients $c_i(t)$. Substituting (6.81) into (6.80) gives

$$\dot{c}_1(t)\boldsymbol{e}_i + \ldots + \dot{c}_n(t)\boldsymbol{e}_n = \boldsymbol{A}[c_1(t)\boldsymbol{e}_1 + \ldots + c_n(t)\boldsymbol{e}_n] \tag{6.82}$$

Since $(\lambda_i, \boldsymbol{e}_i)$ are **spectral pairs** (that is, eigenvalue–eigenvector pairs) for the matrix $\boldsymbol{A}$,

$$\boldsymbol{A}\boldsymbol{e}_i = \lambda_i \boldsymbol{e}_i \quad (i = 1, 2, \ldots, n)$$

(6.82) may be written as

$$[\dot{c}_1(t) - \lambda_1 c_1(t)]\boldsymbol{e}_1 + \ldots + [\dot{c}_n(t) - \lambda_n c_n(t)]\boldsymbol{e}_n = 0 \quad \textbf{(6.83)}$$

Because the eigenvectors $\boldsymbol{e}_i$ are linearly independent, it follows from (6.83) that the system (6.80) is completely represented by the set of uncoupled differential equations

$$\dot{c}_i(t) - \lambda_i c_i(t) = 0 \quad (i = 1, 2, \ldots, n) \quad \textbf{(6.84)}$$

with solutions of the form

$$c_i(t) = \mathrm{e}^{\lambda_i(t-t_0)} c_i(t_0)$$

Then, using (6.81), the system response is

$$\boldsymbol{x}(t) = \sum_{i=1}^{n} c_i(t_0)\,\mathrm{e}^{\lambda_i(t-t_0)}\boldsymbol{e}_i \quad \textbf{(6.85)}$$

Using the given information about the initial state,

$$\boldsymbol{x}(t_0) = \sum_{i=1}^{n} c_i(t_0)\,\boldsymbol{e}_i \quad \textbf{(6.86)}$$

so that the constants $c_i(t_0)$ may be found from the given initial state using the **reciprocal basis vectors** $\boldsymbol{r}_i$ $(i = 1, 2, \ldots, n)$ defined by

$$\boldsymbol{r}_i^{\mathrm{T}}\boldsymbol{e}_j = \delta_{ij}$$

where δ_{ij} is the Kronecker delta. Taking the scalar product of both sides of (6.86) with $\boldsymbol{r}_k$, we have

$$\boldsymbol{r}_k^{\mathrm{T}}\boldsymbol{x}(t_0) = \sum_{i=1}^{n} c_i(t_0)\boldsymbol{r}_k^{\mathrm{T}}\boldsymbol{e}_i = c_k(t_0) \quad (k = 1, 2, \ldots, n)$$

which on substituting in (6.85) gives the system response

$$\boldsymbol{x}(t) = \sum_{i=1}^{n} \boldsymbol{r}_i^{\mathrm{T}}\boldsymbol{x}(t_0)\,\mathrm{e}^{\lambda_i(t-t_0)}\boldsymbol{e}_i \quad \textbf{(6.87)}$$

which is referred to as the **spectral** or **modal form** of the response. The terms $\boldsymbol{r}_i^{\mathrm{T}}\boldsymbol{x}(t_0)\,\mathrm{e}^{\lambda_i(t-t_0)}\boldsymbol{e}_i$ are called the **modes** of the system. Thus, provided that the system matrix $\boldsymbol{A}$ has n linearly independent eigenvectors, this approach has the advantage of enabling us to break down the general system response into the sum of its simple modal responses. The amount of excitation of each mode, represented by $\boldsymbol{r}_i^{\mathrm{T}}\boldsymbol{x}(t_0)$, is dependent only on the initial conditions, so if, for example, the initial state $\boldsymbol{x}(t_0)$ is parallel to the ith eigenvector $\boldsymbol{e}_i$ then only the ith mode will be excited.

It should be noted that if a pair of eigenvalues λ_1, λ_2 are complex conjugates then the modes associated with $\mathrm{e}^{\lambda_1(t-t_0)}$ and $\mathrm{e}^{\lambda_2(t-t_0)}$ cannot be separated from each other. The combined motion takes place in a plane determined by the corresponding eigenvectors $\boldsymbol{e}_1$ and $\boldsymbol{e}_2$ and is oscillatory.

By retaining only the dominant modes, the spectral representation may be used to approximate high-order systems by lower-order ones.

EXAMPLE 6.39 Obtain in spectral form the response of the second-order system

$$\begin{bmatrix} \dot{x}_1 \\ \dot{x}_2 \end{bmatrix} = \begin{bmatrix} -2 & 1 \\ 1 & -1 \end{bmatrix} \begin{bmatrix} x_1 \\ x_2 \end{bmatrix}, \quad x(0) = \begin{bmatrix} 1 \\ 2 \end{bmatrix}$$

and sketch the trajectory.

Solution The eigenvalues of the matrix

$$\boldsymbol{A} = \begin{bmatrix} -2 & 1 \\ 1 & -2 \end{bmatrix}$$

are determined by

$$|\boldsymbol{A} - \lambda \boldsymbol{I}| = \lambda^2 + 4\lambda + 3 = 0$$

that is,

$$\lambda_1 = -1, \quad \lambda_2 = -3$$

with corresponding eigenvectors

$$\boldsymbol{e}_1 = [1 \quad 1]^{\mathrm{T}}, \quad \boldsymbol{e}_2 = [1 \quad -1]^{\mathrm{T}}$$

Denoting the reciprocal basis vectors by

$$\boldsymbol{r}_1 = [r_{11} \quad r_{12}]^{\mathrm{T}}, \quad \boldsymbol{r}_2 = [r_{21} \quad r_{22}]^{\mathrm{T}}$$

and using the relationships

$$\boldsymbol{r}_i^{\mathrm{T}} \boldsymbol{e}_j = \delta_{ij} \quad (i, j = 1, 2)$$

we have

$$\boldsymbol{r}_1^{\mathrm{T}} \boldsymbol{e}_1 = r_{11} + r_{12} = 1, \quad \boldsymbol{r}_1^{\mathrm{T}} \boldsymbol{e}_2 = r_{11} - r_{12} = 0$$

giving

$$r_{11} = \tfrac{1}{2}, \quad r_{12} = \tfrac{1}{2}, \quad \boldsymbol{r}_1 = [\tfrac{1}{2} \quad \tfrac{1}{2}]^{\mathrm{T}}$$

and

$$\boldsymbol{r}_2^{\mathrm{T}} \boldsymbol{e}_2 = r_{21} + r_{22} = 0, \qquad \boldsymbol{r}_2^{\mathrm{T}} \boldsymbol{e}_2 = r_{21} - r_{22} = 1$$

giving

$$r_{21} = \tfrac{1}{2}, \quad r_{22} = -\tfrac{1}{2}, \quad r_2 = [\tfrac{1}{2} \quad -\tfrac{1}{2}]^{\mathrm{T}}$$

Thus

$$\boldsymbol{r}_1^{\mathrm{T}} \boldsymbol{x}(0) = \tfrac{1}{2} + 1 = \tfrac{3}{2}, \quad \boldsymbol{r}_2^{\mathrm{T}} \boldsymbol{x}(0) = \tfrac{1}{2} - 1 = -\tfrac{1}{2}$$

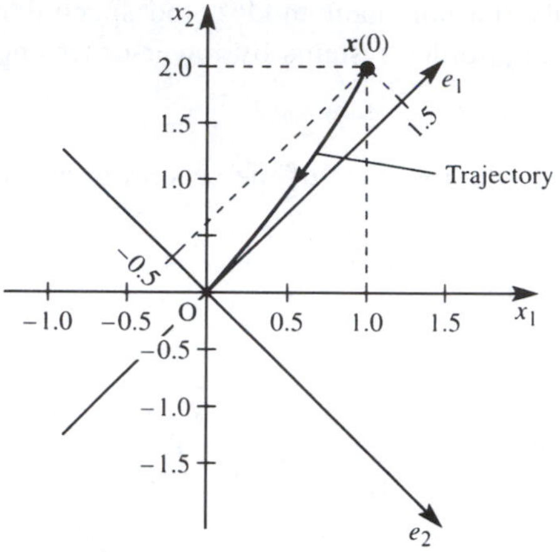

Figure 6.11 Trajectory for Example 6.39.

so that, from (6.87), the system response is

$$x(t) = \sum_{i=1}^{2} r_i^T x(0)\,e^{\lambda_i t} e_i = r_1^T x(0)\,e^{\lambda_1 t} e_1 + r_2^T x(0)\,e^{\lambda_2 t} e_2$$

That is,

$$x(t) = \tfrac{3}{2} e^{-t} e_1 - \tfrac{1}{2} e^{-3t} e_2$$

which is in the required spectral form.

To plot the response, we first draw axes corresponding to the eigenvectors e_1 and e_2, as shown in Figure 6.11. Taking these as coordinate axes, we are at the point $(\frac{3}{2}, -\frac{1}{2})$ at time $t = 0$. As t increases, the movement along the direction of e_2 is much faster than that in the direction of e_1, since e^{-3t} decreases more rapidly than e^{-t}. We can therefore guess the trajectory, without plotting, as sketched in Figure 6.11.

We can proceed in an analogous manner to obtain the spectral representation of the response to the forced system

$$\dot{x}(t) = \boldsymbol{A}x(t) + \boldsymbol{B}u(t)$$

with $x(t_0)$ given. Making the same assumption regarding the linear independence of the eigenvectors e_i $(i = 1, 2, \dots, n)$ of the matrix $\boldsymbol{A}$, the vector $\boldsymbol{B}u(t)$ may also be written as a linear combination of the form

$$\boldsymbol{B}u(t) = \sum_{i=1}^{n} \beta_i(t) e_i \tag{6.88}$$

so that, corresponding to (6.83), we have

$$[\dot{c}_1(t) - \lambda_1 c_1(t) - \beta_1(t)]e_1 + \dots + [\dot{c}_n(t) - \lambda_n c_n(t) - \beta_n(t)]e_n = 0$$

As a consequence of the linear independence of the eigenvectors $\boldsymbol{e}_i$ this leads to the set of uncoupled differential equations

$$\dot{c}_i(t) - \lambda_i c_i(t) - \beta_i(t) = 0 \quad (i = 1, 2, \ldots, n)$$

which, using (6.65), have corresponding solutions

$$c_i(t) = \mathrm{e}^{\lambda_i(t-t_0)} c_i(t_0) + \int_{t_0}^{t} \mathrm{e}^{\lambda_i(t-\tau)} \beta_i(\tau)\,\mathrm{d}\tau \tag{6.89}$$

As for $c_i(t_0)$, the reciprocal basis vectors $\boldsymbol{r}_i$ may be used to obtain the coefficients $\beta_i(\tau)$. Taking the scalar product of both sides of (6.88) with $\boldsymbol{r}_k$ and using the relationships $\boldsymbol{r}_i^{\mathrm{T}}\boldsymbol{e}_j = \delta_{ij}$, we have

$$\boldsymbol{r}_k^{\mathrm{T}}\boldsymbol{Bu}(t) = \beta_k(t) \quad (k = 1, 2, \ldots, n)$$

Thus, from (6.89),

$$c_i(t) = \mathrm{e}^{\lambda_i(t-t_0)} \boldsymbol{r}_i^{\mathrm{T}}\boldsymbol{x}(t_0) + \int_{t_0}^{t} \mathrm{e}^{\lambda_i(t-\tau)} \boldsymbol{r}_i^{\mathrm{T}}\boldsymbol{Bu}(\tau)\,\mathrm{d}\tau$$

giving the spectral form of the system response as

$$\boldsymbol{x}(t) = \sum_{i=1}^{n} c_i(t)\boldsymbol{e}_i$$

6.9.8 Canonical representation

Consider the state-space representation given in (6.56), namely

$$\dot{\boldsymbol{x}} = \boldsymbol{Ax} + \boldsymbol{Bu} \tag{6.56a}$$

$$\boldsymbol{y} = \boldsymbol{Cx} + \boldsymbol{Du} \tag{6.56b}$$

Applying the transformation

$$\boldsymbol{x} = \boldsymbol{Tz}$$

where $\boldsymbol{T}$ is a non-singular matrix, leads to

$$\boldsymbol{T}\dot{\boldsymbol{z}} = \boldsymbol{ATz} + \boldsymbol{Bu}$$

$$\boldsymbol{y} = \boldsymbol{CTz} + \boldsymbol{Du}$$

which may be written in the form

$$\dot{\boldsymbol{z}} = \tilde{\boldsymbol{A}}\boldsymbol{z} = \tilde{\boldsymbol{B}}\boldsymbol{u} \tag{6.90a}$$

$$\boldsymbol{y} = \tilde{\boldsymbol{C}}\boldsymbol{z} = \tilde{\boldsymbol{D}}\boldsymbol{u} \tag{6.90b}$$

where $\boldsymbol{z}$ is now a state vector and

$$\tilde{\boldsymbol{A}} = \boldsymbol{T}^{-1}\boldsymbol{AT}, \qquad \tilde{\boldsymbol{B}} = \boldsymbol{T}^{-1}\boldsymbol{B}, \qquad \tilde{\boldsymbol{C}} = \boldsymbol{CT}, \qquad \tilde{\boldsymbol{D}} = \boldsymbol{D}$$

From (6.79), the input–output transfer-function matrix corresponding to (6.90) is

$$\begin{aligned}
\boldsymbol{G}_1(s) &= \tilde{\boldsymbol{C}}(s\boldsymbol{I} - \tilde{\boldsymbol{A}})^{-1}\tilde{\boldsymbol{B}} + \tilde{\boldsymbol{D}} \\
&= \boldsymbol{CT}(s\boldsymbol{I} - \boldsymbol{T}^{-1}\boldsymbol{AT})^{-1}\boldsymbol{T}^{-1}\boldsymbol{B} + \boldsymbol{D} \\
&= \boldsymbol{CT}(s\boldsymbol{T}^{-1}\boldsymbol{IT} - \boldsymbol{T}^{-1}\boldsymbol{AT})^{-1}\boldsymbol{T}^{-1}\boldsymbol{B} + \boldsymbol{D} \\
&= \boldsymbol{CT}[\boldsymbol{T}^{-1}(s\boldsymbol{I} - \boldsymbol{A})\boldsymbol{T}]^{-1}\boldsymbol{T}^{-1}\boldsymbol{B} + \boldsymbol{D} \\
&= \boldsymbol{CT}[\boldsymbol{T}^{-1}(s\boldsymbol{I} - \boldsymbol{A})^{-1}\boldsymbol{T}]\boldsymbol{T}^{-1}\boldsymbol{B} + \boldsymbol{D} \quad \text{(using the commutative property)} \\
&= \boldsymbol{C}(s\boldsymbol{I} - \boldsymbol{A})^{-1}\boldsymbol{B} + \boldsymbol{D} \\
&= \boldsymbol{G}(s)
\end{aligned}$$

where $\boldsymbol{G}(s)$ is the transfer-function matrix corresponding to (6.56).

Thus the system input–output relationship is unchanged by the transformation, and the linear systems (6.56) and (6.90) are said to be **equivalent**. By the transformation the intrinsic properties of the system, such as stability, controllability and observability, which are of interest to the engineer, are preserved, and there is merit in seeking a transformation leading to a system that is more easily analysed.

Since the transformation matrix $\boldsymbol{T}$ can be arbitrarily chosen, an infinite number of equivalent systems exist. Of particular interest is the case when $\boldsymbol{T}$ is taken to be the modal matrix $\boldsymbol{M}$ of the system matrix $\boldsymbol{A}$; that is

$$\boldsymbol{T} = \boldsymbol{M} = [e_1 \quad e_2 \quad \dots \quad e_n]$$

where e_i $(i = 1, 2, \dots, n)$ are the eigenvectors of the matrix $\boldsymbol{A}$. Under the assumption that the n eigenvalues are distinct,

$$\tilde{\boldsymbol{A}} = \boldsymbol{M}^{-1}\boldsymbol{AM} = \Lambda, \qquad \text{the spectral matrix of } \boldsymbol{A}$$

$$\tilde{\boldsymbol{B}} = \boldsymbol{M}^{-1}\boldsymbol{B}$$

$$\tilde{\boldsymbol{C}} = \boldsymbol{CM}, \qquad \tilde{\boldsymbol{D}} = \boldsymbol{D}$$

so that (6.90) becomes

$$\dot{z} = \Lambda z + \boldsymbol{M}^{-1}\boldsymbol{B}u \tag{6.91a}$$

$$y = \boldsymbol{CM}z + \boldsymbol{D}u \tag{6.91b}$$

Equation (6.91a) constitutes a system of uncoupled linear differential equations

$$\dot{z}_i = \lambda_i z_i + \boldsymbol{b}_i^{\mathrm{T}}\boldsymbol{u} \qquad (i = 1, 2, \dots, n) \tag{6.92}$$

where $z = (z_1, z_2, \dots, z_n)^{\mathrm{T}}$ and $\boldsymbol{b}_i^{\mathrm{T}}$ is the ith row of the matrix $\boldsymbol{M}^{-1}\boldsymbol{B}$. Thus, by reducing (6.56) to the equivalent form (6.91) using the transformation $\boldsymbol{x} = \boldsymbol{M}z$, the modes of the system have been uncoupled, with the new state variables z_i $(i = 1, 2, \dots, n)$ being associated with the ith mode only. The representation (6.91) is called the **normal** or **canonical representation** of the system equations.

From (6.65), the solution of (6.92) is

$$z_i = e^{\lambda_i(t-t_0)}x(t_0) + \int_{t_0}^{t} e^{\lambda_i(t-\tau)} \boldsymbol{b}_i^{\mathrm{T}}\boldsymbol{u}(\tau)\,d\tau \quad (i = 1, \ldots, n)$$

so that the solution of (6.91a) may be written as

$$\boldsymbol{z}(t) = e^{\boldsymbol{\Lambda}(t-t_0)}\boldsymbol{z}(t_0) + \int_{t_0}^{t} e^{\boldsymbol{\Lambda}(t-\tau)}\boldsymbol{M}^{-1}\boldsymbol{B}\boldsymbol{u}(\tau)\,d\tau \tag{6.93}$$

where

$$e^{\boldsymbol{\Lambda}(t-t_0)} = \begin{bmatrix} e^{\lambda_1(t-t_0)} & & 0 \\ & \ddots & \\ 0 & & e^{\lambda_n(t-t_0)} \end{bmatrix}$$

In terms of the original state vector $\boldsymbol{x}(t)$, (6.93) becomes

$$\boldsymbol{x}(t) = \boldsymbol{M}\boldsymbol{z} = \boldsymbol{M}\,e^{\boldsymbol{\Lambda}(t-t_0)}\boldsymbol{M}^{-1}\boldsymbol{x}(t_0) + \int_{t_0}^{t} \boldsymbol{M}\,e^{\boldsymbol{\Lambda}(t-\tau)}\boldsymbol{M}^{-1}\boldsymbol{B}\boldsymbol{u}(\tau)\,d\tau \tag{6.94}$$

and the system response is then obtained from (6.56b) as

$$\boldsymbol{y}(t) = \boldsymbol{C}\boldsymbol{x}(t) + \boldsymbol{D}\boldsymbol{u}(t)$$

By comparing the response (6.94) with that in (6.68), we note that the transition matrix may be written as

$$\boldsymbol{\Phi}(t, t_0) = e^{\boldsymbol{A}(t-t_0)} = \boldsymbol{M}\,e^{\boldsymbol{\Lambda}(t-t_0)}\boldsymbol{M}^{-1}$$

The representation (6.91) may be used to readily infer some system properties. If the system is stable then each mode must be stable, so, from (6.94), each λ_i $(i = 1, 2, \ldots, n)$ must have a negative real part. If, for example, the jth row of the matrix $\boldsymbol{M}^{-1}\boldsymbol{B}$ is zero then, from (6.92), $\dot{z}_j = \lambda_j z_j + 0$, so the input $\boldsymbol{u}(t)$ has as influence on the jth mode of the system, and the mode is said to be **uncontrollable**. A system is said to be **controllable** if all of its modes are controllable.

If the jth column of the matrix $\boldsymbol{CM}$ is zero then, from (6.91b), the response y is independent of z_j, so it is not possible to use information about the output to identify z_j. The state z_j is then said to be **unobservable**, and the overall system is not **observable**.

EXAMPLE 6.40

A third-order system is characterized by the state-space model

$$\dot{\boldsymbol{x}} = \begin{bmatrix} 0 & 1 & 0 \\ 0 & 0 & 1 \\ 0 & -5 & -6 \end{bmatrix}\boldsymbol{x} + \begin{bmatrix} 1 \\ -3 \\ 18 \end{bmatrix}u, \qquad y = [1 \quad 0 \quad 0]\boldsymbol{x}$$

where $\boldsymbol{x} = [x_1 \quad x_2 \quad x_3]^{\mathrm{T}}$. Obtain the equivalent canonical representation of the model and then obtain the response of the system to a unit step $u(t) = H(t)$ given that initially $\boldsymbol{x}(0) = [1 \quad 1 \quad 0]^{\mathrm{T}}$.

Solution The eigenvalues of the matrix

$$\boldsymbol{A} = \begin{bmatrix} 0 & 1 & 0 \\ 0 & 0 & 1 \\ 0 & -5 & -6 \end{bmatrix}$$

are determined by

$$|\boldsymbol{A} - \lambda \boldsymbol{I}| = \begin{vmatrix} -\lambda & 1 & 0 \\ 0 & -\lambda & 1 \\ 0 & -5 & -6-\lambda \end{vmatrix} = 0$$

that is,

$$\lambda(\lambda^2 + 6\lambda + 5) = 0$$

giving $\lambda_1 = 0$, $\lambda_2 = -1$ and $\lambda_3 = -5$, with corresponding eigenvectors

$$\boldsymbol{e}_1 = [1 \quad 0 \quad 0]^{\mathrm{T}}, \quad \boldsymbol{e}_2 = [1 \quad -1 \quad 1]^{\mathrm{T}}, \quad \boldsymbol{e}_3 = [1 \quad -5 \quad 25]^{\mathrm{T}}$$

The corresponding modal and spectral matrices are

$$\boldsymbol{M} = \begin{bmatrix} 1 & 1 & 1 \\ 0 & -1 & -5 \\ 0 & 1 & 25 \end{bmatrix}, \quad \boldsymbol{\Lambda} = \begin{bmatrix} 0 & 0 & 0 \\ 0 & -1 & 0 \\ 0 & 0 & -5 \end{bmatrix}$$

and the inverse modal matrix is determined to be

$$\boldsymbol{M}^{-1} = \frac{1}{20}\begin{bmatrix} 20 & 25 & 4 \\ 0 & -25 & -5 \\ 0 & 1 & 1 \end{bmatrix}$$

In this case $\boldsymbol{B} = [1 \quad -3 \quad 18]^{\mathrm{T}}$, so

$$\boldsymbol{M}^{-1}\boldsymbol{B} = \frac{1}{20}\begin{bmatrix} 20 & 25 & 4 \\ 0 & -25 & -5 \\ 0 & 1 & 1 \end{bmatrix}\begin{bmatrix} 1 \\ -3 \\ 18 \end{bmatrix} = \frac{1}{20}\begin{bmatrix} 20 \\ -15 \\ 15 \end{bmatrix} = \begin{bmatrix} 1 \\ -\frac{3}{4} \\ \frac{3}{4} \end{bmatrix}$$

Likewise, $\boldsymbol{C} = [1 \quad 0 \quad 0]$, giving

$$\boldsymbol{CM} = [1 \quad 0 \quad 0]\begin{bmatrix} 1 & 1 & 1 \\ 0 & -1 & -5 \\ 0 & 1 & 25 \end{bmatrix} = [1 \quad 1 \quad 1]$$

Thus, from (6.91), the equivalent canonical state-space representation is

$$\dot{z} = \begin{bmatrix} \dot{z}_1 \\ \dot{z}_2 \\ \dot{z}_3 \end{bmatrix} = \begin{bmatrix} 0 & 0 & 0 \\ 0 & -1 & 0 \\ 0 & 0 & -5 \end{bmatrix} \begin{bmatrix} z_1 \\ z_2 \\ z_3 \end{bmatrix} + \begin{bmatrix} 1 \\ -\frac{3}{4} \\ \frac{3}{4} \end{bmatrix} u \tag{6.95a}$$

$$y = [1 \quad 1 \quad 1] \begin{bmatrix} z_1 \\ z_2 \\ z_3 \end{bmatrix} \tag{6.95b}$$

When $u(t) = H(t)$, from (6.93) the solution of (6.95a) is

$$z = \begin{bmatrix} e^{0t} & 0 & 0 \\ 0 & e^{-t} & 0 \\ 0 & 0 & e^{-5t} \end{bmatrix} z(0) + \int_0^t \begin{bmatrix} 1 & 0 & 0 \\ 0 & e^{-(t-\tau)} & 0 \\ 0 & 0 & e^{-5(t-\tau)} \end{bmatrix} \begin{bmatrix} 1 \\ -\frac{3}{4} \\ \frac{3}{4} \end{bmatrix} 1\,d\tau$$

where

$$z(0) = M^{-1}x(0) = \frac{1}{20} \begin{bmatrix} 20 & 24 & 4 \\ 0 & -25 & -5 \\ 0 & 1 & 1 \end{bmatrix} \begin{bmatrix} 1 \\ 1 \\ 0 \end{bmatrix} = \begin{bmatrix} \frac{44}{20} \\ -\frac{25}{20} \\ \frac{1}{20} \end{bmatrix}$$

leading to

$$z = \begin{bmatrix} 1 & 0 & 0 \\ 0 & e^{-t} & 0 \\ 0 & 0 & e^{-5t} \end{bmatrix} \begin{bmatrix} \frac{11}{5} \\ -\frac{5}{4} \\ \frac{1}{20} \end{bmatrix} + \int_0^t \begin{bmatrix} 1 \\ -\frac{3}{4} e^{-(t-\tau)} \\ \frac{3}{4} e^{-5(t-\tau)} \end{bmatrix} d\tau$$

$$= \begin{bmatrix} \frac{11}{5} \\ -\frac{5}{4} e^{-t} \\ \frac{1}{20} e^{-5t} \end{bmatrix} + \begin{bmatrix} t \\ -\frac{3}{4} + \frac{3}{4} e^{-t} \\ \frac{3}{20} - \frac{3}{20} e^{-5t} \end{bmatrix} = \begin{bmatrix} t + \frac{11}{5} \\ -\frac{3}{4} - \frac{1}{2} e^{-t} \\ \frac{3}{20} - \frac{1}{10} e^{-5t} \end{bmatrix}$$

Then, from (6.95b),

$$\begin{aligned} y &= z_1 + z_2 + z_3 = (t + \tfrac{11}{5}) + (-\tfrac{3}{4} - \tfrac{1}{2} e^{-t}) + (\tfrac{3}{20} - \tfrac{1}{10} e^{-5t}) \\ &= t + \tfrac{8}{5} - \tfrac{1}{2} e^{-t} - \tfrac{1}{10} e^{-5t} \end{aligned}$$

If we drop the assumption that the eigenvalues of $\boldsymbol{A}$ are distinct then $\tilde{\boldsymbol{A}} = \boldsymbol{M}^{-1}\boldsymbol{A}\boldsymbol{M}$ is no longer diagonal, but may be represented by the corresponding Jordan canonical form $\boldsymbol{J}$ with $\boldsymbol{M}$ being made up of both eigenvectors and generalized eigenvectors of $\boldsymbol{A}$. The equivalent canonical form in this case will be

$$\dot{z} = Jz + M^{-1}Bu$$

$$y = CMz + Du$$

with the solution corresponding to (6.93) being

$$x(t) = M\,e^{J(t-t_0)}M^{-1}x(t_0) + \int_{t_0}^{t} M\,e^{J(t-\tau)}M^{-1}Bu(\tau)\,d\tau$$

6.9.9 Exercises

58 Obtain in spectral form the response of the unforced second-order system

$$\dot{x}(t) = \begin{bmatrix} \dot{x}_1(t) \\ \dot{x}_2(t) \end{bmatrix} = \begin{bmatrix} -\frac{3}{2} & \frac{3}{4} \\ 1 & -\frac{5}{2} \end{bmatrix} x(t),$$

$$x(0) = \begin{bmatrix} 2 \\ 4 \end{bmatrix}$$

Using the eigenvectors as the frame of reference, sketch the trajectory.

59 Using the spectral form of the solution given in (6.87), solve the second-order system

$$\dot{x}(t) = \begin{bmatrix} -2 & 2 \\ 2 & -5 \end{bmatrix} x(t), \qquad x(0) = \begin{bmatrix} 2 \\ 3 \end{bmatrix}$$

and sketch the trajectory.

60 Repeat Exercise 59 for the system

$$\dot{x}(t) = \begin{bmatrix} 0 & -4 \\ 2 & -4 \end{bmatrix} x(t), \qquad x(0) = \begin{bmatrix} 1 \\ 2 \end{bmatrix}$$

61 Determine the equivalent canonical representation of the third-order system

$$\dot{x} = \begin{bmatrix} 1 & 1 & -2 \\ -1 & 2 & 1 \\ 0 & 1 & -1 \end{bmatrix} x + \begin{bmatrix} -1 \\ 1 \\ -1 \end{bmatrix} u$$

$$y = [-2 \quad 1 \quad 0]x$$

62 The solution of a third-order linear system is given by

$$x = \alpha_0 e^{-t} e_0 + \alpha_1 e^{-2t} e_1 + \alpha_2 e^{-3t} e_2$$

where e_0, e_1 and e_2 are linearly independent vectors having values

$$e_0 = [1 \quad 1 \quad 0]^T, \qquad e_1 = [0 \quad 1 \quad 1]^T,$$
$$e_2 = [1 \quad 2 \quad 3]^T$$

Initially, at time $t = 0$ the system state is $x(0) = [1 \quad 1 \quad 1]^T$. Find α_0, α_1 and α_2 using the reciprocal basis method.

63 Obtain the eigenvalues and eigenvectors of the matrix

$$A = \begin{bmatrix} 5 & 4 \\ 1 & 2 \end{bmatrix}$$

Using a suitable transformation $x(t) = Mz(t)$, reduce $\dot{x}(t) = Ax(t)$ to the canonical form $\dot{z}(t) = \Lambda z(t)$, where Λ is the spectral matrix of A. Solve the decoupled canonical form for z, and hence solve for $x(t)$ given that $x(0) = [1 \quad 4]^T$.

64 A second-order system is governed by the state equation

$$\dot{x}(t) = \begin{bmatrix} 3 & 4 \\ 2 & 1 \end{bmatrix} x(t) + \begin{bmatrix} 0 & 1 \\ 1 & 1 \end{bmatrix} u(t) \quad (t \geq 0)$$

Using a suitable transformation $x(t) = Mz(t)$, reduce this to the canonical form

$$\dot{z}(t) = \Lambda z(t) + Bu(t)$$

where Λ is the spectral matrix of

$$\begin{bmatrix} 3 & 4 \\ 2 & 1 \end{bmatrix}$$

and B is a suitable 2×2 matrix.

For the input $u(t) = [4 \quad 3]^T$ solve the decoupled canonical form for z, and hence solve for $x(t)$ given that $x(0) = [1 \quad 2]^T$. Compare the answer with that for Exercises 53 and 54.

65 A third-order single-input–single-output system is characterized by the transfer-function model

$$\frac{Y(s)}{U(s)} = \frac{3s^2 + 2s + 1}{s^3 + 6s^2 + 11s + 6}$$

where $Y(s)$ and $U(s)$ are the Laplace transforms of the system output and input respectively. Express the system model in the state-space form

$$\dot{x} = Ax + bu \quad \textbf{(6.96a)}$$

$$y = c^{\mathrm{T}}x \quad \textbf{(6.96b)}$$

where A is in the companion form. By making a suitable transformation $x = Mz$, reduce the state-space model to its canonical form, and comment on the stability, controllability and observability of the system.

Given that

(i) a necessary and sufficient condition for the system (6.96) to be controllable is that the rank of the **Kalman matrix** $[b \;\; Ab \;\; A^2b \;\; \ldots \;\; A^{n-1}b]$ be the same as the order of A, and

(ii) a necessary and sufficient condition for it to be observable is that the rank of the Kalman matrix $[c \;\; A^{\mathrm{T}}c \;\; (A^{\mathrm{T}})^2c \;\; \ldots \;\; (A^{\mathrm{T}})^{n-1}c]$ be the same as the order of A,

evaluate the ranks of the relevant Kalman matrices to confirm your earlier conclusions on the controllability and observability of the given system.

66 Repeat Exercise 65 for the system characterized by the transfer-function model

$$\frac{s^2 + 3s + 5}{s^3 + 6s^2 + 5s}$$

6.10 Discrete-time systems

In Chapter 3 we considered the role of z transforms in providing a frequency-domain approach to the modelling of discrete-time systems, the analysis and design of which is a subject of increasing importance in many branches of engineering. In this section we consider the time-domain state-space model representation for such systems.

6.10.1 State-space model

Consider the nth-order linear time-invariant discrete-time system modelled by the difference equation

$$y_{k+n} + a_{n-1}y_{k+n-1} + a_{n-2}y_{k+n-2} + \ldots + a_0 y_k = b_0 u_k \quad \textbf{(6.97)}$$

which corresponds to (3.32), with $b_i = 0$ $(i > 0)$. Recall that $\{y_k\}$ is the output sequence, with general term y_k, and $\{u_k\}$ the input sequence, with general term u_k. Following the procedure of Section 6.8.1, we introduce state variables $x_1(k)$, $x_2(k)$, $\ldots$, $x_n(k)$ for the system, defined by

$$x_1(k) = y_k, \quad x_2(k) = y_{k+1}, \quad \ldots, \quad x_n(k) = y_{k+n-1} \quad \textbf{(6.98)}$$

Note that we have used the notation $x_i(k)$ rather than the suffix notation $x_{i,k}$ for clarity. When needed, we shall adopt the same convention for the input term and write $u(k)$ for u_k in the interests of consistency. We now define the state vector corresponding to this choice of state variables as $\boldsymbol{x}(k) = [x_1(k) \quad x_2(k) \quad \dots \quad x_n(k)]^{\mathrm{T}}$. Examining the system of equations (6.98), we see that

$$
\begin{aligned}
x_1(k+1) &= y_{k+1} = x_2(k) \\
x_2(k+1) &= y_{k+2} = x_3(k) \\
&\vdots \\
x_{n-1}(k+1) &= y_{k+n-1} = x_n(k) \\
x_n(k+1) &= y_{k+n} \\
&= -a_{n-1}y_{k+n-1} - a_{n-2}y_{k+n-2} - \dots - a_0 y_k + b_0 u_k \\
&= -a_{n-1}x_n(k) - a_{n-2}x_{n-1}(k) - \dots - a_0 x_1(k) + b_0 u(k)
\end{aligned}
$$

using the alternative notation for u_k.

We can now write the system in the vector–matrix form

$$
\boldsymbol{x}(k+1) = \begin{bmatrix} x_1(k+1) \\ x_2(k+1) \\ \vdots \\ \\ x_n(k+1) \end{bmatrix} = \begin{bmatrix} 0 & 1 & 0 & 0 & \dots & 0 \\ 0 & 0 & 1 & 0 & \dots & 0 \\ \vdots & \vdots & \vdots & \vdots & & \vdots \\ 0 & 0 & 0 & 0 & \dots & 1 \\ -a_0 & -a_1 & -a_2 & -a_3 & \dots & -a_{n-1} \end{bmatrix} \begin{bmatrix} x_1(k) \\ x_2(k) \\ \vdots \\ \\ x_n(k) \end{bmatrix} + \begin{bmatrix} 0 \\ 0 \\ \vdots \\ \\ b_0 \end{bmatrix} u(k)
\tag{6.99}
$$

which corresponds to (6.46) for a continuous-time system. Again, we can write this more concisely as

$$
\boldsymbol{x}(k+1) = \boldsymbol{A}\boldsymbol{x}(k) + \boldsymbol{b}u(k) \tag{6.100}
$$

where $\boldsymbol{A}$ and $\boldsymbol{b}$ are defined as in (6.99). The output of the system is the sequence $\{y_k\}$, and the general term $y_k = x_1(k)$ can be recovered from the state vector $\boldsymbol{x}(k)$ as

$$
y(k) = x_1(k) = [1 \quad 0 \quad 0 \quad \dots \quad 0]\boldsymbol{x}(k) = \boldsymbol{c}^{\mathrm{T}}\boldsymbol{x}(k) \tag{6.101}
$$

As in the continuous-time case, it may be that the output of the system is a combination of the state and the input sequence $\{u(k)\}$, in which case (6.101) becomes

$$
y(k) = \boldsymbol{c}^{\mathrm{T}}\boldsymbol{x}(k) + du(k) \tag{6.102}
$$

Equations (6.100) and (6.102) constitute the state-space representation of the system, and we immediately note the similarity with (6.50a, b) derived for continuous-time systems. Likewise, for the multi-input–multi-output case the discrete-time state-space model corresponding to (6.56) is

$$
\boldsymbol{x}(k+1) = \boldsymbol{A}\boldsymbol{x}(k) + \boldsymbol{B}\boldsymbol{u}(k) \tag{6.103a}
$$

$$
\boldsymbol{y}(k) = \boldsymbol{C}\boldsymbol{x}(k) + \boldsymbol{D}\boldsymbol{u}(k) \tag{6.103b}
$$

EXAMPLE 6.41 Determine the state-space representation of the system modelled by the difference equation

$$y_{k+2} + 0.2y_{k+1} + 0.3y_k = u_k \tag{6.104}$$

Solution We choose as state variables

$$x_1(k) = y_k, \qquad x_2(k) = y_{k+1}$$

Thus

$$x_1(k+1) = x_2(k)$$

and from (6.104),

$$x_2(k+1) = -0.3x_1(k) - 0.2x_2(k) + u(k)$$

The state-space representation is then

$$\boldsymbol{x}(k+1) = \boldsymbol{A}\boldsymbol{x}(k) + \boldsymbol{b}u(k), \qquad y(k) = \boldsymbol{c}^{\mathrm{T}}\boldsymbol{x}(k)$$

with

$$\boldsymbol{A} = \begin{bmatrix} 0 & 1 \\ -0.3 & -0.2 \end{bmatrix}, \quad \boldsymbol{b} = \begin{bmatrix} 0 \\ 1 \end{bmatrix}, \quad \boldsymbol{c}^{\mathrm{T}} = [1 \quad 0]$$

We notice, from reference to Section 3.6.1, that the procedure used in Example 6.41 for establishing the state-space form of the system corresponds to labelling the output of each delay block in the system as a state variable. In the absence of any reason for an alternative choice, this is the logical approach. Section 3.6.1 also gives a clue towards a method of obtaining the state-space representation for systems described by the more general form of (3.32) with $m > 0$. Example 3.19 illustrates such a system, with z transfer function

$$G(z) = \frac{z-1}{z^2 + 3z + 2}$$

The block diagram for this system is shown in Figure 3.9(c) and reproduced for convenience in Figure 6.12. We choose as state variables the outputs from each delay block, it being immaterial whether we start from the left- or the right-hand

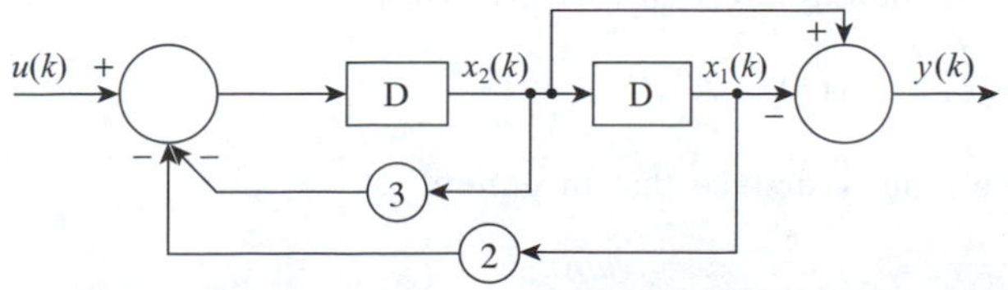

Figure 6.12 Block diagram of system with transfer function $G(z) = (z - 1)/(z^2 + 3z + 2)$.

side of the diagram (obviously, different representations will be obtained depending on the choice we make, but the different forms will yield identical information on the system). Choosing to start on the right-hand side (that is, with $x_1(k)$ the output of the right-hand delay block and $x_2(k)$ that of the left-hand block), we obtain

$$x_1(k+1) = x_2(k)$$

$$x_2(k+1) = -3x_2(k) - 2x_1(k) + u(k)$$

with the system output given by

$$y(k) = -x_1(k) + x_2(k)$$

Thus the state-space form corresponding to our choice of state variables is

$$\boldsymbol{x}(k+1) = \boldsymbol{A}\boldsymbol{x}(k) + \boldsymbol{b}u(k), \qquad y(k) = \boldsymbol{c}^{\mathrm{T}}\boldsymbol{x}(k)$$

with

$$\boldsymbol{A} = \begin{bmatrix} 0 & 1 \\ -2 & -3 \end{bmatrix}, \quad \boldsymbol{b} = \begin{bmatrix} 0 \\ 1 \end{bmatrix}, \quad \boldsymbol{c}^{\mathrm{T}} = [-1 \quad 1]$$

We notice that, in contrast with the system of Example 6.41, the row vector $\boldsymbol{c}^{\mathrm{T}} = [-1 \quad 1]$ now combines contributions from both state variables to form the output $y(k)$.

6.10.2 Solution of the discrete-time state equation

As in Section 6.9.1 for continuous-time systems, we first consider the unforced or homogeneous case

$$\boldsymbol{x}(k+1) = \boldsymbol{A}\boldsymbol{x}(k) \tag{6.105}$$

in which the input $\boldsymbol{u}(k)$ is zero for all time instants k. Taking $k = 0$ in (6.105) gives

$$\boldsymbol{x}(1) = \boldsymbol{A}\boldsymbol{x}(0)$$

Likewise, taking $k = 1$ in (6.105) gives

$$\boldsymbol{x}(2) = \boldsymbol{A}\boldsymbol{x}(1) = \boldsymbol{A}^2\boldsymbol{x}(0)$$

and we readily deduce that in general

$$\boldsymbol{x}(k) = \boldsymbol{A}^k\boldsymbol{x}(0) \quad (k \geqslant 0) \tag{6.106}$$

Equation (6.106) represents the solution of (6.105), and is analogous to (6.67) for the continuous-time case. We define the **transition matrix $\boldsymbol{\Phi}(k)$** of the discrete-time system (6.105) by

$$\boldsymbol{\Phi}(k) = \boldsymbol{A}^k$$

and it is the unique matrix satisfying

$$\boldsymbol{\Phi}(k+1) = \boldsymbol{A}\boldsymbol{\Phi}(k), \qquad \boldsymbol{\Phi}(0) = \boldsymbol{I}$$

where $\boldsymbol{I}$ is the identity matrix.

Since $\boldsymbol{A}$ is a constant matrix, the methods discussed in Section 6.7 are applicable for evaluating the transition matrix. From (6.34a),

$$\boldsymbol{A}^k = \alpha_0(k)\boldsymbol{I} + \alpha_1(k)\boldsymbol{A} + \alpha_2(k)\boldsymbol{A}^2 + \ldots + \alpha_{n-1}(k)\boldsymbol{A}^{n-1} \tag{6.107}$$

where, using (6.34b), the $\alpha_i(k)$ ($k = 0, \ldots, n-1$) are obtained by solving simultaneously the n equations

$$\lambda_j^k = \alpha_0(k) + \alpha_1(k)\lambda_j + \alpha_2(k)\lambda_j^2 + \ldots + \alpha_{n-1}(k)\lambda_j^{n-1} \tag{6.108}$$

where λ_j ($j = 1, 2, \ldots, n$) are the eigenvalues of $\boldsymbol{A}$. As in Section 6.7, if $\boldsymbol{A}$ has repeated eigenvalues then derivatives of λ^k with respect to λ will have to be used. The method for determining $\boldsymbol{A}^k$ is thus very similar to that used for evaluating $e^{\boldsymbol{A}t}$ in Section 6.9.3.

EXAMPLE 6.42 Obtain the response of the second-order unforced discrete-time system

$$\boldsymbol{x}(k+1) = \begin{bmatrix} x_1(k) \\ x_2(k) \end{bmatrix} = \begin{bmatrix} \frac{1}{2} & 0 \\ -1 & \frac{1}{3} \end{bmatrix} \boldsymbol{x}(k)$$

subject to $\boldsymbol{x}(0) = [1 \quad 1]^{\mathrm{T}}$.

Solution In this case the system matrix is

$$\boldsymbol{A} = \begin{bmatrix} \frac{1}{2} & 0 \\ -1 & \frac{1}{3} \end{bmatrix}$$

having eigenvalues $\lambda_1 = \frac{1}{2}$ and $\lambda_2 = \frac{1}{3}$. Since $\boldsymbol{A}$ is a 2×2 matrix, it follows from (6.107) that

$$\boldsymbol{A}^k = \alpha_0(k)\boldsymbol{I} + \alpha_1(k)\boldsymbol{A}$$

with $\alpha_0(k)$ and $\alpha_1(k)$ given from (6.108),

$$\lambda_j^k = \alpha_0(k) + \alpha_1(k)\lambda_j \quad (j = 1, 2)$$

Solving the resulting two equations

$$(\tfrac{1}{2})^k = \alpha_0(k) + (\tfrac{1}{2})\alpha_1(k), \quad (\tfrac{1}{3})^k = \alpha_0(k) + (\tfrac{1}{3})\alpha_1(k)$$

for $\alpha_0(k)$ and $\alpha_1(k)$ gives

$$\alpha_0(k) = 3(\tfrac{1}{3})^k - 2(\tfrac{1}{2})^k, \quad \alpha_1(k) = 6[(\tfrac{1}{2})^k - (\tfrac{1}{3})^k]$$

Thus the transition matrix is

$$\boldsymbol{\Phi}(k) = \boldsymbol{A}^k = \begin{bmatrix} (\tfrac{1}{2})^k & 0 \\ 6[(\tfrac{1}{3})^k - (\tfrac{1}{2})^k] & (\tfrac{1}{3})^k \end{bmatrix}$$

Note that $\boldsymbol{\Phi}(0) = \boldsymbol{I}$, as required.

Then from (6.106) the solution of the unforced system is

$$\boldsymbol{x}(k+1) = \begin{bmatrix} (\tfrac{1}{2})^k & 0 \\ 6[(\tfrac{1}{3})^k - (\tfrac{1}{2})^k] & (\tfrac{1}{3})^k \end{bmatrix} \begin{bmatrix} 1 \\ 1 \end{bmatrix} = \begin{bmatrix} (\tfrac{1}{2})^k \\ 7(\tfrac{1}{3})^k - 6(\tfrac{1}{2})^k \end{bmatrix}$$

Having determined the solution of the unforced system, it can be shown that the solution of the state equation (6.103a) for the forced system with input $\boldsymbol{u}(k)$, analogous to the solution given in (6.68) for the continuous-time system

$$\dot{\boldsymbol{x}} = \boldsymbol{A}\boldsymbol{x} + \boldsymbol{B}\boldsymbol{u}$$

is

$$\boldsymbol{x}(k) = \boldsymbol{A}^k \boldsymbol{x}(0) + \sum_{j=0}^{k-1} \boldsymbol{A}^{k-j-1}\boldsymbol{B}\boldsymbol{u}(j) \tag{6.109}$$

Having obtained the solution of the state equation, the system output or response $\boldsymbol{y}(k)$ is obtained from (6.103b) as

$$\boldsymbol{y}(k) = \boldsymbol{C}\boldsymbol{A}^k \boldsymbol{x}(0) + \boldsymbol{C}\sum_{j=0}^{k-1} \boldsymbol{A}^{k-j-1}\boldsymbol{B}\boldsymbol{u}(j) + \boldsymbol{D}\boldsymbol{u}(k) \tag{6.110}$$

In Section 6.9.5 we saw how the Laplace transform could be used to solve the state-space equations in the case of continuous-time systems. In a similar manner, z transforms can be used to solve the equations for discrete-time systems.

Defining $\mathscr{Z}\{\boldsymbol{x}(k)\} = \boldsymbol{X}(z)$ and $\mathscr{Z}\{\boldsymbol{u}(k)\} = \boldsymbol{U}(z)$ and taking z transforms throughout in the equation

$$\boldsymbol{x}(k+1) = \boldsymbol{A}\boldsymbol{x}(k) + \boldsymbol{B}\boldsymbol{u}(k)$$

gives

$$z\boldsymbol{X}(z) - z\boldsymbol{x}(0) = \boldsymbol{A}\boldsymbol{X}(z) + \boldsymbol{B}\boldsymbol{U}(z)$$

which, on rearranging gives

$$(z\boldsymbol{I} - \boldsymbol{A})X(z) = z\boldsymbol{x}(0) + \boldsymbol{B}U(z)$$

where $\boldsymbol{I}$ is the identity matrix. Premultiplying by $(z\boldsymbol{I} - \boldsymbol{A})^{-1}$ gives

$$X(z) = z(z\boldsymbol{I} - \boldsymbol{A})^{-1}\boldsymbol{x}(0) + (z\boldsymbol{I} - \boldsymbol{A})^{-1}\boldsymbol{B}U(z) \tag{6.111}$$

Taking inverse z transforms gives the response as

$$\boldsymbol{x}(k) = \mathscr{Z}^{-1}\{X(z)\} = \mathscr{Z}^{-1}\{z(z\boldsymbol{I} - \boldsymbol{A})^{-1}\}\boldsymbol{x}(0) + \mathscr{Z}^{-1}\{(z\boldsymbol{I} - \boldsymbol{A})^{-1}\boldsymbol{B}U(z)\} \tag{6.112}$$

which corresponds to (6.75) in the continuous-time case.

On comparing the solution (6.112) with that given in (6.109), we see that the transition matrix $\boldsymbol{\Phi}(t) = \boldsymbol{A}^k$ may also be written in the form

$$\boldsymbol{\Phi}(t) = \boldsymbol{A}^k = \mathscr{Z}^{-1}\{z(z\boldsymbol{I} - \boldsymbol{A})^{-1}\}$$

This is readily confirmed from (6.111), since on expanding $z(z\boldsymbol{I} - \boldsymbol{A})^{-1}$ by the binomial theorem, we have

$$z(z\boldsymbol{I} - \boldsymbol{A})^{-1} = \boldsymbol{I} + \frac{\boldsymbol{A}}{z} + \frac{\boldsymbol{A}^2}{z^2} + \dots + \frac{\boldsymbol{A}^r}{z^r} + \dots$$

$$= \sum_{r=0}^{\infty} \frac{\boldsymbol{A}^r}{z^r} = \mathscr{Z}\{\boldsymbol{A}^k\}$$

EXAMPLE 6.43 Using the z-transform approach, obtain an expression for the state $\boldsymbol{x}(k)$ of the system characterized by the state equation

$$\boldsymbol{x}(k+1) = \begin{bmatrix} 2 & 5 \\ -3 & -6 \end{bmatrix}\boldsymbol{x}(k) + \begin{bmatrix} 1 \\ 1 \end{bmatrix}u(k) \quad (k \geqslant 0)$$

when the input is the unit step function

$$u(k) = \begin{cases} 0 & (k < 0) \\ 1 & (k \geqslant 0) \end{cases}$$

and subject to the initial condition $\boldsymbol{x}(0) = [1 \quad -1]^{\mathrm{T}}$.

Solution In this case

$$\boldsymbol{A} = \begin{bmatrix} 2 & 5 \\ -3 & -6 \end{bmatrix} \quad \text{so} \quad z\boldsymbol{I} - \boldsymbol{A} = \begin{bmatrix} z-2 & -5 \\ 3 & z+6 \end{bmatrix}$$

giving

$$(z\boldsymbol{I} - \boldsymbol{A})^{-1} = \frac{1}{(z+1)(z+3)}\begin{bmatrix} z+6 & 5 \\ -3 & z-2 \end{bmatrix}$$

$$= \begin{bmatrix} \dfrac{\frac{5}{2}}{z+1} - \dfrac{\frac{3}{2}}{z+3} & \dfrac{\frac{5}{2}}{z+1} - \dfrac{\frac{5}{2}}{z+3} \\ \dfrac{-\frac{3}{2}}{z+1} + \dfrac{\frac{3}{2}}{z+3} & \dfrac{-\frac{3}{2}}{z+1} + \dfrac{\frac{5}{2}}{z+3} \end{bmatrix}$$

Then

$$\mathscr{Z}^{-1}\{z(z\boldsymbol{I} - \boldsymbol{A})^{-1}\} = \mathscr{Z}^{-1}\begin{bmatrix} \frac{5}{2}\dfrac{z}{z+1} - \frac{3}{2}\dfrac{z}{z+3} & \frac{5}{2}\dfrac{z}{z+1} - \frac{5}{2}\dfrac{z}{z+3} \\ -\frac{3}{2}\dfrac{z}{z+1} + \frac{3}{2}\dfrac{z}{z+3} & -\frac{3}{2}\dfrac{z}{z+1} + \frac{5}{2}\dfrac{z}{z+3} \end{bmatrix}$$

$$= \begin{bmatrix} \frac{5}{2}(-1)^k - \frac{3}{2}(-3)^k & \frac{5}{2}(-1)^k - \frac{5}{2}(-3)^k \\ -\frac{3}{2}(-1)^k + \frac{3}{2}(-3)^k & -\frac{3}{2}(-1)^k + \frac{5}{2}(-3)^k \end{bmatrix}$$

so that, with $\boldsymbol{x}(0) = [1 \quad -1]^{\mathrm{T}}$, the first term in the solution (6.112) becomes

$$\mathscr{Z}^{-1}\{z(z\boldsymbol{I} - \boldsymbol{A})^{-1}\}\boldsymbol{x}(0) = \begin{bmatrix} (-3)^k \\ -(-3)^k \end{bmatrix} \tag{6.113}$$

Since $U(z) = \mathscr{Z}\{u(k)\} = z/(z-1)$,

$$(z\boldsymbol{I} - \boldsymbol{A})^{-1}\boldsymbol{B}U(z) = \frac{1}{(z+1)(z+3)}\begin{bmatrix} z+6 & 5 \\ -3 & z-2 \end{bmatrix}\begin{bmatrix} 1 \\ 1 \end{bmatrix}\frac{z}{z-1}$$

$$= \frac{z}{(z-1)(z+1)(z+3)}\begin{bmatrix} z+11 \\ z-5 \end{bmatrix}$$

$$= \begin{bmatrix} \frac{3}{2}\dfrac{z}{z-1} - \frac{5}{2}\dfrac{z}{z+1} + \dfrac{z}{z+3} \\ -\frac{1}{2}\dfrac{z}{z-1} + \frac{3}{2}\dfrac{z}{z+1} - \dfrac{z}{z+3} \end{bmatrix}$$

so that the second term in the solution (6.112) becomes

$$\mathscr{Z}^{-1}\{(z\boldsymbol{I} - \boldsymbol{A})^{-1}\boldsymbol{B}U(z)\} = \begin{bmatrix} \frac{3}{2} - \frac{5}{2}(-1)^k + (-3)^k \\ -\frac{1}{2} + \frac{3}{2}(-1)^k - (-3)^k \end{bmatrix} \tag{6.114}$$

Combining (6.113) and (6.114), the response $\boldsymbol{x}(k)$ is given by

$$\boldsymbol{x}(k) = \begin{bmatrix} \frac{3}{2} - \frac{5}{2}(-1)^k + 2(-3)^k \\ -\frac{1}{2} + \frac{3}{2}(-1)^k - 2(-3)^k \end{bmatrix}$$

6.10.3 Exercises

67 Use z transforms to determine $\boldsymbol{A}^k$ for the matrices

(a) $\begin{bmatrix} 0 & 1 \\ 4 & 0 \end{bmatrix}$ (b) $\begin{bmatrix} -1 & 3 \\ 3 & -1 \end{bmatrix}$ (c) $\begin{bmatrix} -1 & 1 \\ 0 & -1 \end{bmatrix}$

68 Solve the discrete-time system specified by

$$x(k+1) = -7x(k) + 4y(k)$$
$$y(k+1) = -8x(k) + y(k)$$

with $x(0) = 1$ and $y(0) = 2$, by writing it in the form $\boldsymbol{x}(k+1) = \boldsymbol{A}\boldsymbol{x}(k)$. Use your answer to calculate $\boldsymbol{x}(1)$ and $\boldsymbol{x}(2)$, and check your answers by calculating $x(1)$, $y(1)$, $x(2)$, $y(2)$ directly from the given difference equations.

69 Using the z-transform approach, obtain an expression for the state $\boldsymbol{x}(k)$ of the system characterized by the state equation

$$\boldsymbol{x}(k+1) = \begin{bmatrix} 0 & 1 \\ -0.16 & -1 \end{bmatrix} \boldsymbol{x}(k) + \begin{bmatrix} 1 \\ 1 \end{bmatrix} u(k)$$

when the input is the unit step function

$$u(k) = \begin{cases} 0 & (k < 0) \\ 1 & (k \geqslant 0) \end{cases}$$

and subject to the initial condition $\boldsymbol{x}(0) = [1 \quad -1]^{\mathrm{T}}$.

70 The difference equation

$$y(k+2) = y(k+1) + y(k)$$

with $y(0) = 0$, and $y(1) = 1$ generates the **Fibonacci sequence** $\{y(k)\}$, which occurs in many practical situations. Taking $x_1(k) = y(k)$ and $x_2(k) = y(k+1)$, express the difference equation in state-space form and hence obtain a general expression for $y(k)$. Show that as $k \to \infty$ the ratio $y(k+1)/y(k)$ tends to the constant $\frac{1}{2}(\sqrt{5} + 1)$. This is the so-called **Golden Ratio**, which has intrigued mathematicians for centuries because of its strong influence on art and architecture. The Golden rectangle, that is, one whose two sides are in this ratio, is one of the most visually satisfying of all geometric forms.

6.11 Engineering application: capacitor microphone

Many smaller portable tape recorders have a capacitor microphone built in, since such a system is simple and robust. It works on the principle that if the distance between the plates of a capacitor changes then the capacitance changes in a known manner, and these changes induce a current in an electric circuit. This current can then be amplified or stored. The basic system is illustrated in Figure 6.13. There

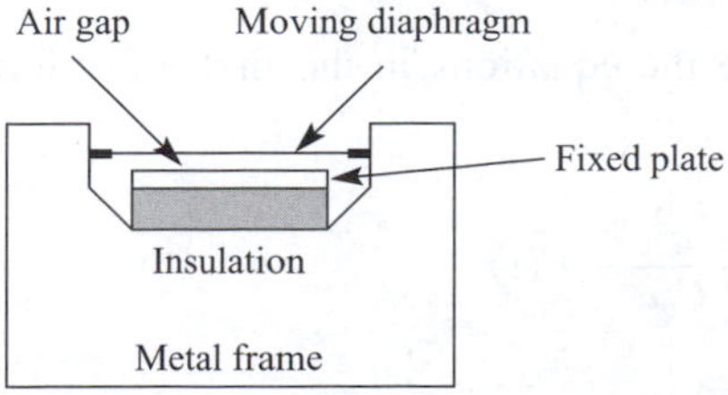

Figure 6.13 Capacitor microphone.

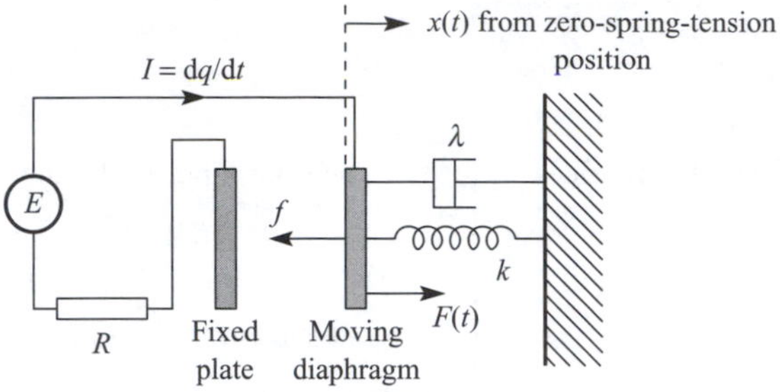

Figure 6.14 Capacitor microphone model.

is a small air gap (about 0.02 mm) between the moving diaphragm and the fixed plate. Sound waves falling on the diaphragm cause vibrations and small variations in the capacitance C; these are certainly sufficiently small that the equations can be *linearized.*

We assume that the diaphragm has mass m and moves as a single unit so that its motion is one-dimensional. The housing of the diaphragm is modelled as a spring-and-dashpot system. The plates are connected through a simple circuit containing a resistance and an imposed steady voltage from a battery. Figure 6.14 illustrates the model. The distance $x(t)$ is measured from the position of zero spring tension, F is the imposed force and f is the force required to hold the moving plate in position against the electrical attraction. The mechanical motion is governed by Newton's equation

$$m\ddot{x} = -kx - \lambda\dot{x} - f + F \tag{6.115}$$

and the electrical circuit equation gives

$$E = RI + \frac{q}{C}, \quad \text{with} \quad \frac{\mathrm{d}q}{\mathrm{d}t} = I \tag{6.116}$$

The variation of capacitance C with x is given by the standard formula

$$C = \frac{C_0 a}{a + x}$$

where a is the equilibrium distance between the plates. The force f is not so obvious, but the following assumption is standard

$$f = \frac{1}{2}q^2\frac{\mathrm{d}}{\mathrm{d}x}\left(\frac{1}{C}\right) = \frac{1}{2}\frac{q^2}{C_0 a}$$

It is convenient to write the equations in the first-order form

$$\dot{x} = v$$

$$m\dot{v} = -kx - \lambda v - \frac{1}{2}\frac{q^2}{C_0 a} + F(t)$$

$$R\dot{q} = -\frac{q(a + x)}{aC_0} + E$$

Furthermore, it is convenient to non-dimensionalize the equations. While it is obvious how to do this for the distance and velocity, for the time and the charge it is less so. There are three natural time scales in the problem: the electrical time $\tau_1 = RC_0$, the spring time $\tau_2^2 = m/k$ and the damping time $\tau_3 = m/\lambda$. Choosing to non-dimensionalize the time with respect to τ_1, the non-dimensionalization of the charge follows:

$$\tau = \frac{t}{\tau_1}, \quad X = \frac{x}{a}, \quad V = \frac{v}{ka/\lambda}, \quad Q = \frac{q}{\sqrt{(2C_0ka^2)}}$$

Then, denoting differentiation with respect to τ by a prime, the equations are

$$X' = \frac{RC_0k}{\lambda}V$$

$$\frac{m}{\lambda RC_0}V' = -X - V - Q^2 + \frac{F}{ka}$$

$$Q' = -Q(1 + X) + \frac{EC_0}{\sqrt{(2C_0ka^2)}}$$

There are four non-dimensional parameters: the external force divided by the spring force gives the first, $G = F/ka$; the electrical force divided by the spring force gives the second, $D^2 = (E^2C_0/2a)/ka$; and the remaining two are

$$A = \frac{RC_0k}{\lambda} = \frac{\tau_1\tau_3}{\tau_2^2}, \qquad B = \frac{m}{\lambda RC_0} = \frac{\tau_3}{\tau_1}$$

The final equations are therefore

$$\left.\begin{aligned} X' &= AV \\ BV' &= -X - V - Q^2 + G \\ Q' &= -Q(1 + X) + D \end{aligned}\right\} \tag{6.117}$$

In equilibrium, with no driving force, $G = 0$ and $V = X' = V' = Q' = 0$, so that

$$\left.\begin{aligned} Q^2 + X &= 0 \\ Q(1 + X) - D &= 0 \end{aligned}\right\} \tag{6.118}$$

or, on eliminating Q,

$$X(1 + X)^2 = -D^2$$

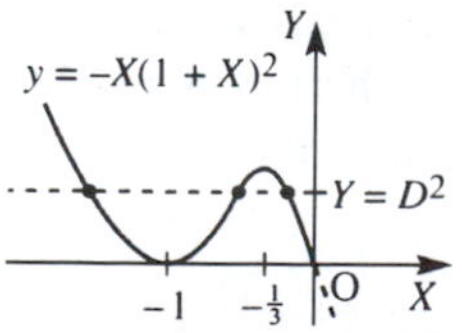

Figure 6.15 Solutions to equations (6.118).

From Figure 6.15, we see that there is always one solution for $X < -1$, or equivalently $x < -a$. The implication of this solution is that the plates have crossed. This is clearly impossible, so the solution is discarded on physical grounds. There are two other solutions if

$$D^2 < \tfrac{1}{3}\left(\tfrac{4}{3}\right)^2 = \tfrac{4}{27}$$

or

$$\frac{E^2 C_0}{2ka^2} < \frac{4}{27} \tag{6.119}$$

We can interpret this statement as saying that the electrical force must not be too strong, and (6.119) gives a precise meaning to what 'too strong' means. There are two physically satisfactory equilibrium solutions $-\frac{1}{3} < X_1 < 0$ and $-1 < X_2 < -\frac{1}{3}$, and the only question left is whether they are stable or unstable.

Stability is determined by small oscillations about the two values X_1 and X_2, where these values satisfy (6.118). Writing

$$X = X_i + \varepsilon, \qquad Q = Q_i + \eta, \qquad V = \theta$$

and substituting into (6.117), neglecting terms in ε^2, y^2, θ^2, $\varepsilon\theta$ and so on, gives

$$\left.\begin{aligned} \varepsilon' &= A\theta \\ B\theta' &= -\varepsilon - \theta - 2Q_i\eta \\ \eta' &= -Q_i\varepsilon - (1 + X_i)\eta \end{aligned}\right\} \tag{6.120}$$

Equations (6.120) are the linearized versions of (6.117) about the equilibrium values. To test for stability, we put $G = 0$ and $\varepsilon = L\,\mathrm{e}^{\alpha\tau}$, $\theta = M\,\mathrm{e}^{\alpha\tau}$, $\eta = N\,\mathrm{e}^{\alpha\tau}$ into (6.120):

$$L\alpha = AM$$

$$BM\alpha = -L - M - 2Q_iN$$

$$N\alpha = -Q_iL - (1 + X_i)N$$

which can be written in the matrix form

$$\alpha\begin{bmatrix} L \\ M \\ N \end{bmatrix} = \begin{bmatrix} 0 & A & 0 \\ -1/B & -1/B & -2Q_i/B \\ -Q_i & 0 & -(1+X_i) \end{bmatrix}\begin{bmatrix} L \\ M \\ N \end{bmatrix}$$

Thus the fundamental stability problem is an eigenvalue problem, a result common to all vibrational stability problems. The equations have non-trivial solutions if

$$0 = \begin{vmatrix} -\alpha & A & B \\ -1/B & -(1/B) - \alpha & -2Q_i/B \\ -Q_i & 0 & -(1+X_i) - \alpha \end{vmatrix}$$

$$= -[B\alpha^3 + (B(1 + X_i) + 1)\alpha^2 + (1 + X_i + A)\alpha + A(1 + X_i - 2Q_i^2)]/B$$

For stability, α must have a negative real part, so that the vibrations damp out, and the Routh–Hurwitz criterion (Section 2.6.2) gives the conditions for this to be the case. Each of the coefficients must be positive, and for the first three

$$B > 0, \qquad B(1 + X_i) + 1 > 0, \qquad 1 + X_i + A > 0$$

are obviously satisfied since $-1 < X_i < 0$. The next condition is

$$A(1 + X_i - 2Q_i^2) > 0$$

which, from (6.118), gives

$$1 + 3X_i > 0, \qquad \text{or} \qquad X_i > -\tfrac{1}{3}$$

Thus the only solution that can possibly be stable is the one for which $X_i > -\frac{1}{3}$; the other solution is unstable. There is one final condition to check,

$$[B(1 + X_i) + 1](1 + X_i + A) - BA(1 + X_i - 2Q_i^2) > 0$$

or

$$B(1 + X_i)^2 + 1 + X_i + A + 2BAQ_2^2 > 0$$

Since all the terms are positive, the solution $X_i > \frac{1}{3}$ is indeed a stable solution.

Having established the stability of one of the positions of the capacitor diaphragm, the next step is to look at the response of the microphone to various inputs. The characteristics can most easily be checked by looking at the frequency response, which is the system response to an individual input $G = b\,e^{j\omega t}$, as the frequency ω varies. This will give information of how the electrical output behaves and for which range of frequencies the response is reasonably flat.

The essential point of this example is to show that a practical vibrational problem gives a stability problem that involves eigenvalues and a response that involves a matrix inversion. The same behaviour is observed for more complicated vibrational problems.

6.12 Engineering application: pole placement

In Chapters 2 and 3 we examined the behaviour of linear systems in both continuous time and discrete time, concentrating on the transform-domain representation, using the Laplace or z transform as appropriate. In this chapter we have examined the same types of system, this time modelled in the form of vector–matrix differential or difference equations. So far we have concentrated on system *analysis*, that is, the question 'Given the system, how does it behave?' In this section we turn our attention briefly to consider the design or synthesis problem, and while it is not possible to produce an exhaustive treatment, it is intended to give the reader an appreciation of the role of mathematics in this task.

6.12.1 Poles and eigenvalues

By now the reader should be convinced that there is an association between system poles as deduced from the system transfer function and the eigenvalues of the system matrix in state-space form. Thus, for example, the system modelled by the second-order differential equation

$$\frac{d^2y}{dt^2} + \frac{1}{2}\frac{dy}{dt} - \frac{1}{2}y = u$$

has transfer function

$$G(s) = \frac{1}{s^2 + \frac{1}{2}s - \frac{1}{2}}$$

The system can also be represented in the state-space form

$$\dot{\boldsymbol{x}} = \boldsymbol{A}\boldsymbol{x} + \boldsymbol{b}u, \qquad y = \boldsymbol{c}^{\mathrm{T}}\boldsymbol{x} \tag{6.121}$$

where

$$\boldsymbol{x} = [x_1 \quad x_2]^{\mathrm{T}}, \qquad \boldsymbol{A} = \begin{bmatrix} 0 & 1 \\ \frac{1}{2} & -\frac{1}{2} \end{bmatrix}, \qquad \boldsymbol{b} = [0 \quad 1]^{\mathrm{T}}, \qquad \boldsymbol{c} = [1 \quad 0]^{\mathrm{T}}$$

It is easy to check that the poles of the transfer function $G(s)$ are at $s = -1$ and $s = \frac{1}{2}$, and that these values are also the eigenvalues of the matrix $\boldsymbol{A}$. Clearly this is an unstable system, with the pole or eigenvalue corresponding to $s = \frac{1}{2}$ located in the right-half of the complex plane. In Section 6.12.2 we examine a method of moving this unstable pole to a new location, thus providing a method of overcoming the stability problem.

6.12.2 The pole placement or eigenvalue location technique

We now examine the possibility of introducing **state feedback** into the system. To do this, we use as system input

$$u = \boldsymbol{k}^{\mathrm{T}}\boldsymbol{x} + u_{\mathrm{ext}}$$

where $\boldsymbol{k} = [k_1 \quad k_2]^{\mathrm{T}}$ and u_{ext} is the external input. The state equation in (6.121) then becomes

$$\dot{\boldsymbol{x}} = \begin{bmatrix} 0 & 1 \\ \frac{1}{2} & -\frac{1}{2} \end{bmatrix}\boldsymbol{x} + \begin{bmatrix} 0 \\ 1 \end{bmatrix}[(k_1x_1 + k_2x_2) + u_{\mathrm{ext}}]$$

That is,

$$\dot{\boldsymbol{x}} = \begin{bmatrix} 0 & 1 \\ k_1 + \frac{1}{2} & k_2 - \frac{1}{2} \end{bmatrix}\boldsymbol{x} + \begin{bmatrix} 0 \\ 1 \end{bmatrix}u_{\mathrm{ext}}$$

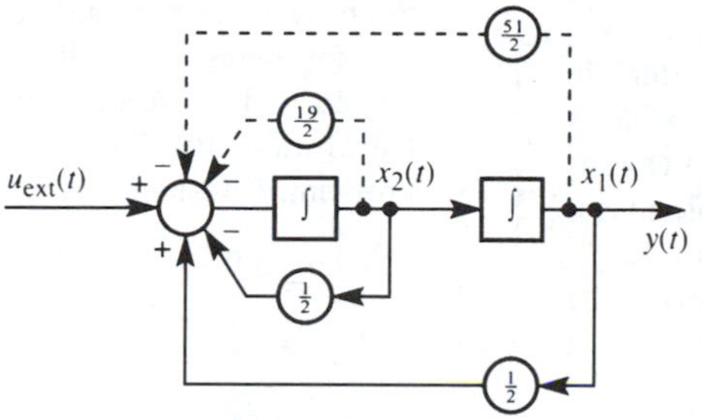

Figure 6.16 Feedback connections for eigenvalue location.

Calculating the characteristic equation of the new system matrix, we find that the eigenvalues are given by the roots of

$$\lambda^2 - (k_2 - \tfrac{1}{2})\lambda - (k_1 + \tfrac{1}{2}) = 0$$

Suppose that we not only wish to stabilize the system, but also wish to improve the response time. This could be achieved if both eigenvalues were located at (say) $\lambda = -5$, which would require the characteristic equation to be

$$\lambda^2 + 10\lambda + 25 = 0$$

In order to make this pole relocation, we should choose

$$-(k_2 - \tfrac{1}{2}) = 10, \qquad -(k_1 + \tfrac{1}{2}) = 25$$

indicating that we take $k_1 = -\frac{51}{2}$ and $k_2 = -\frac{19}{2}$. Figure 6.16 shows the original system and the additional state-feedback connections as dotted lines. We see that for this example at least, it is possible to locate the system poles or eigenvalues wherever we please in the complex plane, by a suitable choice of the vector $\boldsymbol{k}$. This corresponds to the choice of feedback gain, and in practical situations we are of course constrained by the need to specify reasonable values for these. Nevertheless, this technique, referred to as **pole placement**, is a powerful method for system control. There are some questions that remain. For example, can we apply the technique to all systems? Also, can it be extended to systems with more than one input? The following exercises will suggest answers to these questions, and help to prepare the reader for the study of specialist texts.

6.12.3 Exercises

71 An unstable system has Laplace transfer function

$$H(s) = \frac{1}{(s + \frac{1}{2})(s - 1)}$$

Make an appropriate choice of state variables to represent this system in the form

$$\dot{\boldsymbol{x}} = \boldsymbol{A}\boldsymbol{x} + \boldsymbol{b}u, \qquad y = \boldsymbol{c}^{\mathrm{T}}\boldsymbol{x}$$

where

$$\boldsymbol{x} = [x_1 \quad x_2]^{\mathrm{T}}, \qquad \boldsymbol{A} = \begin{bmatrix} 0 & 1 \\ \frac{1}{2} & \frac{1}{2} \end{bmatrix}$$

$$b = [0 \quad 1]^T, \qquad c = [1 \quad 0]^T$$

This particular form of the state-space model in which $\boldsymbol{A}$ takes the companion form and $\boldsymbol{b}$ has a single 1 in the last row is called the **control canonical form** of the system equations, and pole placement is particularly straightforward in this case.

Find a state-variable feedback control of the form $u = \boldsymbol{k}^T\boldsymbol{x}$ that will relocate both system poles at $s = -4$, thus stabilizing the system.

72 Find the control canonical form of the state-space equations for the system characterized by the transfer function

$$G(s) = \frac{2}{(s+1)(s+\frac{1}{4})}$$

Calculate or (better) simulate the step response of the system, and find a control law that relocates both poles at $s = -5$. Calculate or simulate the step response of the new system. How do the two responses differ?

73 The technique for pole placement can be adapted to multi-input systems in certain cases. Consider the system

$$\dot{\boldsymbol{x}} = \boldsymbol{A}\boldsymbol{x} + \boldsymbol{B}\boldsymbol{u}, \qquad y = \boldsymbol{c}^T\boldsymbol{x}$$

where

$$\boldsymbol{x} = [x_1 \quad x_2]^T, \qquad \boldsymbol{u} = [u_1 \quad u_2]^T$$

$$\boldsymbol{A} = \begin{bmatrix} 0 & 1 \\ 6 & 1 \end{bmatrix}, \quad \boldsymbol{B} = \begin{bmatrix} 1 & 0 \\ 1 & 1 \end{bmatrix}, \quad \boldsymbol{c} = [1 \quad 0]^T$$

Writing $\boldsymbol{B}\boldsymbol{u} = \boldsymbol{b}_1u_1 + \boldsymbol{b}_2u_2$, where $\boldsymbol{b}_1 = [1 \quad 1]^T$ and $\boldsymbol{b}_2 = [0 \quad 1]^T$, enables us to work with each input separately. As a first step, use only the input u_1 to relocate both the system poles at $s = -5$. Secondly, use input u_2 only to achieve the same result. Note that we can use either or both inputs to obtain any pole locations we choose, subject of course to physical constraints on the size of the feedback gains.

74 The bad news is that it is not always possible to use the procedure described in Exercise 73. In the first place, it assumes that a full knowledge of the state vector $\boldsymbol{x}(t)$ is available. This may not always be the case; however, in many systems this problem can be overcome by the use of an **observer**. For details, a specialist text on control should be consulted.

There are also circumstances in which the system itself does not permit the use of the technique. Such systems are said to be **uncontrollable**, and the following example, which is more fully discussed in J. G. Reed, *Linear System Fundamentals* (McGraw-Hill, Tokyo, 1983) demonstrates the problem. Consider the system

$$\dot{\boldsymbol{x}} = \begin{bmatrix} 0 & -2 \\ 1 & -3 \end{bmatrix}\boldsymbol{x} + \begin{bmatrix} 2 \\ 1 \end{bmatrix}u$$

with

$$y = [0 \quad 1]\boldsymbol{x}$$

Find the system poles and attempt to relocate both of them, at, say, $s = -2$. It will be seen that no gain vector $\boldsymbol{k}$ can be found to achieve this. Calculating the system transfer function gives a clue to the problem, but Exercise 75 shows how the problem could have been seen from the state-space form of the system.

75 In Exercise 65 it was stated that the system

$$\dot{\boldsymbol{x}} = \boldsymbol{A}\boldsymbol{x} + \boldsymbol{b}u$$
$$y = \boldsymbol{c}^T\boldsymbol{x}$$

where $\boldsymbol{A}$ is an $n \times n$ matrix, is controllable provided that the Kalman matrix

$$\boldsymbol{M} = [\boldsymbol{b} \quad \boldsymbol{A}\boldsymbol{b} \quad \boldsymbol{A}^2\boldsymbol{b} \quad \dots \quad \boldsymbol{A}^{n-1}\boldsymbol{b}]$$

is of rank n. This condition must be satisfied if we are to be able to use the procedure for pole placement. Calculate the Kalman controllability matrix for the system in Exercise 74 and confirm that it has rank less than $n = 2$. Verify that the system of Exercise 71 satisfies the controllability condition.

76 We have noted that when the system equations are expressed in control canonical form, the calculations for pole relocation are particularly easy. The following technique shows how to transform controllable systems into this form. Given the system

$$\dot{\boldsymbol{x}} = \boldsymbol{A}\boldsymbol{x} + \boldsymbol{b}u, \qquad y = \boldsymbol{c}^T\boldsymbol{x}$$

calculate the Kalman controllability matrix $\boldsymbol{M}$, defined in Exercise 75, and its inverse $\boldsymbol{M}^{-1}$. Note that this will only exist for controllable systems. Set $\boldsymbol{v}^T$ as the last row of $\boldsymbol{M}^{-1}$ and form the transformation matrix

$$\boldsymbol{T} = \begin{bmatrix} \boldsymbol{v}^T \\ \boldsymbol{v}^T\boldsymbol{A} \\ \vdots \\ \boldsymbol{v}^T\boldsymbol{A}^{n-1} \end{bmatrix}$$

A transformation of state is now made by introducing the new state vector $z(t) = Tx(t)$, and the resulting system will be in control canonical form. To illustrate the technique, carry out the procedure for the system defined by

$$\dot{x} = \begin{bmatrix} 8 & -2 \\ 35 & -9 \end{bmatrix} x + \begin{bmatrix} 1 \\ 4 \end{bmatrix} u$$

and show that this leads to the system

$$\dot{z} = \begin{bmatrix} 0 & 1 \\ 2 & -1 \end{bmatrix} z + \begin{bmatrix} 0 \\ 1 \end{bmatrix} u$$

Finally, check that the two system matrices have the same eigenvalues, and show that this will always be the case.

6.13 Review exercises (1–22)

1 Obtain the eigenvalues and corresponding eigenvectors of the matrices

(a) $\begin{bmatrix} -1 & 6 & 12 \\ 0 & -13 & 30 \\ 0 & -9 & 20 \end{bmatrix}$ (b) $\begin{bmatrix} 2 & 0 & 1 \\ -1 & 4 & -1 \\ -1 & 2 & 0 \end{bmatrix}$

(c) $\begin{bmatrix} 1 & -1 & 0 \\ -1 & 2 & -1 \\ 0 & -1 & 1 \end{bmatrix}$

2 Find the principal stress values (eigenvalues) and the corresponding principal stress directions (eigenvectors) for the stress matrix

$$T = \begin{bmatrix} 3 & 2 & 1 \\ 2 & 3 & 1 \\ 1 & 1 & 4 \end{bmatrix}$$

Verify that the principal stress directions are mutually orthogonal.

3 Find the values of b and c for which the matrix

$$A = \begin{bmatrix} 2 & -1 & 0 \\ -1 & 3 & b \\ 0 & b & c \end{bmatrix}$$

has $[1 \quad 0 \quad 1]^T$ as an eigenvector. For these values of b and c calculate all the eigenvalues and corresponding eigenvectors of the matrix A.

4 Use Gerschgorin's theorem to show that the largest-modulus eigenvalue λ_1 of the matrix

$$A = \begin{bmatrix} 4 & -1 & 0 \\ -1 & 4 & -1 \\ 0 & -1 & 4 \end{bmatrix}$$

is such that $2 \leqslant |\lambda_1| \leqslant 6$.

Use the power method, with starting vector $x^{(0)} = [-1 \quad 1 \quad -1]^T$ to find λ_1 correct to one decimal place.

5 (a) Using the power method find the dominant eigenvalue and the corresponding eigenvector of the matrix

$$A = \begin{bmatrix} 2 & 1 & 1 \\ 1 & 2.5 & 1 \\ 1 & 1 & 3 \end{bmatrix}$$

starting with an initial vector $[1 \quad 1 \quad 1]^T$ and working to three decimal places.

(b) Given that another eigenvalue A is 1.19 correct to two decimal places, find the value of the third eigenvalue using a property of matrices.

(c) Having determined all the eigenvalues of A, indicate which of these can be obtained by using the power method on the following matrices: (i) A^{-1}; (ii) $A - 3I$.

6 Consider the differential equations

$$\frac{dx}{dt} = 4x + y + z$$

$$\frac{dy}{dt} = 2x + 5y + 4z$$

$$\frac{dz}{dt} = -x - y$$

Show that if it is assumed that there are solutions of the form $x = \alpha e^{\lambda t}$, $y = \beta e^{\lambda t}$ and $z = \gamma e^{\lambda t}$ then the system of equations can be transformed into the eigenvalue problem

$$\begin{bmatrix} 4 & 1 & 1 \\ 2 & 5 & 4 \\ -1 & -1 & 0 \end{bmatrix}\begin{bmatrix} \alpha \\ \beta \\ \gamma \end{bmatrix} = \lambda \begin{bmatrix} \alpha \\ \beta \\ \gamma \end{bmatrix}$$

Show that the eigenvalues for this problem are 5, 3 and 1, and find the eigenvectors corresponding to the smallest eigenvalue.

7 Find the eigenvalues and corresponding eigenvectors for the matrix

$$\boldsymbol{A} = \begin{bmatrix} 8 & -8 & -2 \\ 4 & -3 & -2 \\ 3 & -4 & 1 \end{bmatrix}$$

Write down the modal matrix $\boldsymbol{M}$ and spectral matrix $\boldsymbol{\Lambda}$ of $\boldsymbol{A}$, and confirm that

$$\boldsymbol{M}^{-1}\boldsymbol{A}\boldsymbol{M} = \boldsymbol{\Lambda}$$

8 Show that the eigenvalues of the symmetric matrix

$$\boldsymbol{A} = \begin{bmatrix} 1 & 0 & -4 \\ 0 & 5 & 4 \\ -4 & 4 & 3 \end{bmatrix}$$

are 9, 3 and −3. Obtain the corresponding eigenvectors in normalized form, and write down the normalized modal matrix $\hat{\boldsymbol{M}}$. Confirm that

$$\hat{\boldsymbol{M}}^{\mathrm{T}}\boldsymbol{A}\hat{\boldsymbol{M}} = \boldsymbol{\Lambda}$$

where $\boldsymbol{\Lambda}$ is the spectral matrix of $\boldsymbol{A}$.

9 In a radioactive series consisting of four different nuclides starting with the parent substance N_1 and ending with the stable product N_4 the amounts of each nuclide present at time t are given by the differential equations model

$$\frac{\mathrm{d}N_1}{\mathrm{d}t} = -6N_1$$

$$\frac{\mathrm{d}N_2}{\mathrm{d}t} = 6N_1 - 4N_2$$

$$\frac{\mathrm{d}N_3}{\mathrm{d}t} = 4N_2 - 2N_3$$

$$\frac{\mathrm{d}N_4}{\mathrm{d}t} = 2N_3$$

Express these in the vector–matrix form

$$\dot{\boldsymbol{N}} = \boldsymbol{A}\boldsymbol{N}$$

where $\boldsymbol{N} = [N_1 \quad N_2 \quad N_3 \quad N_4]^{\mathrm{T}}$. Find the eigenvalues and corresponding eigenvectors of $\boldsymbol{A}$. Using the spectral form of the solution, determine $N_4(t)$ given that at time $t = 0$, $N_t = C$ and $N_2 = N_3 = N_4 = 0$.

10 (a) Given

$$\boldsymbol{A} = \begin{bmatrix} 2 & 0 \\ 1 & 1 \end{bmatrix}$$

use the Cayley–Hamilton theorem to find

(i) $\boldsymbol{A}^7 - 3\boldsymbol{A}^6 + \boldsymbol{A}^4 + 3\boldsymbol{A}^3 - 2\boldsymbol{A}^2 + 3\boldsymbol{I}$

(ii) $\boldsymbol{A}^k$, where $k > 0$ is an integer

(b) Using the Cayley–Hamilton theorem, find $\mathrm{e}^{\boldsymbol{A}t}$ when

$$\boldsymbol{A} = \begin{bmatrix} 0 & 1 \\ 0 & -2 \end{bmatrix}$$

11 Show that the matrix

$$\boldsymbol{A} = \begin{bmatrix} 1 & 2 & 3 \\ 0 & 1 & 4 \\ 0 & 0 & 1 \end{bmatrix}$$

has an eigenvalue $\lambda = 1$ with algebraic multiplicity 3. By considering the rank of a suitable matrix, show that there is only one corresponding linearly independent eigenvector $\boldsymbol{e}_1$. Obtain the eigenvector $\boldsymbol{e}_1$ and two further generalized eigenvectors. Write down the corresponding modal matrix $\boldsymbol{M}$ and confirm that $\boldsymbol{M}^{-1}\boldsymbol{A}\boldsymbol{M} = \boldsymbol{J}$, where $\boldsymbol{J}$ is the appropriate Jordan matrix. (*Hint*: In this example care must be taken in applying the procedure to evaluate the generalized eigenvectors to ensure that the triad of vectors takes the form $\{\boldsymbol{T}^2\boldsymbol{\omega}, \boldsymbol{T}\boldsymbol{\omega}, \boldsymbol{\omega}\}$, where $\boldsymbol{T} = \boldsymbol{A} - \lambda\boldsymbol{I}$, with $\boldsymbol{T}^2\boldsymbol{\omega} = \boldsymbol{e}_1$.)

12 The equations of motion of three equal masses connected by springs of equal stiffness are

$$\ddot{x} = -2x + y$$

$$\ddot{y} = x - 2y + z$$

$$\ddot{z} = y - 2z$$

Show that for normal modes of oscillation

$$x = X\cos\omega t, \qquad y = Y\cos\omega t, \qquad z = Z\cos\omega t$$

to exist then the condition on $\lambda = \omega^2$ is

$$\begin{vmatrix} \lambda-2 & 1 & 0 \\ 1 & \lambda-2 & 1 \\ 0 & 1 & \lambda-2 \end{vmatrix} = 0$$

Find the three values of λ that satisfy this condition, and find the ratios $X : Y : Z$ in each case.

13 Classify the following quadratic forms:

(a) $2x^2 + y^2 + 2z^2 - 2xy - 2yz$

(b) $3x^2 + 7y^2 + 2z^2 - 4xy - 4xz$

(c) $16x^2 + 36y^2 + 17z^2 + 32xy + 32xz + 16yz$

(d) $-21x^2 + 30xy - 12xz - 11y^2 + 8yz - 2z^2$

(e) $-x^2 - 3y^2 - 5z^2 + 2xy + 2xz + 2yz$

14 Show that $\boldsymbol{e}_1 = [1 \quad 2 \quad 3]^T$ is an eigenvector of the matrix

$$\boldsymbol{A} = \begin{bmatrix} \frac{7}{2} & -\frac{1}{2} & -\frac{1}{2} \\ 4 & -1 & 0 \\ -\frac{3}{2} & \frac{3}{2} & \frac{1}{2} \end{bmatrix}$$

and find its corresponding eigenvalue. Find the other two eigenvalues and their corresponding eigenvectors.

Write down in spectral form the general solution of the system of differential equations

$$2\frac{dx}{dt} = 7x - y - z$$

$$\frac{dy}{dt} = 4x - y$$

$$2\frac{dz}{dt} = -3x + 3y + z$$

Hence show that if $x = 2$, $y = 4$ and $z = 6$ when $t = 0$ then the solution is

$$x = 2e^t, \qquad y = 4e^t, \qquad z = 6e^t$$

15 A continuous-time system is specified in state-space form as

$$\dot{\boldsymbol{x}}(t) = \boldsymbol{A}\boldsymbol{x}(t) + \boldsymbol{b}u(t)$$

$$y(t) = \boldsymbol{c}^T\boldsymbol{x}(t)$$

where

$$\boldsymbol{A} = \begin{bmatrix} 0 & 6 \\ -1 & -5 \end{bmatrix}, \qquad \boldsymbol{b} = \begin{bmatrix} 0 \\ 1 \end{bmatrix}, \qquad \boldsymbol{c} = \begin{bmatrix} 1 \\ 1 \end{bmatrix}$$

(a) Draw a block diagram to represent the system.

(b) Using Laplace transformations, show that the state transition matrix is given by

$$e^{\boldsymbol{A}t} = \begin{bmatrix} 3e^{-2t} - 2e^{-3t} & 6e^{-2t} - 6e^{-3t} \\ e^{-3t} - e^{-2t} & 3e^{-3t} - 2e^{-2t} \end{bmatrix}$$

(c) Calculate the impulse response of the system, and determine the response $y(t)$ of the system to an input $u(t) = 1$ $(t \geq 0)$, subject to the initial state $\boldsymbol{x}(0) = [1 \quad 0]^T$.

16 Show that the eigenvalues of the matrix

$$\boldsymbol{A} = \begin{bmatrix} 1 & 1 & -2 \\ -1 & 2 & 1 \\ 0 & 1 & -1 \end{bmatrix}$$

are 2, 1 and −1, and find the corresponding eigenvectors. Write down the modal matrix $\boldsymbol{M}$ and spectral matrix $\boldsymbol{\Lambda}$ of $\boldsymbol{A}$, and verify that $\boldsymbol{M\Lambda} = \boldsymbol{AM}$.

Deduce that the system of difference equations

$$\boldsymbol{x}(k + 1) = \boldsymbol{A}\boldsymbol{x}(k)$$

where $\boldsymbol{x}(k) = [x_1(k) \quad x_2(k) \quad x_3(k)]^T$, has a solution

$$\boldsymbol{x}(k) = \boldsymbol{M}\boldsymbol{y}(k)$$

where $\boldsymbol{y}(k) = \boldsymbol{\Lambda}^k\boldsymbol{y}(0)$. Find this solution, given $\boldsymbol{x}(0) = [1 \quad 0 \quad 0]^T$.

17 A linear time-invariant system $(\boldsymbol{A}, \boldsymbol{b}, \boldsymbol{c})$ is modelled by the state-space equations

$$\dot{\boldsymbol{x}}(t) = \boldsymbol{A}\boldsymbol{x}(t) + \boldsymbol{b}u(t)$$

$$y(t) = \boldsymbol{c}^T\boldsymbol{x}(t)$$

where $\boldsymbol{x}(t)$ is the n-dimensional state vector, and $u(t)$ and $y(t)$ are the system input and output respectively. Given that the system matrix $\boldsymbol{A}$ has n distinct non-zero eigenvalues, show that the system equations may be reduced to the canonical form

$$\dot{\boldsymbol{\xi}}(t) = \boldsymbol{\Lambda}\boldsymbol{\xi}(t) + \boldsymbol{b}_1u(t)$$

$$y(t) = \boldsymbol{c}_1^T\boldsymbol{\xi}(t)$$

where $\boldsymbol{\Lambda}$ is a diagonal matrix. What properties of this canonical form determine the controllability and observability of $(\boldsymbol{A}, \boldsymbol{b}, \boldsymbol{c})$?

Reduce to canonical form the system $(\boldsymbol{A}, \boldsymbol{b}, \boldsymbol{c})$ having

$$\boldsymbol{A} = \begin{bmatrix} 1 & 1 & -2 \\ -1 & 2 & 1 \\ 0 & 1 & -1 \end{bmatrix}$$

$$b = \begin{bmatrix} -1 \\ 1 \\ -1 \end{bmatrix} \qquad c = \begin{bmatrix} -2 \\ 1 \\ 0 \end{bmatrix}$$

and comment on its stability, controllability and observability by considering the ranks of the appropriate Kalman matrices $[\boldsymbol{b} \quad \boldsymbol{A}\boldsymbol{b} \quad \boldsymbol{A}^2\boldsymbol{b}]$ and $[\boldsymbol{c} \quad \boldsymbol{A}^T\boldsymbol{c} \quad (\boldsymbol{A}^T)^2\boldsymbol{c}]$.

18 A single-input–single-output system is represented in state-space form, using the usual notation, as

$$\dot{\boldsymbol{x}}(t) = \boldsymbol{A}\boldsymbol{x}(t) + \boldsymbol{b}u(t)$$

$$y(t) = \boldsymbol{c}^T\boldsymbol{x}(t)$$

For

$$\boldsymbol{A} = \begin{bmatrix} -2 & -1 \\ 2 & 0 \end{bmatrix}, \qquad \boldsymbol{b} = \begin{bmatrix} 1 \\ 0 \end{bmatrix}, \qquad \boldsymbol{c} = \begin{bmatrix} 1 \\ 1 \end{bmatrix}$$

show that

$$e^{At} = \begin{bmatrix} e^{-t}(\cos t - \sin t) & -e^{-t}\sin t \\ 2\,e^{-t}\sin t & e^{-t}(\cos t + \sin t) \end{bmatrix}$$

and find $\boldsymbol{x}(t)$ given the $\boldsymbol{x}(0) = 0$ and $u(t) = 1$ $(t \geq 0)$.

Show that the Laplace transfer function of the system is

$$H(s) = \frac{Y(s)}{U(s)} = \boldsymbol{c}(s\boldsymbol{I} - \boldsymbol{A})^{-1}\boldsymbol{b}$$

and find $H(s)$ for this system. What is the system impulse response?

19 The system shown in Figure 6.17 is a realization of a discrete-time system. Show that, with state variables $x_1(k)$ and $x_2(k)$ as shown, the system may be represented as

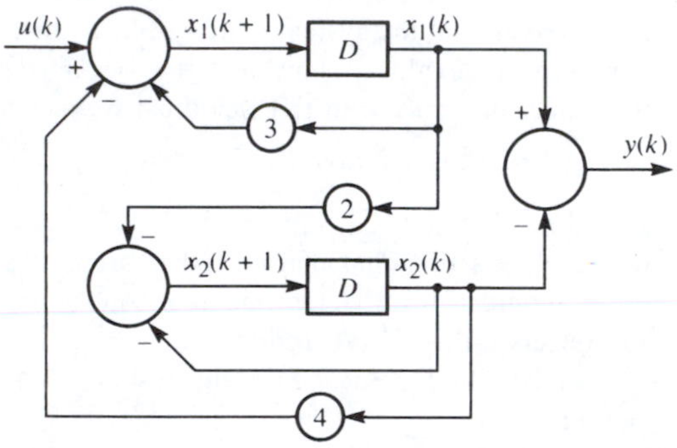

Figure 6.17 Discrete-time system of Review exercise 19.

$$\boldsymbol{x}(k+1) = \boldsymbol{A}\boldsymbol{x}(k) + \boldsymbol{b}u(k)$$
$$y(k) = \boldsymbol{c}^{\mathrm{T}}\boldsymbol{x}(k)$$

where

$$\boldsymbol{A} = \begin{bmatrix} -3 & -4 \\ -2 & -1 \end{bmatrix}, \quad \boldsymbol{b} = \begin{bmatrix} 1 \\ 0 \end{bmatrix}, \quad \boldsymbol{c} = \begin{bmatrix} 1 \\ -1 \end{bmatrix}$$

Calculate the z transfer function of the system, $D(z)$, where

$$D(z) = \boldsymbol{c}(z\boldsymbol{I} - \boldsymbol{A})^{-1}\boldsymbol{b}$$

Reduce the system to control canonical form by the following means:

(i) calculate the controllability matrix $\boldsymbol{M}_c$, where $\boldsymbol{M}_c = [\boldsymbol{b} \quad \boldsymbol{A}\boldsymbol{b}]$ is the matrix with columns $\boldsymbol{b}$ and $\boldsymbol{A}\boldsymbol{b}$;
(ii) show that rank $(\boldsymbol{M}_c) = 2$, and calculate $\boldsymbol{M}_c^{-1}$;
(iii) write down the vector $\boldsymbol{v}^{\mathrm{T}}$ corresponding to the last row of $\boldsymbol{M}_c^{-1}$;
(iv) form the matrix $\boldsymbol{T} = [\boldsymbol{v}^{\mathrm{T}} \quad \boldsymbol{v}^{\mathrm{T}}\boldsymbol{A}]^{\mathrm{T}}$, the matrix with rows $\boldsymbol{v}^{\mathrm{T}}$ and $\boldsymbol{v}^{\mathrm{T}}\boldsymbol{A}$;
(v) calculate $\boldsymbol{T}^{-1}$ and using this matrix $\boldsymbol{T}$, show that the transformation $\boldsymbol{z}(k) = \boldsymbol{T}\boldsymbol{x}(k)$ produces the system

$$\begin{aligned} \boldsymbol{z}(k+1) &= \boldsymbol{T}\boldsymbol{A}\boldsymbol{T}^{-1}\boldsymbol{z}(k) + \boldsymbol{T}\boldsymbol{b}u(k) \\ &= \boldsymbol{C}\boldsymbol{z}(k) + \boldsymbol{b}_c u(k) \end{aligned}$$

where $\boldsymbol{C}$ is of the form

$$\begin{bmatrix} 0 & 1 \\ -\alpha & -\beta \end{bmatrix}$$

and $\boldsymbol{b}_c = [0 \quad 1]^{\mathrm{T}}$. Calculate α and β, and comment on the values obtained in relation to the transfer function $D(z)$.

20 The behaviour of an unforced mechanical system is governed by the differential equation

$$\dot{\boldsymbol{x}}(t) = \begin{bmatrix} 5 & 2 & -1 \\ 3 & 6 & -9 \\ 1 & 1 & 1 \end{bmatrix}\boldsymbol{x}(t), \quad \boldsymbol{x}(0) = \begin{bmatrix} 0 \\ 1 \\ 0 \end{bmatrix}$$

(a) Show that the eigenvalues of the system matrix are 6, 3, 3 and that there is only one linearly independent eigenvector corresponding to the eigenvalue 3. Obtain the eigenvectors corresponding to the eigenvalues 6 and 3 and a further generalized eigenvector for the eigenvalue 3.
(b) Write down a generalized modal matrix $\boldsymbol{M}$ and confirm that

$$\boldsymbol{A}\boldsymbol{M} = \boldsymbol{M}\boldsymbol{J}$$

for an appropriate Jordan matrix $\boldsymbol{J}$.
(c) Using the result

$$\boldsymbol{x}(t) = \boldsymbol{M}e^{\boldsymbol{J}t}\boldsymbol{M}^{-1}\boldsymbol{x}(0)$$

obtain the solution to the given differential equation.

21 A controllable linear plant that can be influenced by one input $u(t)$ is modelled by the differential equation

$$\dot{\boldsymbol{x}}(t) = \boldsymbol{A}\boldsymbol{x}(t) + \boldsymbol{b}u(t)$$

where $\boldsymbol{x}(t) = [x_1(t) \quad x_2(t) \quad \dots \quad x_n(t)]^{\mathrm{T}}$ is the state vector, $\boldsymbol{A}$ is a constant matrix with distinct real eigenvalues $\lambda_1, \lambda_2, \dots, \lambda_n$ and $\boldsymbol{b} = [b_1 \quad b_2 \quad \dots b_n]^{\mathrm{T}}$ is a constant vector. By the application of the feedback control

$$u(t) = K\boldsymbol{v}_K^{\mathrm{T}}\boldsymbol{x}(t)$$

where $\boldsymbol{v}_K$ is the eigenvector of $\boldsymbol{A}^{\mathrm{T}}$ corresponding to the eigenvalue λ_K of $\boldsymbol{A}^{\mathrm{T}}$ (and hence of $\boldsymbol{A}$), the eigenvalue λ_K can be changed to a new real value

ρ_K without altering the other eigenvalues. To achieve this, the feedback gain K is chosen as

$$K = \frac{\rho_K - \lambda_K}{p_K}$$

where $p_K = \boldsymbol{v}_K^T\boldsymbol{b}$.

Show that the system represented by

$$\dot{\boldsymbol{x}}(t) = \begin{bmatrix} 1 & 2 & 0 \\ 0 & -1 & 0 \\ -3 & -3 & -2 \end{bmatrix} \boldsymbol{x}(t) + \begin{bmatrix} 0 \\ 1 \\ 0 \end{bmatrix} u(t)$$

is controllable, and find the eigenvalues and corresponding eigenvectors of the system matrix. Deduce that the system is unstable in the absence of control, and determine a control law that will relocate the eigenvalue corresponding to the unstable mode at the new value -5.

22 (Extended problem) Many vibrational systems are modelled by the vector–matrix differential equation

$$\ddot{\boldsymbol{x}}(t) = \boldsymbol{A}\boldsymbol{x}(t) \tag{1}$$

where $\boldsymbol{A}$ is a constant $n \times n$ matrix and $\boldsymbol{x}(t) = [x_1(t) \quad x_2(t) \ \ldots \ x_n(t)]^{\mathrm{T}}$. By substituting $\boldsymbol{x} = \mathrm{e}^{\lambda t}\boldsymbol{u}$, show that

$$\lambda^2\boldsymbol{u} = \boldsymbol{A}\boldsymbol{u} \tag{2}$$

and that non-trivial solutions for $\boldsymbol{u}$ exist provided that

$$|\boldsymbol{A} - \lambda^2\boldsymbol{I}| = 0 \tag{3}$$

Let $\lambda_1^2, \lambda_2^2, \ldots, \lambda_n^2$ be the solutions of (3) and $\boldsymbol{u}_1, \boldsymbol{u}_2, \ldots, \boldsymbol{u}_n$ the corresponding solutions of (2). Define $\boldsymbol{M}$ to be the matrix having $\boldsymbol{u}_1, \boldsymbol{u}_2, \ldots, \boldsymbol{u}_n$ as its columns and $\boldsymbol{S}$ to be the diagonal matrix having $\lambda_1^2, \lambda_2^2, \ldots, \lambda_n^2$ as its diagonal elements. By applying the transformation $\boldsymbol{x}(t) = \boldsymbol{M}\boldsymbol{q}(t)$, where $\boldsymbol{q}(t) = [q_1(t) \quad q_2(t) \quad \ldots \quad q_n(t)]^{\mathrm{T}}$, to (1), show that

$$\ddot{\boldsymbol{q}} = \boldsymbol{S}\boldsymbol{q} \tag{4}$$

and deduce that (4) has solutions of the form

$$q_i = C_i \sin(\omega_i t + \alpha_i) \tag{5}$$

where c_i and α_i are arbitrary constants and $\lambda_i = \mathrm{j}\omega_i$, with $\mathrm{j} = \sqrt{(-1)}$.

The solutions λ_i^2 of (3) define the **natural frequencies** ω_i of the system. The corresponding solutions q_i given in (5) are called the **normal modes** of the system. The general solution of (1) is then obtained using $\boldsymbol{x}(t) = \boldsymbol{M}\boldsymbol{q}(t)$.

A mass–spring vibrating system is governed by the differential equations

$$\ddot{x}_1(t) = -3x_1(t) + 2x_2(t)$$

$$\ddot{x}_2(t) = x_1(t) - 2x_2(t)$$

with $x_1(0) = 1$ and $x_2(0) = \dot{x}_1(0) = \dot{x}_2(0) = 2$. Determine the natural frequencies and the corresponding normal modes of the system. Hence obtain the general displacement $x_1(t)$ and $x_2(t)$ at time $t \geqslant 0$. Plot graphs of both the normal modes and the general solutions.

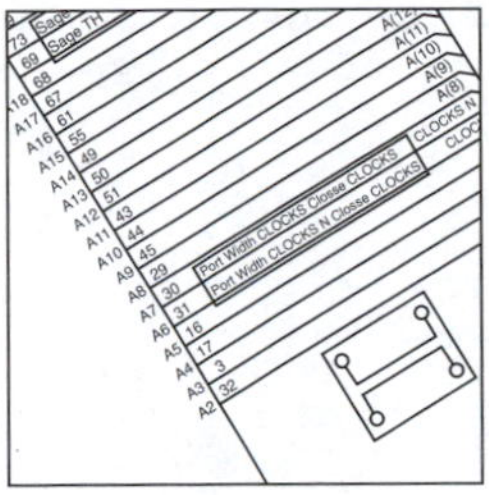

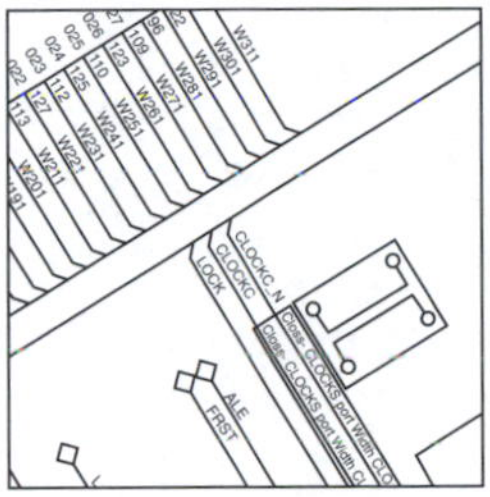

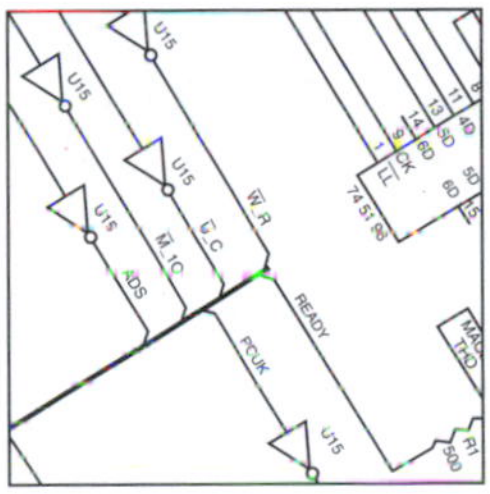

7

Vector Calculus

CONTENTS

7.1 Introduction

In many applications we use functions of the space variable $\boldsymbol{r} = x\boldsymbol{i} + y\boldsymbol{j} + z\boldsymbol{k}$ as models for quantities that vary from point to point in three-dimensional space. There are two types of such functions. There are **scalar point functions**, which model scalar quantities like the temperature at a point in a body, and **vector point functions**, which model vector quantities like the velocity of the flow at a point in a liquid. We can express this more formally in the following way. For each scalar point function f we have a **rule**, $u = f(\boldsymbol{r})$, which assigns to each point with coordinate $\boldsymbol{r}$ in the **domain** of the function a unique real number u. For vector point functions the rule $\boldsymbol{v} = \boldsymbol{F}(\boldsymbol{r})$ assigns to each $\boldsymbol{r}$ a unique vector $\boldsymbol{v}$ in the range of the function. Vector calculus was designed to measure the variation of such functions with respect to the space variable $\boldsymbol{r}$. That development made use of the ideas about vectors (components, addition, subtraction, scalar and vector products) described in Chapter 4 of *Modern Engineering Mathematics* and summarized here in Figure 7.1.

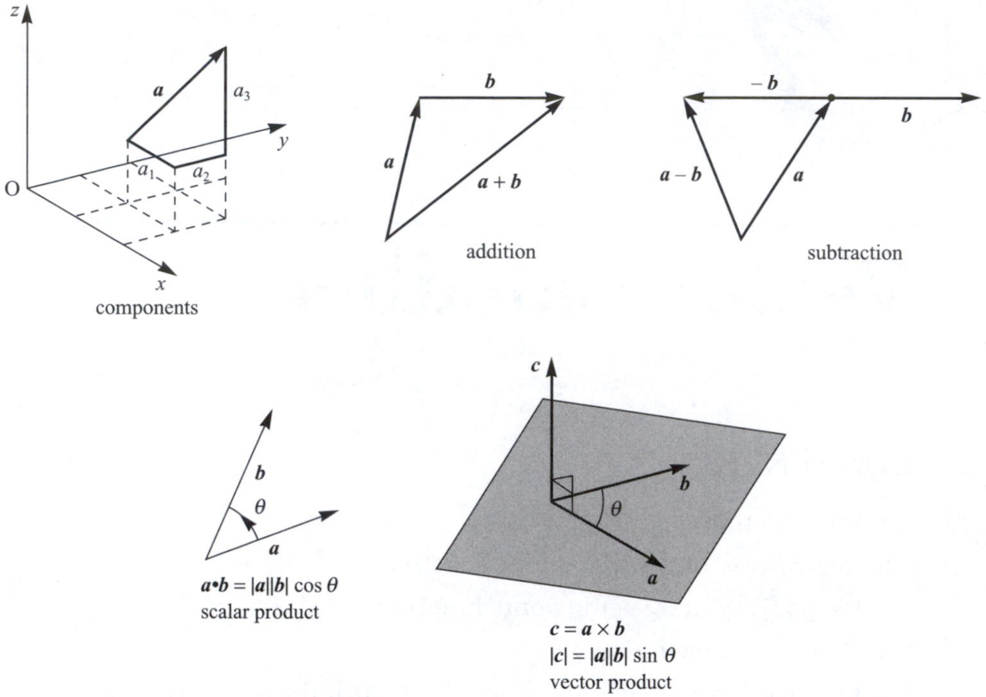

Figure 7.1 Elementary vector algebra.

In component form if $\boldsymbol{a} = (a_1, a_2, a_3)$ and $\boldsymbol{b} = (b_1, b_2, b_3)$ then

$$\boldsymbol{a} \pm \boldsymbol{b} = (a_1 \pm b_1, a_2 \pm b_2, a_3 \pm b_3)$$

$$\boldsymbol{a} \cdot \boldsymbol{b} = (a_1b_1 + a_2b_2 + a_3b_3)$$

$$\boldsymbol{a} \times \boldsymbol{b} = \begin{vmatrix} \boldsymbol{i} & \boldsymbol{j} & \boldsymbol{k} \\ a_1 & a_2 & a_3 \\ b_1 & b_2 & b_3 \end{vmatrix}$$

$$= (a_2b_3 - b_2a_3, b_1a_3 - a_1b_3, a_1b_2 - b_1a_2)$$

The recent development of computer packages for the modelling of engineering problems involving vector quantities has relieved designers of much tedious analysis and computation. To be able to use those packages effectively, however, designers need a good understanding of the mathematical tools they bring to their tasks. It is on that basic understanding that this chapter focuses.

7.1.1 Basic concepts

We can picture a scalar point function $f(\boldsymbol{r})$ by means of its level surfaces $f(\boldsymbol{r})$ = constant. For example, the level surfaces of $f(\boldsymbol{r}) = 2x + 2y - z$ are planes

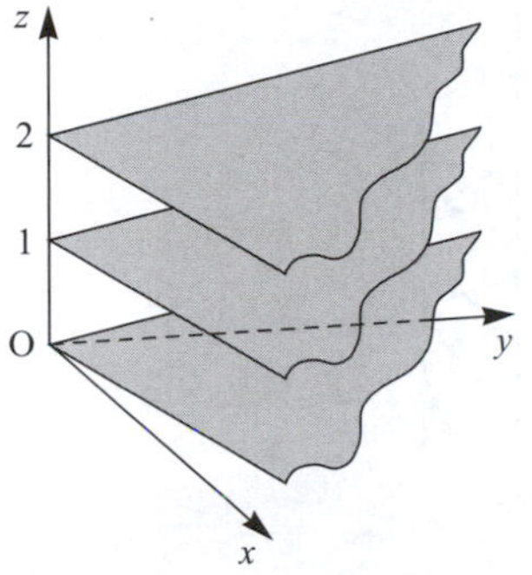

Figure 7.2 Level surfaces of $f(\boldsymbol{r}) = (2, 2, -1) \cdot \boldsymbol{r} = 2x + 2y - z$.

parallel to the plane $z = 2x + 2y$, as shown in Figure 7.2. On the level surface the function value does not change, so the rate of change of the function will be zero along any line drawn on the level surface. An alternative name for a scalar point function is **scalar field**. This is in contrast to the vector point function (or **vector field**). We picture a vector field by its field (or flow) lines. A field line is a curve in space represented by the position vector $\boldsymbol{r}(t)$ such that at each point of the curve its tangent is parallel to the vector field. Thus the field lines of $\boldsymbol{F}(\boldsymbol{r})$ are given by the differential equation

$$\frac{d\boldsymbol{r}}{dt} = \boldsymbol{F}(\boldsymbol{r}), \quad \text{where } \boldsymbol{r}(t_0) = \boldsymbol{r}_0$$

and $\boldsymbol{r}_0$ is the point on the line corresponding to $t = t_0$. This vector equation represents the three simultaneous ordinary differential equations

$$\frac{dx}{dt} = P(x, y, z),$$

$$\frac{dy}{dt} = Q(x, y, z),$$

$$\frac{dz}{dt} = R(x, y, z)$$

where $\boldsymbol{F} = (P, Q, R)$.

Modern computer algebra packages make it easier to draw both the level surfaces of scalar functions and the field lines of vector functions, but to underline the basic ideas we shall consider two simple examples.

EXAMPLE 7.1

Sketch

(a) the level surfaces of the scalar point function $f(\boldsymbol{r}) = z\,e^{-xy}$;

(b) the field lines of the vector point function $\boldsymbol{F}(\boldsymbol{r}) = (y, -x, 1)$.

Solution (a) Consider the level surface given by $f(\boldsymbol{r}) = c$, where c is a number. Then $z\,e^{-xy} = c$ and so $z = c\,e^{xy}$. For c, x and y all positive we can easily sketch part of

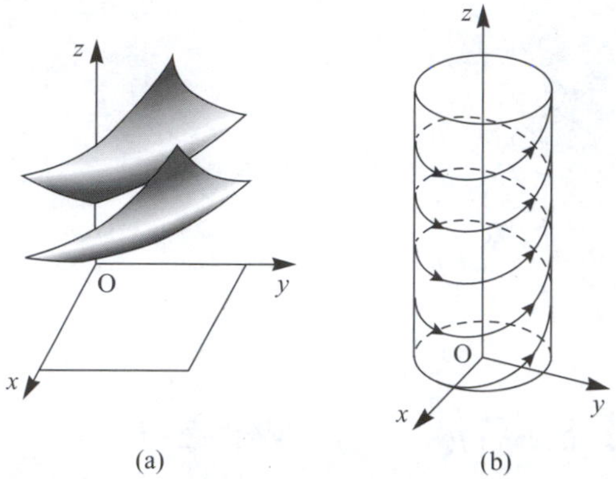

Figure 7.3 (a) Level surfaces of $f(\boldsymbol{r}) = z\,\mathrm{e}^{-xy}$; (b) field lines of $\boldsymbol{F}(\boldsymbol{r}) = (y, -x, 1)$.

the surface as shown in Figure 7.3(a), from which we can deduce the appearance of the whole family of level surfaces.

(b) For the function $\boldsymbol{F}(\boldsymbol{r}) = (y, -x, 1)$ the field lines are given by

$$\frac{\mathrm{d}\boldsymbol{r}}{\mathrm{d}t} = (y, -x, 1)$$

that is, by the simultaneous differential equations

$$\frac{\mathrm{d}x}{\mathrm{d}t} = y, \quad \frac{\mathrm{d}y}{\mathrm{d}t} = -x, \quad \frac{\mathrm{d}z}{\mathrm{d}t} = 1$$

The general solution of these simultaneous equations is

$$x(t) = A\cos t + B\sin t, \quad y(t) = B\cos t - A\sin t, \quad z(t) = t + C$$

where A, B and C are arbitrary constants. Considering, in particular, the field line that passes through (1, 0, 0), we determine the parametric equation

$$(x(t), y(t), z(t)) = (\cos t, -\sin t, t)$$

This represents a circular helix as shown in Figure 7.3(b), from which we can deduce the appearance of the whole family of flow lines.

To investigate the properties of scalar and vector fields further we need to use the calculus of several variables. Here we shall describe the basic ideas and definitions needed for vector calculus. A fuller treatment is given in Chapter 9 of *Modern Engineering Mathematics*.

Given a function $f(x)$ of a single variable x, we measure its rate of change (or gradient) by its derivative with respect to x. This is

$$\frac{\mathrm{d}f}{\mathrm{d}x} = f'(x) = \lim_{\Delta x \to 0} \frac{f(x + \Delta x) - f(x)}{\Delta x}$$

However, a function $f(x, y, z)$ of three independent variables x, y and z does not have a unique rate of change. The value of the latter depends on the direction in which it is measured. The rate of change of the function $f(x, y, z)$ in the x direction is given by its **partial derivative** with respect to x, namely

$$\frac{\partial f}{\partial x} = \lim_{\Delta x \to 0} \frac{f(x + \Delta x, y, z) - f(x, y, z)}{\Delta x}$$

This measures the rate of change of $f(x, y, z)$ with respect to x when y and z are held constant. We can calculate such partial derivatives by differentiating $f(x, y, z)$ with respect to x, treating y and z as constants. Similarly,

$$\frac{\partial f}{\partial y} = \lim_{\Delta y \to 0} \frac{f(x, y + \Delta y, z) - f(x, y, z)}{\Delta y}$$

and

$$\frac{\partial f}{\partial z} = \lim_{\Delta z \to 0} \frac{f(x, y, z + \Delta z) - f(x, y, z)}{\Delta z}$$

define the partial derivatives of $f(x, y, z)$ with respect to y and z respectively.

For conciseness we sometimes use a suffix notation to denote partial derivatives, for example writing f_x for $\partial f/\partial x$. The rules for partial differentiation are essentially the same as for ordinary differentiation, but it must always be remembered which variables are being held constant.

Higher-order partial derivatives may be defined in a similar manner, with, for example,

$$\frac{\partial^2 f}{\partial x^2} = \frac{\partial}{\partial x}\left(\frac{\partial f}{\partial x}\right) = f_{xx}$$

$$\frac{\partial^2 f}{\partial y \partial x} = \frac{\partial}{\partial y}\left(\frac{\partial f}{\partial x}\right) = f_{xy}$$

$$\frac{\partial^3 f}{\partial z \partial y \partial x} = \frac{\partial}{\partial z}\left(\frac{\partial^2 f}{\partial y \partial x}\right) = f_{xyz}$$

EXAMPLE 7.2

Find the first partial derivatives of the functions $f(x, y, z)$ with formula (a) $x + 2y + z^3$, (b) $x^2(y + 2z)$ and (c) $(x + y)/(z^3 + x)$.

Solution

(a) $f(x, y, z) = x + 2y + z^3$. To obtain f_x, we differentiate $f(x, y, z)$ with respect to x, keeping y and z constant. Thus $f_x = 1$, since the derivative of a constant $(2y + z^3)$ with respect to x is zero. Similarly, $f_y = 2$ and $f_z = 3z^2$.

(b) $f(x, y, z) = x^2(y + 2z)$. Here we use the same idea: when we differentiate with respect to one variable, we treat the other two as constants. Thus

$$\frac{\partial}{\partial x}[x^2(y + 2z)] = (y + 2z)\frac{\partial}{\partial x}(x^2) = 2x(y + 2z)$$

$$\frac{\partial}{\partial y}[x^2(y + 2z)] = x^2\frac{\partial}{\partial y}(y + 2z) = x^2(1) = x^2$$

$$\frac{\partial}{\partial z}[x^2(y + 2z)] = x^2\frac{\partial}{\partial z}(y + 2z) = x^2(2) = 2x^2$$

(c) $f(x, y, z) = (x + y)/(z^3 + x)$. Here we use the same idea, together with basic rules from ordinary differentiation:

$$\frac{\partial f}{\partial x} = \frac{(1)(z^3 + x) - (x + y)(1)}{(z^3 + x)^2} \qquad \text{(quotient rule)}$$

$$= \frac{z^3 - y}{(z^3 + x)^2}$$

$$\frac{\partial f}{\partial y} = \frac{1}{z^3 + x}$$

$$\frac{\partial f}{\partial z} = \frac{-3z^2(x + y)}{(z^3 + x)^2} \qquad \text{(chain rule)}$$

In Example 7.2 we used the **chain** (or **composite-function**) **rule** of ordinary differentiation

$$\frac{\mathrm{d}f}{\mathrm{d}x} = \frac{\mathrm{d}f}{\mathrm{d}u}\frac{\mathrm{d}u}{\mathrm{d}x}$$

to obtain the partial derivative $\partial f/\partial z$. The multivariable calculus form of the chain rule is a little more complicated. If the variables u, v and w are defined in terms of x, y and z then the partial derivative of $f(u, v, w)$ with respect to x is

$$\frac{\partial f}{\partial x} = \frac{\partial f}{\partial u}\frac{\partial u}{\partial x} + \frac{\partial f}{\partial v}\frac{\partial v}{\partial x} + \frac{\partial f}{\partial w}\frac{\partial w}{\partial x}$$

with similar expressions for $\partial f/\partial y$ and $\partial f/\partial z$.

EXAMPLE 7.3

Find $\partial T/\partial r$ and $\partial T/\partial \theta$ when

$$T(x, y) = x^3 - xy + y^3$$

and

$$x = r\cos\theta \quad \text{and} \quad y = r\sin\theta$$

Solution By the chain rule,

$$\frac{\partial T}{\partial r} = \frac{\partial T}{\partial x}\frac{\partial x}{\partial r} + \frac{\partial T}{\partial y}\frac{\partial y}{\partial r}$$

In this example

$$\frac{\partial T}{\partial x} = 3x^2 - y \quad \text{and} \quad \frac{\partial T}{\partial y} = -x + 3y^2$$

and

$$\frac{\partial x}{\partial r} = \cos\theta \quad \text{and} \quad \frac{\partial y}{\partial r} = \sin\theta$$

so that

$$\frac{\partial T}{\partial r} = (3x^2 - y)\cos\theta + (-x + 3y^2)\sin\theta$$

Substituting for x and y in terms of r and θ gives

$$\frac{\partial T}{\partial r} = 3r^2(\cos^3\theta + \sin^3\theta) - 2r\cos\theta\sin\theta$$

Similarly,

$$\begin{aligned}\frac{\partial T}{\partial \theta} &= (3x^2 - y)(-r\sin\theta) + (-x + 3y^2)r\cos\theta \\ &= 3r^3(\sin\theta - \cos\theta)\cos\theta\sin\theta + r^2(\sin^2\theta - \cos^2\theta)\end{aligned}$$

EXAMPLE 7.4 Find $\mathrm{d}H/\mathrm{d}t$ when

$$H(t) = \sin(3x - y)$$

and

$$x = 2t^2 - 3 \text{ and } y = \tfrac{1}{2}t^2 - 5t + 1$$

Solution We note that x and y are functions of t only, so that the chain rule becomes

$$\frac{\mathrm{d}H}{\mathrm{d}t} = \frac{\partial H}{\partial x}\frac{\mathrm{d}x}{\mathrm{d}t} + \frac{\partial H}{\partial y}\frac{\mathrm{d}y}{\mathrm{d}t}$$

Note the mixture of partial and ordinary derivatives. H is a function of the one variable t, but its dependence is expressed through the two variables x and y.

Substituting for the derivatives involved, we have

$$\begin{aligned}\frac{\mathrm{d}H}{\mathrm{d}t} &= 3[\cos(3x - y)]4t - [\cos(3x - y)](t - 5) \\ &= (11t + 5)\cos(3x - y) \\ &= (11t + 5)\cos(\tfrac{11}{2}t^2 + 5t - 10)\end{aligned}$$

EXAMPLE 7.5 A scalar point function $f(\boldsymbol{r})$ can be expressed in terms of rectangular cartesian coordinates (x, y, z) or in terms of spherical polar coordinates (r, θ, ϕ), where

$$x = r \sin\theta \cos\phi, \quad y = r \sin\theta \sin\phi, \quad z = r \cos\theta$$

as shown in Figure 7.4. Find $\partial f/\partial x$ in terms of the partial derivatives of the function with respect to r, θ and ϕ.

Solution Using the chain rule, we have

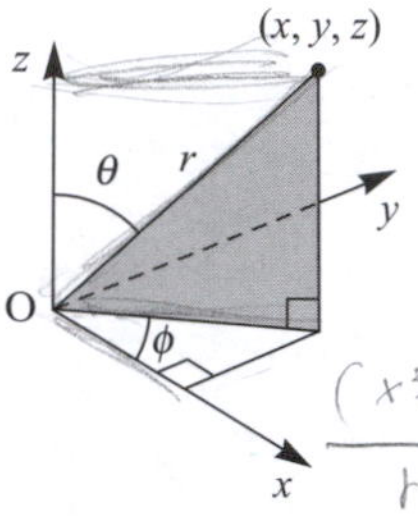

Figure 7.4 Spherical polar coordinates.

$$\frac{\partial f}{\partial x} = \frac{\partial f}{\partial r}\frac{\partial r}{\partial x} + \frac{\partial f}{\partial \theta}\frac{\partial \theta}{\partial x} + \frac{\partial f}{\partial \phi}\frac{\partial \phi}{\partial x}$$

From Figure 7.4, $r^2 = x^2 + y^2 + z^2$, $\tan\phi = y/x$ and $\tan\theta = (x^2 + y^2)^{1/2}/z$, so that

$$\frac{\partial r}{\partial x} = \frac{x}{r} = \sin\theta\cos\phi$$

$$\frac{\partial \phi}{\partial x} = \frac{\partial}{\partial x}\left(\tan^{-1}\frac{y}{x}\right) = -\frac{y}{x^2 + y^2} = -\frac{\sin\phi}{r\sin\theta}$$

$$\frac{\partial \theta}{\partial x} = \frac{\partial}{\partial x}\left\{\tan^{-1}\left[\frac{(x^2 + y^2)^{1/2}}{z}\right]\right\} = \frac{xz}{(x^2 + y^2 + z^2)(x^2 + y^2)^{1/2}}$$

$$= \frac{\cos\phi\cos\theta}{r}$$

Thus

$$\frac{\partial f}{\partial x} = \sin\theta\cos\phi\frac{\partial f}{\partial r} - \frac{\sin\phi}{r\sin\theta}\frac{\partial f}{\partial \phi} + \frac{\cos\phi\cos\theta}{r}\frac{\partial f}{\partial \theta}$$

7.1.2 Exercises

1 Sketch the contours (in two dimensions) of the scalar functions

(a) $f(x, y) = \ln(x^2 + y^2 - 1)$

(b) $f(x, y) = \tan^{-1}[y/(1 + x)]$

2 Sketch the flow lines (in two dimensions) of the vector functions

(a) $\boldsymbol{F}(x, y) = y\boldsymbol{i} + (6x^2 - 4x)\boldsymbol{j}$

(b) $\boldsymbol{F}(x, y) = y\boldsymbol{i} + (\frac{1}{6}x^3 - x)\boldsymbol{j}$

where $\boldsymbol{i}$ and $\boldsymbol{j}$ are unit vectors in the direction of the x and y axes respectively.

3 Sketch the level surfaces of the functions

(a) $f(\boldsymbol{r}) = z - xy$ (b) $f(\boldsymbol{r}) = z - e^{-xy}$

4 Sketch the field lines of the functions

(a) $\boldsymbol{F}(\boldsymbol{r}) = (xy, y^2 + 1, z)$

(b) $\boldsymbol{F}(\boldsymbol{r}) = (yz, zx, xy)$

5 Find all the first and second partial derivatives of the functions

(a) $f(\boldsymbol{r}) = xyz - x^2 + y - z$ (b) $f(\boldsymbol{r}) = x^2yz^3$

(c) $f(\boldsymbol{r}) = z\tan^{-1}(y/x)$

6 Find $\mathrm{d}f/\mathrm{d}t$, where

(a) $f(\mathbf{r}) = x^2 + y^2 - z$, and $x = t^3 - 1$, $y = 2t$, $z = 1/(t-1)$

(b) $f(\mathbf{r}) = xyz$, and $x = \mathrm{e}^{-t}\sin t$, $y = \mathrm{e}^{-t}\cos t$, $z = t$

7 Find $\partial f/\partial y$ and $\partial f/\partial z$ in terms of the partial derivatives of f with respect to spherical polar coordinates (r, θ, ϕ) (see Example 7.5).

8 Show that if $u(\mathbf{r}) = f(r)$, where $r^2 = x^2 + y^2 + z^2$, as usual, and

$$\frac{\partial^2 u}{\partial x^2} + \frac{\partial^2 u}{\partial y^2} + \frac{\partial^2 u}{\partial z^2} = 0$$

then

$$\frac{\mathrm{d}^2 f}{\mathrm{d}r^2} + \frac{2}{r}\frac{\mathrm{d}f}{\mathrm{d}r} = 0$$

Hence find the general form for $f(r)$.

9 Show that

$$V(x, y, z) = \frac{1}{z}\exp\left(-\frac{x^2 + y^2}{4z}\right)$$

satisfies the differential equation

$$\frac{\partial^2 V}{\partial x^2} + \frac{\partial^2 V}{\partial y^2} = \frac{\partial V}{\partial z}$$

10 Verify that $V(x, y, z) = \sin 3x \cos 4y \cosh 5z$ satisfies the differential equation

$$\frac{\partial^2 V}{\partial x^2} + \frac{\partial^2 V}{\partial y^2} + \frac{\partial^2 V}{\partial z^2} = 0$$

7.1.3 Transformations

Example 7.3 may be viewed as an example of *transformation of coordinates*. For example, consider the transformation or mapping from the (x, y) plane to the (s, t) plane defined by

$$s = s(x, y), \quad t = t(x, y) \tag{7.1}$$

Then a function $u = f(x, y)$ of x and y becomes a function $u = F(s, t)$ of s and t under the transformation, and the partial derivatives are related by

$$\left.\begin{aligned}\frac{\partial u}{\partial x} &= \frac{\partial u}{\partial s}\frac{\partial s}{\partial x} + \frac{\partial u}{\partial t}\frac{\partial t}{\partial x}\\ \frac{\partial u}{\partial y} &= \frac{\partial u}{\partial s}\frac{\partial s}{\partial y} + \frac{\partial u}{\partial t}\frac{\partial t}{\partial y}\end{aligned}\right\} \tag{7.2}$$

In matrix notation this becomes

$$\begin{bmatrix}\dfrac{\partial u}{\partial x}\\[2ex] \dfrac{\partial u}{\partial y}\end{bmatrix} = \begin{bmatrix}\dfrac{\partial s}{\partial x} & \dfrac{\partial t}{\partial x}\\[2ex] \dfrac{\partial s}{\partial y} & \dfrac{\partial t}{\partial y}\end{bmatrix}\begin{bmatrix}\dfrac{\partial u}{\partial s}\\[2ex] \dfrac{\partial u}{\partial t}\end{bmatrix} \tag{7.3}$$

The determinant of the matrix of the transformation is called the **Jacobian** of the transformation defined by (7.1) and is abbreviated to

$$\frac{\partial(s, t)}{\partial(x, y)} \quad \text{or simply to} \quad J$$

so that

$$J = \frac{\partial(s, t)}{\partial(x, y)} = \begin{vmatrix} \dfrac{\partial s}{\partial x} & \dfrac{\partial t}{\partial x} \\ \dfrac{\partial s}{\partial y} & \dfrac{\partial t}{\partial y} \end{vmatrix} = \begin{vmatrix} s_x & t_x \\ s_y & t_y \end{vmatrix} \tag{7.4}$$

The matrix itself is referred to as the **Jacobian matrix**. The Jacobian plays an important role in various applications of mathematics in engineering, particularly in implementing changes in variables in multiple integrals, as considered later in this chapter.

As indicated earlier, (7.1) define a transformation of the (x, y) plane to the (s, t) plane and give the coordinates of a point in the (s, t) plane corresponding to a point in the (x, y) plane. If we solve (7.1) for x and y, we obtain

$$x = X(s, t), \quad y = Y(s, t) \tag{7.5}$$

which represent a transformation of the (s, t) plane into the (x, y) plane. This is called the inverse transformation of the transformation defined by (7.1), and, analogously to (7.2), we can relate the partial derivatives by

$$\left.\begin{aligned} \frac{\partial u}{\partial s} &= \frac{\partial u}{\partial x}\frac{\partial x}{\partial s} + \frac{\partial u}{\partial y}\frac{\partial y}{\partial s} \\ \frac{\partial u}{\partial t} &= \frac{\partial u}{\partial x}\frac{\partial x}{\partial t} + \frac{\partial u}{\partial y}\frac{\partial y}{\partial t} \end{aligned}\right\} \tag{7.6}$$

The Jacobian of the inverse transformation (7.5) is

$$J_1 = \frac{\partial(x, y)}{\partial(s, t)} = \begin{vmatrix} x_s & y_s \\ x_t & y_t \end{vmatrix}$$

and, provided $J \neq 0$, it is always true that $J_1 = J^{-1}$ or

$$\frac{\partial(x, y)}{\partial(s, t)}\frac{\partial(s, t)}{\partial(x, y)} = 1$$

If $J = 0$ then the variables s and t defined by (7.1) are functionally dependent; that is, a relationship of the form $f(s, t) = 0$ exists. This implies a non-unique correspondence between points in the (x, y) and (s, t) planes.

EXAMPLE 7.6

(a) Obtain the Jacobian J of the transformation

$$s = 2x + y, \quad t = x - 2y$$

(b) Determine the inverse transformation of the above transformation and obtain its Jacobian J_1. Confirm that $J_1 = J^{-1}$.

Solution (a) Using (7.4), the Jacobian of the transformation is

$$J = \frac{\partial(s, t)}{\partial(x, y)} = \begin{vmatrix} 2 & 1 \\ 1 & -2 \end{vmatrix} = -5$$

(b) Solving the pair of equations in the transformation for x and y gives the inverse transformation as

$$x = \tfrac{1}{5}(2s + t), \quad y = \tfrac{1}{5}(s - 2t)$$

The Jacobian of this inverse transformation is

$$J_1 = \frac{\partial(x, y)}{\partial(s, t)} = \begin{vmatrix} \frac{2}{5} & \frac{1}{5} \\ \frac{1}{5} & -\frac{2}{5} \end{vmatrix} = -\tfrac{1}{5}$$

confirming that $J_1 = J^{-1}$.

$-\frac{4}{25} - \frac{1}{25} = -\frac{5}{25} = \left(-\frac{1}{5}\right)$

EXAMPLE 7.7

Show that the variables x and y given by

$$x = \frac{s+t}{s}, \quad y = \frac{s+t}{t} \tag{7.7}$$

are functionally dependent, and obtain the relationship $f(x, y) = 0$.

Solution

The Jacobian of the transformation (7.7) is

$$J = \frac{\partial(x, y)}{\partial(s, t)} = \begin{vmatrix} x_s & y_s \\ x_t & y_t \end{vmatrix} = \begin{vmatrix} -\dfrac{t}{s^2} & \dfrac{1}{t} \\ \dfrac{1}{s} & -\dfrac{s}{t^2} \end{vmatrix} = \frac{1}{st} - \frac{1}{st} = 0$$

Since $J = 0$, the variables x and y are functionally related.

Rearranging (7.7), we have

$$x = 1 + \frac{t}{s}, \quad y = \frac{s}{t} + 1$$

so that

$$(x - 1)(y - 1) = \frac{t}{s}\frac{s}{t} = 1$$

giving the functional relationship as

$$xy - (x + y) = 0$$

The definition of a Jacobian is not restricted to functions of two variables, and it is readily extendable to functions of many variables. For example, for functions of three variables, if

$$u = U(x, y, z), \quad v = V(x, y, z), \quad w = W(x, y, z) \tag{7.8}$$

represents a transformation in three dimensions from the variables x, y, z to the variables u, v, w then the corresponding Jacobian is

$$J = \frac{\partial(u, v, w)}{\partial(x, y, z)} = \begin{vmatrix} u_x & v_x & w_x \\ u_y & v_y & w_y \\ u_z & v_z & w_z \end{vmatrix}$$

Again, if $J = 0$, it follows that there exists a functional relationship $f(u, v, w) = 0$ between the variables u, v and w defined by (7.8).

7.1.4 Exercises

11 Show that if $x + y = u$ and $y = uv$, then

$$\frac{\partial(x, y)}{\partial(u, v)} = u$$

12 Show that, if $x + y + z = u$, $y + z = uv$ and $z = uvw$, then

$$\frac{\partial(x, y, z)}{\partial(u, v, w)} = u^2v$$

13 If $x = e^u \cos v$ and $y = e^u \sin v$, obtain the two Jacobians

$$\frac{\partial(x, y)}{\partial(u, v)} \quad \text{and} \quad \frac{\partial(u, v)}{\partial(x, y)}$$

and verify that they are mutual inverses.

14 Find the values of the constant parameter λ for which the functions

$$u = \cos x \cos y - \lambda \sin x \sin y$$

$$v = \sin x \cos y + \lambda \cos x \sin y$$

are functionally dependent.

15 Find the value of the constant K for which

$$u = Kx^2 + 4y^2 + z^2$$

$$v = 3x + 2y + z$$

$$w = 2yz + 3zx + 6xy$$

are functionally related, and obtain the corresponding relation.

16 Show that, if $u = g(x, y)$ and $v = h(x, y)$, then

$$\frac{\partial x}{\partial u} = \frac{\partial v}{\partial y}\Big/J \quad \frac{\partial x}{\partial v} = -\frac{\partial u}{\partial y}\Big/J$$

$$\frac{\partial y}{\partial u} = -\frac{\partial v}{\partial x}\Big/J \quad \frac{\partial y}{\partial v} = \frac{\partial u}{\partial x}\Big/J$$

where in each case

$$J = \frac{\partial(u, v)}{\partial(x, y)}$$

17 Use the results of Question 16 to obtain the partial derivatives

$$\frac{\partial x}{\partial u}, \frac{\partial x}{\partial v}, \frac{\partial y}{\partial u}, \frac{\partial y}{\partial v}$$

where

$$u = e^x \cos y \quad \text{and} \quad v = e^{-x} \sin y$$

7.1.5 The total differential

Consider a function $u = f(x, y)$ of two variables x and y. Let Δx and Δy be increments in the values of x and y. Then the corresponding increment in u is given by

$$\Delta u = f(x + \Delta x, y + \Delta y) - f(x, y)$$

We re-write this as two terms: one showing the change in u due to the change in x, and the other showing the change in u due to the change in y. Thus

$$\Delta u = [f(x + \Delta x, y + \Delta y) - f(x, y + \Delta y)] + [f(x, y + \Delta y) - f(x, y)]$$

Dividing the first bracketed term by Δx and the second by Δy gives

$$\Delta u = \frac{f(x+\Delta x, y+\Delta y)-f(x, y+\Delta x)}{\Delta x}\Delta x + \frac{f(x, y+\Delta y)-f(x, y)}{\Delta y}\Delta y$$

From the definition of the partial derivative, we may approximate this expression by

$$\Delta u \approx \frac{\partial f}{\partial x}\Delta x + \frac{\partial f}{\partial y}\Delta y$$

We define the **differential** du by the equation

$$\mathrm{d}u = \frac{\partial f}{\partial x}\Delta x + \frac{\partial f}{\partial y}\Delta y \qquad \textbf{(7.9)}$$

By setting $f(x, y) = f_1(x, y) = x$ and $f(x, y) = f_2(x, y) = y$ in turn in (7.9), we see that

$$\mathrm{d}x = \frac{\partial f_1}{\partial x}\Delta x + \frac{\partial f_1}{\partial y}\Delta y = \Delta x \quad \text{and} \quad \mathrm{d}y = \Delta y$$

so that for the independent variables increments and differentials are equal. For the dependent variable we have

$$\mathrm{d}u = \frac{\partial f}{\partial x}\mathrm{d}x + \frac{\partial f}{\partial y}\mathrm{d}y \qquad \textbf{(7.10)}$$

We see that the differential du is an approximation to the change Δu in $u = f(x, y)$ resulting from small changes Δx and Δy in the independent variables x and y; that is,

$$\Delta u \approx \mathrm{d}u = \frac{\partial f}{\partial x}\mathrm{d}x + \frac{\partial f}{\partial y}\mathrm{d}y = \frac{\partial f}{\partial x}\Delta x + \frac{\partial f}{\partial y}\Delta y \qquad \textbf{(7.11)}$$

a result illustrated in Figure 7.5.

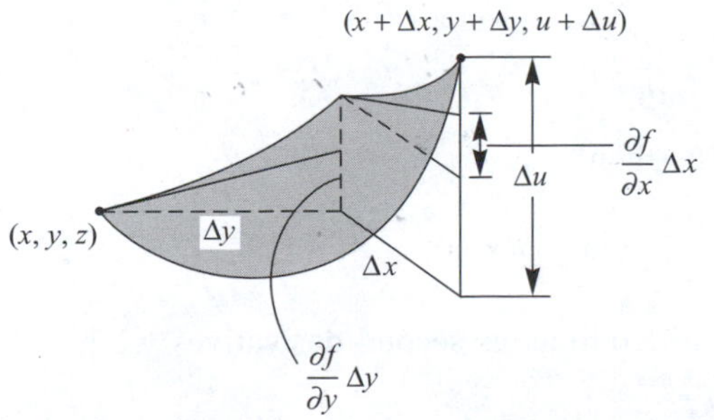

Figure 7.5

This extends to functions of as many variables as we please, provided that the partial derivatives exist. For example, for a function of three variables (x, y, z) defined by $u = f(x, y, z)$ we have

$$\Delta u \approx \mathrm{d}u = \frac{\partial f}{\partial x}\mathrm{d}x + \frac{\partial f}{\partial y}\mathrm{d}y + \frac{\partial f}{\partial z}\mathrm{d}z$$
$$= \frac{\partial f}{\partial x}\Delta x + \frac{\partial f}{\partial y}\Delta y + \frac{\partial f}{\partial z}\Delta z$$

The differential of a function of several variables is often called a **total differential**, emphasizing that it shows the variation of the function with respect to small changes in *all* the independent variables.

EXAMPLE 7.8 Find the total differential of $u(x, y) = x^y$.

Solution Taking partial derivatives we have

$$\frac{\partial u}{\partial x} = yx^{y-1} \quad \text{and} \quad \frac{\partial u}{\partial y} = x^y \ln x$$

Hence, using (7.10),

$$\mathrm{d}u = yx^{y-1}\,\mathrm{d}x + x^y \ln x\,\mathrm{d}y$$

Differentials sometimes arise naturally when modelling practical problems. When this occurs, it is often possible to analyse the problem further by testing to see if the expression in which the differentials occur is a total differential. Consider the equation

$$P(x, y)\,\mathrm{d}x + Q(x, y)\,\mathrm{d}y = 0$$

connecting x, y and their differentials. The left-hand side of this equation is said to be an **exact differential** if there is a function $f(x, y)$ such that

$$\mathrm{d}f = P(x, y)\,\mathrm{d}x + Q(x, y)\,\mathrm{d}y$$

Now we know that

$$\mathrm{d}f = \frac{\partial f}{\partial x}\mathrm{d}x + \frac{\partial f}{\partial y}\mathrm{d}y$$

so if $f(x, y)$ exists then

$$P(x, y) = \frac{\partial f}{\partial x} \quad \text{and} \quad Q(x, y) = \frac{\partial f}{\partial y}$$

For functions with continuous second derivatives we have

$$\frac{\partial^2 f}{\partial x \partial y} = \frac{\partial^2 f}{\partial y \partial x}$$

Thus if $f(x, y)$ exists then

$$\frac{\partial P}{\partial y} = \frac{\partial Q}{\partial x} \tag{7.12}$$

This gives us a test for the existence of $f(x, y)$, but does not tell us how to find it! The technique for finding $f(x, y)$ is shown in Example 7.9.

EXAMPLE 7.9 Show that

$$(6x + 9y + 11)\mathrm{d}x + (9x - 4y + 3)\mathrm{d}y$$

is an exact differential and find the relationship between y and x given

$$\frac{\mathrm{d}y}{\mathrm{d}x} = -\frac{6x + 9y + 11}{9x - 4y + 3}$$

and the condition $y = 1$ when $x = 0$.

Solution In this example

$$P(x, y) = 6x + 9y + 11 \quad \text{and} \quad Q(x, y) = 9x - 4y + 3$$

First we test whether the expression is an exact differential. In this example

$$\frac{\partial P}{\partial y} = 9 \quad \text{and} \quad \frac{\partial Q}{\partial x} = 9$$

so from (7.12), we have an exact differential. Thus we know that there is a function $f(x, y)$ such that

$$\frac{\partial f}{\partial x} = 6x + 9y + 11 \quad \text{and} \quad \frac{\partial f}{\partial y} = 9x - 4y + 3 \tag{7.13a, b}$$

Integrating (7.13a) with respect to x, keeping y constant (that is, reversing the partial differentiation process) we have

$$f(x, y) = 3x^2 + 9xy + 11x + g(y) \tag{7.14}$$

Note that the 'constant' of integration is a function of y. You can check that this expression for $f(x, y)$ is correct by differentiating it partially with respect to x. But we also know from (7.13b) the partial derivative of $f(x, y)$ with respect to y, and this enables us to find $g'(y)$. Differentiating (7.14) partially with respect to y and equating it to (7.13b), we have

$$\frac{\partial f}{\partial y} = 9x + \frac{\mathrm{d}g}{\mathrm{d}y} = 9x - 4y + 3$$

(Note that since g is a function of y only we use $\mathrm{d}g/\mathrm{d}y$ rather than $\partial g/\partial y$.) Thus

$$\frac{\mathrm{d}g}{\mathrm{d}y} = -4y + 3$$

so, on integrating,

$$g(y) = -2y^2 + 3y + C$$

Substituting back into (7.13b) gives

$$f(x, y) = 3x^2 + 9xy + 11x - 2y^2 + 3y + C$$

Now we are given that

$$\frac{dy}{dx} = -\frac{6x + 9y + 11}{9x - 4y + 3}$$

which implies that

$$(6x + 9y + 11)dx + (9x - 4y + 3)dy = 0$$

which in turn implies that

$$3x^2 + 9xy + 11x - 2y^2 + 3y + C = 0$$

The arbitrary constant C is fixed by applying the given condition $y = 1$ when $x = 0$, giving $C = -1$. Thus x and y satisfy the equation

$$3x^2 + 9xy + 11x - 2y^2 + 3y = 1$$

7.1.6 Exercises

18 Determine which of the following are exact differentials of a function, and find, where appropriate, the corresponding function.

(a) $(y^2 + 2xy + 1)dx + (2xy + x^2)dy$

(b) $(2xy^2 + 3y\cos 3x)dx + (2x^2y + \sin 3x)dy$

(c) $(6xy - y^2)dx + (2x\,e^y - x^2)dy$

(d) $(z^3 - 3y)dx + (12y^2 - 3x)dy + 3xz^2dz$

19 Find the value of the constant λ such that

$$(y\cos x + \lambda\cos y)dx + (x\sin y + \sin x + y)dy$$

is the exact differential of a function $f(x, y)$. Find the corresponding function $f(x, y)$ that also satisfies the condition $f(0, 1) = 0$.

20 Show that the differential

$$g(x, y) = (10x^2 + 6xy + 6y^2)dx + (9x^2 + 4xy + 15y^2)dy$$

is not exact, but that a constant m can be chosen so that

$$(2x + 3y)^m g(x, y)$$

is equal to dz, the exact differential of a function $z = f(x, y)$. Find $f(x, y)$.

7.2 Derivatives of a scalar point function

In many practical problems it is necessary to measure the rate of change of a scalar point function. For example, in heat transfer problems we need to know the rate of change of temperature from point to point, because that determines the rate at which heat flows. Similarly, if we are investigating the electric field due to static charges, we need to know the variation of the electric potential from point to point. To determine such information, the ideas of calculus were extended to vector quantities. The first development of this was the concept of the gradient of a scalar point function.

7.2.1 The gradient of a scalar point function

We described in Section 7.1.1 how the gradient of a scalar field depended on the direction along which its rate of change was measured. We now explore this idea further. Consider the rate of change of the function $f(\boldsymbol{r})$ at the point (x, y, z) in the direction of the unit vector (l, m, n). To find this, we need to evaluate the limit

$$\lim_{\Delta r \to 0} \frac{f(\boldsymbol{r} + \Delta \boldsymbol{r}) - f(\boldsymbol{r})}{\Delta r}$$

where $\Delta \boldsymbol{r}$ is in the direction of (l, m, n). In terms of coordinates, this means

$$\begin{aligned} \boldsymbol{r} + \Delta \boldsymbol{r} &= \boldsymbol{r} + \Delta r(l, m, n) \\ &= (x + \Delta x, y + \Delta y, z + \Delta z) \end{aligned}$$

so that

$$\Delta x = l\Delta r, \quad \Delta y = m\Delta r, \quad \Delta z = n\Delta r$$

Thus we have to consider the limit

$$\lim_{\Delta r \to 0} \frac{f(x + l\Delta r, y + m\Delta r, z + n\Delta r) - f(x, y, z)}{\Delta r}$$

We can rewrite this as

$$\begin{aligned} &\lim_{\Delta r \to 0} \left[\frac{f(x + l\Delta r, y + m\Delta r, z + n\Delta r) - f(x, y + m\Delta r, z + n\Delta r)}{l\Delta r}\right] l \\ &+ \lim_{\Delta r \to 0} \left[\frac{f(x, y + m\Delta r, z + n\Delta r) - f(x, y, z + n\Delta r)}{m\Delta r}\right] m \\ &+ \lim_{\Delta r \to 0} \left[\frac{f(x, y, z + n\Delta r) - f(x, y, z)}{n\Delta r}\right] n \end{aligned}$$

Evaluating the limits, remembering that $\Delta x = l\Delta r$ and so on, we find that the rate of change of $f(\boldsymbol{r})$ in the direction of the unit vector (l, m, n) is

$$\frac{\partial f}{\partial x} l + \frac{\partial f}{\partial y} m + \frac{\partial f}{\partial z} n = \left(\frac{\partial f}{\partial x}, \frac{\partial f}{\partial y}, \frac{\partial f}{\partial z}\right) \cdot (l, m, n)$$

The vector

$$\left(\frac{\partial f}{\partial x}, \frac{\partial f}{\partial y}, \frac{\partial f}{\partial z}\right)$$

is called the **gradient** of the scalar point function $f(x, y, z)$, and is denoted by grad f or by ∇f, where ∇ is the vector operator

$$\nabla = \boldsymbol{i}\frac{\partial}{\partial x} + \boldsymbol{j}\frac{\partial}{\partial y} + \boldsymbol{k}\frac{\partial}{\partial z}$$

where $\boldsymbol{i}$, $\boldsymbol{j}$ and $\boldsymbol{k}$ are the usual triad of unit vectors.

The symbol ∇ is called 'del' or sometimes 'nabla'. Then

$$\text{grad } f = \nabla f = \frac{\partial f}{\partial x}\boldsymbol{i} + \frac{\partial f}{\partial y}\boldsymbol{j} + \frac{\partial f}{\partial z}\boldsymbol{k} \equiv \left(\frac{\partial f}{\partial x}, \frac{\partial f}{\partial y}, \frac{\partial f}{\partial z}\right) \tag{7.15}$$

Thus we can calculate the rate of change of $f(x, y, z)$ along any direction we please. If $\hat{\boldsymbol{u}}$ is the unit vector in that direction then

$$(\text{grad } f) \cdot \hat{\boldsymbol{u}}$$

gives the required **directional derivative**, that is, the rate of change of $f(x, y, z)$ in the direction of $\hat{\boldsymbol{u}}$. Remembering that $\boldsymbol{a} \cdot \boldsymbol{b} = |a||b| \cos \theta$, where θ is the angle between the two vectors, it follows that the rate of change of $f(x, y, z)$ is zero along directions perpendicular to grad f and is maximum along the direction parallel to grad f. Furthermore, grad f acts along the normal direction to the level surface of $f(x, y, z)$. We can see this by considering the level surfaces of the function corresponding to c and $c + \Delta c$, as shown in Figure 7.6(a). In going from P on the surface $f(\boldsymbol{r}) = c$ to any point Q on $f(\boldsymbol{r}) = c + \Delta c$, the increase in f is the same whatever point Q is chosen, but the distance PQ will be smallest, and hence the rate of change of $f(x, y, z)$ greatest, when Q lies on the normal $\hat{\boldsymbol{n}}$ to the surface at P. Thus grad f at P is in the direction of the outward normal $\hat{\boldsymbol{n}}$ to the surface $f(\boldsymbol{r}) = u$, and represents in magnitude and direction the greatest rate of increase of $f(x, y, z)$ with distance (Figure 7.6(b)). It is frequently written as

$$\text{grad } f = \frac{\partial f}{\partial n}\hat{\boldsymbol{n}}$$

where $\partial f/\partial n$ is referred to as the normal derivative to the surface $f(\boldsymbol{r}) = c$.

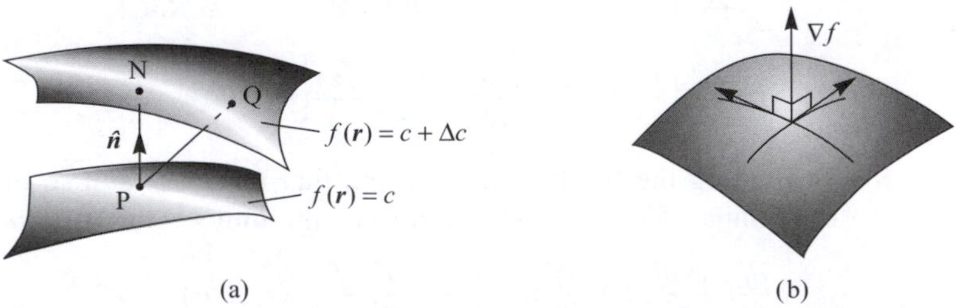

Figure 7.6 (a) Adjacent level surfaces of $f(\boldsymbol{r})$; (b) grad f acts normally to the surface $f(\boldsymbol{r}) = c$.

EXAMPLE 7.10 Find grad f for $f(\boldsymbol{r}) = 3x^2 + 2y^2 + z^2$ at the point (1, 2, 3). Hence calculate

(a) the directional derivative of $f(\boldsymbol{r})$ at (1, 2, 3) in the direction of the unit vector $\frac{1}{3}(2, 2, 1)$;

(b) the maximum rate of change of the function at (1, 2, 3) and its direction.

Solution Since $\partial f/\partial x = 6x$, $\partial f/\partial y = 4y$ and $\partial f/\partial z = 2z$, we have from (7.15) that

$$\text{grad } f = \nabla f = 6x\boldsymbol{i} + 4y\boldsymbol{j} + 2z\boldsymbol{k}$$

At the point (1, 2, 3)

$$\text{grad } f = 6\boldsymbol{i} + 8\boldsymbol{j} + 6\boldsymbol{k}$$

Thus the directional derivative of $f(\boldsymbol{r})$ at (1, 2, 3) in the direction of the unit vector $(\frac{2}{3}, \frac{2}{3}, \frac{1}{3})$ is

$$(6\boldsymbol{i} + 8\boldsymbol{j} + 6\boldsymbol{k}) \cdot (\tfrac{2}{3}\boldsymbol{i} + \tfrac{2}{3}\boldsymbol{j} + \tfrac{1}{3}\boldsymbol{k}) = \tfrac{34}{3}$$

The maximum rate of change of $f(\boldsymbol{r})$ at (1, 2, 3) occurs along the direction parallel to grad f at (1, 2, 3), that is, parallel to (6, 8, 6). The unit vector in that direction is $(3, 4, 3)/\sqrt{34}$ and the maximum rate of change of $f(\boldsymbol{r})$ is $2\sqrt{34}$.

If a surface in three dimensions is specified by the equation $f(x, y, z) = c$, or equivalently $f(\boldsymbol{r}) = c$, then grad f is a vector perpendicular to that surface. This enables us to calculate the normal vector at any point on the surface, and consequently to find the equation of the tangent plane at that point.

EXAMPLE 7.11 A paraboloid of revolution has equation $2z = x^2 + y^2$. Find the unit normal vector to the surface at the point (1, 3, 5). Hence obtain the equation of the normal and the tangent plane to the surface at that point.

Solution A vector normal to the surface $2z = x^2 + y^2$ is given by

$$\text{grad}\,(x^2 + y^2 - 2z) = 2x\boldsymbol{i} + 2y\boldsymbol{j} - 2\boldsymbol{k}$$

At the point (1, 3, 5) the vector has the value $2\boldsymbol{i} + 6\boldsymbol{j} - 2\boldsymbol{k}$. Thus the normal unit vector at the point (1, 3, 5) is $(\boldsymbol{i} + 3\boldsymbol{j} - \boldsymbol{k})/\sqrt{11}$. The equation of the line through (1, 3, 5) in the direction of this normal is

$$\frac{x-1}{1} = \frac{y-3}{3} = \frac{z-5}{-1}$$

and the equation of the tangent plane is

$$(1)(x - 1) + (3)(y - 3) + (-1)(z - 5) = 0$$

which simplifies to $x + 3y - z = 5$ (see Figure 7.7).

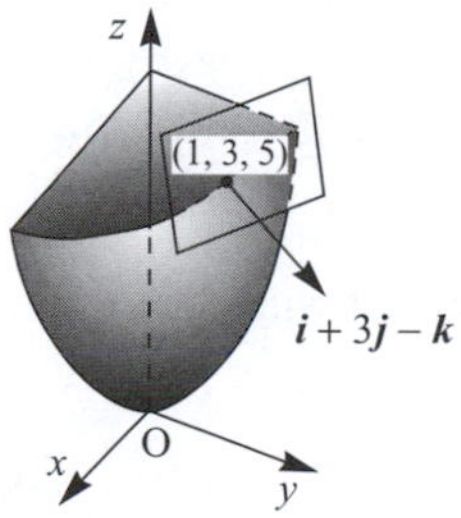

Figure 7.7 Tangent plane at (1, 3, 5) to the paraboloid $2z = x^2 + y^2$.

The concept of the gradient of a scalar field occurs in many applications. The simplest, perhaps, is when $f(\mathbf{r})$ represents the potential in an electric field due to static charges. Then the electric force is in the direction of the greatest decrease of the potential. Its magnitude is equal to that rate of decrease, so that the force is given by $-\text{grad}\, f$.

7.2.2 Exercises

21 Find grad f for $f(\mathbf{r}) = x^2yz^2$ at the point (1, 2, 3). Hence calculate

(a) the directional derivative of $f(\mathbf{r})$ at (1, 2, 3) in the direction of the vector (−2, 3, −6);
(b) the maximum rate of change of the function at (1, 2, 3) and its direction.

22 Find ∇f where $f(\mathbf{r})$ is

(a) $x^2 + y^2 - z$ (b) $z \tan^{-1}(y/x)$
(c) $e^{-x-y+z}/\sqrt{(x^3 + y^2)}$
(d) $xyz \sin\{\pi(x + y + z)\}$

23 Find the directional derivative of $f(\mathbf{r}) = x^2 + y^2 - z$ at the point (1, 1, 2) in the direction of the vector (4, 4, −2).

24 Find a unit normal to the surface $xy^2 - 3xz = -5$ at the point (1, −2, 3).

25 If $\mathbf{r}$ is the usual position vector $\mathbf{r} = x\mathbf{i} + y\mathbf{j} + z\mathbf{k}$, with $|\mathbf{r}| = r$, evaluate

(a) ∇r (b) $\nabla\left(\dfrac{1}{r}\right)$

26 If $\nabla\phi = (2xy + z^2)\mathbf{i} + (x^2 + z)\mathbf{j} + (y + 2xz)\mathbf{k}$, find a possible value for ϕ.

27 Given the scalar function of position

$$\phi(x, y, z) = x^2y - 3xyz + z^3$$

find the value of grad ϕ at the point (3, 1, 2). Also find the directional derivative of ϕ at this point in the direction of the vector (3, −2, 6), that is, in the direction $3\mathbf{i} - 2\mathbf{j} + 6\mathbf{k}$.

28 Find the angle between the surfaces $x^2 + y^2 + z^2 = 9$ and $z = x^2 + y^2 - 3$ at the point (2, −1, 2).

29 Find the equations of the tangent plane and normal line to the surfaces

(a) $x^2 + 2y^2 + 3z^2 = 6$ at (1, 1, 1)
(b) $2x^2 + y^2 - z^2 = -3$ at (1, 2, 3)
(c) $x^2 + y^2 - z = 1$ at (1, 2, 4).

30 (Spherical polar coordinates) When a function $f(\mathbf{r})$ is specified in polar coordinates, it is usual to express grad f in terms of the partial derivatives of f with respect to r, θ and ϕ and the unit vectors $\mathbf{u}_r$, $\mathbf{u}_\theta$ and $\mathbf{u}_\phi$ in the directions of increasing r, θ and ϕ as shown in Figure 7.8. Working from first principles, show that

$$\nabla f = \text{grad}\, f = \frac{\partial f}{\partial r}\mathbf{u}_r + \frac{1}{r}\frac{\partial f}{\partial \theta}\mathbf{u}_\theta + \frac{1}{r\sin\theta}\frac{\partial f}{\partial \phi}\mathbf{u}_\phi$$

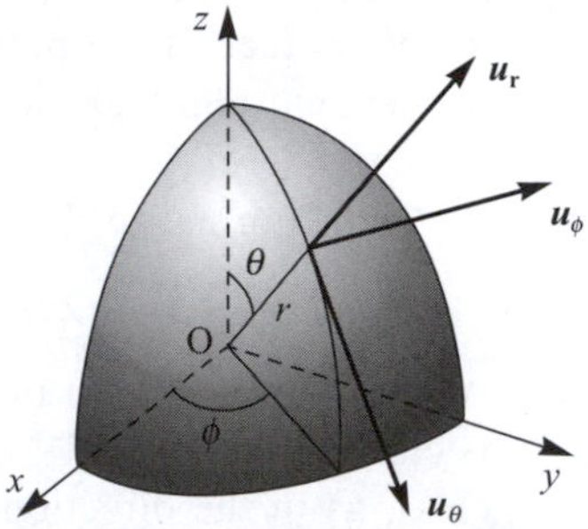

Figure 7.8 Unit vectors associated with spherical polar coordinates.

7.3 Derivatives of a vector point function

When we come to consider the rate of change of a vector point function $\boldsymbol{F}(\boldsymbol{r})$, we see that there are two ways of combining the vector operator ∇ with the vector $\boldsymbol{F}$. Thus we have two cases to consider, namely

$$\nabla \cdot \boldsymbol{F} \quad \text{and} \quad \nabla \times \boldsymbol{F}$$

that is, the scalar product and vector product respectively. Both of these 'derivatives' have physical meanings, as we shall discover in the following sections. Roughly, if we picture a vector field as a fluid flow then at every point in the flow we need to measure the rate at which the field is flowing away from that point and also the amount of spin possessed by the particles of the fluid at that point. The two 'derivatives' given formally above provide these measures.

7.3.1 Divergence of a vector field

Consider the steady motion of a fluid in a region R such that a particle of fluid instantaneously at the point $\boldsymbol{r}$ with coordinates (x, y, z) has a velocity $\boldsymbol{v}(\boldsymbol{r})$ that is independent of time. To measure the flow away from this point in the fluid, we surround the point by an 'elementary' cuboid of side $(2\Delta x) \times (2\Delta y) \times (2\Delta z)$, as shown in Figure 7.9, and calculate the average flow out of the cuboid per unit volume.

The flow out of the cuboid is the sum of the flows across each of its six faces. Representing the velocity of the fluid at (x, y, z) by $\boldsymbol{v}$, the flow out of the face ABCD is given approximately by

$$\boldsymbol{i} \cdot \boldsymbol{v}(x + \Delta x, y, z)(4\Delta y \Delta z)$$

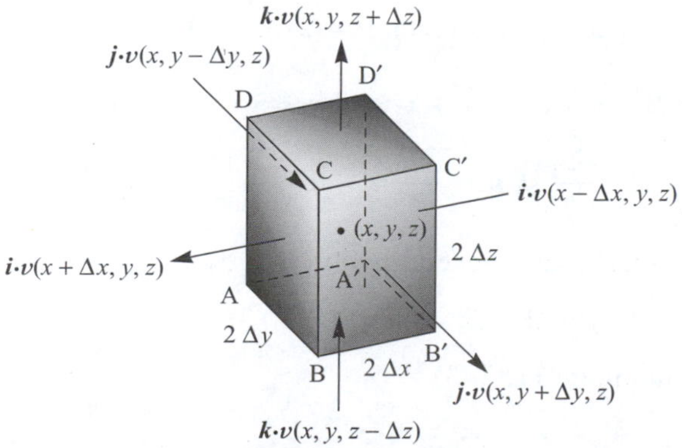

Figure 7.9 Flow out of a cuboid.

The flow out of the face A′B′C′D′ is given approximately by

$$-\boldsymbol{i}\cdot\boldsymbol{v}(x-\Delta x, y, z)(4\Delta y\Delta z)$$

There are similar expressions for the remaining four faces of the cuboid, so that the total flow out of the latter is

$$\begin{aligned}&\boldsymbol{i}\cdot[\boldsymbol{v}(x+\Delta x, y, z)-\boldsymbol{v}(x-\Delta x, y, z)](4\Delta y\Delta z)\\&\quad+\boldsymbol{j}\cdot[\boldsymbol{v}(x, y+\Delta y, z)-\boldsymbol{v}(x, y-\Delta y, z)](4\Delta x\Delta z)\\&\quad+\boldsymbol{k}\cdot[\boldsymbol{v}(x, y, z+\Delta z)-\boldsymbol{v}(x, y, z-\Delta z)](4\Delta x\Delta y)\end{aligned}$$

Dividing by the volume $8\Delta x\Delta y\Delta z$, and proceeding to the limit as $\Delta x, \Delta y, \Delta z \to 0$, we see that the flow away from the point (x, y, z) per unit time is given by

$$\boldsymbol{i}\cdot\frac{\partial \boldsymbol{v}}{\partial x}+\boldsymbol{j}\cdot\frac{\partial \boldsymbol{v}}{\partial y}+\boldsymbol{k}\cdot\frac{\partial \boldsymbol{v}}{\partial z}$$

This may be rewritten as

$$\left(\boldsymbol{i}\frac{\partial}{\partial x}+\boldsymbol{j}\frac{\partial}{\partial y}+\boldsymbol{k}\frac{\partial}{\partial z}\right)\cdot\boldsymbol{v}$$

or simply as $\nabla\cdot\boldsymbol{v}$. Thus we see that the flow away from this point is given by the scalar product of the vector operator ∇ with the velocity vector $\boldsymbol{v}$. This is called the **divergence** of the vector $\boldsymbol{v}$, and is written as $\text{div}\,\boldsymbol{v}$. In terms of components,

$$\begin{aligned}\text{div}\,\boldsymbol{v}=\nabla\cdot\boldsymbol{v}&=\left(\boldsymbol{i}\frac{\partial}{\partial x}+\boldsymbol{j}\frac{\partial}{\partial y}+\boldsymbol{k}\frac{\partial}{\partial z}\right)\cdot(\boldsymbol{i}v_1+\boldsymbol{j}v_2+\boldsymbol{k}v_3)\\&=\frac{\partial v_1}{\partial x}+\frac{\partial v_2}{\partial y}+\frac{\partial v_3}{\partial z}\end{aligned} \tag{7.16}$$

When $\boldsymbol{v}$ is specified in this way, it is easy to compute its divergence. Note that the divergence of a vector field is a scalar quantity.

EXAMPLE 7.12 Find the divergence of the vector $\boldsymbol{v}=(2x-y^2, 3z+x^2, 4y-z^2)$ at the point (1, 2, 3).

Solution Here $v_1 = 2x - y^2$, $v_2 = 3z + x^2$ and $v_3 = 4y - z^2$, so that

$$\frac{\partial v_1}{\partial x}=2,\quad \frac{\partial v_2}{\partial y}=0,\quad \frac{\partial v_3}{\partial z}=-2z$$

Thus from (7.16), at a general point (x, y, z),

$$\text{div}\,\boldsymbol{v}=\nabla\cdot\boldsymbol{v}=2-2z$$

so that at the point (1, 2, 3)

$$\nabla\cdot\boldsymbol{v}=-4$$

A more general way of defining the divergence of a vector field $\boldsymbol{F}(\boldsymbol{r})$ at the point $\boldsymbol{r}$ is to enclose the point in an elementary volume ΔV and find the flow or flux out of ΔV per unit volume. Thus

$$\operatorname{div} \boldsymbol{F} = \nabla \cdot \boldsymbol{F} = \lim_{\Delta V \to 0} \frac{\text{flow out of } \Delta V}{\Delta V}$$

A non-zero divergence at a point in a fluid measures the rate, per unit volume, at which the fluid is flowing away from or towards that point. That implies that either the density of the fluid is changing at the point or there is a source or sink of fluid there. In the case of a non-material vector field, for example in heat transfer, a non-zero divergence indicates a point of generation or absorption. When the divergence is everywhere zero, the flow entering any element of the space is exactly balanced by the outflow. This implies that the lines of flow of the field $\boldsymbol{F}(\boldsymbol{r})$ where $\operatorname{div} \boldsymbol{F} = 0$ must either form closed curves or finish at boundaries or extend to infinity. Vectors satisfying this condition are sometimes termed **solenoidal**.

7.3.2 Exercises

31 Find $\operatorname{div} \boldsymbol{v}$ where

(a) $\boldsymbol{v}(\boldsymbol{r}) = 3x^2y\boldsymbol{i} + z\boldsymbol{j} + x^2\boldsymbol{k}$

(b) $\boldsymbol{v}(\boldsymbol{r}) = (3x + y)\boldsymbol{i} + (2z + x)\boldsymbol{j} + (z - 2y)\boldsymbol{k}$

32 If $\boldsymbol{F} = (2xy^2 + z^2)\boldsymbol{i} + (3x^2z^2 - y^2z^3)\boldsymbol{j} + (yz^2 - xz^3)\boldsymbol{k}$, calculate $\operatorname{div} \boldsymbol{f}$ at the point $(-1, 2, 3)$.

33 Find $\nabla(\boldsymbol{a} \cdot \boldsymbol{r})$, $(\boldsymbol{a} \cdot \nabla)\boldsymbol{r}$ and $\boldsymbol{a}(\nabla \cdot \boldsymbol{r})$, where $\boldsymbol{a}$ is a constant vector and, as usual, $\boldsymbol{r}$ is the position vector $\boldsymbol{r} = (x, y, z)$.

34 The vector $\boldsymbol{v}$ is defined by $\boldsymbol{v} = \boldsymbol{r}r^{-1}$, where $\boldsymbol{r} = (x, y, z)$ and $r = |\boldsymbol{r}|$. Show that

$$\nabla(\nabla \cdot \boldsymbol{v}) \equiv \operatorname{grad} \operatorname{div} \boldsymbol{v} = -\frac{2}{r^3}\boldsymbol{r}$$

35 Find the value of the constant λ such that the vector field defined by

$$\boldsymbol{F} = (2x^2y^2 + z^2)\boldsymbol{i} + (3xy^3 - x^2z)\boldsymbol{j} + (\lambda xy^2z + xy)\boldsymbol{k}$$

is solenoidal.

36 (Spherical polar coordinates) Using the notation introduced in Exercise 30, show, working from first principles, that

$$\nabla \cdot \boldsymbol{v} = \operatorname{div} \boldsymbol{v} = \frac{1}{r^2}\frac{\partial}{\partial r}(r^2 v_r) + \frac{1}{r \sin\theta}\frac{\partial}{\partial\theta}(v_\theta \sin\theta) + \frac{1}{r\sin\theta}\frac{\partial}{\partial\phi}(v_\phi)$$

where $\boldsymbol{v} = v_r\boldsymbol{u}_r + v_\theta\boldsymbol{u}_\theta + v_\phi\boldsymbol{u}_\phi$.

37 A force field $\boldsymbol{F}$, defined by the inverse square law, is given by

$$\boldsymbol{F} = \boldsymbol{r}/r^3$$

Show that $\nabla \cdot \boldsymbol{F} = 0$.

7.3.3 Curl of a vector field

It is clear from observations (for example by watching the movements of marked corks on water) that many fluid flows involve rotational motion of the fluid particles. Complete determination of this motion requires knowledge of the axis of

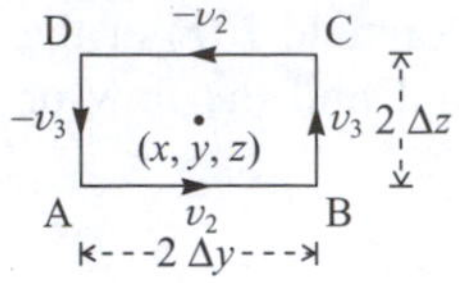

Figure 7.10 Flow around a rectangle.

rotation, the rate of rotation and its sense (clockwise or anticlockwise). The measure of rotation is thus a vector quantity, which we shall find by calculating its x, y and z components separately. Consider the vector field $\boldsymbol{v}(\boldsymbol{r})$. To find the flow around an axis in the x direction at the point $\boldsymbol{r}$, we take an elementary rectangle surrounding $\boldsymbol{r}$ perpendicular to the x direction, as shown in Figure 7.10.

To measure the circulation around the point $\boldsymbol{r}$ about an axis parallel to the x direction, we calculate the flow around the elementary rectangle ABCD and divide by its area, giving

$$[v_2(x, y^*, z - \Delta z)(2\Delta y) + v_3(x, y + \Delta y, z^*)(2\Delta z)$$
$$- v_2(x, \tilde{y}, z + \Delta z)(2\Delta y) - v_3(x, y - \Delta y, \tilde{z})(2\Delta z)]/(4\Delta y \Delta z)$$

where $y^*, \tilde{y} \in (y - \Delta y, y + \Delta y)$, $z^*, \tilde{z} \in (z - \Delta z, z + \Delta z)$ and $\boldsymbol{v} = v_1\boldsymbol{i} + v_2\boldsymbol{j} + v_3\boldsymbol{k}$. Rearranging, we obtain

$$-[v_2(x, \tilde{y}, z + \Delta z) - v_2(x, y^*, z - \Delta z)]/(2\Delta z)$$
$$+[v_3(x, y + \Delta y, z^*) - v_3(x, y - \Delta y, \tilde{z})]/(2\Delta y)$$

Proceeding to the limit as $\Delta y \Delta z \to 0$, we obtain the x component of this vector as

$$\frac{\partial v_3}{\partial y} - \frac{\partial v_2}{\partial z}$$

By similar arguments, we obtain the y and z components as

$$\frac{\partial v_1}{\partial z} - \frac{\partial v_3}{\partial x}, \quad \frac{\partial v_2}{\partial x} - \frac{\partial v_1}{\partial y}$$

respectively.

The vector measuring the rotation about a point in the fluid is called the **curl** of $\boldsymbol{v}$:

$$\text{curl } \boldsymbol{v} = \left(\frac{\partial v_3}{\partial y} - \frac{\partial v_2}{\partial z}\right)\boldsymbol{i} + \left(\frac{\partial v_1}{\partial z} - \frac{\partial v_3}{\partial x}\right)\boldsymbol{j} + \left(\frac{\partial v_2}{\partial x} - \frac{\partial v_1}{\partial y}\right)\boldsymbol{k}$$

$$= \left(\frac{\partial v_3}{\partial y} - \frac{\partial v_2}{\partial z}, \frac{\partial v_1}{\partial z} - \frac{\partial v_3}{\partial x}, \frac{\partial v_2}{\partial x} - \frac{\partial v_1}{\partial y}\right) \tag{7.17}$$

It may be written formally as

$$\text{curl } \boldsymbol{v} = \begin{vmatrix} \boldsymbol{i} & \boldsymbol{j} & \boldsymbol{k} \\ \dfrac{\partial}{\partial x} & \dfrac{\partial}{\partial y} & \dfrac{\partial}{\partial z} \\ v_1 & v_2 & v_3 \end{vmatrix} \tag{7.18}$$

or more compactly as

$$\text{curl } \boldsymbol{v} = \nabla \times \boldsymbol{v}$$

EXAMPLE 7.13 Find the curl of the vector $v = (2x - y^2, 3z + x^2, 4y - z^2)$ at the point (1, 2, 3).

Solution Here $v_1 = 2x - y^2$, $v_2 = 3z + x^2$, $v_3 = 4y - z^2$, so that

$$\text{curl } v = \begin{vmatrix} i & j & k \\ \dfrac{\partial}{\partial x} & \dfrac{\partial}{\partial y} & \dfrac{\partial}{\partial z} \\ 2x - y^2 & 3z + x^2 & 4y - z^2 \end{vmatrix}$$

$$= i\left[\frac{\partial}{\partial y}(4y - z^2) - \frac{\partial}{\partial z}(3z + x^2)\right]$$

$$- j\left[\frac{\partial}{\partial x}(4y - z^2) - \frac{\partial}{\partial z}(2x - y^2)\right]$$

$$+ k\left[\frac{\partial}{\partial x}(3z + x^2) - \frac{\partial}{\partial y}(2x - y^2)\right]$$

$$= i(4 - 3) - j(0 - 0) + k(2x + 2y) = i + 2(x + y)k$$

Thus, at the point (1, 2, 3), $\nabla \times v = (1, 0, 6)$.

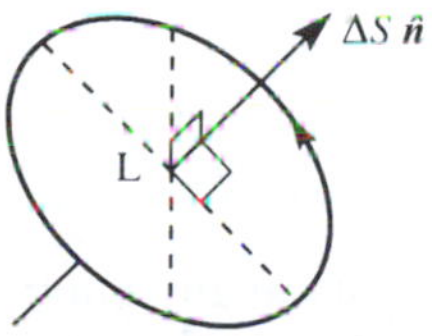

Figure 7.11 Circulation around the element ΔS.

More generally, the component of the curl of a vector field $F(r)$ in the direction of the unit vector $\hat{n}$ at a point L is found by enclosing L by an elementary area ΔS that is perpendicular to $\hat{n}$, as in Figure 7.11, and calculating the flow around ΔS per unit area. Thus

$$(\text{curl } F) \cdot \hat{n} = \lim_{\Delta S \to 0} \frac{\text{flow round } \Delta S}{\Delta S}$$

Another way of visualizing the meaning of the curl of a vector is to consider the motion of a rigid body. We can describe such motion by specifying the angular velocity ω of the body about an axis OA, where O is a fixed point in the body together with the translational (linear) velocity v of O itself. Then at any point P in the body the velocity u is given by

$$u = v + \omega \times r$$

as shown in Figure 7.12. Here v and ω are independent of (x, y, z). Thus

$$\text{curl } u = \text{curl } v + \text{curl } (\omega \times r) = 0 + \text{curl } (\omega \times r)$$

The vector $\omega \times r$ is given by

$$\omega \times r = (\omega_1, \omega_2, \omega_3) \times (x, y, z)$$

$$= (\omega_2 z - \omega_3 y)i + (\omega_3 x - \omega_1 z)j + (\omega_1 y - \omega_2 x)k$$

and

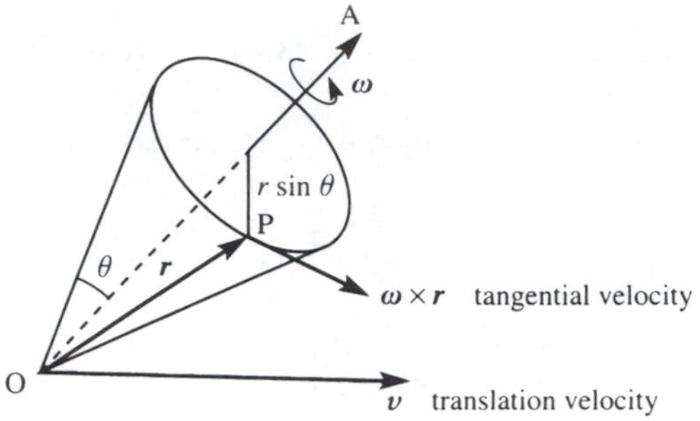

Figure 7.12 Rotation of a rigid body.

$$\operatorname{curl}(\boldsymbol{\omega}\times\boldsymbol{r}) = \begin{vmatrix} \boldsymbol{i} & \boldsymbol{j} & \boldsymbol{k} \\ \dfrac{\partial}{\partial x} & \dfrac{\partial}{\partial y} & \dfrac{\partial}{\partial z} \\ \omega_2 z - \omega_3 y & \omega_3 x - \omega_1 z & \omega_1 y - \omega_2 x \end{vmatrix}$$

$$= 2\omega_1\boldsymbol{i} + 2\omega_2\boldsymbol{j} + 2\omega_3\boldsymbol{k} = 2\boldsymbol{\omega}$$

Thus

$$\operatorname{curl}\boldsymbol{u} = 2\boldsymbol{\omega}$$

that is,

$$\boldsymbol{\omega} = \tfrac{1}{2}\operatorname{curl}\boldsymbol{u}$$

Hence when any rigid body is in motion, the curl of its linear velocity at any point is twice its angular velocity in magnitude and has the same direction.

Applying this result to the motion of a fluid, we can see by regarding particles of the fluid as miniature bodies that when the curl of the velocity is zero there is no rotation of the particle, and the motion is said to be **curl-free** or **irrotational**. When the curl is non-zero, the motion is **rotational**.

7.3.4 Exercises

38 Find $\boldsymbol{u} = \operatorname{curl}\boldsymbol{v}$ when $\boldsymbol{v} = (3xz^2, -yz, x + 2z)$.

39 A vector field is defined by $\boldsymbol{v} = (yz, xz, xy)$. Show that $\operatorname{curl}\boldsymbol{v} = \boldsymbol{0}$.

40 Show that if $\boldsymbol{v} = (2x + yz, 2y + zx, 2z + xy)$ then $\operatorname{curl}\boldsymbol{v} = \boldsymbol{0}$, and find $f(\boldsymbol{r})$ such that $\boldsymbol{v} = \operatorname{grad} f$.

41 By evaluating each term separately, verify the identity

$$\nabla \times (f\boldsymbol{v}) = f(\nabla \times \boldsymbol{v}) + (\nabla f) \times \boldsymbol{v}$$

for $f(\boldsymbol{r}) = x^3 - y$ and $\boldsymbol{v}(\boldsymbol{r}) = (z, 0, -x)$.

42 Find constants a, b and c such that the vector field defined by

$$\boldsymbol{F} = (4xy + az^3)\boldsymbol{i} + (bx^2 + 3z)\boldsymbol{j} + (6xz^2 + cy)\boldsymbol{k}$$

is irrotational. With these values of a, b and c, determine a scalar function $\phi(x, y, z)$ such that $\boldsymbol{F} = \nabla\phi$.

43 If $\boldsymbol{v} = -y\boldsymbol{i} + x\boldsymbol{j} + xyz\boldsymbol{k}$ is the velocity vector of a fluid, find the local value of the angular velocity at the point (1, 3, 2).

44 If the velocity of a fluid at the point (x, y, z) is given by

$$\boldsymbol{v} = (ax + by)\boldsymbol{i} + (cx + dy)\boldsymbol{j}$$

find the conditions on the constants a, b, c and d in order that

$$\text{div}\,\boldsymbol{v} = 0, \quad \text{curl}\,\boldsymbol{v} = 0$$

Verify that in this case

$$\boldsymbol{v} = \tfrac{1}{2}\,\text{grad}\,(ax^2 + 2bxy - ay^2)$$

45 (Spherical polar coordinates) Using the notation introduced in Exercise 30, show that

$$\nabla \times \boldsymbol{v} = \text{curl}\,\boldsymbol{v}$$

$$= \frac{1}{r^2 \sin\theta}\begin{vmatrix} \boldsymbol{u}_r & r\boldsymbol{u}_\theta & r\sin\boldsymbol{u}_\phi \\ \dfrac{\partial}{\partial r} & \dfrac{\partial}{\partial \theta} & \dfrac{\partial}{\partial \phi} \\ v_r & rv_\theta & r\sin v_\phi \end{vmatrix}$$

7.3.5 Further properties of the vector operator ∇

So far we have used the vector operator in three ways:

$$\nabla f = \text{grad}\,f = \frac{\partial f}{\partial x}\boldsymbol{i} + \frac{\partial f}{\partial y}\boldsymbol{j} + \frac{\partial f}{\partial z}\boldsymbol{k}, \qquad f(\boldsymbol{r}) \text{ a scalar field}$$

$$\nabla\cdot\boldsymbol{F} = \text{div}\,\boldsymbol{F} = \frac{\partial f_1}{\partial x} + \frac{\partial f_2}{\partial y} + \frac{\partial f_3}{\partial z}, \qquad \boldsymbol{F}(\boldsymbol{r}) \text{ a vector field}$$

$$\nabla\times\boldsymbol{F} = \text{curl}\,\boldsymbol{F}$$

$$= \left(\frac{\partial f_3}{\partial y} - \frac{\partial f_2}{\partial z}\right)\boldsymbol{i} + \left(\frac{\partial f_1}{\partial z} - \frac{\partial f_3}{\partial x}\right)\boldsymbol{j} + \left(\frac{\partial f_2}{\partial x} - \frac{\partial f_1}{\partial y}\right)\boldsymbol{k}, \qquad \boldsymbol{F}(\boldsymbol{r}) \text{ a vector field}$$

A further application is in determining the directional derivative of a vector field:

$$\boldsymbol{a}\cdot\nabla\boldsymbol{F} = \left(a_1\frac{\partial}{\partial x} + a_2\frac{\partial}{\partial y} + a_3\frac{\partial}{\partial z}\right)\boldsymbol{F}$$

$$= \left(a_1\frac{\partial f_1}{\partial x} + a_2\frac{\partial f_1}{\partial y} + a_3\frac{\partial f_1}{\partial z}\right)\boldsymbol{i} + \left(a_1\frac{\partial f_2}{\partial x} + a_2\frac{\partial f_2}{\partial y} + a_3\frac{\partial f_2}{\partial z}\right)\boldsymbol{j}$$

$$+ \left(a_1\frac{\partial f_3}{\partial x} + a_2\frac{\partial f_3}{\partial y} + a_3\frac{\partial f_3}{\partial z}\right)\boldsymbol{k}$$

The ordinary rules of differentiation carry over to this vector differential operator, but they have to be applied with care, using the rules of vector algebra. For non-orthogonal coordinate systems a specialist textbook should be consulted. Thus for scalar fields $f(\boldsymbol{r})$, $g(\boldsymbol{r})$ and vector fields $\boldsymbol{u}(\boldsymbol{r})$, $\boldsymbol{v}(\boldsymbol{r})$ we have

$$\nabla[f(g(\boldsymbol{r}))] = \frac{\mathrm{d}f}{\mathrm{d}g}\nabla g \qquad \textbf{(7.19a)}$$

$$\nabla[f(\boldsymbol{r})g(\boldsymbol{r})] = g(\boldsymbol{r})\nabla f(\boldsymbol{r}) + f(\boldsymbol{r})\nabla g(\boldsymbol{r}) \qquad \textbf{(7.19b)}$$

$$\nabla[\boldsymbol{u}(\boldsymbol{r})\cdot\boldsymbol{v}(\boldsymbol{r})] = \boldsymbol{v}\times(\nabla\times\boldsymbol{u}) + \boldsymbol{u}\times(\nabla\times\boldsymbol{v}) + (\boldsymbol{v}\cdot\nabla)\boldsymbol{u} + (\boldsymbol{u}\cdot\nabla)\boldsymbol{v} \quad (7.19c)$$

$$\nabla\cdot[f(\boldsymbol{r})\boldsymbol{u}(\boldsymbol{r})] = \boldsymbol{u}\cdot\nabla f + f\nabla\cdot\boldsymbol{u} \quad (7.19d)$$

$$\nabla\times[f(\boldsymbol{r})\boldsymbol{u}(\boldsymbol{r})] = (\nabla f)\times\boldsymbol{u} + f\nabla\times\boldsymbol{u} \quad (7.19e)$$

$$\nabla\cdot[\boldsymbol{u}(\boldsymbol{r})\times\boldsymbol{v}(\boldsymbol{r})] = \boldsymbol{v}\cdot(\nabla\times\boldsymbol{u}) - \boldsymbol{u}\cdot(\nabla\times\boldsymbol{v}) \quad (7.19f)$$

$$\nabla\times[\boldsymbol{u}(\boldsymbol{r})\times\boldsymbol{v}(\boldsymbol{r})] = (\boldsymbol{v}\cdot\nabla)\boldsymbol{u} - \boldsymbol{v}(\nabla\cdot\boldsymbol{u}) - (\boldsymbol{u}\cdot\nabla)\boldsymbol{v} + \boldsymbol{u}(\nabla\cdot\boldsymbol{v}) \quad (7.19g)$$

Higher-order derivatives can also be formed, giving the following:

$$\text{div}\,[\text{grad}\, f(\boldsymbol{r})] = \nabla\cdot\nabla f = \frac{\partial^2 f}{\partial x^2} + \frac{\partial^2 f}{\partial y^2} + \frac{\partial^2 f}{\partial z^2} = \nabla^2 f \quad (7.20)$$

where ∇^2 is called the **Laplacian operator** (sometimes denoted by $\triangle$);

$$\text{curl}\,[\text{grad}\, f(\boldsymbol{r})] = \nabla\times\nabla f(\boldsymbol{r}) \equiv 0 \quad (7.21)$$

since

$$\begin{aligned}\nabla\times\nabla f &= \left(\frac{\partial^2 f}{\partial y\partial z} - \frac{\partial^2 f}{\partial z\partial y}\right)\boldsymbol{i} + \left(\frac{\partial^2 f}{\partial z\partial x} - \frac{\partial^2 f}{\partial x\partial z}\right)\boldsymbol{j} + \left(\frac{\partial^2 f}{\partial x\partial y} - \frac{\partial^2 f}{\partial y\partial x}\right)\boldsymbol{k}\\ &= 0\end{aligned}$$

when all second-order derivatives of $f(\boldsymbol{r})$ are continuous;

$$\text{div}\,[\text{curl}\,\boldsymbol{v}(\boldsymbol{r})] = \nabla\cdot(\nabla\times\boldsymbol{v}) \equiv 0 \quad (7.22)$$

since

$$\frac{\partial}{\partial x}\left(\frac{\partial v_3}{\partial y} - \frac{\partial v_2}{\partial z}\right) + \frac{\partial}{\partial y}\left(\frac{\partial v_1}{\partial z} - \frac{\partial v_3}{\partial x}\right) + \frac{\partial}{\partial z}\left(\frac{\partial v_2}{\partial x} - \frac{\partial v_1}{\partial y}\right) = 0$$

$$\text{grad}\,(\text{div}\,\boldsymbol{v}) = \nabla(\nabla\cdot\boldsymbol{v}) = \left(\boldsymbol{i}\frac{\partial}{\partial x} + \boldsymbol{j}\frac{\partial}{\partial y} + \boldsymbol{k}\frac{\partial}{\partial z}\right)\left(\frac{\partial v_1}{\partial x} + \frac{\partial v_2}{\partial y} + \frac{\partial v_3}{\partial z}\right) \quad (7.23)$$

$$\nabla^2\boldsymbol{v} = \left(\frac{\partial^2}{\partial x^2} + \frac{\partial^2}{\partial y^2} + \frac{\partial^2}{\partial z^2}\right)(v_1\boldsymbol{i} + v_2\boldsymbol{j} + v_3\boldsymbol{k}) \quad (7.24)$$

$$\text{curl}\,[\text{curl}\,\boldsymbol{v}(\boldsymbol{r})] = \nabla\times(\nabla\times\boldsymbol{v}) = \nabla(\nabla\cdot\boldsymbol{v}) - \nabla^2\boldsymbol{v} \quad (7.25)$$

EXAMPLE 7.14 Verify that $\nabla\times(\nabla\times\boldsymbol{v}) = \nabla(\nabla\cdot\boldsymbol{v}) - \nabla^2\boldsymbol{v}$ for the vector field $\boldsymbol{v} = (3xz^2, -yz, x + 2z)$.

Solution

$$\nabla\times\boldsymbol{v} = \begin{vmatrix}\boldsymbol{i} & \boldsymbol{j} & \boldsymbol{k}\\ \dfrac{\partial}{\partial x} & \dfrac{\partial}{\partial y} & \dfrac{\partial}{\partial z}\\ 3xz^2 & -yz & x+2z\end{vmatrix} = (y,\ 6xz - 1,\ 0)$$

$$\nabla \times (\nabla \times \boldsymbol{v}) = \begin{vmatrix} \boldsymbol{i} & \boldsymbol{j} & \boldsymbol{k} \\ \dfrac{\partial}{\partial x} & \dfrac{\partial}{\partial y} & \dfrac{\partial}{\partial z} \\ y & 6xz - 1 & 0 \end{vmatrix} = (-6x, 0, 6z - 1)$$

$$\begin{aligned} \nabla \cdot \boldsymbol{v} &= \frac{\partial}{\partial x}(3xz^2) + \frac{\partial}{\partial y}(-yz) + \frac{\partial}{\partial z}(x + 2z) \\ &= 3z^2 - z + 2 \\ \nabla(\nabla \cdot \boldsymbol{v}) &= (0, 0, 6z - 1) \\ \nabla^2 \boldsymbol{v} &= (\nabla^2(3xz^2), \nabla^2(-yz), \nabla^2(x + 2z)) \\ &= (6x, 0, 0) \end{aligned}$$

Thus

$$\begin{aligned} \nabla(\nabla \cdot \boldsymbol{v}) - \nabla^2 \boldsymbol{v} &= (-6x, 0, 6z - 1) \\ &= \nabla \times (\nabla \times \boldsymbol{v}) \end{aligned}$$

Similar verifications for other identities are suggested in Exercises 7.3.6.

EXAMPLE 7.15 Maxwell's equations in free space may be written, in Gaussian units, as

$$\operatorname{div} \boldsymbol{H} = 0, \quad \operatorname{div} \boldsymbol{E} = 0$$

$$\nabla \times \boldsymbol{H} = \frac{1}{c}\frac{\partial \boldsymbol{E}}{\partial t}, \quad \nabla \times \boldsymbol{E} = -\frac{1}{c}\frac{\partial \boldsymbol{H}}{\partial t}$$

where c is the velocity of light (assumed constant). Show that these equations are satisfied by

$$\boldsymbol{H} = \frac{1}{c}\frac{\partial}{\partial t}\operatorname{grad}\phi \times \boldsymbol{k}, \quad \boldsymbol{E} = -\boldsymbol{k}\frac{1}{c^2}\frac{\partial^2 \phi}{\partial t^2} + \frac{\partial}{\partial z}\operatorname{grad}\phi$$

where ϕ satisfies

$$\nabla^2 \phi = \frac{1}{c^2}\frac{\partial^2 \phi}{\partial t^2}$$

and $\boldsymbol{k}$ is a unit vector along the z axis.

Solution (a) $\boldsymbol{H} = \dfrac{1}{c}\dfrac{\partial}{\partial t}\operatorname{grad}\phi \times \boldsymbol{k}$

gives

$$\operatorname{div} \boldsymbol{H} = \frac{1}{c}\frac{\partial}{\partial t}\operatorname{div}(\operatorname{grad}\phi \times \boldsymbol{k})$$

$$= \frac{1}{c}\frac{\partial}{\partial t}[\boldsymbol{k}\cdot\operatorname{curl}(\operatorname{grad}\phi) - (\operatorname{grad}\phi)\cdot\operatorname{curl}\boldsymbol{k}], \quad \text{from (7.19f)}$$

By (7.21), curl (grad ϕ) = 0, and since $\boldsymbol{k}$ is a constant vector, curl $\boldsymbol{k}$ = 0, so that

$$\operatorname{div}\boldsymbol{H} = 0$$

(b) $$\boldsymbol{E} = -\frac{\boldsymbol{k}}{c^2}\frac{\partial^2\phi}{\partial t^2} + \frac{\partial}{\partial z}\operatorname{grad}\phi$$

gives

$$\operatorname{div}\boldsymbol{E} = -\frac{1}{c^2}\operatorname{div}\left(\boldsymbol{k}\frac{\partial^2\phi}{\partial t^2}\right) + \frac{\partial}{\partial z}\operatorname{div}\operatorname{grad}\phi$$

$$= -\frac{1}{c^2}\frac{\partial}{\partial z}\left(\frac{\partial^2\phi}{\partial t^2}\right) + \frac{\partial}{\partial z}(\nabla^2\phi), \quad \text{by (7.20)}$$

$$= \frac{\partial}{\partial z}\left(\nabla^2\phi - \frac{1}{c^2}\frac{\partial^2\phi}{\partial t^2}\right)$$

and since $\nabla^2\phi = (1/c^2)\partial^2\phi/\partial t^2$, we have

$$\operatorname{div}\boldsymbol{E} = 0$$

(c) $$\operatorname{curl}\boldsymbol{H} = \frac{1}{c}\frac{\partial}{\partial t}\operatorname{curl}(\operatorname{grad}\phi\times\boldsymbol{k})$$

$$= \frac{1}{c}\frac{\partial}{\partial t}[(\boldsymbol{k}\cdot\nabla)\operatorname{grad}\phi$$

$$-\boldsymbol{k}(\operatorname{div}\operatorname{grad}\phi) - (\operatorname{grad}\phi\cdot\nabla)\boldsymbol{k} + \operatorname{grad}\phi(\nabla\cdot\boldsymbol{k})], \quad \text{from (7.19g)}$$

$$= \frac{1}{c}\frac{\partial}{\partial t}\left(\frac{\partial}{\partial z}\operatorname{grad}\phi - \boldsymbol{k}\nabla^2\phi\right), \quad \text{since } \boldsymbol{k} \text{ is a constant vector}$$

$$= \frac{1}{c}\frac{\partial \boldsymbol{E}}{\partial t}$$

(d) $$\operatorname{curl}\boldsymbol{E} = -\frac{1}{c^2}\operatorname{curl}\left(\boldsymbol{k}\frac{\partial^2\phi}{\partial t^2}\right) + \frac{\partial}{\partial z}\operatorname{curl}\operatorname{grad}\phi$$

$$= -\frac{1}{c^2}\begin{bmatrix}\boldsymbol{i} & \boldsymbol{j} & \boldsymbol{k}\\ \dfrac{\partial}{\partial x} & \dfrac{\partial}{\partial y} & \dfrac{\partial}{\partial z}\\ 0 & 0 & \dfrac{\partial^2\phi}{\partial t^2}\end{bmatrix}, \quad \text{since curl grad } \phi = 0 \text{ by (7.21)}$$

$$= -\frac{1}{c^2}\left(\boldsymbol{i}\frac{\partial^3\phi}{\partial y\,\partial t^2} - \boldsymbol{j}\frac{\partial^3\phi}{\partial x\,\partial t^2}\right)$$

Also,

$$\begin{aligned}\frac{\partial \boldsymbol{H}}{\partial t} &= \frac{1}{c}\frac{\partial^2}{\partial t^2}\text{grad } \phi \times \boldsymbol{k}\\ &= \frac{1}{c}\frac{\partial^2}{\partial t^2}(\text{grad } \phi \times \boldsymbol{k}), \quad \text{since } \boldsymbol{k} \text{ is a constant vector}\\ &= \frac{1}{c}\frac{\partial^2}{\partial t^2}\left[\left(\frac{\partial \phi}{\partial x}\boldsymbol{i} + \frac{\partial \phi}{\partial y}\boldsymbol{j} + \frac{\partial \phi}{\partial z}\boldsymbol{k}\right) \times \boldsymbol{k}\right]\\ &= \frac{1}{c}\frac{\partial^2}{\partial t^2}\left(\boldsymbol{i}\frac{\partial \phi}{\partial y} - \boldsymbol{j}\frac{\partial \phi}{\partial x}\right) = \frac{1}{c}\left(\boldsymbol{i}\frac{\partial^3 \phi}{\partial y \partial t^2} - \boldsymbol{j}\frac{\partial^3 \phi}{\partial x \partial t^2}\right)\end{aligned}$$

so that we have

$$\nabla \times \boldsymbol{E} = -\frac{1}{c}\frac{\partial \boldsymbol{H}}{\partial t}$$

7.3.6 Exercises

46 Show that if g is a function of $\boldsymbol{r} = (x, y, z)$ then

$$\text{grad } g = \frac{1}{r}\frac{\mathrm{d}g}{\mathrm{d}r}\boldsymbol{r}$$

Deduce that if $\boldsymbol{u}$ is a vector field then

$$\text{div }[(\boldsymbol{u} \times \boldsymbol{r})g] = (\boldsymbol{r} \cdot \text{curl } \boldsymbol{u})g$$

47 For $\phi(x, y, z) = x^2y^2z^3$ and $\boldsymbol{F}(x, y, z) = x^2y\boldsymbol{i} + xy^2z\boldsymbol{j} - yz^2\boldsymbol{k}$ determine

(a) $\nabla^2\phi$ (b) grad div $\boldsymbol{F}$ (c) curl curl $\boldsymbol{F}$

48 Show that if $\boldsymbol{a}$ is a constant vector and $\boldsymbol{r}$ is the position vector $\boldsymbol{r} = (x, y, z)$ then

$$\text{div }\{\text{grad }[(\boldsymbol{r} \cdot \boldsymbol{r})(\boldsymbol{r} \cdot \boldsymbol{a})]\} = 10(\boldsymbol{r} \cdot \boldsymbol{a})$$

49 Verify the identity

$$\nabla^2\boldsymbol{v} = \text{grad div } \boldsymbol{v} - \text{curl curl } \boldsymbol{v}$$

for the vector field $\boldsymbol{v} = x^2y(x\boldsymbol{i} + y\boldsymbol{j} + z\boldsymbol{k})$.

50 Verify, by calculating each term separately, the identities

$$\text{div }(\boldsymbol{u} \times \boldsymbol{v}) = \boldsymbol{v} \cdot \text{curl } \boldsymbol{u} - \boldsymbol{u} \cdot \text{curl } \boldsymbol{v}$$

$$\text{curl }(\boldsymbol{u} \times \boldsymbol{v}) = \boldsymbol{u} \text{ div } \boldsymbol{v} - \boldsymbol{v} \text{ div } \boldsymbol{u} + (\boldsymbol{v} \cdot \nabla)\boldsymbol{u} - (\boldsymbol{u} \cdot \nabla)\boldsymbol{v}$$

when $\boldsymbol{u} = xy\boldsymbol{j} + xz\boldsymbol{k}$ and $\boldsymbol{v} = xy\boldsymbol{i} + yz\boldsymbol{k}$.

51 If $\boldsymbol{r}$ is the usual position vector $\boldsymbol{r} = (x, y, z)$, show that

(a) $\text{div grad}\left(\frac{1}{r}\right) = 0$

(b) $\text{curl}\left[\boldsymbol{k} \times \text{grad}\left(\frac{1}{r}\right)\right] + \text{grad}\left[\boldsymbol{k} \cdot \text{grad}\left(\frac{1}{r}\right)\right] = 0$

52 If $\boldsymbol{A}$ is a constant vector and $\boldsymbol{r}$ is the position vector $\boldsymbol{r} = (x, y, z)$, show that

(a) $\text{grad}\left(\frac{\boldsymbol{A} \cdot \boldsymbol{r}}{r^3}\right) = \frac{\boldsymbol{A}}{r^3} - 3\frac{(\boldsymbol{A} \cdot \boldsymbol{r})}{r^5}\boldsymbol{r}$

(b) $\text{curl}\left(\frac{\boldsymbol{A} \times \boldsymbol{r}}{r^3}\right) = \frac{2\boldsymbol{A}}{r^3} + \frac{3}{r^5}(\boldsymbol{A} \times \boldsymbol{r}) \times \boldsymbol{r}$

53 If $\boldsymbol{r}$ is the position vector $\boldsymbol{r} = (x, y, z)$, and $\boldsymbol{a}$ and $\boldsymbol{b}$ are constant vectors, show that

(a) $\nabla \times \boldsymbol{r} = 0$

(b) $(\boldsymbol{a} \cdot \nabla)\boldsymbol{r} = \boldsymbol{a}$

(c) $\nabla \times [(\boldsymbol{a} \cdot \boldsymbol{r})\boldsymbol{b} - (\boldsymbol{b} \cdot \boldsymbol{r})\boldsymbol{a}] = 2(\boldsymbol{a} \times \boldsymbol{b})$

(d) $\nabla \cdot [(\boldsymbol{a} \cdot \boldsymbol{r})\boldsymbol{b} - (\boldsymbol{b} \cdot \boldsymbol{r})\boldsymbol{a}] = 0$

54 By evaluating $\nabla \cdot (\nabla f)$, show that the Laplacian in spherical polar coordinates (see Exercise 30) is given by

$$\nabla^2 f = \frac{1}{r^2}\frac{\partial}{\partial r}\left(r^2\frac{\partial f}{\partial r}\right) + \frac{1}{r^2 \sin^2\theta}\frac{\partial^2 f}{\partial \phi^2} + \frac{1}{r^2 \sin\theta}\frac{\partial}{\partial \theta}\left(\sin\theta\frac{\partial f}{\partial \theta}\right)$$

55 Show that Maxwell's equations in free space, namely

$$\operatorname{div} \boldsymbol{H} = 0, \quad \operatorname{div} \boldsymbol{E} = 0$$

$$\nabla \times \boldsymbol{H} = \frac{1}{c}\frac{\partial \boldsymbol{E}}{\partial t}, \quad \nabla \times \boldsymbol{E} = -\frac{1}{c}\frac{\partial \boldsymbol{H}}{\partial t}$$

are satisfied by

$$\boldsymbol{H} = \frac{1}{c}\operatorname{curl}\frac{\partial \boldsymbol{Z}}{\partial t}$$

$$\boldsymbol{E} = \operatorname{curl}\operatorname{curl}\boldsymbol{Z}$$

where the Hertzian vector $\boldsymbol{Z}$ satisfies

$$\nabla^2\boldsymbol{Z} = \frac{1}{c}\frac{\partial^2 \boldsymbol{Z}}{\partial t^2}$$

7.4 Topics in integration

In the previous sections we saw how the idea of the differentiation of a function of a single variable is generalized to include scalar and vector point functions. We now turn to the inverse process of integration. The fundamental idea of an integral is that of summing all the constituent parts that make a whole. More formally, we define the integral of a function $f(x)$ by

$$\int_a^b f(x)\,\mathrm{d}x = \lim_{\substack{n\to\infty \\ \text{all } \Delta x_i \to 0}} \sum_{i=1}^{n} f(\tilde{x}_i)\Delta x_i$$

where $a = x_0 < x_1 < x_2 < \ldots < x_{n-1} < x_n = b$, $\Delta x_i = x_i - x_{i-1}$ and $x_{i-1} \leqslant \tilde{x}_i \leqslant x_i$. Geometrically, we can interpret this integral as the area between the graph $y = f(x)$, the x axis and the lines $x = a$ and $x = b$, as illustrated in Figure 7.13.

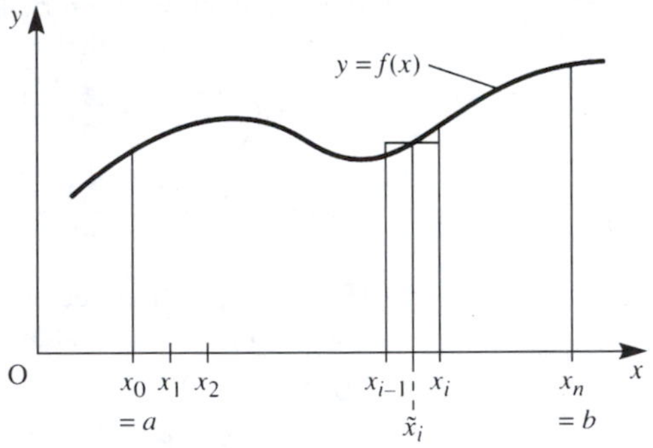

Figure 7.13 Definite integral as an area.

7.4.1 Line integrals

Consider the integral

$$\int_b^a f(x, y)\,\mathrm{d}x, \quad \text{where } y = g(x)$$

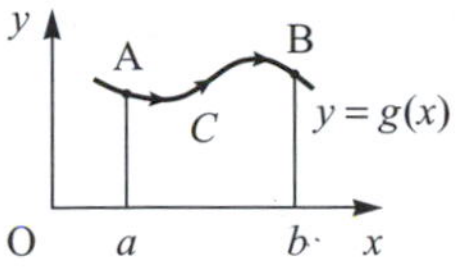

Figure 7.14 Integral along a curve.

This can be evaluated in the usual way by first substituting for y in terms of x in the integrand and then performing the integration

$$\int_a^b f(x, g(x))\,dx$$

Clearly the value of the integral will, in general, depend on the function $y = g(x)$. It may be interpreted as evaluating the integral $\int_a^b f(x, y)\,dx$ along the curve $y = g(x)$, as shown in Figure 7.14. Note, however, that the integral is *not* represented in this case by the area under the curve. This type of integral is called a **line integral**.

There are many different types of such integrals; for example

$$\underset{C}{\int_A^B} f(x, y)\,dx, \quad \underset{C}{\int_A^B} f(x, y)\,ds, \quad \underset{C}{\int_{t_1}^{t_2}} f(x, y)\,dt, \quad \underset{C}{\int_A^B} [f_1(x, y)\,dx + f_2(x, y)\,dy]$$

Here the letter under the integral sign indicates that the integral is evaluated along the curve (or **path**) C. This path is not restricted to two dimensions, and may be in as many dimensions as we please. It is normal to omit the points A and B, since they are usually implicit in the specification of C.

EXAMPLE 7.16

Evaluate $\int_C xy\,dx$ from A(1, 0) to B(0, 1) along the curve C that is the portion of $x^2 + y^2 = 1$ in the first quadrant.

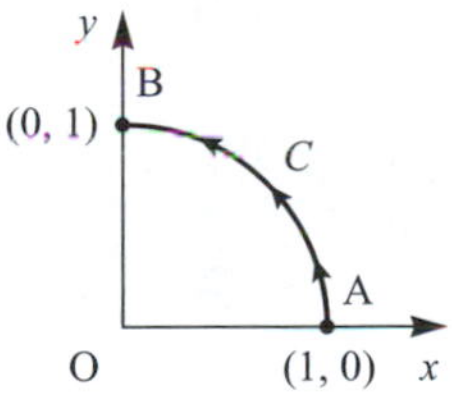

Figure 7.15

Solution The curve C is the first quadrant of the unit circle as shown in Figure 7.15. On the curve, $y = \sqrt{(1 - x^2)}$, so that

$$\int_C xy\,dx = \int_1^0 x\sqrt{(1 - x^2)}\,dx = [-\tfrac{1}{2}\tfrac{2}{3}(1 - x^2)^{3/2}]_1^0 = -\tfrac{1}{3}$$

EXAMPLE 7.17

Evaluate the integral

$$I = \int_C [(x^2 + 2y)\,dx + (x + y^2)\,dy]$$

from A(0, 1) to B(2, 3) along the curve C defined by $y = x + 1$.

Solution The curve C is the straight line $y = x + 1$ from the point A(0, 1) to the point B(2, 3). In this case we can eliminate either x or y. Using

$$y = x + 1 \quad \text{and} \quad \mathrm{d}y = \mathrm{d}x$$

we have, on eliminating y,

$$I = \int_{x=0}^{x=2} \{[x^2 + 2(x+1)]\,\mathrm{d}x + [x + (x+1)^2]\,\mathrm{d}x\}$$

$$= \int_0^2 (2x^2 + 5x + 3)\,\mathrm{d}x = [\tfrac{2}{3}x^3 + \tfrac{5}{2}x^2 + 3x]_0^2 = \tfrac{64}{3}$$

In many practical problems line integrals involving vectors occur. Let $P(\boldsymbol{r})$ be a point on a curve C in three dimensions, and let $\boldsymbol{t}$ be the unit tangent vector at P in the sense of the integration (that is, in the sense of increasing arclength s), as indicated in Figure 7.16. Then $\boldsymbol{t}\,\mathrm{d}s$ is the vector element of arc at P, and

$$\boldsymbol{t}\,\mathrm{d}s = \left[\frac{\mathrm{d}x}{\mathrm{d}s}\boldsymbol{i} + \frac{\mathrm{d}y}{\mathrm{d}s}\boldsymbol{j} + \frac{\mathrm{d}z}{\mathrm{d}s}\boldsymbol{k}\right]\mathrm{d}s = \mathrm{d}x\,\boldsymbol{i} + \mathrm{d}y\,\boldsymbol{j} + \mathrm{d}z\,\boldsymbol{k} = \mathrm{d}\boldsymbol{r}$$

If $f_1(x, y, z)$, $f_2(x, y, z)$ and $f_3(x, y, z)$ are the scalar components of a vector field $\boldsymbol{F}(\boldsymbol{r})$ then

$$\int_C [f_1(x, y, z)\,\mathrm{d}x + f_2(x, y, z)\,\mathrm{d}y + f_3(x, y, z)\,\mathrm{d}z]$$

$$= \int_C \left[f_1(x, y, z)\frac{\mathrm{d}x}{\mathrm{d}s}\,\mathrm{d}s + f_2(x, y, z)\frac{\mathrm{d}y}{\mathrm{d}s}\,\mathrm{d}s + f_3(x, y, z)\frac{\mathrm{d}z}{\mathrm{d}s}\,\mathrm{d}s\right]$$

$$= \int_C \boldsymbol{F}\cdot\boldsymbol{t}\,\mathrm{d}s$$

$$= \int_C \boldsymbol{F}\cdot\mathrm{d}\boldsymbol{r}$$

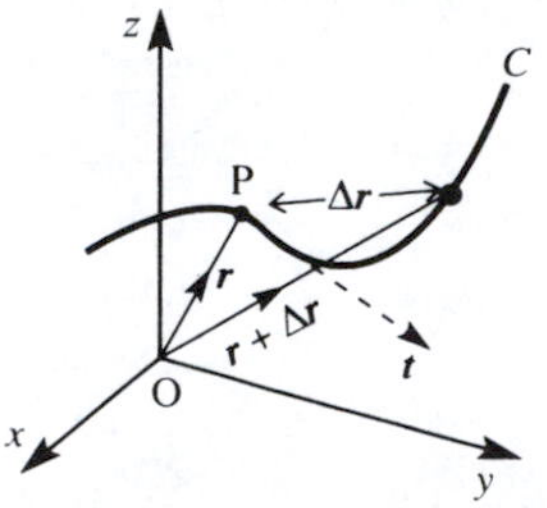

Figure 7.16 Element of arclength.

Thus, given a vector field $\boldsymbol{F}(\boldsymbol{r})$, we can evaluate line integrals of the form $\int_C \boldsymbol{F} \cdot \mathrm{d}\boldsymbol{r}$. In order to make it clear that we are integrating along a curve, the line integral is sometimes written as $\int_C \boldsymbol{F} \cdot \mathrm{d}\boldsymbol{s}$, where $\mathrm{d}\boldsymbol{s} = \mathrm{d}\boldsymbol{r}$ (some authors use $\mathrm{d}\boldsymbol{l}$ instead of $\mathrm{d}\boldsymbol{s}$ in order to avoid confusion with $\mathrm{d}\boldsymbol{S}$, the element of surface area). In a similar manner we can evaluate line integrals of the form $\int_C \boldsymbol{F} \times \mathrm{d}\boldsymbol{r}$.

EXAMPLE 7.18

Calculate (a) $\int_C \boldsymbol{F} \cdot \mathrm{d}\boldsymbol{r}$ and (b) $\int_C \boldsymbol{F} \times \mathrm{d}\boldsymbol{r}$, where C is the part of the spiral $\boldsymbol{r} = (a\cos\theta, a\sin\theta, a\theta)$ corresponding to $0 \leqslant \theta \leqslant \frac{1}{2}\pi$, and $\boldsymbol{F} = r^2\boldsymbol{i}$.

Solution

The curve C is illustrated in Figure 7.17.

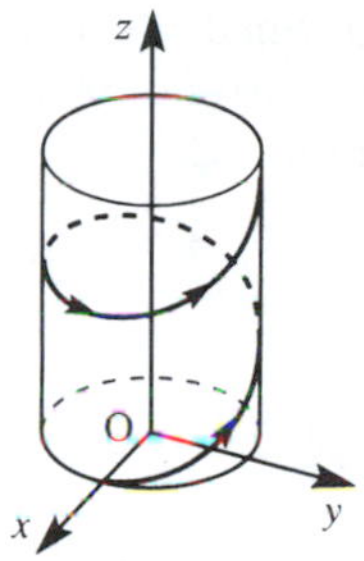

Figure 7.17 The spiral $\boldsymbol{r} = (a\cos\theta, a\sin\theta, a\theta)$.

(a) Since $\boldsymbol{r} = a\cos\theta\boldsymbol{i} + a\sin\theta\boldsymbol{j} + a\theta\boldsymbol{k}$,

$$\mathrm{d}\boldsymbol{r} = -a\sin\theta\,\mathrm{d}\theta\boldsymbol{i} + a\cos\theta\,\mathrm{d}\theta\boldsymbol{j} + a\,\mathrm{d}\theta\boldsymbol{k}$$

so that

$$\boldsymbol{F} \cdot \mathrm{d}\boldsymbol{r} = r^2\boldsymbol{i} \cdot (-a\sin\theta\,\mathrm{d}\theta\boldsymbol{i} + a\cos\theta\,\mathrm{d}\theta\boldsymbol{j} + a\,\mathrm{d}\theta\boldsymbol{k})$$

$$= -ar^2\sin\theta\,\mathrm{d}\theta$$

$$= -a^3(\cos^2\theta + \sin^2\theta + \theta^2)\sin\theta\,\mathrm{d}\theta = -a^3(1 + \theta^2)\sin\theta\,\mathrm{d}\theta$$

since $r = |\boldsymbol{r}| = \sqrt{(a^2\cos^2\theta + a^2\sin^2\theta + a^2\theta^2)}$. Thus,

$$\int_C \boldsymbol{F} \cdot \mathrm{d}\boldsymbol{r} = -a^3\int_0^{\pi/2} (1 + \theta^2)\sin\theta\,\mathrm{d}\theta$$

$$= -a^3[\cos\theta + 2\theta\sin\theta - \theta^2\cos\theta]_0^{\pi/2}, \text{ using integration by parts}$$

$$= -a^3(\pi - 1)$$

(b)
$$\boldsymbol{F} \times \mathrm{d}\boldsymbol{r} = \begin{vmatrix} \boldsymbol{i} & \boldsymbol{j} & \boldsymbol{k} \\ r^2 & 0 & 0 \\ -a\sin\theta\,\mathrm{d}\theta & a\cos\theta\,\mathrm{d}\theta & a\,\mathrm{d}\theta \end{vmatrix}$$

$$= -ar^2\,\mathrm{d}\theta\boldsymbol{j} + ar^2\cos\theta\,\mathrm{d}\theta\boldsymbol{k}$$

$$= -a^3(1 + \theta^2)\,\mathrm{d}\theta\boldsymbol{j} + a^3(1 + \theta^2)\cos\theta\,\mathrm{d}\theta\boldsymbol{k}$$

so that

$$\int_C \boldsymbol{F} \times \mathrm{d}\boldsymbol{r} = -\boldsymbol{j}a^3\int_0^{\pi/2} (1 + \theta^2)\,\mathrm{d}\theta + \boldsymbol{k}a^3\int_0^{\pi/2} (1 + \theta^2)\cos\theta\,\mathrm{d}\theta$$

$$= -\frac{\pi a^3}{24}(12 + \pi^2)\boldsymbol{j} + \frac{a^3}{4}(\pi^2 - 4)\boldsymbol{k}$$

The work done as the point of application of a force $\boldsymbol{F}$ moves along a given path C as illustrated in Figure 7.18 can be expressed as a line integral. The work

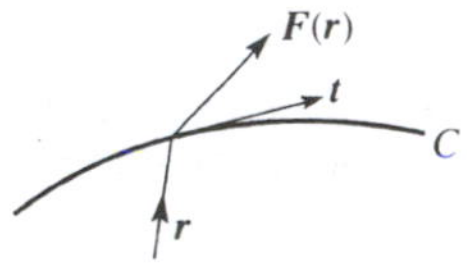

Figure 7.18 Work done by a force $\boldsymbol{F}$.

done as the point of application moves from P($\boldsymbol{r}$) to P′($\boldsymbol{r}$ + d$\boldsymbol{r}$), where $\overrightarrow{PP'} = \mathrm{d}\boldsymbol{r}$, is $\mathrm{d}W = |\mathrm{d}\boldsymbol{r}|\,|\boldsymbol{F}|\cos\theta = \boldsymbol{F}\cdot\mathrm{d}\boldsymbol{r}$. Hence the total work done as P goes from A to B is

$$W = \int_C \boldsymbol{F}\cdot\mathrm{d}\boldsymbol{r}$$

In general, W depends on the path chosen. If, however, $\boldsymbol{F}(\boldsymbol{r})$ is such that $\boldsymbol{F}(\boldsymbol{r})\cdot\mathrm{d}\boldsymbol{r}$ is an exact differential, say $-\mathrm{d}U$, then $W = \int_C -\mathrm{d}U = U_A - U_B$, which depends only on A and B and is the same for all paths C joining A and B. Such a force is a **conservative** force, and $U(\boldsymbol{r})$ is its potential energy, with $\boldsymbol{F}(\boldsymbol{r}) = -\mathrm{grad}\,U$. Forces that do not have this property are said to be **dissipative** or **non-conservative**.

Similarly, if $\boldsymbol{v}(\boldsymbol{r})$ represents the velocity field of a fluid then $\oint_C \boldsymbol{v}\cdot\mathrm{d}\boldsymbol{r}$ is the flow around the closed curve C in unit time. This is sometimes termed the **net circulation integral** of $\boldsymbol{v}$. If $\oint_C \boldsymbol{v}\cdot\mathrm{d}\boldsymbol{r} = 0$ then the fluid is curl-free or irrotational, and in this case $\boldsymbol{v}$ has a potential function $\phi(\boldsymbol{r})$ such that $\boldsymbol{v} = -\mathrm{grad}\,\phi$.

7.4.2 Exercises

56 Evaluate $\int y\,\mathrm{d}s$ along the parabola $y = 2\sqrt{x}$ from A(3, 2√3) to B(24, 4√6).

57 Evaluate $\int_A^B [2xy\,\mathrm{d}x + (x^2 - y^2)\,\mathrm{d}y]$ along the arc of the circle $x^2 + y^2 = 1$ in the first quadrant from A(1, 0) to B(0, 1).

58 Evaluate the integral $\int_C \boldsymbol{V}\cdot\mathrm{d}\boldsymbol{r}$, where $\boldsymbol{V} = (2yz + 3x^2, y^2 + 4xz, 2z^2 + 6xy)$, and C is the curve with parametric equations $x = t^3$, $y = t^2$, $z = t$ joining the points (0, 0, 0) and (1, 1, 1).

59 If $\boldsymbol{A} = (2y + 3)\boldsymbol{i} + xz\boldsymbol{j} + (yz - x)\boldsymbol{k}$, evaluate $\int_C \boldsymbol{A}\cdot\mathrm{d}\boldsymbol{r}$ along the following paths C:

(a) $x = 2t^2$, $y = t$, $z = t^3$ from $t = 0$ to $t = 1$;
(b) the straight lines from (0, 0, 0) to (0, 0, 1), then to (0, 1, 1) and then to (2, 1, 1);
(c) the straight line joining (0, 0, 0) to (2, 1, 1).

60 Prove that $\boldsymbol{F} = (y^2\cos x + z^3)\boldsymbol{i} + (2y\sin x - 4)\boldsymbol{j} + (3xz^2 + z)\boldsymbol{k}$ is a conservative force field. Hence find the work done in moving an object in this field from (0, 1, −1) to (π/2, −1, 2).

61 Find the work done in moving a particle in the force field $\boldsymbol{F} = 3x^2\boldsymbol{i} + (2xz - y)\boldsymbol{j} + z\boldsymbol{k}$ along

(a) the curve defined by $x^2 = 4y$, $3x^3 = 8z$ from $x = 0$ to $x = 2$;
(b) the straight line from (0, 0, 0) to (2, 1, 3).

Does this mean that $\boldsymbol{F}$ is a conservative force? Give reasons for your answer.

62 Prove that the vector field $\boldsymbol{F} = (3x^2 - y, 2yz^2 - x, 2y^2z)$ is conservative, but not solenoidal. Hence evaluate the scalar line integral $\int_C \boldsymbol{F}\cdot\mathrm{d}\boldsymbol{r}$ along any curve C joining the point (0, 0, 0) to the point (1, 2, 3).

63 If $\boldsymbol{F} = xy\boldsymbol{i} - z\boldsymbol{j} + x^2\boldsymbol{k}$ and C is the curve $x = t^2$, $y = 2t$, $z = t^3$ from $t = 0$ to $t = 1$, evaluate the vector line integral $\int_C \boldsymbol{F}\times\mathrm{d}\boldsymbol{r}$.

64 If $\boldsymbol{A} = (3x + y, -x, y - z)$ and $\boldsymbol{B} = (2, -3, 1)$ evaluate the line integral $\oint_C (\boldsymbol{A}\times\boldsymbol{B})\times\mathrm{d}\boldsymbol{r}$ around the circle in the (x, y) plane having centre at the origin and radius 2, traversed in the positive direction.

7.4.3 Double integrals

In the introduction to Section 7.4 we defined the definite integral of a function $f(x)$ of one variable by the limit

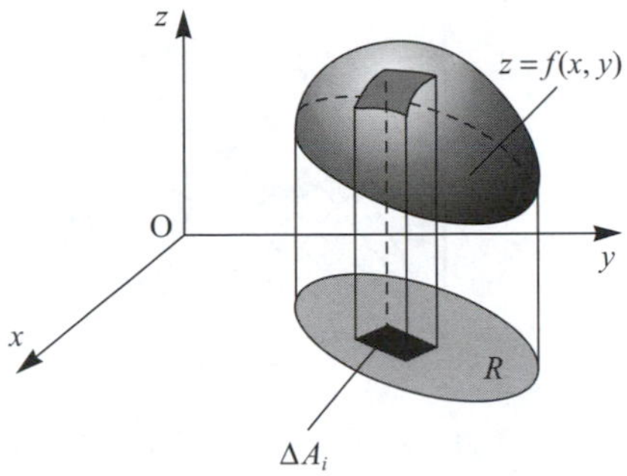

Figure 7.19 Volume as an integral.

$$\int_a^b f(x)\,\mathrm{d}x = \lim_{\substack{n\to\infty \\ \text{all } \Delta x_i \to 0}} \sum_{i=1}^{n} f(\tilde{x}_i)\,\Delta x_i$$

where $a = x_0 < x_1 < x^2 < \ldots < x_n = b$, $\Delta x_i = x_i - x_{i-1}$ and $x_{i-1} \leqslant \tilde{x}_i \leqslant x_i$. This integral is represented by the area between the curve $y = f(x)$ and the x axis and between $x = a$ and $x = b$, as shown in Figure 7.13.

Now consider $z = f(x, y)$ and a region R of the (x, y) plane, as shown in Figure 7.19. Define the integral of $f(x, y)$ over the region R by the limit

$$\iint_R f(x, y)\,\mathrm{d}A = \lim_{\substack{n\to\infty \\ \text{all } \Delta A_i \to 0}} \sum_{i=1}^{n} f(\tilde{x}_i, \tilde{y}_i)\,\Delta A_i$$

where ΔA_i $(i = 1, \ldots, n)$ is a partition of R into n elements of area ΔA_i and $(\tilde{x}_i, \tilde{y}_i)$ is a point in ΔA_i. Now $z = f(x, y)$ represents a surface, and so $f(\tilde{x}_i, \tilde{y}_i)\,\Delta A_i = \tilde{z}_i\,\Delta A_i$ is the volume between $z = 0$ and $z = \tilde{z}_i$ on the base ΔA_i. The integral $\iint_R f(x, y)\,\mathrm{d}A$ is the limit of the sum of all such volumes, and so it is the volume under the surface $z = f(x, y)$ above the region R.

The partition of R into elementary areas can be achieved using grid lines parallel to the x and y axes as shown in Figure 7.20. Then $\Delta A_i = \Delta x_i\,\Delta y_i$, and we can write

$$\iint_R f(x, y)\,\mathrm{d}A = \iint_R f(x, y)\,\mathrm{d}x\mathrm{d}y = \lim_{n\to\infty} \sum_{i=1}^{n} f(\tilde{x}_i, \tilde{y}_i)\Delta x_i\,\Delta y_i$$

Other partitions may be chosen, for example a polar grid as in Figure 7.21. Then the element of area is $(r_i\,\Delta\theta_i)\Delta r_i = \Delta A_i$ and

$$\iint_R f(x, y)\,\mathrm{d}A = \iint_R f(r\cos\theta, r\sin\theta)r\,\mathrm{d}r\,\mathrm{d}\theta \qquad \textbf{(7.26)}$$

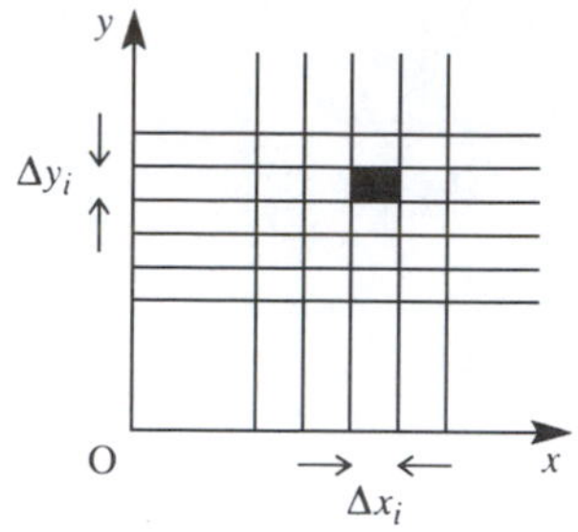

Figure 7.20 A possible grid for the partition of R.

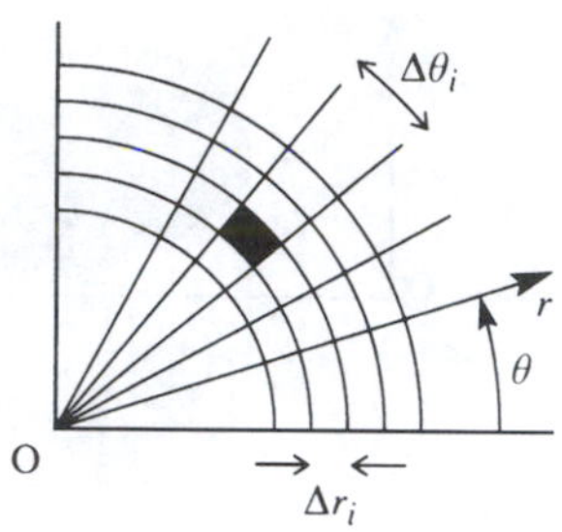

Figure 7.21 Another possible grid for the partition of R.

The expression for ΔA is more complicated when the grid lines do not intersect at right angles; we shall discuss this case in Section 7.4.5.

We can evaluate integrals of the type $\iint_R f(x, y)\,dx\,dy$ as repeated single integrals in x and y. Consequently, they are usually called **double integrals**.

Consider the region R shown in Figure 7.22, with boundary ACBD. Let the curve ACB be given by $y = g_1(x)$ and the curve ADB by $y = g_2(x)$. Then we can evaluate $\iint_R f(x, y)\,dx\,dy$ by summing for y first over the Δy_i, holding x constant ($x = \tilde{x}_i$, say), from $y = g_1(x_i)$ to $y = g_2(x_i)$, and then summing all such strips from A to B, that is, from $x = a$ to $x = b$. Thus we may write

$$\iint_R f(x, y)\,dA = \lim_{\substack{n\to\infty \\ \text{all } \Delta x_i,\, \Delta y_j \to 0}} \sum_{i=1}^{n_2} \left[\sum_{j=1}^{n_1} f(\tilde{x}_i, y_j)\Delta y_j \right] \Delta x_i \quad (n = \min(n_1, n_2))$$

$$= \int_a^b \left[\int_{y=g_1(x)}^{y=g_2(x)} f(x, y)\,dy \right] dx$$

Here the integral inside the brackets is evaluated first, integrating with respect to y, keeping the value of x fixed, and then the result of this integration is integrated with respect to x.

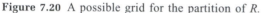

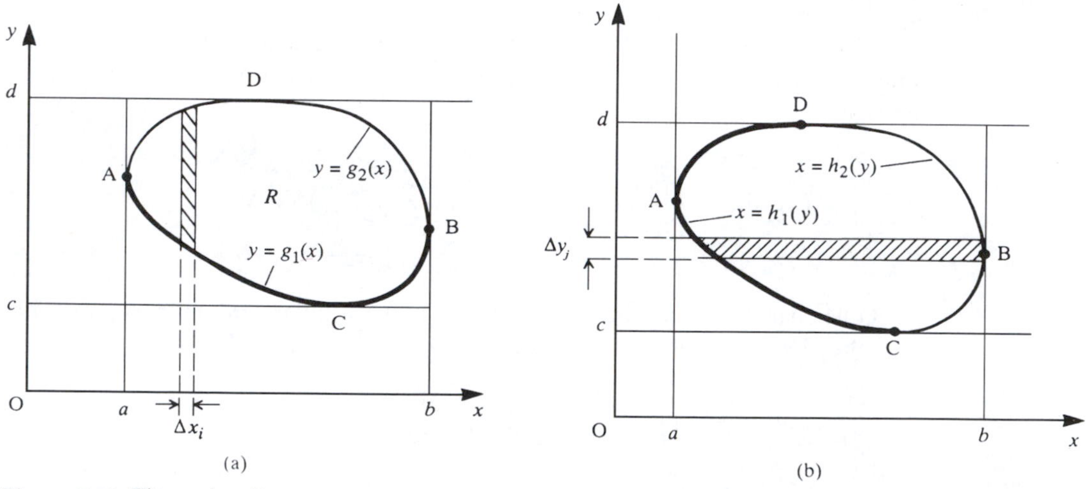

Figure 7.22 The region R.

Alternatively, we can sum for x first and then y. If the curve CAD is represented by $x = h_1(y)$ and the curve CBD by $x = h_2(y)$, we can write the integral as

$$\iint_R f(x, y)\,dA = \lim_{\substack{n\to\infty \\ \text{all } \Delta y_j,\ \Delta x_i \to 0}} \sum_{j=1}^{n_1} \left[\sum_{i=1}^{n_2} f(x_i, \tilde{y}_j)\,\Delta x_i \right] \Delta y_j \qquad (n = \min(n_1, n_2))$$

$$= \int_c^d \left[\int_{x=h_1(y)}^{x=h_2(y)} f(x, y)\,dx \right] dy$$

If the double integral exists then these two results are equal, and in going from one to the other we have changed the order of integration. Notice that the limits of integration are also changed in the process. Often, when evaluating an integral analytically, it is easier to perform the evaluation one way rather than the other.

EXAMPLE 7.19 Evaluate $\iint_R (x^2 + y^2)\,dA$ over the triangle with vertices at (0, 0), (2, 0) and (1, 1).

Solution The domain of integration is shown in Figure 7.23(a). The triangle is bounded by the lines $y = 0$, $y = x$ and $y = 2 - x$.

(a) Integrating with respect to x first, as indicated in Figure 7.23(b), gives

$$\iint_R (x^2 + y^2)\,dA = \int_0^1 \int_{x=y}^{x=2-y} (x^2 + y^2)\,dx\,dy$$

$$= \int_0^1 \left[\tfrac{1}{3}x^3 + y^2 x \right]_{x=y}^{x=2-y} dy$$

$$= \int_0^1 \left[\tfrac{8}{3} - 4y + 4y^2 - \tfrac{8}{3}y^3 \right] dy = \tfrac{4}{3}$$

(b) Integrating with respect to y first, as indicated in Figure 7.23(c), gives

$$\iint_R (x^2 + y^2)\,dA = \int_0^1 \int_{y=0}^{y=x} (x^2 + y^2)\,dy\,dx + \int_1^2 \int_{y=0}^{y=2-x} (x^2 + y^2)\,dy\,dx$$

Note that because the upper boundary of the region R has different equations for it along different parts, the integral has to be split up into convenient subintegrals. Evaluating the integrals we have

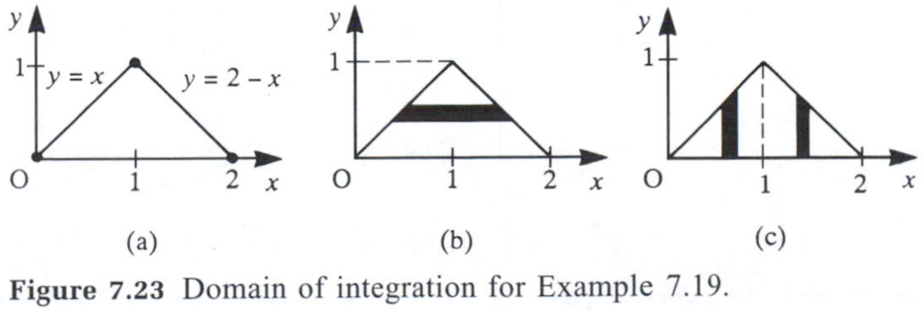

Figure 7.23 Domain of integration for Example 7.19.

$$\int_0^1 \int_{y=0}^{y=x} (x^2+y^2)\,dy\,dx = \int_0^1 [x^2y+\tfrac{1}{3}y^3]_{y=0}^{y=x}\,dx = \int_0^1 \tfrac{4}{3}x^3\,dx = \tfrac{1}{3}$$

$$\int_1^2 \int_{y=0}^{y=2-x} (x^2+y^2)\,dy\,dx = \int_1^2 [x^2y+\tfrac{1}{3}y^3]_{y=0}^{y=2-x}\,dx$$

$$= \int_1^2 (\tfrac{8}{3}-4x+4x^2-\tfrac{4}{3}x^3)\,dx = 1$$

Thus

$$\iint_R (x^2+y^2)\,dA = \tfrac{1}{3}+1 = \tfrac{4}{3}, \quad \text{as before}$$

Clearly, in this example it is easier to integrate with respect to x first.

EXAMPLE 7.20 Evaluate $\iint_R (x + 2y)^{-1/2}\,dA$ over the region $x - 2y \leqslant 1$ and $x \geqslant y^2 + 1$.

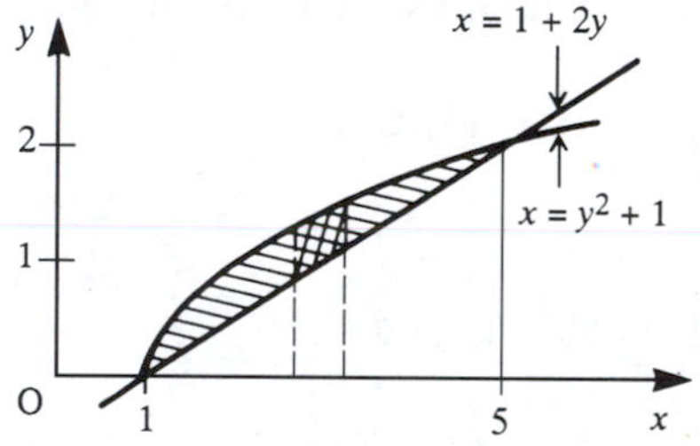

Figure 7.24 Domain of integration for Example 7.20.

Solution The bounding curves intersect where $2y + 1 = y^2 + 1$, which gives $y = 0$ (with $x = 1$) and $y = 2$ (with $x = 5$). The region R is shown in Figure 7.24. In this example we choose to take x first because the formula for the boundary is easier to deal with: $x = y^2 + 1$ rather than $y = (x - 1)^{1/2}$. Thus we obtain

$$\iint_R (x+2y)^{-1/2}\,dA = \int_0^2 \int_{y^2+1}^{2y+1} (x+2y)^{-1/2}\,dx\,dy$$

$$= \int_0^2 [2(x+2y)^{1/2}]_{x=y^2+1}^{x=2y+1}\,dy$$

$$= \int_0^2 [2(4y+1)^{1/2} - 2(y+1)]\,dy$$

$$= [\tfrac{1}{3}(4y+1)^{3/2} - y^2 - 2y]_0^2 = \tfrac{2}{3}$$

As indicated earlier, the evaluation of integrals over a domain R is not restricted to the use of rectangular cartesian coordinates (x, y). Example 7.21 shows how polar coordinates can be used in some cases to simplify the analytical process.

EXAMPLE 7.21 Evaluate $\iint_R x^2y\,\mathrm{d}A$, where R is the region $x^2 + y^2 \leqslant 1$.

Solution The fact that the domain of integration is a circle suggests that polar coordinates are a natural choice for the integration process. Then, from (7.26), $x = r\cos\theta$, $y = r\sin\theta$ and $\mathrm{d}A = r\,\mathrm{d}\theta\,\mathrm{d}r$, and the integral becomes

$$\iint_R x^2y\,\mathrm{d}A = \int_{r=0}^{1}\int_{\theta=0}^{2\pi} r^2\cos^2\theta\; r\sin\theta\; r\,\mathrm{d}\theta\,\mathrm{d}r$$

$$= \int_{r=0}^{1}\int_{\theta=0}^{2\pi} r^4\cos^2\theta\sin\theta\,\mathrm{d}\theta\,\mathrm{d}r$$

Note that in this example the integration is such that we can separate the variables r and θ and write

$$\iint_R x^2y\,\mathrm{d}A = \int_{r=0}^{1} r^4\int_{\theta=0}^{2\pi} \cos^2\theta\sin\theta\,\mathrm{d}\theta\,\mathrm{d}r$$

Furthermore, since the limits of integration with respect to θ do not involve r, we can write

$$\iint_R x^2y\,\mathrm{d}A = \int_{r=0}^{1} r^4\,\mathrm{d}r\int_{\theta=0}^{2\pi} \cos^2\theta\sin\theta\,\mathrm{d}\theta$$

and the double integral in this case reduces to a product of integrals. Thus we obtain

$$\iint_R x^2y\,\mathrm{d}A = [\tfrac{1}{5}r^5]_0^1[-\tfrac{1}{3}\cos^2\theta]_0^{2\pi} = 0$$

7.4.4 Exercises

65 Evaluate the following:

(a) $\displaystyle\int_0^3\int_1^2 xy(x+y)\,\mathrm{d}y\,\mathrm{d}x$ (b) $\displaystyle\int_2^3\int_1^5 x^2y\,\mathrm{d}y\,\mathrm{d}x$

(c) $\displaystyle\int_{-1}^1\int_{-2}^2 (2x^2+y^2)\,\mathrm{d}y\,\mathrm{d}x$

66 Evaluate

$$\iint \frac{x^2}{y}\,\mathrm{d}x\,\mathrm{d}y$$

over the rectangle bounded by the lines $x = 0$, $x = 2$, $y = 1$ and $y = 2$. Check your answer by integrating in the reverse order.

67 Evaluate $\iint (x^2 + y^2)\,dx\,dy$ over the region for which $x \geqslant 0$, $y \geqslant 0$ and $x + y \leqslant 1$.

68 Sketch the domain of integration and evaluate

(a) $\int_1^2 dx \int_x^{2x} \frac{dy}{x^2+y^2}$ (b) $\int_0^1 dx \int_0^{1-x} (x^2+y^2)\,dy$

(c) $\int_0^1 dx \int_{\sqrt{(x-x^2)}}^{\sqrt{(1-x^2)}} \frac{1}{\sqrt{(1-x^2-y^2)}}\,dy$

69 Evaluate $\iint \sin \frac{1}{2}\pi(x + y)\,dx\,dy$ over the triangle whose vertices are (0, 0), (2, 1), (1, 2).

70 Sketch the domains of integration of the double integrals

(a) $\int_0^1 dx \int_x^1 \frac{xy\,dy}{\sqrt{(1+y^4)}}$

(b) $\int_0^{\pi/2} dy \int_0^y (\cos 2y)\sqrt{(1-k^2\sin^2 x)}\,dx$

Change the order of integration, and hence evaluate the integrals.

71 Evaluate

$$\int_0^1 dy \int_{\sqrt{y}}^1 \frac{dx}{\sqrt{[y(1+x^2)]}}$$

72 Sketch the domain of integration of the double integral

$$\int_0^1 \int_0^{\sqrt{(x-x^2)}} \frac{x}{\sqrt{(x^2+y^2)}}\,dy\,dx$$

Express the integral in polar coordinates, and hence show that its value is $\frac{1}{3}$.

73 Sketch the domain of integration of the double integral

$$\int_0^1 dx \int_0^{\sqrt{(1-x^2)}} \frac{x+y}{\sqrt{(x^2+y^2)}}\,dy$$

and evaluate the integral.

74 Evaluate

$$\iint \frac{x+y}{x^2+y^2+a^2}\,dx\,dy$$

over the portion of the first quadrant lying inside the circle $x^2 + y^2 = a^2$.

75 By using polar coordinates, evaluate the double integral

$$\iint \frac{x^2-y^2}{x^2+y^2}\,dx\,dy$$

over the region in the first quadrant bounded by the arc of the parabola $y^2 = 4(1 - x)$ and the coordinate axes.

76 By transforming to polar coordinates, show that the double integral

$$\iint \frac{(x^2+y^2)^2}{(xy)^2}\,dx\,dy$$

taken over the area common to the two circles $x^2 + y^2 = ax$ and $x^2 + y^2 = by$ is ab.

7.4.5 Green's theorem in a plane

This theorem shows the relationship between line integrals and double integrals, and will also provide a justification for the general change of variables in a double integral.

Consider a simple closed curve, C, enclosing the region A as shown in Figure 7.25. If $P(x, y)$ and $Q(x, y)$ are continuous functions with continuous partial derivatives then

$$\oint_C (P\,dx + Q\,dy) = \iint_A \left(\frac{\partial Q}{\partial x} - \frac{\partial P}{\partial y}\right) dx\,dy \qquad (7.27)$$

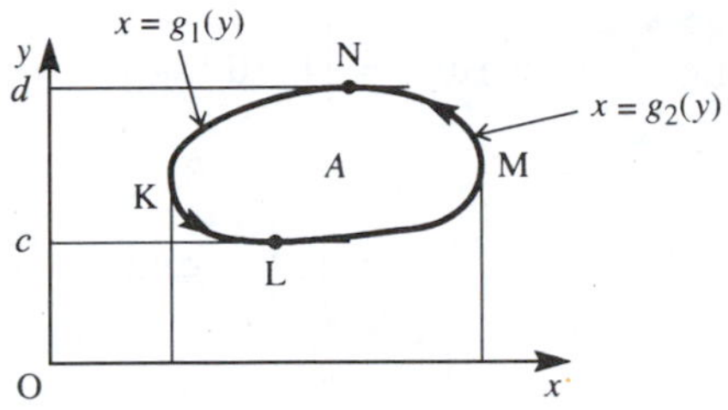

Figure 7.25 Green's theorem.

where C is traversed in the positive sense (that is, so that the bounded area is always on the left). This result is called **Green's theorem** in a plane.

The proof of this result is straightforward. Consider the first term on the right-hand side. Then, with reference to Figure 7.25,

$$\iint_R \frac{\partial Q}{\partial x}\,\mathrm{d}x\,\mathrm{d}y = \int_c^d \left[\int_{g_1(y)}^{g_2(y)} \frac{\partial Q}{\partial x}\,\mathrm{d}x\right]\mathrm{d}y$$

$$= \int_c^d [\,Q(g_2(y), y) - Q(g_1(y), y)\,]\,\mathrm{d}y$$

$$= \int_{\text{LMN}} Q(x, y)\,\mathrm{d}y - \int_{\text{LKN}} Q(x, y)\,\mathrm{d}y$$

$$= \int_{\text{LMNKL}} Q(x, y)\,\mathrm{d}y = \oint_C Q(x, y)\,\mathrm{d}y$$

Similarly,

$$-\iint_A \frac{\partial P}{\partial y}\,\mathrm{d}x\,\mathrm{d}y = \oint_C P(x, y)\,\mathrm{d}x$$

and hence

$$\iint_A \left(\frac{\partial Q}{\partial x} - \frac{\partial P}{\partial y}\right)\mathrm{d}x\,\mathrm{d}y = \oint_C [P(x, y)\,\mathrm{d}x + Q(x, y)\,\mathrm{d}y]$$

An elementary application is shown in Example 7.22.

EXAMPLE 7.22

Evaluate $\oint [2x(x + y)\,\mathrm{d}x + (x^2 + xy + y^2)\,\mathrm{d}y]$ around the square with vertices at (0, 0), (1, 0), (1, 1) and (0, 1) illustrated in Figure 7.26.

Solution Here $P(x, y) = 2x(x + y)$ and $Q(x, y) = x^2 + xy + y^2$, so that $\partial P/\partial y = 2x$, $\partial Q/\partial x = 2x + y$ and $\partial Q/\partial x - \partial P/\partial y = y$. Thus the line integral transforms into an easy double integral

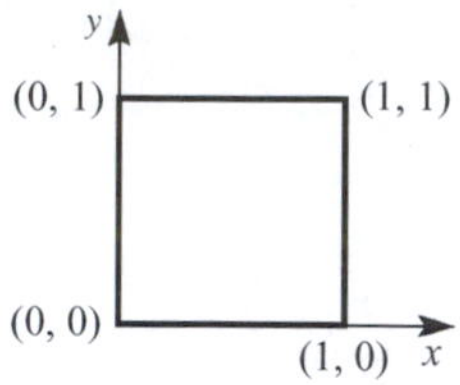

Figure 7.26 Path of integration for Example 7.22.

$$\oint_C [\,2x(x+y)\,\mathrm{d}x + (x^2+xy+y^2)\,\mathrm{d}y\,] = \iint_A y\,\mathrm{d}x\,\mathrm{d}y$$

$$= \int_0^1\!\!\int_0^1 y\,\mathrm{d}x\,\mathrm{d}y$$

$$= \int_0^1 y\,\mathrm{d}y \int_0^1 \mathrm{d}x = \tfrac{1}{2}$$

It follows immediately from Green's theorem (7.27) that the area A enclosed by the closed curve C is given by

$$A = \oint_C x\,\mathrm{d}y \equiv \iint_A 1\,\mathrm{d}x\,\mathrm{d}y = -\oint_C y\,\mathrm{d}x \equiv \iint_A 1\,\mathrm{d}x\,\mathrm{d}y = \tfrac{1}{2}\oint_C (-y\,\mathrm{d}x + x\,\mathrm{d}y)$$

Suppose that under a transformation of coordinates $x = x(u, v)$ and $y = y(u, v)$, the curve becomes C', enclosing an area A'. Then

$$A' = \oint_{C'} u\,\mathrm{d}v \equiv \iint_{A'} \mathrm{d}u\,\mathrm{d}v = \oint_C u\left(\frac{\partial v}{\partial x}\mathrm{d}x + \frac{\partial v}{\partial y}\mathrm{d}y\right)$$

$$= \iint_A \left[\frac{\partial}{\partial x}\left(u\frac{\partial v}{\partial y}\right) - \frac{\partial}{\partial y}\left(u\frac{\partial v}{\partial x}\right)\right]\mathrm{d}x\,\mathrm{d}y = \iint_A \left(\frac{\partial u}{\partial x}\frac{\partial v}{\partial y} - \frac{\partial u}{\partial y}\frac{\partial v}{\partial x}\right)\mathrm{d}x\,\mathrm{d}y$$

This implies that the element of area $\mathrm{d}u\,\mathrm{d}v$ is equivalent to the element

$$\left|\left(\frac{\partial u}{\partial x}\frac{\partial v}{\partial y} - \frac{\partial u}{\partial y}\frac{\partial v}{\partial x}\right)\right|\mathrm{d}x\,\mathrm{d}y$$

Here the modulus sign is introduced to preserve the orientation of the curve under the mapping. Similarly, we may prove that

$$\mathrm{d}x\,\mathrm{d}y = \left|\frac{\partial(x, y)}{\partial(u, v)}\right|\mathrm{d}u\,\mathrm{d}v \qquad \textbf{(7.28)}$$

where $\partial(x, y)/\partial(u, v)$ is the **Jacobian**

$$\frac{\partial x}{\partial u}\frac{\partial y}{\partial v} - \frac{\partial x}{\partial v}\frac{\partial y}{\partial x} = J(x, y)$$

This enables us to make a general change of coordinates in a double integral:

$$\iint_A f(x, y)\,\mathrm{d}x\,\mathrm{d}y = \iint_{A'} f(x(u, v), y(u, v))|J|\,\mathrm{d}u\,\mathrm{d}v \qquad \textbf{(7.29)}$$

where A' is the region in the (u, v) plane corresponding to A in the (x, y) plane.

Note that the above discussion confirms the result

$$\frac{\partial(u, v)}{\partial(x, y)} = \left[\frac{\partial(x, y)}{\partial(u, v)}\right]^{-1}$$

as shown in Section 7.1.3. Using (7.29), the result (7.26) when using polar coordinates is readily confirmed.

EXAMPLE 7.23

Evaluate $\iint xy\,dx\,dy$ over the region in $x \geqslant 0$, $y \geqslant 0$ bounded by $y = x^2 + 4$, $y = x^2$, $y = 6 - x^2$ and $y = 12 - x^2$.

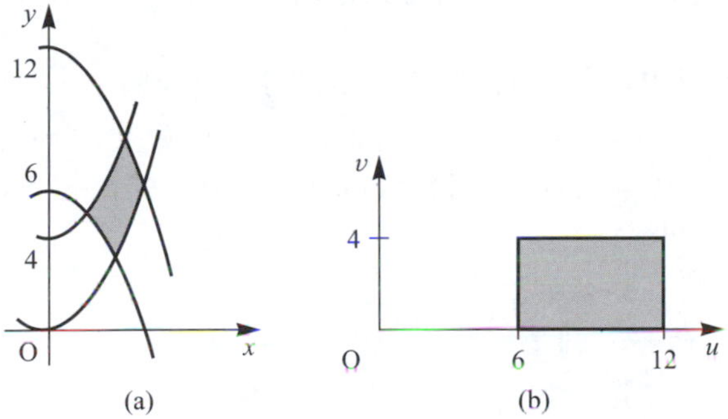

Figure 7.27 Domain of integration for Example 7.23: (a) in the (x, y) plane; (b) in the (u, v) plane.

Solution The domain of integration is shown in Figure 7.27(a). The bounding curves can be rewritten as $y - x^2 = 4$, $y - x^2 = 0$, $y + x^2 = 6$ and $y + x^2 = 12$, so that a natural change of coordinates is to set

$$u = y + x^2, \qquad v = y - x^2$$

Under this transformation, the region of integration becomes the rectangle $6 \leqslant u \leqslant 12$, $0 \leqslant v \leqslant 4$, as shown in Figure 7.27(b). Thus since

$$J(x, y) = \frac{\partial(x, y)}{\partial(u, v)} = \left[\frac{\partial(u, v)}{\partial(x, y)}\right]^{-1} = \frac{1}{4x}$$

the integral simplifies to

$$\iint_A xy\,dx\,dy = \iint_{A'} xy\frac{1}{4x}\,du\,dv$$

Hence

$$\iint_A xy\,dx\,dy = \tfrac{1}{4}\iint_{A'} y\,du\,dv = \tfrac{1}{8}\iint_{A'} (u+v)\,du\,dv, \quad \text{since } y = (u+v)/2$$

$$= \tfrac{1}{8}\int_0^4 dv \int_6^{12} (u+v)\,du = 33$$

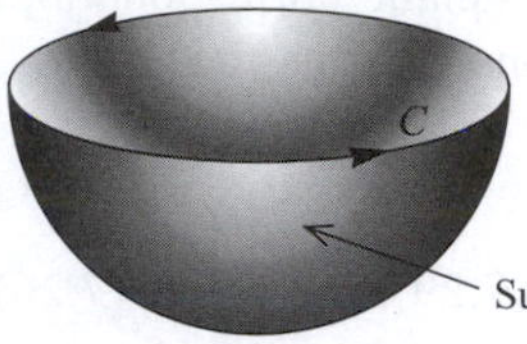

Figure 7.28 Three-dimensional generalization of Green's theorem.

We remark in passing that Green's theorem in a plane may be generalized to three dimensions. Note that the result (7.27) may be written as

$$\oint_C (P, Q, 0) \cdot d\boldsymbol{r} = \iint_A \text{curl}\,[(P, Q, 0)] \cdot \boldsymbol{k}\, dx\, dy$$

For a general surface S with bounding curve C as shown in Figure 7.28 this identity becomes

$$\oint_C \boldsymbol{F}(\boldsymbol{r}) \cdot d\boldsymbol{r} = \iint_S \text{curl}\, \boldsymbol{F}(\boldsymbol{r}) \cdot d\boldsymbol{S}$$

where $d\boldsymbol{S} = \hat{\boldsymbol{n}}\, dS$ is the vector element of surface area and $\hat{\boldsymbol{n}}$ is a unit vector along the normal. This generalization is called Stokes' theorem, and will be discussed in Section 7.4.12 after we have formally introduced the concept of a surface integral.

7.4.6 Exercises

77 Evaluate the line integral

$$\oint_C [\sin y\, dx + (x - \cos y)\, dy]$$

taken in the anticlockwise sense, where C is the perimeter of the triangle formed by the lines

$$y = \tfrac{1}{2}\pi x, \quad y = \tfrac{1}{2}\pi, \quad x = 0$$

Verify your answer using Green's theorem in a plane.

78 Use Green's theorem in a plane to evaluate

$$\oint_C [(xy^2 - y)\, dx + (x + y^2)\, dy]$$

as a double integral, where C is the triangle with vertices at (0, 0), (2, 0) and (2, 2) and is traversed in the anticlockwise direction.

79 Evaluate the line integral

$$I = \oint_C (xy\, dx + x\, dy)$$

where C is the closed curve consisting of $y = x^2$ from $x = 0$ to $x = 1$ and $y = \sqrt{x}$ from $x = 1$ to $x = 0$. Confirm your answer by applying Green's theorem in the plane and evaluating I as a double integral.

80 Use Green's theorem in a plane to evaluate the line integral

$$\oint_C [(e^x - 3y^2)\,dx + (e^y + 4x^2)\,dy]$$

where C is the circle $x^2 + y^2 = 4$. (*Hint*: use polar coordinates to evaluate the double integral.)

81 Evaluate

$$\int_0^a dx \int_x^{2a-x} \frac{y-x}{4a^2+(y+x)^2}\,dy$$

using the transformation of coordinates $u = x + y$. $v = x - y$.

82 Using the transformation

$$x + y = u, \quad \frac{y}{x} = v$$

show that

$$\int_0^1 dy \int_y^{2-y} \frac{x+y}{x^2} e^{x+y}\,dx = \int_0^2 du \int_0^1 e^u\,dv = e^2 - 1$$

7.4.7 Surface integrals

The extensions of the idea of an integral to line and double integrals are not the only generalizations that can be made. We can also extend the idea to integration over a general surface S. Two types of such integrals occur:

(a) $$\iint_S f(x, y, z)\,dS$$

(b) $$\iint_S \boldsymbol{F}(\boldsymbol{r}) \cdot \hat{\boldsymbol{n}}\,dS = \iint_S \boldsymbol{F}(\boldsymbol{r}) \cdot d\boldsymbol{S}$$

In case (a) we have a scalar field $f(\boldsymbol{r})$ and in case (b) a vector field $\boldsymbol{F}(\boldsymbol{r})$. Note that $d\boldsymbol{S} = \hat{\boldsymbol{n}}\,dS$ is the vector element of area, where $\hat{\boldsymbol{n}}$ is the unit vector outward drawn normal to the element dS.

In general, the surface S can be described in terms of two parameters, u and v say, so that on S

$$\boldsymbol{r} = \boldsymbol{r}(u, v) = (x(u, v), y(u, v), z(u, v))$$

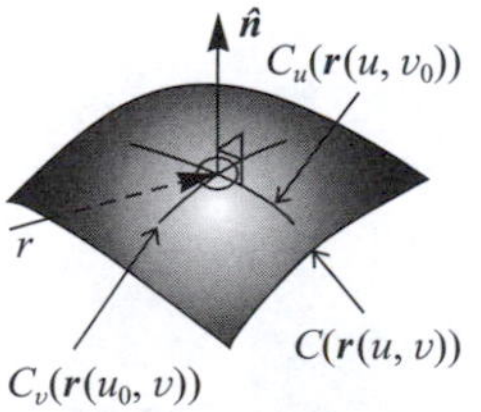

Figure 7.29 Parametric curves on a surface

The surface S can be specified by a scalar point function $C(\boldsymbol{r}) = c$, where c is a constant. Curves may be drawn on that surface, and in particular if we fix the value of one of the two parameters u and v then we obtain two families of curves. On one, $C_u(\boldsymbol{r}(u, v_0))$, the value of u varies while v is fixed, and on the other, $C_v(\boldsymbol{r}(u_0, v))$, the value of v varies while u is fixed, as shown in Figure 7.29. Then as indicated in Figure 7.29, the vector element of area $d\boldsymbol{S}$ is given by

$$d\boldsymbol{S} = \frac{\partial \boldsymbol{r}}{\partial u}\,du \times \frac{\partial \boldsymbol{r}}{\partial v}\,dv = \frac{\partial \boldsymbol{r}}{\partial u} \times \frac{\partial \boldsymbol{r}}{\partial v}\,du\,dv$$

$$= \left(\frac{\partial x}{\partial u}, \frac{\partial y}{\partial u}, \frac{\partial z}{\partial u}\right) \times \left(\frac{\partial x}{\partial v}, \frac{\partial y}{\partial v}, \frac{\partial z}{\partial v}\right) du\,dv = (J_1\boldsymbol{i} + J_2\boldsymbol{j} + J_3\boldsymbol{k})\,du\,dv$$

where

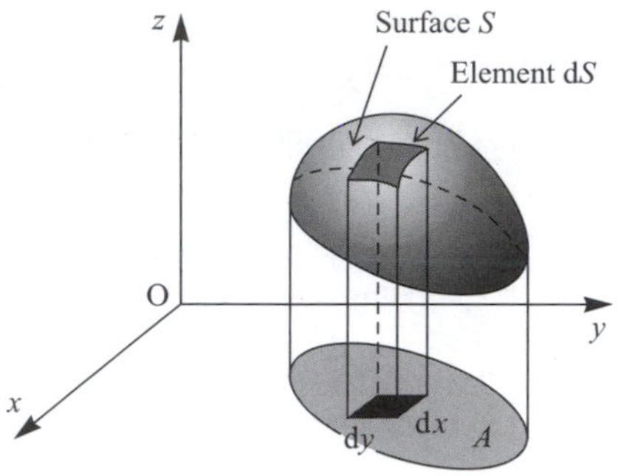

Figure 7.30 A surface described by $z = z(x, y)$.

$$J_1 = \frac{\partial y}{\partial u}\frac{\partial z}{\partial v} - \frac{\partial y}{\partial v}\frac{\partial z}{\partial u}, \quad J_2 = \frac{\partial z}{\partial u}\frac{\partial x}{\partial v} - \frac{\partial z}{\partial v}\frac{\partial x}{\partial u}, \quad J_3 = \frac{\partial x}{\partial u}\frac{\partial y}{\partial v} - \frac{\partial x}{\partial v}\frac{\partial y}{\partial u} \tag{7.30}$$

Hence

$$\iint_S \boldsymbol{F}(\boldsymbol{r}) \cdot \mathrm{d}\boldsymbol{S} = \iint_A (PJ_1 + QJ_2 + RJ_3)\, \mathrm{d}u\, \mathrm{d}v$$

$$\iint_S f(x, y, z)\, \mathrm{d}S = \iint_A f(u, v)\sqrt{(J_1^2 + J_2^2 + J_3^2)}\mathrm{d}u\, \mathrm{d}v$$

where $\boldsymbol{F}(\boldsymbol{r}) = (P, Q, R)$ and A is the region of the (u, v) plane corresponding to S. Here, of course, the terms in the integrands have to be expressed in terms of u and v.

In particular, u and v can be chosen as any two of x, y and z. For example, if $z = z(x, y)$ describes a surface as in Figure 7.30 then

$$\boldsymbol{r} = (x, y, z(x, y))$$

with x and y as independent variables. This gives

$$J_1 = -\frac{\partial z}{\partial x}, \quad J_2 = -\frac{\partial z}{\partial y}, \quad J_3 = 1$$

and

$$\iint_S \boldsymbol{F}(\boldsymbol{r}) \cdot \mathrm{d}\boldsymbol{S} = \iint_A \left(-P\frac{\partial z}{\partial x} - Q\frac{\partial z}{\partial y} + R\right) \mathrm{d}x\, \mathrm{d}y \tag{7.31a}$$

$$\iint_S f(x, y, z)\, \mathrm{d}S = \iint_A f(x, y, z(x, y)) \sqrt{\left[1 + \left(\frac{\partial z}{\partial x}\right)^2 + \left(\frac{\partial z}{\partial y}\right)^2\right]} \mathrm{d}x\, \mathrm{d}y \tag{7.31b}$$

EXAMPLE 7.24 Evaluate the surface integral

$$\iint_S (x + y + z)\,\mathrm{d}S$$

where S is the portion of the sphere $x^2 + y^2 + z^2 = 1$ that lies in the first quadrant.

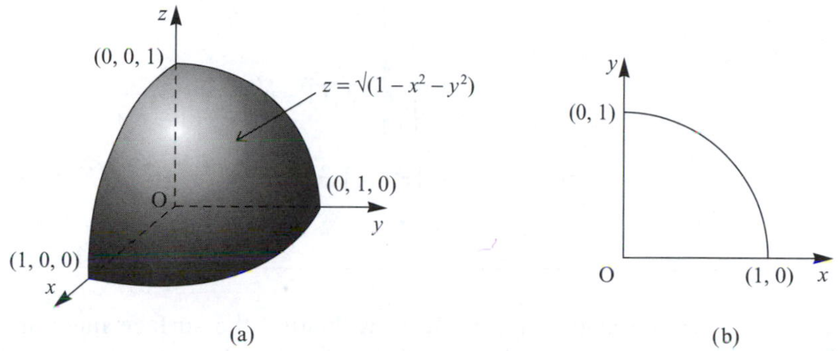

Figure 7.31 (a) Surface S for Example 7.24; (b) quadrant of a circle in the (x, y) plane.

Solution The surface S is illustrated in Figure 7.31(a). Taking

$$z = \sqrt{(1 - x^2 - y^2)}$$

we have

$$\frac{\partial z}{\partial x} = \frac{-x}{\sqrt{(1 - x^2 - y^2)}}, \quad \frac{\partial z}{\partial y} = \frac{-y}{\sqrt{(1 - x^2 - y^2)}}$$

giving

$$\sqrt{\left[1 + \left(\frac{\partial z}{\partial x}\right)^2 + \left(\frac{\partial z}{\partial y}\right)^2\right]} = \sqrt{\left[\frac{x^2 + y^2 + (1 - x^2 - y^2)}{(1 - x^2 - y^2)}\right]} = \frac{1}{\sqrt{(1 - x^2 - y^2)}}$$

Using (7.17) then gives

$$\iint_S (x + y + z)\,\mathrm{d}S = \iint_A [x + y + \sqrt{(1 - x^2 - y^2)}]\frac{1}{\sqrt{(1 - x^2 - y^2)}}\,\mathrm{d}x\,\mathrm{d}y$$

where A is the quadrant of a circle in the (x, y) plane illustrated in Figure 7.31(b).

Transforming to polar coordinates

$$x = r\cos\theta, \quad y = r\sin\theta, \quad \mathrm{d}x\,\mathrm{d}y = r\,\mathrm{d}r\,\mathrm{d}\theta$$

gives

$$\iint_S (x+y+z)\,dS = \int_0^{\pi/2}\int_0^1 \left[\frac{r(\sin\theta+\cos\theta)}{\sqrt{(1-r^2)}}+1\right] r\,dr\,d\theta$$

$$= \int_0^1 \left[\frac{r^2(-\cos\theta+\sin\theta)}{\sqrt{(1-r^2)}}+r\right]_0^{\pi/2} dr$$

$$= \int_0^1 \left[\frac{2r^2}{\sqrt{(1-r^2)}}+\frac{1}{2}r\pi\right] dr$$

$$= \int_0^1 \left[-2\sqrt{(1-r^2)}+\frac{2}{\sqrt{(1-r^2)}}+\frac{1}{2}r\pi\right] dr$$

$$= [-r\sqrt{(1-r^2)}-\sin^{-1}r+2\sin^{-1}r+\tfrac{1}{4}r^2\pi]_0^1$$

$$= \tfrac{3}{4}\pi$$

An alternative approach to evaluating the surface integral in Example 7.24 is to evaluate it directly over the surface of the sphere using spherical polar coordinates. As illustrated in Figure 7.32, on the surface of a sphere of radius a we have

$$x = a\sin\theta\cos\phi, \quad y = a\sin\theta\sin\phi$$

$$z = a\cos\theta, \quad dS = a^2\sin\theta\,d\theta\,d\phi$$

In the sphere of Example 7.24 the radius $a = 1$, so that

$$\iint_S (x+y+z)\,dS = \int_0^{\pi/2}\int_0^{\pi/2} (\sin\theta\cos\phi+\sin\theta\sin\phi+\cos\theta)\sin\theta\,d\theta\,d\phi$$

$$= \int_0^{\pi/2} [\tfrac{1}{4}\pi\cos\phi+\tfrac{1}{4}\pi\sin\phi+\tfrac{1}{2}]\,d\phi = \tfrac{3}{4}\pi$$

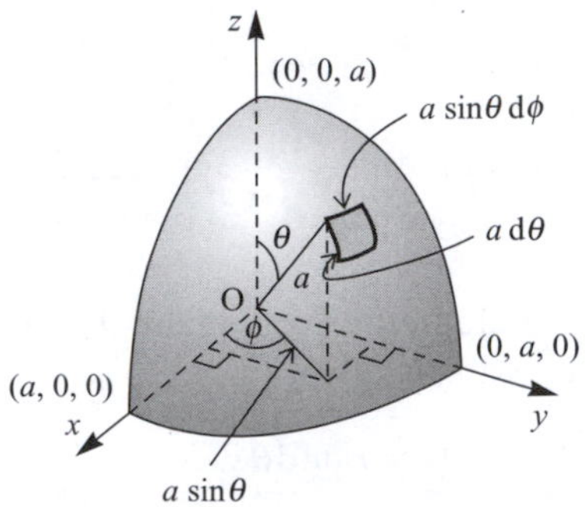

Figure 7.32 Surface element in spherical polar coordinates.

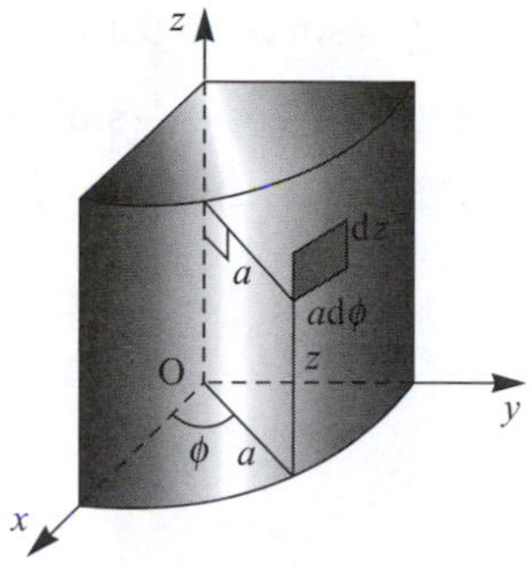

Figure 7.33 Surface element in cylindrical polar coordinates.

In a similar manner, when evaluating surface integrals over the surface of a cylinder of radius a, we have, as illustrated in Figure 7.33,

$$x = a\cos\phi, \quad y = a\sin\phi, \quad z = z, \quad dS = a\,dz\,d\phi$$

EXAMPLE 7.25 Find the surface area of the torus shown in Figure 7.34(a) formed by rotating a circle of radius b about an axis distance a from its centre.

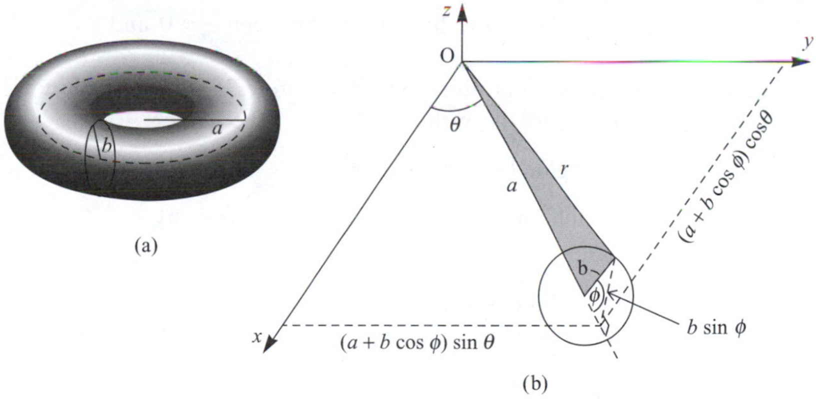

Figure 7.34 (a) Torus of Example 7.25; (b) position vector of a point on the surface of the torus.

Solution From Figure 7.34(b), the position vector $\boldsymbol{r}$ of a point on the surface is given by

$$\boldsymbol{r} = (a + b\cos\phi)\cos\theta\boldsymbol{i} + (a + b\cos\phi)\sin\theta\boldsymbol{j} + b\sin\phi\boldsymbol{k}$$

(Notice that θ and ϕ are not the angles used for spherical polar coordinates.) Thus using 7.16,

$$J_1 = (a + b\cos\phi)\cos\theta(b\cos\phi) - (-b\sin\phi\sin\theta)(0)$$

$$J_2 = (0)(-b\sin\phi\cos\theta) - (b\cos\phi)(a + b\cos\phi)(-\sin\theta)$$

$$J_3 = -(a + b\cos\phi)\sin\theta(-b\sin\phi\sin\theta) - (-b\sin\phi\cos\theta)(a + b\cos\phi)\cos\theta$$

Simplifying, we obtain

$$J_1 = b(a + b\cos\phi)\cos\theta\cos\phi$$

$$J_2 = b(a + b\cos\phi)\sin\theta\cos\phi$$

$$J_3 = b(a + b\cos\phi)\sin\phi$$

and the surface area is given by

$$\begin{aligned} S &= \int_0^{2\pi}\int_0^{2\pi} \sqrt{(J_1^2 + J_2^2 + J_3^2)}\,\mathrm{d}\theta\,\mathrm{d}\phi \\ &= \int_0^{2\pi}\int_0^{2\pi} b(a + b\cos\phi)\,\mathrm{d}\theta\,\mathrm{d}\phi \\ &= 4\pi^2 ab \end{aligned}$$

Thus the surface area of the torus is the product of the circumferences of the two circles that generate it.

EXAMPLE 7.26 Evaluate $\iint_S \boldsymbol{V}\cdot\mathrm{d}\boldsymbol{S}$, where $\boldsymbol{V} = z\boldsymbol{i} + x\boldsymbol{j} - 3y^2z\boldsymbol{k}$ and S is the surface of the cylinder $x^2 + y^2 = 16$ in the first octant between $z = 0$ and $z = 5$.

Solution The surface S is illustrated in Figure 7.35. From Section 7.2.1, the outward normal to the surface is in the direction of the vector

$$\boldsymbol{n} = \text{grad}\,(x^2 + y^2 - 16) = 2x\boldsymbol{i} + 2y\boldsymbol{j}$$

so that the unit outward normal $\hat{\boldsymbol{n}}$ is given by

$$\hat{\boldsymbol{n}} = \frac{2x\boldsymbol{i} + 2y\boldsymbol{j}}{2\sqrt{(x^2 + y^2)}}$$

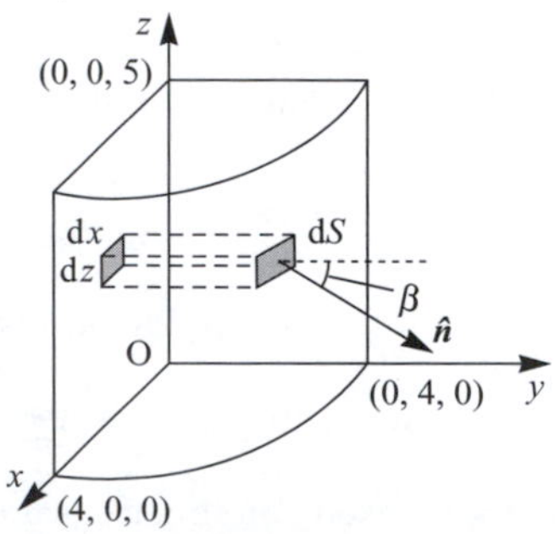

Figure 7.35 Surface S for Example 7.26.

Since on the surface $x^2 + y^2 = 16$,

$$\hat{\boldsymbol{n}} = \tfrac{1}{4}(x\boldsymbol{i} + y\boldsymbol{j})$$

giving

$$\mathrm{d}\boldsymbol{S} = \mathrm{d}S\hat{\boldsymbol{n}} = \tfrac{1}{4}\mathrm{d}S(x\boldsymbol{i} + y\boldsymbol{j})$$

Projecting the element of surface $\mathrm{d}S$ onto the (x, z) plane as illustrated in Figure 7.35, the area $\mathrm{d}x\,\mathrm{d}z$ of the projected element is given by

$$\mathrm{d}x\,\mathrm{d}z = \mathrm{d}S\cos\beta$$

where β is the angle between the normal $\hat{\boldsymbol{n}}$ to the surface element and the normal $\boldsymbol{j}$ to the (x, z) plane. Thus

$$\mathrm{d}x\,\mathrm{d}z = \mathrm{d}S|\hat{\boldsymbol{n}}\cdot\boldsymbol{j}| = \tfrac{1}{4}\mathrm{d}S\,|(x\boldsymbol{i} + y\boldsymbol{j})\cdot\boldsymbol{j}| = \tfrac{1}{4}\mathrm{d}S\,y$$

giving

$$\mathrm{d}S = \frac{4}{y}\,\mathrm{d}x\,\mathrm{d}z$$

Also,

$$\boldsymbol{V}\cdot\mathrm{d}\boldsymbol{S} = \boldsymbol{V}\cdot\hat{\boldsymbol{n}}\,\mathrm{d}S = (z\boldsymbol{i} + x\boldsymbol{j} - 3y^2z\boldsymbol{k})\cdot\left(\frac{x\boldsymbol{i} + y\boldsymbol{j}}{4}\right)\frac{4}{y}\mathrm{d}x\,\mathrm{d}z = \frac{xz + xy}{y}\,\mathrm{d}x\,\mathrm{d}z$$

so that

$$\iint_S \boldsymbol{V}\cdot\mathrm{d}\boldsymbol{S} = \iint_A \frac{xz + xy}{y}\,\mathrm{d}x\,\mathrm{d}z$$

where A is the rectangular region in the (x, z) plane bounded by $0 \leqslant x \leqslant 4$, $0 \leqslant z \leqslant 5$. Noting that the integrand is still evaluated on the surface, we can write $y = \sqrt{(16 - x^2)}$, so that

$$\iint_S \boldsymbol{V}\cdot\mathrm{d}\boldsymbol{S} = \int_0^4\int_0^5\left[x + \frac{xz}{\sqrt{(16 - x^2)}}\right]\mathrm{d}z\,\mathrm{d}x$$

$$= \int_0^4\left[xz + \frac{xz^2}{2\sqrt{(16 - x^2)}}\right]_0^5\mathrm{d}x = \int_0^4\left[5x + \frac{25x}{2\sqrt{(16 - x^2)}}\right]\mathrm{d}x$$

$$= [\tfrac{5}{2}x^2 - \tfrac{25}{2}\sqrt{(16 - x^2)}]_0^4 = 90$$

An alternative approach in this case is to evaluate $\frac{1}{4}\iint_S (xz + xy)\,\mathrm{d}S$ directly over the surface using cylindrical polar coordinates. This is left as Exercise 90 below.

7.4.8 Exercises

83 Evaluate the area of the surface $z = 2 - x^2 - y^2$ lying above the (x, y) plane. (*Hint*: Use polar coordinates to evaluate the double integral.)

84 Evaluate

(a) $\iint_S (x^2 + y^2)\,dS$, where S is the surface area of the plane $2x + y + 2z = 6$ cut off by the planes $z = 0$, $z = 2$, $y = 0$, $y = 3$;
(b) $\iint_S z\,dS$, where S is the surface area of the hemisphere $x^2 + y^2 + z^2 = 1$ $(z > 0)$ cut off by the cylinder $x^2 - x + y^2 = 0$.

85 Evaluate $\iint_S \boldsymbol{v} \cdot d\boldsymbol{S}$, where

(a) $\boldsymbol{v} = (xy, -x^2, x + z)$ and S is the part of the plane $2x + 2y + z = 6$ included in the first octant;
(b) $\boldsymbol{v} = (3y, 2x^2, z^3)$ and S is the surface of the cylinder $x^2 + y^2 = 1$, $0 < z < 1$.

86 Show that $\iint_S z^2\,dS = \frac{1}{2}\pi$, where S is the surface of the sphere $x^2 + y^2 + z^2 = 1$, $z \geqslant 0$.

87 Evaluate the surface integral $\iint_S U(x, y, z)\,dS$, where S is the surface of the paraboloid $z = 2 - (x^2 + y^2)$ above the (x, y) plane and $U(x, y, z)$ is given by

(a) 1 (b) $x^2 + y^2$ (c) z

Give a physical interpretation in each case.

88 Determine the surface area of the plane $2x + y + 2z = 16$ cut off by $x = 0$, $y = 0$ and $x^2 + y^2 = 64$.

89 Show that the area of that portion of the surface of the paraboloid $x^2 + y^2 = 4z$ included between the planes $z = 1$ and $z = 3$ is $\frac{16}{3}\pi(4 - \sqrt{2})$.

90 Evaluate the surface integral in Example 7.26 using cylindrical polar coordinates.

91 If $\boldsymbol{F} = y\boldsymbol{i} + (x - 2xz)\boldsymbol{j} - xy\boldsymbol{k}$, evaluate the surface integral $\iint_S (\text{curl}\,\boldsymbol{F}) \cdot d\boldsymbol{S}$, where S is the surface of the sphere $x^2 + y^2 + z^2 = a^2$, $z \geqslant 0$.

7.4.9 Volume integrals

In Section 7.4.7 we defined the integral of a function over a curved surface in three dimensions. This idea can be extended to define the integral of a function of three variables through a region T of three-dimensional space by the limit

$$\iiint_T f(x, y, z)\,dV = \lim_{\substack{n \to \infty \\ \text{all } \Delta V_i \to 0}} \sum_{i=1}^{n} f(\tilde{x}_i, \tilde{y}_i, \tilde{z}_i)\Delta V_i$$

where $\Delta V_i (i = 1, \dots, n)$ is a partition of T into n elements of volume, and $(\tilde{x}_i, \tilde{y}_i, \tilde{z}_i)$ is a point in ΔV_i as illustrated in Figure 7.36.

In terms of rectangular cartesian coordinates the triple integral can, as illustrated in Figure 7.37, be written as

$$\iiint_T f(x, y, z)\,dV = \int_a^b dx \int_{g_1(x)}^{g_2(x)} dy \int_{h_1(x, y)}^{h_2(x, y)} f(x, y, z)\,dz \qquad \textbf{(7.32)}$$

Note that there are six different orders in which the integration in (7.32) can be carried out.

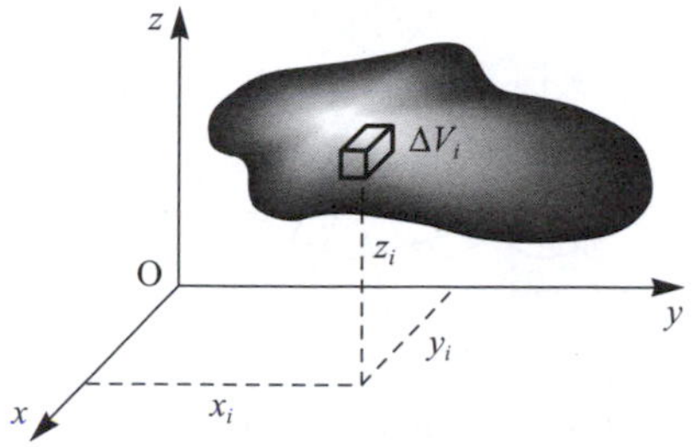

Figure 7.36 Partition of region T into volume elements ΔV_i.

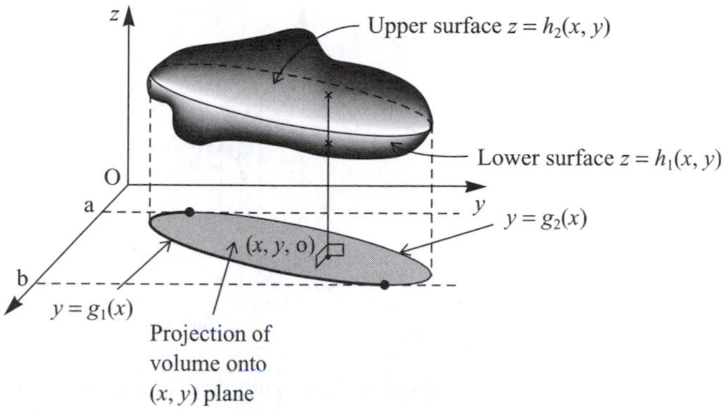

Figure 7.37 The volume integral in terms of rectangular cartesian coordinates.

As we saw for double integrals in (7.28), the expression for the element of volume $dV = dx\,dy\,dz$ under the transformation $x = x(u, v, w)$, $y = y(u, v, w)$, $z = z(u, v, w)$ may be obtained using the Jacobian

$$J = \frac{\partial(x, y, z)}{\partial(u, v, w)} = \begin{vmatrix} \dfrac{\partial x}{\partial u} & \dfrac{\partial y}{\partial u} & \dfrac{\partial z}{\partial u} \\ \dfrac{\partial x}{\partial v} & \dfrac{\partial y}{\partial v} & \dfrac{\partial z}{\partial v} \\ \dfrac{\partial x}{\partial w} & \dfrac{\partial y}{\partial w} & \dfrac{\partial z}{\partial w} \end{vmatrix}$$

as

$$dV = dx\,dy\,dz = |J|\,du\,dv\,dw \tag{7.33}$$

For example, in the case of cylindrical polar coordinates

$$x = \rho\cos\phi, \quad y = \rho\sin\phi, \quad z = z$$

$$J = \rho\begin{vmatrix} \cos\phi & \sin\phi & 0 \\ -\sin\phi & \cos\phi & 0 \\ 0 & 0 & 1 \end{vmatrix} = \rho$$

so that

$$dV = \rho\,d\rho\,d\phi\,dz \tag{7.34}$$

a result illustrated in Figure 7.38.

Similarly, for spherical polar coordinates (r, θ, ϕ)

$$x = r\sin\theta\cos\phi, \quad y = r\sin\theta\sin\phi, \quad z = r\cos\theta$$

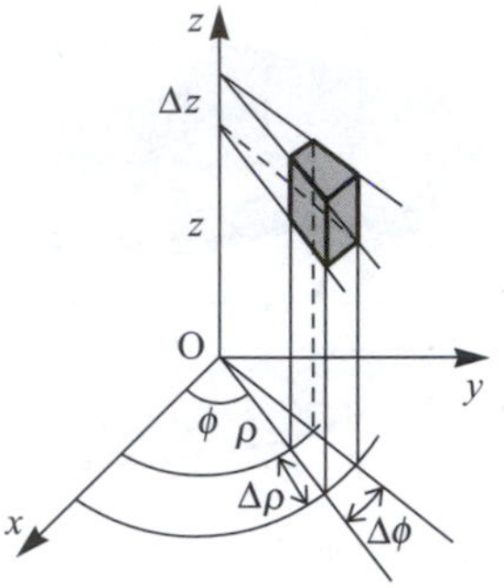

Figure 7.38 Volume element in cylindrical polar coordinates.

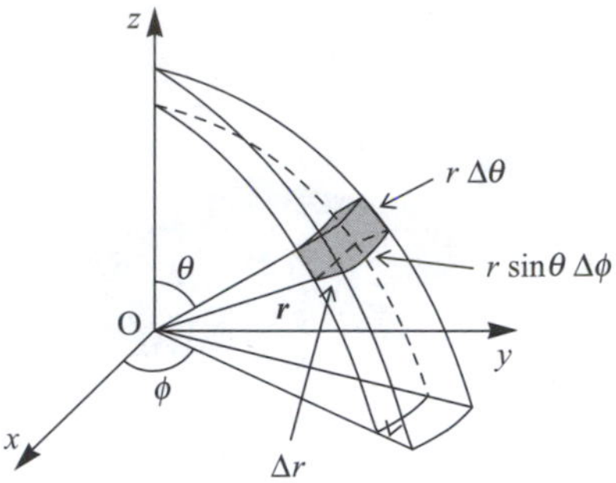

Figure 7.39 Volume element in spherical polar coordinates.

$$J = \begin{vmatrix} \sin\theta\cos\phi & \sin\theta\sin\phi & \cos\theta \\ r\cos\theta\cos\phi & r\cos\theta\sin\phi & -r\sin\theta \\ -r\sin\theta\sin\phi & r\sin\theta\cos\phi & 0 \end{vmatrix} = r^2\sin\theta$$

so that

$$\boxed{\mathrm{d}V = r^2\sin\theta\,\mathrm{d}r\,\mathrm{d}\theta\,\mathrm{d}\phi} \qquad \textbf{(7.35)}$$

a result illustrated in Figure 7.39.

EXAMPLE 7.27

Find the volume and the coordinates of the centroid of the tetrahedron defined by $x \geqslant 0$, $y \geqslant 0$, $z \geqslant 0$ and $x + y + z \leqslant 1$.

Solution The tetrahedron is shown in Figure 7.40. Its volume is

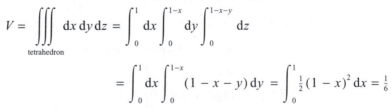

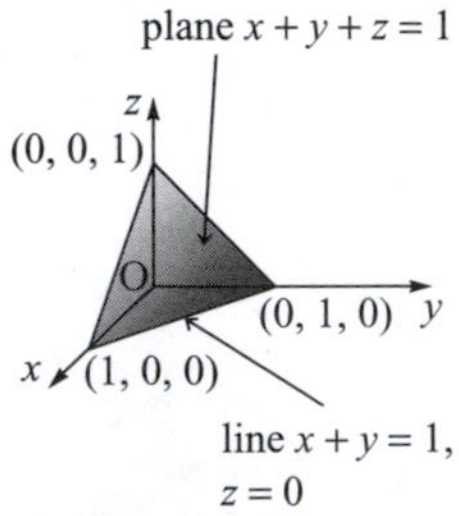

Figure 7.40 Tetrahedron for Example 7.27.

$$V = \iiint\limits_{\text{tetrahedron}} \mathrm{d}x\,\mathrm{d}y\,\mathrm{d}z = \int_0^1 \mathrm{d}x \int_0^{1-x} \mathrm{d}y \int_0^{1-x-y} \mathrm{d}z$$

$$= \int_0^1 \mathrm{d}x \int_0^{1-x} (1 - x - y)\,\mathrm{d}y = \int_0^1 \tfrac{1}{2}(1 - x)^2\,\mathrm{d}x = \tfrac{1}{6}$$

Let the coordinates of the centroid be $(\bar{x}, \bar{y}, \bar{z})$; then

$$\bar{x}V = \iiint\limits_{\text{tetrahedron}} x\,\mathrm{d}V = \iiint\limits_{\text{tetrahedron}} x\,\mathrm{d}x\,\mathrm{d}y\,\mathrm{d}z$$

$$= \int_0^1 \mathrm{d}x \int_0^{1-x} \mathrm{d}y \int_0^{1-x-y} x\,\mathrm{d}z = \int_0^1 \tfrac{1}{2}x(1 - x)^2\,\mathrm{d}x = \tfrac{1}{24}$$

Hence $\bar{x} = \frac{1}{4}$, and by symmetry $\bar{y} = \bar{z} = \frac{1}{4}$.

EXAMPLE 7.28 Find the moment of inertia of a uniform sphere of mass M and radius a about a diameter.

Solution A sphere of radius a has volume $\frac{4}{3}\pi a^3$, so that its density is $3M/4\pi a^3$. Then the moment of inertia of the sphere about the z axis is

$$I = \frac{3M}{4\pi a^3} \iiint_{\text{sphere}} (x^2 + y^2)\,\mathrm{d}x\,\mathrm{d}y\,\mathrm{d}z$$

In this example it is natural to use spherical polar coordinates, so that

$$I = \frac{3M}{4\pi a^3} \iiint_{\text{sphere}} (r^2 \sin^2\theta)\, r^2 \sin\theta\,\mathrm{d}r\,\mathrm{d}\theta\,\mathrm{d}\phi$$

$$= \frac{3M}{4\pi a^3} \int_0^a r^4\,\mathrm{d}r \int_0^{\pi} \sin^3\theta\,\mathrm{d}\theta \int_0^{2\pi} \mathrm{d}\phi = \frac{3M}{4\pi a^3}\left(\frac{1}{5}a^5\right)\left(\frac{4}{3}\right)(2\pi)$$

$$= \tfrac{2}{5}Ma^2$$

7.4.10 Exercises

92 Evaluate the triple integrals

(a) $\displaystyle\int_0^1 \mathrm{d}x \int_0^2 \mathrm{d}y \int_1^3 x^2yz\,\mathrm{d}z$

(b) $\displaystyle\int_0^2\int_1^3\int_2^4 xyz^2\,\mathrm{d}z\,\mathrm{d}y\,\mathrm{d}x$

93 Show that

$$\int_{-1}^1 \mathrm{d}z \int_0^2 \mathrm{d}x \int_{x-z}^{x+z} (x + y + z)\,\mathrm{d}y = 0$$

94 Evaluate $\iiint \sin(x + y + z)\,\mathrm{d}x\,\mathrm{d}y\,\mathrm{d}z$ over the portion of the positive octant cut off by the plane $x + y + z = \pi$.

95 Evaluate $\iiint_V xyz\,\mathrm{d}x\,\mathrm{d}y\,\mathrm{d}z$, where V is the region bounded by the planes $x = 0$, $y = 0$, $z = 0$ and $x + y + z = 1$.

96 Sketch the region contained between the parabolic cylinders $y = x^2$ and $x = y^2$ and the planes $z = 0$ and $x + y + z = 2$. Show that the volume of the region may be expressed as the triple integral

$$\int_0^1\int_{x^2}^{\sqrt{x}}\int_0^{2-x-y} \mathrm{d}z\,\mathrm{d}y\,\mathrm{d}x$$

and evaluate it.

97 Use spherical polar coordinates to evaluate

$$\iiint_V x(x^2 + y^2 + z^2)\,\mathrm{d}x\,\mathrm{d}y\,\mathrm{d}z$$

where V is the region in the first octant lying within the sphere $x^2 + y^2 + z^2 = 1$.

98 Evaluate $\iiint x^2y^2z^2(x + y + z)\,\mathrm{d}x\,\mathrm{d}y\,\mathrm{d}z$ throughout the region defined by $x + y + z \leqslant 1$, $x \geqslant 0$, $y \geqslant 0$, $z \geqslant 0$.

99 Show that if $x + y + z = u$, $y + z = uv$ and $z = uvw$ then

$$\frac{\partial(x, y, z)}{\partial(u, v, w)} = u^2v$$

Hence evaluate the triple integral

$$\iiint_V \exp[-(x+y+z)^3]\,\mathrm{d}x\,\mathrm{d}y\,\mathrm{d}z$$

where V is the volume of the tetrahedron bounded by the planes $x = 0$, $y = 0$, $z = 0$ and $x + y + z = 1$.

100 Evaluate $\iiint_V yz\,\mathrm{d}x\,\mathrm{d}y\,\mathrm{d}z$ taken throughout the prism with sides parallel to the z axis, whose base is the triangle with vertices at (0, 0, 0), (1, 0, 0), (0, 1, 0) and whose top is the triangle with vertices at (0, 0, 2), (1, 0, 1), (0, 1, 1). Find also the position of the centroid of this prism.

101 Evaluate $\iiint z\,\mathrm{d}x\,\mathrm{d}y\,\mathrm{d}z$ throughout the region defined by $x^2 + y^2 \leqslant z^2$, $x^2 + y^2 + z^2 \leqslant 1$, $z > 0$.

102 Using spherical polar coordinates, evaluate $\iiint x\,\mathrm{d}x\,\mathrm{d}y\,\mathrm{d}z$ throughout the positive octant of the sphere $x^2 + y^2 + z^2 = a^2$.

7.4.11 Gauss's divergence theorem

In the same way that Green's theorem relates surface and line integrals, Gauss's theorem relates surface and volume integrals.

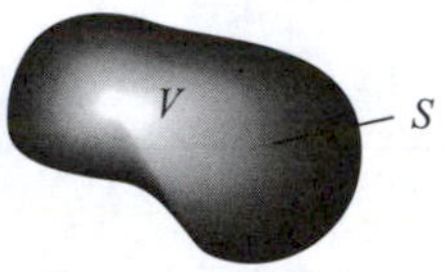

Figure 7.41 Closed volume V with surface S.

Consider the closed volume V with surface area S shown in Figure 7.41. The surface integral $\iint_S \boldsymbol{F}\cdot\mathrm{d}\boldsymbol{S}$ may be interpreted as the flow of a liquid with velocity field $\boldsymbol{F}(\boldsymbol{r})$ out of the volume V. In Section 7.3.1 we saw that the divergence of $\boldsymbol{F}$ could be expressed as

$$\operatorname{div}\boldsymbol{F} = \nabla\cdot\boldsymbol{F} = \lim_{\Delta V\to 0}\frac{\text{flow out of }\Delta V}{\Delta V}$$

In terms of differentials, this may be written

$$\operatorname{div}\boldsymbol{F}\,\mathrm{d}V = \text{flow out of }\mathrm{d}V$$

Consider now a partition of the volume V given by ΔV_i ($i = 1, \dots, n$). Then the total flow out of V is the sum of the flows out of each ΔV_i. That is,

$$\iint_S \boldsymbol{F}\cdot\mathrm{d}\boldsymbol{S} = \lim_{n\to\infty}\sum_{i=1}^{n}(\text{flow out of }\Delta V_i) = \lim_{n\to\infty}\sum_{i=1}^{n}(\operatorname{div}\boldsymbol{F}\Delta V_i)$$

giving

$$\iint_S \boldsymbol{F}\cdot\mathrm{d}\boldsymbol{S} = \iiint_V \operatorname{div}\boldsymbol{F}\,\mathrm{d}V \qquad \textbf{(7.36)}$$

This result is known as the **divergence theorem** or **Gauss's theorem**. It enables us to convert surface integrals into volume integrals, and often simplifies their evaluation.

EXAMPLE 7.29 A vector field $\boldsymbol{F}(\boldsymbol{r})$ is given by

$$\boldsymbol{F}(\boldsymbol{r}) = x^3y\boldsymbol{i} + x^2y^2\boldsymbol{j} + x^2yz\boldsymbol{k}$$

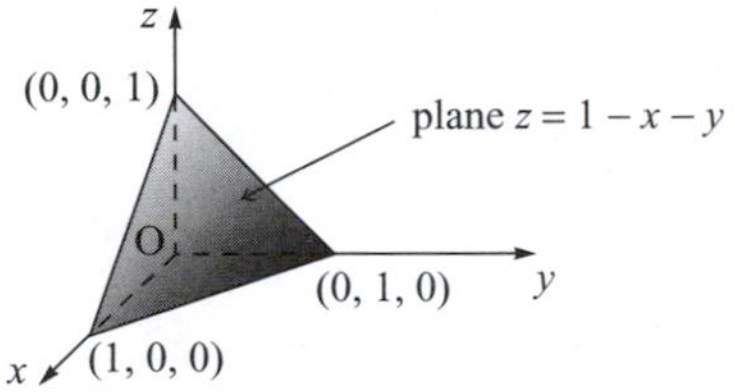

Figure 7.42 Region V and surface S for Example 7.29.

Find $\iint_S \boldsymbol{F} \cdot \mathrm{d}\boldsymbol{S}$, where S is the surface of the region in the first octant for which $x + y + z \leqslant 1$.

Solution We begin by sketching the region V enclosed by S, as shown in Figure 7.42. It is clear that evaluating the surface integral directly will be rather clumsy, involving four separate integrals (one over each of the four surfaces). It is simpler in this case to transform it into a volume integral using the divergence theorem (7.36):

$$\iint_S \boldsymbol{F} \cdot \mathrm{d}\boldsymbol{S} = \iiint_V \operatorname{div} \boldsymbol{F} \, \mathrm{d}V$$

Here

$$\operatorname{div} \boldsymbol{F} = 3x^2y + 2x^2y + x^2y = 6x^2y$$

and we obtain

$$\iint_S \boldsymbol{F} \cdot \mathrm{d}\boldsymbol{S} = \int_0^1 \mathrm{d}x \int_0^{1-x} \mathrm{d}y \int_0^{1-x-y} 6x^2y \, \mathrm{d}z$$

$$= 6\int_0^1 x^2 \, \mathrm{d}x \int_0^{1-x} y \, \mathrm{d}y \int_0^{1-x-y} \mathrm{d}z \qquad \text{(see Example 7.27)}$$

$$= 6\int_0^1 x^2 \, \mathrm{d}x \int_0^{1-x} [(1-x)y - y^2] \, \mathrm{d}y$$

$$= \int_0^1 x^2(1-x)^3 \, \mathrm{d}x = \tfrac{1}{60}$$

EXAMPLE 7.30 Verify the divergence theorem

$$\iint_S \boldsymbol{F} \cdot \mathrm{d}\boldsymbol{S} = \iiint_V \operatorname{div} \boldsymbol{F} \, \mathrm{d}V$$

when $\boldsymbol{F} = 2xz\boldsymbol{i} + yz\boldsymbol{j} + z^2\boldsymbol{k}$ and V is the volume enclosed by the upper hemisphere $x^2 + y^2 + z^2 = a^2$, $z \geqslant 0$.

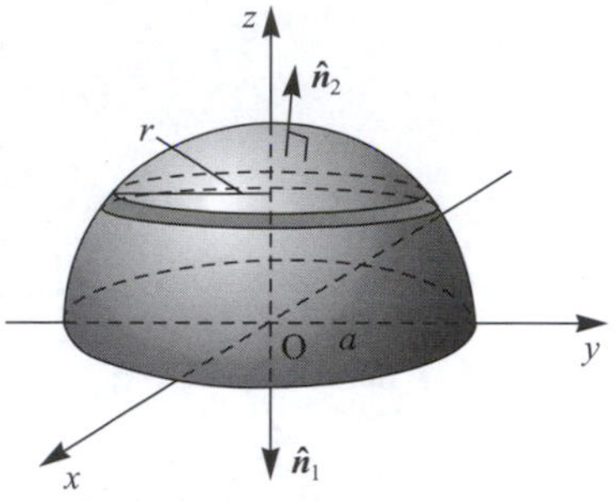

Figure 7.43 Hemisphere for Example 7.30.

Solution The volume V and surface S of the hemisphere are illustrated in Figure 7.43. Note that since the theorem relates to a closed volume, the surface S consists of the flat circular base in the (x, y) plane as well as the hemispherical surface. In this case

$$\text{div}\,\boldsymbol{F} = 2z + z + 2z = 5z$$

so that the volume integral is readily evaluated as

$$\iiint_V 5z\,\mathrm{d}x\,\mathrm{d}y\,\mathrm{d}z = \int_0^a 5z\pi r^2\,\mathrm{d}z = \int_0^a 5\pi z(a^2 - z^2)\mathrm{d}z = \tfrac{5}{4}\pi a^4$$

Considering the surface integral

$$\iint_S \boldsymbol{F}\cdot\mathrm{d}\boldsymbol{S} = \iint_{\text{circular base}} \boldsymbol{F}\cdot\hat{\boldsymbol{n}}_1\,\mathrm{d}S + \iint_{\text{hemisphere}} \boldsymbol{F}\cdot\hat{\boldsymbol{n}}_2\,\mathrm{d}S$$

The unit normal to the base is clearly $\hat{\boldsymbol{n}}_1 = -\boldsymbol{k}$, so

$$\boldsymbol{F}\cdot\hat{\boldsymbol{n}}_1 = -z^2$$

giving

$$\iint_{\text{circular base}} \boldsymbol{F}\cdot\hat{\boldsymbol{n}}_1\,\mathrm{d}S = 0$$

since $z = 0$ on this surface.

The hemispherical surface is given by

$$f(x, y, z) = x^2 + y^2 + z^2 - a^2 = 0$$

so the outward unit normal $\hat{\boldsymbol{n}}_2$ is

$$\hat{\boldsymbol{n}}_2 = \frac{\nabla f}{|\nabla f|} = \frac{2x\boldsymbol{i} + 2y\boldsymbol{j} + 2z\boldsymbol{k}}{2\sqrt{(x^2 + y^2 + z^2)}}$$

Since $x^2 + y^2 + z^2 = a^2$ on the surface,

$$\hat{\boldsymbol{n}}_2 = \frac{x}{a}\boldsymbol{i} + \frac{y}{a}\boldsymbol{j} + \frac{z}{a}\boldsymbol{k}$$

giving

$$F \cdot \hat{n}_2 = \frac{2x^2z}{a} + \frac{y^2z}{a} + \frac{z^3}{a}$$

Hence

$$\iint_{\text{hemisphere}} F \cdot \hat{n}_2 \, dS = \iint_{\text{hemisphere}} \frac{z}{a}(x^2 + a^2)\, dS$$

since $x^2 + y^2 + z^2 = a^2$ on the surface. Transforming to spherical polar coordinates,

$$x = a \sin\theta \cos\phi, \quad z = a \cos\theta, \quad dS = a^2 \sin\theta \, d\theta \, d\phi$$

the surface integral becomes

$$\iint_{\text{hemisphere}} F \cdot \hat{n}_2 \, dS = a^4 \int_0^{2\pi} \int_0^{\pi/2} (\sin\theta \cos\theta + \sin^3\theta \cos\theta \cos^2\phi)\, d\theta \, d\phi$$

$$= a^4 \int_0^{2\pi} [\tfrac{1}{2}\sin^2\theta + \tfrac{1}{4}\sin^4\theta \cos^2\phi]_0^{\pi/2} \, d\phi$$

$$= a^4 \int_0^{2\pi} [\tfrac{1}{2} + \tfrac{1}{4}\cos^2\phi]\, d\phi = \tfrac{5}{4}\pi a^4$$

thus confirming that

$$\iint_S F \cdot dS = \iiint_V \operatorname{div} F \, dV$$

7.4.12 Stokes' theorem

Stokes' theorem is the generalization of Green's theorem, and relates line integrals in three dimensions with surface integrals. At the end of Section 7.3.3 we saw that the curl of the vector F could be expressed in the form

$$\text{curl}\, F \cdot \hat{n} = \lim_{\Delta S \to 0} \frac{\text{flow round } \Delta S}{\Delta S}$$

In terms of differentials, this becomes

$$\text{curl}\, F \cdot dS = \text{flow round } dS$$

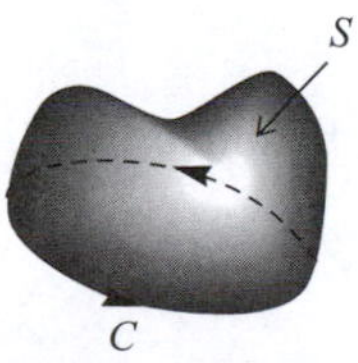

Figure 7.44 Surface S bounded by curve C.

Consider the surface S shown in Figure 7.44, bounded by the curve C. Then the line integral $\oint_c F \cdot dr$ can be interpreted as the total flow of a fluid with velocity field F around the curve C. Partitioning the surface S into elements $\Delta S_i (i = 1, \ldots, n)$, we can write

$$\oint_C \boldsymbol{F} \cdot \mathrm{d}\boldsymbol{r} = \lim_{n \to \infty} \sum_{i=1}^{n} (\text{flow round } \Delta S_i) = \lim_{n \to \infty} \sum_{i=1}^{n} (\text{curl } \boldsymbol{F} \cdot \Delta \boldsymbol{S}_i)$$

so that

$$\oint_C \boldsymbol{F} \cdot \mathrm{d}\boldsymbol{r} = \iint_S (\text{curl } \boldsymbol{F}) \cdot \mathrm{d}\boldsymbol{S} \tag{7.37}$$

This result is known as **Stokes' theorem**. It provides a condition for a line integral to be independent of its path of integration. For, if the integral $\int_A^B \boldsymbol{F} \cdot \mathrm{d}\boldsymbol{r}$ is independent of the path of integration then

$$\int_{C_1} \boldsymbol{F} \cdot \mathrm{d}\boldsymbol{r} = \int_{C_2} \boldsymbol{F} \cdot \mathrm{d}\boldsymbol{r}$$

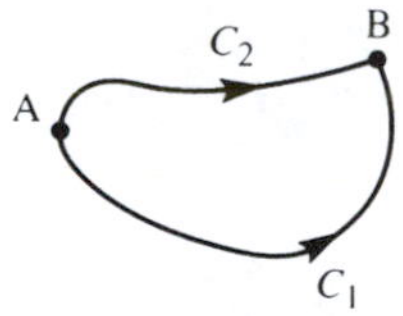

Figure 7.45 Two paths, C_1 and C_2, joining points A and B.

where C_1 and C_2 are two different paths joining A and B as shown in Figure 7.45. Since

$$\int_{C_1} \boldsymbol{F} \cdot \mathrm{d}\boldsymbol{r} = -\int_{-C_2} \boldsymbol{F} \cdot \mathrm{d}\boldsymbol{r}$$

where $-C_2$ is the path C_2 traversed in the opposite direction, we have

$$\int_{C_1} \boldsymbol{F} \cdot \mathrm{d}\boldsymbol{r} + \int_{-C_2} \boldsymbol{F} \cdot \mathrm{d}\boldsymbol{r} = 0$$

That is,

$$\oint_C \boldsymbol{F} \cdot \mathrm{d}\boldsymbol{r} = 0$$

where C is the combined, closed curve formed from C_1 and $-C_2$. Stokes' theorem implies that if $\oint_C \boldsymbol{F} \cdot \mathrm{d}\boldsymbol{r} = 0$ then

$$\iint_S (\text{curl } \boldsymbol{F}) \cdot \mathrm{d}\boldsymbol{S} = 0$$

for any surface S bounded by C. Since this is true for all surfaces bounded by C, we deduce that the integrand must be zero, that is, curl $\boldsymbol{F} = \boldsymbol{0}$. Writing $\boldsymbol{F} = (F_1, F_2, F_3)$, we then have that

$$\boldsymbol{F} \cdot \mathrm{d}\boldsymbol{r} = F_1\,\mathrm{d}x + F_2\,\mathrm{d}y + F_3\,\mathrm{d}z$$

is an exact differential if curl $\boldsymbol{F} = \boldsymbol{0}$, that is, if

$$\frac{\partial F_1}{\partial z} = \frac{\partial F_3}{\partial x}, \quad \frac{\partial F_1}{\partial y} = \frac{\partial F_2}{\partial x}, \quad \frac{\partial F_2}{\partial z} = \frac{\partial F_3}{\partial y}$$

Thus there is a function $f(x, y, z) = f(\boldsymbol{r})$ such that

$$F_1 = \frac{\partial f}{\partial x}, \quad F_2 = \frac{\partial f}{\partial y}, \quad F_3 = \frac{\partial f}{\partial z},$$

that is, such that $\boldsymbol{F}(\boldsymbol{r}) = \text{grad}\, f$.

When $\boldsymbol{F}(\boldsymbol{r})$ represents a field of force, the field is said to be **conservative** (since it conserves rather than dissipates energy). When $\boldsymbol{F}(\boldsymbol{r})$ represents a velocity field for a fluid, the field is said to be curl-free or **irrotational**.

EXAMPLE 7.31

Verify Stokes' theorem for $\boldsymbol{F} = (2x - y)\boldsymbol{i} - yz^2\boldsymbol{j} - y^2z\boldsymbol{k}$, where S is the upper half of the sphere $x^2 + y^2 + z^2 = 1$ and C is its boundary.

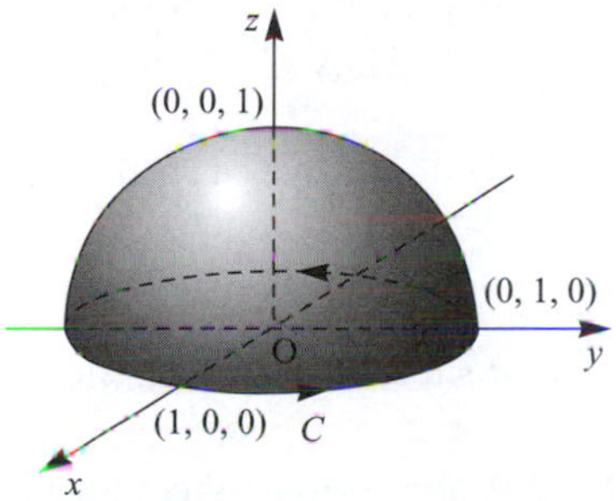

Figure 7.46 Hemispherical surface and boundary for Example 7.31.

Solution The surface and boundary involved are illustrated in Figure 7.46. We are required to show that

$$\oint_C \boldsymbol{F} \cdot \mathrm{d}\boldsymbol{r} = \iint_S \text{curl}\, \boldsymbol{F} \cdot \mathrm{d}\boldsymbol{S}$$

Since C is a circle of unit radius in the (x, y) plane, to evaluate $\oint_C \boldsymbol{F} \cdot \mathrm{d}\boldsymbol{r}$, we take

$$x = \cos\phi, \quad y = \sin\phi$$

so that

$$\boldsymbol{r} = \cos\phi\boldsymbol{i} + \sin\phi\boldsymbol{j}$$

giving

$$\mathrm{d}\boldsymbol{r} = -\sin\phi\, \mathrm{d}\phi\boldsymbol{i} + \cos\phi\, \mathrm{d}\phi\boldsymbol{j}$$

Also, on the boundary C, $z = 0$, so that

$$\boldsymbol{F} = (2x - y)\boldsymbol{i} = (2\cos\phi - \sin\phi)\boldsymbol{i}$$

Thus

$$\oint_C \boldsymbol{F} \cdot \mathrm{d}\boldsymbol{r} = \int_0^{2\pi} (2 \cos \phi - \sin \phi)\boldsymbol{i} \cdot (-\sin \phi \boldsymbol{i} + \cos \phi \boldsymbol{j})\,\mathrm{d}\phi$$

$$= \int_0^{2\pi} (-2 \sin \phi \cos \phi + \sin^2 \phi)\,\mathrm{d}\phi$$

$$= \int_0^{2\pi} [-\sin 2\phi + \tfrac{1}{2}(1 + \cos 2\phi)]\,\mathrm{d}\phi = \pi$$

$$\text{curl}\, \boldsymbol{F} = \begin{vmatrix} \boldsymbol{i} & \boldsymbol{j} & \boldsymbol{k} \\ \dfrac{\partial}{\partial x} & \dfrac{\partial}{\partial y} & \dfrac{\partial}{\partial z} \\ 2x - y & -yz^2 & -y^2 z \end{vmatrix} = \boldsymbol{k}$$

The unit outward drawn normal at a point (x, y, z) on the hemisphere is given by $(x\boldsymbol{i} + y\boldsymbol{j} + z\boldsymbol{k})$, since $x^2 + y^2 + z^2 = 1$. Thus

$$\iint_S \text{curl}\, \boldsymbol{F} \cdot \mathrm{d}\boldsymbol{S} = \iint_S \boldsymbol{k} \cdot (x\boldsymbol{i} + y\boldsymbol{j} + z\boldsymbol{k})\,\mathrm{d}S = \iint_S z\,\mathrm{d}S$$

$$= \int_0^{2\pi} \int_0^{\pi/2} \cos\theta \sin\theta \,\mathrm{d}\theta\,\mathrm{d}\phi = 2\pi[\tfrac{1}{2} \sin^2 \theta]_0^{\pi/2} = \pi$$

Hence $\oint_C \boldsymbol{F} \cdot \mathrm{d}\boldsymbol{r} = \iint_S (\text{curl}\, \boldsymbol{F}) \cdot \mathrm{d}\boldsymbol{S}$, and Stokes' theorem is verified.

7.4.13 Exercises

103 Evaluate $\iint_S \boldsymbol{F} \cdot \mathrm{d}\boldsymbol{S}$, where $\boldsymbol{F} = (4xz, -y^2, yz)$ and S is the surface of the cube bounded by the planes $x = 0$, $x = 1$, $y = 0$, $y = 1$, $z = 0$ and $z = 1$.

104 Use the divergence theorem to evaluate the surface integral $\iint_S \boldsymbol{F} \cdot \mathrm{d}\boldsymbol{S}$, where $\boldsymbol{F} = xz\boldsymbol{i} + yz\boldsymbol{j} + z^2\boldsymbol{k}$ and S is the closed surface of the hemisphere $x^2 + y^2 + z^2 = 4$, $z > 0$. (Note that you are not required to verify the theorem.)

105 Verify the divergence theorem

$$\iint_S \boldsymbol{F} \cdot \mathrm{d}\boldsymbol{S} = \iiint_V \text{div}\, \boldsymbol{F}\,\mathrm{d}V$$

for $\boldsymbol{F} = 4x\boldsymbol{i} - 2y^2\boldsymbol{j} + z^2\boldsymbol{k}$ over the region bounded by $x^2 + y^2 = 4$, $z = 0$ and $z = 3$.

106 Prove that

$$\iiint_V (\text{grad}\, \phi) \cdot (\text{curl}\, \boldsymbol{F})\,\mathrm{d}V = \iint_S (\boldsymbol{F} \times \text{grad}\, \phi) \cdot \mathrm{d}\boldsymbol{S}$$

107 Verify the divergence theorem for $\boldsymbol{F} = (xy + y^2)\boldsymbol{i} + x^2y\boldsymbol{j}$ and the volume V in the first octant bounded by $x = 0$, $y = 0$, $z = 0$, $z = 1$ and $x^2 + y^2 = 4$.

108 Use Stokes' theorem to show that the value of the line integral $\int_A^B \boldsymbol{F} \cdot \mathrm{d}\boldsymbol{r}$ for

$$\boldsymbol{F} = (36xz + 6y \cos x,\ 3 + 6 \sin x + z \sin y,\ 18x^2 - \cos y)$$

is independent of the path joining the points A and B.

109 Use Stokes' theorem to evaluate the line integral $\oint_C \boldsymbol{A} \cdot \mathrm{d}\boldsymbol{r}$, where $\boldsymbol{A} = -y\boldsymbol{i} + x\boldsymbol{j}$ and C is the boundary of the ellipse $x^2/a^2 + y^2/b^2 = 1$, $z = 0$.

110 Verify Stokes' theorem by evaluating both sides of

$$\iint_S (\text{curl}\,\boldsymbol{F}) \cdot \mathrm{d}\boldsymbol{S} = \oint_C \boldsymbol{F} \cdot \mathrm{d}\boldsymbol{r}$$

where $\boldsymbol{F} = (2x - y)\boldsymbol{i} - yz^2\boldsymbol{j} - y^2z\boldsymbol{k}$ and S is the curved surface of the hemisphere $x^2 + y^2 + z^2 = 16$, $z \geqslant 0$.

111 By applying Stokes' theorem to the function $\boldsymbol{a}f(\boldsymbol{r})$, where $\boldsymbol{a}$ is a constant, deduce that

$$\iint_S (\boldsymbol{n} \times \text{grad}\,f)\,\mathrm{d}S = \int_C f(\boldsymbol{r})\mathrm{d}\boldsymbol{r}$$

Verify this result for the function $f(\boldsymbol{r}) = 3xy^2$ and the rectangle in the plane $z = 0$ bounded by the lines $x = 0$, $x = 1$, $y = 0$ and $y = 2$.

112 Verify Stokes' theorem for $\boldsymbol{F} = (2y + z, x - z, y - x)$ for the part of $x^2 + y^2 + z^2 = 1$ lying in the positive octant.

7.5 Engineering application: streamlines in fluid dynamics

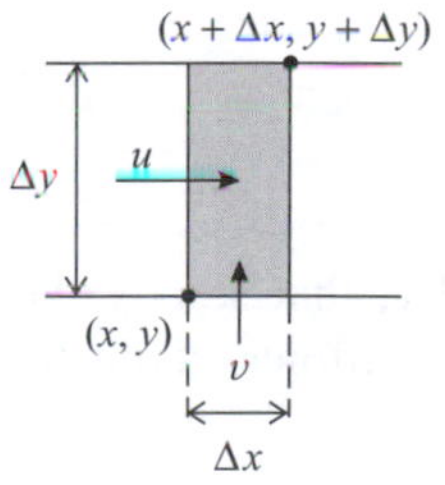

Figure 7.47

As we mentioned in Section 7.1.5, differentials often occur in mathematical modelling of practical problems. An example occurs in fluid dynamics. Consider the case of steady-state incompressible fluid flow in two dimensions. Using rectangular cartesian coordinates (x, y) to describe a point in the fluid, let u and v be the velocities of the fluid in the x and y directions respectively. Then by considering the flow in and flow out of a small rectangle, as shown in Figure 7.47, per unit time, we obtain a differential relationship between $u(x, y)$ and $v(x, y)$ that models the fact that no fluid is lost or gained in the rectangle; that is, the fluid is conserved.

The velocity of the fluid $\boldsymbol{q}$ is a vector point function. The values of its components u and v depend on the spatial coordinates x and y. The flow into the small rectangle in unit time is

$$u(x, \bar{y})\Delta y + v(\bar{x}, y)\Delta x$$

where $\bar{x}$ lies between x and $x + \Delta x$, and $\bar{y}$ lies between y and $y + \Delta y$. Similarly, the flow out of the rectangle is

$$u(x + \Delta x, \tilde{y})\Delta y + v(\tilde{x}, y + \Delta y)\Delta x$$

where $\tilde{x}$ lies between x and $x + \Delta x$ and $\tilde{y}$ lies between y and $y + \Delta y$. Because no fluid is created or destroyed within the rectangle, we may equate these two expressions, giving

$$u(x, \bar{y})\Delta y + v(\bar{x}, y)\Delta x = u(x + \Delta x, \tilde{y})\Delta y + v(\tilde{x}, y + \Delta y)\Delta x$$

Rearranging, we have

$$\frac{u(x + \Delta x, \tilde{y}) - u(x, \bar{y})}{\Delta x} + \frac{v(\tilde{x}, y + \Delta y) - v(\bar{x}, y)}{\Delta y} = 0$$

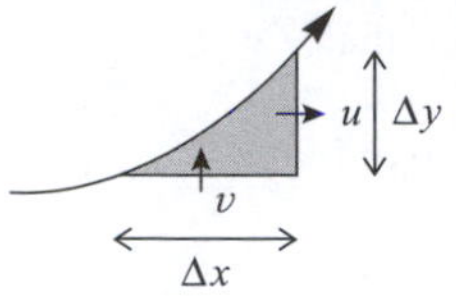

Figure 7.48

Letting $\Delta x \to 0$ and $\Delta y \to 0$ gives the **continuity equation**

$$\frac{\partial u}{\partial x} + \frac{\partial v}{\partial y} = 0$$

The fluid actually flows along paths called **streamlines** so that there is no flow across a streamline. Thus from Figure 7.48 we deduce that

$$v\,\Delta x = u\,\Delta y$$

and hence

$$v\,\mathrm{d}x - u\,\mathrm{d}y = 0$$

The condition for this expression to be an exact differential is

$$\frac{\partial}{\partial y}(v) = \frac{\partial}{\partial x}(-u)$$

or

$$\frac{\partial u}{\partial x} + \frac{\partial v}{\partial y} = 0$$

This is satisfied for incompressible flow since it is just the continuity equation, so that we deduce that there is a function $\psi(x, y)$, called the **stream function**, such that

$$v = \frac{\partial \psi}{\partial x} \quad \text{and} \quad u = -\frac{\partial \psi}{\partial y}$$

It follows that if we are given u and v, as functions of x and y, that satisfy the continuity equation then we can find the equations of the streamlines given by $\psi(x, y) = \text{constant}$.

EXAMPLE 7.32 Find the stream function $\psi(x, y)$ for the incompressible flow that is such that the velocity $\boldsymbol{q}$ at the point (x, y) is

$$(-y/(x^2 + y^2), x/(x^2 + y^2))$$

Solution From the definition of the stream function, we have

$$u(x, y) = -\frac{\partial \psi}{\partial y} \quad \text{and} \quad v(x, y) = \frac{\partial \psi}{\partial x}$$

provided that

$$\frac{\partial u}{\partial x} + \frac{\partial v}{\partial y} = 0$$

Here we have

$$u = \frac{-y}{x^2 + y^2} \quad \text{and} \quad v = \frac{x}{x^2 + y^2}$$

so that

$$\frac{\partial u}{\partial x} = \frac{2xy}{(x^2 + y^2)^2} \quad \text{and} \quad \frac{\partial v}{\partial y} = -\frac{2yx}{(x^2 + y^2)^2}$$

confirming that

$$\frac{\partial u}{\partial x} + \frac{\partial v}{\partial y} = 0$$

Integrating

$$\frac{\partial \psi}{\partial y} = -u(x, y) = \frac{y}{x^2 + y^2}$$

with respect to y, keeping x constant, gives

$$\psi(x, y) = \tfrac{1}{2} \ln (x^2 + y^2) + g(x)$$

Differentiating partially with respect to x gives

$$\frac{\partial \psi}{\partial x} = \frac{x}{x^2 + y^2} + \frac{dg}{dx}$$

Since it is known that

$$\frac{\partial \psi}{\partial x} = v(x, y) = \frac{x}{x^2 + y^2}$$

we have

$$\frac{dg}{dx} = 0$$

which on integrating gives

$$g(x) = C$$

where C is a constant. Substituting back into the expression obtained for $\psi(x, y)$, we have

$$\psi(x, y) = \tfrac{1}{2} \ln (x^2 + y^2) + C$$

A streamline of the flow is given by the equation $\psi(x, y) = k$, where k is a constant. After a little manipulation this gives

$$x^2 + y^2 = a^2 \quad \text{and} \quad \ln a = k - C$$

and the corresponding streamlines are shown in Figure 7.49. This is an example of a **vortex**.

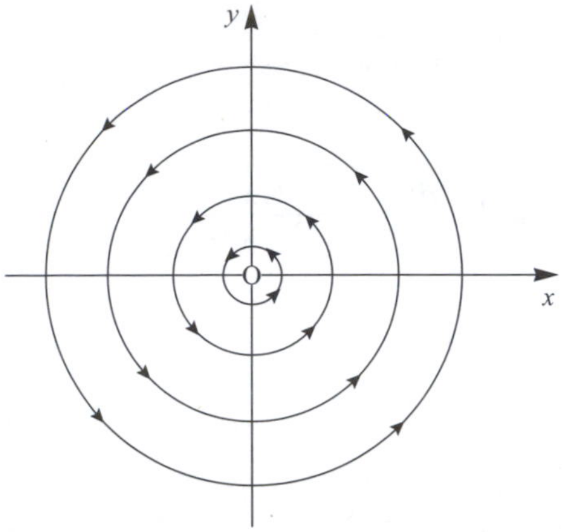

Figure 7.49

7.6 Engineering application: heat transfer

In modelling heat transfer problems we make use of three experimental laws.

(1) Heat flows from hot regions to cold regions of a body.

(2) The rate at which heat flows through a plane section drawn in a body is proportional to its area and to the temperature gradient normal to the section.

(3) The quantity of heat in a body is proportional to its mass and to its temperature.

In the simplest case we consider heat transfer in a medium for which the constants of proportionality in the above laws are independent of direction. Such a medium is called **thermally isentropic**. For any arbitrary region within such a medium we can obtain an equation that models such heat flows. The total amount $Q(t)$ of heat within the region V is

$$Q(t) = \iiint_V c\rho u(\boldsymbol{r}, t)\,\mathrm{d}V$$

where c is the specific heat of the medium, ρ is the density and $u(\boldsymbol{r}, t)$ is the temperature at the point $\boldsymbol{r}$ at time t. Heat flows out of the region through its bounding surface S. The experimental laws (1) and (2) above imply that the rate at which heat flows across an element $\Delta\boldsymbol{S}$ of that surface is $-k\boldsymbol{\nabla}u\cdot\Delta\boldsymbol{S}$, where k is the thermal conductivity of the medium. (The minus sign indicates that heat flows from hot regions to cold.) Thus the rate at which heat flows across the whole surface of the region is given by

$$\iint_S (-k\nabla u) \cdot \mathrm{d}\boldsymbol{S} = -k \iint_S \nabla u \cdot \mathrm{d}\boldsymbol{S}$$

Using Gauss's theorem, we deduce that the rate at which heat flows out of the region is

$$-k \iiint_V \nabla^2 u \, \mathrm{d}V$$

If there are no sources or sinks of heat within the region, this must equal the rate at which the region loses heat, $-\mathrm{d}Q/\mathrm{d}t$. Therefore

$$-\frac{\mathrm{d}}{\mathrm{d}t}\left[\iiint_V c\rho u(\boldsymbol{r}, t)\,\mathrm{d}V\right] = -k \iiint_V \nabla^2 u \,\mathrm{d}V$$

Since

$$\frac{\mathrm{d}}{\mathrm{d}t}\iiint_V u(\boldsymbol{r}, t)\,\mathrm{d}V = \iiint_V \frac{\partial u}{\partial t}\,\mathrm{d}V$$

this implies that

$$\iiint_V \left(k\nabla^2 u - c\rho\frac{\partial u}{\partial t}\right)\mathrm{d}V = 0$$

This models the situation for any arbitrarily chosen region V. The arbitrariness in the choice of V implies that the value of the integral is independent of V and that the integrand is equal to zero. Thus

$$\nabla^2 u = \frac{c\rho}{k}\frac{\partial u}{\partial t}$$

The quantity $k/c\rho$ is termed the **thermal diffusivity** of the medium and is usually denoted by the Greek letter kappa, κ. The differential equation models heat flow within a medium. Its solution depends on the initial temperature distribution $u(\boldsymbol{r}, 0)$ and on the conditions pertaining at the boundary of the region. Methods for solving this equation are discussed in Chapter 9. This differential equation also occurs as a model for water percolation through a dam, for neutron transport in reactors and in charge transfer within charge-coupled devices. We shall now proceed to obtain its solution in a very special case.

EXAMPLE 7.33

A large slab of material has an initial temperature distribution such that one half is at $-u_0$ and the other at $+u_0$. Obtain a mathematical model for this situation and solve it, stating explicitly the assumptions that are made.

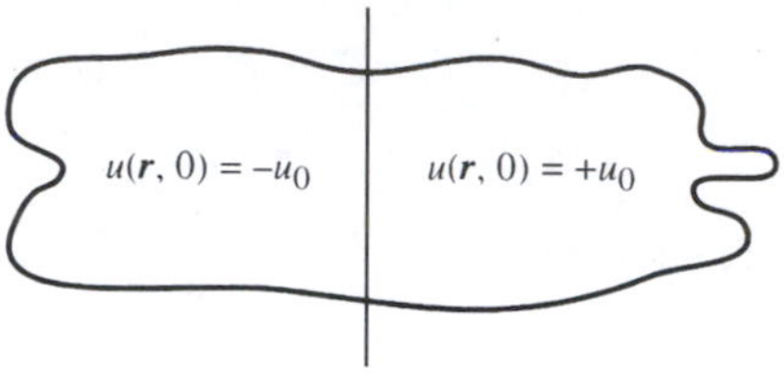

Figure 7.50 Region for Example 7.33.

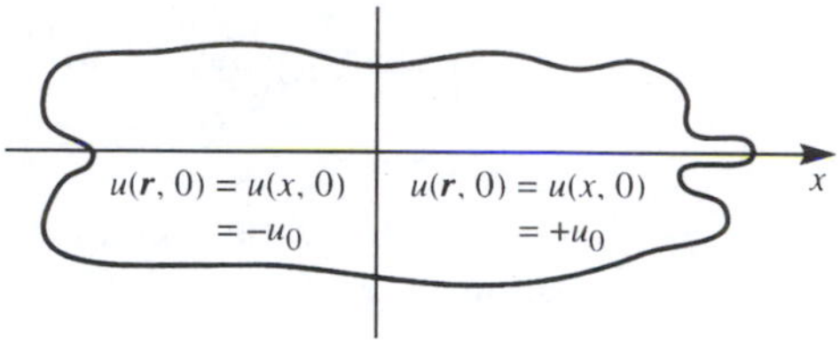

Figure 7.51 Coordinate system for Example 7.33.

Solution When a problem is stated in such vague terms, it is difficult to know what approximations and simplifications may be reasonably made. Since we are dealing with heat transfer, we know that for an isentropic medium the temperature distribution satisfies the equation

$$\nabla^2 u = \frac{1}{\kappa}\frac{\partial u}{\partial t}$$

throughout the medium. We know that the region we are studying is divided so that at $t = 0$ the temperature in one part is $-u_0$ while that in the other is $+u_0$, as illustrated in Figure 7.50. We can deduce from this figure that the subsequent temperature at a point in the medium depends only on the perpendicular distance of the point from the dividing plane. We choose a coordinate system so that its origin lies on the dividing plane and the x axis is perpendicular to it, as shown in Figure 7.51. Then the differential equation simplifies, since $u(\boldsymbol{r}, t)$ is independent of y and z, and we have

$$\frac{\partial^2 u}{\partial x^2} = \frac{1}{\kappa}\frac{\partial u}{\partial t} \quad \text{with} \quad u(x, 0) = \begin{cases} -u_0 & (x < 0) \\ +u_0 & (x \geqslant 0) \end{cases}$$

Thinking about the physical problem also provides us with some further information. The heat flows from the hot region to the cold until (eventually) the temperature is uniform throughout the medium. In this case that terminal temperature is zero, since initially half the medium is at temperature $+u_0$ and the other half at $-u_0$. So we know that $u(x, t) \to 0$ as $t \to \infty$. We also deduce from the initial temperature distribution that $-u_0 \leqslant u(x, t) \leqslant u_0$ for all x and t, since there are no extra sources or sinks of heat in the medium. Summarizing, we have

$$\frac{\partial^2 u}{\partial x^2} = \frac{1}{\kappa}\frac{\partial u}{\partial t} \quad (-\infty < x < \infty, t \geqslant 0) \quad \text{with} \quad \begin{aligned} &u(x, 0) = \begin{cases} -u_0 & (x < 0) \\ +u_0 & (x \geqslant 0) \end{cases} \\ &u(x, t) \quad \text{bounded for all } x \\ &u(x, t) \to 0 \quad \text{as} \quad t \to \infty \end{aligned}$$

There are many approaches to solving this problem (see Chapter 9). One is to investigate the effect of changing the scale of the independent variables x and t. Setting $x = \lambda X$ and $t = \mu T$, where λ and μ are positive constants, the problem becomes

$$\mu \frac{\partial^2 U}{\partial X^2} = \frac{\lambda^2}{\kappa} \frac{\partial U}{\partial T}$$

with $U(X, T) = u(x, t)$ and $U(X, 0) = u_0 \operatorname{sgn} X$. Choosing $\mu = \lambda^2$, we see that

$$\frac{\partial^2 U}{\partial X^2} = \frac{1}{\kappa} \frac{\partial U}{\partial T}, \quad \text{with} \quad U(X, 0) = u_0 \operatorname{sgn} X$$

which implies that the solution $u(x, t)$ of the original equation is also a solution of the scaled equation. Thus

$$u(x, t) = u(\lambda x, \lambda^2 t)$$

which suggests that we should look for a solution expressed in terms of a new variable s that is proportional to the ratio of x to $\sqrt{t}$. Setting $s = ax/\sqrt{t}$, we seek a solution as a function of s:

$$u(x, t) = u_0 f(s)$$

This reduces the partial differential equation for u to an ordinary differential equation for f, since

$$\frac{\partial u}{\partial x} = \frac{a u_0}{\sqrt{t}} \frac{\mathrm{d}f}{\mathrm{d}s}, \quad \frac{\partial^2 u}{\partial x^2} = \frac{a^2 u_0}{t} \frac{\mathrm{d}^2 f}{\mathrm{d}s^2}, \quad \frac{\partial u}{\partial t} = -\frac{1}{2} \frac{a x u_0}{t\sqrt{t}} \frac{\mathrm{d}f}{\mathrm{d}s}$$

Thus the differential equation is transformed into

$$\frac{a^2}{t} \frac{\mathrm{d}^2 f}{\mathrm{d}s^2} = -\frac{ax}{2\kappa t\sqrt{t}} \frac{\mathrm{d}f}{\mathrm{d}s}$$

giving

$$a^2 \frac{\mathrm{d}^2 f}{\mathrm{d}s^2} = -\frac{s}{2\kappa} \frac{\mathrm{d}f}{\mathrm{d}s}$$

Choosing the constant a such that $a^2 = 1/(4\kappa)$ reduces this to the equation

$$\frac{\mathrm{d}^2 f}{\mathrm{d}s^2} = -2s \frac{\mathrm{d}f}{\mathrm{d}s}$$

The initial condition is transformed into two conditions, since for $x < 0$, $s \to -\infty$ as $t \to 0$ and for $x > 0$, $s \to +\infty$ as $t \to 0$. So we have

$$f(s) \to 1 \quad \text{as} \quad s \to \infty$$

$$f(s) \to -1 \quad \text{as} \quad s \to -\infty$$

Integrating the differential equation once gives

$$\frac{\mathrm{d}f}{\mathrm{d}s} = A\,\mathrm{e}^{-s^2}, \quad \text{where } A \text{ is a constant}$$

and integrating a second time gives

$$f(s) = B + A \int e^{-s^2} \, ds$$

The integral occurring here is one that frequently arises in heat transfer problems, and is given a special name. We define the **error function**, erf(x), by the integral

$$\operatorname{erf}(x) = \frac{2}{\sqrt{\pi}} \int_0^x e^{-z^2} \, dz$$

Its name derives from the fact that it is associated with the normal distribution, which is a common model for the distribution of experimental errors. This is a well-tabulated function, and has the property that erf(x) $\to 1$ as $x \to \infty$.

Writing the solution obtained above in terms of the error function, we have

$$f(s) = A \operatorname{erf}(s) + B$$

Letting $s \to \infty$ and $s \to -\infty$ gives two equations for A and B:

$$1 = A + B$$

$$-1 = -A + B$$

from which we deduce $A = 1$ and $B = 0$. Thus

$$f(s) = \operatorname{erf}(s)$$

so that

$$u(x, t) = u_0 \operatorname{erf}\left(\frac{x}{2\sqrt{t}}\right) = \frac{2u_0}{\sqrt{\pi}} \int_0^{x/2\sqrt{t}} e^{-z^2} \, dz$$

7.7 Review exercises (1–20)

1 Show that $u(x, y) = x^n f(t)$, $t = y/x$, satisfies the differential equations

(a) $x \dfrac{\partial u}{\partial x} + y \dfrac{\partial u}{\partial y} = nu$

(b) $x^2 \dfrac{\partial^2 u}{\partial x^2} + 2xy \dfrac{\partial^2 u}{\partial x \partial y} + y^2 \dfrac{\partial^2 u}{\partial y^2} = n(n-1)u$

Verify these results for the function $u(x, y) = x^4 + y^4 + 16x^2y^2$.

2 Find the values of the numbers a and b such that the change of variables $u = x + ay$, $v = x + by$ transforms the differential equation

$$9 \frac{\partial^2 f}{\partial x^2} - 9 \frac{\partial^2 f}{\partial x \partial y} + 2 \frac{\partial^2 f}{\partial y^2} = 0$$

into

$$\frac{\partial^2 f}{\partial u \partial v} = 0$$

Hence deduce that the general solution of the equation is given by

$$u(x, y) = f(x + 3y) + g(x + \tfrac{3}{2}y)$$

where f and g are arbitrary functions.

Find the solution of the differential equation that satisfies the conditions

$$u(x, 0) = \sin x, \quad \frac{\partial u(x, 0)}{\partial y} = 3 \cos x$$

3 A differential $P(x, y, z)\,dx + Q(x, y, z)\,dy + R(x, y, z)\,dz$ is exact if there is a function $f(x, y, z)$ such that

$$P(x, y, z)\,dx + Q(x, y, z)\,dy + R(x, y, z)\,dz = \nabla f \cdot (dx, dy, dz)$$

Show that this implies $\nabla \times (P, Q, R) = 0$. Deduce that curl grad $f = 0$.

4 Find grad f, plot some level curves f = constant and indicate grad f by arrows at some points on the level curves for $f(\boldsymbol{r})$ given by

(a) xy (b) $x/(x^2 + y^2)$

5 Show that if $\boldsymbol{\omega}$ is a constant vector then

(a) grad $(\boldsymbol{\omega} \cdot \boldsymbol{r}) = \boldsymbol{\omega}$

(b) curl $(\boldsymbol{\omega} \times \boldsymbol{r}) = 2\boldsymbol{\omega}$

6 (a) Prove that if $f(\boldsymbol{r})$ is a scalar point function then

curl grad $f = 0$

(b) Prove that if $\boldsymbol{v} = \text{grad}\,[zf(\boldsymbol{r})] + \alpha f(\boldsymbol{r})\boldsymbol{k}$ and $\nabla^2 f = 0$, where α is a constant and f is a scalar point function, then

$$\text{div}\,\boldsymbol{v} = (2 + \alpha)\frac{\partial f}{\partial z}, \quad \nabla^2 \boldsymbol{v} = \text{grad}\left(2\frac{\partial f}{\partial z}\right)$$

7 Show that if $\boldsymbol{F} = (x^2 - y^2 + x)\boldsymbol{i} - (2xy + y)\boldsymbol{j}$, then curl $\boldsymbol{F} = 0$, and find $f(\boldsymbol{r})$ such that $\boldsymbol{F}$ = grad f.

Verify that

$$\int_{(1,2)}^{(2,1)} \boldsymbol{F} \cdot d\boldsymbol{r} = [f(\boldsymbol{r})]_{(1,2)}^{(2,1)}$$

8 A force $\boldsymbol{F}$ acts on a particle that is moving in two dimensions along the semicircle $x = 1 - \cos\theta$, $y = \sin\theta$ $(0 \leqslant \theta \leqslant \pi)$. Find the work done when

(a) $\boldsymbol{F} = \sqrt{(x^2 + y^2)}\,\boldsymbol{i}$

(b) $\boldsymbol{F} = \sqrt{(x^2 + y^2)}\,\hat{\boldsymbol{n}}$

$\hat{\boldsymbol{n}}$ being the unit vector tangential to the path.

9 A force $\boldsymbol{F} = (xy, -y, 1)$ acts on a particle as it moves along the straight line from (0, 0, 0) to (1, 1, 1). Calculate the work done.

10 The force $\boldsymbol{F}$ per unit length of a conducting wire carrying a current $\boldsymbol{I}$ in a magnetic field $\boldsymbol{B}$ is $\boldsymbol{F} = \boldsymbol{I} \times \boldsymbol{B}$. Find the force acting on a circuit whose shape is given by $x = \sin\theta$, $y = \cos\theta$, $z = \sin\frac{1}{2}\theta$, when current I flows in it and when it lies in a magnetic field $\boldsymbol{B} = x\boldsymbol{i} - y\boldsymbol{j} + \boldsymbol{k}$.

11 The velocity $\boldsymbol{v}$ at the point (x, y) in a two-dimensional fluid flow is given by $\boldsymbol{v} = (y\boldsymbol{i} - x\boldsymbol{j})/(x^2 + y^2)$. Find the net circulation around the square $x = \pm 1$, $y = \pm 1$.

12 A metal plate has its boundary defined by $x = 0$, $y = x^2/c$ and $y = c$. The density at the point (x, y) is kxy (per unit area). Find the moment of inertia of the plate about an axis through (0, 0) and perpendicular to the plate.

13 A right circular cone of height h and base radius a is cut into two pieces along a plane parallel to and distance c from the axis of the cone. Find the volume of the smaller piece.

14 The axes of two circular cylinders of radius a intersect at right angles. Show that the volume common to both cylinders may be expressed as the triple integral

$$8\int_0^a dy \int_0^{\sqrt{(a^2-y^2)}} dx \int_0^{\sqrt{(a^2-y^2)}} dz$$

and hence evaluate it.

15 The elastic energy of a volume V of material is $q^2V/2EI$, where q is its stress and E and I are constants. Find the elastic energy of a cylindrical volume of radius r and length l in which the stress varies directly as the distance from its axis, being zero at the axis and q_0 at the outer surface.

16 The velocity of a fluid at the point (x, y, z) has components $(3x^2y, xy^2, 0)$. Find the flow rate out of the triangular prism bounded by $z = 0$, $z = 1$, $x = 0$, $y = 0$ and $x + y = 1$.

17 An electrostatic field has components $(2xy, -y^2, x + y)$ at the point (x, y, z). Find the total flux out of the sphere $x^2 + y^2 + z^2 = a^2$.

18 Verify Stokes' theorem

$$\oint_C \boldsymbol{F} \cdot d\boldsymbol{r} = \iint_S (\text{curl}\,\boldsymbol{F}) \cdot d\boldsymbol{S}$$

where $\boldsymbol{F} = (x^2 + y - 4, 3xy, 2xz + z^2)$ and S is the surface of the hemisphere $x^2 + y^2 + z^2 = 16$ above the (x, y) plane.

19 Use the divergence theorem to evaluate the surface integral

$$\iint_S \boldsymbol{a} \cdot \mathrm{d}\boldsymbol{S}$$

where $\boldsymbol{a} = x\boldsymbol{i} + y\boldsymbol{j} - 2z\boldsymbol{k}$ and S is the surface of the sphere $x^2 + y^2 + z^2 = a^2$ *above* the (x, y) plane.

20 Evaluate the volume integral

$$\iiint_V xyz \, \mathrm{d}V$$

where V denotes the wedge-shaped region bounded in the positive octant by the four planes $x = 0$, $y = 0$, $y = 1 - x$ and $z = 2 - x$.

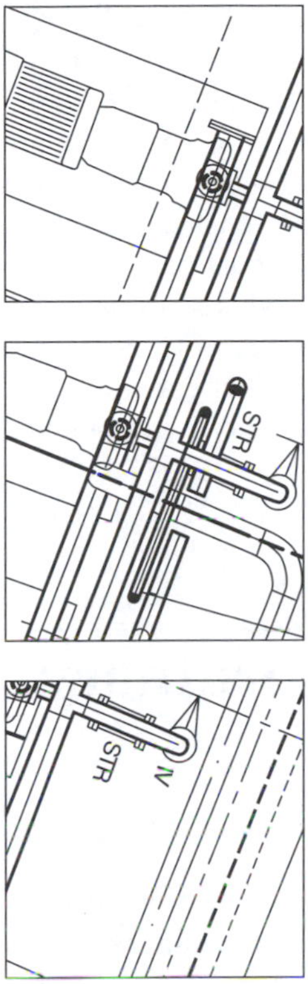

8

Numerical Solution of Ordinary Differential Equations

CONTENT

8.1 Introduction

Frequently the equations which express mathematical models in both engineering analysis and engineering design involve derivatives and integrals of the models' variables. Equations involving derivatives are called **differential equations** and those which include integrals or both integrals and derivatives are called **integral equations** or **integro-differential equations**. Generally integral and integro-differential equations are more difficult to deal with than purely differential ones.

There are many methods and techniques for the analytical solution of elementary ordinary differential equations. The most common of these are covered in most first-level books on engineering mathematics (e.g. *Modern Engineering Mathematics*). However, many differential equations of interest to engineers are not amenable to analytical solution and in these cases we must resort to numerical solutions. Numerical solutions have many disadvantages (it is, for instance, much less obvious how changes of parameters or coefficients in the equations affect the solutions) so an analytical solution is generally more useful where one is available. In this chapter we will develop some basic concepts relating to numerical methods for solving ordinary differential equations and, in the latter parts, indicate some of the more advanced methods which are available for the more complex and difficult problems.

8.2 Numerical solution of first-order ordinary differential equations

In a book such as this we cannot hope to cover all of the many numerical techniques which have been developed for dealing with ordinary differential equations so we will concentrate on presenting a selection of methods which illustrate the main strands of the theory. In so doing we will meet the main theoretical tools and unifying concepts of the area.

8.2.1 A simple solution method: Euler's method

For a first-order differential equation $\mathrm{d}x/\mathrm{d}t = f(t, x)$ we can define a **direction field**. The direction field is that two-dimensional vector field in which the vector at any point (t, x) has the gradient $dx/\mathrm{d}t$. More precisely, it is the field

$$\frac{f(t,x)}{\sqrt{[1+f(t,x)^2]}}[1, f(t, x)]$$

For instance Figure 8.1 shows the direction field of the differential equation $\mathrm{d}x/\mathrm{d}t = x(1 - x)t$.

Since a solution of a differential equation is a function $x(t)$ which has the property $\mathrm{d}x/\mathrm{d}t = f(t, x)$ at all points (t, x) the solutions of the differential equation are curves in the t–x plane to which the direction field lines are tangential at every point. For instance, the curves shown in Figure 8.2 are solutions of the differential equation

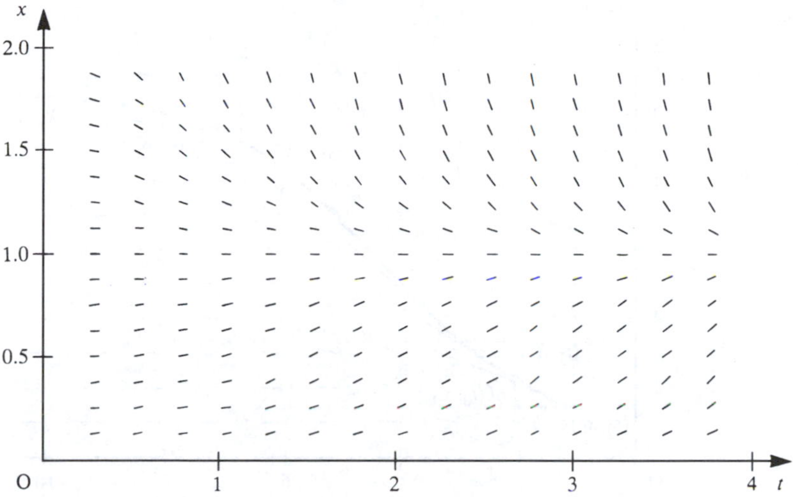

Figure 8.1 The direction field for the equation $dx/dt = x(1 - x)t$.

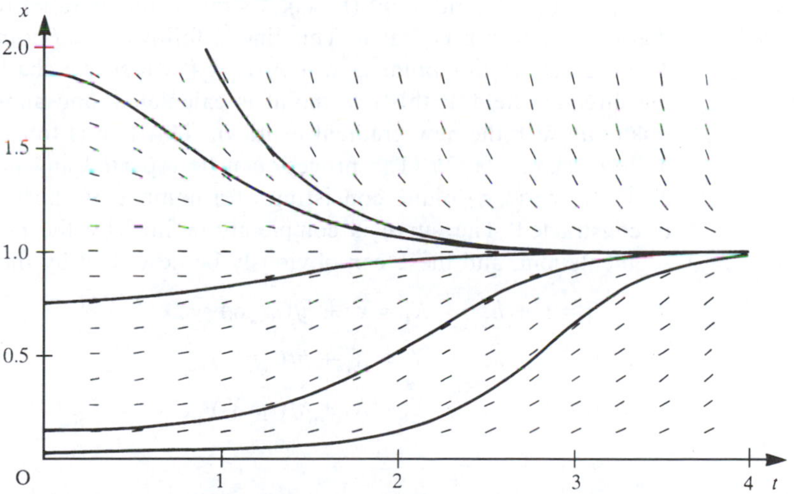

Figure 8.2 Solutions of $dx/dt = x(1 - x)t$ superimposed on its direction field.

$$\frac{dx}{dt} = x(1 - x)t$$

This immediately suggests that a curve representing a solution can be obtained by sketching on the direction field a curve that is always tangential to the lines of the direction field. In Figure 8.3 a way of systematically constructing an approximation to such a curve is shown.

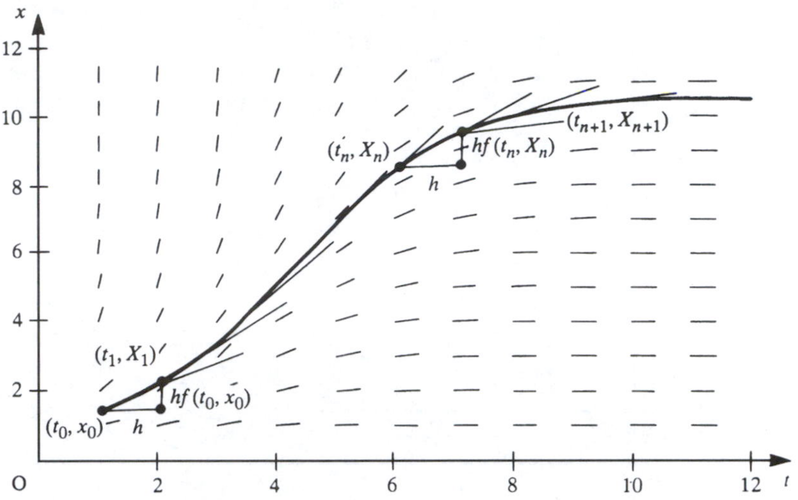

Figure 8.3 The construction of a numerical solution of the equation $dx/dt = f(t, x)$.

Starting at some point (t_0, x_0), a straight line parallel to the direction field at that point, $f(t_0, x_0)$, is drawn. This line is followed to a point with abscissa $t_0 + h$. The ordinate at this point is $x_0 + hf(t_0, x_0)$, which we shall call X_1. The value of the direction field at this new point is calculated, and another straight line from this point with the new gradient is drawn. This line is followed as far as the point with abscissa $t_0 + 2h$. The process can be repeated any number of times, and a curve in the (t, x) plane consisting of a number of short straight line segments is constructed. The curve is completely defined by the points at which the line segments join, and these can obviously be described by the equations

$$
\begin{aligned}
t_1 &= t_0 + h, & X_1 &= x_0 + hf(t_0, x_0) \\
t_2 &= t_1 + h, & X_2 &= X_1 + hf(t_1, X_1) \\
t_3 &= t_2 + h, & X_3 &= X_2 + hf(t_2, X_2) \\
&\vdots & &\vdots \\
t_{n+1} &= t_n + h, & X_{n+1} &= X_n + hf(t_n, X_n)
\end{aligned}
$$

These define, mathematically, the simplest method for integrating first-order differential equations. It is called **Euler's method**. Solutions are constructed step by step, starting from some given starting point (t_0, x_0). For a given t_0 each different x_0 will give rise to a different solution curve. These curves are all solutions of the differential equation, but each corresponds to a different initial condition.

The solution curves constructed using this method are obviously not exact solutions but only approximations to solutions, because they are only tangential to the direction field at certain points. Between these points, the curves are only approximately tangential to the direction field. Intuitively, we expect that, as the distance for which we follow each straight line segment is reduced, the curve we

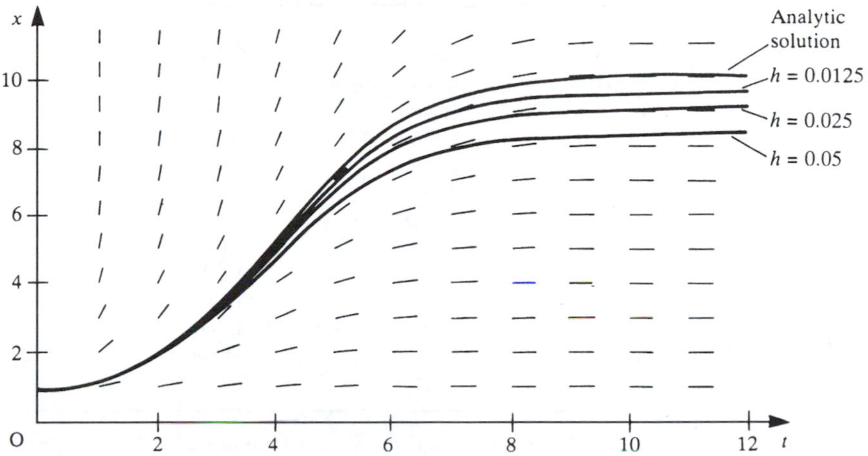

Figure 8.4 The Euler-method solutions of $dx/dt = x^2 t e^{-t}$ for $h = 0.05$, 0.025 and 0.0125.

are constructing will become a better and better approximation to the exact solution. The increment h in the independent variable t along each straight-line segment is called the **step size** used in the solution. In Figure 8.4 three approximate solutions of the initial-value problem

$$\frac{dx}{dt} = x^2 t e^{-t}, \quad x(0) = 0.91 \tag{8.1}$$

for step sizes $h = 0.05$, 0.025 and 0.0125 are shown. These steps are sufficiently small that the curves, despite being composed of a series of short straight lines, give the illusion of being smooth curves. The equation (8.1) actually has an analytical solution, which can be obtained by separation:

$$x = \frac{1}{(1 + t)e^{-t} + C}$$

The analytical solution to the initial-value problem is also shown in Figure 8.4 for comparison. It can be seen that, as we expect intuitively, the smaller the step size the more closely the numerical solution approximates the analytical solution.

EXAMPLE 8.1

The function $x(t)$ satisfies the differential equation

$$\frac{dx}{dt} = \frac{x + t}{xt}$$

and the initial condition $x(1) = 2$. Use Euler's method to obtain an approximation to the value of $x(2)$ using a step size of $h = 0.1$.

t	X	$X+t$	Xt	$h\dfrac{X+t}{Xt}$
1.0000	2.0000	3.0000	2.0000	0.1500
1.1000	2.1500	3.2500	2.3650	0.1374
1.2000	2.2874	3.4874	2.7449	0.1271
1.3000	2.4145	3.7145	3.1388	0.1183
1.4000	2.5328	3.9328	3.5459	0.1109
1.5000	2.6437	4.1437	3.9656	0.1045
1.6000	2.7482	4.3482	4.3971	0.0989
1.7000	2.8471	4.5471	4.8400	0.0939
1.8000	2.9410	4.7410	5.2939	0.0896
1.9000	3.0306	4.9306	5.7581	0.0856
2.0000	3.1162			

Figure 8.5 Computational results for Example 8.1.

Solution The solution is obtained step by step as set out in Figure 8.5. The approximation $X(2) = 3.1162$ results.

8.2.2 Analysing Euler's method

We have introduced Euler's method via an intuitive argument from a geometrical understanding of the problem. Euler's method can be seen in another light – as an application of Taylor series. The Taylor series expansion for a function $x(t)$ gives

$$x(t+h) = x(t) + h\frac{dx}{dt}(t) + \frac{h^2}{2!}\frac{d^2x}{dt^2}(t) + \frac{h^3}{3!}\frac{d^3x}{dt^3}(t) + \dots \qquad (8.2)$$

Using this formula, we could, in theory, given the value of $x(t)$ and all the derivatives of x at t, compute the value of $x(t+h)$ for any given h. If we choose a small value for h then the Taylor series truncated after a finite number of terms will provide a good approximation to the value of $x(t+h)$. Euler's method can be interpreted as using the Taylor series truncated after the second term as an approximation to the value of $x(t+h)$.

In order to distinguish between the exact solution of a differential equation and a numerical approximation to the exact solution (and it should be appreciated that all numerical solutions, however accurate, are only approximations to the exact solution), we shall now make explicit the convention that we used in the last section. The exact solution of a differential equation will be denoted by a lower-case letter and a numerical approximation to the exact solution by the corresponding capital letter. Thus, truncating the Taylor series, we write

$$X(t + h) = x(t) + h\frac{\mathrm{d}x}{\mathrm{d}t}(t) = x(t) + hf(t, x) \tag{8.3}$$

Applying this truncated Taylor series, starting at the point (t_0, x_0) and denoting $t_0 + nh$ by t_n, we obtain

$$X(t_1) = X(t_0 + h) = x(t_0) + hf(t_0, x_0)$$

$$X(t_2) = X(t_1 + h) = X(t_1) + hf(t_1, X_1)$$

$$X(t_3) = X(t_2 + h) = X(t_2) + hf(t_2, X_2)$$

and so on

which is just the Euler-method formula obtained in Section 8.2.1. As an additional abbreviated notation, we shall adopt the convention that $x(t_0 + nh)$ is denoted by x_n, $X(t_0 + nh)$ by X_n, $f(t_n, x_n)$ by f_n, and $f(t_n, X_n)$ by F_n. Hence we may express the Euler method, in general terms, as the recursive rule

$$X_0 = x_0$$

$$X_{n+1} = X_n + hF_n \quad (n \geqslant 0)$$

The advantage of viewing Euler's method as an application of Taylor series in this way is that it gives us a clue to obtaining more accurate methods for the numerical solution of differential equations. It also enables us to analyse in more detail how accurate the Euler method may be expected to be. Using the order notation we can abbreviate (8.2) to

$$x(t + h) = x(t) + hf(t, x) + O(h^2)$$

and, combining this with (8.3), we see that

$$X(t + h) = x(t + h) + O(h^2) \tag{8.4}$$

(Note that in obtaining this result we have used the fact that signs are irrelevant in determining the order of terms; that is, $-O(h^p) = O(h^p)$.) Equation (8.4) expresses the fact that at each step of the Euler process the value of $X(t + h)$ obtained has an error of order h^2, or, to put it another way, the formula used is accurate as far as terms of order h. For this reason Euler's method is known as a **first-order method**. The exact size of the error is, as we intuitively expected, dependent on the size of h, and decreases as h decreases. Since the error is of order h^2, we expect that halving h, for instance, will reduce the error at each step by a factor of four.

This does not, unfortunately, mean that the error in the solution of the initial value problem is reduced by a factor of four. To understand why this is so, we argue as follows. Starting from the point (t_0, x_0) and using Euler's method with a step size h to obtain a value of $X(t_0 + 4)$, say, requires $4/h$ steps. At each step an error of order h^2 is incurred. The total error in the value of $X(t_0 + 4)$ will be the sum of the errors incurred at each step, and so will be $4/h$ times the value of a typical step error. Hence the total error is of the order of $(4/h)O(h^2)$; that is, the total error is $O(h)$. From this argument we should expect that if we compare solutions of a differential equation obtained using Euler's method with different step sizes, halving the step size will halve the error in the solution. Examination of Figure 8.4 confirms that this expectation is roughly correct in the case of the solutions presented there.

EXAMPLE 8.2

Let X_a denote the approximation to the solution of the initial-value problem

$$\frac{dx}{dt} = \frac{x^2}{t+1}, \quad x(0) = 1$$

obtained using Euler's method with a step size $h = 0.1$, and X_b that obtained using a step size of $h = 0.05$. Compute the values of $X_a(t)$ and $X_b(t)$ for $t = 0.1, 0.2, \dots, 1.0$. Compare these values with the values of $x(t)$, the exact solution of the problem. Compute the ratio of the errors in X_a and X_b.

Solution The exact solution, which may be obtained by separation, is

$$x = \frac{1}{1 - \ln(t+1)}$$

The numerical solutions X_a and X_b and their errors are shown in Figure 8.6. Of course, in this figure the values of X_a are recorded at every step whereas those of X_b are only recorded at alternate steps.

Again, the final column of Figure 8.6 shows that our expectations about the effects of halving the step size when using Euler's method to solve a differential equation are confirmed. The ratio of the errors is not, of course, exactly one-half, because there are some higher-order terms in the errors, which we have ignored.

t	X_a	X_b	$x(t)$	$\lvert x - X_a\rvert$	$\lvert x - X_b\rvert$	$\dfrac{\lvert x - X_b\rvert}{\lvert x - X_a\rvert}$
0.000 00	1.000 00	1.000 00	1.000 00			
0.100 00	1.100 00	1.102 50	1.105 35	0.005 35	0.002 85	0.53
0.200 00	1.210 00	1.216 03	1.222 97	0.012 97	0.006 95	0.54
0.300 00	1.332 01	1.342 94	1.355 68	0.023 67	0.012 75	0.54
0.400 00	1.468 49	1.486 17	1.507 10	0.038 61	0.020 92	0.54
0.500 00	1.622 52	1.649 52	1.681 99	0.059 47	0.032 47	0.55
0.600 00	1.798 03	1.837 91	1.886 81	0.088 78	0.048 90	0.55
0.700 00	2.000 08	2.057 92	2.130 51	0.130 42	0.072 59	0.56
0.800 00	2.235 40	2.318 57	2.425 93	0.190 53	0.107 36	0.56
0.900 00	2.513 01	2.632 51	2.792 16	0.279 15	0.159 65	0.57
1.000 00	2.845 39	3.018 05	3.258 89	0.413 50	0.240 84	0.58

Figure 8.6 Computational results for Example 8.2.

8.2.3 Using numerical methods to solve engineering problems

In Example 8.2 the errors in the values of X_a and X_b are quite large (up to about 14% in the worst case). While carrying out computations with large errors such as these is quite useful for illustrating the mathematical properties of computational methods, in engineering computations we usually need to keep errors very much smaller. Exactly how small they must be is largely a matter of engineering judgement. The engineer must decide how accurately a result is needed for a given engineering purpose. It is then up to that engineer to use the mathematical techniques and knowledge available to carry out the computations to the desired accuracy. The engineering decision about the required accuracy will usually be based on the use that is to be made of the result. If, for instance, a preliminary design study is being carried out then a relatively approximate answer will often suffice, whereas for final design work much more accurate answers will normally be required. It must be appreciated that demanding greater accuracy than is actually needed for the engineering purpose in hand will usually carry a penalty in time, effort or cost.

Let us imagine that, for the problem posed in Example 8.2, we had decided we needed the value of $x(1)$ accurate to 1%. In the cases in which we should normally resort to numerical solution we should not have the analytical solution available, so we must ignore that solution. We shall suppose then that we had obtained the values of $X_a(1)$ and $X_b(1)$ and wanted to predict the step size we should need to use to obtain a better approximation to $x(1)$ accurate to 1%. Knowing that the error in $X_b(1)$ should be approximately one-half the error in $X_a(1)$ suggests that the error in $X_b(1)$ will be roughly the same as the difference between the errors in $X_a(1)$ and $X_b(1)$, which is the same as the difference between $X_a(1)$ and $X_b(1)$; that is, 0.172 66. One percent of $X_b(1)$ is roughly 0.03, that is, roughly one-sixth of the error in $X_b(1)$. Hence we expect that a step size roughly one-sixth of that used to obtain X_b will suffice; that is, a step size $h = 0.008\,33$. In practice, of course, we shall round to a more convenient non-recurring decimal quantity such as $h = 0.008$. This procedure is closely related to the Aitken extrapolation procedure sometimes used for estimating limits of convergent sequences and series.

EXAMPLE 8.3

Compute an approximation $X(1)$ to the value of $x(1)$ satisfying the initial-value problem

$$\frac{dx}{dt} = \frac{x^2}{t+1}, \quad x(0) = 1$$

by using Euler's method with a step size $h = 0.008$.

Solution It is worth commenting here that the calculations performed in Example 8.2 could reasonably be carried out on any hand-held calculator, but this new calculation requires 125 steps. To do this is on the boundaries of what might reasonably be done on a hand-held calculator, and is more suited to a micro- or minicomputer. Repeating the calculation with a step size $h = 0.008$ produces the result $X(1) = 3.213\,91$.

We had estimated from the evidence available (that is, values of $X(1)$ obtained using step sizes $h = 0.1$ and 0.05) that the step size $h = 0.008$ should provide a value of $X(1)$ accurate to approximately 1%. Comparison of the value we have just computed with the exact solution shows that it is actually in error by approximately 1.4%. This does not quite meet the target of 1% that we set ourselves. This example therefore serves, first, to illustrate how, given two approximations to $x(1)$ derived using Euler's method with different step sizes, we can estimate the step size needed to compute an approximation within a desired accuracy, and, secondly, to emphasize that the estimate of the appropriate step size is only an *estimate*, and will not *guarantee* an approximate solution to the problem meeting the desired accuracy criterion. If we had been more conservative and rounded the estimated step size down to, say, 0.005, we should have obtained $X(1) = 3.230\,43$, which is in error by only 0.9% and would have met the required accuracy criterion.

Since we have mentioned in Example 8.3 the use of computers to undertake the repetitious calculations involved in the numerical solution of differential equations, it is also worth commenting briefly on the writing of computer programs to implement those numerical solution methods. While it is perfectly possible to write informal, unstructured programs to implement algorithms such as Euler's method, a little attention to planning and structuring a program well will usually be amply rewarded – particularly in terms of the reduced probability of introducing 'bugs'. Another reason for careful structuring is that, in this way, parts of programs can often be written in fairly general terms and can be re-used later for other problems. The two pseudocode algorithms in Figures 8.7 and 8.8 will both produce the table of results in Example 8.2. The pseudocode program of Figure 8.7 is very specific to the problem posed, whereas that of Figure 8.8 is more general, better structured, and more expressive of the structure of mathematical problems. It is generally better to aim at the style of Figure 8.8.

```
x1 ← 1
x2 ← 1
write(keyb, 0, 1, 1, 1)
for i is 1 to 10 do
  x1 ← x1 + 0.1*x1*x1/((i−1)*0.1 + 1)
  x2 ← x2 + 0.05*x2*x2/((i−1)*0.1 + 1)
  x2 ← x2 + 0.05*x2*x2/((i−1)*0.1 + 1.05)
  x ← 1/(1 − ln(i*0.1 + 1))
  write(keyb,0.1*i,x1,x2,x,x − x1,x − x2,(x − x2)/(x − x1))
endfor
```

Figure 8.7 A poorly structured algorithm for Example 8.2.

```
initial_time ← 0
final_time ← 1
initial_x ← 1
step ← 0.1
t ← initial_time
x1 ← initial_x
x2 ← initial_x
h1 ← step
h2 ← step/2
write(keyb,initial_time,x1,x2,initial_x)
repeat
  euler(t,x1,h1,1 → x1)
  euler(t,x2,h2,2 → x2)
  t ← t + step
  x ← exact_solution(t,initial_time,initial_x)
  write(keyb,t,x1,x2,x,abs(x - x1),abs(x - x2),abs((x - x2)/(x - x1)))
until t ≥ final_time

procedure euler(t_old,x_old,step,number → x_new)
  temp_x ← x_old
  for i is 0 to number -1 do
    temp_x ← temp_x + step*derivative(t_old + step*i,temp_x)
  endfor
  x_new ← temp_x
endprocedure

procedure derivative(t,x → derivative)
  derivative ← x*x/(t + 1)
endprocedure

procedure exact_solution(t,t0,x0 → exact_solution)
  c ← ln(t0 + 1) + 1/x0
  exact_solution ← 1/(c - ln(t + 1))
endprocedure
```

Figure 8.8 A better structured algorithm for Example 8.2.

8.2.4 Exercises

(Questions marked with a dagger (†) are intended to be solved with the assistance of a computer.)

1 Find the value of $X(1)$ for the initial-value problem

$$\frac{dx}{dt} = \frac{x}{2(t+1)}, \quad x(0.5) = 1$$

using Euler's method with step size $h = 0.1$.

2 Find the value of $X(0.5)$ for the initial-value problem

$$\frac{dx}{dt} = \frac{4-t}{t+x}, \quad x(0) = 1$$

using Euler's method with step size $h = 0.05$.

†3 Denote the Euler-method solution of the initial-value problem

$$\frac{dx}{dt} = \frac{xt}{t^2+2}, \quad x(1) = 2$$

using step size $h = 0.1$ by $X_a(t)$, and that using $h = 0.05$ by $X_b(t)$. Find the values of $X_a(2)$ and $X_b(2)$. Estimate the error in the value of $X_b(2)$, and suggest a value of step size that would provide a value of $X(2)$ accurate to 0.1%. Find the value of $X(2)$ using this step size. Find the exact solution of the initial-value problem, and determine the actual magnitude of the errors in $X_a(2)$, $X_b(2)$ and your final value of $X(2)$.

†**4** Denote the Euler-method solution of the initial-value problem

$$\frac{dx}{dt} = \frac{1}{xt}, \quad x(1) = 1$$

using step size $h = 0.1$ by $X_a(t)$, and that using $h = 0.05$ by $X_b(t)$. Find the values of $X_a(2)$ and $X_b(2)$. Estimate the error in the value of $X_b(2)$, and suggest a value of step size that would provide a value of $X(2)$ accurate to 0.2%. Find the value of $X(2)$ using this step size. Find the exact solution of the initial-value problem, and determine the actual magnitude of the errors in $X_a(2)$, $X_b(2)$ and your final value of $X(2)$.

†**5** Denote the Euler-method solution of the initial-value problem

$$\frac{dx}{dt} = \frac{1}{\ln x}, \quad x(1) = 1.2$$

using step size $h = 0.05$ by $X_a(t)$, and that using $h = 0.025$ by $X_b(t)$. Find the values of $X_a(1.5)$ and $X_b(1.5)$. Estimate the error in the value of $X_b(1.5)$, and suggest a value of step size that would provide a value of $X(1.5)$ accurate to 0.25%. Find the value of $X(1.5)$ using this step size. Find the exact solution of the initial-value problem, and determine the actual magnitude of the errors in $X_a(1.5)$, $X_b(1.5)$ and your final value of $X(1.5)$.

8.2.5 More accurate solution methods: multistep methods

In Section 8.2.2 we discovered that using Euler's method to solve a differential equation is essentially equivalent to using a Taylor series expansion of a function truncated after two terms. Since, by so doing, we are ignoring terms $O(h^2)$, an error of this order is introduced at each step in the solution. Could we not derive a method for calculating approximate solutions of differential equations which, by using more terms of the Taylor series, provides greater accuracy than Euler's method? We can – but there are some disadvantages in so doing, and various methods have to be used to overcome these.

Let us first consider a Taylor series expansion with the first three terms written explicitly. This gives

$$x(t + h) = x(t) + h\frac{dx}{dt}(t) + \frac{h^2}{2!}\frac{d^2x}{dt^2}(t) + O(h^3) \tag{8.5}$$

Substituting $f(t, x)$ for dx/dt, we obtain

$$x(t + h) = x(t) + hf(t, x) + \frac{h^2}{2!}\frac{df}{dt}(t, x) + O(h^3)$$

Dropping the $O(h^3)$ terms provides an approximation

$$X(t + h) = x(t) + hf(t, x) + \frac{h^2}{2!}\frac{df}{dt}(t, x)$$

such that

$$X(t + h) = x(t + h) + O(h^3)$$

in other words, a numerical approximation method which has an error at each step that is not of order h^2 like the Euler method but rather of order h^3. The corresponding general numerical scheme is

$$X_{n+1} = X_n + hF_n + \frac{h^2}{2}\frac{dF_n}{dt} \tag{8.6}$$

The application of the formula (8.3) in Euler's method was straightforward because an expression for $f(t, x)$ was provided by the differential equation itself. To apply (8.6) as it stands requires an analytical expression for df/dt so that dF_n/dt may be computed. This may be relatively straightforward to provide – or it may be quite complicated. In any event, for computer applications, the necessity of providing the derivative of a function has been a considerable disadvantage because, in the past, computer programs have not been able to handle analytical calculations but only numerical ones. This, of course, is beginning to change as more capable symbolic manipulation (computer algebra) packages become available, but the ability to carry out numerical computations efficiently without access to computer algebra will remain important for some time to come.

Fortunately, there are ways to work around this difficulty. One such method hinges on the observation that it is just as valid to write down Taylor series expansions for negative increments as for positive ones. The Taylor series expansion of $x(t - h)$ is

$$x(t-h) = x(t) - h\frac{dx}{dt}(t) + \frac{h^2}{2!}\frac{d^2x}{dt^2}(t) - \frac{h^3}{3!}\frac{d^3x}{dt^3}(t) + \dots$$

If we write only the first three terms explicitly, we have

$$x(t-h) = x(t) - h\frac{dx}{dt}(t) + \frac{h^2}{2!}\frac{d^2x}{dt^2}(t) + O(h^3)$$

or, rearranging the equation,

$$\frac{h^2}{2!}\frac{d^2x}{dt^2}(t) = x(t-h) - x(t) + h\frac{dx}{dt}(t) + O(h^3)$$

Substituting this into (8.5), we obtain

$$x(t+h) = x(t) + h\frac{dx}{dt}(t) + \left[x(t-h) - x(t) + h\frac{dx}{dt}(t) + O(h^3)\right] + O(h^3)$$

That is,

$$x(t+h) = x(t-h) + 2h\frac{dx}{dt}(t) + O(h^3)$$

or, substituting $f(t, x)$ for dx/dt,

$$x(t+h) = x(t-h) + 2hf(t, x) + O(h^3) \tag{8.7}$$

Alternatively, we could write down the Taylor series expansion of the function dx/dt with an increment of $-h$:

$$\frac{dx}{dt}(t-h) = \frac{dx}{dt}(t) - h\frac{d^2x(t)}{dt^2}(t) + \frac{h^2}{2!}\frac{d^3x}{dt^3}(t) - O(h^3)$$

Writing only the first two terms explicitly and rearranging gives

$$h\frac{d^2x}{dt^2}(t) = \frac{dx}{dt}(t) - \frac{dx}{dt}(t-h) + O(h^2)$$

and substituting this into (8.5) gives

$$x(t+h) = x(t) + h\frac{dx}{dt}(t) + \frac{h}{2}\left[\frac{dx}{dt}(t) - \frac{dx}{dt}(t-h) + O(h^2)\right] + O(h^3)$$

That is,

$$x(t+h) = x(t) + \frac{h}{2}\left[3\frac{dx}{dt}(t) - \frac{dx}{dt}(t-h)\right] + O(h^3)$$

or, substituting $f(t, x)$ for dx/dt,

$$x(t+h) = x(t) + \tfrac{1}{2}h[3f(t, x(t)) - f(t-h, x(t-h))] + O(h^3) \qquad \textbf{(8.8)}$$

Equations (8.5), (8.7) and (8.8) each give an expression for $x(t+h)$ in which all terms up to those in h^2 have been made explicit. In the same way as, by ignoring terms of $O(h^3)$ in (8.5), the numerical scheme (8.6) can be obtained, (8.7) and (8.8) give rise to the numerical schemes

$$X_{n+1} = X_{n-1} + 2hF_n \qquad \textbf{(8.9)}$$

and

$$X_{n+1} = X_n + \tfrac{1}{2}h(3F_n - F_{n-1}) \qquad \textbf{(8.10)}$$

respectively. Each of these alternative schemes, like (8.6), incurs an error $O(h^3)$ at each step.

The advantage of (8.9) or (8.10) over (8.6) arises because the derivative of $f(t, x)$ in (8.5) has been replaced in (8.7) by the value of the function x at the previous time, $x(t-h)$, and in (8.8) by the value of the function f at time $t-h$. This is reflected in (8.9) and (8.10) by the presence of the terms in X_{n-1} and F_{n-1} respectively and the absence of the term in dF_n/dt. The elimination of the derivative of the function $f(t, x)$ from the numerical scheme is an advantage, but it is not without its penalties. In both (8.9) and (8.10) the value of X_{n+1} depends not only on the values of X_n and F_n but also on the value of one or the other at t_{n-1}. This is chiefly a problem when starting the computation. In the case of the Euler scheme the first step took the form

$$X_1 = X_0 + hF_0$$

In the case of (8.9) and (8.10) the first step would seem to take the forms

$$X_1 = X_{-1} + 2hF_0$$

and

$$X_1 = X_0 + \tfrac{1}{2}h(3F_0 - F_{-1})$$

respectively. The value of X_{-1} in the first case and F_{-1} in the second is not normally available. The resolution of this difficulty is usually to use some other method to start the computation, and, when the value of X_1, and therefore also the value of F_1, is available, change to (8.9) or (8.10). The first step using (8.9) or (8.10) therefore involves

$$X_2 = X_0 + 2hF_1$$

or

$$X_2 = X_1 + \tfrac{1}{2}h(3F_1 - F_0)$$

Methods like (8.9) and (8.10) that involve the values of the dependent variable or its derivative at more than one value of the independent variable are called **multi-step methods**. These all share the problem that we have just noted of difficulties in deciding how to start the computation. We shall return to this problem of starting multistep methods in Section 8.2.7.

EXAMPLE 8.4

Solve the initial-value problem

$$\frac{dx}{dt} = \frac{x^2}{t+1}, \quad x(0) = 1$$

posed in Example 8.2 using the scheme (8.10) with a step size $h = 0.1$. Compute the values of $X(t)$ for $t = 0.1, 0.2, \ldots, 1.0$ and compare them with the values of the exact solution $x(t)$.

Solution We shall assume that the value of $X(0.1)$ has been computed using some other method and has been found to be 1.105 35. The computation therefore starts with the calculation of the values of F_1, F_0 and hence X_2. The results of the computation are shown in Figure 8.9.

t	X_n	F_n	$\frac{1}{2}h(3F_n - F_{n-1})$	$x(t)$	$\lvert x - X_n\rvert$
0.000 00	1.000 00	1.000 00			
0.100 00	1.105 35	1.110 73	0.116 61	1.105 35	0.000 00
0.200 00	1.221 96	1.244 32	0.131 11	1.222 97	0.001 01
0.300 00	1.353 07	1.408 31	0.149 03	1.355 68	0.002 61
0.400 00	1.502 10	1.611 65	0.171 33	1.507 10	0.004 99
0.500 00	1.673 44	1.866 92	0.199 46	1.681 99	0.008 55
0.600 00	1.872 89	2.192 33	0.235 50	1.886 81	0.013 91
0.700 00	2.108 39	2.614 90	0.282 62	2.130 51	0.022 11
0.800 00	2.391 01	3.176 08	0.345 67	2.425 93	0.034 92
0.900 00	2.736 68	3.941 80	0.432 47	2.792 16	0.055 48
1.000 00	3.169 14			3.258 89	0.089 75

Figure 8.9 Computational results for Example 8.4.

It is instructive to compare the values of X computed in Example 8.4 with those computed in Example 8.2. Since the method we are using here is a second-order method, the error at each step should be $O(h^3)$ rather than the $O(h^2)$ error of the Euler method. We are using the same step size as for the solution X_a of Example 8.2, so the errors should be correspondingly smaller. Because in this case we know the exact solution of the differential equation, we can compute the errors. Examination of the results shows that they are indeed much smaller than those of the Euler method, and also considerably smaller than the errors in the Euler method solution X_b which used step size $h = 0.05$, half the step size used here.

In fact, some numerical experimentation (which we will not describe in detail) reveals that to achieve a similarly low level of errors, the Euler method requires a step size $h = 0.016$, and therefore 63 steps are required to find the value of $X(1)$. The second-order method of (8.10) requires only 10 steps to find $X(1)$ to a similar accuracy. Thus the solution of a problem to a given accuracy using a second-order method can be achieved in a much shorter computer processing time than using a first-order method. When very large calculations are involved or simple calculations are repeated very many times, such savings are very important.

How do we choose between methods of equal accuracy such as (8.9) and (8.10)? Numerical methods for the solution of differential equations have other properties apart from accuracy. One important property is **stability**. Some methods have the ability to introduce gross errors into the numerical approximation to the exact solution of a problem. The sources of these gross errors are the so-called **parasitic solutions** of the numerical process, which do not correspond to solutions of the differential equation. The analysis of this behaviour is beyond the scope of this book, but methods that are susceptible to it are intrinsically less useful than those that are not. The method of (8.9) can show unstable behaviour, as demonstrated in Example 8.5.

EXAMPLE 8.5

Let X_a denote the approximation to the solution of the initial-value problem

$$\frac{dx}{dt} = -3x + 2e^{-t}, \quad x(0) = 2$$

obtained using the method defined by (8.9), and X_b that obtained using the method defined by (8.10), both with step size $h = 0.1$. Compute the values of $X_a(t)$ and $X_b(t)$ for $t = 0.1, 0.2, \ldots, 2.0$. Compare these with the values of $x(t)$, the exact solution of the problem. In order to overcome the difficulty of starting the processes, assume that the value $X(0.1) = 1.645\,66$ has been obtained by another method.

Solution The exact solution of the problem, which is a linear equation and so may be solved by the integrating-factor method, is

$$x = e^{-t} + e^{-3t}$$

The numerical solutions X_a and X_b and their errors are shown in Figure 8.10. It can be seen that X_a exhibits an unexpected oscillatory behaviour, leading to large

t	X_a	X_b	$x(t)$	$x - X_a$	$x - X_b$
0.000 00	2.000 00	2.000 00	2.000 00		
0.100 00	1.645 66	1.645 66	1.645 66	0.000 00	0.000 00
0.200 00	1.374 54	1.376 56	1.367 54	−0.007 00	−0.009 02
0.300 00	1.148 42	1.159 09	1.147 39	−0.001 04	−0.011 70
0.400 00	0.981 82	0.984 36	0.971 51	−0.010 30	−0.012 84
0.500 00	0.827 46	0.842 27	0.829 66	0.002 20	−0.012 61
0.600 00	0.727 95	0.725 83	0.714 11	−0.013 84	−0.011 72
0.700 00	0.610 22	0.629 54	0.619 04	0.008 83	−0.010 50
0.800 00	0.560 45	0.549 22	0.540 05	−0.020 41	−0.009 17
0.900 00	0.453 68	0.481 64	0.473 78	0.020 10	−0.007 86
1.000 00	0.450 88	0.424 32	0.417 67	−0.033 21	−0.006 66
1.100 00	0.330 30	0.375 33	0.369 75	0.039 45	−0.005 58
1.200 00	0.385 84	0.333 15	0.328 52	−0.057 33	−0.004 64
1.300 00	0.219 27	0.296 60	0.292 77	0.073 50	−0.003 83
1.400 00	0.363 29	0.264 75	0.261 59	−0.101 70	−0.003 15
1.500 00	0.099 93	0.236 83	0.234 24	0.134 31	−0.002 59
1.600 00	0.392 59	0.212 25	0.210 13	−0.182 46	−0.002 12
1.700 00	−0.054 86	0.190 52	0.188 78	0.243 64	−0.001 73
1.800 00	0.498 57	0.171 24	0.169 82	−0.328 76	−0.001 42
1.900 00	−0.287 88	0.154 08	0.152 91	0.440 80	−0.001 16
2.000 00	0.731 13	0.138 77	0.137 81	−0.593 32	−0.000 96

Figure 8.10 Computational results for Example 8.5.

errors in the solution. This is typical of the type of instability from which the scheme (8.9) and those like it are known to suffer. The scheme defined by (8.9) is not unstable for all differential equations, but only for a certain class. The possibility of instability in numerical schemes is one that should always be borne in mind, and the intelligent user is always critical of the results of numerical work and alert for signs of this type of problem.

In this section we have seen how, starting from the Taylor series for a function, schemes of a higher order of accuracy than Euler's method can be constructed. We have constructed two second-order schemes. The principle of this technique can be extended to produce schemes of yet higher orders. They will obviously introduce more values of X_m or F_m (where $m = n - 2, n - 3, \dots$). The scheme (8.10) is, in fact, a member of a family of schemes known as the **Adams–Bashforth formulae**. The first few members of this family are

$$X_{n+1} = X_n + hF_n$$

$$X_{n+1} = X_n + \tfrac{1}{2}h(3F_n - F_{n-1})$$

$$X_{n+1} = X_n + \tfrac{1}{12}h(23F_n - 16F_{n-1} + 5F_{n-2})$$

$$X_{n+1} = X_n + \tfrac{1}{24}h(55F_n - 59F_{n-1} + 37F_{n-2} - 9F_{n-3})$$

The formulae represent first-, second-, third- and fourth-order methods respectively. The first-order Adams–Bashforth formula is just the Euler method, the second-order one is the scheme we introduced as (8.10), while the third- and fourth-order formulae are extensions of the principle we have just introduced. Obviously all of these require special methods to start the process in the absence of values of X_{-1}, F_{-1}, X_{-2}, F_{-2} and so on.

8.2.6 Local and global truncation errors

In Section 8.2.2 we argued intuitively that, although the Euler method introduces an error $O(h^2)$ at each step, it yields an $O(h)$ error in the value of the dependent variable corresponding to a given value of the independent variable. What is the equivalent result for the second-order methods we have introduced in Section 8.2.5? We shall answer this question with a slightly more general analysis that will also be useful to us in succeeding sections.

First let us define two types of error. The **local error** in a method for integrating a differential equation is the error introduced at each step. Thus if the method is defined by

$$X_{n+1} = g(h, t_n, X_n, t_{n-1}, X_{n-1}, \ldots)$$

and analysis shows us that

$$x_{n+1} = g(h, t_n, x_n, t_{n-1}, x_{n-1}, \ldots) + O(h^{p+1})$$

then we say that the local error in the method is of order $p + 1$ or that the method is a pth-order method.

The **global error** of an integration method is the error in the value of $X(t_0 + a)$ obtained by using that method to advance the required number of steps from a known value of $x(t_0)$. Using a pth-order method, the first step introduces an error $O(h^{p+1})$. The next step takes the approximation X_1 and derives an estimate X_2 of x_2 that introduces a further error $O(h^{p+1})$. The number of steps needed to calculate the value $X(t_0 + a)$ is a/h. Hence we have

$$X(t_0 + a) = x(t_0 + a) + \frac{a}{h}O(h^{p+1})$$

Dividing a quantity that is $O(h^r)$ by h produces a quantity that is $O(h^{r-1})$, so we must have

$$X(t_0 + a) = x(t_0 + a) + O(h^p)$$

In other words, the global error produced by a method that has a local error $O(h^{p+1})$ is $O(h^p)$. As we saw in Example 8.2, halving the step size for a calculation using Euler's method produces errors that are roughly half as big. This is consistent with the global error being $O(h)$. Since the local error of the Euler method is $O(h^2)$, this is as we should expect. Let us now repeat Example 8.2 using the second-order Adams–Bashforth method, (8.10).

EXAMPLE 8.6 Let X_a denote the approximation to the solution of the initial-value problem

$$\frac{dx}{dt} = \frac{x^2}{t+1}, \quad x(0) = 1$$

obtained using the second-order Adams–Bashforth method with a step size $h = 0.1$, and X_b that obtained using a step size of $h = 0.05$. Compute the values of $X_a(t)$ and $X_b(t)$ for $t = 0.1, 0.2, \ldots, 1.0$. Compare these values with the values of $x(t)$, the exact solution of the problem. Compute the ratio of the errors in X_a and X_b. In order to start the process, assume that the values $X(-0.1) = 0.904\,68$ and $X(-0.05) = 0.951\,21$ have already been obtained by another method.

t	X_a	X_b	$x(t)$	$\lvert x - X_a \rvert$	$\lvert x - X_b \rvert$	$\dfrac{\lvert x - X_b \rvert}{\lvert x - X_a \rvert}$
0.000 00	1.000 00	1.000 00	1.000 00			
0.100 00	1.104 53	1.105 12	1.105 35	0.000 82	0.000 23	0.28
0.200 00	1.220 89	1.222 39	1.222 97	0.002 08	0.000 58	0.28
0.300 00	1.351 76	1.354 59	1.355 68	0.003 92	0.001 09	0.28
0.400 00	1.500 49	1.505 25	1.507 10	0.006 61	0.001 85	0.28
0.500 00	1.671 44	1.679 03	1.681 99	0.010 55	0.002 96	0.28
0.600 00	1.870 40	1.882 17	1.886 81	0.016 40	0.004 64	0.28
0.700 00	2.105 25	2.123 31	2.130 51	0.025 25	0.007 20	0.29
0.800 00	2.387 00	2.414 70	2.425 93	0.038 93	0.011 23	0.29
0.900 00	2.731 45	2.774 40	2.792 16	0.060 70	0.017 76	0.29
1.000 00	3.162 20	3.230 07	3.258 89	0.096 70	0.028 82	0.30

Figure 8.11 Computational results for Example 8.6.

Solution The exact solution was given in Example 8.2. The numerical solutions X_a and X_b and their errors are shown in Figure 8.11.

Because the method is second-order, we expect the global error to vary like h^2. Theoretically, then, the error in the solution X_b should be one-quarter that in X_a. We see that this expectation is approximately borne out in practice.

Just as previously we outlined how, for the Euler method, we could estimate from two solutions of the differential equation the step size that would suffice to compute a solution to any required accuracy, so we can do the same in a more general way. If we use a pth-order method to compute two estimates $X_a(t_0 + a)$ and $X_b(t_0 + a)$ of $x(t_0 + a)$ using step sizes h and $\frac{1}{2}h$ then, because the global error of the process is $O(h^p)$ we expect the error in $X_a(t_0 + a)$ to be roughly 2^p times that in $X_b(t_0 + a)$. Hence the error in $X_b(t_0 + a)$ may be estimated to be

$$\frac{\lvert X_a(t_0 + a) - X_b(t_0 + a) \rvert}{2^p - 1}$$

If the desired error, which may be expressed in absolute terms or may be derived from a desired maximum percentage error, is ε then the factor k, say, by which the error in $X_b(t_0 + a)$ must be reduced is

$$k = \frac{|X_a(t_0 + a) - X_b(t_0 + a)|}{\varepsilon(2^p - 1)}$$

Since reducing the step size by a factor of q will, for a pth-order error, reduce the error by a factor of q^p, the factor by which step size must be reduced in order to meet the error criterion is the pth root of k. The step size used to compute X_b is $\frac{1}{2}h$, so finally we estimate the required step size as

$$\frac{h}{2}\left(\frac{\varepsilon(2^p - 1)}{|X_a(t_0 + a) - X_b(t_0 + a)|}\right)^{1/p} \tag{8.11}$$

This technique of estimating the error in a numerical approximation of an unknown quantity by comparing two approximations of that unknown quantity whose order of accuracy is known is an example of the application of **Richardson extrapolation**.

EXAMPLE 8.7

Estimate the step size required to compute an estimate of $x(1)$ accurate to 2dp for the initial-value problem in Example 8.6 given the values $X_a(1) = 3.162\,20$ and $X_b(1) = 3.230\,07$ obtained using step sizes $h = 0.1$ and 0.05 respectively.

Solution For the result to be accurate to 2dp the error must be less than 0.005. The estimates $X_a(1)$ and $X_b(1)$ were obtained using a second-order process, so, applying (8.11), with $\varepsilon = 0.005$, $\frac{1}{2}h = 0.05$ and $p = 2$, we have

$$h = 0.05\left(\frac{0.015}{|3.162\,20 - 3.230\,07|}\right)^{1/2} = 0.0235$$

In a real engineering problem what we would usually do is round this down to say 0.02 and recompute $X(1)$ using step sizes $h = 0.04$ and 0.02. These two new estimates of $X(1)$ could then be used to estimate again the error in the value of $X(1)$ and confirm that the desired error criterion had been met.

8.2.7 More accurate solution methods: predictor–corrector methods

In Section 8.2.5 we showed how the third term in the Taylor series expansion

$$x(t + h) = x(t) + h\frac{dx}{dt}(t) + \frac{h^2}{2!}\frac{d^2x}{dt^2}(t) + O(h^3) \tag{8.12}$$

could be replaced by either $x(t - h)$ or $(dx/dt)(t - h)$. These are not the only possibilities. By using appropriate Taylor series expansions, we could replace the term with other values of $x(t)$ or dx/dt. For instance, expanding the function $x(t - 2h)$ about $x(t)$ gives rise to

$$x(t - 2h) = x(t) - 2h\frac{dx}{dt}(t) + 2h^2\frac{d^2x}{dt^2}(t) + O(h^3) \tag{8.13}$$

and eliminating the second-derivative term between (8.12) and (8.13) gives

$$x(t + h) = \frac{3}{4}x(t) + \frac{1}{4}x(t - 2h) + \frac{3}{2}h\frac{dx}{dt}(t) + O(h^3)$$

which, in turn, would give rise to the integration scheme

$$X_{n+1} = \tfrac{3}{4}X_n + \tfrac{1}{4}X_{n-2} + \tfrac{3}{2}hF_n$$

Such a scheme, however, would not seem to offer any advantages to compensate for the added difficulties caused by a two-step scheme using non-consecutive values of X.

The one alternative possibility that does offer some gains is using the value of $(dx/dt)(t + h)$. Writing the Taylor series expansion of $(dx/dt)(t + h)$ yields

$$\frac{dx}{dt}(t+h) = \frac{dx}{dt}(t) + h\frac{d^2x}{dt^2}(t) + O(h^2)$$

and eliminating the second derivative between this and (8.12) gives

$$x(t + h) = x(t) + \frac{h}{2}\left[\frac{dx}{dt}(t) + \frac{dx}{dt}(t + h)\right] + O(h^3) \tag{8.14}$$

leading to the integration scheme

$$X_{n+1} = X_n + \tfrac{1}{2}h(F_n + F_{n+1}) \tag{8.15}$$

This, like (8.9) and (8.10), is a second-order scheme. It has the problem that, in order to calculate X_{n+1}, the value of F_{n+1} is needed, which, in its turn, requires that the value of X_{n+1} be known. This seems to be a circular argument!

One way to work around this problem and turn (8.15) into a usable scheme is to start by working out a rough value of X_{n+1}, use that to compute a value of F_{n+1}, and then use (8.15) to compute a more accurate value of X_{n+1}. Such a process can be derived as follows. We know that

$$x(t + h) = x(t) + h\frac{dx}{dt}(t) + O(h^2)$$

Let

$$\hat{x}(t + h) = x(t) + h\frac{dx}{dt}(t) \tag{8.16}$$

then

$$x(t + h) = \hat{x}(t + h) + O(h^2)$$

or, using the subscript notation defined above,

$$x_{n+1} = \hat{x}_{n+1} + O(h^2)$$

Thus

$$\begin{aligned}\frac{\mathrm{d}x_{n+1}}{\mathrm{d}t} &= f(t_{n+1}, x_{n+1}) \\ &= f(t_{n+1}, \hat{x}_{n+1} + O(h^2)) \\ &= f(t_{n+1}, \hat{x}_{n+1}) + O(h^2)\frac{\partial f}{\partial x}(t_{n+1}, \hat{x}_{n+1}) + O(h^4) \\ &= f(t_{n+1}, \hat{x}_{n+1}) + O(h^2) \qquad \textbf{(8.17)}\end{aligned}$$

In the subscript notation (8.14) is

$$x_{n+1} = x_n + \tfrac{1}{2}h(f(t_n, x_n) + f(t_{n+1}, x_{n+1})) + O(h^3)$$

Substituting (8.17) into this gives

$$x_{n+1} = x_n + \tfrac{1}{2}h(f(t_n, x_n) + f(t_{n+1}, \hat{x}_{n+1}) + O(h^2)) + O(h^3)$$

That is,

$$x_{n+1} = x_n + \tfrac{1}{2}h(f(t_n, x_n) + f(t_{n+1}, \hat{x}_{n+1})) + O(h^3) \qquad \textbf{(8.18)}$$

Equation (8.18) together with (8.16) forms the basis of what is known as a **predictor–corrector method**, which is defined by the following scheme:

(1) compute the 'predicted' value of X_{n+1}, call it $\hat{X}_{n+1}$, from

$$\hat{X}_{n+1} = X_n + hf(t_n, X_n) \qquad \textbf{(8.19a)}$$

(2) compute the 'corrected' value of X_{n+1} from

$$X_{n+1} = X_n + \tfrac{1}{2}h(f(t_n, X_n) + f(t_{n+1}, \hat{X}_{n+1})) \qquad \textbf{(8.19b)}$$

This predictor–corrector scheme, as demonstrated by (8.18), is a second-order method. It has the advantage over (8.9) and (8.10) of requiring only the value of X_n, not X_{n-1} or F_{n-1}. On the other hand, each step requires two evaluations of the function $f(t, x)$, and so the method is less efficient computationally.

EXAMPLE 8.8

Solve the initial value problem

$$\frac{\mathrm{d}x}{\mathrm{d}t} = \frac{x^2}{t+1}, \quad x(0) = 1$$

t	X_n	$f(t_n, X_n)$	$\hat{X}_{n+1}$	$f(t_{n+1}, \hat{X}_{n+1})$	$x(t)$	$\lvert x - X_n \rvert$
0.000 00	1.000 00	1.000 00	1.100 00	1.100 00	1.000 00	0.000 00
0.100 00	1.105 00	1.110 02	1.216 00	1.232 22	1.105 35	0.000 35
0.200 00	1.222 11	1.244 63	1.346 58	1.394 82	1.222 97	0.000 86
0.300 00	1.354 08	1.410 42	1.495 13	1.596 72	1.355 68	0.001 60
0.400 00	1.504 44	1.616 67	1.666 11	1.850 61	1.507 10	0.002 65
0.500 00	1.677 81	1.876 69	1.865 47	2.175 00	1.681 99	0.004 18
0.600 00	1.880 39	2.209 92	2.101 38	2.597 53	1.886 81	0.006 42
0.700 00	2.120 76	2.645 67	2.385 33	3.161 00	2.130 51	0.009 75
0.800 00	2.411 10	3.229 66	2.734 06	3.934 26	2.425 93	0.014 83
0.900 00	2.769 29	4.036 30	3.172 92	5.033 72	2.792 16	0.022 87
1.000 00	3.222 79				3.258 89	0.036 10

Figure 8.12 Computational results for Example 8.8.

posed in Example 8.2 using the second-order predictor–corrector scheme with a step size $h = 0.1$. Compute the values of $X(t)$ for $t = 0.1, 0.2, \dots, 1.0$ and compare them with the values of the exact solution $x(t)$.

Solution The exact solution was given in Example 8.2. The computation proceeds as set out in Figure 8.12.

Comparison of the result of Example 8.8 with those of Examples 8.2 and 8.6 shows that, as we should expect, the predictor–corrector scheme produces results of considerably higher accuracy than the Euler method and of comparable (though slightly better) accuracy to the second-order Adams–Bashforth scheme. We also expect the scheme to have a global error $O(h^2)$, and, in the spirit of Examples 8.2 and 8.6, we confirm this in Example 8.9.

EXAMPLE 8.9 Let X_a denote the approximation to the solution of the initial-value problem

$$\frac{dx}{dt} = \frac{x^2}{t+1}, \quad x(0) = 1$$

obtained using the second-order predictor–corrector method with a step size $h = 0.1$, and X_b that obtained using $h = 0.05$. Compute the values of $X_a(t)$ and $X_b(t)$ for $t = 0.1, 0.2, \dots, 1.0$. Compare these with the values of $x(t)$, the exact solution of the problem. Compute the ratio of the errors in X_a and X_b.

Solution The numerical solutions X_a and X_b and their errors are shown in Figure 8.13. The ratio of the errors confirms that the error behaves roughly $O(h^2)$.

t	X_a	X_b	$x(t)$	$\lvert x - X_a\rvert$	$\lvert x - X_b\rvert$	$\dfrac{\lvert x - X_b\rvert}{\lvert x - X_a\rvert}$
0.000 00	1.000 00	1.000 00	1.000 00			
0.100 00	1.105 00	1.105 26	1.105 35	0.000 35	0.000 09	0.27
0.200 00	1.222 11	1.222 74	1.222 97	0.000 86	0.000 23	0.27
0.300 00	1.354 08	1.355 25	1.355 68	0.001 60	0.000 43	0.27
0.400 00	1.504 44	1.506 38	1.507 10	0.002 65	0.000 72	0.27
0.500 00	1.677 81	1.680 86	1.681 99	0.004 18	0.001 13	0.27
0.600 00	1.880 39	1.885 07	1.886 81	0.006 42	0.001 73	0.27
0.700 00	2.120 76	2.127 87	2.130 51	0.009 75	0.002 64	0.27
0.800 00	2.411 10	2.421 90	2.425 93	0.014 83	0.004 03	0.27
0.900 00	2.769 29	2.785 92	2.792 16	0.022 87	0.006 24	0.27
1.000 00	3.222 79	3.248 98	3.258 89	0.036 10	0.009 91	0.27

Figure 8.13 Computational results for Example 8.9.

In Section 8.2.5 we mentioned the difficulties that multistep methods introduce with respect to starting the computation. We now have a second-order method that does not need values of X_{n-1} or earlier. Obviously we can use this method just as it stands, but we then pay the penalty, in computer processing time, of the extra evaluation of $f(t, x)$ at each step of the process. An alternative scheme is to use the second-order predictor–corrector for the first step and then, because the appropriate function values are now available, change to the second-order Adams–Bashforth scheme – or even, if the problem is one for which the scheme given by (8.9) (which is called the **central difference scheme**) is stable, to that process. In this way we create a hybrid process that retains the $O(h^2)$ convergence and simultaneously minimizes the computational load.

The principles by which we derive (8.14) and so the integration scheme (8.15) can be extended to produce higher-order schemes. Such schemes are called the **Adams–Moulton formulae** and are as follows.

$$X_{n+1} = X_n + hF_{n+1}$$
$$X_{n+1} = X_n + \tfrac{1}{2}h(F_{n+1} + F_n)$$
$$X_{n+1} = X_n + \tfrac{1}{12}h(5F_{n+1} + 8F_n - F_{n-1})$$
$$X_{n+1} = X_n + \tfrac{1}{24}h(9F_{n+1} + 19F_n - 5F_{n-1} + F_{n-2})$$

These are first-, second-, third- and fourth-order formulae respectively. They are all like the one we derived in this section in that the value of F_{n+1} is required in order to compute the value of X_{n+1}. They are therefore usually used as corrector formulae in predictor–corrector schemes. The most common way to do this is to use the $(p - 1)$th-order Adams–Bashforth formula as predictor, with the pth-order Adams–Moulton formula as corrector. This combination can be shown to always produce a scheme of pth order. The predictor–corrector scheme we have derived in this section is of this form, with $p = 2$. Of course, for $p > 2$ the predictor–corrector

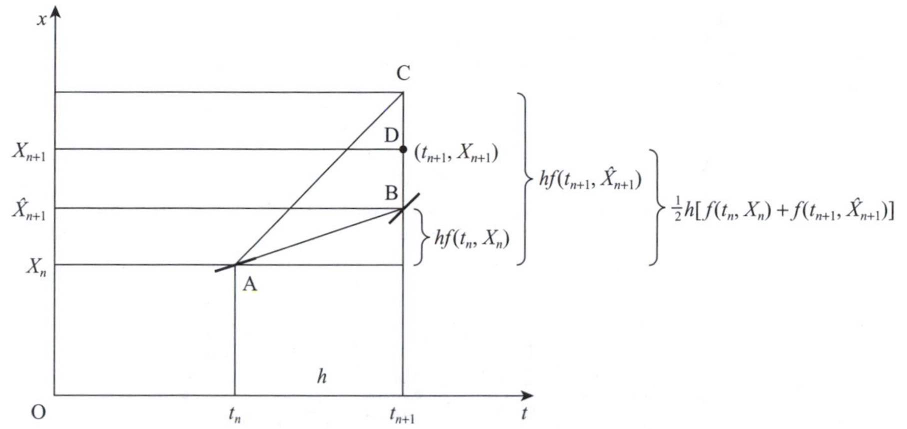

Figure 8.14 A geometrical interpretation of the second-order predictor–corrector method.

formula produced is no longer self-starting, and other means have to be found to produce the first few values of X. We shall return to this topic in the next section.

8.2.8 More accurate solution methods: Runge–Kutta methods

Another class of higher-order methods comprises the Runge–Kutta methods. The mathematical derivation of these methods is quite complicated and beyond the scope of this book. However, their general principle can be explained informally by a graphical argument. Figure 8.14 shows a geometrical interpretation of the second-order predictor–corrector method introduced in the last section. Starting at the point (t_n, X_n), point A in the diagram, the predicted value $\hat{X}_{n+1}$ is calculated. The line AB has gradient $f(t_n, X_n)$, so the ordinate of the point B is the predicted value $\hat{X}_{n+1}$. The line AC in the diagram has gradient $f(t_{n+1}, \hat{X}_{n+1})$, the gradient of the direction field of the equation at point B, so point C has ordinate $X_n + hf(t_{n+1}, \hat{X}_{n+1})$. The midpoint of the line BC, point D, has ordinate $X_n + \frac{1}{2}h(f(t_n, X_n) + f(t_{n+1}, \hat{X}_{n+1}))$, which is the value of X_{n+1} given by the corrector formula. Geometrically speaking, the predictor–corrector scheme can be viewed as the process of calculating the gradient of the direction field of the equation at points A and B and then assuming that the average gradient of the solution over the interval (t_n, t_{n+1}) is reasonably well estimated by the average of the gradients at these two points. The Euler method, of course, is equivalent to assuming that the gradient at point A is a good estimate of the average gradient of the solution over the interval (t_n, t_{n+1}). Given this insight, it is unsurprising that the error performance of the predictor–corrector method is superior to that of the Euler method.

Runge–Kutta methods extend this principle by using the gradient at several points in the interval (t_n, t_{n+1}) to estimate the average gradient of the solution over the interval. The most commonly used Runge–Kutta method is a fourth-order one which can be expressed as follows:

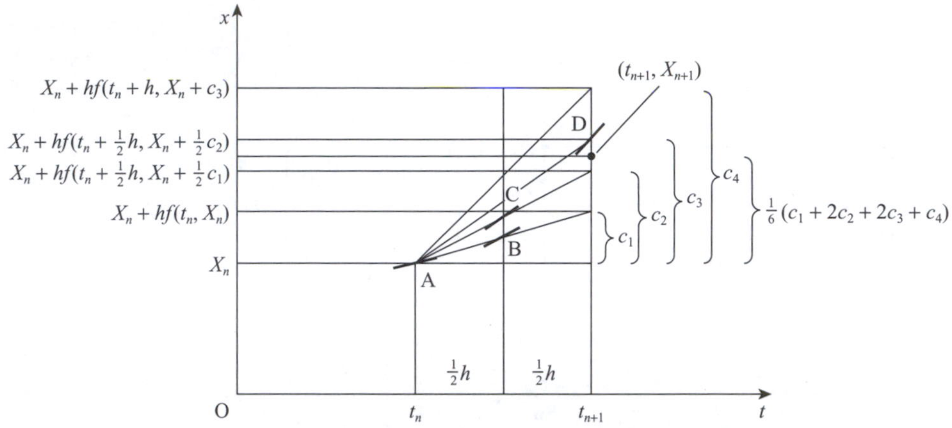

Figure 8.15 A geometrical interpretation of the fourth-order Runge–Kutta method.

$$c_1 = hf(t_n, X_n) \tag{8.20a}$$

$$c_2 = hf(t_n + \tfrac{1}{2}h, X_n + \tfrac{1}{2}c_1) \tag{8.20b}$$

$$c_3 = hf(t_n + \tfrac{1}{2}h, X_n + \tfrac{1}{2}c_2) \tag{8.20c}$$

$$c_4 = hf(t_n + h, X_n + c_3) \tag{8.20d}$$

$$X_{n+1} = X_n + \tfrac{1}{6}(c_1 + 2c_2 + 2c_3 + c_4) \tag{8.20e}$$

Geometrically, this can be understood as the process shown in Figure 8.15. The line AB has the same gradient as the equation's direction field at point A. The ordinate of this line at $t_n + \frac{1}{2}h$ defines point B. The line AC has gradient equal to the direction of the direction field at point B. This line defines point C. Finally, a line AD, with gradient equal to the direction of the direction field at point C, defines point D. The average gradient of the solution over the interval (t_n, t_{n+1}) is then estimated from a weighted average of the gradients at points A, B, C and D. It is intuitively acceptable that such a process is likely to give a highly accurate estimate of the average gradient over the interval.

As was said before, the mathematical proof that the process defined by (8.20a–e) is a fourth-order process is beyond the scope of this text. It is interesting to note that the predictor–corrector method defined by (8.19a, b) could also be expressed as

$$c_1 = hf(t_n, X_n)$$

$$c_2 = hf(t_n + h, X_n + c_1)$$

$$X_{n+1} = X_n + \tfrac{1}{2}(c_1 + c_2)$$

This is also of the form of a Runge–Kutta method (the second-order Runge–Kutta method), so we find that the second-order Runge–Kutta method and the second-order Adams–Bashforth/Adams–Moulton predictor–corrector are, in fact, equivalent processes.

EXAMPLE 8.10 Let X_a denote the approximation to the solution of the initial-value problem

$$\frac{dx}{dt} = \frac{x^2}{t+1}, \quad x(0) = 1$$

obtained using the fourth-order Runge–Kutta method with a step size $h = 0.1$, and X_b that obtained using $h = 0.05$. Compute the values of $X_a(t)$ and $X_b(t)$ for $t = 0.1$, $0.2, \ldots, 1.0$. Compare these with the values of $x(t)$, the exact solution of the problem. Compute the ratio of the errors in X_a and X_b.

t	X_a	X_b	$x(t)$	$\lvert x - X_a\rvert \times 10^3$	$\lvert x - X_b\rvert \times 10^3$	$\frac{\lvert x-X_b\rvert}{\lvert x-X_a\rvert}$
0.000 00	1.000 000 0	1.000 000 0	1.000 000 0			
0.100 00	1.105 350 7	1.105 351 2	1.105 351 2	0.000 55	0.000 04	0.0682
0.200 00	1.222 973 3	1.222 974 5	1.222 974 6	0.001 33	0.000 09	0.0680
0.300 00	1.355 680 2	1.355 682 5	1.355 682 7	0.002 46	0.000 17	0.0679
0.400 00	1.507 091 8	1.507 095 7	1.507 095 9	0.004 10	0.000 28	0.0678
0.500 00	1.681 980 5	1.681 986 6	1.681 987 1	0.006 53	0.000 44	0.0678
0.600 00	1.886 795 2	1.886 804 7	1.886 805 4	0.010 20	0.000 69	0.0677
0.700 00	2.130 491 5	2.130 506 4	2.130 507 4	0.015 92	0.001 08	0.0677
0.800 00	2.425 903 1	2.425 926 6	2.425 928 3	0.025 19	0.001 71	0.0677
0.900 00	2.792 115 5	2.792 153 7	2.792 156 5	0.041 03	0.002 78	0.0677
1.000 00	3.258 821 4	3.258 886 6	3.258 891 4	0.069 94	0.004 74	0.0678

Figure 8.16 Computational results for Example 8.10.

Solution The exact solution was given in Example 8.2. The numerical solutions X_a and X_b and their errors are presented in Figure 8.16.

This example shows, first, that the Runge–Kutta scheme, being a fourth-order scheme, has considerably smaller errors, in absolute terms, than any of the other methods we have met so far (note that Figure 8.16 does not give raw errors but errors times 1000!) and, second, that the expectation we have that the global error should be $O(h^4)$ is roughly borne out in practice (the ratio of $|x - X_a|$ to $|x - X_b|$ is roughly 16 : 1).

Runge–Kutta schemes are single-step methods in the sense that they only require the value of X_n, not the value of X at any steps prior to that. They are therefore entirely self-starting, unlike the predictor–corrector and other multistep methods. On the other hand, Runge–Kutta methods proceed by effectively creating substeps within each step. Therefore they require more evaluations of the function

$f(t, x)$ at each step than multistep methods of equivalent order of accuracy. For this reason, they are computationally less efficient. Because they are self-starting, however, Runge–Kutta methods can be used to start the process for multistep methods. An example of an efficient scheme that consistently has a fourth-order local error is as follows. Start by taking two steps using the fourth-order Runge–Kutta method. At this point values of X_0, X_1 and X_2 are available, so, to achieve computational efficiency, change to the three-step fourth-order predictor–corrector consisting of the third-order Adams–Bashforth/fourth-order Adams–Moulton pair.

8.2.9 Exercises

(Note that Questions 6–13 may be attempted using a hand-held calculator, particularly if it is of the programmable variety. The arithmetic will, however, be found to be tedious, and the use of computer assistance is recommended if the maximum benefit is to be obtained from completing these questions. The questions marked with a dagger (†) are intended to be solved with the aid of a computer.)

6 Using the second-order Adams–Bashforth method (start the process with a single step using the second-order predictor–corrector method),

(a) compute an estimate of $x(0.5)$ for the initial-value problem

$$\frac{dx}{dt} = x^2 \sin t - x, \quad x(0) = 0.2$$

using step size $h = 0.1$;
(b) compute an estimate of $x(1.2)$ for the initial-value problem

$$\frac{dx}{dt} = x^2 e^{tx}, \quad x(0.5) = 0.5$$

using step size $h = 0.1$.

7 Using the third-order Adams–Bashforth method (start the process with two second-order predictor–corrector method steps) compute an estimate of $x(0.5)$ for the initial-value problem

$$\frac{dx}{dt} = \sqrt{(x^2 + 2t)}, \quad x(0) = 1$$

using step size $h = 0.1$.

8 Using the second-order predictor–corrector method,

(a) compute an estimate of $x(0.5)$ for the initial-value problem

$$\frac{dx}{dt} = (2t + x)\sin 2t, \quad x(0) = 0.5$$

using step size $h = 0.05$;
(b) compute an estimate of $x(1)$ for the initial-value problem

$$\frac{dx}{dt} = -\frac{1 + x}{\sin(t + 1)}, \quad x(0) = -2$$

using step size $h = 0.1$.

9 Write down the first three terms of the Taylor series expansions of the functions

$$\frac{dx}{dt}(t - h) \quad \text{and} \quad \frac{dx}{dt}(t - 2h)$$

about $x(t)$. Use these two equations to eliminate

$$\frac{d^2x}{dt^2}(t) \quad \text{and} \quad \frac{d^3x}{dt^3}(t)$$

from the Taylor series expansion of the function $x(t + h)$ about $x(t)$. Show that the resulting formula for $x(t + h)$ is the third member of the Adams–Bashforth family, and hence confirm that this Adams–Bashforth method is a third-order method.

10 Write down the first three terms of the Taylor series expansions of the functions

$$\frac{dx}{dt}(t + h) \quad \text{and} \quad \frac{dx}{dt}(t - h)$$

about $x(t)$. Use these two equations to eliminate

$$\frac{d^2x}{dt^2}(t) \quad \text{and} \quad \frac{d^3x}{dt^3}(t)$$

from the Taylor series expansion of the function $x(t + h)$ about $x(t)$. Show that the resulting formula for $x(t + h)$ is the third member of the Adams–Moulton family, and hence confirm that this Adams–Moulton method is a third-order method.

11 Write down the first four terms of the Taylor series expansion of the function $x(t - h)$ about $x(t)$, and the first three terms of the expansion of the function

$$\frac{dx}{dt}(t - h)$$

about $x(t)$. Use these two equations to eliminate

$$\frac{d^2x}{dt^2}(t) \quad \text{and} \quad \frac{d^3x}{dt^3}(t)$$

from the Taylor series expansion of the function $x(t + h)$ about $x(t)$. Show that the resulting formula is

$$X_{n+1} = -4X_n + 5X_{n-1} + h(4F_n + 2F_{n-1}) + O(h^4)$$

Show that this method is a linear combination of the second-order Adams–Bashforth method and the central difference method (that is, the scheme based on (8.9)). What do you think, in view of this, might be its disadvantages?

12 Using the third-order Adams–Bashforth–Moulton predictor–corrector method (that is, the second-order Adams–Bashforth formula as predictor and the third-order Adams–Moulton formula as corrector), compute an estimate of $x(0.5)$ for the initial-value problem

$$\frac{dx}{dt} = x^2 + t^2, \quad x(0.3) = 0.1$$

using step size $h = 0.05$. (You will need to employ another method for the first step to start this scheme – use the fourth-order Runge–Kutta method).

13 Using the fourth-order Runge–Kutta method,

(a) compute an estimate of $x(0.75)$ for the initial-value problem

$$\frac{dx}{dt} = x + t + xt, \quad x(0) = 1$$

using step size $h = 0.15$;
(b) compute an estimate of $x(2)$ for the initial-value problem

$$\frac{dx}{dt} = \frac{1}{x + t}, \quad x(1) = 2$$

using step size $h = 0.1$.

†14 Consider the initial-value problem

$$\frac{dx}{dt} = x^2 + t^{3/2}, \quad x(0) = -1$$

(a) Compute estimates of $x(2)$ using the second-order Adams–Bashforth scheme (using the second-order predictor–corrector to start the computation) with step sizes $h = 0.2$ and 0.1. From these two estimates of $x(2)$ estimate what step size would be needed to compute an estimate of $x(2)$ accurate to 3dp. Compute $X(2)$, first using your estimated step size and second using half your estimated step size. Does the required accuracy appear to have been achieved?
(b) Compute estimates of $x(2)$ using the second-order predictor–corrector scheme with step sizes $h = 0.2$ and 0.1. From these two estimates of $x(2)$ estimate what step size would be needed with this scheme to compute an estimate of $x(2)$ accurate to 3dp. Compute $X(2)$, first using your estimated step size and second using half your estimated step size. Does the required accuracy appear to have been achieved?
(c) Compute estimates of $x(2)$ using the fourth-order Runge–Kutta scheme with step sizes $h = 0.4$ and 0.2. From these two estimates of $x(2)$ estimate what step size would be needed to compute an estimate of $x(2)$ accurate to 5 dp. Compute $X(3)$, first using your estimated step size and second using half your estimated step size. Does the required accuracy appear to have been achieved?

†15 For the initial-value problem

$$\frac{dx}{dt} = x^2 e^{-t}, \quad x(1) = 1$$

find, by any method, an estimate, accurate to 5 dp, of the value of $x(3)$.

8.2.10 Stiff equations

There is a class of differential equations, known as **stiff differential equations**, that are apt to be somewhat troublesome to solve numerically. It is beyond the scope of this text to explore the topic of stiff equations in any great detail. It is, however, important to be aware of the possibility of difficulties from this source and to be able to recognize the sort of equations that are likely to be stiff. In that spirit we shall present a very informal treatment of stiff equations and the sort of troubles that they cause. Example 8.11 shows the sort of behaviour that is typical of stiff differential equations.

EXAMPLE 8.11 The equation

$$\frac{dx}{dt} = 1 - x, \quad x(0) = 2 \tag{8.21}$$

has analytical solution $x = 1 + e^{-t}$. The equation

$$\frac{dx}{dt} = 50(1 - x) + 50\,e^{-t}, \quad x(0) = 2 \tag{8.22}$$

has analytical solution $x = 1 + \frac{1}{49}(50\,e^{-t} - e^{-50t})$. The two solutions are shown in Figure 8.17.

Suppose that it were not possible to solve the two equations analytically and that numerical solutions must be sought. The form of the two solutions shown in Figure 8.17 is not very different, and it might be supposed (at least naively) that

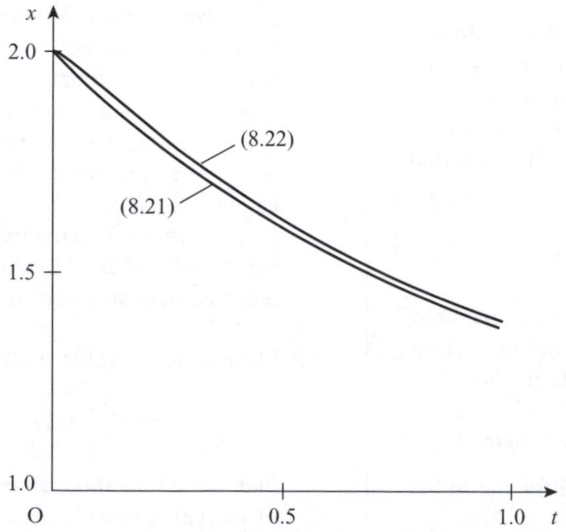

Figure 8.17 The analytical solutions of (8.21) and (8.22).

t	X_a	$\|X_a - x_a\|$	X_b	$\|X_b - x_b\|$	*Ratio of errors*
0.000 00	2.000 00	0.000 000	2.000 00	0.000 000	
0.100 00	1.904 84	0.000 002	1.923 15	0.000 017	11.264 68
0.200 00	1.818 73	0.000 003	1.835 47	0.000 028	10.022 19
0.300 00	1.740 82	0.000 004	1.755 96	0.000 026	6.864 34
0.400 00	1.670 32	0.000 005	1.684 02	0.000 023	5.150 07
0.500 00	1.606 54	0.000 005	1.618 93	0.000 021	4.120 06
0.600 00	1.548 82	0.000 006	1.560 03	0.000 019	3.433 38
0.700 00	1.496 59	0.000 006	1.506 74	0.000 017	2.942 90
0.800 00	1.449 34	0.000 006	1.458 51	0.000 016	2.575 03
0.900 00	1.406 58	0.000 006	1.414 88	0.000 014	2.288 92
1.000 00	1.367 89	0.000 006	1.375 40	0.000 013	2.060 02

Figure 8.18 Computational results for Example 8.11; $h = 0.01$.

t	X_a	$\|X_a - x_a\|$	X_b	$\|X_b - x_b\|$	*Ratio of errors*
0.000 00	2.000 00	0.000 000	2.000 00	0.000 000	
0.100 00	1.904 85	0.000 010	1.922 04	0.001 123	116.951 24
0.200 00	1.818 75	0.000 017	1.835 67	0.000 231	13.270 10
0.300 00	1.740 84	0.000 024	1.756 25	0.000 317	13.438 84
0.400 00	1.670 35	0.000 028	1.684 30	0.000 296	10.384 39
0.500 00	1.606 56	0.000 032	1.619 18	0.000 268	8.328 98
0.600 00	1.548 85	0.000 035	1.560 25	0.000 243	6.942 36
0.700 00	1.496 62	0.000 037	1.506 94	0.000 220	5.950 68
0.800 00	1.449 37	0.000 038	1.458 70	0.000 199	5.206 82
0.900 00	1.406 61	0.000 039	1.415 05	0.000 180	4.628 26
1.000 00	1.367 92	0.000 039	1.375 55	0.000 163	4.165 42

Figure 8.19 Computational results for Example 8.11; $h = 0.025$.

the numerical solution of the two equations would present similar problems. This, however, is far from the case.

Figure 8.18 shows the results of solving the two equations using the second-order predictor–corrector method with step size $h = 0.01$. The numerical and exact solutions of (8.21) are denoted by X_a and x_a respectively, and those of (8.22) by X_b and x_b. The third and fifth columns give the errors in the numerical solutions (compared with the exact solutions), and the last column gives the ratio of the errors. The solution X_a is seen to be considerably more accurate than X_b using the same step size.

Figure 8.19 is similar to Figure 8.18, but with a step size $h = 0.025$. As we might expect, the error in the solution X_a is larger by a factor of roughly six (the global error of the second-order predictor–corrector method is $O(h^2)$). The errors in X_b, however, are larger by more than the expected factor, as is evidenced by the increase in the ratio of the error in X_b to that in X_a.

t	X_a	$\|X_a - x_a\|$	X_b	$\|X_b - x_b\|$
0.000 00	2.000 00	0.000 000	2.000 00	0.000 000
0.100 00	1.904 88	0.000 039	1.873 43	0.049 740
0.200 00	1.818 80	0.000 071	1.707 36	0.128 075
0.300 00	1.740 91	0.000 096	1.421 02	0.334 914
0.400 00	1.670 44	0.000 116	0.802 59	0.881 408
0.500 00	1.606 66	0.000 131	−0.705 87	2.324 778
0.600 00	1.548 95	0.000 142	−4.576 42	6.136 434
0.700 00	1.496 74	0.000 150	−14.695 10	16.201 818
0.800 00	1.449 48	0.000 156	−41.322 43	42.780 932
0.900 00	1.406 73	0.000 158	−111.551 73	112.966 595
1.000 00	1.368 04	0.000 159	−296.925 40	298.300 783

Figure 8.20 Computational results for Example 8.11; $h = 0.05$.

Figure 8.20 shows the results obtained using a step size $h = 0.05$. The errors in X_a are again larger by about the factor expected (25 when compared with Figure 8.18). The solution X_b, however, shows little relationship to the exact solution x_b – so little that the error at $t = 1$ is over 20 000% of the exact solution. Obviously a numerical method that causes such large errors to accumulate is not at all satisfactory.

In Section 8.2.5 we met the idea that some numerical methods can, when applied to some classes of differential equation, show instability. What has happened here is, of course, that the predictor–corrector method is showing instability when used to solve (8.22) with a step size larger than some critical limit. Unfortunately the same behaviour is also manifest by the other methods that we have already come across – the problem lies with the equation (8.22), which is an example of a stiff differential equation.

The typical pattern with stiff differential equations is that, in order to avoid instability, the step size used to solve the equation using normal numerical methods must be very small when compared with the interval over which the equation is to be solved. In other words, the number of steps to be taken is very large and the solution is costly in time and computing resources. Essentially, stiff equations are equations whose solution contains terms involving widely varying time scales. That (8.22) is of this type is evidenced by the presence of terms in both e^{-t} and e^{-50t} in the analytical solution. In order to solve such equations accurately, a step must be chosen that is small enough to cope with the shortest time scale. If the solution is required for times comparable to the long time scales, this can mean that very large numbers of steps are needed and the computer processing time needed to solve the problem becomes prohibitive. In Example 8.11 the time scale of the rapidly varying and the more slowly varying components of the solution differed by only a factor of 50. It is not unusual, in the physical problems arising from engineering investigations, to find time scales differing by three or more orders of magnitude; that is, factors of 1000 or more. In these cases the problems caused are proportionately amplified. Fortunately a number of numerical methods that are particularly efficient at solving stiff differential equations have been developed. Among these are the **backward differentiation methods**. It is beyond the scope of this text to treat these in any detail. The interested reader can find a comprehensive review of numerical methods for stiff differential equations in D. Byrne and A. C. Hindmarsh, Stiff ODE solvers: a review of current and coming attractions, *Journal of Computational Physics*, **70**, 1–62 (1987).

From the engineering point of view, the implication of the existence of stiff equations is that engineers must be aware of the possibility of meeting such equations and also of the nature of the difficulty for the numerical methods – the widely varying time scales inherent in the problem. It is probably easier to recognize that an engineering problem is likely to give rise to a stiff equation or equations because of the physical nature of the problem than it is to recognize a stiff equation in its abstract form isolated from the engineering context from which it arose. As is often the case, a judicious combination of mathematical reasoning and engineering intuition is more powerful than either approach in isolation.

8.2.11 Computer software libraries and the 'state of the art'

In the last few sections we have built up some basic methods for the integration of first-order ordinary differential equations. These methods, particularly the more sophisticated ones – the fourth-order Runge–Kutta and the predictor–corrector methods – suffice for many of the problems arising in engineering practice. However, for more demanding problems – demanding in terms of the scale of the problem or because the problem is characterized by ill behaviour of some form – there exist more sophisticated methods than those we are able to present in this book.

All the methods that we have presented in the last few sections use a fixed step size. Among the more sophisticated methods to which we have just alluded are some that use a variable step size. In Section 8.2.6 we showed how Richardson extrapolation can be used to estimate the size of the error in a numerical solution and, furthermore, to estimate the step size that should be used in order to compute a solution of a differential equation to some desired accuracy. The principle of the variable-step methods is that a running check is kept of the estimated error in the solution being computed. The error may be estimated by a formula derived along principles similar to that of Richardson extrapolation. This running estimate of the error is used to predict, at any point in the computation, how large a step can be taken while still computing a solution within any given error bound specified by the user. The step size used in the solution can be altered accordingly. If the error is approaching the limits of what is acceptable then the step size can be reduced; if it is very much smaller than that which can be tolerated then the step size may be increased in the interests of speedy and efficient computing. For multistep methods the change of step size can lead to quite complicated formulae or procedures. As an alternative, or in addition, to a change of step size, changes can be made in the order of the integration formula used. When increased accuracy is required, instead of reducing the step size, the order of the integration method can be increased, and vice versa. Implementations of the best of these more sophisticated schemes are readily available in packaged software libraries.

The availability of complex and sophisticated 'state of the art' methods is not the only argument for the use of software libraries. It is a good engineering principle that, if an engineer wishes to design and construct a reliable engineering artefact, tried and proven components of known reliability and performance characteristics should be used. This principle can also be extended to engineering software. It is almost always both more efficient, in terms of expenditure of time and intellectual energy, and more reliable, in terms of elimination of bugs and unwanted side-effects, to use software from a known and proven source than to write programs from scratch.

For both the foregoing reasons, when reliable libraries of numerical analysis subprograms are available, their use is strongly recommended. Among the best known of such libraries, in the United Kingdom at least, is the NAG library which is distributed by Numerical Algorithms Group Ltd of Oxford. The section of that library dealing with the numerical solution of ordinary differential equations contains (in the Mk 12 issue) 56 routines. This does not mean that 56 different numerical methods for differential equations are included – most of the routines relate to slightly different ways of using a small number of basic methods. The three methods underlying most of the main routines for initial-value problems are Runge–Kutta–Merson (a development of the Runge–Kutta methods we described in Section 8.2.8), Adams methods (based on the Adams–Bashforth and Adams–Moulton methods described in Section 8.2.7) and backward-differentiation formulae (methods specially designed to work efficiently with stiff systems of equations such as those we described in Section 8.2.10). By choosing appropriate routines from a library such as the NAG library, the engineer is assured that the routines are, as far as is practicable, 'bug-free', that they are efficient and reliable in their operation and that the algorithms used have been chosen and developed by expert numerical analysts from amongst the best available 'state of the art' methods. The reader is warned, however, that not all software libraries maintain such uniformly high standards in all these respects as the NAG library.

It is tempting to believe that the use of software libraries solves all the problems of numerical analysis that an engineering user is likely to meet. Faced with a problem for which analytical methods fail, the engineer need merely thumb through the index to some numerical analysis software library until a method for solving the type of problem currently faced is found. Unfortunately such undiscerning use of packaged software will almost certainly, sooner or later, lead to a gross failure of some sort. If the user is fortunate, the software will be sophisticated enough to detect that the problem posed is outside its capabilities and to return an error message to that effect. If the user is less fortunate, the writer of the software will have assumed that all users are competent numerical analysts and will not have provided any checks against use of the software outside its range of applicability. In that case the software may produce seemingly valid answers while giving no indication of any potential problem. Under such circumstances the undiscerning user of engineering software is on the verge of committing a major engineering blunder. From such circumstances result failed bridges and crashed aircraft! It has been the objective of these sections on the numerical solution of differential equations both to equip readers with numerical methods suitable for the less demanding problems that will arise in their engineering careers and to give them sufficient understanding of the basics of this branch of numerical analysis that they may become discriminating, intelligent and wary users of packaged software and other aids to numerical computing.

8.3 Numerical solution of second- and higher-order differential equations

Obviously, the classes of second- and higher-order differential equations that can be solved analytically, while representing an important subset of the totality of such equations, are relatively restricted. Just as for first-order equations, those for which no analytical solution exists can still be solved by numerical means. The numerical solution of second- and higher-order equations does not, in fact, need any significant new mathematical theory or technique.

8.3.1 Numerical solution of coupled first-order equations

In Section 8.2 we met various methods for the numerical solution of equations of the form

$$\frac{dx}{dt} = f(t, x)$$

that is, first-order differential equations involving a single dependent variable and a single independent variable. However it is possible to have sets of coupled first-order equations, each involving the same independent variable but with more than one dependent variable. An example of these types of equation is

$$\frac{dx}{dt} = x - y^2 + xt \quad \textbf{(8.23a)}$$

$$\frac{dy}{dt} = 2x^2 + xy - t \quad \textbf{(8.23b)}$$

This is a pair of differential equations in the dependent variables x and y with the independent variable t. The derivative of each of the dependent variables depends not only on itself and on the independent variable t, but also on the other dependent variable. Neither of the equations can be solved in isolation or independently of the other – both must be solved simultaneously, or side by side. A pair of coupled differential equations such as (8.23) may be characterized as

$$\frac{dx}{dt} = f_1(t, x, y) \tag{8.24a}$$

$$\frac{dy}{dt} = f_2(t, x, y) \tag{8.24b}$$

For a set of p such equations it is convenient to denote the dependent variables not by $x, y, z, \ldots$ but by $x_1, x_2, x_3, \ldots, x_p$ and the set of equations by

$$\frac{dx_i}{dt} = f_i(t, x_1, x_2, \ldots, x_p) \quad (i = 1, 2, \ldots, p)$$

or equivalently, using vector notation,

$$\frac{d}{dt}[\boldsymbol{x}] = \boldsymbol{f}(t, \boldsymbol{x})$$

where $\boldsymbol{x}(t)$ is a vector function of t given by

$$\boldsymbol{x}(t) = [x_1(t) \quad x_2(t) \quad \ldots \quad x_p(t)]^{\mathrm{T}}$$

$\boldsymbol{f}(t, \boldsymbol{x})$ is a vector-valued function of the scalar variable t and the vector variable $\boldsymbol{x}$.

The Euler method for the solution of a single differential equation takes the form

$$X_{n+1} = X_n + hf(t_n, X_n)$$

If we were to try to apply this method to (8.24a), we should obtain

$$X_{n+1} = X_n + hf_1(t_n, X_n, Y_n)$$

In other words, the value of X_{n+1} depends not only on t_n and X_n but also on Y_n. In the same way, we would obtain

$$Y_{n+1} = Y_n + hf_2(t_n, X_n, Y_n)$$

for Y_{n+1}. In practice, this means that to solve two simultaneous differential equations, we must advance the solution of both equations simultaneously in the manner shown in Example 8.12.

EXAMPLE 8.12

Find the value of $X(1.4)$ satisfying the following initial value problem:

$$\frac{dx}{dt} = x - y^2 + xt, \quad x(1) = 0.5$$

$$\frac{dy}{dt} = 2x^2 + xy - t, \quad y(1) = 1.2$$

using the Euler method with time step $h = 0.1$.

Solution The right-hand sides of the two equations will be denoted by $f_1(t, x, y)$ and $f_2(t, x, y)$ respectively, so

$$f_1(t, x, y) = x - y^2 + xt \quad \text{and} \quad f_2(t, x, y) = 2x^2 + xy - t$$

The initial condition is imposed at $t = 1$, so t_n will denote $1 + nh$, X_n will denote $X(1 + nh)$, and Y_n will denote $Y(1 + nh)$. Then we have

$$\begin{aligned} X_1 &= x_0 + hf_1(t_0, x_0, y_0) & Y_1 &= y_0 + hf_2(t_0, x_0, y_0) \\ &= 0.5 + 0.1f_1(1, 0.5, 1.2) & &= 1.2 + 0.1f_2(1, 0.5, 1.2) \\ &= 0.4560 & &= 1.2100 \end{aligned}$$

for the first step. The next step is therefore

$$\begin{aligned} X_2 &= X_1 + hf_1(t_1, X_1, Y_1) & Y_2 &= Y_1 + hf_2(t_1, X_1, Y_1) \\ &= 0.4560 & &= 1.2100 \\ &\quad + 0.1f_1(1.1, 0.4560, 1.2100) & &\quad + 0.1f_2(1.1, 0.4560, 1.2100) \\ &= 0.4054 & &= 1.1968 \end{aligned}$$

and the third step is

$$\begin{aligned} X_3 &= 0.4054 & Y_3 &= 1.1968 \\ &\quad + 0.1f_1(1.2, 0.4054, 1.1968) & &\quad + 0.1f_2(1.2, 0.4054, 1.1968) \\ &= 0.3513 & &= 1.1581 \end{aligned}$$

Finally, we obtain

$$\begin{aligned} X_4 &= 0.3513 + 0.1f_1(1.3, 0.3513, 1.1581) \\ &= 0.2980 \end{aligned}$$

Hence we have $X(1.4) = 0.2980$.

The principle of solving the two equations side by side extends in exactly the same way to the solution of more than two simultaneous equations and to the solution of simultaneous differential equations by methods other than the Euler method.

EXAMPLE 8.13 Find the value of $X(1.4)$ satisfying the following initial-value problem:

$$\frac{dx}{dt} = x - y^2 + xt, \quad x(1) = 0.5$$

$$\frac{dy}{dt} = 2x^2 + xy - t \quad y(1) = 1.2$$

using the second-order predictor–corrector method with time step $h = 0.1$.

Solution First step:

predictor

$$\hat{X}_1 = x_0 + hf_1(t_0, x_0, y_0) = 0.4560 \qquad \hat{Y}_1 = y_0 + hf_2(t_0, x_0, y_0) = 1.2100$$

corrector

$$\begin{aligned} X_1 &= x_0 + \tfrac{1}{2}h(f_1(t_0, x_0, y_0) + f_1(t_1, \hat{X}_1, \hat{Y}_1)) \\ &= 0.5 + 0.05(f_1(1, 0.5, 1.2) + f_1(1.1, 0.456, 1.21)) \\ &= 0.4527 \end{aligned}$$

$$\begin{aligned} Y_1 &= y_0 + \tfrac{1}{2}h(f_2(t_0, x_0, y_0) + f_2(t_1, \hat{X}_1, \hat{Y}_1)) \\ &= 1.2 + 0.05(f_2(1, 0.5, 1.2) + f_2(1.1, 0.456, 1.21)) \\ &= 1.1984 \end{aligned}$$

Second step:

predictor

$$\hat{X}_2 = X_1 + hf_1(t_1, X_1, Y_1) = 0.4042 \qquad \hat{Y}_2 = Y_1 + hf_2(t_1, X_1, Y_1) = 1.1836$$

corrector

$$\begin{aligned} X_2 &= X_1 + \tfrac{1}{2}h(f_1(t_1, X_1, Y_1) + f_1(t_2, \hat{X}_2, \hat{Y}_2)) \\ &= 0.4527 + 0.05(f_1(1.1, 0.4527, 1.1984) + f_1(1.2, 0.4042, 1.1836)) \\ &= 0.4028 \end{aligned}$$

$$\begin{aligned} Y_2 &= Y_1 + \tfrac{1}{2}h(f_2(t_1, X_1, Y_1) + f_2(t_2, \hat{X}_2, \hat{Y}_2)) \\ &= 1.1984 + 0.05(f_2(1.1, 0.4527, 1.1984) + f_2(1.2, 0.4042, 1.1836)) \\ &= 1.1713 \end{aligned}$$

Third step:

predictor

$$\hat{X}_3 = X_2 + hf_1(t_2, X_2, Y_2) = 0.3542 \qquad \hat{Y}_3 = Y_2 + hf_2(t_2, X_2, Y_2) = 1.1309$$

corrector

$$\begin{aligned} X_3 &= X_2 + \tfrac{1}{2}h(f_1(t_2, X_2, Y_2) + f_1(t_3, \hat{X}_3, \hat{Y}_3)) \\ &= 0.4028 + 0.05(f_1(1.2, 0.4028, 1.1713) + f_1(1.3, 0.3542, 1.1309)) \\ &= 0.3553 \end{aligned}$$

$$\begin{aligned} Y_3 &= Y_2 + \tfrac{1}{2}h(f_2(t_2, X_2, Y_2) + f_2(t_3, \hat{X}_3, \hat{Y}_3)) \\ &= 1.1713 + 0.05(f_2(1.2, 0.4028, 1.1713) + f_2(1.3, 0.3542, 1.1309)) \\ &= 1.1186 \end{aligned}$$

Fourth step:

predictor

$$\hat{X}_4 = X_3 + hf_1(t_3, X_3, Y_3) \qquad \hat{Y}_4 = Y_3 + hf_2(t_3, X_3, Y_3)$$
$$= 0.3119 \qquad\qquad = 1.0536$$

corrector

$$X_4 = X_3 + \tfrac{1}{2}h(f_1(t_3, X_3, Y_3) + f_1(t_4, \hat{X}_4, \hat{Y}_4))$$
$$= 0.3553 + 0.05(f_1(1.3, 0.3553, 1.1186) + f_1(1.4, 0.3119, 1.0536))$$

Hence finally we have $X(1.4) = 0.3155$.

It should be obvious from Example 8.13 that the main drawback of extending the methods we already have at our disposal to sets of differential equations is the additional labour and tedium of the computations. Intrinsically, the computations are no more difficult, merely much more laborious – a prime example of a problem ripe for computerization.

8.3.2 State-space representation of higher-order systems

The solution of differential equation initial-value problems of order greater than one can be reduced to the solution of a set of first-order differential equations using the state-space representation introduced in Section 6.8. This is achieved by a simple transformation, illustrated by Example 8.14.

EXAMPLE 8.14 The initial-value problem

$$\frac{d^2x}{dt^2} + x^2t\frac{dx}{dt} - xt^2 = \tfrac{1}{2}t^2, \quad x(0) = 1.2, \quad \frac{dx}{dt}(0) = 0.8$$

can be transformed into two coupled first-order differential equations by introducing an additional variable

$$y = \frac{dx}{dt}$$

With this definition, we have

$$\frac{d^2x}{dt^2} = \frac{dy}{dt}$$

and so the differential equation becomes

$$\frac{dy}{dt} + x^2ty - xt^2 = \frac{1}{2}t^2$$

Thus the original differential equation can be replaced by a pair of coupled first-order differential equations, together with initial conditions:

$$\frac{dx}{dt} = y, \quad x(0) = 1.2$$

$$\frac{dy}{dt} = -x^2ty + xt^2 + \frac{1}{2}t^2, \quad y(0) = 0.8$$

This process can be extended to transform a pth-order initial-value problem into a set of p first-order equations, each with an initial condition. Once the original equation has been transformed in this way, its solution by numerical methods is just the same as if it had been a set of coupled equations in the first place.

EXAMPLE 8.15 Find the value of $X(0.2)$ satisfying the initial-value problem

$$\frac{d^3x}{dt^3} + xt\frac{d^2x}{dt^2} + t\frac{dx}{dt} - t^2x = 0, \quad x(0) = 1, \quad \frac{dx}{dt}(0) = 0.5, \quad \frac{d^2x}{dt^2}(0) = -0.2$$

using the fourth-order Runge–Kutta scheme with step size $h = 0.05$.

Solution Since this is a third-order equation, we need to introduce two new variables:

$$y = \frac{dx}{dt} \quad \text{and} \quad z = \frac{dy}{dt} = \frac{d^2x}{dt^2}$$

Then the equation is transformed into a set of three first-order differential equations

$$\frac{dx}{dt} = y \qquad x(0) = 1$$

$$\frac{dy}{dt} = z \qquad y(0) = 0.5$$

$$\frac{dz}{dt} = -xtz - ty + t^2x \quad z(0) = -0.2$$

Applied to the set of differential equations

$$\frac{dx}{dt} = f_1(t, x, y, z)$$

$$\frac{dy}{dt} = f_2(t, x, y, z)$$

$$\frac{dz}{dt} = f_3(t, x, y, z)$$

the Runge–Kutta scheme is of the form

$$c_{11} = hf_1(t_n, X_n, Y_n, Z_n)$$

$$c_{21} = hf_2(t_n, X_n, Y_n, Z_n)$$

$$c_{31} = hf_3(t_n, X_n, Y_n, Z_n)$$

$$c_{12} = hf_1(t_n + \tfrac{1}{2}h, X_n + \tfrac{1}{2}c_{11}, Y_n + \tfrac{1}{2}c_{21}, Z_n + \tfrac{1}{2}c_{31})$$

$$c_{22} = hf_2(t_n + \tfrac{1}{2}h, X_n + \tfrac{1}{2}c_{11}, Y_n + \tfrac{1}{2}c_{21}, Z_n + \tfrac{1}{2}c_{31})$$

$$c_{32} = hf_3(t_n + \tfrac{1}{2}h, X_n + \tfrac{1}{2}c_{11}, Y_n + \tfrac{1}{2}c_{21}, Z_n + \tfrac{1}{2}c_{31})$$

$$c_{13} = hf_1(t_n + \tfrac{1}{2}h, X_n + \tfrac{1}{2}c_{12}, Y_n + \tfrac{1}{2}c_{22}, Z_n + \tfrac{1}{2}c_{32})$$

$$c_{23} = hf_2(t_n + \tfrac{1}{2}h, X_n + \tfrac{1}{2}c_{12}, Y_n + \tfrac{1}{2}c_{22}, Z_n + \tfrac{1}{2}c_{32})$$

$$c_{33} = hf_3(t_n + \tfrac{1}{2}h, X_n + \tfrac{1}{2}c_{12}, Y_n + \tfrac{1}{2}c_{22}, Z_n + \tfrac{1}{2}c_{32})$$

$$c_{14} = hf_1(t_n + h, X_n + c_{13}, Y_n + c_{23}, Z_n + c_{33})$$

$$c_{24} = hf_2(t_n + h, X_n + c_{13}, Y_n + c_{23}, Z_n + c_{33})$$

$$c_{34} = hf_3(t_n + h, X_n + c_{13}, Y_n + c_{23}, Z_n + c_{33})$$

$$X_{n+1} = X_n + \tfrac{1}{6}(c_{11} + 2c_{12} + 2c_{13} + c_{14})$$

$$Y_{n+1} = Y_n + \tfrac{1}{6}(c_{21} + 2c_{22} + 2c_{23} + c_{24})$$

$$Z_{n+1} = Z_n + \tfrac{1}{6}(c_{31} + 2c_{32} + 2c_{33} + c_{34})$$

Note that each of the four substeps of the Runge–Kutta scheme must be carried out in parallel on each of the equations, since the intermediate values for all the independent variables are needed in the next substep for each variable; for instance, the computation of c_{13} requires not only the value of c_{12} but also the values of c_{22} and c_{32}. The first step of the computation in this case proceeds thus:

$$X_0 = x_0 = 1 \qquad Y_0 = y_0 = 0.5 \qquad Z_0 = z_0 = -0.2$$

$$c_{11} = hf_1(t_0, X_0, Y_0, Z_0) = hY_0 = 0.025\,000$$

$$c_{21} = hf_2(t_0, X_0, Y_0, Z_0) = hZ_0 = -0.010\,000$$

$$c_{31} = hf_3(t_0, X_0, Y_0, Z_0) = h(-X_0t_0Z_0 - t_0Y_0 + t_0^2X_0) = 0.000\,000$$

$$\begin{aligned} c_{12} &= hf_1(t_0 + \tfrac{1}{2}h, X_0 + \tfrac{1}{2}c_{11}, Y_0 + \tfrac{1}{2}c_{21}, Z_0 + \tfrac{1}{2}c_{31}) \\ &= h(Y_0 + \tfrac{1}{2}c_{21}) \\ &= 0.024\,750 \end{aligned}$$

$$\begin{aligned} c_{22} &= hf_2(t_0 + \tfrac{1}{2}h, X_0 + \tfrac{1}{2}c_{11}, Y_0 + \tfrac{1}{2}c_{21}, Z_0 + \tfrac{1}{2}c_{31}) \\ &= h(Z_0 + \tfrac{1}{2}c_{31}) \\ &= -0.010\,000 \end{aligned}$$

$$\begin{aligned} c_{32} &= hf_3(t_0 + \tfrac{1}{2}h, X_0 + \tfrac{1}{2}c_{11}, Y_0 + \tfrac{1}{2}c_{21}, Z_0 + \tfrac{1}{2}c_{31}) \\ &= h(-(X_0 + \tfrac{1}{2}c_{11})(t_0 + \tfrac{1}{2}h)(Z_0 + \tfrac{1}{2}c_{31}) \\ &\quad - (t_0 + \tfrac{1}{2}h)(Y_0 + \tfrac{1}{2}c_{21}) + (t_0 + \tfrac{1}{2}h)^2(X_0 + \tfrac{1}{2}c_{11})) \\ &= -0.000\,334 \end{aligned}$$

$$\begin{aligned} c_{13} &= hf_1(t_0 + \tfrac{1}{2}h, X_0 + \tfrac{1}{2}c_{12}, Y_0 + \tfrac{1}{2}c_{22}, Z_0 + \tfrac{1}{2}c_{32}) \\ &= h(Y_0 + \tfrac{1}{2}c_{22}) \\ &= 0.024\,750 \end{aligned}$$

$$\begin{aligned} c_{23} &= hf_2(t_0 + \tfrac{1}{2}h, X_0 + \tfrac{1}{2}c_{12}, Y_0 + \tfrac{1}{2}c_{22}, Z_0 + \tfrac{1}{2}c_{32}) \\ &= h(Z_0 + \tfrac{1}{2}c_{32}) \\ &= -0.010\,008 \end{aligned}$$

$$\begin{aligned} c_{33} &= hf_3(t_0 + \tfrac{1}{2}h, X_0 + \tfrac{1}{2}c_{12}, Y_0 + \tfrac{1}{2}c_{22}, Z_0 + \tfrac{1}{2}c_{32}) \\ &= h(-(X_0 + \tfrac{1}{2}c_{12})(t_0 + \tfrac{1}{2}h)(Z_0 + \tfrac{1}{2}c_{32}) \\ &\quad - (t_0 + \tfrac{1}{2}h)(Y_0 + \tfrac{1}{2}c_{22}) + (t_0 + \tfrac{1}{2}h)^2(X_0 + \tfrac{1}{2}c_{12})) \\ &= -0.000\,334 \end{aligned}$$

$$\begin{aligned} c_{14} &= hf_1(t_0 + h, X_0 + c_{13}, Y_0 + c_{23}, Z_0 + c_{33}) \\ &= h(Y_0 + c_{23}) \\ &= 0.024\,499 \end{aligned}$$

$$\begin{aligned} c_{24} &= hf_2(t_0 + h, X_0 + c_{13}, Y_0 + c_{23}, Z_0 + c_{33}) \\ &= h(Z_0 + c_{33}) \\ &= -0.010\,016 \end{aligned}$$

$$\begin{aligned} c_{34} &= hf_3(t_0 + h, X_0 + c_{13}, Y_0 + c_{23}, Z_0 + c_{33}) \\ &= h(-(X_0 + c_{13})(t_0 + h)(Z_0 + c_{33}) \\ &\quad - (t_0 + h)(Y_0 + c_{23}) + (t_0 + h)^2(X_0 + c_{13})) \\ &= -0.000\,584 \end{aligned}$$

$$X_1 = 1.024\,750, \quad Y_1 = 0.489\,994, \quad Z_1 = -0.200\,320$$

The second and subsequent steps are similar – we shall not present the details of the computations. It should be obvious by now that computations like these are sufficiently tedious to justify the effort of writing a computer program to carry out the actual arithmetic. The essential point for the reader to grasp is not the mechanics, but rather the principle whereby methods for the solution of first-order differential equations can be extended to the solution of sets of equations and hence to higher-order equations.

8.3.3 Exercises

16 Transform the following initial-value problems into sets of first-order differential equations with appropriate initial conditions:

(a) $\dfrac{d^2x}{dt^2} + 6(x^2 - t)\dfrac{dx}{dt} - 4xt = 0$

$x(0) = 1, \quad \dfrac{dx}{dt}(0) = 2$

(b) $\dfrac{d^2x}{dt^2} + 4(x^2 - t^2)^{1/2} = 0$

$x(1) = 2, \quad \dfrac{dx}{dt}(1) = 0.5$

(c) $\dfrac{d^2x}{dt^2} - \sin\left(\dfrac{dx}{dt}\right) + 4x = 0$

$x(0) = 0, \quad \dfrac{dx}{dt}(0) = 0$

(d) $\dfrac{d^3x}{dt^3} + t\dfrac{d^2x}{dt^2} + 6\,e^t\dfrac{dx}{dt} - x^2t = e^{2t}$

$x(0) = 1, \quad \dfrac{dx}{dt}(0) = 2, \quad \dfrac{d^2x}{dt^2}(0) = 0$

(e) $\dfrac{d^3x}{dt^3} + t\dfrac{d^2x}{dt^2} + x^2 = \sin t$

$x(1) = 1, \quad \dfrac{dx}{dt}(1) = 0, \quad \dfrac{d^2x}{dt^2}(1) = -2$

(f) $\left(\dfrac{d^3x}{dt^3}\right)^{1/2} + t\dfrac{d^2x}{dt^2} + x^2t^2 = 0$

$x(2) = 0, \quad \dfrac{dx}{dt}(2) = 0, \quad \dfrac{d^2x}{dt^2}(2) = 2$

(g) $\dfrac{d^4x}{dt^4} + x\dfrac{d^2x}{dt^2} + x^2 = \ln t, \quad x(0) = 0, \quad \dfrac{dx}{dt}(0) = 0,$

$\dfrac{d^2x}{dt^2}(0) = 4, \quad \dfrac{d^3x}{dt^3}(0) = -3$

(h) $\dfrac{d^4x}{dt^4} + \left(\dfrac{dx}{dt} - 1\right)t\dfrac{d^3x}{dt^3} + \dfrac{dx}{dt} - (xt)^{1/2}$

$= t^2 + 4t - 5$

$x(0) = a, \quad \dfrac{dx}{dt}(0) = 0, \quad \dfrac{d^2x}{dt^2}(0) = b, \quad \dfrac{d^3x}{dt^3}(0) = 0$

17 Find the value of $X(0.3)$ for the initial-value problem

$$\frac{d^2x}{dt^2} + x^2\frac{dx}{dt} + x = \sin t, \quad x(0) = 0, \quad \frac{dx}{dt}(0) = 1$$

using the Euler method with step size $h = 0.1$.

18 The second-order Adams–Bashforth method for the integration of a single first-order differential equation

$$\frac{dx}{dt} = f(t, x)$$

is

$$X_{n+1} = X_n + \tfrac{1}{2}h[3f(t_n, X_n) - f(t_{n-1}, X_{n-1})]$$

Write down the appropriate equations for applying the same method to the solution of the pair of differential equations

$$\frac{dx}{dt} = f_1(t, x, y), \quad \frac{dy}{dt} = f_2(t, x, y)$$

Hence find the value of $X(0.3)$ for the initial-value problem

$$\frac{d^2x}{dt^2} + x^2\frac{dx}{dt} + x = \sin t, \quad x(0) = 0, \quad \frac{dx}{dt}(0) = 1$$

using this Adams–Bashforth method with step size $h = 0.1$. Use the second-order predictor–corrector method for the first step to start the computation.

19 Use the second-order predictor–corrector method (that is, the first-order Adams–Bashforth formula

as predictor and the second-order Adams–Moulton formula as corrector) to compute an approximation $X(0.65)$ to the solution $x(0.65)$ of the initial-value problem

$$\frac{d^3x}{dt^3} + (x - t)\frac{d^2x}{dt^2} + \left(\frac{dx}{dt}\right)^2 - x^2 = 0$$

$$x(0.5) = -1, \quad \frac{dx}{dt}(0.5) = 1, \quad \frac{d^2x}{dt^2}(0.5) = 2$$

using a step size $h = 0.05$.

†20 Write a computer program to solve the initial-value problem

$$\frac{d^2x}{dt^2} + x^2\frac{dx}{dt} + x = \sin t \quad x(0) = 0, \quad \frac{dx}{dt}(0) = 1$$

using the fourth-order Runge–Kutta method. Use your program to find the value of $X(1.6)$ using step sizes $h = 0.4$ and 0.2. Estimate the accuracy of your value of $X(1.6)$ and estimate the step size that would be necessary to obtain a value of $X(1.6)$ accurate to 6dp.

†21 Write a computer program to solve the initial-value problem

$$\frac{d^3x}{dt^3} + (x - t)\frac{d^2x}{dt^2} + \left(\frac{dx}{dt}\right)^2 - x^2 = 0$$

$$x(0.5) = -1, \quad \frac{dx}{dt}(0.5) = 1, \quad \frac{d^2x}{dt^2}(0.5) = 2$$

using the third-order predictor–corrector method (that is, the second-order Adams–Bashforth formula as predictor with the third-order Adams–Moulton as corrector). Use the fourth-order Runge–Kutta method to overcome the starting problem with this process. Use your program to find the value of $X(2.2)$ using step sizes $h = 0.1$ and 0.05. Estimate the accuracy of your value of $X(2.2)$ and estimate the step size that would be necessary to obtain a value of $X(2.2)$ accurate to 6dp.

8.3.4 Boundary-value problems

Because first-order ordinary differential equations only have one boundary condition, that condition can always be treated as an initial condition. Once we turn to second- and higher-order differential equations, there are, at least for fully determined problems, two or more boundary conditions. If the boundary conditions are all imposed at the same point then the problem is an initial-value problem and can be solved by the methods we have already described. The problems that have been used as illustrations in Sections 8.3.1 and 8.3.2 were all initial-value problems. Boundary-value problems are somewhat more difficult to solve than initial-value problems.

To illustrate the difficulties of boundary-value problems, let us consider second-order differential equations. These have two boundary conditions. If they are both imposed at the same point (and so are initial conditions), the conditions will usually be a value of the dependent variable and of its derivative, for instance a problem like

$$\mathrm{L}[x(t)] = f(t), \quad x(a) = p, \quad \frac{dx}{dt}(a) = q$$

where L is some differential operator. Occasionally, a mixed boundary condition such as

$$Cx(a) + D\frac{dx}{dt}(a) = p$$

will arise. Provided that a second boundary condition on x or $\mathrm{d}x/\mathrm{d}t$ is imposed at the same point, this causes no difficulty, since the boundary conditions can be decoupled, that is, solved to give values of $x(a)$ and $(\mathrm{d}x/\mathrm{d}t)(a)$, before the problem is solved.

If the two boundary conditions are imposed at different points then they could consist of two values of the dependent variable, the value of the dependent variable at one boundary and its derivative at the other, or even linear combinations of the values of the dependent variable and its derivative. For instance, we may have

$$\mathrm{L}[x(t)] = f(t), \quad x(a) = p, \quad x(b) = q$$

or

$$\mathrm{L}[x(t)] = f(t), \quad \frac{\mathrm{d}x}{\mathrm{d}t}(a) = p, \quad x(b) = q$$

or

$$\mathrm{L}[x(t)] = f(t), \quad x(a) = p, \quad \frac{\mathrm{d}x}{\mathrm{d}t}(b) = q$$

or even such systems as

$$\mathrm{L}[x(t)] = f(t), \quad x(a) = p, \quad Ax(b) + B\frac{\mathrm{d}x}{\mathrm{d}t}(b) = q$$

The increased range of possibilities introduced by boundary-value problems almost inevitably increases the problems which may arise in their solution. For instance it may at first sight seem that it should also be possible to solve problems with boundary conditions consisting of the derivative at both boundaries, such as

$$\mathrm{L}[x(t)] = f(t), \quad \frac{\mathrm{d}x}{\mathrm{d}t}(a) = p, \quad \frac{\mathrm{d}x}{\mathrm{d}t}(b) = q$$

Things are unfortunately not that simple – as Example 8.16 shows.

EXAMPLE 8.16 Solve the boundary-value problem

$$\frac{\mathrm{d}^2x}{\mathrm{d}t^2} = 4, \quad \frac{\mathrm{d}x}{\mathrm{d}t}(0) = p, \quad \frac{\mathrm{d}x}{\mathrm{d}t}(1) = q$$

Solution Integrating twice easily yields the general solution

$$x = 2t^2 + At + B$$

The boundary conditions then impose

$$A = p \quad \text{and} \quad 4 + A = q$$

It is obviously not possible to find a value of A satisfying both these equations unless $q = p + 4$. In any event, whether or not p and q satisfy this relation, it is not possible to determine the constant B.

Example 8.16 illustrates the fact that if derivative boundary conditions are to be applied, a supplementary compatibility condition is needed. In addition, there may be a residual uncertainty in the solution. The complete analysis of what types of boundary conditions are allowable for two-point boundary-value problems is beyond the scope of this book. Differential equations of order higher than two increase the range of possibilities even further and introduce further complexities into the determination of what boundary conditions are allowable and valid.

8.3.5 The method of shooting

One obvious way of solving two-point boundary-value problems is a form of systematic trial and error in which the boundary-value problem is replaced by an initial-value problem with initial values given at one of the two boundary points. The initial-value problem can be solved by an appropriate numerical technique and the value of whatever function is involved in the boundary condition at the second boundary point determined. The initial values are then adjusted and another initial-value problem solved. This process is repeated until a solution is found with the appropriate value at the second boundary point.

As an illustration, we shall consider a second-order boundary-value problem of the form

$$\mathrm{L}[x] = f(t), \quad x(a) = p, \quad x(b) = q \tag{8.25}$$

The related initial-value problem

$$\mathrm{L}[x] = f(t), \quad x(a) = p, \quad \frac{\mathrm{d}x}{\mathrm{d}t}(a) = 0 \tag{8.26}$$

could be solved as described in Section 8.3.2. Suppose that doing this results in an approximate solution of (8.26) denoted by X_1. In the same way, denote the solution of the problem

$$\mathrm{L}[x] = f(t), \quad x(a) = p, \quad \frac{\mathrm{d}x}{\mathrm{d}t}(a) = 1 \tag{8.27}$$

by X_2. We now have a situation as shown in Figure 8.21. The values of the two solutions at the point $t = b$ are $X_1(b)$ and $X_2(b)$. The original boundary-value problem (8.25) requires a value q at b. Since q is roughly three-quarters of the way between $X_1(b)$ and $X_2(b)$, we should intuitively expect that solving the initial-value problem

$$\mathrm{L}[x] = f(t), \quad x(a) = p, \quad \frac{\mathrm{d}x}{\mathrm{d}t}(a) = 0.75 \tag{8.28}$$

will produce a solution with $X(b)$ much closer to q. What we have done, of course, is to assume that $X(b)$ varies continuously and roughly in proportion to $(\mathrm{d}x/\mathrm{d}t)(a)$ and then to use linear interpolation to estimate a better value of $(\mathrm{d}x/\mathrm{d}t)(a)$. It is

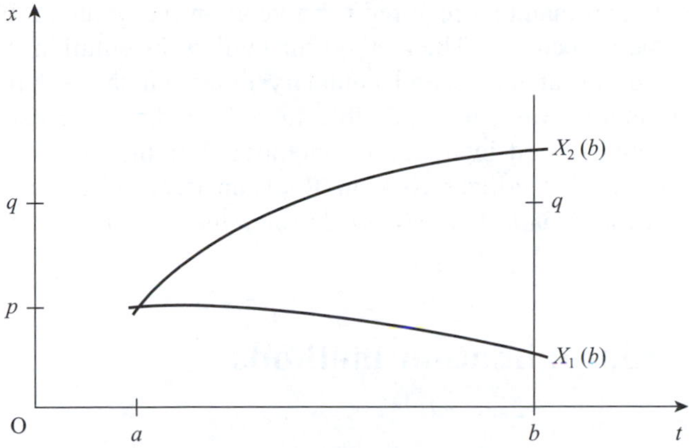

Figure 8.21 The solution of a differential equation by the method of shooting: initial trials.

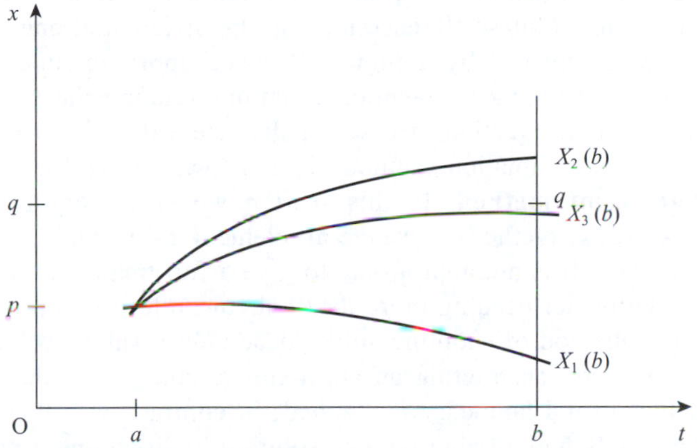

Figure 8.22 The solution of a differential equation by the method of shooting: first refinement.

unlikely, of course, that $X(b)$ will vary exactly linearly with $(\mathrm{d}x/\mathrm{d}t)(a)$ so the solution of (8.28), call it X_3, will be something like that shown in Figure 8.22. The process of linear interpolation to estimate a value of $(\mathrm{d}x/\mathrm{d}t)(a)$ and the subsequent solution of the resulting initial-value problem can be repeated until a solution is found with a value of $X(b)$ as close to q as may be required. This method of solution is known, by an obvious analogy with the bracketing method employed by artillerymen to find their targets, as the **method of shooting**. Shooting is not restricted to solving two-point boundary-value problems in which the two boundary values are values of the dependent variable. Problems involving boundary values on the derivatives can be solved in an analogous manner.

The solution of a two-point boundary-value problem by the method of shooting involves repeatedly solving a similar initial-value problem. It is therefore obvious that the amount of computation required to obtain a solution to a two-point boundary-value problem by this method is certain to be an order of magnitude or more greater than that required to solve an initial-value problem of the same order to the same accuracy. The method for finding the solution that satisfies the boundary condition at the second boundary point which we have just described used linear interpolation. It is possible to reduce the computation required by using more sophisticated interpolation methods. For instance, a version of the method of shooting that utilizes Newton–Raphson iteration is described in R. D. Milne, *Applied Functional Analysis, An Introductory Treatment* (Pitman, London, 1979).

8.3.6 Function approximation methods

The method of shooting is not the only way of solving boundary-value problems numerically. Other methods include various **finite-difference techniques** and a set of methods that can be collectively characterized as **function approximation methods**. In a finite-difference method the differential operator of the differential equation is replaced by a finite-difference approximation to the operator. This leads to a set of linear algebraic equations relating the values of the solution to the differential equation at a set of discrete values of the independent variable. Function approximation methods include various **collocation methods** and the **finite-element method**. In this section we shall very briefly outline function approximation methods and give an elementary example of the use of a collocation method. It is not appropriate to give a full treatment of these methods in this book; the reader needing more detail should refer to more advanced texts.

The method of shooting solves a boundary-value problem by starting at one boundary and constructing an approximate solution to the problem step by step until the second boundary is reached. In contrast with this, function approximation methods find an approximate solution by assuming a particular type or form of function for the solution over the whole range of the problem. This function (usually referred to as the **trial function**) is then substituted into the differential equation and its boundary conditions. Trial functions always contain some unknown parameters, and, once the function has been substituted into the differential equation, some criterion can be used to assign values to these initially unknown parameters in such a way as to make the trial function as close an approximation as possible to the solution of the boundary-value problem.

Unless a very fortuitous choice of trial function is made, it is unlikely that it will be possible to make the function chosen satisfy the differential equation exactly. If, for instance, a trial function depending on some parameters $p_1, p_2, \dots$ and denoted by $X(t; p_1, p_2, \dots)$ is to be used to obtain an approximate solution to the differential equation $\mathrm{L}[x(t)] = 0$ then substituting this function into the differential equation results in a function

$$\mathrm{L}[X(t; p_1, p_2, \dots)] = \eta(t; p_1, p_2, \dots)$$

which is called the **residual** of the equation. Intuitively, it seems likely that making this residual as small as possible will result in a good approximation to the solution of the equation. But what does making a function as small as possible mean? The most common approaches are to make the residual zero at some discrete set of points distributed over the range of the independent variable – this gives rise to collocation methods – or to minimize, in some way, some measure of the overall size of the residual (for instance the integral of the square of the residual) – this is commonly used in finite-element methods.

Thus, for instance, to solve the boundary-value problem

$$\mathrm{L}[x(t)] = 0, \quad x(a) = q, \quad x(b) = r \tag{8.29}$$

we should assume that the trial function $X(t)$, an approximation to $x(t)$, takes some form such as

$$X(t) = \sum_{i=1}^{n} p_i f_i(t) \tag{8.30}$$

where $\{p_i : i = 1, 2, \dots, n\}$ is the set of parameters that are to be determined and $\{f_i(t) : i = 1, 2, \dots, n\}$ is some set of functions of t. Substituting the approximation (8.30) into the original problem (8.29) gives

$$\mathrm{L}\left[\sum_{i=1}^{n} p_i f_i(t)\right] = \eta(t) \tag{8.31a}$$

$$\sum_{i=1}^{n} p_i f_i(a) = q \tag{8.31b}$$

and

$$\sum_{i=1}^{n} p_i f_i(b) = r \tag{8.31c}$$

Equations (8.31b, c) express the requirement that the approximation chosen will satisfy the boundary conditions of the problem. The function $\eta(t)$ in (8.31a) is the residual of the problem. Since (8.31b, c) impose two conditions on the choice of the parameters $p_1, p_2, \dots, p_n$ we need another $n - 2$ conditions to determine all the p_i. For a collocation solution this is done by choosing $n - 2$ values of t such that $a < t_1 < t_2 < \dots < t_{n-2} < b$ and making $\eta(t_k) = 0$ for $k = 1, 2, \dots, n - 2$. Thus we have the n equations

$$\mathrm{L}\left[\sum_{i=1}^{n} p_i f_i(t_k)\right] = 0 \quad (k = 1, 2, \dots, n-2) \tag{8.32a}$$

$$\sum_{i=1}^{n} p_i f_i(a) = q \tag{8.32b}$$

$$\sum_{i=1}^{n} p_i f_i(b) = r \tag{8.32c}$$

for the n unknown parameters $p_1, p_2, \ldots, p_n$. In general, these equations will be nonlinear in the p_i, but if the operator L is a linear operator then they may be rewritten as

$$\sum_{i=1}^{n} p_i L[f_i(t_k)] = 0 \quad (k = 1, 2, \ldots, n-2) \tag{8.33a}$$

$$\sum_{i=1}^{n} p_i f_i(a) = q \tag{8.33b}$$

$$\sum_{i=1}^{n} p_i f_i(b) = r \tag{8.33c}$$

and are linear in the p_i. They therefore constitute a matrix equation for the p_i:

$$\begin{bmatrix} \mathrm{L}[f_1(t_1)] & \mathrm{L}[f_2(t_1)] & \ldots & \mathrm{L}[f_n(t_1)] \\ \mathrm{L}[f_1(t_2)] & \mathrm{L}[f_2(t_2)] & \ldots & \mathrm{L}[f_n(t_2)] \\ \mathrm{L}[f_1(t_3)] & \mathrm{L}[f_2(t_3)] & \ldots & \mathrm{L}[f_n(t_3)] \\ \vdots & \vdots & \ldots & \vdots \\ \mathrm{L}[f_1(t_{n-2})] & \mathrm{L}[f_2(t_{n-2})] & \ldots & \mathrm{L}[f_n(t_{n-2})] \\ f_1(a) & f_2(a) & \ldots & f_n(a) \\ f_1(b) & f_2(b) & \ldots & f_n(b) \end{bmatrix} \begin{bmatrix} p_1 \\ p_2 \\ p_3 \\ \vdots \\ p_{n-2} \\ p_{n-1} \\ p_n \end{bmatrix} = \begin{bmatrix} 0 \\ 0 \\ 0 \\ \vdots \\ 0 \\ q \\ r \end{bmatrix} \tag{8.34}$$

This matrix equation can, of course, be solved by any of the standard methods of linear algebra. If the operator L is nonlinear then (8.32) cannot be expressed in the form (8.33). The equations (8.32) may still be solved for the coefficients p_i, but the solution of nonlinear equations is, in general, a much more difficult task than the solution of linear ones.

The choice of the functions $f_i(t)$ and the collocation points t_k greatly affect the accuracy and speed of convergence of the solution. (The speed of convergence in this context is usually measured by the number of terms it is necessary to take in the approximation (8.30) in order to achieve a solution with a specified accuracy.) Example 8.17 shows a simple application of collocation methods to the solution of a second-order boundary-value problem.

EXAMPLE 8.17

Solve the boundary-value problem

$$\frac{\mathrm{d}^2x}{\mathrm{d}t^2} + \mathrm{e}^t\frac{\mathrm{d}x}{\mathrm{d}t} + x = 0, \quad x(0) = 0, \quad x(2) = 1 \tag{8.35}$$

using a collocation method with

$$X_n(t) = \sum_{i=1}^{n} p_i t^{i-1}$$

Solution The differential operator in this case is linear, so we may construct the matrix equation equivalent to (8.34). With the given approximation, we have

$$\mathrm{L}[f_i(t)] = \mathrm{L}[t^{i-1}] = \begin{cases} [(i-1)(i-2+t\mathrm{e}^t)+t^2]t^{i-3} & (i \geqslant 3) \\ \mathrm{e}^t + t & (i = 2) \\ 1 & (i = 1) \end{cases}$$

We shall choose the collocation points to be equally spaced over the interior of the interval [0, 2]. Thus, for $n = 5$ say, we need three collocation points, which would be 0.5, 1.0 and 1.5. We should therefore obtain the matrix equation

$$\begin{bmatrix} \mathrm{L}[f_1(0.5)] & \mathrm{L}[f_2(0.5)] & \mathrm{L}[f_3(0.5)] & \mathrm{L}[f_4(0.5)] & \mathrm{L}[f_5(0.5)] \\ \mathrm{L}[f_1(1.0)] & \mathrm{L}[f_2(1.0)] & \mathrm{L}[f_3(1.0)] & \mathrm{L}[f_4(1.0)] & \mathrm{L}[f_5(1.0)] \\ \mathrm{L}[f_1(1.5)] & \mathrm{L}[f_2(1.5)] & \mathrm{L}[f_3(1.5)] & \mathrm{L}[f_4(1.5)] & \mathrm{L}[f_5(1.5)] \\ 1 & 0 & 0 & 0 & 0 \\ 1 & 2 & 4 & 8 & 16 \end{bmatrix} \begin{bmatrix} p_1 \\ p_2 \\ p_3 \\ p_4 \\ p_5 \end{bmatrix} = \begin{bmatrix} 0 \\ 0 \\ 0 \\ 0 \\ 1 \end{bmatrix}$$

Computing the numerical values of the matrix elements yields the matrix equation

$$\begin{bmatrix} 1.000 & 2.149 & 3.899 & 4.362 & 3.887 \\ 1.000 & 3.718 & 8.437 & 15.155 & 23.873 \\ 1.000 & 5.982 & 17.695 & 42.626 & 92.565 \\ 1.000 & 0.000 & 0.000 & 0.000 & 0.000 \\ 1.000 & 2.000 & 4.000 & 8.000 & 16.000 \end{bmatrix} \begin{bmatrix} p_1 \\ p_2 \\ p_3 \\ p_4 \\ p_5 \end{bmatrix} = \begin{bmatrix} 0.000 \\ 0.000 \\ 0.000 \\ 0.000 \\ 1.000 \end{bmatrix}$$

whose solution is

$$\boldsymbol{p} = [0.000 \quad 2.636 \quad -1.912 \quad 0.402 \quad 0.010]^{\mathrm{T}}$$

Figure 8.23 shows the solutions X_4, X_5, X_6 and X_7. As we should intuitively expect, taking more terms in the approximation for $x(t)$ causes the successive approximations to converge. In Figure 8.24 the approximations X_5 and X_6 are

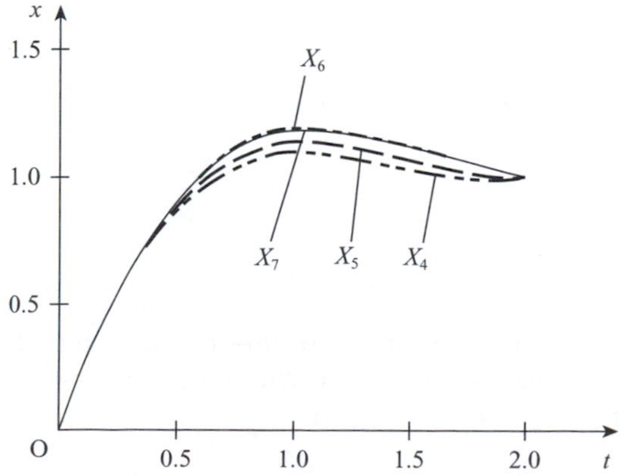

Figure 8.23 A collocation solution of (8.35).

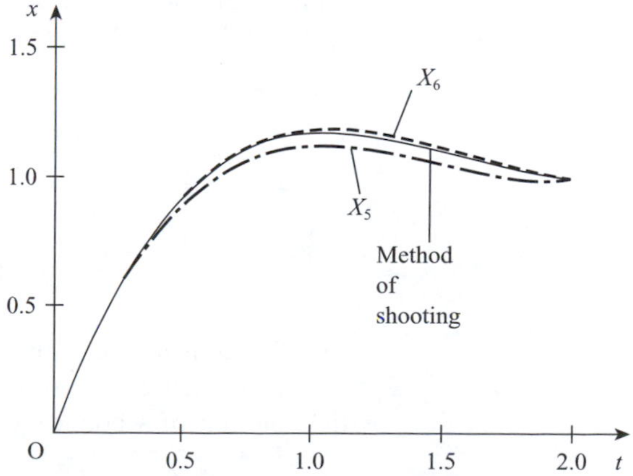

Figure 8.24 Comparison of the collocation solutions with the solution by the method of shooting.

compared with a solution to the problem (8.35) obtained by the method of shooting using a second-order Runge–Kutta integration method. The step size used for the method-of-shooting solution was estimated, using the technique introduced in Section 8.2.6, to yield a solution accurate to better than 3.5×10^{-3}. On this graph the solution X_7 was indistinguishable from the method-of-shooting solution.

Although Example 8.17 gave reasonably good accuracy from a relatively small number of terms in the function $X_n(t)$, difficulties do arise with collocation methods when straightforward power-series approximations like this are used. It is more normal to use some form of orthogonal polynomials, such as Tchebyshev or Legendre polynomials, for the $f_i(t)$. In appropriate cases $f_i(t) = \sin it$ and $\cos it$ are also used. The reader is referred to more advanced texts for details of these functions and their use in collocation methods.

Although they are rather more commonly used for problems involving partial differential equations, finite-element methods may also be used for ordinary differential equation boundary-value problems. The essential difference between finite-element methods and collocation methods of the type described in Example 8.17 lies in the type of functions used to approximate the dependent variable. Finite-element methods use **functions with localized support**. By this, we mean functions that are zero over large parts of the range of the independent variable and only have a non-zero value for some restricted part of the range. A complete approximation to the dependent variable may be constructed from a linear sum of such functions, the coefficients in the linear sum providing the parameters of the function approximation.

EXAMPLE 8.18

A typical simple set of functions with localized support that are often used in the finite-element method are the 'witch's hat' functions. For a one-dimensional boundary-value problem, such as (8.29), the range $[a, b]$ of the independent variable is divided into a number of subranges $[t_0, t_1], [t_1, t_2], \dots, [t_{n-1}, t_n]$ with $t_0 = a$ and $t_n = b$. We then define functions

$$f_i(t) = \begin{cases} \dfrac{t - t_{i-1}}{t_i - t_{i-1}} & (t \in [t_{i-1}, t_i]) \\[2ex] \dfrac{t_{i+1} - t}{t_{i+1} - t_i} & (t \in [t_i, t_{i+1}]) \\[2ex] 0 & (t \notin [t_{i-1}, t_{i+1}]) \end{cases}$$

The function $f_i(t)$ has support (that is, its value is non-zero) only on the interval $[t_{i-1}, t_{i+1}]$. Figure 8.25 shows the form of the functions $f_i(t)$. An approximation to the solution of a boundary-value problem can be formed as

$$X(t) = \sum_{k=0}^{n} p_k f_k(t) \tag{8.36}$$

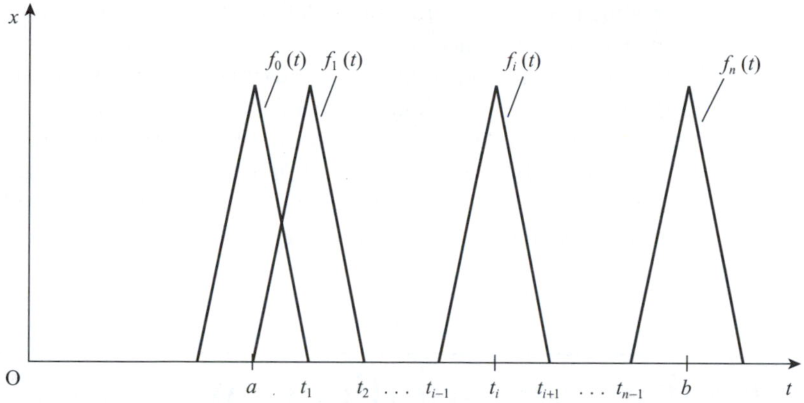

Figure 8.25 The 'witch's hat' functions.

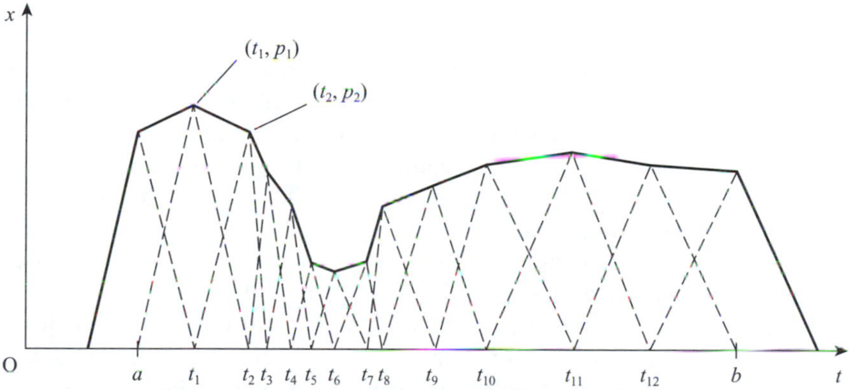

Figure 8.26 The construction of a continuous piecewise-linear approximation function from 'witch's hat' functions.

This equation defines a function that is piecewise-linear and continuous on the range $[a, b]$ as illustrated in Figure 8.26.

The finite-element method provides a general framework for using functions with localized support to construct an approximation to the whole solution. One advantage of using such functions is that the user can, to a considerable extent, tailor the approximation used to the properties of the physical problem. If the problem is expected to give rise to very rapid changes in some region then more functions with local support in that area can be used. In regions where the solution is expected to change relatively slowly fewer functions may be used. In Figure 8.26, for instance, the division of the interval $[a, b]$ into subregions is shown as being finer near t_4 and coarser near t_{11}. This property of functions with local support gives the finite-element method considerable advantages over collocation methods (which use functions defined over the whole range of the problem) and over finite-difference methods.

Just as for the function approximation method illustrated in Example 8.17, the finite-element method requires that some criterion be chosen for determining the values of the unknown parameters in the approximation (8.36). A variety of criteria are commonly used, but we shall not describe these in detail in this section. The use of the finite-element method for obtaining numerical solutions of partial differential equations is described in Section 9.6.

8.4 Engineering application: oscillations of a pendulum

The simple pendulum has been used for hundreds of years as a timing device. A pendulum clock, using either a falling weight or a clockwork spring device to provide motive power, relies on the natural periodic oscillations of a pendulum to ensure good timekeeping. Generally we assume that the period of a pendulum is constant regardless of its amplitude. But this is only true for infinitesimally small amplitude oscillations. In reality the period of a pendulum's oscillations depends on its amplitude. In this section we will use our knowledge of numerical analysis to assist in an investigation of this relationship.

Figure 8.27 shows a simple rigid pendulum mounted on a frictionless pivot swinging in a single plane. By resolving forces in the tangential direction we have, following the classical analysis of such pendula,

$$ma\frac{\mathrm{d}^2\theta}{\mathrm{d}t^2} = -mg\sin\theta$$

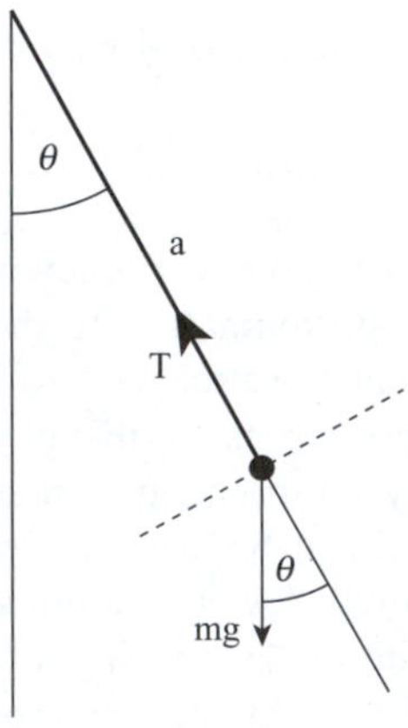

Figure 8.27 A simple pendulum.

that is,

$$\frac{d^2\theta}{dt^2} + \frac{g}{a}\sin\theta = 0 \tag{8.37}$$

For small oscillations of the pendulum we can use the approximation $\sin\theta \approx \theta$ so the equation becomes

$$\frac{d^2\theta}{dt^2} + \frac{g}{a}\theta = 0 \tag{8.38}$$

which is, of course, the simple harmonic motion equation with solutions

$$\theta = A\cos\left(\sqrt{\frac{g}{a}}t\right) + B\sin\left(\sqrt{\frac{g}{a}}t\right)$$

Hence the period of the oscillations is $2\pi\sqrt{(a/g)}$ and is independent of the amplitude of the oscillations.

In reality, of course, the amplitude of the oscillations may not be small enough for the linear approximation $\sin\theta \approx \theta$ to be valid, so it would be useful to be able to solve (8.37). Equation (8.37) is nonlinear so its solution is rather more problematical than (8.38). We will solve the equation numerically. In order to make the solution a little more transparent we will scale it so that the period of the oscillations of the linear approximation (8.38) is unity. This is achieved by setting $t = 2\pi\sqrt{(a/g)}\,\tau$. Equation (8.37) then becomes

$$\frac{d^2\theta}{d\tau^2} + 4\pi^2\sin\theta = 0 \tag{8.39}$$

For an initial amplitude of 30°, the pseudocode algorithm shown in Figure 8.28, which implements the fourth-order Runge–Kutta method described in Section 8.2.8, produces the results $\Theta(6.0) = 23.965\,834$ using a time step of 0.05 and $\Theta(6.0) = 24.018\,659$ with a step of 0.025. Using Richardson extrapolation (see Section 8.2.6) we can predict that the time step needed to achieve 5 sf accuracy (i.e. an error less than 5×10^{-6}) with this fourth-order method is

$$\left[\frac{0.000\,005 \times (2^4 - 1)}{|23.965\,834 - 24.018\,659|}\right]^{1/4} \times 0.025 = 0.0049$$

Repeating the calculation with time steps 0.01 and 0.005 gives $\Theta(6.0) = 24.021\,872\,7$ and $\Theta(6.0) = 24.021\,948\,1$ for which Richardson extrapolation implies an error of 5×10^{-6} as predicted.

As a check we can draw the graph of $|\Theta_{0.01}(\tau) - \Theta_{0.005}(\tau)|/15$, shown in Figure 8.29. This confirms that the error grows as the solution advances and that the maximum error is around 7.5×10^{-6}.

What we actually wanted is an estimate of the period of the oscillations. The most satisfactory way to determine this is to find the interval between the times of successive zero crossings. The time of a zero crossing can be estimated by

```
tol ← 0.000 01
t_start ← 0
t_end ← 6
write(vdu,'Enter amplitude => ')
read(keyb, x0)
x_start ← pi*x0/180
v_start ← 0
write(vdu,'Enter stepsize => ')
read(keyb,h)
write(vdu,t_start,'   ',deg(x_start))
t ← t_start
x ← x_start
v ← v_start
repeat
  rk4(x,v,h → xn,vn)
  x ← xn
  v ← vn
  t ← t+h
until abs(t - t_end) < tol
write(vdu,t, '   ',deg(x))

procedure rk4(x,v,h → xn,vn)
      c11 ← h*f1(x,v)
      c21 ← h*f2(x,v)
      c12 ← h*f1(x + c11/2,v + c21/2)
      c22 ← h*f2(x + c11/2,v + c21/2)
      c13 ← h*f1(x + c12/2,v + c22/2)
      c23 ← h*f2(x + c12/2,v + c22/2)
      c14 ← h*f1(x + c13,v + c23)
      c24 ← h*f2(x + c13,v + c23)
      xn ← x + (c11 + 2*(c12 + c13) + c14)/6
      vn ← v + (c21 + 2*(c22 + c23) + c24)/6
endprocedure

procedure f1(x,v → f1)
  f1 ← v
endprocedure

procedure f2(x,v → f2)
  f2 ← -4*pi*pi*sin (x)
endprocedure

procedure deg(x → deg)
  deg ← 180*x/pi
endprocedure
```

Figure 8.28 A pseudocode algorithm for solving the nonlinear pendulum equation (8.39).

linear interpolation between the data points produced in numerical solution of the differential equation. At a zero crossing the successive values of Θ have the opposite sign. Figure 8.30 shows a modified version of the main part of the algorithm of Figure 8.28. This version determines the times of successive positive to negative zero crossings and the differences between them.

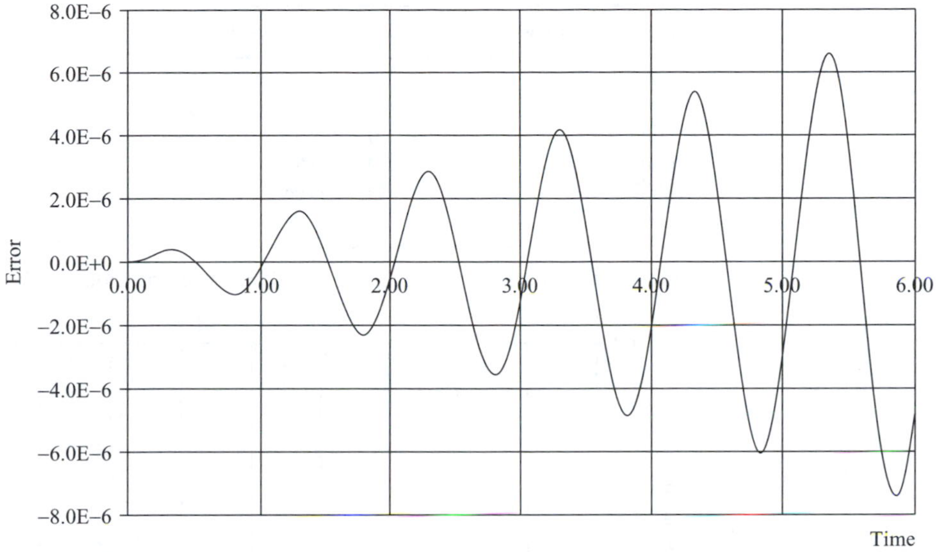

Figure 8.29 Error in solution of equation (8.39) using algorithm (8.28) with $h = 0.005$.

```
tol ← 0.000 01
t_start ← 0
t_end ← 6
write(vdu,'Enter amplitude => ')
read(keyb,x0)
x_start ← pi*x0/180
v_start ← 0
write(vdu,'Enter stepsize => ')
read(keyb,h)
write(vdu,t_start,'   ',deg(x_start))
t ← t_start
x ← x_start
v ← v_start
t_previous_cross ← t_start
repeat
  rk4(x,v,h → xn,vn)
  if(xn*x < 0) and (x > 0) then
    t_cross ← (t*xn – (t + h)*x)/(xn-x)
    write(vdu,t_cross,'   ',t_cross – t_previous_cross)
    t_previous_cross ← t_cross
  endif
  x ← xn
  v ← vn
  t ← t+h
until abs(t – t_end) < tol
```

Figure 8.30 Modification of pseudocode algorithm to find the period of oscillations of equation (8.39).

Time of crossing	Period of last cycle
0.254 352 13	
1.271 761 06	1.017 408 93
2.289 169 73	1.017 408 67
3.306 578 68	1.017 408 95
4.323 987 34	1.017 408 66
5.341 396 30	1.017 408 96

Figure 8.31 Periods of successive oscillations of equation (8.39), $\Theta_0 = 30^\circ$, $h = 0.005$.

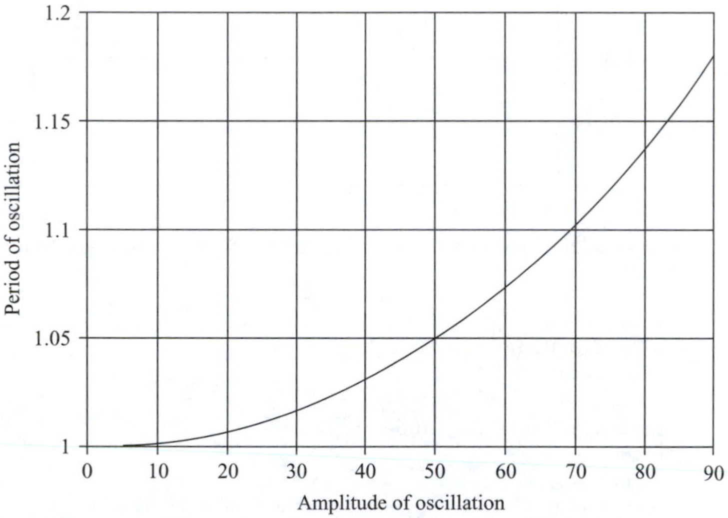

Figure 8.32 Variation of period of oscillations of equation (8.39) with amplitude.

Figure 8.31 shows some results from a program based on the algorithm of Figure 8.30; it is evident that the period has been determined to 6 sf accuracy. Figure 8.32 has been compiled from similar results for other amplitudes of oscillation.

Some spring-powered pendulum clocks are observed to behave in a counter-intuitive way – as the spring winds down the clock gains time where most people intuitively expect it to run more slowly and hence lose time. Figure 8.32 explains this phenomenon. The reason is that, in a spring-powered clock, the spring, acting through the escapement mechanism, exerts forces on the pendulum which, over each cycle of oscillation of the pendulum, result in the application of a tiny net impulse. The result is that just sufficient work is done on the pendulum to overcome the effects of bearing friction, air resistance and any other dissipative effects, and to keep the pendulum swinging with constant amplitude. But, as the spring unwinds the force available is reduced and the impulse gets smaller. The result is that, as the clock winds down, the amplitude of oscillation of the pendulum decreases slightly. Figure 8.32 shows that as the amplitude decreases the period also decreases. Since the period of the pendulum controls the speed of the clock, the clock runs faster as the period decreases! Of course, as the clock winds down even further, the spring reaches a point where it is no longer capable of applying a sufficient impulse to overcome the dissipative forces, the pendulum ceases swinging and the clock finally stops.

8.5 Engineering application: heating of an electrical fuse

The electrical fuse is a simple device for protecting an electrical apparatus or circuit from overload and possible damage after the failure of one or more components in the apparatus. A fuse is usually a short length of thin wire through which the electrical current powering the apparatus flows. If the apparatus fails in such a way as to draw a dangerously increased current, the fuse wire heats up and eventually melts thus disconnecting the apparatus from the power source. In order to design fuses which will not fail during normal use but which will operate reliably and rapidly in abnormal circumstances we must understand the heating of a thin wire carrying an electrical current.

The equation governing the heat generation and dissipation in a wire carrying an electrical current can be formulated as

$$-k\pi r^2 \frac{d^2T}{dx^2} + 2\pi rh(T - T_e)^\alpha = I^2 \frac{\rho}{\pi r^2} \tag{8.40}$$

where T is the temperature of the fuse wire, x is the distance along the wire, k is the thermal conductivity of the material of which the wire is composed, r is the radius of the wire, h is the convective heat transfer coefficient from the surface of the wire, T_e is the ambient temperature of the fuse's surroundings, α is an empirical constant with a value around 1.25, I is the current in the wire and ρ is the resistivity of the wire. Equation (8.40) expresses the balance, in the steady state, between heat generation and heat loss. The first term of the equation represents the transfer of heat along the wire by conduction, the second term is the loss of heat from the surface of the wire by convection and the third term is the generation of heat in the wire by the electrical current.

Taking $\theta = (T - T_e)$ and dividing by $k\pi r^2$, (8.40) can be expressed as

$$\frac{d^2\theta}{dx^2} - \frac{2h}{kr}\theta^\alpha = -\frac{\rho I^2}{k\pi^2 r^4} \tag{8.41}$$

Letting the length of the fuse be $2a$ and scaling the space variable, x, by setting $x = 2aX$, (8.41) becomes

$$\frac{d^2\theta}{dX^2} - \frac{8a^2h}{kr}\theta^\alpha = -\frac{4a^2\rho I^2}{k\pi^2 r^4}$$

```
rho ← 16e-8
kappa ← 63
r ← 5e-4
a ← le-2
hh ← le2
i ← 20
pconst ← 8*hh*a*a/(kappa*r)
qconst ← 4*a*a*rho*i*i/(kappa*pi*pi*r*r*r*r)
tol ← le-5
x_start ← 0.0
x_end ← 1.0
theta_start ← 0.0
write(vdu,'Enter stepsize -->')
read(keyb,h)
write(vdu,'Enter lower limit -->')
read(keyb,theta_dash_low)
write(vdu,'Enter upper limit -->')
read(keyb,theta_dash_high)
desolve(x_start,x_end,h,theta_start,theta_dash_low → th,ql)
desolve(x_start,x_end,h,theta_start,theta_dash_high → th,qh)
repeat
  theta_dash_new ← (qh*theta_dash_low – ql*theta_dash_high)/(qh – ql)
  desolve (x_start,x_end,h,theta_start,theta_dash_new → th,qn)
  if ql*qn>0 then
    ql ← qn
    theta_dash_low ← theta_dash_new
  else
    qh ← qn
    theta_dash_high ← theta_dash_new
  endif
until abs(qn) < tol
write(vdu,th,qn)

procedure desolve(x_0,x_end,h,v1_0,v2_0 → v1_f,v2_f)
  x ← x_0
  v1_o ← v1_0
  v2_o ← v2_0
  rk4(x,v1_o,v2_o,h → v1,v2)
  x ← x+h
  repeat
    pc3(x,v1_o,v2_o,v1,v2,h, → 1_n,v2_n)
    v1_o ← v1
    v2_o ← v2
    v1 ← v1_n
    v2 ← v2_n
    x ← x+h
  until abs(x – x_end) < tol
  v1_f ← v1
  v2_f ← v2
endprocedure
```

Figure 8.33 Pseudocode algorithm for solving equation (8.42).

The boundary conditions are that the two ends of the wire, which are in contact with the electrical terminals in the fuse unit, are kept at some fixed temperature (we will assume that this temperature is the same as T_e). In addition, the fuse has symmetry about its midpoint $x = a$. Hence we may express the complete differential equation problem as

$$\frac{d^2\theta}{dX^2} - \frac{8a^2h}{kr}\theta^{\alpha} = -\frac{4a^2\rho I^2}{k\pi^2 r^4}, \quad \theta(0) = 0, \quad \frac{d\theta}{dX}(1) = 0 \tag{8.42}$$

Equation (8.42) is a nonlinear second-order ordinary differential equation. There is no straightforward analytical technique for tackling it so we must use numerical means. The problem is a boundary-value problem so we must use either the method of shooting or some function approximation method. Figure 8.33 shows a pseudocode algorithm for this problem and Figure 8.34 gives the supporting procedures. The procedure desolve assumes initial conditions of the form $\theta(0) = 0$, $d\theta/dX(0) = \theta_0'$ and solves the differential equation using the third-order

```
procedure rk4 (x,v1,v2,h → v1n,v2n)
  c11 ← h*f1(x,v1,v2)
  c21 ← h*f2(x,v1,v2)
  c12 ← h*f1(x + h/2,v1 + c11/2,v2 + c21/2)
  c22 ← h*f2(x + h/2,v1 + c11/2,v2 + c21/2)
  c13 ← h*f1(x + h/2,v1 + c12/2,v2 + c22/2)
  c23 ← h*f2(x + h/2,v1 + c12/2,v2 + c22/2)
  c14 ← h*f1(x + h,v1 + c13,v2 + c23)
  c24 ← h*f2(x + h,v1 + c13,v2 + c23)
  v1n ← v1 + (c11 + 2*(c12 + c13) + c14)/6
  v2n ← v2 + (c21 + 2*(c22 + c23) + c24)/6
endprocedure

procedure pc3(x, v1_o,v2_o,v1,v2,h → v1_n,v2_n)
  v1_p ← v1 + h*(3*f1(x,v1,v2) – f1(x – h,v1_o,v2_o))/2
  v2_p ← v2 + h*(3*f2(x,v1,v2) – f2(x – h,v1_o,v2_o))/2
  v1_n ← v1 + h*(5*f1(x + h,v1_p,v2_p)
            + 8*f1(x,v1,v2) – f1(x – h,v1_o,v2_o))/12
  v2_n ← v2 + h*(5*f2(x + h,v1_p, v2_p)
            + 8*f2(x,v1,v2) – f2(x – h,v1_o,v2_o))/12
endprocedure

procedure f1(x, theta,theta_dash → f1)
  f1 ← theta_dash;
endprocedure

procedure f2(x,theta,theta_dash → f2)
  if theta < tol then
    f2 ← –qconst
  else
    f2 ← pconst*exp(ln (theta)*1.25) – qconst
  endif
endprocedure
```

Figure 8.34 Subsidiary procedures for pseudocode algorithm for solving equation (8.42).

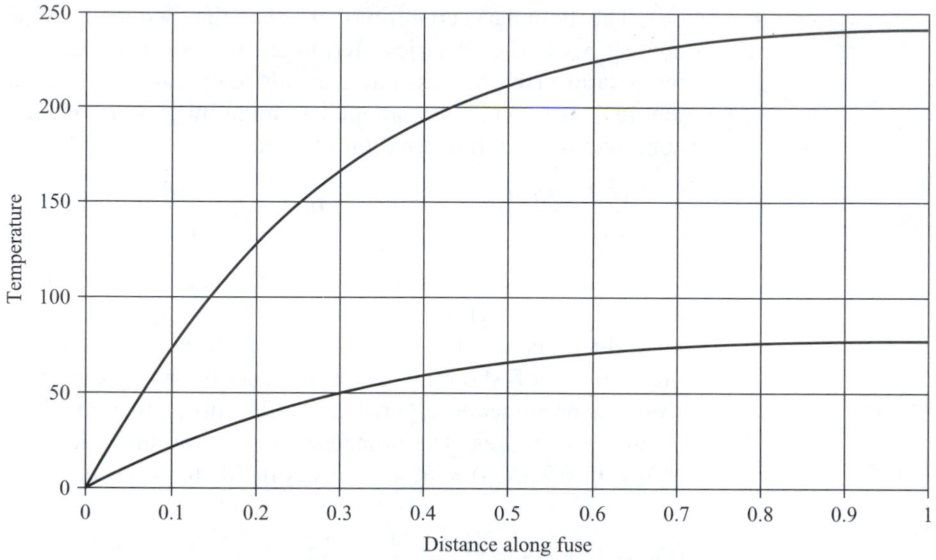

Figure 8.35 Comparison of temperatures in a fuse wire carrying 20 amps and 40 amps.

predictor-corrector method (with a single fourth-order Runge–Kutta step to start the multistep process). The main program uses the method of *regula falsa* to iterate from two starting values of θ_0' which bracket that value of θ_0' corresponding to $d\theta/dX(1) = 0$ which we seek.

Figure 8.35 shows the result of computations using a program based on the algorithm in Figure 8.33. Taking the values of the physical constants as $h = 100\,\mathrm{W\,m^{-2}\,K^{-1}}$, $a = 0.01\,\mathrm{m}$, $k = 63\,\mathrm{W\,m^{-1}\,K^{-1}}$, $\rho = 16 \times 10^{-8}\,\Omega\,\mathrm{m}$ and $r = 5 \times 10^{-4}\,\mathrm{m}$, and taking I as 20 amps and 40 amps gives the lower and upper curves in Figure 8.35 respectively.

Evidently at 20 amps the operating temperature of the middle part of the wire is about 77° above the ambient temperature. If the current increases to 40 amps the temperature increases to about 245° above ambient – just above the melting point of tin! The procedure could obviously be used to design and validate appropriate dimensions (length and diameter) for fuses made from a variety of metals for a variety of applications and rated currents.

8.6 Review exercises (1–10)

1 Solve the differential equation

$$\frac{dx}{dt} = \sqrt{\frac{xt}{x^2 + t^2}}, \quad x(0) = 1$$

to find the value of $X(0.4)$ using the Euler method with steps of size 0.1 and 0.05. By comparing the two estimates of $x(0.4)$ estimate the accuracy of the better of the two values which you have obtained and also the step size you would need to use in order to calculate an estimate of $x(0.4)$ accurate to two decimal places.

2 Solve the differential equation

$$\frac{dx}{dt} = \sin(t^2), \quad x(0) = 2$$

to find the value of $X(0.25)$ using the Euler method with steps of size 0.05 and 0.025. By comparing the two estimates of $x(0.25)$ estimate the accuracy of the better of the two values which you have obtained and also the step size you would need to use in order to calculate an estimate of $x(0.25)$ accurate to 3 decimal places.

†3 Let X_1, X_2 and X_3 denote the estimates of the function $x(t)$ satisfying the differential equation

$$\frac{dx}{dt} = \sqrt{(xt + t)}, \quad x(1) = 2$$

which are calculated using the second-order predictor–corrector method with steps of 0.1, 0.05 and 0.025 respectively. Compute $X_1(1.2)$, $X_2(1.2)$ and $X_3(1.2)$. Show that the ratio of $|X_2 - X_1|$ and $|X_3 - X_2|$ should tend to 4 : 1 as the step size tends to zero. Do your computations bear out this expectation?

†4 Compute the solution of the differential equation

$$\frac{dx}{dt} = e^{-xt}, \quad x(0) = 5$$

for $x = 0$ to 2 using the fourth-order Runge–Kutta method with step sizes of 0.2, 0.1 and 0.05. Estimate the accuracy of the most accurate of your three solutions.

5 In a thick cylinder subjected to internal pressure the radial pressure $p(r)$ at distance r from the axis of the cylinder is given by

$$p + r\frac{dp}{dr} = 2a - p$$

where a is a constant (which depends on the geometry of the cylinder).

If the stress has magnitude p_0 at the inner wall, $r = r_0$, and may be neglected at the outer wall, $r = r_1$, show that

$$p(r) = \frac{p_0 r_0^2}{r_1^2 - r_0^2}\left(\frac{r_1^2}{r^2} - 1\right)$$

If $r_0 = 1$, $r_1 = 2$ and $p_0 = 1$, compare the value of $p(1.5)$ obtained from this analytic solution with the numerical value obtained using the fourth-order Runge–Kutta method with step size $h = 0.5$. (*NB*: with these values of r_0, r_1 and p_0, $a = -1/3$).

†6 Find the values of $X(t)$ for t up to 2 where $X(t)$ is the solution of the differential equation problem

$$\frac{d^3x}{dt^3} + \left(\frac{d^2x}{dt^2}\right)^2 + 4\left(\frac{dx}{dt}\right)^2 - tx = \sin t,$$

$$x(1) = 0.2, \quad \frac{dx}{dt}(1) = 1, \quad \frac{d^2x}{dt^2}(1) = 0$$

using the Euler method with steps of 0.025. Repeat the computation with a step size of 0.0125. Hence estimate the accuracy of the value of $X(2)$ given by your solution.

†7 Find the solution of the differential equation problem

$$\frac{d^2x}{dt^2} + (x^2 - 1)\frac{dx}{dt} + 40x = 0,$$

$$x(0) = 0.02, \quad \frac{dx}{dt}(0) = 0$$

using the second-order predictor–corrector method. Hence find an estimate of the value of $x(4)$ accurate to 4 decimal places.

†8 Find the solution of the differential equation problem

$$\frac{d^3x}{dt^3} + \left|\frac{d^2x}{dt^2}\right|^{\frac{1}{2}} + 4\left(\frac{dx}{dt}\right)^3 - tx = \sin t,$$

$$x(1) = -1, \quad \frac{dx}{dt}(1) = 1, \quad \frac{d^2x}{dt^2}(1) = 2$$

using the fourth-order Runge–Kutta method. Hence find an estimate of the value of $x(2.5)$ accurate to 4 decimal places.

†9 (Extended, open-ended problem.) The second-order, nonlinear, ordinary differential equation

$$\frac{d^2x}{dt^2} + \mu(x^2 - 1)\frac{dx}{dt} + \lambda^2 x = 0$$

governs the oscillations of the Van der Pol oscillator. By scaling the time variable the equation can be reduced to

$$\frac{d^2x}{dt^2} + \mu(x^2 - 1)\frac{dx}{dt} + (2\pi)^2 x = 0$$

Investigate the properties of the Van der Pol oscillator. In particular show that the oscillator shows limit cycle behaviour (that is, the oscillations tend to a form which is independent of the initial conditions and depends only on the parameter μ). Determine the dependence of the limit cycle period on μ.

†**10** (Extended, open-ended problem.) The equation of simple harmonic motion

$$\frac{d^2x}{dt^2} + \lambda^2 x = 0$$

is generally used to model the undamped oscillations of a mass supported on the end of a linear spring (i.e. a spring whose tension is strictly proportional to its extension). Most real springs are actually nonlinear because as their extension or compression increases their stiffness changes. This can be modelled by the equation

$$\frac{d^2x}{dt^2} + 4\pi^2(1 + \beta x^2)x = 0$$

For a 'hard' spring stiffness increases with displacement ($\beta > 0$) and a soft spring's stiffness decreases ($\beta > 0$). Investigate the oscillations of a mass supported by a hard or soft spring. In particular determine the connection between the frequency of the oscillations and their amplitude.

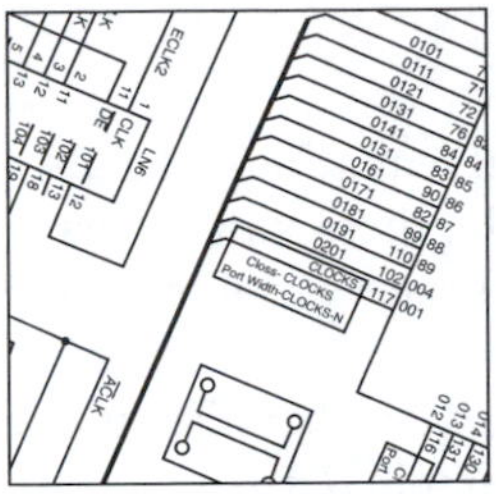
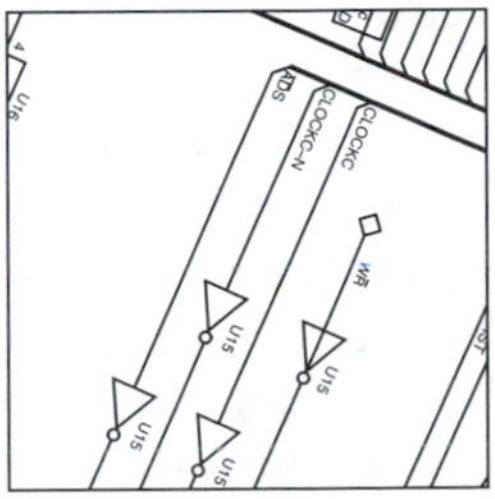
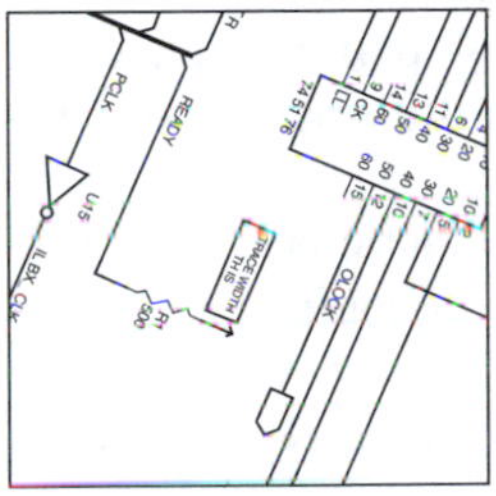

9

Partial Differential Equations

CONTENTS

9.1 Introduction

In Chapter 8 we considered the role of ordinary differential equations in engineering. However, many physical processes fundamental to science and engineering are governed by **partial differential equations**, that is, equations involving partial derivatives. The most familiar of these processes are heat conduction and wave propagation. To describe such phenomena, we make assumptions about gradients (for instance the Fourier law that heat flow is proportional to temperature gradient) and we write down balance equations; partial differential equations are thus produced in a natural way. Unless the situation is very simple, there will be many independent variables, for example a time variable t and a space variable x, and the differential equations *must* involve partial derivatives.

The application of partial differential equations is much wider than the simple situations already mentioned. Maxwell's equations (see Example 7.15) comprise a set of partial differential equations that form the basis of electromagnetic theory, and are fundamental to electrical engineers and physicists. The equations of fluid flow are partial differential equations, and are widely used in aeronautical engineering, acoustics, the study of groundwater flows in civil engineering, the development of most fluid handling devices used in mechanical engineering and in investigating flame and combustion processes in chemical engineering. Quantum mechanics is yet another theory governed by a partial differential equation, the Schrödinger equation, which forms the basis of much of physics, chemistry and electronic engineering. Stress analysis is important in large areas of civil and mechanical engineering, and again requires a complicated set of partial differential equations. This is by no means an exhaustive list, but it does illustrate the importance of partial differential equations and their solution.

One of the major difficulties with partial differential equations is that it is extremely difficult to illustrate their solutions geometrically, in contrast to single-variable problems, where a simple curve can be used. For instance, the temperature in a room, particularly if it is time-varying, is not at all easy to draw or visualize, but such information is of crucial importance to a heating engineer. A second basic problem with partial differential equations is that it is intrinsically more difficult to solve them or even to decide whether a solution exists. The driving force of most physical systems that can be modelled by partial differential equations is determined by either what happens on the boundary of the region under consideration or how the system is started at zero time. Boundaries, therefore, play a very significant role, and we shall see that a problem can have a solution for one set of boundary conditions but not for another. Finding *a* solution to a partial differential equation is often quite straightforward but finding *the* solution that fits the boundary conditions is very difficult.

The solution of partial differential equations has been greatly eased by the use of computers, which have allowed the rapid numerical solution of problems that would otherwise have been intractable. Such methods have generally been integrated into this chapter, since they are now one of the standard techniques available. However, the finite-element method is considered separately, since it is more complicated, and requires a lot of careful thought and work (the section dealing with it can be omitted on a first reading). The finite-element method originated in stress analysis in civil engineering work, but has now spread into most areas where complicated boundaries are encountered.

There are three basic types of equation that appear in most areas of science and engineering, and it is essential to understand their solutions before any progress can be made on more complicated sets of equations, nonlinear equations or equations with variable coefficients.

9.2 General discussion

The three basic types of equation are referred to as the **wave equation**, the **heat-conduction** or **diffusion equation** and the **Laplace equation**. In this section we briefly discuss the formulation of these three basic forms, and then consider each in more detail in later sections. The various sections will concentrate on finding and understanding solutions of the three types of equations in simple regions. The treatment of advanced methods, more complicated equations and other regions will be left to more comprehensive books on partial differential equations (see, for example, Dimitri D. Vvedensky, *Partial Differential Equations with Mathematica*. Addison-Wesley, 1993).

9.2.1 Wave equation

$$\frac{1}{c^2}\frac{\partial^2 u}{\partial t^2} = \frac{\partial^2 u}{\partial x^2} + \frac{\partial^2 u}{\partial y^2} + \frac{\partial^2 u}{\partial z^2} = \nabla^2 u \qquad (9.1)$$

Many phenomena that involve propagation of a signal require the wave equation (9.1) to be solved in the appropriate number of space dimensions. Perhaps the simplest, in one space dimension, is the vibration of a taut string stretched to a uniform tension T between two fixed points as illustrated in Figure 9.1(a), where u is the displacement, x is measured along the equilibrium position of the string and t is time. Applying Newton's law of motion to an element Δs of the string (Figure 9.1b), for motion in the u direction, we have

net force in u direction = mass element × acceleration in u direction

that is,

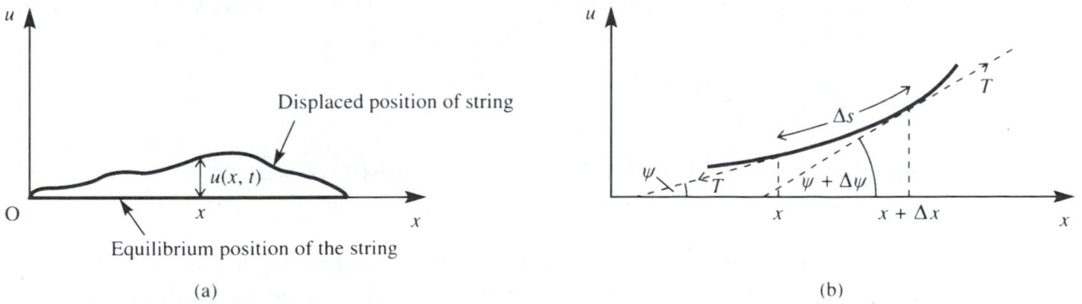

Figure 9.1 Displacement of an element of a taut string.

$$T\sin(\psi+\Delta\psi) - T\sin\psi = \rho\Delta s\frac{\partial^2 u}{\partial t^2} \tag{9.2}$$

where ρ is the mass per unit length of the string. Neglecting terms quadratic in small quantities and using the Taylor series expansions

$$\cos\Delta\psi = 1 + O(\Delta\psi^2) \simeq 1,\ \sin\Delta\psi = \Delta\psi + O(\Delta\psi^3) \simeq \Delta\psi$$

and the expression

$$\Delta s = \sqrt{\left[1+\left(\frac{\partial u}{\partial x}\right)^2\right]}\Delta x \simeq \Delta x$$

for the arclength, (9.2) becomes

$$T\sin\psi + T\cos\psi\Delta\psi - T\sin\psi = \rho\Delta x\frac{\partial^2 u}{\partial t^2}$$

or

$$T\cos\psi\frac{\Delta\psi}{\Delta x} = \rho\Delta x\frac{\partial^2 u}{\partial t^2}$$

which in the limit as $\Delta x \to 0$ becomes

$$T\cos\psi\frac{\partial\psi}{\partial x} = \rho\frac{\partial^2 u}{\partial t^2} \tag{9.3}$$

Again assuming that ψ itself is small for small oscillations of the string, we have $\cos\psi \simeq 1$, and the gradient of the string

$$\frac{\partial u}{\partial x} = \tan\psi \simeq \psi$$

and hence from (9.3) we obtain

$$T\frac{\partial^2 u}{\partial x^2} = \rho\frac{\partial^2 u}{\partial t^2}$$

Thus the displacement of the string satisfies the one-dimensional wave equation

$$\frac{1}{c^2}\frac{\partial^2 u}{\partial t^2} = \frac{\partial^2 u}{\partial x^2} \tag{9.4}$$

and the propagation of the disturbance in the string is given by a solution of this equation where $c^2 = T/\rho$.

By considering the theory of small displacements of a compressible fluid, sound waves can likewise be shown to propagate according to (9.1). The one-dimensional form (9.4) will model the propagation of sound in an organ pipe, while the spherically symmetric version of (9.1) will give a solution for waves emanating from an explosion. Because it is known that most wave phenomena satisfy the wave equation, it is reasonable, from a physical standpoint, that the propagation of electromagnetic waves will also satisfy (9.1). A careful analysis of Maxwell's equations in free space is required to show this result (see Example 7.15). We could give further examples of physical phenomena that have (9.1) as a basic equation, but we have described enough here to establish its importance and the need to look at methods of solution. An aspect of the wave equation that is not often discussed is its bad behaviour. Any discontinuities in a variable or its derivatives will, according to the wave equation, propagate with time. An obvious physical manifestation of this is a shock wave. When an aircraft breaks the sound barrier, a shock is produced and the sonic boom can be heard many miles away. How the shock is produced is a complicated nonlinear effect, but once it has been produced it propagates according to the wave equation.

EXAMPLE 9.1

Show that

$$u = u_0 \sin\left(\frac{\pi x}{L}\right)\cos\left(\frac{\pi ct}{L}\right)$$

satisfies the one-dimensional wave equation and the conditions

(a) a given initial displacement $u(x, 0) = u_0 \sin(\pi x/L)$, and

(b) zero initial velocity, $\partial u(x, 0)/\partial t = 0$.

Solution Clearly the condition (a) is satisfied by inspection. If we now partially differentiate u with respect to t,

$$\frac{\partial u}{\partial t} = -\frac{u_0 \pi c}{L}\sin\left(\frac{\pi x}{L}\right)\sin\left(\frac{\pi ct}{L}\right)$$

so that at $t = 0$ we have $\partial u/\partial t = 0$ and (b) is satisfied.

It remains to show that (9.4) is also satisfied. Using the standard subscript notation for partial derivatives,

$$u_{xx} = \frac{\partial^2 u}{\partial x^2} = -\frac{u_0 \pi^2}{L^2}\sin\left(\frac{\pi x}{L}\right)\cos\left(\frac{\pi ct}{L}\right)$$

$$u_{tt} = \frac{\partial^2 u}{\partial t^2} = -\frac{u_0 \pi^2 c^2}{L^2}\sin\left(\frac{\pi x}{L}\right)\cos\left(\frac{\pi ct}{L}\right)$$

so that the equation is indeed satisfied.

This solution corresponds physically to the fundamental mode of vibration of a taut string plucked at its centre.

EXAMPLE 9.2

Verify that the function

$$u = a\exp\left[-\left(\frac{x}{h} - \frac{ct}{h}\right)^2\right]$$

satisfies the wave equation (9.4). Sketch the graphs of the solution u against x at $t = 0$, $t = 2h/c$ and $t = 4h/c$.

Solution Evaluate the partial derivatives as

$$u_x = \frac{-2a(x-ct)}{h^2}\exp\left[-\left(\frac{x}{h}-\frac{ct}{h}\right)^2\right]$$

$$u_t = \frac{2ac(x-ct)}{h^2}\exp\left[-\left(\frac{x}{h}-\frac{ct}{h}\right)^2\right]$$

and

$$u_{xx} = \frac{-2a}{h^2}\exp\left[-\left(\frac{x}{h}-\frac{ct}{h}\right)^2\right] + \frac{4a(x-ct)^2}{h^4}\exp\left[-\left(\frac{x}{h}-\frac{ct}{h}\right)^2\right]$$

$$u_{tt} = \frac{-2ac^2}{h^2}\exp\left[-\left(\frac{x}{h}-\frac{ct}{h}\right)^2\right] + \frac{4a(x-ct)^2c^2}{h^4}\exp\left[-\left(\frac{x}{h}-\frac{ct}{h}\right)^2\right]$$

Clearly (9.4) is satisfied by these second derivatives.

The curves of u against x are plotted in Figure 9.2, and show a wave initially centred at the origin moving with a constant speed c to the right.

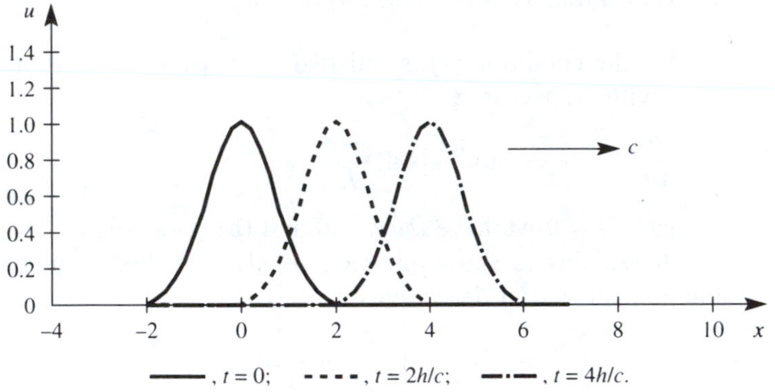

Figure 9.2 Propagating wave in Example 9.2.

9.2.2 Heat-conduction or diffusion equation

$$\frac{1}{\kappa}\frac{\partial u}{\partial t} = \nabla^2 u \tag{9.5}$$

This equation arises most commonly when heat is transferred from a hot area to a cold one by conduction, when the temperature satisfies (9.5).

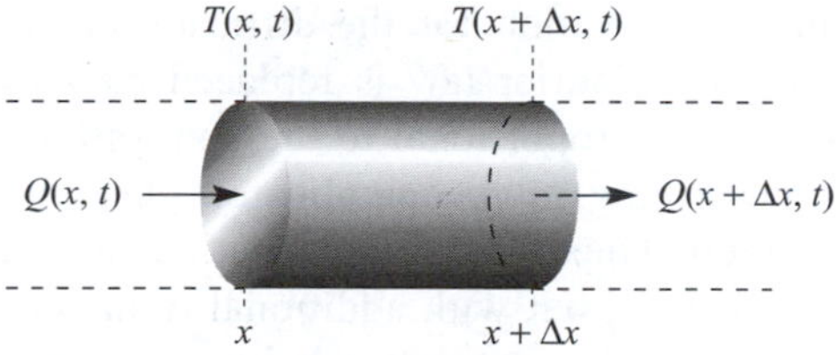

Figure 9.3 Heat flow in an element.

In Section 7.6 a full derivation of the equation (9.5) is made. Here we shall investigate the one-dimensional version in the context of the heat flow along a thin bar. The bar is assumed to have a uniform cross-sectional area and an insulated outer surface through which no heat is lost. It is also assumed that, at any cross-section x = constant, the temperature $T(x, t)$ is uniform. Consider an element of the bar from x to $x + \Delta x$, where x is measured along the length of the bar, as illustrated in Figure 9.3. An amount of heat $Q(x, t)$ per unit time per unit area enters the left-hand face and an amount $Q(x + \Delta x, t)$ leaves the right-hand face of the element. The net increase per unit cross-sectional area in unit time is

$$Q(x, t) - Q(x + \Delta x, t)$$

If c is the specific heat of the bar and ρ is its density then the amount of heat in the element is $c\rho T\Delta x$. The net increase in heat in the element in unit time is

$$c\rho\frac{\partial T}{\partial t}\Delta x$$

and is equated to the net amount entering. Thus

$$c\rho\frac{\partial T}{\partial t}\Delta x = Q(x, t) - Q(x + \Delta x, t)$$

which in the limit as $\Delta x \to 0$ gives

$$c\rho\frac{\partial T}{\partial t} = -\frac{\partial Q}{\partial x} \tag{9.6}$$

The Fourier law for the conduction of heat states that the heat transferred across unit area is proportional to the temperature gradient. Thus

$$Q = -k\frac{\partial T}{\partial x}$$

where k is the thermal conductivity and the minus sign takes into account the fact that heat flows from hot to cold. Substitution for Q in (9.6) gives the **one-dimensional heat equation**

$$\frac{\partial T}{\partial t} = \kappa\frac{\partial^2 T}{\partial x^2} \tag{9.7}$$

where $\kappa = k/c\rho$ is called the **thermal diffusivity**.

An entirely similar derivation for the diffusion equation can be made. The only difference is that the Fourier law is replaced by Fick's law that the diffusional flow of a material is proportional to the concentration gradient.

The equations describing more complicated phenomena, such as the time-dependent electromagnetic equations or the equations of fluid mechanics, have the same basic structure as (9.5), but with additional terms or with coupling to other equations of the same type. We certainly need to know how to solve (9.5) before even contemplating solving these more complex versions.

An essential feature of the heat-conduction equation is that, given a long enough time and assuming that there are no time varying inputs, the temperature will eventually settle down to a steady state. Thus the final solution is independent of time, and hence will satisfy $\partial u/\partial t = 0$ or $\nabla^2 u = 0$. The transient behaviour tells how this solution is approached from its given starting value. Physically it is reasonable that any initial temperature, however complicated, will move to a smooth final solution, and we should not expect the severe difficulties with discontinuities that occur with the wave equation. Exactly how initial discontinuities are treated in a numerical solution, however, can affect the accuracy in the early development of the solution.

EXAMPLE 9.3 Show that

$$T = T_\infty + (T_m - T_\infty)\,\mathrm{e}^{-U(x-Ut)/\kappa} \quad (x \geqslant Ut)$$

satisfies the one-dimensional heat-conduction equation (9.7), together with the boundary conditions $T \to T_\infty$ as $x \to \infty$ and $T = T_m$ at $x = Ut$.

Solution The second term vanishes as $x \to \infty$, for any fixed t, and hence $T \to T_\infty$. When $x = Ut$, the exponential term is unity, so the T_∞s cancel and $T = T_m$. Hence the two boundary conditions are satisfied. Checking both sides of the heat-conduction equation (9.7),

$$\frac{1}{\kappa}\frac{\partial T}{\partial t} = \frac{1}{\kappa}(T_m - T_\infty)\frac{U^2}{\kappa}\,\mathrm{e}^{-U(x-Ut)/\kappa}$$

$$\frac{\partial^2 T}{\partial x^2} = (T_m - T_\infty)\frac{U^2}{\kappa^2}\,\mathrm{e}^{-U(x-Ut)/\kappa}$$

which are obviously equal, so that the equation is satisfied.

The example models a block of material being melted at a temperature T_m, with the melting boundary having constant speed U, and with a steady temperature T_∞ at great distances. An application of this model would be a heat shield on a re-entry capsule ablated by frictional heating.

EXAMPLE 9.4 Show that the function

$$T = \frac{1}{\sqrt{t}} \exp\left(-\frac{x^2}{4\kappa t}\right)$$

satisfies the one-dimensional heat-conduction equation (9.7). Plot T against x for various times t, and comment.

Solution We first calculate the partial derivatives

$$\frac{\partial T}{\partial t} = -\frac{1}{2}\frac{1}{t^{3/2}} \exp\left(\frac{-x^2}{4\kappa t}\right) + \frac{1}{\sqrt{t}}\frac{-x^2}{4\kappa}\frac{-1}{t^2} \exp\left(\frac{-x^2}{4\kappa t}\right)$$

$$\frac{\partial T}{\partial x} = \frac{1}{\sqrt{t}}\frac{-2x}{4\kappa t} \exp\left(\frac{-x^2}{4\kappa t}\right)$$

and

$$\frac{\partial^2 T}{\partial x^2} = \frac{-1}{2\kappa t^{3/2}} \exp\left(\frac{-x^2}{4\kappa t}\right) + \frac{-x}{2\kappa t^{3/2}}\frac{-2x}{4\kappa t} \exp\left(\frac{-x^2}{4\kappa t}\right)$$

It is easily checked that (9.7) is satisfied except at the time $t = 0$, where T is not properly defined. The graph of reduced temperature $T/\sqrt{(4\kappa)}$ against distance x at various times $t = L^2/4\kappa$ can be seen in Figure 9.4. Physically, the problem corresponds to a very hot weld being applied instantaneously to the bar. The initial temperature 'spike' at $x = 0$ is seen to spread out as time progresses, and, as expected from the physical interpretation, T tends to zero for all x as the time becomes large.

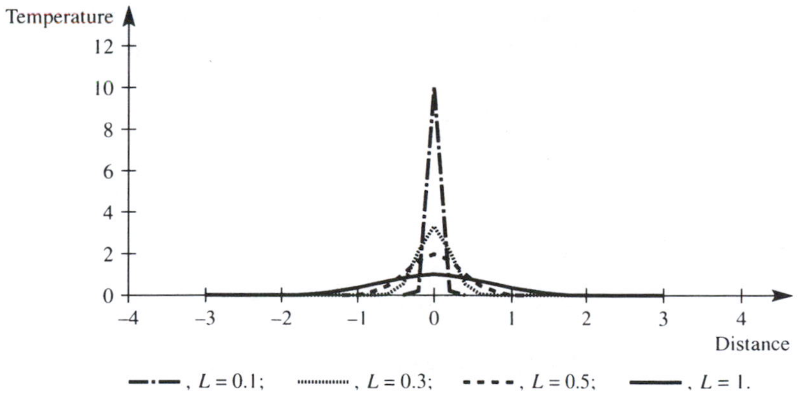

Figure 9.4 Solution of the heat-conduction equation starting from an initial spike in Example 9.4.

9.2.3 Laplace equation

$$\nabla^2 u = 0 \tag{9.8}$$

The simplest physical interpretation of this equation has already been mentioned, namely as the steady-state heat equation. So, for example, the two-dimensional Laplace equation

$$\frac{\partial^2 T}{\partial x^2} + \frac{\partial^2 T}{\partial y^2} = 0$$

could represent the steady-state distribution of temperature over a thin rectangular plate in the (x, y) plane.

Heat transfer is well understood intuitively, and good guesses at steady-state solutions can usually be made. Perhaps less commonly understood is the case of the electrostatic potential in a uniform dielectric, which also satisfies the Laplace equation. Working out the electrical behaviour of a capacitor that is charged in a certain manner simply implies solving (9.8) subject to appropriate boundary conditions. Possibly the least obvious, but extremely important, application of the Laplace equation is in inviscid, irrotational fluid mechanics. To a large extent, subsonic aerodynamics is based on (9.8) as an approximate model. The lift and drag on an aerofoil in a fluid stream can be evaluated accurately from suitable solutions of this equation. It is only close to the aerofoil that viscous and rotational effects become important.

The Laplace equation is a 'smoother' in the sense that it irons out peaks and troughs. Physically, the steady-state heat-conduction context tells us that if a particular point has a higher temperature than neighbouring points then heat will flow from hot to cold until the 'hot spot' is eliminated. Thus there are no interior points at which the solution u of (9.8) is smaller or larger than all of its neighbours. This result can be confirmed mathematically, and establishes that smooth solutions are obtained.

EXAMPLE 9.5 Show that

$$u = x^4 - 2x^3y - 6x^2y^2 + 2xy^3 + y^4$$

satisfies the Laplace equation.

Solution Differentiating

$$u_x = 4x^3 - 6x^2y - 12xy^2 + 2y^3, \quad u_y = -2x^3 - 12x^2y + 6xy^2 + 4y^3$$

$$u_{xx} = 12x^2 - 12xy - 12y^2, \quad u_{yy} = -12x^2 + 12xy + 12y^2$$

so clearly

$$u_{xx} + u_{yy} = 0$$

and the two-dimensional Laplace equation is satisfied.

EXAMPLE 9.6 Show that the function

$$\psi = Uy\left(1 - \frac{a^2}{x^2 + y^2}\right)$$

satisfies the Laplace equation, and sketch the curves ψ = constant.

Solution First calculate the partial derivatives:

$$\psi_x = \frac{2xyUa^2}{(x^2 + y^2)^2}$$

$$\psi_y = U - \frac{Ua^2}{x^2 + y^2} + \frac{2y^2Ua^2}{(x^2 + y^2)^2}$$

$$\psi_{xx} = \frac{2yUa^2}{(x^2 + y^2)^2} - \frac{8x^2yUa^2}{(x^2 + y^2)^3}$$

$$\psi_{yy} = \frac{2yUa^2}{(x^2 + y^2)^2} + \frac{4yUa^2}{(x^2 + y^2)^2} - \frac{8y^3Ua^2}{(x^2 + y^2)^3}$$

Substituting into (9.8) gives

$$\nabla^2\psi = \frac{8yUa^2}{(x^2 + y^2)^2} - \frac{8y(x^2 + y^2)Ua^2}{(x^2 + y^2)^3} = 0$$

and hence the Laplace equation is satisfied.

Secondly, to sketch the contours, we note that $\psi = 0$ on $y = 0$ and on the circle $x^2 + y^2 = a^2$. On keeping $y = y_0$ and letting $x \to \pm\infty$, the second term vanishes, so the curves tend to $\psi = Uy_0$. Figure 9.5 shows the solution, which

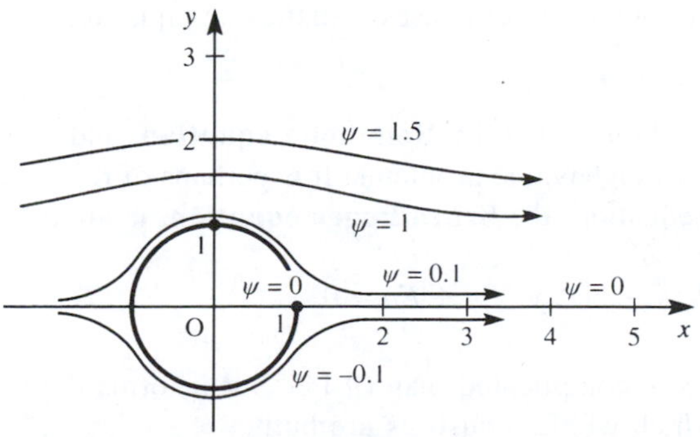

Figure 9.5 Streamlines for flow past a cylinder of radius 1, from the Laplace equation in Example 9.6.

corresponds physically to the flow of an inviscid, irrotational fluid past a cylinder placed in a uniform stream.

9.2.4 Other and related equations

We discussed in Section 9.1 applications in science and engineering. Many such applications are governed by equations that are closely related to the three basic equations discussed above. For example, consider the equations of slow, steady, viscous flow in two dimensions, which take the form

$$\left.\begin{aligned} &\frac{\partial p}{\partial x} = \frac{1}{\mathcal{R}}\nabla^2 u, \quad \frac{\partial p}{\partial y} = \frac{1}{\mathcal{R}}\nabla^2 v \\ &\frac{\partial u}{\partial x} + \frac{\partial v}{\partial y} = 0, \end{aligned}\right\} \tag{9.9}$$

where u, v and p are the non-dimensional velocities and pressure, and $\mathcal{R}$ is the Reynolds number. The system has a familiar look about it, and indeed a little simple manipulation gives $\nabla^2 p = 0$, so that the pressure satisfies the Laplace equation. If p can be calculated then $\partial p/\partial x$ and $\partial p/\partial y$ are known, so we have equations of the form

$$\nabla^2 u = f(x, y) \tag{9.10}$$

This equation is called the **Poisson equation**, and is clearly closely related to the Laplace equation. It can be interpreted physically as steady heat conduction with heat sources in the region. A careful study of the solution of the Laplace equation is required before either (9.10) or (9.9) can be attacked.

If there is good knowledge about the time behaviour of the wave or diffusion equation then we can often obtain important information from them without solving the full equations. For instance, if we put a periodic time dependence $u = e^{j\alpha t}v(x, y, z)$ into (9.1), or if we put an exponentially decaying solution $u = e^{-\beta t}v(x, y, z)$ into (9.5) then the variable v, in both cases, satisfies an equation of the form

$$\nabla^2 v + \lambda v = 0 \tag{9.11}$$

Equation (9.11) is called the **Helmholtz equation**, and plays an important role in the solution of eigenvalue problems. It is perhaps of relevance that the best studied eigenvalue equation, the **Schrödinger equation**, is almost the same, namely

$$\frac{h^2}{8\pi^2 m}\nabla^2 u - V(x, y, z)u + Eu = 0$$

It is a bit more complicated than (9.11), but it forms the basis of quantum mechanics, on which whole industries are built.

The three classical partial differential equations discussed above are all of second order. First-order equations are rarely of much interest in applications in science and engineering. There is, however, a well-developed theory that is much more comprehensive than for second order. One example will give an idea of the power of the theory, but it is left to specialist books to describe the detail.

EXAMPLE 9.7 Find a general solution to the equation

$$\frac{\partial u}{\partial t} + x\frac{\partial u}{\partial x} - y\frac{\partial u}{\partial y} = 0$$

and show that u is constant on the curves xy = constant.

Solution The theory of first-order equations indicates that to solve the equation

$$a\frac{\partial u}{\partial t} + b\frac{\partial u}{\partial x} + c\frac{\partial u}{\partial y} = p$$

we should look at solutions of the ordinary differential equations

$$\frac{\mathrm{d}t}{a} = \frac{\mathrm{d}x}{b} = \frac{\mathrm{d}y}{c} = \frac{\mathrm{d}u}{p}$$

In the current problem these equations are

$$\frac{\mathrm{d}t}{1} = \frac{\mathrm{d}x}{x} = \frac{\mathrm{d}y}{-y} = \frac{\mathrm{d}u}{0}$$

which have a solution $u = C$, $x = A\,\mathrm{e}^t$ and $y = B\,\mathrm{e}^{-t}$, where A, B and C are arbitrary constants.

The variable t can be eliminated to give u = constant on the rectangular hyperbolas $xy = D$. Provided u is known, from the initial conditions, at one point of the curve $xy = D$ then it is known on the curve for all time.

To check the solution independently look for a solution

$$u = g(t, xy)$$

Taking $Y = xy$ and putting this form into the equation gives

$$0 = u_t + xu_x - yu_y = g_t + xyg_Y - yxg_Y$$

and hence $g_t = 0$. Thus g is independent of t explicitly, and u is only a function of the combination xy. If xy = constant then u is fixed on this curve.

A physical interpretation of this result can be given as follows. If an impurity with concentration u is being transported by a fluid with velocity given by $v = (x, -y)$ then, for the case when diffusion can be neglected, the impurity concentration satisfies the equation of the problem. Essentially the solution tells us that the impurities are transported along the hyperbolas xy = constant.

So far, all of the equations that we have considered are *linear*, since they have not included any quadratic (or higher) terms in u or its derivatives. As soon as we move from linear to *nonlinear* problems, a whole new crop of theoretical and computational difficulties arises. Very few such equations can be solved analytically, and devising computational schemes is not easy. Even worse, mathematicians cannot always tell whether or not a solution even exists. An act of faith is usually made by scientists and engineers that their problem is modelled correctly and therefore there must be a mathematical solution reflecting the physics. Often the faith is well founded, but modelling is an imperfect art and there are many things that can go wrong. It may be thought that nonlinear problems do not occur in practice, but this is certainly not the case. For some phenomena, like the behaviour of thermionic valves or avalanche semiconductors or pulsed lasers, it is the nonlinearity that produces the desired effects. Other situations arise where the nonlinearity of the system may or may not be important. For instance, the full steady two-dimensional fluid equations are

$$\left.\begin{aligned} u\frac{\partial u}{\partial x}+v\frac{\partial u}{\partial y} &= -\frac{\partial p}{\partial x}+\frac{1}{\mathcal{R}}\nabla^2 u \\ u\frac{\partial v}{\partial x}+v\frac{\partial v}{\partial y} &= -\frac{\partial p}{\partial y}+\frac{1}{\mathcal{R}}\nabla^2 v \\ \frac{\partial u}{\partial x}+\frac{\partial v}{\partial y} &= 0 \end{aligned}\right\} \tag{9.12}$$

where u, v, p and $\mathcal{R}$ are defined as for (9.9). These equations are *nonlinear* because of the presence of quadratic terms such as $u\,\partial u/\partial x$. It can be seen that (9.12) reduce to (9.9) for slow flow when quadratic terms are neglected. While (9.9) would be applicable to the flow of molten glass, we would need the full equations (9.12) to look at flow close to an aerofoil. Indeed, as $\mathcal{R}$ becomes large, the flow becomes turbulent, that is, unstable, and the applicability of these equations comes into question.

In each of the examples in this section, a solution has been given; it has been checked that the solution satisfies the appropriate partial differential equation. In no case has the boundary condition been part of the specification of the problem, although in several cases boundary conditions were checked. In the next sections the boundary conditions are given as part of the set up of the example. This is the natural way that a physical problem is specified and it proves to be a much tougher proposition.

9.2.5 Exercises

1 Find the possible values of a and b in the expression

$$u = \cos at \sin bx$$

such that it satisfies the wave equation

$$\frac{1}{c^2}\frac{\partial^2 u}{\partial t^2} = \frac{\partial^2 u}{\partial x^2}$$

2 Taking

$$u = f(x + \alpha t)$$

where f is any function, find the values of α that will ensure that u satisfies the wave equation

$$\frac{1}{c^2}\frac{\partial^2 u}{\partial t^2} = \frac{\partial^2 u}{\partial x^2}$$

3 Find all the possible solutions of the heat-conduction equation

$$\frac{1}{\kappa}\frac{\partial u}{\partial t} = \frac{\partial^2 u}{\partial x^2}$$

of the form

$$u(x, t) = e^{\alpha t}V(x)$$

4 Find the values of the constant n for which

$$V = r^n(3\cos^2\theta - 1)$$

satisfies the Laplace equation (in spherical polar coordinates and independent of ϕ)

$$\frac{\partial}{\partial r}\left(r^2\frac{\partial V}{\partial r}\right) + \frac{1}{\sin\theta}\frac{\partial}{\partial\theta}\left(\sin\theta\frac{\partial V}{\partial\theta}\right) = 0$$

for all values of the variables r and θ.

5 Show that $f(x, y) = x^2y^2 + g(x/y)$ satisfies the partial differential equation

$$x\frac{\partial f}{\partial x} + y\frac{\partial f}{\partial y} = 4x^2y^2$$

for any arbitrary function g. It is given that $f = t^2$ on the line with parametric equation $x = 1 - t$, $y = t$; find the function g.

6 The transmission-line equations represent the flow of current along a long, leaky wire such as a trans-Atlantic cable. The equations take the form

$$-\frac{\partial I}{\partial x} = gv + c\frac{\partial v}{\partial t}$$

$$-\frac{\partial v}{\partial x} = rI + L\frac{\partial I}{\partial t}$$

where g, c, r and L are constants and I and v are the current and voltage respectively.

(a) Show that when $r = g = 0$, the equations reduce to the wave equation.
(b) Show that when $L = 0$, the equations reduce to a heat-conduction equation with a forcing term. Write $W = v\,e^{gt/c}$ to reduce to the normal form of the equation.
(c) Put $a = \frac{1}{2}(r/L + g/c)$ and then $w = v\,e^{at}$. Show that when $rc = gL$, w satisfies the wave equation.

7 If $V = x^3 + axy^2$, where a is a constant, show that

$$x\frac{\partial V}{\partial x} + y\frac{\partial V}{\partial y} = 3V$$

Find the value of a if V is to satisfy the equation

$$\frac{\partial^2 V}{\partial x^2} + \frac{\partial^2 V}{\partial y^2} = 0$$

Taking this value of a, show that if $u = r^3V$, where $r^2 = x^2 + y^2$, then

$$\frac{\partial^2 u}{\partial x^2} + \frac{\partial^2 u}{\partial y^2} = 27rV$$

8 Show that $u = e^{-kt}\cos mx\cos nt$ is a solution of the equation

$$c^2\frac{\partial^2 u}{\partial x^2} = \frac{\partial^2 u}{\partial t^2} + 2k\frac{\partial u}{\partial t}$$

provided that the constants k, m, n and c are related by the equation $n^2 + k^2 = c^2m^2$.

9 The **telegraph equation** has the form

$$\frac{\partial^2\phi}{\partial x^2} = \frac{1}{c^2}\left(\frac{\partial^2\phi}{\partial t^2} + k\frac{\partial\phi}{\partial t}\right)$$

where c^2 is the speed of light and k is usually small. Given that $\Phi(x, t)$ is a solution of the wave equation

$$\frac{\partial^2\Phi}{\partial x^2} = \frac{1}{c^2}\frac{\partial^2\Phi}{\partial t^2}$$

show that $\phi(x, t) = \Phi(x, t)\,e^{-kt/2}$ is a solution of the telegraph equation, if terms of order k^2 can be neglected.

10 Show that if f is a function of x only then

$$u = f(x)\sin(ay + b)$$

where a and b are constants, is a solution of the partial differential equation

$$\frac{\partial^2 u}{\partial y^2} = \frac{\partial^2 u}{\partial x^2} - 2a\frac{\partial u}{\partial x}$$

provided that $f(x)$ satisfies the ordinary differential equation

$$\frac{d^2 f}{dx^2} - 2a\frac{df}{dx} + a^2 f = 0$$

Hence show that

$$u = (A + Bx)\,e^{ax}\sin(ay + b)$$

where A and B are arbitrary constants, is a solution of the partial differential equation.

9.3 Solution of the wave equation

In this section we consider methods of solving the wave equation introduced in Section 9.2.1.

9.3.1 d'Alembert solution and characteristics

A classical solution of the one-dimensional wave equation

$$\frac{1}{c^2}\frac{\partial^2 u}{\partial t^2} = \frac{\partial^2 u}{\partial x^2} \tag{9.4}$$

is obtained by changing the axes to reduce the equation to a particularly simple form. Let

$$r = x + ct, \quad s = x - ct$$

Then, using the chain rule procedure for transformation of coordinates (see Section 7.1.1),

$$u_{xx} = u_{rr} + 2u_{rs} + u_{ss}$$

$$u_{tt} = c^2(u_{rr} - 2u_{rs} + u_{ss})$$

so that the wave equation (9.4) becomes

$$4c^2 u_{rs} = 0$$

This equation can now be integrated once with respect to s to give

$$u_r = \frac{\partial u}{\partial r} = \theta(r)$$

where θ is an arbitrary function of r. Now, integrating with respect to r, we obtain

$$u = f(r) + g(s)$$

which, on substituting for r and s, gives the solution of the wave equation (9.4) as

$$u = f(x + ct) + g(x - ct) \tag{9.13}$$

where f and g are *arbitrary functions* and f is just the integral of the arbitrary function θ.

The solution (9.13) is one of the few cases where the general solution of a partial differential equation can be found. However, finding the precise form of the arbitrary functions f and g that satisfy given initial data is not always easy. The initial conditions must give just enough information to evaluate f and g, which are functions of the *single* variables $r = x + ct$ and $s = x - ct$ respectively.

In Example 9.2 we have already seen a simple example of a wave of this type. We first deduced that a function of $x - ct$ satisfied the wave equation, and then showed in Figure 9.2 that it represented a wave travelling in the x direction with velocity c.

The next example is similar.

EXAMPLE 9.8

Check that $u = 1/[1 + (x + ct)^2]$ satisfies the wave equation (9.4) and show that it represents a travelling wave in the $-x$ direction.

Solution Differentiating partially with respect to x and t

$$u_x = \frac{-2(x+ct)}{[1+(x+ct)^2]^2}, \quad u_{xx} = \frac{2[-1+3(x+ct)^2]}{[1+(x+ct)^2]^3}$$

$$u_t = \frac{-2c(x+ct)}{[1+(x+ct)^2]^2}, \quad u_{tt} = \frac{2c^2[-1+3(x+ct)^2]}{[1+(x+ct)^2]^3}$$

and the wave equation is satisfied. Plots of the function u against x for various values of ct are shown in Figure 9.6. The same curve can be seen to be just translated to the left.

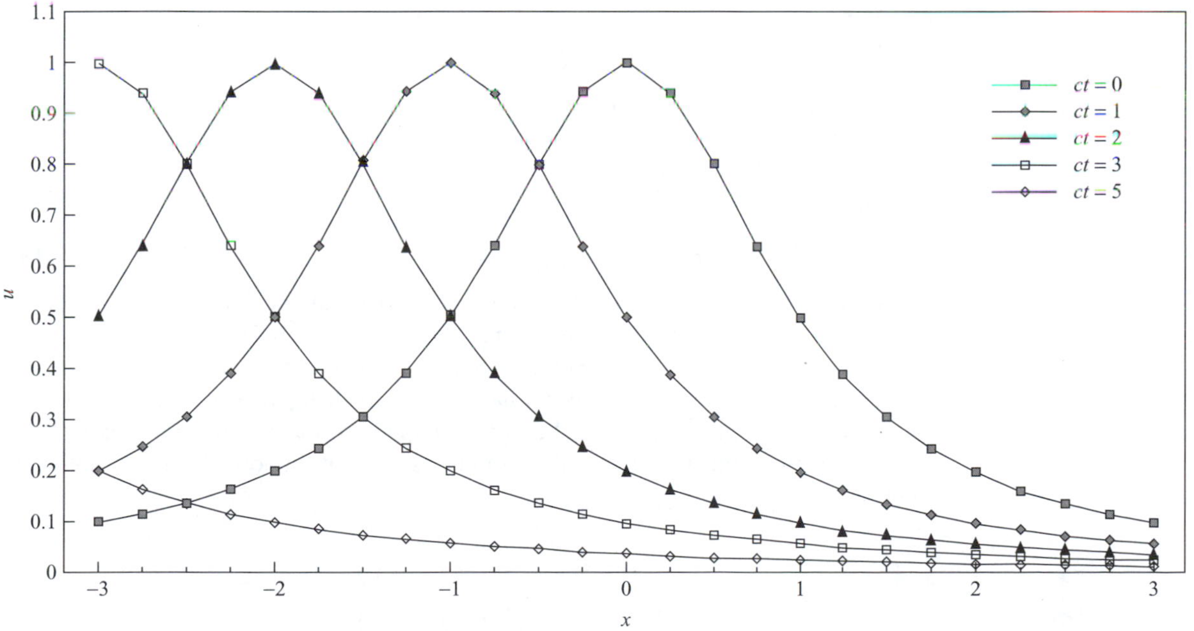

Figure 9.6 Solution to Example 9.8 showing u against x for various values of ct.

In Example 9.9 we attempt the more difficult task of fitting initial conditions to the solution.

EXAMPLE 9.9 Solve the wave equation (9.4) subject to the conditions

(a) zero initial velocity, $\partial u(x, 0)/\partial t = 0$ for all x, and

(b) an initial displacement given by

$$u(x, 0) = F(x) = \begin{cases} 1 - x & (0 \leqslant x \leqslant 1) \\ 1 + x & (-1 \leqslant x \leqslant 0) \\ 0 & \text{otherwise} \end{cases}$$

Solution This example corresponds physically to an infinite string initially at rest, and displaced as in Figure 9.7, which is then released.

From (9.13) we have a solution of the wave equation as

$$u = f(x + ct) + g(x - ct)$$

We now fit the given boundary data. Condition (a) gives

$$0 = cf'(x) - cg'(x) \quad \text{for all } x$$

so that

$$f(x) - g(x) = K = \text{an arbitrary constant}$$

and thus

$$u = f(x + ct) + f(x - ct) - K$$

Similarly, condition (b) gives

$$F(x) = 2f(x) - K$$

so that

$$u = \tfrac{1}{2}F(x + ct) + \tfrac{1}{2}F(x - ct) \tag{9.14}$$

We now have the solution to the equation in terms of the function F defined in condition (b). (Note that the same is true for any function F.)

The solution is plotted in Figure 9.8 as u against x for given times. It may be observed from this example that we have two **travelling waves**, one propagating to the right and one to the left. The initial shape is propagated exactly, except for a factor of two, and the shape discontinuities are not smoothed out, as noted in Section 9.2.1.

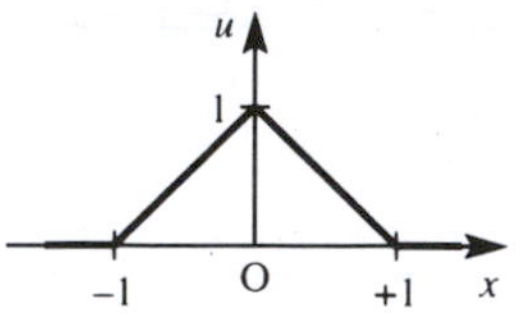

Figure 9.7 Initial displacement in Example 9.9.

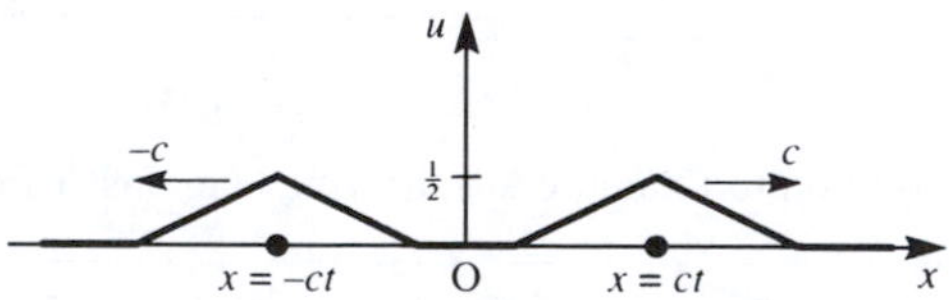

Figure 9.8 Solution to Example 9.9 showing two waves propagating in the $+x$ and $-x$ directions with velocity c.

The analysis in Example 9.9 can be extended to solve the wave equation subject to the general conditions

(a) an initial velocity, $\partial u(x, 0)/\partial t = G(x)$, and

(b) an initial displacement, $u(x, 0) = F(x)$ for all x.

Condition (a) gives, from (9.13)

$$G(x) = c[f'(x) - g'(x)]$$

so that

$$c[f(x) - g(x)] = \int_0^x G(x)\,\mathrm{d}x + Kc$$

Condition (b) gives

$$f(x) + g(x) = F(x)$$

and we can solve for $f(x)$ and $g(x)$ as

$$f(x) = \frac{1}{2}F(x) + \frac{1}{2c}\int_0^x G(x)\,\mathrm{d}x + \frac{1}{2}K$$

$$g(x) = \frac{1}{2}F(x) - \frac{1}{2c}\int_0^x G(x)\,\mathrm{d}x - \frac{1}{2}K$$

The solution thus becomes

$$u = \frac{1}{2}[F(x+ct) + F(x-ct)] + \frac{1}{2c}\int_{x-ct}^{x+ct} G(z)\,\mathrm{d}z \qquad \textbf{(9.15)}$$

which is commonly called the **d'Alembert solution**. As in Examples 9.2, 9.8 and 9.9, it gives rise to waves propagating in the $+x$ and $-x$ directions.

As mentioned in Section 9.1, a major difficulty is to illustrate the solution of a partial differential equation in a simple way. Figure 9.8 is a 'snapshot' at a particular time t, and if we wish to look at the solution over all (x, t) then we have to draw u as a function of the two variables x and t. With some difficulty, we can draw the solution to Example 9.9 in a three-dimensional diagram as in Figure 9.9, but for any higher-dimensional problem such a diagram is clearly impossible. The 'snapshot' in Figure 9.8 corresponds to a plane slice parallel to the (u, x) plane.

The idea of using an (x, t) plane is a very useful one for the wave equation, since the solution

$$u = f(x + ct) + g(x - ct)$$

gives a representation by **characteristics**. If we plot the lines $x + ct$ = constant and $x - ct$ = constant as in Figure 9.10 then we see that the line AP has equation $x - ct = x_0$ and the line BP has equation $x + ct = x_1$. Thus

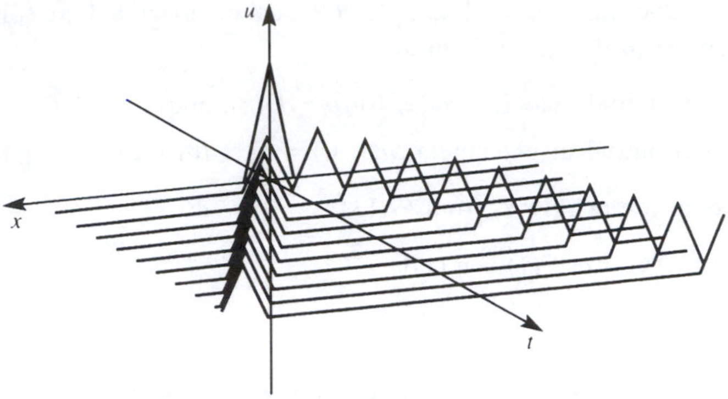

Figure 9.9 Solution to Example 9.9 in (x, t, u) space.

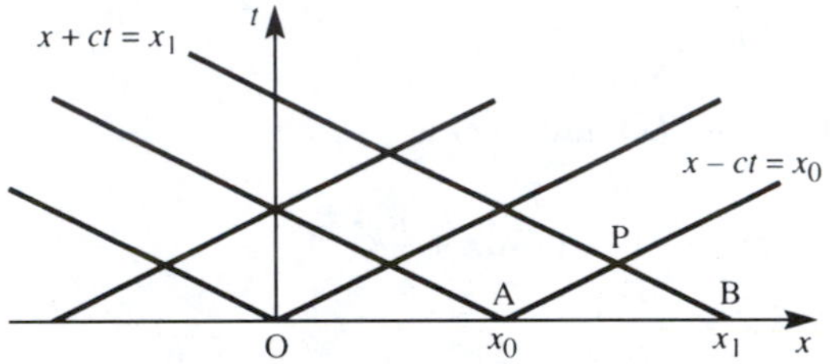

Figure 9.10 Characteristics $x + ct =$ constant and $x - ct =$ constant.

$$\text{on the whole of AP} \quad g(x - ct) = g(x_0)$$

$$\text{on the whole of BP} \quad f(x + ct) = f(x_1)$$

Thus g takes a constant value on AP and f takes a constant value on BP. If we can calculate f and g on the initial line $t = 0$ then we know the value of u at P, namely

$$u(\mathrm{P}) = f(x_1) + g(x_0) \tag{9.16}$$

Since P is an arbitrary point, the solution at any point would be known. The essential problem is to calculate $f(x)$ and $g(x)$ on the line $t = 0$.

Typical conditions on $t = 0$ are

(a) $u(x, 0) = F(x)$, and

(b) $\partial u(x, 0)/\partial t = G(x)$,

which specify the initial position and velocity of the system. Now

$$c\frac{\partial u}{\partial x} + \frac{\partial u}{\partial t} = cf'(x + ct) + cg'(x - ct) + cf'(x + ct) - cg'(x - ct)$$

$$= 2cf'(x + ct)$$

and similarly

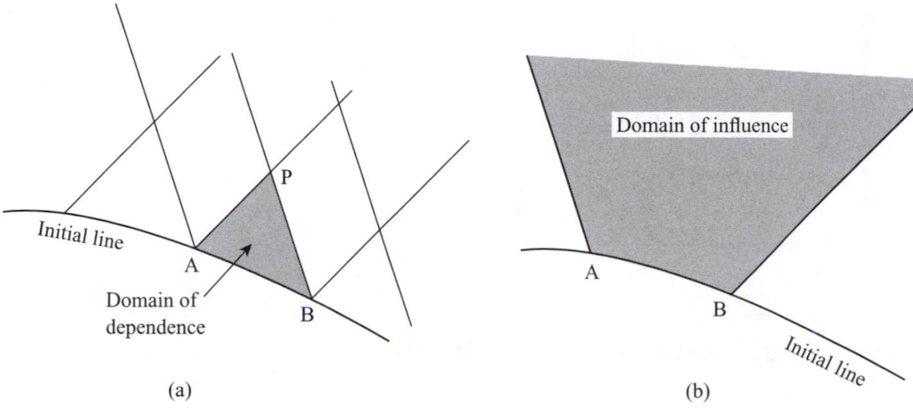

Figure 9.11 Characteristics showing (a) the domain of dependence and (b) the domain of influence.

$$c\frac{\partial u}{\partial x} - \frac{\partial u}{\partial t} = 2cg'(x - ct)$$

On $t = 0$ we know that $\partial u/\partial x = F'(x)$ and $\partial u/\partial t = G(x)$, so we can deduce that

$$cF'(x) + G(x) = 2cf'(x)$$

$$cF'(x) - G(x) = 2cg'(x)$$

Since F and G are given, we can compute

$$f'(x) = \tfrac{1}{2}[F'(x) + G(x)/c]$$

$$g'(x) = \tfrac{1}{2}[F'(x) - G(x)/c]$$

and hence $f(x)$ and $g(x)$ can be computed by straightforward integration.

This method is essentially the same as the d'Alembert method, but it concentrates on calculating $f(x)$ and $g(x)$ on the initial line and then constructing the solution at P by the characteristics AP and BP. The method gives great insight into the behaviour of the solution of such equations, but it is not an easy technique to use in practice. Perhaps the best that can be obtained from characteristics is an idea of how the solution depends on the initial data. In Figure 9.11 the characteristics emanating from the initial line are drawn. To evaluate the solution at P, we must have information on the section of the initial line AB, and the rest of the initial line is irrelevant to the solution at P. This is called the **domain of dependence**. The section of the initial line AB has a **domain of influence** determined by the characteristics through the points A and B. The data on AB cannot influence the solution outside the shaded region in Figure 9.11(b).

EXAMPLE 9.10

Use characteristics to compute the solution of the one-dimensional wave equation (9.4), with speed $c = 1$, given the initial conditions that, for all x and $t = 0$, (a) $u = \exp(-|x|)$ and (b) $\partial u/\partial t = 0$.

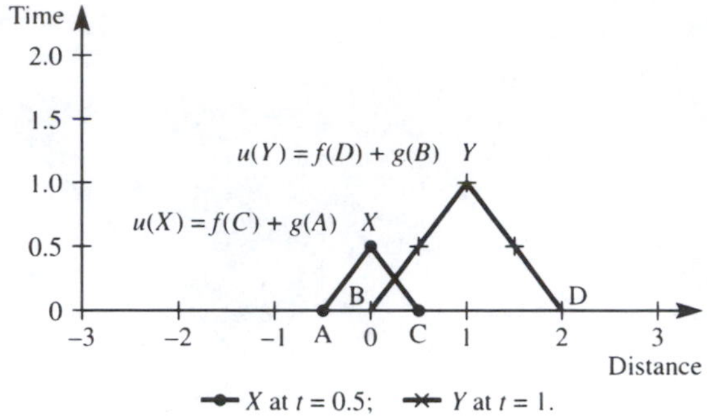

Figure 9.12 Characteristic solution of Example 9.10.

Solution Physically, this problem corresponds to an infinite string released from rest with the initial shape given by condition (a).

We know that the solution takes the form

$$u = f(x + t) + g(x - t)$$

and from the general analysis we have

$$G(x) = 0, \quad F(x) = \exp(-|x|)$$

giving

$$f'(x) = \tfrac{1}{2}F'(x), \quad g'(x) = \tfrac{1}{2}F'(x)$$

Hence we can integrate to give $f(x)$ and $g(x)$ on $t = 0$ as

$$f(x) = \tfrac{1}{2}\exp(-|x|), \quad g(x) = \tfrac{1}{2}\exp(-|x|)$$

We choose to compute u at the next time step $t = 0.5$. Looking at Figure 9.12, we see that the characteristics meet on the line $t = 0.5$ in the (x, t) plane. Taking a few typical points, and according to (9.16),

$$u(0, 0.5) = f(0.5 + 0) + g(0 - 0.5) = 0.3033 + 0.3033 = 0.6066$$

$$u(0.5, 0.5) = f(0.5 + 0.5) + g(0.5 - 0.5) = 0.1839 + 0.5 = 0.6839$$

$$u(1, 0.5) = f(1 + 0.5) + g(1 - 0.5) = 0.1116 + 0.3033 = 0.4149$$

In a similar manner, all the points on the line $t = 0.5$ can be computed. On the line $t = 1$, that is, the next time step, the characteristics are traced back to the initial line and then computed from the appropriate combination of $f(x)$ and $g(x)$. Figure 9.13 is the table of values obtained in the interval $-3 < x < 3$ for a few time steps, and the results are plotted in Figure 9.14. The two waves move to $+\infty$ and to $-\infty$, as expected. Note that $F(x)$ is not differentiable at $x = 0$ so the partial differential equation is not properly satisfied at this point. However, it may also be noted from Figure 9.14 that the cusp is transmitted by the wave, as indicated in Section 9.2.1.

x	$f(x)$	$g(x)$	$u(x, 0)$	$u(x, 0.5)$	$u(x, 1)$	$u(x, 1.5)$	$u(x, 2)$
−3.0000	0.0249	0.0249	0.0498	0.0561	0.0768	0.1171	0.1873
−2.5000	0.0410	0.0410	0.0821	0.0926	0.1267	0.1931	0.3088
−2.0000	0.0677	0.0677	0.1353	0.1526	0.2088	0.3184	0.5092
−1.5000	0.1116	0.1116	0.2231	0.2516	0.3443	0.5249	0.3184
−1.0000	0.1839	0.1839	0.3679	0.4148	0.5677	0.3443	0.2088
−0.5000	0.3033	0.3033	0.6065	0.6839	0.4148	0.2516	0.1526
0.0000	0.5000	0.5000	1.0000	0.6065	0.3679	0.2231	0.1353
0.5000	0.3033	0.3033	0.6065	0.6839	0.4148	0.2516	0.1526
1.0000	0.1839	0.1839	0.3679	0.4148	0.5677	0.3443	0.2088
1.5000	0.1116	0.1116	0.2231	0.2516	0.3443	0.5249	0.3184
2.0000	0.0677	0.0677	0.1353	0.1526	0.2088	0.3184	0.5092
2.5000	0.0410	0.0410	0.0821	0.0926	0.1267	0.1931	0.3088
3.0000	0.0249	0.0249	0.0498	0.0561	0.0768	0.1171	0.1873

Figure 9.13 Computed data for Example 9.10.

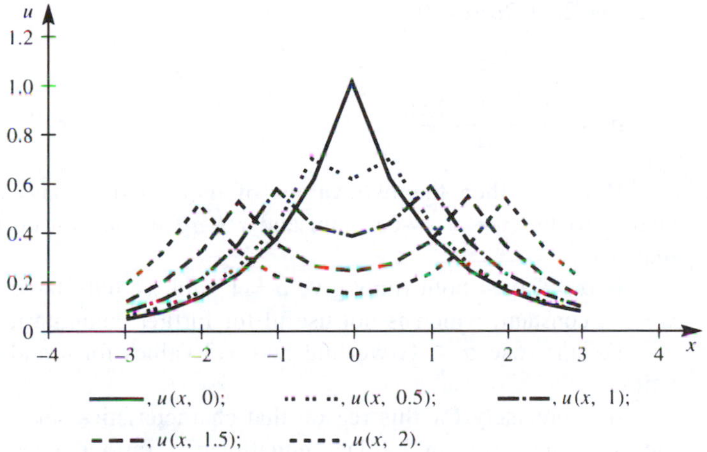

Figure 9.14 Plot of the data from Figure 9.13.

When the initial line comprises only part of the x axis, as for example when a *finite* string is plucked, the evaluation of the arbitrary functions f and g in equation (9.13) and the use of characteristics is no longer straightforward. We do not have a d'Alembert type of solution; a great deal of thought and care is needed.

The idea of characteristics can be applied to more general classes of second-order partial differential equations. Example 9.11 illustrates a case of a constant-coefficient equation.

EXAMPLE 9.11 Find the characteristics of the equation

$$0 = u_{xx} + 2u_{xt} + 2\alpha u_{tt}$$

Study the case when $\alpha = \frac{3}{8}$ and the solution satisfies the boundary conditions

(a) $\partial u(x, 0)/\partial t = 0$ for $x \geqslant 0$

(b) $u(x, 0) = F(x) = \begin{cases} 1 & (0 < x < 1) \\ 0 & (x \geqslant 1) \end{cases}$

(c) $u(0, t) = 0$ for all t

(d) $\partial u(0, t)/\partial x = 0$ for all t

Solution Since the coefficients of the equation are constants, we know that the characteristics are straight lines, so we look for solutions of the form

$$u = u(x + at)$$

Putting $z = x + at$ and writing $u' = du/dz$ and so on, we obtain

$$0 = u_{xx} + 2u_{xt} + 2\alpha u_{tt} = (1 + 2a + 2a^2\alpha)u''$$

Hence for a solution we require

$$1 + 2a + 2a^2\alpha = 0$$

or

$$a = \frac{-1 \pm \surd(1 - 2\alpha)}{2\alpha}$$

If $\alpha > \frac{1}{2}$ then the two values of a (a_1 and a_2 say) are complex, and the characteristics $x + a_1t$ = constant and $x + a_2t$ = constant do not make sense in the real plane.

If $\alpha = \frac{1}{2}$ then both roots give $a = 1$, and we only have a single characteristic $x + t$ = constant, which is not useful for further computation.

For the case $\alpha < \frac{1}{2}$, we find two real values for a and two sets of characteristics.

It is precisely for this reason that characteristics serve no useful purpose for the heat-conduction or Laplace equations. A further discussion can be found in Section 9.7 after the formal classification of equations has been completed.

Take the case $\alpha = \frac{3}{8}$; then we obtain $a_1 = -2$ and $a_2 = -\frac{2}{3}$, so the solution has the form

$$u = f(x - 2t) + g(x - \tfrac{2}{3}t)$$

where f and g are arbitrary functions and the characteristics are the straight lines $x - 2t$ = constant and $x - \frac{2}{3}t$ = constant.

The boundary conditions given in the problem are a little more complicated than in the d'Alembert solution. Conditions (a) and (b) give

$$\left.\begin{aligned} 0 = \frac{\partial u(x, 0)}{\partial t} &= -2f'(x) - \frac{2}{3}g'(x) \\ F(x) = u(x, 0) &= f(x) + g(x) \end{aligned}\right\} \quad (x \geqslant 0)$$

Taking $f(0) = g(0) = 0$, we can integrate the first of these expressions and then solve for $f(x)$ and $g(x)$ on the line $t = 0$ as

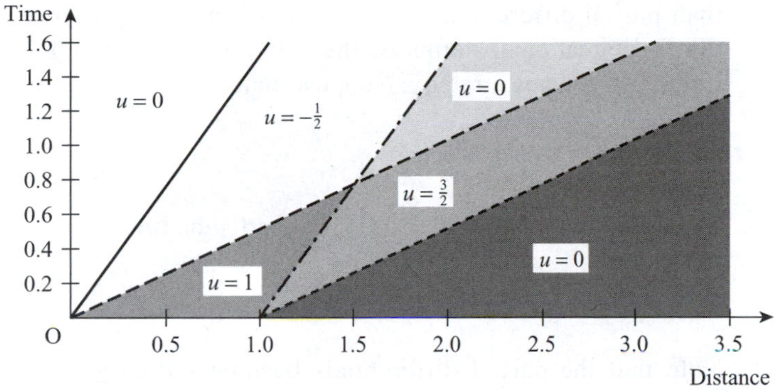

Figure 9.15 Characteristic solution of Example 9.11. The solution u takes the constant values shown in the six regions of the first quadrant.

$$f(x) = -\tfrac{1}{2}F(x), \quad g(x) = \tfrac{3}{2}F(x) \quad (x \geqslant 0)$$

Conditions (c) and (d) say that $u(0, t) = 0$, and $\partial u(0, t)/\partial x = 0$. Thus on the line $x = 0$ we deduce

$$f(z) = g(z) = 0 \quad (z < 0)$$

We can now construct the solution by characteristics. Figure 9.15 illustrates this solution. Because $f(x)$ and $g(x)$ are constant along the respective characteristics, we deduce $u(A) = 0$, $u(B) = -\tfrac{1}{2}$, $u(C) = 1$, $u(D) = 0$, $u(E) = \tfrac{3}{2}$, $u(F) = 0$ at typical points in the six regions that divide up the first quadrant of the (x, t) plane.

For non-constant-coefficient equations the characteristics are not usually straight lines, which causes computational difficulties. In particular, there are some fundamental problems when characteristics of the same family intersect. The solution loses its uniqueness, and 'shocks' can be generated. The classical wave equation (9.4) will propagate these shocks, but it requires 'curved characteristics' to generate them.

9.3.2 Separated solutions

A method of considerable importance is the **method of separation of variables**. The basis of the method is to attempt to look for solutions $u(x, y)$ of a partial differential equation as a product of functions of single variables

$$u(x, y) = X(x)Y(y)$$

The advantage of this approach is that it is sometimes possible to find X and Y as solutions of *ordinary* differential equations. These are very much easier to solve than partial differential equations, and it may be possible to build up solutions of the full equation in terms of the solutions for X and Y. A simple example illustrates the general strategy. Suppose that we wish to solve

$$\frac{\partial u}{\partial x} + \frac{\partial u}{\partial y} = 0$$

Then we should write $u = X(x)Y(y)$ and substitute

$$Y\frac{dX}{dx} + X\frac{dY}{dy} = 0, \quad \text{or} \quad \frac{1}{X}\frac{dX}{dx} = -\frac{1}{Y}\frac{dY}{dy}$$

Note that the partial differentials become ordinary differentials, since the functions are just functions of a single variable. Now

$$\text{LHS} = \frac{1}{X}\frac{dX}{dx} = \text{a function of } x \text{ only}$$

$$\text{RHS} = -\frac{1}{Y}\frac{dY}{dy} = \text{a function of } y \text{ only}$$

Since LHS = RHS for *all* x and y, the only way that this can be achieved is for each side to be a *constant*. We thus have two ordinary differential equations

$$\frac{1}{X}\frac{dX}{dx} = \lambda, \quad -\frac{1}{Y}\frac{dY}{dy} = \lambda$$

These equations can be solved easily as

$$X = B\,e^{\lambda x}, \quad Y = C\,e^{-\lambda y}$$

and thus the solution of the original partial differential equation is

$$u(x, y) = X(x)Y(y) = A\,e^{\lambda(x-y)}$$

where $A = BC$. The constants A and λ are arbitrary. The crucial question is whether the boundary conditions imposed by the problem can be satisfied by a sum of solutions of this type.

The method of separation of variables can be a very powerful technique, and we shall see it used on all three of the basic partial differential equations. It should be noted, however, that all equations do not have separated solutions, and even when they can be obtained it is not always possible to satisfy the boundary conditions with such solutions.

In the case of the heat-conduction equation and the wave equation, the form of one of the functions in the separated solution is dictated by the physics of the problem. We shall see that the separation technique becomes a little simpler when such physical arguments are used. However, for the Laplace equation there is no help from the physics, so the method just described needs to be applied.

In most wave equation problems we are looking for either a travelling-wave solution as in Section 9.3.1 or for periodic solutions, as a result of plucking a violin string for instance. It therefore seems natural to look for specific solutions that have periodicity built into them. These will not be general solutions, but they will be seen to be useful for a whole class of problems. The essential mathematical simplicity of the method comes from only having to solve ordinary differential equations.

The above argument suggests that we seek solutions of the wave equation

$$\frac{1}{c^2}\frac{\partial^2 u}{\partial t^2} = \frac{\partial^2 u}{\partial x^2} \tag{9.4}$$

of the form either

$$u = \sin(c\lambda t)v(x) \tag{9.17a}$$

or

$$u = \cos(c\lambda t)v(x) \tag{9.17b}$$

both of which when substituted into (9.4) give the ordinary differential equation

$$\frac{d^2v}{dx^2} = -\lambda^2 v$$

This is a simple-harmonic equation with solutions $v = \sin\lambda x$ or $v = \cos\lambda x$. We can thus build up a general solution of (9.4) from linear multiples of the four basic solutions

$$u_1 = \cos\lambda ct\sin\lambda x \tag{9.18a}$$

$$u_2 = \cos\lambda ct\cos\lambda x \tag{9.18b}$$

$$u_3 = \sin\lambda ct\sin\lambda x \tag{9.18c}$$

$$u_4 = \sin\lambda ct\cos\lambda x \tag{9.18d}$$

and try to satisfy the boundary conditions using appropriate linear combinations of solutions of this type. We saw an example of such a solution in Example 9.1.

EXAMPLE 9.12

Solve the wave equation (9.4) for the vibration of a string stretched between the points $x = 0$ and $x = l$ and subject to the boundary conditions

(a) $u(0, t) = 0 \quad (t \geqslant 0)$ (fixed at the end $x = 0$);

(b) $u(l, t) = 0 \quad (t \geqslant 0)$ (fixed at the end $x = l$);

(c) $\partial u(x, 0)/\partial t = 0 \quad (0 \leqslant x \leqslant l)$ (with zero initial velocity);

(d) $u(x, 0) = F(x)$ (given initial displacement).

Consider the two cases

(i) $F(x) = \sin(\pi x/l) + \frac{1}{4}\sin(3\pi x/l)$

(ii) $F(x) = \begin{cases} x & (0 \leqslant x \leqslant \frac{1}{2}l) \\ l - x & (\frac{1}{2}l \leqslant x \leqslant l) \end{cases}$

Solution Clearly, we are solving the problem of a stretched string, held at its ends $x = 0$ and $x = l$ and released from rest.

By inspection, we see that the solutions (9.18b, d) cannot satisfy condition (a). We see that condition (b) is satisfied by the solutions (9.18a, c), provided that

$$\sin \lambda l = 0, \quad \text{or} \quad \lambda l = n\pi \quad (n = 1, 2, 3, \ldots)$$

It may be noted that only specific values of λ in (9.18) give permissible solutions. Thus the string can only vibrate with given frequencies, $nc/2l$. The solution (9.18) appropriate to this problem takes the form either

$$u = \cos\left(\frac{nc\pi t}{l}\right)\sin\left(\frac{n\pi x}{l}\right) \tag{9.19a}$$

or

$$u = \sin\left(\frac{nc\pi t}{l}\right)\sin\left(\frac{n\pi x}{l}\right) \tag{9.19b}$$

($n = 1, 2, 3, \ldots$). To satisfy condition (c) for all x, we must choose the solution (9.19a) and omit (9.19b). Clearly, it is not possible to satisfy the initial condition (d) with (9.19a). However, because the wave equation is linear, any *sum* of such solutions is also a solution. Thus we build up a solution

$$u = \sum_{n=1}^{\infty} b_n \cos\left(\frac{nc\pi t}{l}\right)\sin\left(\frac{n\pi x}{l}\right) \tag{9.20}$$

Case (i)

The initial condition (d) for $u(x, 0)$ gives

$$\sum_{n=1}^{\infty} b_n \sin\left(\frac{n\pi x}{l}\right) = \sin\left(\frac{\pi x}{l}\right) + \frac{1}{4}\sin\left(\frac{3\pi x}{l}\right)$$

and the values of b_n can be evaluated by inspection as

$$b_1 = 1,\ b_2 = 0,\ b_3 = \tfrac{1}{4},\ b_4 = b_5 = \ldots = 0$$

The full solution is therefore

$$u = \cos\left(\frac{\pi ct}{l}\right)\sin\left(\frac{\pi x}{l}\right) + \frac{1}{4}\cos\left(\frac{3\pi ct}{l}\right)\sin\left(\frac{3\pi x}{l}\right)$$

The solution is illustrated in Figure 9.16.

Case (ii)

The condition (d) for $u(x, 0)$ simply gives

$$\sum_{n=1}^{\infty} b_n \sin\left(\frac{n\pi x}{l}\right) = f(x) = \begin{cases} x & (0 \leqslant x \leqslant \frac{1}{2}l) \\ l - x & (\frac{1}{2}l \leqslant x \leqslant l) \end{cases}$$

and thus to determine b_n we must find the Fourier sine series expansion of the function $f(x)$ over the finite interval $0 \leqslant x \leqslant l$. We have from (4.33) that

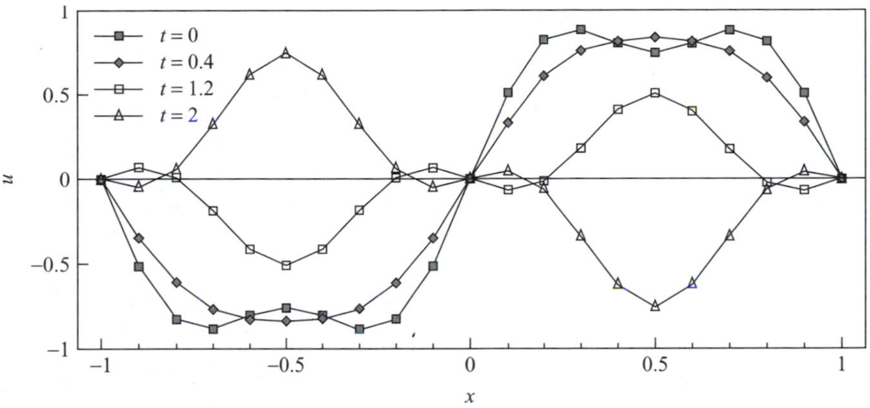

Figure 9.16 Sketch of the solution to Example 9.12 (i) with $c = \frac{1}{3}$ and $l = 1$.

$$\begin{aligned}
b_n &= \frac{2}{l}\int_0^l f(x)\sin\left(\frac{n\pi x}{l}\right)dx \\
&= \frac{2}{l}\int_0^{l/2} x\sin\left(\frac{n\pi x}{l}\right)dx + \frac{2}{l}\int_{l/2}^{l} (l-x)\sin\left(\frac{n\pi x}{l}\right)dx \\
&= \frac{4l}{\pi^2 n^2}\sin\left(\frac{1}{2}n\pi\right) \quad (n = 1, 2, 3, \dots)
\end{aligned}$$

The complete solution of the wave equation in this case is therefore

$$u(x, t) = \frac{4l}{\pi^2}\sum_{n=1}^{\infty}\frac{1}{n^2}\sin\left(\frac{1}{2}n\pi\right)\cos\left(\frac{nc\pi t}{l}\right)\sin\left(\frac{n\pi x}{l}\right) \tag{9.21}$$

or

$$u(x, t) = \frac{4l}{\pi^2}\left[\cos\left(\frac{c\pi t}{l}\right)\sin\left(\frac{\pi x}{l}\right) - \frac{1}{9}\cos\left(\frac{3c\pi t}{l}\right)\sin\left(\frac{3\pi x}{l}\right) + \frac{1}{25}\cos\left(\frac{5c\pi t}{l}\right)\sin\left(\frac{5\pi x}{l}\right) + \dots\right]$$

The complete solution to Example 9.12 Case (ii) gives some very useful information. We see that all the even 'harmonics' have disappeared from the solution and the amplitudes of the harmonics decrease like $1/n^2$. A beautiful theory of musical instruments can be built up from such solutions. We see that for different instruments different harmonics are important and have different amplitudes. It is this that gives an instrument its characteristic sound.

A sensible question that we can ask is whether we can use the sum of the series in (9.21) to plot u. In Chapter 4 we saw that, at discontinuities in particular, Fourier series can be very slow to converge, so that, although (9.21) is a complete

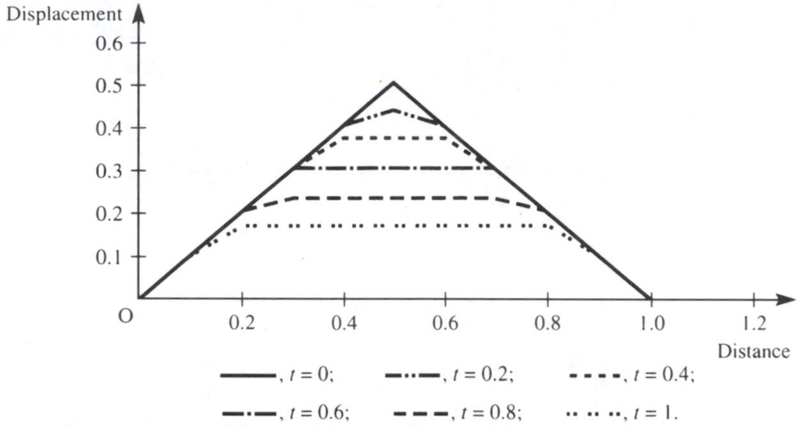

Figure 9.17 Solution of Example 9.12 (ii) with $c = \frac{1}{3}$ and $l = 1$.

solution, does it provide us with any useful information? In the present case there is no particular problem, but the general comment should be noted. The solution (9.21) is plotted in Figure 9.17 with $l = 1$ and $c = \frac{1}{3}$. We note that even with 10 terms $u(0.5, 0) = 0.4899$ instead of the correct value 0.5, so there is a 2% error in the calculated value. Perhaps the most pertinent comment that we can make is that a good number of terms in the series are required to obtain a solution, and exact solutions may not be as useful as we might expect.

Separated solutions depend on judicious use of the known solutions (9.18) of the wave equation to fit the boundary conditions. Although it is not always possible to solve any particular problem using separated solutions, the idea is sufficiently straightforward that it is always worth a try. The extension to other equations and coordinate systems is possible, and some of these will be discussed in Sections 9.4 and 9.5.

EXAMPLE 9.13 Solve the wave equation (9.4) for vibrations in an organ pipe subject to the boundary conditions

(a) $u(0, t) = 0 \quad (t \geqslant 0)$ (the end $x = 0$ is closed);

(b) $\partial u(l, t)/\partial x = 0 \quad (t \geqslant 0)$ (the end $x = l$ is open);

(c) $u(x, 0) = 0 \quad (0 \leqslant x \leqslant l)$ (the pipe is initially undisturbed);

(d) $\partial u(x, 0)/\partial t = v = \text{constant} \quad (0 \leqslant x \leqslant l)$ (the pipe is given an initial uniform blow).

Solution From the solution (9.18), we deduce from condition (a) that solutions (9.18b, d) must be omitted, and similarly from condition (c) that solution (9.18a) is not useful. We are left with the solution (9.18c) to satisfy the boundary condition (b). This can only be satisfied if

$$\cos \lambda l = 0, \quad \text{or} \quad \lambda l = (n + \tfrac{1}{2})\pi \quad (n = 0, 1, 2, \ldots)$$

Thus we obtain solutions of the form

$$u = b_n \sin\left[\frac{(n+\frac{1}{2})\pi ct}{l}\right]\sin\left[\frac{(n+\frac{1}{2})\pi x}{l}\right] \quad (n = 0, 1, 2, \dots)$$

giving a general solution

$$u = \sum_{n=0}^{\infty} b_n \sin\left[\frac{(n+\frac{1}{2})\pi ct}{l}\right]\sin\left[\frac{(n+\frac{1}{2})\pi x}{l}\right]$$

The condition (d) gives

$$v = \sum_{n=0}^{\infty} b_n \frac{(n+\frac{1}{2})\pi c}{l}\sin\left[\frac{(n+\frac{1}{2})\pi x}{l}\right]$$

which, on using (4.33) to obtain the coefficients of the Fourier sine series expansion of the constant v over the finite interval $0 \leqslant x \leqslant l$, gives

$$b_n = \frac{2v}{(n+\frac{1}{2})\pi}\frac{l}{(n+\frac{1}{2})\pi c} = \frac{8lv}{\pi^2 c}\frac{1}{(2n+1)^2}$$

Our complete solution of the wave equation is therefore

$$u = \frac{8lv}{\pi^2 c}\sum_{n=0}^{\infty}\frac{1}{(2n+1)^2}\sin\left[\left(n+\frac{1}{2}\right)\pi\frac{ct}{l}\right]\sin\left[\left(n+\frac{1}{2}\right)\pi\frac{x}{l}\right]$$

or,

$$u = \frac{8lv}{\pi^2 c}\left[\sin\left(\frac{\pi ct}{2l}\right)\sin\left(\frac{\pi x}{2l}\right) + \frac{1}{9}\sin\left(\frac{3\pi ct}{2l}\right)\sin\left(\frac{3\pi x}{2l}\right)\right.$$
$$\left. + \frac{1}{25}\sin\left(\frac{5\pi ct}{2l}\right)\sin\left(\frac{5\pi x}{2l}\right) + \dots\right]$$

It would be instructive to compute this solution and compare it with Figure 9.17, which corresponds to the solution of Example 9.12.

9.3.3 Laplace transform solution

For linear problems that are time-varying from 0 to ∞, as in the case of the wave equation, Laplace transforms provide a formal method of solution. The only difficulty is whether the final inversion can be performed.

First we obtain the Laplace transforms of the partial derivatives

$$\frac{\partial u}{\partial x}, \quad \frac{\partial u}{\partial t}, \quad \frac{\partial^2 u}{\partial x^2}, \quad \frac{\partial^2 u}{\partial t^2}$$

of the function $u(x, t)$, $t \geqslant 0$. Using the same procedure as that used to obtain the Laplace transform of standard derivatives in Section 2.3.1, we have the following.

(a) $$\mathscr{L}\left\{\frac{\partial u}{\partial x}\right\} = \int_0^\infty e^{-st}\frac{\partial u}{\partial x}\,dt = \frac{d}{dx}\int_0^\infty e^{-st}u(x, t)\,dt$$

using Leibniz' rule (see *Modern Engineering Mathematics*) for differentiation under an integral sign. Noting that

$$\mathscr{L}\{u(x, t)\} = U(x, s) = \int_0^\infty e^{-st}u(x, t)\,dt$$

we have

$$\mathscr{L}\left\{\frac{\partial u}{\partial x}\right\} = \frac{d}{dx}U(x, s) \tag{9.22}$$

(b) Writing $y(x, t) = \partial u/\partial x$, repeated application of the result (9.22) gives

$$\mathscr{L}\left\{\frac{\partial y}{\partial x}\right\} = \frac{d}{dx}\mathscr{L}\{y(x, t)\} = \frac{d}{dx}\left(\frac{d}{dx}U(x, s)\right)$$

so that

$$\mathscr{L}\left\{\frac{\partial^2 u}{\partial x^2}\right\} = \frac{d^2U(x, s)}{dx^2} \tag{9.23}$$

(c) $$\mathscr{L}\left\{\frac{\partial u}{\partial t}\right\} = \int_0^\infty e^{-st}\frac{\partial u}{\partial t}\,dt$$

$$= [e^{-st}u(x, t)]_0^\infty + s\int_0^\infty e^{-st}u(x, t)\,dt = [0 - u(x, 0)] + sU(x, s)$$

so that

$$\mathscr{L}\left\{\frac{\partial u}{\partial t}\right\} = sU(x, s) - u(x, 0) \tag{9.24}$$

where we have assumed that $u(x, t)$ is of exponential order.

(d) Writing $v(x, t) = \partial u/\partial t$, repeated application of (9.24) gives

$$\mathscr{L}\left\{\frac{\partial v}{\partial t}\right\} = sV(x, s) - v(x, 0)$$

$$= s[sU(x, s) - u(x, 0)] - v(x, 0)$$

so that

$$\mathscr{L}\left\{\frac{\partial^2 u}{\partial t^2}\right\} = s^2U(x, s) - su(x, 0) - u_t(x, 0) \tag{9.25}$$

where $u_t(x, 0)$ denotes the value of $\partial u/\partial t$ at $t = 0$.

Let us now return to consider the wave equation (9.4)

$$c^2\frac{\partial^2 u}{\partial x^2} = \frac{\partial^2 u}{\partial t^2}$$

subject to the boundary conditions $u(x, 0) = f(x)$ and $\partial u(x, 0)/\partial t = g(x)$. Taking Laplace transforms on both sides of (9.4) and using the results (9.23) and (9.25) gives

$$c^2\frac{\mathrm{d}^2 U(x, s)}{\mathrm{d}x^2} = s^2 U(x, s) - g(x) - sf(x) \qquad \textbf{(9.26)}$$

The problem has thus been reduced to an ordinary differential equation in $U(x, s)$ of a straightforward type. It can be solved for given conditions at the ends of the x range, and the solution can then be inverted to give $u(x, t)$.

EXAMPLE 9.14 Solve the wave equation (9.4) for a semi-infinite string by Laplace transforms, given that

(a) $u(x, 0) = 0$ $(x \geqslant 0)$ (string initially undisturbed);

(b) $\partial u(x, 0)/\partial t = x\,\mathrm{e}^{-x/a}$ $(x \geqslant 0)$ (string given an initial velocity);

(c) $u(0, t) = 0$ $(t \geqslant 0)$ (string held at $x = 0$);

(d) $u(x, t) \to 0$ as $x \to \infty$ for $t \geqslant 0$ (string held at infinity).

Solution Using conditions (a) and (b) and substituting for $f(x)$ and $g(x)$ in the result (9.26), the transformed equation in this case is

$$c^2\frac{\mathrm{d}^2}{\mathrm{d}x^2}U(x, s) = s^2 U(x, s) - x\,\mathrm{e}^{-x/a}$$

By seeking a particular integral of the form

$$U = \alpha x\,\mathrm{e}^{-x/a} + \beta\,\mathrm{e}^{-x/a}$$

we obtain a solution of the differential equation as

$$U(x, s) = A\,\mathrm{e}^{sx/c} + B\,\mathrm{e}^{-sx/c} - \frac{\mathrm{e}^{-x/a}}{c^2/a^2 - s^2}\left[x + \frac{2c^2/a}{c^2/a^2 - s^2}\right]$$

where A and B are arbitrary constants.

Transforming the given boundary conditions (c) and (d), we have $U(0, s) = 0$ and $U(x, s) \to 0$ as $x \to \infty$, which can be used to determine A and B. From the second condition $A = 0$, and the first condition then gives

$$B = \frac{2c^2/a}{(c^2/a^2 - s^2)^2}$$

so that the solution becomes

$$U(x, s) = \frac{2c^2/a}{(c^2/a^2 - s^2)^2}\,\mathrm{e}^{-sx/c} - \frac{\mathrm{e}^{-x/a}}{(c^2/a^2 - s^2)}\left[x + \frac{2c^2/a}{(c^2/a^2 - s^2)}\right]$$

Fortunately in this case these transforms can be inverted from tables of Laplace transforms.

Using the second shift theorem (2.45) together with the Laplace transform pairs

$$\mathscr{L}\{\sinh \omega t\} = \frac{\omega}{s^2 - \omega^2}, \quad \mathscr{L}\{\cosh \omega t\} = \frac{s}{s^2 - \omega^2}$$

$$\mathscr{L}\left\{\frac{\omega t \cosh \omega t - \sinh \omega t}{2\omega^3}\right\} = \frac{1}{(s^2 - \omega^2)^2}$$

we obtain the solution as

$$u = \frac{a}{c}\left[(ct - x)\cosh\left(\frac{ct - x}{a}\right)H(ct - x) - ct\,e^{-x/a}\cosh\left(\frac{ct}{a}\right)\right]$$

$$+ \frac{a}{c}\left[e^{-x/a}(x + a)\sinh\left(\frac{ct}{a}\right) - a\sinh\left(\frac{ct - x}{a}\right)H(ct - x)\right]$$

where $H(t)$ is the Heaviside step function defined in Section 2.5.1.

9.3.4 Exercises

11 Solve the wave equation

$$\frac{\partial^2 u}{\partial x^2} = \frac{1}{c^2}\frac{\partial^2 u}{\partial t^2}$$

subject to the initial conditions

(a) $u(x, 0) = \sin x \quad$ (all x)

(b) $\dfrac{\partial u}{\partial t}(x, 0) = 0 \quad$ (all x)

Use both the d'Alembert solution and the separation of variables method and show that they both give the same result.

12 The spherically symmetric version of the wave equation (9.4) takes the form

$$\frac{1}{c^2}\frac{\partial^2 u}{\partial t^2} = \frac{\partial^2 u}{\partial r^2} + \frac{2}{r}\frac{\partial u}{\partial r}$$

Show, by putting $v = ru$, that it has a solution

$$ru = f(ct - r) + g(ct + r)$$

Interpret the terms as spherical waves.

13 Using the trigonometric identity

$$\sin A \cos B = \tfrac{1}{2}\sin(A - B) + \tfrac{1}{2}\sin(A + B)$$

rewrite the solution (9.21) to Example 9.12 as a progressive wave.

14 Solve the wave equation

$$\frac{\partial^2 u}{\partial x^2} = \frac{1}{c^2}\frac{\partial^2 u}{\partial t^2}$$

subject to the initial conditions

(a) $u(x, 0) = 0 \quad$ (all x)

(b) $\dfrac{\partial u}{\partial t}(x, 0) = x\,e^{-x^2} \quad$ (all x)

15 Find the solutions to the wave equation (9.4) subject to the boundary conditions

(a) $\partial u(x, 0)/\partial t = 0$ for all x

(b) $u = \begin{cases} 1 - x & (0 \leqslant x \leqslant 1) \\ 1 + x & (-1 \leqslant x \leqslant 0) \\ 0 & (|x| \geqslant 1) \end{cases}$ at $t = 0$

using d'Alembert's method. Compare with Example 9.9.

16 Compute the characteristics of the equation

$$3u_{xx} + 6u_{xy} + u_{yy} = 0$$

†17 Use characteristics to compute the solution of the wave equation (9.4), with speed $c = 1$, given the initial conditions that for all x and $t = 0$

(a) $u = 0$ (b) $\partial u/\partial t = \exp(-|x|)$

Use a time step of 0.5 to compute (on a spreadsheet or other package) the first four steps over the range $-3 < x < 3$.

18 Using the separated solution approach of Section 9.3.2, obtain a series solution of the wave equation

$$\frac{\partial^2 u}{\partial x^2} = \frac{1}{c^2}\frac{\partial^2 u}{\partial t^2}$$

subject to the boundary conditions

(a) $u(0, t) = 0 \quad (t > 0)$

(b) $\partial u(x, 0)/\partial t = 0 \quad (0 < x < \pi)$

(c) $u(\pi, t) = 0 \quad (t > 0)$

(d) $u(x, 0) = \pi x - x^2 \quad (0 < x < \pi)$

19 The end at $x = 0$ of an infinitely long string, initially at rest along the x axis, undergoes a periodic displacement $a \sin \omega t$, for $t > 0$, transverse to the x axis. The displacement $u(x, t)$ of any point on the string at any time is given by the solution of the wave equation

$$\frac{\partial^2 u}{\partial t^2} = c^2\frac{\partial^2 u}{\partial x^2} \quad (x > 0, t > 0)$$

subject to the boundary conditions

(a) $u(x, 0) = 0 \quad (x > 0)$

(b) $\partial u(x, 0)/\partial t = 0 \quad (x > 0)$

(c) $u(0, t) = a \sin \omega t \quad (t > 0)$

(d) $|u(x, t)| < L, \quad L$ constant

where the last condition specifies that the displacement is bounded.

Using the Laplace transform method, show that the displacement is given by

$$u(x, t) = a \sin\left[\omega\left(t - \frac{x}{c}\right)\right] H\left(t - \frac{x}{c}\right)$$

where $H(t)$ is the Heaviside step function.

Plot a graph of $u(x, t)$, and discuss.

20 The function $u(r, t)$ satisfies the partial differential equation

$$\frac{\partial^2 u}{\partial r^2} + \frac{2}{r}\frac{\partial u}{\partial r} = \frac{1}{c^2}\frac{\partial^2 u}{\partial t^2}$$

where c is a positive constant. Show that this equation has a solution of the form

$$u = \frac{g(r)}{r}\cos \omega t$$

where ω is a constant and g satisfies

$$\frac{d^2 g}{dr^2} + \frac{\omega^2 g}{c^2} = 0$$

Show that, if u satisfies the conditions

$$u(a, t) = \beta \cos \omega t$$

$$u(b, t) = 0$$

then the solution is

$$u(r, t) = \frac{\beta a \cos \omega t}{r}\frac{\sin[\omega(b - r)/c]}{\sin[\omega(b - a)/c]}$$

9.3.5 Numerical solution

For all but the simplest problems, we have to find a numerical solution. In Section 9.3.1 we saw that characteristics give a possible numerical way of working by extending the solution away from the initial line. While this method is possible, it is difficult to program except for the simplest problems, where other methods would be preferred anyway. In particular, when characteristics are curved it becomes difficult to keep track of the solution front. However, calculus methods suffer because they cannot cope with discontinuities, so that, should these occur, the methods described in this section will tend to 'smear out' the shocks. Characteristics provide one of the few methods that will trap the shocks when we use the fact that the latter are propagated along the characteristics.

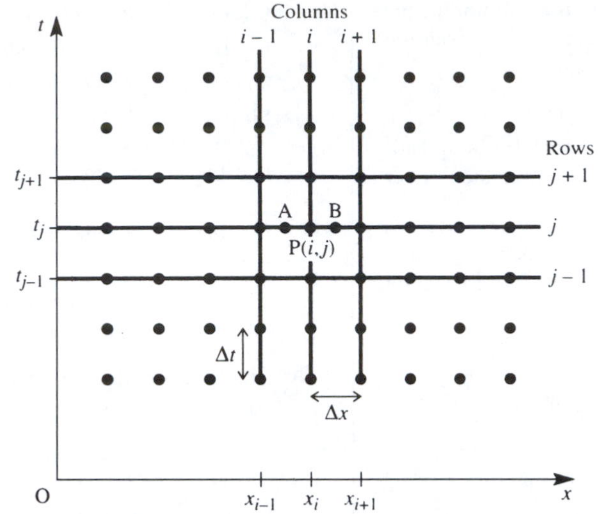

Figure 9.18 Mesh points for a numerical solution of the wave equation.

In Section 10.6 of *Modern Engineering Mathematics* the numerical solution of ordinary differential equations was studied in some detail. The basis of the methods was to construct approximations to differentials in terms of values of the required function at discrete points. The commonest approximation was discussed in Section 8.4.1 of *Modern Engineering Mathematics*, namely the 'chord approximation'

$$\frac{\mathrm{d}f(a)}{\mathrm{d}x} \simeq \frac{f(a+h)-f(a-h)}{2h}$$

and for the second derivative

$$\frac{\mathrm{d}^2f(a)}{\mathrm{d}x^2} \simeq \frac{f(a+h)-2f(a)+f(a-h)}{h^2}$$

The justification of these approximations and the computation of the errors involved depend on the Taylor expansions of the functions. In partial differentiation the approximations are the same except that there is a partial derivative in both x and t for the function $u(x, t)$.

Figure 9.18 illustrates a mesh of points, or **nodes**, with spacing Δx in the x direction and Δt in the t direction. Each node is specified by a pair of integers (i, j), so that the coordinates of the nodal points take the form

$$x_i = a + i\Delta x, \quad t_j = b + j\Delta t$$

and a and b specify the origin chosen. The mesh points or nodes lie on the intersection of the **rows** (j = constant) and **columns** (i = constant).

The approximations are applied to a typical point P, with discretized coordinates (i, j), and with increments $\Delta x = x_{i+1} - x_i$ and $\Delta t = t_{j+1} - t_j$, which are taken to be uniform through the mesh. We know that at the points A and B we can approximate

$$\left(\frac{\partial u}{\partial x}\right)_A \simeq \frac{u(i,j)-u(i-1,j)}{\Delta x}$$

$$\left(\frac{\partial u}{\partial x}\right)_B \simeq \frac{u(i+1,j)-u(i,j)}{\Delta x}$$

so that the second derivative at P has the numerical form

$$\frac{\partial^2 u}{\partial x^2} = \frac{(\partial u/\partial x)_B-(\partial u/\partial x)_A}{\Delta x} = \frac{u(i+1,j)-2u(i,j)+u(i-1,j)}{\Delta x^2}$$

Similarly,

$$\frac{\partial^2 u}{\partial t^2} = \frac{u(i,j+1)-2u(i,j)+u(i,j-1)}{\Delta t^2}$$

Thus the wave equation $\partial^2 u/\partial t^2 = c^2\partial^2 u/\partial x^2$ becomes

$$\frac{u(i,j+1)-2u(i,j)+u(i,j-1)}{\Delta t^2} = c^2\frac{u(i+1,j)-2u(i,j)+u(i-1,j)}{\Delta x^2}$$

which can be rearranged as

$$u(i,j+1) = 2u(i,j)-u(i,j-1) + \lambda^2[u(i+1,j)-2u(i,j)+u(i-1,j)] \tag{9.27}$$

where

$$\lambda = c\,\Delta t/\Delta x$$

Equation (9.27) is a **finite-difference representation** of the wave equation, and provided that u is known on rows $j-1$ and j then $u(i, j+1)$ can be computed on row $j+1$ from (9.27) and thus the solution continued. On the zeroth row the boundary conditions $u(x, 0) = f(x)$ and $\partial u(x, 0)/\partial t = g(x)$ are known, so that $f_i = u(i, 0)$ and $g_i = \partial u(i, 0)/\partial t$ are also known at each node on this row, and these are used to start the process off. From Figure 9.19, we see that

$$g_i = \frac{\partial u}{\partial t} = \frac{u(i, 1)-u(i,-1)}{2\Delta t} \tag{9.28}$$

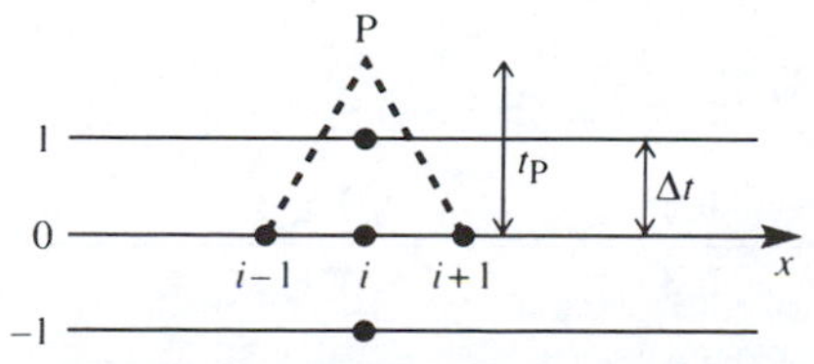

Figure 9.19 The first rows of mesh points in a numerical solution of the wave equation.

Now (9.27) with $j = 0$ becomes

$$u(i, 1) = 2u(i, 0) - u(i, -1) + \lambda^2 [u(i + 1, 0) - 2u(i, 0) + u(i - 1, 0)]$$

Since $u(i, 0) = f_i$ and $u(i, -1) = u(i, 1) - 2\Delta t\, g_i$, (9.27) now takes the form

$$u(i, 1) = (1 - \lambda^2) f_i + \tfrac{1}{2}\lambda^2 (f_{i+1} + f_{i-1}) + \Delta t\, g_i \qquad \textbf{(9.29)}$$

Thus the basic strategy is to compute row zero from $u(i, 0) = f_i$, evaluate row one from (9.29), and then march forward for general row j by (9.27).

EXAMPLE 9.15 Solve the wave equation $\partial^2 u/\partial t^2 = c^2 \partial^2 u/\partial x^2$ numerically with the conditions

(a) $u(x, 0) = \sin(\pi x)$ $(0 \leqslant x \leqslant 1)$ (initial displacement);

(b) $\partial u(x, 0)/\partial t = 0$ $(0 \leqslant x \leqslant 1)$ (initially at rest);

(c) $u(0, t) = u(1, t) = 0$ $(t \geqslant 0)$ (the two ends held fixed).

Use the values $c = 1$, $\Delta x = 0.25$, $\Delta t = 0.1$.

Solution Note that $\lambda^2 = 0.16$. The values at $t = 0$ are given by condition (a)

x	0	0.25	0.5	0.75	1
u	0	0.7071	1	0.7071	0

The values at $t = 0.1$ (or $j = 1$) are computed from (9.29) with $f_i = \sin(\pi x)$

$$u(i, 1) = 0.84 f_i + 0.08(f_{i+1} + f_{i-1})$$

and give

x	0	0.25	0.5	0.75	1
u	0	0.674	0.9531	0.674	0

The first two rows are now complete, so formula (9.27) can be used for each of the subsequent times, for $t = 0.2$ (or $j = 2$)

$$u(i, 2) = 2u(i, 1) - u(i, 0) + 0.16[u(i + 1, 1) - 2u(i, 1) + u(i - 1, 1)]$$

which gives

x	0	0.25	0.5	0.75	1
u	0	0.5777	0.8169	0.5777	0

and for $t = 0.3$ (or $j = 3$)

x	0	0.25	0.5	0.75	1
u	0	0.4272	0.6042	0.4272	0

and so on.

This problem has an exact solution so the results can be compared with $u(x, t) = \sin(\pi x)\cos(\pi t)$.

EXAMPLE 9.16 Solve the wave equation $\partial^2 u/\partial t^2 = c^2\, \partial^2 u/\partial x^2$ for a semi-infinite string, given the initial conditions

(a) $u(x, 0) = x \exp[-5(x-1)^2]$ $(x \geq 0)$ (string given an initial displacement);

(b) $\partial u(x, 0)/\partial t = 0$ $(x \geq 0)$ (string at rest initially);

(c) $u(0, t) = 0$ $(t \geq 0)$ (string held at the point $x = 0$).

Solution Since $g_i = 0$ in (9.29), only the one parameter λ needs to be specified. Figure 9.20 shows the solution of u over eight time steps with $\lambda = 0.5$. It can be seen that the solution splits into two waves, one moving in the $+x$ direction and the other in the $-x$ direction. At a given time $t = 0.8/c$, the u values are presented in the table shown in Figure 9.21 for various values of λ. We see that for $\lambda < 1$ the solution is reasonably consistent, and we have errors of a few per cent. However, for $\lambda = 2$ the solution looks very suspect.

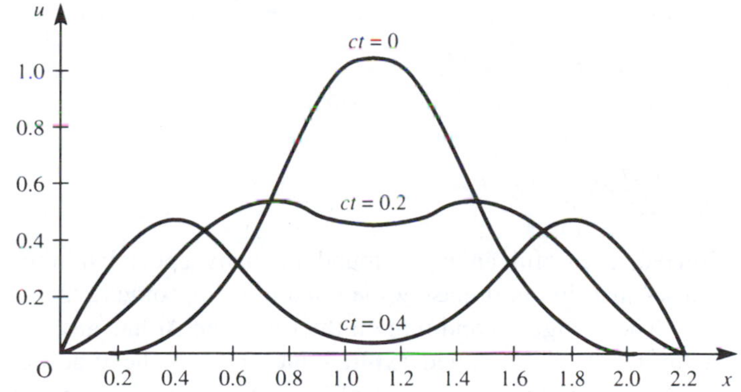

Figure 9.20 Solution of Example 9.16 with $\Delta x = 0.2$, $\lambda = 0.5$ for successive values of ct.

x	0	0.2	0.4	0.6	0.8	1.0
$u(\lambda = 0.25)$	0	0.3451	0.4674	0.3368	0.1353	0.0236
$u(\lambda = 0.5)$	0	0.3487	0.4665	0.3318	0.1340	0.0272
$u(\lambda = 1)$	0	0.3652	0.4582	0.3105	0.1322	0.0408
$u(\lambda = 2)$	0	0.1078	0.3571	0.6334	0.5742	0.2749

Figure 9.21 Table of values of u for a numerical solution of Example 9.16 with $\Delta x = 0.2$ and $ct = 0.8$.

A further two time steps gives, at $ct = 1.6$, the solution

x	0	0.2	0.4	0.6	0.8	1
$u(\lambda = 2)$	0	−3.12	21.75	−10.25	−34.70	32.72

Clearly the solution has gone wild!

Looking back to Figure 9.19, we can attempt an explanation for the apparent divergence of the solution in Example 9.16. The characteristics through the points $(x_{i-1}, 0)$ and $(x_{i+1}, 0)$ are

$$x_{i-1} = x - ct$$

$$x_{i+1} = x + ct$$

which can be solved to give, at the point P,

$$x_P = \tfrac{1}{2}(x_{i+1} + x_{i-1})$$

$$ct_P = \tfrac{1}{2}(x_{i+1} - x_{i-1}) = \Delta x$$

Recalling the work done on characteristics, we should require the new point to be *inside* the domain of dependence defined by the interval (x_{i-1}, x_{i+1}). Hence we require

$$t_P \geqslant \Delta t$$

so

$$\frac{c\Delta t}{\Delta x} \leqslant 1 \tag{9.30}$$

Indeed, a careful analysis, found in many specialist numerical analysis books, shows that this is precisely the condition for convergence of the method.

The stringent condition on the time step Δt has always been considered to be a limitation on so-called **explicit methods** of the type described here, but such methods have the great merit of being very simple to program. As computers get faster, the very short time step is becoming less of a problem, and vector or array processors allow nodes to be dealt with simultaneously, thus making such methods even more competitive.

There are, however, clear advantages in the stability of calculations if an **implicit method** is used. In Figure 9.18 the approximation to u_{xx} may be formed by the average of the approximations from rows $j + 1$ and $j - 1$. Thus

$$\begin{aligned}&[u(i, j + 1) - 2u(i, j) + u(i, j - 1)]/c^2\,\Delta t^2\\ &= \tfrac{1}{2}[u(i + 1, j + 1) - 2u(i, j + 1) + u(i - 1, j + 1)\\ &\quad + u(i + 1, j - 1) - 2u(i, j - 1) + u(i - 1, j - 1)]/\Delta x^2\end{aligned}$$

Assuming that u is known on rows j and $j - 1$, we can rearrange the equation into the convenient form

$$-\lambda^2 u(i+1, j+1) + 2(1+\lambda^2)u(i, j+1) - \lambda^2 u(i-1, j+1)$$
$$= 4u(i, j) + \lambda^2 u(i+1, j-1) - 2(1+\lambda^2)u(i, j-1) + \lambda^2 u(i-1, j-1) \quad \textbf{(9.31)}$$

The right-hand side of (9.31) is known, since it depends only on rows j and $j-1$. The unknowns on row $j+1$ appear on the left-hand side. The equations must now be solved simultaneously using the Thomas algorithm for a tridiagonal matrix (described in Section 5.5.2 of *Modern Engineering Mathematics*). This algorithm is very rapid and requires little storage. It can be shown that the method will proceed satisfactorily for any λ, so that the time step is unrestricted. The evaluation of rows 0 and 1 is the same as for the explicit method, so this can reduce the accuracy, and clearly the algorithm needs a finite x region to allow the matrix inversion.

EXAMPLE 9.17 Solve the wave equation $\partial^2 u/\partial t^2 = c^2 \partial^2 u/\partial x^2$ by an implicit method given

(a) $u(0, t) = 0 \quad (t \geq 0)$ (fixed at $x = 0$);

(b) $u(1, t) = 0 \quad (t \geq 0)$ (fixed at $x = 1$);

(c) $\partial u(x, 0)/\partial t = 0 \quad (0 \leq x \leq 1)$ (zero initial velocity);

(d) $u(x, 0) = \begin{cases} 1 & (x = \frac{1}{4}) \\ 0 & \text{otherwise} \end{cases}$ (displaced at the one point $x = \frac{1}{4}$).

Compare the solutions at a fixed time for various λ.

Solution Here we have a wave equation solved for a string stretched between two points and displaced at a single point.

The numerical solution shows the expected behaviour of a wave splitting into two waves, one moving in the $-x$ direction and the other in the $+x$ direction. The waves are reflected from the ends, and eventually give a complicated wave shape.

The computations were performed with $\Delta x = 0.125$ and various λ or Δt, with $\lambda = c\,\Delta t/\Delta x$. The values of u are given at the same time, $T = \Delta x/c$, for various λ:

x	0	0.125	0.25	0.375	0.5	0.625	0.75	0.875	1
$u(\lambda = 0.2)$	0	0.3394	0.2432	0.3412	0.0352	0.0019	0.0001	0	0
$u(\lambda = 0.1)$	0	0.3479	0.2297	0.3493	0.0344	0.0014	0	0	0
$u(\lambda = 0.05)$	0	0.3506	0.2254	0.3519	0.0341	0.0013	0	0	0
$u(\lambda = 0.025)$	0	0.3514	0.2243	0.3526	0.0340	0.0014	0	0	0

Although the method converges for all λ, the accuracy still requires a small λ (or time step), but the value $\lambda = 0.05$ certainly gives an accuracy of less than 1%. It may be noted that at the chosen value of T the wave has split but has not progressed far enough to be reflected from the end $x = 1$.

The methods described in this section all extend to higher dimensions, and some to nonlinear problems. The work involved is correspondingly greater of course.

9.3.6 Exercises

†21 Use an explicit method to solve the wave equation $\partial^2 u/\partial t^2 = c^2\partial^2 u/\partial x^2$ for the boundary conditions

(a) $u(0, t) = 0 \quad (t \geqslant 0)$
(b) $u(1, t) = 0 \quad (t \geqslant 0)$
(c) $u(x, 0) = 0 \quad (0 \leqslant x \leqslant 1)$
(d) $\dfrac{\partial u(x, 0)}{\partial t} = \begin{cases} x & (0 \leqslant x \leqslant \frac{1}{2}) \\ 1 - x & (\frac{1}{2} \leqslant x \leqslant 1) \end{cases}$

Use $\Delta x = \Delta t = \frac{1}{4}$ and study the behaviour for a variety of values of λ for the first three time steps. Compare your result with the implicit version in (9.31).

†22 An oscillator is started at the end of a tube, and oscillations propagate according to the wave equation. The displacement $u(x, t)$ satisfies

$$\frac{\partial^2 u}{\partial x^2} = \frac{1}{c^2}\frac{\partial^2 u}{\partial t^2}$$

in $0 < x < l$, for $t > 0$, with the boundary conditions

(a) $u(0, t) = a \sin \omega t, \quad u(l, t) = 0 \quad (t > 0)$

(b) $u(x, 0) = \dfrac{\partial u(x, 0)}{\partial t} = 0 \quad (0 \leqslant x \leqslant l)$

where c, a and ω are real positive constants. Show that the solution of the partial differential equation is

$$u(x, t) = \frac{a \sin \omega t \sin [\omega(l - x)/c]}{\sin (\omega l/c)} + \sum_{n=1}^{\infty} \frac{2lac\omega}{\omega^2 l^2 - n^2\pi^2 c^2} \sin\left(\frac{n\pi x}{l}\right) \sin\left(\frac{n\pi ct}{l}\right)$$

provided that $\omega l/\pi c$ is not an integer.

Compare this solution with one computed using the explicit numerical method. Use $l = 1$, $c = 1$, $\omega = \frac{1}{2}\pi$, $\Delta x = 0.2$ and $\Delta t = 0.02$ to evaluate $u(x, 0.06)$.

†23 Solve the equation

$$c^2\frac{\partial^2 u}{\partial x^2} + 2 = \frac{\partial^2 u}{\partial t^2}$$

numerically, subject to the conditions

$$u = x(1 - x), \quad \frac{\partial u}{\partial t} = 0 \quad \text{for all } x \text{ at } t = 0$$

Use

(a) an explicit method with $\Delta x = \Delta t = 0.2$ and $\lambda = 0.5$;
(b) an implicit method with $\Delta x = \Delta t = 0.2$ and $\lambda = 0.5$.

†24 Solve the equation

$$c^2\frac{\partial^2 u}{\partial x^2} + 2 = \frac{\partial^2 u}{\partial t^2}$$

numerically, subject to the conditions

$$u = x(1 - x), \quad \frac{\partial u}{\partial t} = 0 \quad (0 < x < 1) \quad \text{at } t = 0$$

$$u = 0 \quad (x = 0, 1) \quad \text{for } t > 0$$

Use

(a) an explicit method with $\Delta x = \Delta t = 0.2$ and $\lambda = 0.5$;
(b) an implicit method with $\Delta x = \Delta t = 0.2$ and $\lambda = 0.5$.

Compare your solution with that in Exercise 23.

9.4 Solution of the heat-conduction/diffusion equation

In this section we consider methods for solving the heat-conduction/diffusion equation introduced in Section 9.2.2.

9.4.1 Separation method

It was with the aim of solving heat-conduction problems that Fourier (*c*. 1800) first used the idea of separation of variables and Fourier series. As indicated in Section 4.1, many mathematicians at the time argued about the validity of his approach, while he continued to solve many practical problems.

In Section 9.2.2 we noted that the heat-conduction equation

$$\frac{1}{\kappa}\frac{\partial u}{\partial t} = \nabla^2 u \tag{9.5}$$

has a steady-state solution U, provided there are no time varying inputs, satisfying

$$\nabla^2 U = 0$$

and appropriate boundary conditions. One useful way to write the general solution is

$$u = U + v$$

where v also satisfies (9.5) and the boundary conditions for $u - U$. Certainly the heat-conduction interpretation supports this idea, and we base our strategy on first finding U and then determining the transient v that takes the solution from its initial to its final state. We note that $v \to 0$ as $t \to \infty$, so that $u \to U$, and an obvious method is to try an exponential decay to zero. Thus, in the one-dimensional form of the heat-conduction equation

$$\frac{1}{\kappa}\frac{\partial v}{\partial t} = \frac{\partial^2 v}{\partial x^2} \tag{9.32}$$

we seek a separated solution of the special type discussed in Section 9.3.2, where the physics indicates a solution

$$v = \mathrm{e}^{-\alpha t} w(x)$$

which on substitution gives

$$-\frac{\alpha}{\kappa} w = \frac{\mathrm{d}^2 w}{\mathrm{d}x^2}$$

Letting $\alpha/\kappa = \lambda^2$, we can solve this simple-harmonic equation to give

$$w = A \sin \lambda x + B \cos \lambda x$$

and hence

$$v = \mathrm{e}^{-\alpha t}(A \sin \lambda x + B \cos \lambda x) \tag{9.33}$$

Taking the hint from Section 9.3.2, we expect in general to take sums of terms like (9.33) to satisfy all the boundary conditions. Thus we build up a solution

$$v = \sum_{n=1}^{\infty} \mathrm{e}^{-\alpha t}(A_n \sin \lambda x + B_n \cos \lambda x) \tag{9.34}$$

EXAMPLE 9.18 Solve the heat-conduction equation $\partial T/\partial t = \kappa\partial^2 T/\partial x^2$ subject to the boundary conditions

(a) $T = 0$ at $x = 0$ and for all $t > 0$ (held at zero temperature);

(b) $\partial T/\partial x = 0$ at $x = l$ and for all $t > 0$ (no heat loss from this end);

(c) $T = T_0 \sin(3\pi x/2l)$ at $t = 0$ and for $0 \leqslant x \leqslant l$ (given initial temperature profile).

Solution We first note that as $t \to \infty$ the solution will be $T = 0$, so the steady-state solution is zero, and so from (9.33) we consider a solution of the form

$$T = e^{-\alpha t}(A \sin \lambda x + B \cos \lambda x) \tag{9.35}$$

In order to satisfy the boundary condition (a), it is clear that it is not possible to include the cosine term, so $B = 0$. To satisfy the condition (b) then requires

$$\frac{\partial T}{\partial x} = A\,e^{-\alpha t}\,\lambda \cos \lambda x = 0 \quad (x = l)$$

so that

$$\cos \lambda l = 0, \quad \text{or} \quad \lambda l = (n + \tfrac{1}{2})\pi \quad (n = 0, 1, 2, 3, \dots)$$

leading to the solution

$$T = A\,e^{-\alpha t} \sin\left[\left(n + \frac{1}{2}\right)\pi\frac{x}{l}\right]$$

We now compare the T from condition (c) with the solution just obtained at time $t = 0$, giving

$$A\,e^{-0} \sin\left[\left(n + \frac{1}{2}\right)\pi\frac{x}{l}\right] = T_0 \sin\left(\frac{3\pi x}{2l}\right)$$

The unknown parameters can now be identified as $n = 1$ and $A = T_0$, and hence the final solution is

$$T = T_0 \exp\left(-\frac{9\kappa\pi^2 t}{4l^2}\right) \sin\left(\frac{3\pi x}{2l}\right)$$

EXAMPLE 9.19 Solve the heat-conduction equation $\partial u/\partial t = \kappa\partial^2 u/\partial t^2$ subject to the boundary conditions

(a) $u(0, t) = 0 \quad (t \geqslant 0)$ (zero temperature at the end $x = 0$);

(b) $u(l, t) = 0 \quad (t \geqslant 0)$ (zero temperature at the end $x = l$);

(c) $u(x, 0) = u_0(\tfrac{1}{2} - x/l) \quad (0 < x < l)$ (a given initial temperature profile).

Solution We are solving the problem of heat conduction in a bar that is held at zero temperature at its ends and with a given initial temperature profile.

It is clear that the final solution as $t \to \infty$ is $u = 0$, so that $U = 0$ is the steady-state solution, and hence from (9.33) we seek a solution of the form

$$u = e^{-\alpha t}(A \sin \lambda x + B \cos \lambda x)$$

subject to the given boundary conditions. The first of these conditions (a) gives $B = 0$, while the second condition (b) gives

$$\sin \lambda l = 0, \quad \text{or} \quad \lambda l = n\pi \quad (n = 1, 2, \dots)$$

Recalling that $\lambda^2 = \alpha/\kappa$, we find solutions of the form

$$u = A\,e^{-\kappa n^2\pi^2 t/l^2} \sin\left(\frac{n\pi x}{l}\right) \quad (n = 1, 2, \dots)$$

Clearly we cannot satisfy (c) from a single solution, so, as indicated in (9.34), we revert to the sum

$$u = \sum_{n=1}^{\infty} A_n\,e^{-\kappa n^2\pi^2 t/l^2} \sin\left(\frac{n\pi x}{l}\right) \tag{9.36}$$

which is also a solution.

Using the boundary condition (c), we then have

$$u_0\left(\frac{1}{2} - \frac{x}{l}\right) = \sum_{n=1}^{\infty} A_n \sin\left(\frac{n\pi x}{l}\right)$$

and hence, by (4.33),

$$\frac{1}{2}lA_n = u_0 \int_0^l \left(\frac{1}{2} - \frac{x}{l}\right) \sin\left(\frac{n\pi x}{l}\right) dx = \begin{cases} 0 & (\text{odd } n) \\ u_0 l/n\pi & (\text{even } n) \end{cases}$$

Note again that we have used a periodic extension of the given function to obtain a Fourier sine series valid over the interval $0 \leqslant x \leqslant l$. Outside the interval $0 \leqslant x \leqslant l$ we have no physical interest in the solution. Substituting back into (9.36) gives as a final solution

$$u = \frac{u_0}{\pi} \sum_{m=1}^{\infty} \frac{1}{m}\,e^{-4\kappa m^2\pi^2 t/l^2} \sin\left(\frac{2m\pi x}{l}\right) \tag{9.37}$$

or, in an expanded form,

$$u = \frac{u_0}{\pi}\left[e^{-4\kappa\pi^2 t/l^2} \sin\left(\frac{2\pi x}{l}\right) + \frac{1}{2}e^{-16\kappa\pi^2 t/l^2} \sin\left(\frac{4\pi x}{l}\right) + \frac{1}{3}e^{-36\kappa\pi^2 t/l} \sin\left(\frac{6\pi x}{l}\right) + \dots\right]$$

In Figure 9.22, u/u_0 is plotted against x/l at successive times $T = t(4\kappa\pi^2/l^2) = 0$, 0.5, 1, 1.5.

Taking successive terms in the series for the values $T = t(4\kappa\pi^2/l^2) = 0.5$ and $x/l = 0.2$, we get

	1 term	2 terms	3 terms	4 terms
u/u_0	0.1836	0.1963	0.1956	0.1953

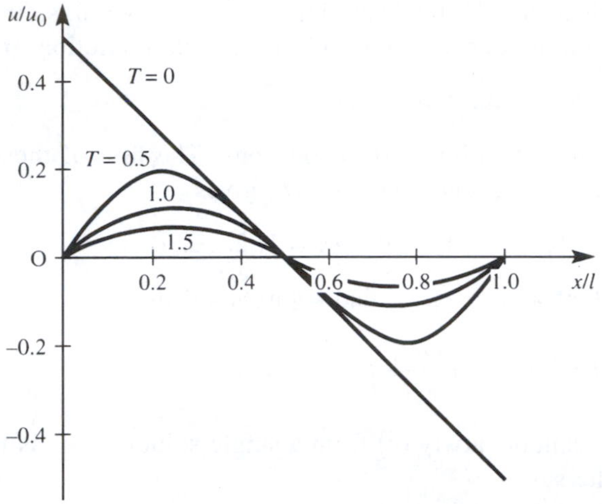

Figure 9.22 Solution of Example 9.19 with $T = t(4\kappa\pi^2/l^2)$.

and we see that three terms of this series would probably be sufficient to give three-figure accuracy. In all such problems some numerical experimentation is required to determine how many terms are required. For small t and x we should expect to need a large number of terms, since the temperature at the end switches from $\frac{1}{2}u_0$ to 0 at time $t = 0$. It is well known, as we saw in Chapter 4, that discontinuities cause convergence difficulties for Fourier series.

It may be noted in Example 9.19 that the initial discontinuity is smoothed out, as we expected from the physical ideas that we outlined in Section 9.2.2.

EXAMPLE 9.20 Solve the heat-conduction equation $\partial u/\partial t = \kappa\partial^2 u/\partial x^2$ in a bar subject to the boundary conditions

(a) $u(0, t) = 0 \quad (t \geqslant 0)$ (the end $x = 0$ is held at zero temperature);

(b) $u(1, t) = 1 \quad (t \geqslant 0)$ (the end $x = 1$ is at temperature 1);

(c) $u(x, 0) = x(2 - x) \quad (0 \leqslant x \leqslant 1)$ (the initial temperature profile is given).

Solution First it is clear that the final steady-state solution is $U = x$, since this satisfies (a) and (b) and also $\nabla^2 U = 0$. Secondly putting $u = U + v$, the new variable v satisfies (9.32), but now the boundary conditions on v are

(a′) $v(0, t) = 0$

(b′) $v(1, t) = 0$

(c′) $v(x, 0) = x(2 - x) - x = x - x^2$

The appropriate solutions in (9.33) can now be selected; the condition (a′) gives $B = 0$, while the condition (b′), $v(1, t) = 0$, gives

$$\sin\lambda = 0, \quad \text{or} \quad \lambda = n\pi \quad (n = 1, 2, \ldots)$$

From (9.34) we then have

$$v = \sum_{n=1}^{\infty} a_n \, e^{-\kappa n^2\pi^2 t} \sin n\pi x$$

and condition (c′) gives

$$x - x^2 = \sum_{n=1}^{\infty} a_n \sin n\pi x$$

Determining the Fourier coefficient using (4.33),

$$\tfrac{1}{2} a_n = \int_0^1 (x - x^2) \sin n\pi x \, dx = \begin{cases} 4/n^3\pi^3 & (\text{odd } n) \\ 0 & (\text{even } n) \end{cases}$$

Thus the complete solution is

$$u = x + \frac{8}{\pi^3} \sum_{n=1}^{\infty} \frac{1}{(2n-1)^3} e^{-\kappa(2n-1)^2\pi^2 t} \sin (2n-1)\pi x$$

or, in expanded form,

$$u = x + \frac{8}{\pi^3}\left[e^{-\kappa\pi^2 t} \sin \pi x + \frac{1}{27} e^{-9\kappa\pi^2 t} \sin 3\pi x + \frac{1}{125} e^{-25\kappa\pi^2 t} \sin 5\pi x + \ldots \right]$$

9.4.2 Laplace transform method

As we saw for the wave equation in Section 9.3.3, Laplace transforms provide an alternative method of solution for the heat-conduction equation. The method has the merit of dealing with the boundary conditions easily, but it suffers from the usual difficulty of performing the final inversion. The following example serves to illustrate these points.

EXAMPLE 9.21 Using the Laplace transform method, solve the heat-conduction equation

$$\frac{\partial u}{\partial t} = \kappa \frac{\partial^2 u}{\partial x^2} \quad (x > 0,\ t > 0)$$

given that $u(x, t)$ remains bounded and satisfies the boundary conditions

(a) $u(x, 0) = 0 \quad (x > 0)$

(b) $u(a, t) = T\delta(t)$

Solution This problem models a semi-infinite insulated bar, coincident with the positive x axis, that is initially at zero temperature and is subjected to an instantaneous heat source (a spot weld) of strength T at the point $x = a > 0$.

Using (9.23) and (9.24) and taking Laplace transforms gives

$$sU(x, s) - u(x, 0) = \kappa\frac{d^2U(x, s)}{dx^2}$$

which, on using condition (a), leads to the ordinary differential equation

$$\frac{d^2U}{dx^2} - \frac{s}{\kappa}U = 0$$

This is readily solved to give

$$U(x, s) = A\,e^{\sqrt{(s/\kappa)}x} + B\,e^{-\sqrt{(s/\kappa)}x}$$

Since the temperature of the bar remains bounded, A must be zero, so that

$$U(x, s) = B\,e^{-\sqrt{(s/\kappa)}x}$$

Condition (b) then gives

$$U(a, s) = \mathscr{L}\{T\delta(t)\} = T = B\,e^{-\sqrt{(s/\kappa)}a}$$

so that

$$B = T\,e^{\sqrt{(s/\kappa)}a}$$

giving

$$U(x, s) = T\,e^{-(x-a)\sqrt{(s/\kappa)}}$$

To find the solution $u(x, t)$, we must invert the Laplace transform. However, in this case, the methods discussed in Chapter 2 do not suffice, and it is necessary to resort to the use of the complex inversion integral, which is dealt with in specialist texts on Laplace transforms (see also Chapter 5, Review exercise 7). Alternatively, we can turn to the extensive tables that exist of Laplace transform pairs, to find that

$$\mathscr{L}^{-1}\{e^{-b\sqrt{s}}\} = \frac{b}{2\sqrt{\pi}}t^{-3/2}\,e^{-b^2/4t}$$

We can then carry out the required inversion to give the solution

$$u(x, t) = \frac{T(x-a)}{2\sqrt{(\pi\kappa)}}t^{-3/2}\,e^{-(x-a)^2/4\kappa t} \qquad (t > 0)$$

It should be noted that the solution of Example 9.21 is not variable separable and could not be obtained from the methods of Section 9.4.1.

9.4.3 Exercises

25 Find the solution to the equation

$$\frac{\partial u}{\partial t} = \kappa\frac{\partial^2 u}{\partial x^2}$$

satisfying the conditions

(a) $\partial u/\partial x = 0$ at $x = 0$ for all t
(b) $u = 0$ at $x = 1$ for all t
(c) $u = a\cos(\pi x)\cos(\frac{1}{2}\pi x)$ for $0 \leqslant x \leqslant 1$ when $t = 0$

26 Use separation of variables to obtain a solution to the heat-conduction equation $\partial u/\partial t = \kappa\partial^2 u/\partial x^2$, given

(a) $\partial u(0, t)/\partial x = 0 \quad (t \geqslant 0)$
(b) $u(l, t) = 0 \quad (t \geqslant 0)$
(c) $u(x, 0) = u_0(\frac{1}{2} - x/l) \quad (0 \leqslant x \leqslant l)$

Compare the solution with that obtained in Example 9.19.

27 The spherically symmetric form of the heat-conduction equation is

$$u_{rr} + \frac{2}{r}u_r = \frac{1}{\kappa}u_t$$

By putting $ru = v$, show that v satisfies the standard one-dimensional heat-conduction equation. What can we expect of a solution as $r \to \infty$?

28 Show that $u(x, t) = t^{\alpha}F(\eta)$, where $\eta = x^2/t$ is a solution of the partial differential equation

$$\frac{\partial u}{\partial t} = \frac{\partial^2 u}{\partial x^2}$$

if F satisfies

$$4\eta\frac{\mathrm{d}^2F}{\mathrm{d}\eta^2} + (2 + \eta)\frac{\mathrm{d}F}{\mathrm{d}\eta} - \alpha F = 0$$

Find non-zero values of α and κ for which $F = \mathrm{e}^{\kappa\eta}$ is a solution.

29 Show that $u(x, t) = f(x)\sin(x - \beta t)$ is a solution of the heat conduction equation

$$\frac{\partial u}{\partial t} = \frac{\partial^2 u}{\partial x^2}$$

provided that f and the constant β are chosen suitably. Give a physical interpretation of the problem in the context of a semi-infinite slab of uniform material occupying the region $x \geqslant 0$.

30 Show that the equation

$$\theta_t = \kappa\theta_{xx} - h(\theta - \theta_0)$$

can be reduced to the standard heat-conduction equation by writing $u = \mathrm{e}^{ht}(\theta - \theta_0)$. How do you interpret the term $h(\theta - \theta_0)$?

31 The voltage v at a time t at a distance x along an electric cable of length L with capacitance and resistance only, satisfies

$$\frac{\partial^2 v}{\partial x^2} = \frac{1}{\kappa}\frac{\partial v}{\partial t}$$

Verify that a form of the solution appropriate to the conditions that $v = v_0$ when $x = 0$, and $v = 0$ when $x = L$, for all values of t, is given by

$$v = v_0\left(1 - \frac{x}{L}\right) + \sum_{n=1}^{\infty} c_n \exp\left(\frac{-\kappa n^2\pi^2 t}{L^2}\right)\sin\left(n\pi\frac{x}{L}\right)$$

where v_0 and the c_n are constants.

Show that if, in addition, $v = 0$ when $t = 0$ for $0 < x < L$,

$$c_n = -\frac{2v_0}{n\pi}$$

32 A uniform bar of length l has its ends maintained at a temperature of 0 °C. Initially, the temperature at any point between the ends of the bar is 10 °C, and, after a time t, the temperature $u(x, t)$ at a distance x from one end of the bar satisfies the one-dimensional heat-conduction equation

$$\frac{\partial^2 u}{\partial x^2} = \frac{1}{\kappa}\frac{\partial u}{\partial t} \quad (x > 0)$$

Write down boundary conditions for the bar and show that the solution corresponding to these conditions is

$$u(x, t) = \frac{20}{\pi}\sum_{n=1}^{\infty}\frac{1}{n}(1 - \cos n\pi)\exp\left(\frac{-\kappa n^2\pi^2 t}{l^2}\right) \times \sin\left(\frac{\pi n x}{l}\right)$$

33 The function $\phi(x, t)$ satisfies the equation

$$\frac{\partial \phi}{\partial t} = a\frac{\partial^2 \phi}{\partial x^2} + b \quad (-h < x < h,\ t > 0)$$

with the boundary conditions

(a) $\phi(-h, t) = \phi(h, t) = 0 \quad (t > 0)$
(b) $\phi(x, 0) = 0 \quad (-h < x < h)$

where a, b and h are positive real constants. Show that the Laplace transform of the solution $\phi(x, t)$ is

$$\frac{b}{s^2}\left\{1 - \frac{\cosh[(s/a)^{1/2}x]}{\cosh[(s/a)^{1/2}h]}\right\}$$

9.4.4 Numerical solution

As for the wave equation, except for the most straightforward problems, we must resort to numerical solutions of the heat-conduction equation. Even when analytical solutions are known, they are not always easy to evaluate because of convergence difficulties near to singularities. They are, of course, crucial in testing the accuracy and efficiency of numerical methods.

We can write the heat-conduction equation

$$\frac{1}{\kappa}\frac{\partial u}{\partial t} = \frac{\partial^2 u}{\partial x^2} \tag{9.38}$$

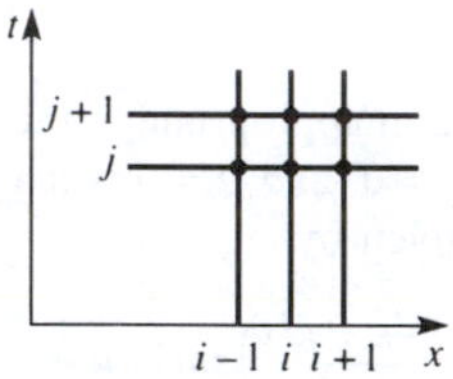

Figure 9.23 Mesh for marching forward the solution of the heat-conduction equation.

in the usual *finite-difference* form, using the notation of Figure 9.23.

We assume that we know the solution up to time step j and we wish to calculate the solution at time step $j + 1$. In Section 9.3.5 we showed how to approximate the second derivative as

$$\frac{\partial^2 u}{\partial x^2} = \frac{u(i+1, j) - 2u(i, j) + u(i-1, j)}{\Delta x^2}$$

To obtain the time derivative, we use the approximation between rows j and $j + 1$:

$$\frac{\partial u}{\partial t} = \frac{u(i, j+1) - u(i, j)}{\Delta t}$$

Putting these into (9.38) gives

$$\frac{u(i, j+1) - u(i, j)}{\kappa \Delta t} = \frac{u(i-1, j) - 2u(i, j) + u(i+1, j)}{\Delta x^2}$$

or, on rearranging,

$$u(i, j+1) = \lambda u(i-1, j) + (1 - 2\lambda)u(i, j) + \lambda u(i+1, j) \tag{9.39}$$

where $\lambda = \kappa\Delta t/\Delta x^2$. Equation (9.39) gives a finite-difference representation of (9.38), and provided that all the values are known on row j, we can then compute u on row $j + 1$ from the simple **explicit formula** (9.39).

First a simple example on a coarse grid.

EXAMPLE 9.22 Use an explicit numerical method to solve the heat conduction equation (9.38) subject to the boundary conditions

(a) $u(0, t) = u(0, 1) = 0 \quad (t \geq 0)$ (both ends held at zero temperature);

(b) $u(x, 0) = \sin(\pi x) \quad (0 \leq x \leq 1)$ (a given initial temperature distribution).

Use the parameters $\Delta t = 0.1$, $\Delta x = 0.25$, $\kappa = 0.1$.

Solution This problem has the exact solution $u = e^{-\pi^2 \kappa t} \sin(\pi x)$, so the accuracy of the numerical solution can be checked easily.

At $t = 0$ (or $j = 0$) the initial values come from the boundary condition (b)

x	0	0.25	0.5	0.75	1
u	0	0.7071	1	0.7071	0

Before proceeding further we first note that $\lambda = 0.16$. At $t = 0.1$ (or $j = 1$) equation (9.39) becomes

$$u(i, 1) = 0.16[u(i-1, 0) + u(i+1, 0)] + 0.68u(i, 0)$$

which, on calculation for the values $x = 0.25, 0.5, 0.75$ or $i = 1, 2, 3$, gives

x	0	0.25	0.5	0.75	1
u	0	0.6408	0.9063	0.6048	0

Note that at the ends, $x = 0$ and 1, the boundary condition (a) $u = 0$ is imposed.

At $t = 0.2$ (or $j = 2$)

$$u(i, 2) = 0.16[u(i-1, 1) + u(i+1, 1)] + 0.68u(i, 1)$$

and performing the calculations

x	0	0.25	0.5	0.75	1
u	0	0.5808	0.8213	0.5808	0

Similarly for $t = 0.3$ (or $j = 3$)

x	0	0.25	0.5	0.75	1
u	0	0.5263	0.7444	0.5263	0
u exact	0	0.5259	0.7437	0.5259	0

and so on.

In the last table the exact values have been included for comparison.

EXAMPLE 9.23 Solve the heat-conduction equation (9.38) subject to the boundary conditions

(a) $\partial u(0, t)/\partial x = 0 \quad (t \geq 0)$ (no heat flow through the end $x = 0$);

(b) $u(1, t) = 1 \quad (t \geq 0)$ (unit temperature held at $x = 1$);

(c) $u(x, 0) = x^2 \quad (0 \leq x \leq 1)$ (a given initial temperature distribution).

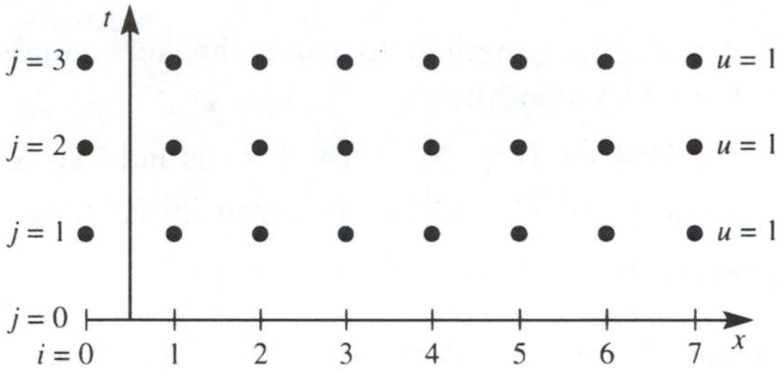

Figure 9.24 Mesh for Example 9.23: $u(7, j) = 1$ and $u(0, j) = u(1, j)$ for all j; $\Delta x = 1/6.5 = 0.1538$.

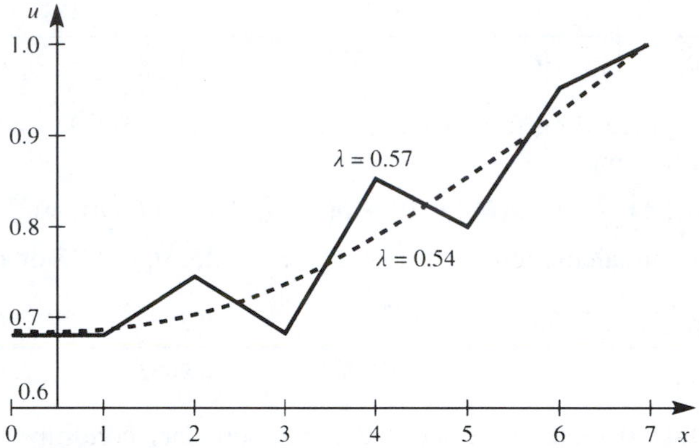

Figure 9.25 Numerical solution of Example 9.13 at time $t = 20\ \Delta x^2/x$, for two values of λ.

Solution To fit the condition (a) most easily, we allow the first mesh space to straddle the t axis as illustrated in Figure 9.24, where six intervals are used in the x direction. The mesh implies that $\Delta x = 1/6.5 = 0.1538$; condition (a) gives $u(0, j) = u(1, j)$ while condition (b) gives $u(7, j) = 1$.

The following table gives values for u at the three times $t = 0$, $0.2\Delta x^2/\kappa$ and $20\Delta x^2/\kappa$, calculated with $\lambda = 0.2$. These results are then compared at $t = 20\Delta x^2/\kappa$, computed with λ taken to be 0.5. It may be noted that there are errors in the third significant figure between the two cases.

i		0	1	2	3	4	5	6	7
$\lambda = 0.2$	$j = 0$	0.0059	0.0059	0.0533	0.1479	0.2899	0.4793	0.7160	1
	$j = 1$	0.0154	0.0154	0.0627	0.1574	0.2994	0.4888	0.7255	1
	$j = 100$	0.6817	0.6817	0.7002	0.7362	0.7874	0.8510	0.9233	1
$\lambda = 0.5$	$j = 40$	0.6850	0.6850	0.7033	0.7389	0.7896	0.8526	0.9241	1

For $\lambda = 0.54$ we can obtain a solution that compares with the solution given, but for $\lambda = 0.6$ the solution diverges wildly. Figure 9.25 shows a plot of the solution near the critical value of λ at a fixed time $t = 20\Delta x^2/\kappa$. As in the table, the solutions are accurate to about 1% for small λ, but when λ gets much above

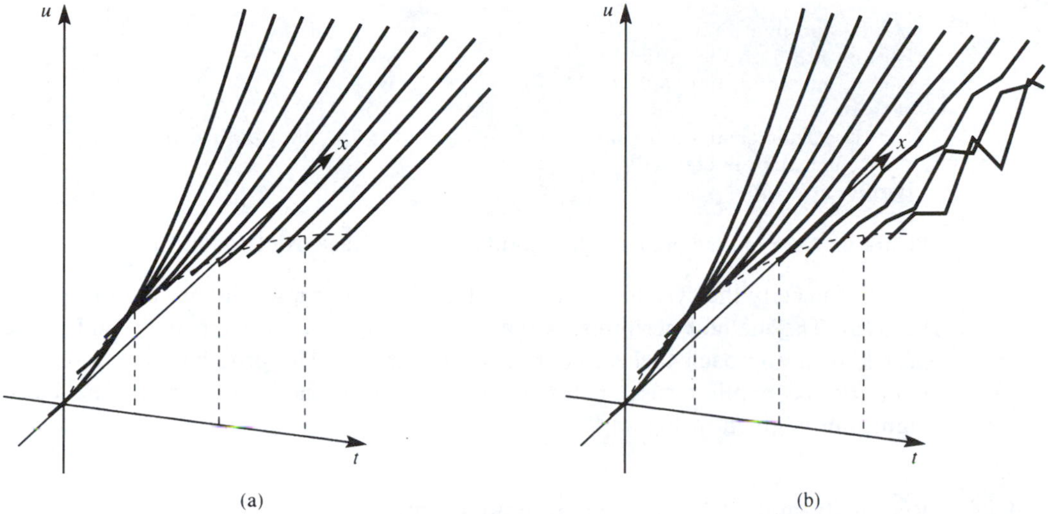

Figure 9.26 Plots of u against x and t from the solution of Example 9.23 with (a) $\lambda = 0.2$; (b) $\lambda = 0.55$.

0.5, oscillations creep in and the solution is meaningless. In Figure 9.26 a further graph illustrates the development of the solution in the two cases $\lambda = 0.2$ and $\lambda = 0.55$. One solution progresses smoothly as time advances, while the other produces oscillations that will eventually lead to divergence.

Comparing Example 9.23 with the numerical solution of the wave equation undertaken in Example 9.16, we observe similar behaviour for the explicit scheme, namely that the method will only converge for small enough time steps or λ. From (9.39) it may be noted that the middle term changes sign at $\lambda = 0.5$, and above this value we might anticipate difficulties. Indeed, some straightforward numerical analysis shows that convergence is certain for $\lambda < 0.5$. It is sufficient here to note that λ must not be too large.

To avoid the limitation on λ, we can again look at an **implicit** formulation of the numerical equations. Returning to Figure 9.23, the idea is to approximate the x derivative by an average of row j and row $j + 1$:

$$u(i, j+1) - u(i, j) = \lambda\{(1-\alpha)[u(i-1, j) - 2u(i, j) + u(i+1, j)] \\ + \alpha[u(i-1, j+1) - 2u(i, j+1) + u(i+1, j+1)]\}$$

where $0 \leqslant \alpha \leqslant 1$ is an averaging parameter. The case $\alpha = 0$ corresponds to the explicit formulation (9.39), while $\alpha = \frac{1}{2}$ is the best known implicit formulation, and constitutes the **Crank–Nicolson method**. With $\alpha = \frac{1}{2}$, we have

$$-\lambda u(i-1, j+1) + 2(1+\lambda)u(i, j+1) - \lambda u(i+1, j+1) \\ = \lambda u(i-1, j) + 2(1-\lambda)u(i, j) + \lambda u(i+1, j) \qquad \textbf{(9.40)}$$

We know the solution on row j, so the right-hand side of (9.40) is known, and the unknowns on row $(j + 1)$ have to be solved for simultaneously.

```
[read λ and final time]
[set up u(i, 0) on the initial line]
[set up end conditions u(0, j), u(N, j)]
repeat
    [Use Thomas algorithm on (9.40) to evaluate u(i, j + 1) for each i]
until [final time is reached]
[write out results]
```

Figure 9.27 An outline pseudocode algorithm for the Crank–Nicolson method.

Fortunately the system is tridiagonal, so the very rapid Thomas algorithm can be used. The method performs extremely well: it converges for all λ, and is the best known approach to heat conduction equations. The programming is harder than for the explicit case, but it is usually worth the extra effort. The basic algorithm is given in Figure 9.27.

EXAMPLE 9.24 Repeat Example 9.22 but with an implicit scheme.

Solution At $t = 0$ (or $j = 0$) the initial values come from the boundary condition and are identical for both the implicit and explicit formulations.

At $t = 0.1$ (or $j = 1$) the three equations of the type (9.40) corresponding to $x = 0.25, 0.5, 0.75$ or $i = 1, 2, 3$ are:

$$-0.16u(0, 1) + 2.32u(1, 1) - 0.16u(2, 1) = 0.16[u(0, 0) + u(2, 0)] + 1.68u(1, 0)$$

$$-0.16u(1, 1) + 2.32u(2, 1) - 0.16u(3, 1) = 0.16[u(1, 0) + u(3, 0)] + 1.68u(2, 0)$$

$$-0.16u(2, 1) + 2.32u(3, 1) - 0.16u(4, 1) = 0.16[u(2, 0) + u(4, 0)] + 1.68u(3, 0)$$

After noting that the end boundary conditions give

$$u(0, 0) = u(0, 1) = u(4, 0) = u(4, 1)$$

and the right-hand sides evaluated from the initial values, the equations can be written in matrix form as

$$\begin{bmatrix} 2.32 & -0.16 & 0 \\ -0.16 & 2.32 & -0.16 \\ 0 & -0.16 & 2.32 \end{bmatrix} \begin{bmatrix} u(1, 1) \\ u(2, 1) \\ u(3, 1) \end{bmatrix} = \begin{bmatrix} 1.348 \\ 1.906 \\ 1.348 \end{bmatrix}$$

The tridiagonal system can be solved to give

x	0	0.25	0.5	0.75	1
u	0	0.6438	0.9105	0.6438	0

For the next time steps the matrix equation is identical, with the j-suffix advanced by 1 at each time step and the right-hand sides re-evaluated from the most recently computed values of u. Subsequent values are

x	0	0.25	0.5	0.75	1
u at $t = 0.2$	0	0.5862	0.829	0.5862	0
u at $t = 0.3$	0	0.5337	0.7547	0.5337	0

EXAMPLE 9.25 Solve the heat-conduction equation (9.38) using the implicit formulation (9.40) for the boundary conditions

(a) $u(0, t) = 0 \quad (t \geqslant 0)$ (the end $x = 0$ is kept at zero temperature);

(b) $u(1, t) = 1 \quad (t \geqslant 0)$ (the end $x = 0$ is kept at unit temperature);

(c) $u(x, 0) = 0 \quad (0 \leqslant x \leqslant 1)$ (initially the bar has zero temperature).

Solution Here a bar is initially at zero temperature. At one end the temperature is raised to the value 1 and kept at that value.

The results of the calculation are presented in Figure 9.28. At time step 0 the temperature distribution is discontinuous. The successive time steps 1, 10, 100 are shown, and the final distribution $u = x$ is labelled ∞.

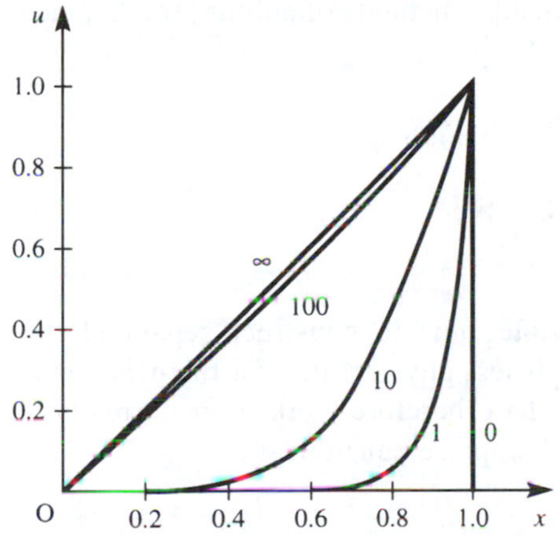

Figure 9.28 Solution of Example 9.25 with $\Delta x = 0.1$, $\lambda = 0.3$ using the Crank–Nicolson scheme.

9.4.5 Exercises

†34 Derive the usual explicit finite-difference representation of the equation

$$\frac{\partial u}{\partial t} = \frac{\partial^2 u}{\partial x^2}$$

Using this scheme with $\Delta t = 0.02$ and $\Delta x = 0.2$, determine an approximate solution of the equation at $t = 0.06$, given that

$$u = x^2 \quad \text{when} \quad t = 0 \quad (0 \leqslant x \leqslant 1)$$

$$u = 0 \quad \text{when} \quad x = 0 \quad (t > 0)$$

$$u = 1 \quad \text{when} \quad x = 1 \quad (t > 0)$$

†35 Use both explicit and implicit numerical formulations to obtain solutions of the heat-conduction equation subject to the boundary conditions

(a) $u(0, t) = 0 \quad (t \geqslant 0)$

(b) $u(1, t) = e^{-t} \quad (t \geqslant 0)$

(c) $u(x, 0) = 0 \quad (0 \leqslant x < 1)$

Compare the two results for $t = 1$.

†36 Given that u satisfies the equation

$$\frac{\partial u}{\partial t} = \frac{\partial^2 u}{\partial x^2}$$

and is subject to the boundary conditions

$$\frac{\partial u}{\partial x} = 1 \quad (x = 0,\ t > 0)$$

$$u = 0 \quad (x = 1,\ t > 0)$$

$$u = x(1 - x) \quad (t = 0,\ 0 \leqslant x \leqslant 1)$$

derive a set of algebraic equations from the implicit formulation in Section 9.4.4 using $\Delta x = 0.2$ and $\Delta t = 0.02$. Use the Thomas algorithm to find the solution at $t = 0.02$.

9.5 Solution of the Laplace equation

In this section we consider methods of solving the Laplace equation introduced in Section 9.2.3.

9.5.1 Separated solutions

It is much less obvious how to construct separated solutions for the Laplace equation, since there is less physical feel for the behaviour except that the solution will be smooth. We shall therefore work more formally, as in Section 9.3.2, and seek a solution of the Laplace equation

$$\frac{\partial^2 u}{\partial x^2} + \frac{\partial^2 u}{\partial y^2} = 0 \tag{9.41}$$

in the form

$$u = X(x)Y(y)$$

which gives on substitution

$$Y\frac{d^2X}{dx^2} + X\frac{d^2Y}{dy^2} = 0$$

or

$$\frac{d^2X}{dx^2}\Big/X = -\frac{d^2Y}{dy^2}\Big/Y = \lambda \tag{9.42}$$

Now $(d^2X/dx^2)/X$ is a function of x only and $-(d^2Y/dy^2)/Y$ is a function of y only. Since they must be equal for *all* x and y, both sides of (9.42) must be a constant, λ. We therefore obtain two equations of simple-harmonic type

$$\frac{d^2X}{dx^2} = \lambda X, \quad \frac{d^2Y}{dy^2} = -\lambda Y$$

The type of solution depends on the sign of λ, and we have a variety of possible solutions:

$$\lambda = -\mu^2 < 0: \quad u = (A \sin \mu x + B \cos \mu x)(C\,e^{\mu y} + D\,e^{-\mu y}) \tag{9.43a}$$

$$\lambda = \mu^2 > 0: \quad u = (A\,e^{\mu x} + B\,e^{-\mu x})(C \sin \mu y + D \cos \mu y) \tag{9.43b}$$

$$\lambda = 0: \quad u = (Ax + B)(Cy + D) \tag{9.43c}$$

where A, B, C and D are arbitrary constants. Using the definitions of the hyperbolic functions, it is sometimes more convenient to express the solution (9.43a) as

$$u = (A \sin \mu x + B \cos \mu x)(C \cosh \mu y + D \sinh \mu y) \tag{9.43d}$$

and (9.43b) as

$$u = (A \sinh \mu x + B \cosh \mu x)(C \sin \mu y + D \cos \mu y) \tag{9.43e}$$

The actual form of the solution depends on the problem in hand, as illustrated in the following examples.

EXAMPLE 9.26 Use the separated solutions (9.43) of the Laplace equation to find the solution to (9.41) satisfying the boundary conditions

$$u(x, 0) = 0 \quad (0 < x < 2)$$

$$u(x, 1) = 0 \quad (0 < x < 2)$$

$$u(0, y) = 0 \quad (0 < y < 1)$$

$$u(2, y) = a \sin 2\pi y \quad (0 < y < 1)$$

Solution To satisfy the first two conditions, we need to choose the separated solutions that include the $\sin \mu y$ terms. Thus we take solution (9.43b)

$$u = (A\,e^{\mu x} + B\,e^{-\mu x})(C \sin \mu y + D \cos \mu y)$$

The first boundary condition gives

$$(A\,e^{\mu x} + B\,e^{-\mu x})D = 0 \quad (0 < x < 2)$$

so that $D = 0$. Thus

$$u = (A'\,e^{\mu x} + B'\,e^{-\mu x}) \sin \mu y$$

where $A' = AC$ and $B' = BC$. The second boundary condition then gives

$$(A'\,e^{\mu x} + B'\,e^{-\mu x}) \sin \mu = 0 \quad (0 < x < 2)$$

so that $\sin \mu = 0$, or $\mu = n\pi$ with n as integer. Thus

$$u = (A'\,e^{n\pi x} + B'\,e^{-n\pi x}) \sin n\pi y$$

From the third boundary condition,

$$(A' + B') \sin n\pi y = 0 \quad (0 < y < 1)$$

so that $B' = -A'$, giving

$$u = A'(e^{n\pi x} - e^{-n\pi x}) \sin n\pi y$$
$$= 2A' \sinh n\pi x \sin n\pi y$$

The final boundary condition then gives

$$2A' \sinh 2n\pi \sin n\pi y = a \sin 2\pi y \quad (0 < y < 1)$$

We must therefore choose $n = 2$, and $a = 2A' \sinh 2n\pi = 2A' \sinh 4\pi$, or $2A' = a/\sinh 4\pi$. The solution is therefore

$$u = a \sin 2\pi y \frac{\sinh 2\pi x}{\sinh 4\pi}$$

EXAMPLE 9.27 Solve the Laplace equation (9.41) for steady heat conduction in the semi-infinite region $0 \leqslant y \leqslant 1$, $x \geqslant 0$ and subject to the boundary conditions

$$\left.\begin{array}{lll}\text{(a)} & u(x, 0) = 0 & (x \geqslant 0) \\ \text{(b)} & u(x, 1) = 0 & (x \geqslant 0) \\ \text{(c)} & u(x, y) \to 0 & \text{as } x \to \infty\end{array}\right\} \text{(temperature kept at zero on two sides and at infinity);}$$

(d) $u(0, y) = 1$ $(0 \leqslant y \leqslant 1)$ (unit temperature on the fourth side).

Solution Clearly from condition (c) we need a solution that is exponential in x, so we take (9.43b):

$$u = (A\,e^{\mu x} + B\,e^{-\mu x})(C \sin \mu y + D \cos \mu y)$$

and since the solution must tend to zero as $x \to \infty$, we have $A = 0$, giving

$$u = e^{-\mu x}(C' \sin \mu y + D' \cos \mu y)$$

where $C' = BC$ and $D' = BD$. Condition (a) then gives $D' = 0$, and (b) gives $\sin \mu = 0$, or $\mu = n\pi$ ($n = 1, 2, \ldots$), so the solution becomes $u = C'\,e^{-n\pi x} \sin n\pi y$ ($n = 1, 2, \ldots$). Because of the linearity of the Laplace equation, we sum over n to obtain the more general solution

$$u = \sum_{n=1}^{\infty} C'_n\,e^{-n\pi x} \sin n\pi y$$

Condition (d) then gives, as before, a classic Fourier series problem

$$1 = \sum_{n=1}^{\infty} C'_n \sin n\pi y \quad (0 \leqslant y \leqslant 1)$$

so that, using (4.33),

$$C'_n = 2\int_0^1 \sin n\pi y\,dy = \begin{cases} 4/n\pi & (\text{odd } n) \\ 0 & (\text{even } n) \end{cases}$$

The complete solution is therefore

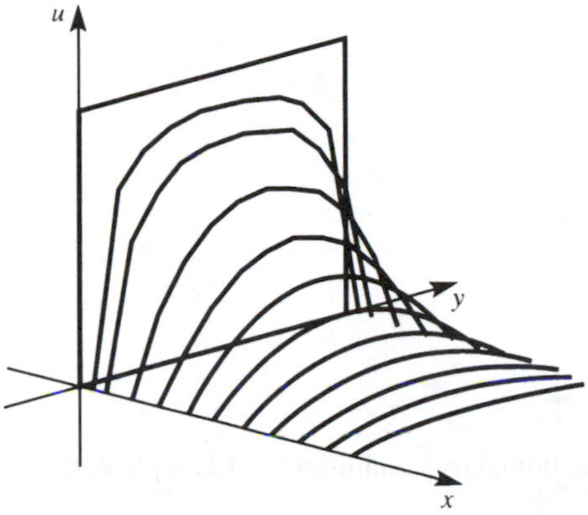

Figure 9.29 Solution of the Laplace equation in Example 9.27.

$$u = \frac{4}{\pi}\sum_{n=1}^{\infty}\frac{1}{2n-1}\,e^{-(2n-1)\pi x}\sin(2n-1)\pi y$$

or, in expanded form,

$$u = \frac{4}{\pi}\left(e^{-\pi x}\sin\pi y + \frac{1}{3}e^{-3\pi x}\sin 3\pi y + \frac{1}{5}e^{-5\pi x}\sin 5\pi y + \dots\right)$$

In Figure 9.29 the solution $u(x, t)$ is plotted in the (x, y) plane. Because of the discontinuity at $x = 0$, for $x = 0.05$ thirty terms of the series were required to compute u to four-figure accuracy, while for $x = 1$, one or two terms were quite sufficient.

It is clear from Example 9.27 that the solutions (9.43) can only be used for rectangular regions. For various cylindrical and spherically symmetric regions separated solutions can be constructed, but they need more complicated Bessel and Legendre functions. There is great merit in calculating exact solutions where we can, since they give significant insight. However, with modern computing techniques it is certainly not necessarily quicker than a straight numerical solution.

EXAMPLE 9.28

Solve the Laplace equation (9.41) in the region $0 \leqslant x \leqslant 1$, $0 \leqslant y \leqslant 2$ with the conditions

(a) $u(x, 0) = x \quad (0 \leqslant x \leqslant 1)$
(b) $u(x, 2) = 0 \quad (0 \leqslant x \leqslant 1)$
(c) $u(0, y) = 0 \quad (0 \leqslant y \leqslant 2)$
(d) $\partial u(1, y)/\partial x = 0 \quad (0 \leqslant y \leqslant 2)$

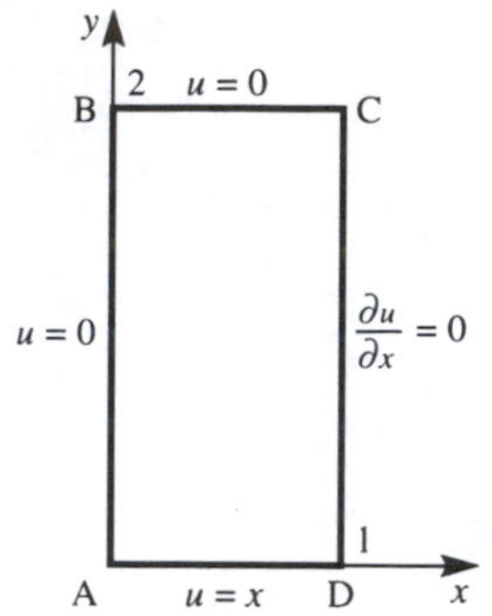

Figure 9.30 Region and boundary conditions for Example 9.28.

Solution The steady heat-conduction interpretation of this problem, looking at Figure 9.30, gives a zero temperature on ABC, an insulated boundary on CD and a linear temperature on AD.

Of the solutions (9.43), we require zeros on AB and zero derivative on CD, so we might expect to use trigonometric solutions in the x direction and exponential (or equivalently sinh and cosh) solutions in the y direction. We therefore take a solution of the form (9.43d):

$$u = (A \sin \mu x + B \cos \mu x)(C \cosh \mu y + D \sinh \mu y)$$

From condition (c), we must take $B = 0$, giving

$$u = (C' \cosh \mu y + D' \sinh \mu y) \sin \mu x$$

where $C' = AC$ and $D' = AD$. Condition (d) then gives $\cos \mu = 0$ or

$$\mu = (n + \tfrac{1}{2})\pi \quad (n = 0, 1, 2, \dots)$$

so the solution becomes

$$u = [C' \cosh (n + \tfrac{1}{2})\pi y + D' \sinh (n + \tfrac{1}{2})\pi y] \sin (n + \tfrac{1}{2})\pi x \quad (n = 0, 1, 2, \dots) \tag{9.44}$$

To satisfy condition (b), it is best to rewrite (9.44) in the equivalent form

$$u = \sin [(n + \tfrac{1}{2})\pi x]\{E \cosh [(n + \tfrac{1}{2})\pi(2 - y)] + F \sinh [(n + \tfrac{1}{2})\pi(2 - y)]\} \quad (n = 0, 1, 2, \dots)$$

We see that (b) now implies $E = 0$, so that our basic solution, summed over all n, is

$$u = \sum_{n=0}^{\infty} F_n \sin [(n + \tfrac{1}{2})\pi x] \sinh [(n + \tfrac{1}{2})\pi(2 - y)]$$

The final condition (a) then gives the standard Fourier series problem

$$x = \sum_{n=0}^{\infty} F_n \sinh [(2n + 1)\pi] \sin [(n + \tfrac{1}{2})\pi x]$$

so that, using (4.33),

$$\frac{1}{2}F_n \sinh(2n+1)\pi = \int_0^1 x \sin\left[\left(n+\frac{1}{2}\right)\pi x\right] dx = \frac{\sin(n+\frac{1}{2})\pi}{\pi^2(n+\frac{1}{2})^2}$$

The solution in expanded form is therefore

$$u = \frac{8}{\pi^2}\left[\sin\frac{1}{2}\pi x \frac{\sinh\frac{1}{2}\pi(2-y)}{\sinh\pi} - \frac{\sin\frac{3}{2}\pi x \sinh\frac{3}{2}\pi(2-y)}{9\sinh 3\pi} + \sin\frac{5}{2}\pi x \frac{\sinh\frac{5}{2}\pi(2-y)}{25\sinh 5\pi} + \dots\right]$$

Curiously, Laplace transform solutions are not natural for the Laplace equation, since there is no obvious semi-infinite parameter. Even in cases like Example 9.27, where we have a semi-infinite region, the Laplace transform in x requires information that is not available.

Another technique for solution of the Laplace equation involves complex variables and was discussed in Section 1.3.2. It is a method that was very widely used in aerodynamics and in electrostatic problems. Since the advent of modern computers, with highly efficient Laplace solvers, the method has fallen somewhat into disuse. It was the cornerstone of all early flight calculations and, for those interested in either the historical context or in the beautiful mathematical theory, the study of complex-variable solutions is essential.

EXAMPLE 9.29

If $f(z) = \phi(x, y) + j\psi(x, y)$ is a complex function of the complex variable $z = x + jy$, verify that ϕ and ψ satisfy the Laplace equation for the case $f(z) = z^2$. Sketch the contours of ϕ = constant and ψ = constant.

Solution

Now

$$f(z) = z^2 = (x^2 - y^2) + j2xy$$

and thus

$$\phi = x^2 - y^2, \quad \psi = 2xy$$

It is trivial to differentiate these functions, and both clearly satisfy the Laplace equation:

$$\nabla^2\phi = 0, \quad \nabla^2\psi = 0$$

The contours of ϕ and ψ are plotted in Figure 9.31. They are both hyperbolas, which intersect at right angles. The usual interpretation of these solutions is as irrotational inviscid fluid flow into a corner.

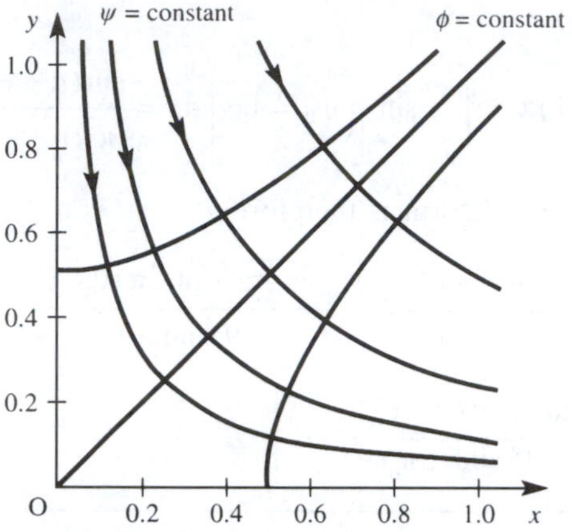

Figure 9.31 Complex-variable solution to the Laplace equation in Example 9.29, showing the streamlines of flow into a corner.

9.5.2 Exercises

37 Use the separated solutions (9.43) to solve the Laplace equation

$$\frac{\partial^2 u}{\partial x^2} + \frac{\partial^2 u}{\partial y^2} = 0$$

in the region $0 \leqslant x \leqslant 1$, $y \geqslant 0$ given the boundary conditions

(a) $u = 0$ on $x = 0$ and $x = 1$ $\quad (y \geqslant 0)$
(b) $u \to 0$ as $y \to \infty$ $\quad (0 \leqslant x \leqslant 1)$
(c) $u = \sin^5(\pi x)$ on $y = 0$ $\quad (0 \leqslant x \leqslant 1)$

[*Note* the identity $\sin^5\theta = \frac{1}{16}(\sin 5\theta - 5\sin 3\theta + 10\sin\theta)$]

38 Solve the Laplace equation $\partial^2 u/\partial x^2 + \partial^2 u/\partial y^2 = 0$ in the region $0 < x < 1$, $0 < y < 1$ subject to the boundary conditions $u(0, y) = 0$, $u(x, 0) = 0$, $u(1, y) = 1$, $u(x, 1) = 1$ by separation methods.

39 (Hadamard example) Show that the Laplace equation

$$\partial^2 u/\partial x^2 + \partial^2 u/\partial y^2 = 0$$

with $u(0, y) = 0$, $u_x(0, y) = (1/n)\sin ny$ $(n > 0)$ has the solution

$$u(x, y) = \frac{1}{n^2}\sinh nx \sin ny$$

Compare this solution, for large n, with the solution to the 'neighbouring' problem, when $u(0, y) = 0$, $u_x(0, y) = 0$, and the solution $u(x, y) = 0$.

40 Show that the function $u(x, y) = e^{-\pi x/2}[y\cos(\pi y/2) - x\sin(\pi y/2)]$ satisfies the Laplace equation

$$\frac{\partial^2 u}{\partial x^2} + \frac{\partial^2 u}{\partial y^2} = 0$$

and the boundary conditions

(a) $u = 0$ on $y = 0$ $\quad (x \geqslant 0)$
(b) $u = -x e^{-\pi x/2}$ on $y = 1$ $\quad (x \geqslant 0)$
(c) $u = y\cos(\pi y/2)$ on $x = 0$ $\quad (0 \leqslant y \leqslant 1)$
(d) $u \to 0$ as $x \to \infty$

41 Show that

$$u(r, \theta) = Br^n \sin n\theta$$

satisfies the Laplace equation in polar coordinates,

$$u_{rr} + \frac{1}{r}u_r + \frac{1}{r^2}u_{\theta\theta} = 0$$

Determine u that is finite for $r \leqslant a$ and given that

$$u(a, \theta) = \sin^3\theta = \tfrac{3}{4}\sin\theta - \tfrac{1}{4}\sin 3\theta$$

42 Verify that

$$u = \frac{-2y}{x^2 + y^2 + 2x + 1}, \quad v = \frac{x^2 + y^2 - 1}{x^2 + y^2 + 2x + 1}$$

both satisfy the Laplace equation, and sketch the curves u = constant and v = constant. Show that

$$u + jv = \frac{j(z-1)}{z+1}$$

where $z = x + jy$.

43 A long bar of square cross-section $0 \leqslant x \leqslant a$, $0 \leqslant y \leqslant a$ has the faces $x = 0$, $x = a$ and $y = 0$ maintained at zero temperature, and the face $y = a$ at a control temperature u_0. Under steady-state conditions the temperature $u(x, y)$ at a point in a cross-section satisfies the Laplace equation

$$\frac{\partial^2 u}{\partial x^2} + \frac{\partial^2 u}{\partial y^2} = 0$$

Write down the boundary conditions for $u(x, y)$, and hence show that $u(x, y)$ is given by

$$u(x, y) = \frac{4u_0}{\pi} \sum_{n=0}^{\infty} \frac{\operatorname{cosech}(2n+1)\pi}{2n+1} \times \sinh\left[(2n+1)\frac{\pi y}{a}\right] \sin\left[(2n+1)\frac{\pi x}{a}\right]$$

44 Heat is flowing steadily in a metal plate whose shape is an infinite rectangle occupying the region $-a < x < a$, $y > 0$ of the (x, y) plane. The temperature at the point (x, y) is denoted by $u(x, y)$. The sides $x = \pm a$ are insulated, the temperature approaches zero as $y \to \infty$, while the side $y = 0$ is maintained at a fixed temperature $-T$ for $-a < x < 0$ and T for $0 < x < a$. It is known that $u(x, y)$ satisfies the Laplace equation

$$\frac{\partial^2 u}{\partial x^2} + \frac{\partial^2 u}{\partial y^2} = 0$$

and the boundary conditions

(a) $u \to 0$ as $y \to \infty$ for all x in $-a < x < a$
(b) $\partial u/\partial x = 0$ when $x = \pm a$
(c) $u(x, 0) = \begin{cases} -T & (-a < x < 0) \\ T & (0 < x < a) \end{cases}$

Using the method of separation, obtain the solution $u(x, y)$ in the form

$$u(x, y) = \frac{4T}{\pi} \sum_{n=0}^{\infty} \frac{1}{2n+1} \exp\left[-\left(n + \frac{1}{2}\right)\frac{\pi y}{a}\right] \times \sin\left[\left(n + \frac{1}{2}\right)\frac{\pi x}{a}\right]$$

45 A thin semicircular plate of radius a has its bounding diameter kept at zero temperature and its curved boundary at a constant temperature T_0. The steady-state temperature $T(r, \theta)$ at a point having polar coordinates (r, θ), referred to the centre of the circle as origin, is given by the Laplace equation

$$\frac{\partial^2 T}{\partial r^2} + \frac{1}{r}\frac{\partial T}{\partial r} + \frac{1}{r^2}\frac{\partial^2 T}{\partial \theta^2} = 0$$

Assuming a separated solution of the form

$$T = R(r)\Theta(\theta)$$

show that

$$T(r, \theta) = \frac{4T_0}{\pi} \sum_{n=0}^{\infty} \frac{(r/a)^{2n+1}}{2n+1} \sin(2n+1)\theta$$

46 The Laplace equation in spherical polar coordinates (r, θ, ϕ) takes the form

$$\frac{\partial}{\partial r}\left(r^2 \frac{\partial V}{\partial r}\right) + \frac{1}{\sin\theta}\frac{\partial}{\partial\theta}\left(\sin\theta \frac{\partial V}{\partial\theta}\right) + \frac{1}{\sin^2\theta}\frac{\partial^2 V}{\partial\phi^2} = 0$$

If V is only a function of r and θ, and V takes the form

$$V = R(r)y(x), \quad \text{where} \quad x = \cos\theta$$

show that

$$\frac{d}{dr}\left(r^2 \frac{dR}{dr}\right) = k(k+1)R$$

$$(1 - x^2)\frac{d^2 y}{dx^2} - 2x\frac{dy}{dx} + k(k+1)y = 0$$

where k is a constant.

The function V satisfies the Laplace equation in the region $a \leqslant r \leqslant b$. On $r = a$, $V = 0$ and on $r = b$, $V = \alpha \sin^2\theta$, where α is a constant. Given that solutions for y are

$$y = \begin{cases} 1 & (k = 0) \\ x & (k = 1) \\ \frac{1}{2}(3x^2 - 1) & (k = 2) \end{cases}$$

find V throughout the region.

9.5.3 Numerical solution

Of the three classical partial differential equations, the Laplace equation proves to be the most difficult to solve. The other two have a natural time variable in them, and it is possible, with a little care, to march forward either by a simple explicit method or by an implicit procedure. In the case of the Laplace equation, information is given around the whole of the boundary of the solution region, so the field variables at *all* mesh points must be solved simultaneously. This in turn leads to a solution by matrix inversion.

The usual numerical approximation for the partial derivatives, discussed in Section 9.3.5, are employed, so that the equation

$$\frac{\partial^2 u}{\partial x^2} + \frac{\partial^2 u}{\partial y} = 0 \tag{9.45}$$

at a typical point, illustrated in Figure 9.32, becomes

$$\frac{u(i+1,j) - 2u(i,j) + u(i-1,j)}{\Delta x^2} + \frac{u(i,j+1) - 2u(i,j) + u(i,j-1)}{\Delta y^2} = 0$$

For the case $\Delta x = \Delta y$ rearranging gives

$$4u(i,j) = u(i+1,j) + u(i-1,j) + u(i,j+1) + u(i,j-1) \tag{9.46}$$

In the typical five-point module (9.46) the increments Δx and Δy are taken to be the same and it is noted that the middle value $u(i, j)$ is the average of its four neighbours. This corresponds to the absence of 'hot spots'. We now examine how (9.46) can be implemented.

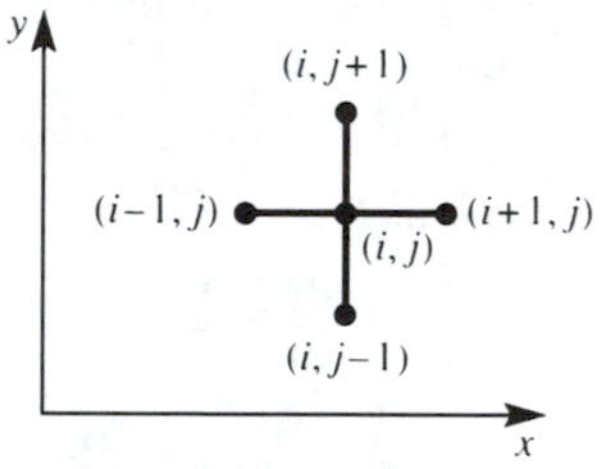

Figure 9.32 Five-point computational module for the Laplace equation.

EXAMPLE 9.30 Solve the Laplace equation (9.45) in the square region $0 \leqslant x \leqslant 1$, $0 \leqslant y \leqslant 1$ with the boundary conditions

(a) $u = 0$ on $x = 0$ (b) $u = 1$ on $x = 1$

(c) $u = 0$ on $y = 0$ (d) $u = 0$ on $y = 1$

Solution For a first solution we take the simplest mesh, illustrated in Figure 9.33(a), which contains only four interior points labelled u_1, u_2, u_3 and u_4. The four equations obtained from (9.46) are

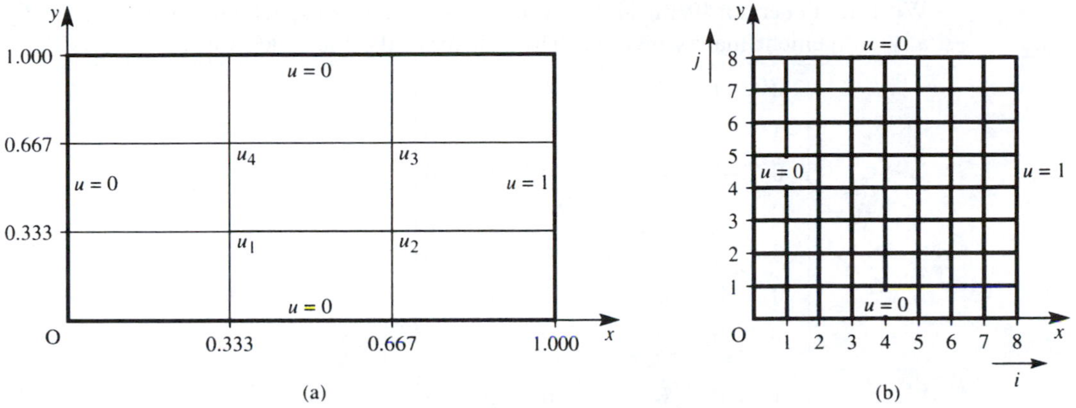

Figure 9.33 Meshes for the solution of the Laplace equation in Example 9.30: (a) a simple mesh containing 4 interior points; (b) a larger mesh with 49 interior points.

$$4u_1 = 0 + 0 + u_2 + u_4$$
$$4u_2 = 0 + 1 + u_3 + u_1$$
$$4u_3 = 1 + 0 + u_4 + u_2$$
$$4u_4 = 0 + 0 + u_1 + u_3$$

which in turn can be written in matrix form as

$$\begin{bmatrix} 4 & -1 & 0 & -1 \\ -1 & 4 & -1 & 0 \\ 0 & -1 & 4 & -1 \\ -1 & 0 & -1 & 4 \end{bmatrix} \begin{bmatrix} u_1 \\ u_2 \\ u_3 \\ u_4 \end{bmatrix} = \begin{bmatrix} 0 \\ 1 \\ 1 \\ 0 \end{bmatrix}$$

This has the solution $u_1 = 0.125$, $u_2 = 0.375$, $u_3 = 0.375$, $u_4 = 0.125$. A larger mesh obtained by dividing the sides up into eight equal parts is indicated in Figure 9.33(b). The equations now take the form

$$\begin{aligned}
4u(1, 1) &= u(2, 1) + 0 \qquad\qquad + u(1, 2) + 0 \\
4u(2, 1) &= u(3, 1) + u(1, 1) + u(2, 2) + 0 \\
&\vdots \\
4u(7, 1) &= 1 \qquad\qquad + u(6, 1) + u(7, 2) + 0 \\
4u(1, 2) &= u(2, 2) + 0 \qquad\qquad + u(1, 3) + u(1, 1) \\
4u(2, 2) &= u(3, 2) + u(1, 2) + u(2, 3) + u(2, 1) \\
&\vdots \\
4u(7, 2) &= 1 \qquad\qquad + u(6, 2) + u(7, 3) + u(7, 1) \\
&\vdots
\end{aligned}$$

We thus generate 49 linear equations in 49 unknowns, which can be solved by any convenient matrix inverter. The matrices take the block form

$$\boldsymbol{A} = \begin{bmatrix} 4 & -1 & 0 & 0 & 0 & 0 & 0 \\ -1 & 4 & -1 & 0 & 0 & 0 & 0 \\ 0 & -1 & 4 & -1 & 0 & 0 & 0 \\ 0 & 0 & -1 & 4 & -1 & 0 & 0 \\ 0 & 0 & 0 & -1 & 4 & -1 & 0 \\ 0 & 0 & 0 & 0 & -1 & 4 & -1 \\ 0 & 0 & 0 & 0 & 0 & -1 & 4 \end{bmatrix}$$

$$\boldsymbol{B} = \begin{bmatrix} -1 & 0 & 0 & 0 & 0 & 0 & 0 \\ 0 & -1 & 0 & 0 & 0 & 0 & 0 \\ 0 & 0 & -1 & 0 & 0 & 0 & 0 \\ 0 & 0 & 0 & -1 & 0 & 0 & 0 \\ 0 & 0 & 0 & 0 & -1 & 0 & 0 \\ 0 & 0 & 0 & 0 & 0 & -1 & 0 \\ 0 & 0 & 0 & 0 & 0 & 0 & -1 \end{bmatrix}$$

$$\boldsymbol{C} = \begin{bmatrix} 0 \\ 0 \\ 0 \\ 0 \\ 0 \\ 0 \\ 1 \end{bmatrix}, \quad \boldsymbol{U}_k = \begin{bmatrix} u(1, k) \\ u(2, k) \\ \\ \vdots \\ \\ \\ u(7, k) \end{bmatrix}$$

so that the equations become

$$\begin{bmatrix} \boldsymbol{A} & \boldsymbol{B} & 0 & 0 & 0 & 0 & 0 \\ \boldsymbol{B} & \boldsymbol{A} & \boldsymbol{B} & 0 & 0 & 0 & 0 \\ 0 & \boldsymbol{B} & \boldsymbol{A} & \boldsymbol{B} & 0 & 0 & 0 \\ 0 & 0 & \boldsymbol{B} & \boldsymbol{A} & \boldsymbol{B} & 0 & 0 \\ 0 & 0 & 0 & \boldsymbol{B} & \boldsymbol{A} & \boldsymbol{B} & 0 \\ 0 & 0 & 0 & 0 & \boldsymbol{B} & \boldsymbol{A} & \boldsymbol{B} \\ 0 & 0 & 0 & 0 & 0 & \boldsymbol{B} & \boldsymbol{A} \end{bmatrix} \begin{bmatrix} \boldsymbol{U}_1 \\ \boldsymbol{U}_2 \\ \boldsymbol{U}_3 \\ \boldsymbol{U}_4 \\ \boldsymbol{U}_5 \\ \boldsymbol{U}_6 \\ \boldsymbol{U}_7 \end{bmatrix} = \begin{bmatrix} \boldsymbol{C} \\ \boldsymbol{C} \\ \boldsymbol{C} \\ \boldsymbol{C} \\ \boldsymbol{C} \\ \boldsymbol{C} \\ \boldsymbol{C} \end{bmatrix} \tag{9.47}$$

The matrix equation (9.47) can be solved by an elimination technique or an iterative method like **successive over-relaxation** (SOR). As indicated in Section 5.5.4 of *Modern Engineering Mathematics*, SOR is the simplest to program, and elimination techniques are best performed by a package from a computer library. For the current problem we present the solution in Figure 9.34, where the cases

j values \ i values	0	1	2	3	4	5	6	7	8
8	0	0	0	0	0	0	0	0	1
7	0	0.017	0.038	0.064	0.103	0.164	0.269	0.483	1
6	0	0.032	0.069	0.117	0.184	0.282	0.431	0.661	1
5	0	0.042	0.089	0.150	0.233	0.350	0.512	0.731	1
4	0	0.045	0.096	0.162	0.250	0.371	0.556	0.749	1
			(0.098)		(0.250)		(0.527)		
3									
2	0		(0.071)		(0.188)		(0.429)		1
1									
0	0		0		0		0		1

Figure 9.34 The solution of Example 9.30. The solution is symmetric about the line $j = 4$; the solution with $\Delta x = 0.125$ is given in the upper half and the solution with $\Delta x = 0.25$ is shown in parentheses in the lower half.

$\Delta x = \frac{1}{8}$ and $\Delta x = \frac{1}{4}$ are both shown. It may be seen from this example that the accuracy of the solution is quite tolerable when the cases $\Delta x = \frac{1}{4}$ and $\Delta x = \frac{1}{8}$ are compared. Note the averaging behaviour of the Laplace equation and observe that the discontinuity in the corner does not spread into the solution. The corner nodes are never used in the numerical calculation, so the discontinuity is avoided.

Because of its simplicity, SOR is an attractive method for solving Laplace-type problems. Equations (9.46) are rewritten with an iteration superscript as

$$u^{n+1}(i, j) = u^n(i, j) + \tfrac{1}{4}w[u^n(i + 1, j) + u^n(i - 1, j) + u^n(i, j + 1) + u^n(i, j - 1) - 4u^n(i, j)] \qquad \textbf{(9.48)}$$

and w is a relaxation factor, discussed in Chapter 5 of *Modern Engineering Mathematics*.

Knowing all the $u(i, j)$ at iteration n, we can use (9.48) to evaluate $u(i, j)$ at iteration $n + 1$. Normally the $u(i, j)$ are over-written in the computer as they are computed, so that some of the ns in the right-hand side of (9.48) become $(n + 1)$s. The order of evaluation of the is and js in (9.48) is critical, but the most obvious methods by rows or columns prove to be satisfactory.

A great deal is known about the optimum relaxation factor w. It is closely related to the value of the maximum eigenvalue of the matrix associated with the problem. For square regions with unit side and with u given on the boundary and equal mesh spacing it can be shown that $w = 2/(1 + \sin \Delta x)$ is the best value. For other problems this is usually used as a starting guess, but numerical experimentation is required to determine an optimum or near-optimum value.

We have only considered u to be given on the boundary, and it is essential to know how to deal with derivative boundary conditions, since these are very common. Let us consider a typical example:

$$\frac{\partial u}{\partial x} = g(y) \quad \text{on } x = 0$$

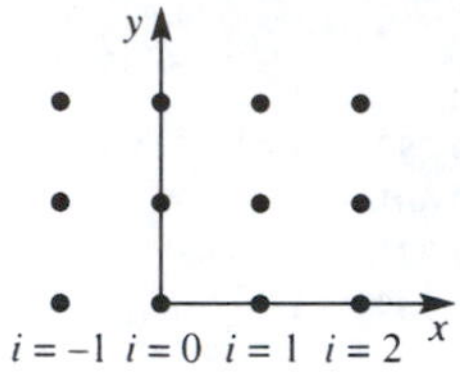

Figure 9.35 Fictitious nodes, $i = -1$, introduced outside the boundary.

We then insert a fictitious line of nodes, as shown in Figure 9.35. Approximately, the boundary condition gives

$$u(1, j) - u(-1, j) = g(y_j)2\Delta x$$

so that

$$u(-1, j) = u(1, j) - 2\Delta x\, g(y_j) \tag{9.49}$$

Equations (9.46) or (9.48) are now solved for $i = 0$ as well as $i > 0$, but at the end of a sweep $u(-1, j)$ will be updated via (9.49).

EXAMPLE 9.31

Solve the Laplace equation (9.45) for steady-state heat conduction in the unit square, given that

(a) $\partial u/\partial x = \frac{1}{2} - y, \quad x = 0$ (steady heat supply on this boundary)

(b) $u = 0$ on $x = 1, y = 0, y = 1$ (zero temperature on the other three sides)

Use $\Delta x = \Delta y = \frac{1}{3}$.

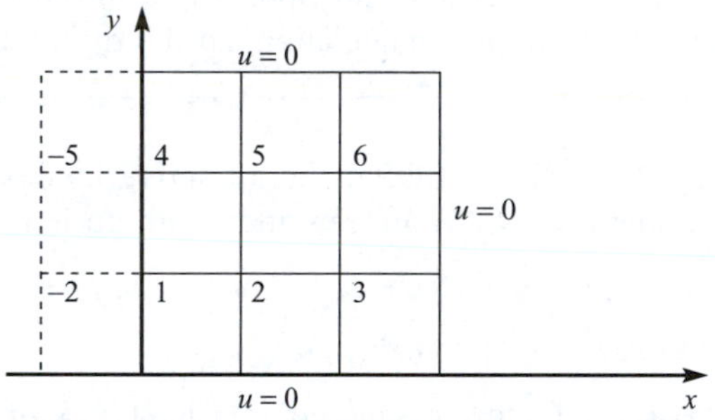

Figure 9.36 The mesh for Example 9.31.

Solution Labelling the six unknown values $u_1, u_2, \ldots, u_6$ as shown in Figure 9.36, equation (9.46) gives

$$4u_1 = 0 + u_2 + u_4 + u_{-2}$$

$$4u_2 = 0 + u_3 + u_5 + u_1$$

$$4u_3 = 0 + 0 + u_6 + u_2$$

$$4u_4 = u_1 + u_5 + 0 + u_{-5}$$

$$4u_5 = u_2 + u_6 + 0 + u_4$$

$$4u_6 = u_3 + 0 + 0 + u_5$$

The values u_{-2} and u_{-5} are evaluated from boundary condition (a)

$$u_5 - u_{-5} = 2h(\tfrac{1}{2} - \tfrac{2}{3}) = -\tfrac{1}{9}, \quad u_2 - u_{-2} = 2h(\tfrac{1}{2} - \tfrac{1}{3}) = \tfrac{1}{9}$$

so the equations become

$$4u_1 = 2u_2 + u_4 - \tfrac{1}{9}$$

$$4u_2 = u_1 + u_3 + u_5$$

$$4u_3 = u_2 + u_6$$

$$4u_4 = u_1 + 2u_5 + \tfrac{1}{9}$$

$$4u_5 = u_2 + u_4 + u_6$$

$$4u_6 = u_3 + u_5$$

Thus there are six linear equations in six unknowns, which can be solved by any convenient method. For instance, SOR as suggested in (9.48) gives the set of equations with iteration counter n and relaxation factor w

$$u_1^{n+1} = u_1^n + \tfrac{w}{4}(2u_2^n + u_4^n - \tfrac{1}{9} - 4u_1^n)$$

$$u_2^{n+1} = u_2^n + \tfrac{w}{4}(u_1^{n+1} + u_3^n + u_5^n - 4u_2^n)$$

$$u_3^{n+1} = u_3^n + \tfrac{w}{4}(u_2^{n+1} + u_6^n - 4u_3^n)$$

$$u_4^{n+1} = u_4^n + \tfrac{w}{4}(u_1^{n+1} + 2u_5^n + \tfrac{1}{9} - 4u_4^n)$$

$$u_5^{n+1} = u_5^n + \tfrac{w}{4}(u_2^{n+1} + u_4^{n+1} + u_6^n - 4u_5^n)$$

$$u_6^{n+1} = u_6^n + \tfrac{w}{4}(u_3^{n+1} + u_5^{n+1} - 4u_6^n)$$

The equations are diagonally dominant so the iterations converge quickly; six significant figures are obtained in 11 iterations with $w = 1$ and at near optimum $w = 1.2$ in 8 iterations.

	u_1	u_2	u_3	u_4	u_5	u_6
$h = \frac{1}{3}$	−0.024 24	−0.005 05	−0.001 01	0.024 24	0.005 05	0.001 01
$h = \frac{1}{6}$	−0.031 21	−0.005 17	−0.000 77	0.031 21	0.005 17	0.000 77
$h = \frac{1}{12}$	−0.033 57	−0.005 22	−0.000 68	0.033 57	0.005 22	0.000 68

The expected symmetry is observed from the solution, physically heat is supplied to the bottom half of the left-hand boundary and an equal amount is extracted from the top half. For comparison of the accuracy, the calculations with $h = \frac{1}{6}$ and $\frac{1}{12}$ have been included in the table.

EXAMPLE 9.32

Solve the Laplace equation (9.45), for steady heat conduction, in the unit square, given that

(a) $\partial u/\partial x = 1$ on $x = 0$ (steady heat supply along this boundary);

(b) $u = y^2$ on $x = 1$

(c) $u = 0$ on $y = 0$

(d) $u = x$ on $y = 1$

(temperature given on three sides).

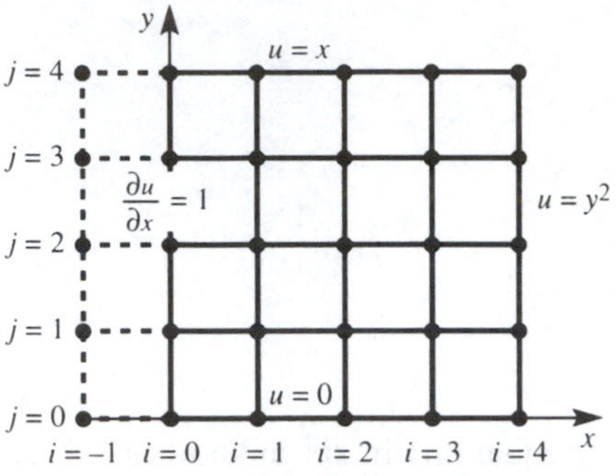

Figure 9.37 The mesh used in Example 9.32.

$j = 4$	0.0000	0.2500	0.5000	0.7500	1.0000
$j = 3$	−0.1190	0.1114	0.3067	0.4644	0.5625
$j = 2$	−0.1987	0.0077	0.1511	0.2384	0.2500
$j = 1$	−0.1912	−0.0330	0.0516	0.0881	0.0625
$j = 0$	0.0000	0.0000	0.0000	0.0000	0.0000
	$i = 0$	$i = 1$	$i = 2$	$i = 3$	$i = 4$

Figure 9.38 Data from the solution of Example 9.32 using a step length 0.25 in each direction.

Solution The region is illustrated in Figure 9.37, with mesh spacing $\Delta x = \Delta y = \frac{1}{4}$. Equation (9.49) just gives $u(-1, j) = u(1, j) - \frac{1}{2}$ for each j. From (9.46), we can write the equations as

$$4u(0, 1) = u(-1, 1) + u(1, 1) + 0 + u(0, 2) = -\tfrac{1}{2} + 2u(1, 1) + u(0, 2)$$

$$4u(1, 1) = u(0, 1) + u(2, 1) + 0 + u(1, 2)$$

$$4u(2, 1) = u(1, 1) + u(3, 1) + 0 + u(2, 2)$$

and so on, and hence obtain 12 equations in 12 unknowns. These can be solved by any convenient method to give a solution as shown in Figure 9.38.

9.5.4 Exercises

†47 Use the five-point difference approximation in (9.46) to solve

$$\frac{\partial^2 u}{\partial x^2} + \frac{\partial^2 u}{\partial y^2} = 0 \quad (0 \leqslant x \leqslant 1, 0 \leqslant y \leqslant 1)$$

where $u(x, 0) = u(0, y) = 0$, $u(x, 1) = x$, $u(1, y) = y(2 - y)$. Find the approximations for $u(\frac{1}{2}, \frac{1}{2})$ for grid sizes $\Delta x = \Delta y = \frac{1}{2}$ and $\Delta x = \Delta y = \frac{1}{4}$.

†48 Use a mesh $\Delta x = \Delta y = \frac{1}{2}$ to solve

$$\frac{\partial^2 u}{\partial x^2} + \frac{\partial^2 u}{\partial y^2} = 0 \quad (0 < x < 1, 0 < y < 1)$$

satisfying $u(0, y) = 0$, $\partial u(x, 0)/\partial y = 0$, $u(1, y) = 1$, $\partial u/\partial y + u = 0$ on $y = 1$.

†**49** A numerical solution is to be determined for the loading of a uniform plate, where the displacement w satisfies the equation

$$\frac{\partial^2 w}{\partial x^2} + \frac{\partial^2 w}{\partial y^2} + 20 = 0$$

and a square mesh of side h is used. Show that, at an interior point 0 with neighbours 1, 2, 3 and 4, the approximation to the equation is

$$4w_0 = w_1 + w_2 + w_3 + w_4 + 20h^2$$

The plate is in the shape of a trapezium whose vertices can be represented by the points (0, 0), (5, 0), (2, 3) and (0, 3). The plate is held on its edges so that on the boundary $w = 0$. Compute the solution for w at the five interior points if h is taken as 1.

†**50** The function $\phi(x, y)$ satisfies the equation

$$\frac{\partial^2 \phi}{\partial x^2} + \frac{\partial^2 \phi}{\partial y^2} = x$$

and the boundary conditions (see Figure 9.39)

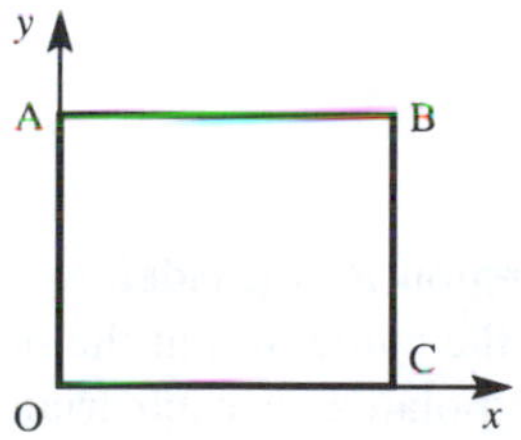

Figure 9.39 Region for Exercise 50.

$$\phi = 3 - y^2 \text{ on OA } (x = 0, 0 \leq y \leq 1)$$

$$\frac{\partial \phi}{\partial y} = -\phi \text{ on AB } (y = 1, 0 < x < 1)$$

$$\phi = 1 \text{ on BC } (x = 1, 0 \leq y \leq 1)$$

$$\phi = 3 - x \text{ on CO } (y = 0, 0 < x < 1)$$

Solve the equation numerically, using a mesh of (a) $h = \frac{1}{2}$ in each direction, (b) $h = \frac{1}{4}$ in each direction.

51 The function $\phi(x, y)$ satisfies the Laplace equation

$$\frac{\partial^2 \phi}{\partial x^2} + \frac{\partial^2 \phi}{\partial y^2} = 0$$

inside the region shown in Figure 9.40. The function ϕ takes the value $\phi = 9x^2$ at all points on the boundary. Making full use of symmetry, formulate a set of finite-difference equations to solve for the nodal values of ϕ on a square grid of side $h = \frac{1}{3}$. Solve for ϕ at the nodal points.

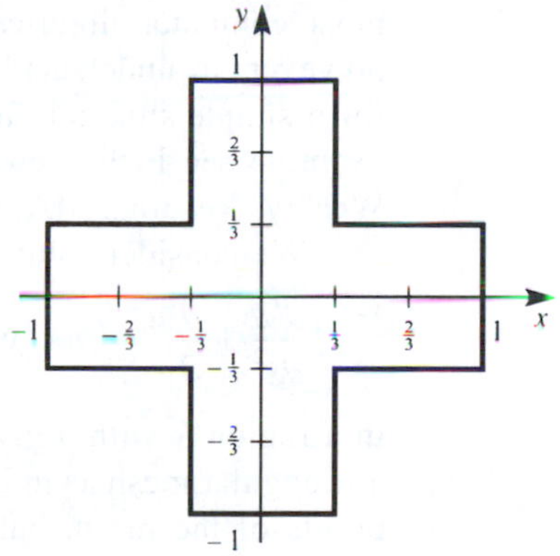

Figure 9.40 Region for Exercise 51.

9.6 Finite elements

In Section 9.5.3 we sought numerical solutions of the Laplace equation, but we noted that only simple geometries could be handled using finite differences. The region described in Exercise 49 is about as difficult as can be treated easily. To adapt methods to awkward regions is not easy, so alternative strategies have been sought. Great advances took place in the 1960s, when civil engineers pioneered the method of finite elements. To solve plate bending problems, they solved the appropriate equations for small patches and then 'stitched' the latter together to form an overall solution. The job is not an easy one, and requires a large amount

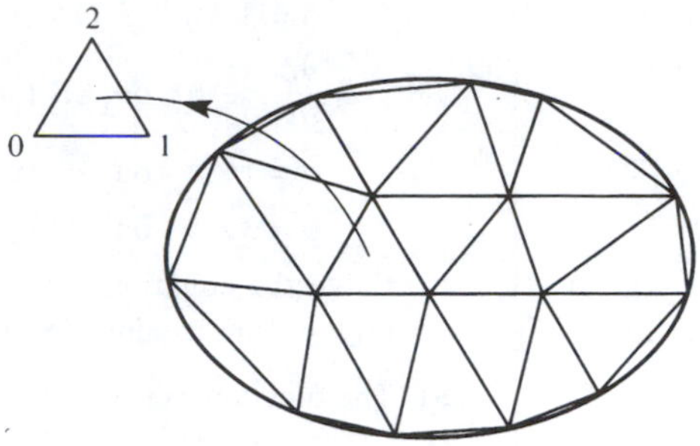

Figure 9.41 Triangular finite-element mesh, with the local numbering of a typical element.

of arithmetic. It was only when large, fast computers became available that the method was viable. This method is now very widely used, and forms the basis of most calculations in stress analysis and, recently, for many fluid flows. It is very adaptable and physically satisfying, but is very difficult to program. This is in contrast to finite differences, which are reasonably easy. In general the advice to anyone employing this technique is to use a finite element 'package', available in most computer libraries, and not to write one's own program. It is important, however, to understand the basis of the method. We shall illustrate this method for a simple situation, but refer to specialist books for details and extensions: for example see P. E. Lewis and J. P. Ward *The Finite Element Method* (Addison-Wesley, Reading, MA, 1991).

We consider solutions to the Poisson equation

$$\frac{\partial^2 u}{\partial x^2} + \frac{\partial^2 u}{\partial y^2} = \rho(x, y) \tag{9.50}$$

in a region R with u given on the boundary of R. The region R is divided up into a triangular mesh as in Figure 9.41. We aim to calculate the value of u at the nodal points of the mesh, but with the function suitably interpolated in each triangle. The simplest situation is obtained if, in a typical triangle, u is approximated as a linear function

$$u = ax + by + c \tag{9.51}$$

taking the values u_0, u_1 and u_2 at the corners. This function can be written explicitly in terms of the functions

$$\begin{aligned}
L_0 &= \begin{vmatrix} x & y & 1 \\ x_1 & y_1 & 1 \\ x_2 & y_2 & 1 \end{vmatrix} \Bigg/ \begin{vmatrix} x_0 & y_0 & 1 \\ x_1 & y_1 & 1 \\ x_2 & y_2 & 1 \end{vmatrix} \\
L_1 &= \begin{vmatrix} x & y & 1 \\ x_2 & y_2 & 1 \\ x_0 & y_0 & 1 \end{vmatrix} \Bigg/ \begin{vmatrix} x_1 & y_1 & 1 \\ x_2 & y_2 & 1 \\ x_0 & y_0 & 1 \end{vmatrix} \\
L_2 &= \begin{vmatrix} x & y & 1 \\ x_0 & y_0 & 1 \\ x_1 & y_1 & 1 \end{vmatrix} \Bigg/ \begin{vmatrix} x_2 & y_2 & 1 \\ x_0 & y_0 & 1 \\ x_1 & y_1 & 1 \end{vmatrix}
\end{aligned} \tag{9.52}$$

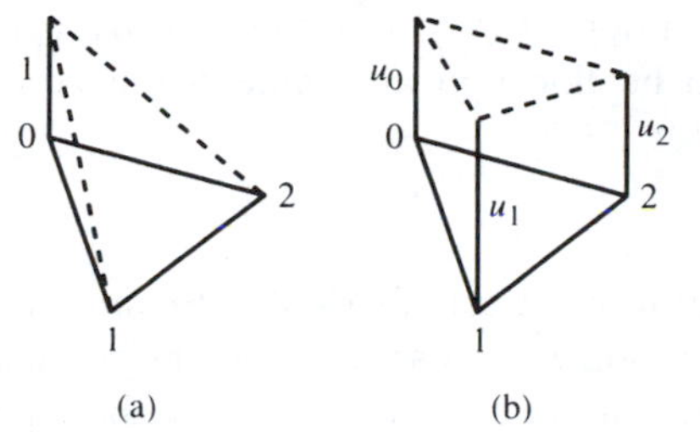

Figure 9.42 (a) The function L_0; (b) u approximated as a linear function in the element.

each of which is taken to be zero outside the triangle with vertices (x_0, y_0), (x_1, y_1) and (x_2, y_2). The denominators are just $2A$, where A is the area of the triangle. The function L_0, illustrated in Figure 9.42(a), takes the values 1 at (x_0, y_0), 0 at (x_1, y_1) and (x_2, y_2); it is linear in the triangle and is taken to be zero elsewhere. The functions L_1 and L_2 behave similarly. The field variable u in the element, denoted by u^e, can now be written as

$$u^e(x, y) = u_0L_0 + u_1L_1 + u_2L_2 \tag{9.53}$$

The situation is illustrated in Figure 9.42(b). Note that $u^e(x, y)$ is a linear function in x and y, $u^e(x_0, y_0) = u_0$, $u^e(x_1, y_1) = u_1$ and $u^e(x_2, y_2) = u_2$, and hence gives an explicit form for the function in (9.51) that has the correct values at the nodes.

EXAMPLE 9.33 Find the linear function that has the values u_0 at (0, 0), u_1 at (1, 1) and u_2 at $(\frac{1}{2}, 1)$.

Solution From (9.52), the functions L_0, L_1 and L_2 are given by

$$L_0 = \begin{vmatrix} x & y & 1 \\ 1 & 1 & 1 \\ \frac{1}{2} & 1 & 1 \end{vmatrix} \Bigg/ \begin{vmatrix} 0 & 0 & 1 \\ 1 & 1 & 1 \\ \frac{1}{2} & 1 & 1 \end{vmatrix} = (\tfrac{1}{2} - \tfrac{1}{2}y)/\tfrac{1}{2} = 1 - y$$

$$L_1 = \begin{vmatrix} x & y & 1 \\ \frac{1}{2} & 1 & 1 \\ 0 & 0 & 1 \end{vmatrix} \Bigg/ \frac{1}{2} = 2x - y$$

$$L_2 = \begin{vmatrix} x & y & 1 \\ 0 & 0 & 1 \\ 1 & 1 & 1 \end{vmatrix} \Bigg/ \frac{1}{2} = 2y - 2x$$

Thus, from (9.53), the required linear function is

$$u = (1 - y)u_0 + (2x - y)u_1 + 2(y - x)u_2$$

or

$$u = x(2u_1 - 2u_2) + y(-u_0 - u_1 + 2u_2) + u_0$$

We build up the solution of (9.50) as the sum over all the elements of the functions constructed to be linear in an element and zero outside the element. Thus

$$u = \sum u^e$$

To be of use, this function must satisfy (9.50) in some approximate sense. The function cannot be differentiated across the element boundaries, since it has discontinuous behaviour. We therefore have to satisfy the equation in an integrated or **'weak' form**.

We use the well-known result that if V is continuous and

$$\iint_R V\phi \, \mathrm{d}x \, \mathrm{d}y = 0 \tag{9.54}$$

for a complete set of functions ϕ (that is, a set of functions that will approximate any continuous function as accurately as desired) then $V \equiv 0$ in R. Using the residual of (9.50) in (9.54) gives

$$0 = \iint_R \left(\frac{\partial^2 u}{\partial x^2} + \frac{\partial^2 u}{\partial y^2} - \rho\right)\phi \, \mathrm{d}x \, \mathrm{d}y$$

$$= \iint_R \left[\frac{\partial}{\partial x}\left(\phi\frac{\partial u}{\partial x}\right) + \frac{\partial}{\partial y}\left(\phi\frac{\partial u}{\partial y}\right) - \frac{\partial u}{\partial x}\frac{\partial \phi}{\partial x} - \frac{\partial u}{\partial y}\frac{\partial \phi}{\partial y} - \rho\phi\right] \mathrm{d}x \, \mathrm{d}y$$

$$= -\iint_R (u_x\phi_x + u_y\phi_y + \rho\phi) \, \mathrm{d}x \, \mathrm{d}y - \int_C (\phi u_y \, \mathrm{d}x - \phi u_x \, \mathrm{d}y)$$

where the final integral is obtained over the boundary C of R using Green's theorem

$$\iint_R \left(\frac{\partial N}{\partial x} - \frac{\partial M}{\partial y}\right) \mathrm{d}x \, \mathrm{d}y = \int_C (M \, \mathrm{d}x + N \, \mathrm{d}y)$$

described in Section 7.4.5.

Choosing ϕ to be zero on the boundary C, we have an integrated form for (9.50) as

$$0 = \iint_R (u_x\phi_x + u_y\phi_y + \rho\phi) \, \mathrm{d}x \, \mathrm{d}y \tag{9.55}$$

It is this integrated or 'weak' form of (9.50) that we shall satisfy. We are therefore satisfying the equation in a global sense over the whole region. In comparison, finite-difference approximations are local to the mesh point. Clearly, we cannot use (9.55) with a complete set of functions ϕ, since there must be infinitely many of them. The best that can be done is to use N test functions ϕ_n ($n = 1, \ldots, N$) when there are N interior nodes. There will then be N equations for the N

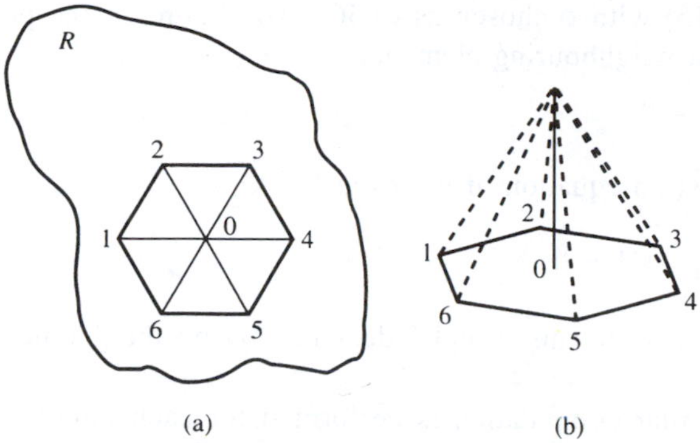

Figure 9.43 (a) a typical node and its neighbours in the region R; (b) the pyramid function used at the typical node.

unknowns u_n ($n = 1, \ldots, N$) at the node points. As $N \to \infty$, the functions ϕ_n must form a complete set, and then the weak form of (9.55) will be satisfied identically rather than approximately. The most popular set of functions ϕ_i is that due to Galerkin, who used the pyramid functions illustrated in Figure 9.43. At a typical point 0 with neighbours 1, 2, 3, ..., m we have

$$\phi = \begin{cases} \left.\begin{array}{ll} 1 & \text{at node } 0 \\ 0 & \text{at nodes } 1, 2, \ldots, m \end{array}\right\} & \begin{array}{l}\text{and piecewise-linear in each} \\ \text{of the neighbouring triangles}\end{array} \\ \text{identically zero outside the neighbouring triangles} \end{cases} \tag{9.56}$$

If there are N nodes in the mesh then there are N pyramid functions of the type (9.56). We substitute each of these functions in turn into (9.55) to satisfy the weak form of our original Poisson equation. Taking a typical node, we see that ϕ is piecewise-linear in the neighbouring triangles. For a typical such triangle 012, ϕ is just the linear function L_0 defined earlier. We substitute $\phi = L_0$ into the right-hand side of (9.55) and use the fact that

$$u = u_0L_0 + u_1L_1 + u_2L_2$$

to obtain the contribution from this particular triangle as

$$I_e = \iint_{\Delta_{012}} \left[\left(u_0\frac{\partial L_0}{\partial x} + u_1\frac{\partial L_1}{\partial x} + u_2\frac{\partial L_2}{\partial x}\right)\frac{\partial L_0}{\partial x}\right.$$
$$\left. + \left(u_0\frac{\partial L_0}{\partial y} + u_1\frac{\partial L_1}{\partial y} + u_2\frac{\partial L_2}{\partial y}\right)\frac{\partial L_0}{\partial y} + \rho_e L_0\right] \mathrm{d}x\,\mathrm{d}y$$

where ρ_e is taken to be constant in the triangle. Since L_i ($i = 0, 1, 2$) are linear, $\partial L_i/\partial x$ and $\partial L_i/\partial y$ are constants, and hence the integrals can be performed explicitly, giving

$$I_e = u_0[(y_1 - y_2)^2 + (x_1 - x_2)^2] + u_1[(y_2 - y_0)(y_1 - y_2) + (x_2 - x_0)(x_1 - x_2)]$$
$$+ u_2[(y_0 - y_1)(y_1 - y_2) + (x_0 - x_1)(x_1 - x_2)] + \tfrac{2}{3}A\rho_e$$

From (9.55) with ϕ chosen as (9.56), we obtain for the point 0 the sum of such terms over neighbouring elements:

$$\sum_{e} I_e = 0$$

This is just an equation of the form

$$\sum_{i=0}^{m} a_i u_i + b = 0$$

where the coefficients a_i and b depend only on the geometry and *not* on the field variables.

A similar computation is performed for each internal node, with ϕ taken to be of the form (9.56). The Poisson equation is linear in the u_i, and since there is one such equation for each internal node, we obtain N equations in the N unknowns u_i ($i = 1, \dots, N$). These form a matrix (called the **stiffness matrix**) equation, which can be solved for the u_i. The general strategy is

(i) calculate all the coefficients;
(ii) assemble the stiffness matrix;
(iii) invert the matrix to obtain the unknowns u_i ($i = 1, 2, 3, \dots, N$);
(iv) calculate any required data from the solution.

EXAMPLE 9.34 Solve the Laplace equation $\partial^2 u/\partial x^2 + \partial^2 u/\partial y^2 = 0$ in the unit square shown in Figure 9.44, subject to the boundary conditions indicated.

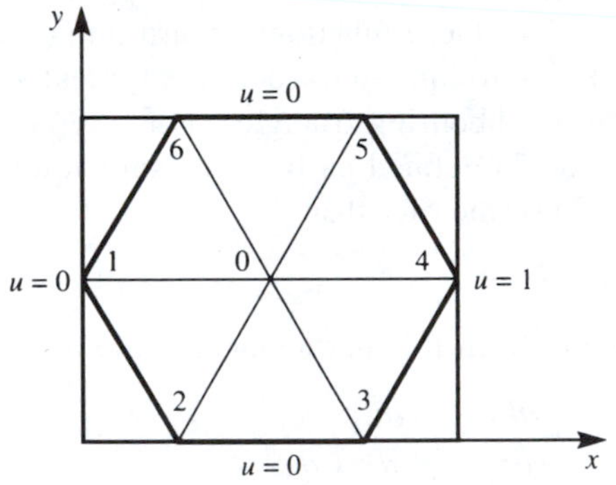

Figure 9.44 Mesh for the finite element solution to Example 9.34.

Solution Here we have a simple rectangular region with $\rho = 0$, and which is the same problem as Example 9.30.

$\Delta 012$	gives a contribution	$0.3125u_0 - 0.1875u_1 - 0.1250u_2$
$\Delta 023$	"	$0.2500u_0 - 0.1250u_2 - 0.1250u_3$
$\Delta 034$	"	$0.3125u_0 - 0.1250u_3 - 0.1875u_4$
$\Delta 045$	"	$0.2500u_0 - 0.1875u_4 - 0.1250u_5$
$\Delta 056$	"	$0.2500u_0 - 0.1250u_5 - 0.1250u_6$
$\Delta 061$	"	$0.3125u_0 - 0.1250u_6 - 0.1875u_1$

Adding all these contributions for the point 0, which is the only unspecified point, gives

$$0 = 1.75u_0 - 0.365u_1 - 0.25u_2 - 0.25u_3 - 0.365u_4 - 0.25u_5 - 0.25u_6$$

so that, knowing $u_1 = u_2 = u_3 = u_5 = u_6 = 0$ and $u_4 = 1$, we obtain $u_0 = 0.2086$. Comparing with Example 9.30, we see that our result is not particularly accurate, which is not surprising, since the mesh chosen here is particularly crude.

It is clear from Example 9.34 that the contributions from each triangle need considerable computational effort and the finite-element method is unsuitable for hand computations.

EXAMPLE 9.35 Solve the Laplace equation in the region shown in Figure 9.45 subject to zero boundary conditions except for the three points indicated. All the triangles are equilateral of side a.

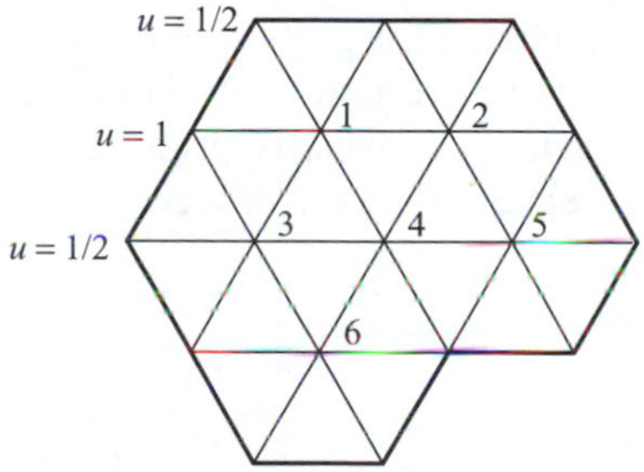

Figure 9.45 Mesh for Example 9.35. The unmarked boundary points are given as $u = 0$.

Solution Note that in the region in Figure 9.45 it would be very difficult to implement a standard finite difference mesh. We now follow the general strategy.

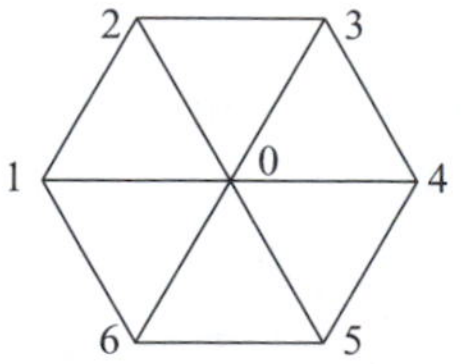

(i) *Calculate the coefficients.* When all the triangles are equilateral, the coefficients are all identical so the amount of computation is greatly reduced. For a typical point

$\Delta 012$	gives a contribution	$\frac{1}{4}a^2(4u_0 - 2u_1 - 2u_2)$
$\Delta 023$	"	$\frac{1}{4}a^2(4u_0 - 2u_2 - 2u_3)$
$\Delta 034$	"	$\frac{1}{4}a^2(4u_0 - 2u_3 - 2u_4)$
$\vdots$		

and hence adding the six contributions gives the total for the typical point

$$a^2(6u_0 - u_1 - u_2 - u_3 - u_4 - u_5 - u_6)$$

(ii) *Assemble the stiffness matrix.* Apply the results in (i) to each of the six active points

$$
\begin{aligned}
6u_1 &= u_3 + u_4 + u_2 + 0 + \tfrac{1}{2} + 1 \\
6u_2 &= u_4 + u_5 + 0 + 0 + 0 + u_1 \\
6u_3 &= 0 + u_6 + u_4 + u_1 + \tfrac{1}{2} + 1 \\
6u_4 &= u_6 + 0 + u_5 + u_2 + u_1 + u_3 \\
6u_5 &= 0 + 0 + 0 + 0 + u_2 + u_4 \\
6u_6 &= 0 + 0 + 0 + u_4 + u_3 + 0
\end{aligned}
$$

and the matrices take the form

$$
\boldsymbol{A} = \underbrace{\begin{bmatrix} 6 & -1 & -1 & -1 & 0 & 0 \\ -1 & 6 & 0 & -1 & -1 & 0 \\ -1 & 0 & 6 & -1 & 0 & -1 \\ -1 & -1 & -1 & 6 & -1 & -1 \\ 0 & -1 & 0 & -1 & 6 & 0 \\ 0 & 0 & -1 & -1 & 0 & 6 \end{bmatrix}}_{\text{stiffness matrix}}, \quad \boldsymbol{b} = \underbrace{\begin{bmatrix} 3/2 \\ 0 \\ 3/2 \\ 0 \\ 0 \\ 0 \end{bmatrix}}_{\text{load vector}}, \quad \boldsymbol{u} = \underbrace{\begin{bmatrix} u_1 \\ u_2 \\ u_3 \\ u_4 \\ u_5 \\ u_6 \end{bmatrix}}_{\text{unknowns}}
$$

(iii) We now need to solve the matrix equation $\boldsymbol{Au} = \boldsymbol{b}$ for the vector $\boldsymbol{u}$. It may be noted that the matrix does not have much 'structure', except that it is diagonally dominant, so a direct inversion is usually preferred unless the dimension of the matrix is very large. The equations were solved using MATLAB to give

$$
\boldsymbol{u} = \begin{bmatrix} 0.3481 \\ 0.0900 \\ 0.3471 \\ 0.1514 \\ 0.0402 \\ 0.0831 \end{bmatrix}
$$

(iv) Calculate any required data from the solution in (iii).

EXAMPLE 9.36 Solve the Laplace equation in the region shown in Figure 9.46 with the given boundary conditions.

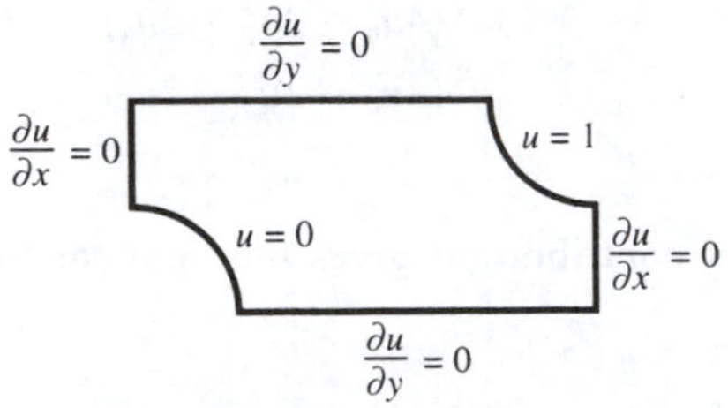

Figure 9.46 The boundary conditions for the solution of Example 9.36.

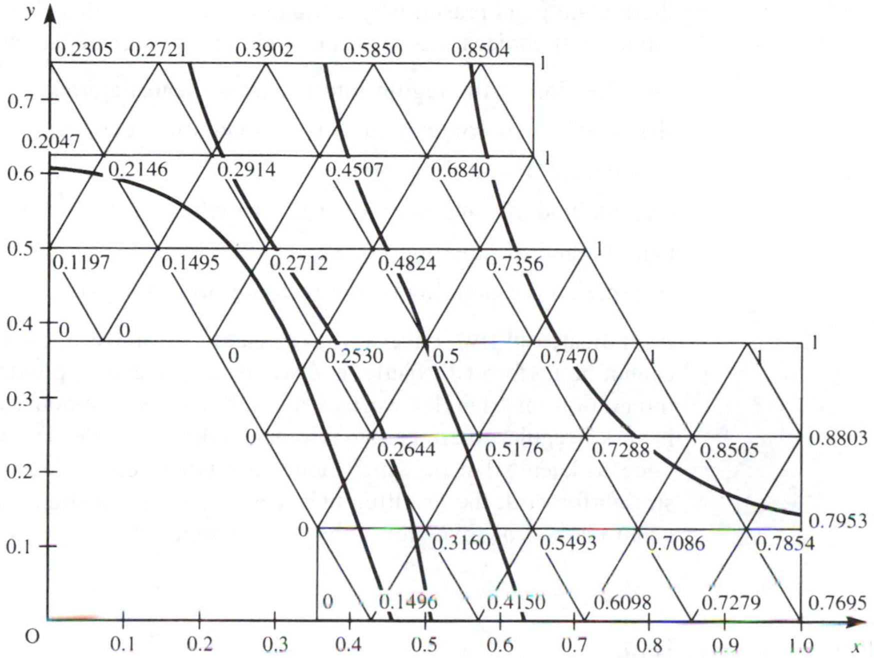

Figure 9.47 The solution of Example 9.36; the nodal values are given and the streamlines drawn.

Solution It should be noted that we have included derivative boundary conditions in this problem. To derive (9.55), we eliminated the boundary integral since $\phi = 0$ on C. In this case $\phi \neq 0$, but $u_y = 0$ on the top and bottom portions and $u_x = 0$ on the left- and right-hand sides, so the line integral is still zero. To calculate the equation for u on these boundaries, fictitious triangles can be added outside the region, and the derivative boundary condition used to extend the solution to the points outside the physical region. This procedure is not necessary, since the weak formulation will satisfy the derivative boundary conditions at least approximately. However, it can improve accuracy and speed up iterative solution methods.

The example chosen is such that the triangles are all equilateral triangles, so the coefficients are relatively easy to compute and, because of the uniformity, there is a sensible ordering to the elements. The solutions at the 33 unknown points are shown in Figure 9.47, and the streamlines in the figure illustrate the 'flow' through the configuration.

In this section no more than the 'flavour' of the finite-element method has been given. The intimate connection with computers makes it difficult to do more than show the complications that occur and give an outline of how they are dealt with.

For many problems a linear approximation is not good enough, for example, in stress analysis for finite deformations, when high derivatives are required. Also, there is no good reason why a triangle is chosen rather than a quadrilateral. There are several choices that must be made at the start of the calculation:

(i) division of the region into triangles, quadrilaterals, . . . ;
(ii) level of approximation, linear, quadratics, cubics, . . . ;
(iii) choice of test function;
(iv) method of integration over elements: exact, Gaussian, . . . ;
(v) method of labelling nodes;
(vi) method of solution of the resulting matrix equation.

As indicated in (iv), once we have abandoned linear approximations, the integrals cannot be performed exactly, and we need to use an approximate method. Gaussian integration for triangles works very well, and is commonly used. When the region is very irregular, there is no obvious labelling of nodes and it is necessary for each node to keep a list of which nodes are neighbours. Because the labelling is not straightforward, the resulting matrices rarely have a simple structure, and the most usual method of inversion is by a full frontal attack with Gaussian elimination.

9.6.1 Exercises

†52 Solve the problem in Exercise 49 using the triangular finite-element mesh shown in Figure 9.48.

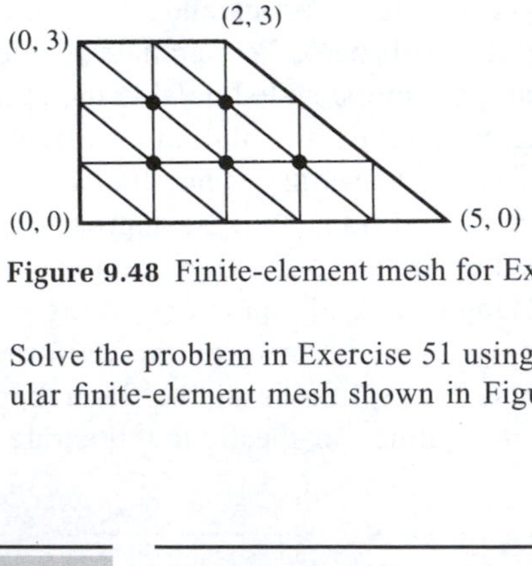

Figure 9.48 Finite-element mesh for Exercise 52.

†53 Solve the problem in Exercise 51 using the triangular finite-element mesh shown in Figure 9.49.

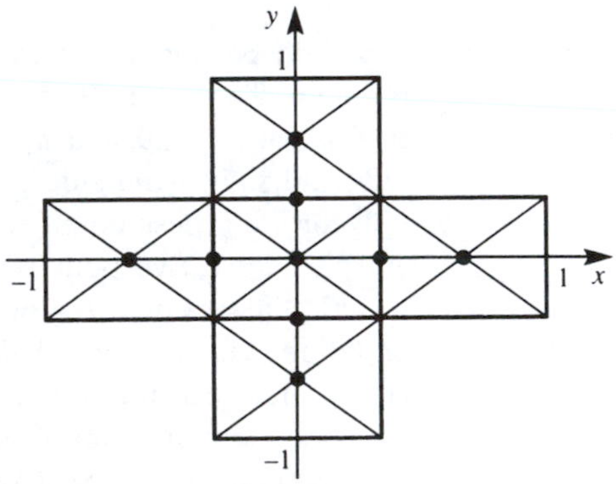

Figure 9.49 Finite-element mesh for Exercise 53.

9.7 General considerations

There are properties of a general nature that can be deduced without reference to any particular partial differential equation. The formal classification of second-order equations and their intimate connection with the appropriateness of boundary conditions will be considered in this section. The much more difficult problems of the existence and uniqueness is left to specialist texts.

9.7.1 Formal classification

In the preceding sections we have discussed in general terms the three classic partial differential equations. We shall now show that second-order equations can be reduced to one of these three types.

Consider the general form of a second-order equation:

$$Au_{xx} + 2Bu_{xy} + Cu_{yy} + Du_x + Eu_y + F = 0 \quad \textbf{(9.57)}$$

where $A, B, C, \ldots$ are constants. If we make a change of variable

$$r = ax + y, \quad s = x + by$$

then the chain rule gives

$$u_{xx} = a^2u_{rr} + 2au_{rs} + u_{ss}$$

$$u_{xy} = au_{rr} + (1 + ab)u_{rs} + bu_{ss}$$

$$u_{yy} = u_{rr} + 2bu_{rs} + b^2u_{ss}$$

Substituting into (9.57) gives

$$u_{rr}(Aa^2 + 2Ba + C) + 2u_{rs}(aA + B + abB + bC) + u_{ss}(A + 2Bb + b^2C)$$
$$+ (aD + E)u_r + (D + Eb)u_s + F = 0$$

If we choose to eliminate the u_{rs} term then we must put

$$a(A + bB) = -(B + bC) \quad \textbf{(9.58)}$$

and we can eliminate a by substitution to obtain

$$(A + 2bB + b^2C)\left[u_{ss} + \frac{AC - B^2}{(A + Bb)^2}u_{rr}\right] + \ldots = 0 \quad \textbf{(9.59)}$$

We can see immediately that the behaviour of (9.59) depends critically on the sign of $AC - B^2$ and this leads to the following classification.

Case 1: AC – B² > 0, *elliptic equations*

On putting $(AC - B^2)/(A + Bb)^2 = \lambda^2$, (9.59) becomes

$$\alpha(u_{ss} + \lambda^2u_{rr}) + \ldots = 0$$

and on further putting $q = r/\lambda$,

$$u_{ss} + u_{qq} + \ldots = 0$$

The second-order terms are just the same as the Laplace operator. Equations such as (9.57) with $AC - B^2 > 0$ are called **elliptic equations**.

Case 2: AC – B² = 0, *parabolic equations*

In this case (9.59) simply becomes

$$u_{ss} + \ldots = 0$$

with only one second-order term surviving. The equation is almost identical to the heat conduction equation. Equations such as (9.57) with $AC - B^2 = 0$ are called **parabolic equations**.

Case 3: AC – B² < 0, *hyperbolic equations*

On putting $(AC - B^2)/(A + Bb)^2 = -\mu^2$, (9.59) becomes

$$\alpha(u_{ss} - \mu^2 u_{rr}) + \ldots = 0$$

and on further putting $t = r/\mu$,

$$u_{ss} - u_{tt} + \ldots = 0$$

which we can identify with the terms of the wave equation. Equations such as (9.57) with $AC - B^2 < 0$ are called **hyperbolic equations**.

Thus we see that simply by changing axes and adjusting length scales, the general equation (9.57) is reduced to one of the three standard types. We therefore have strong reasons for studying the three classical equations very closely. An example illustrates the process.

EXAMPLE 9.37 Discuss the behaviour of the equation

$$u_{xx} + 2u_{xy} + 2\alpha u_{yy} = 0$$

for various values of the constant α.

Solution In the notation of (9.57), $A = 1$, $B = 1$ and $C = 2\alpha$, so from (9.58)

$$a = -\frac{1 + 2\alpha b}{1 + \mathrm{b}}$$

and the change of variables $r = ax + y$, $s = x + by$ gives

$$u_{ss} + \frac{2\alpha - 1}{(1 + b)^2} u_{rr} = 0$$

Thus if $\alpha > \frac{1}{2}$ and $q = r(1 + b)/\surd(2\alpha - 1)$, we have the elliptic equation

$$u_{ss} + u_{qq} = 0$$

If $\alpha = \frac{1}{2}$, we have the parabolic equation

$$u_{ss} = 0$$

If $\alpha < \frac{1}{2}$ and $t = r(1 + b)/\surd(1 - 2\alpha)$, we have the hyperbolic equation

$$u_{ss} - u_{tt} = 0$$

In (9.57) the assumption was that A, B, C, ... were constants. Certainly for many problems this is not the case, and A, B, ... are functions of x and y, and possibly u also. Therefore the analysis described does not hold globally for variable-coefficient equations. However, we can follow the same analysis at each point of the region under consideration. If the equation at *every* point is of one type, say elliptic, then we call the equation elliptic. There are good physical problems where its type can change. One of the best known examples is for transonic flow, where the equation is of the form

$$\left(1 - \frac{u^2}{c^2}\right)\psi_{xx} - \frac{2uv}{c^2}\psi_{xy} + \left(1 - \frac{v^2}{c^2}\right)\psi_{yy} + f(\psi) = 0$$

where u and v are the velocity components and c is a constant. We calculate

$$AC - B^2 = \left(1 - \frac{u^2}{c^2}\right)\left(1 - \frac{v^2}{c^2}\right) - \left(\frac{uv}{c^2}\right)^2 = 1 - \frac{u^2 + v^2}{c^2} = 1 - \frac{q^2}{c^2}$$

If we put $q/c = M$, the Mach number, then for $M > 1$ the flow is hyperbolic and supersonic, while for $M < 1$ it is elliptic and subsonic. It is easy to appreciate that transonic flows are very difficult to compute, since different boundary conditions and techniques are required on the subsonic and supersonic sides.

9.7.2 Boundary conditions

In the preceding sections we chose natural boundary conditions for the three classical partial differential equations. We can formalize these ideas a bit further and look at appropriate boundary conditions and the consequences of choosing inappropriate conditions. We shall confine ourselves to two-variable situations, but it is possible to extend the theory to problems with more variables.

Suppose that we are trying to obtain the solution $u(x, y)$ to a partial differential equation in a region R with boundary C. The commonest boundary conditions involve u or the normal derivative $\partial u/\partial n$ on C. The normal derivative (which is discussed in Section 7.2.1) at a point P on C is the rate of change of u with respect to the variable n along the line that is normal to C at P. The three conditions that are found to occur most regularly are

Cauchy conditions

$$u \text{ and } \frac{\partial u}{\partial n} \text{ given on } C$$

Dirichlet conditions

$$u \text{ given on } C$$

Neumann conditions

$$\frac{\partial u}{\partial n} \text{ given on } C$$

It is common for different conditions to apply to different parts of the boundary C. A boundary is said to be **closed** if conditions are specified on the whole of it, or **open** if conditions are only specified on part of it. The boundary can of course include infinity; conditions at infinity are specified if the boundary is closed or unspecified if it is open.

The natural conditions for the *wave equation* are Cauchy conditions on an open boundary. The d'Alembert solution (9.15) in the (x, t) plane in Section 9.3.1 corresponds to u and $\partial u/\partial t$ given on the open boundary, $t = 0$. Physically these conditions correspond to a given displacement and velocity at time $t = 0$, However the vibrations of a finite string, say a violin string, will be given by mixed conditions (Figure 9.50a). On the initial line $0 \leqslant x \leqslant l$, $t = 0$ (Figure 9.50b) Cauchy conditions will hold, with both u and $\partial u/\partial t$ given. The ends of the string are held fixed, so we have Dirichlet conditions $u = 0$ on $x = 0$, $t \geqslant 0$ and $u = 0$ on $x = l$, $t \geqslant 0$.

Figure 9.50 is typical of the hyperbolic-type equations in the two variables x and t that arise in wave propagation problems. For the second-order system

$$Au_{xx} + 2Bu_{xy} + Cu_{yy} = 0$$

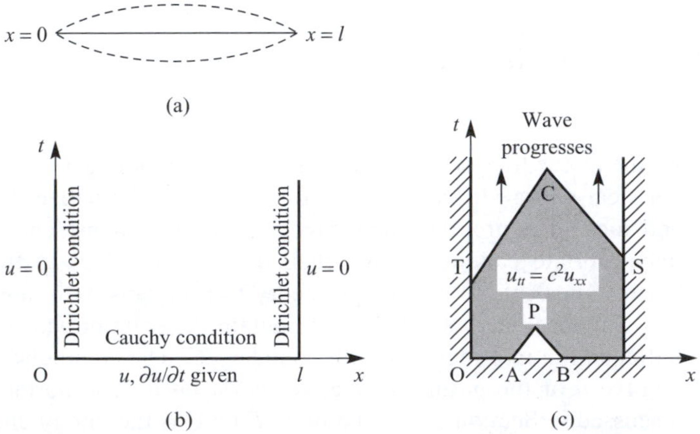

Figure 9.50 (a) A vibrating string fixed at its ends $x = 0$ and $x = l$; (b) the corresponding region and boundary conditions in the (x, t) plane; (c) wave moving forward with time.

the characteristics are defined by

$$\frac{dy}{dx} = \frac{B \pm \sqrt{(B^2 - AC)}}{A} \tag{9.60}$$

For a hyperbolic equation, $B^2 - AC > 0$, so there are two characteristics, which for constant A, B and C are straight lines. Each of the characteristics carries one piece of information from the boundary into the solution region. This is illustrated in Figure 9.50(c), where the solution at P is completely determined from the information on AB. The pair of characteristics then allows us to push the solution further into the region using the characteristics TC and SC. It is clear that Cauchy or initial data is required on the line $x = 0$ and that a hyperbolic equation will be an initial-value-type problem.

There is no reason why the boundaries cannot be at infinity – an extremely long string can sensibly be modelled in this way. Care at such infinite boundaries must be taken, since the modelling of what happens there is not always obvious; certainly it requires thought.

We have mentioned the commonest boundary conditions, but it is possible to conceive of others. However, such conditions do not always give a unique solution; a physical example will illustrate this point.

Consider the problem in Example 9.1, which has the solution

$$u = u_0 \sin\left(\frac{\pi x}{L}\right) \cos\left(\frac{\pi ct}{L}\right)$$

Suppose that a photograph of the string is taken at the times $t = L/2c$ and $t = 3L/2c$. Can the solution then be constructed from these two photographs? At the two times the string has the same shape $u = 0$; that is, the string is in its resting position. One possible solution is therefore that the string has not moved. We know that a non-zero solution to Example 9.1 is possible, so we have two solutions to our problem, and we have lost uniqueness. Specifying the displacement at two successive times is not a sensible set of boundary conditions. Although we have stated an extreme case for clarity, the same problem is there for any T, and non-unique solutions can occur if incorrect boundary conditions are imposed.

The *boundary conditions* for the heat-conduction equation (9.7) for the one-dimensional case are given by specifying u or the normal derivative $\partial u/\partial n$ on a curve C in the (x, t) plane, that is, *Dirichlet* or *Neumann conditions* respectively. Because there is only one time derivative in (9.7), we need only specify one function at $t = 0$ (say), rather than two as in the wave equation. In the simplest one-dimensional problem, at time $t = 0$ the temperature in a bar is given, $u(x, 0) = f(x)$, and at the ends some temperature condition is satisfied for all time. Typical conditions might be $u(0, t) = 0$, so that the end $x = 0$ is kept at zero temperature, and $\partial u(L, t)/\partial x = 0$, which implies no heat loss from the end $x = L$. The situation is illustrated in Figure 9.51. It is clear that, no matter what the starting temperature $f(x)$ is, the solution must tend to $u = 0$ as the final solution.

In the case of a parabolic equation we have $B^2 - AC = 0$, so the characteristics in (9.60) coalesce. Imagine that there are two characteristics very close together.

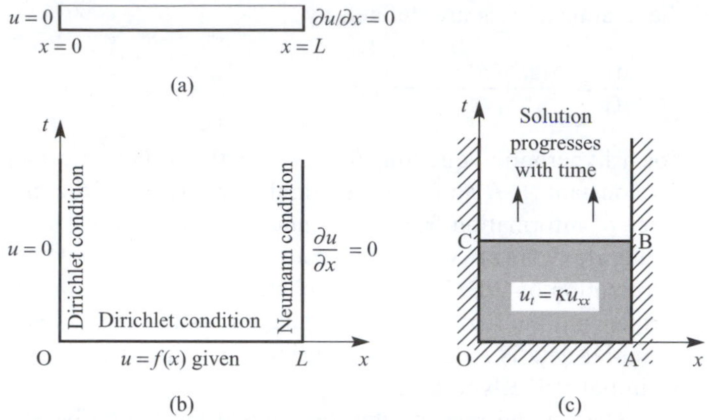

Figure 9.51 (a) A heated bar with a temperature $u = 0$ at $x = 0$ and insulated, $\partial u/\partial x = 0$, at $x = L$; (b) the corresponding region and boundary conditions in the (x, t) plane. (c) Solution can be computed at successive times.

Then the information on the boundaries will propagate a long way, since the two lines will meet 'close to infinity'. We should therefore expect that information on the initial line would propagate forward in time, and because there is only one characteristic that one piece of information on the boundary curve would be sufficient. Figure 9.51(c) illustrates the situation, with the solution on CB being determined by a single boundary condition on each of CO, OA and AB.

Again, as with the wave equation, there is no reason why the bar cannot be of infinite length, at least in a mathematical idealization, so that the initial curve C can include infinite parts. The conditions at infinity are usually quite clear and cause little difficulty.

An interesting feature is that it is very difficult to integrate the heat-conduction equation backwards in time. Suppose we are given a temperature distribution at time $t = T$ and seek the initial distribution at $t = 0$ that produces such a distribution of temperature at $t = T$. If there is an exact solution then the problem can be solved, but it is unstable in the sense that small changes at $t = T$ can lead to huge changes at $t = 0$. Consider, for instance, the solution to the heat conduction equation with $\kappa = 0.5$ in the following two situations:

Given $u = 0$ on $x = 0$ and 1, and at $t = 5$

$$u(x, 5) = \sin(\pi x)\,\mathrm{e}^{-2.5\pi^2}$$

find $u(x, 0)$.

The solution is just one of the separated solutions in (9.33), namely

$$u(x, t) = \sin(\pi x)\,\mathrm{e}^{-0.5\pi^2 t}$$

At $t = 5$ $\quad u < 2 \times 10^{-11}$
At $t = 0$ $\quad u = \sin(\pi x)$

Given $u = 0$ on $x = 0$ and 1, and at $t = 5$

$$u(x, 5) = \sin(2\pi x)\,\mathrm{e}^{-10\pi^2}$$

find $u(x, 0)$.

The solution is just one of the separated solutions in (9.33), namely

$$u(x, t) = \sin(2\pi x)\,\mathrm{e}^{-2\pi^2 t}$$

At $t = 5$ $\quad u < 1.4 \times 10^{-43}$
At $t = 0$ $\quad u = \sin(2\pi x)$

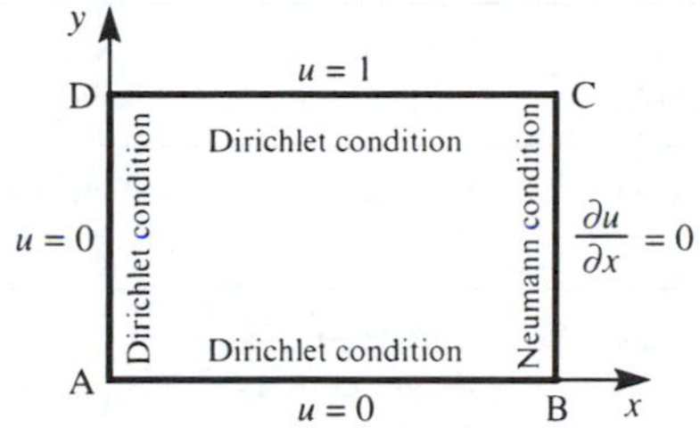

Figure 9.52 Typical boundary conditions for the Laplace equation in a rectangular plate.

The two conditions at $t = 5$ differ by a very small amount ($< 10^{-11}$) but, integrating backwards to $t = 0$, the two solutions are significantly different. Although this analysis is physically artificial it indicates why integrating backwards in time is unstable. This phenomenon, in particular, leads to almost insuperable difficulties when a numerical solution is sought, since errors are inherent in any numerical method. Such a situation applies, for instance, when a space capsule is required to have a specified temperature distribution on reaching its final orbit. The designer wants to know an initial temperature distribution that will achieve this end.

The boundary conditions most relevant to the Laplace equation are *Dirichlet* or *Neumann conditions*. These specify respectively u or the normal derivative $\partial u/\partial n$ on a closed physical boundary. One condition around the whole boundary, which may include an infinite part, is sufficient for this equation. Typically, on a rectangular plate as shown in Figure 9.52, the temperature is maintained at 1 on CD, at 0 on AB and AD, and there is no heat loss from CB.

Figure 9.52 is typical of an elliptic equation where we have conditions on a closed boundary. In (9.60) we have the condition that $B^2 - AC < 0$, so the characteristics associated with the solution do not make sense in the real plane. There is no 'time' in elliptic problems; such problems are concerned with steady-state behaviour and not propagation with time. We are dealing with a fundamentally different situation from the hyperbolic and parabolic cases. The interpretation of the idea of characteristics is unclear physically, and does not prove to be a useful direction to explore, although advanced theoretical treatments do use the concept.

It is possible to solve the Laplace equation with other boundary conditions, for instance Cauchy conditions u and $\partial u/\partial x$ on the y axis. However, an example due to Hadamard (see Exercise 39) shows that the solution is unstable in the sense that small changes in the boundary conditions cause large changes in the solution. This type of problem is not well posed, and should not occur in a physical situation; however, mistakes are made and this type of behaviour should be carefully noted.

Figure 9.53 gives in tabular form a summary of the appropriate boundary conditions for these problems.

Data	Boundary	$\nabla^2 u = u_{tt}$ Hyperbolic	$\nabla^2 u = 0$ Elliptic	$\nabla^2 u = u_t$ Parabolic
Dirichlet or Neumann	Open	Insufficient data	Insufficient data	Unique, stable solution for $t > 0$
	Closed	Not unique	Unique, stable (to an arbitrary constant in the Neumann case)	Overspecified
Cauchy	Open	Unique, stable	Solution may exist, but is unstable	Overspecified
	Closed	Overspecified	Overspecified	Overspecified

Figure 9.53 Appropriateness of boundary conditions to the three classical partial differential equations (adapted from P. M. Morse and H. Feshbach, *Methods of Theoretical Physics*, Volume I. McGraw-Hill, New York, 1953).

9.7.3 Exercises

54 Determine the type of each of the following partial differential equations, and reduce them to the standard form by change of axes:

(a) $u_{xx} + 2u_{xy} + u_{yy} = 0$

(b) $u_{xx} + 2u_{xy} + 5u_{yy} + 3u_x + u = 0$

(c) $3u_{xx} - 5u_{xy} - 2u_{yy} = 0$

55 Find the general solution of the equation Exercise 54(c).

56 Use the change of variable $u = x + y$, $v = x - y$ to transform the partial differential equation

$$\frac{\partial^2 f}{\partial x^2} - 2\frac{\partial^2 f}{\partial x \partial y} + \frac{\partial^2 f}{\partial y^2} = 0 \tag{9.61}$$

to

$$\frac{\partial^2 f}{\partial v^2} = 0$$

Hence compute the general solution of equation (9.61) as

$$f = (x - y)F(x + y) + G(x + y)$$

where F and G are arbitrary functions.

57 Establish the nature of the Tricomi equation

$$yu_{xx} + u_{yy} = 0$$

in the regions (a) $y > 0$, (b) $y = 0$ and (c) $y < 0$. Use (9.60) to determine the characteristics of the equation where they are real.

58 Verify that the function $f = [Ax^3 + (B/x^2)]y(1 - y^2)$, with A and B constants, satisfies the partial differential equation

$$x^2\frac{\partial^2 f}{\partial x^2} + (1 - y^2)\frac{\partial^2 f}{\partial y^2} = 0$$

In which regions is the equation elliptic, parabolic and hyperbolic?

59 Determine the nature of the equation

$$2q\frac{\partial^2 v}{\partial p^2} + 4p\frac{\partial^2 v}{\partial p \partial q} + 2q\frac{\partial^2 v}{\partial q^2} + 2\frac{\partial v}{\partial q} = 0$$

Show that if $p = \frac{1}{2}(x^2 - y^2)$ and $q = \frac{1}{2}(x^2 + y^2)$, the equation reduces to the Laplace equation in x and y.

60 Show that the equation

$$x^2\frac{\partial^2 u}{\partial x^2} - y^2\frac{\partial^2 u}{\partial y^2} = 0$$

is hyperbolic. Sketch the domain of dependence and range of influence from the characteristics.

9.8 Engineering application: wave propagation under a moving load

A wide range of practical problems can be studied under the general heading of moving loads. Cable cars that carry passengers, buckets that remove spoil to waste tips, and cable cranes are very obvious examples, while electric train pantographs on overhead wires are perhaps less obvious. Extending the problem to beams opens up a whole range of new problems, such as trains going over bridges, gantry cranes and the like. An excellent general discussion and wide range of applications is given by L. Fryba, *Vibration of Solids and Structures under Moving Loads* (Noordhoff, Groningen, 1973), and *Initial Value Problems, Fourier Series, Overhead Wires, Partial Differential Equations of Applied Mathematics, Open University Mathematics Unit* M321, 5, 6 and 7 (Milton Keynes, 1974) treats pantographs on overhead wires.

A straightforward linearized theory of a cable tightly stretched between fixed supports provides important information about the behaviour of such systems. Certainly, if such a theory could not be solved then more complicated problems involving large deformations, slack cables or beams would be beyond reach. The basic assumptions are

(a) the deflections from the horizontal are small compared with the length of the cable;

(b) deflections due to the weight of the cable itself are neglected;

(c) the horizontal tension in the cable is so large compared with the perturbations caused by the load that it may be regarded as constant.

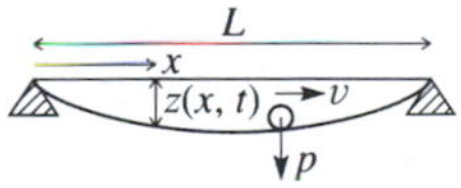

Figure 9.54 Moving load across a taut wire.

Figure 9.54 shows the situation under study and the coordinate system used.

Because the problem is one of small deflections, the basic equation is the wave equation with a forcing term from the moving load:

$$-c^2\frac{\partial^2 z}{\partial x^2}+\frac{\partial^2 z}{\partial t^2}=p(x, t)$$

Since the two ends are fixed, we have

$$z(0, t)=z(l, t)=0 \quad (t \geqslant 0) \tag{9.62}$$

A trolley is assumed to start at $x = 0$, with the cable initially at rest; that is

$$z(x, 0)=\frac{\partial}{\partial t}z(x, 0)=0 \quad (0 \leqslant x \leqslant l) \tag{9.63}$$

It remains to specify the forcing function $p(x, t)$ due to the moving load. We use the simplest assumption of the delta function and step function, as defined in Section 2.5, namely

$$-c^2\frac{\partial^2 z}{\partial x^2} + \frac{\partial^2 z}{\partial t^2} = P\delta\left(t - \frac{x}{v}\right)H(l - x) \qquad \textbf{(9.64)}$$

The delta function models the impulse of the trolley at time t at distance x, while the step function switches off the forcing function when the trolley reaches the end $x = l$.

There are several ways of solving this equation, but here the Laplace transform method will be used. Taking the transform of (9.64) using (9.23) and (9.25) together with the initial condition (9.63), we obtain the ordinary differential equation

$$-c^2Z'' + s^2Z = P\,\mathrm{e}^{-sx/v}H(l - x)$$

Since we have no interest in the case $x > l$, the final term can be omitted, since it is just 1 if $x < l$ and 0 if $x > l$. It is now straightforward to solve this equation as

$$Z = A\,\mathrm{e}^{sx/c} + B\,\mathrm{e}^{-sx/c} + \frac{P\,\mathrm{e}^{-sx/v}}{s^2(1 - c^2/v^2)}$$

Before evaluating A and B, it is clear that the speed $v = c$ causes problems, since the third term is then infinite, and the solution is not valid for this case. The solution is going to depend on whether the trolley speed is subcritical $v < c$ or supercritical $v > c$.

Equation (9.62) gives $Z(0, s) = Z(l, s) = 0$ for the boundary conditions, so that A and B can now be evaluated from

$$0 = A + B + \frac{P}{s^2(1 - c^2/v^2)}$$

$$0 = A\,\mathrm{e}^{sl/c} + B\,\mathrm{e}^{-sl/c} + \frac{P\,\mathrm{e}^{-sl/v}}{s^2(1 - c^2/v^2)}$$

Some straightforward algebra gives A and B, and hence Z, as

$$Z = \frac{P}{s^2(1 - c^2/v^2)}\left\{\mathrm{e}^{-sx/v} - \mathrm{e}^{-sl/v}\frac{\sinh(sx/c)}{\sinh(sl/c)} + \frac{\sinh[s(x - l)/c]}{\sinh(sl/c)}\right\}$$

It is easy to check that the two boundary conditions at $x = 0$ and $x = l$ are satisfied. As with all transform solutions, the main question is whether the inversion can be performed. Fortunately the three terms can be found in tables of transforms, to give

$$\begin{aligned}\frac{z}{P}\left(1 - \frac{c^2}{v^2}\right) &= \left(t - \frac{x}{v}\right)H\left(t - \frac{x}{v}\right)\\ &\quad - H\left(t - \frac{l}{v}\right)\left\{\frac{x}{l}\left(t - \frac{l}{v}\right) + \frac{2l}{c\pi^2}\sum_{n=1}^{\infty}\frac{(-1)^n}{n^2}\sin\frac{n\pi x}{l}\sin\left[\left(\frac{n\pi c}{l}\right)\left(t - \frac{l}{v}\right)\right]\right\}\\ &\quad + \left\{\left(\frac{x - l}{l}\right)t + \frac{2l}{c\pi^2}\sum_{n=1}^{\infty}\frac{(-1)^n}{n^2}\sin\left[\frac{n\pi(x - l)}{l}\right]\sin\left[\frac{n\pi ct}{l}\right]\right\}\end{aligned} \qquad \textbf{(9.65)}$$

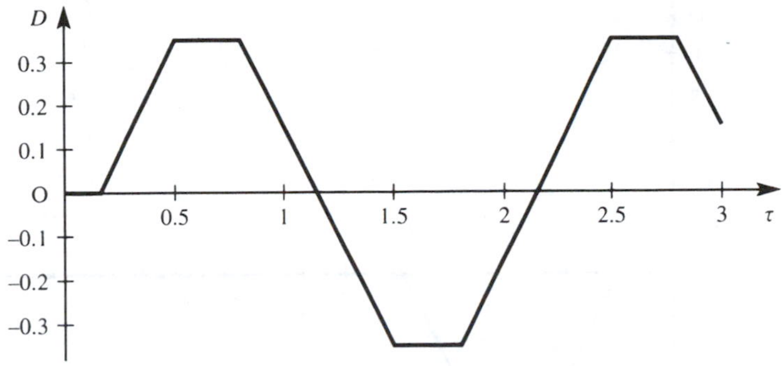

Figure 9.55 Solution of the moving-load problem for $x = \frac{1}{2}l$ and $\lambda = 0.3$; the supercritical case.

The three terms can be identified immediately. The first is the displacement caused by the trolley moving with speed v and hitting the value x after a time $t = x/v$; the second term only appears for $t > l/v$, and gives the reflected wave from $x = l$; while the third term is the wave caused by the trolley disturbance propagating in the cable with wave speed c.

To look a little more closely at the solution (9.65), we shall consider the case $x = \frac{1}{2}l$. Thus the motion of the midpoint will be considered as a function of time. Plotting such waves is easier in non-dimensional form, so we first rewrite (9.65) in terms of

$$D = \frac{cz}{Pl}\left(1 - \frac{c^2}{v^2}\right), \quad \tau = \frac{ct}{l}, \quad \lambda = \frac{c}{v}$$

so that D is the non-dimensional displacement, τ is the non-dimensional time, with $\tau = 1$ corresponding to the time for the wave to propagate the length of the cable, and λ is the ratio of wave speed to trolley speed. The second step is then to take $x = \frac{1}{2}l$ to give

$$\begin{aligned} D &= (\tau - \tfrac{1}{2}\lambda)H(\tau - \tfrac{1}{2}\lambda) \\ &\quad - H(\tau - \lambda)\left\{\frac{1}{2}(\tau - \lambda) - \frac{2}{\pi^2}\left[\sin\pi(\tau - \lambda) - \frac{1}{9}\sin 3\pi(\tau - \lambda)\right.\right. \\ &\quad \left.\left. + \frac{1}{25}\sin 5\pi(\tau - \lambda)\ldots\right]\right\} - \frac{1}{2}\tau + \frac{2}{\pi^2}\left(\sin\pi\tau - \frac{1}{9}\sin 3\pi\tau + \frac{1}{25}\sin 5\pi\tau\ldots\right) \end{aligned}$$

In Figure 9.55 the supercritical case, $\lambda = 0.3$, is displayed. It may be noted that the three terms 'switch on' at times $\tau = 0.15$, $\tau = 0.8$ and $\tau = 0.5$ respectively, corresponding physically to the trolley hitting, the reflected wave arriving and the initial wave arriving. The motion is subsequently periodic, as indicated in the figure.

Similarly, the subcritical case $\lambda = 3$ is shown in Figure 9.56. Here the switches are at $\tau = 1.5$, 2.5 and 0.5 respectively for the three terms, with the same interpretation as above. Because of the very odd choice of parameter λ, only one pulse is seen at the centre point, with the terms subsequently cancelling.

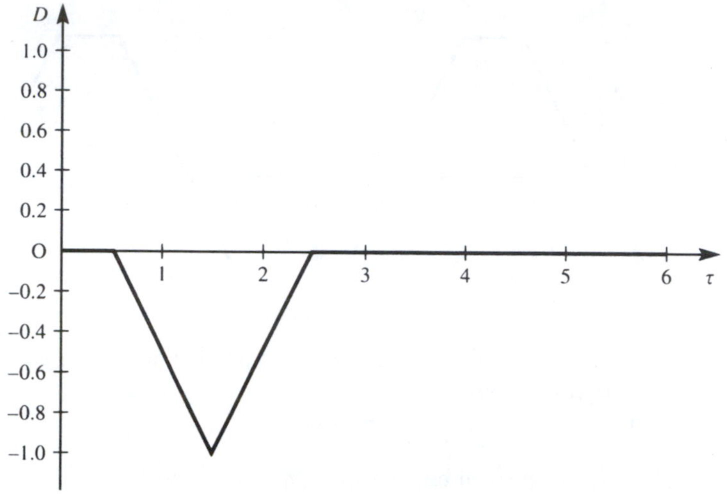

Figure 9.56 Solution of the moving-load problem for $x = \frac{1}{2}l$ and $\lambda = 3$; the subcritical case.

While the model illustrates many of the obvious properties of wave propagation, it clearly has its limitations. The discontinuous behaviour in the gradient of the displacement looks unrealistic, and the absence of damping means that oscillations once started continue for ever. It is clear that more subtle modelling of the phenomenon is required to make the solutions realistic, but the general behaviour of the solution would still be followed.

9.9 Engineering application: blood-flow model

A problem of considerable interest is how to deal with the flow of a fluid through a tube with distensible walls and hence variable cross-section. An obvious application is to the flow of blood in a blood vessel. The full Navier–Stokes equations for viscous flow are difficult to solve and the distensible wall, where boundary conditions are not clear, makes for an impossible problem. An alternative, simpler and more heuristic approach is possible and some useful solutions can be deduced. The work is based on a paper by A. Singer (*Bulletin of Mathematical Biophysics* **31** (1969) 453–70), where more details can be found about the practical application to blood flows.

The assumptions required to set up the model are as follows:

(1) the flow is one-dimensional;

(2) the flow is incompressible and laminar;

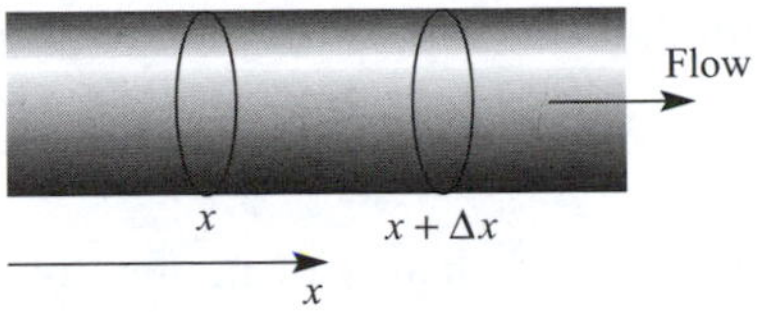

Figure 9.57 An element of the flexible tube in the blood-flow problem.

(3) the flow is slow, so that all quadratic terms can be neglected;
(4) the resistance to flow is assumed to be proportional to the velocity;
(5) there is a leakage through the walls that is proportional to the pressure;
(6) the cross-sectional area S is a function of the pressure only.

We take the situation illustrated in Figure 9.57, and we denote the pressure by p, the velocity by v, the time by t and the axial distance along the tube by x. The first equation that we derive is a continuity equation, which states that in a time Δt the fluid that comes into the element must leave the element:

$$\underbrace{(S_{t+\Delta t}}_{\text{volume after}} - \underbrace{S_t)\Delta x}_{\text{volume before}} = \underbrace{-(vS)_{x+\Delta x}\Delta t}_{\text{volume out of right-hand end}} + \underbrace{(vS)_x\Delta t}_{\text{volume into left-hand end}} - \underbrace{gpS\,\Delta x\Delta t}_{\text{leakage}}$$

The proportionality constant g is the leakage per unit volume of tube per unit time. The equation can be rewritten as

$$\frac{S_{t+\Delta t} - S_t}{\Delta t} + \frac{(vS)_{x+\Delta x} - (vS)_x}{\Delta x} + gpS = 0$$

or

$$\frac{\partial S}{\partial t} + \frac{\partial}{\partial x}(vS) + gpS = 0 \tag{9.66}$$

A second equation is required to evaluate v, and this comes from Newton's law that the force is proportional to the rate of change of momentum. The force in the x direction acting on the element in Figure 9.57 is

$$\text{force} = \underbrace{(pS)_x}_{\text{pressure force on left-hand end}} - \underbrace{(pS)_{x+\Delta x}}_{\text{pressure force on right-hand end}} - \underbrace{vrS\,\Delta x}_{\text{resistance}}$$

where r is the resistance per unit length per unit cross-section per unit time, and is the proportionality constant in assumption (4). The change in the momentum in time Δt is more difficult to compute because of the convection due to the moving fluid. However, these effects only involve second-order terms, and hence can be omitted by assumption (3). The calculation is straightforward under this assumption, so that

$$\Delta M = \text{change in momentum} = \underbrace{\rho(v_{t+\Delta t})S\,\Delta x}_{\text{momentum before}} - \underbrace{\rho(v_t)S\,\Delta x}_{\text{momentum after}}$$

where ρ is the density of the fluid. Thus

$$\frac{\partial M}{\partial t} = \rho S \frac{\partial v}{\partial t} \Delta x$$

Putting the force equal to the rate of change of momentum, we obtain, on taking the limit as $\Delta x \to 0$,

$$\rho S \frac{\partial v}{\partial t} = -\frac{\partial (pS)}{\partial x} - vrS \tag{9.67}$$

Now assumption (6) gives $S = S(p)$, so that

$$\frac{1}{S}\frac{\partial S}{\partial x} = \frac{1}{S}\frac{\mathrm{d}S}{\mathrm{d}p}\frac{\partial p}{\partial x}$$

$$\frac{1}{S}\frac{\partial S}{\partial t} = \frac{1}{S}\frac{\mathrm{d}S}{\mathrm{d}p}\frac{\partial p}{\partial t}$$

We define $c = (1/S)\,\mathrm{d}S/\mathrm{d}p$ as the **distensibility** of the tube, that is, the change in S per unit area per unit change in p. Equations (9.66) and (9.67) become

$$cp_t + v_x + cp_x v + gp = 0$$

$$\rho v_t + p_x + cp_x p + rv = 0$$

and since the term vp_x and pp_x can be neglected by assumption (3), we arrive at our final equations

$$\left.\begin{aligned} cp_t + v_x + gp &= 0 \\ \rho v_t + p_x + rv &= 0 \end{aligned}\right\} \tag{9.68}$$

These are the linearized flow equations, and are identical with the **transmission line** equations describing the flow of electricity down a long, leaky wire such as a transatlantic cable (see Exercise 6).

We can now look at special cases that will prove to be very informative about the various terms in the equation.

Case (i): c = constant, $r = g = 0$

This case corresponds to constant distensibility, which in turn gives $S = A\,\mathrm{e}^{cp}$, since S must satisfy $c = (1/S)\,\mathrm{d}S/\mathrm{d}p$. Thus we have made a specific assumption about how S depends on p. The $r = g = 0$ implies the absence of resistance and leakage. Eliminating p between the two equations in (9.68) gives

$$v_{xx} = (c\rho)v_{tt}$$

which is just the wave equation. We know that any pulse will propagate perfectly with a velocity $u = 1/\sqrt{(c\rho)}$. The assumption in the problem is that the tube is one-dimensional and has no branches. Clearly a heart pulse will propagate to the nearest branch, but there will then be reflection and a complicated behaviour near the branch. In long arteries like the femoral artery the theory can be checked for its validity.

Case (ii): S = constant

Here we are considering a rigid tube where the cross-sectional area does not vary, and hence $c = 0$. Eliminating p between the two equations in (9.68) gives

$$\rho v_t - \frac{1}{g} v_{xx} + rv = 0$$

Substituting $v = V\mathrm{e}^{-rt/\rho}$, we have

$$\rho\left(V_t - \frac{rV}{\rho}\right) - \frac{V_{xx}}{g} + rV = 0$$

so that

$$V_t = \frac{1}{\rho g} V_{xx}$$

which is just the diffusion equation. The solution for this rigid-tube case is therefore a damped, diffusion solution. Typically, if we start with a delta-function pulse at the origin then it can be checked that the solution is

$$v = A\frac{\mathrm{e}^{-rt/\rho}}{t^{1/2}}\mathrm{e}^{-\rho g x^2/4t}$$

where A is a constant. This solution is plotted in Figure 9.58, and shows the rapid damping. Such a pulse would be most unlikely to propagate far enough for blood to reach the whole of the system.

The two cases considered are extremes, but, just from the analysis performed, some conclusions can be drawn. If there is no distensibility then pulses will not propagate but will just diffuse through the system. We conclude that to move blood through the system with a series of pulses is not possible with rigid blood vessels, and we need flexible walls. Certainly for older people with hardening of the arteries, a major problem is to pump blood round the whole system, and this fact is confirmed by the mathematics.

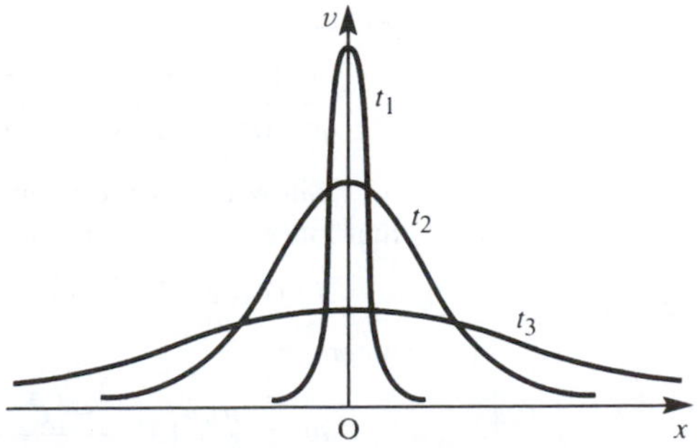

Figure 9.58 Development of the solution to the blood-flow problem from a delta function for successive times t_1, t_2 and t_3.

The actual situation is somewhere between the two cases cited, but there are no simple solutions for such cases except for the 'balanced line' case when $cr = g\rho$ (see Exercise 6 and Review exercise 20). Singer solves the equations numerically for data appropriate to a dog aorta, and compares his results with experiment. Although the agreement is good, there are problems, since there appears to be a residual pressure after each pulse. The overall pressure would therefore build up to levels that are clearly not acceptable.

9.10 Review exercises (1–21)

1 A uniform string is stretched along the x axis and its ends fixed at the points $x = 0$ and $x = a$. The string at the point $x = b$ $(0 < b < a)$ is drawn aside through a small displacement ε perpendicular to the x axis, and released from rest at time $t = 0$. By solving the one-dimensional wave equation, show that at any subsequent time t the transverse displacement y is given by

$$y = \frac{2\varepsilon a^2}{\pi^2 b(a-b)} \sum_{n=1}^{\infty} \frac{1}{n^2} \sin\left(\frac{n\pi b}{a}\right) \sin\left(\frac{n\pi x}{a}\right) \times \cos\left(\frac{n\pi ct}{a}\right)$$

where c is the transverse wave velocity in the string.

2 The function $\phi(x, t)$ satisfies the wave equation

$$\frac{\partial^2 \phi}{\partial x^2} = \frac{\partial^2 \phi}{\partial t^2} \quad (t > 0,\ 0 \leqslant x \leqslant l)$$

and the conditions

$$\phi(x, 0) = x^2 \quad (0 \leqslant x \leqslant l)$$

$$\frac{\partial \phi}{\partial t}(x, 0) = 0 \quad (0 \leqslant x \leqslant l)$$

$$\phi(0, t) = 0 \quad (t > 0)$$

$$\frac{\partial \phi}{\partial x}(l, t) = 2l \quad (t > 0)$$

Show that the Laplace transform of the solution is

$$\frac{2}{s^3} + \frac{x^2}{s} - \frac{2}{s^3}\frac{\cosh s(x-l)}{\cosh sl}$$

Using tables of Laplace transforms, deduce that the solution of the wave equation is

$$2xl - \frac{32l^2}{\pi^3} \sum_{n=0}^{\infty} \frac{(-1)^n}{(2n+1)^3} \cos\left[\frac{(2n+1)(x-l)\pi}{2l}\right] \times \cos\left[\frac{(2n+1)\pi t}{2l}\right]$$

3 The damped vibrations of a stretched string are governed by the equation

$$\frac{1}{c^2}\frac{\partial^2 y}{\partial t^2} + \frac{1}{c^2 \tau}\frac{\partial y}{\partial t} = \frac{\partial^2 y}{\partial x^2} \tag{9.69}$$

where $y(x, t)$ is the transverse deflection, t is the time, x is the position coordinate along the string, and c and τ are positive constants. A taut elastic string, $0 \leqslant x \leqslant l$, is fixed at its end points so that $y(0) = y(l) = 0$. Show that separation of variable solutions of (9.69) satisfying these boundary conditions are of the form

$$y_n(x, t) = T_n(t) \sin\left(\frac{n\pi x}{l}\right) \quad (n = 1, 2, \dots)$$

where

$$\frac{1}{c^2}\frac{\mathrm{d}^2 T_n}{\mathrm{d}t^2} + \frac{1}{c^2 \tau}\frac{\mathrm{d}T_n}{\mathrm{d}t} + \frac{n^2\pi^2 T_n}{l^2} = 0$$

Show that if the parameters c, τ and l are such that $2\pi c\tau > l$, the solutions for T_n are all of the form

$$T_n(t) = \mathrm{e}^{-t/2\tau}(a_n \cos \omega_n t + b_n \sin \omega_n t)$$

where

$$\omega_n = \frac{n\pi c}{l}\left(1 - \frac{l^2}{4\pi^2 n^2 c^2 \tau^2}\right)^{1/2}$$

and a_n and b_n are constants.

Hence find the general solution of (9.69) satisfying the given boundary conditions.

Given the initial conditions $y(x, 0) = 4\sin(3\pi x/l)$ and $(\partial y/\partial t)_{t=0} = 0$, find $y(x, t)$.

4 A thin uniform beam OA of length l is clamped horizontally at both ends. For small transverse vibrations of the beam the displacement $u(x, t)$ at time t at a distance x from O satisfies the equation

$$\frac{\partial^4 u}{\partial x^4} + \frac{1}{a^2}\frac{\partial^2 u}{\partial t^2} = 0$$

where a is a constant. The restriction that the beam is clamped horizontally gives the boundary conditions

$$u = 0, \quad \frac{\partial u}{\partial x} = 0 \quad (x = 0, l)$$

Show that for periodic solutions of the type

$$u(x, t) = V(x)\sin(\omega t + \varepsilon)$$

where ω and ε are constants, to exist, V must satisfy an equation of the form

$$\frac{d^4 V}{dx^4} = \alpha^4 V \qquad \textbf{(9.70)}$$

where $\alpha^4 = (\omega/a)^2$, and the boundary conditions

$$V(0) = V'(0) = V(l) = V'(l) = 0$$

Verify that

$$V = A\cosh\alpha x + B\cos\alpha x + C\sinh\alpha x + D\sin\alpha x$$

where A, B, C and D are constants, satisfies (9.70), and show that this function satisfies the boundary conditions provided that

$$B = -A, \quad D = -C$$

and α is a root of

$$\cos\alpha l\cosh\alpha l = 1$$

5 In a uniform bar of length l the temperature $\theta(x, t)$ at a distance x from one end satisfies the equation

$$\frac{\partial^2\theta}{\partial x^2} = a^2\frac{\partial\theta}{\partial t}$$

where a is a constant. The end $x = l$ is kept at zero temperature and the other end $x = 0$ is perfectly insulated, so that

$$\theta(l, t) = 0, \quad \frac{\partial\theta}{\partial t}(0, t) = 0 \quad (t > 0)$$

Using the method of separation of variables, show that if initially the temperature in the bar is $\theta(x, 0) = f(x)$ then subsequently the temperature is

$$\theta(x, t) = \sum_{n=0}^{\infty} A_{2n+1}\cos\left[\frac{(2n+1)\pi x}{2l}\right] \times \exp\left[-\frac{(2n+1)^2\pi^2 t}{4a^2 l^2}\right]$$

where

$$A_{2n+1} = \frac{2}{l}\int_0^l f(x)\cos\left[\frac{(2n+1)\pi x}{2l}\right]dx$$

Given $\theta(x, 0) = \theta_0(l - x)$, where θ_0 is a constant, determine the subsequent temperature in the bar.

6 Prove that if $z = x/\sqrt{t}$ and $\phi(x, t) = f(z)$ satisfies the heat-conduction equation

$$\kappa\frac{\partial^2\phi}{\partial x^2} = \frac{\partial\phi}{\partial t} \qquad \textbf{(9.71)}$$

then $f(z)$ must be of the form

$$f(z) = A\,\text{erf}\left(\frac{z}{2\sqrt{\kappa}}\right) + B$$

where A and B are constants and the **error function** is defined as

$$\text{erf}(\xi) = \frac{2}{\sqrt{\pi}}\int_0^{\xi} e^{-u^2}\,du$$

A heat-conducting solid occupies the semi-infinite region $x \geqslant 0$. At time $t = 0$ the temperature everywhere in the solid has the value T_0. The temperature at the surface, $x = 0$, is suddenly raised, at $t = 0$, to the constant value $T_0 + \phi_0$ and is then maintained at this temperature. Assuming that the temperature field in the solid has the form

$$T = \phi(x, t) + T_0$$

where ϕ satisfies (9.71) in $x > 0$, find the solution of this problem.

†7 Use the explicit method and the Crank–Nicolson formula to solve the heat-conduction equation

$$\frac{\partial^2\phi}{\partial x^2} = \frac{\partial\phi}{\partial t}$$

given that ϕ satisfies the conditions

$$\phi = 1 \quad (0 \leqslant x \leqslant 1, t = 0)$$

$$\frac{\partial\phi}{\partial x} = \begin{cases} \phi & (x = 0) \\ -\phi & (x = 1) \end{cases} \quad (t \geqslant 0)$$

Compute $\phi(x, t)$ at $x = 0, 0.2, 0.4, 0.6, 0.8, 1$ when $t = 0.004$ and $t = 0.008$.

8 An infinitely long bar of square cross-section has faces $x = 0$, $x = a$, $y = 0$, $y = a$. The bar is made of heat-conducting material, and under steady-state conditions the temperature T satisfies the Laplace equation

$$\frac{\partial^2 T}{\partial x^2} + \frac{\partial^2 T}{\partial y^2} = 0$$

All the faces except $y = 0$ are kept at zero temperature, while the temperature in the face $y = 0$ is given by $T(x, 0) = x(a - x)$. Show that the temperature distribution in the bar is

$$\frac{8a^2}{\pi^3} \sum_{r=0}^{\infty} \frac{\sin[(2r+1)\pi x/a] \sinh[(2r+1)\pi(a-y)/a]}{(2r+1)^3 \sinh(2r+1)\pi}$$

†9 (Harder) The function ϕ satisfies the Poisson equation

$$\frac{\partial^2 \phi}{\partial x^2} + \frac{\partial^2 \phi}{\partial y^2} = xy^2$$

inside the area bounded by the parabola $y^2 = x$ and the line $x = 2$. The function ϕ is given at all points on the boundary as $\phi = 1$. By using a square grid of side $\frac{1}{2}$ and making full use of symmetry, formulate a set of finite-difference equations for the unknown values ϕ, and solve.

10 A semi-infinite region of incompressible fluid of density ρ and viscosity μ is bounded by a plane wall in the plane $z = 0$ and extends throughout the region $z \geqslant 0$. The wall executes oscillations in its own plane so that its velocity at time t is $U \cos \omega t$. No pressure gradients or body forces are operative. It can be shown that the velocity of the liquid satisfies the equation

$$\frac{\partial u}{\partial t} = v \frac{\partial^2 u}{\partial z^2}$$

where $v = \mu/\rho$. Establish that an appropriate solution of the equation is

$$u = U e^{-\alpha z} \cos(\omega t - \alpha z)$$

where $\alpha = \sqrt{(\omega/2v)}$.

11 Determine the value of the constant k so that

$$U = t^k e^{-r^2/4t}$$

satisfies the partial differential equation

$$\frac{1}{r^2} \frac{\partial}{\partial r}\left(r^2 \frac{\partial U}{\partial r}\right) = \frac{\partial U}{\partial t}$$

Sketch the solution for successive values of t.

12 The function $z(x, y)$ satisfies

$$\frac{\partial z}{\partial x} + \frac{\partial z}{\partial y} = 0$$

with the boundary conditions

$$z = 2x \quad \text{when } y = -x \quad (x > 0)$$

Find the unique solution for z and the region in which this solution holds.

13 The function $\phi(x, y)$ satisfies the Laplace equation

$$\frac{\partial^2 \phi}{\partial x^2} + \frac{\partial^2 \phi}{\partial y^2} = 0$$

in the region $0 < x < \pi$, $0 < y$, and also the boundary conditions

$$\phi \to 0 \quad \text{as } y \to \infty$$

$$\phi(0, y) = \phi(\pi, y) = 0$$

Show that an appropriate separation of variables solution is

$$\phi = \sum_{n=1}^{\infty} c_n \sin(nx) e^{-ny}$$

Show that if further

$$\phi(x, 0) = x(\pi - x)$$

then $c_{2m} = 0$ while the odd coefficients are given by

$$c_{2m+1} = \frac{8}{\pi(2m+1)^3}$$

14 The boundary-value problem associated with the torsion of a prism of rectangular cross-section $-a \leqslant x \leqslant a$, $-b \leqslant y \leqslant b$ entails the solution of

$$\frac{\partial^2 \chi}{\partial x^2} + \frac{\partial^2 \chi}{\partial y^2} = -2$$

subject to $\chi = 0$ on the boundary. Show that the differential equation and the boundary conditions on $x = \pm a$ are satisfied by a solution of the form

$$\chi = a^2 - x^2 + \sum_{n=0}^{m} A_{2n+1} \cosh\left[\frac{(2n+1)\pi y}{2a}\right] \times \cos\left[\frac{(2n+1)\pi x}{2a}\right]$$

From the condition $\chi = 0$ on the boundaries $y = \pm b$, evaluate the coefficients A_{2n+1}.

15 When $0 < x < 1$ and $t > 0$ the function $u(x, t)$ satisfies the wave equation

$$\frac{\partial^2 u}{\partial x^2} = \frac{\partial^2 u}{\partial t^2}$$

and is also subject to the following boundary conditions:

(a) $u(0, t) = u(1, t) = 0$ for all $t > 0$

(b) $\dfrac{\partial u}{\partial t}(x, 0) = 0 \quad (0 < x < 1)$

(c) $u(x, 0) = 1 - x \quad (0 < x < 1)$

Use the separation method to find the solution for $u(x, t)$ that is valid for $0 < x < 1$ and $t > 0$.

16 The excess porewater pressure $u(z, t)$ in an infinite layer of clay satisfies the diffusion equation

$$\frac{\partial u}{\partial t} = c\frac{\partial^2 u}{\partial z^2} \quad (t > 0,\ 0 < z < h)$$

where t is the time in minutes, z is the vertical height in metres from the base of the clay layer and c is the coefficient of consolidation. There is complete drainage at the top and bottom of the clay layer, which is of thickness h. The distribution of excess porewater pressure $u(z, t)$ is A at $t = 0$ where A is a constant. Show that

$$u(z, t) = \frac{4A}{\pi}\sum_{n=0}^{\infty}\frac{\sin[(2n+1)\pi z/h]}{2n+1}\,e^{-c\pi^2(2n+1)^2 t/h^2}$$

17 By seeking a separated solution of the form $\phi = X(x)T(t)$, find a solution to the telegraph equation

$$\frac{\partial^2\phi}{\partial x^2} = \frac{1}{c^2}\left(\frac{\partial^2\phi}{\partial t^2} + K\frac{\partial\phi}{\partial t}\right)$$

satisfying the conditions

(a) $\phi = A\cos px$ for all values of x and for $t = 0$ for the case when $c^2p^2 > \frac{1}{4}K^2$;

(b) $\phi = A$ and $\partial\phi/\partial t = -\frac{1}{2}AK$ for $x = 0$ and $t = 0$.

18 For the two-dimensional flow of an incompressible fluid the continuity equation may be expressed as

$$\frac{\partial}{\partial r}(rv_r) + \frac{\partial v_\theta}{\partial\theta} = 0$$

where r and θ are polar coordinates in a plane parallel to the flow, and v_r and v_θ are the respective velocity components. Show that there exists a stream function ψ such that

$$v_r = \frac{1}{r}\frac{\partial\psi}{\partial\theta}$$

$$v_\theta = -\frac{\partial\psi}{\partial r}$$

satisfy the continuity equation.

Take

$$\psi = Ur\sin\theta - \frac{Ua^2}{r}\sin\theta$$

and interpret the solution physically.

19 (An extended problem) Section 9.8 looked at wave propagation caused by moving loads on cables. For loads on beams a similar analysis models such problems as trains going over bridges or loads moving on gantry cranes. Use a similar analysis for the beam equation

$$a^2\frac{\partial^4 u}{\partial x^4} + \frac{\partial^2 u}{\partial t^2} = p(x, t)$$

20 (An extended problem) In the blood-flow model in Section 9.8 consider the following cases:

(a) S = constant, $g = 0$ for a pulsating flow

$v = v_0\,e^{j\omega t}$ at $x = 0$ for all t

(b) S = constant, $r = 0$ for a pulsating flow

$v = v_0\,e^{j\omega t}$ at $x = 0$ for all t

(c) the balanced-line case when $g\rho = rc$. Show that $v = e^{-gt/c}U$ gives

$$\frac{\partial^2 U}{\partial t^2} = \frac{1}{c\rho}\frac{\partial^2 U}{\partial x^2}$$

Solve the equation and interpret your solution.

21 (An extended problem) Fluid flows steadily in the two-dimensional channel shown in Figure 9.59. The temperature $\theta = \theta(x, t)$ depends only on the distance x along the channel and the time t. The fluid flows at a constant rate so that an amount L crosses any given section in unit time. The specific heat of the fluid is a constant c, and the heat H in a length δx with cross-section S is therefore

$$H = c(S\,\delta x)\theta$$

Heat is transferred through the walls of the channel, AB and DC, at a rate proportional to the temperature difference between the inside and outside. Heat conduction in the x direction is neglected. Show that the heat balance in the element ABCD leads to the equation

$$cS\frac{\partial\theta}{\partial t} = -Lc\frac{\partial\theta}{\partial x} + k_1(\theta'' - \theta) + k_2(\theta' - \theta)$$

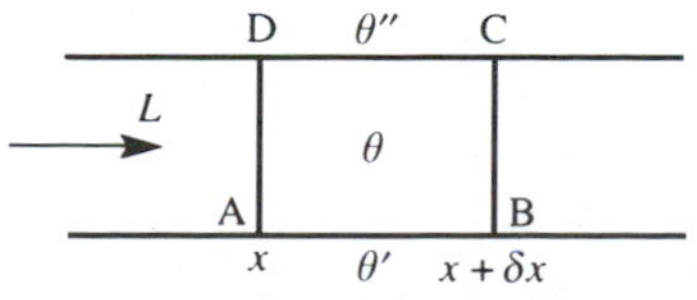

Figure 9.59 An element of the channel in Review exercise 21.

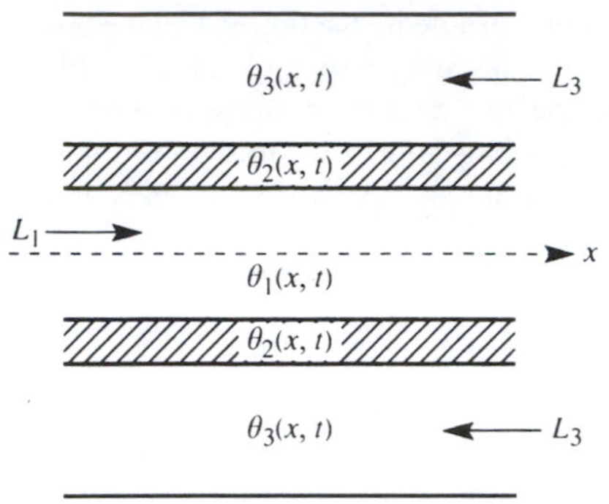

Figure 9.60 Heat-exchanger configuration in Review exercise 21.

This type of analysis can now be applied to the long heat exchanger illustrated in Figure 9.60. The configuration is considered to be two-dimensional and symmetric with respect to the x axis; in the inner region the flow is to the right, while in the outer regions it is to the left. The regions are separated by metal walls in which similar assumptions to the above are made, except of course there is no fluid flow.

Set up the equations of the system in the form

$$c_1S_1\frac{\partial\theta_1}{\partial t} = -L_1c_1\frac{\partial\theta_1}{\partial x} + 2k_1(\theta_2 - \theta_1)$$

$$c_2S_2\frac{\partial\theta_2}{\partial t} = -k_1(\theta_1 - \theta_2) + k_2(\theta_3 - \theta_2)$$

$$c_3S_3\frac{\partial\theta_3}{\partial t} = L_3c_3\frac{\partial\theta_3}{\partial x} + k_2(\theta_2 - \theta_3)$$

where the assumption is made that there is no heat flow through the outside lagged walls. Solve the steady-state equations and fit the arbitrary constants to the conditions that at the inlet ($x = 0$) the fluid enters the inner region at a given temperature $\theta_1 = T_1$, while at the outlet ($x \to \infty$) the fluid in the outer regions enters at a given temperature $\theta_3 = T_3$. Find flow rates that ensure that this situation is possible, and discuss the implications of any results obtained.

Discuss the assumptions made in setting up this problem, the limitations imposed by the assumptions, possible applications of this type of analysis, and extensions of the work, for example a time-dependent solution.

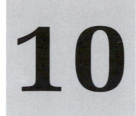

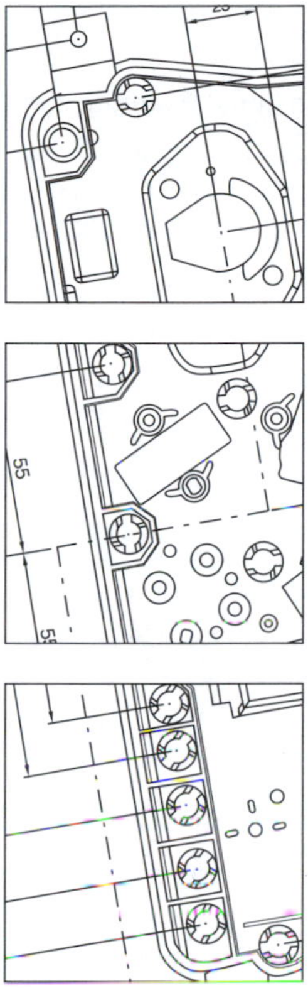

10

Optimization

CONTENTS

10.1 Introduction

The need to get the 'best' out of a system is a very strong motivation in much of engineering. A typical problem may be to obtain the maximum amount of product or to minimize the cost of a process or to find a configuration that gives maximum strength. Sometimes what is 'best' is easy to define, but frequently the problem is not so clear cut, and a lot of thought is required to reach an appropriate function to optimize. In most cases there are very severe and natural constraints operating: the problem may be one of maximizing the amount of product, subject to the supply of materials; or it may be minimizing the cost of production, with constraints due to safety standards. Indeed, much of modern optimization is concerned with constraints and how to deal with them.

We have seen in Chapter 9 of *Modern Engineering Mathematics* how to obtain the maximum and minimum of a function of many variables. However, the methods described there founder very quickly because it becomes impossible to solve the resulting equations analytically. A simple one-dimensional example soon shows that a numerical solution is required.

EXAMPLE 10.1 Find the positive x value that maximizes the function

$$y = \frac{\tanh x}{1 + x}$$

Solution Equating the derivative to zero gives

$$\frac{dy}{dx} = 0 = \frac{(1 + x)\operatorname{sech}^2 x - \tanh x}{(1 + x)^2}$$

so that we need to solve

$$1 + x = \tfrac{1}{2}\sinh 2x$$

which has no simple positive solutions that can be obtained analytically.

To solve such problems, a set of numerical algorithms was developed during the 1960s as fast computers became available to perform the large amounts of arithmetic required. These algorithms will be described in Section 10.4. Perhaps the main stimulus for this development came from the space industries, where small percentage savings, achieved by doing some mathematics, could save vast amounts of money. The ideas were quickly taken up by 'expensive' areas of engineering, such as the chemical and steel industries and aircraft production.

The idea of dealing with constraints is not new: Lagrange developed the theory of **equality**-constrained optimization around the 1800s. However, it was not until the 1940s that **inequality** constraints were studied with any seriousness. The use of Lagrange multipliers for equality constraints was also introduced in Chapter 9 of *Modern Engineering Mathematics*, and will be looked at again in more detail in Section 10.3 below. The only work on inequality constraints will be for linear programming problems in Section 10.2. Where inequality constraints are nonlinear, the problems become very difficult to deal with, and are the province of specialist books on optimization. Linear programming, however, is much more straightforward, and the basic simplex algorithm has been spectacularly successful – so successful in fact that many workers try to force their problems to be linear when they are clearly not. The computer scientist's maxim GIGO ('garbage in, garbage out') is very applicable to people who try to fit the problem to the mathematics rather than the mathematics to the problem!

Before considering detailed methods of solution of optimization problems, we shall look at a few examples. Let us first revisit an extended form of the milk carton problem considered in Example 9.4 (and illustrated in Figure 9.9) of *Modern Engineering Mathematics*.

EXAMPLE 10.2 A milk carton is designed from a sheet of waxed cardboard as illustrated in Figure 10.1, where a 5 mm overlap has been allowed.

It is to contain 2 pints of milk, and we require the minimum surface area for the carton.

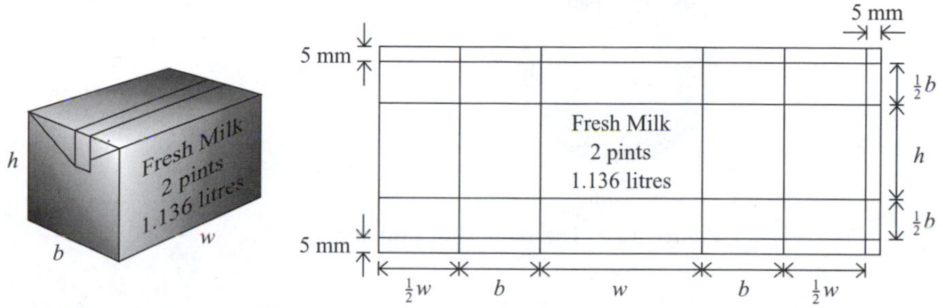

Figure 10.1 Waxed cardboard milk container opened up, with measurements in millimetres and with a 5 mm overlap.

Solution The only difference between this example and Example 9.4 of *Modern Engineering Mathematics* is that we no longer insist on a square cross-section. The total area in square millimetres is

$$A = (2b + 2w + 5)(h + b + 10)$$

and the volume of the two-pint container is

$$\text{volume} = hbw = 1\,136\,000\ \text{mm}^3$$

We first note that a constraint, the given volume, occurs naturally in the problem. Because of its simplicity, we can eliminate w from the constraint to give

$$A = (h + b + 10)\left(\frac{2\,272\,000}{hb} + 2b + 5\right)$$

Following the standard minimization procedure and equating partial derivatives to zero gives

$$\frac{\partial A}{\partial h} = \frac{2\,272\,000}{hb} + 2b + 5 - (h + b + 10)\frac{2\,272\,000}{h^2 b} = 0$$

$$\frac{\partial A}{\partial b} = \frac{2\,272\,000}{hb} + 2b + 5 - (h + b + 10)\left(\frac{-2\,272\,000}{hb^2} + 2\right) = 0$$

We therefore have two highly nonlinear equations in the two unknowns h and b, which cannot be solved without resorting to numerical techniques. We shall return to this problem later in Examples 10.11 and 10.17 to see how a practical solution can be obtained.

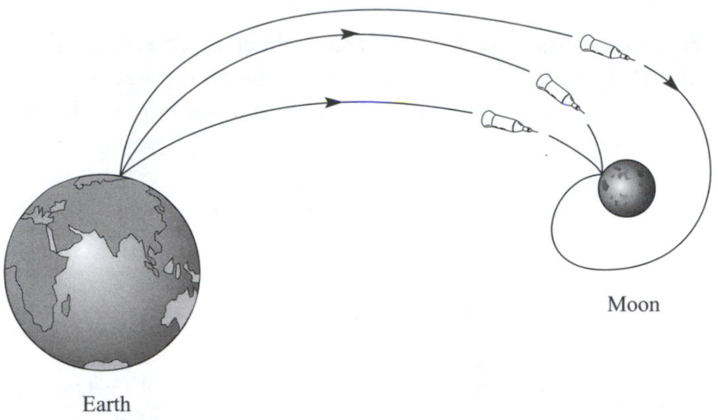

Figure 10.2 The moonshot problem.

Most practical optimization problems come from very expensive projects where savings of a few per cent can be very significant. Laying natural gas or water pipe networks are typical examples. Without considering the expense of installing compressors, the problem is to minimize the capital cost. This cost is directly related to the weight of the pipe, subject to constraints imposed by pressure-drop limitations, which in turn depend on the pipe diameter in a nonlinear way. Adding the compressors imposes further costs and constraints.

Heat exchangers provide an example of a system where we try to remove heat. We design the flow rates, the pipe sizes and pipe spacing to maximize the heat transferred. A related heating problem might be the design of an industrial furnace. It is required that the energy consumption be minimized subject to constraints on the heat flow and the maintenance of various temperatures.

A final example, the moonshot problem, illustrates a large-scale, very complicated problem that stimulated much of the recent developments in optimization (see Figure 10.2). Which path from a point on the Earth to a point on the Moon should be chosen to minimize the weight of fuel carried by a rocket? The complicated relation between the weight of fuel, the mechanical equations of the rocket and the path must be established before it is possible to proceed to obtain the optimum. The numerous constraints on the strengths of materials, the maximum tolerable acceleration etc. add to the difficulty of the problem.

In the problems discussed we have assumed that an optimum exists at a point, and we have asked for the mathematical conditions that must hold. The other way round is much more difficult. Given that the appropriate conditions hold, does an optimum exist, and if so what type of optimum is it? For many simple finite-dimensional problems these conditions are known, but may not be very simple to apply. To serve as a reminder, the condition $f'(0) = 0$ is a necessary condition for a maximum to exist for the differentiable function $f(x)$ at $x = 0$. It is not sufficient, however, as can be seen from the three functions $f_1(x) = x^2$, $f_2(x) = x^3$ and $f_3(x) = -x^2$, which have respectively a minimum, a point of inflection and a maximum at the origin. In many dimensions the difficulties are similar, but much more complicated.

10.2 Linear programming

10.2.1 Introduction

In Section 10.1 it was indicated that constraints are very important in most applications. When all functions are linear, there is an extremely efficient algorithm, developed by Danzig in the 1940s, which will be described for the **linear programming (LP)** problem.

We shall start by posing a particular problem and looking at a simple graphical solution.

EXAMPLE 10.3 A manufacturing company makes two circuit boards R1 and R2, constructed as follows:

R1 comprises 3 resistors, 1 capacitor, 2 transistors and 2 inductances;
R2 comprises 4 resistors, 2 capacitors and 3 transistors.

The available stocks for a day's production are 2400 resistors, 900 capacitors, 1600 transistors and 1200 inductances. It is required to calculate how many R1 and how many R2 the company should produce daily in order to maximize its overall profits, knowing that it can make a profit on an R1 circuit board of 5p and on an R2 circuit board of 9p.

Solution If the company produces daily x of type R1 and y of type R2 then its stock limitations give

$$3x + 4y \leqslant 2400 \tag{10.1a}$$

$$x + 2y \leqslant 900 \tag{10.1b}$$

$$2x + 3y \leqslant 1600 \tag{10.1c}$$

$$2x \leqslant 1200 \tag{10.1d}$$

$$x \geqslant 0, \quad y \geqslant 0$$

and it makes a profit z given by

$$z = 5x + 9y \tag{10.2}$$

These inequalities are plotted on a diagram as in Figure 10.3(a). The shaded region defines the area for which *all* the inequalities are satisfied, and is called the **feasible region**. The lines of constant profit z = constant, defined by (10.2), are plotted as 'dashed' lines in Figure 10.3(b). It is clear from the geometry that the largest possible value of z that intersects the feasible region is at S with $x = 500$,

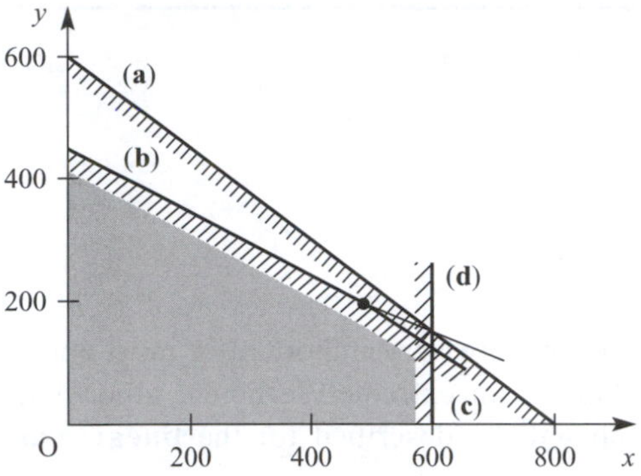

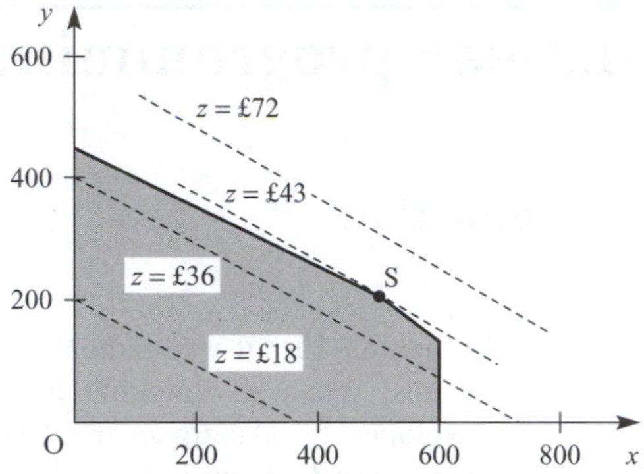

(a) (b)

Figure 10.3 (a) Feasible region for the circuit board manufacture problem. (b) Lines of constant z show that S (500, 200) gives the optimum.

	Available	Used	Left over
Resistors	2400	2300	100
Capacitors	900	900	0
Transistors	1600	1600	0
Inductances	1200	1000	200

Figure 10.4 Table of stock usage.

$y = 200$, and this gives the optimal solution. At this point we can analyse the usage of the stocks as in Figure 10.4 and note that a profit of £43 has been made.

Example 10.3 has encapsulated much of the LP method, and we shall try to extract the maximum amount of information from this example. The graphical method will only work if the problem has two variables, so we need to consider how to translate the geometry into an algebraic form that will work with any number of variables. Although we shall concentrate in this chapter on small problems in order to illustrate the methods, in practical problems there can be hundreds of variables and constraints. Large problems bring further difficulties that will not be considered here; for instance, how a large amount of information can be input into a computer accurately or how large data sets are handled in the computer.

From Figure 10.3 it can be seen that the solutions must be at a 'corner' of the feasible region, other than in the exceptional case when the profit line $z =$ constant is parallel to one of the constraints. This follows through into many-dimensional problems, so that it is *only necessary to inspect the corners of the feasible region*. The **simplex method**, described in Section 10.2.3, uses this fact and selects a starting corner, chooses the neighbouring corner that increases z the most, and then repeats the process until no improvement is possible. The method writes the equations into a standard form; it then automates the choice of corner and finally reprocesses the equations back to the standard form again.

Once a solution has been obtained, it may be observed from Figures 10.3 and 10.4 that the **binding constraints** (b) and (c) intersect at S and are satisfied identically, so that all the stocks are used, while the **non-binding constraints** (a) and (d) leave some stock unused. It can also be seen from Figure 10.3 that the constraint (a) is redundant since it does not intersect the feasible region. These might appear obvious comments, but they prove to be useful and relevant observations when a sensitivity analysis is performed. Such an analysis asks whether or not the solution changes as the stocks vary or the costs vary, or the coefficients are changed. In practice, parameters vary over a period, and we wish to know whether a new calculation must be performed or whether the solution that we have already obtained can be used.

There are a whole series of special cases that we must consider:

(a) Does a feasible region exist? If the system is modelled correctly, a feasible region always exists for a sensible problem; but if an error is made then it is easy to eliminate the region completely. For instance, in (10.1d) putting – 1200 instead of 1200 would give an empty region, and this must be detected by any program.

(b) Is the solution degenerate? If the profit lines z = constant are parallel to any one of the final binding constraints then any feasible part of the constraint gives a solution.

(c) Can we get unbounded regions and solutions? Again it is very easy to construct problems where the regions are unbounded and finite solutions may or may not exist. Just maximizing (10.2) subject only to the single constraint (10.1d) provides an unbounded region. Interpreting this situation, the only constraint that we have is on the inductances. If this were true, we could make infinitely many circuits of type R2 and make an infinite profit!

10.2.2 Simplex algorithm: an example

We now need to convert the ideas of Section 10.2.1 into a useful algebraic algorithm. There is a whole array of technical terms that are used in LP, and they will be introduced as we reconsider Example 10.3 to develop the solution method. The first step is to introduce **slack variables** r, s, t and u into (10.1) to make the inequality constraints into equality constraints.

If we are given x and y in the feasible region, the variables r, s, t and u provide a measure of how much 'slack' is available before all the corresponding resource is used up, so

$$3x + 4y + r = 2400 \qquad \textbf{(10.3a)}$$

$$x + 2y + s = 900 \qquad \textbf{(10.3b)}$$

$$2x + 3y + t = 1600 \qquad \textbf{(10.3c)}$$

$$2x + u = 1200 \qquad \textbf{(10.3d)}$$

where x, y, r, s, t and u are now all greater than or equal to zero. We now have more variables than equations, and this enables us to construct a **feasible basic solution** by inspection:

$$\underbrace{x = y = 0;}_{\text{non-basic variables}} \quad \underbrace{r = 2400, \quad s = 900, \quad t = 1600, \quad u = 1200}_{\text{basic variables}}$$

with 4 **basic variables** (the same number as constraints, which are non-zero) and 2 **non-basic variables** (the remainder of the variables, which are zero).

The algebraic equivalent of moving to a neighbouring corner is to increase one of the non-basic variables from zero to its largest possible value. From the profit function given in (10.2), we have

$$z = 5x + 9y$$

Currently z has the value zero, and it seems sensible to change y, since the coefficient of y is larger; this will increase z the most. So keep $x = 0$ in (10.3) and increase y to its maximum value in each case: either

(a) change y to 600 and reduce r to zero, or

(b) change y to 450 and reduce s to zero, or

(c) change y to $533\frac{1}{3}$ and reduce t to zero, or

(d) note that there is no effect on changing y.

Choose option (b), since increasing y above 450 will make s negative, which would then violate the condition that all variables must be positive. Interchange s and y between the set of basic and non-basic variables and rewrite in the same form as (10.3). This is achieved by solving for y from (10.3b) and substituting to give

$$x - 2s + r = 600 \qquad \textbf{(10.4a)}$$

$$\tfrac{1}{2}x + \tfrac{1}{2}s + y = 450 \qquad \textbf{(10.4b)}$$

$$\tfrac{1}{2}x - \tfrac{3}{2}s + t = 250 \qquad \textbf{(10.4c)}$$

$$2x + u = 1200 \qquad \textbf{(10.4d)}$$

and, from (10.2),

$$z = 4050 + \tfrac{1}{2}x - \tfrac{9}{2}s \qquad \textbf{(10.4e)}$$

The problem is now reduced to exactly the same form as (10.3), and the same procedure can be applied. The non-basic variables are $x = s = 0$, and the basic variables are $r = 600$, $y = 450$, $t = 250$ and $u = 1200$, and z has increased its value from 0 to 4050.

Now only x can be increased, since increasing the other non-basic variable, s, would decrease z. Increasing x to 500 in (10.4c) and reducing t to 0 is the best that can be done. Using (10.4c) to write $x = 3s - 2t + 500$, we now eliminate x from the other equations to give

$$s - 2t + r \qquad\qquad = 100 \tag{10.5a}$$

$$2s - t \quad + y \qquad = 200 \tag{10.5b}$$

$$-3s + 2t \qquad + x \quad = 500 \tag{10.5c}$$

$$6s - 4t \qquad\qquad + u = 200 \tag{10.5d}$$

and

$$z = 4300 - 3s - t \tag{10.5e}$$

We now have the final solution, since increasing s or t can only decrease z. Thus we have $x = 500$, $y = 200$, which is in agreement with the previous graphical solution, the maximum profit is $z = 4300$ as before, and the amounts left over in Figure 10.4 are just the 100 and 200 appearing on the right-hand sides of (10.5a, d).

We have just described the essentials of the **simplex algorithm**, although the method of working may have appeared a little haphazard. It can be tidied up and formalized by writing the whole system in **tableau form**. Equations (10.3) are written with the basic variables in the left-hand column, the coefficients in the equations placed in the appropriate array element and the objective function z placed in the first row with minus signs inserted.

		Non-basic variables		Basic variables				
		x	y	r	s	t	u	Solution
Objective function z		−5	−9	0	0	0	0	0
Basic variables	r	3	4	1	0	0	0	2400
	s	1	2	0	1	0	0	900
	t	2	3	0	0	1	0	1600
	u	2	0	0	0	0	1	1200

The current solution can easily be read from the tableau. The basic variables in the left-hand column are equal to the values in the solution column, so $r = 2400$, $s = 900$, $t = 1600$ and $u = 1200$. The remaining non-basic variables are zero, namely $x = y = 0$. The profit z is read similarly as the entry in the solution column, namely $z = 0$. The negative signs in the z row ensure that z remains positive in the subsequent manipulation. It should be noted that a 4×4 unit matrix (shown shaded) occurs in the tableau in the basic variable columns, with zeros occurring above in the z row. This standard display is always the starting place for the simplex method, with the only possible complication being that the columns of the unit matrix might be shuffled around. The algorithm can now be performed in a series of steps:

Step 1

Choose the most negative entry in the z row and mark that column (the y column in this case).

Step 2

Evaluate the ratios of the solution column and the *positive* entries in the y column, choose the smallest of these and mark that row (the s row in this case).

	x	y	r	s	t	u	Solution	
z	−5	−9	0	0	0	0	0	Ratios
r	3	4	1	0	0	0	2400	2400/4 = 600
s	1	(2)	0	1	0	0	900	900/2 = 450
t	2	3	0	0	1	0	1600	1600/3 = $533\frac{1}{3}$
u	2	0	0	0	0	1	1200	–

Step 3

Change the marked basic variable in the left-hand column to the marked non-basic variable in the top row (in this case s changes to y in the left-hand column).

Step 4

Make the pivot (the element in the position where the marked row and column cross) 1 by dividing through. In this case we divide the row elements by 2. These series of steps lead to the tableau

	x	y	r	s	t	u	Solution
z	−5	−9	0	0	0	0	0
r	3	4	1	0	0	0	2400
y	$\frac{1}{2}$	1	0	$\frac{1}{2}$	0	0	450
t	2	2	0	0	1	0	1600
u	2	0	0	0	0	1	1200

Step 5

Clear the y column by subtracting an appropriate multiple of the y row (this is just Gaussian elimination); for example (z row) + 9 × (y row), (r row) − 4 × (y row) and so on. This leads to the tableau

	x	y	r	s	t	u	Solution
z	$-\frac{1}{2}$	0	0	$\frac{9}{2}$	0	0	4050
r	1	0	1	−2	0	0	600
y	$\frac{1}{2}$	1	0	$\frac{1}{2}$	0	0	450
t	$\frac{1}{2}$	0	0	$-\frac{3}{2}$	1	0	250
u	2	0	0	0	0	1	1200

This tableau can now easily be recognized as equations (10.4). It may be noted that the unit matrix appears in the tableau again, with the columns permuted, and the z row has zero entries in the basic variable columns.

The tableau is in exactly the standard form, and is ready for reapplication of the five given steps. *Steps 1 and 2* give the tableau

	x	y	r	s	t	u	Solution	
z	$-\frac{1}{2}$	0	0	$\frac{9}{2}$	0	0	4050	Ratios
r	1	0	1	−2	0	0	600	600
y	$\frac{1}{2}$	1	0	$\frac{1}{2}$	0	0	450	900
t	($\frac{1}{2}$)	0	0	$-\frac{3}{2}$	1	0	250	500
u	2	0	0	0	0	1	1200	600

Steps 3, 4 and 5 then produce a final tableau

	x	y	r	s	t	u	Solution
z	0	0	0	3	1	0	4300
r	0	0	1	1	−2	0	100
y	0	1	0	2	−1	0	200
x	1	0	0	−3	2	0	500
u	0	0	0	6	−4	1	200

All the entries in the z row are now positive, so the optimum is achieved. The solution is read from the tableau directly; the left-hand column equals the right-hand column, giving $z = 4300$, $r = 100$, $y = 200$, $x = 500$ and $u = 200$, which is in agreement with the solution obtained in Example 10.3.

10.2.3 Simplex algorithm: general theory

We can now generalize the problem to the standard form of finding the **maximum** of the objective function

$$z = c_1x_1 + c_2x_2 + \ldots + c_nx_n$$

subject to the constraints

$$\left.\begin{array}{l} a_{11}x_1 + a_{12}x_2 + \ldots + a_{1n}x_n \leqslant b_1 \\ a_{21}x_1 + a_{22}x_2 + \ldots + a_{2n}x_n \leqslant b_2 \\ \quad\vdots \qquad\quad \vdots \qquad\qquad\quad \vdots \qquad\quad \vdots \\ a_{m1}x_1 + a_{m2}x_2 + \ldots + a_{mn}x_n \leqslant b_m \end{array}\right\} \qquad (10.6)$$

by the **simplex algorithm**, where the $b_1, b_2, \ldots, b_m$ are all positive. By introducing the slack variables $x_{n+1}, \ldots, x_{n+m} \geqslant 0$ we convert (10.6) into the **standard tableau**

	x_1	x_2	$\ldots$	x_n	x_{n+1}	x_{n+2}	$\ldots$	x_{n+m}	Solution
z	$-c_1$	$-c_2$	$\ldots$	$-c_n$	0	0	$\ldots$	0	0
x_{n+1}	a_{11}	a_{12}	$\ldots$	a_{1n}	1	0	$\ldots$	0	b_1
x_{n+2}	a_{21}	a_{22}	$\ldots$	a_{2n}	0	1	$\ldots$	0	b_2
$\vdots$	$\vdots$	$\vdots$		$\vdots$	$\vdots$	$\vdots$		$\vdots$	$\vdots$
x_{n+m}	a_{m1}	a_{m2}	$\ldots$	a_{mn}	0	0	$\ldots$	1	b_m

Any subsequent tableau takes this general form, with an $m \times m$ unit matrix in the basic variables columns. As noted in the previous example, the basic variables change, so the left-hand column will have m entries, which can be any of the variables, $x_1, \ldots, x_{m+n}$, and the unit-matrix columns are usually not in the above neat form but are permuted.

The five basic steps in the algorithm follow quite generally:

Step 1

Choose the most negative value in the z row, say $-c_i$. If all the entries are positive then the maximum has been achieved. (Identify column i)

Step 2

Evaluate $b_1/a_{1i}, b_2/a_{2i}, \ldots, b_m/a_{mi}$ for all *positive* a_{ki}. Select the minimum of these numbers, say b_j/a_{ji}. (Identify row j)

Step 3

Replace x_{n+j} by x_i in the basic variables in the left-hand column. (Change the basis)

Step 4

In row j replace a_{jk} by a_{jk}/a_{ji} for $k = 1, \ldots, n + m + 1$. (Note that the first row and the final column are treated as part of the tableau for computation purposes, $-c_p = a_{0p}$, $b_q = a_{q(n+m+1)}$.) (Make pivot = 1)

Step 5

In all other rows, $l \neq j$, replace a_{lk} by $a_{lk} - a_{li}a_{jk}$ for all $k = 1, \ldots, m + n + 1$ and for each row $l = 0, \ldots, m$ $(l \neq j)$. (Gaussian elimination)

The algorithm is then repeated until at Step 1 the maximum is achieved. The method provides an extremely efficient way of searching through the corners of the feasible region. To inspect all corners would require the computation of $\binom{m+n}{m}$ points, while the simplex algorithm reduces this very considerably, often down to something of the order of $m + n$.

Several checks should be made at the completion of each cycle, since it may be possible to identify an exceptional case. Perhaps the most complicated of the exceptions is when one of the $b_i = 0$ during the calculation, implying that one of the basic variables is zero. This can be a temporary effect, in which case the problem goes away at the next iteration, or it may be permanent, and that basic variable is indeed zero in the optimal solution. The best that may be said, other than going into sophisticated techniques found in specialist books on LP, is that problems are possible and the computation should be watched carefully. The solution can get into a cycle that cannot be broken.

A second exception, that should be noted carefully, occurs when one of the $c_i = 0$ for a *non-basic* variable in the optimal tableau. The normal simplex algorithm can then change the solution without changing the z row by selecting this i column at Step 1. Because $c_i = 0$, Step 5 is never used on the z row at all. This case corresponds to a degenerate solution with many **alternative solutions** to the problem, and geometrically the profit function is parallel to one of the constraints.

The third exception occurs at Step 2 when all the $a_{1i}, a_{2i}, \ldots, a_{mi}$ in the optimal column are zero or negative and it becomes impossible to identify a row to continue the method. The region in this case is **unbounded**, and a careful look at the original problem is required to decide whether this is reasonable, since it may still be possible to get a solution to such a problem.

EXAMPLE 10.4 Find the maximum of

$$z = 5x_1 + 4x_2 + 6x_3$$

subject to

$$4x_1 + x_2 + x_3 \leqslant 19$$

$$3x_1 + 4x_2 + 6x_3 \leqslant 30$$

$$2x_1 + 4x_2 + x_3 \leqslant 25$$

$$x_1 + x_2 + 2x_3 \leqslant 15$$

$$x_1, x_2, x_3 \geqslant 0$$

Solution The example cannot be solved graphically, since it has three variables, but is in a correct form for the simplex algorithm. The initial tableau gives the solution $x_4 = 19$, $x_5 = 30$, $x_6 = 25$, $x_7 = 15$ and non-basic variables $x_1 = x_2 = x_3 = 0$.

	x_1	x_2	x_3	x_4	x_5	x_6	x_7	Solution	
z	−5	−4	−6	0	0	0	0	0	Ratios
x_4	4	1	1	1	0	0	0	19	19/1 = 19
x_5	3	4	(6)	0	1	0	0	30	30/6 = 5
x_6	2	4	1	0	0	1	0	25	25/1 = 25
x_7	1	1	2	0	0	0	1	15	15/2 = 7.5

In the initial tableau the pivot is identified, and x_5 is removed from the basic variable column and replaced by x_3. The pivot is made equal to unity by dividing the x_3 row by 6. The other entries in the x_3 column are then made zero by the Gaussian elimination in Step 5. This gives the tableau

	x_1	x_2	x_3	x_4	x_5	x_6	x_7	Solution	
z	−2	0	0	0	1	0	0	30	Ratios
x_4	($\frac{7}{2}$)	$\frac{1}{3}$	0	1	$-\frac{1}{6}$	0	0	14	4
x_3	$\frac{1}{2}$	$\frac{2}{3}$	1	0	$\frac{1}{6}$	0	0	5	10
x_6	$\frac{3}{2}$	$\frac{10}{3}$	0	0	$-\frac{1}{6}$	1	0	20	13.3
x_7	0	$-\frac{1}{3}$	0	0	$-\frac{1}{3}$	0	1	5	–

The process is then repeated and the pivot is again found, x_4 is replaced by x_1 in the first column, and the next tableau is constructed by following the remaining steps of the simplex algorithm, giving the tableau

	x_1	x_2	x_3	x_4	x_5	x_6	x_7	Solution
z	0	$\frac{4}{21}$	0	$\frac{4}{7}$	$\frac{19}{21}$	0	0	38
x_1	1	$\frac{2}{21}$	0	$\frac{2}{7}$	$-\frac{1}{21}$	0	0	4
x_3	0	$\frac{13}{21}$	1	$-\frac{1}{7}$	$\frac{4}{21}$	0	0	3
x_6	0	$\frac{67}{21}$	0	$-\frac{3}{7}$	$-\frac{2}{21}$	1	0	14
x_7	0	$-\frac{1}{3}$	0	0	$-\frac{1}{3}$	0	1	5

Thus the solution is now optimal, and gives $x_1 = 4$, $x_2 = 0$, $x_3 = 3$, and $z = 38$ as the maximum value. Note that the first two constraints are binding, that is satisfied exactly, while the other two are not. This can easily be deduced by looking at the slack variables in the initial tableau. We have $x_4 = x_5 = 0$, corresponding to the first two constraints, and $x_6 \neq 0$, $x_7 \neq 0$ for the last two constraints.

EXAMPLE 10.5

A firm has two plants, P1 and P2, that can produce a particular chemical. The product is made from three constituents, A, B and C. In a given period there are 36 000 litres of A, 30 000 litres of B and 12 000 litres of C available. Plant P1 requires the constituents A, B, C to be mixed in the ratio 4:2:1 respectively, and the manufacturer makes a profit of £1.50 per litre of product; plant P2 requires the ratio 3:3:1, and gives a profit of £1 per litre of product.

Determine how production should be allocated to each plant to maximize the profits, and how much of A, B and C remain.

There is a major breakdown in the supply of chemical C, so that only 8000 litres are available in the given period. How should production be changed to maximize the profits, how much has profit been reduced, and how much of A, B and C remain?

Solution

For each 1000 litres produced in plant P1, $\frac{4}{7} \times 1000$ will be constituent A, $\frac{2}{7} \times 1000$ will be B and $\frac{1}{7} \times 1000$ will be C. For each 1000 litres produced in plant P2, $\frac{3}{7} \times 1000$ will be constituent A, $\frac{3}{7} \times 1000$ will be B and $\frac{1}{7} \times 1000$ will be C. Thus, taking the three constituents in turn and letting x_1 and x_2 represent respectively the amount (in 1000 litre units) produced in plants P1 and P2, we obtain

$$\begin{aligned} \tfrac{4}{7}x_1 + \tfrac{3}{7}x_2 &\leqslant 36 & \qquad 4x_1 + 3x_2 &\leqslant 252 \\ \tfrac{2}{7}x_1 + \tfrac{3}{7}x_2 &\leqslant 30 \quad \text{or} & 2x_1 + 3x_2 &\leqslant 210 \\ \tfrac{1}{7}x_1 + \tfrac{1}{7}x_2 &\leqslant 12 & x_1 + x_2 &\leqslant 84 \end{aligned}$$

$$x_1, x_2 \geqslant 0$$

and the profit

$$z = 1.5x_1 + x_2$$

We can immediately construct the initial tableau

	x_1	x_2	x_3	x_4	x_5	Solution	
z	−1.5	−1	0	0	0	0	Ratios
x_3	(4)	3	1	0	0	252	63
x_4	2	3	0	1	0	210	105
x_5	1	1	0	0	1	84	84

The pivot has been found, and hence we introduce x_1 into the basis and construct the next tableau following the steps of the simplex algorithm:

	x_1	x_2	x_3	x_4	x_5	Solution
z	0	$\frac{1}{8}$	$\frac{3}{8}$	0	0	94.5
x_1	1	$\frac{3}{4}$	$\frac{1}{4}$	0	0	63
x_4	0	$\frac{3}{2}$	$-\frac{1}{2}$	1	0	84
x_5	0	$\frac{1}{4}$	$-\frac{1}{4}$	0	1	21

The z row is all positive, and hence we can immediately read off the solution (multiply by 1000 to re-establish proper costs)

$$x_1 = 63\,000, \quad x_2 = 0, \quad z = £94\,500$$

and only plant P1 is utilized. From the initial tableau we see that since $x_3 = 0$, there are zero litres of A remaining; $x_4 = 84$, so that we have $(84/7) \times 1000 = 12\,000$ litres of B remaining; and $x_5 = 21$, so that $(21/7) \times 1000 = 3000$ litres of C remain.

After the breakdown, the 12 000 litres of C are reduced to 8000 litres, so that the first tableau becomes

	x_1	x_2	x_3	x_4	x_5	Solution	
z	−1.5	−1	0	0	0	0	Ratios
x_3	4	3	1	0	0	252	63
x_4	2	3	0	1	0	210	105
x_5	(1)	1	0	0	1	56	56

We note that we have a different pivot, and hence we expect a different solution. The next tableau is derived in the usual way, giving

	x_1	x_2	x_3	x_4	x_5	Solution
z	0	0.5	0	0	1.5	84
x_3	0	−1	1	0	−4	28
x_4	0	1	0	1	−2	98
x_1	1	1	0	0	1	56

The tableau is again optimal, so

$$x_1 = 56\,000, \quad x_2 = 0, \quad z = £84\,000$$

The profit is thus reduced by £10 500 by the breakdown, but still only plant P1 is used. The remaining amounts of A, B and C can be checked to be 4000, 14 000 and zero litres respectively.

Since this problem has only two variables, it would be instructive to check these results using the graphical method.

EXAMPLE 10.6 Find the maximum of

$$z = 4x_1 + 2x_2 + 4x_3$$

subject to

$$3x_1 + x_2 + 2x_3 \leqslant 320$$

$$x_1 + x_2 + x_3 \leqslant 100$$

$$2x_1 + x_2 + 2x_3 \leqslant 200$$

Solution

	x_1	x_2	x_3	x_4	x_5	x_6	Solution	
z	−4	−2	−4	0	0	0	0	Ratios
x_4	3	1	2	1	0	0	320	160
x_5	1	1	(1)	0	1	0	100	100
x_6	2	1	2	0	0	1	200	100

Note that in the above tableau there is some arbitrariness in the choices in both Steps 1 and 2. In Step 1 the column is chosen arbitrarily between the x_1 and x_3 columns. From the ratios, the x_5 row is selected from the x_5 and x_6 rows at Step 2, which both have equal ratios. Steps 3–5 are then followed to give the tableau

	x_1	x_2	x_3	x_4	x_5	x_6	Solution
z	0	2	0	0	4	0	400
x_4	1	−1	0	1	−2	0	120
x_3	(1)	1	1	0	1	0	100
x_6	0	−1	0	0	−2	1	0

Although this is the optimal solution with $x_1 = x_2 = 0$, $x_3 = 100$ and $z = 400$, we have $c_1 = 0$ in the z row. Since x_1 is a non-basic variable, there is degeneracy. If we follow through the algorithm, choosing the first column at Step 1, we obtain an equally optimal solution in the following tableau. Replace x_3 by x_1 in the basic variables, and subtract the x_1 row from the x_4 row:

	x_1	x_2	x_3	x_4	x_5	x_6	Solution
z	0	2	0	0	4	0	400
x_4	0	−2	−1	1	−3	0	20
x_1	1	1	1	0	1	0	100
x_6	0	−1	0	0	−2	1	0

This solution gives $x_1 = 100$, $x_2 = x_3 = 0$ and $z = 400$ once more. It can easily be deduced that $x_1 = 100(1 - \alpha)$, $x_2 = 0$, $x_3 = 100\alpha$ is an optimal solution for any $0 \leqslant \alpha \leqslant 1$ with $z = 400$. We could have observed this fact geometrically, since z is just a multiple of the left-hand side of the last constraint.

10.2.4 Exercises

1 Use the graphical method to find the maximum value of

$$f = 4x + 5y$$

subject to

$$3x + 7y \leqslant 10$$
$$2x + y \leqslant 3$$
$$x, y \geqslant 0$$

2 Find $x_1, x_2, x_3, x_4 \geqslant 0$ that maximize

$$f = 6x_1 + x_2 + 2x_3 + 4x_4$$

subject to

$$2x_1 + x_2 + x_4 \leqslant 3$$
$$x_1 + x_3 + x_4 \leqslant 4$$
$$x_1 + x_2 + 3x_3 + 2x_4 \leqslant 10$$

3 A manufacturer produces two types of cupboard, which are constructed from chipboard and oak veneer that both come in standard widths. The first type requires 4 m of chipboard and 5 m of oak veneer, takes 5 h of labour to produce and gives a profit of £24 per unit. The second type requires 5 m of chipboard and 2 m of oak veneer, takes 3 h of labour to produce and gives a £12 profit per unit.

On a weekly basis there are 400 m of chipboard available, 200 m of oak veneer and a maximum of 250 h of labour. Write this problem in a linear programming form. Use the simplex method to determine how many cupboards of each type should be made to maximize profits. How much profit is made? Which of the scarce resources remain unused? Show that the amount of oak veneer available can be reduced to 175 m without affecting the basis. What is the new solution, and by how much is the profit reduced?

4 A factory manufactures nails and screws. The profit yield is 2p per kg nails and 3p per kg screws. Three units of labour are required to manufacture 1 kg nails and 6 units to make 1 kg screws. Twenty-four units of labour are available. Two units of raw material are needed to make 1 kg nails and 1 unit for 1 kg screws. Determine the manufacturing policy that yields maximum profit from 10 units of raw material.

5 A manufacturer makes two types of cylinder, CYL1 and CYL2. Three materials, M1, M2 and M3, are required for the manufacture of each cylinder. The following information is provided:

	Quantities of materials required		
	M1	M2	M3
CYL1	1	1	2
CYL2	5	2	1

Quantities of materials available		
M1	M2	M3
45	21	24

£4 profit is made on one CYL1 and £3 profit on one CYL2. How many of each cylinder should the manufacturer make in order to maximize profit?

6 Find the optimal solution of the following LP problem: maximize

$$z = kx_1 + 20x_2$$

subject to

$$x_1 + 2x_2 \leqslant 20$$
$$3x_1 + x_2 \leqslant 25$$
$$x_1, x_2 \geqslant 0$$

where k is a positive parameter representing variable profitability. Use both the simplex method and the graphical method, and interpret the results geometrically.

7 Use the simplex method to solve the following problem: maximize

$$2x_1 + x_2 + 4x_3 + x_4$$

subject to

$$\begin{aligned} 2x_1 + x_3 &\leqslant 3 \\ x_1 + 3x_3 + x_4 &\leqslant 4 \\ 4x_2 + x_3 + x_4 &\leqslant 3 \\ x_1, x_2, x_3, x_4 &\geqslant 0 \end{aligned}$$

8 The Yorkshire Clothing Company makes two styles of jacket, the 'York' and the 'Wetherby'. The York requires 3 m of cloth and 3 h of labour, and makes a profit of £25. The Wetherby needs 4 m of cloth and 2 h of labour, and makes a profit of £30. The Yorkshire has 400 m of cloth available and 300 h of labour available each week. Advise the company on the number of each style it should produce in order to maximize profits.

The company is prepared to buy more cloth to increase its profits, but it will not employ any more labour. Under this revised policy, is there a strategy that will increase its profits?

9 A publisher has three books available for printing, B1, B2 and B3. The books require varying amounts of paper, and the total paper supplies are limited:

	B1	B2	B3	Total units available
Units of paper required per 1000 copies	3	2	1	60
Profit per 1000 copies	£900	£800	£300	

Books B1 and B2 are similar in content, and the total combined market for these two books is estimated to be at most 15 000 copies. Determine how many copies of each book should be printed to maximize the overall profit.

10 Euroflight is considering the purchase of new aircraft. Long-range aircraft cost £4 million each, medium-range £2 million each and short-range £1 million each, and Euroflight has £60 million to invest. The estimated profit from each type of aircraft is £0.4 million, £0.3 million and £0.15 million respectively. The company has trained pilots for at most a total 25 aircraft. Maintenance facilities are limited to a maximum of the equivalent of 30 short-range aircraft. Long-range aircraft need twice as much maintenance as short-range ones, and medium-range 1.5 times as much. Set this up as a linear programming problem, and solve it. Aircraft can only be bought in integer numbers, so estimate how many of each type should be bought.

10.2.5 Two-phase method

The previous section only dealt with '≤' constraints and did not consider '≥' constraints. These prove to be much more troublesome, since there is no obvious initial feasible solution, and Phase 1 of the **two-phase method** is solely concerned with getting such a solution. Once this has been obtained, we then move to Phase 2. This is the standard simplex method, starting from the solution just obtained from Phase 1. A simple example will illustrate the problems involved and the basic ideas of the two-phase method.

EXAMPLE 10.7 Find the maximum of

$$z = x + y$$

subject to

$$\begin{aligned} -x + 2y &\leqslant 6 \\ x \quad &\leqslant 4 \\ 2x + y &\geqslant 4 \\ x, y &\geqslant 0 \end{aligned}$$

Solution The region defined by the constraints is shown in Figure 10.5. It is clear from the figure that the origin is *not* in the feasible region and that $x = 4$, $y = 5$ gives the optimal solution. We have already appreciated that the graphical method is only useful for two-dimensional problems, and we must explore how the simplex method copes with this problem.

Add in the positive variables r, s and t to give

$$\begin{aligned} -x + 2y + r \quad\quad &= 6 \\ x \quad\quad + s \quad &= 4 \\ 2x + y \quad\quad - t &= 4 \end{aligned}$$

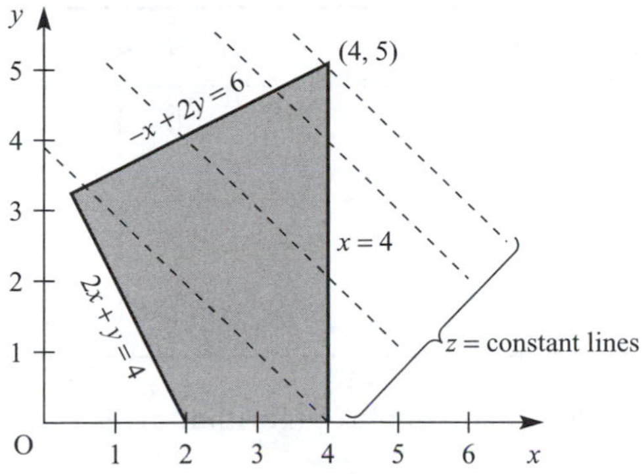

Figure 10.5 The feasible region.

The r and s are the usual slack variables. Because we must subtract t to take away the surplus, t is called a **surplus variable**. The obvious solution $x = y = 0$, $r = 6$, $s = 4$, $t = -4$ does *not* satisfy the condition that all variables be positive. The algebra is saying that the origin is not in the feasible region. Because the simplex method works so well, the last equation is forced into standard form by adding in yet another variable, u, called an **artificial variable**, to give

$$2x + y - t + u = 4$$

Now we have a feasible solution $x = y = t = 0$, $r = 6$, $s = 4$, $u = 4$, but not to the problem we originally stated. As the term 'artificial variable' implies, we wish to get rid of u and then reduce the problem back to our original one at a feasible corner. The variable u can be eliminated by forcing it to zero and this can be done by entering **Phase 1** with a new cost function

$$z' = -u$$

We see that if we can maximize z' then this is at $u = 0$, and our Phase 1 will be complete. The simplex tableau for Phase 1 then takes the form

	x	y	r	s	t	u	Solution
z	−1	−1	0	0	0	0	0
z'	0	0	0	0	0	1	0
r	−1	2	1	0	0	0	6
s	1	0	0	1	0	0	4
u	2	1	0	0	−1	1	4

where z has been included for the elimination but does *not* enter the optimization: only the z' row is considered in Phase 1. It may be observed that the tableau is not of standard form, since u is a basic variable and the (z', u) entry is non-zero. This must be remedied by subtracting the u row from the z' row to give the standard-form tableau

	x	y	r	s	t	u	Solution
z	−1	−1	0	0	0	0	0
z'	−2	−1	0	0	1	0	−4
r	−1	2	1	0	0	0	6
s	1	0	0	1	0	0	4
u	(2)	1	0	0	−1	1	4

Manipulation using the usual simplex algorithm gives the tableau

	x	y	r	s	t	u	Solution
z	0	$-\frac{1}{2}$	0	0	$-\frac{1}{2}$	$\frac{1}{2}$	2
z'	0	0	0	0	0	1	0
r	0	$\frac{5}{2}$	1	0	$-\frac{1}{2}$	$\frac{1}{2}$	8
s	0	$-\frac{1}{2}$	0	1	$\frac{1}{2}$	$-\frac{1}{2}$	2
x	1	$\frac{1}{2}$	0	0	$-\frac{1}{2}$	$\frac{1}{2}$	2

At this stage $z' = 0$ and $u = 0$, so that we have driven u out of the problem, and the z' row and u column can now be deleted. The Phase 1 solution gives $x = 2$, $y = 0$, which can be observed to be a corner of the feasible region in Figure 10.5.

We now enter **Phase 2**, with the z' row and u column deleted, and perform the usual sequence of steps. The initial tableau is

	x	y	r	s	t	Solution
z	0	$-\frac{1}{2}$	0	0	$-\frac{1}{2}$	2
r	0	($\frac{5}{2}$)	1	0	$-\frac{1}{2}$	8
s	0	$-\frac{1}{2}$	0	1	$\frac{1}{2}$	2
x	1	$\frac{1}{2}$	0	0	$-\frac{1}{2}$	2

Two further cycles are required, leading sequentially to the following two tableaux:

	x	y	r	s	t	Solution
z	0	0	$\frac{1}{5}$	0	$-\frac{3}{5}$	$\frac{18}{5}$
y	0	1	$\frac{2}{5}$	0	$-\frac{1}{5}$	$\frac{16}{5}$
s	0	0	$\frac{1}{5}$	1	($\frac{2}{5}$)	$\frac{18}{5}$
x	1	0	$-\frac{1}{5}$	0	$-\frac{2}{5}$	$\frac{2}{5}$

	x	y	r	s	t	Solution
z	0	0	$\frac{1}{2}$	$\frac{3}{2}$	0	9
y	0	1	$\frac{1}{2}$	$\frac{1}{2}$	0	5
t	0	0	$\frac{1}{2}$	$\frac{5}{2}$	1	9
x	1	0	0	1	0	4

We now have an optimum solution, since all the z row entries are non-negative with $x = 4$, $y = 5$ and objective function $z = 9$ in agreement with the graphical solution.

The general **two-phase strategy** is then as follows:

Phase 1

(a) Introduce slack and surplus variables.
(b) Introduce artificial variables alongside the surplus variables, say $x_p, \dots, x_q$.
(c) Write the artificial cost function

$$z' = -x_p - x_{p+1} \dots -x_q$$

(d) Subtract rows $x_p, x_{p+1}, \dots, x_q$ from the cost function z' to ensure there are zeros in the entries in the z' row corresponding to the basic variables.
(e) Use the standard simplex method to maximize z' (keeping the z row as an extra row) until $z' = 0$ and

$$x_p = x_{p+1} = \dots = x_q = 0$$

Phase 2

(a) Eliminate the z' row and artificial columns $x_p, \dots, x_q$.
(b) Use the standard simplex method to maximize the objective function z.

There are other approaches to obtaining an initial feasible basic solution, but Phase 1 of the two-phase method gives an efficient way of obtaining a starting point. Geometrically, it uses the simplex method to search the non-feasible vertices until it is driven to a vertex in the feasible region.

EXAMPLE 10.8 Use the two-phase method to solve the following LP problem: maximize

$$z = 4x_1 + \tfrac{1}{2}x_2 + x_3$$

subject to

$$x_1 + 2x_2 + 3x_3 \geq 2$$
$$2x_1 + x_2 + x_3 \leq 5$$
$$x_1, x_2, x_3 \geq 0$$

Solution

Phase 1 Introduce a surplus variable x_4 and a corresponding artificial variable x_5 into the first inequality. A slack variable x_6 is required for the second inequality. The artificial cost is just

$$z' = -x_5$$

and we can construct the initial tableau

	x_1	x_2	x_3	x_4	x_5	x_6	Solution
z	−4	$-\frac{1}{2}$	−1	0	0	0	0
z'	0	0	0	0	1	0	0
x_5	1	2	3	−1	1	0	2
x_6	2	1	1	0	0	1	5

We subtract the x_5 row from the z' row to eliminate the 1 from the (z', x_5) element, giving the tableau

	x_1	x_2	x_3	x_4	x_5	x_6	Solution
z	−4	$-\frac{1}{2}$	−1	0	0	0	0
z'	−1	−2	−3	1	0	0	−2
x_5	1	2	(3)	−1	1	0	2
x_6	2	1	1	0	0	1	5

We now apply the steps of the simplex algorithm to give the tableau

	x_1	x_2	x_3	x_4	x_5	x_6	Solution
z	$-\frac{11}{3}$	$\frac{1}{6}$	0	$-\frac{1}{3}$	$\frac{1}{3}$	0	$\frac{2}{3}$
z'	0	0	0	0	1	0	0
x_3	$\frac{1}{3}$	$\frac{2}{3}$	1	$-\frac{1}{3}$	$\frac{1}{3}$	0	$\frac{2}{3}$
x_6	$\frac{5}{3}$	$\frac{1}{3}$	0	$\frac{1}{3}$	$-\frac{1}{3}$	1	$\frac{13}{3}$

Since $z' = 0$ and the artificial variable x_5 has been driven into the non-basic variables, phase 1 ends.

Phase 2 The z' row and the x_5 column are now deleted, and the following sequence of tableaux constructed following the rules of the simplex algorithm:

	x_1	x_2	x_3	x_4	x_6	Solution
z	$-\frac{11}{3}$	$\frac{1}{6}$	0	$-\frac{1}{3}$	0	$\frac{2}{3}$
x_3	$\frac{1}{3}$ (circled)	$\frac{2}{3}$	1	$-\frac{1}{3}$	0	$\frac{2}{3}$
x_6	$\frac{5}{3}$	$\frac{1}{3}$	0	$\frac{1}{3}$	1	$\frac{13}{3}$

	x_1	x_2	x_3	x_4	x_6	Solution
z	0	$\frac{15}{2}$	11	−4	0	8
x_1	1	2	3	−1	0	2
x_6	0	−3	−5	2 (circled)	1	1

	x_1	x_2	x_3	x_4	x_6	Solution
z	0	$\frac{3}{2}$	1	0	2	10
x_1	1	$\frac{1}{2}$	$\frac{1}{2}$	0	$\frac{1}{2}$	$\frac{5}{2}$
x_4	0	$-\frac{3}{2}$	$-\frac{5}{2}$	1	$\frac{1}{2}$	$\frac{1}{2}$

The solution is now optimal, with $x_1 = \frac{5}{2}$, $x_2 = x_3 = 0$ and $z = 10$. Note that the first inequality is not binding, since $x_4 \neq 0$, while the second inequality is binding.

EXAMPLE 10.9 Three ores, A, B and C, are blended to form 100 kg of alloy; the percentage contents and the costs are as follows:

Ore	A	B	C
Iron	70	60	0
Lead	20	10	40
Copper	10	30	60
Cost (£ kg^{-1})	3000	2000	1000

The alloy must contain at least 20% iron, at least 25% lead but less than 48% copper. Find the blend of ores that minimizes the cost of the alloy.

Solution Let x_1, x_2 and x_3 be the weights (kg) of ores A, B and C respectively in the 100 kg of alloy. The constraints give

$$\begin{aligned} \text{iron} \quad & 0.7x_1 + 0.6x_2 \geqslant 20 \\ \text{lead} \quad & 0.2x_1 + 0.1x_2 + 0.4x_3 \geqslant 25 \\ \text{copper} \quad & 0.1x_1 + 0.3x_2 + 0.6x_3 \leqslant 48 \end{aligned}$$

and to make the 100 kg of alloy,

$$x_1 + x_2 + x_3 = 100$$

The cost is

$$3000x_1 + 2000x_2 + 1000x_3$$

which is to be *minimized.*

To reduce the problem to standard form, we change the problem to a *maximization* of

$$z = -3000x_1 - 2000x_2 - 1000x_3$$

For the inequality constraints we use surplus variables x_4 and x_5 and a slack variable x_6. We require two artificial variables x_7 and x_8 alongside the surplus variables. Thus the inequalities become

$$\begin{aligned} 0.7x_1 + 0.6x_2 \qquad - x_4 \qquad + x_7 &= 20 \\ 0.2x_1 + 0.1x_2 + 0.4x_3 \qquad - x_5 \qquad + x_8 &= 25 \\ 0.1x_1 + 0.3x_2 + 0.6x_3 \qquad + x_6 &= 48 \end{aligned}$$

To deal with the equality constraint, we introduce a further artificial variable x_9:

$$x_1 + x_2 + x_3 + x_9 = 100$$

We must drive x_9 to zero, to ensure that the equality holds, so it is essential to put x_9 into the artificial cost function. (Note that this is the standard way of dealing with an equality constraint.) We first enter *Phase 1*. Steps (a)–(c) of Phase 1 of the two-phase method give the initial tableau

	x_1	x_2	x_3	x_4	x_5	x_6	x_7	x_8	x_9	Solution
z	3000	2000	1000	0	0	0	0	0	0	0
z'	0	0	0	0	0	0	1	1	1	0
x_7	0.7	0.6	0	−1	0	0	1	0	0	20
x_8	0.2	0.1	0.4	0	−1	0	0	1	0	25
x_6	0.1	0.3	0.6	0	0	1	0	0	0	48
x_9	1	1	1	0	0	0	0	0	1	100

It is necessary to remove the 1s from the z' row in the basic variable columns x_7, x_8 and x_9. Following (d) of the general strategy, we replace the z' row by (z' row) − (x_7 row) − (x_8 row) − (x_9 row) to give the tableau

	x_1	x_2	x_3	x_4	x_5	x_6	x_7	x_8	x_9	Solution
z	3000	2000	1000	0	0	0	0	0	0	0
z'	−1.9	−1.7	−1.4	1	1	0	0	0	0	−145
x_7	0.7	0.6	0	−1	0	0	1	0	0	20
x_8	0.2	0.1	0.4	0	−1	0	0	1	0	25
x_6	0.1	0.3	0.6	0	0	1	0	0	0	48
x_9	1	1	1	0	0	0	0	0	1	100

Several tableaux need to be completed to drive z' to zero and complete Phase 1, with the final tableau being

	x_1	x_2	x_3	x_4	x_5	x_6	x_7	x_8	x_9	Solution
z	0	−2000	0	0	−10 000	0	0	10 000	−1000	−250 000
z'	0	0	0	0	0	0	1	1	1	0
x_1	1	1.5	0	0	5	0	0	−5	2	75
x_3	0	−0.5	1	0	−5	0	0	5	−1	25
x_6	0	0.45	0	0	2.5	1	0	−2.5	0.4	25.5
x_4	0	0.45	0	1	3.5	0	−1	−3.5	1.4	32.5

Removing the artificial variables and the z' row gives the tableau

	x_1	x_2	x_3	x_4	x_5	x_6	Solution
z	0	−2000	0	0	−10 000	0	−250 000
x_1	1	1.5	0	0	5	0	75
x_3	0	−0.5	1	0	−5	0	25
x_6	0	0.45	0	0	2.5	1	25.5
x_4	0	0.45	0	1	3.5	0	32.5

The algorithm is now ready for *Phase 2*, since a sensible feasible basic solution is available. The standard procedure leads, after many cycles, to the final tableau

	x_1	x_2	x_3	x_4	x_5	x_6	Solution
z	333.3	0	0	0	0	3333.3	−140 000
x_4	0.3	0	0	1	0	−2	4
x_3	−0.67	0	1	0	0	3.33	60
x_2	1.67	1	0	0	0	−3.33	40
x_5	−0.3	0	0	0	1	1	3

and the solution can be read off as

$$x_1 = 0, \quad x_2 = 40, \quad x_3 = 60$$

with the cost minimized at £140 000. (Note that the cost in the tableau is negative, since the original problem is a minimization problem). It may be noted that $x_6 = 0$, so the copper constraint is binding while the iron constraint gives 4% more than the required minimum and the lead constraint gives 3% more than the required minimum.

Except for simple, illustrative examples, the amount of computational work in the two-phase strategy is heavy and requires the use of a computer package. Even for the comparatively simple Example 10.9, many tableaux were required in the solution.

10.2.6 Exercises

11 Use the simplex method to find positive values of x_1 and x_2 that minimize

$$f = 10x_1 + x_2$$

subject to

$$4x_1 + x_2 \leqslant 32$$
$$2x_1 + x_2 \geqslant 12$$
$$2x_1 - x_2 \leqslant 4$$
$$-2x_1 + x_2 \leqslant 8$$

Sketch the points obtained by the simplex method on a graph, indicating how the points progress through Phases 1 and 2 to the solution.

12 Solve the following LP problem: minimize

$$2x_1 + 7x_2 + 4x_3 + 5x_4$$

subject to

$$x_1 - x_3 - x_4 \geqslant 0$$
$$x_2 + x_3 \geqslant 2$$
$$x_1, x_2, x_3, x_4 \geqslant 0$$

13 A trucking company requires antifreeze that contains at least 50% of pure glycol and at least 5% of anticorrosive additive. The company can buy three products, A, B and C, whose constituents and costs are as follows:

	A	B	C
% glycol	65	25	80
% additive	10	3	0
Cost (£/litre)	1.8	0.9	1.5

What blend will provide the required antifreeze solution at minimum cost? What is the cost of 100 litres of solution?

14 A builder is constructing three different styles of house on an estate, and is deciding which styles to erect in the next phase of building. There are 40 plots of equal size, and the different styles require 1, 2 and 2 plots respectively. The builder anticipates shortages of two materials, and estimates the requirements and supplies (in appropriate units) to be as follows:

	Requirements			
	Style 1	Style 2	Style 3	Total supply
Facing stone	1	2	5	58
Weather boarding	3	2	1	72

The local authority insists that there be at least 5 more houses of style 2 than style 1. If the profits on the houses are £1000, £1500 and £2500 for styles 1, 2 and 3 respectively, find how many of each style the builder should construct to maximize the total profit.

15 The Footsie company produces boots and shoes. If no boots are made, the company can produce a maximum of 250 pairs of shoes in a day. Each pair of boots takes twice as long to make as each pair of shoes. The maximum daily sales of boots and shoes are 200, but 25 pairs of boots must be produced to satisfy an important customer. The profits per pair of boots and shoes are £8 and £5 respectively. Determine the daily production plan to maximize profits. Use the two-phase method to obtain the solution, and verify your result with a graphical solution.

16 In Exercise 9 there is an additional union agreement that at least 50 000 books must be printed. Does the solution change? If so, calculate the new optimum strategy.

17 A manufacturer produces three types of carpeting, C1, C2 and C3. Two of the raw materials, M1 and M2, are in short supply. The following table gives the supplies of M1 and M2 available (in 1000s of kg), the quantities of M1 and M2 required for each 1000 m^2 of carpet, and the profits made from each type of carpet (in £1000s per 1000 m^2):

	Quantities required		
	M1	M2	Profit
C1	1	1	2
C2	1	1	3
C3	1	0	0
Quantities available	5	4	

Carpet of type C3 is non-profit-making, but is included in the range in order to enable the company to satisfy its customers. The company has policies that require that, if x_1, x_2 and x_3 1000s of m^2 of C1, C2 and C3 respectively are made then

$$x_1 \geqslant 1$$

and

$$x_1 - x_2 + x_3 \geqslant 2$$

How much carpet of each type should the company manufacture in order to satisfy the constraints and maximize profits?

10.3 Lagrange multipliers

10.3.1 Equality constraints

In Section 10.2 we looked at the situation where all the functions were linear. As soon as functions become nonlinear, the problems become very much more difficult. This is generally the case in most of mathematics, and is certainly true in optimization problems.

In Section 9.8.4 of *Modern Engineering Mathematics* it was shown how to use Lagrange multipliers to solve the problem of the optimization of a nonlinear function of many variables subject to **equality constraints**. For the general problem it was shown that the necessary conditions for the extremum of

$$f(x_1, x_2, \ldots, x_n)$$

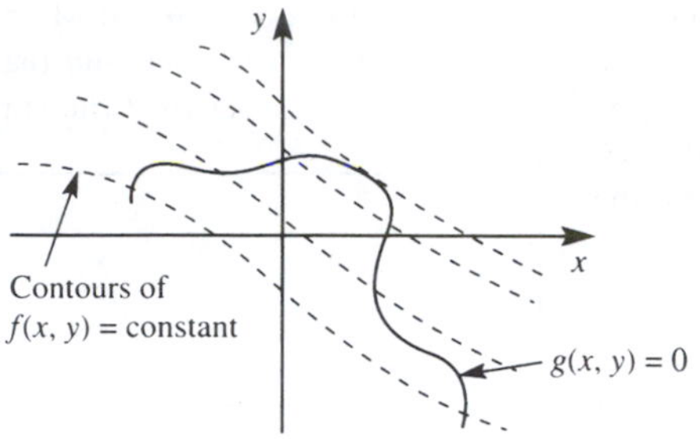

Figure 10.6 Lagrange multiplier problem.

subject to

$$g_i(x_1, x_2, \ldots, x_n) = 0 \quad (i = 1, 2, \ldots, m\ (m < n))$$

are

$$\frac{\partial f}{\partial x_k} + \lambda_1 \frac{\partial g_1}{\partial x_k} + \lambda_2 \frac{\partial g_2}{\partial x_k} + \ldots + \lambda_m \frac{\partial g_m}{\partial x_k} = 0 \quad (k = 1, 2, \ldots, n) \qquad \textbf{(10.7)}$$

where $\lambda_1, \lambda_2, \ldots, \lambda_m$ are the **Lagrange multipliers**. These n equations must be solved together with the m constraints $g_i = 0$. Thus there are $n + m$ equations in the $n + m$ unknowns $x_1, x_2, \ldots, x_n$ and $\lambda_1, \lambda_2, \ldots, \lambda_m$.

For a two-variable problem of finding the extremum of

$$f(x, y) \quad \text{subject to the single constraint} \quad g(x, y) = 0$$

the problem is illustrated geometrically in Figure 10.6. We are looking for the maximum, say, of the function $f(x, y)$, but *only* considering those points in the plane that lie on the curve $g(x, y) = 0$. The mathematical conditions look much simpler of course, as

$$\left.\begin{aligned} f_x + \lambda g_x &= 0 \\ f_y + \lambda g_y &= 0 \\ g &= 0 \end{aligned}\right\} \qquad \textbf{(10.8)}$$

There are two comments that should be noted. First, the method fails if $g_x = g_y = 0$ at the solution point. Such points are called **singular points**: fortunately they are rare, but their existence should be noted. Secondly, sufficient conditions for a maximum, a minimum or a saddle point can be derived, but they are difficult to apply and not very useful. Example 10.12 will illustrate an intuitive approach to sufficiency.

A few examples should be enough to remind the reader of the problems involved and to show the techniques required to solve the equations.

EXAMPLE 10.10 Fermat stated in 1661 that 'Light travels along the shortest path'. Find the path that joins the eye to an object when they are in separate media (Figure 10.7).

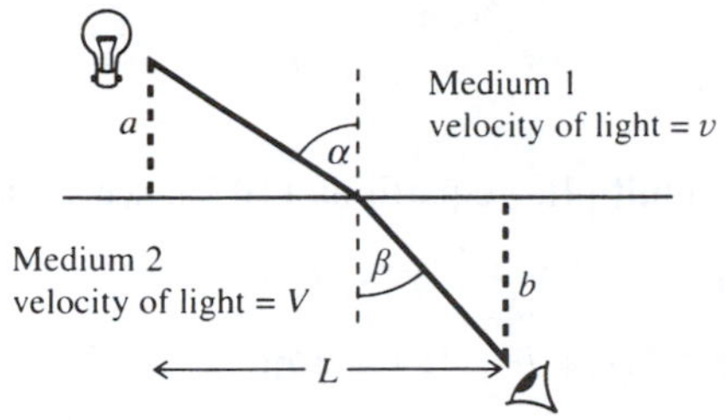

Figure 10.7 Fermat's shortest-path problem.

Solution The velocities of light in the two media are v and V. The time of transit of light is given by the geometry as

$$T = \frac{a}{v \cos \alpha} + \frac{b}{V \cos \beta}$$

and is then subject to the geometrical constraint that

$$L = a \tan \alpha + b \tan \beta$$

Applying (10.8),

$$0 = \frac{\partial T}{\partial \alpha} + \lambda \frac{\partial g}{\partial \alpha} = \frac{a}{v} \sec \alpha \tan \alpha + \lambda a \sec^2 \alpha$$

$$0 = \frac{\partial T}{\partial \beta} + \lambda \frac{\partial g}{\partial \beta} = \frac{b}{V} \sec \beta \tan \beta + \lambda b \sec^2 \beta$$

These give as the only solution

$$\sin \alpha = -\lambda v, \quad \sin \beta = -\lambda V$$

or

$$\frac{\sin \alpha}{\sin \beta} = \frac{v}{V} = \mu$$

which is known as **Snell's law**.

In Example 10.10, to obtain Snell's law, the solution of the equation was quite straightforward, but it is rarely so easy. Frequently it is technically the most difficult task, and it is easy to miss solutions. We return to the 'milk carton' problem discussed earlier to illustrate the point.

EXAMPLE 10.11 Find the minimum area of the milk carton problem stated in Example 10.2 and illustrated in Figure 10.1.

Solution Taking measurements in millimetres, the basic mathematical problem is to minimize

$$A = (2b + 2w + 5)(h + b + 10)$$

subject to

$$hbw = 1\,136\,000$$

Applying the Lagrange multiplier equations (10.7), we obtain

$$0 = (2b + 2w + 5) \qquad\qquad\qquad\qquad + \lambda bw$$

$$0 = (2b + 2w + 5) + 2(h + b + 10) + \lambda hw$$

$$0 = \qquad\qquad\qquad\qquad 2(h + b + 10) + \lambda hb$$

giving four equations in the four unknowns h, b, w and λ. If we eliminate λ and w from these equations, we are left with the same equations that we derived in Example 10.2, which have no simple analytical solutions.

The only way to proceed further with Example 10.11 is by a numerical solution. Thus, even with simple problems such as this one, we need a numerical algorithm, and in most realistic problems in science and engineering we encounter similar severe computational difficulties. It is often a problem even to write down the function or the constraints explicitly. Such functions frequently emerge as numbers from a complicated computer program. It is therefore essential to look for efficient numerical algorithms to optimize such functions. This will be the substance of Section 10.4.

A final example shows the situation where there are more than two variables involved. An indication will be given of the difficulties of *proving* that the point obtained is a maximum.

EXAMPLE 10.12 A hopper is to be made from a cylindrical portion connected to a conical portion as indicated in Figure 10.8. It is required to find the maximum volume subject to a given surface area.

Solution We can compute the volume of the hopper as

$$V = \pi R^2 L + \tfrac{1}{3}\pi R^3 \tan\alpha$$

and find its maximum subject to the surface area being given as

$$A = 2\pi RL + \pi R^2 \sec\alpha$$

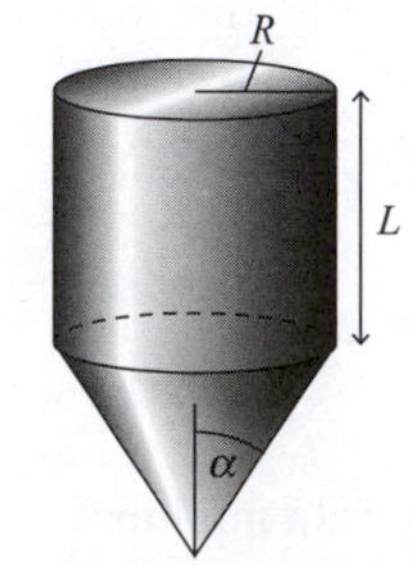

Figure 10.8 Hopper.

Applying (10.7) with the appropriate variables, we obtain

$$0 = \frac{\partial V}{\partial R} + \lambda\frac{\partial g}{\partial R} = 2\pi RL + \pi R^2 \tan\alpha + \lambda(2\pi L + 2\pi R\sec\alpha)$$

$$0 = \frac{\partial V}{\partial L} + \lambda\frac{\partial g}{\partial L} = \pi R^2 + \lambda 2\pi R$$

$$0 = \frac{\partial V}{\partial \alpha} + \lambda\frac{\partial g}{\partial \alpha} = \frac{1}{3}\pi R^3 \sec^2\alpha + \lambda\pi R^2 \sec\alpha\tan\alpha$$

First, λ can be easily evaluated as $\lambda = -\frac{1}{2}R$. The last of the above three equations becomes

$$\pi R^2 \sec^2 \alpha(\tfrac{1}{3}R + \lambda \sin \alpha) = 0$$

and hence $\sin \alpha = \frac{2}{3}$. Since $0 \leqslant \alpha \leqslant \frac{1}{2}\pi$, we have $\alpha = 0.730$ rad or 41.8°. A little further algebraic manipulation between the constraints and the above equations gives $R^2 = A/\pi\sqrt{5}$, so that $R = 0.377A^{1/2}$ and $V = 0.126A^{3/2}$.

Normally it would be assumed that this is a maximum for the volume on intuitive or geometrical grounds. To prove this rigorously, we take small variations around the suspected maximum and show that the volume is larger than all its neighbours. For simplicity let $R^2 = (A/\pi\sqrt{5})(1+ \delta)$, $\sec \alpha = (3/\sqrt{5})(1 + \varepsilon)$, evaluate L from the constraint $g = 0$ and then calculate V to *second order* in ε and δ by Taylor's theorem. Some careful algebra gives

$$V = \frac{A^{3/2}}{3\pi^{1/2}5^{1/4}}\left(1 - \frac{1}{8}\delta^2 - \frac{9}{16}\varepsilon^2\right)$$

This shows that for any non-zero values of ε and δ we obtain a smaller volume, and hence we have proved that we have found a maximum.

It should be reiterated that the major problem lies in solving the Lagrange multiplier equations and not in writing them down. This is typical, and supports the need for good numerical algorithms to solve such problems; they only have to be marginally more difficult than Example 10.12 to become impossible to manipulate analytically.

10.3.2 Inequality constraints

Although we do not intend to consider them in any detail here, for reference we shall state the basic extension of (10.7) to the case of **inequality constraints**. Kuhn and Tucker proved the following result in the 1940s.

To maximize the function

$$f(x_1, \ldots, x_n)$$

subject to

$$g_i(x_1, \ldots, x_n) \leqslant 0 \ (i = 1, \ldots, m)$$

the equivalent conditions to (10.7) are

$$\frac{\partial f}{\partial x_k} - \lambda_1 \frac{\partial g_1}{\partial x_k} - \ldots - \lambda_m \frac{\partial g_m}{\partial x_k} = 0 \quad (k = 1, \ldots, n)$$

$$\left.\begin{aligned} \lambda_i g_i &= 0 \\ \lambda_i &\geq 0 \\ g_i &\leq 0 \end{aligned}\right\} \quad (i = 1, \ldots, m)$$

The equation $\lambda_i g_i = 0$ gives two alternative conclusions for each constraint, either

(a) $g_i = 0$, in which case the constraint is 'active' and the corresponding $\lambda_i > 0$, or

(b) $\lambda_i = 0$ and $g_i < 0$, so that the optimum is away from this constraint and the Lagrange multiplier is not necessary.

Implementation of the Kuhn–Tucker result is not very easy, even though in principle it looks straightforward. There are so many cases to check that it becomes very susceptible to error.

10.3.3 Exercises

(Questions marked with a dagger (†) are intended to be solved with the assistance of a computer.)

18 Find the optimum of $f = xy^2z$ subject to $x + 2y + 3z = 6$ using a Lagrange multiplier method.

19 Show that the stationary points of $f = x^2 + y^2 + z^2$ subject to $x + y - z = 0$ and $yz + 2zx - 2xy = 1$ are given by the solution of the equations

$$0 = 2x + \lambda + (2z - 2y)\mu$$
$$0 = 2y + \lambda + (z - 2x)\mu$$
$$0 = 2z - \lambda + (y + 2x)\mu$$

Add the last of these two equations, and show that either $\mu = -2$ or $y + z = 0$. Hence deduce the stationary points.

20 A rectangular box without a lid is to be made. It is required to maximize the volume for a given surface area. Find the dimensions of the box when the total surface area is A.

21 Find the shortest distance from the origin to the ellipse

$$\frac{x^2}{a^2} + \frac{y^2}{b^2} = 1 \quad (a < b)$$

22 Determine the lengths of the sides of a rectangle with maximum area that can be inscribed within the ellipse

$$\frac{x^2}{a^2} + \frac{y^2}{b^2} = 1$$

†23 (Harder) The lowest frequency of vibration, α, of an elastic plate can be computed by minimizing

$$I[\omega] = \iint_R (\nabla^2\omega)^2 \mathrm{d}x\, \mathrm{d}y$$

subject to

$$\iint_R \omega^2 \mathrm{d}x\, \mathrm{d}y = 1$$

over all functions $\omega(x, y)$, where R is the region of the plate in the (x, y) plane. If R is the square region $|x| \leq 1$, $|y| \leq 1$ and the plate is clamped at its edges, use the approximation

$$\omega = A \cos^2 \tfrac{1}{2}\pi x \cos^2 \tfrac{1}{2}\pi y$$

to show that $I[\omega_{\min}] = \alpha^2 = \frac{8}{9}\pi^4$.

Use the improved approximation

$$\omega = \cos^2 \tfrac{1}{2}\pi x \cos^2 \tfrac{1}{2}\pi y (A + B \cos \tfrac{1}{2}\pi x \cos \tfrac{1}{2}\pi y)$$

to get a better estimate of α.

(*Note*: $\int_{-1}^{1} \cos^{2n} \frac{1}{2}\pi z\, \mathrm{d}z = (2n)!/(n!)^2 2^{2n-1}$ for non-negative integers n. Preferably use an algebraic symbolic manipulator, for example MAPLE, to evaluate the differentials and integrals.)

24 Use the Kuhn–Tucker criteria to find the minimum of

$$2x_1^2 + x_2^2 + 2x_1x_2$$

subject to

$$x_1 - x_2 \leq \alpha$$

where α is a parameter. Find the critical value of α at which the nature of the solution changes. Sketch the situation geometrically to illustrate the change.

10.4 Hill climbing

10.4.1 Single-variable search

Most practical problems give calculations that cannot be performed explicitly, and need a numerical technique. Typical cases are those in Examples 10.1 and 10.11, where the final equations cannot be solved analytically and we need to resort to numerical methods. This is not an uncommon situation, and **hill climbing methods** were devised, mainly in the 1960s, to cope with just such problems.

In many engineering problems the functions that we are trying to optimize cannot be written down explicitly. Take for example a vibrational problem where the frequencies of vibration are calculated from an eigenvalue problem. These frequencies will depend on the parameters of the physical system, and it may be necessary to make the largest frequency as low as possible. To illustrate this idea, suppose that the eigenvalues come from the equation

$$\begin{vmatrix} a-\lambda & -1 & 0 \\ -1 & -\lambda & -1 \\ 0 & -1 & a^2-\lambda \end{vmatrix} = 0$$

where there is just one parameter a. In this case the mathematical problem is to find $\min_a(\lambda_{max})$. We note that there is no explicit formula for λ_{max} as a simple function of a, and our function is the result of a solution of the determinantal equation. For some values of a the function can be calculated easily, $\lambda_{max}(-1) = \sqrt{3}$, $\lambda_{max}(0) = \sqrt{2}$ and $\lambda_{max}(1) = 2$, but any other value requires a considerable amount of work. Calculating the derivative of the function with respect to a is too difficult even to contemplate.

In the determinant example we have a function of a single variable, but in most problems there are many variables. One of the commonest methods of attack is to obtain the maximum or minimum as a *sequence* of single-variable searches. We choose a direction and search in this direction until we have found the optimum of the function in the chosen direction. We then select a new direction and repeat the process. For this to be a successful method, we need to be able to perform single-variable searches very efficiently. This section therefore deals with single-variable problems, and only then in Section 10.4.3 are multivariable techniques discussed. In deciding on a strategy for solution, one crucial point is whether derivatives can or cannot be calculated. In the eigenvalue problem, calculation of the derivative is difficult, and would probably not be attempted. If the derivative can be obtained, however, more information is available, and any numerical method can be speeded up considerably. With the increase in sophistication of computers, this is becoming a less important consideration, since a good numerical approximation to the derivative is usually quite satisfactory.

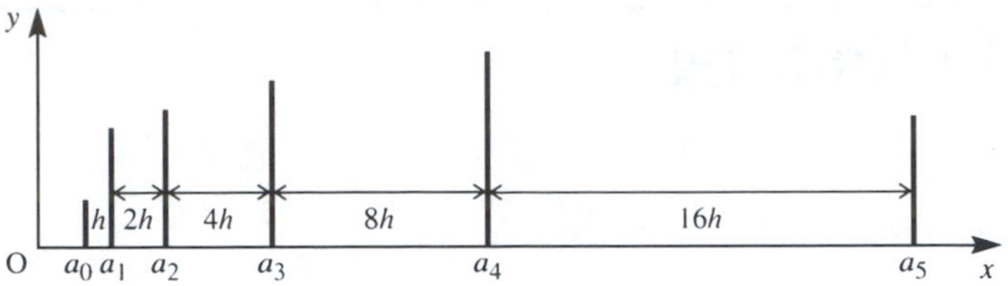

Figure 10.9 Bracketing procedure.

```
(a) {Derivative not known}
    read (keyb, a, h, n)
    it ← 0
    old val ← f(a)
    a ← a + h; val ← f(a)
    repeat
      old old val ← old val; old val ← val
      it ← it + 1; h ← 2*h; a ← a + h
      val ← f(a)
    until (((val < old val) and (old old val < old val)) or (it > n))
    if it > n then write (vdu, no max)
      else write (vdu, 'bracket given by', a – 3h, a)
    end if

(b) {Derivative known} read (keyb, a, h, n)
    it ← 0; deriv ← fdash(a); a ← a + h
    if fdash(a) < 0 then stop else
      repeat
        it ← it + 1
        deriv ← fdash(a)
        h = 2*h; a ← a + h
      until (deriv < 0) or (it > n)
      if (it > n) then write(vdu, 'no max')
      else write (vdu, 'bracket given by', a – 3h/2, a – h).
      endif
    endif
```

Figure 10.10 Pseudocode for obtaining a bracket for the maximum of $f(x)$. Function segments for $f(a)$ and $fdash(a)$ are declared elsewhere.

The basic problem is to determine the maximum of a function $y = f(x)$ that is difficult to evaluate and for which the derivative may or may not be available. The task is performed in two stages: in Stage 1 we bracket the maximum by obtaining x_1 and x_2 such that $x_1 \leqslant x_{max} \leqslant x_2$ as described in Figures 10.9 and 10.10, and in Stage 2 we devise a method that iterates to the maximum to any desired accuracy, as in Figures 10.11 and 10.12.

A **bracket** is most quickly achieved by starting at a given point, taking a step in the increasing direction and proceeding in this direction, doubling the step length at each step until the bracket is obtained (Figure 10.9). The basic idea is summarized in Figure 10.10, which gives a pseudocode procedure for this technique. It is written as a 'stand-alone' segment, but it would normally be incorporated in a more general program. It is assumed that sensible values for a and h have been chosen, but in a working program great care has to be taken. A great deal of effort is required to cope with inappropriate choices and to prevent the program aborting.

The algorithm works very efficiently provided that appropriate safeguards are included, but it is not foolproof. The maximum number of steps chosen, n, is usually 10, and the initial value of h is small compared with the overall dimension of the problem under consideration.

EXAMPLE 10.13 Find a bracket for the first maximum of $f(x) = x \sin x$ using the algorithm in Figure 10.10.

Solution Choose $a = 0$ and $h = 0.01$; the algorithm then gives

a	0	0.01	0.02	0.04	0.08	0.16	0.32	0.64	1.28	2.56	5.12
f	0	0.000	0.000	0.002	0.006	0.025	0.101	0.382	1.226	1.406	−4.701
f'	0	0.020	0.040	0.080	0.160	0.317	0.618	1.111	1.325	−1.590	—

If the derivative is *not* used then the bracket is $1.28 \leqslant x \leqslant 5.12$.
If the derivative is used then the bracket is $1.28 \leqslant x \leqslant 2.56$.

Stage 2 of the calculation is to use the bracket just obtained and then iterate to an accurate maximum. A simple and efficient approach is to use a **polynomial approximation** to estimate the maximum and then choose the 'best' points to repeat the calculation.

If it is assumed that *no derivative* is available then a bracket is known from the algorithm of Figure 10.10, so that x_1, x_2, x_3 and the corresponding f_1, f_2, f_3, with $f_1 < f_2$ and $f_3 < f_2$, are given. The *quadratic polynomial* through these points can be written down immediately: it is just the **Lagrange interpolation formula** that was discussed in Section 2.4 of *Modern Engineering Mathematics*. It can easily be checked that the quadratic which passes through the required points is given by

$$f \simeq F = \frac{(x - x_1)(x - x_2)}{(x_3 - x_1)(x_3 - x_2)} f_3 + \frac{(x - x_2)(x - x_3)}{(x_1 - x_2)(x_1 - x_3)} f_1 + \frac{(x - x_3)(x - x_1)}{(x_2 - x_3)(x_2 - x_1)} f_2 \qquad \textbf{(10.9)}$$

Then $F' = 0$ at the point x^*, which is given, after a little algebra, by

```
read (keyb, x1, x2, x3, f1, f2, f3, eps)
repeat
  {compute xstar from (10.10)}
  fstar ← f(xstar)
  if fstar > f2
    then
      if xstar < x2 then x3 ← x2 else x1 ← x2
      endif
      x2 ← xstar
    else
      if xstar < x2 then x1 ← xstar else x3 ← xstar
      endif
  endif
until (abs(x2 - xstar) < eps)
write (vdu, 'best value', 'x = ', xstar, 'f = ', fstar)
```

(a)

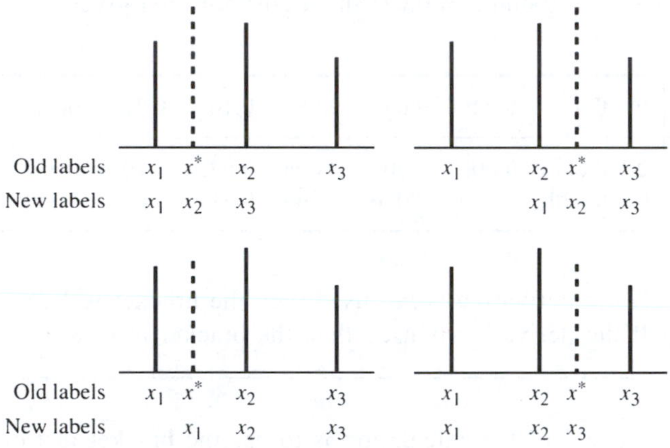

(b)

Figure 10.11 (a) Pseudocode for the quadratic approximation algorithm; the function segment for $f(x)$ is declared elsewhere; (b) diagrams corresponding to the four cases considered in the program.

$$x^* = \frac{(x_2^2 - x_3^2)f_1 + (x_3^2 - x_1^2)f_2 + (x_1^2 - x_2^2)f_3}{2[(x_2 - x_3)f_1 + (x_3 - x_1)f_2 + (x_1 - x_2)f_3]} \tag{10.10}$$

A pseudocode procedure for the algorithm that uses this new x^* and f^* is given in Figure 10.11, where the *best* three values are chosen for the next iteration.

The method works exceptionally well, but again it is not totally foolproof, and remedial checks need to be put into a working program. The stopping criterion used in the program is based on the difference between the x values. This is not always necessarily the best one to use. The criterion is very problem-dependent, and requires thought and numerical experimentation.

EXAMPLE 10.14 Find the first maximum of $f(x) = x \sin x$ given the values from Example 10.13, namely $x_1 = 1.28$, $x_2 = 2.56$, $x_3 = 5.12$, $f_1 = 1.226$, $f_2 = 1.406$ and $f_3 = -4.701$.

Solution From (10.10), $x^* = 2.03$ and $f^* = 1.820$, so for the next iteration choose

$$x_1 = 1.28, \quad x_2 = 2.03, \quad x_3 = 2.56$$

$$f_1 = 1.226, \quad f_2 = 1.820, \quad f_3 = 1.406$$

From (10.10), $x^* = 1.98$ and $f^* = 1.816$, so for the next iteration choose

$$x_1 = 1.98, \quad x_2 = 2.03, \quad x_3 = 2.56$$

$$f_1 = 1.816, \quad f_2 = 1.820, \quad f_3 = 1.406$$

From (10.10), $x^* = 2.027$ and $f^* = 1.820$, so the method has almost converged.

When the *derivative* is available, a better approximating polynomial than (10.9) can be used, since x_1, f_1, f_1' (> 0), x_2, f_2, and f_2' (< 0) are known from the bracketing algorithm, and this data can be fitted to a *cubic polynomial*

$$f \simeq F = ax^3 + bx^2 + cx + d$$

$$F' = 3ax^2 + 2bx + c$$

In this case fitting the values just gives the matrix equation

$$\begin{bmatrix} f_1 \\ f_2 \\ f_1' \\ f_2' \end{bmatrix} = \begin{bmatrix} x_1^3 & x_1^2 & x_1 & 1 \\ x_2^3 & x_2^2 & x_2 & 1 \\ 3x_1^2 & 2x_1 & 1 & 0 \\ 3x_2^2 & 2x_2 & 1 & 0 \end{bmatrix} \begin{bmatrix} a \\ b \\ c \\ d \end{bmatrix} \qquad \textbf{(10.11)}$$

and the maximum of F is given by $F' = 0$, so that

$$x^* = \frac{-b \pm (b^2 - 3ac)^{1/2}}{3a} \qquad \textbf{(10.12)}$$

and the negative sign is chosen (the positive sign for a minimum problem) to ensure that $x_1 < x^* < x_2$.

A simple algorithm uses these results to choose the appropriate bracket for the next iteration. The algorithm is illustrated in Figure 10.12, and is an efficient iterative way of evaluating the maximum. Unfortunately it is very easy to make errors in a hand computation, and many people prefer the quadratic algorithm for this purpose. On a computer, however, the cubic approximation method is almost universally used.

```
read (keyb, x1, x2, f1, f2, fdash1, fdash2, eps)
  repeat at
    {evaluate a,b,c from (10.11)}
    {calculate xstar from (10.12)}
    fstar ← f(xstar); fdashstar ← fdash(xstar)
  if fdashstar > 0        then x1 ← xstar; f1 ← fstar;
    fdash1 ← fdashstar else x2 ← xstar; f2 ← fstar;
    fdash2 ← fdashstar
  endif
  until (abs(f dashstar) , eps)
write (vdu, 'max at', xstar; 'f = ', fstar)
```

(a)

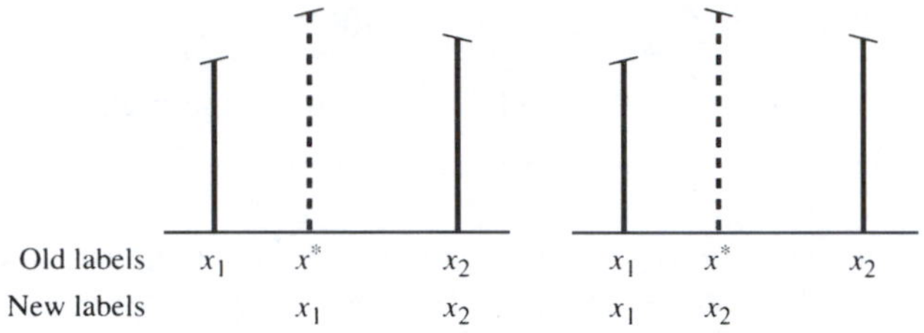

(b)

Figure 10.12 (a) Pseudocode for the cubic approximation algorithm for the maximum of $f(x)$; the function segments $f(x)$ and $fdash(x)$ are declared elsewhere; (b) diagrams corresponding to the two cases considered in the program.

EXAMPLE 10.15

Find the first maximum of $f(x) = x \sin x$ given the bracket values from Example 10.13, namely $x_1 = 1.28$, $f_1 = 1.226$, $f_1' = 1.325$, and $x_2 = 2.56$, $f_2 = 1.406$, $f_2' = -1.590$.

Solution Solving (10.11) gives $a = -0.3328$, $b = 0.7783$ and $c = 0.9683$, and hence, from (10.12),

$$x^* = 2.036, \quad f^* = 1.820, \quad f^{*\prime} = -0.0184$$

Thus x_2, f_2 and f_2' are replaced by x^*, f^* and $f^{*\prime}$, and x_1, f_1 and f_1' are retained. Equation (10.11) is now recomputed and solved to give

$$a = -0.4635, \quad b = 1.417, \quad c = -0.0242$$

Hence, from (10.12),

$$x^* = 2.030, \quad f^* = 1.820, \quad f^{*\prime} = -0.0021$$

Further iterations may be performed to get an even more accurate value. Comparison with Example 10.14 shows that both the quadratic and cubic algorithms work well for this function.

10.4.2 Exercises

†**25** (a) Find a bracket for the minimum of the function $f(x) = x + 1/x^2$. Start at $x = 0.1$ and $h = 0.2$.
(b) Use two cycles of the quadratic approximation to obtain an estimate of the minimum.
(c) Use one cycle of the cubic approximation to obtain an estimate of the minimum.

†**26** The function

$$f = \frac{\sin x}{1+x^2}$$

has been computed as follows:

x	-0.5	0	1	3
f	-0.3825	0	0.4207	0.0141
f'	0.3952	1	-0.1506	-0.1075

Compare the brackets, obtained from the bracketing procedure, to be used in calculating the maximum of the function: (a) without using the derivatives, and (b) using the derivatives. Use these brackets to perform one iteration of each of the quadratic and the cubic algorithms. Compare the values obtained from the two calculations.

†**27** Starting with the bracket $1 < x < 3$, determine an approximation to the maximum of the function

$$f(x) = x(e^{-x} - e^{-2x})$$

(a) using two iterations of the quadratic algorithm, and
(b) using two iterations of the cubic algorithm.

†**28** (Harder) Use the quadratic algorithm to obtain an estimate of the value of x that gives the minimum value to the largest root of the eigenvalue equation (that is, $\min_x [\lambda_{max}(x)]$)

$$\begin{vmatrix} x-\lambda & -1 & 0 \\ -1 & -\lambda & -1 \\ 0 & -1 & x^2-\lambda \end{vmatrix} = 0$$

Use the bracket given by $x = 1$ and $x = -1$.

29 Show that if $x_1 = x_2 - h$ and $x_3 = x_2 + h$ then (10.10) reduces to

$$x^* = x_2 + \frac{1}{2}h\frac{f_1 - f_3}{f_1 - 2f_2 + f_3}$$

(*Note*: This provides a better hand computation method than the Lagrange interpolation approach. The best x value is chosen and h is replaced by $\frac{1}{2}h$ after each step.) Show that this formula is a numerical form of the Newton–Raphson method applied to the equation $f'(x) = 0$.

10.4.3 Simple multivariable searches

As indicated in Section 10.4.1, many multivariable search methods use a sequence of single-variable searches to achieve a maximum. The fundamental question is how to choose a sensible direction in which to search for the top of the hill with a very limited amount of local information. The problem can be visualized when no derivatives are available as sitting in a dense fog and trying to get to the top of the hill with only an altimeter available. If derivatives are available then the fog has lifted a little, and we can now see a few feet around, so that at least we can see which is the uphill direction. The criteria for choosing a direction are (a) an easy choice of direction and (b) one that gives an efficient climbing method. This is not a simple task, and although the methods described in this section are rarely used nowadays, they do provide the basis for more advanced methods.

Modern methods are difficult to program, not because the basic method is difficult but because of the vast amount of remedial action that must be taken to prevent the program failing when something goes wrong. The general advice here is to understand the basic idea behind a method and then to implement it using a program from a reliable software library such as NAG (distributed by Numerical Algorithms Group Ltd of Oxford, UK).

Perhaps the most obvious method of choosing a search direction is to use the locally steepest direction. For the function $f(x_1, x_2, \dots, x_n)$ this is known to be in the gradient direction $\boldsymbol{G} = \text{grad } f = [\partial f/\partial x_1 \quad \partial f/\partial x_2 \quad \dots \quad \partial f/\partial x_n]$. If the derivatives are not available then in current practice they are evaluated numerically.

The gradient direction can be easily proved to give the maximum change. From a given point $(a_1, \dots, a_n)$, we proceed in the direction $(h_1, h_2, \dots, h_n)$ with given step length h so that $h_1^2 + h_2^2 + \dots + h_n^2 = h^2$. We then require

$$\max F = f(a_1 + h_1, \dots, a_n + h_n) - f(a_1, \dots, a_n)$$

subject to the constraint

$$h_1^2 + \dots + h_n^2 = h^2$$

The problem is one of Lagrange multipliers, so

$$0 = \frac{\partial F}{\partial h_i} - \lambda 2h_i = \frac{\partial f}{\partial h_i} - 2\lambda h_i \quad (i = 1, \dots, n)$$

Thus

$$\frac{\partial f}{\partial h_i} = \frac{\partial f}{\partial x_i} = 2\lambda h_i \quad (i = 1, \dots, n)$$

and hence

$$[h_1, h_2, \dots, h_n] \text{ is proportional to } \left[\frac{\partial f}{\partial x_1}, \dots, \frac{\partial f}{\partial x_n}\right] = \boldsymbol{G}$$

The method of **steepest ascent** (or **steepest descent**, for minima) then proceeds from the point $\boldsymbol{a}$ by choosing the gradient direction $\boldsymbol{G}$ (or $-\boldsymbol{G}$ for minima) for a search direction. We therefore need to find

$$\max_t \{g(t) = f(a_1 + tG_1, a_2 + tG_2, \dots, a_n + tG_n)\} \tag{10.13}$$

Since (10.13) is a function of a single variable t, the methods of Section 10.4.1 are appropriate, and we should expect the cubic algorithm to be used in the optimization. Once the best available point in the search direction has been found, $\boldsymbol{a} + t_{\max} \boldsymbol{G}$, the new gradient direction is computed and the whole process is repeated. The algorithm is fairly straightforward, and is summarized in Figure 10.13.

The steepest-ascent (or descent) method has the great advantage of being simple and secure, but it has the disadvantage of being very slow, particularly near to the optimum. It is rarely used nowadays, but does form the basis of the hill climbing methods described in Section 10.4.5.

```
read (keyb, a1, . . . , an)
  repeat
    {evaluate f(a) and G1 = ∂f/∂x1, G2 = ∂f/∂x2, . . . }
    {maximize g(t) in (10.13) by the cubic algorithm
      to give a new point a ← a + tmaxG}
  until {G = 0}
```

Figure 10.13 Steepest-ascent algorithm.

EXAMPLE 10.16

Find the maximum of the function

$$f(x_1, x_2) = -(x_1 - 1)^4 - (x_1 - x_2)^2$$

by the steepest-ascent method, starting at the point (0, 0).

Solution It is clear that (1, 1) gives the maximum, but this example is used to illustrate the basic method. The gradient is easily calculated from the partial derivatives

$$\frac{\partial f}{\partial x_1} = -4(x_1 - 1)^3 - 2(x_1 - x_2), \quad \frac{\partial f}{\partial x_2} = 2(x_1 - x_2)$$

Cycle 1: At the point (0, 0), $f = -1$, $\boldsymbol{G} = [4 \quad 0]$, the search direction is $x_1 = 4t$, $x_2 = 0$, and we require

$$\max_t \{g(t) = -(4t - 1)^4 - 16t^2\}$$

This can be calculated as $t_{max} = 0.102\,56$, so that we can start Cycle 2.

Cycle 2: Here $\boldsymbol{x} = (0.410\,25, 0)$, $f = -0.259$ and $\boldsymbol{G} = [0 \quad 0.8205]$. The next search is in the direction $x_1 = 0.410\,25$, $x_2 = 0.8205t$, and we require

$$\max_t \{g(t) = -0.1210 - (0.410\,25 - 0.8205t)^2\}$$

This has the obvious solution $t_{max} = \frac{1}{2}$, so that we can move to Cycle 3.

Cycle 3: Here $\boldsymbol{x} = (0.410\,25, 0.410\,25)$, $f = -0.1210$, and $\boldsymbol{G} = [0.8205 \quad 0]$. The calculation can be continued until $\boldsymbol{G} = 0$ to the required accuracy.

There are one or two points to note from this calculation. The function value is steadily increasing, which is a good feature of the method, but after the first few iterations the method progresses in a large number of very small steps. The successive search directions are parallel to the axes, and hence are perpendicular to each other. This perpendicularity is just a restatement of the known result that the gradient vector is perpendicular to the contours (see Section 7.2.1). In Example 10.16 the function $g(t)$ is written down explicitly for clarity. In practice on a computer this would not be done, since once the search direction has been established, x_1 and x_2 are known functions of t only, and by the chain rule we have

$$\frac{dg}{dt} = \frac{\partial f}{\partial x_1}\frac{dx_1}{dt} + \frac{\partial f}{\partial x_2}\frac{dx_2}{dt} = \frac{\partial f}{\partial x_1}G_1 + \frac{\partial f}{\partial x_2}G_2$$

Since x_1 and x_2 are known, both $\partial f/\partial x_1$ and $\partial f/\partial x_2$ can be calculated, and $\boldsymbol{G} = [G_1 \quad G_2]$ is the known search direction; therefore $\mathrm{d}g/\mathrm{d}t$ is computed without explicitly writing down the function g.

The major criticism of the steepest-ascent method is that it is slow to converge, and so the question arises as to how it can be speeded up. It is well known that Newton–Raphson methods (see Section 9.5.8 of *Modern Engineering Mathematics*) converge very rapidly, so the same basic idea is tried for these problems. It is convenient to employ matrix notation, and indeed most multidimensional optimization methods are written in matrix form. This gives a compact notation, and arrays appear naturally in computer languages.

Taylor's theorem (see Section 9.8.1 of *Modern Engineering Mathematics*) can be written in matrix form to second order as

$$f(a_1 + h_1, \ldots, a_n + h_n) \simeq f(a_1, \ldots, a_n) + \boldsymbol{h}^{\mathrm{T}}\boldsymbol{G} + \tfrac{1}{2}\boldsymbol{h}^{T}\boldsymbol{J}\boldsymbol{h} \tag{10.14}$$

where

$$\boldsymbol{h} = \begin{bmatrix} h_1 \\ \vdots \\ h_n \end{bmatrix}, \quad \boldsymbol{G} = \begin{bmatrix} \dfrac{\partial f}{\partial x_1} \\ \vdots \\ \dfrac{\partial f}{\partial x_n} \end{bmatrix}, \quad \boldsymbol{J} = \begin{bmatrix} \dfrac{\partial^2 f}{\partial x_1^2} & \cdots & \dfrac{\partial^2 f}{\partial x_1\,\partial x_n} \\ \vdots & & \vdots \\ \dfrac{\partial^2 f}{\partial x_n\,\partial x_1} & \cdots & \dfrac{\partial^2 f}{\partial x_n^2} \end{bmatrix}$$

The form (10.14) can be verified by multiplying out the matrices and comparing with the standard form of Taylor's theorem. The vector $\boldsymbol{G}$ is just the gradient vector and $\boldsymbol{J}$ is an $n \times n$ symmetric matrix of second derivatives called the **Hessian** (or Jacobian) **matrix**.

The **Newton method** takes (10.14) as an approximation to f, finds the maximum (or minimum) of the quadratic approximation, and uses this best value to start the cycle again.

The optimum of (10.14) is given by $\partial f/\partial h_i = 0$ $(i = 1, 2, \ldots, n)$. The first of these conditions gives

$$0 = \frac{\partial f}{\partial h_1} = [1 \quad 0 \quad 0 \quad \ldots \quad 0]\boldsymbol{G} + \frac{1}{2}[1 \quad 0 \quad \ldots \quad 0]\boldsymbol{J}\boldsymbol{h} + \frac{1}{2}\boldsymbol{h}^{\mathrm{T}}\boldsymbol{J}\begin{bmatrix} 1 \\ 0 \\ 0 \\ \vdots \\ 0 \end{bmatrix} \tag{10.15}$$

Noting that, since $\boldsymbol{J}$ is symmetric, for any vectors $\boldsymbol{a}$ and $\boldsymbol{b}$ we have

$$\boldsymbol{a}^{\mathrm{T}}\boldsymbol{J}\boldsymbol{b} = (\boldsymbol{a}^{\mathrm{T}}\boldsymbol{J}\boldsymbol{b})^{\mathrm{T}} = \boldsymbol{b}^{\mathrm{T}}\boldsymbol{J}\boldsymbol{a}$$

and hence (10.15) can be written as

$$0 = [1 \quad 0 \quad 0 \quad \ldots \quad 0]\,(\boldsymbol{G} + \boldsymbol{J}\boldsymbol{h})$$

Similarly for the other components we obtain

```
repeat
    {a_i known, calculate G_i, J_i}
    {evaluate a_{i+1} = a_i - J_i^{-1} G_i}
  until {sufficient accuracy}
```

Figure 10.14 Newton algorithm.

$$0 = [0 \quad 1 \quad 0 \quad \ldots \quad 0]\,(\boldsymbol{G} + \boldsymbol{Jh})$$

$$\vdots \qquad\qquad\qquad\qquad \vdots$$

$$0 = [0 \quad 0 \quad 0 \quad \ldots \quad 1]\,(\boldsymbol{G} + \boldsymbol{Jh})$$

The only way to satisfy this set of equations is to have

$$\boldsymbol{G} + \boldsymbol{Jh} = 0$$

and hence

$$\boldsymbol{h} = -\boldsymbol{J}^{-1}\boldsymbol{G}$$

The basic algorithm is now very straightforward, as indicated in Figure 10.14.

When Newton's algorithm of Figure 10.14 converges, it does so very rapidly and satisfies our request for a fast method. Unfortunately it is very unreliable, particularly for higher-dimensional problems, unless the starting value is close to the maximum. The reason for this is fairly clear. The analysis given only uses the necessary condition for a maximum, but it would apply equally well to a minimum or saddle point. In multidimensional problems saddle points are abundant, and the usual failure of the Newton method is that it proceeds towards a distant saddle point and then diverges.

EXAMPLE 10.17 Use Newton's method to find the maximum of

$$A = (h + b + 10)\left(\frac{2\,272\,000}{hb} + 2b + 5\right)$$

Solution Here we have returned to the 'milk carton' problem of Example 10.2. We can calculate

$$\frac{\partial A}{\partial h} = \left(\frac{2\,272\,000}{hb} + 2b + 5\right) - (h + b + 10)\frac{2\,272\,000}{h^2 b}$$

$$\frac{\partial A}{\partial b} = \left(\frac{2\,272\,000}{hb} + 2b + 5\right) + (h + b + 10)\left(-\frac{2\,272\,000}{hb^2} + 2\right)$$

$$\frac{\partial^2 A}{\partial h^2} = 2(b + 10)\frac{2\,272\,000}{h^3 b}$$

$$\frac{\partial^2 A}{\partial b\,\partial h} = 2 + \frac{10 \times 2\,272\,000}{h^2 b^2}$$

$$\frac{\partial^2 A}{\partial b^2} = 4 + \frac{2 \times 2\,272\,000(h + 10)}{hb^3}$$

Iteration 1

$$\boldsymbol{a} = \begin{bmatrix} 100 \\ 100 \end{bmatrix}, \quad \boldsymbol{G} = \begin{bmatrix} -44.92 \\ 375.1 \end{bmatrix}, \quad \boldsymbol{J} = \begin{bmatrix} 4.998 & 2.227 \\ 2.227 & 8.998 \end{bmatrix}$$

$$\boldsymbol{a}_{\text{new}} = \begin{bmatrix} 100 \\ 100 \end{bmatrix} - \begin{bmatrix} 0.2249 & -0.0557 \\ -0.0557 & 0.1249 \end{bmatrix} \begin{bmatrix} -44.92 \\ 375.1 \end{bmatrix} = \begin{bmatrix} 131 \\ 50.6 \end{bmatrix}$$

Iteration 2

$$\boldsymbol{a} = \begin{bmatrix} 131 \\ 50.6 \end{bmatrix}, \quad \boldsymbol{G} = \begin{bmatrix} -52.3 \\ -465.7 \end{bmatrix}, \quad \boldsymbol{J} = \begin{bmatrix} 2.421 & 2.517 \\ 2.517 & 41.75 \end{bmatrix}$$

$$\boldsymbol{a}_{\text{new}} = \begin{bmatrix} 131 \\ 50.6 \end{bmatrix} - \begin{bmatrix} -0.4407 & -0.0266 \\ -0.0266 & 0.0255 \end{bmatrix} \begin{bmatrix} -52.3 \\ 465.7 \end{bmatrix} = \begin{bmatrix} 141.7 \\ 61.1 \end{bmatrix}$$

Iteration 3

$$\boldsymbol{a} = \begin{bmatrix} 141.7 \\ 61.1 \end{bmatrix}, \quad \boldsymbol{G} = \begin{bmatrix} -4.493 \\ -98.76 \end{bmatrix}, \quad \boldsymbol{J} = \begin{bmatrix} -1.858 & -2.303 \\ 2.303 & 25.33 \end{bmatrix}$$

$$\boldsymbol{a}_{\text{new}} = \begin{bmatrix} 141.7 \\ 61.1 \end{bmatrix} - \begin{bmatrix} 0.6067 & -0.0552 \\ -0.0552 & 0.0445 \end{bmatrix} \begin{bmatrix} -4.493 \\ -98.76 \end{bmatrix} = \begin{bmatrix} 139 \\ 65.2 \end{bmatrix}$$

The iterations converge very rapidly; $h = 139$, $b = 65$ is not far from the solution, and gives $A = 827\ \text{cm}^2$.

10.4.4 Exercises

30 Follow the first two complete cycles in the steepest-descent algorithm for finding the minimum of the function

$$f(x_1, x_2) = \boldsymbol{x}^{\text{T}} \begin{bmatrix} 1 \\ 0 \end{bmatrix} + \frac{1}{2}\boldsymbol{x}^{\text{T}} \begin{bmatrix} 1 & -1 \\ -1 & 2 \end{bmatrix} \boldsymbol{x}$$

starting at $\boldsymbol{a} = \begin{bmatrix} 1 \\ 1 \end{bmatrix}$.

31 Show that the function

$$f(x, y) = 2(x + y)^2 + (x - y)^2 + 3x + 2y$$

has a minimum at the point $(-\frac{7}{16}, -\frac{3}{16})$.

Starting at the point (0, 0), use one iteration of the steepest-descent algorithm to determine an approximation to the minimum point. Show that one iteration of Newton's method yields the minimum point from *any* starting point.

†32 Use one complete cycle of the steepest-ascent method to find the maximum of the function

$$f(x, y, z) = -(x - y + z)^2 - (2x + z - 2)^2 - z^4$$

starting from (2, 2, 2).

†33 Minimize the function

$$f(x, y, z) = (x - y + z)^2 + (2x + z - 2)^2 + z^4$$

by Newton's method. Complete one cycle of the iteration, starting at (2, 2, 2).

10.4.5 Advanced multivariable searches

To overcome the problems of evaluating second derivatives, which are rarely available, and of the unreliability of the Newton method, but to use its speed of convergence, several methods were produced in the early 1960s. Two have survived and are now the methods currently available in most program libraries. **Conjugate-gradient methods** give one approach, but these will not be described here. We shall look at a method due to Davidon, commonly called **DFP** (after Davidon, Fletcher and Powell, who developed the method) or **quasi-Newton methods**. There is a whole class of such methods, but only one will be studied.

The basic idea of the DFP method is to look for the *minimum* of the function $f(x_1, \ldots, x_n)$ with gradient $\boldsymbol{G} = [\partial f/\partial x_1 \quad \partial f/\partial x_2 \quad \ldots \quad \partial f/\partial x_n]^{\mathrm{T}}$ by iterating with a matrix $\boldsymbol{H}_i$, which will be updated at each iteration, so that

$$\boldsymbol{a}_{i+1} = \boldsymbol{a}_i - \lambda \boldsymbol{H}_i \boldsymbol{G}_i \tag{10.16}$$

The reliable but slow steepest-descent method chooses $\boldsymbol{H}_i = \boldsymbol{I}$, the unit matrix, and $\lambda = \lambda_{\min}$, while the less reliable but fast Newton method chooses $\boldsymbol{H}_i = \boldsymbol{J}_i^{-1}$ and $\lambda = 1$. Thus the idea is to compute a sequence of $\boldsymbol{H}_i$ so that $\boldsymbol{H}_0 = \boldsymbol{I}$ and $\boldsymbol{H}_i \to \boldsymbol{J}^{-1}$ as the minimum is approached. The basic analysis required to implement this scheme is quite difficult and beyond the scope of this book, so only the briefest of outlines will be given. For a quadratic function

$$f = c + \boldsymbol{x}_i^{\mathrm{T}} \boldsymbol{G} + \tfrac{1}{2} \boldsymbol{x}_i^{\mathrm{T}} \boldsymbol{J} \boldsymbol{x}_i$$

at two successive points the gradient is given by

$$\boldsymbol{G}_i = \boldsymbol{G} + \boldsymbol{J}\boldsymbol{x}_i$$

$$\boldsymbol{G}_{i+1} = \boldsymbol{G} + \boldsymbol{J}\boldsymbol{x}_{i+1}$$

so, subtracting,

$$\boldsymbol{G}_{i+1} - \boldsymbol{G}_i = \boldsymbol{J}(\boldsymbol{x}_{i+1} - \boldsymbol{x}_i)$$

Writing $\boldsymbol{h}_i = \boldsymbol{x}_{i+1} - \boldsymbol{x}_i$ and $\boldsymbol{y}_i = \boldsymbol{G}_{i+1} - \boldsymbol{G}_i$ and working on the assumption that $\boldsymbol{H}_i\boldsymbol{J} = \boldsymbol{I}$ since we require $\boldsymbol{H}_i \to \boldsymbol{J}^{-1}$, we obtain

$$\boldsymbol{H}_i \boldsymbol{y}_i = \boldsymbol{H}_i \boldsymbol{J} \boldsymbol{h}_i = \boldsymbol{h}_i \tag{10.17}$$

It is found to be impossible to satisfy (10.17) unless an exact solution is known, so the next best thing is to satisfy (10.17) one step behind as

$$\boldsymbol{H}_{i+1} \boldsymbol{y}_i = \boldsymbol{h}_i \tag{10.18}$$

There is a whole class of matrices that satisfy the key equation (10.18) but only the original Davidon matrix (and still one of the best) is quoted here, namely

$$\boldsymbol{H}_{i+1} = \boldsymbol{H}_i - \frac{\boldsymbol{H}_i \boldsymbol{y}_i \boldsymbol{y}_i^{\mathrm{T}} \boldsymbol{H}_i}{\boldsymbol{y}_i^{\mathrm{T}} \boldsymbol{H}_i \boldsymbol{y}_i} + \frac{\boldsymbol{h}_i \boldsymbol{h}_i^{\mathrm{T}}}{\boldsymbol{h}_i^{\mathrm{T}} \boldsymbol{y}_i} \tag{10.19}$$

```
read {initial values x0, H0 = I}
     {calculate the gradient G0 and f0}
repeat
     {Find min f(xi - lHiGi), by the cubic algorithm}
           λ
     {Put xi+1 = xi - lminHiGi}
     {Calculate fi+1, Gi+1 and hence hi = xi+1 - xi
       and yi = Gi+1 - Gi}
     {Update Hi to Hi+1 by (10.19)}
until {sufficient accuracy}
```

Figure 10.15 DFP algorithm for the minimum of $f(x_1, x_2, \ldots, x_n)$.

It can be shown that for a quadratic function this sequence of $\boldsymbol{H}$s produces $\boldsymbol{J}^{-1}$ in n iterations, where n is the dimension of the problem. The basic algorithm is described in Figure 10.15 for the minimum of a general function $f(x_1, \ldots, x_n)$ with gradient $\boldsymbol{G} = [\partial f/\partial x_1 \quad \ldots \quad \partial f/\partial x_n]^{\mathrm{T}}$.

This method was a major breakthrough in the early 1960s, and is still one of the best and most reliable available. Proofs of convergence and computational experience are available in advanced texts on optimization. To repeat a word of warning: these programs are very long and tedious to write because of the large amount of checking and remedial work that has to be inserted to prevent the program stopping. Such programs are available in software libraries, and these should be used.

EXAMPLE 10.18 Use the DFP method to find the minimum of

$$f(x, y) = (x - y - 1)^4 + (2x + y - 5)^2$$

starting at (0; 0).

Solution Note that only the first derivatives

$$\boldsymbol{G} = \begin{bmatrix} 4(x-y-1)^3 + 4(2x+y-5) \\ -4(x-y-1)^3 + 2(2x+y-5) \end{bmatrix}$$

are required in this method.

Iteration 1

$$\boldsymbol{a}_0 = \begin{bmatrix} 0 \\ 0 \end{bmatrix}, \quad \boldsymbol{G}_0 = \begin{bmatrix} -24 \\ -6 \end{bmatrix}, \quad \boldsymbol{H}_0 = \begin{bmatrix} 1 & 0 \\ 0 & 1 \end{bmatrix}, \quad f_0 = 26$$

and we search in the direction

$$\boldsymbol{a} = \begin{bmatrix} 0 \\ 0 \end{bmatrix} - \lambda \begin{bmatrix} 1 & 0 \\ 0 & 1 \end{bmatrix} \begin{bmatrix} -24 \\ -6 \end{bmatrix} = \begin{bmatrix} 24\lambda \\ 6\lambda \end{bmatrix}$$

for a minimum of f, that is, we compute

$$\min_{\lambda}\{(18\lambda-1)^4+(54\lambda-5)^2\}$$

The cubic algorithm gives $\lambda_{\min} = 0.089\,72$, so

$$\boldsymbol{a}_1 = \begin{bmatrix} 2.153 \\ 0.538 \end{bmatrix}, \quad \boldsymbol{G}_1 = \begin{bmatrix} 0.306 \\ -1.241 \end{bmatrix}$$

$$\boldsymbol{h}_0 = \begin{bmatrix} 2.153 \\ 0.538 \end{bmatrix}, \quad \boldsymbol{y}_0 = \begin{bmatrix} 24.31 \\ 4.76 \end{bmatrix}$$

Thus

$$\boldsymbol{H}_1 = \begin{bmatrix} 1 & 0 \\ 0 & 1 \end{bmatrix} - \frac{1}{613.4}\begin{bmatrix} 24.31 \\ 4.76 \end{bmatrix}[\,24.31 \quad 4.76\,] + \frac{1}{54.89}\begin{bmatrix} 2.153 \\ 0.538 \end{bmatrix}[\,2.153 \quad 0.538\,]$$

$$= \begin{bmatrix} 0.1214 & -0.1675 \\ -0.1675 & 0.9684 \end{bmatrix}$$

Iteration 2

$$\boldsymbol{a}_1 = \begin{bmatrix} 2.153 \\ 0.538 \end{bmatrix}, \quad \boldsymbol{G}_1 = \begin{bmatrix} 0.306 \\ -1.241 \end{bmatrix}, \quad \boldsymbol{H}_1 = \begin{bmatrix} 0.1214 & -0.1675 \\ -0.1675 & 0.9684 \end{bmatrix}, \quad f_1 = 0.167$$

We now search in the direction

$$\boldsymbol{a} = \begin{bmatrix} 2.153 \\ 0.538 \end{bmatrix} - \lambda \begin{bmatrix} 0.1214 & -0.1675 \\ -0.1675 & 0.9684 \end{bmatrix}\begin{bmatrix} 0.306 \\ -1.241 \end{bmatrix} = \begin{bmatrix} 2.153 - 0.2449\lambda \\ 0.538 + 1.2525\lambda \end{bmatrix}$$

and the cubic algorithm gives $\lambda_{\min} = 0.272$. We can now compute the next point and its gradient as

$$\boldsymbol{a}_2 = \begin{bmatrix} 2.086 \\ 0.879 \end{bmatrix}, \quad \boldsymbol{G}_2 = \begin{bmatrix} 0.2427 \\ 0.0678 \end{bmatrix}$$

and then calculate the vectors

$$\boldsymbol{h}_1 = \begin{bmatrix} -0.0666 \\ 0.3407 \end{bmatrix}, \quad \boldsymbol{y}_1 = \begin{bmatrix} -0.0636 \\ 1.3083 \end{bmatrix}, \quad \boldsymbol{H}_1\boldsymbol{y}_1 = \begin{bmatrix} -0.2269 \\ 1.2776 \end{bmatrix}$$

The new Davidon matrix is calculated from (10.19) as

$$\boldsymbol{H}_2 = \begin{bmatrix} 0.1007 & -0.0460 \\ -0.0460 & 0.2582 \end{bmatrix}$$

Iteration 3

$$\boldsymbol{a}_2 = \begin{bmatrix} 2.086 \\ 0.879 \end{bmatrix}, \quad \boldsymbol{G}_2 = \begin{bmatrix} 0.2427 \\ 0.0678 \end{bmatrix}, \quad \boldsymbol{H}_2 = \begin{bmatrix} 0.1007 & -0.0460 \\ -0.0460 & 0.2582 \end{bmatrix}, \quad f = 0.0044$$

The iterations continue until convergence at $x = 2$ and $y = 1$.

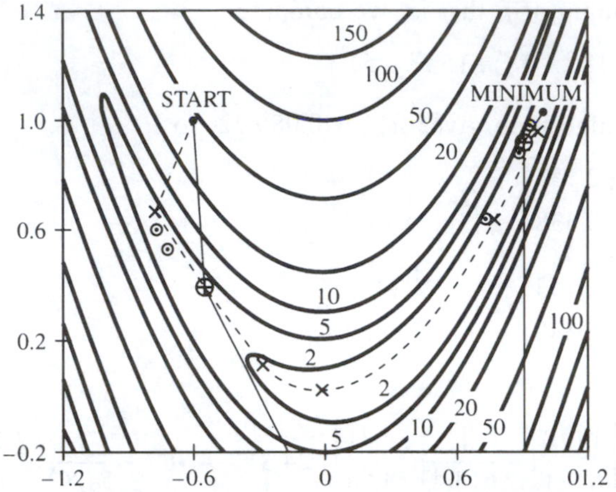

Figure 10.16 Minimum of the function $f = 100(y^2 - x)^2 + (1 - x)^2$: ⊕—⊕, Newton's method at iterations 1, 2 (off diagram), 3, 6; × _ _ ×, DFP method at iterations 1, 2, 5, 10, 15; ⊙ . . . ⊙, Steepest descent at iterations 10, 20, 21, 30, 40.

To compare the three methods considered above, we apply them to the same function

$$f(x, y) = 100(y^2 - x)^2 + (1 - x)^2$$

By plotting the contours of the function as in Figure 10.16 and superimposing the paths followed by the three methods, we can compare their performance. We start from (−0.6, 1). The Newton and DFP methods converge to 5 figures of accuracy in 6 and 18 iterations respectively, while the steepest-descent method has not converged in 40 iterations. The following table gives some indication of the way the methods progress for this example:

Iteration	5	10	15	20
Newton	(1, 0.998)	Converged		
DFP	(0.012, 1.026)	(0.791, 0.608)	(0.992, 0.982)	Converged
Steepest-descent	(−0.788, 0.632)	(−0.768, 0.587)	–	(−0.718, 0.513)

A major development since the 1970s has been in devising modifications that avoid the line searches. It was found that the latter were very time-consuming, so there was great pressure to avoid them. The searches were replaced by a single step in the search direction and, provided that the function value is reduced by a sufficient amount, this is accepted for recalculation of $\boldsymbol{H}$. To be efficient, the method requires a variant of the DFP updating, and it needs more care when failures occur; it is, however, one of the most competitive methods available.

The early development of numerical optimization algorithms led to very distinct methods for cases when derivatives were or were not available. As computers developed in speed, this difference became less necessary, since derivatives could be calculated very rapidly by a numerical method. The only difference between the two cases in modern programs is that the quadratic or cubic algorithms are used for the line searches in the two cases. A DFP-type algorithm or a conjugate-gradient method (mentioned earlier) is then used to iterate to the minimum.

Adapting methods to deal with constraints is of intense interest in practice, and has been a major thrust in the subject. For fully nonlinear problems with nonlinear constraints the practical difficulties are very severe, and the production of efficient, robust programs is at a premium. If the constraints are all linear then the problem is easier, and methods of projection onto 'active' constraints reduce it to an unconstrained problem of the type discussed in this section. When the function and the constraints are all linear we have the linear programming problem of Section 10.2.

10.4.6 Exercises

†**34** Minimize the following functions by the DFP method, completing two cycles:

(a) $f(x, y) = (x - y)^2 + 4(x - 1)^2$, starting at (2, 2);
(b) $f(x, y, z) = (x - y + z)^2 + (2x + z - 2)^2 + z^4$, starting at (0, 0, 0).

†**35** Show that the updating formula

$$\boldsymbol{H}_{i+1} = \boldsymbol{H}_i + \frac{(\boldsymbol{h}_i - \boldsymbol{H}_i\boldsymbol{y}_i)(\boldsymbol{h}_i - \boldsymbol{H}_i\boldsymbol{y}_i)^{\mathrm{T}}}{(\boldsymbol{h}_i - \boldsymbol{H}_i\boldsymbol{y}_i)^{\mathrm{T}}\boldsymbol{y}_i}$$

satisfies (10.18). Follow the DFP method through for two cycles, but using this update for $\boldsymbol{H}$, on the functions

(a) $x^2 + 2y^2$, starting at (1, 2);
(b) $x^2 + y^2 + 2z^4$, starting at (1, 2, 3).

36 Show that the update formula (with suffixes suppressed)

$$\boldsymbol{H}' = \boldsymbol{H} + \boldsymbol{v}\boldsymbol{p}^{\mathrm{T}} - \boldsymbol{H}\boldsymbol{u}\boldsymbol{q}^{\mathrm{T}}$$

satisfies the basic quasi-Newton equation (10.18)

$$\boldsymbol{H}'\boldsymbol{u} = \boldsymbol{v}$$

where $\boldsymbol{p}$ and $\boldsymbol{q}$ are vectors satisfying

$$\boldsymbol{p}^{\mathrm{T}}\boldsymbol{u} = 1, \quad \boldsymbol{q}^{\mathrm{T}}\boldsymbol{u} = 1$$

but are otherwise arbitrary.

By making a suitable choice of α, β and α', β' in the expressions

$$\boldsymbol{p} = \alpha\boldsymbol{v} + \beta\boldsymbol{H}\boldsymbol{u}$$

$$\boldsymbol{q} = \alpha'\boldsymbol{v} + \beta'\boldsymbol{H}\boldsymbol{u}$$

show that the Davidon formula (10.19) and the formula in Exercise 35 can be obtained.

10.5 Engineering application: chemical processing plant

A chemical processing plant consists of a main processing unit and two recovery units. Chemicals A and B are fed into the plant, and produce a maximum output of 100 t day^{-1} of material C. The effluent stream is rich in chemical B, which can be recovered from the primary and secondary recovery units. The total recovered,

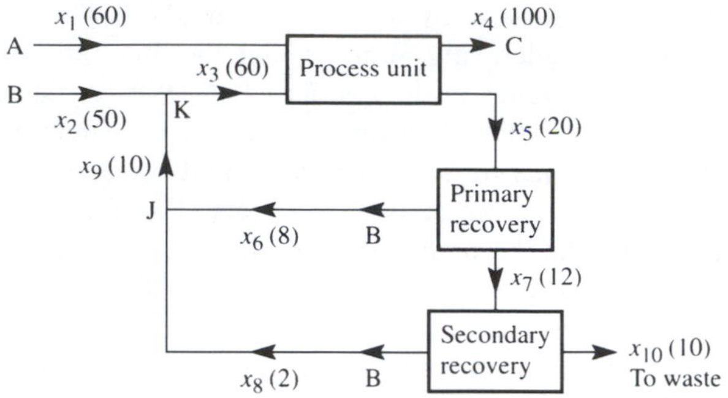

Figure 10.17 Schematic diagram of a chemical processing plant. The numbers in parentheses give the maximum flow.

when at full throughput, is 10 t day^{-1} of pure B, and it is fed back into the incoming stream of chemical B. The process is illustrated in Figure 10.17; the numbers in parentheses indicate the maximum flow (in t day^{-1}) that can be sent down the pipes, and the x_i indicate the actual flow (in t day^{-1}).

The chemistry of the process implies that the chemicals must be mixed in given ratios. For the present system it is found that

$$x_1{:}x_3 = 1{:}1, \quad x_1{:}x_4 = 3{:}5, \quad x_1{:}x_5 = 3{:}1, \quad x_5{:}x_7 = 5{:}3$$

$$x_5{:}x_6 = 5{:}2, \quad x_7{:}x_8 = 6{:}1, \quad x_7{:}x_{10} = 6{:}5$$

must be maintained for any flow through the system. Chemical A costs £100 t^{-1}, chemical B costs £120 t^{-1} and chemical C sells for £220 t^{-1}. The running costs are as follows:

	Variable costs	Fixed costs
Process unit	£70 t^{-1} of product	£500 day^{-1}
Primary recovery	£30 t^{-1} of input	£200 day^{-1}
Secondary recovery	£40 t^{-1} of input	£100 day^{-1}
Disposal of waste	£30 t^{-1}	
Indirect cost		£400 day^{-1}

It is required to find the most profitable operating policy that can be achieved.

The profit can be written down for a day's production as

$$\begin{aligned} z = {} & -(\text{fixed costs}) + (\text{profit from sale of C}) - (\text{costs of chemicals A and B}) \\ & - (\text{process unit costs}) - (\text{primary unit costs}) - (\text{secondary unit costs}) \\ & - (\text{waste product costs}) \\ = {} & -1200 + (220x_4) - (100x_1 + 120x_2) \\ & - (70x_4) - (30x_5) - (40x_7) \\ & - 30x_{10} \end{aligned}$$

$$z = -1200 - 100x_1 - 120x_2 + 150x_4 - 30x_5 - 40x_7 - 30x_{10}$$

The constraints on the flow, given by the maximum throughput, are

$$x_1 \leqslant 60, \quad x_2 \leqslant 50, \quad x_3 \leqslant 60, \quad x_4 \leqslant 100, \quad x_5 \leqslant 20, \quad x_6 \leqslant 8, \quad x_7 \leqslant 12,$$
$$x_8 \leqslant 2, \quad x_9 \leqslant 10, \quad x_{10} \leqslant 10$$

The constraints on the chemistry given by the fixed ratios can be written in a convenient form as

$$x_1 - x_3 = 0, \quad 5x_1 - 3x_4 = 0, \quad x_1 - 3x_5 = 0, \quad 3x_5 - 5x_7 = 0, \quad 2x_5 - 5x_6 = 0,$$
$$x_7 - 6x_8 = 0, \quad 5x_7 - 6x_{10} = 0$$

Finally, at the junctions J and K, continuity (what flows in equals what flows out) gives

$$x_6 + x_8 - x_9 = 0, \quad x_2 + x_9 - x_3 = 0$$

The problem thus has 10 variables, 10 inequality constraints and 9 equality constraints. The choice is whether to use the equality constraints to eliminate some of the variables or just to treat the 19 constraints directly by LP. The equations are sufficiently simple to solve for the variables as

$$x_2 = \tfrac{5}{6}x_1, \quad x_3 = x_1, \quad x_4 = \tfrac{5}{3}x_1, \quad x_5 = \tfrac{1}{3}x_1, \quad x_6 = \tfrac{2}{15}x_1, \quad x_7 = \tfrac{1}{5}x_1, \quad x_8 = \tfrac{1}{30}x_1,$$
$$x_9 = \tfrac{2}{15}x_1, \quad x_{10} = \tfrac{1}{6}x_1, \quad z = 27x_1 - 1200$$

Thus x_1 must be as large as possible, that is, at the value 60, giving a maximum profit of £420 day^{-1}. We must check that all the constraints are satisfied, and indeed this is the case. It is easily seen that each variable reaches its maximum possible value indicated in Figure 10.17.

When we look at variations on the problem, it becomes less clear whether to eliminate or just to use LP directly on the modified equations. For instance a very sensible question is whether it is worth using the primary or secondary recovery units. We can consider this question by allowing a portion to go to waste (at the same cost given previously), as indicated in Figure 10.18.

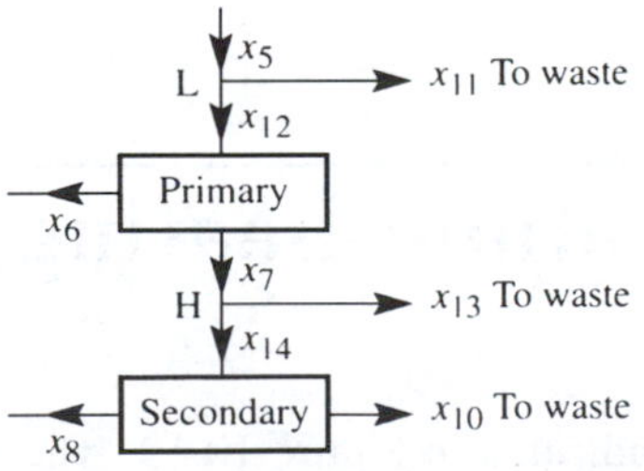

Figure 10.18 Modification to the chemical processing plant.

We add to the previous continuity equations similar equations for the junctions L and H:

$$x_5 - x_{11} - x_{12} = 0$$

$$x_7 - x_{13} - x_{14} = 0$$

The inputs to the primary and secondary units are now x_{12} and x_{14}, so we need to modify the fixed ratio chemical constraint as follows:

replace $3x_5 - 5x_7 = 0$ by $3x_{12} - 5x_7 = 0$

replace $2x_5 - 5x_6 = 0$ by $2x_{12} - 5x_6 = 0$

replace $x_7 - 6x_8 = 0$ by $x_{14} - 6x_8 = 0$

replace $5x_7 - 6x_{10} = 0$ by $5x_{14} - 6x_{10} = 0$

In the cost function x_5 is replaced by x_{12}, and x_7 by x_{14}, and the additional waste costs $(-30x_{11} - 30x_{13})$ must be added:

$$z = -1200 + 150x_4 - 100x_1 - 120x_2$$
$$- 30x_{12} - 30x_{11} - 40x_{14} - 30x_{13} - 30x_{10}$$

We now have three free variables, so we shall certainly need an LP approach. An LP package was used to obtain the solution

$x_1 = 57.69$, $x_2 = 50$, $x_3 = 57.69$, $x_4 = 96.15$

$x_5 = 19.23$, $x_6 = 7.69$, $x_7 = 11.54$, $x_8 = 0$

$x_9 = 7.69$, $x_{10} = 0$, $x_{11} = 0$, $x_{12} = 19.23$

$x_{13} = 11.54$, $x_{14} = 0$

and the profit is £530.77 day^{-1}. It can be seen that $x_{11} = 0$, so nothing is sent to waste before the primary recovery unit; but $x_{14} = 0$, so that all the material from the primary to the secondary goes to waste, and the secondary unit is bypassed. The effect of this strategy is to increase the profit by about 20%.

There are many other variations that can be considered for this model. For instance, pumps often go wrong, so it is important to investigate what happens if the maximum flows are reduced or even cut completely. Once the basic program has been set up, such variations are quite straightforward to implement.

10.6 Engineering application: heating fin

A heating fin is of the shape indicated in Figure 10.19, where the wall temperature is T_1, the ambient temperature is T_0 and within the fin the value $T = T(x)$ is

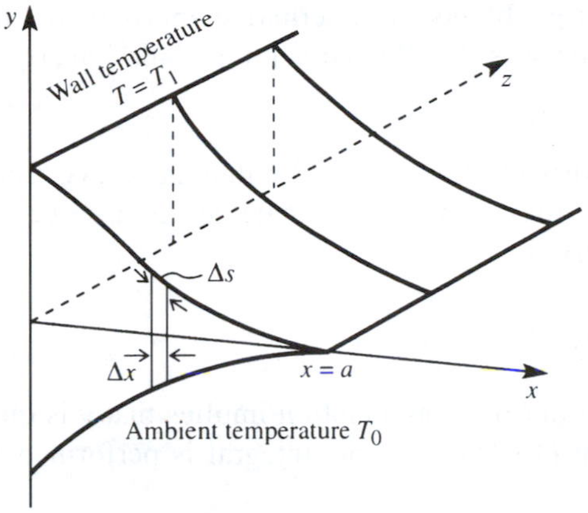

Figure 10.19 Heating fin.

assumed to depend only on x and to be independent of y and z. Heat is transferred by conduction along the fin, which has thermal conductivity k, and heat is transferred to the outside according to Newton's law of cooling, with surface heat-transfer coefficient h. Considering an area of the fin of unit width in the z direction and height $2y$, the heat transferred by conduction in the x direction is $k2y\,\mathrm{d}T/\mathrm{d}x$. The net transfer through the element illustrated is $(\mathrm{d}/\mathrm{d}x)(k2y\,\mathrm{d}T/\mathrm{d}x)$. The total heat lost through a surface element of unit width in the z direction and length $\Delta s = [1 + (\mathrm{d}y/\mathrm{d}x)^2]^{1/2}\,\Delta x$ along the surface is $h(T - T_0)\Delta s$. Since there are two surfaces, we can write the heat-transfer equation as

$$2\frac{\mathrm{d}}{\mathrm{d}x}\left[ky\frac{\mathrm{d}}{\mathrm{d}x}(T-T_0)\right] = 2h(T-T_0)\left[1+\left(\frac{\mathrm{d}y}{\mathrm{d}x}\right)^2\right]^{1/2}$$

Provided that $\mathrm{d}y/\mathrm{d}x$ is not too large, $(\mathrm{d}y/\mathrm{d}x)^2$ can be neglected, giving

$$\frac{\mathrm{d}}{\mathrm{d}x}\left[y\frac{\mathrm{d}}{\mathrm{d}x}(T-T_0)\right] = \frac{h}{k}(T-T_0) \qquad \textbf{(10.20)}$$

The mass of the fin is given, so that its cross-sectional area is known, and hence

$$\int_0^a y\,\mathrm{d}x = \tfrac{1}{2}A \qquad \textbf{(10.21)}$$

Finally, we wish to maximize the heat transfer, so we require the maximum of

$$I = 2h\int_0^a (T-T_0)\,\mathrm{d}x \qquad \textbf{(10.22)}$$

over all possible functions $y(x)$.

The problem involves choosing a function $y(x)$ that satisfies (10.21) and then solving (10.20) for $T - T_0$. This is then substituted into (10.22) and the integral evaluated. Out of all possible such functions y, we choose the one that maximizes I. The scheme outlined is extremely difficult and belongs to a class called **variational problems**. An alternative approximate method must be sought. The *assumption* made is that the temperature falls linearly with x as

$$T - T_0 = (T_1 - T_0)(1 - \alpha x)$$

We also assume that $y = 0$ at $x = a$. We thus have two free parameters, α and a, which we can use to give an approximate solution. Given $T - T_0$, y can be computed from (10.20) as

$$y = \frac{h}{k\alpha}\left(a - \frac{1}{2}\alpha a^2 - x + \frac{1}{2}\alpha x^2\right)$$

It can be seen that our basic assumption implies that y is quadratic in x. To satisfy the area constraint (10.21), a simple integral is performed to give

$$\frac{1}{2}A = \frac{h}{k\alpha}a^2\left(\frac{1}{2} - \frac{1}{3}\alpha a\right)$$

which gives a relation between α and a. The function I is now integrated as

$$I = 2h(T_1 - T_0)\int_0^a (1 - \alpha x)\,\mathrm{d}x$$

or

$$\frac{I}{2h(T_1 - T_0)} = a - \frac{1}{2}\alpha a^2$$

Thus the very difficult problem has been reduced to a Lagrange multiplier problem of maximizing

$$f = a - \tfrac{1}{2}\alpha a^2$$

subject to

$$g = 0 = \frac{a^2}{2\alpha} - \frac{1}{3}a^3 - S^3 \tag{10.23}$$

where $S^3 = kA/2h$. The Lagrange multiplier analysis gives the equations

$$\frac{\partial(f - \lambda g)}{\partial a} = 1 - \alpha a - \lambda\left(\frac{a}{\alpha} - a^2\right) = 0$$

$$\frac{\partial(f - \lambda g)}{\partial \alpha} = -\frac{1}{2}a^2 + \lambda\frac{a^2}{2\alpha^2} = 0$$

Hence

$$\lambda = \alpha^2$$

$$1 - 2\alpha a + \alpha^2 a^2 = 0$$

so that $a\alpha = 1$. Substituting back into the constraint (10.23) gives

$$S^3 = \frac{1}{6}a^3, \quad \text{or} \quad a = \left(\frac{3kA}{h}\right)^{1/3}$$

and therefore

$$T - T_0 = (T_1 - T_0)\left(1 - \frac{x}{a}\right)$$

so that

$$\frac{y}{a} = \frac{ha}{2k}\left(1 - \frac{x}{a}\right)^2$$

Thus, given the physical parameters k, h and A, the 'best' shape can be derived. This model shows how a very difficult mathematical problem in optimization can be reduced to a much more straightforward one by an appropriate choice of test functions for $T - T_0$.

10.7 Review exercises (1–26)

1 Use the simplex method to find the maximum of the function

$$F = 12x_1 + 8x_2$$

subject to the constraints

$$x_1 + x_2 \leqslant 350$$
$$2x_1 + x_2 \leqslant 600$$
$$x_1 + 3x_2 \leqslant 900$$
$$x_1, x_2 \geqslant 0$$

Check your results with a graphical solution.

2 A motor manufacturer makes a 'standard', a 'super' and a 'deluxe' version of a particular model of car. It is found that each week two of the materials are in limited supply: that of chromium trim being limited to 1600 m per week and that of soundproofing material to 1500 m^2 per week. The quantity of each of these materials required by a car of each type is as follows:

	Standard	Super	Deluxe
Chromium trim (m)	10	20	30
Soundproofing (m^2)	10	15	20

All other materials are in unlimited supply.

The manufacturer knows that any number of standard models can be sold, but it is estimated that the combined market for the super and deluxe versions is limited to 50 models per week. In addition there is a contractual obligation to supply a total of 70 cars (of any type) each week.

The profits on a standard, super and deluxe model are £100, £300 and £400 respectively. Assuming that the facilities to manufacture any number of cars are available, how many of each model should be made to maximize the weekly profit, and what is that profit?

3 A manufacturer makes three types of sailboard and is trying to decide how many of each to make in a given week. There are 400 h of labour available, and the three types of sailboard require respectively 10, 20 and 30 h of labour to construct. A shortage of fibreglass and of resin coating is anticipated. The quantities required by each type of sailboard are as follows:

	Type 1	Type 2	Type 3	Total supplies
Fibreglass (kg)	5	10	25	290
Resin coating (litres)	3	2	1	72

If the profits on types 1, 2 and 3 are £10, £15 and £25 respectively, how many of each type should the manufacturer make to maximize profit?

4 A poor student lives on bread and cheese. The bread contains 1000 calories and 25 g protein in each kilogram, and the cheese has 2000 calories and 100 g of protein per kg. To maintain a good diet, the student requires at least 3000 calories and 100 g of protein per day. Bread costs 60p per kg and cheese 180p per kg. Find the minimum cost of bread and cheese needed per day to maintain the diet.

5 Use Lagrange multipliers to find the maximum and minimum distances from the origin to the point P lying on the curve

$$x^2 - xy + y^2 = 1$$

6 A solid body of volume V and surface area S is formed by joining together two cubes of different sizes so that every point on one side of the smaller cube is in contact with the larger cube. If $S = 7\text{ m}^2$, find the maximum and/or minimum values of V for which both cubes have non-zero volumes.

7 Find the maximum distance from the point (1, 0, 0) to the surface represented by

$$2x + y^2 + z = 8$$

8 Find the local extrema of the function

$$F(x, y, z) = x + 2y + 3z$$

subject to the constraint

$$x^2 + y^2 + z^2 = 14$$

Obtain also the *global* maximum and minimum values of F in the region

$$x \geqslant 0, \quad y \geqslant 0, \quad z \geqslant 0, \quad x^2 + y^2 + z^2 \leqslant 14$$

9 A triangle with sides a, b, c has given perimeter $2s$. Recall that the area of the triangle, A, is given by the formula

$$A^2 = s(s - a)(s - b)(s - c)$$

(i) If a is given, use Lagrange multipliers to find the values of b and c that make the area a maximum.
(ii) If a, b, c are unrestricted use Lagrange multipliers to find the values that make the area a maximum.

10 A nuclear reactor is in the form of a circular cylinder of radius r and height h. According to the theory of nuclear diffusion, the restriction

$$\left(\frac{a}{r}\right)^2 + \left(\frac{\pi}{h}\right)^2 = b$$

applies, where a and b are constants. Use Lagrange multipliers to find the values of r and h that make the volume of the reactor a maximum.

†11 According to lubrication theory, the lift on a pad bearing, where fluid flows in the narrow gap between a pad and a fixed piece of machinery, is given by

$$F = A\frac{1}{(k-1)^2}\left(\ln k - 2\frac{k-1}{k+1}\right)$$

where A is a constant, $k = h_1/h_2 > 1$ and h_1 and h_2 are the gap widths at the front and back of the pad. Find the value of k that makes F a maximum by using the bracket/quadratic approximation technique.

†12 A cylindrical can of radius R (cm) and height H (cm) is to be made with volume 1000 cm^3. The cost of making the can is proportional to

(amount of metal) × (machine factor)

where the amount of metal is proportional to the surface area of the can (including the two ends) and the machine factor is given by $1 + [1 - (H/4R)]^2$ and reflects the difficulty of machining the can. Show that the cost is

$$2\left(\frac{1000}{R} + \pi R^2\right)\left[1 + \left(1 - \frac{1000}{4\pi R^3}\right)^2\right]$$

Find a bracket for R and use the quadratic algorithm to estimate the radius that minimizes the cost.

†13 Use the quadratic approximation method to obtain a first estimate of the minimum of the function

$$f(x) = 1 - t + t^2$$

where t is the non-negative root of

$$t^2 + tx - (1 - x^2) = 0$$

Start with the interval $0 \leqslant x \leqslant 1$ and note that for $x = 0.5$, $t = 0.6514$ and $f = 0.7729$.

†14 In Figure 10.20 the disc rotates at a constant angular velocity, so $\theta = \alpha t$. The subsequent movement of

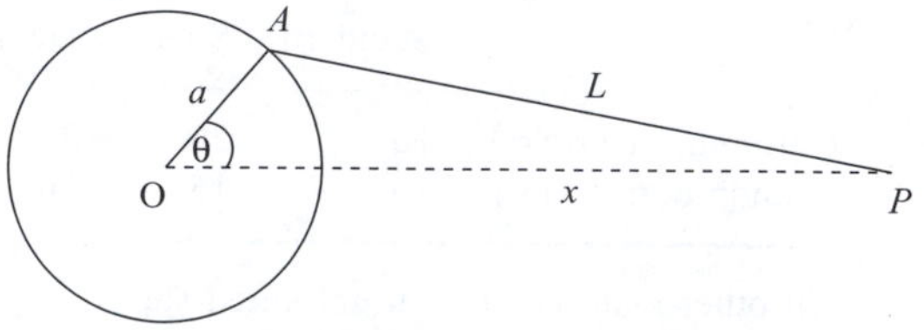

Figure 10.20 Disc and slider in Review exercise 14.

the slider P gives $x = x(t)$. If $L/a = \lambda$ show that the velocity, v, of the slider is given by

$$\frac{v}{\alpha a} = -\sin\theta\left[1 + \frac{\cos\theta}{\sqrt{(\lambda^2 - \sin^2\theta)}}\right]$$

Use the bracket and quadratic approximation technique to evaluate the maximum and minimum velocities of the slider in the case $\lambda = 3$.

†15 A trucking company estimates that the cost of running a truck is

$$0.02\left(2v^{\frac{1}{4}} + \frac{v}{10}\right)$$

pounds per mile at a constant speed v. The driver earns £5 per hour. Find the cost for a journey of D miles. What speed is recommended to minimize the cost?

†16 Use (a) the steepest-descent method, (b) the Newton method and (c) the DFP method to find the position of the minimum of the function

$$f(x, y) = (x - y)^2 + \tfrac{1}{16}(x + y + 1)^4$$

starting at (0, 0). Perform two cycles of each method, and compare your results.

†17 A compound pendulum consists of a rectangular laminar with a heavy particle embedded in it, as illustrated in Figure 10.21. For small oscillations about the equilibrium, $\theta = \alpha$, and putting $\theta = \alpha + \varepsilon$, the equation of motion is

$$\ddot{\varepsilon} = -\mu^2\varepsilon$$

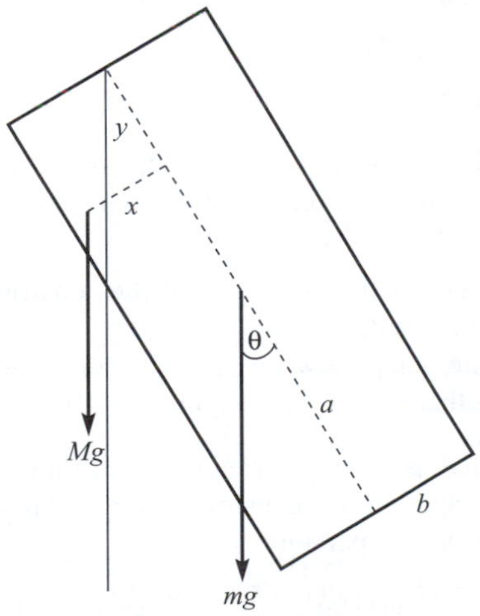

Figure 10.21 Compound pendulum in Review exercise 17.

where the period of the oscillations μ is given, after a substantial calculation, by

$$\frac{a\mu^2}{g} = \frac{\sqrt{[(\lambda + Y)^2 + X^2]}}{X^2 + Y^2 + \lambda(\frac{4}{3} + k^2)}$$

with

$$X = x/a, \quad Y = y/a, \quad k = b/a, \quad \lambda = m/M$$

Explore this expression for maximum and minimum values in the region $|X| \leqslant k$, $Y \leqslant 2$, take the case $\lambda = \frac{1}{2}$ and $k = \frac{1}{4}$.

†18 A method called **Partan** uses the notation $\boldsymbol{D}^{(i)}$ for the gradient of f evaluated at $\boldsymbol{x} = \boldsymbol{x}^{(i)}$. The iteration scheme for evaluating the minimum of f using Partan is

$$\boldsymbol{x}^{(2)} = \boldsymbol{x}^{(1)} - \mu_1\boldsymbol{D}^{(1)}$$

$$\left.\begin{aligned} \boldsymbol{z}^{(i)} &= \boldsymbol{x}^{(i)} - \mu_i\boldsymbol{D}^{(i)} \\ \boldsymbol{x}^{(i+1)} &= \boldsymbol{z}^{(i)} + \lambda_i(\boldsymbol{z}^{(i)} - \boldsymbol{x}^{(i-1)}) \end{aligned}\right\} \quad (i = 2, 3, \dots)$$

where μ_i and λ_i are chosen by optimum line searches. Sketch the progress of this method up to the point $\boldsymbol{x}^{(4)}$ for a scalar function of two variables $f(x_1, x_2)$.

Illustrate the use of Partan and the method of steepest descent on the quadratic function

$$f = (x_1 - x_2)^2 + (x_1 - 1)^2$$

starting at (0, 0).

†19 The Newton method described in Section 10.4.3 often fails to converge. One way of overcoming this problem is to restrict the step length at each iteration. Given that $\boldsymbol{x} = \boldsymbol{a} + \boldsymbol{h}$, this can be implemented by constraining $\boldsymbol{h}$ to have length L, where

$$\boldsymbol{h}^{\mathrm{T}}\boldsymbol{h} = L^2$$

Use a Lagrange multiplier λ to show that the result gives

$$\boldsymbol{x}_{\text{new}} = \boldsymbol{x}_{\text{old}} - (\boldsymbol{J} + \lambda\boldsymbol{I})^{-1}\boldsymbol{G}$$

The algorithm is then implemented by successively using $\lambda = 0, 1, 10, 100, \dots$ until a reduction in the function f is obtained.

Starting at $\boldsymbol{x} = [1 \quad 1]^{\mathrm{T}}$, perform one complete step of the modified algorithm on the function

$$f = x^2 + y^2 - x^2y$$

†20 It is required to use a Newton method to minimize a function F that is a sum of squares

$$F(x_1, x_2, \dots, x_n) = f_1^2(x_1, \dots, x_n) + \dots + f_m^2(x_1, \dots, x_n)$$

where the derivatives of $f_1, f_2, \ldots, f_m$ are available.

In matrix notation we define

$$\boldsymbol{x} = \begin{bmatrix} x_1 \\ \vdots \\ x_n \end{bmatrix}, \quad \boldsymbol{a} = \begin{bmatrix} a_1 \\ \vdots \\ a_n \end{bmatrix}, \quad \boldsymbol{f} = \begin{bmatrix} f_1 \\ \vdots \\ f_m \end{bmatrix}$$

and the matrix of derivatives as

$$\boldsymbol{J} = \begin{bmatrix} \frac{\partial f_1}{\partial x_1} & \cdots & \frac{\partial f_1}{\partial x_n} \\ \vdots & & \vdots \\ \frac{\partial f_m}{\partial x_1} & \cdots & \frac{\partial f_m}{\partial x_n} \end{bmatrix}$$

The functions $f_1, \ldots, f_m$ are replaced by their linear approximation about the point $\boldsymbol{x} = \boldsymbol{a}$, derived from Taylor's theorem. Show that the minimum of F is obtained at

$$\boldsymbol{x} = \boldsymbol{a} - (\boldsymbol{J}^{\mathrm{T}}\boldsymbol{J})^{-1}\boldsymbol{J}^{\mathrm{T}}\boldsymbol{f}$$

where all the terms on the right-hand side are evaluated at $\boldsymbol{x} = \boldsymbol{a}$.

Construct an algorithm to iterate to the minimum based on this idea.

Use the method on the functions

(a) $F = (x - y)^2 + \frac{1}{16}(x + y + 1)^4$, starting at $x = 0$, $y = 0$;

(b) $F = \left(\dfrac{1}{x+y}\right)^2 + \left(\dfrac{x}{1+2x+y}\right)^2$, starting at $x = 1$, $y = 1$.

†21 It is known from experience that a curve of the form

$$y = 1/(a + bx)$$

should give a good fit to experimental data in the form of a set of points

$$(x_i, y_i) \quad (i = 1, 2, \ldots, p)$$

It is required to calculate a and b by a best least-squares fit, and thus to minimize

$$F(a, b) = \sum_{i=1}^{p}\left(y_i - \frac{1}{a + bx_i}\right)^2$$

Use the least-squares algorithm described in Review exercise 20 to fit the function to the data points

$$(0, 1) \quad (1, 0.6), \quad (2, 0.3), \quad (3, 0.2)$$

22 A quadratic function $f(x_1, x_2, \ldots, x_n)$ with a unique minimum is given in matrix form as

$$f = c + \boldsymbol{b}^{\mathrm{T}}\boldsymbol{x} + \tfrac{1}{2}\boldsymbol{x}^{\mathrm{T}}\boldsymbol{A}\boldsymbol{x}$$

Show that a search in the direction

$$\boldsymbol{x} = \boldsymbol{a} + \lambda\boldsymbol{d}$$

produces the minimum at

$$\lambda_{\min} = \frac{-(\boldsymbol{b} + \boldsymbol{A}\boldsymbol{a})^{\mathrm{T}}\boldsymbol{d}}{\boldsymbol{d}^{\mathrm{T}}\boldsymbol{A}\boldsymbol{d}}$$

23 Complete two complete cycles of the steepest-descent algorithm for the function

$$f(x, y) = (1 - x)^2 + (x - y)^2$$

starting at $x = 0$, $y = 0$. Use Review exercise 22 for the minimization in the search directions.

Show that the minimum is obtained in a single iteration of the Newton method.

†24 (Harder) It is required to solve the differential equation

$$yy'' - y'^2 + y' = 0$$

with the boundary conditions $y(0) = 1$ and $y(1) = 3$, by a shooting method. The equation is solved for the initial conditions

$$y(0) = 1, \quad y'(0) = \alpha$$

by any suitable method (for example by a Runge–Kutta method). With this solution, calculate

$$F(\alpha) = y(1)$$

and then try to drive F to the value 3 by minimization of

$$[F(\alpha) - 3]^2$$

In this example illustrate the method by using the exact solution

$$y = (1 + b)\,\mathrm{e}^{x/b} - b$$

for the forward integration.

25 (An extended problem) In the chemical processing plant model in Section 10.5 consider the profits when

(a) the primary pump is faulty and the constraint $x_5 \leqslant 12$ is imposed;

(b) the waste pump between primary and secondary fails so that $x_{13} = 0$ (see Figure 10.18).

26 (An extended problem) Extend the heating fin analysis in Section 10.6, using a higher approximation to the temperature:

$$T - T_0 = (T_1 - T_0)(1 - \alpha x - \beta x^2)$$

Compare the shape of the fin with that given in the text, and compute the heat transferred in each case.

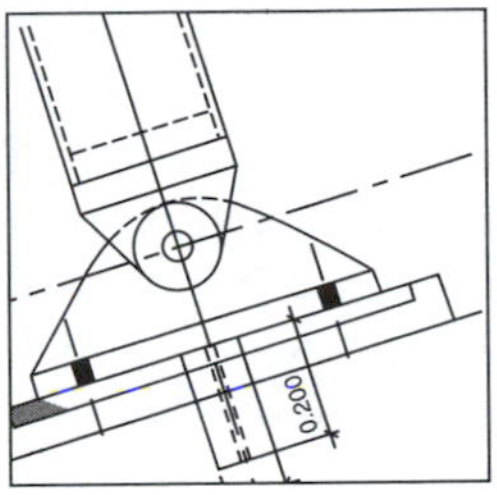

11

Applied Probability and Statistics

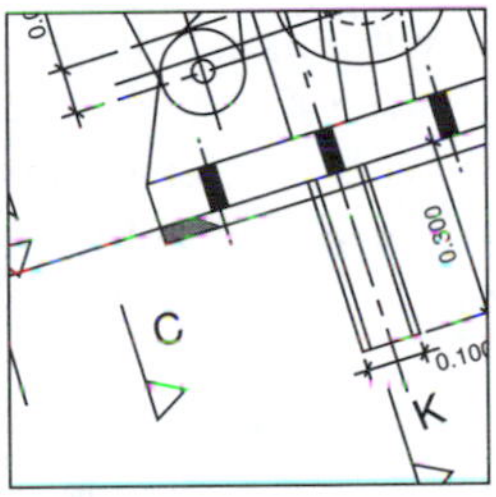

CONTENTS

11.1 Introduction

Applications of probability and statistics in engineering are very far-reaching. Data from experiments have to be analysed and conclusions drawn, decisions have to be made, production and distribution have to be organized and monitored, and quality has to be controlled. In all of these activities probability and statistics have at least a supporting role to play, and sometimes a central one.

The distinction between applied probability and statistics is blurred, but essentially it is this: **applied probability** is about mathematical modelling of a situation that involves random uncertainty, whereas **statistics** is the business of handling data and drawing conclusions, and can be regarded as a branch of applied probability. Most of this chapter is about statistics, but Section 11.10 on queueing theory is applied probability.

When applying statistical methods to a practical problem, the most visible activity is the processing of data, using either a hand calculator or, increasingly often, a computer statistical package. Either way, a formula or standard procedure from a textbook is being applied to the data. The relative ease and obviousness of this activity sometimes leads to a false sense that there is nothing more to it. On the contrary, the handling of the data (by whatever means) is quite superficial compared with the essential task of trying to understand both the problem at hand and the assumptions upon which the various statistical procedures are based. If the wrong procedure is chosen, a wrong conclusion may be drawn.

It is, unfortunately, all too easy to use a formula while overlooking its theoretical basis, which largely determines its applicability. It is true that there are some statistical methods that continue to work reliably even where the assumptions upon which they are based do not hold (such methods are called **robust**), but it is unwise to rely too heavily upon this and even worse to be unaware of the assumptions at all.

The conclusions of a statistical analysis are often expressed in a qualified way such as 'We can be 95% sure that . . .'. At first this seems vague and inadequate. Perhaps a decision has to be made, but the statistical conclusion is not expressed simply as 'yes' or 'no'. A statistical analysis is rather like a legal case in which the witness is required to tell 'the whole truth and nothing but the truth'. In the present context 'the whole truth' means that the statistician must glean as much information from the data as is possible until nothing but pure randomness remains. 'Nothing but the truth' means that the statistician must not state the conclusion with any greater degree of certainty or confidence than is justified by the analysis. In fact there is a practical compromise between truth and precision that will be explained in Section 11.3.3. The result of all this is that the decision-maker is aided by the analysis but not pre-empted by it.

In this chapter we shall first review the basic theory of probability and then cover some applications that are beneficial in engineering and many other fields: the statistics of means, proportions and correlation, linear regression and goodness-of-fit testing, queueing theory and quality control.

11.2 Review of basic probability theory

This section contains an overview of the basic theory used in the remainder of this chapter. No attempt is made to explain or justify the ideas or results: a full account can be found in Chapter 12 of *Modern Engineering Mathematics* or elsewhere. For the same reason there are no examples or exercises. In the process of reviewing the basic theory, this section also establishes the pattern of notation used throughout the chapter, which follows standard conventions as far as possible. No reader should embark on this chapter without having a fairly thorough understanding of the material in this section.

11.2.1 The rules of probability

We associate a probability $P(A)$ with an **event** A, which in general is a subset of a **sample space** S. The usual set-theoretic operations apply to the events (subsets) in S, and there are corresponding rules that must be satisfied by the probabilities.

Complement rule

$$P(S - A) = 1 - P(A)$$

The **complement** of an event A is often written as $\bar{A}$.

Addition rule

$$P(A \cup B) = P(A) + P(B) - P(A \cap B)$$

For **disjoint** events, $A \cap B = \varnothing$, and the addition rule takes the simple form

$$P(A \cup B) = P(A) + P(B)$$

Product rule

$$P(A \cap B) = P(A)P(B \mid A)$$

This is actually the definition of the **conditional probability** $P(B \mid A)$ of B given A. If A and B are **independent** then the product rule takes the simple form

$$P(A \cap B) = P(A)P(B)$$

11.2.2 Random variables

A **random variable** has a sample space of possible numerical values together with a **distribution** of probabilities. Random variables can be either **discrete** or **continuous**. For a discrete random variable (X, say) the possible values can be written as a list $\{v_1, v_2, v_3, \dots\}$ with corresponding probabilities $P(X = v_1)$, $P(X = v_2)$, $P(X = v_3)$, . . . The **mean** of X is then defined as

$$\mu_X = \sum_k v_k P(X = v_k)$$

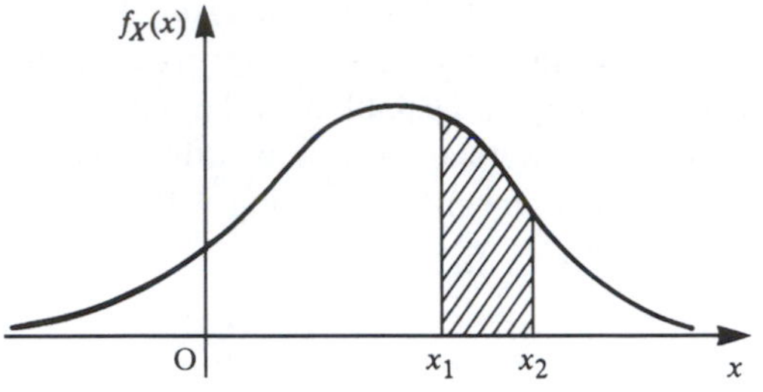

Figure 11.1 Probability of interval from density function.

(sum over all possible values), and is a measure of the central location of the distribution. The **variance** of X is defined as

$$\text{Var}(X) = \sigma_X^2 = \sum_k (v_k - \mu_X)^2 P(X = v_k)$$

and is a measure of dispersion of the distribution about the mean. The symbols μ and σ^2 are conventional for these quantities. In general, the **expected value** of a function $h(X)$ of X is defined as

$$E\{h(X)\} = \sum_k h(v_k) P(X = v_k)$$

of which the mean and variance are special cases. The **standard deviation** σ_X is the square root of the variance.

For a **continuous** random variable (X, say), there is a **probability density function** $f_X(x)$ and a **cumulative distribution function** $F_X(x)$. The cumulative distribution function is defined as

$$F_X(x) = P(X \leqslant x)$$

and is the indefinite integral of the density function:

$$F_X(x) = \int_{-\infty}^{x} f_X(t)\,\mathrm{d}t$$

These functions determine the probabilities of events for the random variable. The probability that the variable X takes a value within the real interval (x_1, x_2) is the area under the density function over that interval, or equivalently the difference in values of the distribution function at its ends:

$$P(x_1 < X < x_2) = \int_{x_1}^{x_2} f_X(t)\,\mathrm{d}t = F_X(x_2) - F_X(x_1)$$

(see Figure 11.1). Note that the events $x_1 < X < x_2$, $x_1 \leqslant X < x_2$, $x_1 < X \leqslant x_2$ and $x_1 \leqslant X \leqslant x_2$ are all equivalent in probability terms for a continuous random variable X because the probability of X being exactly equal to either x_1 or x_2 is zero. The mean and variance of X, and the expected value of a function $h(X)$, are defined in terms of the density function by

$$\mu_X = \int_{-\infty}^{\infty} x f_X(x)\,dx$$

$$\text{Var}(X) = \sigma_X^2 = \int_{-\infty}^{\infty} (x - \mu_X)^2 f_X(x)\,dx$$

$$E\{h(X)\} = \int_{-\infty}^{\infty} h(x) f_X(x)\,dx$$

These definitions assume that the random variable is defined for values of x from $-\infty$ to ∞. If the random variable is defined in general for values in some real interval, say (a, b), then the domain of integration can be restricted to that interval, or alternatively the density function can be defined to be zero outside that interval.

Just as events can be independent (and then obey a simple product rule of probabilities), so can random variables be independent. We shall consider this in more detail in Section 11.4. Means and variances of random variables (whether discrete or continuous) have the following important properties (X and Y are random variables, and c is an arbitrary constant):

$$E(cX) = cE(X) = c\mu_X$$

$$\text{Var}(cX) = c^2\text{Var}(X) = c^2\sigma_X^2$$

$$E(X + c) = E(X) + c = \mu_X + c$$

$$\text{Var}(X + c) = \text{Var}(X) = \sigma_X^2$$

$$E(X + Y) = E(X) + E(Y) = \mu_X + \mu_Y$$

(this applies whether or not X and Y are independent),

$$\text{Var}(X + Y) = \text{Var}(X) + \text{Var}(Y) = \sigma_X^2 + \sigma_Y^2$$

(this applies only when X and Y are independent).

It is also useful to note that $\text{Var}(X) = E(X^2) - [E(X)]^2$.

11.2.3 The Bernoulli, binomial and Poisson distributions

The simplest example of a discrete distribution is the **Bernoulli distribution**. This has just two values: $X = 1$ with probability p and $X = 0$ with probability $1 - p$, from which the mean and variance are p and $p(1 - p)$ respectively.

The binomial and Poisson distributions are families of discrete distributions whose probabilities are generated by formulae, and which arise in many real situations. The **binomial distribution** governs the number (X, say) of 'successes' in n independent 'trials', with a probability p of 'success' at each trial:

$$P(X = k) = \binom{n}{k} p^k (1-p)^{n-k}$$

where the range of possible values (k) is $\{0, 1, 2, \ldots, n\}$. The binomial distribution can be thought of as the sum of n independent Bernoulli random variables. This distribution (more properly, *family* of distributions) has two **parameters**, n and p. In terms of these parameters, the mean and variance are

$$\mu_X = np$$

$$\sigma_X^2 = np(1 - p)$$

The **Poisson distribution** is defined as

$$P(X = k) = \frac{\lambda^k e^{-\lambda}}{k!}$$

where the range of possible values (k) is the set of non-negative integers $\{0, 1, 2, \ldots\}$. This has mean and variance both equal to the single parameter λ (see Section 11.7.1), and, by setting $\lambda = np$, provides a useful approximation to the binomial distribution that works when n is large and p is small (see Section 10.7.2). As a guide, the approximation can be used when $n \geqslant 25$ and $p \leqslant 0.1$. The Poisson distribution has many other uses, as will be seen in Section 11.10.

11.2.4 The normal distribution

This is a family of continuous distributions with probability density function given by

$$f_X(x) = \frac{1}{\sigma_X \sqrt{(2\pi)}} \exp\left[-\frac{1}{2}\left(\frac{x - \mu_X}{\sigma_X}\right)^2\right]$$

for $-\infty < x < +\infty$, where the parameters μ_X and σ_X are the mean and standard deviation of the distribution. It is conventional to denote the fact that a random variable X has a normal distribution by

$$X \sim N(\mu_X, \sigma_X^2)$$

The **standard normal distribution** is a special case with zero mean and unit variance, often denoted by Z:

$$Z \sim N(0, 1)$$

Tables of the standard normal cumulative distribution function

$$\Phi(z) = P(Z \leqslant z) = \frac{1}{\sqrt{(2\pi)}} \int_{-\infty}^{z} e^{-t^2/2} \, dt$$

are widely available (see for example Figure 11.2). These tables can be used for probability calculations involving arbitrary normal random variables. For example, if $X \sim N(\mu_X, \sigma_X^2)$ then

z	.00	.01	.02	.03	.04	.05	.06	.07	.08	.09
.0	.5000	.5040	.5080	.5120	.5160	.5199	.5239	.5279	.5319	.5359
.1	.5398	.5438	.5478	.5517	.5557	.5596	.5636	.5675	.5714	.5753
.2	.5793	.5832	.5871	.5910	.5948	.5987	.6026	.6064	.6103	.6141
.3	.6179	.6217	.6255	.6293	.6331	.6368	.6406	.6443	.6480	.6517
.4	.6554	.6591	.6628	.6664	.6700	.6736	.6772	.6808	.6844	.6879
.5	.6915	.6950	.6985	.7019	.7054	.7088	.7123	.7157	.7190	.7224
.6	.7257	.7291	.7324	.7357	.7389	.7422	.7454	.7486	.7517	.7549
.7	.7580	.7611	.7642	.7673	.7704	.7734	.7764	.7794	.7823	.7852
.8	.7881	.7910	.7939	.7967	.7995	.8023	.8051	.8078	.8106	.8133
.9	.8159	.8186	.8212	.8238	.8264	.8289	.8315	.8340	.8365	.8389
1.0	.8413	.8438	.8461	.8485	.8508	.8531	.8554	.8577	.8599	.8621
1.1	.8643	.8665	.8686	.8708	.8729	.8749	.8770	.8790	.8810	.8830
1.2	.8849	.8869	.8888	.8907	.8925	.8944	.8962	.8980	.8997	.9015
1.3	.9032	.9049	.9066	.9082	.9099	.9115	.9131	.9147	.9162	.9177
1.4	.9192	.9207	.9222	.9236	.9251	.9265	.9279	.9292	.9306	.9319
1.5	.9332	.9345	.9357	.9370	.9382	.9394	.9406	.9418	.9429	.9441
1.6	.9452	.9463	.9474	.9484	.9495	.9505	.9515	.9525	.9535	.9545
1.7	.9554	.9564	.9573	.9582	.9591	.9599	.9608	.9616	.9625	.9633
1.8	.9641	.9649	.9656	.9664	.9671	.9678	.9686	.9693	.9699	.9706
1.9	.9713	.9719	.9726	.9732	.9738	.9744	.9750	.9756	.9761	.9767
2.0	.9772	.9778	.9783	.9788	.9793	.9798	.9803	.9808	.9812	.9817
2.1	.9821	.9826	.9830	.9834	.9838	.9842	.9846	.9850	.9854	.9857
2.2	.9861	.9864	.9868	.9871	.9875	.9878	.9881	.9884	.9887	.9890
2.3	.9893	.9896	.9898	.9901	.9904	.9906	.9909	.9911	.9913	.9916
2.4	.9918	.9920	.9922	.9925	.9927	.9929	.9931	.9932	.9934	.9936
2.5	.9938	.9940	.9941	.9943	.9945	.9946	.9948	.9949	.9951	.9952
2.6	.9953	.9955	.9956	.9957	.9959	.9960	.9961	.9962	.9963	.9964
2.7	.9965	.9966	.9967	.9968	.9969	.9970	.9971	.9972	.9973	.9974
2.8	.9974	.9975	.9976	.9977	.9977	.9978	.9979	.9979	.9980	.9981
2.9	.9981	.9982	.9982	.9983	.9984	.9984	.9985	.9985	.9986	.9986
3.0	.9987	.9987	.9987	.9988	.9988	.9989	.9989	.9989	.9990	.9990
3.1	.9990	.9991	.9991	.9991	.9992	.9992	.9992	.9992	.9993	.9993
3.2	.9993	.9993	.9994	.9994	.9994	.9994	.9994	.9995	.9995	.9995
3.3	.9995	.9995	.9995	.9996	.9996	.9996	.9996	.9996	.9996	.9997
3.4	.9997	.9997	.9997	.9997	.9997	.9997	.9997	.9997	.9997	.9998

z	1.282	1.645	1.960	2.326	2.576	3.090	3.291	3.891	4.417
$\Phi(z)$	.90	.95	.975	.99	.995	.999	.9995	.999 95	.999 995
$2[1-\Phi(z)]$	.20	.10	.05	.02	.01	.002	.001	.000 1	.000 01

Figure 11.2 Table of the standard normal cumulative distribution function $\Phi(z)$.

$$P(X \leqslant a) = P\left(\frac{X-\mu_X}{\sigma_X} \leqslant \frac{a-\mu_X}{\sigma_X}\right) = \Phi\left(\frac{a-\mu_X}{\sigma_X}\right)$$

The key result for applications of the normal distribution is the **central limit theorem**: if $\{X_1, X_2, X_3, \ldots, X_n\}$ are independent and identically distributed random variables (the distribution being arbitrary), each with mean μ_X and variance σ_X^2, and if

$$W_n = \frac{X_1 + \ldots + X_n}{n}, \quad Z_n = \frac{X_1 + \ldots + X_n - n\mu_X}{\sigma_X \sqrt{n}}$$

then, as $n \to \infty$, the distributions of W_n and Z_n tend to $W_n \sim N(\mu_X, \sigma_X^2/n)$ and $Z_n \sim N(0, 1)$ respectively. Loosely speaking, the sum of independent identically distributed random variables tends to a normal distribution.

This theorem is proved in Section 11.7.3, and in the key to many statistical processes, some of which are described in Section 11.3. One corollary is that the normal distribution can be used to approximate the binomial distribution when n is sufficiently large: if X is binomial with parameters n and p then the approximating distribution (by equating the means and variances) is $Y \sim N(np, np(1-p))$. This is explained (together with the important **continuity correction**) in Section 12.5.5 of *Modern Engineering Mathematics*, and the approximation is used as follows:

$$P(X \leqslant k) \simeq \Phi\left(\frac{k+0.5-np}{\sqrt{[np(1-p)]}}\right)$$

$$P(X = k) \simeq \Phi\left(\frac{k+0.5-np}{\sqrt{[np(1-p)]}}\right) - \Phi\left(\frac{k-0.5-np}{\sqrt{[np(1-p)]}}\right)$$

As a guide, the approximation can be used when $n \geqslant 25$ and $0.1 \leqslant p \leqslant 0.9$.

11.2.5 Sample measures

It is conventional to denote a random variable by an upper-case letter (X, say), and an actual observed value of it by the corresponding lower-case letter (x, say). An observed value x will be one of the set of possible values (sample space) for the random variable, which for a discrete random variable may be written as a list of the form $\{v_1, v_2, v_3, \ldots\}$. It is possible to observe a random variable many times (say n times) and obtain a series of values. In this case we assume that the random variable X refers to a **population** (whose characteristics may be unknown), and the series of random variables $\{X_1, X_2, \ldots, X_n\}$ as a **sample**. Each X_i is assumed to have the characteristics of the population, so they all have the same distribution. The actual series of values $\{x_1, x_2, \ldots, x_n\}$ consists of data upon which we can work, but it is useful to define certain sample measures in terms of the random variables $\{X_1, X_2, \ldots, X_n\}$ in order to interpret the data. Principal among these measures are the **sample average** and **sample variance**, defined as

$$\bar{X} = \frac{1}{n}\sum_{i=1}^{n} X_i, \quad S_X^2 = \frac{1}{n}\sum_{i=1}^{n}(X_i - \bar{X})^2$$

respectively, and it is useful to note that the sample variance is the average of the squares minus the square of the average:

$$S_X^2 = \overline{X^2} - (\bar{X})^2$$

We shall also need the following alternative definition of sample variance in Section 11.3.5:

$$S_{X,n-1}^2 = \frac{1}{n-1}\sum_{i=1}^{n}(X_i - \bar{X})^2$$

We can use the properties of means and variances (summarized in Section 11.2.2) to find the mean and variance of the sample average as follows:

$$E(\bar{X}) = \frac{1}{n}E(X_1 + \ldots + X_n) = \frac{1}{n}[E(X_1) + \ldots + E(X_n)]$$

$$= \frac{n\mu_X}{n} = \mu_X$$

$$\mathrm{Var}(\bar{X}) = \frac{1}{n^2}\mathrm{Var}(X_1 + \ldots + X_n) = \frac{1}{n^2}[\mathrm{Var}(X_1) + \ldots + \mathrm{Var}(X_n)]$$

$$= \frac{n\sigma_X^2}{n^2} = \frac{\sigma_X^2}{n}$$

Here we are assuming that the population mean and variance are μ_X and σ_X^2 respectively (which may be unknown values in practice), and that the observations of the random variables X_i are *independent*, a very important requirement in statistics.

11.3 Estimating parameters

11.3.1 Interval estimates and hypothesis tests

The first step in statistics is to take some data from an experiment and make inferences about the values of certain parameters. Such parameters could be the mean and variance of a population, or the correlation between two variables for a population. The data are never sufficient to determine the values exactly, but two kinds of inferences can be made:

(a) a range of values can be quoted, within which it is believed with high probability that the population parameter value lies, or

(b) a decision can be made as to whether or not the data are compatible with a particular value of the parameter.

The first of these is called **interval estimation**, and provides an assessment of the value that is rather more honest than merely quoting a single number derived from the sample data, which may be more or less uncertain depending upon the sample size. The second approach is called **hypothesis testing** and allows a value of particular interest to be assessed. These two approaches are usually covered in separate chapters in introductory textbooks on statistics, but they are closely related and are often used in conjunction with each other. Tests of simple hypotheses about parameter values will therefore be covered here within the context of interval estimation.

11.3.2 Distribution of the sample average

Suppose that a clearly identified population has a numerical characteristic with an unknown mean value, such as the mean lifetime for a kind of electronic component or the mean salary for a job category. A natural way to estimate this unknown mean is to take a sample from the population, measure the appropriate characteristic, and find the average value. If the sample size is n and the measured values are $\{x_1, x_2, \ldots, x_n\}$ then the average value

$$\bar{x} = \frac{1}{n}\sum_{i=1}^{n} x_i$$

is a reasonable estimate of the population mean μ_X provided that the sample is *representative* and *independent*, and the size n is sufficiently large.

We can be more precise about how useful this estimate is if we treat the sample average as a random variable. Now we have a sample $\{X_1, X_2, \ldots, X_n\}$ with average

$$\bar{X} = \frac{1}{n}\sum_{i=1}^{n} X_i$$

and the mean and variance of $\bar{X}$ are given by

$$E(\bar{X}) = \mu_X, \quad \text{Var}(\bar{X}) = \frac{\sigma_X^2}{n}$$

(see Section 11.2.5). This shows that the expected value of the average is indeed equal to the population mean, and that the variance is smaller for larger samples. However, we can go further. The central limit theorem (Section 11.2.4) tells us that sums of identical random variables tend to have a normal distribution regardless of the distribution of the variables themselves. The only requirement is that a sufficient number of variables contribute to the sum (the actual number required depends very much on the shape of the underlying distribution).

The sample average is a sum of random variables, and therefore has (approximately) a normal distribution for a sufficiently large sample:

$$\bar{X} \sim N(\mu_X, \sigma_X^2/n)$$

This allows us to use a general method of inference concerning means instead of a separate method for each underlying distribution – even if this were known, which is usually not the case. In practice, a sample size of 25 or more is usually sufficient for the normal approximation.

EXAMPLE 11.1 For all children taking an examination, the mean mark was 60%, with a standard deviation of 8%. A particular class of 30 children achieved an average of 63%. Is this unusual?

Solution The average of 63% is higher than the mean, but not by very much. We do not know the true distribution of marks, but the sample average has (approximately) a normal distribution. We can test the idea that this particular class result is a fluke by reducing the sample average to a standard normal in the manner described in Section 11.2.4 and checking its value against the table of the cumulative distribution function $\Phi(z)$ (Figure 11.2):

$$P(\bar{X} \geqslant 63) = P\left(\frac{\bar{X} - 60}{8/\sqrt{30}} \geqslant \frac{63 - 60}{8/\sqrt{30}}\right) = 1 - \Phi(2.054) = 0.020$$

It is unlikely (one chance in 50) that an average as high as this could occur by chance, assuming that the ability of the class is typical. Figure 11.3 illustrates that the result is towards the tail of the distribution. It therefore seems that this class is unusually successful.

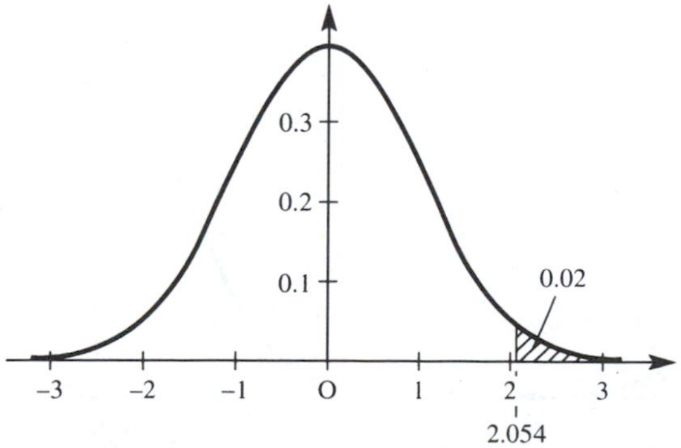

Figure 11.3 Normal density function for Example 11.1.

11.3.3 Confidence interval for the mean

A useful notation will be introduced here. For the standard normal distribution, define z_α to be the point on the z axis for which the area under the density function to its right is equal to α:

$$P(Z > z_\alpha) = \alpha$$

or equivalently

$$\Phi(z_\alpha) = 1 - \alpha$$

(see Figure 11.4a). From the standard normal table we have $z_{0.05} = 1.645$ and $z_{0.025} = 1.96$. By symmetry

$$P(-z_{\alpha/2} < Z < z_{\alpha/2}) = 1 - \alpha$$

(see Figure 11.4b). Assuming normality of the sample average, we have

$$P\left(-z_{\alpha/2} < \frac{\bar{X} - \mu_X}{\sigma_X/\sqrt{n}} < z_{\alpha/2}\right) = 1 - \alpha$$

which, after multiplying through the inequality by $\sigma_X/\sqrt{n}$ and changing the sign, gives

$$P\left(-z_{\alpha/2}\frac{\sigma_X}{\sqrt{n}} < \mu_X - \bar{X} < z_{\alpha/2}\frac{\sigma_X}{\sqrt{n}}\right) = 1 - \alpha$$

so that

$$P\left(\bar{X} - z_{\alpha/2}\frac{\sigma_X}{\sqrt{n}} < \mu_X < \bar{X} + z_{\alpha/2}\frac{\sigma_X}{\sqrt{n}}\right) = 1 - \alpha$$

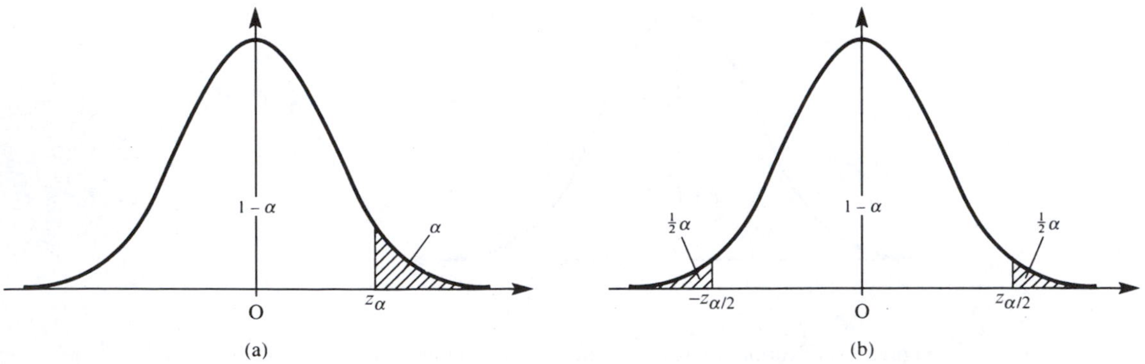

Figure 11.4 Normal density functions with (a) z_α and (b) $z_{\alpha/2}$.

Assume for now that the standard deviation of X is known (it is actually very rare for σ_X to be known when μ_X is unknown, but we shall discuss this case first for simplicity and later consider the more general situation where both μ_X and σ_X are unknown).

The interval defined by $(\bar{X} \pm z_{\alpha/2}\sigma_X/\sqrt{n})$ is called a **100(1 − α)% confidence interval for the mean**, with variance known. If a value for α is specified, the upper and lower limits of this interval can be calculated from the sample average. The probability is $1 - \alpha$ that the true mean lies between them.

EXAMPLE 11.2 The temperature (in degrees Celsius) at ten points chosen at random in a large building is measured, giving the following list of readings:

$$\{18^\circ, 16.5^\circ, 17.5^\circ, 18^\circ, 19.5^\circ, 16.5^\circ, 18^\circ, 17^\circ, 19^\circ, 17.5^\circ\}$$

The standard deviation of temperature through the building is known from past experience to be 1°C. Find a 90% confidence interval for the mean temperature in the building.

Solution The average of the ten readings is 17.75°C, and, using $z_{0.05} = 1.645$, the 90% confidence interval is

$$(17.75 \pm 1.645(1/\sqrt{10})) = (17.1, 18.3)$$

The confidence interval is used to indicate the degree of uncertainty in the sample average. The simplicity of the calculation is deceptive because the idea is very important and easily misunderstood. It is not the mean that is random but rather the interval that would enclose it $100(1 - \alpha)$% of the times the experiment is performed. It is tempting to think of the interval as fixed by the experiment and the mean as a random variable that has a probability $1 - \alpha$ of lying within it, but this is not correct.

Typical values of α are 0.1, 0.05 and 0.01, giving 90%, 95% and 99% confidence intervals respectively. The value chosen is a compromise between truth and precision, as illustrated in Figure 11.5. A statement saying that the mean lies

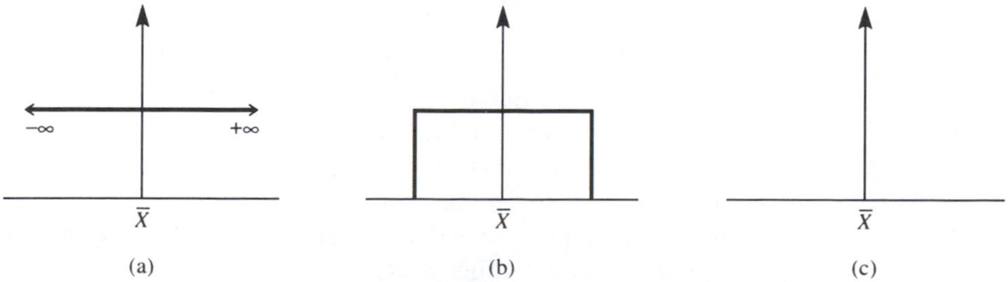

Figure 11.5 Confidence intervals: (a) infinite interval; (b) finite interval; (c) point value.

within the interval $(-\infty, \infty)$ is 100% true (certain to be the case), but totally uninformative because of its total imprecision. None of the possible values is ruled out. On the other hand, saying that the mean equals the exact value given by the sample average is maximally precise, but again of limited value because the statement is false – or rather the probability of its truth is zero. A statement quoting a finite interval for the mean has a probability of being true, chosen to be quite high, and at the same time it rules out most of the possible values and therefore is highly informative. The higher the probability of truth, the lower the informativeness, and vice versa.

The width of the interval also depends on the sample size. A larger experiment yields a more precise result. If figures for the confidence $1 - \alpha$ and precision (width of the interval) are specified in advance then the sample size can be chosen sufficiently large to satisfy these requirements. In some experimental situations (for example destructive testing) there are incentives to keep sample sizes as small as possible. The experimenter must weigh up these conflicting objectives and design the experiment accordingly.

EXAMPLE 11.3 A machine fills cartons of liquid; the mean fill is adjustable but the dial on the gauge is not very accurate. The standard deviation of the quantity of fill is 6 ml. A sample of 30 cartons gave a measured average content of 570 ml. Find 90% and 95% confidence intervals for the mean.

Solution Using $\alpha = 0.05$ and $z_{0.025} = 1.96$, the 95% confidence interval is

$$(570 \pm 1.960(6/\surd 30)) = (567.8, 572.1)$$

Likewise, using $\alpha = 0.1$ and $z_{0.05} = 1.645$, the 90% confidence interval is

$$(570 \pm 1.645(6/\surd 30)) = (568.2, 571.8)$$

As expected, the 95% interval is slightly wider.

11.3.4 Testing simple hypotheses

As explained in Section 11.3.1, the testing of hypotheses about parameter values is complementary to the estimation process involving an interval. A 'simple' hypothesis is one that specifies a particular value for the parameter, as opposed to an interval, and it is this type that we shall consider. The following remarks apply generally to parameter hypothesis testing, but will be directed in particular to hypotheses concerning means.

There are two kinds of errors that can occur when testing hypotheses:

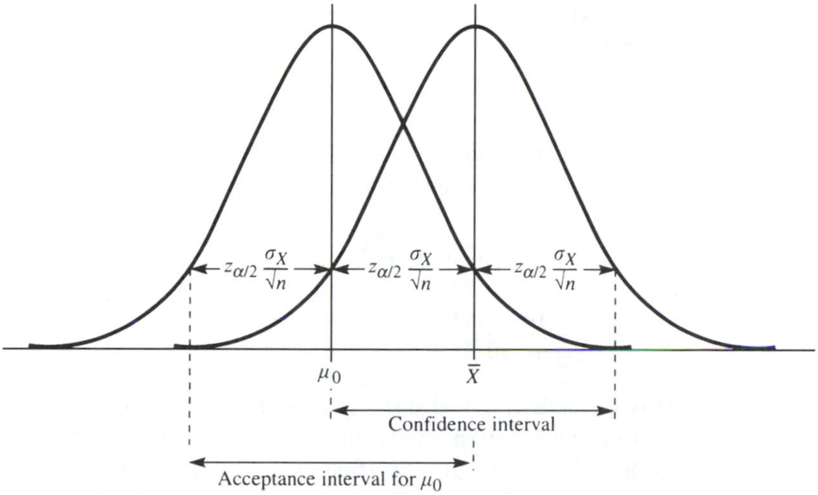

Figure 11.6 Confidence interval and hypothesis test.

(a) a true hypothesis can be rejected (this is usually referred to as a **type I error**), or

(b) a false hypothesis can be accepted (this is usually called a **type II error**).

In reality, all hypotheses that prescribe particular values for parameters are false, but they may be approximately true and rejection may be the result of an experimental fluke. This is the sense in which a type I error can occur. Any hypothesis will be rejected if the sample size is large enough. Acceptance really means that there is insufficient evidence to reject the hypothesis, but this is not an entirely negative view because if the hypothesis has survived the test then it has some degree of dependability.

Normally a simple hypothesis is tested by evaluating a **test statistic**, a quantity that depends upon the sample and leads to rejection of the hypothesized parameter value if its magnitude exceeds a certain threshold. If the hypothesized mean is μ_0 then the test statistic for the mean is

$$Z = \frac{\bar{X} - \mu_0}{\sigma_X / \sqrt{n}}$$

with the hypothesis 'rejected at significance level α' if $|Z| > z_{\alpha/2}$.

The **significance level** can be regarded as the probability of false rejection, an error of type I. If the hypothesis is true then Z has a standard normal distribution and the probability that it will exceed $z_{\alpha/2}$ in magnitude is α. If Z does exceed this value then either the hypothesis is wrong or else a rare event has occurred. It is easy to show that the test statistic lies on this threshold (for significance level α) exactly when the hypothesized value lies at one or other extreme of the $100(1 - \alpha)\%$ confidence interval (see Figure 11.6). An alternative way to test the hypothesis is therefore to see whether or not the value lies within the confidence interval.

EXAMPLE 11.4 For the situation described in Example 11.3 test the hypothesis that the mean fill of liquid is 568 ml (one imperial pint).

Solution The value of the test statistic is

$$Z = \frac{570 - 568}{6/\sqrt{30}} = 1.83$$

This exceeds $z_{0.05} = 1.645$ (10% significance), but is less than $z_{0.025} = 1.96$ (5% significance). Alternatively, the quoted figure lies within the 95% confidence interval but outside the 90% confidence interval. Either way, the hypothesis is rejected at the 10% significance level but accepted at the 5% level. If the actual mean is 568 ml then there is less than one chance in 10 (but more than one in 20) that a result as extreme as 570 ml will be obtained. It looks as though the true mean is larger than the intended value, but the evidence is not particularly strong. The probability of false rejection (type I error) is somewhere between 5% and 10%, which is small but not negligible.

Examples 11.3 and 11.4 set the pattern for the interpretation and use of confidence intervals. We shall now see how to apply these ideas more generally.

11.3.5 Other confidence intervals and tests concerning means

Mean when variance is unknown

With the basic ideas of interval estimation and hypothesis testing established, it is relatively easy to cover other cases. The first and most obvious is to remove the assumption that the variance is known. If the sample size is large then there is essentially no problem, because the sample standard deviation $S_{X,n-1}$ can be used in place of σ_X in the confidence interval, where

$$S_{X,n-1}^2 = \frac{1}{n-1}\sum_{i=1}^{n}(X_i - \bar{X})^2$$

This definition was introduced in Section 11.2.5. Note that the sum is divided by $n - 1$ rather than n. For a large sample this makes little difference, but for a small sample this form must be used because the 't distribution' requires it.

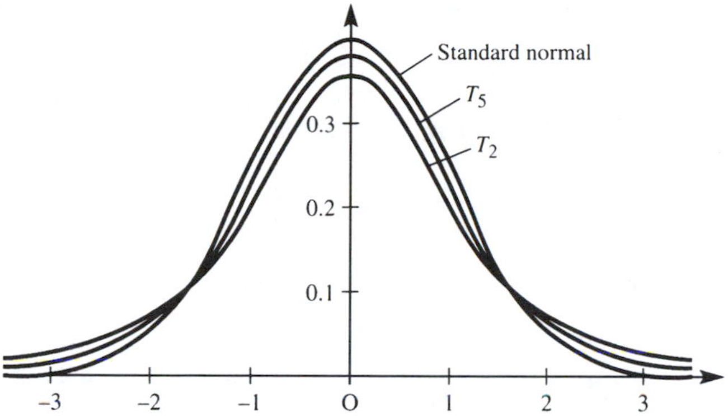

Figure 11.7 Density functions of T_n and z.

Suppose that the sample size is small, say less than 25. Using $S_{X,n-1}$ in place of σ_X adds an extra uncertainty because this estimate is itself subject to error. Furthermore, the central limit theorem cannot be relied upon to ensure that the sample average has a normal distribution. We have to assume that the data themselves are normal. In this situation the random variable

$$T_n = \frac{\bar{X} - \mu_X}{S_{X,n-1}/\sqrt{n}}$$

has a ***t* distribution** with parameter $n - 1$. This distribution resembles the normal distribution, as can be seen in Figure 11.7, which shows the density functions of T_2 and T_5 together with that of the standard normal distribution. In fact T_n tends to the standard normal distribution as $n \to \infty$. The parameter of the t distribution (whose value here is one less than the size of the sample) is usually called the **number of degrees of freedom**.

Defining $t_{\alpha,n-1}$ by

$$P(T_n > t_{\alpha,n-1}) = \alpha$$

(by analogy with z_α), we can derive a $100(1 - \alpha)\%$ confidence interval for the mean by the method used in Section 11.3.3:

$$\left(\bar{X} \pm t_{\alpha/2,n-1} \frac{S_{X,n-1}}{\sqrt{n}}\right)$$

This takes explicit account of the uncertainty caused by the use of $S_{X,n-1}$ in place of σ_X. Values of $t_{\alpha,n-1}$ for typical values of α can be read directly from the table of the t distribution, an example of which is shown in Figure 11.8. To obtain a test statistic for an assumed mean μ_0, simply replace μ_X by μ_0 in the definition of T_n.

n	$\alpha = 0.10$	$\alpha = 0.05$	$\alpha = 0.025$	$\alpha = 0.01$	$\alpha = 0.005$	n
1	3.078	6.314	12.706	31.821	63.657	1
2	1.886	2.920	4.303	6.965	9.925	2
3	1.638	2.353	3.182	4.541	5.841	3
4	1.533	2.132	2.776	3.747	4.604	4
5	1.476	2.015	2.571	3.365	4.032	5
6	1.440	1.943	2.447	3.143	3.707	6
7	1.415	1.895	2.365	2.998	3.499	7
8	1.397	1.860	2.306	2.896	3.355	8
9	1.383	1.833	2.262	2.821	3.250	9
10	1.372	1.812	2.228	2.764	3.169	10
11	1.363	1.796	2.201	2.718	3.106	11
12	1.356	1.782	2.179	2.681	3.055	12
13	1.350	1.771	2.160	2.650	3.012	13
14	1.345	1.761	2.145	2.624	2.977	14
15	1.341	1.753	2.131	2.602	2.947	15
16	1.337	1.746	2.120	2.583	2.921	16
17	1.333	1.740	2.110	2.567	2.898	17
18	1.330	1.734	2.101	2.552	2.878	18
19	1.328	1.729	2.093	2.539	2.861	19
20	1.325	1.725	2.086	2.528	2.845	20
21	1.323	1.721	2.080	2.518	2.831	21
22	1.321	1.717	2.074	2.508	2.819	22
23	1.319	1.714	2.069	2.500	2.807	23
24	1.318	1.711	2.064	2.492	2.797	24
25	1.316	1.708	2.060	2.485	2.787	25
26	1.315	1.706	2.056	2.479	2.779	26
27	1.314	1.703	2.052	2.473	2.771	27
28	1.313	1.701	2.048	2.467	2.763	28
29	1.311	1.699	2.045	2.462	2.756	29
∞	1.282	1.645	1.960	2.326	2.576	∞

Figure 11.8 Table of the t distribution $t_{\alpha,n}$. (Based on Table 12 of *Biometrika Tables for Statisticians*, Volume 1. Cambridge University Press, 1954. By permission of the *Biometrika* trustees.)

EXAMPLE 11.5 The measured lifetimes of a sample of 20 electronic components gave an average of 1250 h, with a sample standard deviation of 96 h. Assuming that the lifetime has a normal distribution, find a 95% confidence interval for the mean lifetime of the population, and test the hypothesis that the mean is 1300 h.

Solution The appropriate figure from the t table is $t_{0.025,19} = 2.093$, so the 95% confidence interval is

$$(1250 \pm 2.093(96)/\sqrt{20}) = (1205,\ 1295)$$

The claim that the mean lifetime is 1300 h is therefore rejected at the 5% significance level. The same conclusion is reached by evaluating

$$T_n = \frac{1250 - 1300}{96/\sqrt{20}} = -2.33$$

which exceeds $t_{0.025,19}$ in magnitude.

Difference between means

Now suppose that we have not just a single sample but two samples from different populations, and that we wish to compare the separate means. Assume also that the variances of the two populations are equal but unknown (the most common situation). Then it can be shown that the $100(1 - \alpha)\%$ confidence interval for the difference $\mu_1 - \mu_2$ between the means is

$$\left(\bar{X}_1 - \bar{X}_2 \pm t_{\alpha/2,n} S_p \sqrt{\left(\frac{1}{n_1} + \frac{1}{n_2}\right)}\right)$$

where $\bar{X}_1$ and $\bar{X}_2$ are the respective sample averages, n_1 and n_2 are the respective sample sizes, S_1^2 and S_2^2 are the respective sample variances (using the '$n - 1$' form as above),

$$S_p^2 = \frac{(n_1 - 1)S_1^2 + (n_2 - 1)S_2^2}{n_1 + n_2 - 2}$$

is a pooled estimate of the unknown variance, and

$$n = n_1 + n_2 - 2$$

is the parameter for the t table. The corresponding test statistic for an assumed difference $d_0 = \mu_1 - \mu_2$ is

$$T_n = \frac{\bar{X}_1 - \bar{X}_2 - d_0}{S_p\sqrt{(1/n_1 + 1/n_2)}}$$

For small samples the populations have to be normal, but for larger samples this is not required and the t-table figure can be replaced by $z_{\alpha/2}$.

EXAMPLE 11.6

Two kinds of a new plastic material are to be compared for strength. From tensile strength measurements of 10 similar pieces of each type, the sample averages and standard deviations were as follows:

$$\bar{X}_1 = 78.3, \quad S_1 = 5.6, \quad \bar{X}_2 = 84.2, \quad S_2 = 6.3$$

Compare the mean strengths, assuming normal data.

Solution The pooled estimate of the standard deviation is 5.960, the t table gives $t_{0.025,18} = 2.101$, and the 95% confidence interval for the difference between means is

$$(78.3 - 84.2 \pm 2.101(5.96)/\sqrt{5}) = (-11.5, -0.3)$$

The difference is significant at the 5% level because zero does not lie within the interval. Also, assuming zero difference gives

$$T_n = \frac{78.3 - 84.2}{5.96/\sqrt{5}} = -2.21$$

which confirms the 5% significance.

It is also possible to set up confidence intervals and tests for the variance σ_X^2, or for comparing two variances for different populations. The process of testing means and variances within and between several populations is called **analysis of variance**. This has many applications, and is well covered in statistics textbooks.

11.3.6 Interval and test for proportion

The ideas of interval estimation do not just apply to means. If probability is interpreted as a long-term proportion (which is one of the common interpretations) then measuring a sample proportion is a way of estimating a probability. The binomial distribution (Section 11.2.3) points the way. We count the number of 'successes', say X, in n 'trials', and estimate the probability p of success at each trial, or the long-term proportion, by the sample proportion

$$\hat{p} = \frac{X}{n}$$

(it is common in statistics to place the 'hat' symbol ˆ over a parameter to denote an estimate of that parameter). This only provides a point estimate. To obtain a confidence interval, we can exploit the normal approximation to the binomial (Section 11.2.4)

$$X \sim N(np, np(1-p))$$

approximately, for large n. Dividing by n preserves normality, so

$$\hat{p} \sim N\left(p, \frac{p(1-p)}{n}\right)$$

Following the argument in Section 11.3.3, we have

$$P\left(p - z_{\alpha/2}\sqrt{\left[\frac{p(1-p)}{n}\right]} < \hat{p} < p + z_{\alpha/2}\sqrt{\left[\frac{p(1-p)}{n}\right]}\right) = 1 - \alpha$$

and, after rearranging the inequality,

$$P\left(\hat{p} - z_{\alpha/2}\sqrt{\left[\frac{p(1-p)}{n}\right]} < p < \hat{p} + z_{\alpha/2}\sqrt{\left[\frac{p(1-p)}{n}\right]}\right) = 1 - \alpha$$

Because p is unknown, we have to make a further approximation by replacing p by $\hat{p}$ inside the square root, to give an approximate $100(1 - \alpha)\%$ confidence interval for p:

$$\left(\hat{p} \pm z_{\alpha/2}\sqrt{\left[\frac{\hat{p}(1-\hat{p})}{n}\right]}\right)$$

The corresponding test statistic for an assumed proportion p_0 is

$$Z = \frac{X - np_0}{\sqrt{[np_0(1-p_0)]}}$$

with $\pm z_{\alpha/2}$ as the rejection points for significance level α.

EXAMPLE 11.7 In an opinion poll conducted with a sample of 1000 people chosen at random, 30% said that they support a certain political party. Find a 95% confidence interval for the actual proportion of the population who support this party.

Solution The required confidence interval is obtained directly as

$$\left(0.3 \pm 1.96\sqrt{\left[\frac{(0.3)(0.7)}{1000}\right]}\right) = (0.27, 0.33)$$

A variation of about 3% either way is therefore to be expected when conducting opinion polls with sample sizes of this order, which is fairly typical, and this figure is often quoted in the news media as an indication of maximum likely error.

A similar argument that also exploits the fact that the difference between two independent normal random variables is also normal leads to the following $100(1 - \alpha)\%$ confidence interval for the difference between two proportions, when $\hat{p}_1$ and $\hat{p}_2$ are the respective sample proportions:

$$\left(\hat{p}_1 - \hat{p}_2 \pm z_{\alpha/2}\sqrt{\left[\frac{\hat{p}_1(1-\hat{p}_1)}{n_1} + \frac{\hat{p}_2(1-\hat{p}_2)}{n_2}\right]}\right)$$

Again it is assumed that n_1 and n_2 are reasonably large. The test statistic for equality of proportions is

$$Z = \frac{\hat{p}_1 - \hat{p}_2}{\sqrt{[\hat{p}(1-\hat{p})(1/n_1 + 1/n_2)]}}$$

where $\hat{p} = (X_1 + X_2)/(n_1 + n_2)$ is a pooled estimate of the proportion.

EXAMPLE 11.8 One hundred samples of an alloy are tested for resistance to fatigue. Half have been prepared using a new process and the other half by a standard process. Of those prepared by the new process, 35 exhibit good fatigue resistance, whereas only 25 of those prepared in the standard way show the same performance. Is the new process better than the standard one?

Solution The proportions of good samples are 0.7 for the new process and 0.5 for the standard one, so a 95% confidence interval for the difference between the true proportions is

$$\left(0.7 - 0.5 \pm 1.96\sqrt{\left[\frac{(0.7)(0.3)}{50} + \frac{(0.5)(0.5)}{50}\right]}\right) = (0.01, 0.39)$$

The pooled estimate of proportion is

$$\hat{p} = (35 + 25)/(50 + 50) = 0.6$$

so that

$$Z = \frac{0.7 - 0.5}{\sqrt{[(0.6)(0.4)/25]}} = 2.04.$$

Both approaches show that the difference is significant at the 5% level. However, it is only just so: if one more sample for the new process had been less fatigue-resistant, the difference would not have been significant at this level. This suggests that the new process is effective – but, despite the apparently large difference in success rates, the evidence is not very strong.

This method only applies to independent sample proportions. It would not be legitimate to apply it, for instance, to a more elaborate version of the opinion poll (Example 11.7) in which respondents can choose between two (or more) political parties or else support neither. Support for one party usually precludes support for another, so the proportions of those interviewed who support the two parties are not independent. More elaborate confidence intervals, based on the multinomial distribution, can handle such situations. This shows how important it is to understand the assumptions upon which statistical methods are based. It would be very easy to look up 'difference between proportions' in an index and apply an inappropriate formula.

11.3.7 Exercises

1 An electrical firm manufactures light bulbs whose lifetime is approximately normally distributed with a standard deviation of 50 h.

(a) If a sample of 30 bulbs has an average life of 780 h, find a 95% confidence interval for the mean lifetime of the population.

(b) How large a sample is needed if we wish to be 95% confident that our sample average will be within 10 h of the population mean?

2 Quantities of a trace impurity in 12 specimens of a new material are measured (in parts per million) as follows:

8.8, 7.1, 7.9, 10.2, 8.9, 7.7, 10.6, 9.4, 9.2, 7.5, 9.0, 8.4

Find a 95% confidence interval for the population mean, assuming that the distribution is normal.

3 A sample of 30 pieces of a semiconductor material gave an average resistivity of 73.2 mΩ m, with a sample standard deviation of 5.4 mΩ m. Obtain a 95% confidence interval for the resistivity of the material, and test the hypothesis that this is 75 mΩ m.

4 The mean weight loss of 16 grinding balls after a certain length of time in mill slurry is 3.42 g, with a standard deviation of 0.68 g. Construct a 99% confidence interval for the true mean weight loss of such grinding balls under the stated conditions.

5 While performing a certain task under simulated weightlessness, the pulse rate of 32 astronaut trainees increased on the average by 26.4 beats per minute, with a standard deviation of 4.28 beats per minute. Construct a 95% confidence interval for the true average increase in the pulse rate of astronaut trainees performing the given task.

6 The quality of a liquid being used in an etching process is monitored automatically by measuring the attenuation of a certain wavelength of light passing through it. The criterion is that when the attenuation reaches 58%, the liquid is declared as 'spent'. Ten samples of the liquid are used until they are judged as 'spent' by the experts. The light attenuation is then measured, and gives an average result of 56%, with a standard deviation of 3%. Is the criterion satisfactory?

7 A fleet car company has to decide between two brands A and B of tyre for its cars. An experiment is conducted using 12 of each brand, run until they wear out. The sample averages and standard deviations of running distance (in km) are respectively 36 300 and 5000 for A, and 39 100 and 6100 for B. Obtain a 95% confidence interval for the difference in means, assuming the distributions to be normal, and test the hypothesis that brand B tyres outrun brand A tyres.

8 A manufacturer claims that the lifetime of a particular electronic component is unaffected by temperature variations within the range 0–60 °C. Two samples of these components were tested, and their measured lifetimes (in hours) recorded as follows:

0 °C: 7250, 6970, 7370, 7910, 6790, 6850, 7280, 7830

60 °C: 7030, 7270, 6510, 6700, 7350, 6770, 6220, 7230

Assuming that the lifetimes have a normal distribution, find 90% and 95% confidence intervals for the difference between the mean lifetimes at the two temperatures, and hence test the manufacturer's claim at the 5% and 10% significance levels.

9 Suppose that out of 540 drivers tested at random, 38 were found to have consumed more than the legal limit of alcohol. Find 90% and 95% confidence intervals for the true proportion of drivers who were over the limit during the time of the tests. Are the results compatible with the hypothesis that this proportion is less than 5%?

10 It is known that approximately one-quarter of all houses in a certain area have inadequate loft insulation. How many houses should be inspected if the difference between the estimated and true proportions having inadequate loft insulation is not to exceed 0.05, with probability 90%? If in fact 200 houses are inspected, and 55 of them have inadequate loft insulation, find a 90% confidence interval for the true proportion.

11 A drug-manufacturer claims that the proportion of patients exhibiting side-effects to their new anti-arthritis drug is at least 8% lower than for the standard brand X. In a controlled experiment 31 out of 100 patients receiving the new drug exhibited side-effects, as did 74 out of 150 patients receiving brand X. Test the manufacturer's claim using 90% and 95% confidence intervals.

12 Suppose that 10 years ago 500 people were working in a factory, and 180 of them were exposed to a material which is now suspected as being carcinogenic. Of those 180, 30 have since developed cancer, whereas 32 of the other workers (who were not exposed) have also since developed cancer. Obtain a 95% confidence interval for the difference between the proportions with cancer among those exposed and not exposed, and assess whether the material should be considered carcinogenic, on this evidence.

11.4 Joint distributions and correlation

Just as it is possible for events to be dependent upon one another in that information to the effect that one has occurred changes the probability of the other, so it is possible for random variables to be associated in value. In this section we show how the degree of dependence between two random variables can be defined and measured.

11.4.1 Joint and marginal distributions

The idea that two variables, each of which is random, can be associated in some way might seem mysterious at first, but can be clarified with some familiar examples. For instance, if one chooses a person at random and measures his or her height and weight, each measurement is a random variable – but we know that taller people also tend to be heavier than shorter people, so the outcomes will be related. On the other hand, a person's birthday and telephone number are not likely to be related in any way. In general, we need a measure of the simultaneous distribution of two random variables.

For two discrete random variables X and Y with possible values $\{u_1, \dots, u_m\}$ and $\{v_1, \dots, v_n\}$ respectively, the **joint distribution** of X and Y is the set of all joint probabilities of the form

$$P(X = u_k \cap Y = v_j\} \quad (k = 1, \dots, m; j = 1, \dots, n)$$

The joint distribution contains all relevant information about the random variables separately, as well as their joint behaviour. To obtain the distribution of one variable, we sum over the possible values of the other:

$$P(X = u_k) = \sum_{j=1}^{n} P(X = u_k \cap Y = v_j) \quad (k = 1, \dots, m)$$

$$P(Y = v_j) = \sum_{k=1}^{m} P(X = u_k \cap Y = v_j) \quad (j = 1, \dots, n)$$

The distributions obtained in this way are called **marginal distributions** of X and Y.

EXAMPLE 11.9 Two textbooks are selected at random from a shelf containing three statistics texts, two mathematics texts and three engineering texts. Denoting the number of books selected in each subject by S, M and E respectively, find (a) the joint distribution of S and M, and (b) the marginal distributions of S, M and E.

	M			
S	0	1	2	Total
0	$\frac{3}{28}$	$\frac{3}{14}$	$\frac{1}{28}$	$\frac{5}{14}$
1	$\frac{9}{28}$	$\frac{3}{14}$		$\frac{15}{28}$
2	$\frac{3}{28}$			$\frac{3}{28}$
Total	$\frac{15}{28}$	$\frac{3}{7}$	$\frac{1}{28}$	1

Figure 11.9 Joint distribution for Example 11.9.

Solution (a) The joint distribution (shown in Figure 11.9) is built up element by element using the addition and product rules of probability as follows:

$$P(S = M = 0) = P(E = 2) = (\tfrac{3}{8})(\tfrac{2}{7}) = \tfrac{3}{28}$$

that is, the probability that the first book is an engineering text (three chances out of eight) times the probability that the second book is also (two remaining chances out of seven). Continuing,

$$\begin{aligned} P(S = 0 \cap M = 1) &= P(M = 1 \cap E = 1) \\ &= (\tfrac{2}{8})(\tfrac{3}{7}) + (\tfrac{3}{8})(\tfrac{2}{7}) = \tfrac{3}{14} \end{aligned}$$

that is, the probability that the first book is a mathematics text and the second an engineering text, plus the (equal) probability of the books being the other way round. The other probabilities are derived similarly.

(b) The marginal distributions of S and M are just the row and column totals as shown in Figure 11.9. The marginal distribution of E can also be derived from the table:

$$P(E = 2) = P(S = M = 0) = \tfrac{3}{28}$$

$$P(E = 1) = P(S = 1 \cap M = 0) + P(S = 0 \cap M = 1) = \tfrac{15}{28}$$

$$P(E = 0) = P(S = 2) + P(S = 1 \cap M = 1) + P(M = 2) = \tfrac{5}{14}$$

This is the same as the marginal distribution of S, which is not surprising, because there are the same numbers of engineering and statistics books on the shelf.

In order to apply these ideas of joint and marginal distributions to continuous random variables, we need to build on the interpretation of the probability density function. The **joint density function** of two continuous random variables X and Y, denoted by $f_{X,Y}(x, y)$, is such that

$$P(x_1 < X < x_2 \quad \text{and} \quad y_1 < Y < y_2) = \int_{x_1}^{x_2}\int_{y_1}^{y_2} f_{X,Y}(x, y)\,dy\,dx$$

for all intervals (x_1, x_2) and (y_1, y_2). This involves a double integral over the two variables x and y. This is necessary because the joint density function must indicate the relative likelihood of every combination of values of X and Y, just as the joint distribution does for discrete random variables. The joint density function is transformed into a probability by integrating over an interval for both variables. The double integral here can be regarded as a pair of single-variable integrations, with the outer variable (x) held constant during the integration with respect to the inner variable (y). In fact the same answer is obtained if the integration is performed the other way around.

The **marginal density functions** for X and Y are obtained from the joint density function in a manner analogous to the discrete case: by integrating over all values of the unwanted variable:

$$f_X(x) = \int_{-\infty}^{\infty} f_{X,Y}(x, y)\,\mathrm{d}y \quad (-\infty < x < \infty)$$

$$f_Y(y) = \int_{-\infty}^{\infty} f_{X,Y}(x, y)\,\mathrm{d}x \quad (-\infty < y < \infty)$$

EXAMPLE 11.10 The joint density function of random variables X and Y is

$$f_{X,Y}(x, y) = \begin{cases} 1 & (0 \leqslant x \leqslant 1,\ cx \leqslant y \leqslant cx + 1) \\ 0 & \text{otherwise} \end{cases}$$

where c is a constant such that $0 \leqslant c \leqslant 1$ (which means that $f_{X,Y}(x, y)$ is unity over the trapezoidal area shown in Figure 11.10 and zero elsewhere). Find the marginal distributions of X and Y. Also find the probability that neither X nor Y exceeds one-half, assuming $c = 1$.

Solution To find the marginal distribution of X, we integrate with respect to y:

$$f_X(x) = \begin{cases} \displaystyle\int_{cx}^{cx+1} \mathrm{d}y = 1 & (0 \leqslant x \leqslant 1) \\ 0 & \text{otherwise} \end{cases}$$

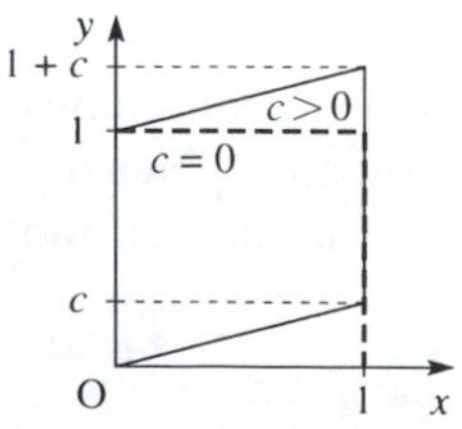

Figure 11.10 Density function for Example 11.10.

The marginal distribution for Y is rather more complicated. Integrating with respect to x and assuming that $0 < c \leqslant 1$,

$$f_Y(y) = \begin{cases} 1 - \dfrac{1}{c}(y-1) & (1 \leqslant y \leqslant 1 + c) \\ 1 & (c \leqslant y \leqslant 1) \\ \dfrac{y}{c} & (0 \leqslant y \leqslant c) \end{cases}$$

(Exercise 15). When $c = 0$, the marginal distribution for Y is the same as that for X. Finally, when $c = 1$,

$$P(X \leqslant \tfrac{1}{2} \quad \text{and} \quad Y \leqslant \tfrac{1}{2}) = \int_0^{1/2}\int_x^{1/2} 1\,\mathrm{d}y\,\mathrm{d}x = \int_0^{1/2} (\tfrac{1}{2} - x)\,\mathrm{d}x = \tfrac{1}{8}$$

Here the inner integral (with respect to y) is performed with x treated as constant, and the resulting function of x is integrated to give the answer.

The definitions of joint and marginal distributions can be extended to any number of random variables.

11.4.2 Independence

The idea of independence of events can be extended to random variables to give us the important case in which no information is shared between them. This is important in experiments where essentially the same quantity is measured repeatedly, either within a single experiment involving repetition or between different experiments. As mentioned before, independence within a sample is one of the properties that qualifies the sample for analysis and conclusion.

> Two random variables X and Y are called **independent** if their joint distribution factorizes into the product of their marginal distributions:
>
> $P(X = u_k \cap Y = v_j) = P(X = u_k)P(Y = v_j)$ in the discrete case
>
> $f_{X,Y}(x, y) = f_X(x)\, f_Y(y)$ in the continuous case

For example, the random variables X and Y in Example 11.10 are independent if and only if $c = 0$.

EXAMPLE 11.11 The assembly of a complex piece of equipment can be divided into two stages. The times (in hours) required for the two stages are random variables (X and Y, say) with density functions e^{-x} and $2\,\mathrm{e}^{-2y}$ respectively. Assuming that the stage assembly times are independent, find the probability that the assembly will be completed within four hours.

Solution The assumption of independence implies that

$$f_{X,Y}(x, y) = f_X(x) f_Y(y) = 2\,\mathrm{e}^{-(x+2y)}$$

If the time for the first stage is x, the total time will not exceed four hours if

$$Y < 4 - x$$

so the required value is

$$P(X + Y < 4) = \int_0^4 \int_0^{4-x} f_{X,Y}(x, y)\,dy\,dx = \int_0^4 \int_0^{4-x} 2\,e^{-(x+2y)}\,dy\,dx$$

$$= \int_0^4 (e^{-x} - e^{-(8-x)})\,dx = 0.964$$

Where random variables are dependent upon one another, it is possible to express this dependence by defining a **conditional distribution** analogous to conditional probability, in terms of the joint distribution (or density function) and the marginal distributions. Instead of pursuing this idea here, we shall consider a numerical measure of dependence that can be estimated from sample data.

11.4.3 Covariance and correlation

The use of mean and variance for a random variable is motivated partly by the difficulty in determining the full probability distribution in many practical cases. The joint distribution of two variables presents even greater difficulties. Since we already have numerical measures of location and dispersion for the variables individually, it seems reasonable to define a measure of association of the two variables that is independent of their separate means and variances so that the new measure provides essentially new information about the variables.

There are four objectives that it seems reasonable for such a measure to satisfy. Its value should

(a) be zero for independent variables,

(b) be non-zero for dependent variables,

(c) indicate the degree of dependence in some well-defined sense, detached from the individual means and variances,

(d) be easy to estimate from sample data.

It is actually rather difficult to satisfy all of these, but the most popular measure of association gets most of the way.

The **covariance** of random variables X and Y, denoted by Cov (X, Y), is defined as

$$\text{Cov}(X, Y) = E\{(X - \mu_X)(Y - \mu_Y)\}$$

$$= \begin{cases} \displaystyle\sum_{k=1}^{m} \sum_{j=1}^{n} (u_k - \mu_X)(v_j - \mu_Y)P(X = u_k \cap Y = v_j) \\ \displaystyle\int_{-\infty}^{\infty} \int_{-\infty}^{\infty} (x - \mu_X)(y - \mu_Y) f_{X,Y}(x, y)\,dx\,dy \end{cases}$$

for discrete and continuous variables respectively. The **correlation** $\rho_{X,Y}$ is the covariance divided by the product of the standard deviations:

$$\rho_{X,Y} = \frac{\text{Cov}(X, Y)}{\sigma_X \sigma_Y}$$

If whenever the random variable X is larger than its mean the random variable Y also tends to be larger than its mean then the product $(X - \mu_X)(Y - \mu_Y)$ will tend to be positive. The same will be true if both variables tend to be smaller than their means simultaneously. The covariance is then positive. A negative covariance implies that the variables tend to move in opposite directions with respect to their means. Both covariance and correlation therefore measure association relative to the mean values of the variables. It turns out that correlation measures association relative to the standard deviations as well.

It should be noted that the variance of a random variable X is the same as the covariance with itself:

$$\text{Var}(X) = \text{Cov}(X, X)$$

Also, by expanding the product within the integral or sum in the definition of covariance, it is easy to show that an alternative expression is

$$\text{Cov}(X, Y) = E(X\,Y) - E(X)E(Y)$$

Although the sign of the covariance indicates the direction of the dependence, its magnitude depends not only on the degree of dependence but also upon the variances of the random variables, so it fails to satisfy the objective (c). In contrast, the correlation is limited in range

$$-1 \leqslant \rho_{X,Y} \leqslant +1$$

and it adopts the limiting values of this range only when the random variables are linearly related:

$$\rho_{X,Y} = \pm 1 \quad \text{if and only if there exist } a, b \text{ such that } Y = aX + b$$

(this is proved in most textbooks on probability theory). The magnitude of the correlation indicates the degree of linear relationship, so that objective (c) is satisfied.

EXAMPLE 11.12 Find the correlation of the random variables S and M in Example 11.9.

Solution The joint and marginal distributions of S and M are shown in Figure 11.9. First we find the expected values of S and S^2 from the marginal distribution, and hence the variance and standard deviation:

$$E(S) = \tfrac{15}{28} + (2)(\tfrac{3}{28}) = \tfrac{21}{28}, \quad E(S^2) = \tfrac{15}{28} + (4)(\tfrac{3}{28}) = \tfrac{27}{28}$$

$$\text{Var}(S) = \tfrac{27}{28} - (\tfrac{21}{28})^2$$

from which

$$\sigma_S = \tfrac{3}{28}\sqrt{35}$$

Next we do the same for M:

$$E(M) = \tfrac{3}{7} + (2)(\tfrac{1}{28}) = \tfrac{1}{2}, \quad E(M^2) = \tfrac{3}{7} + (4)(\tfrac{1}{28}) = \tfrac{4}{7}$$

$$\text{Var}(M) = \tfrac{4}{7} - \tfrac{1}{4} = \tfrac{9}{28}$$

from which

$$\sigma_M = \tfrac{3}{2}\sqrt{\tfrac{1}{7}}$$

All products of S and M are zero except when both are equal to one, so the expected value of the product is

$$E(SM) = \tfrac{3}{14}$$

The correlation now follows easily:

$$\rho_{S,M} = \frac{E(SM) - E(S)E(M)}{\sigma_S \sigma_M} = \frac{\tfrac{3}{14} - (\tfrac{21}{28})(\tfrac{1}{2})}{(\tfrac{3}{28}\sqrt{35})(\tfrac{3}{2}\sqrt{\tfrac{1}{7}})} = -\frac{1}{\sqrt{5}}$$

The correlation is negative because if there are more statistics books in the selection then there will tend to be fewer mathematics books, and vice versa.

EXAMPLE 11.13 Find the correlation of the random variables X and Y in Example 11.10.

Solution Proceeding as in Example 11.12, we have for X

$$E(X) = \int_0^1 x\,dx = \tfrac{1}{2}$$

$$E(X^2) = \int_0^1 x^2\,dx = \tfrac{1}{3}$$

so that $\text{Var}(X) = E(X^2) - [E(X)]^2 = \tfrac{1}{12}$. Also, for Y

$$E(Y) = \int_0^c \frac{y^2}{c}\,dy + \int_c^1 y\,dy + \int_1^{1+c} y\left[1 - \frac{1}{c}(y-1)\right]dy$$

$$= \tfrac{1}{2}(1 + c) \quad \text{after simplification}$$

$$E(Y^2) = \int_0^c \frac{y^3}{c}\,dy + \int_c^1 y^2\,dy + \int_1^{1+c} y^2\left[1 - \frac{1}{c}(y-1)\right]dy$$

$$= \tfrac{1}{3}(1 + c^2) + \tfrac{1}{2}c \quad \text{after simplification}$$

so that $\text{Var}(Y) = E(Y^2) - [E(Y)]^2 = \tfrac{1}{12}(1 + c^2)$. For the expected value of the product we have

$$E(XY) = \int_0^1 \int_{cx}^{cx+1} xy\,dy\,dx = \frac{1}{2}\int_0^1 x(1 + 2cx)\,dx = \frac{1}{4} + \frac{1}{3}c$$

Finally, the correlation between X and Y is

$$\rho_{X,Y} = \frac{E(XY) - E(X)E(Y)}{\sqrt{[\mathrm{Var}(X)\,\mathrm{Var}(Y)]}} = \frac{\frac{1}{4} + \frac{1}{3}c - \frac{1}{4}(1+c)}{\frac{1}{12}\sqrt{(1+c^2)}} = \frac{c}{\sqrt{(1+c^2)}}$$

Note that in fact the result of Example 11.13 holds for any value of c, and not just for the range $0 \leqslant c \leqslant 1$ assumed in Example 11.10. As the value of c increases (positive or negative), the correlation increases also, but its magnitude never exceeds one. It is also clear that if X and Y are independent then $c = 0$ and the correlation is zero. Refer to Figure 11.10 for a geometrical interpretation. When $c = 0$, the sample space is a square within which all points are equally likely, so there is no association between the variables. As c increases (positive or negative), the sample space becomes more elongated as the variables become more tightly coupled to one another.

The general relationship between independence and correlation is expressed as follows: if the random variables X and Y are independent then their correlation is zero. This is easily shown as follows for continuous random variables (or by a similar argument for discrete random variables). First we have

$$f_{X,Y}(x, y) = f_X(x)\, f_Y(y)$$

and then

$$\int_{-\infty}^{\infty}\int_{-\infty}^{\infty} (x-\mu_X)(y-\mu_Y)\, f_X(x)\, f_Y(y)\, \mathrm{d}x\, \mathrm{d}y$$

$$= \int_{-\infty}^{\infty} (x-\mu_X)\, f_X(x)\, \mathrm{d}x \int_{-\infty}^{\infty} (y-\mu_Y)\, f_Y(y)\, \mathrm{d}y$$

$$= (\mu_X - \mu_X)(\mu_Y - \mu_Y) = 0$$

Unfortunately, the converse does not hold: zero correlation does not imply independence. In general, correlation is a measure of linear dependence, and may be zero or very small for variables that are dependent in a *nonlinear* way (see Exercise 14). Objective (a) is satisfied, therefore, but not objective (b) in general.

Another problem with correlation is that a non-zero value does not imply the presence of a causal relationship between the variables or the phenomena that they measure. Correlation can be 'spurious', deriving from some third variable that may be unrecognized at the time. For example, among the economic statistics that are gathered together from many countries and published, there are figures for the number of telephones per head of population, birth rate, and the gross domestic product per capita (GDP). It turns out that there is a large negative correlation between number of telephones per head and the birth rate, but no-one would suggest that telephones have any direct application in birth control. The GDP is a measure of wealth, and there is a large positive correlation between this and the number of telephones per head, and a large negative correlation between GDP and birth rate, both for quite genuine reasons. The correlation between telephones and birth rate is therefore spurious, and a more sophisticated measure called the **partial correlation** can be used to eliminate the third variable (provided that it is recognized and measured).

We have considered all the objectives except (d); that this is satisfied is shown in Section 11.4.4.

11.4.4 Sample correlation

There are two kinds of situations where we take samples of values of two random variables X and Y. First we might be interested in the same property for two different populations. Perhaps there is evidence that the mean values are different, so we take samples of each and compare them. This situation was discussed in Section 11.3.5. The second kind involves two different properties for the same population. It is to this situation that correlation applies. We take a single sample from the population and measure the pair of random variables (X_i, Y_i) for each $i = 1, \ldots, n$.

For a sample $\{(X_1, Y_1), \ldots, (X_n, Y_n)\}$ the **sample correlation coefficient** is defined as

$$r_{X,Y} = \frac{\frac{1}{n}\sum_{i=1}^{n}[(X_i-\bar{X})(Y_i-\bar{Y})]}{S_X S_Y}$$

Like the true correlation, the sample correlation is limited in value to the range $[-1, 1]$ and $r_{X,Y} = \pm 1$ when (and only when) all of the points lie along a line. Figure 11.11 contains four typical **scatter diagrams** of samples plotted on the (x, y) plane, with an indication of the correlation for each one. The range of behaviour is shown from independence (a) through imperfect correlation (b) and (c) to a perfect linear relationship (d).

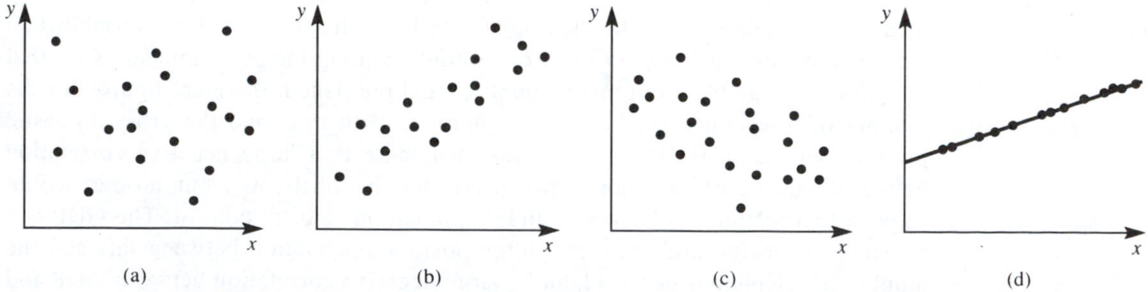

Figure 11.11 Scatter plots for two random variables: (a) $\rho_{X,Y} = 0$; (b) $\rho_{X,Y} > 0$; (c) $\rho_{X,Y} < 0$; (d) $\rho_{X,Y} = 1$.

```
{ Program to compute sample correlation.
x(k) and y(k) are the arrays of data,
n is the sample size,
xbar and ybar are the sample averages,
sx and sy are the sample standard deviations,
rxy is the sample correlation,
Mx, My, Qx, Qy, Qxy hold running totals. }

Mx ← 0; My ← 0
Qx ← 0; Qy ← 0
Qxy ← 0
for k is 1 to n do
  diffx ← x(k) − Mx
  diffy ← y(k) − My
  Mx ← ((k − 1)*Mx + x(k))/k
  My ← ((k − 1)*My + y(k))/k
  Qx ← Qx + (1 − 1/k)*diffx*diffx
  Qy ← Qy + (1 − 1/k)*diffy*diffy
  Qxy ← Qxy + (1 − 1/k)*diffx*diffy
endfor
xbar ← Mx; ybar ← My
sx ← sqr(Qx/n); sy ← sqr(Qy/n)
rxy ← Qxy/(n*sx*sy)
```

Figure 11.12 Pseudocode listing for sample correlation.

By expanding the product within the outer bracket in the numerator, it is easy to show that an alternative expression is

$$r_{X,Y} = \frac{\overline{XY} - (\bar{X})(\bar{Y})}{S_X S_Y}$$

which is useful for hand calculation. For computer calculation the best method involves the successive sums of products:

$$Q_{XY,k} = \sum_{i=1}^{k} (X_i - M_{X,k})(Y_i - M_{Y,k})$$

where

$$M_{X,k} = \frac{1}{k}\sum_{i=1}^{k} X_i, \quad M_{Y,k} = \frac{1}{k}\sum_{i=1}^{k} Y_i$$

and then $r_{X,Y} = Q_{XY,n}/nS_X S_Y$. The pseudocode listing in Figure 11.12 exploits the recurrence relation

$$Q_{XY,k} = Q_{XY,k-1} + \left(1 - \frac{1}{k}\right)(X_k - M_{X,k-1})(Y_k - M_{Y,k-1})$$

which allows for a single pass through the data with no loss of accuracy.

EXAMPLE 11.14 A material used in the construction industry contains an impurity suspected of having an adverse effect upon the material's performance in resisting long-term operational stresses. Percentages of impurity and performance indexes for 22 specimens of this material are as follows:

% Impurity X_i	4.4	5.5	4.2	3.0	4.5	4.9	4.6	5.0	4.7	5.1	4.4
Performance Y_i	12	14	18	35	23	29	16	12	18	21	27
% Impurity X_i	4.1	4.9	4.7	5.0	4.6	3.6	4.9	5.1	4.8	5.2	5.2
Performance Y_i	13	19	22	20	16	27	21	13	18	17	11

Find the sample correlation coefficient.

Solution The following quantities are easily obtained from the data:

$$\bar{X} = 4.6545, \quad S_X = 0.550\,81, \quad \bar{Y} = 19.1818, \quad S_Y = 6.0350, \quad \overline{XY} = 87.3591$$

(Note that it is advisable to record these results to several significant digits in order to avoid losing precision when calculating the difference within the numerator of $r_{X,Y}$.) The sample correlation is then $r_{X,Y} = -0.58$, the negative value suggesting that the impurity has an adverse affect upon performance. It remains to be seen whether this is statistically significant.

11.4.5 Interval and test for correlation

Correlation is more difficult to deal with than mean and proportion, but for normal random variables X and Y with a true correlation $\rho_{X,Y}$ the sample statistic

$$Z = \frac{\sqrt{(n-3)}}{2} \ln\left[\frac{(1+r_{X,Y})(1-\rho_{X,Y})}{(1-r_{X,Y})(1+\rho_{X,Y})}\right]$$

is approximately standard normal for large n. This can be used directly as a test statistic for an assumed value of $\rho_{X,Y}$. Alternatively, an approximate $100(1-\alpha)\%$ confidence interval for $\rho_{X,Y}$ can be derived:

$$\left(\frac{1+r-c(1-r)}{1+r+c(1-r)}, \frac{1+r-(1-r)/c}{1+r+(1-r)/c}\right)$$

where

$$c = \exp\left[\frac{2z_{\alpha/2}}{\sqrt{(n-3)}}\right]$$

(the subscripts X and Y have been dropped from $r_{X,Y}$ in this formula).

EXAMPLE 11.15 For the data in Example 11.14 find 95% and 99% confidence intervals for the true correlation between percentage of impurity and performance index, and test the hypothesis that these are independent.

Solution The sample correlation (from the 22 specimens) was found in Example 11.14 to be -0.58. For the 95% confidence interval the constant $c = 2.458$ and the interval itself is $(-0.80, -0.21)$. Similarly, the 99% confidence interval is $(-0.85, -0.07)$. Assuming $\rho_{XY} = 0$, the value of the test statistic is $Z = -2.89$, which exceeds $z_{0.005} = 2.576$ in magnitude. Either way, we can be more than 99% confident that the impurity has an adverse effect upon performance.

11.4.6 Rank correlation

As has been previously emphasized the correlation only works as a measure of dependence if

(1) n is reasonably large,

(2) X and Y are *numerical* characteristics,

(3) the dependence is *linear*, and

(4) X and Y each have a *normal* distribution.

There is an alternative form of sample correlation, which has greater applicability, requiring only that

(1) n is reasonably large,

(2) X and Y are *rankable* characteristics, and

(3) the dependence is *monotonic* (that is, always in the same direction, which may be forward or inverse, but not necessarily linear).

The variables X and Y can have any distribution. For a set of data $X_1, \ldots, X_n$, a **rank** of 1 is assigned to the smallest value, 2 to the next-smallest and so on up to a rank of n assigned to the largest. This applies wherever the values are distinct. Tied values are given the mean of the ranks they would receive if slightly different. The following is an example:

X_i	8	3	5	8	1	9	6	5	3	5	7	2
Rank	10.5	3.5	6	10.5	1	12	8	6	3.5	6	9	2

The **Spearman rank correlation coefficient** r_S for data $(X_1, Y_1), \ldots, (X_n, Y_n)$ is the correlation of the ranks of X_i and Y_i, where the data $X_1, \ldots, X_n$ and $Y_1, \ldots, Y_n$ are ranked separately. If the number of tied values is small compared with n then

$$r_S \simeq 1 - \frac{6}{n(n^2 - 1)} \sum_{i=1}^{n} d_i^2$$

where d_i is the difference between the rank of X_i and that of Y_i. The value of r_S always lies in the interval $[-1, 1]$, and adopts its extreme values only when the rankings precisely match (forwards or in reverse).

To test for dependence, special tables must be used for small samples ($n < 20$), but for larger samples the test statistic

$$Z = r_S\sqrt{(n - 1)}$$

is approximately standard normal.

EXAMPLE 11.16 Find and test the rank correlation for the data in Example 11.14.

Solution The data with their ranks are as follows:

X_i	4.4	5.5	4.2	3.0	4.5	4.9	4.6	5.0	4.7	5.1	4.4
Rank	5.5	22	4	1	7	14	8.5	16.5	10.5	18.5	5.5
Y_i	12	14	18	35	23	29	16	12	18	21	27
Rank	2.5	6	11	22	18	21	7.5	2.5	11	15.5	19.5

X_i	4.1	4.9	4.7	5.0	4.6	3.6	4.9	5.1	4.8	5.2	5.2
Rank	3	14	10.5	16.5	8.5	2	14	18.5	12	20.5	20.5
Y_i	13	19	22	20	16	27	21	13	18	17	11
Rank	4.5	13	17	14	7.5	19.5	15.5	4.5	11	9	1

From this, the rank correlation is $r_S = -0.361$, and $Z = -1.66$, which exceeds $z_{0.05} = 1.645$ and is therefore just significant at the 10% level. If the approximate formula is used, the sum of squares of differences is 2398, so

$$r_S \simeq 1 - \frac{(6)(2398)}{(22)(483)} = -0.354$$

and $Z = -1.62$, which is just short of significance.

These results show that the rank correlation is a more conservative test than the sample correlation $r_{X,Y}$, in that a larger sample tends to be needed before the hypothesis of independence is rejected. A price has to be paid for the wider applicability of the method.

11.4.7 Exercises

13 Suppose that the random variables X and Y have the following joint distribution:

	X		
Y	1	2	3
1	0	0.17	0.08
2	0.20	0.11	0
3	0.14	0.25	0.05

Find (a) the marginal distributions of X and Y, (b) $P(Y = 3 \mid X = 2)$, and (c) the mean, variance and correlation coefficient of X and Y.

14 Consider the random variable X with density function

$$f_X(x) = \begin{cases} 1 & (-\frac{1}{2} < x < \frac{1}{2}) \\ 0 & \text{otherwise} \end{cases}$$

Show that the covariance of X and X^2 is zero. (This shows that zero covariance does not imply independence, because obviously X^2 is dependent on X.)

15 The joint density function of random variables X and Y is

$$f_{X,Y}(x, y) = \begin{cases} 1 & (0 \leqslant x \leqslant 1;\ cx \leqslant y \leqslant cx + 1) \\ 0 & \text{otherwise} \end{cases}$$

where c is a constant such that $0 \leqslant c \leqslant 1$. Find the marginal density function for Y (see Example 11.10).

16 Let the random variables X and Y have joint density function given by

$$f_{X,Y}(x, y) = \begin{cases} c(1 - y) & (0 \leqslant x \leqslant y \leqslant 1) \\ 0 & \text{otherwise} \end{cases}$$

Find (a) the value of the constant c, (b) $P(x < \frac{3}{4}, y > \frac{1}{2})$, and (c) the marginal density functions for X and Y.

17 Let the random variables X and Y represent the lifetimes (in hundreds of hours) of two types of components used in an electronic system. The joint density function is given by

$$f_{X,Y}(x, y) = \begin{cases} \frac{1}{8} x\,e^{-(x+y)/2} & (x > 0, y > 0) \\ 0 & \text{otherwise} \end{cases}$$

Find (a) the probability that two components (one of each type) will each last longer than 100 h, and (b) the probability that a component of the second type (Y) will have a lifetime in excess of 200 h.

18 The ball and socket of a joint are separately moulded and then assembled together. The diameter of the ball is a random variable X between 29.8 and 30.3 mm, all values being equally likely. The internal diameter of the socket is a random variable Y between 30.1 and 30.6 mm, again with all values equally likely. The condition for an acceptable fit is that $0 \leqslant Y - X \leqslant 0.6$ mm. Find the probability of this condition being satisfied, assuming that the random variables are independent.

19 The following are the measured heights and weights of eight people:

Height (cm)	182.8	162.5	175.2	185.4	170.1	167.6	177.8	172.7
Weight (kg)	86.1	58.3	83.0	92.4	60.2	69.3	83.6	72.7

Find the sample correlation coefficient.

20 The number of minutes it took 10 mechanics to assemble a piece of machinery in the morning (X) and in the late afternoon (Y) were measured, with the following results:

X	11.1	10.3	12.0	15.1	13.7	18.5	17.3	14.2	14.8	15.3
Y	10.9	14.2	13.8	21.5	13.2	21.1	16.4	19.3	17.4	19.0

Find the sample correlation coefficient.

21 If the sample correlation between resistance and failure time for 30 overloaded resistors is 0.7, find a 95% confidence interval for the true correlation.

22 Find a 95% confidence interval for correlation between height and weight using the data in Exercise 19.

23 Marks obtained by 20 students taking examinations in mathematics and computer studies were as follows:

Math.	45	77	43	64	58	64	58	54	71	45
	57	52	67	57	54	54	61	58	55	42
Comp.	64	67	47	75	42	65	58	42	70	44
	44	67	49	70	51	58	37	60	42	36

Find the sample correlation coefficient and the 90% and 95% confidence intervals. Hence test the hypothesis that the two marks are independent at the 5% and 10% significance levels. Also find and test the rank correlation.

11.5 Regression

A procedure that is very familiar to engineers is that of drawing a good straight line through a set of points on a graph. When calibrating a measuring instrument, for example, known inputs are applied, the readings are noted and plotted, a straight line is drawn as close to the points as possible (there are bound to be small errors, so they will not all lie on the line), and the graph is then used to interpret the readings for unknown inputs. It is possible to draw the line by eye, but there is a better way, which involves calculating the slope and intercept of the line from the data. The given line then minimizes the total squared error for the data points. This procedure (which for historical reasons is called **regression**) can be applied in general to pairs of random variables.

Computer packages are very often used to carry out the regression calculations and display the results. This is of special value when the data tend to follow a curve and various nonlinear models are tried and compared (see Section 11.5.4).

11.5.1 The method of least squares

The correlation was introduced in Section 11.4.3 as a way of measuring the dependence between random variables. Subsequently, we have seen how the correlation can be estimated and the dependence tested using sample data. We can take the idea of correlation between variables (say X and Y) a stage further by assuming that the sample pairs $\{(X_1, Y_1), \ldots, (X_n, Y_n)\}$ satisfy a linear relationship of the form

$$Y_i = a + bX_i + E_i \quad (i = 1, \ldots, n)$$

where a and b are unknown coefficients and the random variables E_i have zero mean and represent residual errors. This assumption is prompted by the scatter diagrams in Figure 11.11, which illustrate how the points may be concentrated around a line. Figure 11.13 shows a typical scatter diagram again, this time with the line drawn in. If we can estimate the coefficients a and b so as to give the best fit, we shall be able to predict the value of Y when the value of X is known.

The least-squares approach is to choose estimates $\hat{a}$ and $\hat{b}$ to minimize the sum of squares of the values E_i:

$$Q = \sum_{i=1}^{n} E_i^2 = \sum_{i=1}^{n} [\, Y_i - (\hat{a} + \hat{b}X_i)\,]^2$$

Equating to zero the partial derivatives of this sum with respect to the two coefficients gives a pair of equations that determine the minimum:

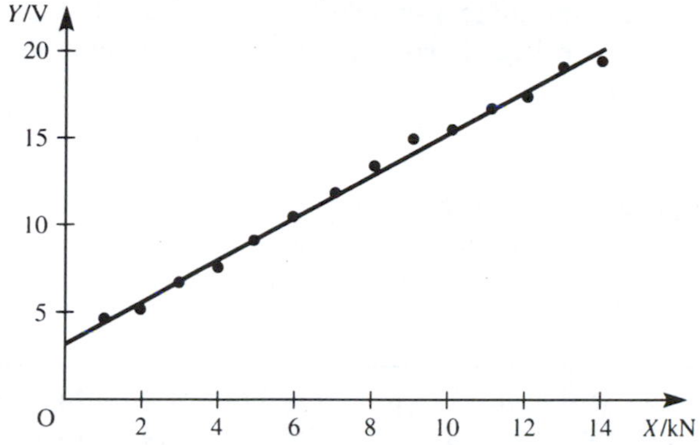

Figure 11.13 Scatter plot with regression line (Example 11.17).

$$\frac{\partial Q}{\partial \hat{a}} = 0 = -2\sum_{i=1}^{n}[Y_i - (\hat{a} + \hat{b}X_i)]$$

$$\frac{\partial Q}{\partial \hat{b}} = 0 = -2\sum_{i=1}^{n}X_i[Y_i - (\hat{a} + \hat{b}X_i)]$$

These can be rewritten as

$$n\hat{a} + (\Sigma_i X_i)\hat{b} = (\Sigma_i Y_i)$$

$$(\Sigma_i X_i)\hat{a} + (\Sigma_i X_i^2)\hat{b} = (\Sigma_i X_i Y_i)$$

(were $\Sigma_i = \Sigma_{i=1}^n$) from which the solution is

$$\hat{b} = \frac{S_{XY}}{S_X^2}, \quad \hat{a} = \bar{Y} - \hat{b}\bar{X}$$

where

$$\bar{X} = \frac{1}{n}\Sigma_i X_i, \quad \bar{Y} = \frac{1}{n}\Sigma_i Y_i$$

and

$$S_{XY} = \frac{1}{n}\Sigma_i (X_i - \bar{X})(Y_i - \bar{Y}) = \overline{XY} - (\bar{X})(\bar{Y})$$

$$S_X^2 = \frac{1}{n}\Sigma_i (X_i - \bar{X})^2 = \overline{X^2} - (\bar{X})^2$$

$$S_Y^2 = \frac{1}{n}\Sigma_i (Y_i - \bar{Y})^2 = \overline{Y^2} - (\bar{Y})^2$$

are the sample variances and covariance.

This process of fitting a straight line through a set of data of the form $\{(X_1, Y_1), \ldots, (X_n, Y_n)\}$ is called **linear regression**, and the coefficients are called **regression coefficients**.

EXAMPLE 11.17 A strain gauge has been bonded to a steel beam, and is being calibrated. The resistance of the strain gauge is converted into a voltage appearing on a meter. Known forces (X, in kN) are applied and voltmeter measurements (Y, in V) are as follows:

X	1	2	3	4	5	6	7	8	9	10	11	12	13	14
Y	4.4	4.9	6.4	7.3	8.8	10.3	11.7	13.2	14.8	15.3	16.5	17.2	18.9	19.3

Fit a regression line through the data and estimate the tension in the beam when the meter reading is 13.8 V.

Solution The following quantities are calculated from the data:

$$\bar{X} = 7.5, \quad S_X = 4.031\,13, \quad \bar{Y} = 12.0714, \quad S_Y = 4.950\,68, \quad \overline{XY} = 110.421$$

(When using a hand calculator to solve linear regression problems, it is advisable to work to at least five or six significant digits during intermediate calculations, because the subtraction in the numerator of $\hat{b}$ often results in the loss of some leading digits.) From these results, $\hat{b} = 1.22$ and $\hat{a} = 2.89$ (Figure 11.13). The estimated value of tension for a reading of $Y = 13.8$ V is given by

$$13.8 = 2.89 + 1.22X$$

from which $X = 8.9$ kN.

Figure 11.14 shows a pseudocode listing for linear regression. The program is very similar to that in Figure 11.12 for the sample correlation, and the link between these will be explained in Section 11.5.3. In addition to the regression coefficients $\hat{a}$ and $\hat{b}$, an estimate of the residual standard deviation is returned, which is explained below.

11.5.2 Normal residuals

The process of fitting a straight line through the data by minimizing the sum of squares of the errors does not involve any statistics as such. However, we often need to test whether the slope of the regression line is significantly different from zero, because this will reveal whether there is any dependence between the random variables. For this purpose we must make the assumption that the errors E_i, called the **residuals**, have a normal distribution:

```
{  Program to compute linear regression coefficients.
x(k) and y(k) are the arrays of data,
n is the sample size,
xbar and ybar are the sample averages,
sx and sy are the sample standard deviations,
bhat is the regression slope result,
ahat is the regression intercept result,
se is the residual standard deviation,
Mx, My, Qx, Qy, Qxy hold running totals.  }

Mx ← 0; My ← 0
Qx ← 0; Qy ← 0
Qxy ← 0
for k is 1 to n do
  diffx ← x(k) - Mx
  diffy ← y(k) - My
  Mx ← ((k - 1)*Mx+x(k))/k
  My ← ((k - 1)*My+y(k))/k
  Qx ← Qx+(1-1/k)*diffx*diffx
  Qy ← Qy+(1-1/k)*diffy*diffy
  Qxy ← Qxy+(1-1/k)*diffx*diffy
endfor
xbar ← Mx; ybar ← My
sx ← sqr(Qx/n); sy ← sqr(Qy/n)
bhat ← Qxy/(n*sx*sx)
ahat ← ybar - bhat*xbar
se ← sqr(sy*sy - bhat*bhat*sx*sx)
```

Figure 11.14 Pseudocode listing for linear regression.

$$E_i \sim N(0, \sigma_E^2)$$

The unknown variance σ_E^2 can be estimated by defining

$$S_E^2 = \frac{1}{n}\sum_{i=1}^{n} E_i^2 = \frac{1}{n}\sum_{i=1}^{n}[Y_i - (\hat{a} + \hat{b}X_i)]^2$$

Using the earlier result that $\hat{a} = \bar{Y} - \hat{b}\bar{X}$ gives a more convenient form:

$$\begin{aligned} S_E^2 &= \frac{1}{n}\sum_i [(Y_i - \bar{Y}) - \hat{b}(X_i - \bar{X})]^2 \\ &= \frac{1}{n}\sum_i [(Y_i - \bar{Y})^2 - 2\hat{b}(X_i - \bar{X})(Y_i - \bar{Y}) + \hat{b}^2(X_i - \bar{X})^2] \\ &= S_Y^2 - 2\hat{b}S_{XY} + \hat{b}^2 S_X^2 \\ &= S_Y^2 - \hat{b}^2 S_X^2 \end{aligned}$$

This result is used in the pseudocode listing (Figure 11.14).

Various confidence intervals are derived in more advanced texts covering linear regression. Here the most useful results will simply be quoted. The $100(1 - \alpha)\%$ confidence interval for the regression slope b is given by

$$\left(\hat{b} \pm t_{\alpha/2,\, n-2} \frac{S_E}{S_X\sqrt{(n-2)}}\right)$$

It is often useful to have an estimate of the mean value of Y for a given value of X, say $X = x$. The point estimate is $\hat{a} + \hat{b}x$, and the $100(1 - \alpha)\%$ confidence interval for this is

$$\left(\hat{a} + \hat{b}x \pm t_{\alpha/2,n-2} S_E \sqrt{\left[\frac{1 + (x - \bar{X})^2/S_X^2}{n-2}\right]}\right)$$

EXAMPLE 11.18 Estimate the residual standard deviation and find a 95% confidence interval for the regression slope for the data in Example 11.17. Also test the hypothesis that the tension in the beam is 10 kN when a voltmeter reading of 15 V is obtained.

Solution Using the results obtained in Example 11.17, the residual standard deviation is

$$S_E = 0.418$$

and, using $t_{0.025,12} = 2.179$, the 95% confidence interval for b is

$$\left(1.22 \pm 2.179 \frac{0.418}{(4.031)\sqrt{12}}\right) = (1.16, 1.29)$$

Obviously the regression slope is significant – but this is not in doubt. To test the hypothesis that the tension is 10 kN, we can use the 95% confidence interval for the corresponding voltage, which is

$$\left(2.89 + 1.22(10) \pm 2.179(0.418)\sqrt{\left[\frac{1 + (10 - 7.5)^2/(4.031)^2}{12}\right]}\right)$$

$$= (14.8, 15.4)$$

The measured value of 15 V lies within this interval, so the hypothesis is accepted at the 5% level. A better way to approach this would be to reverse the regression (use force as the Y variable and voltage as the X variable), so that a confidence interval for the tension in the beam for a given voltage could be obtained and the assumed value tested. For the present data this gives (9.6, 10.1) at 95%, so again the hypothesis is accepted (Exercise 25).

11.5.3 Regression and correlation

Both regression and correlation are statistical methods for measuring the linear dependence of one random variable upon another, so it is not surprising that there is a connection between them. From the definition of the sample correlation $r_{X,Y}$ (Section 11.4.4) and the result for the regression slope $\hat{b}$, it follows immediately that

$$r_{X,Y} = \frac{S_{XY}}{S_X S_Y} = \frac{\hat{b}S_X}{S_Y}$$

Another expression for the residual variance is then

$$S_E^2 = S_Y^2 - \left(\frac{S_Y r_{X,Y}}{S_X}\right)^2 S_X^2 = S_Y^2(1 - r_{X,Y}^2)$$

This result has an important interpretation. S_Y^2 is the total variation in the Y values, and S_E^2 is the residual variation after the regression line has been identified, so $r_{X,Y}^2$ is the proportion of the total variation in the Y values that is accounted for by the regression on X: informally, it represents how closely the points are clustered about the line. This is a measure of goodness of fit that is especially useful when the dependence between X and Y is nonlinear and different models are to be compared.

11.5.4 Nonlinear regression

Sometimes the dependence between two random variables is nonlinear, and this shows clearly in the scatter plot; see for instance Figure 11.15. Fitting a straight line through the data would hardly be appropriate. Instead, various models of the dependence can be assumed and tested. In each case the value of $r_{X,Y}^2$ indicates the success of the model in capturing the dependence. One form of nonlinear regression model involves a quadratic or higher-degree polynomial:

$$Y_i = a_0 + a_1X_i + a_2X_i^2 + E_i \quad (i = 1, \ldots, n)$$

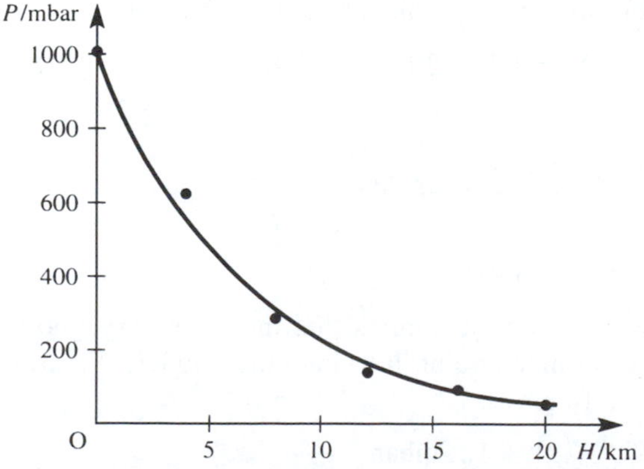

Figure 11.15 Nonlinear regression (Example 11.19).

The three coefficients a_0, a_1 and a_2 can be identified by a multivariate regression method that is beyond the scope of this text. A simpler approach is to try models of the form

$$Y_i = aX_i^b F_i \quad (i = 1, \dots, n)$$

or

$$Y_i = F_i \exp(a + bX_i) \quad (i = 1, \dots, n)$$

where each F_i is a residual multiplicative error. On taking logarithms, these models reduce to the standard linear form:

$$\ln Y_i = \ln a + b \ln X_i + E_i \quad (i = 1, \dots, n)$$

or

$$\ln Y_i = a + bX_i + E_i \quad (i = 1, \dots, n)$$

which can be solved by the usual method.

EXAMPLE 11.19 The following data for atmospheric pressure (P, in mbar) at various heights (H, in km) have been obtained:

Height H	0	4	8	12	16	20
Pressure P	1012	621	286	141	104	54

The relationship between height and pressure is believed to be of the form

$$P = e^{a+bH}$$

where a and b are constants. Fit and assess a model of this form and predict the atmospheric pressure at a height of 14 km.

Solution Taking logarithms and setting $Y = \ln P$ gives

$$Y_i = a + bH_i + E_i$$

for which the following results are easily obtained:

$$\bar{H} = 10, \quad S_H = 6.831\,30, \quad \bar{Y} = 5.431\,52, \quad S_Y = 1.016\,38, \quad \overline{HY} = 47.4081$$

Hence

$$\hat{b} = \frac{\overline{HY} - (\bar{H})(\bar{Y})}{S_H^2} = -0.148$$

$$\hat{a} = \bar{Y} - \hat{b}\bar{H} = 6.91$$

Also, $r_{H,Y}^2 = 0.99$, which implies that the fit is very good (Figure 11.15). In this case there is not much point in trying other models. Finally, the predicted pressure at a height of 14 km is

$$P = e^{6.91-0.148(14)} = 126\,\text{mbar}$$

11.5.5 Exercises

24 Measured deflections (in mm) of a structure under a load (in kg) were recorded as follows:

Load X	1	2	3	4	5	6	7	8	9	10	11	12
Deflection Y	16	35	45	74	86	96	106	124	134	156	164	182

Draw a scatter plot of the data. Find the linear regression coefficients and predict the deflection for a load of 15 kg.

25 Using the data in Example 11.17, carry out a regression of force against voltage, and obtain a 95% confidence interval for the tension in the beam when the voltmeter reads 15 V, as described in Example 11.18.

26 Weekly advertising expenditures X_i and sales Y_i for a company are as follows (in units of £100):

X_i	40	20	25	20	30	50	40	20	50	40	25	50
Y_i	385	400	395	365	475	440	490	420	560	525	480	510

(a) Fit a regression line and predict the sales for an advertising expenditure of £6000.
(b) Estimate the residual standard deviation and find a 95% confidence interval for the regression slope. Hence test the hypothesis that the sales do not depend upon advertising expenditure.
(c) Find a 95% confidence interval for the mean sales when advertising expenditure is £6000.

27 A machine that can be run at different speeds produces articles, of which a certain number are defective. The number of defective items produced per hour depends upon machine speed (in rev s^{-1}) as indicated in the following experimental run:

Speed	8	9	10	11	12	13	14	15
Defectives per hour	7	12	13	13	13	16	14	18

Find the regression line for number of defectives against speed, and a 90% confidence interval for the mean number of defectives per hour when the speed is 14 rev s^{-1}.

28 Sometimes it is required that the regression line passes through the origin, in which case the only regression coefficient is the slope of the line. Use the least-squares procedure to show that the estimate of the slope is then

$$\hat{b} = (\Sigma_i X_i Y_i)/\Sigma_i X_i^2$$

29 A series of measurements of voltage across and current through a resistor produced the following results:

Voltage (V)	1	2	3	4	5	6	7	8	9	10	11	12
Current (mA)	6	18	27	30	42	48	58	69	74	81	94	99

Estimate the resistance, using the result of the previous exercise.

30 The pressure P of a gas corresponding to various volumes V was recorded as follows:

V (cm^3)	50	60	70	90	100
P (kg cm^{-2})	64.7	51.3	40.5	25.9	7.8

The ideal gas law is given by the equation

$$PV^\lambda = C$$

where λ and C are constants. By taking logarithms and using the least-squares method, estimate λ and C from the data and predict P when $V = 80\ cm^3$.

31 The following data show the unit costs of producing certain electronic components and the number of units produced:

Lot size X_i	50	100	250	500	1000	2000	5000
Unit cost Y_i	108	65	21	13	4	2.2	1

Fit a model of the form $Y = aX^b$ and predict the unit cost for a lot size of 300.

11.6 Goodness-of-fit tests

The common classes of distributions, especially the binomial, Poisson and normal distributions, which often govern the data in experimental contexts, are used as the basis for statistical methods of estimation and testing. A question that naturally arises is whether or not a given set of data actually follows an assumed distribution. If it does then the statistical methods can be used with confidence. If not then some alternative should be considered. The general procedure used for testing this can also be used to test for dependence between two variables.

11.6.1 Chi-square distribution and test

No set of data will follow an assumed distribution exactly, but there is a general method for testing a distribution as a statistical hypothesis. If the hypothesis is accepted then it is reasonable to use the distribution as an approximation to reality.

First the data must be partitioned into classes. If the data consist of observations from a discrete distribution then they will be in classes already, but it may be appropriate to combine some classes if the numbers of observations are small. For each class the number of observations that would be expected to occur under the assumed distribution can be worked out. The following quantity acts as a test statistic for comparing the observed and expected class numbers:

$$\chi^2 = \sum_{k=1}^{m} \frac{(f_k - e_k)^2}{e_k}$$

where f_k is the number of observations in the kth class, e_k is the expected number in the kth class and m is the number of classes. Clearly, χ^2 is a non-negative quantity whose magnitude indicates the extent of the discrepancy between the histogram of data and the assumed distribution. For small samples the histogram is erratic and the comparison invalid, but for large samples the histogram should approximate the true distribution. It can be shown that for a large sample the random variable χ^2 has a **'chi-square' distribution**. This class of distributions is widely used in statistics, and a typical chi-square probability density function is shown in Figure 11.16. We are interested in particular in the value of $\chi^2_{\alpha,n}$ to the right of which the area under the density function curve is α, where n is the (single) parameter of the distribution. These values are extensively tabulated; a typical table is shown in Figure 11.17.

The hypothesis of the assumed distribution is rejected if

$$\chi^2 > \chi^2_{\alpha,m-t-1}$$

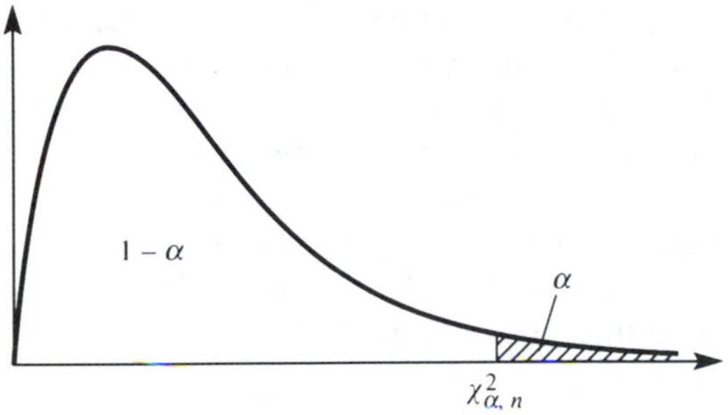

Figure 11.16 The chi-square distribution with $\chi^2_{\alpha,n}$.

n	$\alpha = 0.05$	$\alpha = 0.025$	$\alpha = 0.01$	$\alpha = 0.005$	n
1	3.841	5.024	6.635	7.879	1
2	5.991	7.378	9.210	10.597	2
3	7.815	9.348	11.345	12.838	3
4	9.488	11.143	13.277	14.860	4
5	11.070	12.832	15.086	16.750	5
6	12.592	14.449	16.812	18.548	6
7	14.067	16.013	18.475	20.278	7
8	15.507	17.535	20.090	21.955	8
9	16.919	19.023	21.666	23.589	9
10	18.307	20.483	23.209	25.188	10
11	19.675	21.920	24.725	26.757	11
12	21.026	23.337	26.217	28.300	12
13	22.362	24.736	27.688	29.819	13
14	23.685	26.119	29.141	31.319	14
15	24.996	27.488	30.578	32.801	15
16	26.296	28.845	32.000	34.267	16
17	27.587	30.191	33.409	35.718	17
18	28.869	31.526	34.805	37.156	18
19	30.144	32.852	36.191	38.582	19
20	31.410	34.170	37.566	39.997	20
21	32.671	35.479	38.932	41.401	21
22	33.924	36.781	40.289	42.796	22
23	35.172	38.076	41.638	44.181	23
24	36.415	39.364	42.980	45.558	24
25	37.652	40.646	44.314	46.928	25
26	38.885	41.923	45.642	48.290	26
27	40.113	43.194	46.963	49.645	27
28	41.337	44.461	48.278	50.993	28
29	42.557	45.722	49.588	52.336	29
30	43.773	46.979	50.892	53.672	30

Figure 11.17 Table of the chi-square distribution $\chi^2_{\alpha,n}$.

where α is the significance level and t is the number of independent parameters estimated from the data and used for computing the e_k values. The significance level is the probability of false rejection, as discussed in Section 11.3.4. The only difference is that there is now no estimation procedure underlying the hypothesis test, which must stand alone. Sometimes the hypothesis is deliberately vague, for example a parameter value may be left unspecified. If the data themselves are used to fix parameter values in the assumed distribution before testing then the test must be strengthened to allow for this in the form of a correction t in the chi-square parameter.

A useful rule of thumb when using this test is that there should be at most a small number (one or two) of classes with an expected number of observations less than five. If necessary, classes in the tails of the distribution can be merged.

EXAMPLE 11.20 A die is tossed 600 times, and the numbers of occurrences of the numbers one to six are recorded respectively as 89, 113, 98, 104, 117 and 79. Is the die fair or biased?

Solution The expected values are $e_k = 100$ for all k, and the test value is $\chi^2 = 10.4$. This is less than $\chi^2_{0.05,5} = 11.07$, so we should expect results as erratic as this at least once in 20 similar experiments. The die may well be biased, but the data are insufficient to justify this conclusion.

EXAMPLE 11.21 The numbers of trucks arriving per hour at a warehouse are counted for each of 500 h. Counts of zero up to eight arrivals are recorded on respectively 52, 151, 130, 102, 45, 12, 5, 1 and 2 occasions. Test the hypothesis that the numbers of arrivals have a Poisson distribution, and estimate how often there will be nine or more arrivals in one hour.

Solution The hypothesis stipulates a Poisson distribution, but without specifying the parameter λ. Since the mean of the Poisson distribution is λ and the average number of arrivals per hour is 2.02 from the data, it is reasonable to assume that $\lambda = 2$. The columns in the table in Figure 11.18 show the observed counts f_k, the Poisson

Trucks	f_k	p_k	e_k	χ^2
0	52	0.1353	67.7	3.63
1	151	0.2707	135.3	1.81
2	130	0.2707	135.3	0.21
3	102	0.1804	90.2	1.54
4	45	0.0902	45.1	0.00
5	12	0.0361	18.0	2.02
6	5	0.0120	6.0	0.17
7	3	0.0046	2.3	0.24
Totals	500	1.0	500	9.62

Figure 11.18 Chi-square calculation for Example 11.21.

probabilities p_k, the expected counts $e_k = 500p_k$ and the individual χ^2 values for each class. The last two classes have been combined because the numbers are so small. One parameter has been estimated from the data, so the total χ^2 value is compared with $\chi^2_{0.05,6} = 12.59$. The Poisson hypothesis is accepted, and on that basis the probability of nine or more trucks arriving in one hour is

$$P(9 \text{ or more}) = 1 - \sum_{k=0}^{8} \frac{2^k e^{-2}}{k!} = 0.000\,237$$

This will occur roughly once in every 4200 h of operation.

Because so many statistical methods assume normal data, it is important to have a test for normality, and the chi-square method can be used (Exercise 36 and Section 11.8.4).

11.6.2 Contingency tables

In Section 11.4.3 the correlation was introduced as a measure of dependence between two random variables. The sample correlation (Section 11.4.4) provides an estimate from data. This measure only applies to numerical random variables, and then only works for linear dependence (Exercise 14). The rank correlation (Section 11.4.6) has more general applicability, but still requires that the data be classified in order of rank. The chi-square testing procedure can be adapted to provide at least an indicator of dependence that has the widest applicability of all.

Suppose that each item in a sample of size n can be separately classified as one of $A_1, \ldots, A_r$ by one criterion, and as one of $B_1, \ldots, B_c$ by another (these may be numerical values, but not necessarily). The number f_{ij} of items in the sample that are classified as 'A_i and B_j' can be counted for each $i = 1, \ldots, r$ and $j = 1, \ldots, c$. The table of these numbers (with r rows and c columns) is called a **contingency table** (Figure 11.19). The question is whether the two criteria are

Class	B_1	...	B_c	Total
A_1	f_{11}	...	f_{1c}	f_{1+}
⋮	⋮		⋮	⋮
A_r	f_{r1}	...	f_{rc}	f_{r+}
Total	f_{+1}	...	f_{+c}	n

Figure 11.19 Contingency table.

independent. If not then some combinations of A_i and B_j will occur significantly more often (and others less often) than would be expected under the assumption of independence. We first have to work out how many would be expected under an assumption of independence.

Let the row and column totals be denoted by

$$f_{i+} = \sum_{j=1}^{c} f_{ij} \quad (i = 1, \ldots, r)$$

$$f_{+j} = \sum_{i=1}^{r} f_{ij} \quad (j = 1, \ldots, c)$$

If the criteria are independent then the joint probability for each combination can be expressed as the product of the separate marginal probabilities:

$$P(A_i \cap B_j) = P(A_i)P(B_j)$$

The chi-square procedure will be used to see how well the data fit this assumption. To test it, we can estimate the marginal probabilities from the row and column totals,

$$P(A_i) \simeq \frac{f_{i+}}{n}, \quad P(B_j) \simeq \frac{f_{+j}}{n}$$

and multiply the product of these by n to obtain the expected number e_{ij} for each combination:

$$e_{ij} = n\frac{f_{i+}}{n}\frac{f_{+j}}{n} = \frac{f_{i+}f_{+j}}{n}$$

The chi-square goodness-of-fit statistic follows from the actual and expected numbers (f_{ij} and e_{ij}) as a sum over all the rows and columns:

$$\chi^2 = \sum_{i=1}^{r}\sum_{j=1}^{c} \frac{(f_{ij} - e_{ij})^2}{e_{ij}}$$

If the value of this is large then the hypothesis of independence is rejected, because the actual and expected counts differ by more than can be attributed to chance. As explained in Section 11.6.1, the largeness is judged with respect to $\chi^2_{\alpha,m-t-1}$ from the chi-square table. The number of classes, m, is the number of rows times the number of columns, rc. The number of independent parameters estimated from the data, t, is the number of independent marginal probabilities $P(A_i)$ and $P(B_j)$:

$$t = (r - 1) + (c - 1)$$

The number is not $r + c$, because the row and column totals must equal one, so when all but one are specified, the last is determined. Finally,

$$m - t - 1 = rc - (r + c - 2) - 1 = (r - 1)(c - 1)$$

The hypothesis of independence is therefore rejected (at significance level α) if

$$\chi^2 > \chi^2_{\alpha,(r-1)(c-1)}$$

EXAMPLE 11.22 An accident inspector makes spot checks on working practices during visits to industrial sites chosen at random. At one large construction site the numbers of accidents occurring per week were counted for a period of three years, and each week was also classified as to whether or not the inspector had visited the site during the previous week. The results are shown in bold print in Figure 11.20. Do visits by the inspector tend to reduce the number of accidents, at least in the short term?

	Number of accidents				
	0	1	2	3	Total
Visit Residual	**20** (13.38) 2.96	**3** (7.08) −1.99	**1** (2.46) −1.07	**0** (1.08) −1.16	24
No visit Residual	**67** (73.62) −2.96	**43** (38.92) 1.99	**15** (13.54) 1.07	**7** (5.92) 1.16	132
Total	87	46	16	7	156

Figure 11.20 Contingency table for Example 11.22.

Solution The respective row and column totals are shown in Figure 11.20, together with the expected numbers e_{ij} in parentheses in each cell. For example, the top left cell has observed number 20, row total 24, column total 87, $n = 156$, and hence the expected number

$$e_{11} = (24)(87)/156 = 13.38$$

The chi-square sum is

$$\chi^2 = \frac{(20 - 13.38)^2}{13.38} + \ldots + \frac{(7 - 5.92)^2}{5.92} = 8.94$$

With two rows, four columns and a significance level of 5%, the appropriate number from the chi-square table is $\chi^2_{0.05,3} = 7.815$. The calculated value exceeds this, and by comparing the observed and expected numbers in the table, it seems clear that the visits by the inspector do tend to reduce the number of accidents. This is not, however, significant at the 2.5% level.

A significant chi-square value does not by itself reveal what part or parts of the table are responsible for the lack of independence. A procedure that is often helpful in this respect is to work out the **adjusted residual** for each cell, defined as

$$d_{ij} = \frac{f_{ij} - e_{ij}}{\sqrt{[e_{ij}(1 - f_{i+}/n)(1 - f_{+j}/n)]}}$$

Under the assumption of independence, these are approximately standard normal, so a significant value for a cell suggests that that cell is partly responsible for the dependence overall. The adjusted residuals for the contingency table in Example 11.22 are shown in Figure 11.20, and support the conclusion that visits by the inspector tend to reduce the number of accidents.

For a useful survey of procedures for analysing contingency tables see B. S. Everitt, *The Analysis of Contingency Tables*, 2nd edn (Chapman & Hall, London, 1992).

11.6.3 Exercises

32 The number of books borrowed from a library that is open five days a week is as follows: Monday 153, Tuesday 108, Wednesday 120, Thursday 114, Friday 145. Test (at 5% significance) whether the number of books borrowed depends on the day of the week.

33 A new process for manufacturing light fibres is being tested. Out of 50 samples, 32 contained no flaws, 12 contained one flaw and 6 contained two flaws. Test the hypothesis that the number of flaws per sample has a Poisson distribution.

34 In an early experiment on the emission of α-particles from a radioactive source, Rutherford obtained the following data on counts of particles during constant time intervals:

Number of particles	0	1	2	3	4	5	6	7	8	9	10	>10
Number of intervals	57	203	383	525	532	408	273	139	45	27	10	6

Test the hypothesis that the number of particles emitted during an interval has a Poisson distribution.

35 A quality control engineer takes daily samples of four television sets coming off an assembly line. In a total of 200 working days he found that on 110 days no sets required adjustments, on 73 days one set requires adjustments, on 16 days two sets and on 1 day three sets. Use these results to test the hypothesis that 10% of sets coming off the assembly line required adjustments, at 5% and 1% significance levels. Also test this using the confidence interval for proportion (Section 11.3.6), using the total number of sets requiring adjustments.

36 Two samples of 100 data have been grouped into classes as shown in Figure 11.21. The sample average and standard deviation in each case were 10.0 and 2.0 respectively.

(a) Draw the histogram for each sample.
(b) Test each sample for normality using the measured parameters.

Class	Sample 1	Sample 2
< 6.5	4	3
6.5– 7.5	6	6
7.5– 8.5	16	16
8.5– 9.5	16	13
9.5–10.5	17	26
10.5–11.5	20	7
11.5–12.5	12	19
12.5–13.5	6	5
> 13.5	3	5

Figure 11.21 Data classification for Exercise 36.

37 Shipments of electronic devices have been received by a firm from three sources: A, B and C. Each device is classified as either perfect, intermediate (imperfect but acceptable), or unacceptable. From source A 89 were perfect, 23 intermediate and 12 unacceptable. Corresponding figures for source B were 62, 12 and 8 respectively, and for source C 119, 30 and 21 respectively. Is there any significant difference in quality between the devices received from the three sources?

38 Cars produced at a factory are chosen at random for a thorough inspection. The number inspected and the number of those that were found to be unsuitable for shipment were counted monthly for one year as follows:

Month	Jan.	Feb.	Mar.	Apr.	May	Jun.
Inspected	450	550	550	400	600	450
Defective	8	14	6	3	7	8
Month	Jul.	Aug.	Sep.	Oct.	Nov.	Dec.
Inspected	450	200	450	600	600	550
Defective	16	5	12	6	15	9

Is there a significant variation in quality through the year?

11.7 Moment generating functions

This section is more difficult than the rest of this chapter, and can be treated as optional during a first reading. It contains the proofs of theoretical results, such as the central limit theorem, that are of great importance, as seen earlier in this chapter. The technique introduced here is the moment generating function, which is a useful tool for finding means and variances of random variables as well as for proving these essential results. The moment generating function also bears a striking resemblance to the Laplace transform considered in Chapter 2.

11.7.1 Definition and simple applications

The **moment generating function** of a random variable X is defined as

$$M_X(t) = E(\mathrm{e}^{tX}) = \begin{cases} \displaystyle\sum_{k=1}^{m} \mathrm{e}^{tv_k} P(X = v_k) & \text{in the discrete case} \\ \displaystyle\int_{-\infty}^{\infty} \mathrm{e}^{tx} f_X(x)\,\mathrm{d}x & \text{in the continuous case} \end{cases}$$

where t is a real variable. This is an example of the expected value of a function $h(X)$ of the random variable X (Section 11.2.2). The moment generating function does not always exist, or it may exist only for certain values of t. In cases where it fails to exist there is an alternative (called the **characteristic function**) which always exists and has similar properties. When the moment generating function does exist, it is unique in the sense that no two distinct distributions can have the same moment generating function.

To see how the moment generating function earns its name, the first step is to differentiate it with respect to t and then let t tend to zero:

$$\frac{\mathrm{d}}{\mathrm{d}t}M_X(t)|_{t=0} = E(X\,\mathrm{e}^{tX})|_{t=0} = E(X) = \mu_X$$

Thus the first derivative gives the mean. Differentiating again gives

$$\frac{\mathrm{d}^2}{\mathrm{d}t^2}M_X(t)|_{t=0} = E(X^2\,\mathrm{e}^{tX})|_{t=0} = E(X^2)$$

From this result we obtain the variance (from Section 11.2.2)

$$\mathrm{Var}(X) = E(X^2) - [E(X)]^2$$

We can summarize these results as follows. If a random variable X has mean μ_X, variance σ_X^2 and moment generating function $M_X(t)$ defined for t in some neighbourhood of zero then

$$\mu_X = M_X^{(1)}(0), \quad \sigma_X^2 = M_X^{(2)}(0) - [M_X^{(1)}(0)]^2$$

where the superscript in parentheses denotes the order of the derivative. Furthermore,

$$E(X^k) = M_X^{(k)}(0) \quad (k = 1, 2, \dots)$$

EXAMPLE 11.23 Show that the mean and variance of the Poisson distribution with parameter λ are both equal to λ (Section 11.2.3).

Solution If X has a Poisson distribution then

$$P(X = k) = \frac{\lambda^k \mathrm{e}^{-\lambda}}{k!} \quad (k = 0, 1, 2, \dots)$$

and the moment generating function is

$$M_X(t) = \sum_{k=0}^{\infty} \mathrm{e}^{kt}\frac{\lambda^k \mathrm{e}^{-\lambda}}{k!} = \mathrm{e}^{-\lambda}\sum_{k=0}^{\infty}\frac{(\lambda\mathrm{e}^t)^k}{k!} = \exp[\lambda(\mathrm{e}^t - 1)]$$

Differentiating this with respect to t gives

$$M_X^{(1)}(t) = \lambda\mathrm{e}^t\exp[\lambda(\mathrm{e}^t - 1)]$$

$$M_X^{(2)}(t) = (\lambda^2\mathrm{e}^{2t} + \lambda\mathrm{e}^t)\exp[\lambda(\mathrm{e}^t - 1)]$$

from which

$$\mu_X = M_X^{(1)}(0) = \lambda$$

$$\sigma_X^2 = M_X^{(2)}(0) - \mu_X^2 = \lambda$$

as expected.

EXAMPLE 11.24 Show that the mean and standard deviation of the exponential distribution

$$f_X(x) = \begin{cases} \lambda e^{-\lambda x} & (x \geqslant 0) \\ 0 & (x < 0) \end{cases}$$

are both equal to $1/\lambda$ (Section 11.10.2).

Solution The moment generating function is

$$M_X(t) = \int_0^\infty \lambda e^{-(\lambda - t)x} \, dx = \frac{\lambda}{\lambda - t}$$

Note that the integral only exists for $t < \lambda$, but we can differentiate it and set $t = 0$ for any positive value of λ:

$$M_X^{(1)}(0) = \lambda^{-1} = \mu_X$$

$$M_X^{(2)}(0) = 2\lambda^{-2}$$

from which $\sigma_X^2 = \lambda^{-2}$.

11.7.2 The Poisson approximation to the binomial

In addition to its utility for finding means and variances, the moment generating function is a very useful theoretical tool. In this section and the next we shall use it to prove two of the most important results in probability theory, but first we need the following general property of moment generating functions.

> Suppose that X and Y are independent random variables with moment generating functions $M_X(t)$ and $M_Y(t)$. Then the moment generating function of their sum is given by
>
> $$M_{X+Y}(t) = M_X(t)M_Y(t)$$

To prove this, we shall assume that X and Y are continuous random variables with a joint density function $f_{X,Y}(x, y)$; however, the proof is essentially the same if either or both are discrete. By definition,

$$M_{X+Y}(t) = E[e^{t(X+Y)}] = \int_{-\infty}^{\infty} \int_{-\infty}^{\infty} e^{t(x+y)} f_{X,Y}(x, y) \, dx \, dy$$

Both factors of the integrand themselves factorize (noting the independence of X and Y) to give

$$M_{X+Y}(t) = \int_{-\infty}^{\infty} \int_{-\infty}^{\infty} [e^{tx} f_X(x)][e^{ty} f_Y(y)] \, dx \, dy$$

The two integrals can now be separated, and the result follows:

$$M_{X+Y}(t) = \int_{-\infty}^{\infty} e^{tx} f_X(x)\,dx \int_{-\infty}^{\infty} e^{ty} f_Y(y)\,dy$$
$$= M_X(x)M_Y(y)$$

It follows that if $\{X_1, \ldots, X_n\}$ are independent and identically distributed random variables, each with moment generating function $M_X(t)$, and if $Z = X_1 + \ldots + X_n$, then

$$M_Z(t) = [M_X(t)]^n$$

We now proceed to the main result. If the random variable Y has a binomial distribution with parameters n and p and the random variable X has a Poisson distribution with parameter $\lambda = np$ then as $n \to \infty$ and $p \to 0$, the distribution of X tends to that of Y.

First let the random variable B have a Bernoulli distribution with parameter p (Section 11.2.3). The moment generating function of B is then

$$M_B(t) = e^t p + (1 - p) = 1 + p(e^t - 1)$$

Since the binomial random variable Y is the sum of n copies of B, it follows that the moment generating function of the binomial distribution is

$$M_Y(t) = [1 + p(e^t - 1)]^n$$

It can be proved that for any real z

$$\lim_{n\to\infty} \left(1 + \frac{z}{n}\right)^n = e^z$$

Now let $z = np(e^t - 1)$ and assume that $\lambda = np$:

$$M_Y(t) = \left(1 + \frac{z}{n}\right)^n \simeq e^z = \exp[\lambda(e^t - 1)] = M_X(t)$$

(the moment generating function for the Poisson distribution was derived in Example 11.23). The uniqueness of the moment generating function implies that the distributions of X and Y must be similar, provided that n is sufficiently large. It is also required that $p \to 0$ as $n \to \infty$, so that z does not grow without bound.

This approximation has many applications; see for example Section 11.9.2.

11.7.3 Proof of the central limit theorem

This theorem is of vital importance in statistics because it tells us that the sample average for a sample of at least moderate size tends to have a normal distribution even when the data themselves do not. The result (Section 11.2.4) is here restated and proved using the moment generating function.

THEOREM 11.1 The central limit theorem

If $\{X_1, \dots, X_n\}$ are independent and identically distributed random variables, each with mean μ_X and variance σ_X^2, and if

$$Z_n = \frac{X_1 + \dots + X_n - n\mu_X}{\sigma_X\sqrt{n}}$$

then the distribution of Z_n tends to the standard normal distribution as $n \to \infty$.

Proof First let the random variables Y_i be defined by

$$Y_i = \frac{X_i - \mu_X}{\sigma_X\sqrt{n}} \quad (i = 1, \dots, n)$$

so that

$$E(Y_i) = 0$$

$$E(Y_i^2) = \frac{1}{n}$$

$$E(Y_i^3) = \frac{c}{n\sqrt{n}}, \quad \text{where } c \text{ is a constant}$$

Expanding the moment generating function $M_Y(t)$ for each Y_i as a Maclaurin series (to four terms) gives

$$M_Y(t) \simeq M_Y(0) + M_Y^{(1)}(0)t + M_Y^{(2)}(0)\frac{t^2}{2!} + M_Y^{(3)}(0)\frac{t^3}{3!}$$

By the results in Section 11.7.1, the coefficients of this series can be replaced by the successive moments $E(Y_i^k)$:

$$M_Y(t) \simeq E(1) + E(Y_i)t + E(Y_i^2)\frac{t^2}{2!} + E(Y_i^3)\frac{t^3}{3!} \simeq 1 + \frac{t^2}{2n} + \frac{ct^3}{6n\sqrt{n}}$$

Now Z_n is the sum of the variables Y_i,

$$Z_n = \sum_{i=1}^{n} Y_i$$

so the moment generating function of Z_n is the nth power of that of Y_i:

$$M_{Z_n}(t) = [M_Y(t)]^n$$

Retaining only the first two terms of a binomial expansion of this gives

$$M_{Z_n}(t) = \left(1 + \frac{t}{2n}\right)^n + n\left(1 + \frac{t^2}{2n}\right)^{n-1}\left(\frac{ct^3}{6n\sqrt{n}}\right) + \dots$$

$$\to e^{t^2/2} \quad \text{as } n \to \infty$$

because all terms except the first will tend to zero.

It only remains to show that this is the moment generating function for the standard normal distribution; see Exercise 43. □

11.7.4 Exercises

39 A continuous random variable X has density function

$$f_X(x) = \begin{cases} cx\mathrm{e}^{-2x} & (x > 0) \\ 0 & (x \leqslant 0) \end{cases}$$

Find the value of the constant c. Derive the moment generating function, and hence find the mean and variance of X.

40 Prove that if $X_1, \dots, X_n$ are independent Poisson random variables with parameters $\lambda_1, \dots, \lambda_n$ then their sum

$$Y = X_1 + \dots + X_n$$

is also Poisson, with parameter $\lambda = \lambda_1 + \dots + \lambda_n$.

41 A factory contains 30 machines each with breakdown probability 0.01 in any one hour and 40 machines each with breakdown probability 0.005 in any one hour. Use the result of Exercise 40, together with the Poisson approximation to the binomial, to find the probability that a total of three or more machine breakdowns will occur in any one hour.

42 A manufacturer has agreed to dispatch small servomechanisms in cartons of 100 to a distributor. The distributor requires that 90% of cartons contain at most one defective servomechanism. Assuming the Poisson approximation to the binomial, obtain an equation for the Poisson parameter λ such that the distributor's requirements are just satisfied. Solve for λ by one of the standard methods for nonlinear equations (approximate solution 0.5), and hence find the required proportion of manufactured servomechanisms that must be satisfactory.

43 Show that the moment generating function of the standard normal distribution is $e^{t^2/2}$. (*Hint*: Complete the square in the exponential.)

11.8 Engineering application: analysis of engine performance data

11.8.1 Introduction

Statistical methods are often used in conjunction with each other. So far in this chapter the examples and exercises have almost always been designed to illustrate the various topics one at a time. This section contains an example of a more

Engine A				Engine B			
A	*T*	*A*	*T*	*B*	*U*	*B*	*U*
27.7	24	24.1	7	24.9	13	24.3	17
24.3	25	23.1	14	21.4	19	24.5	16
23.7	18	23.4	16	24.1	18	26.1	18
22.1	15	23.1	9	27.5	19	27.7	14
21.8	19	24.1	14	27.5	21	24.3	19
24.7	16	28.6	23	25.7	17	26.1	5
23.4	17	20.2	14	24.9	17	24.0	17
21.6	14	25.7	18	23.3	19	24.9	18
24.5	18	24.6	18	22.5	21	26.7	23
26.1	20	24.0	12	28.5	12	27.3	28
24.8	15	24.9	18	25.9	17	23.9	18
23.7	15	21.9	20	26.9	13	23.1	10
25.0	22	25.1	16	27.7	17	25.5	25
26.9	18	25.7	16	25.4	23	24.9	22
23.7	19	23.5	11	25.3	30	25.9	16

Figure 11.22 Data for engine case study.

extended problem to which several topics are relevant, and correspondingly there are several stages to the analysis.

The background to the problem is this. Suppose that the fuel consumption of a car engine is tested by measuring the time that the engine runs at constant speed on a litre of standard fuel. Two prototype engines, A and B, are being compared for fuel consumption. For each engine a series of tests is performed in various ambient temperatures, which are also recorded. Figure 11.22 contains the data. There are 30 observations each for the four random variables concerned:

A, running time in minutes for engine A;

T, ambient temperature in degrees Celsius for engine A;

B, running time in minutes for engine B;

U, ambient temperature in degrees Celsius for engine B.

The histograms for the running times are compared in Figure 11.23(a) and those for the temperatures in Figure 11.23(b). The overall profile of temperatures is very similar for the two series of tests, differing only in the number of unusually high or low figures encountered. The profiles of running times appear to differ rather more markedly. It is clear that displaying the data in this way is useful, but some analysis will have to be done in order to determine whether the differences are significant.

When planning the analysis, it is as well to consider the questions to which most interest attaches. Do the mean running times for the two engines differ? Does the running time depend on temperature? If so, and if there is a difference in the temperatures for the test series on engines A and B, can this account for any apparent difference in fuel consumption? More particularly, are the data normal? This has a bearing on the methods used, and hence on the conclusions drawn.

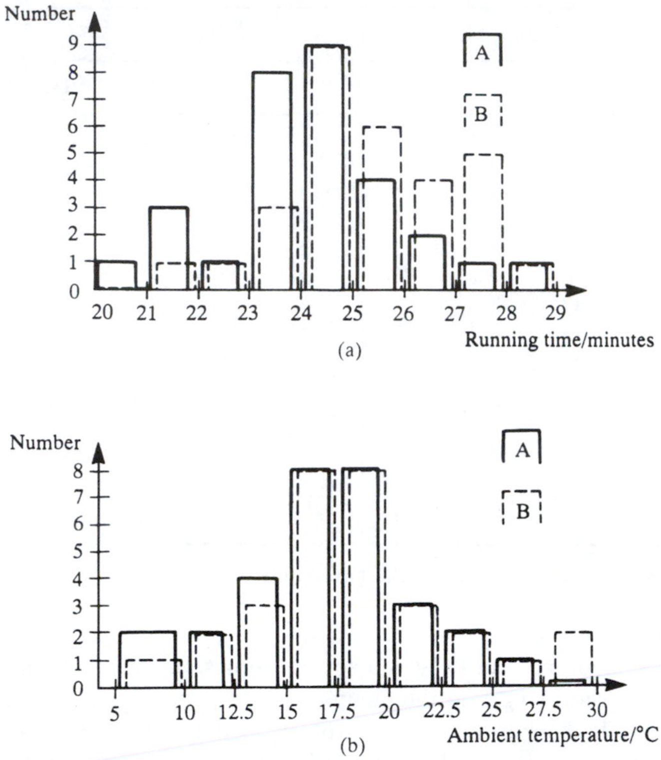

Figure 11.23 Histograms of engine data: (a) running times; (b) temperatures.

11.8.2 Difference in mean running times and temperatures

The sample averages and both versions of standard deviation for the data in Figure 10.22 are as follows:

$$\begin{aligned}
\bar{A} &= 24.20, & \bar{T} &= 16.70, & \bar{B} &= 25.36, & \bar{U} &= 18.07\\
S_A &= 1.761, & S_T &= 4.001, & S_B &= 1.657, & S_U &= 4.932\\
S_{A,n-1} &= 1.791, & S_{T,n-1} &= 4.070, & S_{B,n-1} &= 1.685, & S_{U,n-1} &= 5.017
\end{aligned}$$

The average running time for engine B is slightly higher than for engine A. The sample standard deviations are very similar, so we can assume that the true standard deviations are equal and use the method for comparing means discussed in Section 11.3.5. The pooled estimate of the variance is

$$S_p^2 = \frac{(n-1)(S_{A,n-1}^2 + S_{B,n-1}^2)}{2(n-1)} = \frac{3.208 + 2.839}{2} = 3.023$$

and the relevant value from the t-table is $t_{0.025,58} = 1.960$. In fact the sample is large enough for the value for the normal distribution to be taken. The 95% confidence interval for the difference $\mu_B - \mu_A$ is

$$(\bar{B} - \bar{A} \pm 1.96S_p/\sqrt{15}) = (0.28, 2.04)$$

Even the 99% confidence interval (0.12, 2.20) does not contain the zero point, so we can be almost certain that the difference in mean running times is significant.

Following the same procedure for the temperatures gives the 95% confidence interval for the difference $\mu_U - \mu_T$ to be (−0.94, 3.68). Superficially, this is not significant – and even if the running times do depend on temperature, the similarity of the two test series in this respect enables this factor to be discounted. If so, the fuel performance for engine B is superior to that for engine A. However, if the temperature sensitivity were very high then even a difference in the average too small to give a significant result by this method could create a misleading difference in the fuel consumption figures. This possibility needs to be examined.

11.8.3 Dependence of running time on temperature

The simplest way to test for dependence is to correlate times and temperatures for each engine. To compute the sample correlations, we need the following additional results from the data:

$$\overline{AT} = 407.28, \quad \overline{BU} = 457.87$$

The sample correlations (Section 11.4.4) of A and T and of B and U are then

$$r_{A,T} = \frac{\overline{AT} - (\bar{A})(\bar{T})}{S_A S_T} = 0.445, \quad r_{B,U} = \frac{\overline{BU} - (\bar{B})(\bar{U})}{S_B S_U} = -0.030$$

and the respective 95% confidence intervals (Section 10.4.5) are

$$(0.10, 0.69), \quad (-0.39, 0.34)$$

This is a quite definitive result: the running time for engine A depends positively upon the ambient temperature, but that for engine B does not. The confidence intervals are based on the assumption that all the variables A, T, B and U are normal. The histograms have this character, and a test for normality will be covered later.

Linear regression also reveals the dependence of running time on temperature. Here we assume that the variables are related by a linear model as follows:

$$\left.\begin{aligned} A_i &= c_A + d_A T_i + E_i \\ B_i &= c_B + d_B U_i + F_i \end{aligned}\right\} \quad (i = 1, \ldots, n)$$

where c_A, d_A, c_B and d_B are constants, and the random variables E_i and F_i represent residual errors. Using the results of the least-squares analysis in Section 11.5.1, for engine A we have

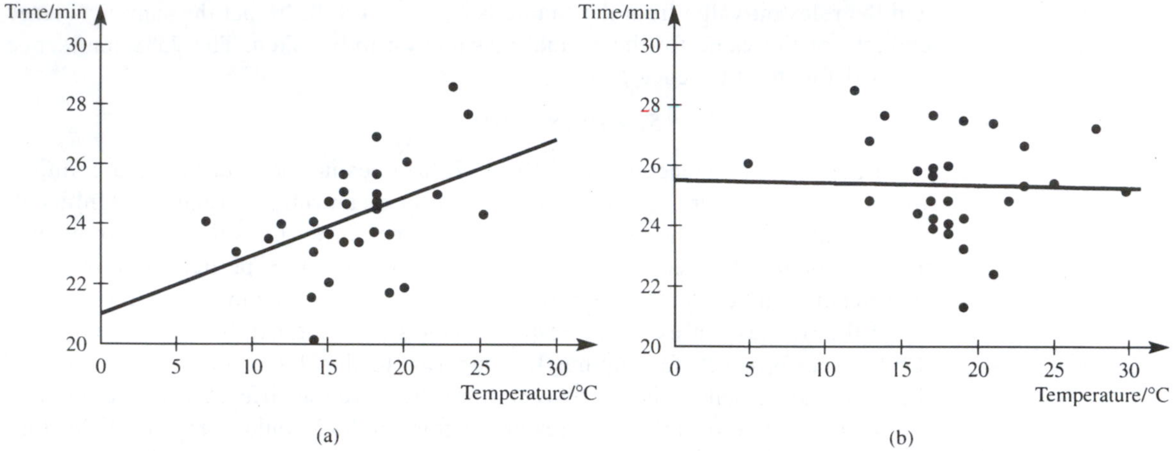

Figure 11.24 Regression of running time against temperature: (a) engine A; (b) engine B.

$$\hat{d}_A = \frac{\overline{AT} - (\bar{A})(\bar{T})}{S_T^2} = 0.196, \quad \hat{c}_A = \bar{A} - \hat{d}_A\bar{T} = 20.9$$

Likewise, for engine B

$$\hat{d}_B = -0.010, \quad \hat{c}_B = 25.5$$

Figure 11.24 contains scatter plots of the data with these regression lines drawn. The points are well scattered about the lines. The residual variances, using the results in Section 11.5.2 and 11.5.3, are

$$S_E^2 = S_A^2 - \hat{d}_A S_T^2 = S_A^2(1 - r_{A,T}^2) = 2.49$$

$$S_F^2 = 2.74$$

As explained in Section 11.5.3, the respective values of r^2 indicate the extent to which the variation in running times is due to the dependence on temperature. For engine B there is virtually no such dependence. For engine A we have $r_{A,T}^2 = 0.198$, so nearly 20% of the variation in running times is accounted for in this way.

If we assume that the residuals E_i and F_i are normal, we can obtain confidence intervals for the regression slopes. The appropriate value from the t table is $t_{\alpha/2,n-2} = t_{0.025,28} = 2.048$, so the 95% confidence interval for d_A is

$$\left(\hat{d}_A \pm 2.048 \frac{S_E}{S_T\sqrt{28}}\right) = (0.04, 0.35)$$

The significance shown here confirms that found for the correlation. The 95% confidence interval for d_B is $(-0.14, 0.12)$.

We now return to the main question. We know that the average running time for engine A was significantly lower than that for engine B. However, we also know that the running time for engine A depends on temperature, and that the average temperature during the test series for engine A was somewhat lower than for engine B. Could this account for the difference?

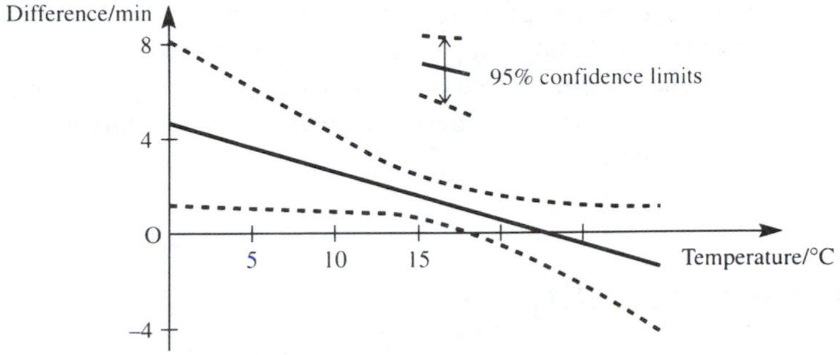

Figure 11.25 Predicted difference in mean running times.

A simple way to test this is to look at the average values. The difference in average temperatures is (18.07 − 16.70) = 1.37 and the regression slope d_A is estimated to be 0.196, so that a deficit in the average running time of (1.37)(0.196) $\simeq$ 0.27 min is predicted on this basis. This cannot account for the actual difference of 1.16 min, and does not even bring the zero point within the 95% confidence interval for the difference in means $\mu_B - \mu_A$.

To examine this more carefully, we can try using the regression slopes to correct all running times to the same temperature. Choosing the average temperature for the series B for this purpose, we can let

$$\left.\begin{aligned} X_i &= A_i + \hat{d}_A(\bar{U} - T_i) \\ Y_i &= B_i + \hat{d}_B(\bar{U} - U_i) \end{aligned}\right\} \quad (i = 1, \ldots, n)$$

The 95% confidence interval for the difference between mean running times, $\mu_Y - \mu_X$, at this temperature is then (0.06, 1.72), using the method in Section 11.3.5. The problem with this is that the estimates of d_A and d_B themselves have rather wide confidence intervals, and it is not satisfactory to adopt a point value to apply the temperature correction.

We need a more direct way to obtain the confidence interval for the difference between mean running times at any particular temperature. This is possible using a more general theory of linear regression, formulated using matrix algebra, which allows for any number of regression coefficients instead of just two; see for example G. A. F. Seber, *Linear Regression Analysis* (Wiley, New York, 1977). Applied to the present problem, estimates of the regression slopes d_A and d_B and intercepts c_A and c_B are obtained simultaneously, with the same values as before, and it now becomes possible to obtain a confidence interval for any linear combination of these four unknowns. In particular, the confidence interval for the difference between mean running times at any temperature t is based on the linear combination

$$(c_B + d_Bt) - (c_A + d_At)$$

Space precludes coverage of the analysis here, but the results can be seen in Figure 11.25. At any temperature below 22.4 °C engine B is predicted to have the advantage over engine A (this temperature being the point at which the regression lines cross). At any temperature below 18.1 °C the 95% confidence interval for the difference is entirely positive, and we should say that engine B has a significant advantage. This is the best comparison of the engines that is possible using the data presented.

11.8.4 Test for normality

The confidence interval statistics are all based on the assumption of normality of the data. Although the sample sizes are reasonably large, so that the central limit theorem can be relied upon to weaken this requirement, it is worth applying a test for normality to see whether there is any clear evidence to the contrary. Here the regression residuals E_i and F_i will be tested using the method described in Section 11.6.1.

Figure 11.26 shows the histogram of all 60 'standardized' residuals. The residuals have zero mean in any case, and are standardized by dividing by the standard deviation so that they can be compared with a standard normal distribution. It is convenient to use intervals of width 0.4, and the comparison is developed in Figure 11.27.

The normal probabilities for each interval are obtained from the standard normal table of the cumulative distribution function $\Phi(z)$, Figure 11.2, taking successive differences:

$$P(z_1 < Z < z_2) = \Phi(z_2) - \Phi(z_1)$$

These probabilities are multiplied by 60 to obtain the expected number in each interval, and the difference between the observed and expected number for each interval is squared and then divided by the expected number to give the contribution to the total chi-square:

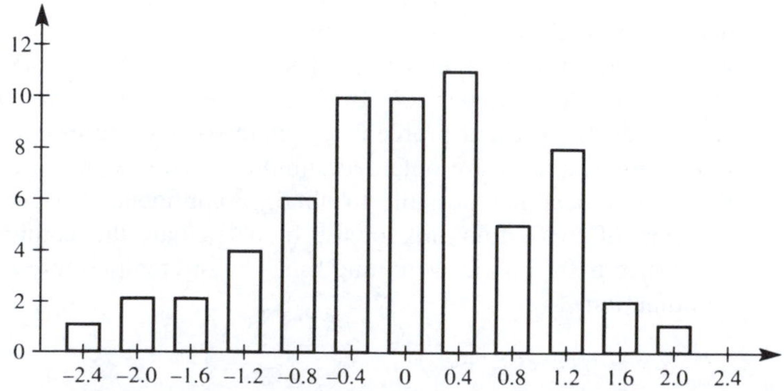

Figure 11.26 Histogram of residuals.

Interval	Observed (f_k)	Probability	Expected (e_k)	Chi-square
$(-\infty, -1.4)$	5	0.0808	4.848	0.005
$(-1.4, -1.0)$	4	0.0779	4.674	0.097
$(-1.0, -0.6)$	6	0.1156	6.936	0.126
$(-0.6, -0.2)$	10	0.1464	8.784	0.168
$(-0.2, +0.2)$	8	0.1586	9.516	0.242
$(+0.2, +0.6)$	11	0.1464	8.784	0.559
$(+0.6, +1.0)$	5	0.1156	6.936	0.540
$(+1.0, +1.4)$	8	0.0779	4.674	2.367
$(+1.4, +\infty)$	3	0.0808	4.848	0.704
Totals	60	1.0	60	4.809

Figure 11.27 Table of the test for normality.

$$\chi^2 = \sum_{k=1}^{m} \frac{(f_k - e_k)^2}{e_k} = 4.81$$

This is small compared with $\chi^2_{0.05,8} = 15.507$, so the hypothesis of normality is accepted.

It is unwise when applying this test in general to have many classes with expected numbers less than five, so the intervals in the tails of the histogram have been merged.

11.8.5 Conclusions

All the questions posed in Section 11.8.1 have now been answered. Engine B has an average running time that is significantly higher than that for engine A, showing that it has the advantage in fuel consumption. However, this statement requires qualification. The running time for engine A depends upon ambient temperature. The temperature difference between the two test series was not significant, and does not account for the difference in average running times. However, engine B will only maintain its fuel advantage up to a certain point. This point has been estimated, but cannot be identified very precisely because of considerable residual scatter in the data. There are many potential sources of this scatter, such as errors in measuring out the fuel, or variations in the quantity and consistency of the engine oil. The scatter has a normal distribution, which justifies the statistics behind the conclusions reached.

11.9 Engineering application: statistical quality control

11.9.1 Introduction

Every manufacturer recognizes the importance of quality, and every manufacturing process involves some variation in the quality of its output, however that is to be measured. Experience shows that tolerating a lack of quality tends to be more costly in the end than promoting a quality approach. It follows that quality control is a major and increasing concern, and methods of statistical quality control are more important than ever. The domain of these methods now extends to the construction and service industries as well as to manufacturing – wherever there is a process that can be monitored in quantitative terms.

Traditionally, quality control involved the accumulation of batches of manufactured items, the testing of samples extracted from these batches, and the acceptance or rejection (with appropriate rectifying action) of these batches depending upon the outcome. The essential problem with this is that it is too late within the process: it is impossible to inspect or test quality into a product. More recently the main concern has been to design the quality into the product or service and to monitor the process to ensure that the standard is maintained, in order to prevent any deficiency. Assurance can then be formally given to the customer that proper procedures are in place.

Control charts play an important role in the implementation of quality. The idea of these is introduced in Section 12.6 in *Modern Engineering Mathematics*, where Shewhart charts for counts of defectives are described. In order for this section to be as self-contained as possible, some of that material is repeated here. This section then covers more powerful control charts and extends the scope of what they monitor.

First note that there are two main alternative measures of quality: **attribute** and **variable**. In attribute measure, regular samples from the process are inspected and for each sample the number that fail according to some criterion is plotted on a chart. In variable measure, regular samples are again taken, but this time the sample average for some numerical measure (such as dimension or lifetime) is plotted.

11.9.2 Shewhart attribute control charts

Figure 11.28 is an example of a Shewhart control chart: a plot of the count of 'defectives' (the number in the sample failing according to some chosen criterion)

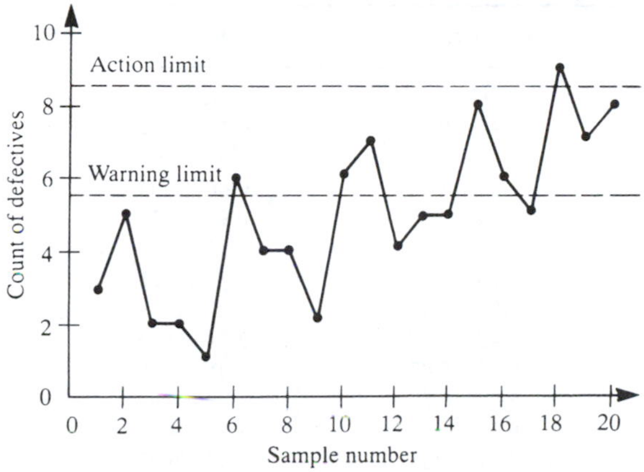

Figure 11.28 Attribute control chart for Example 11.25.

against sample number. It is assumed that a small (specified) proportion of 'defective' items in the process is permitted. Also shown on the chart are two limits on the count of defectives, corresponding to probabilities of one in 40 and one in 1000 of a sample count falling outside if the process is 'in control', that is, conforming to the specification. These are called **warning** and **action limits** respectively, and are denoted by c_W and c_A.

Any sample point falling outside the action limit would normally result in the process being suspended and the problem corrected. Roughly one in 40 sample points will fall outside the warning limit purely by chance, but if this occurs repeatedly or if there is a clear trend upwards in the counts of defectives then action may well be taken before the action limit itself is crossed.

To obtain the warning and action limits, we use the Poisson approximation to the binomial. If the acceptable proportion of defective items is p, usually small, and the sample size is n then for a process in control the defective count C, say, will be a binomial random variable with parameters n and p. Provided that n is not too small, the Poisson approximation can be used (Section 11.7.2):

$$P(C \geqslant c) \simeq \sum_{k=c}^{\infty} \frac{(np)^k \mathrm{e}^{-np}}{k!}$$

Equating this to $\frac{1}{40}$ and then to $\frac{1}{1000}$ gives equations that can be solved for the warning limit c_W and the action limit c_A respectively, in terms of the product np. This is the basis for the table shown in Figure 11.29, which enables c_W and c_A to be read directly from the value of np.

EXAMPLE 11.25 Regular samples of 50 are taken from a process making electronic components, for which an acceptable proportion of defectives is 5%. Successive counts of defectives in each sample are as follows:

c_W or c_A	np for c_W	np for c_A
1.5	<0.44	<0.13
2.5	0.44– 0.87	0.13– 0.32
3.5	0.87– 1.38	0.32– 0.60
4.5	1.38– 1.94	0.60– 0.94
5.5	1.94– 2.53	0.94– 1.33
6.5	2.53– 3.16	1.33– 1.77
7.5	3.16– 3.81	1.77– 2.23
8.5	3.81– 4.48	2.23– 2.73
9.5	4.48– 5.17	2.73– 3.25
10.5	5.17– 5.87	3.25– 3.79
11.5	5.87– 6.59	3.79– 4.35
12.5	6.59– 7.31	4.35– 4.93
13.5	7.31– 8.05	4.93– 5.52
14.5	8.05– 8.80	5.52– 6.12
15.5	8.80– 9.55	6.12– 6.74
16.5	9.55–10.31	6.74– 7.37
17.5	10.31–11.08	7.37– 8.01
18.5	11.08–11.85	8.01– 8.66
19.5	11.85–12.63	8.66– 9.31
20.5	12.63–13.42	9.31– 9.98
21.5	13.42–14.21	9.98–10.65
22.5	14.21–15.00	10.65–11.33
23.5	15.00–15.80	11.33–12.02
24.5	15.80–16.61	12.02–12.71
25.5	16.61–17.41	12.71–13.41
26.5	17.41–18.23	13.41–14.11
27.5	18.23–19.04	14.11–14.82
28.5	19.04–19.86	14.82–15.53
29.5	19.86–20.68	15.53–16.25
30.5		16.25–16.98
31.5		16.98–17.70
32.5		17.70–18.44
33.5		18.44–19.17
34.5		19.17–19.91
35.5		19.91–20.66

Figure 11.29 Shewhart attribute control limits: n is sample size, p is probability of defect, c_W is warning limit and c_A is action limit.

Sample	1	2	3	4	5	6	7	8	9	10	11	12	13	14	15	16	17	18	19	20
Count	3	5	2	2	1	6	4	4	2	6	7	4	5	5	8	6	5	9	7	8

At what point would the decision be taken to stop and correct the process?

Solution The control chart is shown in Figure 11.28. From $np = 2.5$ and Figure 11.29 we have the warning limit $c_W = 5.5$ and the action limit $c_A = 8.5$. The half-integer values are to avoid ambiguity when the count lies on a limit. There are warnings at samples 6, 10, 11, 15 and 16 before the action limit is crossed at sample 18.

Strictly, the decision should be taken at that point, but the probability of two consecutive warnings is less than one in 1600 by the product rule of probabilities, which would justify taking action after sample 11.

An alternative practice (especially popular in the USA) is to dispense with the warning limit and to set the action limit (called the **upper control limit, UCL**) at three standard deviations above the mean. Because the count of defectives is binomial with mean np and variance $np(1-p)$, this means that

$$\text{UCL} = np + 3\sqrt{[np(1-p)]}$$

EXAMPLE 11.26 Find the UCL and apply it to the data in Example 11.25.

Solution From $n = 50$ and $p = 0.05$ we infer that UCL = 7.1, which is between the warning limit c_W and the action limit c_A in Example 11.25. The decision to correct the process would be taken after the 15th sample, the first to exceed the UCL.

11.9.3 Shewhart variable control charts

Suppose now that the appropriate assessment of quality involves measurement on a continuous scale rather than success or failure under some criterion. This arises whenever some dimension of the output is critical for applications. Again we take samples, but this time we measure this critical dimension and average the results. The Shewhart chart for this variable measure is a plot of successive sample averages against sample number.

The warning and action limits c_W and c_A on a Shewhart chart are those points for which the probabilities of a false alarm (where the result exceeds the limit even though the process is in control) are one in 40 and one in 1000 respectively. For variable measure the critical quantity can be either too high or too low, so the sample average must be tested in each direction with the stated probability of exceedance for each limit. It follows that the limits are determined by

$$P(\bar{X} > \mu_X + c_W) = P(\bar{X} < \mu_X - c_W) = \tfrac{1}{40}$$

$$P(\bar{X} > \mu_X + c_A) = P(\bar{X} < \mu_X - c_A) = \tfrac{1}{1000}$$

where $\bar{X}$ is the sample average and μ_X the design mean.

Provided that the sample size n is not too small, the central limit theorem allows the sample average to be assumed normal (Section 11.3.2),

$$\bar{X} \sim N(\mu_X, \sigma_X^2/n)$$

and the normal distribution table (Figure 11.2) then gives

$$c_W = \frac{1.96\sigma_X}{\sqrt{n}}, \quad c_A = \frac{3.09\sigma_X}{\sqrt{n}}$$

EXAMPLE 11.27 Measurements of sulphur dioxide concentration (in μg m^{-3}) in the air are taken daily at five locations, and successive average readings are as follows:

64.2, 56.9, 57.7, 67.9, 61.7, 59.7, 55.6, 63.7
58.3, 66.4, 67.2, 65.2, 63.1, 67.6, 64.1, 66.7

It is suspected that the mean increased during that time. Assuming normal data with a long-term mean of 60.0 and standard deviation of 8.0, investigate whether an increase occurred.

Solution From $n = 5$ and $\sigma_X = 8$ we have $c_W = 7.0$ and $c_A = 11.1$ (Figure 11.30). The warning limit is 67.0, which is exceeded by sample numbers 4, 11 and 14. The action limit is 71.1, which is not exceeded. The readings are suspiciously high – but not sufficiently so for the conclusion to be justified.

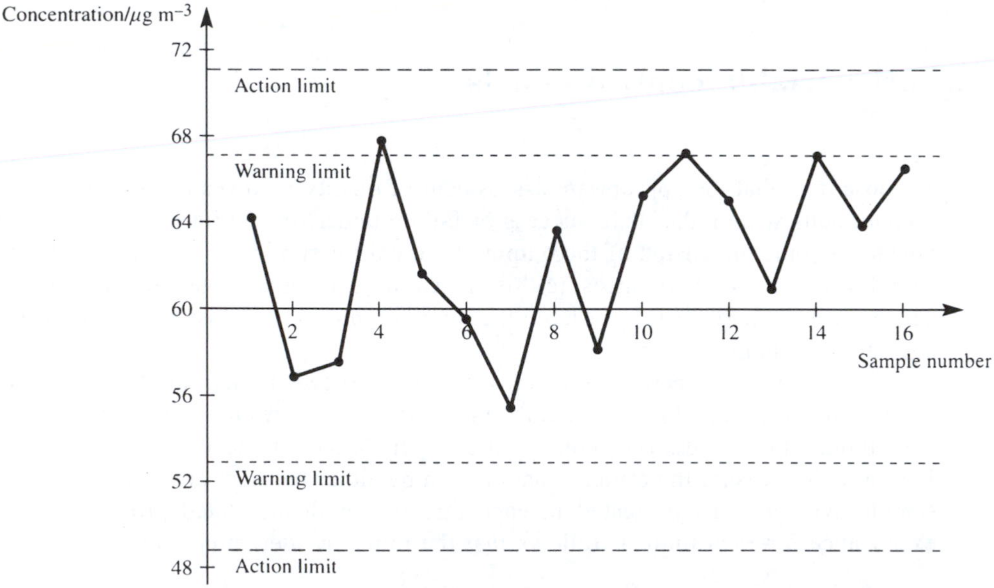

Figure 11.30 Variable control chart for Example 11.27.

As discussed in Section 11.9.2, the practice in the USA is somewhat different: there are no warning limits, only action limits at three standard deviations on either side of the mean. For a variable chart this allows a deviation from the mean of at most $3\sigma_X/\sqrt{n}$, which is very close to the action limit usually used in the UK.

11.9.4 Cusum control charts

The main concern in designing a control chart is to achieve the best compromise between speedy detection of a fault on the one hand and avoidance of a proliferation of false alarms on the other. If the chart is too sensitive, it will lead to a large number of unnecessary shutdowns. The Shewhart charts, on the other hand, are rather conservative in that they are slow to indicate a slight but genuine shift in performance away from the design level. This derives from the fact that each sample point is judged independently and may well lie inside the action limits, whereas the cumulative evidence over several samples might justify an earlier decision. Rather informal methods involving repeated warnings and trends are used, but it is preferable to employ a more powerful control chart. The **cumulative sum (cusum)** chart achieves this, and is easily implemented on a small computer.

Suppose that we have a sequence $\{Y_1, Y_2, \dots\}$ of observations, which may be either counts of defectives or sample averages. From this a new sequence $\{S_0, S_1, \dots\}$ is obtained by setting

$$S_0 = 0,$$

$$S_m = \max\{0, S_{m-1} + Y_m - r\} \quad (m = 1, 2, \dots)$$

where r is a constant 'reference value'. This gives a cumulative sum of values of $Y_m - r$, which is reset to zero whenever it goes negative. The out-of-control decision is made when

$$S_m > h$$

where h is a constant 'decision interval'. This will detect an increasing mean; a separate but similar procedure can be used to detect a decreasing mean. Values of r and h for both attribute and variable types of control can be obtained from tables such as those in J. Murdoch, *Control Charts* (Macmillan, London, 1979), from which the attribute table in Figure 11.31 has been extracted. For variable

np	r	h	np	r	h
0.22	1	1.5	2.35	4	4.5
0.39	1	2.5	2.60	4	5.5
0.51	2	1.5	2.95	5	4.5
0.62	1	4.5	3.24	5	5.5
0.69	1	5.5	3.89	6	5.5
0.79	2	2.5	4.16	6	6.5
0.86	3	1.5	5.32	7	8.5
1.05	2	3.5	6.07	8	8.5
1.21	3	2.5	7.04	9	9.5
1.52	3	3.5	8.01	10	10.5
1.96	3	5.5	9.00	11	11.5
2.16	5	2.5	10.00	12	12.5

Figure 11.31 Cusum attribute chart control data.

measure (with process design mean μ_X and standard deviation σ_X) the following are often used:

$$r = \mu_X + \frac{\sigma_X}{2\sqrt{n}}, \quad h = 5\frac{\sigma_X}{\sqrt{n}}$$

EXAMPLE 11.28 Regular samples of 50 are taken from a process making electronic components, for which an acceptable proportion of defectives is 5%. Successive counts of defectives in each sample are as follows:

Sample	1	2	3	4	5	6	7	8	9	10	11	12	13	14	15	16	17	18	19	20
Count	3	5	2	2	1	6	4	4	2	6	7	4	5	5	8	6	5	9	7	8

At what point would the decision be taken to stop and correct the process?

Solution The acceptable proportion of defectives is $p = 0.05$ and the regular sample size is $n = 50$. From the table in Figure 11.31, with $np = 2.5$ the nearest figures for reference value and decision interval are $r = 4$ and $h = 5.5$. The following shows the cusum S_m below each count of defectives Y_m, and the cusum is also plotted in Figure 11.32:

Count	3	5	2	2	1	6	4	4	2	6	7	4	5	5	8	6	5	9	7	8
Cusum	0	1	0	0	0	2	2	2	0	2	5	5	6	7	11	13	14	19	22	26

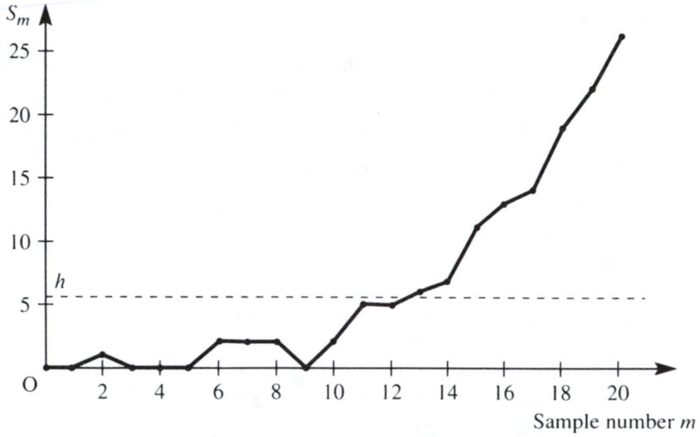

Figure 11.32 Cusum control chart for Example 11.28.

For example,

$$S_{13} = S_{12} + Y_{13} - r = 5 + 5 - 4 = 6$$

and because this exceeds $h = 5.5$, the decision to take action would be made after the 13th sample. This result can be compared with that of a Shewhart chart applied to the same data (Example 11.25), which suggests that action should be taken after 18 samples.

EXAMPLE 11.29 Construct a cusum chart for the sulphur dioxide data in Example 11.27.

Solution From $\mu_X = 60$, $\sigma_X = 8$ and $n = 5$ we have

$$r = 60 + \frac{8}{2\sqrt{5}} = 61.8, \quad h = 5\frac{8}{\sqrt{5}} = 17.9$$

The following table shows the sample average $\bar{X}_m$ and cusum S_m for $1 \leqslant m \leqslant 16$, and the cusum is also plotted in Figure 11.33:

Average	64.2	56.9	57.7	67.9	61.7	59.7	55.6	63.7
Cusum	2.4	0	0	6.1	6.0	3.9	0	1.9
Average	58.3	66.4	67.2	65.2	63.1	67.6	64.1	66.7
Cusum	0	4.6	10.0	13.4	14.7	20.5	22.8	27.7

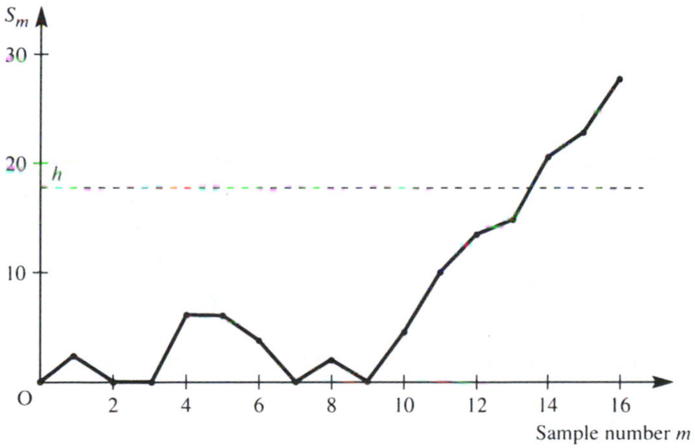

Figure 11.33 Cusum control chart for Example 11.29.

Because $S_{14} = 20.5$ exceeds $h = 17.9$, this chart suggests that the SO_2 concentration did increase during the experiment, a stronger result than that obtained from the Shewhart chart in Example 11.27.

It can be shown that the cusum method will usually detect an out-of-control condition (involving a slight process shift) much sooner than the strict Shewhart method, but with essentially the same risk of a false alarm. For instance, the cusum method leads to a decision after 13 samples in Example 11.28 compared with 18 samples in Example 11.25 for the same data. The measure used to compare the two methods is the **average run length (ARL)**, which is the mean number of samples required to detect an increase in proportion of defectives (or process average) to some specified level. It has been shown that the ARL for the Shewhart chart can be up to four times that for the cusum chart (J. Murdoch, *Control Charts*, Macmillan, London, 1979).

11.9.5 Moving-average control charts

The cusum chart shows that the way to avoid the relative insensitivity of the Shewhart chart is to allow the evidence of a shift in performance to accumulate over several samples. There are also **moving-average control charts**, which are based upon a weighted sum of a number of observations. The best of these, which is very similar to the cusum chart in operation, is the **geometric moving-average (GMA) chart**. This will be described here for variable measure, but it also works for attribute measure (Exercise 52).

Suppose that the successive sample averages are $\bar{X}_1, \bar{X}_2, \ldots$, each from a sample of size n. Also suppose that the design mean and variance are μ_X and σ_X^2. Then the GMA is the new sequence given by

$$S_0 = \mu_X$$

$$S_m = r\bar{X}_m + (1 - r)S_{m-1} \quad (m = 1, 2, \ldots)$$

where $0 < r \leqslant 1$ is a constant. The statistical properties of this sequence are simpler than for the cusum sequence. First, by successively substituting for S_{m-1}, S_{m-2} and so on, we can express S_m directly in terms of the sample averages:

$$S_m = r\sum_{i=0}^{m-1}[(1-r)^i\bar{X}_{m-i}] + (1-r)^m\mu_X$$

Then, using the summation formula

$$\sum_{i=0}^{m-1}x^i = \frac{1-x^m}{1-x} \quad (|x| < 1)$$

it is easy to show (Exercise 53) that the mean and variance of S_m are

$$\mu_{S_m} = E(S_m) = \mu_X$$

$$\sigma_{S_m}^2 = \mathrm{Var}(S_m) = \frac{r}{2-r}[1-(1-r)^{2m}]\frac{\sigma_X^2}{n}$$

After the first few samples the variance of S_m tends to a constant value:

$$\sigma_{S_m}^2 \to \left(\frac{r}{2-r}\right)\frac{\sigma_X^2}{n} \quad \text{as } m \to \infty$$

If US practice is followed then the upper and lower control limits can be set at $(\mu_X \pm 3\sigma_{S_m})$. If UK practice is followed then, from the approximate normality of the sample averages and the fact that sums of normal random variables are also normal, it follows that S_m is approximately normal, so the warning and action limits can be set at $(\mu_X \pm 1.96\sigma_{S_m})$ and $(\mu_X \pm 3.09\sigma_{S_m})$ respectively (although the warning limits now have less significance).

It remains to choose a value for r. If we set $r = 1$, the whole approach reduces to the standard Shewhart charts. Small values of r (say around 0.2) lead to early recognition of small shifts of process mean, but if r is too small, a large shift may remain undetected for some time.

EXAMPLE 11.30 Construct GMA charts for the sulphur dioxide data in Example 11.27, using $r = 0.2$ and $r = 0.4$.

Solution The control charts can be seen in Figure 11.34. Clearly the warning and action limits converge fairly quickly to constant values, so little is lost by using those values in practice. The warning limit is exceeded from sample 11 for both values of r. The action limit is exceeded from sample 14 for $r = 0.2$ (as for the cusum chart in Example 11.29) and at sample 16 for $r = 0.4$.

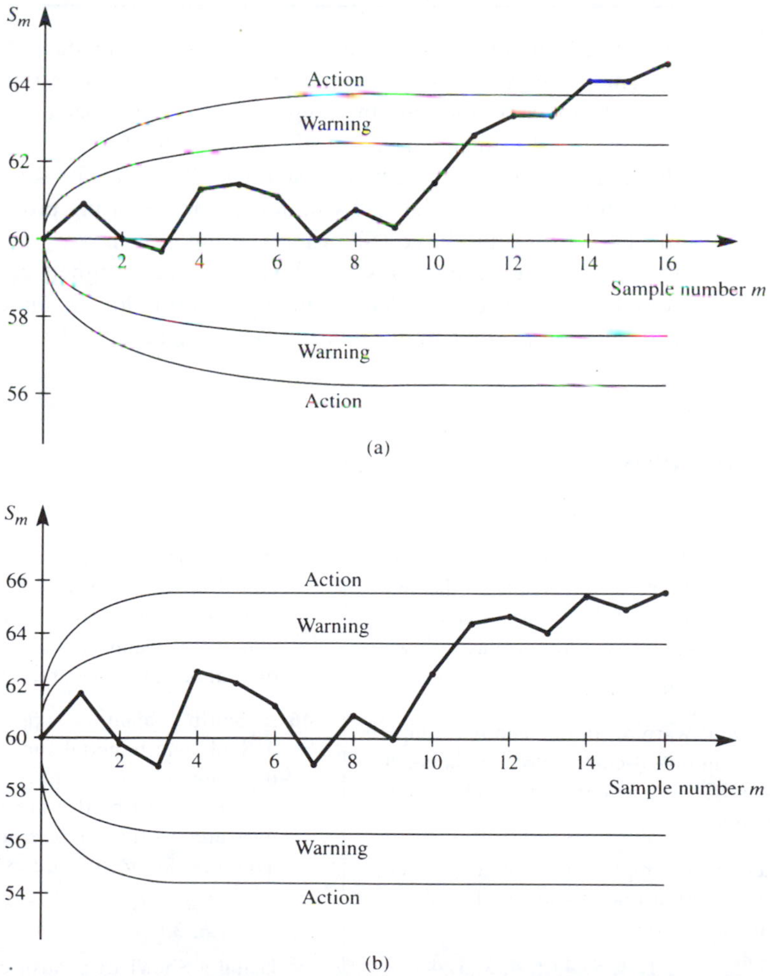

Figure 11.34 Moving-average control chart for Example 11.30: (a) $r = 0.2$; (b) $r = 0.4$.

11.9.6 Range charts

The **sample range** is defined as the difference between the largest and smallest values in the sample. The range has two functions in quality control where the quality is of the variable rather than the attribute type. First, if the data are normal then the range (R, say) provides an estimate $\hat{\sigma}$ of the standard deviation σ by

$$\hat{\sigma} = R/d$$

where d is a constant that depends upon the sample size n as follows:

n	2	3	4	5	6	7	8	9	10	11	12
d	1.128	1.693	2.059	2.326	2.534	2.704	2.847	2.970	3.078	3.173	3.258

It is clearly quicker to evaluate this than the sample standard deviation S, and for the small samples typically used in quality control the estimate is almost as good.

The other reason why the range is important is because the quality of production can vary in dispersion as well as (or instead of) in mean. Control charts for the range R are more commonly used than charts for the sample standard deviation S when monitoring variability within the manufacturing process, and all three types of chart discussed above (Shewhart, cusum and moving-average) can be applied to the range. **Range charts** (or **R charts**) are designed using tables that can be found in specialized books on quality control, for example D. C. Montgomery, *Introduction to Statistical Quality Control*, 2nd edn (Wiley, New York, 1991).

11.9.7 Exercises

44 It is intended that 90% of electronic devices emerging from a machine should pass a simple on-the-spot quality test. The numbers of defectives among samples of 50 taken by successive shifts are as follows:

5, 8, 11, 5, 6, 4, 9, 7, 12, 9, 10, 14

Find the action and warning limits, and the sample number at which an out-of-control decision is taken. Also find the UCL (US practice) and the sample number for action.

45 Thirty-two successive samples of 100 castings each, taken from a production line, contained the following numbers of defectives

3, 3, 5, 3, 5, 0, 3, 1, 3, 5, 4, 2, 4, 3, 5, 4
3, 4, 5, 6, 5, 6, 4, 4, 7, 5, 4, 8, 5, 6, 6, 7

If the proportion that are defective is to be maintained at 0.02, use the Shewhart method (both UK and US standards) to indicate whether this proportion is being maintained, and if not then give the number of samples after which action should be taken.

46 A bottling plant is supposed to fill bottles with 568 ml (one imperial pint) of liquid. The standard deviation of the quantity of fill is 3 ml. Regular samples of 10 bottles are taken and their contents measured. After subtracting 568 from the sample averages, the results are as follows:

−0.2, 1.3, 2.1, 0.3, −0.8, 1.7, 1.3, 0.6, 2.5, 1.4, 1.6, 3.0

Using a Shewhart control chart, determine whether the mean fill requires readjustment.

47 Average reverse-current readings (in nA) for samples of 10 transistors taken at half-hour intervals are as follows:

12.8, 11.2, 13.4, 12.1, 13.6, 13.9, 12.3, 12.9, 13.8, 13.1, 12.9, 14.0, 13.7, 13.4, 14.2, 13.1, 14.0, 14.0, 15.1, 14.3

The standard deviation is 3 nA. At what point, if any, does the Shewhart control method indicate that the reverse current has increased from its design value of 12 nA?

48 Using the data in Exercise 46, apply (a) a cusum control chart and (b) a moving-average control chart with $r = 0.3$.

49 Using the data in Exercise 47, apply (a) a cusum control chart and (b) a moving-average control chart with $r = 0.3$.

50 Apply a cusum control chart to the data in Exercise 44.

51 Apply a cusum control chart to the data in Exercise 45.

52 The diameters of the castings in Exercise 45 are also important. Twelve of each sample of 100 were taken, and their diameters measured and averaged. The differences (in mm) between the successive averages and the design mean diameter of 125 mm were as follows:

0.1, 0.3, −0.2, 0.4, 0.1, 0.0, 0.2, −0.1, 0.2, 0.4, 0.5, 0.1, 0.4, 0.6, 0.3, 0.4, 0.3, 0.6, 0.5, 0.4, 0.2, 0.3, 0.5, 0.7, 0.3, 0.1, 0.6, 0.5, 0.6, 0.7, 0.4, 0.5

Use (a) Shewhart, (b) cusum and (c) moving-average (with $r = 0.2$) control methods to test for an increase in actual mean diameter, assuming a standard deviation of 1 mm.

53 Prove that the mean and variance of the geometric moving-average S_m defined in Section 11.9.5 for variable measure are given by

$$E(S_m) = \mu_X$$

$$\sigma_{S_m}^2 = \text{Var}(S_m) = \frac{r}{2-r}[1-(1-r)^{2m}]\frac{\sigma_X^2}{n}$$

54 Suppose that the moving-average control chart is to be applied to the counts of defectives in attribute quality control. Find the mean and variance of S_m in terms of the sample size n, the design proportion of defectives p and the coefficient r. Following US practice, set the upper control limit at three standard deviations above the mean, and apply the method to the data in Example 11.28, using $r = 0.2$.

55 The design diameter of a moulded plastic component is 6.00 cm, with a standard deviation of 0.2 cm. The following data consist of successive averages of samples of 10 components:

6.04, 6.12, 5.99, 6.02, 6.04, 6.11, 5.97, 6.06, 6.05, 6.06, 6.17, 6.03, 6.13, 6.05, 6.17, 5.97, 6.07, 6.14, 6.03, 5.99, 6.10, 6.01, 5.96, 6.12, 6.02, 6.20, 6.11, 5.98, 6.02, 6.12

After how many samples do the Shewhart, cusum and moving-average (with $r = 0.2$) control methods indicate that action is needed?

11.10 Poisson processes and the theory of queues

Probability theory is often applied to the analysis and simulation of systems, and this can be a valuable aid to design and control. This section, which is therefore applied probability rather than statistics, will illustrate how this progresses from an initial mathematical model through the analysis stage to a simulation.

11.10.1 Typical queueing problems

Queues are everywhere: in banks and ticket offices, at airports and seaports, traffic intersections and hospitals, and in computer and communication networks. Somebody has to decide on the level of service facilities. The problem, in essence, is that it is costly to keep customers waiting for a long time, but it is also costly to provide enough service facilities so that no customer ever has to wait at all. Queues of trucks, aeroplanes or ships may be costly because of the space they occupy or the lost earnings during the idle time. Queues of people may be costly because of lost productivity or because people will often go elsewhere in preference to joining a long queue. Queues of jobs or packets in computer networks are costly in loss of time-efficiency. Service facilities are costly in capital, staffing and maintenance. Probabilistic modelling, often combined with simulation, allows performance evaluation for queues and networks, which can be of great value in preparing the ground for design decisions.

The mathematical model of a simple queueing system is based on the situation shown in Figure 11.35. **Customers** join the queue at random times that are independent of each other – the **inter-arrival time** (between successive arrivals) is a random variable. When a **service channel** is free, the next customer to be served is selected from the queue in a manner determined by the **service discipline**. After being served the customer departs from the queueing system. The **service time** for each customer is another random variable. The distributions of inter-arrival time and service time are usually assumed to take one of a number of standard patterns. The commonest assumption about service discipline is that the next customer to be served is the one who has been queueing the longest time (first in, first out).

The queueing system may be regarded from either a static or a dynamic viewpoint. Dynamically, the system might start from an initial state of emptiness and build up with varying rates of arrivals and varying numbers of service channels depending upon queue length. This is hard to deal with mathematically, but can be treated by computer simulation. Useful information about queues can, however, be obtained from the static viewpoint, in which the rate at which arrivals

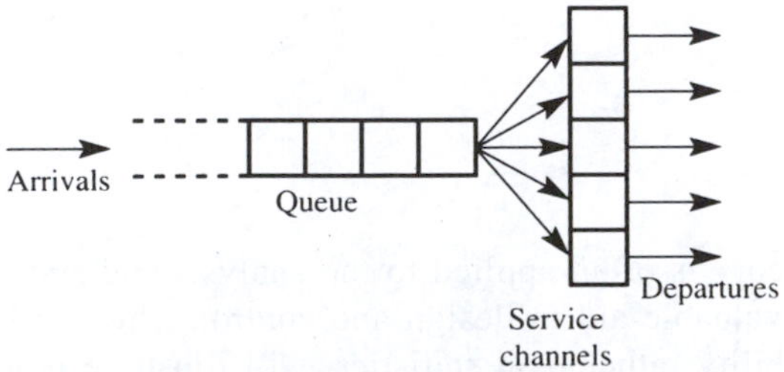

Figure 11.35 A typical queueing system.

occur is constant, as is the number of service channels, and the system is assumed to have been in operation sufficiently long to have reached a steady state. At any time the queue length will be a random variable, but the distribution of queue length is then independent of time.

We need to find the distributions of queue length and of waiting time for the customer, and how these vary with the number of service channels. Costs can be worked out from these results.

11.10.2 Poisson processes

Consider the arrivals process for a queueing system. We shall assume that the customers join the queue at random times that are independent of each other. Other assumptions about the pattern of arrivals would give different results, but this is the most common one. We can therefore think of the arrivals as a stream of events occurring at random along a time axis, as depicted in Figure 11.36. The inter-arrival time T, say, will be a continuous random variable with probability density function $f_T(t)$ and cumulative distribution function $F_T(t)$.

One way to formulate the assumption of independent random arrivals is to assert that at any moment the distribution of the time until the next arrival is independent of the time elapsed since the previous arrival (because arrivals are 'blind' to each other). This is known as the **memoryless property**, and can be expressed as

$$P(T \leqslant t + h \,|\, T > t) = P(T \leqslant h) \quad (t, h \geqslant 0)$$

where t denotes the actual time since the previous arrival and h denotes a possible time until the next arrival. Using the definition of conditional probability (Section 11.2.1), we can write this in terms of the distribution function $F_T(t)$ as

$$\frac{P(t < T \leqslant t+h)}{1-P(T \leqslant t)} = \frac{F_T(t+h)-F_T(t)}{1-F_T(t)} = F_T(h)$$

Rearranging at the second equality and then dividing through by h gives

$$\frac{1}{h}[F_T(t+h)-F_T(t)] = \frac{F_T(h)}{h}[1-F_T(t)]$$

Letting $h \to 0$, we obtain a first-order linear differential equation for $F_T(t)$:

$$\frac{\mathrm{d}}{\mathrm{d}t}F_T(t) = \lambda[1-F_T(t)]$$

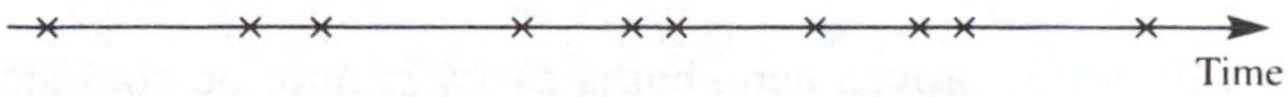

Figure 11.36 Random events (×) on a time axis.

where

$$\lambda = \lim_{h \to 0} \frac{F_T(h)}{h}$$

With the initial condition $F_T(0) = 0$ (because inter-event times must be positive), the solution is

$$F_T(t) = 1 - e^{-\lambda t} \quad (t \geqslant 0)$$

and hence the probability density function is

$$f_T(t) = \frac{d}{dt} F_T(t) = \lambda e^{-\lambda t} \quad (t \geqslant 0)$$

This is the density function of an **exponential distribution with parameter λ**, and it follows that the mean time between arrivals is $1/\lambda$ (see Section 11.7.1). The parameter λ is the **rate of arrivals** (number per unit time).

EXAMPLE 11.31 A factory contains 30 machines of a particular type, each of which breaks down every 100 operating hours on average. It is suspected that the breakdowns are not independent. The operating time intervals between 10 consecutive breakdowns (of any machine) are measured and the shortest such interval is only six minutes. Does this lend support to the suspicion of non-independent breakdowns?

Solution Collectively, the machines break down at the rate of 30/100 or 0.3 per hour. If the breakdowns are independent then the interval between successive breakdowns will have an exponential distribution with parameter 0.3. The probability that such an interval will exceed six minutes is

$$P(\text{interval} > 0.1) = \int_{0.1}^{\infty} 0.3\, e^{-0.3t}\, dt = e^{-0.3(0.1)} = 0.9704$$

and the probability that all nine intervals (between 10 breakdowns) will exceed this time is $(0.9704)^9 = 0.763$. Hence the probability that the shortest interval will be six minutes or less is one minus this, or 0.237. This is quite likely to have happened by chance, so it does not support the suspicion of non-independent intervals.

The assumption of independent random arrivals therefore leads to a particular distribution of inter-arrival time, parametrized by the rate of arrivals. Two further conclusions also emerge. First, the number of arrivals that occur during a fixed interval of length H has a Poisson distribution with parameter λH:

$$P(k \text{ arrivals during interval of length } H) = \frac{(\lambda H)^k e^{-\lambda H}}{k!} \quad (k = 0, 1, 2, \ldots)$$

This will not be proved here, but is easily seen to be consistent with an exponential distribution of inter-arrival time T because

$$F_T(t) = P(T \leqslant t) = 1 - P(T > t)$$
$$= 1 - P \text{ (no event during interval of length } t)$$
$$= 1 - e^{-\lambda t}$$

using the Poisson distribution. Because of this distribution, events conforming to these assumptions are known as a **Poisson process**.

The other conclusion is that the probability that an arrival occurs during a short interval of length h is equal to $\lambda h + O(h^2)$, regardless of the history of the process. Suppose that a time t has elapsed since the previous arrival, and consider a short interval of length h starting from that point:

$$P(\text{arrival during } (t, t+h)) = P(T \leqslant t + h \mid T > t) = F_T(h)$$
$$= 1 - e^{-\lambda h} = \lambda h + O(h^2)$$

using the memoryless property and the expansion of $e^{\lambda h}$ to first order. Furthermore, the probability of more than one arrival during a short interval of length h is $O(h^2)$.

EXAMPLE 11.32 A computer receives on average 60 batch jobs per day. They arrive at a constant rate throughout the day and independently of each other. Find the probability that more than four jobs will arrive in any one hour.

Solution The assumptions for a Poisson process hold, so the number of jobs arriving in one hour is a Poisson random variable with parameter $\lambda H = 60/24$. Hence

$$P(\text{more than four jobs}) = 1 - P(0 \text{ or } 1 \text{ or } 2 \text{ or } 3 \text{ or } 4 \text{ jobs})$$
$$= 1 - e^{-\lambda H}\left[1 + \lambda H + \frac{(\lambda H)^2}{2!} + \frac{(\lambda H)^3}{3!} + \frac{(\lambda H)^4}{4!}\right]$$
$$= 0.109$$

11.10.3 Single service channel queue

Consider a queueing system with a Poisson arrival process with mean rate λ per unit time, and a single service channel. The behaviour of the queueing system depends not only on the arrival process but also upon the distribution of service times. A common assumption here is that the service time distribution (like that of inter-arrival time) is exponential. Thus the probability density function of service time S is

$$f_S(s) = \mu e^{-\mu s} \quad (s \geqslant 0)$$

Unlike the inter-arrival time distribution in Section 11.10.2, this is not based on an assumption of independence or the memoryless property, but simply on the fact that in many queueing situations most customers are served quickly but a few take a lot longer, and the form of the distribution conforms with this fact. This assumption is therefore on much weaker ground than that for the arrival time distribution. The parameter μ is the mean number of customers served in unit time (with no idle periods), and the mean service time is $1/\mu$. With this service distribution, the probability that a customer in the service channel will have departed after a short time h is equal to $\mu h + O(h^2)$, independent of the time already spent in the service channel.

Distribution of the number of customers in the system

We can now derive the distribution of the number of customers in the queueing system. Considering the system as a whole (queue plus service channel), the number of customers in the system at time t is a random variable. Let $p_n(t)$ be the distribution of this random variable:

$$p_n(t) = P(n \text{ customers in the system at time } t) \quad (n = 0, 1, 2, \ldots)$$

Consider the time $t + h$, where h is small. The probability of more than one arrival or more than one departure during this time is $O(h^2)$, and will be ignored. There are four ways in which there can be n (assumed greater than zero) customers in the system at that time:

(1) there are n in the system at t, and no arrival or departure by $t + h$; the probability of this is given by

$$p_n(t)(1 - \lambda h)(1 - \mu h) + O(h^2) = p_n(t)(1 - \lambda h - \mu h) + O(h^2)$$

(2) there are n in the system at t, and one arrival and one departure by $t + h$; the probability is given by

$$p_n(t)(\lambda h)(\mu h) + O(h^2) = O(h^2)$$

(3) there are $n - 1$ in the system at t, and one arrival but no departure by $t + h$; the probability is given by

$$p_{n-1}(t)(\lambda h)(1 - \mu h) + O(h^2) = p_{n-1}(t)(\lambda h) + O(h^2)$$

(4) there are $n + 1$ in the system at t, and no arrivals but one departure by $t + h$; the probability is given by

$$p_{n+1}(t)(1 - \lambda h)(\mu h) + O(h^2) = p_{n+1}(t)(\mu h) + O(h^2)$$

Summing the probabilities of these mutually exclusive events gives the probability of n customers in the system at time $t + h$ as

$$\begin{aligned} p_n(t + h) = {} & p_n(t)(1 - \lambda h - \mu h) + p_{n-1}(t)(\lambda h) \\ & + p_{n+1}(t)(\mu h) + O(h^2) \quad (n = 1, 2, \ldots) \end{aligned} \tag{11.1}$$

Similarly, there are two ways in which the system can be empty ($n = 0$) at time $t + h$: empty at t and no arrival before $t + h$, or one customer at t who departs before $t + h$. This gives

$$p_0(t + h) = p_0(t)(1 - \lambda h) + p_1(t)(\mu h) + O(h^2) \tag{11.2}$$

Rearranging Equations (11.1) and (11.2) and taking the limit as $h \to 0$, we obtain

$$\begin{aligned}\frac{\mathrm{d}}{\mathrm{d}t}p_n(t) &= \lim_{h\to 0}\frac{1}{h}[p_n(t+h)-p_n(t)]\\ &= -(\lambda + \mu)p_n(t) + \lambda p_{n-1}(t) + \mu p_{n+1}(t) \quad (n = 1, 2, \dots)\\ \frac{\mathrm{d}}{\mathrm{d}t}p_0(t) &= -\lambda p_0(t) + \mu p_1(t)\end{aligned}$$

This is a rather complex set of recursive differential equations for the probabilities $p_n(t)$. If we assume that the arrival and service parameters λ and μ are constant and that the system has been in operation for a long time then the distribution will not depend upon t; the derivatives therefore vanish, and we are left with the following algebraic equations for the *steady-state* distribution p_n:

$$0 = -(\lambda + \mu)p_n + \lambda p_{n-1} + \mu p_{n+1} \quad (n = 1, 2, \dots)$$

$$0 = -\lambda p_0 + \mu p_1$$

Defining the ratio of arrival and service parameters λ and μ as $\rho = \lambda/\mu$ and dividing through by μ, we have

$$p_{n+1} = (1 + \rho)p_n - \rho p_{n-1} \quad (n = 1, 2, \dots)$$

$$p_1 = \rho p_0$$

To solve these, we first assume that $p_n = \rho^n p_0$. Clearly this works for $n = 0$ and $n = 1$. Substituting,

$$p_{n+1} = (1 + \rho)\rho^n p_0 - \rho\rho^{n-1}p_0 = \rho^{n+1}p_0$$

so the assumed form holds for $n + 1$, and therefore for all n by induction. It remains only to identify p_0 from the fact that the distribution must sum to unity over $n = 0, 1, 2, \dots$:

$$1 = p_0\sum_{n=0}^{\infty}\rho^n = \frac{p_0}{1-\rho}$$

Hence $p_0 = 1 - \rho$ and

$$p_n = (1 - \rho)\rho^n \quad (n = 0, 1, 2, \dots)$$

This is known as the **geometric distribution**, and is a discrete version of the exponential distribution (Figure 11.37). Note that this result requires that $\rho < 1$, or equivalently $\lambda < \mu$. If this condition fails to hold, the arrival rate swamps the capacity of the service channel, the queue gets longer and longer, and no steady-state condition exists.

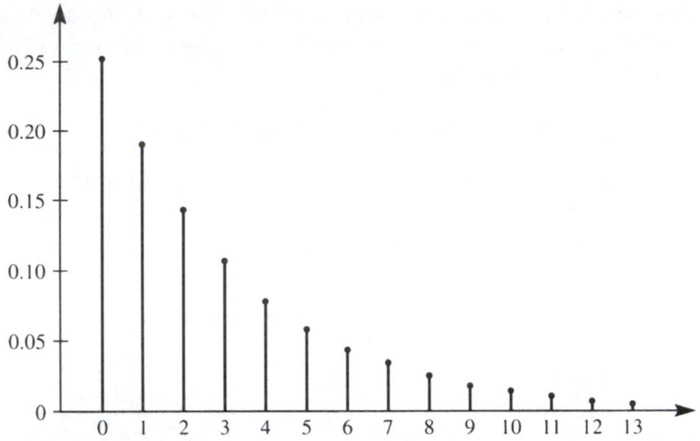

Figure 11.37 Geometric distribution (with $\rho = 0.75$).

Queue length and waiting time

The queue length distribution now follows easily:

$$P(\text{queue empty}) = p_0 + p_1 = 1 - \rho^2$$

$$P(n \text{ in queue}) = P(n + 1 \text{ in system}) = (1 - \rho)\rho^{n+1} \quad (n = 1, 2, \ldots)$$

Denoting the mean numbers of customers in the system and in the queue by N_S and N_Q respectively,

$$N_S = \sum_{n=0}^{\infty} np_n = \frac{\rho}{1-\rho}, \quad N_Q = \sum_{n=1}^{\infty} (n-1)p_n = \frac{\rho^2}{1-\rho}$$

(Exercise 58). Since in the steady state the mean time between departures must equal the mean time between arrivals ($1/\lambda$), it is plausible that the mean total time in the system for each customer, W_S say, is given by

$$W_S = \text{mean number in system} \times \text{mean time between departures}$$

$$= \frac{\rho/\lambda}{1-\rho} = \frac{1}{\mu - \lambda}$$

The mean waiting time in the queue, W_Q say, is then

$$W_Q = \text{mean time in system} - \text{mean service time}$$

$$= W_S - \frac{1}{\mu} = \frac{\rho}{\mu - \lambda}$$

These results for W_S and W_Q can be derived more formally from the respective waiting time distributions. For example, the distribution of total time in the system can be shown to be exponential with parameter $\mu - \lambda$, and the waiting time in the queue can be expressed as

$$P(\text{waiting time in queue} \geqslant t) = \rho\, e^{-(\mu-\lambda)t} \quad (t > 0)$$

EXAMPLE 11.33 If customers in a shop arrive at a single check-out point at the rate of 30 per hour and if the service times have an exponential distribution, what mean service time will ensure that 80% of customers do not have to wait more than five minutes in the queue and what will be the mean queue length?

Solution With $\lambda = 0.5$ and $t = 5$, the queue waiting time gives

$$0.2 = \rho\, e^{-(\mu-\lambda)t}$$

that is,

$$0.2\mu = 0.5\, e^{2.5-5\mu}$$

This is a nonlinear equation for μ, which may be solved by standard methods to give $\mu = 0.743$. The mean service time is therefore $1/\mu$ or 1.35 min, and the mean queue length is

$$N_Q = \frac{\rho^2}{1-\rho} = 1.39, \quad \text{using } \rho = \frac{\lambda}{\mu} = 0.673$$

EXAMPLE 11.34 Handling equipment is to be installed at an unloading bay in a factory. An average of 20 trucks arrive during each 10 h working day, and these must be unloaded. The following three schemes are being considered:

Scheme	Fixed cost/ £ per day	Operating cost/ £ per hour	Mean handling rate/ trucks per hour
A	90	45	3
B	190	50	4
C	450	60	6

Truck waiting time is costed at £30 per hour. Assuming an exponential distribution of truck unloading time, find the best scheme.

Solution Viewing this as a queueing problem, we have

λ = arrival rate per hour = 2.0

μ = unloading rate per hour

mean waiting time for each truck = $1/(\mu - \lambda)$

Hence the mean delay cost per truck is

$$\frac{30}{\mu - 2}$$

and the mean delay cost per day is

$$\frac{20 \times 30}{\mu - 2}$$

The proportion of time that the equipment is running is equal to the probability that the system is not empty (the **utilization**), which is

$$1 - p_0 = \frac{\lambda}{\mu} = \rho$$

Hence the mean operating cost per day is 10ρ times operating cost per hour. The total cost per day (in £) is the sum of the fixed, operating and delay costs, as follows:

Scheme	μ	ρ	Fixed	Operating	Delay	Total
A	3	0.6667	90	300	600	990
B	4	0.5	190	250	300	740
C	6	0.3333	450	200	150	800

Hence scheme B minimizes the total cost.

11.10.4 Queues with multiple service channels

For the case where there are c service channels, all with an exponential service time distribution with parameter μ, a line of argument similar to that in Section 11.10.3 can be found in many textbooks on queueing theory. In particular, it can be shown that the distribution p_n of the number of customers in the system is

$$p_n = \begin{cases} \dfrac{\rho^n}{n!} p_0 & (0 \leqslant n \leqslant c) \\[2ex] \dfrac{\rho^n}{c^{n-c} c!} p_0 & (n > c) \end{cases}$$

where $\rho = \lambda/\mu$ and

$$p_0 = \left[\sum_{n=0}^{c-1} \frac{\rho^n}{n!} + \frac{\rho^c}{(c-1)!(c-\rho)} \right]^{-1}$$

The mean numbers in the queue and in the system are

$$N_Q = \frac{\rho^{c+1}}{(c-1)!(c-\rho)^2}p_0, \quad N_S = N_Q + \rho$$

and the mean waiting times in the queue and in the system are

$$W_Q = \frac{N_Q}{\lambda}, \quad W_S = W_Q + \frac{1}{\mu}$$

EXAMPLE 11.35 For the unloading bay problem in Example 11.34 a fourth option would be to install two sets of equipment under scheme A (there is space available to do this). The fixed costs would then double but the operating costs per bay would be the same. Evaluate this possibility.

Solution With two bays under scheme A, we have $\lambda = 2$, $\mu = 3$ and $c = 2$, so that $\rho = \frac{2}{3}$, and the probability that the system is empty at any time is

$$p_0 = \left(1 + \rho + \frac{\rho^2}{2-\rho}\right)^{-1} = \frac{1}{2}$$

The probabilities of one truck (one bay occupied) and of two or more trucks (both bays occupied) are then

$$P_1 = \rho p_0 = \tfrac{1}{3}$$

$$P(\text{two or more trucks}) = 1 - \tfrac{1}{2} - \tfrac{1}{3} = \tfrac{1}{6}$$

The total operating cost per day is the operating cost for when one or other bay is working (£45 per hour) plus that for when both bays are working (£90 per hour), which is

$$10[\tfrac{1}{3}(45) + \tfrac{1}{6}(90)] = 300$$

The mean number in the queue is

$$\frac{\rho^3}{(2-\rho)^2}p_0 = 0.08333$$

so that the mean total time in the system for each truck is

$$\tfrac{1}{2}(0.08333) + \tfrac{1}{3} = 0.375$$

Multiplying by the cost per hour and the number of trucks gives the delay cost per day:

$$20(30)(0.375) = 225$$

The total cost per day of this scheme is therefore

$$2(90) + 300 + 225 = £705$$

This is less than the £740 under scheme B, the best of the single-bay options.

11.10.5 Queueing system simulation

The assumption that the service time distribution is exponential, which underlies the results in Sections 11.10.3 and 11.10.4, is often unrealistic. It is known that it leads to predicted waiting times that tend to be pessimistic, as a result of which costs based on these predictions are often overestimated. Theoretical results for other service distributions exist (see for example E. Page, *Queueing Theory in OR*. Butterworth, London, 1972), but it is often instructive to simulate a queueing system and find the various answers numerically. It is then easy to vary the arrival and service distributions, and the transient (non-steady-state) behaviour of the system also reveals itself.

Figure 11.38 shows a pseudocode listing of a single-channel queueing system simulation, which is easily modified to cope with multiple channels. Each event consists of either an arrival or a departure. The variables next_arrival and next_departure are used to represent the time to the next arrival and the time to the next departure respectively, and the type of the next event is determined by whichever is the smaller. New arrival and departure times are returned by the functions arrival_time and departure_time, which generate values from appropriate exponential distributions. The distribution of the number of customers in the system is built up as an array, normalized at the end of the simulation, from which the mean and other results can be obtained.

```
{  Procedure to simulate a single-channel queueing system.
time is the running time elapsed,
limit is the simulation length,
number is the number of items in the system,
maximum is the maximum number allowed in the system,
system[i] is the distribution (array) of times with i in the
   system, assumed initialized to zero,
mean is the mean number in the system,
infinity contains a very large number,
arrival_rate is the arrival distribution parameter,
service_rate is the service distribution parameter,
next_arrival is the time to the next arrival,
next_departure is the time to the next departure,
rnd() is a function assumed to return a uniform random (0,1)
   number.
The simulation starts with the system empty,
limit, maximum, arrival_rate and service_rate must be set.   }

time ← 0
next_arrival ← arrival_time(arrival_rate)
next_departure ← infinity
number ← 0
repeat
  if next_arrival < next_departure then
    time ← time + next_arrival
```

Figure 11.38 Pseudocode listing for queueing system simulation.

```
    system[number] ← system[number] + next_arrival
    arrival()
  else
    time ← time + next_departure
    system[number] ← system[number] + next_departure
    departure()
  endif
until time > limit
mean ← 0
for i is 0 to maximum do
  system[i] ← system[i]/time
  mean ← mean + i* system[i]
endfor
********************
procedure arrival()
{  procedure to handle an arrival,
changes the values of next_arrival, next_departure, and number  }
  if number = 0 then
    next_departure ← departure_time(service_rate)
  else
    next_departure ← next_departure - next_arrival
  endif
  number ← number + 1
  if number = maximum then
    next_arrival ← infinity
  else
    next_arrival ← arrival_time(arrival_rate)
  endif
endprocedure
********************
procedure departure()
{  procedure to handle a departure,
changes the values of next_arrival, next_departure, and number  }
  if number = maximum then
    next_arrival ← arrival_time(arrival_rate)
  else
    next_arrival ← next_arrival - next_departure
  endif
  number ← number - 1
  if number = 0 then
    next_departure ← infinity
  else
    next_departure ← departure_time(service_rate)
  endif
endprocedure
********************
function arrival_time(arrival_rate)
{  function to generate a new arrival time  }
  U ← rnd()
  return( - (log(U))/arrival_rate)
endfunction
function departure_time(service_rate)
{  function to generate a new departure time  }
  U ← rnd()
  return ( - (log/U))/service_rate)
endfunction
```

Figure 11.38 continued.

What is typically found from such a simulation (with Poisson arrivals and an exponential service distribution) is that there is good agreement with the predicted results as long as $\rho = \lambda/\mu$ is small, but the results become more erratic as $\rho \to 1$. It takes a very long time for the distribution to reach its steady-state form when the value of ρ is close to unity. In that situation the theoretical steady-state results may be of limited value.

11.10.6 Exercises

56 A sea area has on average 15 gales annually, evenly distributed throughout the year. Assuming that the gales occur independently, find the probability that more than two gales will occur in any one month.

57 Suppose that the average number of telephone calls arriving at a switchboard is 30 per hour, and that they arrive independently. What is the probability that no calls will arrive in a three-minute period? What is the probability that more than five calls will arrive in a five-minute period?

58 Show that for a single-channel queue with Poisson arrivals and exponential service time distribution the mean numbers of customers in the system and in the queue are

$$N_S = \frac{\rho}{1-\rho}, \quad N_Q = \frac{\rho^2}{1-\rho}$$

where ρ is the ratio of arrival and service rates. (*Hint*: Differentiate the equation

$$\sum_{n=0}^{\infty} \rho^n = \frac{1}{1-\rho}$$

with respect to ρ.)

59 Patients arrive at the casualty department of a hospital at random, with a mean arrival rate of three per hour. The department is served by one doctor, who spends on average 15 minutes with each patient, actual consulting times being exponentially distributed. Find

(a) the proportion of time that the doctor is idle;

(b) the mean number of patients waiting to see the doctor;

(c) the probability of there being more than three patients waiting;

(d) the mean waiting time for patients;

(e) the probability of a patient having to wait longer than one hour.

60 A small company operates a cleaning and re-catering service for passenger aircraft at an international airport. Aircraft arrive requiring this service at a mean rate of λ per hour, and arrive independently of each other. They are serviced one at a time, with an exponential distribution of service time. The cost for each aircraft on the ground is put at c_1 per hour, and the cost of servicing the planes at a rate μ is $c_2\mu$ per hour. Prove that the service rate that minimizes the total cost per hour is

$$\mu = \lambda + \sqrt{\left(\frac{c_1\lambda}{c_2}\right)}$$

61 The machines in a factory break down in a Poisson pattern at an average rate of three per hour during the eight-hour working day. The company has two service options, each involving an exponential service time distribution. Option A would cost £20 per hour, and the mean repair time would be 15 min. Option B would cost £40 per hour, with a mean repair time of 12 min. If machine idle time is costed at £60 per hour, which option should be adopted?

62 Ships arrive independently at a port at a mean rate of one every three hours. The time a ship occupies a berth for unloading and loading has an exponential distribution with a mean of 12 hours. If the mean delay to ships waiting for berths is to be kept below six hours, how many berths should there be at the port?

63 In a self-service store the arrival process is Poisson, with on average one customer arriving every 30 s. A single cashier can serve customers every 48 s on average, with an exponential distribution of service time. The store managers wish to minimize the mean waiting time for customers. To do this, they can either double the service rate by providing an additional server to pack the customer's goods (at a single cash desk) or else provide a second cash desk. Which option is preferable?

11.11 Bayes' theorem and its applications

To end this chapter, we return to the foundations of probability and inference. The definition of conditional probability is fundamental to the subject, and from it there follows the theorem of Bayes, which has far-reaching implications.

11.11.1 Derivation and simple examples

The definition in Section 11.2.1 of the conditional probability of an event B given that another event A occurs can be rewritten as

$$P(A \cap B) = P(B \mid A)P(A)$$

If A and B are interchanged then this becomes

$$P(A \cap B) = P(A \mid B)P(B)$$

The left-hand sides are equal, so we can equate the right-hand sides and rearrange, giving

$$P(A \mid B) = \frac{P(B \mid A)P(A)}{P(B)}$$

Now suppose that B is known to have occurred, and that this can only happen if one of the mutually exclusive events

$$\{A_1, \ldots, A_n\}, \quad A_i \cap A_j = \varnothing \quad (i \neq j)$$

has also occurred, but which one is not known. The relevance of the various events A_i to the occurrence of B is expressed by the conditional probabilities $P(B \mid A_i)$. Suppose that the probabilities $P(A_i)$ are also known. The examples below will show that this is a common situation, and we should like to work out the conditional probabilities

$$P(A_i \mid B) = \frac{P(B \mid A_i)P(A_i)}{P(B)}$$

To find the denominator, we sum from 1 to n:

$$\sum_{i=1}^{n} P(A_i \mid B) = 1 = \frac{1}{P(B)} \sum_{i=1}^{n} P(B \mid A_i)P(A_i)$$

The sum is equal to 1 by virtue of the assumption that B could not have occurred without one of the A_i occurring. We therefore obtain a formula for $P(B)$:

$$P(B) = \sum_{i=1}^{n} P(B \mid A_i)P(A_i)$$

which is sometimes called the **rule of total probability**. Hence we have the following theorem.

THEOREM 11.2 **Bayes' theorem**

If $\{A_1, \ldots, A_n\}$ are mutually exclusive events, one of which must occur given that another event B occurs, then

$$P(A_i \mid B) = \frac{P(B \mid A_i)P(A_i)}{\sum_{j=1}^{n} P(B \mid A_j)P(A_j)} \quad (i = 1, \ldots, n)$$

□

EXAMPLE 11.36 Three machines produce similar car parts. Machine A produces 40% of the total output, and machines B and C produce 25% and 35% respectively. The proportions of the output from each machine that do not conform to the specification are 10% for A, 5% for B and 1% for C. What proportion of those parts that do not conform to the specification are produced by machine A?

Solution Let D represent the event that a particular part is defective. Then, by the rule of total probability, the overall proportion of defective parts is

$$P(D) = P(D \mid A)P(A) + P(D \mid B)P(B) + P(D \mid C)P(C)$$

$$= (0.1)(0.4) + (0.05)(0.25) + (0.01)(0.35) = 0.056$$

Using Bayes' theorem,

$$P(A \mid D) = \frac{P(D \mid A)P(A)}{P(D)} = \frac{(0.1)(0.4)}{0.056} = 0.714$$

so that machine A produces 71.4% of the defective parts.

EXAMPLE 11.37 Suppose that 0.1% of the people in a certain area have a disease D and that a mass screening test is used to detect cases. The test gives either a positive or a negative result for each person. Ideally, the test would always give a positive result for a person who has D, and would never do so for a person who has not. In practice the test gives a positive result with probability 99.9% for a person who has D, and with probability 0.2% for a person who has not. What is the probability that a person for whom the test is positive actually has the disease?

Solution Let T represent the event that the test gives a positive result. Then the proportion of positives is

$$P(T) = P(T \mid D)P(D) + P(T \mid \bar{D})P(\bar{D})$$

$$= (0.999)(0.001) + (0.002)(0.999) \simeq 0.003$$

and the desired result is

$$P(D \mid T) = \frac{P(T \mid D)P(D)}{P(T)} = \frac{(0.999)(0.001)}{0.003} = \tfrac{1}{3}$$

Despite the high basic reliability of the test, only one-third of those people receiving a positive result actually have the disease. This is because of the low incidence of the disease in the population, which means that a positive result is twice as likely to be a false alarm as it is to be correct.

In connection with Example 11.37, it might be wondered why the reliability of the test was quoted in the problem in terms of

$P(\text{positive result} \mid \text{disease})$ and $P(\text{positive result} \mid \text{no disease})$

instead of the seemingly more useful

$P(\text{disease} \mid \text{positive result})$ and $P(\text{disease} \mid \text{negative result})$

The reason is that the latter figures are contaminated, in a sense, by the incidence of the disease in the population. The figures quoted for reliability are intrinsic to the test, and may be used anywhere the disease occurs, regardless of the level of incidence.

11.11.2 Applications in probabilistic inference

The scope for applications of Bayes' theorem can be widened considerably if we assume that the calculus of probability can be applied not just to events as subsets of a sample space but also to more general statements about the world. Events are essentially statements about facts that may be true on some occasions and false on others. Scientific theories and hypotheses are much deeper statements, which have great explanatory and predictive power, and which are not so much true or false as gaining or lacking in evidence. One way to assess the extent to which some evidence E supports a hypothesis H is in terms of the conditional probability $P(H \mid E)$. The relative frequency interpretation of probability does not normally apply in this situation, so a subjective interpretation is adopted. The quantity $P(H \mid E)$ is regarded as a **degree of belief** in hypothesis H on the basis of evidence E. In an attempt to render the theory as objective as possible, the rules of probability are strictly applied, and an inference mechanism based on Bayes' theorem is employed.

Suppose that there are in fact two competing hypotheses H_1 and H_2. Let X represent all background information and evidence relevant to the two hypotheses. The probabilities $P(H_1 \mid X)$ and $P(H_1 \mid X \cap E)$ are called the **prior** and **posterior probabilities** of H_1, where E is a new piece of evidence. Similarly, there are prior and posterior probabilities of H_2. Applying Bayes' theorem to both H_1 and H_2 and cancelling the common denominator $P(E)$ gives

$$\frac{P(H_1 \mid X \cap E)}{P(H_2 \mid X \cap E)} = \frac{P(E \mid H_1 \cap X)}{P(E \mid H_2 \cap X)} \frac{P(H_1 \mid X)}{P(H_2 \mid X)}$$

The left-hand side and the second factor on the right-hand side are called the **posterior odds** and **prior odds** respectively, favouring H_1 over H_2. The first factor on the right-hand side is called the **likelihood ratio**, and it measures how much more likely it is that the evidence event E would occur if the hypothesis H_1 were true than if H_2 were true. The new evidence E therefore 'updates' the odds, and the process can be repeated as often as desired, provided that the likelihood ratios can be calculated.

EXAMPLE 11.38 From experience it is known that when a particular type of single-board microcomputer fails, this is twice as likely to be caused by a short on the serial interface (H_1) as by a faulty memory circuit (H_2). The standard diagnostic test is to measure the voltage at a certain point on the board, and from experience it is also known that a drop in voltage there occurs nine times out of ten when the memory circuit is faulty but only once in six occasions of an interface short. How does the observed drop in voltage (E) affect the assessment of the cause of failure?

Solution The prior odds are two to one in favour of H_1, and the likelihood ratio is (1/6)/(9/10), so the posterior odds are given by

$$\frac{P(H_1 \mid E)}{P(H_2 \mid E)} = (\tfrac{10}{54})(2) = 0.370$$

The evidence turns the odds around to about 2.7 to one in favour of H_2.

EXAMPLE 11.39 An oil company is prospecting for oil in a certain area, and is conducting a series of seismic experiments. It is known from past experience that if oil is present in the rock strata below then there is on average one chance in three that a characteristic pattern will appear on the trace recorded by the seismic detector after a test. If oil is absent then the pattern can still appear, but is less likely, appearing only once in four tests on average. After 150 tests in the area the pattern has been seen on 48 occasions. Assuming prior odds of 3:1 against the presence of oil, find the updated odds. Also find the 90% confidence interval for the true probability of the pattern appearing after a test, and hence consider whether oil is present or not.

Solution Let H_1 and H_2 represent the hypotheses that oil is present and that it is absent respectively. There were effectively 150 pieces of evidence gathered, and the odds need to be multiplied by the likelihood ratio for each. Each time the pattern is present the likelihood ratio is

$$\frac{P(\text{pattern} \mid H_1)}{P(\text{pattern} \mid H_2)} = \tfrac{1}{3}/\tfrac{1}{4} = \tfrac{4}{3}$$

and each time it is absent the likelihood ratio is

$$\frac{P(\text{no pattern} \mid H_1)}{P(\text{no pattern} \mid H_2)} = \tfrac{2}{3}/\tfrac{3}{4} = \tfrac{8}{9}$$

The updated odds, letting E represent the total evidence, become

$$\frac{P(H_1|E)}{P(H_2|E)} = (\tfrac{4}{3})^{48}(\tfrac{8}{9})^{102}(\tfrac{1}{3}) = 2.01$$

The odds that there is oil present are therefore raised to 2:1 in favour.

Confidence intervals for proportions were covered in Section 11.3.6. The proportion of tests for which the pattern was observed is 48/150 or 0.32, so the 90% confidence interval for the probability of appearance is

$$\left(0.32 \pm 1.645\sqrt{\left[\frac{(0.32)(0.68)}{150}\right]}\right) = (0.26, 0.38)$$

The hypothesis that oil is absent is not compatible with this, because the pattern should then appear with probability 0.25, whereas the hypothesis that oil is present is fully compatible.

For the problem in Example 11.38 it is conceivable that there could be enough repetitions for the relative frequency interpretation to be placed on the probabilities of the two hypotheses. In contrast, in Example 11.39 the probability of the presence or absence of oil is not well suited to a frequency interpretation, but the subjective interpretation is available.

Example 11.39 also provides a contrast between the 'Bayesian' and 'classical' inference approaches. The classical confidence interval appears to lead to a definite result: H_1 is true and H_2 is false. This definiteness is misleading, because it is possible (although not likely) that the opposite is the case, but the evidence supports one hypothesis more than the other. The Bayesian approach has the merit of indicating this relative support quantitatively.

One area where Bayesian inference is very important is in decision support and expert systems. In classical decision theory Bayesian inference is used to update the probabilities of various possible outcomes of a decision, as further information becomes available. This allows an entire programme of decisions and their consequences to be planned (see D. V. Lindley, *Making Decisions*, 2nd edn. Wiley, London, 1985). Expert systems often involve a process of reasoning from evidence to hypothesis with a Bayesian treatment of uncertainty (see for example R. Forsyth, ed., *Expert Systems, Principles and Case Studies*. Chapman & Hall, London, 1984).

11.11.3 Exercises

64 An explosion at a construction site could have occurred as a result of (a) static electricity, (b) malfunctioning of equipment, (c) carelessness or (d) sabotage. It is estimated that such an explosion would occur with probability 0.25 as a result of (a), 0.20 as a result of (b), 0.40 as a result of (c) and 0.75 as a result of (d). It is also judged that the prior probabilities of the four causes of the explosion are (a) 0.20, (b) 0.40, (c) 0.25, (d) 0.15. Find the posterior probabilities and hence the most likely cause of the explosion.

65 Three marksmen (A, B and C) fire at a target. Their success rates at hitting the target are 60% for A, 50% for B and 40% for C. If each marksman fires one shot at the target and two bullets hit it, then which is more probable: that C hit the target, or did not?

66 An accident has occurred on a busy highway between city A, of 100 000 people, and city B, of 200 000 people. It is known only that the victim is from one of the two cities and that his name is Smith. A check of the records reveals that 10% of city A's population is named Smith and 5% of city B's population has that name. The police want to know where to start looking for relatives of the victim. What is the probability that the victim is from city A?

67 In a certain community, 8% of all adults over 50 have diabetes. If a health service in this community correctly diagnoses 95% of all persons with diabetes as having the disease, and incorrectly diagnoses 2% of all persons without diabetes as having the disease, find the probabilities that

(a) the community health service will diagnose an adult over 50 as having diabetes,

(b) a person over 50 diagnosed by the health service as having diabetes actually has the disease.

68 A stockbroker correctly identifies a stock as being a good one 60% of the time and correctly identifies a stock as being a bad one 80% of the time. A stock has a 50% chance of being good. Find the probability that a stock is good if

(a) the stockbroker identifies it as good,

(b) k out of n stockbrokers of equal ability independently identify it as good.

69 On a communications channel, one of three sequences of letters can be transmitted: AAAA, BBBB and CCCC, where the prior probabilities of the sequences are 0.3, 0.4 and 0.3 respectively. It is known from the noise in the channel that the probability of correct reception of a transmitted letter is 0.6, and the probability of incorrect reception of the other two letters is 0.2 for each. It is assumed that the letters are distorted independently of each other. Find the most probable transmitted sequence if ABCA is received.

70 The number of accidents per day occurring at a road junction was recorded over a period of 100 days. There were no accidents on 84 days, one accident on 12 days, and two accidents on four days. One hypothesis is that the number of accidents per day has a Poisson distribution with parameter λ (unspecified), and another is that the distribution is binomial with parameters $n = 3$ and p (unspecified). Use the average number of accidents per day to identify the unspecified parameters and compare the hypotheses assuming that the binomial is initially thought to be twice as likely as the Poisson.

71 The following **multinomial distribution** is a generalization of the binomial distribution. Suppose that there are k distinct possible outcomes of an experiment, with probabilities $p_1, \ldots, p_k$, and that the experiment is repeated n times. The probability of obtaining a number n_1 of occurrences of the first possible outcome, n_2 of the second, and so on up to n_k of the kth is

$$P(n_1, \ldots, n_k) = \frac{n!}{n_1! \ldots n_k!}(p_1)^{n_1} \ldots (p_k)^{n_k}$$

Suppose now that there are two competing hypotheses H_1 and H_2. H_1 asserts that the probabilities are $p_1, \ldots, p_k$ as above, and H_2 asserts that they are $q_1, \ldots, q_k$. Prove that the logarithm of the likelihood ratio is

$$\ln\left[\frac{P(n_1, \ldots, n_k \mid H_1)}{P(n_1, \ldots, n_k \mid H_2)}\right] = \sum_{i=1}^{k} n_i \ln\left(\frac{p_i}{q_i}\right)$$

72 According to the design specification, of the components produced by a machine, 92% should have no defect, 5% should have defect A alone, 2% should have defect B alone and 1% should have both defects. Call this hypothesis H_1. The user suspects that the machine is producing more components (say a proportion p_B) with defect B alone, and also more components (say a proportion p_{AB}) with both defects, but is satisfied that 5% have defect A alone. Call this hypothesis H_2. Of a sample of 1000 components, 912 had no defects, 45 had A alone,

27 had B alone and 16 had both. Using the multinomial distribution (as in Exercise 71), maximize $\ln P(912, 45, 27, 16 \mid H_2)$ with respect to p_B and p_{AB}, and find the posterior odds assuming prior odds of 5:1 in favour of H_1.

73 It is suggested that higher-priced cars are assembled with greater care than lower-priced cars. To investigate this, a large luxury model A and a compact hatchback B were compared for defects when they arrived at the dealer's showroom. All cars were manufactured by the same company. The numbers of defects for several of each model were recorded:

A: {5, 4, 3, 5, 3, 4}

B: {8, 6, 8, 9, 5}

The number of defects in each car can be assumed to be governed by a Poisson distribution with parameter λ. Compare the hypothesis H_1 that $\lambda_A \neq \lambda_B$ with H_2 that $\lambda_A = \lambda_B = \lambda$, using the average numbers of defects to identify the λ values and assuming no initial preference between the hypotheses.

11.12 Review exercises (1–8)

1 The amplitude d of vibration of a damped pendulum is expected to diminish by

$$d = d_0 e^{-\lambda t}$$

Successive amplitudes are measured from a trace as follows:

t	1.01	2.04	3.12	4.09	5.22	6.30	7.35	8.39	9.44	10.50
d	2.46	1.75	1.26	0.94	0.90	0.79	0.52	0.49	0.31	0.21

Find a 95% confidence interval for the damping coefficient λ.

2 Successive masses of 1 kg were hung from a wire, and the position of a mark at its lower end was measured as follows:

Load/kg	0	1	2	3	4	5	6	7
Position/cm	6.12	6.20	6.26	6.32	6.37	6.44	6.50	6.57

It is expected that the extension Y is related to the force X by

$$Y = LX/EA$$

where $L = 101.4$ cm is the length, $A = 1.62 \times 10^{-5}$ cm^2 is the area and E is the Young's modulus of the material. Find a 95% confidence interval for the Young's modulus.

3 The table in Figure 11.39 gives the intervals, in hours, between arrivals of cargo ships at a port during a period of six weeks. It is helpful to the port authorities to know whether the times of arrival are random or whether they show any regularity. Fit an exponential distribution to the data and test for goodness-of-fit.

4 When large amounts of data are processed, there is a danger of transcription errors occurring (for example a decimal point in the wrong place), which could bias the results. One way to avoid this is to test

6.8	2.1	1.0	28.1	5.8	19.7	2.9	16.3	10.7	25.3	12.5	1.6	3.0	9.9	15.9
21.3	9.1	6.9	5.6	2.0	2.2	10.2	6.5	6.8	42.5	2.9	7.3	3.1	2.6	1.0
3.8	14.7	3.8	13.9	2.9	4.1	22.7	5.8	7.6	6.4	11.3	51.6	15.6	2.6	7.6
1.2	0.7	1.9	1.8	0.7	0.4	72.0	10.7	8.3	15.1	3.6	6.0	0.1	3.1	12.9
2.2	17.6	3.6	2.4	3.2	0.4	4.4	17.1	7.1	10.1	18.8	3.4	0.2	4.9	12.9
1.8	22.4	11.6	4.2	18.0	3.0	16.2	6.8	3.7	13.6	15.7	0.7	2.7	18.8	29.8
4.9	6.8	10.7	0.9	2.4	3.8	9.0	8.8	4.8	0.3	4.6	4.9	6.1	33.0	6.5

Figure 11.39 Time interval data for Review exercise 3.

for **outliers** in the data. Suppose that $X_1, \ldots, X_n$ are independent exponential random variables, each with a common parameter λ. Let the random variable Y be the largest of these divided by the sum:

$$Y = X_{\max} \Big/ \sum_i X_i$$

It can be shown (V. Barnett and T. Lewis, *Outliers in Statistical Data*. Wiley, Chichester, 1978) that the distribution function of Y is given by

$$F_Y(y) = \sum_{k=0}^{[1/y]} (-1)^k \binom{n}{k} (1 - ky)^{n-1}$$

$$\left(\frac{1}{n} \leqslant y \leqslant 1 \right)$$

where $[1/y]$ denotes the integer part of $1/y$. For the data in the Review exercise 3 (Figure 11.39) test the largest value to see whether it is reasonable to expect such a value if the data truly have an exponential distribution. Find 95% confidence intervals for the mean inter-arrival time with this value respectively included and excluded from the data.

5 A surgeon has to decide whether or not to perform an operation on a patient suspected of suffering from a rare disease. If the patient has the disease, he has a 50:50 chance of recovering after the operation but only a one in 20 chance of survival if the operation is not performed. On the other hand, there is a one in five chance that a patient who has not got the disease would die as a result of the operation. How will the decision depend upon the surgeon's assessment of the probability p that the patient has the disease? (*Hint*: Use $P(B \mid A) = P(B \mid A \cap C)P(C) + P(B \mid A \cap \bar{C})P(\bar{C})$, where A and C are independent.)

6 A factory contains 200 machines, each of which becomes misaligned on average every 200 h of operation, the misalignments occurring at random and independently of each other and of other machines. To detect the misalignments, a quality control chart will be followed for each machine, based on one sample of output per machine per hour. Two options have been worked out: option A would cost £1 per hour per machine, whereas option B would cost £1.50 per hour per machine. The control charts differ in their average run lengths (ARL) to a signal of action required. Option A (Shewhart) has an ARL of 20 for a misaligned machine, but will also generate false alarms with an ARL of 1000 for a well-adjusted machine. Option B (cusum) has an ARL of four for a misaligned machine and an ARL of 750 for a well-adjusted machine.

When a control chart signals action required, the machine will be shut down and will join a queue of machines awaiting servicing. A single server will operate, with a mean service time of 30 min and standard deviation of 15 min, regardless of whether the machine was actually misaligned. This is all that is known of the service time distribution, but use can be made of the **Pollaczek–Khintchine formula**, which applies to single-channel queues with arbitrary service distributions:

$$N_S = \rho + \frac{(\lambda \sigma_S)^2 + \rho^2}{2(1 - \rho)}$$

(the notation is as in Section 11.10.3, with σ_S the standard deviation of service time).

During the time that a machine is in the queue and being serviced, its lost production is costed at £200 per hour. In addition, if the machine is found to have been misaligned then its output for the previous several hours (given on average by the ARL) must be examined and if necessary rectified, at a cost of £10 per production hour.

Find the total cost per hour for each option, and hence decide which control scheme should be implemented.

7 A transmission channel for binary data connects a source to a receiver. The source emits a 0 with probability α and a 1 with probability $1 - \alpha$, each symbol independent of every other. The noise in the channel causes some bits to be interpreted incorrectly. The probability that a bit will be inverted is p (whether a 0 or a 1, the channel is 'symmetric').

(a) Using Bayes' theorem, express the four probabilities that the source symbol is a 0 or a 1 given that the received symbol is a 0 or a 1.

(b) If p is small and the receiver chooses to deliver whichever source symbol is the more likely given the received symbol, find the conditions on α such that the source symbol is assumed to be the same as the received symbol.

8 If discrete random variables X and Y can take possible values $\{u_1, \ldots, u_m\}$ and $\{v_1, \ldots, v_n\}$ respectively, with joint distribution $P(u_k, v_j)$ (see Section 11.4.1), the **mutual information** between X and Y is defined as

$$I(X;Y) = \sum_{k=1}^{m}\sum_{j=1}^{n} P(u_k, v_j)\log_2 \frac{P(u_k, v_j)}{P(X = u_k)P(Y = v_j)}$$

Show that for the binary symmetric transmission channel referred to in Review exercise 7, if X is the source symbol, Y the received symbol and $\alpha = \frac{1}{2}$ then

$$I(X;\ Y) = 1 + p\log_2 p + (1 - p)\log_2(1 - p)$$

The interpretation of this quantity is that it measures (in 'bits') the average amount of information received for each bit of data transmitted. Show that $I(X;\ Y) = 0$ when $p = \frac{1}{2}$ and that $I(X;\ Y) \to 1$ as $p \to 0$ and as $p \to 1$. Interpret this result.

Answers to Exercises

CHAPTER 1

Exercises

1 (a) $y = \frac{5}{2}x + \frac{5}{4}$ (b) $y = \frac{1}{4}x - \frac{3}{4}$

2 $z = 2, \frac{1}{2}\pi$

3 $u = 6v$

6 Semi-infinite strip $v > 0$, $|u| < 1$

7 (a) $u = v\sqrt{3} - 4$
(b) $v = -u\sqrt{3}$
(c) $(u + 1)^2 + (v - \sqrt{3})^2 = 4$
(d) $u^2 + v^2 = 8$

8 (a) $\alpha = \frac{1}{5}(-2 + \mathrm{j})$, $\beta = \frac{3}{5}(1 + 2\mathrm{j})$
(b) $u + 2v < 3$
(c) $(5u - 3)^2 + (5v - 6)^2 < 20$
(d) $\frac{3}{10}(1 + 3\mathrm{j})$

9 Interior of circle, centre $(0, -1/2c)$, radius $1/2c$; half-plane $v < 0$; region outside the circle, centre $(0, -1/2c)$, radius $1/2c$)

10 Circle, centre $(\frac{1}{2}, -\frac{2}{3})$, radius $\frac{7}{6}$

11 $\mathrm{Re}(w) = 1/2a$, half-plane $\mathrm{Re}(w) > 1/2a$

12 $w = \dfrac{z + 1}{\mathrm{j}z - \mathrm{j}}$,
$\mathrm{Re}(z) = \mathrm{const}(k)$ to circles
$u^2 + \left(v - \dfrac{k}{1 - k}\right)^2 = \dfrac{1}{(1 - k)^2}$ plus $v = -1$ $(k = 1)$
$\mathrm{Im}(z) = \mathrm{const}(l)$ to circles $\left(u + \dfrac{1}{l}\right)^2 + (v + 1)^2 = \dfrac{1}{l^2}$
plus $u = 0$ $(l = 0)$

13 (a) $1 + \mathrm{j}, \mathrm{j}, \infty$
(b) $|w| > \sqrt{2}$
(c) $v = 0$, $(u - 1)^2 + v^2 = 1$
(d) $\pm 2^{1/4}\,\mathrm{e}^{\mathrm{j}\pi/8}$

14 Segment of the imaginary axis $|v| \geqslant 1$

15 (a) Upper segment of the circle, centre $(\frac{2}{3}, -\frac{2}{3})$, radius $\frac{1}{3}\sqrt{5}$, cut off by the line $u - 3v = 1$

16 Circle, centre $(\frac{5}{3}, 0)$, radius $\frac{4}{3}$

17 $z_0 = \mathrm{j}$, $\theta_0 = \pi$

18 $|w - 1| < 1$; $\left|w - \frac{4}{3}\right| > \frac{2}{3}$

19 $w = \mathrm{e}^{\mathrm{j}\theta_0}\dfrac{z - z_0}{z_0^* z - 1}$, where θ_0 is any real number

20 Region enclosed between the inverted parabola $v = 2 - (u^2/8)$ and the real axis

21 $u = 0$, $2mu = (1 - m^2)v$

23 $u = x + \dfrac{x}{x^2 + y^2}$, $v = y - \dfrac{y}{x^2 + y^2}$; $v = 0$; ellipses, $u^2 + v^2 = r^2$ and $x^2 + y^2 = r^2$, r large

24 (a) $\mathrm{e}^z(z + 1)$ (b) $4\cos 4z$ (d) $-2\sin 2z$

25 $a = -1$, $b = 1$
$w = z^2 + \mathrm{j}z^2$, $\mathrm{d}w/\mathrm{d}z = 2(1 + \mathrm{j})z$

26 $v = 2y + x^2 - y^2$

27 $\mathrm{e}^x(x \sin y + y \cos y)$, $z\,\mathrm{e}^z$

28 $\cos x \sinh y$, $\sin z$

29 (a) $x^4 - 6x^2y^2 + y^4 = \beta$
(b) $2\,\mathrm{e}^{-x}\sin y + x^2 - y^2 = \beta$

30 (a) $(x^2 - y^2)\cos 2x - 2xy \sin 2y + \mathrm{j}[2xy \cos 2x + (x^2 - y^2)\sin 2y]$
(b) $\sin 2x \cosh 2y + \mathrm{j}\cos 2x \sinh 2y$

31 $u = \cos^{-1}\{2y^2\{x^2 + y^2 - 1 + \sqrt{[(x^2 + y^2 - 1)^2 + 4y^2]}\}\}$
$v = \sinh^{-1}\sqrt{\{\frac{1}{2}(x^2 + y^2 - 1) + \frac{1}{2}\sqrt{[(x^2 + y^2 - 1)^2 + 4y^2]}\}}$

33 (a) 0
(b) 3, 4
(c) $\frac{1}{2}$, $\frac{1}{4}(-1 + \mathrm{j}\sqrt{3})$, $\frac{1}{4}(-1 - \mathrm{j}\sqrt{3})$

34 $z = \pm\mathrm{j}$

35 (a) region outside unit circle
(b) $1 \leqslant u^2 + v^2 < \mathrm{e}^2$, $0 \leqslant v \leqslant u \tan 1$
(c) outside unit circle, u and v of opposite sign

36

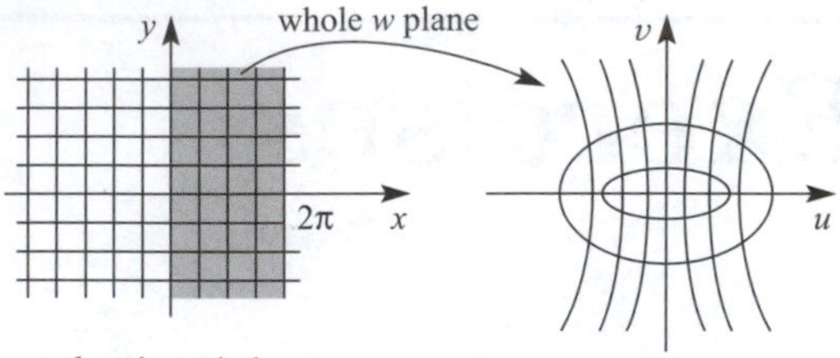

$x = k \rightarrow$ hyperbola
$y = k \rightarrow$ ellipse

37 $4a$, ellipse centred at origin, since axes are $\dfrac{b}{a^2 + b^2}$ and $\dfrac{b}{|b^2 - a^2|}$

38 (a) $j + z - jz^2 - z^3 + jz^4 + \ldots$
(b) $\dfrac{1}{z} + \dfrac{j}{z^2} - \dfrac{1}{z^3} - \dfrac{j}{z^4} + \dfrac{1}{z^5} + \ldots$
(c) $1 - (z - 1 - j) + (z - 1 - j)^2 - (z - 1 - j)^3 + \ldots$

39 (a) $1 - 2z^2 + 3z^4 - 4z^6 + \ldots$
(b) $1 - 3z^2 + 6z^4 - 10z^6 + \ldots$

40 (a) $\frac{1}{2} - \frac{1}{4}(z - 1) + \frac{1}{8}(z - 1)^2 - \frac{1}{16}(z - 1)^3$; 2
(b) $\frac{1}{4} - \frac{1}{16}(z - 2j)^2 + \frac{1}{64}(z - 2j)^4 - \frac{1}{256}(z - 2j)^6$; 2
(c) $-\frac{1}{2}j + \frac{1}{2}(1 + j)(z - 1 - j) + \frac{3}{4}(z - 1 - j)^2 + \frac{1}{2}(j - 1)(z - 1 - j)^3$; $\sqrt{2}$

41 $1 - z + z^3 + \ldots$

42 1, 1, $\sqrt{5}$; f is singular at $z = j$

43 $z + \frac{1}{3}z^2 + \frac{2}{15}z^5 + \ldots$; $\frac{1}{2}\pi$

44 (a) $\dfrac{1}{z} + 2 + 3z + 4z^2 + \ldots\ (0 < |z| < 1)$
(b) $\dfrac{1}{(z-1)^2} - \dfrac{1}{z-1} + 1 - (z - 1) + (z - 1)^2 - \ldots$
$(0 < |z - 1| < 1)$

45 (a) $\ldots + \dfrac{1}{5!z^3} - \dfrac{1}{3!z} + z$
(b) $z - \dfrac{1}{3!z} + \dfrac{1}{5!z^3} - \ldots$ 952
(c) $a^2 \sin \dfrac{1}{a} + zf'(a) + \ldots$

46 (a) $+\frac{1}{2}z + \frac{3}{4}z^2 + \frac{7}{8}z^3 + \frac{15}{16}z^4 + \ldots$
(b) $\ldots - \dfrac{1}{z^2} - \dfrac{1}{z} - 1 - \dfrac{1}{2} - \dfrac{1}{8}z^3 - \ldots$
(c) $\dfrac{1}{z} + \dfrac{3}{z^2} + \dfrac{7}{z^3} + \dfrac{15}{z^4} + \ldots$
(d) $\dfrac{1}{z-1} + \dfrac{2}{(z-1)^2} + \dfrac{2}{(z-1)^3} + \ldots$
(e) $-1 + \dfrac{2}{z-2} + (z - 2) - (z - 2)^2 + (z - 2)^3 + (z - 2)^4 - \ldots$

47 (a) $z = 0$, double pole
(b) $z = j$, simple pole; $z = -j$, double pole
(c) $z = \pm 1, \pm j$, simple poles
(d) $z = jn\pi$ (n an integer), simple poles
(e) $z = \pm j\pi$, simple poles
(f) $z = 1$, essential singularity
(g) Simple zero at $z = 1$ and simple poles at $z = \pm j$
(h) Simple zero at $z = -j$, simple pole at $z = 3$ and a pole of order 3 at $z = -2$
(i) Simple poles at $z = 2 + j$, $2 - j$ and a pole of order 2 at $z = 0$

48 (a) $\dfrac{z}{2!} - \dfrac{z^3}{4!} + \dfrac{z^5}{5!} - \ldots$ (removable singularity)
(b) $\dfrac{1}{z^3} + \dfrac{1}{z} + \dfrac{z}{2!} + \dfrac{z^3}{3!} + \dfrac{z^5}{4!} + \dfrac{z^7}{5!} + \ldots$ (pole of order 3)
(c) $\dfrac{1}{z} + \dfrac{1}{2!z^3} + \dfrac{1}{4!z^5} - \ldots$ (essential singularity)
(d) $\tan^{-1} 2 + \dfrac{2}{5}z - \dfrac{6}{25}z^2 + \ldots$ (analytic point)

50 (a) Simple poles at $z = -1, 2$; residues $\frac{1}{3}, \frac{5}{3}$
(b) Simple pole at $z = 1$, double pole at $z = 0$; residues -1, 1
(c) Simple poles at $z = 1, 3j, -3j$; residues $\frac{1}{2}$, $\frac{5}{12}(3 - j)$, $\frac{5}{12}(3 + j)$
(d) Simple poles at $z = 0, 2j, -2j$; residues $-\frac{1}{4}$, $-\frac{3}{8} + \frac{3}{4}j$, $-\frac{3}{8} - \frac{3}{4}j$
(e) Pole of order 5 at $z = 1$, residue 19
(f) Pole of order 2 at $z = 1$, residue 4
(g) Simple pole at $z = -3$, double pole at $z = 1$; residues $-\frac{1}{8}, \frac{1}{8}$
(h) Simple poles at $z = 0, -2, -1$; residues $\frac{3}{2}, -\frac{5}{2}$, 1

51 (a) 1 (simple pole)
(b) $-\frac{1}{12}(3 + j\sqrt{3}) \sin [\frac{1}{2}(1 + j\sqrt{3})]$ (simple pole)
(c) $\frac{1}{4}(1 + j)\sqrt{2}$ (simple pole)
(d) $-\pi$ (simple pole) (e) $-j\frac{1}{4}$ (double pole)

52 (a) $-\frac{1}{2}$ (triple pole) (b) $-\frac{14}{25}$ (double pole)
(c) $e^{n\pi}$ (double pole)

53 $-\frac{44}{3} - j\frac{8}{3}$, all cases

54 0, all cases

56 $0,\ 2\pi\text{j}$

57 $\frac{4}{5}\pi\text{j},\ \frac{12}{5}\pi\text{j}$

58 $\frac{4}{17}\pi(9+\text{j}2)$, 0

59 (a) $-\frac{3}{8}\pi\text{j}$ (b) 0

60 (a) 0 (b) $2\pi\text{j}$

61 (a) $-\frac{4}{9}\pi\text{j}$ (b) $2\pi\text{j}$

62 $z = \text{j},\ -\frac{3}{10}\text{j}$; $z = -\text{j},\ \frac{3}{10}\text{j}$; $z = \text{j}\sqrt{6},\ \frac{2}{15}\text{j}\sqrt{6}$; $z = -\text{j}\sqrt{6},\ -\frac{2}{15}\text{j}\sqrt{6}$

(a) 0, (b) $\frac{3}{5}\pi$, (c) 0

63 (a) 0 (b) 0

64 (a) $2\pi\text{j},\ \frac{5}{2}\pi\text{j}$

(b) $\frac{2}{25}\pi(25-\text{j}39)$

(c) $0,\ \frac{19}{108}\pi\text{j},\ -\frac{19}{108}\pi\text{j}$

(d) $0-\frac{487}{162}\pi\text{j},\ -3\pi\text{j}$

65 (a) $2\pi/\sqrt{3}$ (b) $\frac{1}{2}\pi$ (c) $\frac{5}{288}\pi$ (d) $\frac{1}{12}\pi$

(e) $\frac{8}{3}\pi$ (f) $\frac{7}{10}\pi$ (g) π (h) $\pi/2\sqrt{2}$

(i) $\frac{1}{2}\pi$ (j) $\pi(1-3/\sqrt{5})$

66 $2axV_0/(x^2+y^2)$

67 (a) (0, 0), (0, 1), (0, 7), (7, 0)

(b) $v = 0$ (c) $u = 0$

68 $H(x, y) = 2y - y^2 + x^2$;

$W = 2z - \text{j}z^2$

70 (a) (0, 0), (1, 0), (−1, 0)

(b) $u = 0$ (c) $v = 0$

1.9 Review exercises

1 (a) 3j (b) 7 + j4 (c) 1 (d) j2

2 $y = 2x$ gives $3u + v = 3$, $u + 2v = 3$ and $3v - u = 1$ respectively

$x + y = 1$ gives $v = 1$, $v - u = 3$ and $u = 1$ respectively

3 (a) $\alpha = -\frac{1}{5}(3+\text{j}4)$, $\beta = 3 + \text{j}$

(b) $13 \leqslant 3u + 4v$

(c) $|w - 3 - \text{j}| \leqslant 1$

(d) $\frac{1}{4}(7-\text{j})$

4 (a) $u^2 + v^2 + u - v = 0$

(b) $u = 3v$

(c) $u^2 + v^2 + u - 2v = 0$

(d) $4(u^2 + v^2) = u$

5 Left hand Right hand

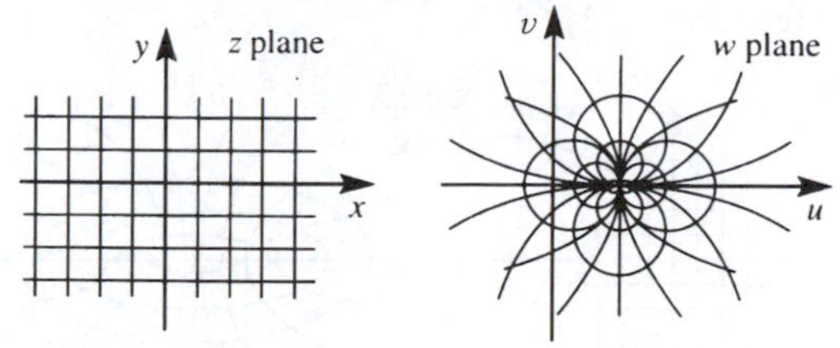

$$x = k \rightarrow \left(u - \frac{k}{k-1}\right)^2 + v^2 = \frac{1}{(k-1)^2}$$

$$y = l \rightarrow (u-1)^2 + \left(v + \frac{1}{l}\right)^2 = \frac{1}{l^2}$$

Fixed points: $1 \pm \sqrt{2}$

6 Fixed points $z = \pm\sqrt{2}/2$

$r = 1 \Rightarrow u = 0$

7 $u = x^3 - 3xy^2,\ v = 3x^2y - y^3$

8 $(z \sin z)$ $v = y \sin x \cosh y + x \cos x \sinh y$

9 $w = 1/z$

10 Ellipse is given by $\dfrac{x^2}{(R + a^2/4k)^2} + \dfrac{y^2}{(R - a^2/4k)^2} = 1$

11 $1 - z^3 + z^6 - z^9 + z^{12} - \dots$;

$1 - 2z^3 + 3z^6 - 4z^9 + \dots$

12 (a) $1 - 2z + 2z^2 - 2z^3$; 1

(b) $\frac{1}{2} - \frac{1}{2}(z-1) + \frac{1}{4}(z-1)^2 - \frac{1}{6}(z-1)^4$; $\sqrt{2}$

(c) $\frac{1}{2}(1+\text{j}) + \frac{1}{2}\text{j}(z-\text{j}) - \frac{1}{4}(1+\text{j})(z-\text{j})^2 - \frac{1}{8}(z-\text{j})^3$; $\sqrt{2}$

13 1, 1, 1, $\frac{1}{2}\sqrt{5}$, $2\sqrt{2}$ respectively

14 (a) $\frac{1}{z} - z + z^3 - z^5 + \dots\ (0 < |z| < 1)$

(b) $\frac{1}{2} - (z-1) + \frac{5}{4}(z-1)^2 + \dots\ (|z-1| < 1)$

15 (a) Taylor series

(b) and (c) are essential singularities, the principal parts are infinite

16 (a) $\frac{1}{2}(\text{e}^{2x}\cos 2y - 1) + \text{j}\frac{1}{2}\text{e}^{2x}\sin 2y$

(b) $\cos 2x \cosh 2y - \text{j} \sin 2x \sinh 2y$

(c)

$$\frac{x \sin x \cosh y + y \cos x \sinh y + \text{j}(x \cos x \sinh y - y \sin x \cosh y)}{x^2 + y^2}$$

(d) $\dfrac{\tan x(1 - \tanh^2 y) + \text{j} \tanh y(1 + \tan^2 x)}{1 + \tan^2 x \tanh^2 y}$

17 (a) Conformal (b) j, $-1-j$
(c) ± 0.465, $\pm j0.465$

18

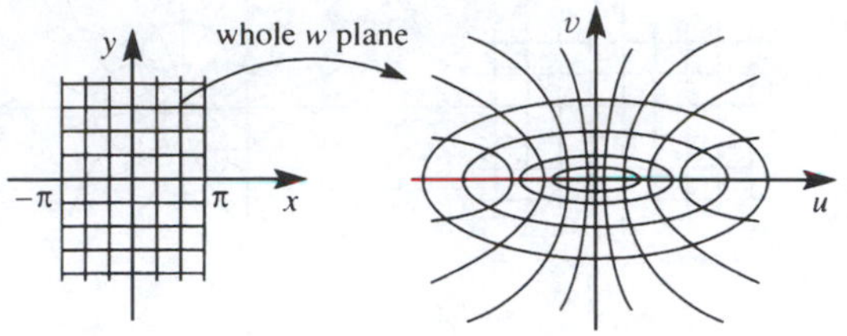

$x = k \to$ hyperbolas, $\dfrac{u^2}{\cos^2 k} - \dfrac{v^2}{\sin^2 k} = 1$

$y = l \to$ ellipses, $\dfrac{u^2}{\cosh^2 l} + \dfrac{v^2}{\sinh^2 l} = 1$

19 (a) Simple pole at $z = 0$
(b) Double poles at $z = 2, 2e^{2\pi j/3}, 2e^{4\pi j/3}$
(c) Simple poles at $z = +1, \pm j$, removable singularity at $z = -1$
(d) Simple poles at $z = \frac{1}{2}(2n+1)\pi j$ $(n = 0, \pm 1, \pm 2, \dots)$
(e) No singularities in finite plane (entire)
(f) Essential singularity at $z = 0$
(g) Essential (non-isolated) singularity at $z = 0$

20 (a) $2e^{-2}$ (b) 0 (c) 0 (d) 0

21 Zeros: ± 1, $-\frac{3}{2} \pm \frac{1}{2}j\sqrt{11}$
Poles: 0, $e^{\pi j/4}$, $e^{3\pi j/4}$, $e^{5\pi j/4}$, $e^{7\pi j/4}$
Residues (respectively) -5, $\dfrac{6+3\sqrt{2}}{4} - j$,
$\dfrac{6-3\sqrt{2}}{4} + \dfrac{j}{4}$, $\dfrac{6-3\sqrt{2}}{4} - \dfrac{j}{-j}$, $\dfrac{6+3\sqrt{2}}{4} + \dfrac{j}{4}$

22 $-204 - 324j$

23 (a) $-\frac{2}{5}\pi j$ (b) 0 (c) (i) 0, (ii) $3\pi j$ (d) 0, 0
(e) $-\pi$ (f) $j\dfrac{\pi}{6}, -\dfrac{4\pi}{3}j$

24 (a) $\frac{7}{50}\pi$ (b) $\frac{1}{8}\pi\sqrt{2}$ (c) $-\dfrac{11\pi}{24}$ (d) $\dfrac{19\pi}{12}$

CHAPTER 2

Exercises

1 (a) $\dfrac{s}{s^2-4}$, Re$(s) > 2$ (b) $\dfrac{2}{s^3}$, Re$(s) > 0$
(c) $\dfrac{3s+1}{s^2}$, Re$(s) > 0$ (d) $\dfrac{1}{(s+1)^2}$, Re$(s) > -1$

2 (a) 5 (b) -3 (c) 0 (d) 3 (e) 2
(f) 0 (g) 0 (h) 0 (i) 2 (j) 3

3 (a) $\dfrac{5s-3}{s^2}$, Re$(s) > 0$
(b) $\dfrac{42}{s^4} - \dfrac{6}{s^2+9}$, Re$(s) > 0$
(c) $\dfrac{3s-2}{s^2} + \dfrac{4s}{s^2+4}$, Re$(s) > 0$
(d) $\dfrac{s}{s^2-9}$, Re$(s) > 3$
(e) $\dfrac{2}{s^2-4}$, Re$(s) > 2$
(f) $\dfrac{5}{s+2} + \dfrac{3}{s} - \dfrac{2s}{s^2+4}$, Re$(s) > 0$
(g) $\dfrac{4}{(s+2)^2}$, Re$(s) > -2$
(h) $\dfrac{4}{s^2+6s+13}$, Re$(s) > -3$
(i) $\dfrac{2}{(s+4)^3}$, Re$(s) > -4$
(j) $\dfrac{36-6s+4s^2-2s^3}{s^4}$, Re$(s) > 0$
(k) $\dfrac{2s+15}{s^2+9}$, Re$(s) > 0$
(l) $\dfrac{s^2-4}{(s^2+4)^2}$, Re$(s) > 0$
(m) $\dfrac{18s^2-54}{(s^2+9)^3}$, Re$(s) > 0$
(n) $\dfrac{2}{s^3} - \dfrac{3s}{s^2+16}$, Re$(s) > 0$
(o) $\dfrac{2}{(s+2)^3} + \dfrac{s+1}{s^2+2s+5} + \dfrac{3}{s}$, Re$(s) > 0$

4 (a) $\frac{1}{4}(e^{-3t} - e^{-7t})$ (b) $-e^{-t} + 2e^{3t}$
(c) $\frac{4}{9} - \frac{1}{3}t - \frac{4}{9}e^{-3t}$ (d) $2\cos 2t + 3\sin 2t$
(e) $\frac{1}{64}(4t - \sin 4t)$ (f) $e^{-2t}(\cos t + 6\sin t)$
(g) $\frac{1}{8}(1 - e^{-2t}\cos 2t + 3e^{-2t}\sin 2t)$
(h) $e^{t} - e^{-t} + 2te^{-t}$
(i) $e^{-t}(\cos 2t + 3\sin 2t)$ (j) $\frac{1}{2}e^{t} - 3e^{2t} + \frac{11}{2}e^{3t}$
(k) $-2e^{-3t} + 2\cos(\sqrt{2}t) - \sqrt{\tfrac{1}{2}}\sin(\sqrt{2}t)$
(l) $\frac{1}{5}e^{t} - \frac{1}{5}e^{-t}(\cos t - 3\sin t)$
(m) $e^{-t}(\cos 2t - \sin 2t)$ (n) $\frac{1}{2}e^{2t} - 2e^{3t} + \frac{3}{2}e^{-4t}$
(o) $-e^{t} + \frac{3}{2}e^{2t} - \frac{1}{2}e^{-2t}$
(p) $4 - \frac{9}{2}\cos t + \frac{1}{2}\cos 3t$

(q) $9\,e^{-2t} - e^{-3t/2}[7\cos(\frac{1}{2}\sqrt{3}t) - \sqrt{3}\sin(\frac{1}{2}\sqrt{3}t)]$

(r) $\frac{1}{9}e^{-t} - \frac{1}{10}e^{-2t} - \frac{1}{90}e^{-t}(\cos 3t + 3\sin 3t)$

5 (a) $x(t) = e^{-2t} + e^{-3t}$

(b) $x(t) = \frac{35}{78}e^{4t/3} - \frac{3}{26}(\cos 2t + \frac{2}{3}\sin 2t)$

(c) $x(t) = \frac{1}{5}(1 - e^{-t}\cos 2t - \frac{1}{2}e^{-t}\sin 2t)$

(d) $y(t) = \frac{1}{25}(12\,e^{-t} + 30t\,e^{-t} - 12\cos 2t + 16\sin 2t)$

(e) $x(t) = -\frac{7}{5}e^{t} + \frac{4}{3}e^{2t} + \frac{1}{15}e^{-4t}$

(f) $x(t) = e^{-2t}(\cos t + \sin t + 3)$

(g) $x(t) = \frac{13}{12}e^{t} - \frac{1}{3}e^{-2t} + \frac{1}{4}e^{-t}(\cos 2t - 3\sin 2t)$

(h) $y(t) = -\frac{2}{3} + t + \frac{2}{3}e^{-t}[\cos(\sqrt{2}t) + \sqrt{\tfrac{1}{2}}\sin(\sqrt{2}t)]$

(i) $x(t) = (\frac{1}{8} + \frac{3}{4}t)e^{-2t} + \frac{1}{2}t^2 e^{-2t} + \frac{3}{8} - \frac{1}{2}t + \frac{1}{4}t^2$

(j) $x(t) = \frac{1}{5} - \frac{1}{5}e^{-2t/3}(\cos\frac{1}{3}t + 2\sin\frac{1}{3}t)$

(k) $x(t) = t\,e^{-4t} - \frac{1}{2}\cos 4t$

(l) $y(t) = e^{-t} + 2t\,e^{-2t/3}$

(m) $x(t) = \frac{5}{4} + \frac{1}{2}t - e^{t} + \frac{5}{12}e^{2t} - \frac{2}{3}e^{-t}$

(n) $x(t) = \frac{9}{20}e^{-t} - \frac{7}{16}\cos t + \frac{25}{16}\sin t - \frac{1}{80}\cos 3t$
$- \frac{3}{80}\sin 3t$

6 (a) $x(t) = \frac{1}{4}(\frac{15}{4}e^{3t} - \frac{11}{4}e^{t} - e^{-2t}),\ y(t) = \frac{1}{8}(3e^{3t} - e^{t})$

(b) $x(t) = 5\sin t + 5\cos t - e^{t} - e^{2t} - 3$
$y(t) = 2\,e^{t} - 5\sin t + e^{2t} - 3$

(c) $x(t) = 3\sin t - 2\cos t + e^{-2t}$
$y(t) = -\frac{7}{2}\sin t + \frac{9}{2}\cos t - \frac{1}{2}e^{-3t}$

(d) $x(t) = \frac{3}{2}e^{t/3} - \frac{1}{2}e^{t},\ y(t) = -1 + \frac{1}{2}e^{t} + \frac{3}{2}e^{t/3}$

(e) $x(t) = 2\,e^{t} + \sin t - 2\cos t$
$y(t) = \cos t - 2\sin t - 2\,e^{t}$

(f) $x(t) = -3 + e^{t} + 3\,e^{-t/3}$
$y(t) = t - 1 - \frac{1}{2}e^{t} + \frac{3}{2}e^{-t/3}$

(g) $x(t) = 2t - e^{t} + e^{-2t},\ y(t) = t - \frac{7}{2} + 3\,e^{t} + \frac{1}{2}e^{-2t}$

(h) $x(t) = 3\cos t + \cos(\sqrt{3}t)$
$y(t) = 3\cos t - \cos(\sqrt{3}t)$

(i) $x(t) = \cos(\sqrt{\tfrac{3}{10}}t) + \frac{3}{4}\cos(\sqrt{6}t)$
$y(t) = \frac{5}{4}\cos(\sqrt{\tfrac{3}{10}}t) - \frac{1}{4}\cos(\sqrt{6}t)$

(j) $x(t) = \frac{1}{3}e^{t} + \frac{2}{3}\cos 2t + \frac{1}{3}\sin 2t$
$y(t) = \frac{2}{3}e^{t} - \frac{2}{3}\cos 2t - \frac{1}{3}\sin 2t$

7 $I_1(s) = \dfrac{E_1(50 + s)s}{(s^2 + 10^4)(s + 100)^2}$

$I_2(s) = \dfrac{Es^2}{(s^2 + 10^4)(s + 100)^2}$

$i_2(t) = E(-\frac{1}{200}e^{-100t} + \frac{1}{2}t\,e^{-100t} + \frac{1}{200}\cos 100t)$

9 $i_1(t) = 20\sqrt{\tfrac{1}{7}}\,e^{-t/2}\sin(\frac{1}{2}\sqrt{7}t)$

10 $x_1(t) = -\frac{3}{10}\cos(\sqrt{3}t) - \frac{7}{10}\cos(\sqrt{13}t)$
$x_2(t) = -\frac{1}{10}\cos(\sqrt{3}t) + \frac{21}{10}\cos(\sqrt{13}t),\ \sqrt{3},\ \sqrt{13}$

13 $f(t) = tH(t) - tH(t-1)$

14 (a) $f(t) = 3t^2 - [3(t-4)^2 + 22(t-4) + 43]H(t-4)$
$- [2(t-6) + 4]H(t-6)$

$F(s) = \dfrac{6}{s^3} - \left(\dfrac{6}{s^3} + \dfrac{22}{s^2} + \dfrac{43}{s}\right)e^{-4s} - \left(\dfrac{2}{s^3} + \dfrac{4}{s}\right)e^{-6s}$

(b) $f(t) = t - 2(t-1)H(t-1) + (t-2)H(t-2)$

$F(s) = \dfrac{1}{s^2} - \dfrac{2}{s^2}e^{-s} + \dfrac{1}{s^2}e^{-2s}$

15 (a) $\frac{1}{6}(t-5)^3 e^{2(t-5)}H(t-5)$

(b) $\frac{3}{2}[e^{-(t-2)} - e^{-3(t-2)}]H(t-2)$

(c) $[t - \cos(t-1) - \sin(t-1)]H(t-1)$

(d) $\sqrt{\tfrac{1}{3}}\,e^{-(t-\pi)/2}\{\sqrt{3}\cos[\frac{1}{2}\sqrt{3}(t-\pi)]$
$+ \sin[\frac{1}{2}\sqrt{3}(t-\pi)]\}H(t-\pi)$

(e) $H(t - \frac{4}{5}\pi)\cos 5t$

(f) $[t - \cos(t-1) - \sin(t-1)]H(t-1)$

16 $x(t) = e^{-t} + (t-1)[1 - H(t-1)]$

17 $x(t) = 2\,e^{-t/2}\cos(\frac{1}{2}\sqrt{3}t) + t - 1 - 2H(t-1)$
$\{t - 2 + e^{-(t-1)/2}\{\cos[\frac{1}{2}\sqrt{3}(t-1)]$
$- \sqrt{\tfrac{1}{3}}\sin[\frac{1}{2}\sqrt{3}(t-1)]\}\}$
$+ H(t-2)\{t - 3 + e^{-(t-2)/2}$
$\{\cos[\frac{1}{2}\sqrt{3}(t-2)] - \sqrt{\tfrac{1}{3}}\sin[\frac{1}{2}\sqrt{3}(t-2)]\}\}$

18 $x(t) = e^{-t} + \frac{1}{10}(\sin t - 3\cos t + 4\,e^{\pi}e^{-2t}$
$- 5\,e^{\pi/2}e^{-t})H(t - \frac{1}{2}\pi)$

19 $f(t) = 3 + 2(t-4)H(t-4)$

$F(s) = \dfrac{3}{s} + \dfrac{2}{s^2}e^{-4s}$

$x(t) = 3 - 2\cos t + 2[t - 4 - \sin(t-4)]H(t-4)$

20 $\theta_0(t) = \frac{3}{10}(1 - e^{-3t}\cos t - 3e^{-3t}\sin t)$
$- \frac{3}{10}[1 - e^{3a}e^{-3t}\cos(t-a)$
$- 3e^{3a}e^{-3t}\sin(t-a)]H(t-a)$

21 $\theta_0(t) = \frac{1}{32}(3 - 2t - 3e^{-4t} - 10t\,e^{-4t})$
$+ \frac{1}{32}[2t - 3 + (2t-1)e^{-4(t-1)}]H(t-1)$

23 $\dfrac{3 - 3e^{-2s} - 6s\,e^{-4s}}{s^2(1 - e^{-4s})}$

24 $\dfrac{K}{T}\dfrac{1}{s^2} - \dfrac{K}{s}\dfrac{e^{-sT}}{1 - e^{-sT}}$

25 (a) $2\delta(t) + 9e^{-2t} - 19e^{-3t}$

(b) $\delta(t) - \frac{5}{2}\sin 2t$

(c) $\delta(t) - e^{-t}(2\cos 2t + \frac{1}{2}\sin 2t)$

26 (a) $x(t) = (\frac{1}{6} - \frac{2}{3}e^{-3t} + \frac{1}{2}e^{-4t})$
$+ (e^{-3(t-2)} - e^{-4(t-2)})H(t-2)$

(b) $x(t) = \frac{1}{2}e^{6\pi}e^{-3t}H(t-2\pi)\sin 2t$

(c) $x(t) = 5e^{-3t} - 4e^{-4t} + (e^{-3(t-3)} - e^{-4(t-3)})H(t-3)$

27 (a) $f'(t) = g'(t) - 43\delta(t-4) - 4\delta(t-6)$

$$g'(t) = \begin{cases} 6t & (0 \le t < 4) \\ 2 & (4 \le t < 6) \\ 0 & (t \ge 6) \end{cases}$$

(b) $$g'(t) = \begin{cases} 1 & (0 \le t < 1) \\ -1 & (1 \le t < 2) \\ 0 & (t \ge 2) \end{cases}$$

(c) $f'(t) = g'(t) + 5\delta(t) - 6\delta(t-2) + 15\delta(t-4)$

$$g'(t) = \begin{cases} 2 & (0 \le t < 2) \\ -3 & (2 \le t < 4) \\ 2t-1 & (t \ge 4) \end{cases}$$

28 $x(t) = -\frac{19}{9}e^{-5t} + \frac{19}{9}e^{-2t} - \frac{4}{3}te^{-2t}$

30 $q(t) = \dfrac{E}{Ln}e^{-\mu t}\sin nt,\ n^2 = \dfrac{1}{LC} - \dfrac{R^2}{4L^2},\ \mu = \dfrac{R}{2L}$

$i(t) = \dfrac{E}{Ln}e^{-\mu t}(n\cos nt - \mu\sin nt)$

31 $y(x) = \dfrac{1}{48EI}[2Mx^4/l + 8W(x - \frac{1}{2}l)^3H(x - \frac{1}{2}l)$
$- 4(M + W)x^3 + (2M + 3W)l^2x]$

32 $y(x) = \dfrac{w(x_2^2 - x_1^2)x^2}{4EI} - \dfrac{w(x_2 - x_1)x^3}{6EI}$
$+ \dfrac{w}{24EI}[(x - x_1)^4H(x - x_1)$
$- (x - x_2)^4H(x - x_2)]$

$y_{max} = wl^4/8EI$

33 $y(x) = \dfrac{W}{EI}[\frac{1}{6}x^3 - \frac{1}{6}(x-b)^3H(x-b) - \frac{1}{2}bx^2]$

$$= \begin{cases} -\dfrac{Wx^2}{6EI}(3b - x) & (0 < x \le b) \\ -\dfrac{Wb^2}{6EI}(3x - b) & (b < x \le l) \end{cases}$$

34 (a) $\dfrac{3s+2}{s^2+2s+5}$

(b) $s^2 + 2s + 5 = 0$, order 2

(c) Poles $-1 \pm j2$; zero $-\frac{2}{3}$

35 $\dfrac{s^2+5s+6}{s^3+5s^2+17s+13}$, $s^3 + 5s^2 + 17s + 13 = 0$
order 3, zeros -3, -2, poles -1, $-2 \pm j3$

36 (a) Marginally stable (b) Unstable
(c) Stable (d) Stable (e) Unstable

37 (a) Unstable
(b) Stable
(c) Marginally stable
(d) Stable
(e) Stable

40 $K > \frac{2}{3}$

41 (a) $3e^{-7t} - 3e^{-8t}$ (b) $\frac{1}{3}e^{-4t}\sin 3t$
(c) $\frac{2}{3}(e^{4t} - e^{-2t})$ (d) $\frac{1}{3}e^{2t}\sin 3t$

42 $\dfrac{s+8}{(s+1)(s+2)(s+4)}$

47 $\frac{2}{7}, \frac{4}{5}$

49 (a) $\frac{1}{54}[2 - e^{-3t}(9t^2 + 6t + 2)]$

(b) $\frac{1}{125}[e^{-3t}(5t+2) + e^{2t}(5t-2)]$

(c) $\frac{1}{16}(4t - 1 + e^{-4t})$

51 $e^{-3t} - e^{-4t}$

$x(t) = \frac{1}{12}A[1 - 4e^{-3t} + 3e^{-4t} - (1 - 4e^{-3(t-T)}$
$+ 3e^{-4(t-T)})H(t-T)]$

52 $e^{-2t}\sin t,\ \frac{1}{5}[1 - e^{-2t}(\cos t + 2\sin t)]$

2.8 Review exercises

1 (a) $x(t) = \cos t + \sin t - e^{-2t}(\cos t + 3\sin t)$

(b) $x(t) = -3 + \frac{13}{7}e^{t} + \frac{15}{7}e^{-2t/5}$

2 (a) $e^{-t} - \frac{1}{2}e^{-2t} - \frac{1}{2}e^{-t}(\cos t + \sin t)$

(b) $i(t) = 2e^{-t} - 2e^{-2t}$
$+ V[e^{-t} - \frac{1}{2}e^{-2t} - \frac{1}{2}e^{-t}(\cos t + \sin t)]$

3 $x(t) = -t + 5\sin t - 2\sin 2t$,
$y(t) = 1 - 2\cos t + \cos 2t$

4 $\frac{1}{5}(\cos t + 2\sin t)$

$e^{-t}[(x_0 - \frac{1}{5})\cos t + (x_1 + x_0 - \frac{3}{5})\sin t]$

$\sqrt{\frac{1}{5}}$, 63.4° lag

6 (a) (i) $\dfrac{s\cos\phi - \omega\sin\phi}{s^2+\omega^2}$

(ii) $\dfrac{s\sin\phi + \omega(\cos\phi + \sin\phi)}{s^2 + 2\omega s + 2\omega^2}$

(b) $\frac{1}{20}(\cos 2t + 2\sin 2t) + \frac{1}{20}e^{-2t}(39\cos 2t + 47\sin 2t)$

7 (a) $e^{-2t}(\cos 3t - 2\sin 3t)$

(b) $y(t) = 2 + 2\sin t - 5e^{-2t}$

8 $x(t) = e^{-8t} + \sin t,\ y(t) = e^{-8t} - \cos t$

9 $q(t) = \frac{1}{500}(5e^{-100t} - 2e^{-200t}) - \frac{1}{500}(3\cos 100t - \sin 100t)$,
current leads by approximately 18.5°

10 $x(t) = \frac{29}{20}e^{-t} + \frac{445}{1212}e^{-t/5} + \frac{1}{3}e^{-2t} - \frac{1}{505}(76\cos 2t - 48\sin 2t)$

11 (a) $\theta = \frac{1}{100}(4e^{-4t} + 10te^{-4t} - 4\cos 2t + 3\sin 2t)$

(b) $i_1 = \frac{1}{7}(e^{4t} + 6e^{-3t}),\ i_2 = \frac{1}{7}(e^{-3t} - e^{4t})$

12 $i = \dfrac{E}{R}[1 - e^{-nt}(\cos nt + \sin nt)]$

13 $i_1 = \dfrac{E(4 - 3e^{-Rt/L} - e^{-3Rt/L})}{6R},\ i_2 \to E/3R$

14 $x_1(t) = \frac{1}{3}[\sin t - 2\sin 2t + \sqrt{3}\sin(\sqrt{3}t)]$

$x_2(t) = \frac{1}{3}[\sin t + \sin 2t - \sqrt{3}\sin(\sqrt{3}t)]$

15 (a) (i) $e^{-t}(\cos 3t + \sin 3t)$

(ii) $e^t - e^{2t} + 2te^t$

(b) $y(t) = \frac{1}{2}e^{-t}(8 + 12t + t^3)$

16 (a) $\frac{5}{2}e^{7t}\sin 2t$

(b) $\dfrac{n^2 i}{Ks(s^2 + 2Ks + n^2)},\ \theta(t) = \dfrac{i}{K}(1 - e^{-Kt}) - ite^{-Kt}$

17 (a) (ii) $e^{-(t-\alpha)}[\cos 2(t-\alpha) - \frac{1}{2}\sin 2(t-\alpha)]H(t-\alpha)$

(b) $y(t) = \frac{1}{10}[e^{-t}(\cos 2t - \frac{1}{2}\sin 2t) + 2\sin t - \cos t] + \frac{1}{10}[e^{-(t-\pi)}(\cos 2t - \frac{1}{2}\sin 2t) + \cos t - 2\sin t]H(t-\pi)$

18 $i(t) = \frac{1}{250}[e^{-40t} - 2H(1 - \frac{1}{2}T)e^{-40(t-T/2)} + 2H(t-T)e^{-40(t-T)} - 2H(t - \frac{3}{2}T)e^{-40(t-3T/2)} + \dots]$
Yes, since time constant is large compared with T

19 $e^{-t}\sin t,\ \frac{1}{2}[1 - e^{-t}(\cos t + \sin t)]$

20 $EI\dfrac{d^4y}{dx^4} = 12 + 12H(x-4) - R\delta(x-4)$,

$y(0) = y'(0) = y(4) = y^{(2)}(5) = y^{(3)}(0) = 0$

$y(x) =$
$$\begin{cases} \frac{1}{2}x^4 - 4.25x^3 + 9x^2 & (0 \leqslant x \leqslant 4) \\ \frac{1}{2}x^4 - 4.25x^3 + 9x^2 + \frac{1}{2}(x-4)^4 - 7.75(x-4)^3 & (4 \leqslant x \leqslant 5) \end{cases}$$

25.5 kN, 18 kN m

21 (a) $f(t) = H(t-1) - H(t-2)$

$x(t) = H(t-1)(1 - e^{-(t-1)}) - H(t-2)(1 - e^{-(t-2)})$

(b) 0, E/R

23 (a) $t - 2 + (t+2)e^{-t}$

(b) $y = t + 2 - 2e^t + 2te^t,\ y(t) = \frac{1}{2}t^2 + y_1$

24 $EIy = -\frac{2}{9}Wlx^2 + \frac{10}{81}Wx^3 - \dfrac{W(x-l)^3}{6}H(x-l)$

$EI\dfrac{d^4y}{dx^4} = -W\delta(x-l) - w[H(x) - H(x-l)]$

25 (a) $x(t) = \frac{1}{6}\{1 + e^{3(t-a)/2}[\sqrt{3}\sin(\frac{1}{2}\sqrt{3}t) - \cos(\frac{1}{2}\sqrt{3}t)]H(t-a)\}$

26 (a) No (b) $\dfrac{1}{s^2 + 2s + (K-3)}$ (d) $K > 3$

27 (a) 4 (b) $\frac{1}{10}$

28 (a) $\dfrac{K}{s^2 + (1 + KK_1)s + K}$

(c) $K = 12.5,\ K_1 = 0.178$

(d) 0.65 s, 2.48 s, 1.86 s

29 (a) $K_2 = M_2\omega^2$

30 (b) Unstable (c) $\beta = 2.5 \times 10^{-5}$, 92 dB

(d) −8 dB, 24°

(e) $K = 10^6,\ \tau_1 = 10^{-6},\ \tau_2 = 10^{-7},\ \tau_3 = 4 \times 10^{-8}$

(f) $s^3 + 36 \times 10^6 s^2 + 285 \times 10^{12}s + 25 \times 10^{18}(1 + 10^7\beta) = 0$

CHAPTER 3

Exercises

1 (a) $\dfrac{4}{4z+1},\ |z| > \frac{1}{4}$ (b) $\dfrac{z}{z-3},\ |z| > 3$

(c) $\dfrac{z}{z+2},\ |z| > 2$ (d) $\dfrac{-z}{z-2},\ |z| > 2$

(e) $3\dfrac{z}{(z-1)^2},\ |z| > 1$

2 $e^{-2\omega kT} \leftrightarrow \dfrac{z}{z - e^{-2\omega T}}$

4 $\dfrac{1}{z^3}\dfrac{2z}{2z-1} = \dfrac{2}{z^2(2z-1)}$

5 (a) $\dfrac{5z}{5z+1}$ (b) $\dfrac{z}{z+1}$

6 $\dfrac{2z}{2z-1}$, $\dfrac{2z}{(2z-1)^2}$

8 (a) $\{e^{-4kT}\} \leftrightarrow \dfrac{z}{z-e^{-4T}}$

(b) $\{\sin kT\} \leftrightarrow \dfrac{z \sin T}{z^2 - 2z\cos T + 1}$

(c) $\{\cos 2kT\} \leftrightarrow \dfrac{z(z-\cos 2T)}{z^2 - 2z\cos 2T + 1}$

11 (a) 1 (b) $(-1)^k$ (c) $(\frac{1}{2})^k$ (d) $\frac{1}{3}(-\frac{1}{3})^k$
(e) j^k (f) $(-j\sqrt{2})^k$
(g) 0 $(k = 0)$, 1 $(k > 0)$
(h) 1 $(k = 0)$, $(-1)^{k+1}$ $(k > 0)$

12 (a) $\frac{1}{3}[1-(-2)^k]$ (b) $\frac{1}{7}[3^k - (-\frac{1}{2})^k]$
(c) $\frac{1}{3} + \frac{1}{6}(-\frac{1}{2})^k$ (d) $\frac{2}{3}(\frac{1}{2})^k + \frac{2}{3}(-1)^{k+1}$
(e) $\sin\frac{1}{2}k\pi$ (f) $2^k \sin\frac{1}{6}k\pi$
(g) $\frac{5}{2}k + \frac{1}{4}(1-3^k)$
(h) $k + 2\sqrt{\frac{1}{3}}\cos(\frac{1}{3}k - \frac{3}{2}\pi)$

13 (a) $\{0, 1, 0, 0, 0, 0, 0, 2\}$
(b) $\{1, 0, 3, 0, 0, 0, 0, 0, 0, -2\}$
(c) $\{0, 5, 0, 1, 3\}$
(d) $\{0, 0, 1, 1\} + \{(-\frac{1}{3})^k\}$
(e) $1(k = 0)$, $\frac{5}{2}(k = 1)$ $\frac{5}{4}(k = 2)$ $-\frac{1}{8}(-\frac{1}{2})^{k-3}(k \geq 3)$
(f) $\begin{cases} 0 & (k = 0) \\ 3 - 2k + 2^{k-1} & (k \geq 1) \end{cases}$
(g) $\begin{cases} 0 & (k = 0) \\ 2 - 2^{k-1} & (k \geq 1) \end{cases}$

14 $y_{k+2} + \frac{1}{2}y_{k+1} = x_k$, $y_{k+2} + \frac{1}{4}y_{k+1} - \frac{1}{5}y_k = x_k$

15 (a) $y_k = k$ (b) $y_k = \frac{3}{10}(9^k) + \frac{17}{10}(-1)^k$
(c) $2^{k-1}\sin\frac{1}{2}k\pi$ (d) $2(-\frac{1}{2})^k + 3^k$

16 (a) $y_k = \frac{2}{5}(-\frac{1}{2})^k - \frac{9}{10}(\frac{1}{3})^k + \frac{1}{2}$
(b) $y_k = \frac{7}{2}(3^k) - 6(2^k) + \frac{5}{2}$
(c) $y_n = \frac{2}{5}(3^n) - \frac{2}{3}(2^n) + \frac{4}{15}(\frac{1}{2})^n$
(d) $y_n = 2(3^{n-1})\sin\frac{1}{6}n\pi + 1$
(e) $y_n = \frac{18}{5}(-\frac{1}{2})^n - \frac{8}{5}(-2)^n - 2n - 1$
(f) $y_n = -\frac{1}{2}[2^n + (-2)^n] + 1 - n$

17 (b) 7, £4841

18 $y_k = 2^k - \frac{1}{2}(3^k) + \frac{1}{2}$

19 As $k \to \infty$, $I_k \to 2G$ as a damped oscillation

21 (a) $\dfrac{1}{z^2 - 3z + 2}$
(b) $\dfrac{z-1}{z^2 - 3z + 1}$
(c) $\dfrac{z+1}{z^3 - z^2 + 2z + 1}$

22

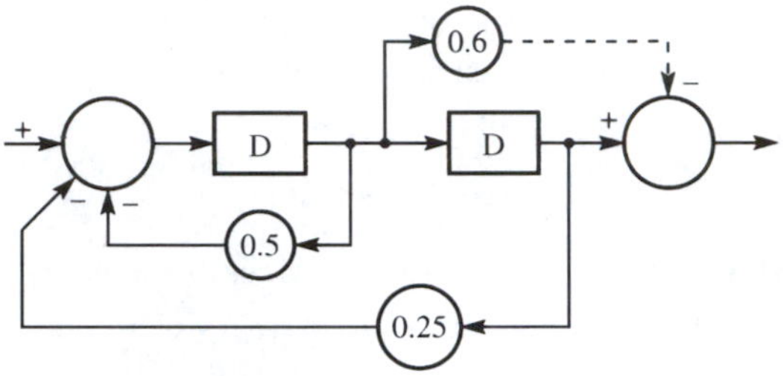

23 (a) $\frac{1}{2}\{(-\frac{1}{4})^k - (-\frac{1}{2})^k\}$
(b) $2(3^k)\sin\frac{1}{6}(k+1)\pi$
(c) $\frac{2}{3}(0.4)^k + \frac{1}{3}(-0.2)^k$
(d) $4^{k+1} + 2^k$

24 $\begin{cases} 0 & (k = 0) \\ 2^{k-1} - 1 & (k \geq 1) \end{cases}$

$\begin{cases} 0 & (k = 0) \\ 2^{k-1} & (k \geq 1) \end{cases}$

25 (a), (b) and (c) are stable; (d) is unstable; (e) is marginally stable

26 $2 - (\frac{1}{2})^k$

27 $y_n = -4(\frac{1}{2})^n + 2(\frac{1}{3})^n + 2(\frac{2}{3})^n$

29 q form:
$(Aq^2 + Bq + C)y_k = \Delta^2(q^2 + 2q + 1)u_k$
δ form:
$[A\Delta^2\delta^2 + (2\Delta A + \Delta B)\delta + (A + B + C)]y_k$
$= \Delta^2(4 + 4\Delta\delta + \Delta^2\delta^2)u_k$
$A = 2\Delta^2 + 6\Delta + 4$
$B = 4\Delta^2 - 8$
$C = 2\Delta^2 - 6\Delta + 4$

30 $\dfrac{1}{s^3 + 2s^2 + 2s + 1}$
$[(\Delta^3 + 4\Delta^2 + 8\Delta + 8)\delta^3 + (6\Delta^2 + 16\Delta + 16)\delta^2$
$+ (12\Delta + 16)\delta + 8]y_k = (2 + T\delta)^3 u_k$

32 $\dfrac{12(z^2 - z)}{(12 + 5\Delta)z^2 + (8\Delta - 12)z - \Delta}$

$\dfrac{12\gamma(1 + \Delta\gamma)}{\Delta(12 + 5\Delta)\gamma^2 + (8\Delta - 12)\gamma + 12}$

3.10 Review exercises

4 $3 + 2k$

5 $\frac{1}{6} + \frac{1}{3}(-2)^k - \frac{1}{2}(-1)^k$

7 $\dfrac{2z}{(z - e)^{3T}} - \dfrac{z}{z - e^{-2T}}$

8 (a) $\left\{\dfrac{1}{a - b}(a^n - b^n)\right\}$

(b) (i) $3^{k-1}k$ (ii) $2\sqrt{\frac{1}{3}}\sin\frac{1}{3}k\pi$

9 $\frac{3}{2} - \frac{1}{2}(-1)^k - 2^k$

10 $(-1)^k$

13 $\frac{1}{2}A[2 - 2(\frac{1}{2})^k - k(\frac{1}{2})^{k-1}]$

CHAPTER 4

Exercises

1 (a) $f(t) = -\dfrac{1}{4}\pi - \dfrac{2}{\pi}\sum_{n=1}^{\infty}\dfrac{\cos(2n-1)t}{(2n-1)^2} + \sum_{n=1}^{\infty}\left[\dfrac{3\sin(2n-1)t}{2n-1} - \dfrac{\sin 2nt}{2n}\right]$

(b) $f(t) = \dfrac{1}{4}\pi + \dfrac{2}{\pi}\sum_{n=1}^{\infty}\dfrac{\cos(2n-1)t}{(2n-1)^2} - \sum_{n=1}^{\infty}\dfrac{\sin nt}{n}$

(c) $f(t) = \dfrac{2}{\pi}\sum_{n=1}^{\infty}\dfrac{\sin nt}{n}$

(d) $f(t) = \dfrac{2}{\pi} + \dfrac{4}{\pi}\sum_{n=1}^{\infty}\dfrac{(-1)^{n+1}\cos 2nt}{4n^2 - 1}$

(e) $f(t) = \dfrac{2}{\pi} + \dfrac{4}{\pi}\sum_{n=1}^{\infty}\dfrac{(-1)^{n+1}\cos nt}{4n^2 - 1}$

(f) $f(t) = \dfrac{1}{2}\pi - \dfrac{4}{\pi}\sum_{n=1}^{\infty}\dfrac{\cos(2n-1)t}{(2n-1)^2}$

(g) $f(t) = -\dfrac{4}{\pi}\sum_{n=1}^{\infty}\dfrac{\cos(2n-1)t}{(2n-1)^2} - \sum_{n=1}^{\infty}\dfrac{\sin 2nt}{n}$

(h) $f(t) = \left(\dfrac{1}{2}\pi + \dfrac{1}{\pi}\sinh\pi\right) + \dfrac{2}{\pi}\sum_{n=1}^{\infty}\left[\dfrac{(-1)^n - 1}{n^2} + \dfrac{(-1)^n\sinh\pi}{n^2 + 1}\right]\cos nt - \dfrac{2}{\pi}\sum_{n=1}^{\infty}\dfrac{n(-1)^n}{n^2 + 1}\sinh\pi\sin nt$

2 $f(t) = \dfrac{1}{3}\pi^2 + 4\sum_{n=1}^{\infty}\dfrac{\cos nt}{n^2}$

Taking $t = \pi$ gives the required result.

3 $q(t) = Q\left[\dfrac{1}{2} - \dfrac{4}{\pi^2}\sum_{n=1}^{\infty}\dfrac{\cos(2n-1)t}{(2n-1)^2}\right]$

4 $f(t) = \dfrac{5}{\pi} + \dfrac{5}{2}\sin t - \dfrac{10}{\pi}\sum_{n=1}^{\infty}\dfrac{\cos 2nt}{4n^2 - 1}$

5 Taking $t = 0$ and $t = \pi$ gives the required answers.

6 $f(t) = \dfrac{1}{4}\pi - \dfrac{2}{\pi}\sum_{n=1}^{\infty}\dfrac{\cos(4n-2)t}{(2n-1)^2}$

Taking $t = 0$ gives the required series.

7 $f(t) = \dfrac{3}{2} + \dfrac{4}{\pi^2}\sum_{n=1}^{\infty}\dfrac{\cos(2n-1)t}{(2n-1)^2}$

Replacing t by $t - \frac{1}{2}\pi$ gives the following sine series of odd harmonics:

$$f\left(t - \frac{1}{2}\pi\right) - \frac{3}{2} = -\frac{4}{\pi^2}\sum_{n=1}^{\infty}\frac{(-1)^n\sin(2n-1)t}{(2n-1)^2}$$

8 $f(t) = \dfrac{2l}{\pi}\sum_{n=1}^{\infty}\dfrac{(-1)^{n+1}}{n}\sin\dfrac{n\pi t}{l}$

9 $f(t) = \dfrac{2K}{\pi}\sum_{n=1}^{\infty}\dfrac{1}{n}\sin\dfrac{n\pi t}{l}$

10 $f(t) = \dfrac{3}{2} + \dfrac{6}{\pi}\sum_{n=1}^{\infty}\dfrac{1}{(2n-1)}\dfrac{\sin(2n-1)\pi t}{5}$

11 $v(t) = \dfrac{A}{\pi}\left(1 + \dfrac{1}{2}\pi\sin\omega t - 2\sum_{n=1}^{\infty}\dfrac{\cos 2n\omega t}{4n^2 - 1}\right)$

12 $f(t) = \dfrac{1}{3}T^2 + \dfrac{4T^2}{\pi^2}\sum_{n=1}^{\infty}\dfrac{(-1)^n}{n^2}\cos\dfrac{n\pi t}{T}$

13 $e(t) = \dfrac{E}{2}\left(1 - \dfrac{2}{\pi}\sum_{n=1}^{\infty}\dfrac{1}{n}\sin\dfrac{2\pi nt}{T}\right)$

15 $f(t) = -\dfrac{8}{\pi^2}\sum_{n=1}^{\infty}\dfrac{1}{(2n-1)^2}\cos(2n-1)\pi t$

16 (a) $f(t) = \frac{2}{3} - \frac{1}{\pi^2}\sum_{n=1}^{\infty}\frac{1}{n^2}\cos 2n\pi t + \frac{1}{\pi}\sum_{n=1}^{\infty}\frac{1}{n}\sin 2n\pi t$

(b) $f(t) = \frac{1}{\pi}\sum_{n=1}^{\infty}\frac{1}{n}\sin 2n\pi t + \frac{2}{\pi}\sum_{n=1}^{\infty}\left[\frac{1}{2n-1} + \frac{4}{\pi^2(2n-1)^3}\right] \times \sin(2n-1)\pi t$

(c) $f(t) = \frac{2}{3} + \frac{4}{\pi^2}\sum_{n=1}^{\infty}\frac{(-1)^{n+1}}{n^2}\cos n\pi t$

17 $f(t) = \frac{1}{6}\pi^2 - \sum_{n=1}^{\infty}\frac{1}{n^2}\cos 2nt$

$f(t) = \frac{8}{\pi}\sum_{n=1}^{\infty}\frac{1}{(2n-1)^3}\sin(2n-1)t$

18 $f(x) = \frac{8a}{\pi^2}\sum_{n=1}^{\infty}\frac{(-1)^{n+1}}{(2n-1)^2}\sin\frac{(2n-1)\pi x}{l}$

19 $f(x) = \frac{2l}{\pi^2}\sum_{n=1}^{\infty}\frac{(-1)^{n+1}}{(2n-1)^2}\sin\frac{2(2n-1)\pi x}{l}$

20 $f(t) = \frac{1}{2}\sin t + \frac{4}{\pi}\sum_{n=1}^{\infty}\frac{n(-1)^{n+1}}{4n^2-1}\sin 2nt$

21 $f(x) = -\frac{1}{2}A - \frac{4A}{\pi^2}\sum_{n=1}^{\infty}\frac{1}{(2n-1)^2}\cos\frac{(2n-1)\pi x}{l}$

22 $T(x) = \frac{8KL^2}{\pi^3}\sum_{n=1}^{\infty}\frac{1}{(2n-1)^3}\sin\frac{(2n-1)\pi x}{L}$

23 $f(t) = \frac{1}{2} + \frac{1}{2}\cos\pi t + \frac{4}{\pi}\sum_{n=1}^{\infty}\frac{1}{4n^2-1}\sin 2n\pi t - \frac{2}{\pi}\sum_{n=1}^{\infty}\frac{1}{2n-1}\sin(2n-1)\pi t$

26 (c) $1 + 4\sum_{n=1}^{\infty}\frac{(-1)^{n+1}}{n}\sin nt$

29 (a) $\frac{1}{6}\pi^2 + \sum_{n=1}^{\infty}\frac{2}{n^2}(-1)^n\cos nt + \sum_{n=1}^{\infty}\frac{1}{\pi}\left[-\frac{\pi^2}{n}(-1)^n + \frac{2}{n^3}(-1)^n - \frac{2}{n^3}\right]\sin nt$

(b) $a_n = 0$

$b_n = \frac{4}{n\pi}\left(\cos n\pi - \cos\frac{1}{2}n\pi\right) + 2\left(\frac{3\pi}{4n^2}\sin\frac{1}{2}n\pi - \frac{\pi^2}{8n}\cos\frac{1}{2}n\pi + \frac{3}{n^3}\cos\frac{1}{2}n\pi - \frac{6}{\pi n^4}\sin\frac{1}{2}n\pi\right),$

$\frac{1}{\pi}\left[\left(\frac{3}{2}\pi^2 - 16\right)\sin t + \frac{1}{8}(32 + \pi^3 - 6\pi)\sin 2t - \frac{1}{3}\left(\frac{32}{9} + \frac{1}{2}\pi^2\right)\sin 3t + \dots\right]$

(c) $-\frac{4}{\pi^2}\sum_{n=1}^{\infty}\frac{\cos(2n-1)\pi t}{(2n-1)^2} + \frac{2}{\pi}\sum_{n=1}^{\infty}\frac{\sin(2n-1)t}{(2n-1)}$

(d) $\frac{1}{4} + \frac{2}{\pi^2}\sum_{n=1}^{\infty}\frac{1}{(2n-1)^2}\cos 2(2n-1)\pi t$

30 $e(t) = 5 + \frac{20}{\pi}\sum_{n=1}^{\infty}\frac{1}{2n-1}\sin(2n-1)100\pi t$

$i_{ss}(t) \simeq 0.008\cos(100\pi t - 1.96) + 0.005\cos(300\pi t - 0.33)$

31 $f(t) = \frac{400}{\pi}\sum_{n=1}^{\infty}\frac{1}{2n-1}\sin(2n-1)t$

$x_{ss}(t) \simeq 0.14\sin(\pi t - 0.1) + 0.379\sin(3\pi t - 2.415) + 0.017\sin(5\pi t - 2.83)$

32 $f(t) = \frac{100}{\pi}\sum_{n=1}^{\infty}\frac{(-1)^{n+1}}{n}\sin 2\pi nt$

$x_{ss}(t) \simeq 0.044\sin(2\pi t - 3.13) - 0.0052\sin(4\pi t - 3.14)$

33 $e(t) = \frac{100}{\pi} + 50\sin 50\pi t - \frac{200}{\pi}\sum_{n=1}^{\infty}\frac{\cos 100\pi nt}{4n^2-1}$

$i_{ss}(t) \simeq 0.78\cos(50\pi t + (-0.17)) - 0.01\sin(100\pi t + (-0.48))$

35 $f(t) = \frac{1}{2} + \sum_{\substack{n=-\infty\\n\neq 0}}^{\infty}\frac{\mathrm{j}}{2n\pi}[(-1)^n - 1]\mathrm{e}^{\mathrm{j}n\pi t/2}$

36 (a) $\frac{3}{4}\pi + \sum_{\substack{n=-\infty\\n\neq 0}}^{\infty}\frac{1}{2\pi}\left\{\frac{\mathrm{j}\pi}{n} - \frac{1}{n^2}[1 + (-1)^n]\right\}\mathrm{e}^{\mathrm{j}nt}$

(b) $\frac{a}{2}\sin\omega t - \sum_{n=-\infty}^{\infty}\frac{a}{2\pi(n^2-1)}[(-1)^n + 1]\,\mathrm{e}^{\mathrm{j}n\omega t}$

(c) $\frac{3}{2} + \sum_{\substack{n=-\infty\\n\neq 0}}^{\infty}\frac{\mathrm{j}}{2\pi n}[1 - (-1)^n]\,\mathrm{e}^{\mathrm{j}nt}$

(d) $\frac{2}{\pi}\sum_{n=-\infty}^{\infty}\frac{1}{1-4n^2}\mathrm{e}^{2\mathrm{j}nt}$

38 (b) (i) 17.74, (ii) 17.95
(c) 18.14; (i) 2.20%, (ii) 1.05%

39 (a) $c_0 = 15,\ c_n = \dfrac{30}{\mathrm{j}n\pi}(1 - \mathrm{e}^{-\mathrm{j}n\pi/2})$

$15,\ \dfrac{30}{\pi}(1-\mathrm{j}),\ -\dfrac{30}{\pi}\mathrm{j},\ -\dfrac{10}{\pi}(1+\mathrm{j}),\ 0,\ \dfrac{6}{\pi}(1-\mathrm{j})$

(b) 15 W, 24.30 W, 12.16 W, 2.70 W, 0.97 W
(c) 60 W
(d) 91.9%

40 0.19, 0.10, 0.0675

41 (c) $c_0 = 0,\ c_1 = \frac{3}{2},\ c_2 = 0,\ c_3 = -\frac{7}{8}$

42 (c) $c_0 = \frac{1}{4},\ c_1 = \frac{1}{2},\ c_2 = \frac{5}{16},\ c_3 = 0$

46 (b) $c_1 = 0,\ c_2 = \surd(2\pi),\ c_3 = 0$, MSE = 0

4.9 Review exercises

1 $f(t) = \dfrac{1}{6}\pi^2 + \displaystyle\sum_{n=1}^{\infty} \dfrac{2}{n^2}(-1)^n \cos nt$

$+ \displaystyle\sum_{n=1}^{\infty} \left[\dfrac{\pi}{2n-1} - \dfrac{4}{\pi(2n-1)^3}\right] \sin(2n-1)t$

$- \displaystyle\sum_{n=1}^{\infty} \dfrac{\pi}{2n} \sin 2nt$

Taking $T = \pi$ gives the required sum

2 $f(t) = \dfrac{1}{9}\pi + \dfrac{2}{\pi}\displaystyle\sum_{n=1}^{\infty} \dfrac{1}{n^2}\left\{\cos \dfrac{1}{3}n\pi - \dfrac{1}{3}[2 + (-1)^n]\right\} \cos nt,$

$\frac{2}{9}\pi$

3 (a) $f(t) = \dfrac{2T}{\pi^2}\displaystyle\sum_{n=1}^{\infty} \dfrac{(-1)^{n+1}}{(2n-1)^2} \sin \dfrac{2(2n-1)\pi t}{T}$

(b) $-\frac{1}{4}T$; (c) Taking $t = \frac{1}{4}T$ gives $S = \frac{1}{8}\pi^2$

5 $f(t) = \dfrac{4}{\pi}\displaystyle\sum_{n=1}^{\infty} \dfrac{(-1)^n \sin(2n-1)t}{(2n-1)^2}$

8 $f(x) = \dfrac{4}{\pi}\displaystyle\sum_{n=1}^{\infty} \dfrac{(-1)^{n+1}}{(2n-1)^2} \sin(2n-1)x$

$f(x) = \dfrac{1}{4}\pi - \dfrac{2}{\pi}\displaystyle\sum_{n=1}^{\infty} \dfrac{\cos 2(2n-1)x}{(2n-1)^2}$

10 (a) $f(t) = \displaystyle\sum_{n=1}^{\infty} \dfrac{2}{n} \sin nt$

(b) $f(t) = \dfrac{1}{2}\pi + \dfrac{4}{\pi}\displaystyle\sum_{n=1}^{\infty} \dfrac{1}{(2n-1)^2} \cos(2n-1)t$

13 (a) $f(t) = \dfrac{1}{2}\pi - \dfrac{4}{\pi}\displaystyle\sum_{n=1}^{\infty} \dfrac{1}{(2n-1)^2} \cos(2n-1)t$

(b) $g(t) = \dfrac{4}{\pi}\displaystyle\sum_{n=1}^{\infty} \dfrac{1}{2n-1} \sin(2n-1)t$

15 (a) $v(t) = \dfrac{10}{\pi} + 5 \sin \dfrac{2\pi t}{T} - \dfrac{20}{\pi}\displaystyle\sum_{n=1}^{\infty} \dfrac{1}{4n^2-1} \cos \dfrac{4n\pi t}{T}$

(b) 2.5 W, 9.01%

16 $g(t) = \dfrac{4}{\pi}\displaystyle\sum_{n=1}^{\infty} \dfrac{1}{2n-1} \sin(2n-1)t$

$f(t) = 1 + g(t)$

18 (b) $\dfrac{\sin \omega t - \omega \cos \omega t}{1 + \omega^2}$ $\quad \dfrac{4}{\pi}\displaystyle\sum_{n=1}^{\infty} \dfrac{\sin \alpha t - \alpha \cos \alpha t}{(2n-1)(1+\alpha^2)}$

$\alpha = (4n-2)\pi/T$

19 (c) $T_0 = 1,\ T_1 = t_1,\ T_2 = 2t^2 - 1,\ T_3 = 4t^3 - 3t$
(d) $\frac{1}{16}T_5 - \frac{5}{8}T_4 + \frac{33}{16}T_3 - \frac{5}{2}T_2 + \frac{95}{5}T_1 - \frac{79}{8}T_0$
(e) $\frac{33}{4}t^3 - 5t^2 + \frac{91}{16}t - \frac{59}{8},\ \frac{11}{16},\ t = -1$

CHAPTER 5

Exercises

1 $\dfrac{2a}{a^2 + \omega^2}$

2 $AT^2\mathrm{j}\omega \operatorname{sinc}^2 \dfrac{\omega T}{2}$

3 $AT \operatorname{sinc}^2 \dfrac{\omega t}{2}$

4 $8K \operatorname{sinc} 2\omega,\ 2K \operatorname{sinc} \omega,\ 2K(4 \operatorname{sinc} 2\omega - \operatorname{sinc} \omega)$

5 $4 \operatorname{sinc} \omega - 4 \operatorname{sinc} 2\omega$

7 $\dfrac{\omega_0}{(a + \mathrm{j}\omega)^2 + \omega_0^2}$

10 $F_s = \dfrac{x}{x^2 + a^2},\ F_c = \dfrac{x}{x^2 + a^2}$

12 $\dfrac{1}{(1 - \omega^2) + 3\mathrm{j}\omega}$

13 $4 \operatorname{sinc} 2\omega - 2 \operatorname{sinc} \omega$

14 $\frac{1}{2}T[\operatorname{sinc} \frac{1}{2}(\omega_0 - \omega)T + \operatorname{sinc} \frac{1}{2}(\omega_0 + \omega)T]$

15 $\frac{1}{2}T\mathrm{e}^{-\mathrm{j}\omega T/2}[\mathrm{e}^{\mathrm{j}\omega_0 T/2} \operatorname{sinc} \frac{1}{2}(\omega - \omega_0)T$
$+ \mathrm{e}^{-\mathrm{j}\omega_0 T/2} \operatorname{sinc} \frac{1}{2}(\omega + \omega_0)T]$

16 $j[\operatorname{sinc}(\omega + 2) - \operatorname{sinc}(\omega - 2)]$

18 $4AT \cos \omega\tau \operatorname{sinc} \omega T$

19 High pass filter

20 $\pi e^{-a|\omega|}$

21 $T[\operatorname{sinc}(\omega - \omega_0)T + \operatorname{sinc}(\omega + \omega_0)T]$

26 $\frac{1}{2}\pi j[\delta(\omega + \omega_0) - \delta(\omega - \omega_0)] - \frac{\omega_0}{\omega_0^2 - \omega^2}$

28 $\{2, 0, 2, 0\}$

29 $\{2, 0, 2, 0\}$

5.9 Review exercises

1 $\frac{\sin \omega}{\omega^2} - \frac{\cos 2\omega}{\omega}$

2 $-\frac{\pi j}{\omega} \operatorname{sinc} 2\omega$

7 (a) $\frac{1}{a-b}(e^{at} - e^{bt})H(t)$
(b) (i) $te^{2t}H(t)$ (ii) $(t - 1 + e^{-t})H(t)$

8 (a) $-\sin \omega_0(t + \frac{1}{4}\pi)$ (b) $\cos \omega_0 t$
(c) $je^{j\omega_0 t}$ (d) $-je^{-j\omega_0 t}$

17 (a) $\frac{a + 2\pi s}{a^2 + 4\pi^2 s^2}$
(b) $\frac{1}{2\pi s}(\sin 2\pi sT - \cos 2\pi sT + 1)$

CHAPTER 6

Exercises

1 (a)

2 $A = \begin{bmatrix} \sqrt{\frac{1}{2}} & \sqrt{\frac{1}{2}} & 0 \\ \sqrt{\frac{1}{2}} & -\sqrt{\frac{1}{2}} & 0 \\ 0 & 0 & 1 \end{bmatrix}$

The transformation rotates the e_1, e_2 plane through $\pi/4$ about the e_3 axis.

3 (a), (c) and (d)

4 The set of all odd quintic polynomials; it has dimension 3.

5 (a) $\lambda^3 - 12\lambda^2 + 40\lambda - 35$
(b) $\lambda^4 - 4\lambda^3 + 2\lambda^2 + 5\lambda + 2$

6 (a) 2, 0; $[1 \quad 1]^T$, $[1 \quad -1]^T$
(b) 4, −1; $[2 \quad 3]^T$, $[1 \quad -1]^T$
(c) 9, 3, −3; $[-1 \quad 2 \quad 2]^T$, $[2 \quad 2 \quad -1]^T$, $[2 \quad -1 \quad +2]^T$
(d) 3, 2, 1; $[2 \quad 2 \quad 1]^T$, $[1 \quad 1 \quad 0]^T$, $[0 \quad -2 \quad 1]^T$
(e) 14, 7, −7; $[2 \quad 6 \quad 3]^T$, $[6 \quad -3 \quad 2]^T$, $[3 \quad 2 \quad -6]^T$
(f) 2, 1, −1; $[-1 \quad 1 \quad 1]^T$, $[1 \quad 0 \quad -1]^T$, $[1 \quad 2 \quad -7]^T$
(g) 5, 3, 1; $[2 \quad 3 \quad -1]^T$, $[1 \quad -1 \quad 0]^T$, $[0 \quad -1 \quad 1]^T$
(h) 4, 3, 1; $[2 \quad -1 \quad -1]^T$, $[2 \quad -1 \quad 0]^T$, $[4 \quad 1 \quad -2]^T$

7 (a) 5, $[1 \quad 1 \quad 1]^T$; 1 (repeated) with two linearly independent eigenvectors, e.g. $[0 \quad 1 \quad 2]^T$, $[1 \quad 0 \quad -1]^T$
(b) −1, $[8 \quad 1 \quad 3]^T$; 2 (repeated) with one linearly independent eigenvector, e.g. $[1 \quad -1 \quad 0]^T$
(c) 1, $[4 \quad 1 \quad -3]^T$; 2 (repeated) with one linearly independent engenvector, e.g. $[3 \quad 1 \quad -2]^T$
(d) 2, $[2 \quad 1 \quad 2]^T$; 1 (repeated) with two linearly independent eigenvectors, e.g. $[0 \quad 2 \quad -1]^T$, $[2 \quad 0 \quad 3]^T$

8 1, $[-3 \quad 1 \quad 1]^T$

9 2, e.g. $[1 \quad 0 \quad 1]^T$, $[0 \quad 1 \quad 1]^T$

12 −6, 3, 2; $[2 \quad 1 \quad 1]^T$, $[-1 \quad 1 \quad 1]^T$, $[0 \quad 1 \quad -1]^T$

13 $[1 \quad -1 \quad 0]^T$

14 8.59, $[0.61 \quad 0.71 \quad 0.35]^T$

15 (a) 3.62, $[0.62 \quad 1 \quad 1]^T$ (b) 7, $[0.25 \quad 0.5 \quad 1]^T$
(c) 2.62, $[1 \quad -0.62 \quad -0.62 \quad 1]^T$

16 $\lambda_1 = 6$; $\lambda_2 = 3$, $\boldsymbol{e}_2 = [1 \quad 1 \quad -1]^T$; $\lambda_3 = 2$, $\boldsymbol{e}_3 = [1 \quad -1 \quad 0]^T$

18 10.132, 4.491, 0.373

19 (b) 0.59

20 5, 2, −1; $[-1 \quad 5 \quad 3]^T$, $[0 \quad 2 \quad 1]^T$, $[1 \quad 0 \quad 0]^T$

21 6, 3, 1; $[1 \quad 2 \quad 0]^T$, $[0 \quad 0 \quad 1]^T$, $[2 \quad -1 \quad 0]^T$

22 18, 9, −9; $[2 \quad 1 \quad 2]^T$, $[1 \quad 2 \quad -2]^T$, $[-2 \quad 2 \quad 1]^T$

23 2, 1, −1; $[1 \quad 3 \quad 1]^T$, $[3 \quad 2 \quad 1]^T$, $[1 \quad 0 \quad 1]^T$

24 −9, 6, 3; $[1 \quad 2 \quad -2]^T$, $[2 \quad 1 \quad 2]^T$, $[-2 \quad 2 \quad 1]^T$
$\boldsymbol{L} = \frac{1}{3}\boldsymbol{M}$, $\boldsymbol{M}$ modal matrix

25 $[0 \quad 0 \quad 1]^T$, $\begin{bmatrix} 2 & 2 & 0 \\ 2 & 5 & 0 \\ 0 & 0 & 3 \end{bmatrix}$

26 $\boldsymbol{J} = \begin{bmatrix} 1 & 1 & 0 \\ 0 & 1 & 1 \\ 0 & 0 & 1 \end{bmatrix}$

27 $\lambda = -2$: $[0 \quad 1 \quad 1 \quad 0]^T$, $[1 \quad 0 \quad 0 \quad 1]^T$
$\lambda = 4$: $[0 \quad 1 \quad -1 \quad 0]^T$, $[6 \quad -1 \quad 0 \quad -6]^T$

$$\boldsymbol{J} = \begin{bmatrix} -2 & 0 & 0 & 0 \\ 0 & -2 & 0 & 0 \\ 0 & 0 & 4 & 1 \\ 0 & 0 & 0 & 4 \end{bmatrix}$$

28 (a) Positive-definite
(b) Positive-semidefinite (c)Indefinite

29 $2a > 1$, $2b^2 < 6a - 3$

30 Positive-semidefinite, eigenvalues 3, 3, 0

31 $k > 2$; when $k = 2$ Q is positive-semidefinite

32 $a > 2$

33 $\lambda > 5$

35 (a) $\begin{bmatrix} 3 & 4 \\ 2 & 3 \end{bmatrix}$ (b) $\begin{bmatrix} 7 & 10 \\ 5 & 7 \end{bmatrix}$ (c) $\begin{bmatrix} 17 & 24 \\ 12 & 17 \end{bmatrix}$

36 (a) $\frac{1}{3}\begin{bmatrix} 2 & -1 \\ -1 & 2 \end{bmatrix}$ (b) $\frac{1}{11}\begin{bmatrix} -2 & 5 & -1 \\ -1 & -3 & 5 \\ 7 & -1 & -2 \end{bmatrix}$

37 $\begin{bmatrix} 47231 & 47342 & 47270 \\ 47342 & 47195 & 47306 \\ 47270 & 47306 & 47267 \end{bmatrix}$

38 (a) $\begin{bmatrix} e^t & 0 \\ te^t & e^t \end{bmatrix}$ (b) $\begin{bmatrix} e^t & 0 \\ e^{2t} - e^t & e^{2t} \end{bmatrix}$

40 (a) $\begin{bmatrix} 2t & 2 \\ -1 & 2t-1 \end{bmatrix}$ (b) $\begin{bmatrix} \frac{10}{3} & 0 \\ \frac{7}{2} & \frac{23}{6} \end{bmatrix}$

41 $\begin{bmatrix} t^4 + 2t^2 + t - 4 & t^3 - t^2 + t - 1 \\ 5t^2 + 5 & 5t - 5 \end{bmatrix}$

42 (a) $\dot{\boldsymbol{x}} = \begin{bmatrix} 0 & 1 & 0 \\ 0 & 0 & 1 \\ -4 & -5 & -4 \end{bmatrix}\boldsymbol{x} + \begin{bmatrix} 0 \\ 0 \\ 1 \end{bmatrix}u,$

$y = [1 \quad 0 \quad 0]\boldsymbol{x}$

(b) $\dot{\boldsymbol{x}} = \begin{bmatrix} 0 & 1 & 0 & 0 \\ 0 & 0 & 1 & 0 \\ 0 & 0 & 0 & 1 \\ 0 & -4 & -2 & 0 \end{bmatrix}\boldsymbol{x} + \begin{bmatrix} 0 \\ 0 \\ 0 \\ 5 \end{bmatrix}u,$

$y = [1 \quad 0 \quad 0 \quad 0]\boldsymbol{x}$

43 (a) $\dot{\boldsymbol{x}} = \begin{bmatrix} 0 & 1 & 0 \\ 0 & 0 & 1 \\ -7 & -5 & -6 \end{bmatrix}\boldsymbol{x} + \begin{bmatrix} 0 \\ 0 \\ 1 \end{bmatrix}u,$

$y = [5 \quad 3 \quad 1]\boldsymbol{x}$

(b) $\dot{\boldsymbol{x}} = \begin{bmatrix} 0 & 1 & 0 \\ 0 & 0 & 1 \\ 0 & -3 & -4 \end{bmatrix}\boldsymbol{x} + \begin{bmatrix} 0 \\ 0 \\ 1 \end{bmatrix}u, \quad y = [2 \quad 3 \quad 1]\boldsymbol{x}$

44 $\dot{\boldsymbol{x}} = \begin{bmatrix} -R_1/L_1 & -R_1/L_1 & -1/L_1 \\ -R_1/L_1 & -(R_1 + R_2)/L_2 & -1/L_2 \\ 1/C & 1/C & 1/C \end{bmatrix}\boldsymbol{x} + \begin{bmatrix} 1/L_1 \\ 1/L_2 \\ 0 \end{bmatrix}u,$

$y = [0 \quad R_2 \quad 0]\boldsymbol{x}$

45 A possible model is

$$\dot{\boldsymbol{x}} = \begin{bmatrix} B(M_1 + M)/MM_1 & 1 & 0 & 0 \\ -(K_1M + KM_1 + KM)/MM_1 & 0 & 1 & 0 \\ -K_1B/MM_1 & 0 & 0 & 1 \\ -K_1K/MM_1 & 0 & 0 & 0 \end{bmatrix}\boldsymbol{x}$$

$$+ \begin{bmatrix} 0 \\ 0 \\ K_1B/MM_1 \\ K_1K_2/MM_1 \end{bmatrix}u, \quad y = [1 \quad 0 \quad 0 \quad 0]\boldsymbol{x}$$

46 $$\begin{bmatrix} \dot{x}_1 \\ \dot{x}_2 \end{bmatrix} = \begin{bmatrix} -\dfrac{(R_1 + R_2 + R_4)}{\alpha C_1} & \dfrac{R_1}{\alpha C_1} \\ \dfrac{R_1}{\alpha C_2} & -\dfrac{R_1 + R_3}{\alpha C_2} \end{bmatrix}\begin{bmatrix} x_1 \\ x_2 \end{bmatrix} + \begin{bmatrix} \dfrac{R_2 + R_4}{\alpha C_1} \\ \dfrac{R_3}{\alpha C_2} \end{bmatrix}u$$

$$\begin{bmatrix} y_1 \\ y_2 \end{bmatrix} = \begin{bmatrix} \dfrac{R_1}{\alpha} & -\dfrac{R_1 + R_3}{\alpha} \\ -\dfrac{R_3}{\alpha}(R_1 + R_2 + R_4) & \dfrac{R_1R_3}{\alpha} \end{bmatrix} + \begin{bmatrix} \dfrac{R_3}{\alpha} \\ \dfrac{R_3}{\alpha}(R_4 + R_2) \end{bmatrix}u,$$

$\alpha = R_1R_3 + (R_1 + R_3)(R_2 + R_4)$
$-2.6 \times 10^2, -1.1 \times 10^2$

48 $\begin{bmatrix} e^t & 0 \\ te^t & e^t \end{bmatrix}$

49 $\begin{bmatrix} e^{-t}(1+t) & te^{-t} \\ -te^{-t} & e^{-t}(1-t) \end{bmatrix}$, $y = e^{-t}(1 + 2t)$

50 $[e^t \quad e^t(t+1)]^T$

51 $x_1 = 2 - 4e^{-2t} + 3e^{-3t}$, $x_2 = 8e^{-2t} - 9e^{-3t}$

52 $x_1 = 4te^{-2t} + e^{-2t} - e^{-t}$, $x_2 = 3e^{-t} - 2e^{-2t} - 4te^{-t}$

53 $\boldsymbol{x}(t) = \begin{bmatrix} -5 + \frac{8}{3}e^{-t} + \frac{10}{3}e^{5t} \\ 3 - \frac{8}{3}e^{-t} + \frac{5}{3}e^{5t} \end{bmatrix}$

54 Same as Exercise 53

55 $x_1 = x_2 = 2e^{-2t} - e^{-3t}$

56 $\frac{15}{4} - \frac{5}{2}t + \frac{9}{4}e^{-2t} - 6e^{-t}$

57 $\boldsymbol{x}(t) = \begin{bmatrix} 4te^{-t} + e^{-2t} \\ -4te^{-t} - 2e^{-2t} + 2e^{-t} \end{bmatrix}$

58 $\boldsymbol{x}(t) = e^t[3 \quad 2]^T - e^{-3t}[1 \quad -2]^T$

59 $\boldsymbol{x}(t) = \frac{1}{5}\begin{bmatrix} 14e^{-t} - 4e^{-6t} \\ 7e^{-t} + 8e^{-6t} \end{bmatrix}$

60 $\boldsymbol{x}(t) = e^{-2t}\{(\cos 2t - \sin 2t)[2 \quad 1]^T - (\cos 2t + \sin 2t)[0 \quad 1]^T\}$

61 $\dot{z} = \begin{bmatrix} 2 & 0 & 0 \\ 0 & 1 & 0 \\ 0 & 0 & -1 \end{bmatrix} z + \begin{bmatrix} \frac{1}{3} \\ 0 \\ -\frac{4}{3} \end{bmatrix} u,$

$y = [1 \quad -4 \quad -2]z$

62 $\alpha_0 = \frac{1}{2}$, $\alpha_1 = -\frac{1}{2}$, $\alpha_2 = \frac{1}{2}$

63 6, 1, $[4 \quad 1]^T$, $[1 \quad -1]^T$

$x_1 = 4e^{6t} - 3e^t$, $x_2 = e^{6t} + 3e^t$

64 Same as Exercise 54

65 $\dot{z} = \begin{bmatrix} -1 & 0 & 0 \\ 0 & -2 & 0 \\ 0 & 0 & -3 \end{bmatrix} z + \begin{bmatrix} \frac{1}{2} \\ -1 \\ \frac{1}{2} \end{bmatrix} u, \quad y = [2 \quad 9 \quad 22]z$

The system is stable, controllable and observable

66 $\dot{z} = \begin{bmatrix} 0 & 0 & 0 \\ 0 & -1 & 0 \\ 0 & 0 & -5 \end{bmatrix} z + \begin{bmatrix} \frac{1}{5} \\ -\frac{1}{4} \\ \frac{1}{20} \end{bmatrix} u, \quad y = [5 \quad 3 \quad 15]z$

The system is marginally stable, controllable and observable

67 (a) $\frac{2^k}{4}\begin{bmatrix} 2 & 1 \\ 4 & 2 \end{bmatrix} + \frac{(-2)^k}{4}\begin{bmatrix} 2 & -1 \\ -4 & 2 \end{bmatrix}$

(b) $\frac{(-4)^k}{2}\begin{bmatrix} 1 & -1 \\ -1 & 1 \end{bmatrix} + 2^{k-1}\begin{bmatrix} 1 & 1 \\ 1 & 1 \end{bmatrix}$

(c) $(-1)^k\begin{bmatrix} 1 & -k \\ 0 & 1 \end{bmatrix}$

68 $x(k) = 5^k(\cos k\theta + \sin k\theta)$, $y(k) = 5^k(2\cos k\theta)$, $\cos\theta = -\frac{3}{5}$

69 $\boldsymbol{x}(k) = \begin{bmatrix} \frac{25}{18} - \frac{17}{6}(-0.2)^k + \frac{22}{9}(-0.8)^k \\ \frac{7}{18} - (3.4/6)(-0.2)^k - (17.6/9)(-0.8)^k \end{bmatrix}$

70 $y(k) = \frac{1}{\sqrt{5}}\left[\left(\frac{1+\sqrt{5}}{2}\right)^k - \left(\frac{1-\sqrt{5}}{2}\right)^k\right]$

71 $u(t) = [-\frac{33}{2} \quad -\frac{17}{2}]\boldsymbol{x}(t) + u_{ext}$

72 $u(t) = [-\frac{99}{4} \quad -\frac{35}{4}]\boldsymbol{x}(t) + u_{ext}$

73 $u(t) = [-\frac{35}{6} \quad -\frac{31}{6}]\boldsymbol{x}(t) + u_{ext}$

$u(t) = [-31 \quad -11]\boldsymbol{x}(t) + u_{ext}$

75 $\boldsymbol{M} = \begin{bmatrix} 2 & -2 \\ 1 & -1 \end{bmatrix}$, rank 1, $\boldsymbol{M} = \begin{bmatrix} 0 & 1 \\ 1 & \frac{1}{2} \end{bmatrix}$, rank 2

6.13 Review exercises

1 (a) 5, 2, −1; $[11 \quad 5 \quad 3]^T$, $[8 \quad 2 \quad 1]^T$, $[1 \quad 0 \quad 0]^T$
(b) 3, 2, 1; $[1 \quad 2 \quad 1]^T$, $[2 \quad 1 \quad 0]^T$, $[1 \quad 0 \quad -1]^T$
(c) 3, 1, 0; $[1 \quad -2 \quad 1]^T$, $[1 \quad 0 \quad -1]^T$, $[1 \quad 1 \quad 1]^T$

2 6, 3, 1; $[1 \quad 1 \quad 1]^T$, $[1 \quad 1 \quad -2]^T$, $[1 \quad -1 \quad 0]^T$

3 $b = 1$, $c = 2$; $\lambda = 2, 4, 1$;
$[1 \quad -2 \quad -1]^T$, $[1 \quad 1 \quad -1]^T$

4 5.4

5 (a) 4.56; $[0.72 \quad 0.84 \quad 1]^T$ (b) 1.75
(c) (i) 1.19 (ii) 1.75

6 $[0 \quad -1 \quad 1]^T$

7 3, 2, 1; $[2 \quad 1 \quad 1]^T$, $[3 \quad 2 \quad 1]^T$, $[4 \quad 3 \quad 2]^T$

8 $[\frac{1}{3} \quad -\frac{2}{3} \quad -\frac{2}{3}]^T$, $[\frac{2}{3} \quad \frac{2}{3} \quad -\frac{1}{3}]^T$, $[\frac{2}{3} \quad -\frac{1}{3} \quad \frac{2}{3}]^T$

9 −6, −4, −2, 0; $[1 \quad -3 \quad 3 \quad -1]^T$, $[0 \quad 1 \quad -2 \quad 1]^T$, $[0 \quad 0 \quad 1 \quad -1]^T$, $[0 \quad 0 \quad 0 \quad 1]^T$;
$C - Ce^{-6t} + 3Ce^{-4t} - 3Ce^{-2t}$

10 (a) (i) $\begin{bmatrix} -29 & 0 \\ -32 & 3 \end{bmatrix}$ (ii) $\begin{bmatrix} 2^k & 0 \\ 2^k - 1 & 1 \end{bmatrix}$

(b) $\begin{bmatrix} 1 & \frac{1}{2}(1 - e^{-2t}) \\ 0 & e^{-2t} \end{bmatrix}$

11 $\boldsymbol{e}_1 = [1 \quad 0 \quad 0]^T$, $\boldsymbol{e}_2^* = [\frac{3}{8} \quad \frac{1}{2} \quad 0]^T$, $\boldsymbol{e}_3^* = [0 \quad 0 \quad \frac{1}{8}]^T$

12 2, $2 \pm \sqrt{2}$; 1:0:−1, 1:$-\sqrt{2}$:1, 1:$\sqrt{2}$:1

13 (a) Positive-semidefinite (b) Positive-definite
(c) Indefinite
(d) Negative-semidefinite
(e) Negative-definite

14 1; 3, $[1 \quad 1 \quad 0]^T$; −1, $[0 \quad -1 \quad 1]^T$

15 (c) $4e^{-2t} - 3e^{-3t}$, $y(t) = 1$ $(t \geqslant 0)$

16 $[1 \quad 3 \quad 1]^T$, $[3 \quad 2 \quad 1]^T$, $[1 \quad 0 \quad 1]^T$

$$\boldsymbol{x}(k) = \begin{bmatrix} \frac{1}{6}(-1)^k - \frac{1}{3}(2^k) + \frac{3}{2} \\ 1 - 2^k \\ \frac{1}{6}(-1)^k - \frac{1}{3}(2^k) + \frac{1}{2} \end{bmatrix}$$

17 $\boldsymbol{\Lambda} = \begin{bmatrix} 2 & 0 & 0 \\ 0 & 1 & 0 \\ 0 & 0 & -1 \end{bmatrix}$, $\boldsymbol{b}_1 = [\frac{1}{3} \quad 0 \quad -\frac{4}{3}]^T$,

$\boldsymbol{c} = [1 \quad -4 \quad -2]^T$
The system is uncontrollable but observable; it is stable

18 $\boldsymbol{x}(t) = \begin{bmatrix} e^{-t}\sin t \\ 1 - e^{-t}(\cos t + \sin t) \end{bmatrix}$

$H(s) = \dfrac{s+2}{(s+1)^2 + 1}$

$e^{-t}(\cos t + \sin t)$

19 $D(z) = \dfrac{z+3}{z^2 + 4z - 5}$ (i) $\boldsymbol{M}_c = \begin{bmatrix} 1 & -3 \\ 0 & -2 \end{bmatrix}$

(ii) $\boldsymbol{M}_c^{-1} = \begin{bmatrix} 1 & -\frac{3}{2} \\ 0 & -\frac{1}{2} \end{bmatrix}$ (iii) $\boldsymbol{v}^T = [0 \quad -\frac{1}{2}]$

(iv) $\boldsymbol{T} = \begin{bmatrix} 0 & -\frac{1}{2} \\ 1 & \frac{1}{2} \end{bmatrix}$ (v) $\boldsymbol{T}^{-1} = \begin{bmatrix} 1 & 1 \\ -2 & 0 \end{bmatrix}$

$\alpha = -5$, $\beta = 4$

20 (a) 6, $[3 \quad 2 \quad 1]^T$; 3, $[1 \quad -1 \quad 0]^T$, $[\frac{1}{3} \quad \frac{1}{3} \quad \frac{1}{3}]^T$

(c) $\dfrac{1}{3}\begin{bmatrix} (-3 + 3t)e^{3t} + 3e^{6t} \\ (1 + 3t)e^{3t} + 2e^{6t} \\ -e^{3t} + e^{6t} \end{bmatrix}$

21 1, −1, −2; $[1 \quad 0 \quad -1]^T$, $[1 \quad -1 \quad 0]^T$, $[0 \quad 0 \quad 1]^T$
$u(t) = -6\{x_1(t) + x_2(t)\}$

22 $x_1 = \frac{5}{3}\cos t + 2\sin t - \frac{2}{3}\cos 2t$
$x_2 = \frac{5}{3}\cos t + 2\sin t + \frac{1}{3}\cos 2t$

CHAPTER 7

Exercises

1 (a) Circles centre (0, 0), $x^2 + y^2 = 1 + e^C$
(b) Straight lines through (−1, 0), $y = (x+1)\tan C$

2 (a) Family of curves $y^2 = 4x^2(x - 1) + C$
(b) Family of curves $y^2 = \frac{1}{12}x^2(x^2 - 12) + C$

3 (a) $z - xy = C$
(b) $xy = \ln(C + z)$

4 (a) $(A\cos(t + B), \tan(t + B), Ce^t)$, curves on hyperbolic cylinders
(b) Curves defined by the intersections of mutually orthogonal hyperbolic cylinders, $x^2 - y^2 = c$, $x^2 - z^2 = k$

5 (a) $f_x = yz - 2x$, $f_y = xz + 1$, $f_z = xy - 1$,
$f_{xx} = -2$, $f_{xy} = z$, $f_{xz} = y$, $f_{yy} = 0$, $f_{yz} = x$, $f_{zz} = 0$
(b) $f_x = 2xyz^3$, $f_y = x^2z^3$, $f_z = 3x^2yz^2$, $f_{xx} = 2yz^3$,
$f_{xy} = 2xz^3$, $f_{xz} = 6xyz^2$, $f_{yy} = 0$, $f_{yz} = 3x^2z^2$, $f_{zz} = 6x^2y$
(c) $f_x = -z/(x^2 + y^2)$, $f_y = zx/(x^2 + y^2)$,
$f_z = \tan^{-1}(y/x)$, $f_{xx} = 2xz/(x^2 + y^2)^2$,
$f_{xy} = 2yz/(x^2 + y^2)^2$, $f_{xz} = -1/(x^2 + y^2)$, $f_{zz} = 0$,
$f_{yy} = -2xyz/(x^2 + y^2)^2$, $f_{yz} = x/(x^2 + y^2)$

6 (a) $2t(3t^6 - 6t^5 + 3t^4 - 3t^3 + 4t^2 + 3t + 4)/(t - 1)^2$
(b) $te^{-2t}(\cos 2t - \sin 2t) + \frac{1}{2}e^{-2t}\sin 2t$

7 $\dfrac{\partial f}{\partial y} = \dfrac{\partial f}{\partial r}\sin\theta\sin\phi + \dfrac{\partial f}{\partial \theta}\dfrac{\sin\phi\cos\theta}{r} + \dfrac{\partial f}{\partial \phi}\dfrac{\cos\phi}{r\sin\theta}$

$\dfrac{\partial f}{\partial z} = \dfrac{\partial f}{\partial r}\cos\theta - \dfrac{\partial f}{\partial \theta}\dfrac{\sin\theta}{r}$

8 $A/r + B$

13 $e^{2u}, \dfrac{1}{x^2 + y^2}$

14 ± 1

15 $9, v^2 = u^2 + 2w$

18 (a) $xy^2 + x^2y + x + c$ (b) $x^2y^2 + y \sin 3x + c$
(c) Not exact (d) $z^3x - 3xy + 4y^3$

19 $-1, y \sin x - x \cos y + \frac{1}{2}(y^2 - 1)$

20 $m = 2$
$8x^5 + 36x^4y + 62x^3y^2 + 63x^2y^3 + 54xy^4 + 27y^5 + c$

21 (a) $\frac{117}{7}$ (b) $39, \frac{1}{13}(12, 3, 4)$

22 (a) $(2x, 2y, -1)$
(b) $(-yz/(x^2 + y^2), xz/(x^2 + y^2), \tan^{-1}(y/x))$
(c) $\dfrac{e^{-x-y+z}}{(x^3 + y^2)^{3/2}}$
$(-x^3 - y^2 - \frac{3}{2}x^2, -x^3 - y^2 - y, x^3 + y^2)$
(d) $(yz \sin \pi(x + y + z) + \pi xyz \cos \pi(x + y + z),$
$xy \sin \pi(x + y + z) + \pi xyz \cos \pi(x + y + z),$
$xz \sin \pi(x + y + z) + \pi xyz \cos \pi(x + y + z))$

23 3

24 $(5\boldsymbol{i} + 4\boldsymbol{j} + 3\boldsymbol{k})/\sqrt{50}$

25 (a) $\boldsymbol{r}/r$ (b) $-\boldsymbol{r}/r^3$

26 $\phi = x^2y + z^2x + zy$

27 $-9\boldsymbol{j} + 3\boldsymbol{k}, \frac{36}{7}$

28 54°25′

29 (a) $x + 2y + 3z = 6, x - 1 = \frac{1}{2}(y - 1) = \frac{1}{3}(z - 1)$
(b) $2x + 2y - 3z = -3,$
$\frac{1}{2}(x - 1) = \frac{1}{2}(y - 2) = \frac{1}{3}(3 - z)$
(c) $2x + 4y - z = 6, \frac{1}{2}(x - 1) = \frac{1}{4}(y - 2) = 4 - z$

31 (a) $6xy$ (b) 4

32 -61

33 $\boldsymbol{a}, \boldsymbol{a}, 3\boldsymbol{a}$

35 -13

38 $(y, 6xz - 1, 0)$

40 $x^2 + y^2 + z^2 + xyz$

42 $a = 2, b = 2, c = 3; \phi = 2x^2y + 2z^3x + 3zy + \text{const}$

43 $2\sqrt{11}$ rad s^{-1}

44 $d = -a, c = b$

47 (a) $2y^2z^3 + 2x^2z^3 + 6x^2y^2z$
(b) $2y(1 + z)\boldsymbol{i} + 2(x + xz - z)\boldsymbol{j} + 2y(x - 1)\boldsymbol{k}$
(c) $2yz\boldsymbol{i} + 2(x - z)\boldsymbol{j} + 2yx\boldsymbol{k}$

56 156

57 0

58 $\frac{61}{12}$

59 (a) $\frac{288}{35}$ (b) 10 (c) 8

60 $15 + 4\pi$

61 (a) 16 (b) 16 (c) No

62 35

63 $-\frac{9}{10}\boldsymbol{i} - \frac{2}{3}\boldsymbol{j} + \frac{7}{5}\boldsymbol{k}$

64 $4\pi(7\boldsymbol{i} + 3\boldsymbol{j})$

65 (a) 24, (b) 76, (c) 16

66 $\frac{8}{3}\ln 2$

67 $\frac{1}{6}$

68 (a) $(\ln 2)\tan^{-1}(\frac{1}{3})$ (b) $\frac{1}{6}$ (c) $\frac{1}{2}\pi - 1$

69 $-4/3\pi^2$

70 (a) $\frac{1}{4}(\sqrt{2} - 1)$ (b) $[(1 - k^2)^{3/2} - 1]/3k^2$

71 $2(\sqrt{2} - 1)$

73 1

74 $2a(1 - \frac{1}{4}\pi)$

75 $\frac{1}{3}(6\pi - 20)$

77 $\frac{1}{4}\pi + 2/\pi - 1$

78 0

79 $\frac{11}{60}$

80 0

81 $\frac{1}{2}a(1 - \frac{1}{4}\pi)$

83 $\frac{13}{3}\pi$

84 (a) $\frac{183}{4}$ (b) $\frac{1}{4}\pi$

85 (a) $\frac{27}{4}$ (b) 0

87 (a) $\frac{13}{3}\pi$ (b) $\frac{149}{30}\pi$ (c) $\frac{37}{10}\pi$

88 24π

90 90

91 0

92 (a) $\frac{8}{3}$ (b) $\frac{448}{3}$

94 $\frac{1}{2}\pi^2 - 2$

95 $\frac{1}{90}$

96 $\frac{11}{30}$

97 $\frac{1}{24}\pi$

98 $\frac{13}{60}$

99 $(1 - e^{-1})/6$

100 $\frac{2}{15}$; $(\frac{5}{16}, \frac{5}{16}, \frac{11}{16})$

101 $\frac{1}{8}\pi$

102 $\frac{1}{16}\pi a^4$

103 $\frac{3}{2}$

104 12π

105 84π

109 $2\pi ab$

110 16π

7.7 Review exercises

2 $\sin(x + 3y)$

7 $\frac{1}{3}x^3 - y^2x + \frac{1}{2}x^2 + \frac{1}{2}y^2 + c$

8 (a) $\frac{8}{3}$ (b) 8

9 $\frac{5}{6}$

10 $-\frac{4}{3}\boldsymbol{j}$

11 0

12 $\frac{13}{80}kc^6$

13 $\frac{2}{3}ha^2\left[\frac{1}{2}\pi - \sin^{-1}\left(\frac{c}{a}\right)\right] - \frac{1}{3}hcl - \frac{hc^3}{3a}\tanh^{-1}\left(\frac{l}{a}\right)$

$l = \sqrt{(a^2 - c^2)}$

14 $\frac{16}{3}a^3$

15 $\pi q_0^2 r^2 l/4EI$

16 $\frac{1}{3}$

17 0

19 0

20 $\frac{13}{240}$

CHAPTER 8

Exercises

1 $X(1) = 1.1571$

2 $X(0.5) = 2.1250$

3 $X_a(2) = 2.811\,489$, $X_b(2) = 2.819\,944$,
$x(t) = 2\sqrt{(2 + t^2)}/\sqrt{3}$

4 $X_a(2) = 1.573\,065$, $X_b(2) = 1.558\,541$,
$x(t) = \sqrt{(1 + 2\ln t)}$

5 $X_a(1.5) = 2.241\,257$, $X_b(1.5) = 2.206\,232$,
$x(t)\ln x(t) - x(t) = t - 1.981\,214$

6 (a) $X(0.5) = 0.1238$ (b) $X(1.2) = 1.3740$

7 $X(0.5) = 1.7460$

8 (a) $X(0.5) = 0.7948$ (b) $X(1) = -1.3511$

12 $X(0.5) = 0.1353$

13 (a) $X(0.75) = 3.2345$ (b) $X(2) = 2.2771$

14 (a) $X_{0.2}(2) = 2.242\,408$, $X_{0.1}(2) = 2.613\,104$
Richardson extrapolation estimates the error as 0.123 565 so a step less than 0.0064 should be used.
(b) $X_{0.2}(2) = 2.788\,158$, $X_{0.1}(2) = 2.863\,456$
Richardson extrapolation estimates the error as 0.025 099 so a step less than 0.014 should be used.
(c) $X_{0.4}(2) = 2.884\,046$, $X_{0.2}(2) = 2.897\,402$
Richardson extrapolation estimates the error as 0.000 890 so a step less than 0.057 should be used.
$x(2) = 2.898\,51$ to 5 dp

15 $X(3) = 1.466\,47$

16 (a) $dx/dt = v$, $x(0) = 1$
$dv/dt = 4xt - 6(x^2 - t)v$, $v(0) = 2$
(b) $dx/dt = v$, $x(1) = 2$
$dv/dt = -4(x^2 - t^2)$, $v(1) = 0.5$
(c) $dx/dt = v$, $x(0) = 0$
$dv/dt = -\sin v - 4x$, $v(0) = 0$
(d) $dx/dt = v$, $x(0) = 1$
$dv/dt = w$, $v(0) = 2$
$dw/dt = e^{2t} + x^2t - 6e^tv - tw$, $w(0) = 0$
(e) $dx/dt = v$, $x(1) = 1$
$dv/dt = w$, $v(1) = 0$
$dw/dt = \sin t - x^2 - tw$, $w(1) = -2$
(f) $dx/dt = v$, $x(2) = 0$
$dv/dt = w$, $v(2) = 0$
$dw/dt = (x^2t^2 + tw)^2$, $w(2) = 2$
(g) $dx/dt = v$, $x(0) = 0$
$dv/dt = w$, $v(0) = 0$

$dw/dt = u, \quad w(0) = 4$
$du/dt = \ln t - x^2 - xw, \quad u(0) = -3$

(h) $dx/dt = v, \quad x(0) = a$
$dv/dt = w, \quad v(0) = 0$
$dw/dt = u, \quad w(0) = b$
$du/dt = t^2 + 4t - 5 + \sqrt{(xt)} - v - (v-1)tu,$
$u(0) = 0$

17 $X(0.3) = 0.299\,90$

18 $X(0.3) = 0.299\,64$

19 $X(0.65) = -0.826\,03$

20 $X_{0.4}(1.6) = 1.220\,254, \quad X_{0.2}(1.6) = 1.220\,055$
Richardson extrapolation estimates the error as 0.000 013 so, to obtain an error less than 5×10^{-7}, a step less than 0.088 should be used.

21 $X_{0.1}(2.2) = 2.923\,350\,36, \quad X_{0.05}(2.2) = 2.925\,417\,56$
Richardson extrapolation estimates the error as 0.000 295 so, to obtain an error less than 5×10^{-7}, a step less than 0.0060 should be used.

Review exercises

1 $X_{0.1}(0.4) = 1.125\,583, \quad X_{0.05}(0.4) = 1.142\,763$
Richardson extrapolation estimates the error as 0.017 180 so, to obtain an error less than 5×10^{-3}, a step less than 0.0146 should be used.

2 $X_{0.05}(0.25) = 2.003\,749, \quad X_{0.025}(0.25) = 2.004\,452$
Richardson extrapolation estimates the error as 0.000 703 so, to obtain an error less than 5×10^{-4}, a step less than 0.0178 should be used.

3 $X_1(1.2) = 2.374\,037, \quad X_2(1.2) = 2.374\,148,$
$X_3(1.2) = 2.374\,176$

4 $X(1) = 5.194\,323$ accurate to 6 dp.

6 $X_{0.025}(2) = 0.847\,035, \quad X_{0.0125}(2) = 0.844\,066$
Richardson extrapolation estimates the error as 0.002 969 so we have $X(2) = 0.84$.

7 $X(4) = 0.1458$ (using step size 0.002)

8 $X(2.5) = -0.6532$ (using step size 0.025)

CHAPTER 9

Exercises

1 $a^2 = b^2c^2$

2 $\alpha = \pm c$

3 For $\alpha = 0$: $V = A + Bx$
For $\alpha > 0$:
$V = A \sinh at + B \cosh at$, where $a^2 = \alpha/\kappa$
or $C\,e^{at} + D\,e^{-at}$
For $\alpha < 0$:
$V = A \cos bt + B \sin bt$, where $b^2 = -\alpha/\kappa$

4 $n = -3, 2$

5 $g(z) = (1 - 2z)/(1 + z)^4$

6 (a) $I_{xx} = (Lc)I_{tt}$
(b) $v_{xx} = (rg)v + (rc)v_t$ and $W_x = (rc)W_{tt}$
(c) $w_{xx} = (Lc)w_{tt}$

7 $a = -3$

11 $u = \sin x \cos ct$

13
$$u = \frac{2l}{\pi^2}\left\{\left[\sin\frac{\pi(x-ct)}{l} - \frac{1}{9}\sin\frac{3\pi(x-ct)}{l} + \frac{1}{25}\sin\frac{5\pi(x-ct)}{l} + \dots\right] + \left[\sin\frac{\pi(x+ct)}{l} - \frac{1}{9}\sin\frac{3\pi(x+ct)}{l} + \frac{1}{25}\sin\frac{5\pi(x+ct)}{l} + \dots\right]\right\}$$

14 $u = \dfrac{a}{\omega c}\sin \omega ct \cos \omega x$

$u = \frac{1}{4c}[\exp\{-(x-ct)^2\} + \exp\{-(x+ct)^2\}]$

15 $u = \frac{1}{2}F(x - ct) + \frac{1}{2}F(x + ct)$, where

$$F(z) = \begin{cases} 1-z & (0 \leqslant z \leqslant 1) \\ 1+z & (-1 \leqslant z \leqslant 0) \\ 0 & (|z| \geqslant 1) \end{cases}$$

16 $x + (-3 - \sqrt{6})y = \text{constant}$
and
$x + (-3 + \sqrt{6})y = \text{constant}$

17

x	$f(x)$	$u(x, 0)$	$u(x, 0.5)$	$u(x, 1)$	$u(x, 1.5)$	$u(x, 2)$
−3.0	0.024 893	0	0.025 943	0.058 509	0.106 010	0.180 570
−2.5	0.041 042	0	0.042 774	0.096 466	0.174 781	0.297 710
−2.0	0.067 667	0	0.070 522	0.159 046	0.288 166	0.490 842
−1.5	0.111 565	0	0.116 272	0.262 222	0.475 106	0.681 635
−1.0	0.183 939	0	0.191 700	0.432 332	0.655 692	0.791 166
−0.5	0.303 265	0	0.316 060	0.585 169	0.748 392	0.847 392
0	0.5	0	0.393 469	0.632 120	0.776 869	0.864 664
0.5	0.696 734	0	0.316 060	0.585 169	0.748 392	0.847 392
1.0	0.816 060	0	0.191 700	0.432 332	0.655 692	0.791 166
1.5	0.888 434	0	0.116 272	0.262 222	0.475 106	0.681 635
2.0	0.932 332	0	0.070 522	0.159 046	0.288 166	0.490 842
2.5	0.958 957	0	0.042 774	0.096 466	0.174 781	0.297 710
3.0	0.975 106	0	0.025 943	0.058 509	0.106 010	0.180 570

18 $u = \dfrac{8}{\pi}\displaystyle\sum_{n=1}^{\infty}\frac{1}{(2n-1)^3}\sin(2n-1)x\cos(2n-1)ct$

21 Explicit with $\lambda = 0.5$

x	$t = 0$	$t = 0.25$	$t = 0.5$	$t = 1$	$t = 1.5$
0	0	0	0	0	0
0.25	0	0.0625	0.125	0.179 687	0.210 937
0.50	0	0.125	0.218 75	0.265 625	0.269 531
0.75	0	0.0625	0.125	0.179 687	0.210 937
1.00	0	0	0	0	0

Implicit with $\lambda = 0.5$

x	$t = 0$	$t = 0.25$	$t = 0.5$	$t = 1$
0	0	0	0	0
0.25	0	0.0625	0.122 45	0.174 07
0.50	0	0.125	0.224 49	0.281 5
0.75	0	0.0625	0.122 45	0.174 07
1.00	0	0	0	0

22 Explicit

x	$t = 0$	$t = 0.02$	$t = 0.04$	$t = 0.06$	$t = 0.08$
0	0	0.031 410	0.062 790	0.094 108	0.125 333
0.2	0	0	0.000 314	0.001 249	0.003 101
0.4	0	0	0	0.000 003	0.000 018
0.6	0	0	0	0	0.000 000
0.8	0	0	0	0	0
1.0	0	0	0	0	0

23 Explicit

x	$t = 0$	$t = 0.2$	$t = 0.4$	$t = 0.6$
0	0	0.03	0.12	0.27
0.2	0.16	0.19	0.28	0.43
0.4	0.24	0.27	0.36	0.51
0.6	0.24	0.27	0.36	0.51
0.8	0.16	0.19	0.28	0.43
1.0	0	0.03	0.12	0.27

24 Explicit

x	$t = 0$	$t = 0.2$	$t = 0.4$	$t = 0.6$
0	0	0	0	0
0.2	0.16	0.19	0.2725	0.388 75
0.4	0.24	0.27	0.36	0.508 125
0.6	0.24	0.27	0.36	0.508 125
0.8	0.16	0.19	0.2725	0.388 75
1.0	0	0	0	0

25 $u = \frac{1}{2}a[\exp(-\frac{9}{4}\kappa\pi^2 t)\cos\frac{3}{2}\pi x + \exp(-\frac{1}{4}\kappa\pi^2 t)\cos\frac{1}{2}\pi x]$

26 $$u = \sum_{n=0}^{\infty} a_n \exp\left[\frac{-\kappa(n+\frac{1}{2})^2\pi^2 t}{l^2}\right]\cos\left[\left(n+\frac{1}{2}\right)\frac{\pi x}{l}\right]$$

where

$$a_n = u_0\left[\frac{8}{(2n+1)^2\pi^2} - \frac{2(-1)^n}{(2n+1)\pi}\right]$$

28 $\alpha = -\frac{1}{2}$, $\kappa = -\frac{1}{4}$

29 $\beta = 2$

30 The term represents heat loss at a rate proportional to the excess temperature over θ_0.

32 $u(0, t) = u(l, t) = 0$ for all t
$u(x, 0) = 10$ for $0 < x < l$

34

x	$t = 0$	$t = 0.02$	$t = 0.04$	$t = 0.06$	$t = 0.08$	$t = 0.1$
0	0	0	0	0	0	0
0.2	0.04	0.08	0.1	0.12	0.135	0.1475
0.4	0.16	0.2	0.24	0.27	0.295	0.315
0.6	0.36	0.4	0.44	0.47	0.495	0.515
0.8	0.64	0.68	0.7	0.72	0.735	0.7475
1.0	1	1	1	1	1	1

35 At $t = 1$ with $\lambda = 0.4$ and $\Delta t = 0.05$
Explicit

x	0	0.2	0.4	0.6	0.8	1.0
u	0	0.1094	0.2104	0.2939	0.3497	0.3679

Implicit

x	0	0.2	0.4	0.6	0.8	1.0
u	0	0.1056	0.2045	0.2883	0.3465	0.3679

36

$$\boldsymbol{A} = \begin{bmatrix} 3 & 1 & 0 & 0 & 0 \\ -0.5 & 3 & -0.5 & 0 & 0 \\ 0 & -0.5 & 3 & -0.5 & 0 \\ 0 & 0 & -0.5 & 3 & -0.5 \\ 0 & 0 & 0 & -0.5 & 3 \end{bmatrix}$$

$$\boldsymbol{b} = \begin{bmatrix} 0.56 \\ 0.28 \\ 0.44 \\ 0.44 \\ 0.28 \end{bmatrix} \quad \boldsymbol{Au} = \boldsymbol{b} \text{ gives } \quad \boldsymbol{u} = \begin{bmatrix} -0.039\,99 \\ 0.120\,036 \\ 0.200\,202 \\ 0.201\,178 \\ 0.126\,863 \end{bmatrix}$$

37 $u = \frac{5}{8}e^{-\pi y}\sin\pi x - \frac{5}{16}e^{-3\pi y}\sin 3\pi x + \frac{1}{16}e^{-5\pi y}\sin 5\pi x$

38 $u = x + \dfrac{2}{\pi}\sum_{n=1}^{\infty} \dfrac{\sin n\pi x}{n \sinh n\pi}$
$\times \{\sinh n\pi y + (-1)^n \sinh n\pi(1-y)\}$

41 $u(r, \theta) = \frac{3}{4} r \sin\theta - \frac{1}{4} r^3 \sin(3\theta)$

42 v = const gives circles with centre $\left(\dfrac{-v}{v-1}, 0\right)$ and radius $1/|v-1|$
u = const gives circles with centre $\left(1, \dfrac{-1}{u}\right)$ and radius $1/|u|$

43 Boundary conditions are $u(0, y) = u(a, y) = 0$, $0 \leqslant y < a$; and $u(x, 0) = 0$, $0 \leqslant x \leqslant a$, $u(x, a) = u_0$, $0 < x < a$

46 $V = \dfrac{2\alpha}{3}\dfrac{\left(1 - \dfrac{a}{r}\right)}{\left(1 - \dfrac{a}{b}\right)} - \dfrac{\alpha}{3}(2 - 3\sin^2\theta)\dfrac{\left(r^2 - \dfrac{a^5}{r^3}\right)}{\left(b^2 - \dfrac{a^5}{b^3}\right)}$

47 For $\Delta x = \Delta y = 0.5$ $u(0.5, 0.5) = 0.3125$
For $\Delta x = \Delta y = 0.25$ $u(0.5, 0.5) = 0.3047$

48 $u(0.5, 0) = 0.4839$, $u(0.5, 0.5) = 0.4677$, $u(0.5, 1) = 0.3871$

49 $u(1, 1) = 10.674$, $u(2, 1) = 12.360$, $u(3, 1) = 8.090$, $u(1, 2) = 10.337$, $u(2, 2) = 10.674$

50 $h = 1/2$ gives $\phi(0.5, 0.5) = 1.8611$ and $\phi(0.5, 1) = 1.3194$
For $h = 1/4$ ϕ is given in the table

y					
1	2	1.601 566 6	1.286 764 7	1.056 521 6	1
0.75	2.4375	1.967 955 1	1.581 801 5	1.257 228 7	1
0.5	2.75	2.266 577 2	1.846 507 3	1.437 466 9	1
0.25	2.9375	2.517 471 5	2.131 433 8	1.693 006 4	1
0.0	3	2.75	2.5	2.25	1
x	0	0.25	0.5	0.75	1

51 $\phi(0, 0) = 1.5909$, $\phi(0, \frac{1}{3}) = 2.0909$, $\phi(0, \frac{2}{3}) = 4.7727$, $\phi(\frac{1}{3}, 0) = 1.0909$, $\phi(\frac{2}{3}, 0) = 0.7727$ and other values can be obtained by symmetry.

54 Parabolic; $r = x - y$ and $s = x + y$ gives $u_{ss} = 0$
Elliptic; $r = -3x + y$ and $s = x + y$ gives
$8(u_{ss} + u_{rr}) - 9u_r + 3u_s + u = 0$
Hyperbolic; $r = 9x + y$ and $s = x + y$ gives
$49u_{rr} - u_{ss} = 0$

55 $u = f(2x + y) + g(x - 3y)$

57 (a) elliptic; (b) parabolic; (c) hyperbolic
For $y < 0$ characteristics are $(-y)^{3/2} \pm \frac{3}{2}x$ = constant

58 elliptic if $|y| < 1$
parabolic if $x = 0$ or $y = \pm 1$
hyperbolic if $|y| > 1$

59 $p > q$ or $p < -q$ then hyperbolic; $p = q$ then parabolic; $-q < p < q$ then elliptic

9.10 Review exercises

3 $y = 4\,\mathrm{e}^{-t/2\tau} \sin\left(3\pi\dfrac{x}{l}\right)\left(\cos\omega_3 t + \dfrac{1}{2\omega_3\tau}\sin\omega_3 t\right)$

where $\omega_3 = 3\pi\dfrac{c}{l}\left(1 - \dfrac{l^2}{36\pi^2 c^2 \tau^2}\right)^{1/2}$

5 $A_{2n+1} = 8\theta_0 l/\pi^2(2n+1)^2$

6 $T = T_0 + \phi_0[1 - \mathrm{erf}(x/2\sqrt{(\kappa t)})]$

7 Explicit

x	$t = 0$	$t = 0.004$	$t = 0.008$
0	1	1	0.96
0.2	1	1	1
0.4	1	1	1
0.6	1	1	1
0.8	1	1	1
1.0	1	1	0.96

Implicit

x	$t = 0$	$t = 0.004$	$t = 0.008$
0	1	0.9103	0.8357
0.2	1	0.9959	0.9851
0.4	1	0.9998	0.9989
0.6	1	0.9998	0.9989
0.8	1	0.9959	0.9851
1.0	1	0.9103	0.8357

9

$y = 1$			1	0.928 592 5
$y = 0.5$		0.987 574 3	0.956 962 1	0.937 999 5
$y = 1$	1	0.984 980 8	0.964 774 6	0.960 193 4
	$x = 0$	$x = 0.5$	$x = 1$	$x = 1.5$

11 $k = -\frac{3}{2}$

12 $z = x - y$, valid in the region $x \geqslant y$

14 $A_{2n+1} = \dfrac{32a^2(-1)^{n+1}}{\pi^3(2n+1)^3 \cosh\left[\dfrac{(2n+1)\pi b}{2a}\right]}$

15 $u(x, t) = \frac{2}{\pi}\sum_{n=1}^{\infty}\frac{1}{n}\sin n\pi x\cos n\pi t$

17 $\phi = A\cos(px)e^{-Kt/2}\cos\omega t$, where $\omega^2 = c^2p^2 - \frac{1}{4}K^2$

18 On $r = a$, $v_r = 0$, so there is no flow through the cylinder $r = a$. As $r \to \infty$, $v_r \to U\cos\theta$ and $v_\theta \to -U\sin\theta$, so the flow is steady at infinity and parallel to the x axis.

CHAPTER 10

Exercises

1 $x = 1, y = 1, f = 9$

2 $x_1 = 1.5, x_2 = 0, x_3 = 2.5, x_4 = 0, f = 14$

3 Original problem:
20 of type 1, 50 of type 2, profit = £1080, 70 m chipboard remain
Revised problem:
5 of type 1, 75 of type 2, profit = £1020, 5 m chipboard remain

4 4 kg nails, 2 kg screws, profit 14 p

5 9 of CYL1, 6 of CYL2 and profit £54

6 For $k \geq 60$: $x_1 = \frac{25}{3}$, $x_2 = 0$, $z = \frac{25}{3}k$
For $60 \geq k \geq 10$: $x_1 = 6$, $x_2 = 7$, $z = 140 + 6k$
For $k \leq 10$: $x_1 = 0$, $x_2 = 10$, $z = 200$

7 $x_1 = 1, x_2 = 0.5, x_3 = 1, x_4 = 0, f = 6.5$

8 LP solution gives $x_1 = 66.67$, $x_2 = 50$, $f = £3166.67$. Profit is improved if more cloth is bought; for example, if the amount of cloth is increased to 410 m then $x_1 = 63.33$, $x_2 = 55$ and $f = £3233.33$.

9 B1, 0; B2, 15 000; B3, 30 000; profit £21 000

10 Long range 15, medium range 0, short range 0, estimated profit £6 million

11 $x_1 = 1, x_2 = 10, f = 20$

12 $x_1 = 2, x_2 = 0, x_3 = 2, x_4 = 0, f = 12$

13 36.63% of A, 44.55% of B, 18.81% of C, profit per 100 litres £134.26

14 6 of style 1, 11 of style 2, 6 of style 3, total profit £37 500

15 Boots 50, shoes 150, profit £1150

16 B1, 0; B2, 10 000; B3, 40 000; profit is down to £20 000

17 $x_1 = 2500\,\text{m}^2$, $x_2 = 1500\,\text{m}^2$, $x_3 = 1000\,\text{m}^2$, profit £9500

18 Several possible optima: (0, 3, 0); $(\frac{3}{2}, \frac{3}{2}, \frac{1}{2})$; $(6 - 3t, 0, t)$ for any t

19 (0, 1, 1); (0, −1, −1); $(2, -1, 1)/\sqrt{7}$; $-(2, -1, 1)/\sqrt{7}$

20 For given surface area S, $b = c = 2a$, where $a^2 = \frac{1}{12}S$ and $V = 4a^3$

21 $x = \pm a$ and $y = 0$

22 $x = a\sqrt{2}$, $y = b\sqrt{2}$, area $= 8ab$

23 $A = 1.33$, $B = 0.2$, $I = 130$

24 For $\alpha \geq 0$ minimum at (0, 0); for $\alpha < 0$ minimum at $(-2\alpha, 3\alpha)/5$

25 (a) Bracket (without using derivatives) $0.7 < x < 3.1$
(b) Iteration 1:

a	0.7	$f(a)$	2.7408
b	1.9	$f(b)$	2.17
c	3.1	$f(c)$	3.2041

Iteration 2:

a	0.7	$f(a)$	2.7408
b	1.7253	$f(b)$	2.0612
c	1.9	$f(c)$	2.17

gives $b = 1.5127$ and $f(b) = 1.9497$
(c)

	Iteration 1		Iteration 2	
x	0.7	3.1	0.7	1.5129
$f(x)$	2.7408	3.2041	2.7408	1.9498
$f'(x)$	−4.8309	0.9329	−4.8309	0.4224

gives $x = 1.1684$ and $f = 1.9009$

26 (a) Iteration 1:

a	0	$f(a)$	0
b	1	$f(b)$	0.420 74
c	3	$f(c)$	0.014 11

Iteration 2:

a	0	$f(a)$	0
b	1	$f(b)$	0.420 74
c	1.5113	$f(c)$	0.303 96

gives $x = 0.989\,79$ and $f = 0.422\,24$

(b)

	Iteration 1		Iteration 2	
x	0	1	0	0.8667
$f(x)$	0	0.4207	0	0.4352
$f'(x)$	1	−0.1506	1	−0.0612

gives $x = 0.8242$, $f = 0.4371$, $f' = -0.0247$

27 (a) Iteration 1:

a	1	$f(a)$	0.232 54
b	1.6667	$f(b)$	0.255 33
c	3	$f(c)$	0.141 93

Iteration 2:

a	1	$f(a)$	0.232 54
b	1.6200	$f(b)$	0.257 15
c	1.6667	$f(c)$	0.255 33

gives $x = 1.4784$ and $f = 0.260\,22$

(b) Iteration 1:

x	1	3
$f(x)$	0.232 54	0.141 93
$f'(x)$	0.135 34	−0.087 18

Iteration 2:

x	1	1.507 7
$f(x)$	0.232 54	0.259 90
$f'(x)$	0.135 34	−0.013 68

gives $x = 1.4462$, $f = 0.260\,35$, $f' = -0.000\,14$

28 $x = 1$, $\lambda_{max} = 2$; $x = 0$, $\lambda_{max} = 1.414\,21$; $x = -1$, $\lambda_{max} = 1.732\,05$. One application of the quadratic algorithm gives $x = -0.148\,26$ and $\lambda_{max} = 1.3854$.

30 Iteration 1:

$$\boldsymbol{a} = \begin{bmatrix} 1 \\ 1 \end{bmatrix}, \quad f = \tfrac{3}{2}, \quad \nabla f = \begin{bmatrix} 1 \\ 1 \end{bmatrix}$$

Iteration 2:

$$\boldsymbol{a} = \begin{bmatrix} -1 \\ -1 \end{bmatrix}, \quad f = -\tfrac{1}{2}, \quad \nabla f = \begin{bmatrix} 1 \\ -1 \end{bmatrix}$$

Iteration 3:

$$\boldsymbol{a} = \begin{bmatrix} -\frac{7}{5} \\ -\frac{3}{5} \end{bmatrix}, \quad f = -0.9, \quad \nabla f = \begin{bmatrix} \frac{1}{5} \\ -\frac{1}{5} \end{bmatrix}$$

31 Steepest descent gives the point (−0.382, −0.255) and $f = 0.828$

32 (1.1905, 2.1619, 0.2191) and $f = -0.9284$

33 (0.333, 1.667, 1.333) and $f = 3.161$

34 (a) After step 1

$$\boldsymbol{a} = \begin{bmatrix} 1.2 \\ 2 \end{bmatrix}, \quad f = 0.8, \quad \boldsymbol{g} = \begin{bmatrix} 0 \\ 1.6 \end{bmatrix}$$

$$\boldsymbol{H} = \begin{bmatrix} 0.1385 & 0.1923 \\ 0.1923 & 0.9615 \end{bmatrix}$$

After step 2 the exact solution $x = 1$, $y = 1$ is obtained

(b) After cycle 1

$$\boldsymbol{a}_1 = \begin{bmatrix} 0.5852 \\ 0 \\ 0.2926 \end{bmatrix}, \quad f = 1.0662, \quad \boldsymbol{g}_1 = \begin{bmatrix} -0.3918 \\ -1.7557 \\ 0.7822 \end{bmatrix}$$

$$\boldsymbol{H}_1 = \begin{bmatrix} 0.3681 & 0.1593 & -0.4047 \\ 0.1593 & 0.9632 & 0.1002 \\ -0.4047 & 0.1002 & 0.7418 \end{bmatrix}$$

After cycle 2

$$\boldsymbol{a}_2 = \begin{bmatrix} 1.0190 \\ 0.9813 \\ -0.0372 \end{bmatrix}, \quad f = 2.999 \times 10^{-6},$$

$$\boldsymbol{g}_2 = \begin{bmatrix} 0.0046 \\ -0.0012 \\ 0.0027 \end{bmatrix}$$

35 (a) After cycle 1

$$\boldsymbol{a} = \begin{bmatrix} 0.485 \\ -0.061 \end{bmatrix}, \quad f = 0.2424, \quad \boldsymbol{g} = \begin{bmatrix} 0.970 \\ -0.242 \end{bmatrix}$$

$$\boldsymbol{H} = \begin{bmatrix} 0.995 & -0.062 \\ -0.062 & 0.258 \end{bmatrix}$$

After cycle 2 the minimum at $x = 0$, $y = 0$ is obtained

(b) After cycle 1

$$a_1 = \begin{bmatrix} 0.970 \\ 1.940 \\ -0.218 \end{bmatrix}, \quad f = 4.711, \quad g_1 = \begin{bmatrix} 1.940 \\ 3.881 \\ -0.083 \end{bmatrix}$$

$$H_1 = \begin{bmatrix} 1 & 0 & -0.0001 \\ 0 & 1 & -0.0003 \\ -0.0001 & -0.0003 & 0.0149 \end{bmatrix}$$

After cycle 2

$$a_2 = \begin{bmatrix} 0 \\ 0 \\ -0.2171 \end{bmatrix}, \quad f = 0.004, \quad g_2 = \begin{bmatrix} 0 \\ 0 \\ -0.0819 \end{bmatrix}$$

$$H_2 = \begin{bmatrix} 0.7 & -0.6 & 0.0003 \\ -0.6 & -0.2 & 0.0005 \\ 0.0003 & 0.0005 & 0.0149 \end{bmatrix}$$

Note that there is considerable rounding error in the calculation of H_2.

10.7 Review exercises

1 $x_1 = 250$, $x_2 = 100$, $F = 3800$

2 Standard 20, super 10, deluxe 40, profit £21 000

3 $x_1 = 22$, $x_2 = 0$, $x_3 = 6$, profit £370

4 2 kg bread and 0.5 kg cheese, cost 210 p

5 Maximum at (1, 1) and (−1, −1), with distance = $\sqrt{2}$
Minimum at $(\sqrt{\frac{1}{3}}, -\sqrt{\frac{1}{3}})$ and $(-\sqrt{\frac{1}{3}}, \sqrt{\frac{1}{3}})$, with distance = $\sqrt{\frac{2}{3}}$

6 Sides are $3\sqrt{\frac{1}{10}}$ and $2\sqrt{\frac{1}{10}}$

7 $(\frac{17}{5}, 0, \frac{6}{5})$, with distance 2.638

8 (1, 2, 3) with $F = 14$, and (−1, −2, −3) with $F = -14$. (1, 2, 3) gives the global maximum and (0, 0, 0) gives the global minimum.

9 (i) $b = c$ (ii) $a = b = c$

10 $h^2 = 3\pi^2/b$, $r^2 = 3a^2/2b$

11 $k = 2.2$

12 Bracket:

R	3.5	5.5	9.5
Cost	1124	704	1418

Quadratic algorithm gives $R = 6.121$ and cost = 802, so $R = 5.5$ still gives the best result.

13 Quadratic algorithm always gives $x = 0.5$ for any intermediate value. However, taking

x	t	f
0.4	0.7381	0.8067
0.5	0.6514	0.7729
1.0	0	1

gives $x = 0.578$, $t = 0.577$, $f = 0.756$

14 Maximum at $\theta = 5$ rad, minimum at $\theta = 1.28$ rad

15 44 mph

16 Both steepest-descent and DFP give convergence at the first minimization to the exact solution $x = -0.5$, $y = -0.5$, $f = 0$. The Newton method gives

$$a_1 = \begin{bmatrix} -0.1667 \\ -0.1667 \end{bmatrix}, \quad f = 0.0123$$

$$a_2 = \begin{bmatrix} -0.2778 \\ -0.2778 \end{bmatrix}, \quad f = 0.002\,44$$

17 Maximum of 0.803 at $X = 0$, $Y = 0.622$, minimum of 0.358 at $X = Y = 0$

18 Partan

$$x_1 = \begin{bmatrix} 0 \\ 0 \end{bmatrix}, \quad f = 1 \qquad x_2 = \begin{bmatrix} 0.5 \\ 0 \end{bmatrix}, \quad f = 0.5$$

$$z_2 = \begin{bmatrix} 0.5 \\ 0.5 \end{bmatrix}, \quad f = 0.25 \qquad x_3 = \begin{bmatrix} 1 \\ 1 \end{bmatrix}, \quad f = 0$$

Steepest-descent

$$x_1 = \begin{bmatrix} 0 \\ 0 \end{bmatrix}, \quad f = 1 \qquad x_2 = \begin{bmatrix} 0.5 \\ 0 \end{bmatrix}, \quad f = 0.5$$

$$x_3 = \begin{bmatrix} 0.5 \\ 0.5 \end{bmatrix}, \quad f = 0.25$$

$$x_4 = \begin{bmatrix} 0.75 \\ 0.5 \end{bmatrix}, \quad f = 0.125$$

19 Start values:

$$a_0 = \begin{bmatrix} 1 \\ 1 \end{bmatrix}, \quad f = 1$$

$\lambda = 0$:

$$a_1 = \begin{bmatrix} 1.5 \\ 1 \end{bmatrix}, \quad f = 1 \quad \text{(no improvement)}$$

$\lambda = 1$:

$$a_1 = \begin{bmatrix} 3 \\ 2 \end{bmatrix}, \quad f = -0.5 \quad \text{(ready for next iteration)}$$

20 (a)

$$a_0 = \begin{bmatrix} 0 \\ 0 \end{bmatrix}, \quad F = 0.0625$$

$$a_1 = \begin{bmatrix} -0.25 \\ -0.25 \end{bmatrix}, \quad F = 0.0039$$

$$a_2 = \begin{bmatrix} -0.375 \\ -0.375 \end{bmatrix}, \quad F = 0.0002$$

(b)

$$a_0 = \begin{bmatrix} 1 \\ 1 \end{bmatrix}, \quad F = 0.3125$$

$$a_1 = \begin{bmatrix} 0.333 \\ 3.667 \end{bmatrix}, \quad F = 0.0664$$

21

$$a_0 = \begin{bmatrix} 1 \\ 0 \end{bmatrix}, \quad F = 1.29$$

$$a_1 = \begin{bmatrix} 1.07 \\ 0.27 \end{bmatrix}, \quad F = 0.239$$

23

$$a_0 = \begin{bmatrix} 0 \\ 0 \end{bmatrix}, \quad f = 1$$

$$a_1 = \begin{bmatrix} 0.5 \\ 0 \end{bmatrix}, \quad f = 0.5$$

$$a_2 = \begin{bmatrix} 0.5 \\ 0.5 \end{bmatrix}, \quad F = 0.25$$

24 Bracket gives

α	$[F(\alpha) - 3]^2$
1.4	0.0776
1.5	0.0029
1.6	0.0369

Quadratic algorithm gives $\alpha^* = 1.518$ and $f = 9 \times 10^{-5}$

CHAPTER 11

Exercises

1 (a) (762, 798) (b) 97

2 (8.05, 9.40)

3 (71.2, 75.2), accept

4 (2.92, 3.92)

5 (24.9, 27.9)

6 95% confidence interval (53.9, 58.1), criterion satisfactory

7 (−1900, 7500), reject

8 90%: (34, 758), 95%: (−45, 837), reject at 10% but accept at 5%

9 90%: (0.052, 0.089), 95%: (0.049, 0.092), reject at 10% but accept at 5%. Test statistic leads to rejection at both 10% and 5% levels, and is more accurate

10 203, (0.223, 0.327)

11 90%: (−0.28, −0.08), 95%: (−0.30, −0.06), accept at 10% but reject at 5%

12 (0.003, 0.130), carcinogenic

13 (a) X: (0.34, 0.53, 0.13), Y: (0.25, 0.31, 0.44)
(b) 0.472, (c) $E(X) = 1.79$, $\text{Var}(X) = 0.426$, $E(Y) = 2.19$, $\text{Var}(Y) = 0.654$, $\rho_{X,Y} = -0.246$

16 (a) 6, (b) 0.484,
(c) $f_X(x) = 6(\frac{1}{2} - x + \frac{1}{2}x^2)$, $f_Y(y) = 6(1 - y)y$

17 (a) 0.552, (b) 0.368

18 0.84

19 0.934

20 0.732

21 (0.45, 0.85)

22 (0.67, 0.99)

23 0.444, 90%: (0.08, 0.70), 95%: (0.00, 0.74), just significant at 5%, rank correlation 0.401, significant at 10%

24 $a = 6.315$, $b = 14.64$, $y = 226$

26 (a) $a = 343.7$, $b = 3.221$, $y = 537$;
(b) (0.46, 5.98), reject; (c) (459, 615)

27 $a = 0.107$, $b = 1.143$, (14.4, 17.8)

29 120 Ω

30 $\lambda = 2.66$, $C = 2.69 \times 10^6$, $P = 22.9$

31 $a = 7533$, $b = -1.059$, $y = 17.9$

32 $\chi^2 = 12.3$, significant at 5%

33 $\chi^2 = 1.35$, accept Poisson

34 $\chi^2 = 12.97$, accept Poisson

35 $\chi^2 = 11.30$, significant at 5% but not at 1%, for proportion 95%: (0.111, 0.159), 99%: (0.104, 0.166), significant at 1%

37 $\chi^2 = 1.30$, not significant

38 $\chi^2 = 20.56$, significant at 5%

39 $c = 4$, $M_X(t) = 4/(t-2)^2$, $E(X) = 1$, $\text{Var}(X) = \frac{1}{2}$

41 0.014

42 0.995

44 Warning 9.5, action 13.5, sample 12, UCL = 11.4, sample 9

45 UK sample 28, US sample 25

46 Action 2.93, sample 12

47 Action 14.9, sample 19 but repeated warnings

48 (a) sample 9, (b) sample 9

49 (a) sample 10, (b) sample 12

50 sample 10

51 sample 16

52 (a) Repeated warnings, (b) sample 15, (c) sample 14

54 sample 11

55 Shewhart, sample 26; cusum, sample 13; moving-average, sample 11

56 0.132

57 0.223, 0.042

59 (a) $\frac{1}{4}$, (b) $2\frac{1}{4}$, (c) 0.237, (d) 45 min, (e) 0.276

61 Mean costs per hour: A, £200; B, £130

62 6

63 Second cash desk

64 Sabotage

65 $P(C|\text{two hits}) = 0.526$

66 $\frac{1}{2}$

67 (a) 0.0944, (b) 0.81

68 (a) $\frac{3}{4}$, (b) $[1 + (\frac{1}{3})^k 2^{n-k}]^{-1}$

69 AAAA

70 1.28:1 in favour of Poisson

72 2.8:1 in favour of H_2

73 12.8:1 in favour of H_1

11.12 Review exercises

1 (0.202, 0.266)

2 $(96.1 \times 10^6, 104.9 \times 10^6)$

3 $\chi^2 = 3.35$ (using class intervals of length 5, with a single class for all values greater than 30), accept exponential

4 Outlier 72 significant at 5%, outlier included (7.36, 11.48); excluded (7.11, 10.53)

5 Operate if $p > \frac{4}{13}$

6 Cost per hour: A, £632.5; B, £603.4

7 (a) $P(\text{input } 0|\text{output } 0) = \dfrac{\bar{p}\alpha}{\bar{p}\alpha + p\bar{\alpha}}$ etc.

(b) $p < \alpha < 1 - p$

Index

Emboldened page references indicate where an entry has been defined in the text.

D

E

M

N

O

P

Q

R

S

T